ANALYTICAL CHEMISTRY

ANALYTICAL CHEMISTRY

A Chemist and Laboratory Technician's Toolkit

BRYAN M. HAM
AIHUI MAHAM

Published by John Wiley & Sons, Inc., Hoboken, New Jersey
Published simultaneously in Canada

Library of Congress Cataloging-in-Publication Data:

Ham, Bryan M.
Analytical chemistry : a chemist and laboratory technician's toolkit / Bryan M. Ham, Aihui Maham.
pages cm
Includes bibliographical references and index.
ISBN 978-1-118-71484-3 (cloth)
1. Chemistry, Analytic–Handbooks, manuals, etc. I. Maham, Aihui, 1974– II. Title.
QD78.H36 2016
543–dc23

2015018014

Cover image courtesy of Bryan M. Ham & Aihui MaHam

Set in 9.5/11.5pt Times by SPi Global, Pondicherry, India

Printed in the United States of America

V10013835_091019

1 2016

DEDICATION

This book is dedicated to the newest, most precious addition to our family: our first baby girl.

CONTENTS

Preface xxiii

Author Biographies xxv

Acknowledgments xxvii

1 Chemist and Technician in the Analytical Laboratory 1

1.1 Introduction—The Analytical Chemist and Technician, 1
1.2 Today's Laboratory Chemist and Technician, 1
1.2.1 Computers in the Laboratory, 1
1.2.2 Laboratory Information Management Systems (LIMS), 1
1.3 ChemTech—The Chemist and Technician Toolkit Companion, 1
1.3.1 Introduction to ChemTech, 1
1.3.1.1 Opening ChemTech, 2
1.3.1.2 Interactive Periodic Table, 2
1.4 Chapter Layout, 2
1.4.1 Glassware, Chemicals, and Safety, 2
1.4.2 Basic Math and Statistics, 2
1.4.3 Graphing and Plotting, 4
1.4.4 Making Laboratory Solutions, 4
1.4.5 Titrimetric Analysis, 4
1.4.6 Electrochemistry, 5
1.4.7 Laboratory Information Management System (or Software) LIMS, 5
1.4.8 Instrumental Analyses—Spectroscopy, 5
1.4.9 Instrumental Analyses—Chromatography, 5
1.4.10 Instrumental Analyses—Mass Spectrometry, 5
1.4.10.1 Mass Analyzers, 5
1.4.10.2 Mass Ionization, 5
1.4.11 Small Molecule and Macromolecule Analysis, 5
1.5 Users of ChemTech, 6

2 Introduction to the Analytical Laboratory 7

2.1 Introduction to the Laboratory, 7
2.1.1 The Scientific Method, 7
2.2 Laboratory Glassware, 7
2.2.1 Volumetric Flasks, 7

2.2.2 Beakers and Erlenmeyer Flasks, 7
2.2.3 Graduated Cylinders, 8
2.2.4 Pipettes, 8
2.2.4.1 Steps for Using Pipette Bulb (a), 8
2.2.4.2 Steps for Using Pipette Bulb (b and c), 10
2.2.4.3 Autopipettes, 11
2.2.5 Evaporating Dishes, 11
2.2.6 Flames and Furnaces in the Laboratory, 11
2.2.6.1 Bunsen Burners, 11
2.2.6.2 Crucibles, 11
2.2.6.3 Ashing Samples, 11
2.2.6.4 Muffle Furnaces, 14
2.2.7 Laboratory Fume Hoods, 14
2.2.8 Drying Ovens, 15
2.2.9 Balances, 15
2.2.10 Refrigerators and Freezers, 16
2.2.11 Test Tubes, 16
2.2.12 Soxhlet Extractions, 16
2.2.13 Vacuum Pumps, 18
2.3 Conclusion, 18

3 Laboratory Safety **19**

3.1 Introduction, 19
3.2 Proper Personal Protection and Appropriate Attire, 19
3.2.1 Proper Eye Protection, 19
3.2.2 Proper Laboratory Coats, 20
3.3 Proper Shoes and Pants, 20
3.4 Laboratory Gloves, 20
3.4.1 Natural Rubber (Latex), 21
3.4.2 Nitrile, 22
3.4.3 Neoprene, 22
3.4.4 Butyl, 22
3.4.5 Polyvinyl Chloride (PVC), 22
3.4.6 Polyvinyl Alcohol (PVA), 22
3.4.7 Viton, 22
3.4.8 Silver Shield/4H, 22
3.5 General Rules to Use Gloves, 22
3.6 Material Safety Data Sheet (MSDS), 22
3.7 Emergency Eye Wash and Face Wash Stations, 23
3.8 Emergency Safety Showers, 24
3.9 Fire Extinguishers, 24
3.9.1 Types of Fires, 24
3.10 Clothing Fire in the Laboratory, 25
3.11 Spill Cleanup Kits, 25
3.12 Chemicals and Solvents, 27
3.13 First Aid Kits, 27
3.14 Gasses and Cylinders, 29
3.15 Sharps Containers and Broken Glass Boxes, 29
3.16 Occupational Safety and Health Administration (OSHA), 29

4 Basic Mathematics in the Laboratory **83**

4.1 Introduction to Basic Math, 83
4.2 Units and Metric System, 83
4.2.1 Introduction to the Metric System, 83
4.2.2 Units of the Metric System, 83
4.2.3 Converting the *SI* Units, 84

4.3 Significant Figures, 84
4.3.1 Significant Figure Rules, 84
4.4 Scientific Calculators, 86
4.4.1 Example Calculator, 86
4.4.2 Window's Calculator, 86
4.4.2.1 Windows' Scientific versus Standard Calculator, 86
4.5 ChemTech Conversion Tool, 89
4.5.1 Using the Conversion Tool, 89
4.5.2 Closing the Conversion Tool, 89
4.6 Chapter Key Concepts, 89
4.7 Chapter Problems, 92

5 Analytical Data Treatment (Statistics) 93

5.1 Errors in the Laboratory, 93
5.1.1 Systematic Errors, 93
5.1.2 Random Errors, 93
5.2 Expressing Absolute and Relative Errors, 94
5.3 Precision, 94
5.3.1 Precision versus Accuracy, 94
5.4 The Normal Distribution Curve, 94
5.4.1 Central Tendency of Data, 95
5.4.1.1 The Arithmetic Mean, 95
5.4.1.2 The Median, 95
5.4.1.3 The Mode, 95
5.4.1.4 Sticking with the Mean, 95
5.5 Precision of Experimental Data, 96
5.5.1 The Range, 96
5.5.2 The Average Deviation, 96
5.5.3 The Standard Deviation, 97
5.5.3.1 Root Mean Square, 97
5.5.3.2 Sample Standard Deviation, 97
5.5.3.3 Comparison of the Three Methods, 97
5.5.3.4 Using the Scientific Calculator, 97
5.5.3.5 Coefficient of Variation, 97
5.6 Normal Distribution Curve of a Sample, 97
5.7 ChemTech Statistical Calculations, 98
5.7.1 Introduction to ChemTech Statistics, 98
5.7.2 ChemTech Chapter 5, 98
5.7.2.1 Entering Data, 100
5.7.2.2 Calculating the Statistics, 100
5.7.2.3 The Results Output, 100
5.7.2.4 Results not Expected, 100
5.7.2.5 Using ChemTech for Large Value Set, 101
5.7.2.6 The Results Page, 101
5.7.2.7 Resetting the Page, 101
5.8 Student's Distribution t Test for Confidence Limits, 101
5.8.1 Accuracy, 101
5.8.2 The Student's t Test, 102
5.8.3 Calculating the Student's t Value, 102
5.8.4 Probability Level, 103
5.8.5 Sulfate Concentration Confidence Limits, 103
5.8.6 Sulfate t Distribution Curve, 103
5.8.7 Determining Types of Error, 103
5.8.7.1 Glucose Content, 104
5.8.8 Determining Error in Methodology, 104
5.8.8.1 Magnesium Primary Standard, 104

5.9 Tests of Significance, 104
5.9.1 Difference in Means, 104
5.9.2 Null Hypothesis, 105
5.10 Treatment of Data Outliers, 105
5.10.1 The Q Test, 105
5.10.2 The T_n Test, 106
5.11 Chapter Key Concepts, 106
5.12 Chapter Problems, 107

6 Plotting and Graphing **109**

6.1 Introduction to Graphing, 109
6.1.1 The Invention of the Graph, 109
6.1.2 Importance of Graphing, 109
6.2 Graph Construction, 109
6.2.1 Axis and Quadrants, 110
6.3 Rectangular Cartesian Coordinate System, 110
6.4 Curve Fitting, 110
6.5 Redrawn Graph Example, 110
6.6 Graphs of Equations, 111
6.6.1 Introduction, 111
6.6.2 Copper Sulfate Data, 111
6.6.3 Plotting the Data, 111
6.6.4 Best Fit Line, 111
6.6.5 Point-Slope Equation of a Line, 112
6.6.6 Finding the Slope (m), 112
6.6.7 Finding the y-Intercept (b), 112
6.6.8 Solving for x, 113
6.6.9 Estimating the Slope and Intercept, 113
6.6.10 Deriving the Equation from the Slope and Intercept, 113
6.7 Least-Squares Method, 114
6.7.1 Plotting Data with Scatter, 114
6.7.2 Linear Regression, 114
6.7.3 Curve Fitting the Data, 114
6.8 Computer-Generated Curves, 115
6.8.1 Using ChemTech to Plot Data, 115
6.8.2 Entering the Data, 115
6.8.3 Plotting the Data, 116
6.8.4 Linear Regression of the Data, 116
6.8.5 Adding the Best Fit Line, 118
6.8.6 Entering a Large Set of Data, 118
6.9 Calculating Concentrations, 119
6.10 Nonlinear Curve Fitting, 119
6.11 Chapter Key Concepts, 123
6.12 Chapter Problems, 124

7 Using Microsoft Excel® in the Laboratory **125**

7.1 Introduction to Excel®, 125
7.2 Opening Excel® in ChemTech, 125
7.3 The Excel® Spreadsheet, 125
7.3.1 Spreadsheet Menus and Quick Access Toolbars, 127
7.4 Graphing in Excel®, 127
7.4.1 Making Column Headings, 127
7.4.2 Entering Data into Columns, 128
7.4.3 Saving the Spreadsheet, 129
7.4.4 Constructing the Graph, 129
7.4.5 The Chart Wizard, 130

7.4.6 The Chart Source Data, 130
7.4.7 Chart Options, 131
7.5 Charts in Excel® 2010, 132
7.6 Complex Charting in Excel® 97-2003, 132
7.6.1 Calcium Atomic Absorption (AAS) Data, 132
7.6.2 Entering Ca Data into Spreadsheet, 135
7.6.3 Average and Standard Deviation, 135
7.6.4 Constructing the Calibration Curve, 135
7.6.5 Entering the Chart Options, 136
7.6.6 Error Bars, 137
7.6.7 Trendline, 138
7.7 Complex Charting in Excel® 2010, 139
7.7.1 Entering the Data, 139
7.7.2 Using the Formula Search Function, 139
7.7.3 Inserting the Chart, 140
7.7.4 Formatting the Chart, 140
7.8 Statistical Analysis Using Excel®, 141
7.8.1 Open and Save Excel® StatExp.xls, 141
7.8.2 Sulfate Data, 141
7.8.3 Excel® Confidence Function, 142
7.8.4 Excel® Student's *t* Test, 142
7.8.4.1 Spreadsheet Calculation I, 142
7.8.4.2 Spreadsheet Calculation II, 143
7.8.5 Excel® Tools Data Analysis, 143
7.8.5.1 Analysis ToolPak, 143
7.8.5.2 ToolPak Functions, 143
7.8.5.3 Data Analysis *t*-Test: Two-Sample Assuming Unequal Variances, 144
7.8.5.4 Analysis ToolPak *F*-test, 145
7.8.5.5 Analysis ToolPak Statistical Summary, 145

8 Making Laboratory Solutions **147**

8.1 Introduction, 147
8.2 Laboratory Reagent Fundamentals, 147
8.3 The Periodic Table, 147
8.3.1 Periodic Table Descriptive Windows, 148
8.4 Calculating Formula Weights, 148
8.5 Calculating the Mole, 148
8.6 Molecular Weight Calculator, 148
8.7 Expressing Concentration, 148
8.7.1 Formal (*F*) Solutions, 149
8.7.1.1 Formal (*F*) Solution Example, 149
8.7.2 Molal (*m*) Solutions, 149
8.7.2.1 Molal (*m*) Solution—Simple Example, 149
8.7.2.2 Molal (*m*) Solution—Complex Example, 149
8.7.3 Molar (*M*) Solutions, 150
8.7.3.1 Molar (*M*) Solution Example, 150
8.7.3.2 Molar (*M*) Solution of K_2CO_3, 151
8.7.4 Normal (*N*) Solutions, 151
8.7.4.1 Normal (*N*) Solution Calculation Example, 152
8.8 The Parts per (PP) Notation, 153
8.9 Computer-Based Solution Calculations, 153
8.9.1 Computer-Based Concentration Calculation—Molarity I, 154
8.9.2 Computer-Based Concentration Calculation—Molarity II, 154
8.9.3 Computer-Based Concentration Calculation—Normality I, 155
8.9.4 Computer-Based Concentration Calculation—Normality II, 156

8.10 Reactions in Solution, 157
8.11 Chapter Key Concepts, 157
8.12 Chapter Problems, 158

9 Acid–Base Theory and Buffer Solutions **159**

9.1 Introduction, 159
9.2 Acids and Bases in Everyday Life, 159
9.3 The Litmus Test, 159
9.4 Early Acid–Base Descriptions, 160
9.5 Brønsted–Lowry Definition, 160
9.6 The Equilibrium Constant, 161
9.7 The Acid Ionization Constant, 161
9.8 Calculating the Hydrogen Ion Concentration, 162
9.9 The Base Ionization Constant, 163
9.9.1 OH^- Ion Concentration Example, 163
9.9.2 Percent Ionization Example, 164
9.10 Ion Product for Water, 164
9.11 The Solubility Product Constant (K_{sp}), 164
9.11.1 Solubility of Silver(I) Thiocyanate, 164
9.11.2 Solubility of Lithium Carbonate, 166
9.12 The pH of a Solution, 166
9.13 Measuring the pH, 167
9.13.1 The Glass Electrode, 167
9.14 Buffered Solutions—Description and Preparing, 168
9.14.1 Le Chatelier's Principle, 169
9.14.2 Titration Curve of a Buffer, 169
9.14.3 Natural Buffer Solutions, 169
9.14.4 Calculating Buffer pH, 170
9.14.5 Buffer pH Calculation I, 170
9.15 ChemTech Buffer Solution Calculator, 170
9.16 Chapter Key Concepts, 171
9.17 Chapter Problems, 172

10 Titration—A Volumetric Method of Analysis **175**

10.1 Introduction, 175
10.2 Reacting Ratios, 175
10.3 The Equivalence Point, 176
10.4 Useful Relationships for Calculations, 176
10.5 Deriving the Titration Equation, 176
10.5.1 Titration Calculation Example, 176
10.6 Titrations in ChemTech, 177
10.6.1 Acid/Base Titrations Using Molar Solutions, 177
10.6.2 Titration Calculation Example, 177
10.7 Acid/Base Titration Endpoint (Equivalence Point), 178
10.8 Acid/Base Titration Midpoint, 179
10.9 Acid/Base Titration Indicators, 180
10.9.1 The Ideal Indicator, 180
10.10 Titrations Using Normal Solutions, 181
10.10.1 Normal Solution Titration Example, 181
10.11 Polyprotic Acid Titration, 181
10.12 ChemTech Calculation of Normal Titrations, 182
10.13 Performing a Titration, 183
10.13.1 Titration Glassware, 183
10.13.2 Titration Steps, 183
10.14 Primary Standards, 184

10.15 Standardization of Sodium Hydroxide, 185
10.15.1 NaOH Titrant Standardization Example, 185
10.16 Conductometric Titrations (Nonaqueous Solutions), 186
10.17 Precipitation Titration (Mohr Method for Halides), 188
10.17.1 Basic Steps in Titration, 188
10.17.2 Important Considerations, 189
10.18 Complex Formation with Back Titration (Volhard Method for Anions), 189
10.18.1 Iron(III) as Indicator, 189
10.18.2 Chloride Titration, 189
10.18.3 The General Calculation, 189
10.18.4 Chloride Titration, 190
10.18.4.1 Volhard Chloride Analysis Example, 190
10.18.4.2 The Titration Steps, 190
10.19 Complex Formation Titration with EDTA for Cations, 190
10.19.1 EDTA–Metal Ion Complex Formation, 191
10.19.2 The Stability Constant, 191
10.19.3 Metal Ions Titrated, 191
10.19.4 Influence of pH, 191
10.19.5 Buffer and Hydroxide Complexation, 192
10.19.6 Visual Indicators, 193
10.20 Chapter Key Concepts, 194
10.21 Chapter Problems, 195

11 Oxidation–Reduction (Redox) Reactions **197**

11.1 Introduction, 197
11.2 Oxidation and Reduction, 197
11.3 The Volt, 198
11.4 The Electrochemical Cell, 198
11.5 Redox Reaction Conventions, 198
11.5.1 Electrode Potential Tables, 198
11.5.2 The Standard Hydrogen Electrode (SHE), 199
11.5.3 The SHE Half-Reaction, 199
11.5.4 Writing the Standard Electrode Potentials, 199
11.5.5 Drawing a Galvanic Cell, 199
11.5.6 Calculating the Cell Potential, 200
11.5.6.1 Iron and Zinc Cell, 200
11.5.6.2 Nickel and Silver Cell, 200
11.6 The Nernst Equation, 200
11.6.1 Nernst Equation Example I, 201
11.6.2 Nernst Equation Example II, 201
11.6.3 Nernst Equation Example III, 201
11.7 Determining Redox Titration Endpoints, 202
11.8 Potentiometric Titrations, 202
11.8.1 Detailed Potentiometer, 202
11.8.2 Half-Reactions, 202
11.8.3 The Nernst Equation, 203
11.8.4 Assumed Reaction Completion, 203
11.8.5 Calculated Potentials of Ce^{4+}, 204
11.9 Visual Indicators Used in Redox Titrations, 204
11.10 Pretitration Oxidation–Reduction, 205
11.10.1 Reducing Agents, 205
11.10.2 Oxidizing Agents, 205
11.11 Ion-Selective Electrodes, 206

11.12 Chapter Key Concepts, 206
11.13 Chapter Problems, 207

12 Laboratory Information Management System (LIMS) **209**

12.1 Introduction, 209
12.2 LIMS Main Menu, 209
12.3 Logging in Samples, 209
12.4 Entering Test Results, 209
12.5 Add or Delete Tests, 211
12.6 Calculations and Curves, 212
12.7 Search Wizards, 214
12.7.1 Searching Archived Samples, 214
12.7.2 General Search, 214
12.7.3 Viewing Current Open Samples, 216
12.8 Approving Samples, 218
12.9 Printing Sample Reports, 220

13 Ultraviolet and Visible (UV/Vis) Spectroscopy **221**

13.1 Introduction to Spectroscopy in the Analytical Laboratory, 221
13.2 The Electromagnetic Spectrum, 221
13.3 Ultraviolet/Visible (UV/Vis) Spectroscopy, 221
13.3.1 Wave and Particle Theory of Light, 222
13.3.2 Light Absorption Transitions, 223
13.3.3 The Color Wheel, 224
13.3.4 Pigments, 224
13.3.5 Inorganic Elemental Analysis, 224
13.3.6 The Azo Dyes, 225
13.3.7 UV-Visible Absorption Spectra, 228
13.3.8 Beer's Law, 228
13.4 UV/Visible Spectrophotometers, 230
13.5 Special Topic (Example)—Spectrophotometric Study of Dye Compounds, 234
13.5.1 Introduction, 234
13.5.2 Experimental Setup for Special Topic Discussion, 235
13.5.3 UV/Vis Study of the Compounds and Complexes, 235
13.6 Chapter Key Concepts, 236
13.7 Chapter Problems, 237

14 Fluorescence Optical Emission Spectroscopy **239**

14.1 Introduction to Fluorescence, 239
14.2 Fluorescence and Phosphorescence Theory, 240
14.2.1 Radiant Energy Absorption, 240
14.2.2 Fluorescence Principle—Jabłoński Diagram, 240
14.2.3 Excitation and Electron Spin States, 240
14.2.3.1 Quantum Numbers, 241
14.2.3.2 Electron Spin States, 241
14.3 Phosphorescence, 241
14.4 Excitation and Emission Spectra, 242
14.5 Rate Constants, 243
14.5.1 Emission Times, 243
14.5.2 Relative Rate Constants (k), 243
14.6 Quantum Yield Rate Constants, 243
14.7 Decay Lifetimes, 244
14.8 Factors Affecting Fluorescence, 244
14.8.1 Excitation Wavelength (Instrumental), 244
14.8.2 Light Source (Instrumental), 244
14.8.3 Filters, Optics, and Detectors (Instrumental), 245

14.8.4 Cuvettes and Cells (Instrumental), 245
14.8.5 Structure (Sample), 246
14.8.5.1 Fluorescein and Beta-(β)-Carotene, 246
14.8.5.2 Diatomic Oxygen Molecular Orbital Diagram, 246
14.8.5.3 Examples of Nonfluorescent and Fluorescent Compounds, 247
14.8.5.4 Other Structural Influences, 247
14.8.5.5 Scattering (Sample), 248
14.9 Quantitative Analysis and Beer–Lambert Law, 248
14.10 Quenching of Fluorescence, 249
14.11 Fluorometric Instrumentation, 249
14.11.1 Spectrofluorometer, 249
14.11.1.1 Light Source, 250
14.11.1.2 Monochromators, 250
14.11.1.3 Photomultiplier tube (PMT), 251
14.11.2 Multidetection Microplate Reader, 252
14.11.3 Digital Fluorescence Microscopy, 252
14.11.3.1 Light Source, 252
14.11.3.2 Filter Cube, 253
14.11.3.3 Objectives and Grating, 253
14.11.3.4 Charged-Coupled Device (CCD), 254
14.12 Special Topic—Flourescence Study of Dye-A007 Complexes, 255
14.13 Chapter Key Concepts, 257
14.14 Chapter Problems, 258

15 Fourier Transform Infrared (FTIR) Spectroscopy **261**

15.1 Introduction, 261
15.2 Basic IR Instrument Design, 261
15.3 The Infrared Spectrum and Molecular Assignment, 263
15.4 FTIR Table Band Assignments, 264
15.5 FTIR Spectrum Example I, 270
15.6 FTIR Spectrum Example II, 270
15.7 FTIR Inorganic Compound Analysis, 271
15.8 Chapter Key Concepts, 271
15.9 Chapter Problems, 273

16 Nuclear Magnetic Resonance (NMR) Spectroscopy **277**

16.1 Introduction, 277
16.2 Frequency and Magnetic Field Strength, 277
16.3 Continuous-Wave NMR, 278
16.4 The NMR Sample Probe, 280
16.5 Pulsed Field Fourier Transform NMR, 280
16.6 Proton NMR Spectra Environmental Effects, 280
16.6.1 Chemical Shift, 281
16.6.2 Spin–Spin Splitting (Coupling), 281
16.6.3 Interpretation of NMR Spectra, 283
16.6.3.1 2-Amino-3-Methyl-Pentanoic Acid, 283
16.6.3.2 Unknown I, 283
16.7 Carbon-13 NMR, 283
16.7.1 Introduction, 283
16.7.2 Carbon-13 Chemical Shift, 284
16.7.3 Carbon-13 Splitting, 286
16.7.4 Finding the Number of Carbons, 286
16.7.5 Carbon-13 NMR Examples, 286
16.8 Special Topic—NMR Characterization of Cholesteryl Phosphate, 287
16.8.1 Synthesis of Cholesteryl Phosphate, 288
16.8.2 Single-Stage and High-Resolution Mass Spectrometry, 288
16.8.3 Proton Nuclear Magnetic Resonance (^{1}H-NMR), 289

16.8.4 Theoretical NMR Spectroscopy, 289
16.8.5 Structure Elucidation, 289
16.9 Chapter Key Concepts, 292
16.10 Chapter Problems, 293
References, 294

17 Atomic Absorption Spectroscopy (AAS) **295**

17.1 Introduction, 295
17.2 Atomic Absorption and Emission Process, 295
17.3 Atomic Absorption and Emission Source, 296
17.4 Source Gases and Flames, 296
17.5 Block Diagram of AAS Instrumentation, 296
17.6 The Light Source, 297
17.7 Interferences in AAS, 299
17.8 Electrothermal Atomization—Graphite Furnace, 299
17.9 Instrumentation, 300
17.10 Flame Atomic Absorption Analytical Methods, 301

18 Atomic Emission Spectroscopy **303**

18.1 Introduction, 303
18.2 Elements in Periodic Table, 303
18.3 The Plasma Torch, 303
18.4 Sample Types, 304
18.5 Sample Introduction, 304
18.6 ICP-OES Instrumentation, 305
18.6.1 Radially Viewed System, 306
18.6.2 Axially Viewed System, 308
18.6.3 Ergonomic Sample Introduction System, 309
18.6.4 Innovative Optical Design, 310
18.6.5 Advanced CID Camera Technology, 310
18.7 ICP-OES Environmental Application Example, 310

19 Atomic Mass Spectrometry **325**

19.1 Introduction, 325
19.2 Low-Resolution ICP-MS, 325
19.2.1 The PerkinElmer NexION® 350 ICP-MS, 325
19.2.2 Interface and Quadrupole Ion Deflector (QID), 325
19.2.3 The Collision/Reaction Cell, 325
19.2.4 Quadrupole Mass Filter, 328
19.3 High-Resolution ICP-MS, 328

20 X-ray Fluorescence (XRF) and X-ray Diffraction (XRD) **333**

20.1 X-Ray Fluorescence Introduction, 333
20.2 X-Ray Fluorescence Theory, 333
20.3 Energy-Dispersive X-Ray Fluorescence (EDXRF), 334
20.3.1 EDXRF Instrumentation, 334
20.3.1.1 Basic Components, 334
20.3.1.2 X-Ray Sources, 334
20.3.1.3 Detectors, 335
20.3.2 Commercial Instrumentation, 337
20.4 Wavelength Dispersive X-Ray Fluorescence (WDXRF), 337
20.4.1 Introduction, 337
20.4.2 WDXRF Instrumentation, 338
20.4.2.1 Simultaneous WDXRF Instrumentation, 338
20.4.2.2 Sequential WDXRF Instrumentation, 340
20.5 Applications of XRF, 341

20.6 X-ray Diffraction (XRD), 342
20.6.1 Introduction, 342
20.6.2 X-Ray Crystallography, 344
20.6.3 Bragg's Law, 345
20.6.4 Diffraction Patterns, 345
20.6.5 The Goniometer, 346
20.6.6 XRD Spectra, 346

21 Chromatography—Introduction and Theory 351

21.1 Preface, 351
21.2 Introduction to Chromatography, 351
21.3 Theory of Chromatography, 351
21.4 The Theoretical Plate Number N, 355
21.5 Resolution R_S, 356
21.6 Rate Theory versus Plate Theory, 357
21.6.1 Multiple Flow Paths or Eddy Diffusion (A Coefficient), 358
21.6.2 Longitudinal (Molecular) Diffusion (B Coefficient), 359
21.6.3 Mass Transfer Resistance between Phases (C_S and C_M Coefficients), 361
21.7 Retention Factor k', 361
References, 362

22 High Performance Liquid Chromatography (HPLC) 363

22.1 HPLC Background, 363
22.2 Design and Components of HPLC, 363
22.2.1 HPLC Pump, 366
22.2.2 HPLC Columns, 368
22.2.2.1 HPLC Column Stationary Phases, 368
22.2.3 HPLC Detectors, 372
22.2.4 HPLC Fraction Collector, 374
22.2.5 Current Commercially Available HPLC Systems, 375
22.2.6 Example of HPLC Analyses, 375
22.2.6.1 HPLC Analysis of Acidic Pesticides, 375

23 Solid-Phase Extraction 381

23.1 Introduction, 381
23.2 Disposable SPE Columns, 381
23.3 SPE Vacuum Manifold, 381
23.4 SPE Procedural Bulletin, 381

24 Plane Chromatography: Paper and Thin-Layer Chromatography 395

24.1 Plane Chromatography, 395
24.2 Thin-Layer Chromatography, 395
24.3 Retardation Factor (R_F) in TLC, 398
24.3.1 Example I, 398
24.3.2 Example II, 398
24.4 Plate Heights (H) and Counts (N) in TLC, 398
24.5 Retention Factor in TLC, 399

25 Gas-Liquid Chromatography 401

25.1 Introduction, 401
25.2 Theory and Principle of GC, 401
25.3 Mobile-Phase Carrier Gasses in GC, 403
25.4 Columns and Stationary Phases, 404
25.5 Gas Chromatograph Injection Port, 406
25.5.1 Injection Port Septa, 407
25.5.1.1 Merlin Microseal, 407

25.5.2 Injection Port Sleeve (Liner), 408
25.5.2.1 Attributes of a Proper Liner, 409
25.5.3 Injection Port Flows, 412
25.5.4 Packed Column Injection Port, 412
25.5.5 Capillary Column Split Injection Port, 414
25.5.6 Capillary Column Splitless Injection Port, 414
25.6 The GC Oven, 415
25.7 GC Programming and Control, 417
25.8 GC Detectors, 418
25.8.1 Flame Ionization Detector (FID), 418
25.8.2 Electron Capture Detector (ECD), 418
25.8.3 Flame Photometric Detector (FPD), 419
25.8.4 Nitrogen Phosphorus Detector (NPD), 419
25.8.5 Thermal Conductivity Detector (TCD), 420

26 Gas Chromatography–Mass Spectrometry (GC–MS) 421

26.1 Introduction, 421
26.2 Electron Ionization (EI), 421
26.3 Electron Ionization (EI)/OE Processes, 422
26.4 Oleamide Fragmentation Pathways: OE $M^{+\bullet}$ by Gas Chromatography/Electron Ionization Mass Spectrometry, 425
26.5 Oleamide Fragmentation Pathways: EE $[M+H]^+$ by ESI/Ion Trap Mass Spectrometry, 426
26.6 Quantitative Analysis by GC/EI–MS, 429
26.7 Chapter Problems, 431
References, 433

27 Special Topics: Strong Cation Exchange Chromatography and Capillary Electrophoresis 435

27.1 Introduction, 435
27.1.1 Overview and Comparison of HPLC and CZE, 435
27.2 Strong Ion Exchange HPLC, 435
27.3 CZE, 435
27.3.1 Electroosmotic Flow (EOF), 436
27.3.2 Applications of CZE, 436
27.4 Binding Constants by Cation Exchange and CZE, 436
27.4.1 Ranking of Binding Constants, 436
27.4.2 Experimental Setup, 436
27.4.3 UV/Vis Study of the Compounds and Complexes, 437
27.4.4 Fluorescence Study of the Dye/A007 Complexes, 438
27.4.5 Computer Modeling of the Complex, 438
27.4.6 Cation Exchange Liquid Chromatography Results, 440
27.4.6.1 Description of HPLC Pseudophase, 441
27.4.7 Capillary Electrophoresis (CE), 441
27.4.7.1 Introduction, 441
27.4.7.2 CE Instrumentation, 441
27.4.7.3 Theory of CE Separation, 441
27.4.7.4 Results of CE Binding Analysis of Dyes and A007, 441
27.4.7.5 Electropherograms of Dye/A007 Complexes, 446
27.5 Comparison of Methods, 446
27.6 Conclusions, 448
References, 448

28 Mass Spectrometry 449

28.1 Definition and Description of Mass Spectrometry, 449
28.2 Basic Design of Mass Analyzer Instrumentation, 449

28.3 Mass Spectrometry of Protein, Metabolite, and Lipid Biomolecules, 451
28.3.1 Proteomics, 451
28.3.2 Metabolomics, 452
28.3.3 Lipidomics, 454
28.4 Fundamental Studies of Biological Compound Interactions, 455
28.5 Mass-to-Charge (*m/z*) Ratio: How the Mass Spectrometer Separates Ions, 457
28.6 Exact Mass versus Nominal Mass, 458
28.7 Mass Accuracy and Resolution, 460
28.8 High-Resolution Mass Measurements, 461
28.9 Rings Plus Double Bonds (*r* + *db*), 463
28.10 The Nitrogen Rule in Mass Spectrometry, 464
28.11 Chapter Problems, 465
References, 465

29 Ionization in Mass Spectrometry **467**

29.1 Ionization Techniques and Sources, 467
29.2 Chemical Ionization (CI), 467
29.2.1 Positive CI, 468
29.2.2 Negative CI, 470
29.3 Atmospheric Pressure Chemical Ionization (APCI), 471
29.4 Electrospray Ionization (ESI), 472
29.5 Nanoelectrospray Ionization (Nano-ESI), 474
29.6 Atmospheric Pressure Photo Ionization (APPI), 477
29.6.1 APPI Mechanism, 478
29.6.2 APPI VUV Lamps, 478
29.6.3 APPI Sources, 478
29.6.4 Comparison of ESI and APPI, 479
29.7 Matrix Assisted Laser Desorption Ionization (MALDI), 483
29.8 FAB, 485
29.8.1 Application of FAB versus EI, 487
29.9 Chapter Problems, 489
References, 489

30 Mass Analyzers in Mass Spectrometry **491**

30.1 Mass Analyzers, 491
30.2 Magnetic and Electric Sector Mass Analyzer, 491
30.3 Time-of-Flight Mass Analyzer (TOF/MS), 496
30.4 Time-of-Flight/Time-of-Flight Mass Analyzer (TOF–TOF/MS), 497
30.5 Quadrupole Mass Filter, 500
30.6 Triple Quadrupole Mass Analyzer (QQQ/MS), 502
30.7 Three-Dimensional Quadrupole Ion Trap Mass Analyzer (QIT/MS), 503
30.8 Linear Quadrupole Ion Trap Mass Analyzer (LTQ/MS), 506
30.9 Quadrupole Time-of-Flight Mass Analyzer (Q-TOF/MS), 507
30.10 Fourier Transform Ion Cyclotron Resonance Mass Analyzer (FTICR/MS), 508
30.10.1 Introduction, 508
30.10.2 FTICR Mass Analyzer, 509
30.10.3 FTICR Trapped Ion Behavior, 509
30.10.4 Cyclotron and Magnetron Ion Motion, 515
30.10.5 Basic Experimental Sequence, 515
30.11 Linear Quadrupole Ion Trap Fourier Transform Mass Analyzer (LTQ–FT/MS), 517
30.12 Linear Quadrupole Ion Trap Orbitrap Mass Analyzer (LTQ–Orbitrap/MS), 518
30.13 Chapter Problems, 527
References, 527

31 Biomolecule Spectral Interpretation: Small Molecules **529**

31.1 Introduction, 529
31.2 Ionization Efficiency of Lipids, 529

31.3 Fatty Acids, 530
31.3.1 Negative Ion Mode Electrospray Behavior of Fatty Acids, 532
31.4 Wax Esters, 537
31.4.1 Oxidized Wax Esters, 538
31.4.2 Oxidation of Monounsaturated Wax Esters by Fenton Reaction, 538
31.5 Sterols, 542
31.5.1 Synthesis of Cholesteryl Phosphate, 542
31.5.2 Single-Stage and High-Resolution Mass Spectrometry, 543
31.5.3 Proton Nuclear Magnetic Resonance (^{1}H-NMR), 543
31.5.4 Theoretical NMR Spectroscopy, 544
31.5.5 Structure Elucidation, 544
31.6 Acylglycerols, 548
31.6.1 Analysis of Monopentadecanoin, 548
31.6.2 Analysis of 1,3-Dipentadecanoin, 548
31.6.3 Analysis of Triheptadecanoin, 550
31.7 ESI-Mass Spectrometry of Phosphorylated Lipids, 551
31.7.1 Electrospray Ionization Behavior of Phosphorylated Lipids, 551
31.7.2 Positive Ion Mode ESI of Phosphorylated Lipids, 553
31.7.3 Negative Ion Mode ESI of Phosphorylated Lipids, 556
31.8 Chapter Problems, 556
References, 557

32 Macromolecule Analysis **559**

32.1 Introduction, 559
32.2 Carbohydrates, 559
32.2.1 Ionization of Oligosaccharides, 561
32.2.2 Carbohydrate Fragmentation, 561
32.2.3 Complex Oligosaccharide Structural Elucidation, 564
32.3 Nucleic Acids, 565
32.3.1 Negative Ion Mode ESI of a Yeast 76-mer $tRNA^{Phe}$, 569
32.3.2 Positive Ion Mode MALDI Analysis, 573
32.4 Chapter Problems, 576
References, 577

33 Biomolecule Spectral Interpretation: Proteins **579**

33.1 Introduction to Proteomics, 579
33.2 Protein Structure and Chemistry, 579
33.3 Bottom-up Proteomics: Mass Spectrometry of Peptides, 580
33.3.1 History and Strategy, 580
33.3.2 Protein Identification through Product Ion Spectra, 584
33.3.3 High-Energy Product Ions, 587
33.3.4 De Novo Sequencing, 587
33.3.5 Electron Capture Dissociation, 589
33.4 Top-Down Proteomics: Mass Spectrometry of Intact Proteins, 590
33.4.1 Background, 590
33.4.2 GP Basicity and Protein Charging, 591
33.4.3 Calculation of Charge State and Molecular Weight, 592
33.4.4 Top-Down Protein Sequencing, 593
33.5 PTM of Proteins, 594
33.5.1 Three Main Types of PTM, 594
33.5.2 Glycosylation of Proteins, 594
33.5.3 Phosphorylation of Proteins, 596
33.5.3.1 Phosphohistidine as PTM, 602
33.5.4 Sulfation of Proteins, 608
33.5.4.1 Glycosaminoglycan Sulfation, 608
33.5.4.2 Tyrosine Sulfation, 609

33.6 Systems Biology and Bioinformatics, 614
33.6.1 Biomarkers in Cancer, 616
33.7 Chapter Problems, 618
References, 619

Appendix I: Chapter Problem Answers **621**

Appendix II: Atomic Weights and Isotopic Compositions **627**

Appendix III: Fundamental Physical Constants **631**

Appendix IV: Redox Half Reactions **633**

Appendix V: Periodic Table of Elements **637**

Appendix VI: Installing and Running Programs **639**

Index **641**

PREFACE

This book is an analytical chemistry book that has two forms: a traditional hardcover book, and an electronic version contained within an analytical chemistry toolkit program. Today's students, technicians, and chemist are all familiar with and daily use a computer. Many textbooks are being converted to online and electronic versions where students study the electronic books using computers or handheld electronics such as Kindles. The book begins with an introduction to the laboratory including safety, glassware, and laboratory basics, and then moves through the fundamentals of analytical techniques such as spectroscopy and chromatography, most common laboratory instrumentation, and examples of laboratory programs such as laboratory information management systems (LIMS). The book also includes a companion teaching, reference, and toolkit program called ChemTech-ToolKit. The ChemTech-ToolKit program contains lesson exercises that stress and review over topics covered in the book, it contains useful calculators, an interactive periodic table, and a copy of all of the chapters of the book that can be opened and read on a computer or handheld electronics such as a Kindle.

The analytical chemist and technician are an invaluable part of the chemistry laboratory. He or she is charged with performing analyses, updating records, taking inventory, documenting projects, keeping the laboratory clean, updating instrumentation and analyses, making sure the laboratory is safe, and giving support to the chief chemist are just a brief description of what the chemist and technician may be charged with. Choosing a career as a chemist or technician in any type of laboratory such as environmental, petroleum, contract, medical, clinical, or biological to name a few examples can be a very rewarding career. Some positions are more demanding than others, but most are challenging. New methods are often needed to be learned or developed. New instrumentation is needed to be set up and used. The chemist and technician also need to keep track of the testing that is being performed and documenting results.

The technicians and chemists in laboratories today are also routinely using computers on a daily basis. Computers are used to control instrumentation and record the data that is being produced. The instrument vendors also often have their own software used for their instruments that the technician, chemist, or laboratory worker needs to learn and use on a routine, daily basis. The personal computer is also used by the chemist and technician on a daily basis for entering laboratory data and writing reports. It is pretty safe to say that most students today have been introduced to the basic operations of the personal computer. A very useful computer program found in most laboratories today is the LIMS. LIMS are used to input data; track sample progress; record sample data such as company, type, and tests needed; and so on. Later in Chapter 12, we are introduced to a LIMS example that we will use to log in samples, input data, search samples, approve samples, and print reports and certificates of analysis (C of A). Also becoming more common in laboratories are electronic laboratory notebooks, and we will take a brief look at using them.

This book is a comprehensive study of analytical chemistry as it pertains to the laboratory analyst and chemist. There are numerous chapters in the book devoted to the basics of analytical chemistry and introductions to the laboratory. The book includes an interactive program called ChemTech-ToolKit for chemists and laboratory technicians. The program acts as a learning aid as we move through the various aspects of analytical chemistry laboratory work and the needed skills to be learned. The ChemTech-ToolKit program has reference tables and an interactive periodic table. It has a link to a LIMS program that we will learn to use. The program also has a review of most of the chapters in this book. The combination of the ChemTech-ToolKit program and the chapters in this book can help to prepare the student for a rewarding career as a chemist or a laboratory technician. The book is also a useful reference for the established chemist or technician already working in the laboratory.

The ChemTech-ToolKit program is a teaching tool designed to equip the chemist, the laboratory analyst, and the technician for a career in the analytical laboratory whether it is a clinical laboratory, an industrial, petrochemical, petroleum, environmental, college or university, or contract laboratory. The book covers the basics of the laboratory including safety, glassware, and balances. Most of the fundamental aspects of laboratory analysis are also covered including titrations, chromatography, and instrumental analyses. Because the ChemTech-ToolKit program contains key points of most of the chapters, ChemTech-ToolKit can be used as a reference and toolkit for the chemist and technician throughout his career. ChemTech-ToolKit progresses from the basics of laboratory fundamentals in safety and glassware through advanced laboratory techniques and analyses. The ChemTech-ToolKit program and the book are a valuable source of reference material also for the industrial or academic researcher in the laboratory. The program and book are in a convenient format for looking up techniques and information in one place for numerous aspects of the laboratory and of samples whether it is the analyst, chemist, or researcher using them. The book and program can also be useful as a supplemental learning source for the college student studying the sciences such as the chemistry, biology, or premed student. The combination of this book with the ChemTech-ToolKit program would be a perfect approach for a community college that is designing a program that would teach and prepare laboratory technicians for an associate degree, and is an undergraduate analytical chemistry textbook.

The book also meets the need of an undergraduate or graduate level analytical chemistry class as all the main topics of analytical chemistry are covered.

Bryan M. Ham
Aihui MaHam

AUTHOR BIOGRAPHIES

Dr. Bryan M. Ham has worked in analytical chemistry laboratories for over 20 years including petrochemical, environmental, foodstuff, and life sciences research. He has published 15 research papers in peer-reviewed journals and two books: *Even Electron Mass Spectrometry with Biomolecule Applications*, John Wiley & Sons, Inc., 2008, and *Proteomics of Biological Systems: Protein Phosphorylation Using Mass Spectrometry Techniques*, John Wiley & Sons, Inc., 2012. He is currently working for the Department of Homeland Security at the US Customs and Border Protection New York Laboratory. He is currently a member of the American Society for Mass Spectrometry (ASMS) and the American Chemical Society (ACS). His research interests include the application of mass spectrometry for biomolecular analysis in the areas of proteomics, lipidomics, and metabolomics and foodstuff chemistry.

Dr. Aihui MaHam is an expert in nanomaterials including the synthesis and characterization of chemical and biological nanosensors. She is also an expert in the field of inorganic materials chemistry and their characterization utilizing methodologies such as SEM, XRD, XRF, and OES. She has published numerous research papers including a recent review in the journal *Small* entitled *Protein-Based Nanomedicine Platforms for Drug Delivery*, which has been cited over 140 times by other researchers. She is currently working for the Department of Homeland Security at the US Customs and Border Protection New York Laboratory. Her current interests include the fundamentals of analytical chemistry and instrumental analyses as applied to modern analytical laboratories.

ACKNOWLEDGMENTS

I would like to acknowledge all those persons whose input, review, and criticisms helped enormously in the early structuring and final content of this book. I would like to include in the acknowledgment each of my past mentors of professors, chief chemists, team leaders, and laboratory managers whose guidance and instruction built with and upon all of the experience that I have gained working in so many areas of analytical chemistry. It was in these different laboratories (petrochemical, contract, environmental, foodstuffs, academic, and government) where the ideas behind this book took shape.

I also include my late parents in the acknowledgment that always supported us in whatever it was we were pursuing.

Finally, and most important of all is the acknowledgment of my wife Dr Aihui MaHam whose consultations, support, reviewing, and invaluable encouragement saw through the entire process of this book from start to finish with an unending presence of which the project would most certainly not have completed to this level without.

1

CHEMIST AND TECHNICIAN IN THE ANALYTICAL LABORATORY

1.1 Introduction—The Analytical Chemist and Technician
1.2 Today's Laboratory Chemist and Technician
  1.2.1 Computers in the Laboratory
  1.2.2 Laboratory Information Management Systems (LIMS)
1.3 ChemTech—The Chemist and Technician Toolkit Companion
  1.3.1 Introduction to ChemTech
1.4 Chapter Layout
  1.4.1 Glassware, Chemicals, and Safety
  1.4.2 Basic Math and Statistics
  1.4.3 Graphing and Plotting
  1.4.4 Making Laboratory Solutions
  1.4.5 Titrimetric Analysis
  1.4.6 Electrochemistry
  1.4.7 Laboratory Information Management System (or Software) LIMS
  1.4.8 Instrumental Analyses—Spectroscopy
  1.4.9 Instrumental Analyses—Chromatography
  1.4.10 Instrumental Analyses—Mass Spectrometry
  1.4.11 Small Molecule and Macromolecule Analysis
1.5 Users of ChemTech

1.1 INTRODUCTION—THE ANALYTICAL CHEMIST AND TECHNICIAN

The analytical chemist and technician are an invaluable part of the chemistry laboratory. He/she is charged with performing analyses, updating records, taking inventory, documenting projects, keeping the laboratory clean, updating instrumentation and analyses, making sure the laboratory is safe, and giving support to the chief chemist. Choosing a career as a chemist or technician in any type of laboratory, such as environmental, petroleum, contract, medical, clinical, or biological to name a few, can be very rewarding. Some positions are more demanding than others, but most are challenging. New methods are often needed to be learned or developed. New instrumentation is needed to be set up and used. The chemist and technician also need to keep track of the testing that is being performed and documenting results.

1.2 TODAY'S LABORATORY CHEMIST AND TECHNICIAN

1.2.1 Computers in the Laboratory

The technicians and chemists in laboratories today are also routinely using computers on a daily basis. Computers are used to control instrumentation and record the data that are being produced. The instrument vendors also often have their own software used for their instruments that the technician, chemist, or laboratory worker needs to learn and use on a routine, daily basis. The personal computer is also used by the chemist and technician on a daily basis for entering laboratory data and writing reports. It is pretty safe to say that most students today have been introduced to the basic operations of the personal computer.

1.2.2 Laboratory Information Management Systems (LIMS)

A very useful computer program found in most laboratories today is the Laboratory Information Management System (LIMS). LIMS is used to input data, track sample progress, record sample data, such as company, type, and tests needed. In Chapter 11, we will be introduced to a LIMS example that we will use to log in samples, input data, search samples, approve samples, and print reports and certificates of analysis (C of A). Also becoming more common in laboratories are electronic laboratory notebooks, and we will take a brief look at using them.

1.3 ChemTech—THE CHEMIST AND TECHNICIAN TOOLKIT COMPANION

1.3.1 Introduction to ChemTech

This textbook is a comprehensive study of analytical chemistry as it pertains to the laboratory analyst and chemist. There are

Analytical Chemistry: A Chemist and Laboratory Technician's Toolkit, First Edition. Bryan M. Ham and Aihui MaHam.

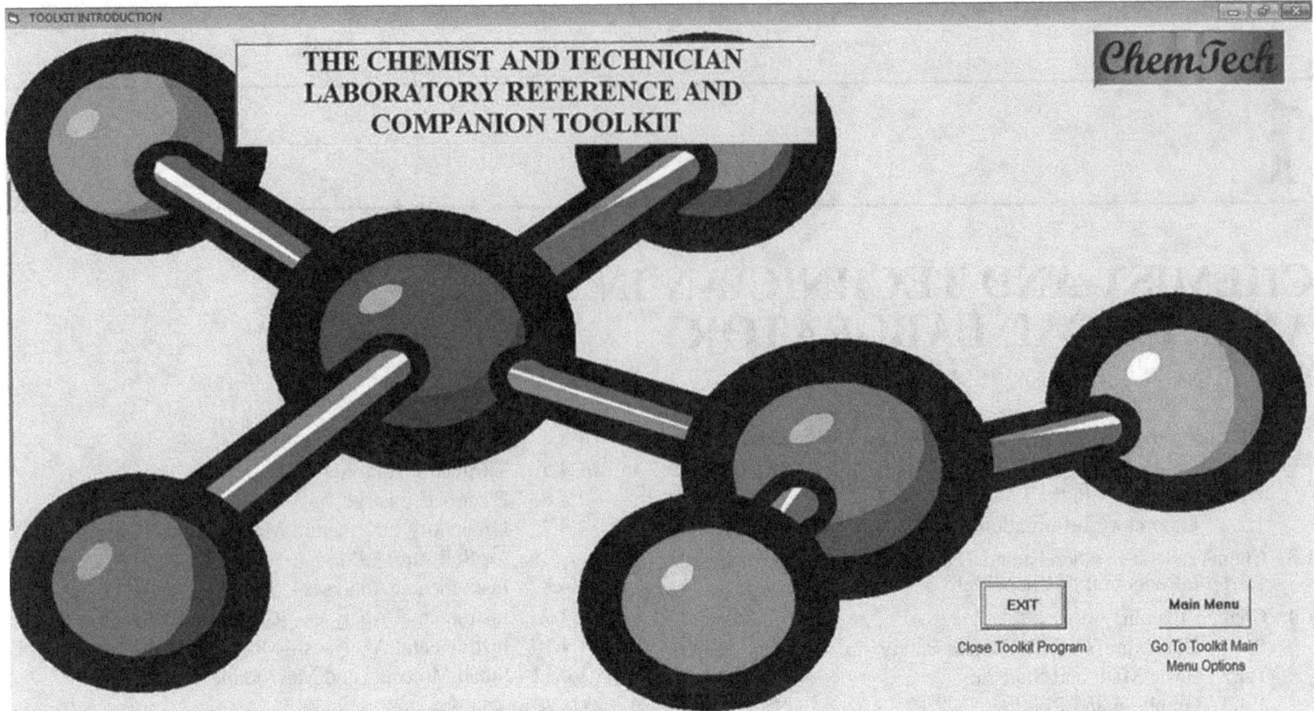

FIGURE 1.1 The ChemTech programs introduction page when it is first opened. The buttons to the bottom right will exit the program or take the user to the main menu of ChemTech.

numerous chapters in the textbook devoted to the basics of analytical chemistry and introductions to the laboratory. This textbook includes an interactive program called ChemTech for chemists and laboratory technicians. The program acts as a learning aid as we move through the various aspects of analytical chemistry, laboratory work, and the needed skills to be learned. The ChemTech program has reference tables and an interactive periodic table. It has a link to a LIMS program that we will learn to use. The program also has a review of most of the chapters in this textbook. The combination of the ChemTech program and the chapters in this textbook can prepare the student for a rewarding career as a chemist or a laboratory technician.

1.3.1.1 Opening ChemTech Let us begin by opening ChemTech by inserting the CD/DVD disk that came with the textbook into your CD/DVD player. The ChemTech program should automatically start, and the main introduction page should look like that found in Figure 1.1. Go down to the bottom right of the page and click on the button "Main Menu." This will open up a page that includes many of the chapters present in the textbook. A link to each chapter is located in the box. Clicking on one of the buttons next to each chapter will open up pages associated with each chapter topic. Located to the right of the page are links to calculators, an interactive periodic table, and some reference tables (Fig. 1.2).

1.3.1.2 Interactive Periodic Table Let us start by clicking the interactive periodic table button to open up the table. Once opened, the page should look like that in Figure 1.3. If you click on an element, a new page will open with facts about the element. Ten of the elements, listed at the bottom of the page, include a rotatable movie of the element. Click on the carbon "C" element to open its fact page. Different facts about the element are listed along with general information about the element. The page should look like the one in Figure 1.4. This is the same for each of the element pages in the periodic table. Also included are pictures of various representations of the elements. Take a few minutes to look at some of the other element information pages. Always remember that you can come back to this page for information about the elements. Click the "Return To Main Menu" button.

1.4 CHAPTER LAYOUT

1.4.1 Glassware, Chemicals, and Safety

Now that we have been introduced to the ChemTech program, we can begin to look at the other chapters. The textbook is designed to take the reader through many of the basic aspects of working in a laboratory. It also teaches the fundamental skills needed of a chemist and technician in science, including basic mathematics. An introduction to laboratory glassware, the layout of laboratories, and instruments used in the laboratory is covered in early chapters. Safety in the laboratory is also covered in an early chapter (see Chapter 3) to give the reader an overview of important aspects of safety in the laboratory.

1.4.2 Basic Math and Statistics

An overview of mathematics used in the laboratory is presented in Chapter 4. The chapter takes the reader through fundamental

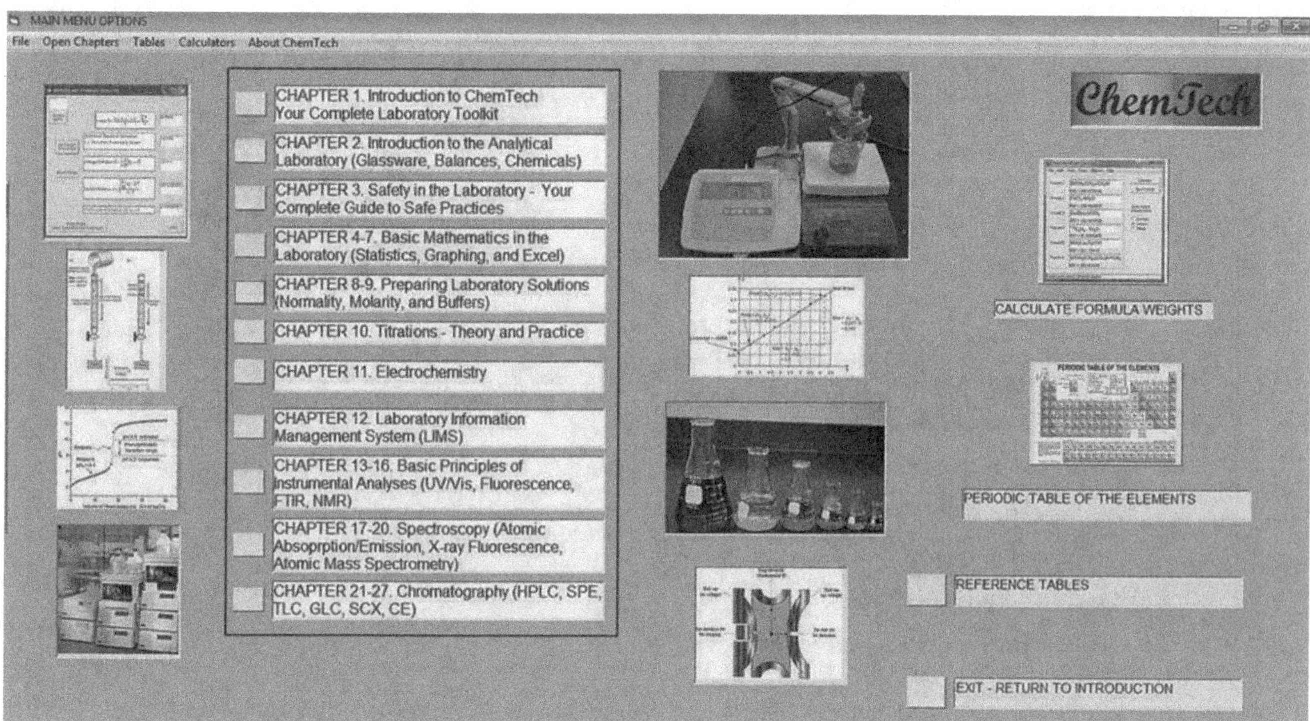

FIGURE 1.2 ChemTech Main Menu Page. The boxed-in area contains links to the various chapters covered in the textbook. To the right are links to calculators, reference tables, and an interactive periodic table.

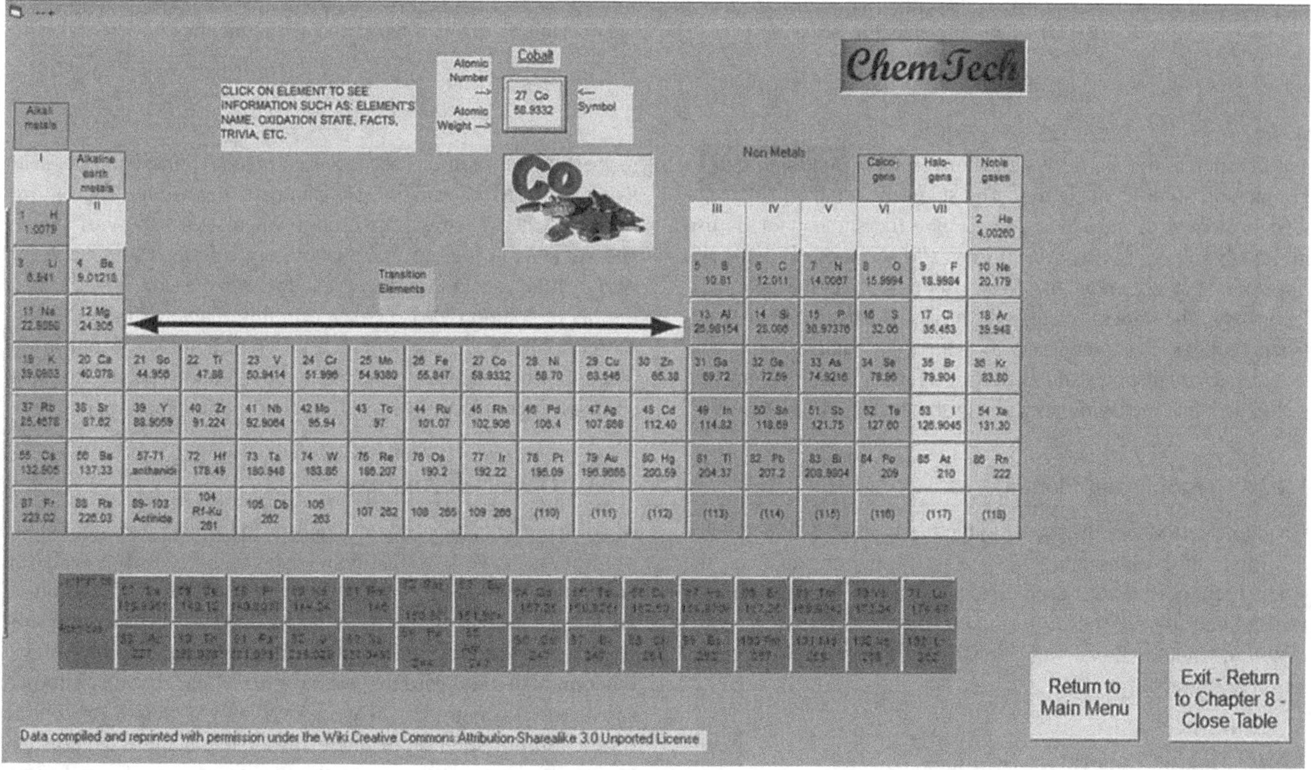

FIGURE 1.3 Interactive periodic table of the elements. Click on an element link to open a page that includes facts, pictures, and movies about the elements.

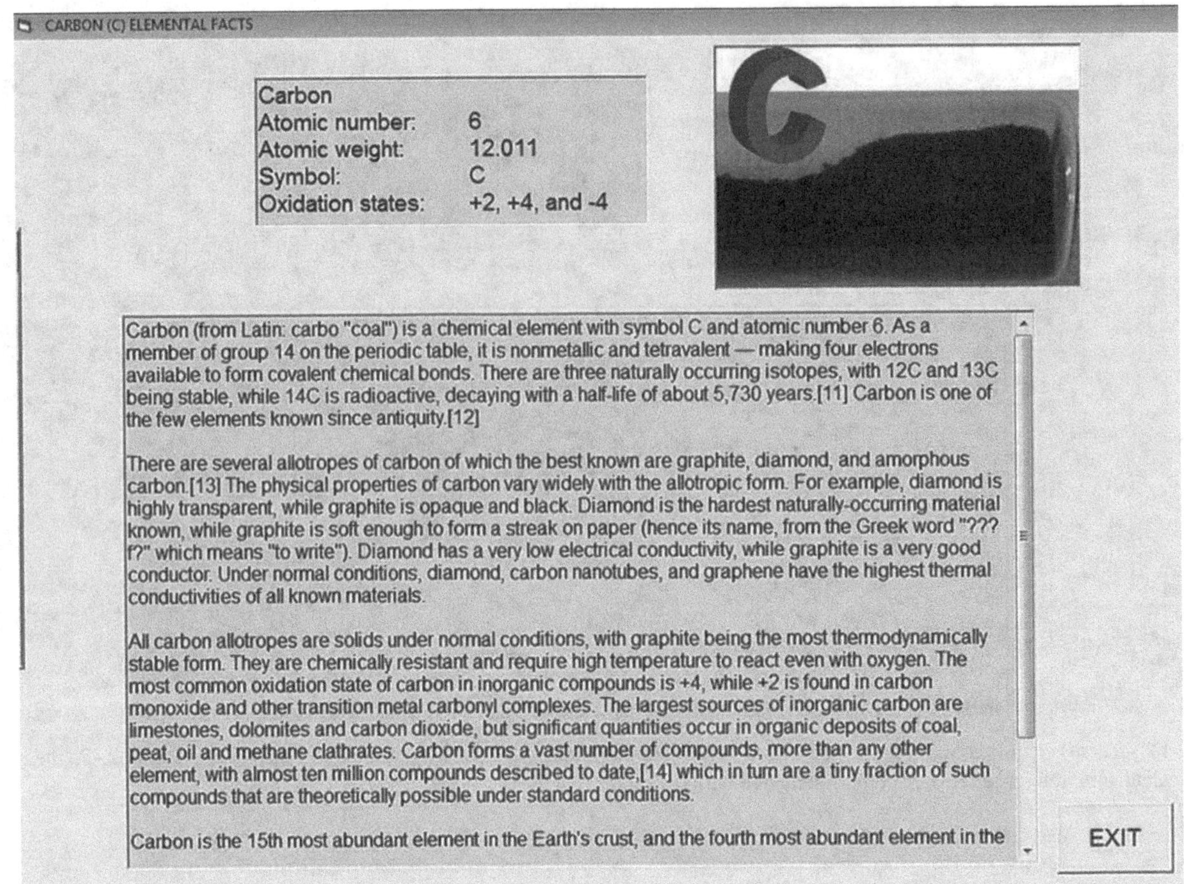

FIGURE 1.4 Carbon element information page. The page contains facts about the element, and pictures.

aspects such as the metric system, conversions such as pounds to grams, and significant figures. The scientific hand calculator is next presented to introduce and prepare the student for its use in the laboratory. The math next moves to statistics in Chapter 5. Statistics is a common and very useful tool in the analytical laboratory. The student is taught the basics of statistics while using ChemTech as a learning aid to help with understanding the associated calculations and concepts. Plotting and graphing are techniques that are used every day in the analytical laboratory.

1.4.3 Graphing and Plotting

Chapter 6 introduces the reader to the basic construction and use of graphs in the laboratory. To aid in the plotting of data, the ChemTech program is used to construct the graphs. Finally, for the math portion of the textbook, Microsoft Excel® is introduced in Chapter 7 as a valuable lab tool for spreadsheet calculation and graphing.

1.4.4 Making Laboratory Solutions

Making solutions in the laboratory is a necessary skill for the chemist and laboratory analyst. There are in fact full time positions that are devoted to making laboratory solutions. We will learn to make solutions of different concentrations and representations such as molarity (M), normality (N), and parts per million (ppm). The program ChemTech will be used to calculate a variety of solutes and solvents needed to make laboratory solutions. In Chapter 9, we will look at acid–base theory, and how to measure and calculate the pH of solutions. Also covered in this chapter is how to make buffer solutions. The ChemTech program is a useful tool in the laboratory for calculating the amount of solutions needed to make some of the most common buffers used in the laboratory.

1.4.5 Titrimetric Analysis

Making buffers and solutions leads us to the analytical technique of titrations. A very useful and widely employed technique in the analytical laboratory is the titrimetric (volumetric) method of analysis. Titrimetric analysis is the process of measuring a substance of unknown concentration in a solution of interest via reaction with a standard that we have made that contains a known substance concentration. If we take a known weight or volume of our solid or solution of interest, we can calculate the concentration of the unknown from the measured use of our known concentration solution. The ChemTech program will be used for a multitude of titrations using various solutions and analyte measurements.

1.4.6 Electrochemistry

Chapter 11 covers the area of electrochemistry in the analytical laboratory including oxidation–reduction reactions and the electrochemical cell. Working with redox equations is covered along with the important Nernst equation. The fundamentals of electrochemistry lead us to redox titrations such as potentiometric titrations.

1.4.7 Laboratory Information Management System (or Software) LIMS

Chapter 12 covers a program that is utilized in most laboratories today, the LIMS. LIMS are used to input data, track sample progress, record sample data such as company, type, and tests needed. In Chapter 12, we will be introduced to a LIMS example that we will use to log in samples, input data, search samples, approve samples, and print reports and certificates of analysis (C of A). Also becoming more common in laboratories are electronic laboratory notebooks that are often coupled with LIMS, and we will take a brief look at using them.

1.4.8 Instrumental Analyses—Spectroscopy

Chapters 13–20 cover a wide range of instrumental analyses that the analyst needs to be introduced to. The analytical laboratory utilizes the phenomenon of the electromagnetic spectrum for an untold number of analyses. Chemists and technicians in the analytical laboratory often make use of the special interaction of molecules with electromagnetic radiation, with the assistance of analytical instrumentation, such as ultraviolet/visible (UV/Vis) spectrophotometers, fluorometers, and Fourier transform infrared spectrometers (FTIR) to measure, identify, and even quantitate compounds of interest. Analysis of metals is also covered in these chapters.

1.4.9 Instrumental Analyses—Chromatography

Chapters 21–27 are dedicated to the area of chromatography. Chromatography is the separation of analyte species using a combination of a mobile phase and a stationary phase. The instrumental techniques covered in these chapters involve different types of chromatography, an extremely useful and quite common technique found in most analytical laboratories that the technician may find himself working in. The chromatographic techniques covered include: Column Liquid Chromatography (LC), High-Performance Liquid Chromatography (HPLC), Solid-Phase Extraction (SPE), Thin-Layer Chromatography (TLC), and Gas-Liquid Chromatography (GC). Chapter 21 starts with the basic theory behind chromatography and then looks in detail at the aforementioned instrumental techniques that include the important components of the chromatography instrumentation. There are illustrative examples throughout the chapters to help the technician in mastering each section followed by a set of problems to be worked at the end of the chapter.

1.4.10 Instrumental Analyses—Mass Spectrometry

Chapters 28–30 cover the more advanced topic of mass spectrometry. We previously looked at the analytical technique of gas chromatography coupled to a single quadrupole mass spectrometer. This is a robust, stable, and well-characterized instrumental analysis that has mostly been automated where the analyst is not called upon for advanced interpretation. The spectra are searched against a library if needed for identifications.

1.4.10.1 Mass Analyzers In these chapters, we will look further at mass spectrometers that are increasingly being used in laboratories today. These include an electric and magnetic sector mass analyzer, a time-of-flight mass analyzer (TOF/MS), a time-of-flight/time-of-flight mass analyzer (TOF–TOF/MS), the hybrid (hybrids are mass analyzers that couple together two separate types of mass analyzers) quadrupole time-of-flight mass analyzer (Q-TOF/MS), a triple quadrupole or linear ion trap mass analyzer (QQQ/MS or LIT/MS), a three-dimensional quadrupole ion trap mass analyzer (QIT/MS), a Fourier transform ion cyclotron mass analyzer (FTICR/MS), and finally the linear ion trap-Orbitrap mass analyzer (IT-Orbitrap/MS). Also included are discussions of the two more recently introduced hybrid mass analyzers in use in laboratories today: the linear quadrupole ion trap Fourier transform mass spectrometer (LTQ-FT/MS) and the linear quadrupole ion trap Orbitrap mass spectrometer (LTQ-Orbitrap/MS).

1.4.10.2 Mass Ionization Also covered are the ionization sources used with the mass spectrometers. Source/ionization systems include electron ionization (EI), electrospray ionization (ESI), chemical ionization (CI), atmospheric pressure chemical ionization (APCI), atmospheric pressure photo ionization (APPI), and matrix-assisted laser desorption ionization (MALDI). These ionization techniques produce ions of analyte molecules (often designated as "M" for molecule), which includes molecular ions $M^{+\bullet}$(from EI), protonated molecules ($[M+H]^+$), deprotonated molecules ($[M–H]^-$), and metal ($[M+metal]^+$, e.g., $[M+Na]^+$) or halide ($[M+halide]^-$, e.g., $[M+Cl]^-$) adduct (all possible from ESI, CI, APCI, APPI, and MALDI).

1.4.11 Small Molecule and Macromolecule Analysis

Chapters 31–33 cover the advanced topics of small molecule analysis, macromolecule analysis, and proteomics. The study of a biological system's compliment of proteins (e.g., from cell, tissue, or a whole organism) at any given state in time has become a major area of focus for research and study in many different fields and applications. In proteomic studies, mass spectrometry can be employed to analyze both the intact, whole protein and the resultant peptides obtained from enzyme-digested proteins. The area of proteomics has been applied to a wide spectrum of physiological samples often based on comparative studies where a specific biological system's protein expression is compared to either another system or the same system under stress. Often in the past the comparison is made using two-dimensional electrophoresis where the gel maps for the two systems are compared looking for changes such as the presence or absence of proteins and the up or down regulation of proteins. Proteins of interest are cut from the gel and identified by mass spectrometry. More recently, proteomics is performed using the advanced instrumentation nano-HPLC/nano-electrospray mass spectrometry in conjunction with bioinformatics software. All will be covered in this chapter.

1.5 USERS OF ChemTech

The ChemTech program is a teaching tool designed to equip the chemist, the laboratory analyst, and the technician for a career in the analytical laboratory whether it is a clinical laboratory, an industrial, petrochemical, petroleum, environmental, college or university, or contract laboratory. This textbook covers the basics of the laboratory, including safety, glassware, and balances. Most of the fundamental aspects of laboratory analysis are also covered including titrations, chromatography, and instrumental analyses. Because the ChemTech program contains key points of most of the chapters, it can be used as a reference and toolkit for the chemist and technician throughout his/her career. ChemTech progresses from the basics of laboratory fundamentals in safety and glassware through advanced laboratory techniques and analyses. The ChemTech program and this textbook are a valuable source of reference material also for the industrial or academic researcher in the laboratory. Both are in a convenient format for looking up techniques and information in one place for numerous aspects of the laboratory and of samples whether it is the analyst, chemist, or researcher using them. They can also be useful as a supplemental learning source for the college student studying the sciences, such as the chemistry, biology, or premed student. The combination of this textbook with the ChemTech program would be a perfect approach for a community college that is designing a program that would teach and prepare laboratory technicians for an associate degree.

2

INTRODUCTION TO THE ANALYTICAL LABORATORY

2.1 Introduction to the Laboratory
 2.1.1 The Scientific Method
2.2 Laboratory Glassware
 2.2.1 Volumetric Flasks
 2.2.2 Beakers and Erlenmeyer Flasks
 2.2.3 Graduated Cylinders
 2.2.4 Pipettes
 2.2.5 Evaporating Dishes
 2.2.6 Flames and Furnaces in the Laboratory
 2.2.7 Laboratory Fume Hoods
 2.2.8 Drying Ovens
 2.2.9 Balances
 2.2.10 Refrigerators and Freezers
 2.2.11 Test Tubes
 2.2.12 Soxhlet Extractions
 2.2.13 Vacuum Pumps
2.3 Conclusion

2.1 INTRODUCTION TO THE LABORATORY

2.1.1 The Scientific Method

Many of us are first introduced to the workings of a simple laboratory as early as the fifth grade. Here, the young students are introduced to the scientific method:

- Ask a Question
- Do Background Research
- Construct a Hypothesis
- Test Your Hypothesis by Doing an Experiment
- Analyze Your Data and Draw a Conclusion
- Communicate Your Results

The experiments usually performed are to produce colors using pH or perhaps observing a visible reaction of acid with shale material (calcium carbonate). In general, the makeup of an analytical laboratory will consist of the basic components of bench space to perform work, sinks for access to water, electrical outlets, and fume hoods to remove harmful vapors. A typical student laboratory is depicted in Figure 2.1 offering a benchtop for performing experiments. Usually located in the middle of the benches are outlets for water and gases such as methane. Methane is used for fueling Bunsen burners or Meker burners used for either heating or burning, which we will look at later in the chapter.

2.2 LABORATORY GLASSWARE

2.2.1 Volumetric Flasks

Glassware is an important and indispensible tool for the analytical laboratory technician and chemist. Taking time to familiarize oneself with the different laboratory glassware and their uses increases safety in the laboratory and also reduces work time and effort. Let us start by looking at an example of commonly used laboratory glassware, the volumetric flask. A schematic of some typical volumetric flasks is shown in Figure 2.2. Note on the neck of the flask there is a line which circumferences the neck. This is the fill line. Bringing the volume up to this point will place into the flask the amount of solvent listed as the flask's size. These flasks are extensively used to make standards, perform dilutions, and adjust volumes.

2.2.2 Beakers and Erlenmeyer Flasks

Some other common glassware used in the analytical laboratory includes beakers and Erlenmeyer flasks, depicted in Figure 2.3. While the beakers and flasks do contain graduations on their sides indicating different volumes, these are only approximate and should not be used for volumetric analysis where exact volumes are needed.

Analytical Chemistry: A Chemist and Laboratory Technician's Toolkit, First Edition. Bryan M. Ham and Aihui MaHam.

FIGURE 2.1 Example of a typical student laboratory. Note the bench space for doing experiments on and sinks in the middle for water and gas outlets.

FIGURE 2.2 Examples of volumetric flasks including 100, 250, 500, and 1000 ml sizes.

2.2.3 Graduated Cylinders

Graduated cylinders and pipettes are used to measure and transfer volumes. Figure 2.4 depicts examples of graduated cylinders commonly used in the laboratory, which also range in sizes from small volume (5 ml) to large volume (1000 ml). Reading the liquid line in a graduated cylinder, a volumetric flask, or a pipette requires the proper visualization of the "meniscus." When a solution is contained within a cylinder such as the neck of a volumetric flask or a graduated cylinder, the liquid due to an electrostatic attraction to the wall will climb slightly up the surface of the cylinder in contact with the liquid and form a small curvature to the surface of the liquid known as the meniscus. The bottom of the meniscus is used to calibrate the glassware and is also where the analyst reads the amount of liquid from. Care must be taken though in reading the meniscus due to an effect known as parallax error. If the analyst looks down at the meniscus while reading it, a slight overestimation of the volume is made. By contrast, if the analyst looks up at the meniscus while reading it, a slight underestimation of the volume is made. The correct way is to look at the meniscus from a point that the eye is level with the meniscus. These three conditions are depicted in Figure 2.5. It can be helpful in reading the meniscus by placing a white card with a black strip on it behind the flask or cylinder and aligning the top of the black strip with the bottom of the meniscus to make the bottom of the meniscus clearer. An example of the use of a white card with a black strip is depicted in Figure 2.6.

2.2.4 Pipettes

Pipettes are used to measure and transfer volumetric amounts of liquids from samples to flasks, or from flask to flask. Pipettes come in many forms and volume transfer accuracies. Class A pipettes are the most accurate pipettes and are usually designated as "to deliver" or TD their contents volumetrically. Class A pipettes are made of glass and come in a broad variety of volumes (e.g., 0.5, 1, 5, and 25 ml). Figure 2.7 depicts some class A pipettes. The flow can be stopped and controlled by placing the finger over the top of the pipette. By adjusting the finger slightly, the flow can be allowed to drain the pipette. Figure 2.8 depicts two types of common pipette bulbs used in the analytical laboratory to draw liquid up into the pipette, and to control the release of the liquid (there are in fact a numerous amount of bulbs available that might be used by the analyst; these are just two examples, but are illustrative in general). Class A pipettes are "to deliver" pipettes; thus, they are allowed to gravity drain. The liquid left in the tip is not "blown" out but has been calibrated as part of the volume, and hence it is left in.

2.2.4.1 Steps for Using Pipette Bulb (a)

1. Insert the pipette into the liquid to be transferred keeping the tip in the liquid at all times while drawing up (if the tip comes out of the liquid while drawing up, the liquid will quickly go into the pipette bulb contaminating it and the liquid).
2. Squeeze the pipette bulb and then place the bulb over the top of the pipette to make a seal.
3. Slowly and gradually release your squeeze on the bulb in small, continuous amounts to draw the liquid into the pipette.
4. Draw the liquid above past the calibration mark and quickly slide your index finger over the top hole of the pipette to trap the liquid inside.
5. Keeping hold with your finger on top, pull the pipette out of the liquid and allow the excess amount of liquid in the pipette to drain down just to the calibration mark.
6. Wipe excess liquid off the outside of the pipette and drain the liquid into the receiver flask.
7. Let the liquid gravity drain out. When finished, touch the pipette tip to the inside glass wall of the receiver flask to remove the last drop.
8. Clean the pipette and store.

(a)

(b)

FIGURE 2.3 Other common glassware used in the analytical laboratory: (a) beakers and (b) Erlenmeyer flasks.

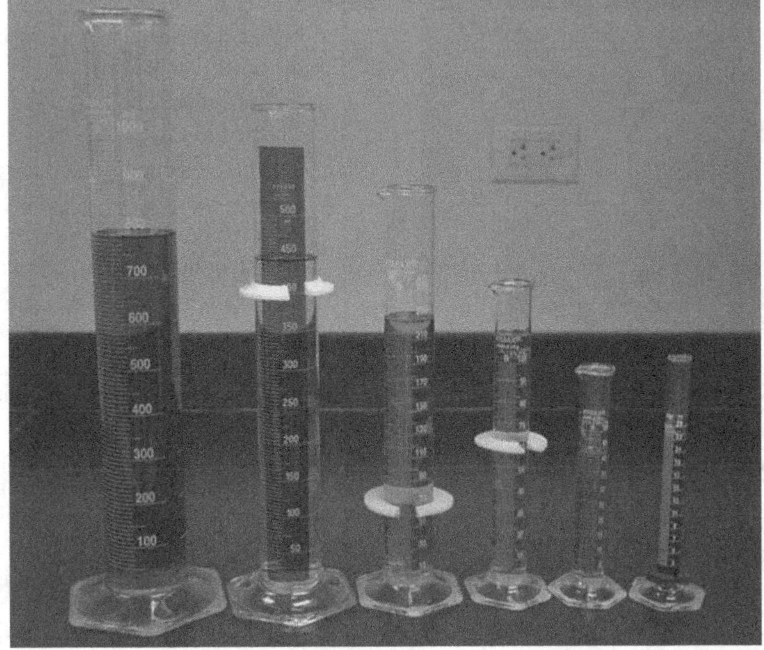

FIGURE 2.4 Examples of graduated cylinders. Pictured are 5, 10, 25, 100, and 500 ml cylinders.

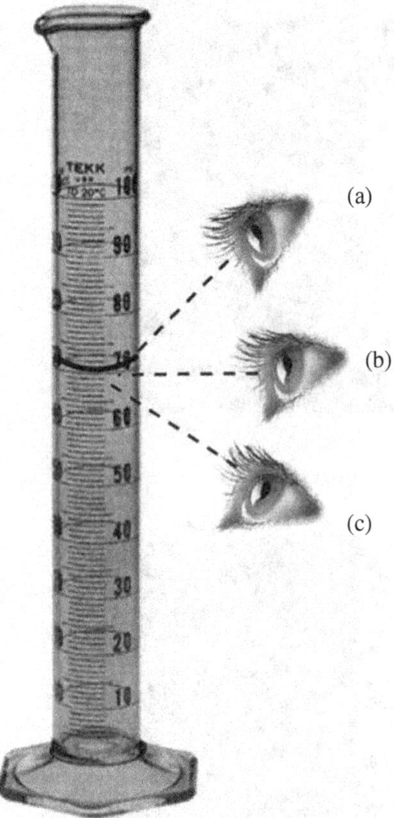

FIGURE 2.5 Proper reading of the meniscus. (a) Reading from above produces an overestimation of the volume. (b) Reading with the eye level to the meniscus is the most accurate volume reading. (c) Reading from below produces an underestimation of the volume.

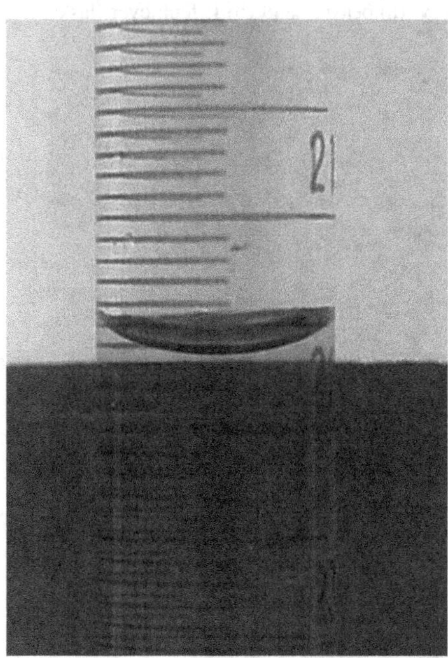

FIGURE 2.6 Example of using a white paper with a black strip for reading volumetric glassware meniscus.

FIGURE 2.7 Class A pipettes with different volumes, and the proper way to hold the pipette with liquid.

2.2.4.2 *Steps for Using Pipette Bulb (b and c)*

1. Press the A valve at the top of the bulb and squeeze the bulb. Let loose of the A valve and the bulb should look like the one to the left-hand side of Figure 2.8(b).
2. Slide the pipette bulb securely onto the top of the pipette as depicted in Figure 2.8(c).
3. Insert the pipette tip into the solution to be transferred.
4. Squeeze the B valve to draw the liquid up into the pipette just to the calibration mark. If passed, the C valve can be used to drain liquid back down to the calibration mark.
5. Squeeze the C valve to drain the liquid into the receiver vessel.

Transferring exact, calibrated volumes using class A pipettes is an important technique and needs to be practiced and mastered by the analyst. The analyst will also use what are known as serological pipettes (also called disposable pipettes) that may be sterile, and usually are "to contain, TC" pipettes where the entire volume is transferred to the receiving flask. Examples of serological pipettes are depicted in Figure 2.9(a). Other common "disposable" pipettes used by the analyst in the analytical laboratory include glass Pasteur pipettes (Fig. 2.9(b) with

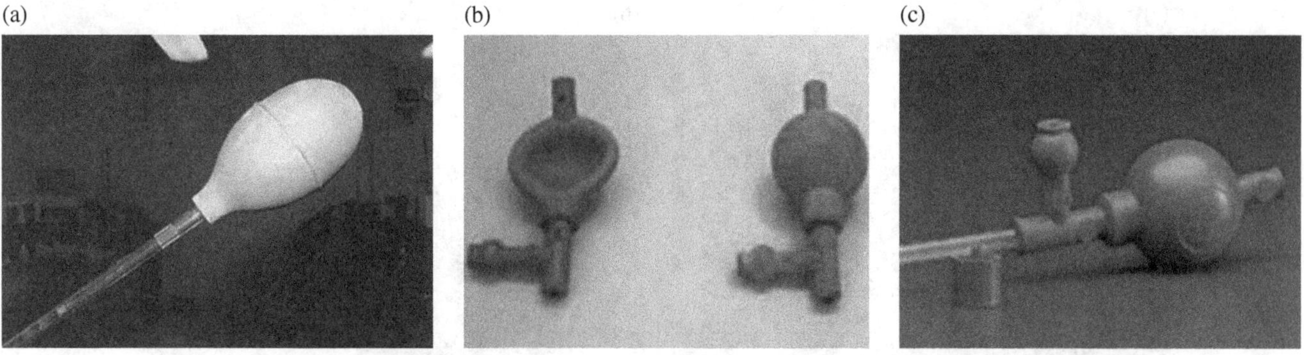

FIGURE 2.8 Common pipette bulbs used to draw liquid up into the pipette.

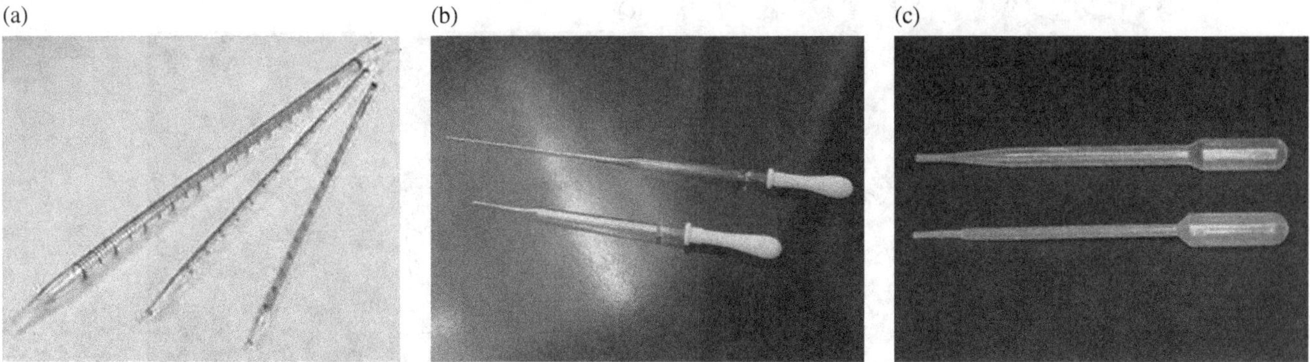

FIGURE 2.9 Examples of common disposable pipettes. (a) Serological pipettes that are "to contain, TC," (b) glass Pasteur pipettes and their bulbs, and (c) plastic transfer pipettes.

associated bulbs) and plastic disposable transfer pipettes shown in Figure 2.9(c).

*2.2.4.3 **Autopipettes*** Finally, many analytical laboratories use autopipettes where a volume can be dialed in or set to draw from a solution, and then transferred. Figure 2.10(a) and (b) depicts examples of autopipettes and their holding rack (Fig. 2.10(a)). These pipettes use disposable plastic tips that are pushed onto the end of the pipette before drawing up solution (shown in Fig. 2.10(c)). Autopipettes are very convenient and work best with aqueous solutions.

2.2.5 Evaporating Dishes

Evaporating dishes, drying dishes, and watch glasses are also commonly used in the analytical laboratory, and are depicted in Figure 2.11. An example of using watch glasses to cover beakers being heated on a hot plate is depicted in Figure 2.12.

2.2.6 Flames and Furnaces in the Laboratory

*2.2.6.1 **Bunsen Burners*** Bunsen burners are commonly used in the laboratory for various analyses; or for heating glassware to bend, or boiling water in a beaker, or burning samples. Figure 2.13 shows different views of a laboratory Bunsen burner. The Bunsen burner is connected with a hose to a natural gas source, usually located in the middle of the bench.

*2.2.6.2 **Crucibles*** Figure 2.14 depicts the use of the Bunsen burner to heat a crucible. The crucible is on a triangle holder with a ring stand. An example of a triangle burner holder is depicted in Figure 2.15. Figure 2.16(a) shows a number of rings on a stand (often called a ring stand) of various sizes. Figure 2.16(b) depicts a ring stand holding a filter funnel. An example of a crucible and laboratory tongs is depicted in Figure 2.17. The tongs are used to grab and hold things in the laboratory, such as a crucible after it has been heated by a Bunsen burner as shown in Figure 2.14.

*2.2.6.3 **Ashing Samples*** Crucibles like the ones shown in these figures are often used to ash a sample. This is where a sample that has been weighed into the crucible is ignited with a flame to burn off water and organic compounds to leave inorganic compounds such as sodium, calcium, and potassium, which is called the ash of the sample. Usually, the sample and crucible are ignited under a fume hood using the Bunsen burner. After the sample no longer burns and looks like black soot, the crucible is placed into a furnace at 550 °C to complete the ashing.

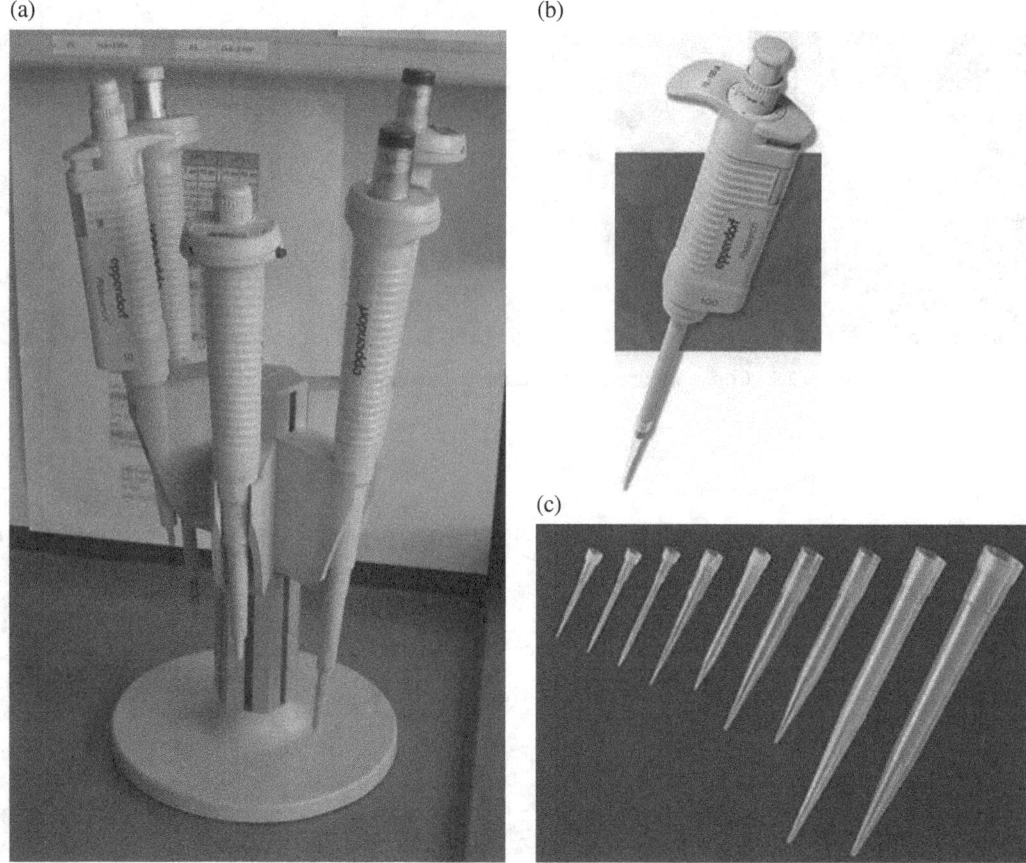

FIGURE 2.10 (a) Example of autopipettes in a holding rack. (b) and (c) Plastic disposable pipette tips.

FIGURE 2.11 Evaporating dishes, drying dishes, and watch glass.

FIGURE 2.12 Example of the use of watch glasses being used to cover beakers being heated on a hot plate.

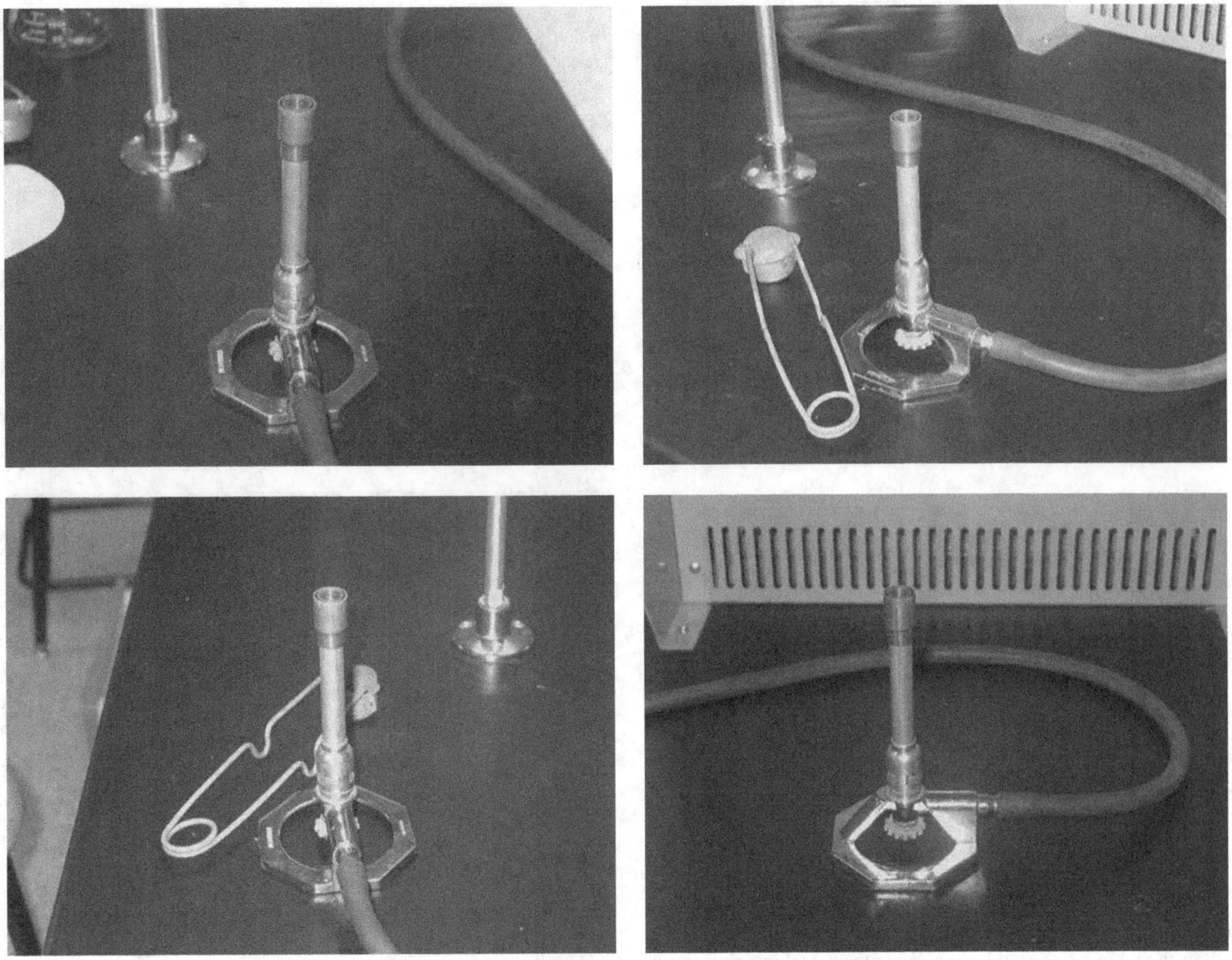

FIGURE 2.13 Some different views of a laboratory Bunsen burner.

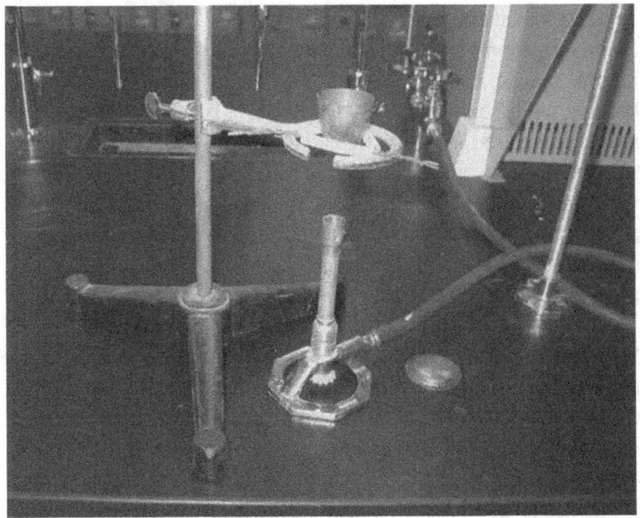

FIGURE 2.14 Bunsen burner heating a crucible.

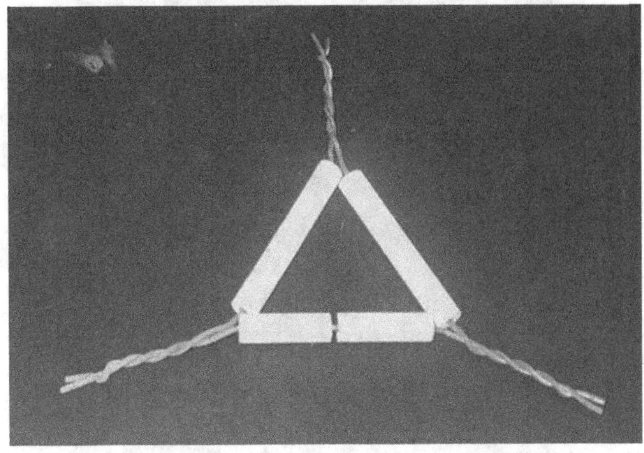

FIGURE 2.15 Triangle burner holder.

(a)

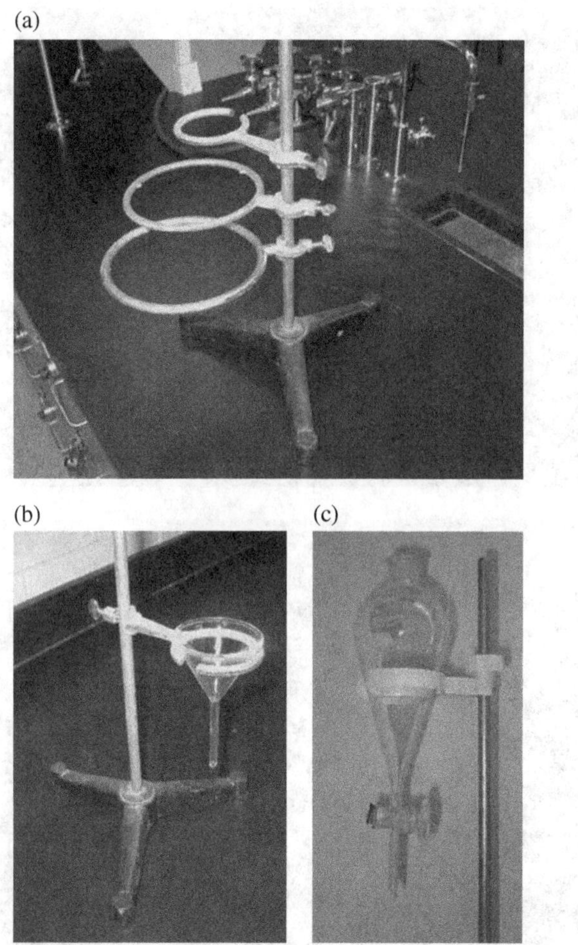

(b) (c)

FIGURE 2.16 (a) Examples of ring stands of different sizes. (b) A filter funnel in a ring stand. (c) A separator funnel in a ring stand.

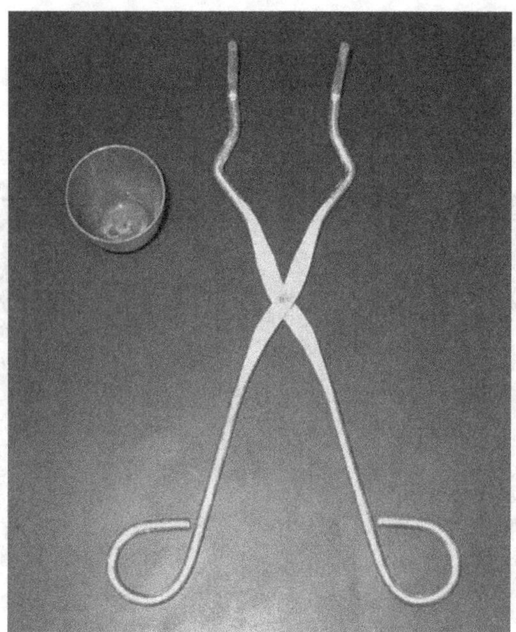

FIGURE 2.17 Example of a crucible and a pair of laboratory tongs. The tongs are used to grasp and hold things like a hot crucible.

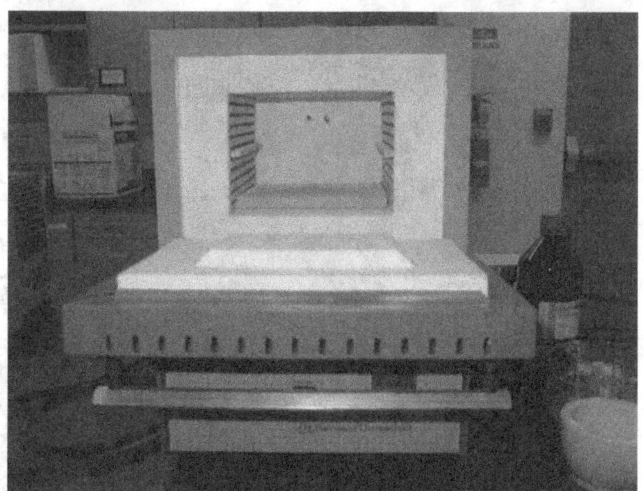

FIGURE 2.18 A laboratory furnace used for ashing samples.

2.2.6.4 Muffle Furnaces An example of a furnace is depicted in Figure 2.18. These are referred to as muffle furnaces and can be used for heating at temperatures up to 1200 °C. A common characterization of a sample is to get the moisture content by drying the sample in a oven at 105 °C, then the organic content by burning, and then ashing the sample at 550 °C in the muffle furnace. Finally, the sample is treated at 900–1000 °C in the furnace to burn off carbonates. The sample can then be reported as having a moisture content, organic content, and ash content, which is composed of metals, such as calcium, sodium, and potassium.

2.2.7 Laboratory Fume Hoods

When working with volatile solvents, burning samples in crucibles, heating solutions producing vapors, or any work that may be evolving fumes, vapors, or smells, the process will be performed within a fume hood. An example of a laboratory fume hood is depicted in Figure 2.19. The fume hood consists of a box area to work in, a front glass hood sash that can be raised and lowered, and a vent system. The vent system is piping that draws the air out of the hood and outside air into the hood. Usually, there is a fan blower motor that is drawing the air out of the hood. There are

FIGURE 2.19 An example of a common laboratory hood. The compartment is positive vented at the top. A blower is connected to the vent pipe at the top of the hood.

FIGURE 2.20 A drying oven located to the left of the laboratory fume hood. These are lower-temperature ovens with temperatures ranging from 40 to 220 °C.

usually gasses that are plumbed into the side of the hood, and a small sink in the back of the hood.

2.2.8 Drying Ovens

Located to the left of the laboratory hood is a laboratory drying oven, depicted in Figure 2.20. These ovens work at low temperatures as compared to the laboratory muffle furnaces. The drying ovens usually operate at temperatures between 40 and 220 °C. The two common types are gravity convection and mechanical convection ovens and vacuum drying ovens. The gravity convection process is as air is heated, it expands and possesses less density (weight per unit volume) than cooler air. Therefore, the heated air rises and displaces the cooler air (the cooler air descends). The method of dry-heat gravity convection produces inconsistent temperatures within the chamber and has a very slow turnover. A mechanical convection oven contains a blower that actively forces heated air throughout all areas of the chamber. The flow created by the blower ensures uniform temperatures and the equal transfer of heat throughout the load. For this reason, the mechanical convection oven is the more efficient of the two processes. The vacuum drying oven uses low atmospheric pressure to aid in the drying process.

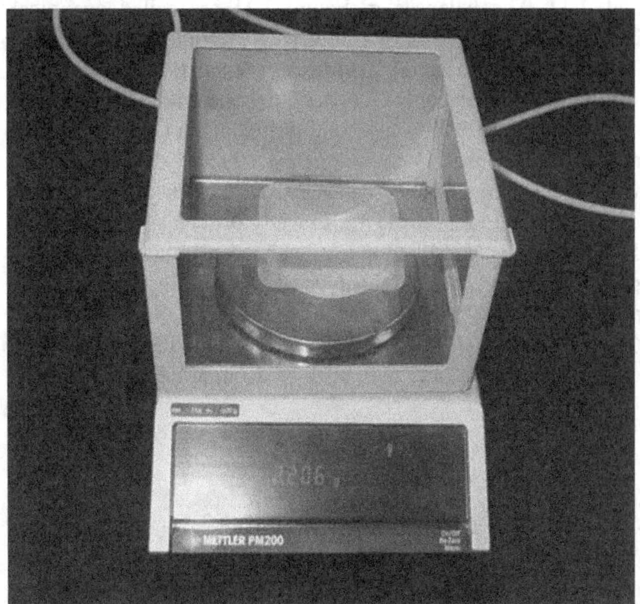

FIGURE 2.21 Examples of analytical balances. Used for precise weighing of amounts up to approximately 220–260 g depending upon the make of the balance.

2.2.9 Balances

Balances used for weighing things in the laboratory are used on a daily basis. Examples of balances are depicted in Figures 2.21

and 2.22. The balances in Figure 2.21 are called analytical balances and are the most common type found in laboratories. These are used to weigh chemicals, glassware, crucibles, and so on, usually between weights of 1 mg and 240 g. The balance shown in Figure 2.22 is a second type called a top loading balance. These are used for weighing things that may be heavier than what needs to be measured on an analytical balance. The weights being weighed range from 1 g to many different amounts from 500 g to 50 kg. Also, the top loading balances often possess less precision (0.1 g) than the analytical. The weight display on an analytical usually goes to four decimal places, or to the 0.1 mg (e.g., 5.3498 g or 29.4 mg). The top loading balance often will display weights to one, two, or three decimal places depending upon the weight.

FIGURE 2.22 A top loading balance used for larger amounts of the item to be weighed. Also used when the precision of an analytical balance is not necessary.

2.2.10 Refrigerators and Freezers

Other common items in the analytical laboratory include refrigerators, such as the one depicted in Figure 2.23, which is actually an explosion proof refrigerator. Standards, samples, and reagents are often stored in laboratory refrigerators as illustrated. Freezers are also commonly used in the laboratory to store standards, samples, and reagents. The temperature of the laboratory refrigerators is usually around 3 to 4 °C, the freezers are at −20, −30, and −80 °C.

2.2.11 Test Tubes

We will continue to look at some common glassware and apparatus that the analyst is likely to work with in the laboratory. Figure 2.24 (a) depicts tests tubes in a tube rack, and a tube holder used when the tube is hot or when heating the tube over a flame. Figure 2.24(b) shows some holders filled with rubber stoppers, and Figure 2.24(c) depicts a holder containing some laboratory thermometers.

2.2.12 Soxhlet Extractions

Figure 2.25 depicts some glassware called Soxhlet apparatus used to extract compounds from a matrix. For example, this is used to extract oil and grease from soil samples. In Figure 2.25(a), there are five Soxhlet apparatus in series. There is a close-up view of the Soxhlet apparatus in Figure 2.25(b). The apparatus comprises a round-bottom reaction flask at the bottom of the setup. The flask is where the extraction solvent is placed and heated. The second stage of the setup consists of the Soxhlet extractor. This is where a large paper thimble is placed that contains the sample. The top of the setup is a condenser where the heated solvent vapors are condensed and fall onto the sample. The solvent will fill up the middle Soxhlet part to a certain level and then drain into the bottom flask taking the extracted compounds with it. This way the extracted compounds are continuously concentrated in the bottom flask. Finally, the bottom flask is removed and connected to a solvent rotator evaporator as shown in Figure 2.26.

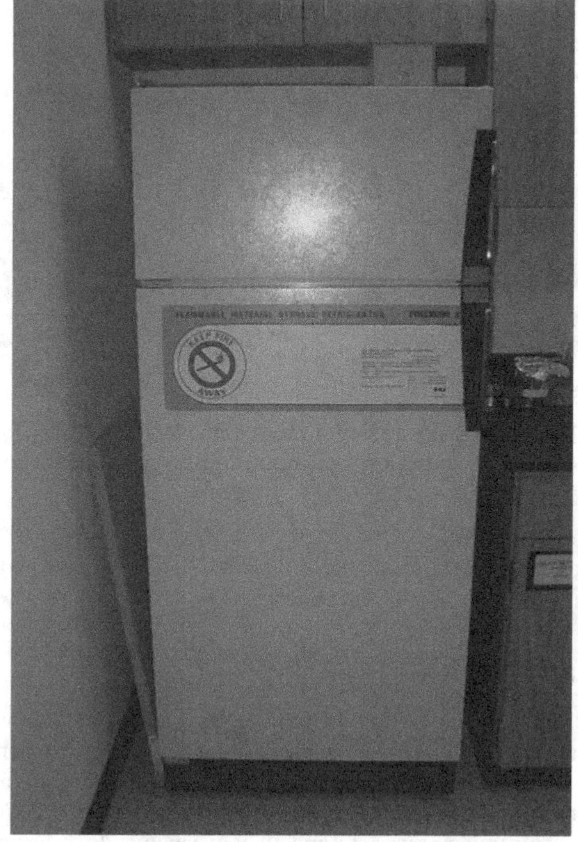

FIGURE 2.23 A laboratory refrigerator used to store standards, samples, and reagents.

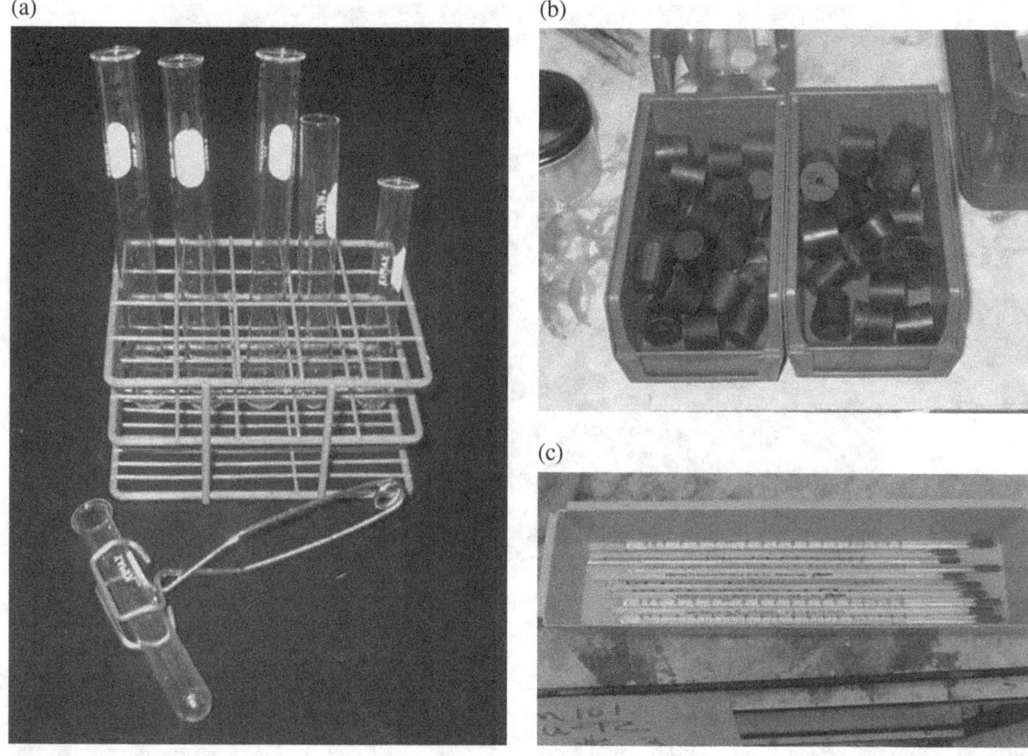

FIGURE 2.24 (a) Test tubes in a test tube rack. Also included is a test tube holder for holding when the tube is hot or being heated. (b) Rubber stoppers. (c) Laboratory thermometers.

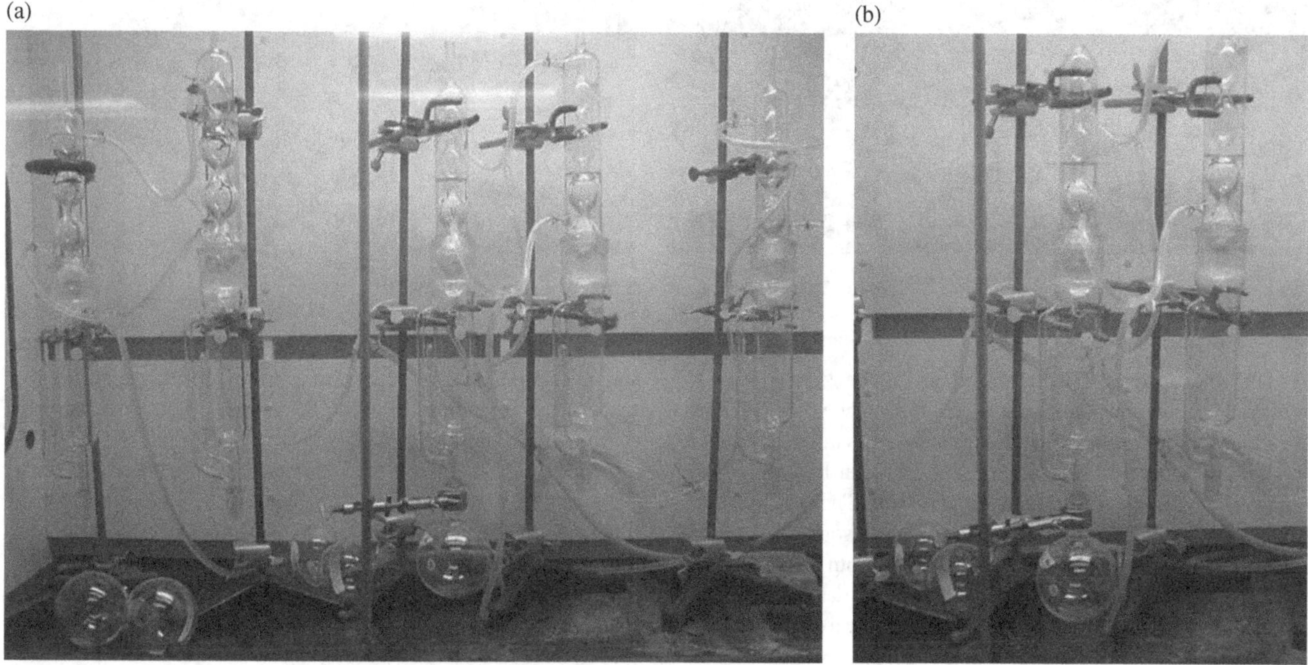

FIGURE 2.25 Soxhlet apparatus used to extract compounds from a matrix such as oil and grease from soil samples. (a) Five Soxhlet apparatus in series. (b) Close-up view of the Soxhlet apparatus composed of a round-bottom reaction flask, a Soxhlet extractor, and a condenser where the heated solvent vapors are condensed and fall onto the sample.

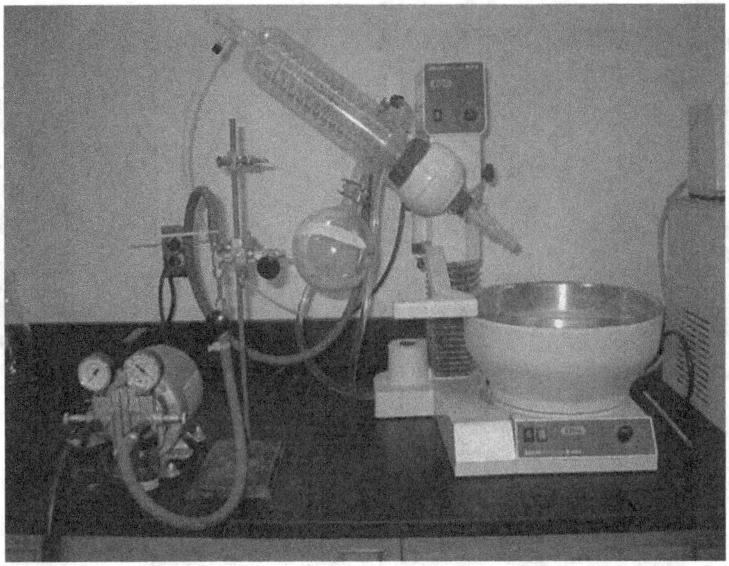

FIGURE 2.26 Rotovap solvent evaporator.

FIGURE 2.27 Examples of laboratory vacuum pumps.

The rotovap evaporator depicted in Figure 2.26 is used to remove large amounts of solvent from a flask such as the one used for the Soxhlet extraction. The flask is attached to the evaporator at the bottom right and lowered into a heated water bath. The flask will rotate in the water bath heating the solvent. The system is also under vacuum to aid in the evaporation of the solvent. The vacuum is created by the small pump located to the bottom left of the system. The evaporator also has a condenser to condense the solvent vapor back to liquid to collect in the solvent receiver flask.

2.2.13 Vacuum Pumps

Vacuum pumps are another common apparatus found in the laboratory. Many instruments require a vacuum pump to keep components under low pressure: mass spectrometers, X-ray fluorescence spectrometers, inductively coupled plasma mass spectrometers, and a few examples of instrumentation that require vacuum pumps. Figure 2.27 depicts a couple of examples of vacuum pumps.

2.3 CONCLUSION

The laboratory is a place where the chemist, technician, and analyst do their work. Like most circumstances of starting to work in a new place, there is a period of becoming familiar with the laboratory where the technician begins to feel comfortable in working. At the beginning, it seems that there is an overwhelming amount of aspects to being a laboratory technician or analyst, but it is obtainable with the proper instruction and training. The following chapters are devoted to training and preparing the technician to begin an interesting, fun, and very rewarding career.

3

LABORATORY SAFETY

3.1 Introduction
3.2 Proper Personal Protection and Appropriate Attire
    3.2.1 Proper Eye Protection
    3.2.2 Proper Laboratory Coats
3.3 Proper Shoes and Pants
3.4 Laboratory Gloves
    3.4.1 Natural Rubber (Latex)
    3.4.2 Nitrile
    3.4.3 Neoprene
    3.4.4 Butyl
    3.4.5 Polyvinyl Chloride (PVC)
    3.4.6 Polyvinyl Alcohol (PVA)
    3.4.7 Viton
    3.4.8 Silver Shield/4H
3.5 General Rules to Use Gloves
3.6 Material Safety Data Sheet (MSDS)
3.7 Emergency Eye Wash and Face Wash Stations
3.8 Emergency Safety Showers
3.9 Fire Extinguishers
    3.9.1 Types of Fires
3.10 Clothing Fire in the Laboratory
3.11 Spill Cleanup Kits
3.12 Chemicals and Solvents
3.13 First Aid Kits
3.14 Gasses and Cylinders
3.15 Sharps Containers and Broken Glass Boxes
3.16 Occupational Safety and Health Administration (OSHA)

3.1 INTRODUCTION

The chemistry laboratory is a very interesting place to work. The chemist, analyst, and technician can have a lot of fulfilling times enjoying the tasks and duties associated with being an analyst. However, the practice of safety in the chemistry laboratory is a vital aspect to keeping a safe, productive, and healthy environment for everyone. All those working in the laboratory need to realize the potential hazards of the laboratory and must know the emergency procedures associated with laboratory accidents. All of the work or experiments performed in the laboratory need to follow the specific safety instructions or procedures in place. While there are specific rules associated with each laboratory, all laboratories follow the basic core safety and emergency policies that are outlined in this chapter. As a good housekeeping practice, all the chemicals must be properly organized (such as alphabetically on shelves or in flame cabinets for flammable liquids), and labeled clearly. The proper and appropriate personal protective equipment (PPE) must be worn at all times in the laboratory. Safety is something that is not only said in meetings and spoken about during training, but must always be forefront in the thoughts and awareness of the laboratory worker, whether a chemist, analyst, or technician.

There are some universal and general laboratory safety rules that must be followed in all chemistry laboratories. In the simplest terms the most basic safety rules to follow include wearing proper PPE, insuring to conduct oneself in a safe manner, and to properly handle samples, chemicals, glassware, and apparatus.

3.2 PROPER PERSONAL PROTECTION AND APPROPRIATE ATTIRE

3.2.1 Proper Eye Protection

Proper eye protection must be worn at all times in the laboratory. Safety goggles are the most effective PPE for protecting the eyes in the laboratory. Goggles are the most effective protection for the eyes from splashed chemicals. Unlike safety glasses, goggles cover, and enclose the eyes completely thus giving the maximum protection from splashed chemicals. Figure 3.1 illustrates proper safety glasses (a) and goggles (b). Prescription normal everyday eye glasses from the eye doctor are not considered proper eye protection. They are not of sufficient strength to protect the eye from a hard contact such as from flying debris. Contact lenses alone are never allowed to be worn in a chemistry laboratory.

Analytical Chemistry: A Chemist and Laboratory Technician's Toolkit, First Edition. Bryan M. Ham and Aihui MaHam.

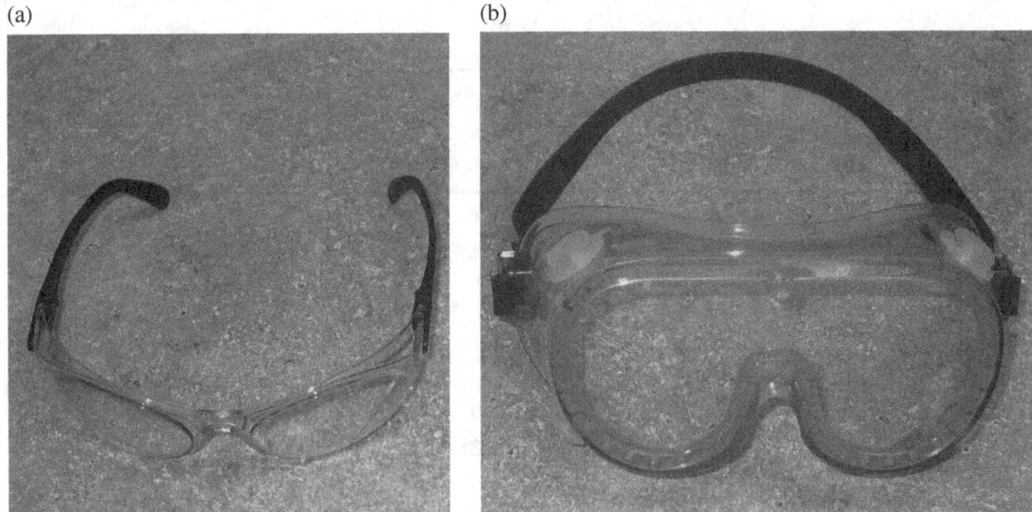

FIGURE 3.1 Examples of proper safety glasses (a) and goggles (b).

FIGURE 3.2 Typical white laboratory coat.

Many different chemical vapors may accumulate and then concentrate under the lenses, eventually causing serious eye damage. Special light such as lasers, ultraviolet light, welding and glassblowing require the use of specialized glasses which are coated to block harmful radiation. Prescription safety glasses are allowed in the laboratory and are often offered as a choice for safety glasses.

3.2.2 Proper Laboratory Coats

Proper laboratory coats should be worn at all times in the laboratory. The laboratory coat is very important to protect against chemical spills, chemical splashes, chemical vapors, cold, heat, and moisture. The coat protects both you and your cloths helping to minimize the contamination of clothes worn by laboratory workers. Laboratory coats are available from chemical supply companies such as VWR International, Laboratory Safety Supply, Fisher Scientific, and Sigma-Aldrich. Figure 3.2 illustrates examples of laboratory coats. Laboratory coats must be worn and stored in the laboratory to prevent spreading chemicals outside the laboratory such as wearing the laboratory coat home or in the car. This can possibly contaminate the car or home with whatever chemicals are on the coat. If corrosive or toxic chemicals contaminate the laboratory coats, the coat should be removed and disposed as a hazardous chemical waste. Usually, the laboratory that the technician is working at will have a system of collecting and sending the dirty laboratory coats out for washing and cleaning.

3.3 PROPER SHOES AND PANTS

Proper shoes and pants must be worn in the laboratory. Closed toe shoes and long pants are appropriate attire for wear in the laboratory. Sandals and open-toed shoes are never allowed in the laboratory. Splashed chemicals or chemical vapors may contact the skin creating a potential health hazard to the analyst.

3.4 LABORATORY GLOVES

Appropriate gloves must be worn at all times in the laboratory. Proper gloves must be worn as long as technicians and analyst are working with chemicals, burning materials, working with extreme temperatures as cold, heat, operating abrasive instruments, and any other hazards. Gloves are commercially available from numerous chemical supply companies. Even though gloves can protect the skin against hazard, gloves can still be punctured allowing contact with chemicals. It is important to realize that gloves can be changed frequently. It is also possible to use dual-layer gloves if mixtures of corrosive chemicals are involved

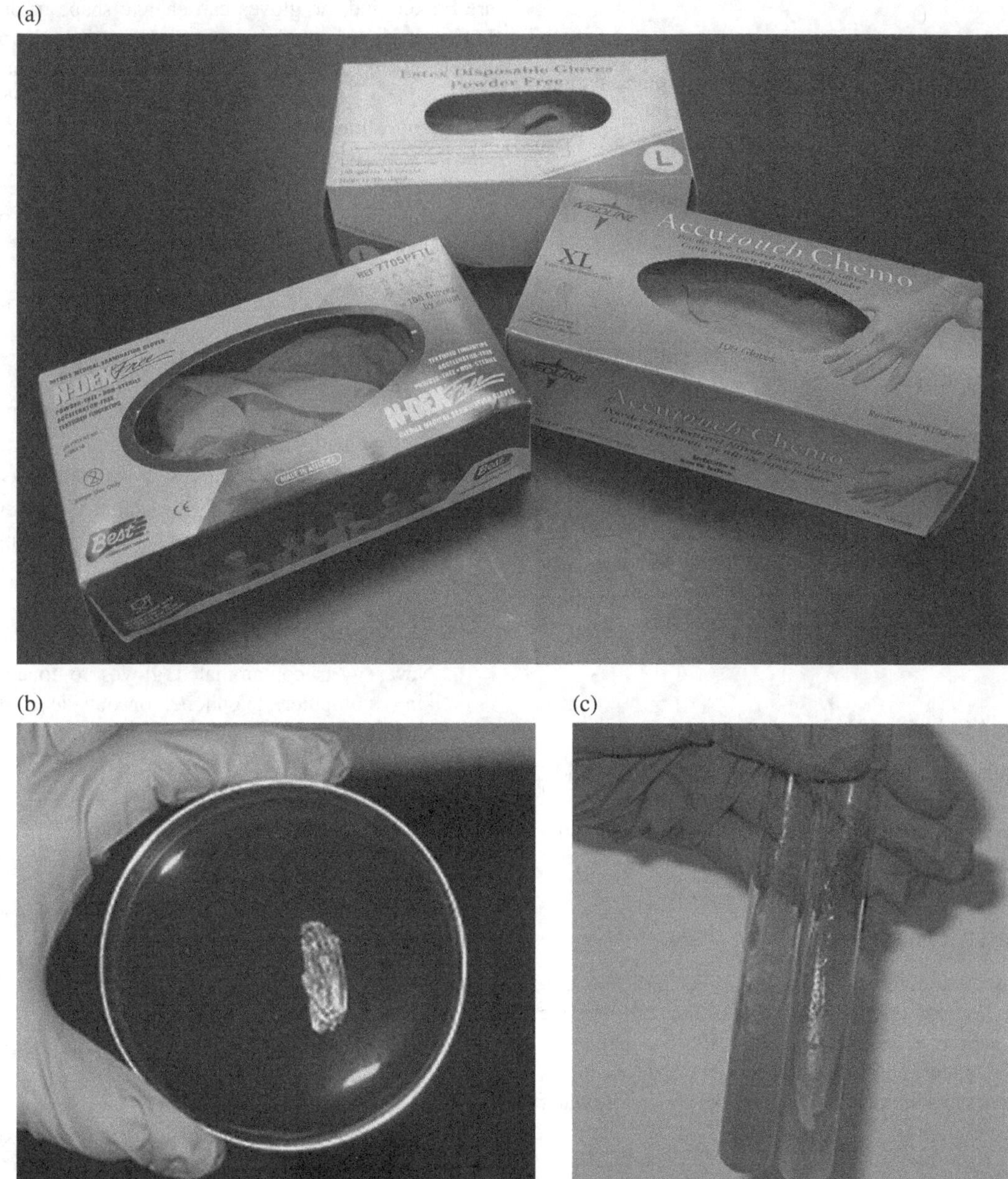

FIGURE 3.3 Common latex and neoprene types of gloves. (a) Glove boxes and (b) and (c) illustrate examples of wearing the gloves to handle laboratory glassware.

or the analyst is operating instruments requiring different toxic chemicals.

Gloves are made of different chemical materials for different protection purposes. Currently, universal glove protection cannot be provided by any one type of gloves material. It is vital to select proper gloves to avoid chemicals contacting the skin. The main concern is the ultimate goal of the protection. It can be against specific chemicals, concern about dexterity of the gloves, or to protect from contamination to the final product. The most common glove materials are both natural and synthetic. Figure 3.3 illustrates common latex and neoprene types of gloves where (a) shows the glove boxes, and (b) and (c) illustrate examples of wearing the gloves to handle laboratory glassware.

The most commonly used ones are listed below:

3.4.1 Natural Rubber (Latex)

Latex is naturally produced rubber that is inherently flexible and resilient. It resists acids, base, alcohols, ketones, and various salts. It can be used to process food, assemble electronics and handle chemicals. However, thin latex gloves cannot always protect the skin while handling chemicals. The disadvantage of latex gloves is allergic reactions by some people due to proteins in the natural rubber. Powder-free latex or synthetic latex gloves are recommended by most Environmental Health and Safety (EH&S) Departments.

3.4.2 Nitrile

Nitrile is a synthetic latex rubber that does not produce a protein allergic reaction in people. Nitrile gloves can replace natural rubber. Nitrile is highly durable to abrasion, puncture, and cut. It provides good resistance to certain solvents and acids (formaldehyde, isopropanol, methanol, and phosphoric acid), oils, vegetable, and animal fats. Nitrile material gloves are not recommended to protect against chloroform or acetone.

3.4.3 Neoprene

A synthetic rubber with oil-resistant properties which are better than those of natural rubber. It can be used in petrochemical plants.

3.4.4 Butyl

A synthetic rubber with best resistance to ketones (such as acetone, M.E.K.), esters, and highly resistant to corrosive acids. It actively reacts with halogen compounds. It can be used in the chemical processing industry, paint, and plastic industries. It is also resistant to pesticide permeation.

3.4.5 Polyvinyl Chloride (PVC)

A plastic polymer that offers high resistance to inorganic acids, aromatics, base, and salts but is not recommended for petrochemical products, halogen compounds, ketone chemicals, and aldehydes chemicals.

3.4.6 Polyvinyl Alcohol (PVA)

A plastic polymer that highly resists strong solvents such as aromatic, aliphatic, ketone, and chlorinated ones that can degrade natural rubber, neoprene, and PVC gloves. It also has high durability to abrasion, snags, punctures, and cuts. But PVA becomes lubricous in aqueous solutions, thus it cannot be used in water based chemical solutions.

3.4.7 Viton

A synthetic fluoroelastomer that is high resistant to most chemicals and solvents, especially aromatic and chlorinated. It is very impermeable to oils, lubricants, most mineral acids, hydraulic fluids, and aqueous solutions. It can be used in aircraft, automobile, and chemical industries.

3.4.8 Silver Shield/4H

A synthetic polymer of five layer laminate of polyehthylene (PE) and ethylene vinyl alcohol (EVOH). It highly resists most toxic chemicals, including aromatics, chlorine, ketones, esters, aliphatics, and alcohols. It can be used in chemical and petrochemical industries, medical laboratory, waste disposal (spill cleanup), photo developing, and hazmat control operations.

All the gloves materials have two properties known as degradation and permeation. Degradation means the gloves material are broken and the gloves can change shape or degrade more. Different glove materials have different degradation times due to contact with chemicals. Permeation means the chemicals break through the gloves materials from the external to the internal surface. Permeation- and degradation-resistance are the most important parameters of various gloves materials. Some other factors are breakthrough time, useful time, chemical concentration, contacting time, work application, inherent nature of chemicals, thermal conditions, and so on.

3.5 GENERAL RULES TO USE GLOVES

1. Select the proper gloves for the proper chemical environment.
2. Before use check the gloves for apparent physical damage.
3. Never reuse disposable gloves after the first use.
4. When a pair of gloves is taken off, try to fold the exterior inside to avoid contacting skin.
5. Once a pair of gloves has been removed, wash hands with soap.
6. Never wear contaminated gloves to touch food, drinking, computer, telephone, or outside of the working environment.

There are other types of gloves such as welding gloves, cut-resistant gloves, mechanics gloves, high to low temperature gloves. These are specialized gloves for specific environments. The authors do not put emphasis on them. If interested, these gloves can be checked with different gloves manufacturers listed above.

3.6 MATERIAL SAFETY DATA SHEET (MSDS)

Material safety data sheet (MSDS) is a sheet providing detailed information of a specific chemical or chemical mixtures regarding the chemical, physical, and safety properties. It is a very important component for the safety environment and workplace. It provides detailed safety procedure for employees and emergency personnel to safely expose to the particular chemical. All MSDS of chemicals should be saved and stored in an easy to access place for professional workers and emergency personnel. Much useful information can be obtained from MSDS such as:

1. Physical and chemical properties (chemical composition, melting point, boiling point, solubility, density, pH, odor and flash point, etc).
2. Health effects.
3. Toxicological information.
4. First aid measures.
5. Stability and reactivity.
6. Handling and storage.
7. Disposal considerations.

SIGMA-ALDRICH

MATERIAL SAFETY DATA SHEET

Date Printed: 07/15/2008
Date Updated: 03/06/2008
Version 1.5

Section 1 - Product and Company Information

Product Name	PETROLEUM ETHER HIGH BOILING, BP 60-80 C
Product Number	77394
Brand	FLUKA
Company	Sigma-Aldrich
Address	3050 Spruce Street SAINT LOUIS MO 63103 US
Technical Phone:	800-325-5832
Fax:	800-325-5052
Emergency Phone:	314-776-6555

Section 2 - Composition/Information on Ingredient

Substance Name	CAS #		SARA 313
PETROLEUM ETHER	101316-46-5		Yes

Ingredient Name	CAS #	Percent	SARA 313
NAPHTHA (PETROLEUM), HYDROTREATED LIGHT	64742-49-0	>= 50 <= 100	No
N-HEXANE	110-54-3	>= 1 <= 5	Yes

Synonyms ligroin

Section 3 - Hazards Identification

EMERGENCY OVERVIEW
Flammable (USA) Extremely Flammable (EU). Harmful.
Irritating to eyes, respiratory system and skin. Harmful: danger of serious damage to health by prolonged exposure through inhalation. Possible risk of impaired fertility. Harmful: may cause lung damage if swallowed. Vapors may cause drowsiness and dizziness.
Target organ(s): Nerves. Kidneys.

HMIS RATING
HEALTH: 2*
FLAMMABILITY: 4
REACTIVITY: 0

NFPA RATING
HEALTH: 2
FLAMMABILITY: 4
REACTIVITY: 0

*additional chronic hazards present.

For additional information on toxicity, please refer to Section 11.

Section 4 - First Aid Measures

FIGURE 3.4 MSDS for petroleum ether.

8. Ecological information.
9. Spill handling procedures.
10. Exposure control/ protective equipment/ personal protection.

In working in chemical environments, safety and health information can be accessible to workers by the MSDS, chemical product labels, and education/training programs. MSDS provides not only the particular substance's chemical and physical properties but also instructions and procedures for the chemical to be used properly and safely as well as point out the potential hazards related to the specific substance. MSDS should be available to all employees that are potentially exposed to chemicals or hazards. This is required by the Occupational Safety and Health Administration (OSHA). MSDS is used for professional workers as well as accessible to fire departments and emergency officials. All chemical products are required to label clearly the following: name of product, name of manufacturer, hazard symbols, hazard risk phrases, proper use of the product, first aid, and reference to the MSDS. The product's label provides brief, clear information of the chemical, where the detailed technical data are contained in MSDS.

An example of an MSDS sheet for petroleum ether is illustrated in Figure 3.4.

3.7 EMERGENCY EYE WASH AND FACE WASH STATIONS

The best treatment for chemical splashes of the eye and face is immediate flushing with copious amounts of water for 15 min using one of the laboratory eyes and face washes. Figure 3.5 is an example of an eye wash located at a laboratory sink. When the wash handle is pushed down the water flows. This is illustrated in Figure 3.6. To flush out the eyes the person leans his face into the stream of water. Eye and Face Washes are equipped with a stay-open ball valve allowing the water to continuously flow while flushing out the eyes. All active eye and face washes should be flushed by laboratory personnel on a weekly basis by allowing the water to flow for 3 min, to remove stagnant water from the pipes. Simply having plastic eye wash bottles is not acceptable as a main eye wash station in the laboratory. They can be used in office settings outside of the laboratory.

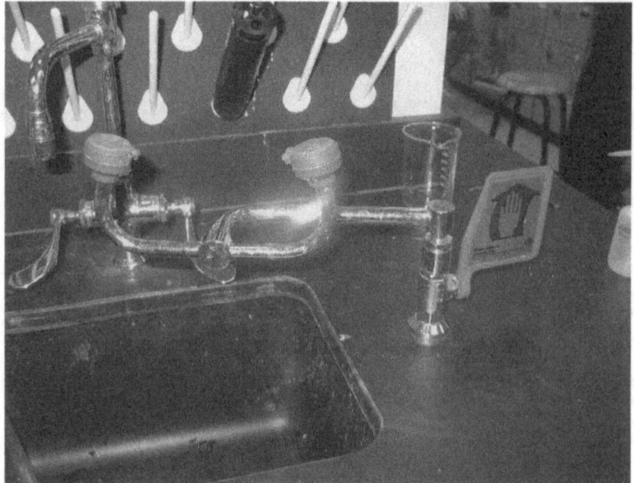

FIGURE 3.5 Eyewash located at the sink in the analytical laboratory.

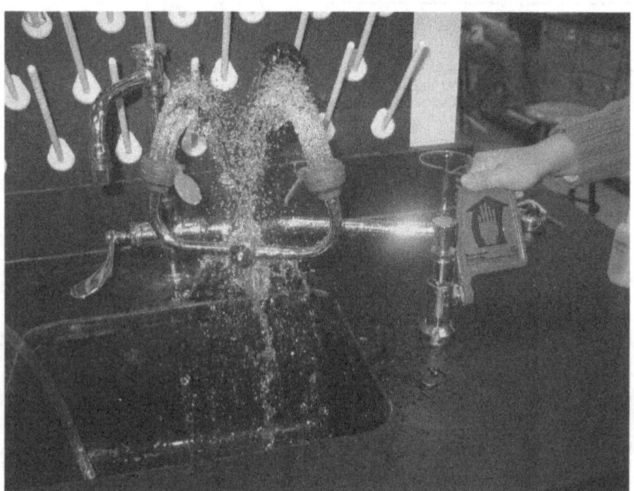

FIGURE 3.6 Eyewash lever has been activated to flush water.

3.8 EMERGENCY SAFETY SHOWERS

Splashes and spills that cover a larger part of the body than the eyes or face are to be washed off using the emergency safety shower. These are showers that spray a shower of water over the entire body. Figure 3.7 illustrates a safety shower that is also equipped with an eye wash. Usually the mechanism to use the safety shower is to stand under the shower head and pull the activation arm down releasing the shower water. Often there will be an alarm that is activated when the shower arm is pulled down alerting others that an emergency has taken place that requires the use of the shower. The area under and around the safety showers must be free from clutter. Items are not to be stored in this area. The shower and areas around it must allow a person to come to the shower without obstruction.

3.9 FIRE EXTINGUISHERS

All laboratories are equipped with different types of fire extinguishers at strategic locations. Fire extinguishers are usually

FIGURE 3.7 Safety shower that is also equipped with an eye wash station.

FIGURE 3.8 The fire triangle listing the three elements required for a fire, fuel, oxygen, and temperature.

located where they are easily accessible and recognizable in the case of an emergency. Fire extinguishers are closely regulated and rated to ensure that they are in proper working order at all times. The elements required to sustain a fire are illustrated in Figure 3.8 in what is often referred to as the fire triangle. The fire triangle elements include fuel to support the fire, oxygen to sustain the fire, and heat to raise the substance to its ignition temperature. If you can take away any one of these you can extinguish the fire.

3.9.1 Types of Fires

Not all fires are the same because the fuel involved creates a different combustion process. The different types of fires require different agents for disrupting the fire triangle in order to extinguish them. The first type of fire is the Class A which is combusted from ordinary materials such as paper, wood, clothing, plastics, and other daily items. The Class A extinguishers

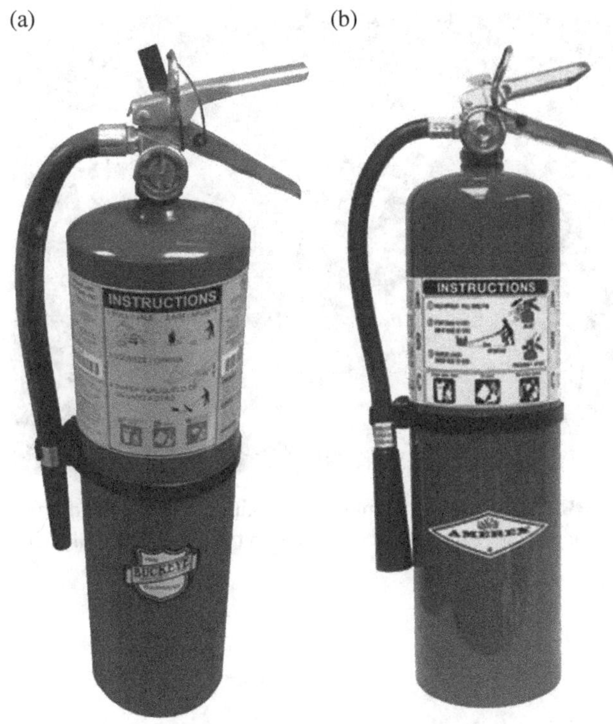

FIGURE 3.9 (a) Class A water filled fire extinguisher. (b) A foam fire extinguisher.

CLASSES OF FIRES	TYPES OF FIRES	PICTURE SYMBOL
A	Wood, paper, cloth, trash & other ordinary materials.	
B	Gasoline, oil, paint and other flammable liquids.	
C	May be used on fires involving live electrical equipment without danger to the operator.	
D	Combustible metals and combustible metal alloys.	D
K	Cooking media (Vegetable or Animal Oils and Fats)	

FIGURE 3.10 A common laboratory placard illustrating the Classes of fires, Types of fires, and Picture symbols for each type.

include water extinguishers and foam extinguishers. Figure 3.9 illustrates (a) a Class A water filled extinguisher, and (b) a foam extinguisher.

Class B types of fires involve flammable liquids such as paint, organic solvents, gasoline, and petroleum oil. Also included in Class B are fires involving flammable gasses such as propane and butane. Cooking oils and grease are not included in Class B. A water filled fire extinguisher cannot be used for a Class B type of fire. The water will splash the liquid that is on fire and can spread it. Carbon dioxide and dry chemical fire extinguishers are used for Class B.

Class C fires involve electrical equipment that is energized. This can be electrical motors, appliances in the house, and transformers. If the electrical power is removed from the apparatus that is on fire, than the Class C fire becomes a Class A or B. Carbon dioxide, dry chemical, and halogenated fire extinguishers are used for Class C fires.

Class D fires cover fires due to metals such as sodium, potassium, and magnesium. Dry powder fire extinguishers are used for Class D fires.

Class K fires cover fires in cooking oils and greases. Wet chemical fire extinguishers are used for Class K fires.

Figure 3.10 illustrates a placard often seen in laboratories listing the Classes of fires, Types of fires, and the Picture symbols for each type.

3.10 CLOTHING FIRE IN THE LABORATORY

Stop, Drop & Roll, and then remove clothing is a well-known action if personal clothing catches fire. The act of rolling on the ground or floor will help to extinguish the flames, especially if you are alone. If someone else is present covering the person with a fire blanket will also extinguish the flames.

3.11 SPILL CLEANUP KITS

All laboratories keep spill cleanup kits that are visible and easily accessible. These kits are used to cleanup small spills that may take place in the laboratory during the course of a work day. Generally, there are three types of spills:

1. Large, uncontrollable spills that involve more than 1 gallon (3.785 l) of substance.
2. Small spills that involve less than 1 gallon (3.785 l) of substance.
3. Minor spills that involve amounts less than 100 ml.

There are a variety of spill kits that are available for use in the laboratory. The most common spill kits are used to cleanup acid spills, caustic spills, organic solvent spills, mercury spills, and formaldehyde spills. Figure 3.11 illustrates a common type of spill kit. The kit includes five agent bottles used for different types of spills according to the type, such as acid, caustic, or solvent. The kit also includes gloves for personal protection, labels for proper identification of spill material, a scoop, and other safety items.

The general steps for cleaning up an accidental spill are as follows:

1. First step is to keep personal safety in mind. Isolate the spill area and make sure you have the proper protection equipment such as gloves, safety glasses, and a respiratory if needed.

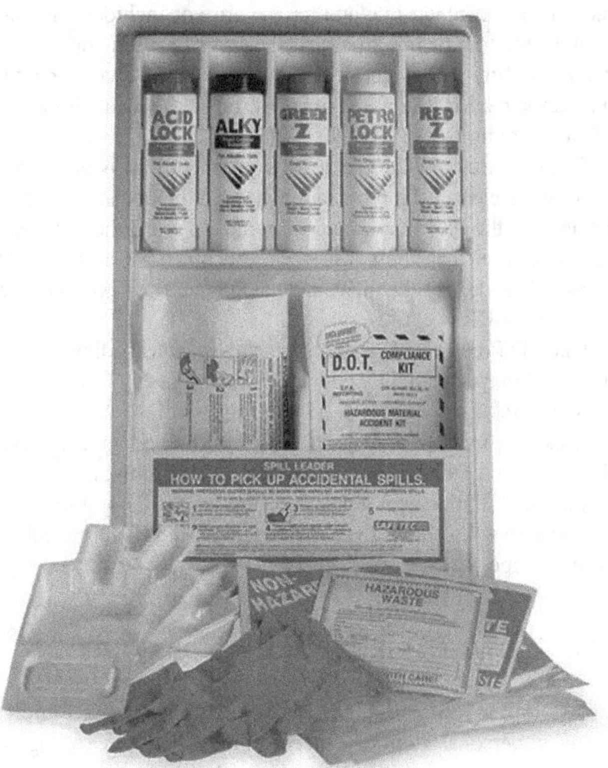

FIGURE 3.11 Common laboratory spill kit including agent bottles for spill types, personal protection gloves, a scoop, and other safety materials.

2. Next, try to identify the spill. The most common are types of spills to identify include acid, caustic, solvent, or formaldehyde.
3. The next step is to select the agent or material to control the spill. Many of the spill kits have bottles in them labeled for acid spills and caustic spills. These bottles are filled with material that will absorb and neutralize the spill. The kits also have absorbent cloths and pillows used to absorb solvent spills.
4. The next step is to treat the spill with the neutralizer/absorbent. Encircle the spill with the proper cleanup agent; then mix the agent into the spill with a spatula or mixing tool supplied in the kit.
5. If the spill is a solvent spill then place the absorbent pillows or towels over the spill to soak it up.
6. The treated spill is then cleaned up by scooping the agent treated spill material into a proper disposal bag that is supplied in the spill kit. The absorbent pillows and blankets that have been used to absorb the solvent spill are also to be placed into the disposal bags supplied with the spill kit. Some spill kits have buckets with lids that are used to dispose of solvent filled absorbent pillows or blankets.
7. The bags for disposal need to be properly labeled as spill cleanup material.
8. Clean up the scraper and scooper that were used to clean up the treated spill by rinsing off.

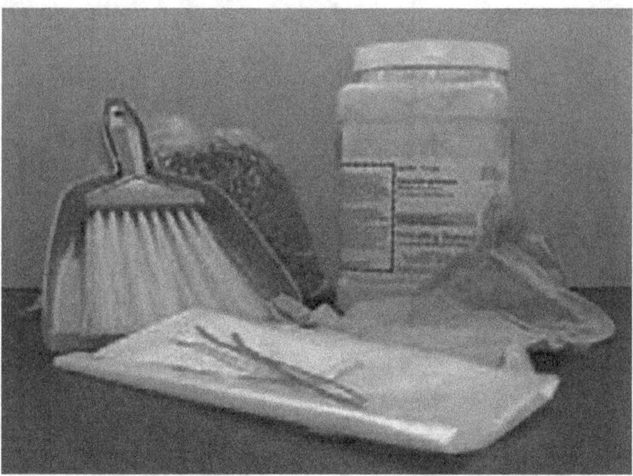

FIGURE 3.12 Example of spill cleanup kit with a scoop and brush used for sweeping up spills that have been treated with a spill agent.

FIGURE 3.13 Example of spill cleanup kit that contains absorbent pillows and disposal buckets.

A second example of an accidental spill kit is illustrated in Figure 3.12. This spill kit shows a good example of a scoop and brush used to sweep up a spill after it has been treated with one of the absorbent agents. Figure 3.13 is a good example of a spill cleanup kit that has absorbent pillows and buckets with lids.

3.12 CHEMICALS AND SOLVENTS

The analytical laboratory will keep and store a wide variety of chemicals and solvents. Dry chemicals and small bottles of nonflammable liquids are often kept on shelves for easy access. Some places will keep chemicals in a stock room. This is often the case in schools, community colleges, and universities. Figure 3.14 is an example of chemicals stored on wooden shelves in a laboratory. The author of the figure points out that the arrangement is not the most proper and even that flammables are stored on the shelves. Storage in this fashion is very susceptible to fires and may result in an accident. Flammable solvents and chemicals are stored in special laboratory flame cabinets. These chemicals are not stored with the other chemicals but are kept in special cabinets. Figure 3.15 illustrates a couple of laboratory flame cabinets used to store flammable solvents and chemicals. Figure 3.16 is another example of chemicals stored on shelves in the laboratory. Some observations are that chemicals are being stored above face level. Many chemicals are stored above face level have the potential for spilling into a worker's face during their removal or placement on the shelf. Some chemicals shown on the upper shelves have ground glass or cork stoppers. One should note the good practice of having a lip on chemical storage shelves to protect from containers falling off from vibrations due to seismic activity, road traffic, building work, and so on.

FIGURE 3.14 Storage of chemicals on wooden shelves. Are any chemicals stored here oxidizers? Note the flammable liquid loading on the right. This shelving arrangement would not last should a fire occur even with fire sprinkler activation. (Reprinted from Journal of Chemical Health and Safety, 15(2), Fred Simmons, David Quigley, Helena Whyte, Janeen Robertson, David Freshwater, Lydia Boada-Clista, J.C. Laul, Chemical storage: Myths vs. reality, 23–30, Copyright (2008), with permission from Elsevier.)

3.13 FIRST AID KITS

First aid kits are always available in laboratories. The kits usually contain Disposable Gloves, Band-Aids, Gauze Bandage, Gauze Pads, and Ice Packs. These kits usually also contain topical creams, liquids, or ointments.

FIGURE 3.15 Examples of laboratory flame cabinets used to store flammable solvents and chemicals.

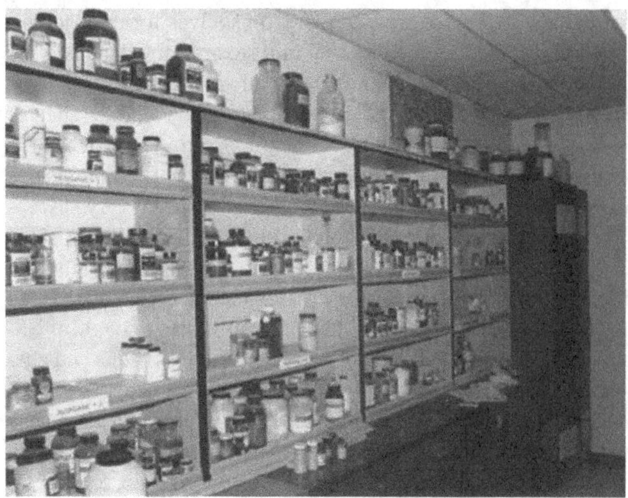

FIGURE 3.16 Chemicals stored above face level. (Reprinted from Journal of Chemical Health and Safety, 15(2), Fred Simmons, David Quigley, Helena Whyte, Janeen Robertson, David Freshwater, Lydia Boada-Clista, J.C. Laul, Chemical storage: Myths vs. reality, 23–30, Copyright (2008), with permission from Elsevier.)

FIGURE 3.17 An example of gas cylinders used in the laboratory.

FIGURE 3.18 Gas cylinders secured with safety chains. (Reprinted with permission under the Wiki Creative Commons Attribution-Sharealike 3.0 Unported License.)

FIGURE 3.19 Gas cylinders in banks as six-packs and eight-packs.

3.14 GASSES AND CYLINDERS

Gasses such as helium, hydrogen, oxygen, air, and nitrogen are commonly used in laboratories usually for supplying gasses to instrumentation. The gasses are contained in gas cylinders. The gas cylinders are metal containers that are cylindrical and come in many sizes. Figure 3.17 illustrates a couple of cylinders that have been plumbed to instrumentation. There are many different cylinder sizes and capacities. For example, typically used in the laboratory is the tallest cylinder at 60 in. that has a capacity of 330 cu ft of gas. The shortest is 18 in with a gas capacity of 22 cu ft. Other sizes include 56 in. (250 cu ft), 48 in. (122 cu ft), 36 in. (80 cu ft), 27 in. (55 cu ft), and 22 in. (40 cu ft). Figure 3.18 shows a set of tanks that is secured with chains. For safety, tanks are never allowed to stand without some type of securing to prevent them from falling. Sometimes gas cylinders are ordered and used a six-packs and eight-packs as illustrated in Figure 3.19. These are banks of the same gas such as six oxygen cylinders. The advantage of this is that the analyst does not need to keep changing the cylinder as it runs out of gas. The time for changing the cylinder has increased by a factor of six or eight.

3.15 SHARPS CONTAINERS AND BROKEN GLASS BOXES

For the disposal of razor blades, hypodermic needles, syringes and other sharp items Sharps containers are used. These are typically red plastic boxes, labeled Sharp container, located throughout the laboratory. Broken glass is disposed of in “Glass Only” boxes. These are usually blue and white thick cardboard boxes that are labeled Glass Only.

3.16 OCCUPATIONAL SAFETY AND HEALTH ADMINISTRATION (OSHA)

The following is a laboratory safety publication from the OSHA (Fig. 3.20).

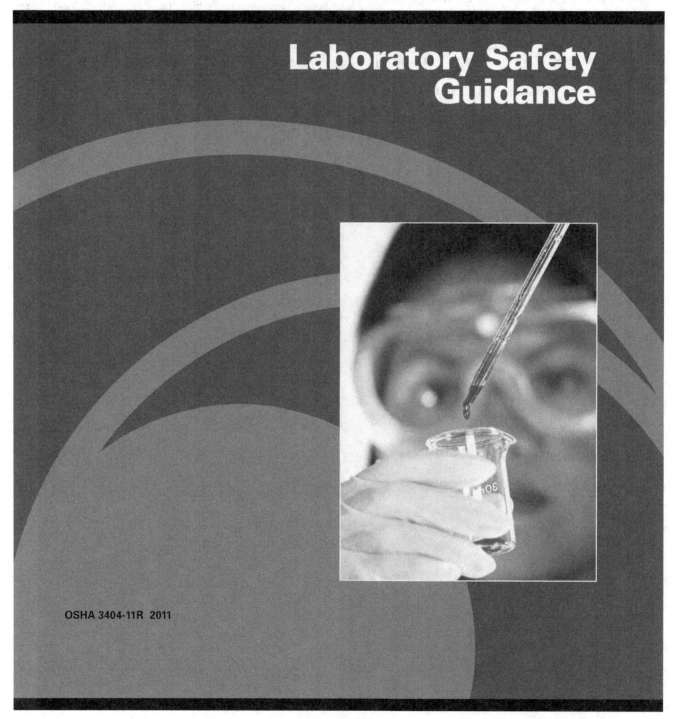

FIGURE 3.20 A laboratory safety publication from the Occupational Safety and Health Administration (OSHA 3404-11R 2011).

Occupational Safety and Health Act of 1970
"To assure safe and healthful working conditions for working men and women; by authorizing enforcement of the standards developed under the Act; by assisting and encouraging the States in their efforts to assure safe and healthful working conditions; by providing for research, information, education, and training in the field of occupational safety and health."

This publication provides a general overview of a particular standards-related topic. This publication does not alter or determine compliance responsibilities which are set forth in OSHA standards, and the *Occupational Safety and Health Act of 1970*. Moreover, because interpretations and enforcement policy may change over time, for additional guidance on OSHA compliance requirements, the reader should consult current administrative interpretations and decisions by the Occupational Safety and Health Review Commission and the courts.

Material contained in this publication is in the public domain and may be reproduced, fully or partially, without permission. Source credit is requested but not required.

This information will be made available to sensory-impaired individuals upon request. Voice phone: (202) 693-1999; teletypewriter (TTY) number: 1-877-889-5627.

FIGURE 3.20 (Continued)

Laboratory Safety Guidance

Occupational Safety and Health Administration
U.S. Department of Labor

OSHA 3404-11R
2011

U.S. Department of Labor
Hilda L. Solis, Secretary of Labor

FIGURE 3.20 (Continued)

This guidance document is not a standard or regulation, and it creates no new legal obligations. It contains recommendations as well as descriptions of mandatory safety and health standards. The recommendations are advisory in nature, informational in content, and are intended to assist employers in providing a safe and healthful workplace. The *Occupational Safety and Health Act* requires employers to comply with safety and health standards and regulations promulgated by OSHA or by a state with an OSHA-approved state plan. In addition, the Act's General Duty Clause, Section 5(a)(1), requires employers to provide their employees with a workplace free from recognized hazards likely to cause death or serious physical harm.

2

FIGURE 3.20 (Continued)

Contents

Introduction 4

OSHA Standards 5

Hierarchy of Controls 8

Chemical Hazards 9

Laboratory Standard 9

Hazard Communication Standard 13

Specific Chemical Hazards 13

Air Contaminants Standard 13

Formaldehyde Standard 14

Latex 15

Chemical Fume Hoods 15

Biological Hazards 15

Biological Agents (other than Bloodborne Pathogens) and Biological Toxins 15

Bloodborne Pathogens 17

Research Animals 19

Biological Safety Cabinets (BSCs) 21

Physical Hazards and Others 21

Ergonomic Hazards 21

Ionizing Radiation 21

Non-ionizing Radiation 22

Noise 23

Safety Hazards 24

Autoclaves and Sterilizers 24

Centrifuges 24

Compressed Gases 24

Cryogens and Dry Ice 25

Electrical 25

Fire 26

Lockout/Tagout 27

Trips, Slips and Falls 28

References 29

Appendices 30

Additional OSHA Information 30

Other Governmental and Non-governmental Agencies Involved in Laboratory Safety 40

Most Common Zoonotic Diseases in Animal Workers 45

Complaints, Emergencies and Further Assistance 46

OSHA Regional Offices 48

FIGURE 3.20 (Continued)

Introduction

More than 500,000 workers are employed in laboratories in the U.S. The laboratory environment can be a hazardous place to work. Laboratory workers are exposed to numerous potential hazards including chemical, biological, physical and radioactive hazards, as well as musculoskeletal stresses. Laboratory safety is governed by numerous local, state and federal regulations. Over the years, OSHA has promulgated rules and published guidance to make laboratories increasingly safe for personnel. This document is intended for supervisors, principal investigators and managers who have the primary responsibility for maintaining laboratories under their supervision as safe, healthy places to work and for ensuring that applicable health, safety and environmental regulations are followed. Worker guidance in the form of Fact Sheets and QuickCards™ is also provided for certain hazards that may be encountered in laboratories. There are several primary OSHA standards that apply to laboratories and these are discussed below. There are also other OSHA standards that apply to various aspects of laboratory activities and these are referred to in this document.

The Occupational Exposure to Hazardous Chemicals in Laboratories standard (29 CFR 1910.1450) was created specifically for non-production laboratories. Additional OSHA standards provide rules that protect workers, including those that who in laboratories, from chemical hazards as well as biological, physical and safety hazards. For those hazards that are not covered by a specific OSHA standard, OSHA often provides guidance on protecting workers from these hazards. This document is designed to make employers aware of the OSHA standards as well as OSHA guidance that is available to protect workers from the diverse hazards encountered in laboratories. The extent of detail on specific hazards provided in this document is dependent on the nature of each hazard and its importance in a laboratory setting. In addition to information on OSHA standards and guidance that deal with laboratory hazards, appendices are provided with information on other governmental and non-governmental agencies that deal with various aspects of laboratory safety.

This Laboratory Safety Guidance booklet deals specifically with laboratories within the jurisdiction of Federal OSHA. There are twenty-five states and two U.S. Territories (Puerto Rico and the Virgin Islands) that have their own OSHA-approved occupational safety and health standards, which may be different from federal standards, but must be at least "as effective as" the federal standards. Contact your local or state OSHA office for further information. More information on OSHA-approved state plans is available at: www.osha.gov/dcsp/osp/index.html.

FIGURE 3.20 (Continued)

OSHA Standards

Section 5(a)(1) of the *Occupational Safety and Health Act of 1970* (OSH Act), the **General Duty Clause**, requires that employers "shall furnish to each of his employees employment and a place of employment which are free from recognized hazards that are causing or likely to cause death or serious physical harm to his employees." Therefore, even if an OSHA standard has not been promulgated that deals with a specific hazard or hazardous operation, protection of workers from all hazards or hazardous operations may be enforceable under section 5(a)(1) of the OSH Act. For example, best practices that are issued by non-regulatory organizations such as the National Institute for Occupational Safety and Health (NIOSH), the Centers for Disease Control and Prevention (CDC), the National Research Council (NRC), and the National Institutes of Health (NIH), can be enforceable under section 5(a)(1).

The principal OSHA standards that apply to all non-production laboratories are listed below. Although this is not a comprehensive list, it includes standards that cover the major hazards that workers are most likely to encounter in their daily tasks. Employers must be fully aware of these standards and must implement all aspects of the standards that apply to specific laboratory work conditions in their facilities.

The Occupational Exposure to Hazardous Chemicals in Laboratories standard (29 CFR 1910.1450), commonly referred to as the Laboratory standard, requires that the employer designate a Chemical Hygiene Officer and have a written Chemical Hygiene Plan (CHP), and actively verify that it remains effective. The CHP must include provisions for worker training, chemical exposure monitoring where appropriate, medical consultation when exposure occurs, criteria for the use of personal protective equipment (PPE) and engineering controls, special precautions for particularly hazardous substances, and a requirement for a Chemical Hygiene Officer responsible for implementation of the CHP. The CHP must be tailored to reflect the specific chemical hazards present in the laboratory where it is to be used. Laboratory personnel must receive training regarding the Laboratory standard, the CHP, and other laboratory safety practices, including exposure detection, physical and health hazards associated with chemicals, and protective measures.

The Hazard Communication standard (29 CFR 1910.1200), sometimes called the HazCom standard, is a set of requirements first issued in 1983 by OSHA. The standard requires evaluating the potential hazards of chemicals, and communicating information concerning those hazards and appropriate protective measures to employees. The standard includes provisions for: developing and maintaining a written hazard communication program for the workplace, including lists of hazardous chemicals present; labeling of containers of chemicals in the workplace, as well as of containers of chemicals being shipped to other workplaces; preparation and distribution of material safety data sheets (MSDSs) to workers and downstream employers; and development and implementation of worker training programs regarding hazards of chemicals and protective measures. This OSHA standard requires manufacturers and importers of hazardous chemicals to provide material safety data sheets to users of the chemicals describing potential hazards and other information. They must also attach hazard warning labels to containers of the chemicals. Employers must make MSDSs available to workers. They must also train their workers in the hazards caused by the chemicals workers are exposed to and the appropriate protective measures that must be used when handling the chemicals.

The Bloodborne Pathogens standard (29 CFR 1910.1030), including changes mandated by the *Needlestick Safety and Prevention Act of 2001*, requires employers to protect workers from infection with human bloodborne pathogens in the workplace. The standard covers all workers with "reasonably anticipated" exposure to blood or other potentially infectious materials (OPIM). It requires that information and training be provided before the worker begins work that may involve occupational exposure to bloodborne pathogens, annually thereafter, and before a worker is offered hepatitis B vaccination. The Bloodborne Pathogens standard also requires advance information and training for all workers in research laboratories who handle human immunodeficiency virus (HIV) or hepatitis B virus (HBV). The standard was issued as a performance standard, which means that the employer must develop a written exposure control plan (ECP) to provide a safe and healthy work environment, but is allowed some flexibility in accomplishing this goal. Among other things, the ECP requires employers to make an exposure determination, establish proce-

FIGURE 3.20 (Continued)

dures for evaluating incidents, and determine a schedule for implementing the standard's requirements, including engineering and work practice controls. The standard also requires employers to provide and pay for appropriate PPE for workers with occupational exposures. Although this standard only applies to bloodborne pathogens, the protective measures in this standard (e.g., ECP, engineering and work practice controls, administrative controls, PPE, housekeeping, training, post-exposure medical follow-up) are the same measures for effectively controlling exposure to other biological agents.

The Personal Protective Equipment (PPE) standard (29 CFR 1910.132) requires that employers provide and pay for PPE and ensure that it is used wherever "hazards of processes or environment, chemical hazards, radiological hazards, or mechanical irritants are encountered in a manner capable of causing injury or impairment in the function of any part of the body through absorption, inhalation or physical contact." [29 CFR 1910.132(a) and 1910.132(h)]. In order to determine whether and what PPE is needed, the employer must "assess the workplace to determine if hazards are present, or are likely to be present, which necessitate the use of [PPE]," 29 CFR 1910.132(d)(1). Based on that assessment, the employer must select appropriate PPE (e.g., protection for eyes, face, head, extremities; protective clothing; respiratory protection; shields and barriers) that will protect the affected worker from the hazard, 29 CFR 1910.132 (d)(1)(i), communicate selection decisions to each affected worker, 29 CFR 1910.132 (d)(1)(ii), and select PPE that properly fits each affected employee, 29 CFR 1910.132(d)(1)(iii). Employers must provide training for workers who are required to use PPE that addresses when and what PPE is necessary, how to wear and care for PPE properly, and the limitations of PPE, 29 CFR 1910.132(f).

The Eye and Face Protection standard (29 CFR 1910.133) requires employers to ensure that each affected worker uses appropriate eye or face protection when exposed to eye or face hazards from flying particles, molten metal, liquid chemicals, acids or caustic liquids, chemical gases or vapors, or potentially injurious light radiation, 29 CFR 1910.133(a).

The Respiratory Protection standard (29 CFR 1910.134) requires that a respirator be provided to each worker when such equipment is necessary to protect the health of such individual. The employer must provide respirators that are appropriate and suitable for the purpose intended, as described in 29 CFR 1910.134(d)(1). The employer is responsible for establishing and maintaining a respiratory protection program, as required by 29 CFR 1910.134(c), that includes, but is not limited to, the following: selection of respirators for use in the workplace; medical evaluations of workers required to use respirators; fit testing for tight-fitting respirators; proper use of respirators during routine and emergency situations; procedures and schedules for cleaning, disinfecting, storing, inspecting, repairing and discarding of respirators; procedures to ensure adequate air quality, quantity, and flow of breathing air for atmosphere-supplying respirators; training of workers in respiratory hazards that they may be exposed to during routine and emergency situations; training of workers in the proper donning and doffing of respirators, and any limitations on their use and maintenance; and regular evaluation of the effectiveness of the program.

The Hand Protection standard (29 CFR 1910.138), requires employers to select and ensure that workers use appropriate hand protection when their hands are exposed to hazards such as those from skin absorption of harmful substances; severe cuts or lacerations; severe abrasions; punctures; chemical burns; thermal burns; and harmful temperature extremes, 29 CFR 1910.138(a). Further, employers must base the selection of the appropriate hand protection on an evaluation of the performance characteristics of the hand protection relative to the task(s) to be performed, conditions present, duration of use, and the hazards and potential hazards identified, 29 CFR 1910.138(b).

The Control of Hazardous Energy standard (29 CFR 1910.147), often called the "Lockout/Tagout" standard, establishes basic requirements for locking and/or tagging out equipment while installation, maintenance, testing, repair, or construction operations are in progress. The primary purpose of the standard is to protect workers from the unexpected energization or startup of machines or equipment, or release of stored energy. The procedures apply to the shutdown of all potential energy sources associated with machines or equipment, including pressures, flows of fluids and gases, electrical power, and radiation.

In addition to the standards listed above, other OSHA standards that pertain to electrical safety

FIGURE 3.20 (Continued)

(29 CFR 1910 Subpart S-Electrical); fire safety (Portable Fire Extinguishers standard, 29 CFR 1910.157); and slips, trips and falls (29 CFR 1910 Subpart D – Walking-Working Surfaces, Subpart E - Means of Egress, and Subpart J - General Environmental Controls) are discussed at pages 25-28. These standards pertain to general industry, as well as laboratories. When laboratory workers are using large analyzers and other equipment, their potential exposure to electrical hazards associated with this equipment must be assessed by employers and appropriate precautions taken. Similarly, worker exposure to wet floors or spills and clutter can lead to slips/trips/falls and other possible injuries and employers must assure that these hazards are minimized. While large laboratory fires are rare, there is the potential for small bench-top fires, especially in laboratories using flammable solvents. It is the responsibility of employers to implement appropriate protective measures to assure the safety of workers.

FIGURE 3.20 (Continued)

Hierarchy of Controls

Occupational safety and health professionals use a framework called the "hierarchy of controls" to select ways of dealing with workplace hazards. The hierarchy of controls prioritizes intervention strategies based on the premise that the best way to control a hazard is to systematically remove it from the workplace, rather than relying on workers to reduce their exposure. The types of measures that may be used to protect laboratory workers, prioritized from the most effective to least effective, are:

- engineering controls;
- administrative controls;
- work practices; and
- personal protective equipment (PPE).

Most employers use a combination of control methods. Employers must evaluate their particular workplace to develop a plan for protecting their workers that may combine both immediate actions as well as longer term solutions. A description of each type of control for non-production laboratories follows.

Engineering controls are those that involve making changes to the work environment to reduce work-related hazards. These types of controls are preferred over all others because they make permanent changes that reduce exposure to hazards and do not rely on worker behavior. By reducing a hazard in the workplace, engineering controls can be the most cost-effective solutions for employers to implement.

Examples include:

- Chemical Fume Hoods; and
- Biological Safety Cabinets (BSCs).

Administrative controls are those that modify workers' work schedules and tasks in ways that minimize their exposure to workplace hazards.

Examples include:

- Developing a Chemical Hygiene Plan; and
- Developing Standard Operating Procedures for chemical handling.

Work practices are procedures for safe and proper work that are used to reduce the duration, frequency or intensity of exposure to a hazard. When defining safe work practice controls, it is a good idea for the employer to ask workers for their suggestions, since they have firsthand experience with the tasks as actually performed. These controls need to be understood and followed by managers, supervisors and workers.

Examples include:

- No mouth pipetting; and
- Chemical substitution where feasible (e.g., selecting a less hazardous chemical for a specific procedure).

Personal Protective Equipment (PPE) is protective gear needed to keep workers safe while performing their jobs. Examples of PPE include respirators (for example, N95), face shields, goggles and disposable gloves. While engineering and administrative controls and proper work practices are considered to be more effective in minimizing exposure to many workplace hazards, the use of PPE is also very important in laboratory settings.

It is important that PPE be:

- Selected based upon the hazard to the worker;
- Properly fitted and in some cases periodically refitted (e.g., respirators);
- Conscientiously and properly worn;
- Regularly maintained and replaced in accord with the manufacturer's specifications;
- Properly removed and disposed of to avoid contamination of self, others or the environment; and
- If reusable, properly removed, cleaned, disinfected and stored.

The following sections of this document are organized based upon classes of hazards, i.e., chemical, biological, physical, safety and other hazards. The organization of these sections and/or subsections may differ somewhat. For instance, OSHA's Laboratory standard is described in greater detail than any other standard in this document. This is because this is the only standard that is specific to laboratories (i.e., non-production laboratories). In all other sections, only those specific aspects of various standards that are considered most relevant to non-production laboratories are discussed. In sections of this document where there are no specific OSHA standards that apply, guidance in the form of Fact Sheets or QuickCards™ may be provided.

OSHA®
Occupational Safety and Health Administration

FIGURE 3.20 (Continued)

Chemical Hazards

Hazardous chemicals present physical and/or health threats to workers in clinical, industrial, and academic laboratories. Laboratory chemicals include cancer-causing agents (carcinogens), toxins (e.g., those affecting the liver, kidney, and nervous system), irritants, corrosives, sensitizers, as well as agents that act on the blood system or damage the lungs, skin, eyes, or mucous membranes. OSHA rules regulate exposures to approximately 400 substances.

Laboratory Standard (29 CFR 1910.1450)

In 1990, OSHA issued the Occupational Exposure to Hazardous Chemicals in Laboratories standard (29 CFR 1910.1450). Commonly known as the Laboratory standard, it was developed to address workplaces where relatively small quantities of hazardous chemicals are used on a non-production basis. However, not all laboratories are covered by the Laboratory standard. For example, most quality control laboratories are not covered under the standard. These laboratories are usually adjuncts of production operations which typically perform repetitive procedures for the purpose of assuring reliability of a product or a process. On the other hand, laboratories that conduct research and development and related analytical work are subject to the requirements of the Laboratory standard, regardless of whether or not they are used only to support manufacturing.

The purpose of the Laboratory standard is to ensure that workers in non-production laboratories are informed about the hazards of chemicals in their workplace and are protected from chemical exposures exceeding allowable levels [i.e., OSHA permissible exposure limits (PELs)] as specified in Table Z of the Air Contaminants standard (29 CFR 1910.1000) and as specified in other substance-specific health standards. The Laboratory standard achieves this protection by establishing safe work practices in laboratories to implement a Chemical Hygiene Plan (CHP).

Scope and Application

The Laboratory standard applies to all individuals engaged in laboratory use of hazardous chemicals. Work with hazardous chemicals outside of laboratories is covered by the Hazard Communication standard (29 CFR 1910.1200). Laboratory uses of chemicals which provide no potential for exposure (e.g., chemically impregnated test media or prepared kits for pregnancy testing) are not covered by the Laboratory standard.

Formaldehyde is one of the most commonly used hazardous chemicals in laboratories. The OSHA Formaldehyde standard (29 CFR 1910.1048) specifically deals with protecting workers from the hazards associated with exposure to this chemical. It should be noted that the scope of the Formaldehyde standard is not affected in most cases by the Laboratory standard. The Laboratory standard specifically does not apply to formaldehyde use in histology, pathology and human or animal anatomy laboratories; however, if formaldehyde is used in other types of laboratories which are covered by the Laboratory standard, the employer must comply with 29 CFR 1910.1450.

Program Description

The Laboratory standard consists of five major elements:

- Hazard identification;
- Chemical Hygiene Plan;
- Information and training;
- Exposure monitoring; and
- Medical consultation and examinations.

Each laboratory covered by the Laboratory standard must appoint a Chemical Hygiene Officer (CHO) to develop and implement a Chemical Hygiene Plan. The CHO is responsible for duties such as monitoring processes, procuring chemicals, helping project directors upgrade facilities, and advising administrators on improved chemical hygiene policies and practices. A worker designated as the CHO must be qualified, by training or experience, to provide technical guidance in developing and implementing the provisions of the CHP.

Hazard Identification

Each laboratory must identify which hazardous chemicals will be encountered by its workers. All containers for chemicals must be clearly labeled. An employer must ensure that workers do not use, store, or allow any other person to use or store, any hazardous substance in his or her laboratory if the container does not meet the labeling requirements outlined in the Hazard Communication standard,

FIGURE 3.20 (Continued)

29 CFR 1910.1200(f)(4). Labels on chemical containers must not be removed or defaced.

Material Safety Data Sheets (MSDSs) for chemicals received by the laboratory must be supplied by the manufacturer, distributor, or importer and must be maintained and readily accessible to laboratory workers. MSDSs are written or printed materials concerning a hazardous chemical. Employers must have an MSDS in the workplace for each hazardous chemical in use.

MSDS sheets must contain:
1. Name of the chemical;
2. Manufacturer's information;
3. Hazardous ingredients/identity information;
4. Physical/chemical characteristics;
5. Fire and explosion hazard data;
6. Reactivity data;
7. Health hazard data;
8. Precautions for safe handling and use; and
9. Control measures.

The United States is participating in the Global Harmonization System of Classifying and Labeling Chemicals (GHS) process and is planning to adopt the GHS in its Hazard Communication standard. The GHS process is designed to improve comprehensibility, and thus the effectiveness of the Hazard Communication standard (HCS), and help to further reduce illnesses and injuries. GHS is a system that defines and classifies the hazards of chemical products, and communicates health and safety information on labels and material safety data sheets (called Safety Data Sheets, or SDSs, in the GHS). The most significant changes to the Hazard Communication standard will include changing terminology: "hazard determination" to "hazard classification" (along with related terms) and "material safety data sheet" to "safety data sheet." The goal is that the same set of rules for classifying hazards, and the same format and content for labels and safety data sheets (SDS) will be adopted and used around the world. An international team of hazard communication experts developed GHS.

The biggest visible impact of the GHS is the appearance of and information required for labels and SDSs. Labels will require signal words, pictograms, precautionary statements and appropriate hazard statements. The GHS system covers all hazardous chemicals and may be adopted to cover chemicals in the workplace, transport, consumer products, and pesticides. SDSs will follow a new 16-section format, containing requirements similar to those identified in the American National Standards Institute (ANSI) Z400 and International Organization for Standardization (ISO) 11014 standards. Information on GHS classification, labels and SDSs is available at: http://www.unece.org/trans/danger/publi/ghs/ghs_welcome_e.html.

Chemical Hygiene Plan (CHP)

The purpose of the CHP is to provide guidelines for prudent practices and procedures for the use of chemicals in the laboratory. The Laboratory standard requires that the CHP set forth procedures, equipment, PPE and work practices capable of protecting workers from the health hazards presented by chemicals used in the laboratory.

The following information must be included in each CHP:

Standard Operating Procedures (SOPs): Prudent laboratory practices which must be followed when working with chemicals in a laboratory. These include general and laboratory-specific procedures for work with hazardous chemicals.

Criteria for Exposure Control Measures: Criteria used by the employer to determine and implement control measures to reduce worker exposure to hazardous chemicals including engineering controls, the use of PPE and hygiene practices.

Adequacy and Proper Functioning of Fume Hoods and other Protective Equipment: Specific measures that must be taken to ensure proper and adequate performance of protective equipment, such as fume hoods.

Information and Training: The employer must provide information and training required to ensure that workers are apprised of the hazards of chemicals in their work areas and related information.

Requirement of Prior Approval of Laboratory Procedures: The circumstances under which certain laboratory procedures or activities require approval from the employer or employer's designee before work is initiated.

Medical Consultations and Examinations: Provisions for medical consultation and examination when exposure to a hazardous chemical has or may have taken place.

FIGURE 3.20 (Continued)

Chemical Hygiene Officer Designation: Identification of the laboratory CHO and outline of his or her role and responsibilities; and, where appropriate, establishment of a Chemical Hygiene Committee.

Particularly Hazardous Substances: Outlines additional worker protections for work with particularly hazardous substances. These include select carcinogens, reproductive toxins, and substances which have a high degree of acute toxicity.

Information and Training

Laboratory workers must be provided with information and training relevant to the hazards of the chemicals present in their laboratory. The training must be provided at the time of initial assignment to a laboratory and prior to assignments involving new exposure situations.

The employer must inform workers about the following:

- The content of the OSHA Laboratory standard and its appendices (the full text must be made available);
- The location and availability of the Chemical Hygiene Plan;
- Permissible exposure limits (PELs) for OSHA-regulated substances, or recommended exposure levels for other hazardous chemicals where there is no applicable standard;
- Signs and symptoms associated with exposure to hazardous chemicals in the laboratory; and
- The location and availability of reference materials on the hazards, safe handling, storage and disposal of hazardous chemicals in the laboratory, including, but not limited to, MSDSs.

Training must include the following:

- Methods and observations used to detect the presence or release of a hazardous chemical. These may include employer monitoring, continuous monitoring devices, and familiarity with the appearance and odor of the chemicals;
- The physical and health hazards of chemicals in the laboratory work area;
- The measures that workers can take to protect themselves from these hazards, including protective equipment, appropriate work practices, and emergency procedures;
- Applicable details of the employer's written Chemical Hygiene Plan;
- Retraining, if necessary.

Exposure Determination

OSHA has established permissible exposure limits (PELs), as specified in 29 CFR 1910, subpart Z, for hundreds of chemical substances. A PEL is the chemical-specific concentration in inhaled air that is intended to represent what the average, healthy worker may be exposed to daily for a lifetime of work without significant adverse health effects. The employer must ensure that workers' exposures to OSHA-regulated substances do not exceed the PEL. However, most of the OSHA PELs were adopted soon after the Agency was first created in 1970 and were based upon scientific studies available at that time. Since science has continued to move forward, in some cases, there may be health data that suggests a hazard to workers below the levels permitted by the OSHA PELs. Other agencies and organizations have developed and updated recommended occupational exposure limits (OELs) for chemicals regulated by OSHA, as well as other chemicals not currently regulated by OSHA. Employers should consult other OELs, in addition to the OSHA PEL, to make a fully informed decision about the potential health risks to workers associated with chemical exposures. The American Conference of Governmental Industrial Hygienists (ACGIH), the American Industrial Hygiene Association (AIHA), the National Institute for Occupational Safety and Health (NIOSH), as well as some chemical manufacturers have established OELs to assess safe exposure limits for various chemicals.

Employers must conduct exposure monitoring, through air sampling, if there is reason to believe that workers may be exposed to chemicals above the action level or, in the absence of an action level, the PEL. Periodic exposure monitoring should be conducted in accord with the provisions of the relevant standard. The employer should notify workers of the results of any monitoring within 15 working days of receiving the results. Some OSHA chemical standards have specific provisions regarding exposure monitoring and worker notification. Employers should consult relevant standards to see if these provisions apply to their workplace.

Medical Consultations and Examinations

Employers must do the following:

- Provide all exposed workers with an opportunity to receive medical attention by a licensed physician, including any follow-up examinations which the examining physician determines to be necessary.

FIGURE 3.20 (Continued)

- Provide an opportunity for a medical consultation by a licensed physician whenever a spill, leak, explosion or other occurrence results in the likelihood that a laboratory worker experienced a hazardous exposure in order to determine whether a medical examination is needed.
- Provide an opportunity for a medical examination by a licensed physician whenever a worker develops signs or symptoms associated with a hazardous chemical to which he or she may have been exposed in the laboratory.
- Establish medical surveillance for a worker as required by the particular standard when exposure monitoring reveals exposure levels routinely exceeding the OSHA action level or, in the absence of an action level, the PEL for an OSHA regulated substance.
- Provide the examining physician with the identity of the hazardous chemical(s) to which the individual may have been exposed, and the conditions under which the exposure may have occurred, including quantitative data, where available, and a description of the signs and symptoms of exposure the worker may be experiencing.
- Provide all medical examinations and consultations without cost to the worker, without loss of pay, and at a reasonable time and place.

The examining physician must complete a written opinion that includes the following information:

- Recommendations for further medical follow-up.
- The results of the medical examination and any associated tests.
- Any medical condition revealed in the course of the examination that may place the individual at increased risk as a result of exposure to a hazardous chemical in the workplace.
- A statement that the worker has been informed of the results of the consultation or medical examination and any medical condition that may require further examination or treatment. However, the written opinion must not reveal specific findings of diagnoses unrelated to occupational exposure.

A copy of the examining physician's written opinion must be provided to the exposed worker.

Recordkeeping

Employers must also maintain an accurate record of exposure monitoring activities and exposure measurements as well as medical consultations and examinations, including medical tests and written opinions. Employers generally must maintain worker exposure records for 30 years and medical records for the duration of the worker's employment plus 30 years, unless one of the exemptions listed in 29 CFR 1910.1020(d)(1)(i)(A)-(C) applies. Such records must be maintained, transferred, and made available, in accord with 29 CFR 1910.1020, to an individual's physician or made available to the worker or his/her designated representative upon request.

Roles and Responsibilities in Implementing the Laboratory Standard

The following are the National Research Council's recommendations concerning the responsibilities of various individuals for chemical hygiene in laboratories.

Chief Executive Officer

- Bears ultimate responsibility for chemical hygiene within the facility.
- Provides continuing support for institutional chemical hygiene.

Chemical Hygiene Officer

- Develops and implements appropriate chemical hygiene policies and practices.
- Monitors procurement, use, and disposal of chemicals used in the lab.
- Ensures that appropriate audits are maintained.
- Helps project directors develop precautions and adequate facilities.
- Knows the current legal requirements concerning regulated substances.
- Seeks ways to improve the chemical hygiene program.

Laboratory Supervisors

- Have overall responsibility for chemical hygiene in the laboratory.
- Ensure that laboratory workers know and follow the chemical hygiene rules.
- Ensure that protective equipment is available and in working order.
- Ensure that appropriate training has been provided.
- Provide regular, formal chemical hygiene and housekeeping inspections, including routine inspections of emergency equipment.
- Know the current legal requirements concerning regulated substances.
- Determine the required levels of PPE and equipment.

FIGURE 3.20 (Continued)

- Ensure that facilities and training for use of any material being ordered are adequate.

Laboratory Workers
- Plan and conduct each operation in accord with the facility's chemical hygiene procedures, including use of PPE and engineering controls, as appropriate.
- Develop good personal chemical hygiene habits.
- Report all accidents and potential chemical exposures immediately.

For more detailed information, OSHA has developed a Safety and Health Topics Page on Laboratories available at: www.osha.gov/SLTC/laboratories/index.html. See the Appendix for other OSHA documents relevant to this topic.

Two OSHA Fact Sheets have been developed to supplement this section. One is entitled **Laboratory Safety – OSHA Laboratory Standard**, and the other is entitled **Laboratory Safety – Chemical Hygiene Plan**; both are available online at www.osha.gov.

Hazard Communication Standard (29 CFR 1910.1200)

This standard is designed to protect against chemical source illnesses and injuries by ensuring that employers and workers are provided with sufficient information to recognize, evaluate and control chemical hazards and take appropriate protective measures.

The steps that employers must take to comply with the requirements of this standard must include, but are not limited to:
- Development and maintenance of a written hazard communication program for the workplace, including lists of hazardous chemicals present;
- Ensuring that containers of chemicals in the workplace, as well as containers of chemicals being shipped to other workplaces, are properly labeled;
- Ensuring that material safety data sheets (MSDSs) for chemicals that workers may be exposed to are made available to workers; and
- Development and implementation of worker training programs regarding hazards of chemicals they may be exposed to and the appropriate protective measures that must be used when handling these chemicals.

This OSHA standard also requires manufacturers and importers of hazardous chemicals to provide MSDSs to users of the chemicals describing potential hazards and other information. They must also attach hazard warning labels to containers of the chemicals. Distributors of hazardous chemicals must also provide MSDSs to employers and other distributors.

An OSHA QuickFacts entitled **Laboratory Safety – Labeling and Transfer of Chemicals** has been developed to supplement this section and is available online at www.osha.gov.

Specific Chemical Hazards

Air Contaminants standard (29 CFR 1910.1000)

The Air Contaminants standard provides rules for protecting workers from airborne exposure to over 400 chemicals. Several of these chemicals are commonly used in laboratories and include: toluene, xylene, and acrylamide. Toluene and xylene are solvents used to fix tissue specimens and rinse stains. They are primarily found in histology, hematology, microbiology and cytology laboratories.

Toluene		
Exposure routes	**Symptoms**	**Target Organs**
Inhalation; Ingestion; Skin and/or eye contact; Skin absorption.	Irritation of eyes, nose; Weakness, exhaustion, confusion, euphoria, headache; Dilated pupils, tearing; Anxiety; Muscle fatigue; Insomnia; Tingling, pricking, or numbness of skin; Dermatitis; Liver, kidney damage.	Eyes; Skin; Respiratory system; Central nervous system; Liver; Kidneys.

FIGURE 3.20 (Continued)

Xylene		
Exposure routes	**Symptoms**	**Target Organs**
Inhalation; Ingestion; Skin and/or eye contact; Skin absorption.	Irritation of eyes, skin, nose, throat; Dizziness, excitement, drowsiness, incoherence, staggering gait; Corneal vacuoliza-tion (cell debris); Anorexia, nausea, vomiting, abdominal pain; Dermatitis.	Eyes; Skin; Respiratory system; Central nervous system; GI tract; Blood; Liver; Kidneys.

Acrylamide is usually found in research laboratories and is used to make polyacrylamide gels for separations of macromolecules (e.g., DNA, proteins).

Acrylamide		
Exposure routes	**Symptoms**	**Target Organs**
Inhalation; Ingestion; Skin and/or eye contact; Skin absorption.	Irritation of eyes, skin; Ataxia (staggering gait), numb limbs, tingling, pricking, or numbness of skin; Muscle weakness; Absence of deep tendon reflex; Hand sweating; Tearing, Drowsiness; Reproductive effects; Potential occupational carcinogen.	Eyes; Skin; Central nervous system; Peripheral nervous system; Reproductive system (in animals: tumors of the lungs, testes, thyroid and adrenal glands).

Employers must do the following to prevent worker exposure:
Implement a written program for chemicals that workers are exposed to and that meet the requirements of the Hazard Communication standard. This program must contain provisions for worker training, warning labels and access to Material Safety Data Sheets (MSDSs).

Formaldehyde standard (29 CFR 1910.1048)

Formaldehyde is used as a fixative and is commonly found in most laboratories. The employer must ensure that no worker is exposed to an airborne concentration of formaldehyde which exceeds 0.75 parts formaldehyde per million parts of air (0.75 ppm) as an 8-hour time weighted average (TWA), 29 CFR 1910.1048(c)(1).

The Hazard Communication standard requires employers to maintain an MSDS, which manufacturers or distributors of formaldehyde are required to provide. The MSDS must be kept in an area that is accessible to workers that may be exposed to formaldehyde.

Formaldehyde		
Exposure routes	**Symptoms**	**Target Organs**
Inhalation; Ingestion; Skin and/or eye contact.	Irritation of eyes, skin, nose, throat, respiratory system; Tearing; Coughing; Wheezing; Dermatitis; Potential occupational nasal carcinogen.	Eyes; Skin; Respiratory system.

Employers must provide the following to workers to prevent exposure:

- Appropriate PPE, 29 CFR 1910.132, 29 CFR 1910.133, and 29 CFR 1910.1048(h).
- Acceptable eyewash facilities within the immediate work area for emergency use, if there is any possibility that a worker's eyes may be splashed with solutions containing 0.1 percent or greater formaldehyde, 29 CFR 1910.1048(i)(3).

FIGURE 3.20 (Continued)

Latex

One of the most common chemicals that laboratory workers are exposed to is latex, a plant protein. The most common cause of latex allergy is direct contact with latex, a natural plant derivative used in making certain disposable gloves and other products. Some healthcare workers have been determined to be latex sensitive, with reactions ranging from localized dermatitis (skin irritation) to immediate, possibly life-threatening reactions. Under OSHA's Personal Protective Equipment standard, 29 CFR 1910.132, the employer must ensure that appropriate personal protective equipment (PPE) is accessible at the worksite or issued to workers. Latex-free gloves, glove liners, powder-free gloves, or other similar alternatives are obtainable and must be readily accessible to those workers who are allergic to latex gloves or other latex-containing PPE, 29 CFR 1910.1030(c)(3)(iii).

Latex allergy should be suspected in workers who develop certain symptoms after latex exposure, including:

- nasal, eye, or sinus irritation
- hives or rash
- difficulty breathing
- coughing
- wheezing
- nausea
- vomiting
- diarrhea

An exposed worker who exhibits these symptoms should be evaluated by a physician or other licensed healthcare professional because further exposure could cause a serious allergic reaction.

Once a worker becomes allergic to latex, special precautions are needed to prevent exposures. Certain medications may reduce the allergic symptoms, but complete latex avoidance is the most effective approach.

Appropriate work practices should be used to reduce the chance of reactions to latex. If a worker must wear latex gloves, oil-based hand creams or lotions (which can cause glove deterioration) should not be used unless they have been shown to reduce latex-related problems and maintain glove barrier protection. After removing latex gloves, workers should wash their hands with a mild soap and dry them thoroughly.

An OSHA QuickFacts entitled **Laboratory Safety – Latex Allergy** has been developed to supplement this section and is available online at www.osha.gov.

Specific Engineering Control - Chemical Fume Hoods

The fume hood is often the primary control device for protecting laboratory workers when working with flammable and/or toxic chemicals. OSHA's Occupational Exposure to Hazardous Chemicals in Laboratories standard, 29 CFR 1910.1450, requires that fume hoods be maintained and function properly when used, 29 CFR 1910.1450(e)(3)(iii).

An OSHA QuickFacts entitled **Laboratory Safety – Chemical Fume Hoods** has been developed to supplement this section and is available online at www.osha.gov.

Biological Hazards

Biological Agents (other than Bloodborne Pathogens) and Biological Toxins

Many laboratory workers encounter daily exposure to biological hazards. These hazards are present in various sources throughout the laboratory such as blood and body fluids, culture specimens, body tissue and cadavers, and laboratory animals, as well as other workers.

A number of OSHA's Safety and Health Topics Pages mentioned below have information on select agents and toxins. These are federally regulated biological agents (e.g., viruses, bacteria, fungi, and prions) and toxins that have the potential to pose a severe threat to public health and safety, to animal or plant health, or to animal or plant products. The agents and toxins that affect animal and plant health are also referred to as high-consequence livestock pathogens and toxins, non-overlap agents and toxins, and listed plant pathogens. Select agents and toxins are defined by lists that appear in sections 73.3 of Title 42 of the Code of Federal Regulations (HHS/CDC *Select Agent Regulations*), sections 121.3 and 121.4 of Title 9 of the Code of Federal Regulations (USDA/APHIS/VS Select Agent Regulations), and section 331.3 of Title 7 of the Code of Federal Regulations (plants - USDA/APHIS/PPQ *Select Agent Regulations*) and Part 121, Title 9, Code of Federal Regulations (animals – USDA/APHIS). Select agents and toxins that are regulated by both HHS/CDC and USDA/APHIS are referred to as "over-

FIGURE 3.20 (Continued)

lap" select agents and toxins (see 42 CFR section 73.4 and 9 CFR 121.4).
Employers may use the list below as a starting point for technical and regulatory information about some of the most virulent and prevalent biological agents and toxins. The OSHA Safety and Health Topics Page entitled Biological Agents can be accessed at: www.osha.gov/SLTC/biologicalagents/index.html.

Anthrax. Anthrax is an acute infectious disease caused by a spore-forming bacterium called *Bacillus anthracis.* It is generally acquired following contact with anthrax-infected animals or anthrax-contaminated animal products. ***Bacillus anthracis*** **is an HHS and USDA select agent.**

Avian Flu. Avian influenza is caused by Influenza A viruses. These viruses normally reside in the intestinal tracts of water fowl and shore birds, where they cause little, if any, disease. However, when they are passed on to domestic birds, such as chickens, they can cause deadly contagious disease, highly pathogenic avian influenza (HPAI). **HPAI viruses are considered USDA/APHIS select agents.**

Botulism. Cases of botulism are usually associated with consumption of preserved foods. However, botulinum toxins are currently among the most common compounds explored by terrorists for use as biological weapons. **Botulinum neurotoxins, the causative agents of botulism, are HHS/CDC select agents.**

Foodborne Disease. Foodborne illnesses are caused by viruses, bacteria, parasites, toxins, metals, and prions (microscopic protein particles). Symptoms range from mild gastroenteritis to life-threatening neurologic, hepatic and renal syndromes.

Hantavirus. Hantaviruses are transmitted to humans from the dried droppings, urine, or saliva of mice and rats. Animal laboratory workers and persons working in infested buildings are at increased risk to this disease.

Legionnaires' Disease. Legionnaires' disease is a bacterial disease commonly associated with water-based aerosols. It is often the result of poorly maintained air conditioning cooling towers and potable water systems.

Molds and Fungi. Molds and fungi produce and release millions of spores small enough to be air-, water-, or insect-borne which may have negative effects on human health including, allergic reactions, asthma, and other respiratory problems.

Plague. The World Health Organization reports 1,000 to 3,000 cases of plague every year. A bioterrorist release of plague could result in a rapid spread of the pneumonic form of the disease, which could have devastating consequences. ***Yersinia pestis,*** **the causative agent of plague, is an HHS/CDC select agent.**

Ricin. Ricin is one of the most toxic and easily produced plant toxins. It has been used in the past as a bioterrorist weapon and remains a serious threat. **Ricin is an HHS/CDC select toxin.**

Severe Acute Respiratory Syndrome (SARS). SARS is an emerging, sometimes fatal, respiratory illness. According to the Centers for Disease Control and Prevention (CDC), the most recent human cases of SARS were reported in China in April 2004 and there is currently no known transmission anywhere in the world.

Smallpox. Smallpox is a highly contagious disease unique to humans. It is estimated that no more than 20 percent of the population has any immunity from previous vaccination. **Variola major virus, the causative agent for smallpox, is an HHS/CDC select agent.**

Tularemia. Tularemia is also known as "rabbit fever" or "deer fly fever" and is extremely infectious. Relatively few bacteria are required to cause the disease, which is why it is an attractive weapon for use in bioterrorism. ***Francisella tularensis,*** **the causative agent for tularemia, is an HHS/CDC select agent.**

Viral Hemorrhagic Fevers (VHFs). Hemorrhagic fever viruses are among the agents identified by the Centers for Disease Control and Prevention (CDC) as the most likely to be used as biological weapons. Many VHFs can cause severe, life-threatening disease with high fatality rates. **Many VHFs are HHS/CDC select agents; for example, Marburg virus, Ebola viruses, and the Crimean-Congo hemorrhagic fever virus.**

An additional OSHA Safety and Health Topics page on Pandemic Influenza has been added in response to the 2009 H1N1 influenza pandemic. It can be accessed at: www.osha.gov/dsg/topics/pandemicflu/index.html.

FIGURE 3.20 (Continued)

Pandemic Influenza. A pandemic is a global disease outbreak. An influenza pandemic occurs when a new influenza virus emerges for which there is little or no immunity in the human population; begins to cause serious illness; and then spreads easily person-to-person worldwide.

The list above does not include all of the biological agents and toxins that may be hazardous to laboratory workers. New agents will be added over time. For agents that may pose a hazard to laboratory workers but are not listed above, consult the CDC web page at: www.cdc.gov. See Appendix for more information on BSL levels.

Material Safety Data Sheets (MSDSs) on Infectious Agents

Although MSDSs for chemical products have been available to workers for many years in the U.S. and other countries, Canada is the only country that has developed MSDSs for infectious agents. These MSDSs were produced by the Canadian Public Health Agency for personnel working in the life sciences as quick safety reference material relating to infectious microorganisms.

These MSDSs on Infectious Agents are organized to contain health hazard information such as infectious dose, viability (including decontamination), medical information, laboratory hazard, recommended precautions, handling information and spill procedures. These MSDSs are available at: www.phac-aspc.gc.ca/msds-ftss.

Bloodborne Pathogens

The OSHA Bloodborne Pathogens (BBP) standard (29 CFR 1910.1030) is designed to protect workers from the health hazards of exposure to bloodborne pathogens. Employers are subject to the BBP standard if they have workers whose jobs put them at reasonable risk of coming into contact with blood or other potentially infectious materials (OPIM). Employers subject to this standard must develop a written Exposure Control Plan, provide training to exposed workers, and comply with other requirements of the standard, including use of Standard Precautions when dealing with blood and OPIM. In 2001, in response to the *Needlestick Safety and Prevention Act,* OSHA revised the Bloodborne Pathogens standard. The revised standard clarifies the need for employers to select safer needle devices and to involve workers in identifying and choosing these devices. The updated standard also requires employers to maintain a log of injuries from contaminated sharps.

OSHA estimates that 5.6 million workers in the healthcare industry and related occupations are at risk of occupational exposure to bloodborne pathogens, including HIV, HBV, HCV, and others. All occupational exposure to blood or OPIM places workers at risk for infection with bloodborne pathogens. OSHA defines blood to mean human blood, human blood components, and products made from human blood. OPIM means: (1) The following human body fluids: semen, vaginal secretions, cerebrospinal fluid, synovial fluid, pleural fluid, pericardial fluid, peritoneal fluid, amniotic fluid, saliva in dental procedures, any body fluid that is visibly contaminated with blood, and all body fluids in situations where it is difficult or impossible to differentiate between body fluids; (2) Any unfixed tissue or organ (other than intact skin) from a human (living or dead); and (3) HIV- or HBV-containing cell or tissue cultures, organ cultures, and HIV- or HBV-containing culture medium or other solutions; and blood, organs, or other tissues from experimental animals infected with HIV or HBV.

The Centers for Disease Control and Prevention (CDC) notes that although more than 200 different diseases can be transmitted from exposure to blood, the most serious infections are hepatitis B virus (HBV), hepatitis C virus (HCV), and human immunodeficiency virus (HIV). Fortunately, the risk of acquiring any of these infections is low. HBV is the most infectious virus of the three viruses listed above. For an unvaccinated healthcare worker, the risk of developing an infection from a single needlestick or a cut exposed to HBV-infected blood ranges from 6-30%. The risk for infection from HCV- and HIV-infected blood under the same circumstances is 1.8 and 0.3 percent, respectively. This means that after a needlestick/cut exposure to HCV-contaminated blood, 98.2% of individuals do not become infected, while after a similar exposure to HIV-contaminated blood, 99.7% of individuals do not become infected. (http://www.cdc.gov/OralHealth/infectioncontrol/faq/bloodborne_exposures.htm).

Many factors influence the risk of becoming infected after a needlestick or cut exposure to HBV-, HCV- or HIV-contaminated blood. These factors include the health status of the individual, the volume of the blood exchanged, the concentration of the virus in the blood, the extent of the cut or the depth of penetration of the needlestick, etc.

FIGURE 3.20 (Continued)

Employers must ensure that workers are trained and prohibited from engaging in the following activities:

- Mouth pipetting/suctioning of blood or OPIM, 29 CFR 1910.1030(d)(2)(xii);
- Eating, drinking, smoking, applying cosmetics or lip balm, or handling contact lenses in work areas where there is a reasonable likelihood of occupational exposure to blood or OPIM, 29 CFR 1910.1030(d)(2)(ix); and
- Storage of food or drink in refrigerators, freezers, shelves, cabinets or on countertops or benchtops where blood or OPIM are present, 29 CFR 1910.1030(d)(2)(x).

Employers must ensure that the following are provided:

- Appropriate PPE for workers if blood or OPIM exposure is anticipated, 29 CFR 1910.1030(d)(3);
  - The type and amount of PPE depends on the anticipated exposure.
    - Gloves must be worn when hand contact with blood, mucous membranes, OPIM, or non-intact skin is anticipated, or when handling contaminated items or surfaces, 29 CFR 1910.1030(d)(3)(ix).
    - Surgical caps or hoods and/or shoe covers or boots must be worn in instances when gross contamination can reasonably be anticipated such as during autopsies or orthopedic surgery, 29 CFR 1910.1030(d)(3)(xii).
- Effective engineering and work practice controls to help remove or isolate exposures to blood and bloodborne pathogens, 29 CFR 1910.1030(d)(2)(i), CPL 02-02-069 (CPL 2-2.69); and
- Hepatitis B vaccination (if not declined by a worker) under the supervision of a physician or other licensed healthcare professional to all workers who have occupational exposure to blood or OPIM, 29 CFR 1910.1030(f)(1)(ii)(A)-(C).

Labels

When any blood, OPIM or infected animals are present in the work area, a hazard warning sign (see graphic) incorporating the universal biohazard symbol, 29 CFR 1910.1030(g)(1)(ii)(A), must be posted on all access doors, 29 CFR 1910.1030(e)(2)(ii)(D).

Engineering Controls and Work Practices for All HIV/HBV Laboratories

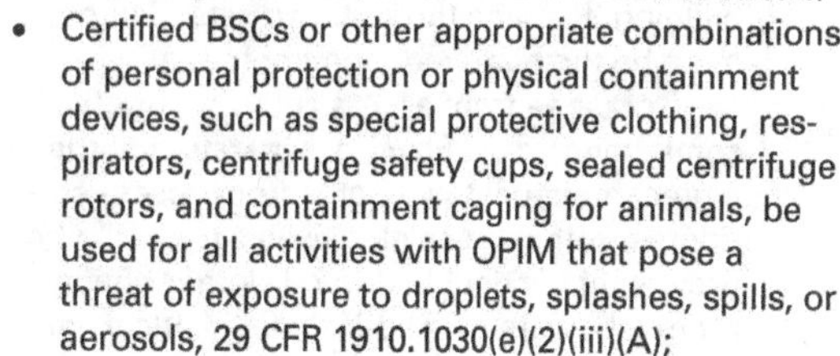

Employers must ensure that:

- All activities involving OPIM are conducted in Biological Safety Cabinets (BSCs) or other physical-containment devices; work with OPIM must not be conducted on the open bench, 29 CFR 1910.1030(e)(2)(ii)(E);
- Certified BSCs or other appropriate combinations of personal protection or physical containment devices, such as special protective clothing, respirators, centrifuge safety cups, sealed centrifuge rotors, and containment caging for animals, be used for all activities with OPIM that pose a threat of exposure to droplets, splashes, spills, or aerosols, 29 CFR 1910.1030(e)(2)(iii)(A);
- Each laboratory contains a facility for hand washing and an eyewash facility which is readily available within the work area, 29 CFR 1910.1030(e)(3)(i); and
- Each work area contains a sink for washing hands and a readily available eyewash facility. The sink must be foot, elbow, or automatically operated and must be located near the exit door of the work area, 29 CFR 1910.1030(e)(4)(iii).

Additional BBP Standard Requirements Apply to HIV and HBV Research Laboratories

Requirements include:

- Waste materials:
  - All regulated waste must either be incinerated or decontaminated by a method such as autoclaving known to effectively destroy bloodborne pathogens, 29 CFR 1910.1030(e)(2)(i); and
  - Contaminated materials that are to be decontaminated at a site away from the work area must be placed in a durable, leakproof, labeled or color-coded container that is closed before being removed from the work area, 29 CFR 1910.1030(e)(2)(ii)(B).
- Access:
  - Laboratory doors must be kept closed when work involving HIV or HBV is in progress, 29 CFR 1910.1030(e)(2)(ii)(A);
  - Access to the production facilities' work area must be limited to authorized persons. Written policies and procedures must be established whereby only persons who have been advised of the potential biohazard, who

FIGURE 3.20 (Continued)

meet any specific entry requirements, and who comply with all entry and exit procedures must be allowed to enter the work areas and animal rooms, 29 CFR 1910.1030(e)(2)(ii)(C);

- Access doors to the production facilities' work area or containment module must be self-closing, 29 CFR 1910.1030(e)(4)(iv);
- Work areas must be separated from areas that are open to unrestricted traffic flow within the building. Passage through two sets of doors must be the basic requirement for entry into the work area from access corridors or other contiguous areas. Physical separation of the high-containment work area from access corridors or other areas or activities may also be provided by a double-doored clothes-change room (showers may be included), airlock, or other access facility that requires passing through two sets of doors before entering the work area, 29 CFR 1910.1030(e)(4)(i); and
- The surfaces of doors, walls, floors and ceilings in the work area must be water-resistant so that they can be easily cleaned. Penetrations in these surfaces must be sealed or capable of being sealed to facilitate decontamination, 29 CFR 1910.1030(e)(4)(ii).

(These requirements **do not apply** to clinical or diagnostic laboratories engaged solely in the analysis of blood, tissue, or organs, 29 CFR 1910.1030(e)(1).)

Research Animals

All procedures on animals should be performed by properly trained personnel. By using safe work practices and appropriate PPE, 29 CFR 1910.132(a), workers can minimize the likelihood that they will be bitten, scratched, and/or exposed to animal body fluids and tissues.

Possible Injuries/Illnesses

The most common work-related health complaints reported by individuals working with small animals are the following:

1. Sprains;
2. Strains;
3. Bites; and
4. Allergies.

Of these injuries, allergies (i.e., exaggerated reactions by the body's immune system) to proteins in small animals' urine, saliva, and dander are the greatest potential health risk. An allergic response may evolve into life-long asthma. Because mice and rats are the animals most frequently used in research studies, there are more reports of allergies to rodents than other laboratory animals. Most workers who develop allergies to laboratory animals will do so within the first twelve months of working with them. Sometimes reactions only occur in workers after they have been handling animals for several years. Initially, the symptoms are present within minutes of the worker's exposure to the animals. Approximately half of allergic workers will have their initial symptoms subside and then recur three or four hours following the exposure.

Employers should adopt the following best practices to reduce allergic responses of workers:

- Eliminate or minimize exposure to the proteins found in animal urine, saliva and dander.
- Limit the chances that workers will inhale or have skin contact with animal proteins by using well-designed air handling and waste management systems.
- Have workers use appropriate PPE (e.g., gloves, gowns, hair covers, respirators) to further minimize their risk of exposure.

Zoonotic Diseases

There are a host of possible infectious agents that can be transferred from animals to humans. These are referred to as zoonotic diseases. The common routes of exposure to infectious agents are inhalation, inoculation, ingestion and contamination of skin and mucous membranes. Inhalation hazards may arise during work practices that can generate aerosols. These include the following: centrifugation, mixing (e.g., blending, vortexing, and sonication), pouring/decanting and spilling/splashing of culture fluids. Inoculation hazards include needlesticks and lacerations from sharp objects. Ingestion hazards include the following: splashes to the mouth, placing contaminated articles/fingers in mouth, consumption of food in the laboratory, and mouth pipetting. Contamination of skin and mucous membranes can occur via splashes or contact with contaminated fomites (e.g., towels, bedclothes, cups, money). Some of the zoonotic diseases that can be acquired from animals are listed below.

Zoonotic Diseases – Wild and Domesticated Animals

Wild rodents and other wild animals may inflict an injury such as a bite or scratch. Workers need to receive training on the correct way to capture and handle any wild animals. While they may carry or shed organisms that may be potentially infectious to humans, the primary health risk to individuals

FIGURE 3.20 (Continued)

working with captured animals is the development of an allergy. The development of disease in the human host often requires a preexisting state that compromises the immune system. Workers who have an immune compromising medical condition or who are taking medications that impair the immune system (steroids, immunosuppressive drugs, or chemotherapy) are at higher risk for contracting a rodent disease.

Wild rodents may act as carriers for viruses such as Hantavirus and lymphocytic choriomeningitis virus (LCMV) depending on where they were captured. Additionally, each rodent species may harbor their own range of bacterial diseases, such as tularemia and plague. These animals may also have biting insect vectors which can act as a potential carrier of disease (mouse to human transmission).

Examples of zoonotic diseases that can be transmitted from wild and domesticated animals to humans are listed in the table at page 45 in the Appendix.

Zoonotic Diseases – Non-human Primates (e.g., monkeys)

It should not be surprising that, given our many similarities, humans and non-human primates are susceptible to similar infectious agents. Because of our differences, the consequences of infection with the same agent often vary considerably. Infection may cause few if any symptoms in one group and may be lethal to the other. Exposures to body fluids from non-human primates should be treated immediately.

In 2003, a report entitled, Occupational Safety and Health in the Care and Use of Non-Human Primates (see References) was published. This report covers topics relevant to facilities in which non-human primates are housed or where non-human primate blood or tissues are handled. The report describes the hazards associated with work involving non-human primates and discusses the components of a successful occupational health and safety program, including hazard identification, risk assessment and management, institutional management of workers after a suspected occupational exposure, applicable safety regulations, and personnel training.

Employers should ensure that workers are trained to adhere to the following good practices to prevent exposure to zoonotic diseases when working with research animals:

- Avoid use of sharps whenever possible. Take extreme care when using a needle and syringe to inject research animals or when using sharps during necropsy procedures. Never remove, recap, bend, break, or clip used needles from disposable syringes. Use safety engineered needles when practical.
- Take extra precautions when handling hoofed animals. Due to the physical hazards of weight and strength of the animal, large hoofed mammals pose additional concerns for workers. Hoofed mammals may resist handling and may require multiple workers to administer medication or perform other functions.
- Keep hands away from mouth, nose and eyes.
- Wear appropriate PPE (i.e., gloves, gowns, face protection) in all areas within the animal facility.
  - A safety specialist may recommend additional precautions, based upon a risk assessment of the work performed.
- Wear tear-resistant gloves to prevent exposure by animal bites. Micro-tears in the gloves may compromise the protection they offer.
- Remove gloves and wash hands after handling animals or tissues derived from them and before leaving areas where animals are kept.
- Use mechanical pipetting devices (no mouth pipetting).
- Never eat, drink, smoke, handle contact lenses, apply cosmetics, or take or apply medicine in areas where research animals are kept.
- Perform procedures carefully to reduce the possibility of creating splashes or aerosols.
- Contain operations that generate hazardous aerosols in BSCs or other ventilated enclosures, such as animal bedding dump stations.
- Wear eye protection.
- Wear head/hair covering to protect against sprays or splashes of potentially infectious fluids.
- Keep doors closed to rooms where research animals are kept.
- Clean all spills immediately.
- Report all incidents and equipment malfunctions to the supervisor.
- Promptly decontaminate work surfaces when procedures are completed and after surfaces are soiled by spills of animal material or waste.
- Properly dispose of animal waste and bedding.
- Workers should report all work-related injuries and illnesses to their supervisor immediately.
- Following a bite by an animal or other injury in which the wound may be contaminated, first aid should be initiated at the work site.

FIGURE 3.20 (Continued)

- Contaminated skin and wounds should be washed thoroughly with soap and water for 15 minutes.
- Contaminated eyes and mucous membranes should be irrigated for 15 minutes using normal saline or water.

• Consult an occupational health physician concerning wound care standard operating procedures (SOPs) for particular animal bites/scratches.

An OSHA QuickCard™ entitled **Laboratory Safety – Working with Small Animals** has been developed to supplement this section and is available online at www.osha.gov.

Specific Engineering Control – Biological Safety Cabinets (BSCs)

Properly maintained BSCs, when used in conjunction with good microbiological techniques, provide an effective containment system for safe manipulation of moderate and high-risk infectious agents [Biosafety Level 2 (BSL 2) and 3 (BSL 3) agents]. BSCs protect laboratory workers and the immediate environment from infectious aerosols generated within the cabinet.

Biosafety Cabinet Certifications

BSCs must be certified when installed, whenever they are moved and at least annually, 29 CFR 1030(e)(2)(iii)(B).

An OSHA Fact Sheet entitled **Laboratory Safety – Biosafety Cabinets (BSCs)** has been developed to supplement this section and is available online at www.osha.gov.

Physical Hazards and Others

Besides exposure to chemicals and biological agents, laboratory workers can also be exposed to a number of physical hazards. Some of the common physical hazards that they may encounter include the following: ergonomic, ionizing radiation, non-ionizing radiation and noise hazards. These hazards are described below in individual sections.

Ergonomic Hazards

Laboratory workers are at risk for repetitive motion injuries during routine laboratory procedures such as pipetting, working at microscopes, operating microtomes, using cell counters and keyboarding at computer workstations. Repetitive motion injuries develop over time and occur when muscles and joints are stressed, tendons are inflamed, nerves are pinched and the flow of blood is restricted. Standing and working in awkward positions in front of laboratory hoods/biological safety cabinets can also present ergonomic problems.

By becoming familiar with how to control laboratory ergonomics-related risk factors, employers can reduce chances for occupational injuries while improving worker comfort, productivity, and job satisfaction. In addition to the general ergonomic guidance, laboratory employers are reminded of some simple adjustments that can be made at the workplace. While there is currently no specific OSHA standard relating to ergonomics in the laboratory workplace, it is recommended that employers provide the information to laboratory workers contained in the new OSHA fact sheet highlighted below.

An OSHA Fact Sheet entitled **Laboratory Safety – Ergonomics for the Prevention of Musculoskeletal Disorders in Laboratories** has been developed to supplement this section and is available online at osha.gov.

Ionizing Radiation

OSHA's Ionizing Radiation standard, 29 CFR 1910. 1096, sets forth the limitations on exposure to radiation from atomic particles. Ionizing radiation sources are found in a wide range of occupational settings, including laboratories. These radiation sources can pose a considerable health risk to affected workers if not properly controlled.

Any laboratory possessing or using radioactive isotopes must be licensed by the Nuclear Regulatory Commission (NRC) and/or by a state agency that has been approved by the NRC, 10 CFR 31.11 and 10 CFR 35.12.

The fundamental objectives of radiation protection measures are: (1) to limit entry of radionuclides into the human body (via ingestion, inhalation, absorption, or through open wounds) to quantities as low as reasonably achievable (ALARA) and always within the established limits; and (2) to limit exposure to external radiation to levels that are within established dose limits and as far below these limits as is reasonably achievable.

FIGURE 3.20 (Continued)

All areas in which radioactive materials are used or stored must conspicuously display the symbol for radiation hazards and access should be restricted to authorized personnel.

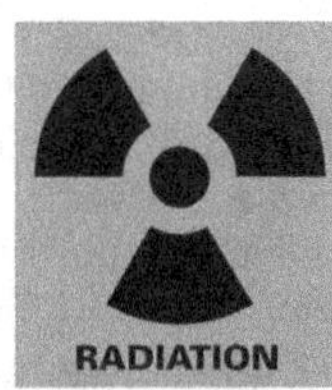

The OSHA Ionizing Radiation standard requires precautionary measures and personnel monitoring for workers who are likely to be exposed to radiation hazards. Personnel monitoring devices (film badges, thermoluminescent dosimeters (TLD), pocket dosimeters, etc.) must be supplied and used if required to measure an individual's radiation exposure from gamma, neutron, energetic beta, and X-ray sources. The standard monitoring device is a clip-on badge or ring badge bearing the individual assignee's name, date of the monitoring period and a unique identification number. The badges are provided, processed and reported through a commercial service company that meets current requirements of the National Institute of Standards and Technology's National Voluntary Laboratory Accreditation Program (NIST NVLAP).

It is important for employers to understand and follow all applicable regulations for the use of isotopes. In institutional settings, it is the responsibility of each institution to ensure compliance with local, state, and federal laws and regulations; to obtain licenses for official use of radioactive substances; and to designate a radiation safety officer (RSO) to oversee and ensure compliance with state and/or NRC requirements. Information on radioactive materials licenses may be obtained from the Department of Public Health from individual states or from the NRC.

The following OSHA Safety and Health Topics Page provides links to technical and regulatory information on the control of occupational hazards from ionizing radiation:
www.osha.gov/SLTC/radiationionizing/index.html.

Non-ionizing Radiation

Non-ionizing radiation is described as a series of energy waves composed of oscillating electric and magnetic fields traveling at the speed of light. Non-ionizing radiation includes the spectrum of ultraviolet (UV), visible light, infrared (IR), microwave (MW), radio frequency (RF), and extremely low frequency (ELF). Lasers commonly operate in the UV, visible, and IR frequencies. Non-ionizing radiation is found in a wide range of occupational settings and can pose a considerable health risk to potentially exposed workers if not properly controlled.

The following OSHA Safety and Health Topics Pages provide links to technical and regulatory information on the control of occupational hazards from non-ionizing radiation and are available at:
www.osha.gov/SLTC/radiation_nonionizing/index.html.

Extremely Low Frequency Radiation (ELF)
Extremely Low Frequency (ELF) radiation at 60 HZ is produced by power lines, electrical wiring, and electrical equipment. Common sources of intense exposure include ELF induction furnaces and high-voltage power lines.

Radiofrequency and Microwave Radiation
Microwave radiation (MW) is absorbed near the skin, while radiofrequency (RF) radiation may be absorbed throughout the body. At high enough intensities both will damage tissue through heating. Sources of RF and MW radiation include radio emitters and cell phones.

Infrared Radiation (IR)
The skin and eyes absorb infrared radiation (IR) as heat. Workers normally notice excessive exposure through heat sensation and pain. Sources of IR radiation include heat lamps and IR lasers.

Visible Light Radiation
The different visible frequencies of the electromagnetic (EM) spectrum are "seen" by our eyes as different colors. Good lighting is conducive to increased production, and may help prevent incidents related to poor lighting conditions. Excessive visible radiation can damage the eyes and skin.

Ultraviolet Radiation (UV)
Ultraviolet radiation (UV) has a high photon energy range and is particularly hazardous because there are usually no immediate symptoms of excessive exposure. Sources of UV radiation in the laboratory include black lights and UV lasers.

Laser Hazards
Lasers typically emit optical (UV, visible light, IR) radiations and are primarily an eye and skin hazard. Common lasers include CO_2 IR laser; helium - neon, neodymium YAG, and ruby visible lasers, and the Nitrogen UV laser.

LASER is an acronym which stands for Light Amplification by Stimulated Emission of Radiation.

FIGURE 3.20 (Continued)

The laser produces an intense, highly directional beam of light. The most common cause of laser-induced tissue damage is thermal in nature, where the tissue proteins are denatured due to the temperature rise following absorption of laser energy.

The human body is vulnerable to the output of certain lasers, and under certain circumstances, exposure can result in damage to the eye and skin. Research relating to injury thresholds of the eye and skin has been carried out in order to understand the biological hazards of laser radiation. It is now widely accepted that the human eye is almost always more vulnerable to injury than human skin.

Noise

OSHA's Occupational Noise Exposure standard, 29 CFR 1910.95, requires employers to develop and implement a hearing conservation program that includes the use of PPE (e.g., hearing protectors), if workers are exposed to a time-weighted average (TWA) of ≥ 85 dBA over an 8-hour work shift. In addition, when workers are exposed to noise levels ≥ 85 dBA, the employer must develop a monitoring program to assess noise levels. The monitoring program must include the following components:

- All continuous, intermittent, and impulsive sound levels from 80-130 dBA must be included in noise measurements, 29 CFR 1910.95(d)(2)(i);
- Instruments used to measure worker noise exposure must be calibrated to ensure measurement accuracy, 29 CFR 1910.95(d)(2)(ii); and
- Monitoring must be repeated whenever a change in production, process, equipment, or controls increases noise exposures, 29 CFR 1910.95(d)(3).

Laboratory workers are exposed to noise from a variety of sources. Operation of large analyzers (e.g., chemistry analyzer), fume hoods, biological safety cabinets, incubators, centrifuges (especially ultracentrifuges), cell washers, sonicators, and stirrer motors, all contribute to the noise level in laboratories. Further sources of noise in laboratories include fans and compressors for cryostats, refrigerators, refrigerated centrifuges, and freezers. As an example, a high-speed refrigerated centrifuge alone can generate noise levels as high as 65 dBA. To provide some further context, a whisper registers approximately 30 dBA; normal conversation about 50 to 60 dBA; a ringing phone 80 dBA and a power mower 90 dBA. If noise levels exceed 80 dBA, people must speak very loudly to be heard, while at noise levels of 85 to 90 dBA, people have to shout.

In order to determine if the noise levels in the laboratory are above the threshold level that damages hearing, the employer must conduct a noise exposure assessment using an approved sound level monitoring device, such as a dosimeter, and measuring an 8-hour TWA exposure. If the noise levels are found to exceed the threshold level, the employer must provide hearing protection at no cost to the workers and train them in the proper use of the protectors. The potential dangers of miscommunicating instructions or laboratory results are obvious, and efforts should be made to improve the design of clinical laboratories and to evaluate new instrumentation with regard to the impact of these factors on worker noise exposure. The employer should evaluate the possibility of relocating equipment to another area or using engineering controls to reduce the noise level below an 8-hour TWA of 85 dBA in order to comply with OSHA's Occupational Noise Exposure standard.

While most laboratories' noise levels do not equal or exceed the 8-hour TWA of 85 dBA, certain accrediting agencies are implementing special emphasis programs on noise reduction in the laboratory. Because noise is becoming more of a concern in the clinical setting, the College of American Pathologists has added evaluation of noise in the laboratory under their general checklist for accreditation (GEN.70824).

Health Effects

Exposure to continuous noise may lead to the following stress-related symptoms:

- Depression;
- Irritability;
- Decreased concentration in the workplace;
- Reduced efficiency and decreased productivity;
- Noise-induced hearing loss;
- Tinnitus (i.e., ringing in the ears); and
- Increased errors in laboratory work.

There are several steps that employers can take to minimize the noise in the laboratory, including:

- Moving noise-producing equipment (e.g., freezers, refrigerators, incubators and centrifuges) from the laboratory to an equipment room;
- Locating compressors for controlled-temperature rooms remotely; and
- Providing acoustical treatment on ceilings and walls.

FIGURE 3.20 (Continued)

An OSHA Fact Sheet entitled **Laboratory Safety – Noise** has been developed to supplement this section and is available online at www.osha.gov.

Safety Hazards

Employers must assess tasks to identify potential worksite hazards and provide and ensure that workers use appropriate personal protective equipment (PPE) as stated in the PPE standard, 29 CFR 1910.132.

Employers must require workers to use appropriate hand protection when hands are exposed to hazards such as sharp instruments and potential thermal burns. Examples of PPE which may be selected include using oven mitts when handling hot items, and steel mesh or cut-resistant gloves when handling or sorting sharp instruments as stated in the Hand Protection standard, 29 CFR 1910.138.

Autoclaves and Sterilizers

Workers should be trained to recognize the potential for exposure to burns or cuts that can occur from handling or sorting hot sterilized items or sharp instruments when removing them from autoclaves/sterilizers or from steam lines that service the autoclaves.

In order to prevent injuries from occurring, employers must train workers to follow good work practices such as those outlined in the QuickCard™ highlighted below.

An OSHA QuickFacts entitled **Laboratory Safety – Autoclaves/Sterilizers** has been developed to supplement this section and is available online at www.osha.gov.

Centrifuges

Centrifuges, due to the high speed at which they operate, have great potential for injuring users if not operated properly. Unbalanced centrifuge rotors can result in injury, even death. Sample container breakage can generate aerosols that may be harmful if inhaled.

The majority of all centrifuge accidents are the result of user error. In order to prevent injuries or exposure to dangerous substances, employers should train workers to follow good work practices such as those outlined in the QuickCard™ highlighted below.

Employers should instruct workers when centrifuging infectious materials that they should wait 10 minutes after the centrifuge rotor has stopped before opening the lid. Workers should also be trained to use appropriate decontamination and cleanup procedures for the materials being centrifuged if a spill occurs and to report all accidents to their supervisor immediately.

An OSHA QuickFacts entitled **Laboratory Safety – Centrifuges** has been developed to supplement this section and is available online at www.osha.gov.

Compressed Gases

According to OSHA's Laboratory standard, a **"compressed gas"** (1) is a gas or mixture of gases in a container having an absolute pressure exceeding 40 pounds per square inch (psi) at 70°F (21.1°C); or (2) is a gas or mixture of gases having an absolute pressure exceeding 104 psi at 130°F (54.4°C) regardless of the pressure at 70°F (21.1°C); or (3) is a liquid having a vapor pressure exceeding 40 psi at 100°F (37.8°C) as determined by ASTM (American Society for Testing and Materials) D-323-72, [29 CFR 1910.1450(c)(1)-(3)].

Within laboratories, compressed gases are usually supplied either through fixed piped gas systems or individual cylinders of gases. Compressed gases can be toxic, flammable, oxidizing, corrosive, or inert. Leakage of any of these gases can be hazardous. Leaking inert gases (e.g., nitrogen) can quickly displace air in a large area creating an oxygen-deficient atmosphere; toxic gases (e.g., can create poison atmospheres; and flammable (oxygen) or reactive gases can result in fire and exploding cylinders. In addition, there are hazards from the pressure of the gas and the physical weight of the cylinder. A gas cylinder falling over can break containers and crush feet. The gas cylinder can itself become a missile if the cylinder valve is broken off. Laboratories must include compressed gases in their inventory of chemicals in their Chemical Hygiene Plan.

Compressed gases contained in cylinders vary in chemical properties, ranging from inert and harmless to toxic and explosive. The high pressure of the gases constitutes a serious hazard in the event that gas cylinders sustain physical damage and/or are exposed to high temperatures.

FIGURE 3.20 (Continued)

Store, handle, and use compressed gases in accord with OSHA's Compressed Gases standard (29 CFR 1910.101) and Pamphlet P-1-1965 from the Compressed Gas Association.

- All cylinders whether empty or full must be stored upright.
- Secure cylinders of compressed gases. Cylinders should never be dropped or allowed to strike each other with force.
- Transport compressed gas cylinders with protective caps in place and do not roll or drag the cylinders.

Cryogens and Dry Ice

Cryogens, substances used to produce very low temperatures [below -153°C (-243°F)], such as liquid nitrogen (LN_2) which has a boiling point of -196°C (-321°F), are commonly used in laboratories. Although not a cryogen, solid carbon dioxide or dry ice which converts directly to carbon dioxide gas at -78°C (-109°F) is also often used in laboratories. Shipments packed with dry ice, samples preserved with liquid nitrogen, and in some cases, techniques that use cryogenic liquids, such as cryogenic grinding of samples, present potential hazards in the laboratory.

Overview of Cryogenic Safety Hazards

The safety hazards associated with the use of cryogenic liquids are categorized as follows:

(1) *Cold contact burns*

Liquid or low-temperature gas from any cryogenic substance will produce effects on the skin similar to a burn.

(2) *Asphyxiation*

Degrees of asphyxia will occur when the oxygen content of the working environment is less than 20.9% by volume. This decrease in oxygen content can be caused by a failure/leak of a cryogenic vessel or transfer line and subsequent vaporization of the cryogen. Effects from oxygen deficiency become noticeable at levels below approximately 18% and sudden death may occur at approximately 6% oxygen content by volume.

(3) *Explosion - Pressure*

Heat flux into the cryogen from the environment will vaporize the liquid and potentially cause pressure buildup in cryogenic containment vessels and transfer lines. Adequate pressure relief should be provided to all parts of a system to permit this routine outgassing and prevent explosion.

(4) *Explosion - Chemical*

Cryogenic fluids with a boiling point below that of liquid oxygen are able to condense oxygen from the atmosphere. Repeated replenishment of the system can thereby cause oxygen to accumulate as an unwanted contaminant. Similar oxygen enrichment may occur where condensed air accumulates on the exterior of cryogenic piping. Violent reactions, e.g., rapid combustion or explosion, may occur if the materials which make contact with the oxygen are combustible.

Employer Responsibility

It is the responsibility of the employer, specifically the supervisor in charge of an apparatus, to ensure that the cryogenic safety hazards are minimized. This will entail (1) a safety analysis and review for all cryogenic facilities, (2) cryogenic safety and operational training for relevant workers, (3) appropriate maintenance of cryogenic systems in their original working order, i.e., the condition in which the system was approved for use, and (4) upkeep of inspection schedules and records.

Employers must train workers to use the appropriate personal protective equipment (PPE)

Whenever handling or transfer of cryogenic fluids might result in exposure to the cold liquid, boil-off gas, or surface, protective clothing must be worn. This includes:

- face shield or safety goggles;
- safety gloves; and
- long-sleeved shirts, lab coats, aprons.

Eye protection is required at all times when working with cryogenic fluids. When pouring a cryogen, working with a wide-mouth Dewar flask or around the exhaust of cold boil-off gas, use of a full face shield is recommended.

Hand protection is required to guard against the hazard of touching cold surfaces. It is recommended that Cryogen Safety Gloves be used by the worker.

An OSHA QuickFacts entitled **Laboratory Safety – Cryogens and Dry Ice** has been developed to supplement this section and is available online at www.osha.gov.

Electrical

In the laboratory, there is the potential for workers to be exposed to electrical hazards including electric shock, electrocutions, fires and explosions. Damaged electrical cords can lead to possible

FIGURE 3.20 (Continued)

shocks or electrocutions. A flexible electrical cord may be damaged by door or window edges, by staples and fastenings, by equipment rolling over it, or simply by aging.

The potential for possible electrocution or electric shock or contact with electrical hazards can result from a number of factors, including the following:

- Faulty electrical equipment/instrumentation or wiring;
- Damaged receptacles and connectors; and
- Unsafe work practices.

Employers are responsible for complying with OSHA's standard 1910 Subpart S-Electrical

Subpart S is comprehensive and addresses electrical safety requirements for the practical safeguarding of workers in their workplaces. This Subpart includes, but is not limited to, these requirements:

- Electrical equipment must be free from recognized hazards, 29 CFR 1910.303(b)(1);
- Listed or labeled equipment must be used or installed in accord with any instructions included in the listing or labeling, 29 CFR 1910.303(b)(2);
- Sufficient access and working space must be provided and maintained around all electrical equipment operating at ≤ 600 volts to permit ready and safe operation and maintenance of such equipment, 29 CFR 1910.303(g)(1);
- Ensure that all electrical service near sources of water is properly grounded.
- Tag out and remove from service all damaged receptacles and portable electrical equipment, 29 CFR 1910.334(a)(2)(ii);
- Repair all damaged receptacles and portable electrical equipment before placing them back into service, 29 CFR 1910.334(a)(2)(ii);
- Ensure that workers are trained not to plug or unplug energized equipment when their hands are wet, 29 CFR 1910.334(a)(5)(i);
- Select and use appropriate work practices, 29 CFR 1910.333; and
- Follow requirements for Hazardous Classified Locations, 29 CFR 1910.307. This section covers the requirements for electric equipment and wiring in locations that are classified based on the properties of the flammable vapors, liquids or gases, or combustible dusts or fibers that may be present therein and the likelihood that a flammable or combustible concentration or quantity is present.

Notes:

- Only "Qualified Persons," as defined by OSHA in 29 CFR 1910.399, are to work on electrical circuits/systems.
- Workers must be trained to know the locations of circuit breaker panels that serve their lab area.

An OSHA QuickFacts entitled **Laboratory Safety – Electrical Hazards** has been developed to supplement this section and is available online at www.osha.gov.

Fire

Fire is the most common serious hazard that one faces in a typical laboratory. While proper procedures and training can minimize the chances of an accidental fire, laboratory workers should still be prepared to deal with a fire emergency should it occur. In dealing with a laboratory fire, all containers of infectious materials should be placed into autoclaves, incubators, refrigerators, or freezers for containment.

Small bench-top fires in laboratory spaces are not uncommon. Large laboratory fires are rare. However, the risk of severe injury or death is significant because fuel load and hazard levels in labs are typically very high. Laboratories, especially those using solvents in any quantity, have the potential for flash fires, explosion, rapid spread of fire, and high toxicity of products of combustion (heat, smoke, and flame).

Employers should ensure that workers are trained to do the following in order to prevent fires

- Plan work. Have a written emergency plan for your space and/or operation.
- Minimize materials. Have present in the immediate work area and use only the minimum quantities necessary for work in progress. Not only does this minimize fire risk, it reduces costs and waste.
- Observe proper housekeeping. Keep work areas uncluttered, and clean frequently. Put unneeded materials back in storage promptly. Keep aisles, doors, and access to emergency equipment unobstructed at all times.
- Observe restrictions on equipment (i.e., keeping solvents only in an explosion-proof refrigerator).
- Keep barriers in place (shields, hood doors, lab doors).

FIGURE 3.20 (Continued)

- Wear proper clothing and personal protective equipment.
- Avoid working alone.
- Store solvents properly in approved flammable liquid storage cabinets.
- Shut door behind you when evacuating.
- Limit open flames use to under fume hoods and only when constantly attended.
- Keep combustibles away from open flames.
- Do not heat solvents using hot plates.
- Remember the "RACE" rule in case of a fire.
  - R= Rescue/remove all occupants
  - A= Activate the alarm system
  - C= Confine the fire by closing doors
  - E= Evacuate/Extinguish

Employers should ensure that workers are trained in the following emergency procedures

- Know what to do. You tend to do under stress what you have practiced or pre-planned. Therefore, planning, practice and drills are essential.
- Know where things are: The nearest fire extinguisher, fire alarm box, exit(s), telephone, emergency shower/eyewash, and first-aid kit, etc.
- Be aware that emergencies are rarely "clean" and will often involve more than one type of problem. For example, an explosion may generate medical, fire, and contamination emergencies simultaneously.
- Train workers and exercise the emergency plan.
- Learn to use the emergency equipment provided.

Employers must be knowledgeable about OSHA's Portable Fire Extinguishers standard, 29 CFR 1910.157, and train workers to be aware of the different fire extinguisher types and how to use them. OSHA's Portable Fire Extinguishers standard, 29 CFR 1910.157, applies to the placement, use, maintenance, and testing of portable fire extinguishers provided for the use of workers. This standard requires that a fire extinguisher be placed within 75 feet for Class A fire risk (ordinary combustibles; usually fuels that burn and leave "ash") and within 50 feet for high-risk Class B fire risk (flammable liquids and gases; in the laboratory many organic solvents and compressed gases are fire hazards).

The two most common types of extinguishers in the chemistry laboratory are pressurized dry chemical (Type BC or ABC) and carbon dioxide. In addition, you may also have a specialized Class D dry powder extinguisher for use on flammable metal fires. Water-filled extinguishers are not acceptable for laboratory use.

Employers should train workers to remember the "PASS" rule for fire extinguishers

PASS summarizes the operation of a fire extinguisher.

P – Pull the pin

A – Aim extinguisher nozzle at the base of the fire

S – Squeeze the trigger while holding the extinguisher upright

S – Sweep the extinguisher from side to side; cover the fire with the spray

Employers should train workers on appropriate procedures in the event of a clothing fire

- If the floor is not on fire, STOP, DROP and ROLL to extinguish the flames or use a fire blanket or a safety shower if not contraindicated (i.e., there are no chemicals or electricity involved).
- If a coworker's clothing catches fire and he/she runs down the hallway in panic, tackle him/her and smother the flames as quickly as possible, using appropriate means that are available (e.g., fire blanket, fire extinguisher).

Lockout/Tagout

Workers performing service or maintenance on equipment may be exposed to injuries from the unexpected energization, startup of the equipment, or release or stored energy in the equipment. OSHA's Control of Hazardous Energy standard, 29 CFR 1910.147, commonly referred to as the "Lockout/Tagout" standard, requires the adoption and implementation of practices and procedures to shut down equipment, isolate it from its energy source(s), and prevent the release of potentially hazardous energy while maintenance and servicing activities are being performed. It contains minimum performance requirements, and definitive criteria for establishing an effective program for the control of hazardous energy. However, employers have the flexibility to develop Lockout/Tagout programs that are suitable for their respective facilities.

This standard establishes basic requirements involved in locking and/or tagging equipment while

LABORATORY SAFETY 27

FIGURE 3.20 (Continued)

installation, maintenance, testing, repair or construction operations are in progress. The primary purpose is to prevent hazardous exposure to personnel and possible equipment damage. The procedures apply to the shutdown of all potential energy sources associated with the equipment. These could include pressures, flows of fluids and gases, electrical power, and radiation. This standard covers the servicing and maintenance of machines and equipment in which the "**unexpected**" energization or startup of the machines or equipment, or release of stored energy could cause injury to workers.

Under the standard, the term "unexpected" also covers situations in which the servicing and/or maintenance is performed during ongoing normal production operations if:

- A worker is required to remove or bypass machine guards or other safety devices, 29 CFR 1910.147(a)2)(ii)(A) or
- A worker is required to place any part of his or her body into a point of operation or into an area on a machine or piece of equipment where work is performed, or into the danger zone associated with the machine's operation, 29 CFR 1910.147(a)(2)(ii)(B).

The Lockout/Tagout standard establishes minimum performance requirements for the control of such hazardous energy.

Maintenance activities can be performed with or without energy present. A probable, underlying cause of many accidents resulting in injury during maintenance is that work is performed without the knowledge that the system, whether energized or not, can produce hazardous energy. Unexpected and unrestricted release of hazardous energy can occur if: (1) all energy sources are not identified; (2) provisions are not made for safe work practices with energy present; or (3) deactivated energy sources are reactivated, mistakenly, intentionally, or accidentally, without the maintenance worker's knowledge.

Problems involving control of hazardous energy require procedural solutions. Employers must adopt such procedural solutions for controlling hazards to ensure worker safety during maintenance. However, such procedures are effective only if strictly enforced. Employers must, therefore, be committed to strict implementation of such procedures.

Trips, Slips and Falls

Worker exposure to wet floors or spills and clutter can lead to slips/trips/falls and other possible injuries. In order to keep workers safe, employers are referred to OSHA standard 29 CFR 1910 Subpart D – Walking-Working Surfaces, Subpart E - Means of Egress, and Subpart J - General environmental controls which states the following:

- Keep floors clean and dry, 29 CFR 1910.22(a)(2). In addition to being a slip hazard, continually wet surfaces promote the growth of mold, fungi, and bacteria that can cause infections.
- Provide warning (caution) signs for wet floor areas, 29 CFR 1910.145(c)(2).
- Where wet processes are used, maintain drainage and provide false floors, platforms, mats, or other dry standing places where practicable, or provide appropriate waterproof footgear, 29 CFR 1910.141(a)(3)(ii).
- The Walking/Working Surfaces standard requires that all employers keep all places of employment clean and orderly and in a sanitary condition, 29 CFR 1910.22(a)(1).
- Keep aisles and passageways clear and in good repair, with no obstruction across or in aisles that could create a hazard, 29 CFR 1910.22(b)(1). Provide floor plugs for equipment, so that power cords need not run across pathways.
- Keep exits free from obstruction. Access to exits must remain clear of obstructions at all times, 29 CFR 1910.37(a)(3).
- Ensure that spills are reported and cleaned up immediately.
- Eliminate cluttered or obstructed work areas.
- Use prudent housekeeping procedures such as using caution signs, cleaning only one side of a passageway at a time, and provide good lighting for all halls and stairwells to help reduce accidents, especially during the night hours.
- Instruct workers to use the handrail on stairs, to avoid undue speed, and to maintain an unobstructed view of the stairs ahead of them even if that means requesting help to manage a bulky load.
- Eliminate uneven floor surfaces.
- Promote safe work practices, even in cramped working spaces.
- Avoid awkward positions, and use equipment that makes lifting easier.

FIGURE 3.20 (Continued)

References

American Chemical Society, Safety in Academic Chemistry Laboratories. 1990. 5th Edition.

Burnett L, Lunn G, Coico R. Biosafety: Guidelines for working with pathogenic and infectious microorganisms. Current Protocols in Microbiology. 2009. 13:1A.1.1.-1A.1.14.

Centers for Disease Control and Prevention (CDC), National Institutes of Health (NIH). Primary Containment for Biohazards: Selection, Installation, and Use of Biological Safety Cabinets. 2007. 3rd Edition.

Centers for Disease Control and Prevention (CDC), National Institutes of Health (NIH). Biosafety Manual. 2007. 5th Edition. Washington, DC: U.S. Government Printing Office.

Centers for Disease Control and Prevention (CDC), Safety Survival Skills II. Laboratory Safety. A Primer on Safe Laboratory Practice and Emergency Response for CDC Workers. 2004. Available at: www.cdc.gov/od/ohs/safety/S2.pdf (Accessed January 7, 2009).

Clinical Laboratory Standards Institute (formerly NCCLS) document GP17-A2. Clinical Laboratory Safety. 2nd Edition. 2004.

Clinical Laboratory Standards Institute (formerly NCCLS) document GP18-A2. Laboratory Design. 2nd Edition. 2007.

Clinical Laboratory Standards Institute (formerly NCCLS) document M29-A3. Protection of Laboratory Workers from Occupationally Acquired Infections. 3rd Edition. 2005.

Committee on Occupational Health and Safety in the Care and Use of Non-human Primates, National Research Council. 2003. Occupational Health and Safety in the Care and Use of Non-human Primates. 2003. The National Academy Press, Washington, D.C.

Darragh AR, Harrison H, Kenny S. Effect of ergonomics intervention on workstations of microscope workers. American Journal of Occupational Therapy. 2008. 62:61-69.

Davis D. Laboratory Safety: A Self Assessment Workbook, ASCP Press, 1st Edition, 2008.

Furr AK. CRC Handbook of Laboratory Safety, 5th Edition, Chemical Rubber Company Press, 2000.

Gile TJ. Ergonomics in the laboratory. Lab Med. 2001. 32:263-267.

Illinois State University. Chemical Hygiene Plan for Chemistry Laboratories: Information and Training, 1995.

Kimman TG, Smit E, Klein MR. Evidence-based biosafety: A review of the principles and effectiveness of microbiological containment. Clinical Microbiology Reviews. 2008. 21:403-425.

National Institute of Occupational Safety and Health, Registry of Toxic Effects of Chemical Substances, (published annually) U.S. Department of Health and Human Services, Occupational Health Guidelines for Chemical Hazards, NIOSH/OSHA.

National Research Council, Prudent Practices in the Laboratory: Handling and Management of Chemical Hazards, National Academy Press, 2011.

Rose S. Clinical Laboratory Safety. J.B. Lippincott. Philadelphia, PA, 1984.

Singh K. Laboratory-acquired infections. Clinical Infectious Diseases. 2009. 49:142-147.

University of Illinois at Urbana-Champaign. UIUC Model Chemical Hygiene Plan, 1999.

University of Nebraska – Lincoln. UNL Environmental Health and Safety. Safe Operating Procedures, 2005-2008.

Vecchio D, Sasco AJ, Cann CI. 2003. Occupational risk in health care and research. American Journal of Industrial Medicine. 43:369-397.

FIGURE 3.20 (Continued)

Appendices

Additional OSHA Information

Chemical Hazards

Laboratory workers may be exposed to a variety of hazardous chemicals on the job. The following OSHA resources provide information on how to prevent or reduce exposure to some of the more common chemicals.

OSHA Standards

The Air Contaminants standard (1910.1000) provides rules for protecting workers from exposure to over 400 chemicals.

- Complete standard
  - **29 CFR 1910.1000**
    http://www.osha.gov/pls/oshaweb/owadisp.show_document?p_table=STANDARDS&p_id=9991
- Hospital eTool
  - ***Laboratories – Common safety and health topics***
    - ***Toluene, Xylene, or Acrylamide Exposure***
      http://www.osha.gov/SLTC/etools/hospital/lab/lab.html#Toulene,Xylene,orAcrylamideExposure

The Ethylene Oxide standard (29 CFR 1910.1047) requires employers to provide workers with protection from occupational exposure to ethylene oxide (EtO).

- Complete standard
  - **29 CFR 1910.1047**
    http://www.osha.gov/pls/oshaweb/owadisp.show_document?p_table=STANDARDS&p_id=10070
- Fact Sheet
  - ***Ethylene Oxide***
    http://www.osha.gov/OshDoc/data_General_Facts/ethylene-oxide-factsheet.pdf
- Booklet
  - ***Ethylene Oxide (EtO): Understanding OSHA's Exposure Monitoring Requirements.***
    OSHA Publication 3325 (2007). http://www.osha.gov/Publications/OSHA_ethylene_oxide.pdf
  - ***Small Business Guide for Ethylene Oxide.*** OSHA Publication 3359 (2009).
    http://www.osha.gov/Publications/ethylene-oxide-final.html
- Safety and Health Topics Page
  - ***Ethylene Oxide***
    http://www.osha.gov/SLTC/ethyleneoxide/index.html

The Formaldehyde standard (29 CFR 1910.1048) requires employers to provide workers with protection from occupational exposure to formaldehyde.

- Complete standard
  - **29 CFR 1910.1048**
    http://www.osha.gov/pls/oshaweb/owadisp.show_document?p_table=STANDARDS&p_id=10075
- Fact Sheet
  - ***Formaldehyde***
    http://www.osha.gov/OshDoc/data_General_Facts/formaldehyde-factsheet.pdf
- Hospital eTool
  - ***Laboratories – Common safety and health topics***
    - ***Formaldehyde Exposure***
      http://www.osha.gov/SLTC/etools/hospital/lab/lab.html#FormaldehydeExposure
- Safety and Health Topics Page
  - ***Formaldehyde***
    http://www.osha.gov/SLTC/formaldehyde/index.html

30

FIGURE 3.20 (Continued)

The Hazard Communication standard (29 CFR 1910.1200) is designed to protect against chemical source illnesses and injuries by ensuring that employers and employees are provided with sufficient information to recognize, evaluate and control chemical hazards and take appropriate protective measures.

In addition to the information provided at page 13 of this document, the following documents are available in either electronic or hard copy formats or both.

- Complete standard
  - **29 CFR 1910.1200**
    http://www.osha.gov/pls/oshaweb/owadisp.show_document?p_table=STANDARDS&p_id=10099
- Brochures
  - ***Chemical Hazard Communication.*** OSHA Publication 3084 (1998).
    http://www.osha.gov/Publications/osha3084.pdf
  - ***Hazard Communication Guidance for Combustible Dusts.*** OSHA Publication 3371 (2009).
    http://www.osha.gov/Publications/osha3371.pdf
  - ***Hazard Communication Guidelines for Compliance.*** OSHA publication 3111 (2000).
    http://www.osha.gov/Publications/osha3111.pdf
- Sample program
  - ***Model Plans and Programs for the OSHA Bloodborne Pathogens and Hazard Communications Standards.*** OSHA Publication 3186 (2003).
    http://www.osha.gov/Publications/osha3186.pdf
- QuickFacts
  - ***Laboratory Safety – Labeling and Transfer of Chemicals.*** OSHA Publication 3410 (2011).
    http://www.osha.gov/Publications/laboratory/OSHAquickfacts-lab-safety-labeling-chemical-transfer.pdf
- Safety and Health Topics Pages
  - ***Hazard Communication: Foundation of Workplace Chemical Safety Programs***
    http://www.osha.gov/dsg/hazcom/MSDSenforcementInitiative.html
  - ***Hazard Communication – HAZCOM Program***
    http://www.osha.gov/dsg/hazcom/solutions.html
  - ***Hazardous Drugs***
    http://www.osha.gov/SLTC/hazardousdrugs/index.html

The Occupational Exposure to Hazardous Chemicals in Laboratories standard (29 CFR 1910.1450), commonly referred to as the Laboratory standard, requires that the employer designate a Chemical Hygiene Officer and have a written Chemical Hygiene Plan (CHP), and actively verify that it remains effective.

In addition to the information provided at page 9 of this document, the following documents are available in either electronic or hard copy formats or both.

- Complete standard
  - **29 CFR 1910.1450**
    http://www.osha.gov/pls/oshaweb/owadisp.show_document?p_table=STANDARDS&p_id=10106
- Fact Sheet
  - ***Laboratory Safety – OSHA Laboratory Standard***
    http://www.osha.gov/Publications/laboratory/OSHAfactsheet-laboratory-safety-osha-lab-standard.pdf
  - ***Laboratory Safety – Chemical Hygiene Plan***
    http://www.osha.gov/Publications/laboratory/OSHAfactsheet-laboratory-safety-chemical-hygiene-plan.pdf
- Hospital eTool
  http://www.osha.gov/SLTC/etools/hospital/lab/lab.html
  - ***Laboratories – Common safety and health topics:***
    - ***Bloodborne Pathogens (BBPs)***
      http://www.osha.gov/SLTC/etools/hospital/lab/lab.html#BloodbornePathogens
    - ***Tuberculosis (TB)*** https://www.osha.gov/SLTC/etools/hospital/lab/lab.html#Tuberculosis
    - ***OSHA Laboratory Standard***
      http://www.osha.gov/SLTC/etools/hospital/lab/lab.html#OSHA_Laboratory_Standard
    - ***Formaldehyde Exposure***
      http://www.osha.gov/SLTC/etools/hospital/lab/lab.html#FormaldehydeExposure
    - ***Toluene, Xylene, or Acrylamide Exposure***
      http://www.osha.gov/SLTC/etools/hospital/lab/lab.html#Toulene,Xylene,orAcrylamideExposure

FIGURE 3.20 (Continued)

- ***Needle Stick and Sharps Injuries***
  http://www.osha.gov/SLTC/etools/hospital/lab/lab.html#NeedlestickInjuries
- ***Work Practices and Behaviors***
  http://www.osha.gov/SLTC/etools/hospital/lab/lab.html#WorkPractices
- ***Engineering Controls***
  http://www.osha.gov/SLTC/etools/hospital/lab/lab.html#EngineeringControls
- ***Morgue***
  http://www.osha.gov/SLTC/etools/hospital/lab/lab.html#Morgue
- ***Latex Allergy***
  http://www.osha.gov/SLTC/etools/hospital/lab/lab.html#LatexAllergy
- ***Slips/Trips/Falls***
  http://www.osha.gov/SLTC/etools/hospital/lab/lab.html#Slips/Trips/Falls
- ***Ergonomics***
  http://www.osha.gov/SLTC/etools/hospital/lab/lab.html#Ergonomics

Additional OSHA Information on Chemical Hazards

Beryllium

- Hazard Information Bulletin
  - ***Preventing Adverse Effects from Exposure to Beryllium in Dental Laboratories.*** (2002).
    http://www.osha.gov/dts/hib/hib_data/hib20020419.html
- Safety and Health Topics Page
  - ***Beryllium***
    http://www.osha.gov/SLTC/beryllium/index.html

Glutaraldehyde

- Booklet
  - ***Best Practices for the Safe Use of Glutaraldehyde in Health Care.*** OSHA Publication 3258-08N, (2006).
    http://www.osha.gov/Publications/glutaraldehyde.pdf
- Hospital eTool
  - ***Glutaraldehyde***
    http://www.osha.gov/SLTC/etools/hospital/hazards/glutaraldehyde/glut.html

Latex

- Safety and Health Information Bulletin
  - ***Potential for Sensitization and Possible Allergic Reaction to Natural Rubber Latex Gloves and other Natural Rubber Products.*** (2008).
    http://www.osha.gov/dts/shib/shib012808.html
- Letters of Interpretation
  - ***Bloodborne Pathogens and the issue of latex allergy and latex hypersensitivity.*** (1995 - 10/23/1995).
    http://www.osha.gov/pls/oshaweb/owadisp.show_document?p_table=INTERPRETATIONS&p_id=21987
  - ***Concern of potential adverse affects from latex by consumers and health care patients with Hevea Natural Rubber Latex Allergy.*** (2004 - 01/29/2004).
    http://www.osha.gov/pls/oshaweb/owadisp.show_document?p_table=INTERPRETATIONS&p_id=24742
  - ***Labeling of Latex.*** (1996 - 01/11/1996).
    http://www.osha.gov/pls/oshaweb/owadisp.show_document?p_table=INTERPRETATIONS&p_id=22040
- Hospital eTool
  - ***Latex Allergy***
    http://www.osha.gov/SLTC/etools/hospital/hazards/latex/latex.html
- Safety and Health Topics Page
  - ***Latex Allergy***
    http://www.osha.gov/SLTC/latexallergy/index.html
- QuickFacts
  - ***Laboratory Safety - Latex Allergy.*** OSHA Publication 3411 (2011).
    http://www.osha.gov/Publications/laboratory/OSHAquickfacts-lab-safety-latex-allergy.pdf

32

FIGURE 3.20 (Continued)

Mercury is commonly found in thermometers, manometers, barometers, gauges, valves, switches, batteries, and high-intensity discharge (HID) lamps. It is also used in amalgams for dentistry, preservatives, heat transfer technology, pigments, catalysts, and lubricating oils.

- Safety and Health Topics Page
  - ***Mercury***
    http://www.osha.gov/SLTC/mercury/index.html

Biological Hazards

The Bloodborne Pathogens standard (29 CFR 1910.1030), including changes mandated by the *Needlestick Safety and Prevention Act of 2001*, requires employers to protect workers from infection from human bloodborne pathogens in the workplace. The standard covers all workers with "reasonably anticipated" exposure to blood or other potentially infectious materials (OPIM).

- Complete standard
  - **29 CFR 1910.1030**
    http://www.osha.gov/pls/oshaweb/owadisp.show_document?p_table=STANDARDS&p_id=10051
- Standard interpretations
  - ***OSHA's standard interpretations for 29 CFR 1910.1030***
    http://www.osha.gov/pls/oshaweb/owasrch.search_form?p_doc_type=INTERPRETATIONS&p_toc_level=3&p_keyvalue=1910.1030&p_status=CURRENT
- Brochure/Sample program
  - ***Model Plans and Programs for the OSHA Bloodborne Pathogens and Hazard Communications Standards.*** OSHA Publication 3186 (2003).
    http://www.osha.gov/Publications/osha3186.html
- Fact Sheets (Accessible through the Safety and Health Topics Page entitled, *Bloodborne Pathogens and Needlestick Prevention*) NOTE: The links provided below are for the old Fact Sheets. All of these have been updated and approved for publication (2010) - please upload the new Fact Sheets
  - ***OSHA's Bloodborne Pathogens Standard***
    http://www.osha.gov/OshDoc/data_BloodborneFacts/bbfact01.pdf
  - ***Protecting Yourself When Handling Contaminated Sharps***
    http://www.osha.gov/OshDoc/data_BloodborneFacts/bbfact02.pdf
  - ***Personal Protective Equipment Reduces Exposure to Bloodborne Pathogens***
    http://www.osha.gov/OshDoc/data_BloodborneFacts/bbfact03.pdf
  - ***Exposure Incidents***
    http://www.osha.gov/OshDoc/data_BloodborneFacts/bbfact04.pdf
  - ***Hepatitis B Vaccination Protection***
    http://www.osha.gov/OshDoc/data_BloodborneFacts/bbfact05.pdf
- Safety and Health Topics Page
  - ***Bloodborne Pathogens and Needlestick Prevention***
    http://www.osha.gov/SLTC/bloodbornepathogens/index.html
- Safety and Health Information Bulletins
  - ***Use of Blunt-Tip Suture Needles to Decrease Percutaneous Injuries to Surgical Personnel.*** (2007).
    http://www.cdc.gov/niosh/docs/2008-101/
  - ***Disposal of Contaminated Needles and Blood Tube Holders Used for Phlebotomy.*** (2003).
    http://www.osha.gov/dts/shib/shib101503.html
  - ***Potential for Occupational Exposure to Bloodborne Pathogens from Cleaning Needles Used in Allergy Testing Procedures.*** (1995).
    http://www.osha.gov/dts/hib/hib_data/hib19950921.html
  - ***Sharps Disposal Containers with Needle Removal Features.*** (1993).
    http://www.osha.gov/dts/hib/hib_data/hib19930312.html
- Hospital eTool
  - ***Bloodborne Pathogens***
    http://www.osha.gov/SLTC/etools/hospital/hazards/bbp/bbp.html

FIGURE 3.20 (Continued)

Additional OSHA Information on Biological Agents

Tuberculosis

- Hospital eTool
  - ***Sample Tuberculosis Exposure Control Plan***
    http://www.osha.gov/SLTC/etools/hospital/hazards/tb/sampleexposurecontrolplan.html
  - ***Tuberculosis***
    http://www.osha.gov/SLTC/etools/hospital/hazards/tb/tb.html
- Safety and Health Topics Page
  - ***Tuberculosis***
    http://www.osha.gov/SLTC/tuberculosis/index.html

Physical Hazards and Others

Ionizing Radiation standard (29 CFR 1910.1096). Ionizing radiation sources may be found in a wide range of occupational settings, including, but not limited to, healthcare facilities, research institutions, nuclear reactors and their support facilities, nuclear weapons production facilities, and other various manufacturing settings. These radiation sources pose considerable health risks to affected workers if not properly controlled. This standard requires employers to conduct a survey of the types of radiation used in the facility, including x-rays, to designate restricted areas to limit worker exposure and to require those working in designated areas to wear personal radiation monitors. In addition, radiation areas and equipment must be labeled and equipped with caution signs.

- Complete standard
  - **29 CFR 1910.1096**
    http://www.osha.gov/pls/oshaweb/owadisp.show_document?p_table=STANDARDS&p_id=10098
- Safety and Health Topics Page
  - ***Ionizing Radiation***
    http://www.osha.gov/SLTC/radiationionizing/index.html
- Hospital eTool
  - ***Radiation Exposure***
    http://www.osha.gov/SLTC/etools/hospital/clinical/radiology/radiology.html#Radiation

Occupational Noise Exposure standard (29 CFR 1910.95). This standard requires employers to have a hearing conservation program in place if workers are exposed to a time-weighted average of 85 decibels (dB) over an 8-hour work shift.

- Complete standard
  - **29 CFR 1910.95**
    http://www.osha.gov/pls/oshaweb/owadisp.show_document?p_table=STANDARDS&p_id=10625
- Safety and Health Topics Page
  - ***Noise and Hearing Conservation***
    http://www.osha.gov/SLTC/noisehearingconservation/index.html
- Fact Sheet
  - **Laboratory Safety – Noise**
    http://www.osha.gov/Publications/laboratory/OSHAfactsheet-laboratory-safety-noise.pdf

Additional OSHA Information on Physical Hazards

Centrifuges

- QuickFacts
  - ***Laboratory Safety – Centrifuges.*** OSHA Publication 3406 (2011).
    http://www.osha.gov/Publications/laboratory/OSHAquickfacts-lab-safety-centrifuges.pdf

Cryogens & Dry Ice

- QuickFacts
  - ***Laboratory Safety – Cryogens & Dry Ice.*** OSHA Publication 3408 (2011).
    http://www.osha.gov/Publications/laboratory/OSHAquickfacts-lab-safety-cryogens-dryice.pdf

Laser hazards

- Safety and Health Information Bulletin
  - ***Hazard of Laser Surgery Smoke*** (1988).
    http://www.osha.gov/dts/hib/hib_data/hib19880411.html

OSHA®
Occupational Safety and Health Administration

FIGURE 3.20 (Continued)

- Hospital eTool
  - ***Laser Hazards***
    http://www.osha.gov/SLTC/etools/hospital/surgical/lasers.html
- Safety and Health Topics Pages
  - ***Laser Hazards***
    http://www.osha.gov/SLTC/laserhazards/index.html
  - ***Laser/Electrosurgery Plume***
    http://www.osha.gov/SLTC/laserelectrosurgeryplume/index.html

Safety Hazards

The Control of Hazardous Energy standard (29 CFR 1910.147), often called the "Lockout/Tagout" standard, establishes basic requirements for locking and/or tagging out equipment while installation, maintenance, testing, repair, or construction operations are in progress. The primary purpose of the standard is to protect workers from the unexpected energization or start-up of machines or equipment, or release of stored energy.

- Complete standard
  - **29 CFR 1910.147**
    http://www.osha.gov/pls/oshaweb/owadisp.show_document?p_table=STANDARDS&p_id=9804
- Booklet
  - ***Control of Hazardous Energy Lockout/Tagout***. OSHA Publication 3120 (2002).
    http://www.osha.gov/Publications/osha3120.pdf
- Safety and Health Topics Page
  - ***Control of Hazardous Energy (Lockout/Tagout)***
    http://www.osha.gov/SLTC/controlhazardousenergy/index.html

Electrical Hazards standards (29 CFR 1910 Subpart S). Wiring deficiencies are one of the hazards most frequently cited by OSHA. OSHA's electrical standards include design requirements for electrical systems and safety-related work practices. If flammable gases are used, special wiring and equipment installation may be required.

- Complete standard
  - **29 CFR 1910 Subpart S**
    http://www.osha.gov/pls/oshaweb/owadisp.show_document?p_table=STANDARDS&p_id=10135
- Booklet
  - ***Controlling Electrical Hazards.*** OSHA Publication 3075 (2002).
    http://www.osha.gov/Publications/osha3075.pdf
- Safety and Health Topics Page
  - ***Electrical***
    http://www.osha.gov/SLTC/electrical/index.html
- Hospital eTool
  - ***Electrical Hazards***
    http://www.osha.gov/SLTC/etools/hospital/hazards/electrical/electrical.html
- QuickFacts
  - ***Laboratory Safety – Electrical Hazards.*** OSHA Publication 3409 (2011).
    http://www.osha.gov/Publications/laboratory/OSHAquickfacts-lab-safety-electrical-hazards.pdf

Fire Prevention Plans standard (29 CFR 1910.39). OSHA recommends that all employers have a Fire Prevention Plan. A plan is mandatory when required by an OSHA standard. Additional fire hazard information is available via OSHA publications and web pages.

- Complete standard
  - **29 CFR 1910.39**
    http://www.osha.gov/pls/oshaweb/owadisp.show_document?p_table=STANDARDS&p_id=12887
- Booklet
  - ***Fire Service Features of Buildings and Fire Protection Systems.*** OSHA Publication 3256 (2006).
    http://www.osha.gov/Publications/osha3256.pdf
- Expert Advisor
  - ***Fire Safety Advisor***
    http://www.osha.gov/dts/osta/oshasoft/softfirex.html

LABORATORY SAFETY 35

FIGURE 3.20 (Continued)

- Fact Sheet
  - ***Fire Safety in the Workplace***
    http://www.osha.gov/OshDoc/data_General_Facts/FireSafetyN.pdf
- Safety and Health Topics Page
  - ***Fire Safety***
    http://www.osha.gov/SLTC/firesafety/index.html
- eTool
  - ***Evacuation Plans and Procedures***
    http://www.osha.gov/SLTC/etools/evacuation/index.html

Additional OSHA Information on Safety Hazards

Compressed gas

- Safety and Health Topics Page
  - ***Compressed Gas and Equipment***
    http://www.osha.gov/SLTC/compressedgasequipment/index.html

Ergonomics

- Fact Sheet
  - ***Laboratory Safety – Ergonomics for the Prevention of Musculoskeletal Disorders***
    http://www.osha.gov/Publications/laboratory/OSHAfactsheet-laboratory-safety-ergonomics.pdf

Engineering Controls

Autoclaves/Sterilizers

- QuickFacts
  - ***Laboratory Safety – Autoclaves/Sterilizers.*** OSHA Publication 3405 (2011).
    http://www.osha.gov/Publications/laboratory/OSHAquickfacts-lab-safety-autoclaves-sterilizers.pdf

Biosafety Cabinets (BSCs)

- Fact Sheet
  - ***Laboratory Safety – Biosafety Cabinets (BSCs).***
    http://www.osha.gov/Publications/laboratory/OSHAfactsheet-laboratory-safety-biosafety-cabinets.pdf

Chemical Fume Hoods

- QuickFacts
  - ***Laboratory Safety – Chemical Fume Hoods.*** OSHA Publication 3407 (2011).
    http://www.osha.gov/Publications/laboratory/OSHAquickfacts-lab-safety-chemical-fume-hoods.pdf

Personal Protective Equipment

The Personal Protective Equipment (PPE) standard (29 CFR 1910.132) requires that employers provide PPE and ensure that it is used wherever "hazards of processes or environment, chemical hazards, radiological hazards, or mechanical irritants [are] encountered in a manner capable of causing injury or impairment in the function of any part of the body through absorption, inhalation or physical contact," 29 CFR 1910.132(a).

- Complete standards
  - **29 CFR 1910 Subpart I**
    http://www.osha.gov/pls/oshaweb/owadisp.show_document?p_table=STANDARDS&p_id=10118
- Fact Sheet
  - ***Personal Protective Equipment***
    http://www.osha.gov/OshDoc/data_General_Facts/ppe-factsheet.pdf
- Brochures/Booklets
  - ***Personal Protective Equipment.*** OSHA Publication 3151 (2003).
    http://www.osha.gov/Publications/osha3151.html
- Safety and Health Topics Page
  - ***Personal Protective Equipment***
    http://www.osha.gov/SLTC/personalprotectiveequipment/index.html

The Eye and Face Protection standard (29 CFR 1910.133) requires that employers ensure that each affected employee uses appropriate eye or face protection when exposed to eye or face hazards from flying

FIGURE 3.20 (Continued)

particles, molten metal, liquid chemicals, acids or caustic liquids, chemical gases or vapors, or potentially injurious light radiation, 29 CFR 1910.133(a).

- Complete standard
  - **29 CFR 1910.133**
    http://www.osha.gov/pls/oshaweb/owadisp.show_document?p_table=STANDARDS&p_id=9778
- eTool
  - ***Eye and Face Protection***
    http://www.osha.gov/SLTC/etools/eyeandface/index.html
- Safety and Health Topics Page
  - ***Eye and Face Protection***
    http://www.osha.gov/SLTC/eyefaceprotection/index.html

The Respiratory Protection standard (29 CFR 1910.134) requires that a respirator be provided to each worker when such equipment is necessary to protect their health. The employer must provide respirators that are appropriate based on the hazards to which the worker is exposed and factors that affect respirator performance and reliability, as described in 29 CFR 1910.134(d)(1).

- Complete standard
  - **29 CFR 1910.134**
    http://www.osha.gov/pls/oshaweb/owadisp.show_document?p_table=STANDARDS&p_id=12716&p_text_version=FALSE
- Guidance Documents
  - ***Respiratory Protection.*** OSHA Publication 3079 (2002).
  - ***Small Entity Compliance Guide for OSHA's Respiratory Protection Standard.*** OSHA Publication 9071 (1999).
    http://www.osha.gov/Publications/secgrev-current.pdf
  - ***Assigned Protection Factors for the Revised Respiratory Protection Standard.*** OSHA Publication 3352 (2009).
    http://www.osha.gov/SLTC/etools/respiratory/index.html
- Fact Sheet
  - ***Respiratory Infection Control: Respirators Versus Surgical Masks***
- eTool
  - ***Respiratory Protection***
    http://www.osha.gov/SLTC/etools/respiratory/index.html
- Safety and Health Topics Page
  - ***Respiratory Protection***
    http://www.osha.gov/SLTC/respiratoryprotection/index.html

The Hand Protection standard (29 CFR 1910.138), requires that employers select and require workers to use appropriate hand protection when their hands are exposed to hazards such as those from skin absorption of harmful substances; severe cuts or lacerations; severe abrasions; punctures; chemical burns; thermal burns; and harmful temperature extremes, 29 CFR 1910.138(a). Further, employers must base the selection of the appropriate hand protection on an evaluation of the performance characteristics of the hand protection relative to the task(s) to be performed, conditions present, duration of use, and the hazards and potential hazards identified, 29 CFR 1910.138(b).

- Complete standard
  - **29 CFR 1910.138**
    http://www.osha.gov/pls/oshaweb/owadisp.show_document?p_table=STANDARDS&p_id=9788

Miscellaneous Information

Emergency Action Plan standard (29 CFR 1910.38). OSHA recommends that all employers have an Emergency Action Plan. A plan is mandatory when required by an OSHA standard. An Emergency Action Plan describes the actions workers should take to ensure their safety in a fire or other emergency situation.

- Complete standard
  - **29 CFR 1910.38**
    http://www.osha.gov/pls/oshaweb/owadisp.show_document?p_table=STANDARDS&p_id=9726

FIGURE 3.20 (Continued)

- Brochures/Booklets
  - ***Principal Emergency Response and Preparedness – Requirements and Guidance.*** OSHA Publication 3122 (2004) http://www.osha.gov/Publications/osha3122.pdf
  - ***How to Plan for Workplace Emergencies and Evacuations.*** OSHA Publication 3088 (2001). http://www.osha.gov/Publications/osha3088.pdf
- QuickFacts
  - ***Laboratory Safety – Working with Small Animals.*** OSHA Publication 3412 (2011). http://www.osha.gov/Publications/laboratory/OSHAquickfacts-lab-safety-working-with-small-animals.pdf
- eTool
  - ***Evacuation Plans and Procedures***
    http://www.osha.gov/SLTC/etools/evacuation/index.html
- eTool
  - ***Emergency Preparedness and Response***
    http://www.osha.gov/SLTC/emergencypreparedness/index.html

Exit Routes standards (29 CFR 1910.34 – 29 CFR 1910.37). All employers must comply with OSHA's requirements for exit routes in the workplace.

- Complete standards
  - **29 CFR 1910.34**
    http://www.osha.gov/pls/oshaweb/owadisp.show_document?p_table=STANDARDS&p_id=12886
  - **29 CFR 1910.35**
    http://www.osha.gov/pls/oshaweb/owadisp.show_document?p_table=STANDARDS&p_id=9723
  - **29 CFR 1910.36**
    http://www.osha.gov/pls/oshaweb/owadisp.show_document?p_table=STANDARDS&p_id=9724
  - **29 CFR 1910.37**
    http://www.osha.gov/pls/oshaweb/owadisp.show_document?p_table=STANDARDS&p_id=9725
- Fact Sheet
  - ***Emergency Exit Routes.*** http://www.osha.gov/OshDoc/data_General_Facts/emergency-exit-routes-factsheet.pdf
- QuickCard™
  - ***Emergency Exit Routes.*** OSHA Publication 3183 (2003).

Medical and First Aid standard (29 CFR 1910.151). OSHA requires employers to provide medical and first-aid personnel and supplies commensurate with the hazards of the workplace. The details of a workplace medical and first-aid program are dependent on the circumstances of each workplace and employer.

- Complete standard
  - **29 CFR 1910.151**
    http://www.osha.gov/pls/oshaweb/owadisp.show_document?p_table=STANDARDS&p_id=9806
- Brochures/Booklets
  - ***Best Practices Guide: Fundamentals of a Workplace First-Aid Program.*** OSHA Publication 3317 (2006) http://www.osha.gov/Publications/OSHA3317first-aid.pdf
- Safety and Health Topics Page
  - ***Medical and First Aid***
    http://www.osha.gov/SLTC/medicalfirstaid/index.html

Recordkeeping standard (29 CFR 1904). OSHA requires most employers to keep records of workplace injuries and illnesses. The employer should first determine if it is exempt from the routine recordkeeping requirements. An employer is not required to keep OSHA injury and illness records (unless asked to do so in writing by OSHA or the Bureau of Labor Statistics) if:

- It had 10 or fewer workers during all of the last calendar year (29 CFR 1904.1); or
- It is engaged in certain low-hazard industries (29 CFR Part 1904, Subpart B, Appendix A). The following types of healthcare facilities are exempt from OSHA's injury and illness recordkeeping requirements, regardless of size:
  - Offices and Clinics of Medical Doctors (SIC 801)
  - Offices and Clinics of Dentists (SIC 802)
  - Offices of Osteopathic Physicians (SIC 803)

OSHA®
Occupational Safety and Health Administration

FIGURE 3.20 (Continued)

- Offices of Other Health Care Practitioners (SIC 804)
- Medical and Dental Laboratories (SIC 807)
- Health and Allied Services, Not Elsewhere Classified (SIC 809)

If an employer does not fall within one of these exemptions, it must comply with OSHA's recordkeeping requirements. Download OSHA's recordkeeping forms or order them from the OSHA Publications Office at www.osha.gov.

For additional information on the Recordkeeping standard, see the following OSHA documents.

- Complete standards
  - ***Recording and reporting occupational injuries and illness.*** 29 CFR 1904
    http://www.osha.gov/pls/oshaweb/owasrch.search_form?p_doc_type=STANDARDS&p_toc_level=1&p_keyvalue=1904
  - ***Recording criteria for needlestick and sharps injuries.*** 29 CFR 1904.8
    http://www.osha.gov/pls/oshaweb/owadisp.show_document?p_table=STANDARDS&p_id=9639
- Standard Interpretations
  - ***Recordkeeping Handbook - The Regulation and Related Interpretations for Recording and Reporting Occupational Injuries and Illnesses.*** OSHA Publication 3245 (2005).
    http://www.osha.gov/recordkeeping/handbook/index.html
- Fact Sheets
  - ***Highlights of OSHA's Recordkeeping Rule***
    http://www.osha.gov/OshDoc/data_RecordkeepingFacts/RKfactsheet1.pdf
  - ***OSHA Recordkeeping Help***
    http://www.osha.gov/OshDoc/data_RecordkeepingFacts/RKfactsheet2.pdf
- Brochures
  - ***Access to Medical and Exposure Records.*** OSHA Publication 3110 (2001).
    http://www.osha.gov/Publications/osha3110.pdf
  - ***RECORDKEEPING - It's new, it's improved, and it's easier....*** OSHA Publication 3169 (2001).
    http://www.osha.gov/Publications/osha3169.pdf
  - ***Recordkeeping Handbook.*** OSHA Publication 3245 (2005).
    http://www.osha.gov/Publications/osha3245.pdf
- OSHA Web Page
  - ***Injury and Illness: Recordkeeping***
    http://www.osha.gov/recordkeeping/index.html

Access to Worker Exposure and Medical Records standard (29 CFR 1910.1020).
This standard requires all employers, regardless of size or industry, to report the work-related death of any worker or hospitalizations of three or more workers. It also requires employers to provide workers, their designated representatives, and OSHA with access to worker exposure and medical records. Employers generally should maintain worker exposure records for 30 years and medical records for the duration of the worker's employment plus 30 years, unless one of the exemptions listed in 29 CFR 1910.1020(d)(1)(i) (A)-(C) applies.

All employers covered by OSHA recordkeeping requirements must post the OSHA Poster (or state plan equivalent) in a prominent location in the workplace. The OSHA Poster can be downloaded or ordered in either English or Spanish.

The following OSHA document provides more detailed information on this standard.

- Booklet
  - ***Access to Medical and Exposure Records.*** OSHA Publication 3110 (2001).
    http://www.osha.gov/Publications/osha3110.pdf

NOTE: If your workplace is in a state operating an OSHA-approved state program, state plan recordkeeping and reporting regulations, although substantially identical to federal ones, may have different exemptions or more stringent or supplemental requirements, such as for reporting of fatalities and catastrophes. Contact your state program directly for additional information.

FIGURE 3.20 (Continued)

Other Governmental and Non-governmental Agencies Involved in Laboratory Safety

U.S. Environmental Protection Agency (EPA)

Microbial Products of Biotechnology: Final Rule (62 FR 17910)

The regulation under which the TSCA Biotechnology Program functions is titled "Microbial Products of Biotechnology; Final Regulation Under the Toxic Substances Control Act" (TSCA), published in the Federal Register on April 11, 1997. This rule was developed under TSCA Section 5, which authorizes the Agency to, among other things, review new chemicals before they are introduced into commerce. Under a 1986 intergovernmental policy statement, intergeneric microorganisms (microorganisms created to contain genetic material from organisms in more than one taxonomic genus) are considered new chemicals under TSCA Section 5. The Biotechnology rule sets forth the manner in which the Agency will review and regulate the use of intergeneric microorganisms in commerce, or commercial research.

Documents relevant to this rule can be found at the following web site:
http://www.epa.gov/oppt/biotech/pubs/biorule.htm.

U.S. Nuclear Regulatory Commission (NRC)

10 CFR 31.11 – General license for use of byproduct material for certain in vitro clinical or laboratory testing. Link at: http://www.nrc.gov/reading-rm/doc-collections/cfr/part031/part031-0011.html.

U.S. Department of Transportation (DOT)

An infectious substance is regulated as a hazardous material under the DOT's Hazardous Materials Regulations (HMR; 49 CFR Parts 171-180). The HMR apply to any material DOT determines is capable of posing an unreasonable risk to health, safety, and property when transported in commerce. An infectious substance must conform to all applicable HMR requirements when offered for transportation or transported by air, highway, rail, or water.

DOT's Pipeline and Hazardous Materials Safety Administration (PHMSA) published a final rule on June 1, 2006, revising the requirements in the HMR applicable to the transportation of infectious substances. The new requirements became effective October 1, 2006. Changes under the new rule apply to parts 171, 172, 173, and 175 of the HMR and include the following:

- New classification system
- New and revised definitions
- Revised marking requirements
- Revised packaging requirements
- New shipping paper requirements
- New security plan requirements
- New carriage by aircraft requirements

A guide to these changes is available at:
http://www.phmsa.dot.gov/staticfiles/PHMSA/DownloadableFiles/Files/Transporting_Infectious_Substances_brochure.pdf.

U.S. Department of Health and Human Services (HHS)

Centers for Disease Control and Prevention (CDC)

Biosafety Levels

Laboratory supervisors are responsible for ensuring that appropriate safety and health precautions are in place in the laboratory. Therefore, for each biosafety level, there are specific supervisory qualifications as assurance that laboratory workers are provided with effective supervision. Various types of specialized controls and equipment are used to provide primary barriers between the microorganism and the laboratory worker. These range from disposable gloves and other PPE to complex biosafety cabinets or other containment devices.

The laboratory director is specifically and primarily responsible for the safe operation of the laboratory. His/her knowledge and judgment are critical in assessing risks and appropriately applying these recommendations. The recommended biosafety level represents those conditions under which the agent can ordinarily be safely handled. Special characteristics of the agents used, the training and experience of personnel, and the nature or function of the laboratory may further influence the director in applying these recommendations.

The U.S. Department of Health and Human Services' (DHHS) Centers for Disease Control and Prevention (CDC) defines four levels of biosafety, which are outlined below. Selection of an appropriate biosafety level for work with a particular agent or animal study (see Animal Facilities) depends upon a number of factors. Some of the most important are the virulence, pathogenicity, biological stability, route of spread, and communicability of the agent; the nature or function of the laboratory; the procedures and manipulations involving the agent; the endemicity (restricted to a locality/region) of the agent; and the availability of effective vaccines or therapeutic measures.

40

OSHA®
Occupational Safety and Health Administration

FIGURE 3.20 (Continued)

CDC Summary of Recommended Biosafety Levels for Infectious Agents				
Biosafety Level	**Agent Characteristics**	**Practices**	**Safety Equipment**	**Facilities (secondary barriers)**
BSL-1	Not known to consistently cause disease in healthy adults	Standard microbiological Practices	None	Open bench top sink
BSL-2	Associated with human disease, hazard from percutaneous injury, ingestion, mucous membrane exposure	Standard microbiological Practices Limited access Biohazard warning signs Sharps precautions Biosafety manual defining any needed waste decontamination or medical surveillance policies.	Class I or II biosafety cabinets (BSCs) or other containment devices used for all agents that cause splashes or aerosols of infectious materials Laboratory coats and gloves Face protection as needed	Open bench top sink Autoclave
BSL-3	Indigenous or exotic agents with potential for aerosol transmission; disease may have serious or lethal consequences	All BSL-2 practices Controlled access Decontamination of all waste Decontamination of laboratory clothing before laundering Baseline serum	Class I or II BSCs or other physical containment devices used for all open manipulations of agents Protective lab clothing and gloves Respiratory protection as needed	Open bench top sink Autoclave Physical separation from access corridors Self-closing, double-door access Exhaust air not recirculated Negative airflow in laboratory
BSL-4	Dangerous/exotic agents which pose high risk of life-threatening disease; aerosol-transmitted lab infections; or related agents with unknown risk of transmission	All BSL-3 practices Clothing change before entering Shower on exit All material decontaminated on exit from facility	All procedures conducted in Class III BSCs, or Class I or II BSCs in combination with full-body, air-supplied, positive pressure personnel suit.	BSL-3 plus: Separate building or isolated zone Dedicated supply and exhaust, vacuum, and decontamination systems Other requirements outlined in the text

LABORATORY SAFETY 41

FIGURE 3.20 (Continued)

NOTE: The following information has been adapted from *Biosafety in Microbiological and Biomedical Laboratories*, 5th Ed. (BMBL, 5th Ed.), which is published jointly by the U.S. Centers for Disease Control and Prevention (CDC) and the National Institutes of Health (NIH), and is available online at www.cdc.gov/od/ohs/biosfty/bmbl5/bmbl5toc.htm. Laboratory workers and supervisors are strongly urged to review this publication directly before engaging in any experimentation.

Biosafety Level 1 (BSL-1)

BSL-1 is appropriate for working with microorganisms that are not known to cause disease in healthy humans. BSL-l practices, safety equipment, and facility design and construction are appropriate for undergraduate and secondary educational training and teaching laboratories, and for other laboratories in which work is done with defined and characterized strains of viable microorganisms not known to consistently cause disease in healthy adult humans. *Bacillus subtilis, Naegleria gruberi*, infectious canine hepatitis virus, and exempt organisms under the *NIH Recombinant DNA Guidelines* (http://www4.od nih.gov/oba/rac/guidelines/guidelines.html) are representative of microorganisms meeting these criteria. Many agents not ordinarily associated with disease processes in humans are, however, opportunistic pathogens and may cause infection in the young, the aged, and immunodeficient or immunosuppressed individuals. Vaccine strains that have undergone multiple in vivo passages should not be considered avirulent simply because they are vaccine strains.

BSL-1 represents a basic level of containment that relies on standard microbiological practices with no special primary or secondary barriers recommended, other than a sink for hand washing.

Biosafety Level 2 (BSL-2)

The facility, containment devices, administrative controls, and practices and procedures that constitute BSL-2 are designed to maximize safe working conditions for laboratory personnel working with agents of moderate risk to personnel and the environment. BSL-2 practices, equipment, and facility design and construction are applicable to clinical, diagnostic, teaching, and other laboratories in which work is done with the broad spectrum of indigenous moderate-risk agents that are present in the community and associated with human disease of varying severity. With good microbiological techniques, these agents can be used safely in activities conducted on the open bench, provided the potential for producing splashes or aerosols is low. Hepatitis B virus, H1V, the salmonellae, and Toxoplasma spp. are representative of microorganisms assigned to this containment level.

Biosafety Level 2 is also appropriate when work is done with any human-derived blood, body fluids, tissues, or primary human cell lines where the presence of an infectious agent may be unknown. Laboratory personnel in the United States working with human-derived materials should refer to the U.S. Occupational Safety and Health Administration (OSHA) *Bloodborne Pathogens Standard* (OSHA 1991), available online at www.osha.gov/pls/oshaweb/owadisp.show_document?p_table=STANDARDS7p_id=1005, for required precautions.

Primary hazards to personnel working with these agents relate to accidental percutaneous or mucous membrane exposures, or ingestion of infectious materials. Extreme caution should be taken with contaminated needles or sharp instruments. Even though organisms routinely manipulated at Biosafety Level 2 are not known to be transmissible by the aerosol route, procedures with aerosol or high splash potential that may increase the risk of such personnel exposure must be conducted in primary containment equipment, or in devices such as a biological safety cabinet (BSC) or safety centrifuge cups. Personal protective equipment (PPE) should be used as appropriate, such as splash shields, face protection, gowns, and gloves.

Secondary barriers such as hand washing sinks and waste decontamination facilities must be available to reduce potential environmental contamination.

Biosafety Level 3 (BSL-3)

BSL-3 is suitable for work with infectious agents which may cause serious or potentially lethal diseases as a result of exposure by the inhalation route. This may apply to clinical, diagnostic, teaching, research, or production facilities in which work is done with indigenous or exotic agents with potential for respiratory transmission, and which may cause serious and potentially lethal infection. *Mycobacterium tuberculosis*, St. Louis encephalitis virus, and *Coxiella burnetti* are representative of the microorganisms assigned to this level. Primary hazards to personnel working with these agents relate to autoinoculation, ingestion, and exposure to infectious aerosols.

At BSL-3, more emphasis is placed on primary and secondary barriers to protect personnel in contiguous

FIGURE 3.20 (Continued)

areas, the community, and the environment from exposure to potentially infectious aerosols. For example, all laboratory manipulations should be performed in a BSC or other enclosed equipment, such as a gas-tight aerosol generation chamber. Secondary barriers for this level include controlled access to the laboratory and ventilation requirements that minimize the release of infectious aerosols from the laboratory.

Biosafety Level 4 (BSL-4)
BSL-4 practices, safety equipment, and facility design and construction are applicable for work with dangerous and exotic agents that pose a high individual risk of life-threatening disease, which may be transmitted via the aerosol route, and for which there is no available vaccine or therapy. Agents with a close or identical antigenic relationship to Biosafety Level 4 agents also should be handled at this level. When sufficient data are obtained, work with these agents may continue at this or at a lower level. Viruses such as Marburg or Congo-Crimean hemorrhagic fever are manipulated at Biosafety Level 4.

The primary hazards to personnel working with Biosafety Level 4 agents are respiratory exposure to infectious aerosols, mucous membrane or broken skin exposure to infectious droplets, and autoinoculation. All manipulations of potentially infectious diagnostic materials, isolates, and naturally or experimentally infected animals pose a high risk of exposure and infection to laboratory personnel, the community, and the environment.

The laboratory worker's complete isolation from aerosolized infectious materials is accomplished primarily by working in a Class III BSC or in a full-body, air-supplied, positive-pressure personnel suit. The BSL-4 facility itself is generally a separate building or completely isolated zone with complex, specialized ventilation requirements and waste management systems to prevent release of viable agents to the environment.

Animal Biosafety Levels
The CDC defines four biosafety levels for activities involving infectious disease work with experimental animals. These combinations of practices, safety equipment, and facilities are designated Animal Biosafety Levels 1, 2, 3, and 4, and provide increasing levels of protection to personnel and the environment.

Protocols using live animals must first be reviewed and approved by an Institutional Animal Care and Use Committee (IACUC) or must conform to governmental regulations regarding the care and use of laboratory animals. Follow all appropriate guidelines for the use and handling of infected animals.

For more information, refer to Section V of the BMBL, 5th Ed., available online at www.cdc.gov/od/ohs/biosafty/bmbl5/bmbl5toc.htm.

National Institutes of Health (NIH)
The NIH Office of Biotechnology Activities (OBA) promotes science, safety, and ethics in biotechnology through advancement of knowledge, enhancement of public understanding, and development of sound public policies. OBA accomplishes its mission through analysis, deliberation, and communication of scientific, medical, ethical, legal, and social issues.

OBA fulfills its mission through four important programs:

- Recombinant DNA (RAC)
- Genetics, Health, Society (SACGHS)
- Dual Use Research (NSABB)
- Clinical Research Policy Analysis and Coordination (CRpac)

Links to each of the programs listed above are provided at the OBA website: http://oba.od.nih.gov/oba/index.html.

National Institute for Occupational Safety and Health (NIOSH)
The NIOSH **Pocket Guide to Chemical Hazards (NPG)** (available at: www.cdc.gov/niosh/npg) provides a source of general industrial hygiene information on several hundred chemicals/classes for workers, employers, and occupational health professionals. While the NPG does not contain an analysis of all pertinent data, it presents key information and data in abbreviated or tabular form for chemicals or substance groupings (e.g., cyanides, fluorides, manganese compounds) that are found in the work environment. The information contained in the NPG should help users recognize and control occupational chemical hazards.

Other Government Web Links for Access to Additional Information Concerning Laboratory Safety
The Animal Plant Health Inspection Service (APHIS), www.usda/aphis.gov

U.S. Department of Agriculture (USDA), www.usda.gov

National Institute for Occupational Safety and Health (NIOSH), www.niosh.gov

LABORATORY SAFETY 43

FIGURE 3.20 (Continued)

U.S. Department of Health and Human Services (DHHS), www.hhs.gov

U.S. Department of Transportation (DOT), www.dot.gov

U.S. Food and Drug Administration (FDA), www.fda.gov

Government Regulatory Agency Web Links

Code of Federal Regulations Search Engine, www.access.gpo.gov/nara/cfr/index.html

Environmental Protection Agency, www.epa.gov

Federal Register Search Engine, www.access.gpo.gov/su_docs/aces/aces140.html

Food and Drug Administration, www.fda.gov

Nuclear Regulatory Commission, www.nrc.gov

Occupational Safety and Health Administration (OSHA), www.osha.gov

Non-governmental Agency Web Links for Access to Additional Information Concerning Laboratory Safety

American Biological Safety Association (ABSA), www.absa.org

College of American Pathologists (CAP), www.cap.org

Institute for Laboratory Animal Research (ILAR), www.dels.nas.edu/ilar_n/ilarhome

National Fire Protection Association (NFPA), www.nfpa.org

Dictionary of Safety Terms

Oregon OSHA Dictionary of Safety Terms - Spanish to English, www.orosha.org/pdf/dictionary/spanish-english.pdf

Oregon OSHA Dictionary of Safety Terms - English to Spanish, www.orosha.org/pdf/dictionary/english-spanish.pdf

44

FIGURE 3.20 (Continued)

Most Common Zoonotic Diseases in Workers

Workers that work with animals may be exposed to a number of zoonotic diseases. Examples of some of the zoonotic diseases that workers may be exposed to are listed in the table below.

Disease	Disease agent	Animals				
		Cats	Dogs	Birds	Farm Animals	Wild Animals
Brucellosis	*Brucella canis*		X			
Campylobacteriosis	*Campylobacter jejuni*	X	X		X	
Cat Scratch Fever	*Bartonella henselae*	X				
Cryptococcosis	*Cryptococcus neoformans* and other species			X		
Hemorrhagic fever with renal syndrome (HFRS) and hantavirus pulmonary syndrome (HPS)	Hantavirus					X
Lymphocytic choriomeningitis	Lymphocytic choriomeningitis virus (LCMV)			X		
Pasteurella pneumonia	*Pasteurella haemolytica*				X	
Histoplasmosis	*Histoplasma capsulatum*			X		
Orf	Poxvirus				X	
Plague	*Yersinia pestis*					X
Q-fever	*Coxiella burnetii*				X	
Rabies	Rabies virus	X	X			
Salmonellosis	*Salmonella enterica* serovar Typhi				X	
Toxoplasmosis	*Toxoplasma gondii*	X				
Tularemia	*Tularemia francisella*					X

FIGURE 3.20 (Continued)

Complaints, Emergencies and Further Assistance

Workers have the right to a safe workplace. The *Occupational Safety and Health Act of 1970* (OSH Act) was passed to prevent workers from being killed or seriously harmed at work. The law requires employers to provide their employees with working conditions that are free of known dangers. Workers may file a complaint to have OSHA inspect their workplace if they believe that their employer is not following OSHA standards or that there are serious hazards. Further, the Act gives complainants the right to request that their names not be revealed to their employers. It is also against the law for an employer to fire, demote, transfer, or discriminate in any way against a worker for filing a complaint or using other OSHA rights.

To report an emergency, file a complaint, or seek OSHA advice, assistance, or products, call (800) 321-OSHA (6742) or contact your nearest OSHA regional, area, or state plan office listed or linked to at the end of this publication. The teletypewriter (TTY) number is (877) 889-5627. You can also file a complaint online by visiting OSHA's website at www.osha.gov. Most complaints submitted online may be resolved informally over the phone or by fax with your employer. Written complaints, that are signed by a worker or their representative and submitted to the closest OSHA office, are more likely to result in an on-site OSHA inspection.

Compliance Assistance Resources

OSHA can provide extensive help through a variety of programs, including free workplace consultations, compliance assistance, voluntary protection programs, strategic partnerships, alliances, and training and education. For more information on any of the programs listed below, visit OSHA's website at www.osha.gov or call 1-800-321-OSHA (6742).

Establishing an Injury and Illness Prevention Program

The key to a safe and healthful work environment is a comprehensive injury and illness prevention program.

Injury and illness prevention programs, known by a variety of names, are universal interventions that can substantially reduce the number and severity of workplace injuries and alleviate the associated financial burdens on U.S. workplaces. Many states have requirements or voluntary guidelines for workplace injury and illness prevention programs. In addition, numerous employers in the United States already manage safety using injury and illness prevention programs, and we believe that all employers can and should do the same. Employers in the construction industry are already required to have a health and safety program. Most successful injury and illness prevention programs are based on a common set of key elements. These include management leadership, worker participation, hazard identification, hazard prevention and control, education and training, and program evaluation and improvement. Visit OSHA's website at http://www.osha.gov/dsg/topics/safetyhealth/index.html for more information and guidance on establishing effective injury and illness prevention programs in the workplace.

Compliance Assistance Specialists

OSHA has compliance assistance specialists throughout the nation who can provide information to employers and workers about OSHA standards, short educational programs on specific hazards or OSHA rights and responsibilities, and information on additional compliance assistance resources. Contact your local OSHA office for more information.

OSHA Consultation Service for Small Employers

The OSHA Consultation Service provides **free assistance** to small employers to help them identify and correct hazards, and to improve their injury and illness prevention programs. Most of these services are delivered on site by state government agencies or universities using well-trained professional staff.

Consultation services are available to private sector employers. Priority is given to small employers with the most hazardous operations or in the most high-hazard industries. These programs are largely funded by OSHA and are delivered at no cost to employers who request help. Consultation services are separate from enforcement activities. To request such services, an employer can phone or write to the OSHA Consultation Program. See the Small Business section of OSHA's website for contact information for the consultation offices in every state.

- **Safety and Health Achievement Recognition Program**
  Under the consultation program, certain exemplary employers may request participation in OSHA's Safety and Health Achievement

FIGURE 3.20 (Continued)

Recognition Program (SHARP). Eligibility for participation includes, but is not limited to, receiving a full-service, comprehensive consultation visit, correcting all identified hazards, and developing an effective injury and illness prevention program.

Cooperative Programs

OSHA offers cooperative programs to help prevent fatalities, injuries and illnesses in the workplace.

- **OSHA's Alliance Program**
  Through the Alliance Program, OSHA works with groups committed to worker safety and health to prevent workplace fatalities, injuries, and illnesses. These groups include businesses, trade or professional organizations, unions, consulates, faith- and community-based organizations, and educational institutions. OSHA and the groups work together to develop compliance assistance tools and resources, share information with workers and employers, and educate workers and employers about their rights and responsibilities.
- **Challenge Program**
  This program helps employers and workers improve their injury and illness prevention programs and implement an effective system to prevent fatalities, injuries and illnesses.
- **OSHA Strategic Partnership Program (OSPP)**
  Partnerships are formalized through tailored agreements designed to encourage, assist and recognize partner efforts to eliminate serious hazards and achieve model workplace safety and health practices.
- **Voluntary Protection Programs (VPP)**
  The VPP recognize employers and workers in private industry and federal agencies who have implemented effective injury and illness prevention programs and maintain injury and illness rates below national Bureau of Labor Statistics averages for their respective industries. In VPP, management, labor, and OSHA work cooperatively and proactively to prevent fatalities, injuries, and illnesses.

OSHA Training Institute Education Centers

The OSHA Training Institute (OTI) Education Centers are a national network of nonprofit organizations authorized by OSHA to conduct occupational safety and health training to private sector workers, supervisors and employers.

Susan Harwood Training and Education Grants

OSHA provides grants to nonprofit organizations to provide worker education and training on serious job hazards and avoidance/prevention strategies.

Information and Publications

OSHA has a variety of educational materials and electronic tools available on its website at www.osha.gov. These include Safety and Health Topics Pages, Safety Fact Sheets, Expert Advisor software, copies of regulations and compliance directives, videos and other information for employers and workers. OSHA's software programs and eTools walk you through safety and health issues and common problems to find the best solutions for your workplace.

OSHA's extensive publications help explain OSHA standards, job hazards, and mitigation strategies and provide assistance in developing injury and illness prevention programs.

For a listing of free publications, visit OSHA's website at www.osha.gov or call 1-800-321-OSHA (6742).

QuickTakes

OSHA's free, twice monthly online newsletter, *QuickTakes*, offers the latest news about OSHA initiatives and products to assist employers and workers in finding and preventing workplace hazards. To sign up for *QuickTakes*, visit OSHA's website at www.osha.gov and click on *QuickTakes* at the top of the page.

Contacting OSHA

To order additional copies of this publication, to get a list of other OSHA publications, to ask questions or to get more information, to contact OSHA's free consultation service, or to file a confidential complaint, contact OSHA at 1-800-321-OSHA (6742), (TTY) 1-877-889-5627 or visit www.osha.gov.

For assistance, contact us.
We are OSHA. We can help.
It's confidential.

LABORATORY SAFETY 47

FIGURE 3.20 (Continued)

OSHA Regional Offices

Region I
Boston Regional Office
(CT*, ME, MA, NH, RI, VT*)
JFK Federal Building, Room E340
Boston, MA 02203
(617) 565-9860 (617) 565-9827 Fax

Region II
New York Regional Office
(NJ*, NY*, PR*, VI*)
201 Varick Street, Room 670
New York, NY 10014
(212) 337-2378 (212) 337-2371 Fax

Region III
Philadelphia Regional Office
(DE, DC, MD*, PA, VA*, WV)
The Curtis Center
170 S. Independence Mall West
Suite 740 West
Philadelphia, PA 19106-3309
(215) 861-4900 (215) 861-4904 Fax

Region IV
Atlanta Regional Office
(AL, FL, GA, KY*, MS, NC*, SC*, TN*)
61 Forsyth Street, SW, Room 6T50
Atlanta, GA 30303
(678) 237-0400 (678) 237-0447 Fax

Region V
Chicago Regional Office
(IL*, IN*, MI*, MN*, OH, WI)
230 South Dearborn Street
Room 3244
Chicago, IL 60604
(312) 353-2220 (312) 353-7774 Fax

Region VI
Dallas Regional Office
(AR, LA, NM*, OK, TX)
525 Griffin Street, Room 602
Dallas, TX 75202
(972) 850-4145 (972) 850-4149 Fax
(972) 850-4150 FSO Fax

Region VII
Kansas City Regional Office
(IA*, KS, MO, NE)
Two Pershing Square Building
2300 Main Street, Suite 1010
Kansas City, MO 64108-2416
(816) 283-8745 (816) 283-0547 Fax

Region VIII
Denver Regional Office
(CO, MT, ND, SD, UT*, WY*)
1999 Broadway, Suite 1690
Denver, CO 80202
(720) 264-6550 (720) 264-6585 Fax

Region IX
San Francisco Regional Office
(AZ*, CA*, HI*, NV*, and American Samoa, Guam and the Northern Mariana Islands)
90 7th Street, Suite 18100
San Francisco, CA 94103
(415) 625-2547 (415) 625-2534 Fax

Region X
Seattle Regional Office
(AK*, ID, OR*, WA*)
300 Fifth Avenue, Suite 1280
Seattle, WA 98104-2397
(206) 757-6700 (206) 757-6705 Fax

*These states and territories operate their own OSHA-approved job safety and health plans and cover state and local government employees as well as private sector employees. The Connecticut, Illinois, New Jersey, New York and Virgin Islands programs cover public employees only. (Private sector workers in these states are covered by Federal OSHA). States with approved programs must have standards that are identical to, or at least as effective as, the Federal OSHA standards.

Note: To get contact information for OSHA area offices, OSHA-approved state plan offices and OSHA consultation projects, please visit us online at www.osha.gov or call us at 1-800-321-OSHA (6742).

48

FIGURE 3.20 (Continued)

FIGURE 3.20 (Continued)

For more information:

U.S. Department of Labor
www.osha.gov (800) 321-OSHA (6742)

FIGURE 3.20 (Continued)

4

BASIC MATHEMATICS IN THE LABORATORY

4.1 Introduction to Basic Math
4.2 Units and Metric System
4.2.1 Introduction to the Metric System
4.2.2 Units of the Metric System
4.2.3 Converting the *SI* Units
4.3 Significant Figures
4.3.1 Significant Figure Rules
4.4 Scientific Calculators
4.4.1 Example Calculator
4.4.2 Window's Calculator
4.5 ChemTech Conversion Tool
4.5.1 Using the Conversion Tool
4.5.2 Closing the Conversion Tool
4.6 Chapter Key Concepts
4.7 Chapter Problems

4.1 INTRODUCTION TO BASIC MATH

The basic use of rudimentary mathematics in the laboratory is an essential skill required by the laboratory technician, the analytical chemist, and the research chemist. We will begin with the most fundamental treatment of data involving adding, subtracting, multiplying, and dividing to calculate sums and averages to start with. In later chapters, we will move on to more complex statistical treatments of laboratory measurement-generated data, including the use of computer spreadsheets, which facilitate and greatly ease the tasks of data manipulation for the analyst. Throughout the chapter, we will also incorporate the use of ChemTech at various stages to work problems that reflect a real-life laboratory setting in relation to fundamental mathematic techniques. As always, ChemTech also serves as a future reference and refresher for the technician in the analytical laboratory.

While the analyst in the laboratory is usually following a prescribed method for a testing procedure, he/she will be required to accurately record results in either a laboratory notebook, laboratory testing sheets, or a laboratory computer-based system (covered in Chapter 11 Laboratory Information Management System). Laboratories today are still a mixture of automated data calculation and manual calculation by the technician; thus, the basic math skills presented in this chapter are essential tools needed.

4.2 UNITS AND METRIC SYSTEM

4.2.1 Introduction to the Metric System

Laboratory measurements are primarily recorded in the *International System* (abbreviated *SI* from the French *Le Système International d'Unités*) standardized units. The SI units were adopted through an international agreement between scientists in 1960, and are now widely adopted by most countries. The most recent SI units (the mole was added in 1971 by the 14th General Conference on Weights and Matters (CGPM)) are listed in Table 4.1. This is a system of units derived from the metric system, a system that has its roots originating from the late eighteenth century (ca 1791) France. Antoine-Laurent Lavoisier, known as the "father of modern chemistry," was instructed along with other contemporary scientists by Louis XVI of France to construct a unified, natural, and universal system of measurement to replace the disparate systems then in use. This effort resulted in the production of the metric system, a system of measurement, which is a base-10 system (all units expressed as a factor of 10).

4.2.2 Units of the Metric System

Interestingly, a unit for volume is not listed among the *SI* units in Table 4.1. The common practice is to use the *liter* (l) in the laboratory as a unit for volumetric designations, and the associated

Analytical Chemistry: A Chemist and Laboratory Technician's Toolkit, First Edition. Bryan M. Ham and Aihui MaHam.

TABLE 4.1 The *International System* (SI) Base Units.

Unit Quantity	Unit Name	Unit Symbol
Length	Meter	m
Mass	Kilogram	kg
Time	Second	s
Amount of material	Mole	mol
Temperature	Kelvin	K
Luminous intensity	Candela	cd
Electric current	Ampere	A

TABLE 4.2 The Metric System.

Symbol	Prefix	Multiplication Factor	
E	exa	10^{18}	1,000,000,000,000,000,000
P	peta	10^{15}	1,000,000,000,000,000
T	tera	10^{12}	1,000,000,000,000
G	giga	10^{9}	1,000,000,000
M	mega	10^{6}	1,000,000
k	kilo	10^{3}	1,000
h	hecta	10^{2}	100
da	deca	10^{1}	10
d	deci	10^{-1}	0.1
c	centi	10^{-2}	0.01
m	milli	10^{-3}	0.001
μ	micro	10^{-6}	0.000,001
n	nano	10^{-9}	0.000,000,001
p	pico	10^{-12}	0.000,000,000,001
f	femto	10^{-15}	0.000,000,000,000,001
a	atto	10^{-18}	0.000,000,000,000,000,001

factors of the 10-based metric system such as the widely used milliliter (ml). Volume is in fact a three-dimensional length measurement, thus it is not considered to be a base unit. For example, one liter is equal to one cubic decimeter ($1\,l = 1\,dm^3$), and one milliliter is equal to one cubic centimeter ($1\,ml = 1\,cm^3$). Table 4.2 lists the most common metric units, including symbols, prefixes, and the associated multiplication factors. Values that are greater than 1 would use deca (da, ×10) to exa (E, $\times 10^{18}$), and values less than 1 would use deci (d, ×0.1) to atto (a, $\times 10^{-18}$). The metric system expresses the *SI* units in factors of 10:

$$1\text{ kilometer} = 10^3\text{m}$$
$$1\text{ hectometer} = 10^2\text{m}$$
$$1\text{ decameter} = 10\text{ m}$$
$$1\text{ decimeter} = 10^{-1}\text{m}$$
$$1\text{ centimeter} = 10^{-2}\text{m}$$
$$1\text{ millimeter} = 10^{-3}\text{m}$$

4.2.3 Converting the *SI* Units

Often it may be required to convert between *SI* units, and between other units such as the English system. Table 4.3 is a listing of the metric system units, including the approximate U.S. equivalent values that can be used as a conversion table. Appendix III gives a more extensive listing of various conversions. Your ChemTech program contains these tables as easy reference to use in the laboratory, a link found in the Main Menu of Chapter 4. When converting values by hand, often called the *unit-factor method*, the *factor-label method*, or *dimensional analysis*, the value to be converted is multiplied by a conversion factor where the unit of the original value is replaced with the desired unit. Let us look at an example to get an idea of how this works.

Example 4.1 Convert 326.8 meters (m) to yards (yd)

$$\text{Conversion factor}: \frac{1.0936\text{ yd}}{\text{m}} \tag{4.1}$$

$$326.8\text{ m} \times \frac{1.0936\text{yd}}{\text{m}} = 357.4\text{ yd} \tag{4.2}$$

Note in the example that multiplying by the conversion factor will result in the cancellation of the original unit replacing it with the desired unit.

4.3 SIGNIFICANT FIGURES

The result to the above example is also expressed in the proper number of significant figures. When performing laboratory calculations, only the proper number of significant figures should be reported in a final result. The definition of significant figures states that only those digits that are known with certainty plus the first uncertain digit are to be reported. The last digit making up a number that is being reported is generally understood to be uncertain by ±1, unless otherwise designated due to a circumstance of known uncertainty. This is illustrated in the following set of numbers which all contain four significant figures: 0.2058, 2.058, 2058, and 2.058×10^5. The 2, 0, and 5 are certain and the 8 is uncertain but significant.

4.3.1 Significant Figure Rules

The following general rules will give a guideline for the proper treatment of numbers in the laboratory:

1. Zeros contained within digits are significant and should be counted, such as 107 contains three significant figures and 1007 contains four. Leading zeros in numbers, such as 0.00124 and 0.124, where both contain three significant figures, should not be counted and only serve to locate the decimal place. Terminal zeros, such as 3.5600, which contains five significant figures, are to be counted, or if they are not significant they should not be reported.
2. When reporting a number with the proper significant figures, the last digit is rounded up by 1 if the digit to its right to be eliminated due to being a nonsignificant digit is greater than 5. If the nonsignificant digit to be eliminated is less than 5, do not round up. If the nonsignificant digit is

TABLE 4.3 Listing of Metric System Units with U.S. Equivalent Approximations.

Metric System

Length

Unit	Abbreviation	Number of Meters	Approximate U.S. Equivalent
Kilometer	km	1,000	0.62 mile
Hectometer	hm	100	328.08 feet
Decameter	dam	10	32.81 feet
Meter	m	1	39.37 inches
Decimeter	dm	0.1	3.94 inches
Centimeter	cm	0.01	0.39 inch
Millimeter	mm	0.001	0.039 inch
Micrometer	μm	0.000001	0.000039 inch

Area

Unit	Abbreviation	Number of Square Meters	Approximate U.S. Equivalent
Square kilometer	sq km *or* km^2	1,000,000	0.3861 square miles
Hectare	ha	10,000	2.47 acres
Are	a	100	119.60 square yards
Square centimeter	sq cm *or* cm^2	0.0001	0.155 square inch

Volume

Unit	Abbreviation	Number of Cubic Meters	Approximate U.S. Equivalent
Cubic meter	m^3	1	1.307 cubic yards
Cubic decimeter	dm^3	0.001	61.023 cubic inches
Cubic centimeter	cu cm *or* cm^3 *also* cc	0.000001	0.061 cubic inch

Capacity

Unit	Abbreviation	Number of Liters	Approximate U.S. Equivalent		
			Cubic	Dry	Liquid
Kiloliter	kl	1,000	1.31 cubic yards		
Hectoliter	hl	100	3.53 cubic feet	2.84 bushels	
Decaliter	dal	10	0.35 cubic foot	1.14 pecks	2.64 gallons
Liter	l	1	61.02 cubic inches	0.908 quart	1.057 quarts
Cubic decimeter	dm^3	1	61.02 cubic inches	0.908 quart	1.057 quarts
Deciliter	dl	0.10	6.1 cubic inches	0.18 pint	0.21 pint
Centiliter	cl	0.01	0.61 cubic inch		0.338 fluid ounce
Milliliter	ml	0.001	0.061 cubic inch		0.27 fluid dram
Microliter	μl	0.000001	0.000061 cubic inch		0.00027 fluid dram

Mass and Weight

Unit	Abbreviation	Number of Grams	Approximate U.S. Equivalent
Metric ton	t	1,000,000	1.102 short tons
Kilogram	kg	1,000	2.2046 pounds
Hectogram	hg	100	3.527 ounces
Decagram	dag	10	0.353 ounce
Gram	g	1	0.035 ounce
Decigram	dg	0.10	1.543 grains
Centigram	cg	0.01	0.154 grain
Milligram	mg	0.001	0.015 grain
Microgram	μg	0.000001	0.000015 grain

5, then round up the last significant digit to the nearest even number. For example, suppose a result is to contain four significant figures, thus 2.3655 would be rounded up to 2.366, 2.3645 would be rounded to 2.364, while 2.3654 is reported as 2.365.

3. When multiplication and division are being performed for an analytical result, the answer will be rounded off to possess the same number of significant figures as that of the number used in the calculation that contains the least number of significant figures. Exact values and definitions

(such as multiplying by 100 to obtain % or pph) do not affect the rounding of the final result. The following examples illustrate this concept:

$$\frac{125.6\,\text{g}}{438.909\,\text{g}} \times 100\% = 28.62\% \qquad (4.3)$$

$$(28.367\,\text{ml}) \times \left(\frac{0.107\,\text{mg}}{\text{ml}}\right) = 3.04\,\text{mg} \qquad (4.4)$$

$$\frac{(0.043351) \times \left(\frac{0.0981\,\text{mol}}{1}\right) \times (1.0 \times 10^6)}{9.2\,\text{mol}} \qquad (4.5)$$

$$= 4.6 \times 10^2\,\text{ppm, or } 460\,\text{ppm (not 462, or 462.2 ppm)}$$

4. For addition and subtraction, only retain the number of significant figures that includes as many digits to the right of the decimal place as that of the number that contains the least digits to the right of the decimal place. For example, $124.508 + 7.12 = 131.63$.
5. In logs, the initial zeros to the right of the decimal point are considered to be significant. The log of 1.03 would be expressed as 0.013 with three significant figures. This is further illustrated with the pH treatment of a hydrogen proton concentration: suppose a solution contained $[H^+] = 2.1 \times 10^{-11}$ M, the pH calculated as $-\log[H^+]$ would equal 10.68 and not 10 or 10.7. The pH is to be expressed with the same number of significant figures to the right of the decimal point as that of the nonexponential significant figures of the pre-log value. In other words, the 2.1 of the H^+ concentration equals two significant figures.

4.4 SCIENTIFIC CALCULATORS

It is quite common in the analytical laboratory today to perform many calculations using handheld calculators. These calculators can perform the most basic functions of addition, subtraction, multiplication, and division making routine calculations by the analyst fast and error free. Handheld calculators are also able to perform more complex calculations such as trigonometric functions, logarithms, and roots that once required the reliance upon multiple tables. Simple statistics are also very easy to perform on the handheld calculator, such as the standard deviation of a number of replicate analyses. It is therefore needed to spend a little time getting familiar with, or reviewing (if the reader is already versed in) the use of a common handheld calculator, the TI-30XA scientific calculator. The TI-30XA scientific calculator is depicted in Figure 4.1 showing the arrangement of the function keys.

4.4.1 Example Calculator

Figures 4.2, 4.3, 4.4, 4.5, 4.6, 4.7, 4.8, and 4.9 are adapted from the Texas Instruments Web site instructions for the

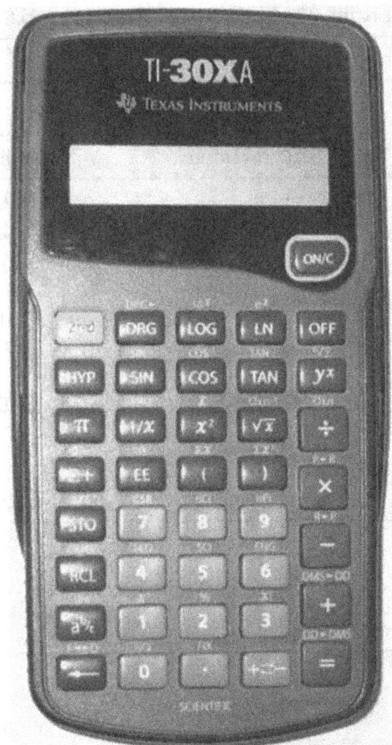

FIGURE 4.1 The handheld TI-30XA scientific calculator. (Reprinted with permission from Texas Instruments.)

TI-30XA scientific calculator. The figures begin with the on and off keys of the calculator and end with the notation used with the calculator.

4.4.2 Window's Calculator

Now that we have reviewed the use of the TI-30XA handheld calculator, let us take a look at a second option that the laboratory technician has in using the Windows-based calculator. To start the calculator, go to "Chapter 4 Main Menu" in ChemTech. Here, as depicted in Figure 4.10, with the click of the first button (the calculator) we will find a computer-based calculator that can be opened and used. Computer calculators such as this can be used in the laboratory the same way that the handheld calculators are used.

4.4.2.1 Windows' Scientific versus Standard Calculator The calculator is depicted in Figure 4.11 showing the two available forms as "Standard" and "Scientific." To switch back and forth between these two versions, go to the drop-down menu in the calculator's toolbar and select the "View" option. In the drop-down menu, you can choose either the standard view or the scientific view. Go ahead and try a few operations on the calculator using the computer keypad and mouse. You will see that the same types of operations can be performed as with the handheld calculators. In general, the Windows-based calculator can also be opened by going to the "Start" button and selecting "All Programs"—"Accessories"—"Calculator," which will open the calculator depicted in Figure 4.11.

Basic Operations

TI-30Xa (battery)

- [ON/C] turns on the TI-30Xa.
- [OFF] turns off the TI-30Xa and clears display, settings, and pending operations, but not memory.
- APD™ (Automatic Power Down™) turns off the TI-30Xa automatically if no key is pressed for about 5 minutes, but does not clear display, settings, pending operations, or memory.

  Note: [ON/C] after APD retrieves display, pending operations, settings, and memory.

TI-30Xa Solar

- To turn on the TI-30Xa Solar, expose the solar panel to light and press [ON/AC]. **Note:** Always press [ON/AC] to clear the calculator because memory and display may contain incorrect numbers.
- To turn off the TI-30Xa Solar, cover the solar panel with the slide case.

2nd Functions

2nd functions are printed above the keys. [2nd] selects the 2nd function of the next key pressed. For example, 2 [2nd] [x³] calculates the cube of 2.

FIGURE 4.2 Basic Operations including turning the calculator on and off, and the second function key. (Reprinted with permission from Texas Instruments.)

Basic Arithmetic

[+] [−] [×] [÷]	60 [+] 5 [×] 12 [=]	120.
[=]	Completes all pending operations. With constant (K), repeats the operation and value.	
[+⇄−]	Changes sign of value just entered. 1 [+] 8 [+⇄−] [+] 12 [=]	5.
[(] [)]	Parenthetical expression (up to 15 open). [=] closes all open parentheses.	
[π]	Pi is calculated with 12 digits (3.14159265359), displayed with 10 digits (3.141592654). 2 [×] [π] [=]	6.283185307

FIGURE 4.3 Basic arithmetic keys. (Reprinted with permission from Texas Instruments.)

Order of Operations

1st	Expressions inside parentheses.
2nd	Single-variable functions that perform the calculation and display the result immediately (square, square root, cube, cube root, trigonometric, factorial, logarithmic, percent, reciprocals, angle conversions).
3rd	Combinations and permutations.
4th	Exponentiation and roots.
5th	Multiplication and division.
6th	Addition and subtraction.
7th	[=] completes all operations.

The TI-30Xa uses AOS™ (Algebraic Operating System). It stores up to 4 pending operations (2 when STAT is displayed).

FIGURE 4.4 Explanation of the order of operations. (Reprinted with permission from Texas Instruments.)

Powers and Roots

[1/x]	8 [1/x] [+] 4 [1/x] [=]	0.375
[x²]	6 [x²] [+] 2 [=]	38.
[√x]	256 [√x] [+] 4 [√x] [=]	18.
[2nd] [x³]	2 [2nd] [x³] [+] 2 [=]	10.
[2nd] [∛x]	8 [2nd] [∛x] [+] 4 [=]	6.
[yˣ]	5 [yˣ] 3 [=]	125.
[2nd] [ˣ√y]	8 [2nd] [ˣ√y] 3 [=]	2.

Logarithmic Functions

[LOG]	15.32 [LOG]	1.185258765
	[+] 12.45 [LOG] [=]	2.280428117
[2nd] [10ˣ]	2 [2nd] [10ˣ] [−] 10 [x²] [=]	0.
[LN]	15.32 [LN]	2.729159164
	[+] 12.45 [LN] [=]	5.250879787
[2nd] [eˣ]	.693 [2nd] [eˣ]	1.999705661
	[+] 1 [=]	2.999705661

(e=2.71828182846)

FIGURE 4.5 Powers, roots, and logarithmic functions. (Reprinted with permission from Texas Instruments.)

One-Variable Statistics

[2nd] [CSR]	Clears all statistical data.
[Σ+]	Enters a data point.
[2nd] [Σ−]	Removes a data point.
[2nd] [FRQ]	Adds or removes multiple occurrences of a data point. Enter data point, press [2nd] [FRQ], enter frequency (1–99), press [Σ+] to add or [2nd] [Σ−] to remove data points.
[2nd] [Σx]	Sum.
[2nd] [Σx²]	Sum of squares.
[2nd] [x̄]	Mean.
[2nd] [σxn]	Population standard deviation (n weighting).
[2nd] [σxn-1]	Sample standard deviation (n-1 weighting).
[2nd] [n]	Number of data points.

Find the sum, mean, population standard deviation, and sample standard deviation for the data set: 45, 55, 55, 55, 60, 80. The last data point is erroneously entered as 8, removed with [2nd] [Σ−], and then correctly entered as 80.

[2nd] [CSR] (if STAT is displayed)		
45 [Σ+]	n=	1
55 [2nd] [FRQ] 3 [Σ+]	n=	4
60 [Σ+]	n=	5
8 [Σ+]	n=	6
8 [2nd] [Σ−]	n=	5
80 [Σ+]	n=	6
[2nd] [Σx] (sum)		350.
[2nd] [x̄] (mean)		58.33333333
[2nd] [σxn] (deviation, n weighting)		10.67187373
[2nd] [σxn-1] (deviation, n-1 weighting)		11.69045194

FIGURE 4.6 Statistical operations on the calculator. (Reprinted with permission from Texas Instruments.)

Example 4.2 Using the conversion tables (Table 4.3 or Appendix III), convert the following units including the proper number of significant figures (you may use either a handheld calculator or the computer calculator found in Chapter 4 Main Menu of ChemTech, Figure 4.11):

1. Convert 5.3 ounces to grams.
2. Convert 0.4 metric tons to pounds.
3. Convert 10 milliliter to ounces.

Answers: (1) 150; (2) 882; (3) 0.338

Check your results against the answers and resolve any problems that may have arisen.

Clearing and Correcting

[ON/C] (battery) [CE/C] (solar)	Clears value (before operation key) and K, but not M1, M2, M3, or STAT.
[ON/C] [ON/C] (battery) [CE/C] [CE/C] (solar)	Clears display, errors, all pending operations and K, but not M1, M2, M3, or STAT.
[OFF] [ON/C] (battery)	Clears display, errors, all pending operations, K, and STAT, but not M1, M2, and M3. Sets DEG angle units, floating-decimal format.
[ON/AC] (solar)	Clears display, errors, all pending operations, K, STAT, M1, M2, and M3. Sets DEG angle units, floating-decimal format.
[←]	Deletes right-most character in display.
0 [STO] n	Clears memory n.
[2nd] [FLO]	Clears SCI or ENG notation.
[2nd] [FIX] [.]	Clears FIX notation.
[2nd] [CSR]	Clears all statistical data.

FIGURE 4.7 Clearing and correcting. (Reprinted with permission from Texas Instruments.)

Memory

The calculator has 3 memories. When a memory contains a number other than 0, M1, M2, or M3 displays. To clear a single memory, press 0 [STO] 1, 0 [STO] 2, or 0 [STO] 3. To clear all 3 memories (solar only), press [ON/AC].

[STO] n	Stores displayed value in memory n, replacing current value.		
	23 [STO] 1	M1	23.
	[+] 2 [=]	M1	25.
[RCL] n	Recalls value in memory n. (continued)		
	[RCL] 1	M1	23.
	[+] 3 [=]	M1	26.
[2nd] [SUM] n	Adds displayed value to memory n. (continued)		
	4 [2nd] [SUM] 1	M1	4.
	[RCL] 1	M1	27.
[2nd] [EXC] n	Exchanges displayed and memory values. (continued)		
	3 [×] 5 [=]	M1	15.
	[2nd] [EXC] 1	M1	27.
	[2nd] [EXC] 1	M1	15.

FIGURE 4.8 Explanation of the calculator's memory. (Reprinted with permission from Texas Instruments.)

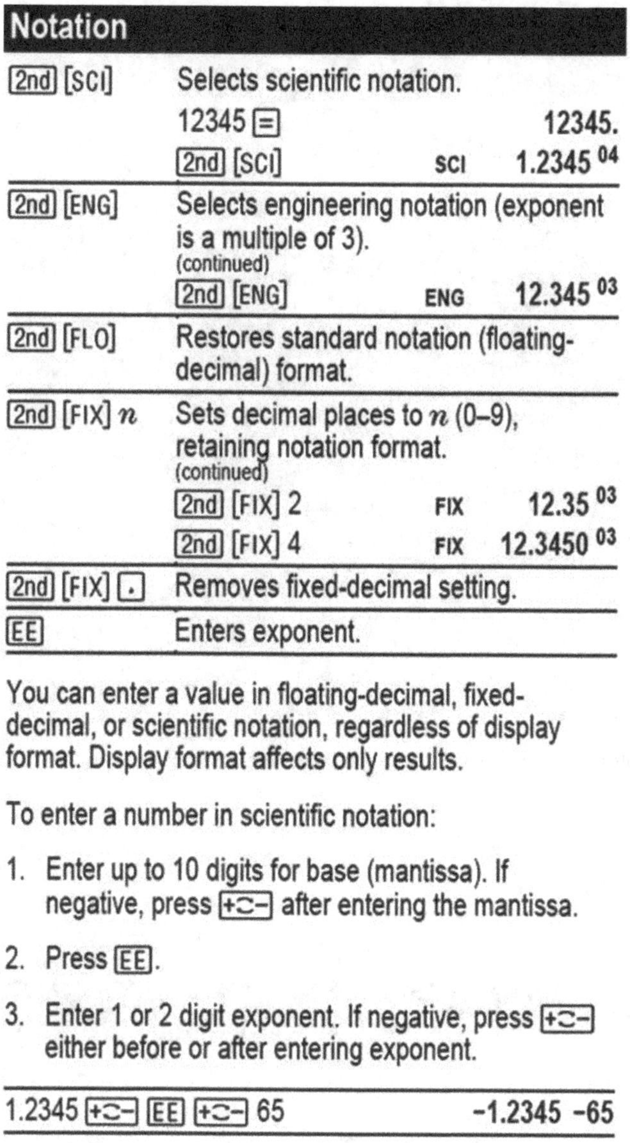

Notation

[2nd] [SCI]	Selects scientific notation.		
	12345 [=]		12345.
	[2nd] [SCI]	SCI	1.2345 04
[2nd] [ENG]	Selects engineering notation (exponent is a multiple of 3). (continued)		
	[2nd] [ENG]	ENG	12.345 03
[2nd] [FLO]	Restores standard notation (floating-decimal) format.		
[2nd] [FIX] *n*	Sets decimal places to *n* (0–9), retaining notation format. (continued)		
	[2nd] [FIX] 2	FIX	12.35 03
	[2nd] [FIX] 4	FIX	12.3450 03
[2nd] [FIX] [.]	Removes fixed-decimal setting.		
[EE]	Enters exponent.		

You can enter a value in floating-decimal, fixed-decimal, or scientific notation, regardless of display format. Display format affects only results.

To enter a number in scientific notation:

1. Enter up to 10 digits for base (mantissa). If negative, press [+/-] after entering the mantissa.
2. Press [EE].
3. Enter 1 or 2 digit exponent. If negative, press [+/-] either before or after entering exponent.

1.2345 [+/-] [EE] [+/-] 65	-1.2345 -65

FIGURE 4.9 The notation used for the calculator. (Reprinted with permission from Texas Instruments.)

4.5 ChemTech CONVERSION TOOL

We will now look at a conversion tool that can be used on the computer in the laboratory using ChemTech to do various conversions. Let us go to "Chapter 4 Main Menu" in ChemTech and click on the first shortcut link labeled "lbs to grams" as depicted in Figure 4.12. This will open up a computer-based conversion tool depicted in Figure 4.13.

4.5.1 Using the Conversion Tool

To use the conversion tool, click on the "Mass" folder tab. In the "Input" window, select the "gram (g)*" option. In the "Output" window, select the "ounce (avdp)(oz)" option. Below the "Input" window options, there is an Input textbox. Type "5" into the Input textbox. You will see that this is immediately converted to "0.176368" in the Output textbox. Here we have converted 5 g to 0.176368 oz. Of course when using these calculators and unit converters we must keep in mind the rules for reporting values with their correct number of significant figures, as discussed in Section 4.2 Significant figures.

4.5.2 Closing the Conversion Tool

To close the unit converter, either go to the main menu "File" selection and press "Exit," or alternatively press the red X button in the top right-hand corner of the Convert window. Exiting the unit converter program will bring you back to the Chapter 4 Main Menu in ChemTech. Clicking any of the other folder tabs in the unit converter will bring up other options for the conversion of units associated with pressure, volume, time, and so on. Use the conversion tool to convert the units found in Problem 4.1.

Example 4.3 Convert 287.65 cubic inches (in.^{3}) to cubic centimeters (cm^3)

$$\text{Conversion factor}: \frac{1\ \text{cm}^3}{0.061\ \text{in.}^3} \tag{4.6}$$

$$287.65\ \text{in.}^3 \times \frac{1\ \text{cm}^3}{0.061\ \text{in.}^3} = 4715.57\ \text{cm}^3 \tag{4.7}$$

Example 4.4 Convert 719.3 grams (g) to pounds (lb)

$$\text{Conversion factor}: \frac{1\ \text{lb}}{453.59\ \text{g}} \tag{4.8}$$

$$719.3\ \text{g} \times \frac{1\ \text{lb}}{453.59\ \text{g}} = 1.6\ \text{lb} \tag{4.9}$$

Example 4.5 Convert 5.74 quarts (qt) to liters (l)

$$\text{Conversion factor}: \frac{1\ \text{l}}{1.0567\ \text{qt}} \tag{4.10}$$

$$5.74\ \text{qt} \times \frac{1\ \text{l}}{1.0567\ \text{qt}} = 5.43\ \text{l} \tag{4.11}$$

4.6 CHAPTER KEY CONCEPTS

4.1 The basic use of rudimentary mathematics in the laboratory is an essential skill required of the laboratory technician, the analytical chemist, and the research chemist.

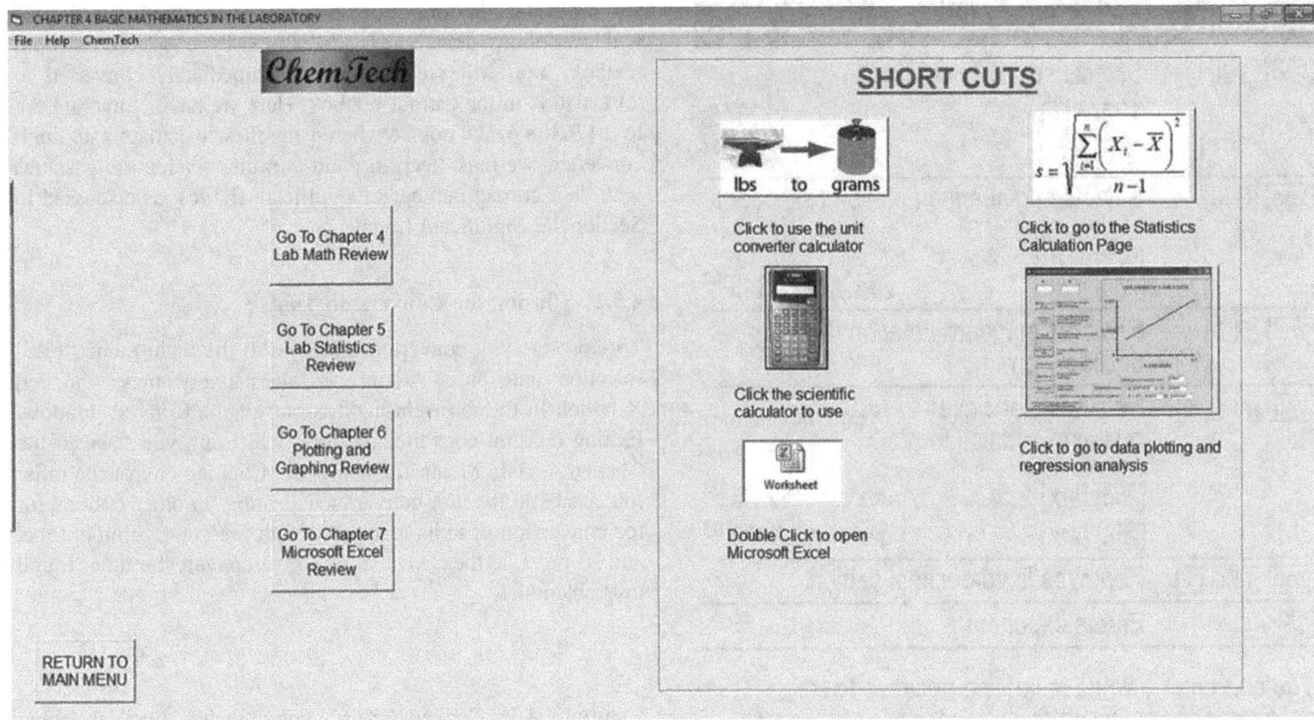

FIGURE 4.10 Chapter 4 Main Menu illustrating the top calculator button to click to open the computer-based calculator.

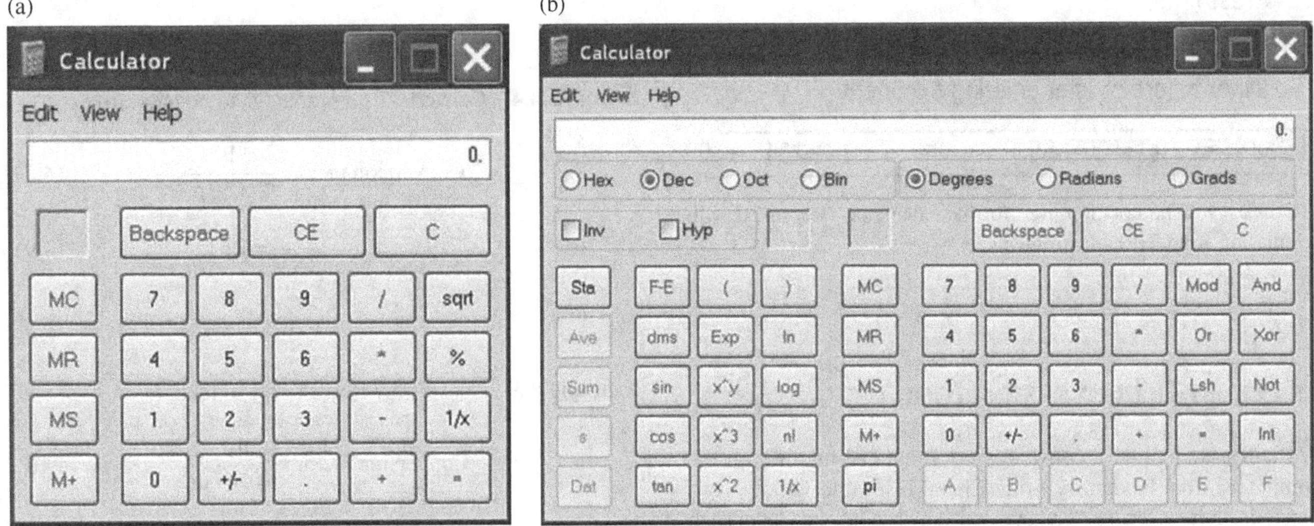

FIGURE 4.11 Windows-based calculator illustrating (a) the standard view and (b) the scientific view.

CHAPTER 4 BASIC MATHEMATICS IN THE LABORATORY
File Help ChemTech
ChemTech
SHORT CUTS
lbs to grams
Click to use the unit converter calculator
Click to go to the Statistics Calculation Page
Go To Chapter 4 Lab Math Review
Go To Chapter 5 Lab Statistics Review
Click to open unit converter
Click the scientific calculator to use
Go To Chapter 6 Plotting and Graphing Review
Worksheet
Click to go to data plotting and regression analysis
Go To Chapter 7 Microsoft Excel Review
Double Click to open Microsoft Excel
RETURN TO MAIN MENU

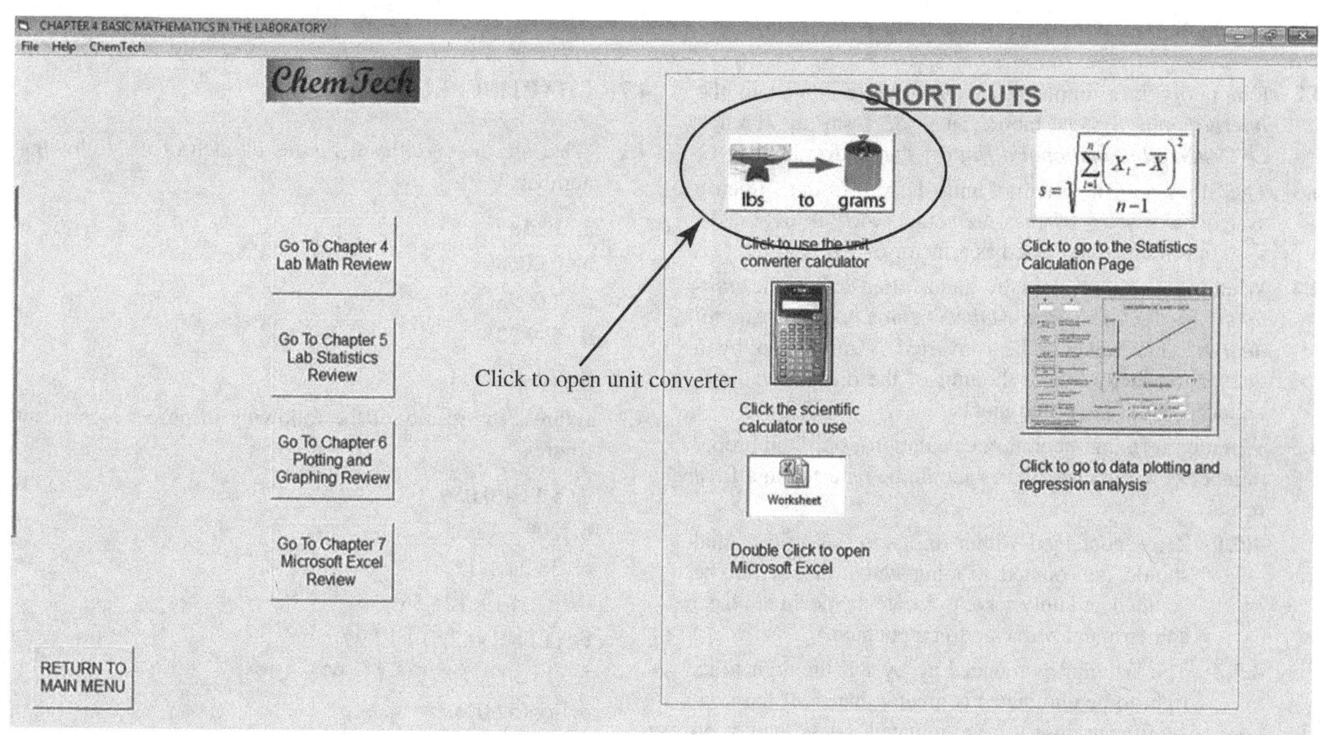

FIGURE 4.12 "Chapter 4 Main Menu" in ChemTech. Click on the first shortcut link labeled "lbs to grams" as illustrated to open up a computer-based conversion tool.

Convert
File Options Help
Pressure Speed Temperature Time Torque Volume Volume - Dry
Acceleration Amt. of Substance Angle Area Computer Concentration
Custom Density Distance Energy Flow Force Light Mass Power
Input
carat
grain (gr)
gram (g) *
kilogram (kg) *
megagram (Mg)
microgram (µg)
milligram (mg)
ounce (avdp) (oz)
ounce (troy) (ozt)
pennyweight
Output
carat
grain (gr)
gram (g)
kilogram (kg)
megagram (Mg)
microgram (µg)
milligram (mg)
ounce (avdp) (oz)
ounce (troy) (ozt)
pennyweight
Input: 5 gram (g) *
Output: 0.176368 ounce (avdp) (oz)

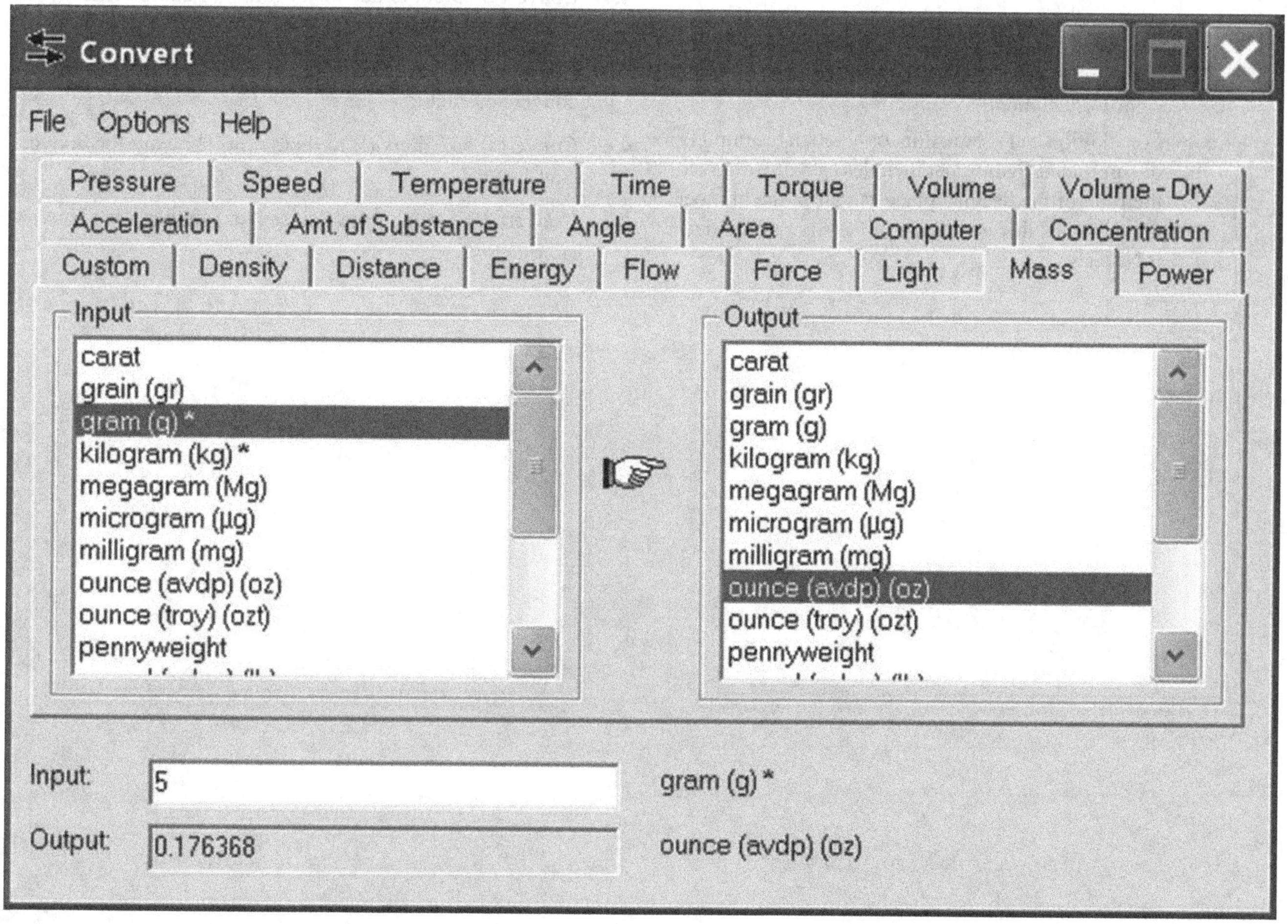

FIGURE 4.13 Example of a computer-based conversion tool.

4.2 Laboratory measurements are primarily recorded in the *International System* (abbreviated *SI* from the French *Le Système International d'Unités*) standardized units.

4.3 The SI units are a system of units derived from the metric system, a system of measurement, which is a base-10 system (all units expressed as a factor of 10).

4.4 When converting values by hand, often called the *unit-factor method*, the *factor-label method*, or *dimensional analysis*, the value to be converted is multiplied by a conversion factor where the unit of the original value is replaced with the desired unit.

4.5 When performing laboratory calculations, only the proper number of significant figures should be reported in a final result.

4.5.1 Zeros contained within digits are significant and should be counted, leading zeros should not be counted and only serve to locate the decimal place, and terminal zeros are to be counted.

4.5.2 The last digit is rounded up by 1 if the digit to its right to be eliminated is greater than 5. If the non-significant digit to be eliminated is less than 5, do not round up. If the nonsignificant digit is 5, then round up the last significant digit to the nearest even number.

4.5.3 When multiplication and division are being performed, the answer will be rounded off to possess the same number of significant figures as that of the number used that contains the least number of significant figures.

4.5.4 For addition and subtraction, only retain the number of significant figures that includes as many digits to the right of the decimal place as that of the number that contains the least digits to the right of the decimal place.

4.7 CHAPTER PROBLEMS

4.1 What are the significant figures of each of the following numbers?

a. 12.58
b. 0.00689
c. 7.0098
d. 589.225
e. 0.127

4.2 Express the results of the following in proper significant figures.

a. $5.11 + 0.056$
b. $209 - 12.35$
c. $38.24 \div 12$
d. 83.181×121.56
e. $(1.58 \times 442.3) + 12.99$
f. $(13.56 \div 1.876) - (5.997 \times 8.12)$
g. $\log(52.01)$
h. $\log(0.001) \times \log(72)$

4.3 Convert 3.40 meters (m) to inches (in.) showing the conversion factor.

4.4 Convert 36.5 cubic decimeter (dm^3) to cubic inches ($in.^3$) showing the conversion factor.

4.5 Convert 2.19 kilograms (kg) to pounds (lb) showing the conversion factor.

4.6 Convert 0.532 liters (l) to quarts (qt) showing the conversion factor.

4.7 Convert 355 grams (g) to ounces (oz) showing the conversion factor.

5

ANALYTICAL DATA TREATMENT (STATISTICS)

5.1 Errors in the Laboratory
5.1.1 Systematic Errors
5.1.2 Random Errors
5.2 Expressing Absolute and Relative Errors
5.3 Precision
5.3.1 Precision versus Accuracy
5.4 The Normal Distribution Curve
5.4.1 Central Tendency of Data
5.5 Precision of Experimental Data
5.5.1 The Range
5.5.2 The Average Deviation
5.5.3 The Standard Deviation
5.6 Normal Distribution Curve of a Sample
5.7 ChemTech Statistical Calculations
5.7.1 Introduction to ChemTech Statistics
5.7.2 ChemTech Chapter 5
5.8 Student's Distribution t Test for Confidence Limits
5.8.1 Accuracy
5.8.2 The Student's t Test
5.8.3 Calculating the Student's t Value
5.8.4 Probability Level
5.8.5 Sulfate Concentration Confidence Limits
5.8.6 Sulfate t Distribution Curve
5.8.7 Determining Types of Error
5.8.8 Determining Error in Methodology
5.9 Tests of Significance
5.9.1 Difference in Means
5.9.2 Null Hypothesis
5.10 Treatment of Data Outliers
5.10.1 The Q Test
5.10.2 The T_n Test
5.11 Chapter Key Concepts
5.12 Chapter Problems

5.1 ERRORS IN THE LABORATORY

Even though the greatest care can be taken by the analyst to obtain and record data as "error free" as possible, error will still always exist. This is something that the analyst cannot escape, but with care in analysis and understanding of the types of error that can be introduced into the system, error can be brought to a minimum. There are two types of error that the analyst needs to understand: systematic (determinate) and random (indeterminate) errors associated with laboratory analysis.

5.1.1 Systematic Errors

Systematic errors are introduced into the analysis by a malfunctioning instrument, an analytical method that contains a defect, a bad calibration, or by the analyst. For example, if the function of the instrument has changed, say the instrument's detector response is suddenly lowered due to faulty electronics, the results obtained will contain this error. Other examples could include: an incorrect calibration of an instrument, perhaps from improper standard preparation which would give a systematic error in the results, or a reagent used in the analysis has changed. These types of errors are correctable and can be eliminated by the analyst if he/she is able to identify the underlying cause of the error and rectify it. We will see in a later section the use of control monitor charting that allows the tracking of trends, which alerts the analyst to systematic errors being introduced into routine measurements.

5.1.2 Random Errors

Because of the fact that there is always some uncertainty to all laboratory measurements, random errors are unavoidable for the analyst. Even the most careful measurements will inherently contain some error; however, the introduced random error will be an equal distribution of positive and negative errors. From this we know that an individual measurement most likely will contain a greater magnitude in error as compared to the average of multiple replicate measurements.

Analytical Chemistry: A Chemist and Laboratory Technician's Toolkit, First Edition. Bryan M. Ham and Aihui MaHam.

5.2 EXPRESSING ABSOLUTE AND RELATIVE ERRORS

Sometimes the analyst may need to express in simple terms the absolute and relative errors of a result. The absolute error of a result X_i is equal to $X_i-\mu$, where μ is the "true value" of the result. The accuracy of a result, which is an expression of the nearness of a value (X_i) or the arithmetic mean ($\bar{X}$) to the true value (μ), is often presented in terms of error such as the absolute error $X_i-\mu$ or $\bar{X}-\mu$, or the relative error, which is equal to the absolute error divided by the true value:

$$\text{Absolute error: } X_i-\mu \text{ or } \bar{X}-\mu \tag{5.1}$$

$$\text{Relative error: pph} = \frac{(X_i-\mu)}{\mu} \times 100; \quad \text{ppt} = \frac{(X_i-\mu)}{\mu} \times 1000 \tag{5.2}$$

The absolute error will retain the units of the measurement while the relative error does not.

5.3 PRECISION

When the analyst is considering the "precision" of a set of measurements, the expression is given as a deviation of a set of measurements from the set's arithmetic mean. From this we see that the precision of a set of results is the agreement of the results among themselves. Again, precision can also be expressed in terms of the absolute deviation or the relative deviation of the set of results. Thus, the absolute deviation will retain the units of the measurement while the relative deviation does not. In the following sections, we will look closer at the estimation of deviation where the most commonly reported expressions are the range, average deviation, and the standard deviation.

5.3.1 Precision versus Accuracy

There is a clear distinction between precision and accuracy when dealing with and interpreting analytical results. Having a high degree of precision does not necessarily indicate a high degree of accuracy. It is possible to have one without the other, to have neither, or to have both. This is illustrated in Figure 5.1 where (a) depicts the instance of a relatively high degree in precision in the analytical measurement with no accuracy. In the target of (a), the results are bunched together indicating that the measurement has a good level of repeatability but the average of these results will be far from the true value which is represented by the center or bulls eye of the target. In Figure 5.1(b), the average of the results will fall close to the true value; however, the repeatability of the measurements contains a large degree of scatter. This instance helps also to illustrate the tendency in analytical testing where the greater number of measurements that are made will help to obtain a more accurate average result, although there is a high level of scatter in the data. Finally, Figure 5.1(c) depicts the case where both precision and accuracy exist in the measured results.

5.4 THE NORMAL DISTRIBUTION CURVE

In analytical measurements, the data collected are an estimate of the true value (μ) that is usually represented by the arithmetic mean ($\bar{X}$) of the data. This is due to only a small set of values being collected that is known as a "sample" of the population while an infinite set of data values would in fact represent the population itself as μ. Let us look at what is known as the *normal distribution curve* (also referred to as the *probability curve* or *error curve*) when an infinite number of data points are obtained and only random errors are present. A normal distribution curve is depicted in Figure 5.2 for an infinite number of replicate measurements for some population mean value μ. The curve in Figure 5.2 possesses a bell or Gaussian symmetrical shape, where the *y*-axis represents the probability of the occurrence of a measurement, and the x-axis represents the distribution of the data centered around the population mean μ. This brings us to two descriptive aspects of analytical data, namely the *central tendency of data* and the precision of the data expressed as the *standard deviation* (σ), which we shall take a closer look at both.

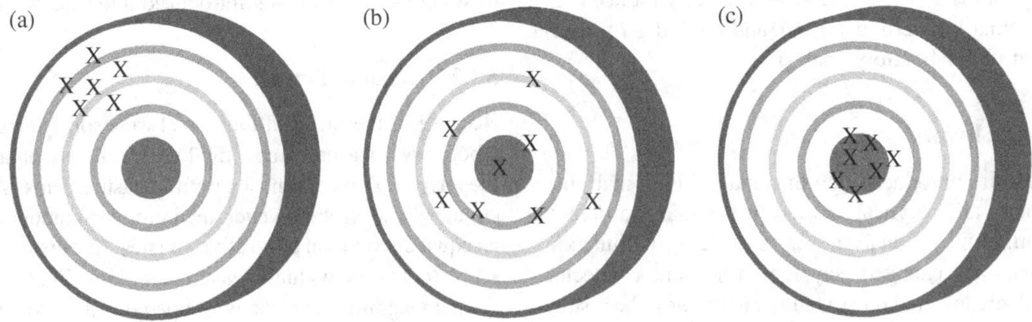

FIGURE 5.1 A target representation of the concepts of precision and accuracy in the treatment of analytical data. (a) A relatively high degree in precision in the analytical measurement with no accuracy. (b) The average of the results will fall close to the true value, however the repeatability of the measurements contains a large degree of scatter. This illustrates the tendency in analytical testing where the greater number of measurements that are made will help to obtain a more accurate average result, even though there is a high level of scatter in the data. (c) Both precision and accuracy exist in the measured results.

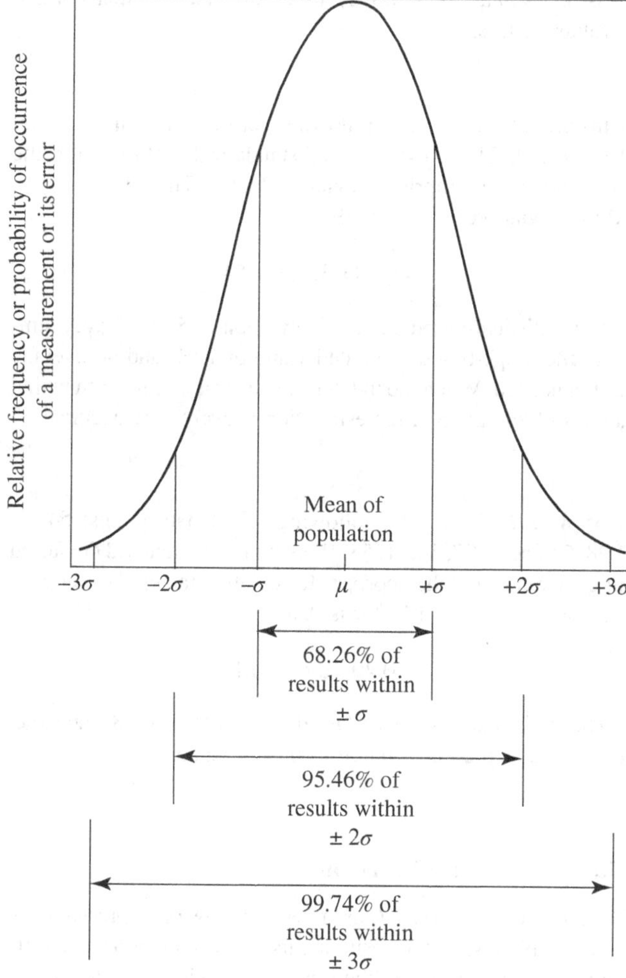

FIGURE 5.2 The normal distribution curve.

5.4.1 Central Tendency of Data

As presented earlier, the true value of a measurement (μ) is usually not known and is estimated using the *mean*, *median*, or *mode* of a set of values. When determining a value such as the chloride concentration [Cl^-] (in molarity, M) of an unknown sample, we are obtaining a number of replicate measurements that are realistic in number (perhaps three to six replicates). Using the aspect of central tendency, a value can be obtained (X) from a sample (limited number of replicate measurements) of the population (which represents an infinite number of replicate measurements) that is an estimate of the population's true value (μ). If our chloride measurement [Cl^-] does not contain systematic error but only random error, then it will have a distribution similar to that of Figure 5.2.

5.4.1.1 The Arithmetic Mean The arithmetic mean is usually chosen in the analytical laboratory to represent the central tendency of a set of replicate data collected. The arithmetic mean ($\bar{X}$), or more commonly referred to as the average value of the sample, is calculated using the following formula:

$$\bar{X}=\frac{X_1+X_2+\cdots+X_n}{n}=\frac{\sum_{i=1}^{n}X_i}{n} \tag{5.3}$$

For example

$$4,\ 4,\ 5,\ 6,\ 7,\ 7 (n=6)$$

$$\bar{X}=\frac{4+4+5+6+7+7}{6}=\frac{33}{6}=5.5$$

$$4,\ 4,\ 5,\ 6,\ 7,\ 7,\ 7 (n=7)$$

$$\bar{X}=\frac{4+4+5+6+7+7+7}{6}=\frac{40}{7}=5.7$$

5.4.1.1.1 Advantage of the Mean The advantage of the arithmetic mean over the use of the median or mode is that the mean uses all values that contribute to the final mean value. This is usually best for a larger set of values and values that do not contain any significant outliers. Significant outliers are data values that clearly possess a significant degree of error, perhaps a mistake was made in reading a burette when collecting that particular data point that is quite different from the others. We will look at a test for determining whether a data point is significantly different and can be removed later using the *Q test*.

5.4.1.2 The Median If a significant outlier is included in the mean calculation, it will skew the estimated value away from the true value. Using the median can be a way to eliminate the effect of a data point that possesses significant error. The median (*M*) is found by arranging the data in order of magnitude and determining the middle value of the data. If the number of data is even, then the median is the average of the two middle values. If the number of data is odd, then the median is the middle value.

For example

$$4,\ 4,\ 5,\ 6,\ 7,\ 7 \quad \text{median}=5.5$$

$$4,\ 4,\ 5,\ 6,\ 7,\ 7,\ 7 \quad \text{median}=6$$

5.4.1.3 The Mode The mode is found as the data value that occurs most frequently in the data. While the mode can be easily found by visual inspection, it is not very useful for a small set of data points.

For example

$$4,\ 4,\ 5,\ 6,\ 7,\ 7 \quad \text{median}=4 \text{ and } 7$$

$$4,\ 4,\ 5,\ 6,\ 7,\ 7,\ 7 \quad \text{median}=7$$

5.4.1.4 Sticking with the Mean In these simple examples, we see that different values are obtained when comparing the mean, median, and mode. For our purposes, we will essentially use the mean value of a set of results. One interesting rule is that the mean of n replicates or results in $(n)^{1/2}$ times more reliable than X_i results. This indicates that X of four results would be two times more reliable than that of one result, X of six results would be 2.45 times, X of eight results would be 2.83, X of ten results would be 3.16, and so on. Clearly the analyst should collect a number of results in order to attempt to estimate the true value of the analysis. However, time and sample size often dictates that only a select number of analyses may be performed, usually at least three.

5.5 PRECISION OF EXPERIMENTAL DATA

We now have our description for the estimation of the central tendency of data as the arithmetic mean. Let us look at the precision of our measurements, which tells the magnitude of the indeterminate error that was associated with our measurements. There are three commonly used calculations in the laboratory, which describe the precision: the *range*, the *average deviation*, and the *standard deviation*.

5.5.1 The Range

The simplest way to estimate the precision, scatter, or indeterminate error of a small set of data points is to use the range of the data. There is a rather straightforward approach have demonstrated a rather straightforward approach using a table of *deviation factors* (k) for data sets that have $n = 2–10$ data points. The range of the data is first obtained by ordering the data from the least value to the greatest value. The range is then calculated as $X_n–X_1$. Then, using the deviation factors listed in Table 5.1, the standard deviation is obtained by multiplying the range by the deviation factor associated for the number of data points within the data set.

$$\text{Standard deviation } (s) = \text{Deviation Factor}(k) \times \text{Range} \quad (5.4)$$

$$s = k \times \text{Range} \quad (5.5)$$

For a small set of data points, the estimation of the standard deviation from the range and deviation factor can be quite similar to the statistical-based calculated standard deviation (s), but much easier to derive. However, as the number of data points approaches infinity, the estimation of the standard deviation from the range will reduce to an efficiency of zero. This approach for precision estimation is mostly used for data sets that contain 10 values or less.

TABLE 5.1 Table of Deviation Factors.

n, number of data points	k, deviation factor
2	0.89
3	0.59
4	0.49
5	0.43
6	0.40
7	0.37
8	0.35
9	0.34
10	0.33

Example 5.1 Using the following set of data points: 20.35, 22.14, 18.99, 21.11, estimate the standard deviation using the range and the appropriate deviation factor. The range is 3.15 and the deviation factor is 0.49.

$$s = 0.49 \times 3.15 = 1.54$$

The statistical-based standard deviation (s) is 1.33, very similar to the range-based estimated value of 1.54, and an average was not needed. When the indeterminate error is smaller than that shown in Example 5.1, the estimation is even more accurate.

Example 5.2 Using the following set of data points: 54.22, 53.98, 54.36, 54.07, 54.31, 53.91, estimate the standard deviation using the range and the appropriate deviation factor. The range is 0.45 and the deviation factor is 0.40.

$$s = 0.40 \times 0.45 = 0.18$$

The statistical-based standard deviation (s) is 0.18, the exact same value as the range-based estimated value.

5.5.2 The Average Deviation

The average deviation ($\bar{d}$) is a second way of expressing the precision of a set of measurements. Its calculation is a little more complex than using the range and deviation factor where the individual deviation of each measurement from the mean is calculated. In the calculation of the average deviation the absolute deviation is taken, thus all deviations are positive. The formula used to calculate the average deviation is

$$\bar{d} = \frac{1}{n}\sum_{i=1}^{n}|X_i - \bar{X}| \quad (5.6)$$

Example 5.3 Using the following table of data values, calculate the average deviation and compare it to the estimated standard deviation derived from the range (Table 5.2).

We first calculate the mean of the values as 31.1. We then calculate the individual deviations listed in the right-hand column. Summing up the individual deviations, we obtain a number of 7.5. Then using Equation 5.7, we calculate the average deviation as

$$\bar{d} = \frac{1}{8}(7.5) = 0.9 \quad (5.7)$$

Using a range value of 3.3 and a deviation factor of 0.35, we can calculate the estimation of the standard deviation as 1.2. This value is similar to that of the average deviation, but slightly different indicating that there is some scatter in the results.

TABLE 5.2 Experimental values.

X_i	$\lvert X_i-\bar{X}\rvert$
32.9	1.8
29.6	1.5
31.5	0.4
32.1	1.0
30.8	0.3
29.9	1.2
30.5	0.6
31.8	0.7
$\bar{X}=31.1$	$\sum\lvert X_i-\bar{X}\rvert=7.5$

TABLE 5.3 Experimental values.

X_i	$(X_i-\bar{X})^2$
32.9	3.2
29.6	2.2
31.5	0.2
32.1	1.0
30.8	0.1
29.9	1.4
30.5	0.4
31.8	0.5
$\bar{X}=31.1$	$\sum(X_i-\bar{X})^2=9.0$

5.5.3 The Standard Deviation

5.5.3.1 ***Root Mean Square*** Calculating the standard deviation (or the root mean square RMS) of a set of data points is a more accurate measure of the dispersion of the data set tested. This is due to the standard deviation being based off of a statistical theoretical basis that is more affected to data points that possess a greater degree of error. For the infinite set of data points making up a population such as the normal distribution curve of Figure 5.2, the standard deviation (σ) is the square root taken of the average square difference between the population mean (μ) and the individual observations:

$$\sigma=\sqrt{\frac{\sum_{i=1}^{n}(X_i-\mu)^2}{n}} \tag{5.8}$$

5.5.3.2 ***Sample Standard Deviation*** When we are calculating the precision of a subset of the population, or a sample of the population in the laboratory, we use an estimation form of the population standard deviation of Equation 5.8, called the sample standard deviation (s):

$$s=\sqrt{\frac{\sum_{i=1}^{n}(X_i-\bar{X})^2}{n-1}} \tag{5.9}$$

Note the difference in Equation 5.9 versus Equation 5.8 where the population mean (μ) has been replaced by the sample mean ($\bar{X}$), and the population number (n) by the sample degrees of freedom (n–1). Let us take our data table from Example 5.3 and calculate the sample standard deviation presented in Equation 5.9.

Example 5.4 Using the following set of data calculate the sample standard deviation and compare it to the average deviation and estimated standard deviation from the range.

We have the same average of the data as before as $\bar{X}=31.1$, but now we have placed into the right-hand column the square of the individual deviations from the sample mean. The summation of the squared deviations is $\sum(X_i-\bar{X})^2=9.0$ (Table 5.3).

Using 7 as our degrees of freedom ($n-1$) we can now use Equation 5.9 to calculate the sample standard deviation:

$$s=\sqrt{\frac{9.0}{7}}=1.1$$

5.5.3.3 ***Comparison of the Three Methods*** Interestingly, we calculated the average deviation to be 0.9, the estimated standard deviation using the range at 1.2, and now the sample standard deviation at 1.1. All three values are very similar and demonstrate that for a small sample set with relatively good precision all three approaches are agreeable. However, most data generated in the analytical laboratory are subjected to the sample standard deviation calculation of Equation 5.9 for precision measurement due to the theoretical nature of the calculation.

5.5.3.4 ***Using the Scientific Calculator*** Also, the handheld calculators used in the laboratory today readily calculate the sample standard deviation with just a few keystrokes needed. The TI-30XA description and exercise in Illustration 5 "Statistical operations on the calculator" gives a straightforward presentation of basic statistical operations on the handheld calculator. The analyst needs to keep in mind though that when we are calculating the sample standard deviation (s) the second function "$\sigma xn-1$" is to be used (Fig. 5.3).

5.5.3.5 ***Coefficient of Variation*** Finally, the variation (or measure of dispersion) can be expressed as a relative standard deviation (RSD), also called *coefficient of variation* (*CV*) where the sample standard deviation (s) is divided by the sample mean ($\bar{X}$), $CV=s/\bar{X}$. The *CV* is often expressed as parts per hundred (pph) when multiplied by 100 (also called a percentage).

5.6 NORMAL DISTRIBUTION CURVE OF A SAMPLE

The normal distribution curve, also called a Gaussian distribution or Bell curve is the shape that is observed when the data points of a system that possess a normal distribution is plotted. Now that we have a calculation for the sample standard deviation, let us take another look at the normal distribution curve as applied to our

One-Variable Statistics

[2nd] [CSR]	Clears all statistical data.
[Σ+]	Enters a data point.
[2nd] [Σ-]	Removes a data point.
[2nd] [FRQ]	Adds or removes multiple occurrences of a data point. Enter data point, press [2nd] [FRQ], enter frequency (1–99), press [Σ+] to add or [2nd] [Σ-] to remove data points.
[2nd] [Σx]	Sum.
[2nd] [Σx²]	Sum of squares.
[2nd] [x̄]	Mean.
[2nd] [σxn]	Population standard deviation (*n* weighting).
[2nd] [σxn-1]	Sample standard deviation (*n*-1 weighting).
[2nd] [n]	Number of data points.

Find the sum, mean, population standard deviation, and sample standard deviation for the data set: 45, 55, 55, 55, 60, 80. The last data point is erroneously entered as 8, removed with [2nd] [Σ-], and then correctly entered as 80.

[2nd] [CSR] (if STAT is displayed)		
45 [Σ+]	n=	1
55 [2nd] [FRQ] 3 [Σ+]	n=	4
60 [Σ+]	n=	5
8 [Σ+]	n=	6
8 [2nd] [Σ-]	n=	5
80 [Σ+]	n=	6
[2nd] [Σx] (sum)		350.
[2nd] [x̄] (mean)		58.33333333
[2nd] [σxn] (deviation, *n* weighting)		10.67187373
[2nd] [σxn-1] (deviation, *n*-1 weighting)		11.69045194

FIGURE 5.3 Statistical operations on the calculator. (Reprinted with permission from Texas Instruments).

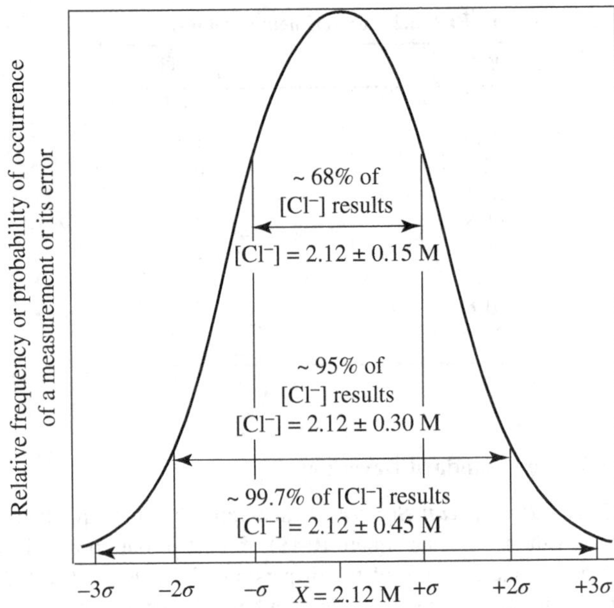

FIGURE 5.4 Normal distribution curve for chloride measurement as $[Cl^-]$. The mean concentration is $\bar{X} = 2.12$ M, with a standard deviation $s = 0.15$.

measurement of chloride concentration as molarity $[Cl^-]$. If the measurement of the chloride concentration has been performed with care taken in the methodology, then systematic error should not contribute to the deviation of the experimental value from the true value leaving only indeterminate random error. Thus, according to the normal distribution curve, if we have determined a mean chloride concentration ($\bar{X}$) of 2.12 M with a standard deviation (s) of 0.15, two-thirds of the measurements (~68%) should be within the range of 2.12 ± 0.15 M. This is depicted in Figure 5.4. Following this, ~95% of the measurements would fall within the range of 2.12 ± 0.30 M, and ~99.7% of the measurements would fall within the range of 2.12 ± 0.45 M.

5.7 ChemTech STATISTICAL CALCULATIONS

5.7.1 Introduction to ChemTech Statistics

At this point we now have tools to calculate the average ($\bar{X}$) of a set of measurements, and three methods for the variation or dispersion of the data as the estimated standard deviation, the average deviation ($\bar{d}$), and the sample standard deviation (s). Again, we are assuming that the error involved in the measurements is attributable to random, indeterminate error associated with all forms of measurements. In this assumption, we have identified and eliminated any systematic, determinate error such as a bad calibration or a faulty instrument. While these values can readily be calculated by hand, when a large set of values are to be processed it becomes quite tedious and often error prone to determine the statistics by hand. There are a number of alternative approaches for the laboratory technician to use to calculate statistics of data collected in the laboratory. These include the scientific calculator as we worked with in Section 5.5.3.4, more advanced statistical programs such as Excel® that we will use later and other computer-based programs such as ChemTech.

5.7.2 ChemTech Chapter 5

Let us open ChemTech and go to the "Chapter 5 Main Menu." Click the button "CALCULATE STATISTICS 1" as depicted in Figure 5.5 to open a statistical calculation page. Here we have a method of calculating the simple statistics of a set of data without the need to calculate averages or deviations by hand. Let us start with a small set of data to input and calculate the statistics. Enter the following data into the upper left data input box of the statistical calculation page as shown in Figure 5.6.

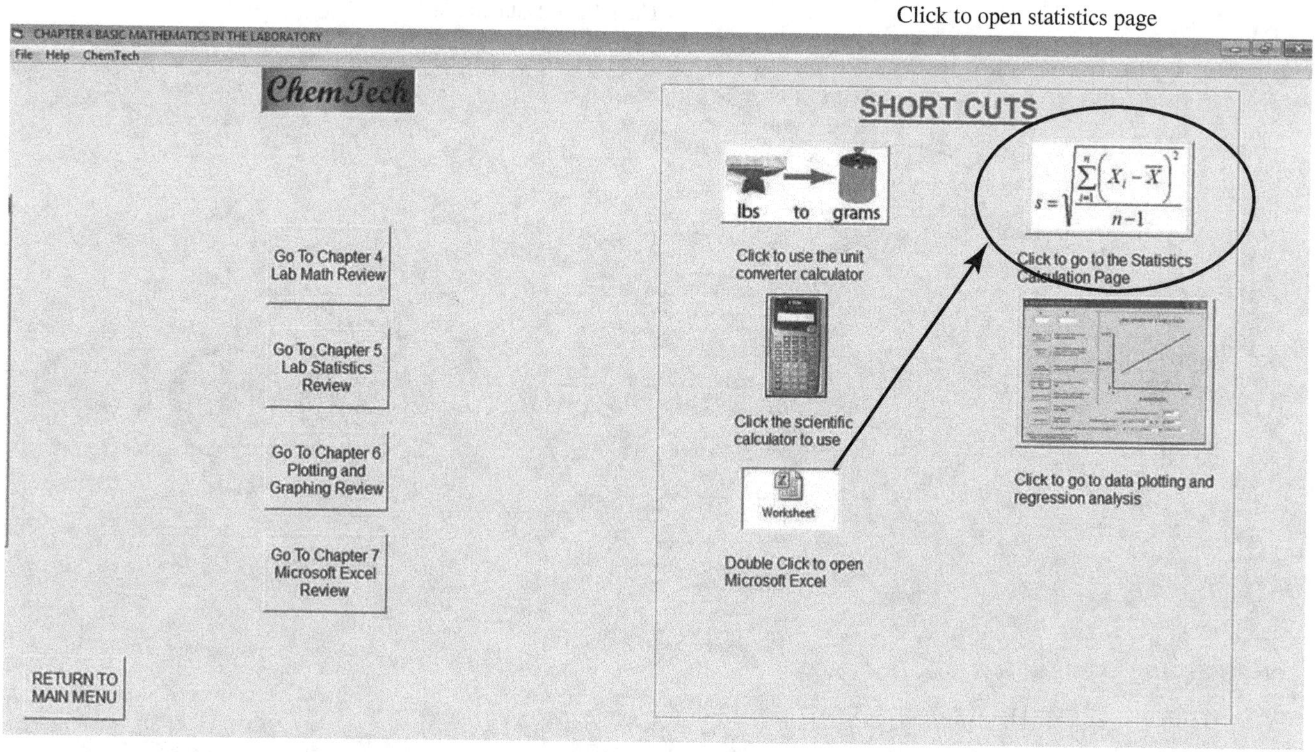

Click to open statistics page

CHAPTER 4 BASIC MATHEMATICS IN THE LABORATORY

File Help ChemTech

ChemTech

SHORT CUTS

Go To Chapter 4 Lab Math Review

Go To Chapter 5 Lab Statistics Review

Go To Chapter 6 Plotting and Graphing Review

Go To Chapter 7 Microsoft Excel Review

RETURN TO MAIN MENU

lbs to grams

Click to use the unit converter calculator

Click the scientific calculator to use

Worksheet

Double Click to open Microsoft Excel

$s = \sqrt{\frac{\sum_{i=1}^{n}(X_i - \overline{X})^2}{n-1}}$

Click to go to the Statistics Calculation Page

Click to go to data plotting and regression analysis

FIGURE 5.5 Chapter 5 Main Menu illustrating the "CALCULATE STATISTICS 1" button to click to open the computer-based statistics calculation page.

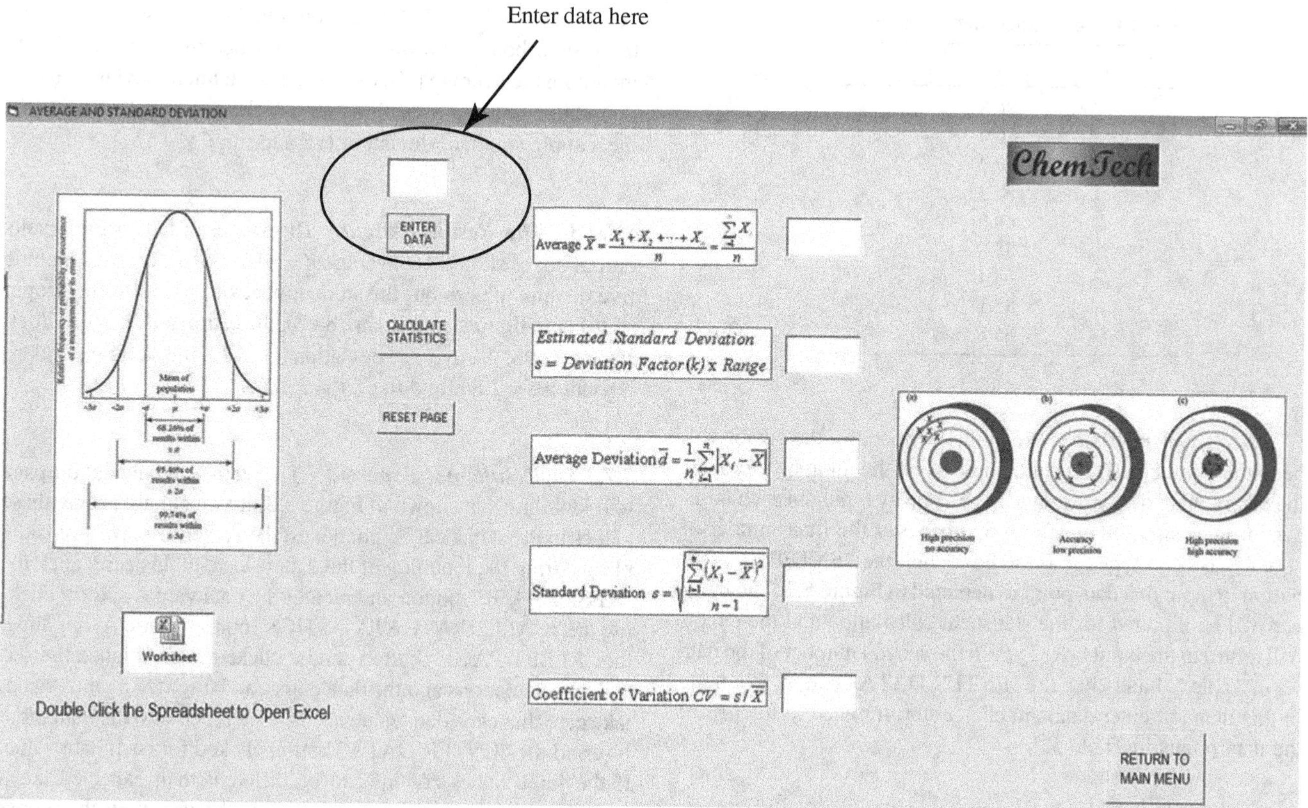

FIGURE 5.6 The "CALCULATE STATISTICS 1" page from ChemTech used for the computer-based statistics calculations.

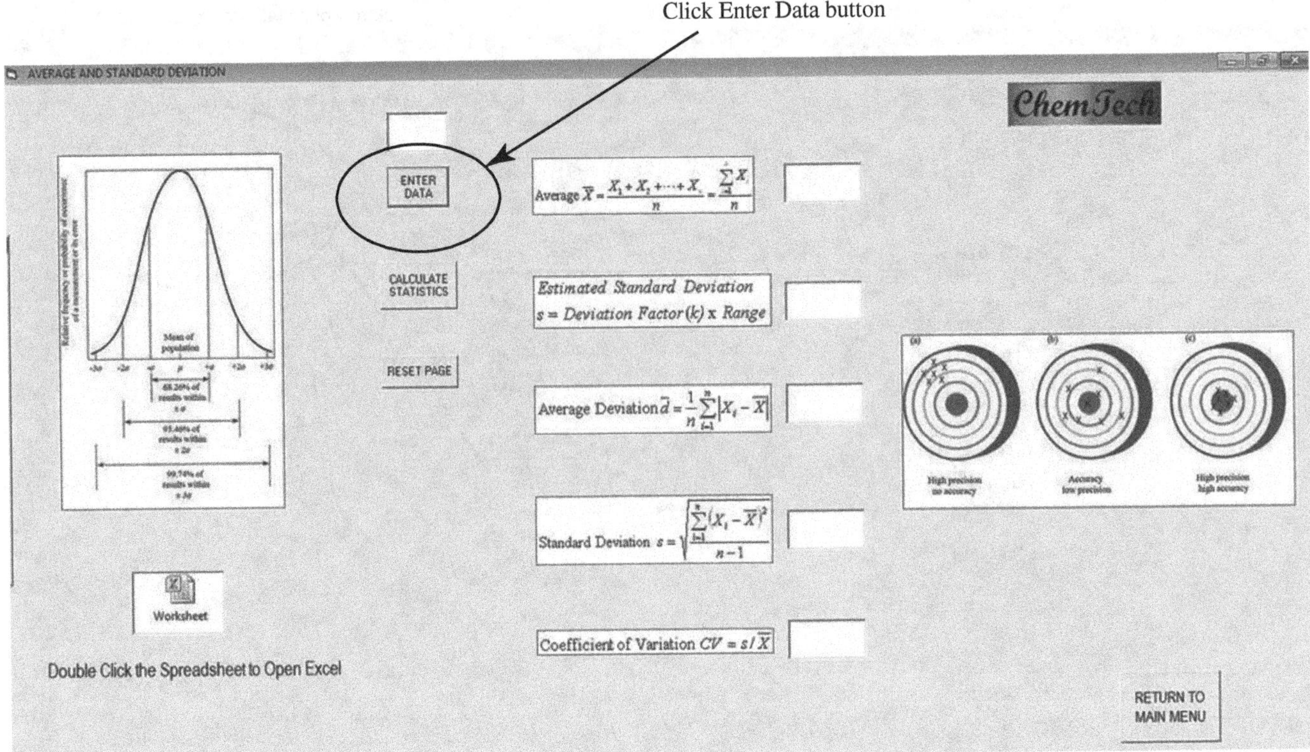

FIGURE 5.7 The "CALCULATE STATISTICS 1" page from ChemTech. Enter data into the top data input box and press the "Enter Data" button to enter each number for statistical calculations.

TABLE 5.4 Experimental values.

X_i
86.3
79.9
87.7
84.2
88.6
83.8
85.4
81.1
85.8

5.7.2.1 ***Entering Data*** When the form opens, the cursor will be in the input box. If it is not, click inside the input box to place the cursor there. Enter the data one by one by typing in each number into the input box. After a number, say the first number of "86.3" has been typed into the box, click the "ENTER DATA" button to input that data point as depicted in Figure 5.7. The number will be inputted for the statistical calculation and the cursor will return to the input box. Type in the second number of the data set in Table 5.4 and click the "ENTER DATA" button. Continue in this manner, enter data and click enter, for each of the rest of the data points in Table 5.4.

5.7.2.2 ***Calculating the Statistics*** After the last data point has been inputted into the statistical program, press the "CALCULATE STATISTICS" button. After pressing the button the results boxes will display the statistics of the inputted data including the average ($\bar{X}$), the estimated standard deviation based upon the range and deviation factor k, the average deviation ($\bar{d}$), the sample standard deviation (s), and the CV.

5.7.2.3 ***The Results Output*** The values listed in the results output boxes should be: average $\bar{X} = 84.75555$ (all results include five decimal places but the student needs to round to the proper significant figures, in this case 84.8), the estimated standard deviation as 2.96, the average deviation $\bar{d} = 2.23$, the sample standard deviation $s = 2.86$, and the $CV = 3.38\%$.

5.7.2.4 ***Results not Expected*** The statistical calculation page will look like that shown in Figure 5.8. If you are not seeing these values in the statistical output boxes, then an error may have taken place during the inputting of the data values. In this case, click the "RESET PAGE" button and reenter the data values and try clicking the "CALCULATE STATISTICS" button again. At any time, the "RESET PAGE" button can be clicked to reinitialize the statistical calculations and the data entry can be started again. Again, take care that each data value needs to be typed into the data input box and the "ENTER DATA" button clicked for each data value. If the final values continue to be different than expected, keep resetting the page and practice inputting the data until the proper results are obtained.

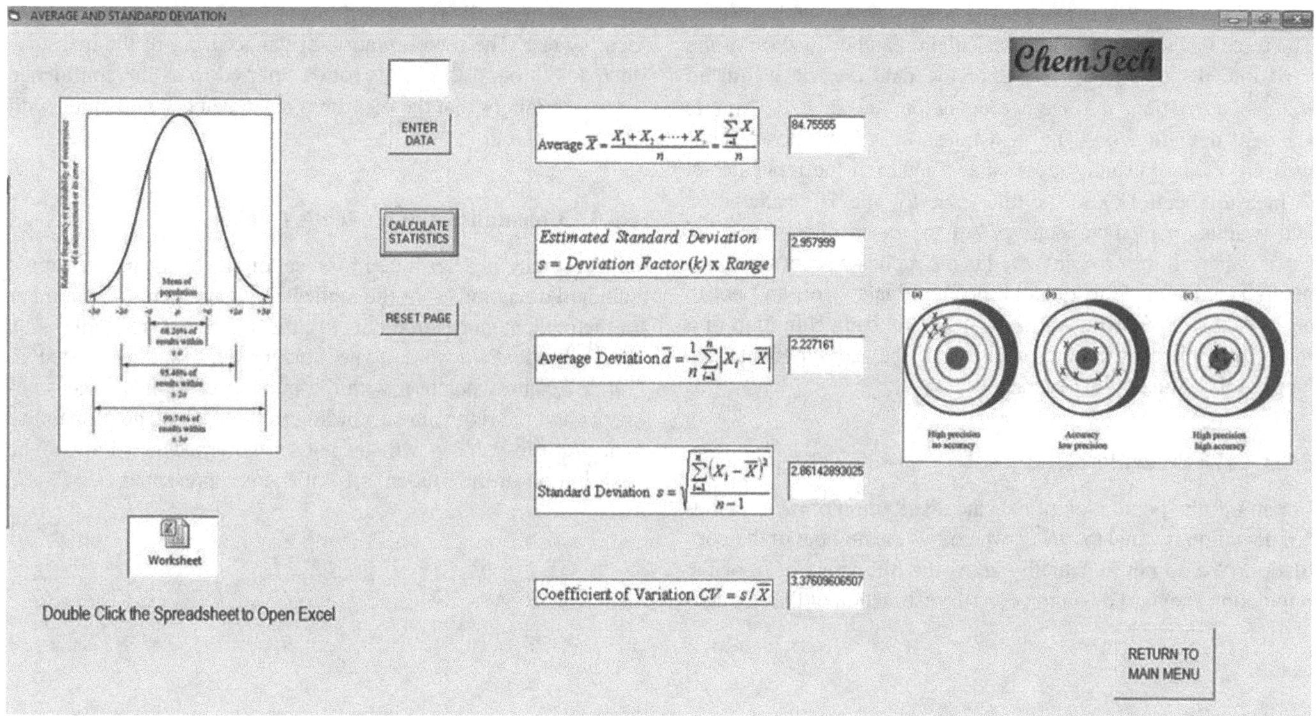

FIGURE 5.8 Results obtained from the data entered from Table 5.4 after clicking the "CALCULATE STATISTICS" button on the statistics page.

TABLE 5.5 Experimental values.

X_i	X_i
1012.358	1003.623
998.5513	1010.899
995.6621	1009.539
1004.815	1015.816
1001.559	1000.893
1014.689	997.6355
1007.158	1005.883
992.8899	1011.691
997.6615	993.1161
1005.891	999.3631

5.7.2.5 ***Using ChemTech for Large Value Set*** The number of data values in Table 5.4 is 9 ($n = 9$). While the statistics could have been calculated by hand using the set of Equations 5.1–5.8, it is very tedious and time consuming. ChemTech can be used in your laboratory at any time to help simplify the calculation of statistics that have been gathered in the laboratory. Let us look at an example of data points ($n = 20$) that would truly be a daunting task to calculate the statistics by hand. If not already open, open ChemTech and go to the Chapter 5 Main Menu page. Click the button "CALCULATE STATISTICS 1" to open a statistical calculation page. Enter the following values in Table 5.5 as we did before. After the last data point has been inputted into the statistical program, press the "CALCULATE STATISTICS" button. After pressing the button the results boxes will again display the statistics of the inputted data including the average ($\bar{X}$), the estimated standard deviation based upon the range and deviation factor k, the average deviation ($\bar{d}$), the sample standard deviation (s), and the CV.

5.7.2.6 ***The Results Page*** The values listed in the results output boxes should be: average $\bar{X} = 1003.985$, the estimated standard deviation as NA, the average deviation $\bar{d} = 5.889$, the sample standard deviation $s = 7.026086$, and the $CV = 0.6998\%$. Note that the reported value for the estimated standard deviation is "NA." This is due to the number of data points being greater that $n = 10$.

5.7.2.7 ***Resetting the Page*** Again as before, if you are not seeing these values in the statistical output boxes, then an error may have taken place during the inputting of the data values. In this case, click the "RESET PAGE" button and reenter the data values and try clicking the "CALCULATE STATISTICS" button again. At any time the "RESET PAGE" button can be clicked to reinitialize the statistical calculations and the data entry can be started again. As before, take care that each data value needs to be typed into the data input box and the "ENTER DATA" button clicked for each data value. If the final values continue to be different than expected, keep resetting the page and practice inputting the data until the proper results are obtained.

5.8 STUDENT'S DISTRIBUTION t TEST FOR CONFIDENCE LIMITS

5.8.1 Accuracy

In the previous sections, we have been looking at ways to calculate and estimate the precision of a set of replicate analyses that was gathered in the laboratory based upon the mean of the results. The three methods used for the calculation and estimation of the variation, dispersion, or scatter of the data are the estimated standard deviation, the average deviation ($\bar{d}$), and the sample standard deviation (s). The next question is how close is our measured value, usually reported as a mean of the replicate set of measurements ($\bar{X}$), to the true value μ? This is a calculation and estimation of the "accuracy" of the mean of the replicate set of values. If the error involved in the replicate set of measurements is attributable to random, indeterminate error, and not to any systematic, determinate error such as a bad calibration or a faulty instrument, then the precision of the measurements can be used to estimate the accuracy.

5.8.2 The Student's *t* Test

For an infinite population of data the distribution of mean standard deviation is equal to $\sigma/N^{1/2}$, where N is the number of the population. We do not in actuality know the true value of μ, or the population standard deviation σ, so the estimation of the standard deviation of the distribution of means is calculated using the Student's t test. The representation of the accuracy of the measurements is through the limits (often referred to as the confidence limits) or ranges that the measured mean value $\bar{X}$ will agree with the true value μ.

5.8.3 Calculating the Student's *t* Value

The accuracy is determined for a set of values using the sample standard deviation (s) of the replicate measurements, the number of the replicate measurements (n), the degrees of freedom (n−1), and a Student's t value. The Student's t values are listed in Table 5.6 in association with the number of replicate measurements and the risk value according to the percentage probability that the true value μ will lie within the associated limits. The distribution of the Student's t statistic is represented as

$$\pm t = \frac{\bar{X}-\mu}{s/\sqrt{n}} \tag{5.10}$$

TABLE 5.6 Percentile Values of the Student's Distribution t_α Used for Calculating Confidence Limits.

n	df (n−1)	$t_{0.995}$	$t_{0.99}$	$t_{0.975}$	$t_{0.95}$	$t_{0.90}$	$t_{0.80}$	$t_{0.75}$	$t_{0.70}$
2	1	63.657	31.821	12.706	6.314	3.078	1.376	1.000	0.727
3	2	9.925	6.965	4.303	2.920	1.886	1.061	0.816	0.617
4	3	5.841	4.541	3.182	2.353	1.638	0.978	0.765	0.584
5	4	4.604	3.747	2.776	2.132	1.533	0.941	0.741	0.569
6	5	4.032	3.365	2.571	2.015	1.476	0.920	0.727	0.559
7	6	3.707	3.143	2.447	1.943	1.440	0.906	0.718	0.553
8	7	3.499	2.998	2.365	1.895	1.415	0.896	0.711	0.549
9	8	3.355	2.896	2.306	1.860	1.397	0.889	0.706	0.546
10	9	3.250	2.821	2.262	1.833	1.383	0.883	0.703	0.543
11	10	3.169	2.764	2.228	1.812	1.372	0.879	0.700	0.542
12	11	3.106	2.718	2.201	1.796	1.363	0.876	0.697	0.540
13	12	3.055	2.681	2.179	1.782	1.356	0.873	0.695	0.539
14	13	3.012	2.650	2.160	1.771	1.350	0.870	0.694	0.538
15	14	2.977	2.624	2.145	1.761	1.345	0.868	0.692	0.537
16	15	2.947	2.602	2.131	1.753	1.341	0.866	0.691	0.536
17	16	2.921	2.583	2.120	1.746	1.337	0.865	0.690	0.535
18	17	2.898	2.567	2.110	1.740	1.333	0.863	0.689	0.534
19	18	2.878	2.552	2.101	1.734	1.330	0.862	0.688	0.534
20	19	2.861	2.539	2.093	1.729	1.328	0.861	0.688	0.533
21	20	2.845	2.528	2.086	1.725	1.325	0.860	0.687	0.533
22	21	2.831	2.518	2.080	1.721	1.323	0.859	0.686	0.532
23	22	2.819	2.508	2.074	1.717	1.321	0.858	0.686	0.532
24	23	2.807	2.500	2.069	1.714	1.319	0.858	0.685	0.532
25	24	2.797	2.492	2.064	1.711	1.318	0.857	0.685	0.531
26	25	2.787	2.485	2.060	1.708	1.316	0.856	0.684	0.531
27	26	2.779	2.479	2.056	1.706	1.315	0.856	0.684	0.531
28	27	2.771	2.473	2.052	1.703	1.314	0.855	0.684	0.531
29	28	2.763	2.467	2.048	1.701	1.313	0.855	0.683	0.530
30	29	2.756	2.462	2.045	1.699	1.311	0.854	0.683	0.530
31	30	2.750	2.457	2.042	1.697	1.310	0.854	0.683	0.530
41	40	2.704	2.423	2.021	1.684	1.303	0.851	0.681	0.529
61	60	2.660	2.390	2.000	1.671	1.296	0.848	0.679	0.527
121	120	2.617	2.358	1.980	1.658	1.289	0.845	0.677	0.526
∞+1	∞	2.576	2.326	1.960	1.645	1.282	0.842	0.674	0.524

Rearranging Equation 5.9, we are able to put the equation into a form that expresses the estimation of the range of values centered around the mean $\bar{X}$ that will include the true value μ.

$$\mu = \bar{X} \pm \frac{ts}{\sqrt{n}} \tag{5.11}$$

To calculate the confidence limits of the mean $\bar{X}$, the following form of the equation is used:

$$\bar{X} \pm \frac{ts}{\sqrt{n}} \tag{5.12}$$

5.8.4 Probability Level

The calculation of the confidence limits is based upon a Student's t distribution curve. The probability (risk value α or percentage probability: 100–100α) that the true value μ will fall within the estimated confidence limits is determined by the t value used from Table 5.6. The t value is associated with the number of replicate measurements and the degree of certainty that is desired. If the number of replicates is $n = 10$, then the degrees of freedom $df = n - 1$ or 9. The Student's t value with a 90% probability level that the true value will fall within the confidence limits range would be 1.833.

5.8.5 Sulfate Concentration Confidence Limits

Example 5.5 A gravimetric measurement of sulfur content as molarity (M) sulfate $[SO_4^{2-}]$ was repeated 10 times with the following resultant concentrations (Table 5.7):

Using ChemTech Chapter 5 "CALCULATE STATISTICS 1" the average sulfate concentration is determined at $\bar{X} = 2.106$M with a sample standard deviation $s = 0.0358$ M. If we choose a factor of risk level at $\alpha = 0.05$ (i.e., confidence of a 95% probability that μ would fall within our limits), then the degrees of freedom is $df = 9$ and the Student's t value is 2.262. Using Equation 5.12, the confidence limits are calculated as

$$\bar{X} \pm \frac{ts}{\sqrt{n}} = 2.106\ \text{M} \pm \frac{(2.262)(0.0358\ \text{M})}{\sqrt{10}} = 2.106\ \text{M} \pm 0.026\ \text{M} \tag{5.13}$$

TABLE 5.7 Experimental values.

X_i
2.125
2.103
2.110
2.131
2.108
2.129
2.107
2.111
2.009
2.130

5.8.6 Sulfate t Distribution Curve

From these calculations, it can be determined that with a probability of 95%, μ will fall within the sulfate concentration $[SO_4^{2-}]$ range of 2.080–2.132 M. A representative t distribution curve, similar to those found in Figure 5.2, can be constructed from these data using $\bar{X} = 2.106$ M, $s = 0.0358$ M, and $\alpha = 0.05$ and is depicted in Figure 5.9. Illustratively, the middle of the curve contains a region or area where there exists a 95% probability that the true sulfate concentration μ will be contained within the range of 2.080–2.132 M. Note that the two ends of the curve contain areas equal to $\alpha/2$ or 0.025 that represents the fraction of the risk that the true value μ lies in an area that is either larger or smaller than the range set by the confidence limits. This area of the curve represents the 5% probability that the true value μ is not contained within the sulfate concentration $[SO_4^{2-}]$ range of 2.080–2.132 M.

5.8.7 Determining Types of Error

There are other different uses of the Student's t test besides the determination of a range of values that will contain the true value to a certain level of confidence. The t values listed in Table 5.6 can also be used to determine whether a testing procedure contains only random error or a systematic error in the analysis exists. Taking the average and standard deviation of a set of replicate analysis, if the calculated t value is greater than that listed in Table 5.6 for the associated degrees of freedom and the level of confidence than it is assumed that a systematic error exists.

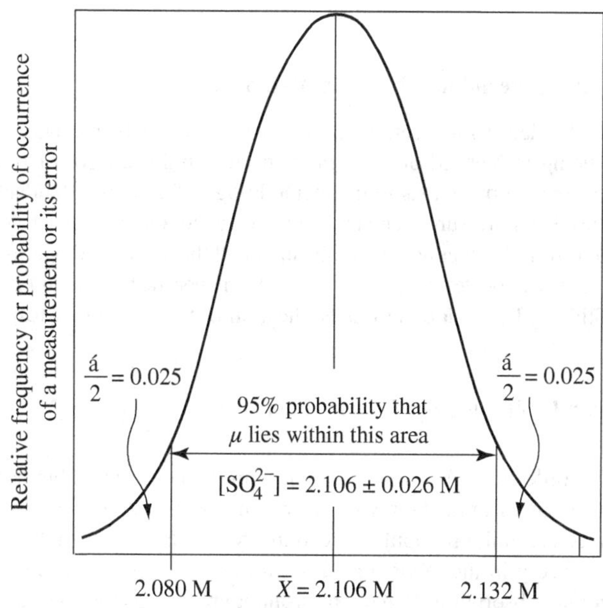

FIGURE 5.9 Student's t distribution curve of the sulfate analysis in example 10.10 with mean sulfate concentration $[SO_4^{2-}]$ of $\bar{X} = 2.106$ M, sample standard deviation $s = 0.0358$ M, and risk factor $\alpha = 0.05$.

TABLE 5.8 Experimental values.

X_i
0.536
0.533
0.541
0.539
0.542
0.544
0.535
0.545
0.531
0.535

5.8.7.1 Glucose Content

Example 5.6 A spectrophotometric measurement of glucose content in plasma as percent glucose (%Glu) was repeated 10 times with the following resultant concentrations (Table 5.8):

Using ChemTech Chapter 5 "CALCULATE STATISTICS 1," the average glucose concentration is determined at $\bar{X} = 0.538\%$ with a sample standard deviation $s = 0.00479\%$. If we choose a factor of risk level at $\alpha = 0.05$ (i.e., confidence of a 95% probability that μ would fall within our limits), then the degrees of freedom is $df = 9$ and the Student's t value is 2.262. Using Equation 5.11, the confidence limits are calculated as

$$\bar{X} \pm \frac{ts}{\sqrt{n}} = 0.538\% \pm \frac{(2.262)(0.00479\%)}{\sqrt{10}} \qquad (5.14)$$
$$= 0.538\% \pm 0.003\%$$

From these calculations, it can be determined that with a probability of 95%, μ will fall within the glucose concentration range of 0.535–0.541%.

5.8.8 Determining Error in Methodology

The Student's t test can be used to determine whether a method is being influenced only by random indeterminate error or if a systematic error exists in the methodology. If a known standard is used and measured repeatedly by the methodology in question, then from the average value obtained and the standard deviation a range can be determined to measure whether the true value falls within it. Let us look at this application of the Student's t test.

5.8.8.1 Magnesium Primary Standard

Example 5.7 A primary magnesium standard was obtained from the manufacturer with a known concentration $\mu = 175.1$ parts per million (ppm). The primary standard was analyzed repeatedly in the laboratory using a methodology set up on an atomic absorption flame spectrophotometer. Using the data obtained and listed in Table 5.9 below; determine whether there is only random error in the measurement, or the possibility of a systematic error existing in the methodology.

TABLE 5.9 Experimental values.

X_i (ppm)
178.3
175.9
176.8
174.4
175.1
173.9
174.7
177.2
175.5
176.6
178.1
177.5

Using ChemTech Chapter 5 "CALCULATE STATISTICS 1," the average magnesium concentration is determined at $\bar{X} = 176.2$ ppm with a sample standard deviation $s = 1.47$ ppm. If we choose a factor of risk level for a one-tailed test at $\alpha = 0.05$ (i.e., confidence of a 95% probability that μ would fall within our limits), then the degrees of freedom is $df = 11$ and the Student's t value is 1.812.

$$\mu = \bar{X} \pm \frac{ts}{\sqrt{n}} = 176.2\ \text{ppm} \pm \frac{(1.812)(1.47\ \text{ppm})}{\sqrt{12}} \qquad (5.15)$$
$$= 176.2\ \text{ppm} \pm 0.77\ \text{ppm}$$

We see that the true concentration value of 175.1 ppm does not in fact lie within our experimental range of 175.4–177.0 ppm. We can therefore conclude that a systematic error may exist in our present methodology. Upon further inspection and evaluation of the system, the technician was able to determine that all of the known standards he/she analyzed now on the atomic absorption spectrophotometer were reading slightly high. The calibration of the instrument had drifted and needed recalibration.

5.9 TESTS OF SIGNIFICANCE

The Student's t test can also be used to determine whether two separate analytical methods have the same mean in the absence of any systematic error. If the two methods accurately measure a given parameter and the variance is due only to inherent random error, then the calculated t value derived from the replicate analysis of the two methods should be less than the t value listed in Table 5.9 for the associated degrees of freedom and desired confidence value.

5.9.1 Difference in Means

Example 5.8 A gas chromatography (GC) measurement of cholesterol content from a corn oil extract is to be compared to a method that has recently been developed on a new high-pressure liquid chromatography (HPLC) instrument installed in the laboratory. Using the data listed in Table 5.10 below,

TABLE 5.10 Experimental values.

X_i (GC, %)	X_i (HPLC, %)
3.51	3.52
3.54	3.56
3.49	3.48
3.47	3.56
3.52	3.46
3.46	3.51
3.53	3.52

determine whether there is a significant difference in the mean of the two methods.

Using ChemTech Chapter 5 "CALCULATE STATISTICS 1" the average cholesterol concentration for the GC method is $\bar{X}_{GC} = 3.50\%$ with a sample standard deviation $s_{GC} = 0.030\%$, and for the new HPLC method $\bar{X}_{HPLC} = 3.52\%$ with a sample standard deviation $s_{HPLC} = 0.028\%$. When we are comparing two method's mean value for a sample measurement, we perform a test of significance for two unknown means and also two unknown standard deviations. Because the measured results are a small set of the population the population standard deviations are not known, and are estimated by the calculated values s_1 and s_2. When this is done, the test statistic is defined by the *two-sample t statistic* illustrated by the following equation:

$$t = \frac{(\bar{x}_1 - \bar{x}_2) - (\mu_1 - \mu_2)}{\sqrt{\frac{s_1^2}{n_1} + \frac{s_2^2}{n_2}}} \tag{5.16}$$

5.9.2 Null Hypothesis

When comparing the means of two sets of data we are performing a statistical hypothesis in order to answer the question of whether the means are the same or different. The assumption that the means are the same is called the *null hypothesis* and is illustrated as H_o, where $\mu_1 - \mu_2 = 0$. The case where the means are not equal is called the alternative hypothesis and is illustrated as H_1, where $\mu_1 - \mu_2 \neq 0$. If we choose a factor of risk level at $\alpha = 0.05$ (i.e., confidence of a 95% probability that μ would fall within our limits), then the degrees of freedom is taken from the sample set with the lowest number of data. In this problem, the $df = 6$ and the associated Student's t value is 2.015. Using Equation 5.15 and setting $\mu_1 - \mu_2 = 0$, the calculated t value for the comparison is

$$t = \frac{(3.50 - 3.52)}{\sqrt{\frac{0.030^2}{7} + \frac{0.028^2}{7}}} \tag{5.17}$$

$$t = -1.2895$$

The absolute value of the calculated t value (1.2895) is less than the Student's t test value (2.015) thus indicating that there is not a significant difference in the means obtained from the two methodology.

To calculate the difference in mean confidence interval, the following relationship is used:

$$(\bar{x}_1 - \bar{x}_2) \pm t* \sqrt{\frac{s_1^2}{n_1} + \frac{s_2^2}{n_2}} \tag{5.18}$$

The $t*$ value used is for a two-tailed test thus the critical value for the particular level of confidence is equal to (1–C)/2. For example, at a level of 95% confidence ($t_{0.95}$), the Student's t value from Table 5.6 would be chosen from the $t_{0.975}$ column with the associated degrees of freedom. We take from Table 5.6, a $t*$ value of 2.571 and calculate the confidence interval as

$$(3.50 - 3.52) \pm 2.571 \sqrt{\frac{0.030^2}{7} + \frac{0.028^2}{7}} \tag{5.19}$$

$$-0.02 \pm 0.040$$

This is a range from −0.06 to 0.02, which includes the value 0 also indicating that there is not a significant difference of the two methods' means at the 0.05 level of confidence.

5.10 TREATMENT OF DATA OUTLIERS

A typical rule of thumb in the laboratory is measurements in "threes." When analyses are replicated it is a good idea to collect at least three values, where three is the lower limit or the least amount of replicates that can be collected for average and standard deviation determination for outlier evaluation. The third value is collected to help support at least one of the other two. It is better to collect as many values as possible for an analysis, however this is often impractical. A compromise in time and effort is to collect three replicate analyses and compare the values. If one of the three is drastically different from the other two there may arise a need to repeat the analysis again. A question arises though as to how does the analyst decide whether a value is drastically different or just slightly different? How different does a value have to be from the other values to denote it as an "outlier" and remove it from the set of replicate results? There are two rather simple approaches for the treatment of small sets of data, consisting usually from $n = 3$ to $n = 10$, to test for the presence of an outlier and a subsequent decision for its removal.

5.10.1 The *Q* Test

The Q test is the simplest approach to the treatment of small data set outliers. The Q test uses the difference between the suspected outlier's nearest neighbor divided by the range of the data set. The experimental ratio value Q_{exp} is then compared to a table of critical values Q_{crit} (Table 5.11) according to the sample size (n) and the level of desired confidence.

When testing for an outlier value it is either the smallest value in the data set or the largest. The associated formulas to use according to either the smallest or largest value for outlier testing are

$$\text{Smallest value } (x_1): Q_{exp} = \frac{x_2 - x_1}{x_n - x_1} \tag{5.20}$$

TABLE 5.11 Critical Values for Rejection Quotient Q.

Number of Observations (n)	90% Confidence Level $Q_{0.90}$	95% Confidence Level $Q_{0.95}$	99% Confidence Level $Q_{0.99}$
3	0.941	0.970	0.994
4	0.765	0.829	0.926
5	0.642	0.710	0.821
6	0.560	0.625	0.740
7	0.507	0.568	0.680
8	0.468	0.526	0.634
9	0.437	0.493	0.598
10	0.412	0.466	0.568

$$\text{Largest value } (x_n): Q_{\text{exp}} = \frac{x_n - x_{n-1}}{x_n - x_1}\sqrt{b^2-4ac} \tag{5.21}$$

If $Q_{\text{exp}} > Q_{\text{crit}}$ (from Table 5.10), then the suspected outlier value can be rejected and removed from the data set. Often, a 90% confidence level ($Q_{0.90}$) is used for evaluating possible outliers from data sets. Let us look at an example of the use of the Q test.

Example 5.9 A replicate analysis of a soil sample for moisture content resulted in the following values: 47.89, 45.76, 46.55, 49.92, 46.13, 47.28, and 45.35 g/m^3. With a confidence value of 90%, can the 49.92 g/m^3 value be removed or not? We will need to use the largest value equation Q test:

$$\text{Largest value } (x_n): Q_{\text{exp}} = \frac{x_n - x_{n-1}}{x_n - x_1} \tag{5.22}$$

$$Q_{\text{exp}} = \frac{49.92-47.89}{49.92-45.35}$$
$$Q_{\text{exp}} = 0.444 \tag{5.23}$$

The Q test critical value for $n=7$ and $Q_{0.90}$ is 0.507. The experimental value does not meet the criteria $Q_{\text{exp}} > Q_{\text{crit}}$; therefore, the value of 49.92 g/m^3 cannot be removed from the data set but must be retained.

5.10.2 The T_n Test

A second approach to testing the validity of rejecting a possible outlier data point is provided by the American Society for Testing Materials (ASTM) that is based upon the average ($\bar{X}$) and standard deviation (s) of the entire data set. The T_n test uses the absolute difference between the suspected outlier and the data set average divided by the standard deviation of the data set. The experimental ratio value T_{exp} is then compared to a table of critical values T_{crit} (Table 5.12) according to the sample size (n) and the level of desired confidence.

When testing for an outlier value it is either the smallest value in the data set or the largest. The associated formula to use according to either the smallest or largest value for outlier testing is

TABLE 5.12 Critical Values for Rejection Quotient T_n.

Number of Observations (n)	95% Confidence Level $T_{0.95}$	97.5% Confidence Level $T_{0.975}$	99% Confidence Level $T_{0.99}$
3	1.15	1.15	1.15
4	1.46	1.48	1.49
5	1.67	1.71	1.75
6	1.82	1.89	1.94
7	1.94	2.02	2.10
8	2.03	2.13	2.22
9	2.11	2.21	2.32
10	2.18	2.29	2.41

$$T_{\text{exp}} = \frac{|x_i - \bar{X}|}{s} \tag{5.24}$$

If $T_{\text{exp}} > T_{\text{crit}}$ (from Table 5.11), then the suspected outlier value can be rejected and removed from the data set. Often, a 95% confidence level ($T_{0.95}$) is used for evaluating possible outliers from data sets. Let us look at an example of the use of the T_n test.

Example 5.10 A replicate analysis of a soybean oil sample for stearic acid content resulted in the following values: 7.856%, 7.801%, 7.925%, 7.455%, 7.884%, 7.919%, and 7.837%. With a confidence value of 95%, can the 7.455% value be removed or not? Using ChemTech Chapter 5 "CALCULATE STATISTICS 1" the average stearic acid content is $\bar{X} = 7.807\%$ with a sample standard deviation $s = 0.1781\%$. Plugging into the critical quotient T_n Equation 5.21 we get a T_{exp} value of

$$T_{\text{exp}} = \frac{|7.455-7.807|}{0.1781}$$
$$T_{\text{exp}} = 1.98 \tag{5.25}$$

The T_n test critical value for $n=7$ and $T_{0.95}$ is 1.94. The experimental value does meet the criteria $T_{\text{exp}} > T_{\text{crit}}$; therefore, the value of 7.455% can be removed from the data set at the 95% confidence level. Note however that the criteria of $T_{\text{exp}} > T_{\text{crit}}$ at the 97.5% confidence level ($T_{\text{crit}} = 2.02$) is not met and the data point would have to be retained.

5.11 CHAPTER KEY CONCEPTS

5.1 There are two types of error that the analyst needs to understand: systematic (determinate) and random (indeterminate) errors associated with laboratory analysis.

5.2 Systematic errors are introduced into the analysis by a malfunctioning instrument, an analytical method that contains a defect, a bad calibration, or by the analyst.

5.3 Random errors are due to the fact that there is always some uncertainty to all laboratory measurements.

5.4 The absolute error of a result X_i is equal to $X_i - \mu$, where μ is the "true value" of the result.

5.5 The relative error is equal to the absolute error divided by the true value.

5.6 The precision of a set of results is the agreement of the results among themselves.

5.7 The accuracy of a result is determined by how close the experimental value is to the true value.

5.8 The *normal distribution curve* (also referred to as the *probability curve* or *error curve*) possesses a bell or Gaussian symmetrical shape, where the y-axis represents the probability of the occurrence of a measurement, and the x-axis represents the distribution of the data centered around the population mean μ.

5.9 The true value of a measurement (μ) is usually not known and is estimated using the *mean*, *median*, or *mode* of a set of values.

5.10 Calculating the standard deviation (or the root mean square, RMS) of a set of data points is a more accurate measure of the dispersion of the data set tested.

5.11 The normal distribution curve, also called a Gaussian distribution or Bell curve is the shape that is observed when the data points of a system that possess a normal distribution is plotted.

5.12 For an infinite population of data, the distribution of mean standard deviation is equal to $\sigma/N^{1/2}$, where N is the number of the population.

5.12 CHAPTER PROBLEMS

5.1 Using the following set of data points: 10.2, 9.8, 10.6, 9.9, 10.3 estimate the standard deviation using the range and the appropriate deviation factor. The range is 0.8 and the deviation factor is 0.43.

5.2 Using the following set of data points: 76.23, 74.21, 78.92, 75.25, 77.29, 74.88 estimate the standard deviation using the range and the appropriate deviation factor. The range is 4.71 and the deviation factor is 0.40.

5.3 Using the following table of data values, calculate the average deviation, and compare it to the estimated standard deviation derived from the range.

X_i
54.3
55.9
53.1
56.7
55.2
54.8
57.3
53.5

5.4 Using the following set of data calculate the sample standard deviation and compare it to the average deviation and estimated standard deviation from the range.

X_i
54.3
55.9
53.1
56.7
55.2
54.8
57.3
53.5

5.5 A gravimetric measurement of phosphorus content as molarity (M) phosphate [PO_4^{3-}] was repeated 10 times. Use the following resultant concentrations to calculate the confidence limits.

X_i
7.131
7.154
7.197
7.121
7.159
7.129
7.107
7.111
7.009
7.130

5.6 A spectrophotometric measurement of protein content in plasma as percent protein (%Pro) was repeated 10 times. Calculate the confidence limits and determine whether μ will fall within the glucose concentration range.

X_i
0.356
0.349
0.351
0.354
0.347
0.344
0.350
0.352
0.355
0.351

5.7 A primary iron standard was obtained from the manufacturer with a known concentration $\mu = 502.6$ ppm. The primary standard was analyzed repeatedly in the laboratory using a methodology set up on an atomic absorption flame spectrophotometer. Using the data obtained and listed below determine whether there is only random error in the measurement, or the possibility of a systematic error existing in the methodology.

X_i (ppm)
502.1
501.8
502.3
501.7
501.5
502.6
502.4
501.6
501.9
502.4
502.7
502.2

5.8 A GC measurement of palmitic acid content from a corn oil extract is to be compared to a method that has recently been developed on a new HPLC instrument installed in the laboratory. Using the data listed below, determine whether there is a significant difference in the mean of the two methods.

X_i (GC, %)	X_i (HPLC, %)
8.51	8.52
8.54	8.56
8.49	8.48
8.47	8.56
8.52	8.46
8.46	8.51
8.53	8.52

5.9 A replicate analysis of a soil sample for moisture content resulted in the following values: 68.59, 62.31, 65.56, 64.13, 67.49, 67.10, and 66.55 g/m^3. With a confidence value of 90%, can the 68.59 g/m^3 value be removed or not using the largest value equation Q test?

5.10 A replicate analysis of a soybean oil sample for stearic acid content resulted in the following values: 3.553%, 3.826%, 3.758%, 3.321%, 3.799%, 3.801%, and 3.692%. With a confidence value of 95%, can the 3.321% value be removed or not using the critical quotient T_n Equation?

6

PLOTTING AND GRAPHING

6.1 Introduction to Graphing
  6.1.1 The Invention of the Graph
  6.1.2 Importance of Graphing
6.2 Graph Construction
  6.2.1 Axis and Quadrants
6.3 Rectangular Cartesian Coordinate System
6.4 Curve Fitting
6.5 Redrawn Graph Example
6.6 Graphs of Equations
  6.6.1 Introduction
  6.6.2 Copper Sulfate Data
  6.6.3 Plotting the Data
  6.6.4 Best Fit Line
  6.6.5 Point-Slope Equation of a Line
  6.6.6 Finding the Slope (*m*)
  6.6.7 Finding the *y*-Intercept (*b*)
  6.6.8 Solving for *x*
  6.6.9 Estimating the Slope and Intercept
  6.6.10 Deriving the Equation from the Slope and Intercept
6.7 Least-Squares Method
  6.7.1 Plotting Data with Scatter
  6.7.2 Linear Regression
  6.7.3 Curve Fitting the Data
6.8 Computer-Generated Curves
  6.8.1 Using ChemTech to Plot Data
  6.8.2 Entering the Data
  6.8.3 Plotting the Data
  6.8.4 Linear Regression of the Data
  6.8.5 Adding the Best Fit Line
  6.8.6 Entering a Large Set of Data
6.9 Calculating Concentrations
6.10 Nonlinear Curve Fitting
6.11 Chapter Key Concepts
6.12 Chapter Problems

6.1 INTRODUCTION TO GRAPHING

6.1.1 The Invention of the Graph

The fundamental idea behind the construction of a two-dimensional graph laid flat and moving up or down on the plane of a page was first developed by two separate mathematicians: Rene Descartes (1596–1650), a French philosopher and mathematician, and Pierre de Fermat (1601–1665), a French jurist and mathematician. It is in Descartes's famous book, *Geometry*, published in 1635 that is generally credited with the introduction to the rectangular Cartesian coordinate system (named after Descartes), a system to represent ordered pairs of numbers (x, y) in different quadrants of a set of axes laid in the plane of the page. Descartes's *Geometry* laid the foundation for modern analytical geometry and introduced concepts in mathematics that was needed for the invention and further development of calculus.

6.1.2 Importance of Graphing

Plotting and graphing data is a very important skill that the laboratory technician needs to posses. Whether you have extensive knowledge in this area, a little familiarity with this, or are a complete beginner we will start at the most basic aspects of plotting data and move into a more in depth study of the important uses of graphs in the analytical laboratory including calibration curves, relationship studies, and control charts.

6.2 GRAPH CONSTRUCTION

The basic construction of a graph is depicted in Figure 6.1, which is typically used in the analytical laboratory for simple scatter diagrams of a linear (and nonlinear) relationship between two variables, such as the concentration of a standard and its response on an analytical instrument.

Analytical Chemistry: A Chemist and Laboratory Technician's Toolkit, First Edition. Bryan M. Ham and Aihui MaHam.

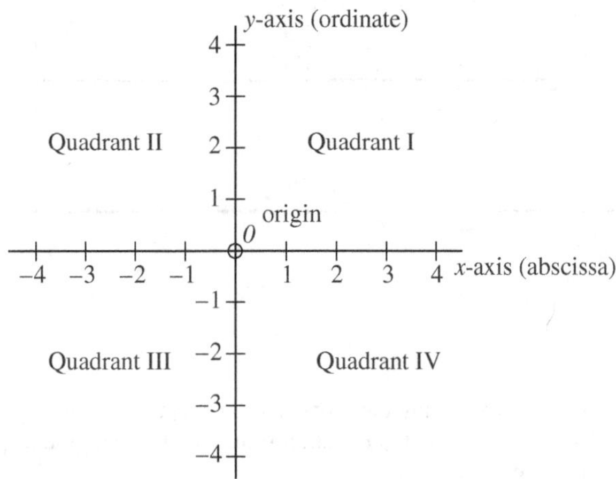

FIGURE 6.1 Basic construction of a graph based upon the rectangular Cartesian coordinate system, typically used in the analytical laboratory. Key components include the vertical *y*-axis (ordinate), the horizontal *x*-axis (abscissa), the origin, and the four quadrants.

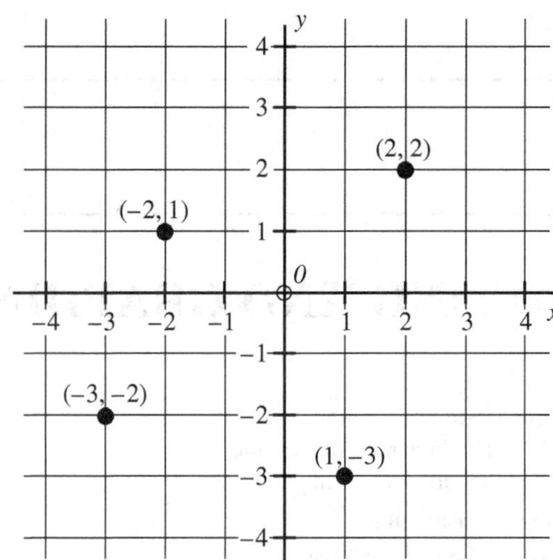

FIGURE 6.2 Rectangular Cartesian coordinate system and point location.

6.2.1 Axis and Quadrants

The graph is made up of the vertical *y*-axis (ordinate) and of the horizontal *x*-axis (abscissa). The fundamental design of a graph is based on the rectangular Cartesian coordinate system where the plane of the paper is divided into four quadrants. In quadrant I, the *x*-axis and *y*-axis values are both positive and increase in magnitude radially out from the mutually 0 (zero) location (known as the origin). In quadrant II, the *x*-axis values are negative and the *y*-axis values are positive. In quadrant III, both the *x*-axis and *y*-axis values are negative. Finally, in quadrant IV, the *x*-axis values are positive while the *y*-axis values are negative.

6.3 RECTANGULAR CARTESIAN COORDINATE SYSTEM

The rectangular Cartesian coordinate system depicted in Figure 6.2 contains a network of gridlines that will make it easier for us to see the location of a particular point within the system. When a point in two-dimensional space within the predefined confines of the system is to be located, its distances from the *x*- and *y*-axes with respect to each other are used to designate the point's location. The *x* coordinate is the distance from the *y*-axis while the *y* coordinate is the distance from the *x*-axis. This is designated as (x, y) such as the (2, 2), the (1, −3), the (−3, −2), and the (−2, 1) locations depicted in Figure 6.2. To locate the point (2, 2) we move to the right in the positive direction 2 units on the *x*-axis and up in the positive direction 2 units on the *y*-axis. This locates us at the coordinate (2, 2) in quadrant I of the rectangular Cartesian coordinate system. Each of the subsequent points listed are obtained in the same manner, such as the (1, −3) coordinate that is located by moving 1 unit to the right in the positive direction on the *x*-axis and down in the negative direction 3 units on the *y*-axis. This point is located in quadrant IV of the rectangular Cartesian coordinate system in Figure 6.2.

6.4 CURVE FITTING

In the analytical laboratory, there are two methods for drawing a graph or plotting data for generating a curve (curve fitting). This can be accomplished either by hand or using a computer (also by some handheld computers or calculators). We will start off with the basic construction of curves using graph paper to illustrate and learn the fundamental premises behind curve fitting. Laboratory graph paper is typically a sheet composed of horizontal and vertical grid lines. This allows the convenient setting up of *x*- and *y*-axes and the plotting of the different values into generally quadrant I. The values and the scale of the *x*- and *y*-axes are generally chosen for the representation of the data in the most meaningful way.

6.5 REDRAWN GRAPH EXAMPLE

Let us look at a specific example, which will help us to visualize a meaningful construction of the graph. Suppose we have the following set of coordinates: (5.1, 325), (12.8, 369), (22.4, 406), and (27.7, 443). It is clear that the magnitude of the two sets of numbers is quite different with the *y* values of much greater magnitude. If a plot is constructed that contains equal axis distributions, the plot may look something like Figure 6.3, which does not represent the data in a meaningful way. The data are contained within only a small portion of the graph. When preparing a plot of data, the axes do not have to be chosen on the same scale and often are drawn to span most of the graph paper. This was not done in Figure 6.3. The graph has been redrawn in Figure 6.4 where the axes scales have been modified to include a smaller region that more closely spans the data values. The *x*-axis spans a region from

0 to 30 with a grid line at every 5 units while the y-axis spans a region from 300 to 460 with a grid line at 20 unit increments. We see that the axis can be chosen to span regions that more clearly illustrate the data being plotted. In this way as we will see later it is easier to extract more information from the plotted results.

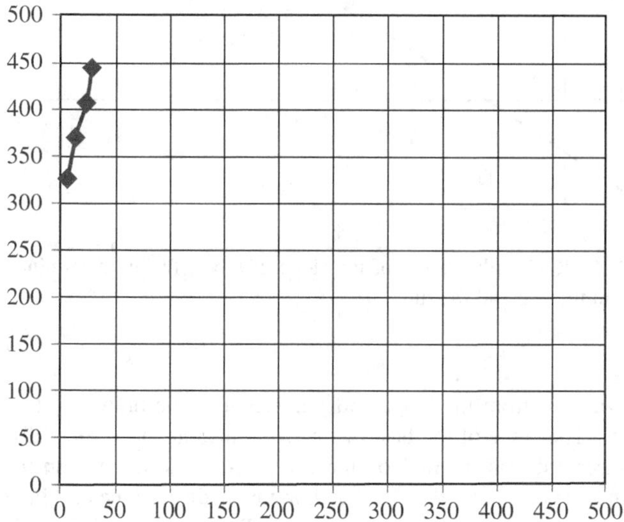

FIGURE 6.3 Plot of the set of coordinates (5.1, 325), (12.8, 369), (22.4, 406), and (27.7, 443) where the formatting of the x-axis and the y-axis has been kept on the same scale from 0 to 500.

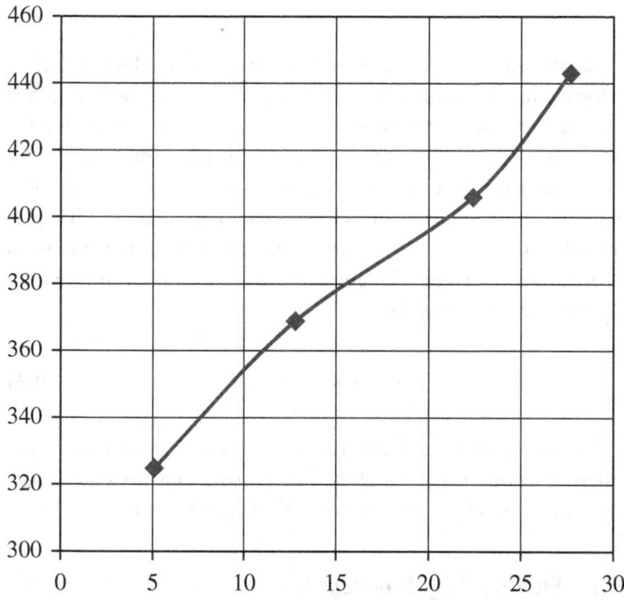

FIGURE 6.4 Plot of the set of coordinates (5.1, 325), (12.8, 369), (22.4, 406), and (27.7, 443) where the formatting of the x-axis and the y-axis has been done to better illustrate the data.

6.6 GRAPHS OF EQUATIONS

6.6.1 Introduction

When data points are plotted, the relationship between them can be used to derive very useful equations in the analytical laboratory. We will use the following example to define a linear relationship between sets of data and their use in setting up a calibration curve.

6.6.2 Copper Sulfate Data

Suppose we have collected the following pairs of data from a spectrophotometric analysis of a series of copper (Cu(II)) standards that we have made (from crushed and dried copper sulfate, $CuSO_4$) in the laboratory to set up a quantitative analysis of a wastewater runoff sample:

6.6.3 Plotting the Data

The data collected on the spectrophotometer listed in Table 6.1 are plotted in Figure 6.5. The data were plotted by choosing x-axis and y-axis scales that would reasonably contain the data. In this case, the x-axis scale ranges from 0 to 4.5, while the y-axis range is from 0 to 0.35. Graph paper is used to construct curves by hand such as the one in this example. Often, when making the axis, different scales will have to be tried until you have one that covers the data ranges in a convenient manner.

6.6.4 Best Fit Line

When constructing and evaluating laboratory data by hand; estimations have to be made. For example, we would like to construct and define a curve from our plotted data in

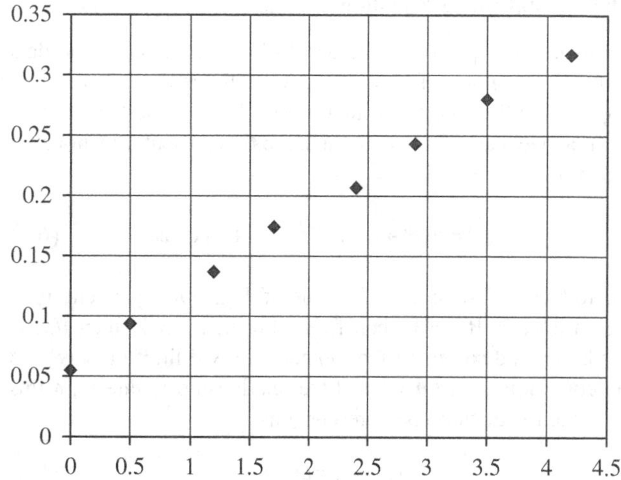

FIGURE 6.5 Plot of the set of concentration values for a series of Cu(II) standards (x-axis) versus their absorbance (arbitrary units, y-axis) collected on a spectrophotometer.

TABLE 6.1 Experimental data.

Cu(II) concentration (ppm) x-axis		0.5	1.2	1.7	2.4	2.9	3.5	4.2
Absorbance (abu) y-axis	0.055	0.094	0.137	0.174	0.207	0.243	0.280	0.317

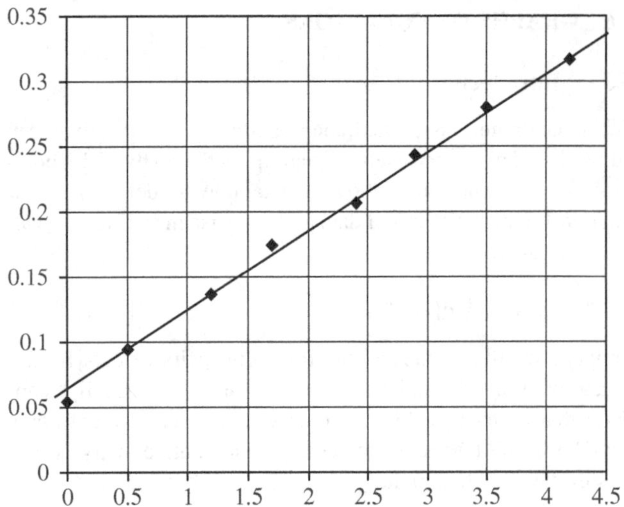

FIGURE 6.6 Plot of the set of concentration values for a series of Cu(II) standards (x-axis) versus their absorbance (arbitrary units, y-axis) collected on a spectrophotometer with a "best fit" line drawn across the data points.

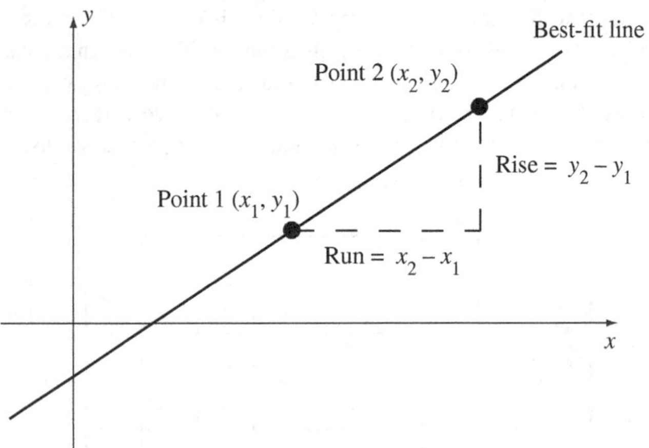

FIGURE 6.7 Illustration of the slope of a best fit line determined from the rise and the run.

Figure 6.5. To do this, we lay a ruler across the data points in a manner where most of the data points are along the ruler. A "best fit" line is then drawn across the data points as depicted in Figure 6.6. Note that not all of the data points touch the line, but our estimation contains most. The best fit line drawn from the experimental data points is next used to extract a curve function that represents the linear relationship between the concentration of the Cu(II) standards and the absorbance measured on the spectrophotometer.

6.6.5 Point-Slope Equation of a Line

To obtain the equation of the best fit line drawn across the data points in the graph of Figure 6.6, we will need to calculate the slope of the line and the y-intercept of the line. Best fit lines as the one we have drawn in Figure 6.6 represent functions of the form

$$\mathrm{f}(x) = \mathrm{a}_n x^n + \mathrm{a}_{n-1} x^{n-1} + \cdots + \mathrm{a}_1 x + \mathrm{a}_0, \tag{6.1}$$

where f(x) is a polynomial function of degree n dependent upon the variable x. If $n = 1$, then f(x) is linear; if $n = 2$, then f(x) is quadratic; and so on. At the moment we will limit ourselves to the condition where $n = 1$ and the relationship is linear. In this case, the relationship is represented as

$$\mathrm{f}(x) = y = \mathrm{a}x + \mathrm{a}_0, \tag{6.2}$$

where y is the dependent variable, x is the independent variable, and a_0 is a constant. The function illustrated in Equation 6.2 is known as the Point-Slope Equation of a line.

6.6.6 Finding the Slope (m)

When we have a curve line as that depicted in Figure 6.6 constructed from experimental data, we can derive the linear function $y = \mathrm{a}x + \mathrm{a}_0$ from the slope and y-intercept of the drawn best fit line. The slope of the best fit line is defined as the vertical displacement from point 1 on the curve (x_1, y_1) to point 2 on the curve (x_2, y_2) divided by the horizontal displacement. Many students learn this concept as "rise divided by run." This is depicted in Figure 6.7 for a best fit drawn line. The equation for the slope (m) of the curve is

$$\text{slope, } m = \frac{y_2 - y_1}{x_2 - x_1} \tag{6.3}$$

Slopes can be either negative or positive but nonzero. A horizontal line contains a zero slope, and a vertical line has an undefined slope. Both of these instances are not useful for our discussion here because we are looking for relationships in the laboratory where a change in the x value corresponds to a change in the y value. If the slope is positive, then the line goes upward to the right. If the slope is negative, then the line goes downward to the right. The general equation for a linear line with slope m is represented as

$$y - y_1 = m(x - x_1) \tag{6.4}$$

and is known as the point-slope form of the equation of a line. If there is a point (x_1, y_1) that the line passes, then any other point will also be included on the line only if $(y - y_1)/(x - x_1) = m$.

6.6.7 Finding the y-Intercept (b)

The special form of Equation 6.4 is derived by taking the point that the line intercepts the y-axis (point (0, b) where b is called the y-intercept) and substituting this point into Equation 6.5 as

$$y - b = m(x - 0) \tag{6.5}$$

$$y = mx + b \tag{6.6}$$

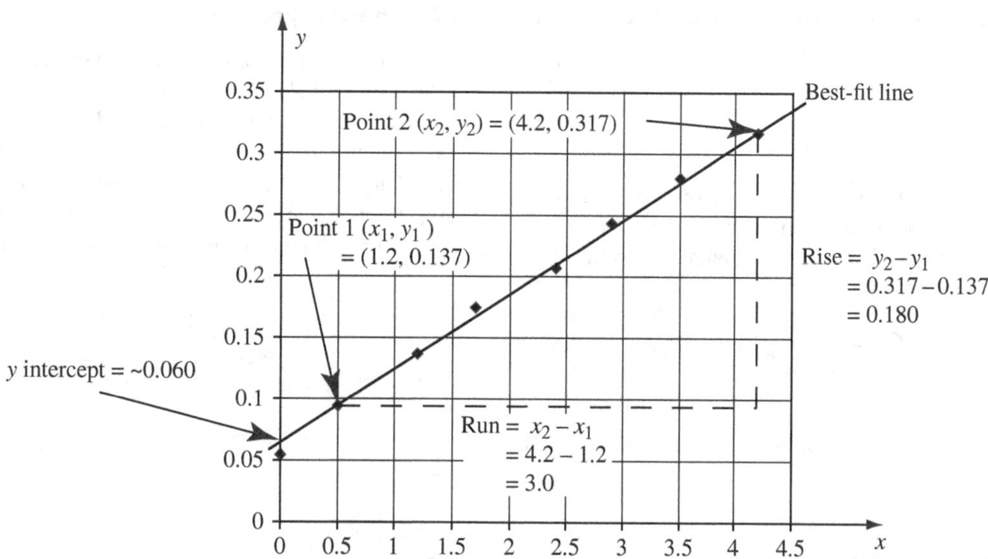

FIGURE 6.8 Plot of the set of concentration values for a series of Cu(II) standards (x-axis) versus their absorbance (arbitrary units, y-axis) collected on a spectrophotometer with a "best fit" line drawn across the data points. The slope and y-intercept can be estimated by hand by choosing two points on the best fit curve and extending the curve through the y-axis.

6.6.8 Solving for x

This equation is called the slope-intercept form of the linear line equation and is the most useful form for the analytical laboratory technician. As written in Equation 6.6, y is the dependent variable and x is the independent variable. Often when we set up a calibration graph, the x-axis is chosen as the variable that we want to calculate for an unknown using the parameter collected in the laboratory whose values are plotted on the y-axis. Thus, an often useful form of the slope-intercept equation is to set x as the dependent variable and y as the independent variable:

$$x = \frac{y}{m} - \frac{b}{m} = \frac{(y-b)}{m} \tag{6.7}$$

As we will see, in this form the analyst can collect the independent parameter y in the laboratory and calculate the dependent parameter x from the equation. This is after a general form of the equation is found from a curve generated in the laboratory.

A summary of the different forms of linear equations is as follows:

Standard Form:	$Ax + By = C$
Horizontal Line:	$y = \text{k}$, k is a constant
Vertical Line:	$x = \text{k}$, k is a constant
Slope-Intercept Form:	$y = mx + b$
Point-Slope Form:	$y - y1 = m(x - x1)$

6.6.9 Estimating the Slope and Intercept

If we look back at our plot in Figure 6.6 of the set of concentration values for a series of Cu(II) standards (x-axis) versus their absorbance (arbitrary units, y-axis) collected on a spectrophotometer with a "best fit" line drawn across the data points. The slope and y-intercept can be estimated by hand by choosing two points on the best fit curve, and extending the curve through the y-axis. This is depicted in Figure 6.8 where the important steps have been labeled on the graph. In the graph, the two points chosen are (1.2, 0.137) for point 1 (x_1, y_1), and (4.2, 0.317) for point 2 (x_2, y_2). The rise is calculated as $y_2 - y_1$ at 0.180, and the run is calculated as $x_2 - x_1$ at 3.0. The slope (m) equates to the ratio of the rise over the run at $0.180/3.0 = 0.06$. The y-intercept (b) is estimated from the best fit line that has been extended through the y-axis at approximately 0.060.

6.6.10 Deriving the Equation from the Slope and Intercept

From the slope and the y-intercept, we are able to derive a best fit equation to predict other unknown concentrations from the curve:

$$y = mx + b \tag{6.6}$$

$$y = 0.06x + 0.060 \tag{6.8}$$

Let us test the equation from the known data in Table 6.1 where a concentration of 2.4 ppm has an absorbance of 0.207 units. If we plug the concentration value of 2.4 ppm for the independent x parameter, we can estimate the associated absorbance at:

$$y = 0.06(2.4) + 0.06 \tag{6.9}$$

$$y = 0.204 \tag{6.10}$$

An absorbance of 0.204 for the standard with concentration of 2.4 ppm is quite close to the known value of 0.207 absorbance units. As presented earlier the point-slope equation can be written as

$$x = \frac{y}{m} - \frac{b}{m} \quad (6.7)$$

which allows the estimation of the concentration of the standard from the absorbance value. Again, using the absorbance of 0.207 for the 2.4 ppm Cu(II) standard, we can estimate the concentration from the best fit curve:

$$x = \frac{0.207}{0.06} - \frac{0.06}{0.06} \quad (6.11)$$

$$x = 2.45\text{ ppm} \quad (6.12)$$

6.7 LEAST-SQUARES METHOD

6.7.1 Plotting Data with Scatter

The error associated with estimating the best fit curve from a set of plotted values by hand as was introduced in the previous section tends to be very subjective according to the technician performing the plotting. In the example of the curve fit data for the Cu(II) concentration calibration spectrophotometric analysis of standards the plot in Figure 6.8 is quite linear and the drawing of the best fit line by hand is not too difficult. There are times when this is not the case and the technician must use a substantial amount of judgment in drawing the best fit line. The plot of a set of experimental data obtained in the laboratory that illustrates this type of situation is depicted in Figure 6.9. The plot in Figure 6.9a depicts the data without a line connecting the points. We can already see that the points do not lie in a straight line. In Figure 6.9b, the data points have been connected with a line but the curve is much too complicated. Finally, in Figure 6.9c a best fit line has been drawn through the data points for a linear curve that more accurately represents the overall trend of the data being plotted.

6.7.2 Linear Regression

To solve this problem, a more mathematical approach is usually employed that makes use of the linear regression approach to derive the best fit line from a calculated slope and y-intercept that uses all of the data pairs in the construction of the calibration curve. The least-squares curve is generated from the line that passes through the centroid of the points $(\bar{X}, \bar{Y})$ according to the following slope (m) and y-intercept (b) equations:

$$m = \frac{\sum_i (X_i - \bar{X})(Y_i - \bar{Y})}{\sum_i (X_i - \bar{X})^2} \quad (6.13)$$

$$b = \bar{Y} - m\bar{X} \quad (6.14)$$

6.7.3 Curve Fitting the Data

Let us take the results we have collected for the following pairs of data from a spectrophotometric analysis of a series of copper (Cu(II)) standards that we have made (from crushed and dried copper sulfate, $CuSO_4$) in the laboratory to calculate the best fit curve using the linear regression formulas of Equations 6.13 and 6.14 (Table 6.2).

TABLE 6.2 Experimental data.

X_i	Y_i	$X_i - \bar{X}$	$(X_i - \bar{X})^2$	$Y_i - \bar{Y}$	$(X_i - \bar{X})(Y_i - \bar{Y})$
0.50	0.094	−1.84	3.38	−0.113	0.208
1.20	0.137	−1.14	1.30	−0.070	0.0798
1.70	0.174	−0.64	0.41	−0.033	0.0211
2.40	0.207	0.06	0.00	0.000	0.000
2.90	0.243	0.56	0.31	0.036	0.0202
3.50	0.280	1.16	1.34	0.073	0.0847
4.20	0.317	1.86	3.46	0.110	0.205
$\bar{X} = 2.34$	$\bar{Y} = 0.207$		$\Sigma = 10.2$		$\Sigma = 0.619$

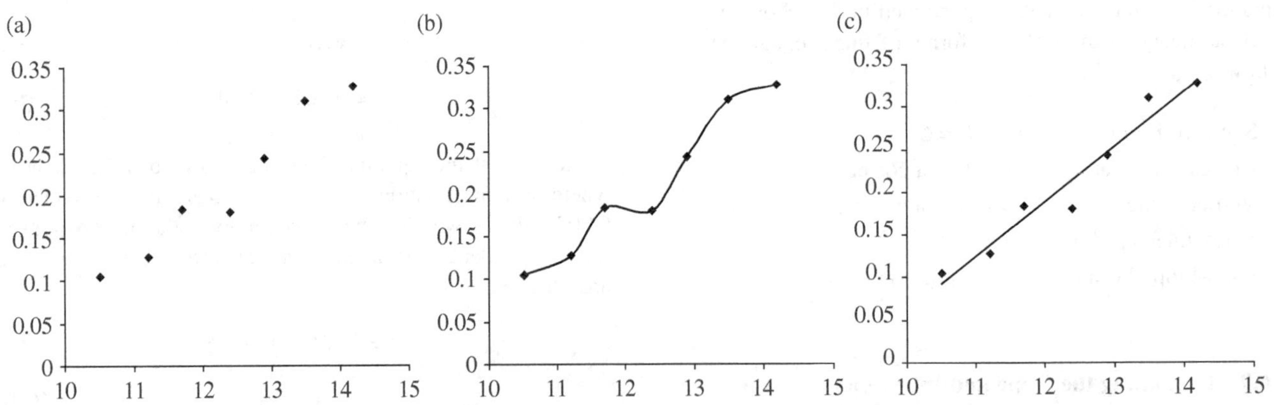

FIGURE 6.9 Plot of experimental data obtained in the laboratory illustrating chart complexity. (a) Data without a line connecting the points. (b) Data points have been connected with a line but the curve is much too complicated. (c) A best fit line drawn through the data points for a linear curve more accurately representing the overall trend of the data being plotted.

$$m=\frac{\sum_i (X_i-\bar{X})(Y_i-\bar{Y})}{\sum_i (X_i-\bar{X})^2} \tag{6.13}$$

$$m=\frac{0.619}{10.2} \tag{6.15}$$

$$m=0.0607 \tag{6.16}$$

$$b=\bar{Y}-m\bar{X} \tag{6.14}$$

$$b=0.207-(0.0607)(2.34) \tag{6.17}$$

$$b=0.0650 \tag{6.18}$$

From our calculated slope of 0.0607 and y-intercept of 0.0650 we are now able to set up a general point-slope equation from the curve data:

$$y=mx+b \tag{6.6}$$

$$y=0.0607x+0.0650 \tag{6.19}$$

This is very similar to the results that we obtained for the hand fit best curve that was generated in Figure 6.8, as is expected.

6.8 COMPUTER-GENERATED CURVES

Typically the analytical laboratory that the chemist or technician will be working in will have computer programs where the analyst may input the experimental data and generate best fit curves and also perform regression analyses for curve fitting. We will now look at the use of some of these programs and how they greatly simplify the tasks of plotting curves and obtaining best fit curves. Your ChemTech program contains a page that will allow us to input data and generate plots of the data along with regression analysis for a best fit curve.

6.8.1 Using ChemTech to Plot Data

ChemTech can be used in your laboratory at any time to help simplify the plotting of data and best fit line curve fitting of data that have been gathered in the laboratory. If not already open, open ChemTech and go to the Chapter 6 Main Menu page. Click the button "PLOT DATA" to open a data graphing and regression calculation page, as depicted in Figure 6.10. The graphing interface page that is now showing is depicted in Figure 6.11. Let us spend a moment to become familiar with the various parts of the page. In the upper left-hand corner are the x–y data input boxes. These will be used to input the x and y data points in which we want to plot. Below the x and y data point input boxes is a button labeled "ENTER X–Y DATA." This will be clicked each time a set of x and y values are being entered. Below this button is the "PLOT X–Y DATA" button, which we will use to plot the data after all of the data points have been entered. To the right is a blank box where the plotted data as a graph is displayed. Let us begin by entering a small set of data and displaying a graph.

6.8.2 Entering the Data

Using the data from Table 6.3, enter each concentration value for the "x" data and each associated absorbance for the "y" data. The

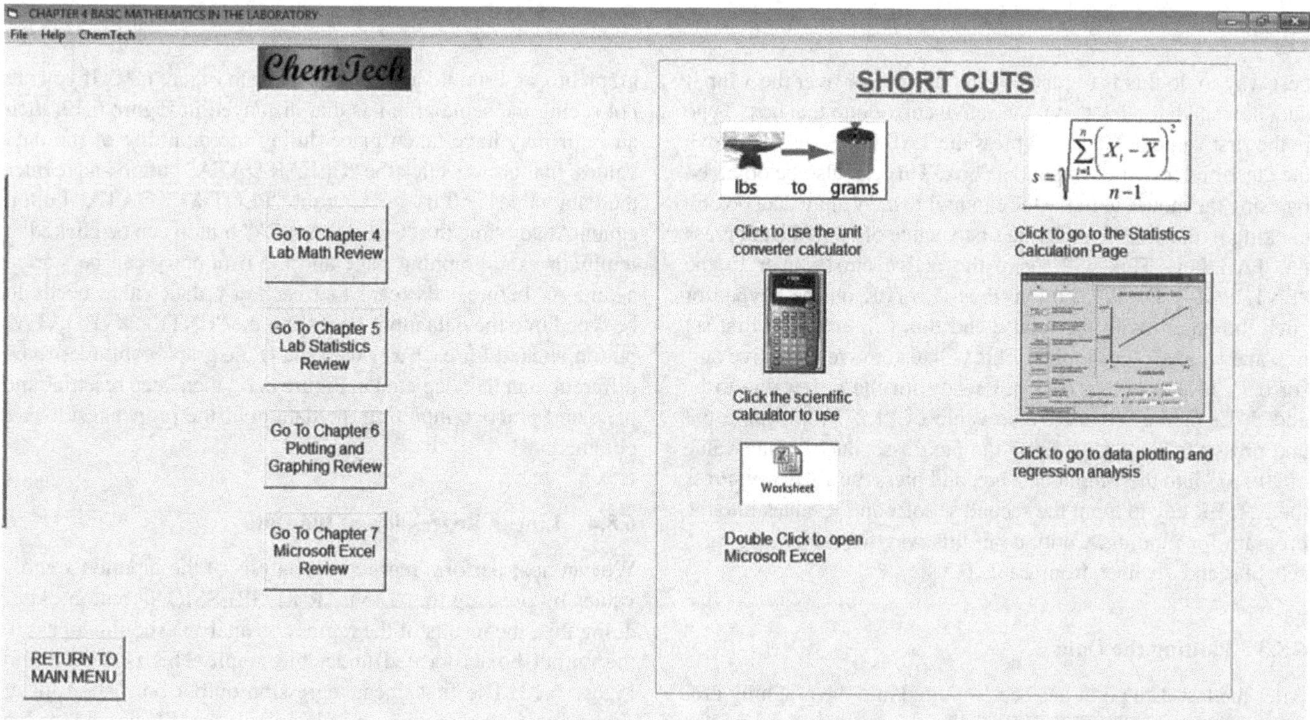

FIGURE 6.10 ChemTech Chapter 6 Main Menu page. Click the button "PLOT DATA" to open a data graphing and regression calculation page.

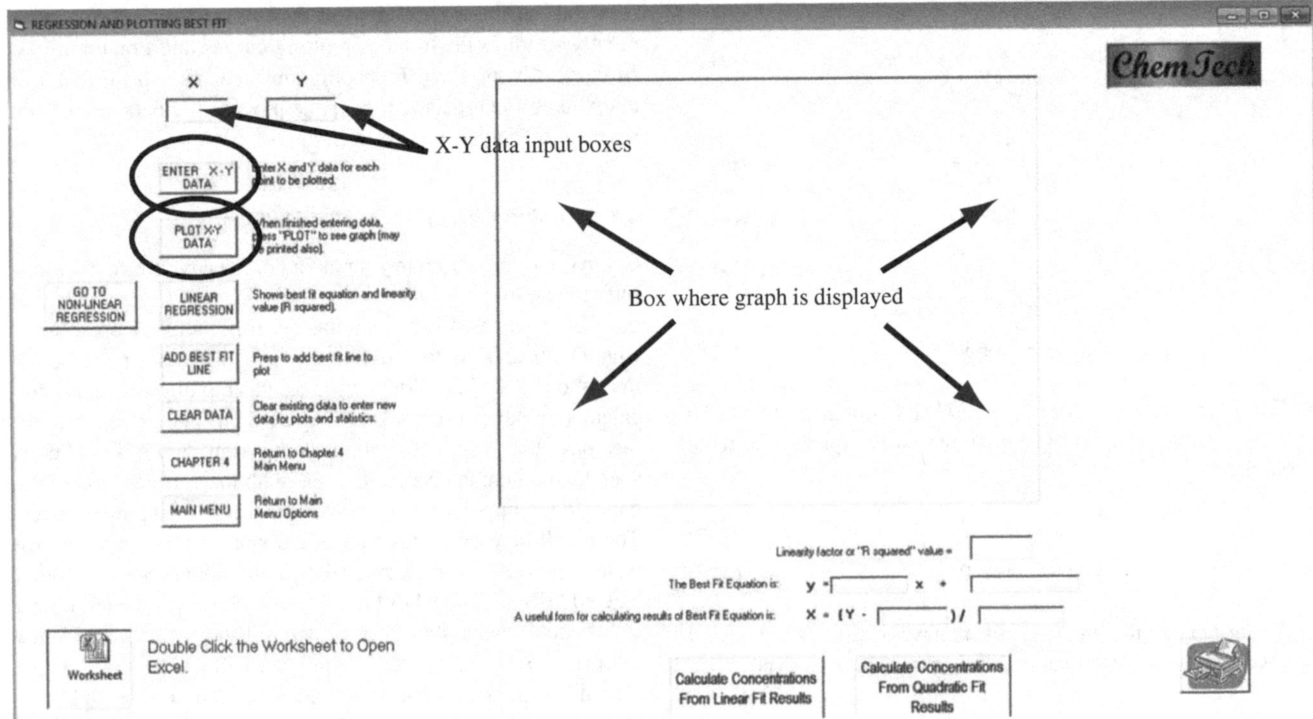

FIGURE 6.11 Data plotting page. The data are entered into the *X* and *Y* input boxes for plotting. After each associated *x* and *y* data points are entered, the "ENTER *X*–*Y* DATA" button is clicked.

TABLE 6.3 Experimental data.

Cu(II) concentration (ppm) *x*-axis	0.5	1.2	1.7	2.4	2.9	3.5	4.2
Absorbance (abu) *y*-axis	0.094	0.137	0.174	0.207	0.243	0.280	0.317

best way to do this is to place the mouse cursor over the *x* input data box and left click to put the active cursor into that box. Type in the first value of "0.5" and press the TAB key. This will move the cursor over to the *y* input data box. This can also be obtained by using the mouse to move the cursor into the *y* input data box by clicking in the box. Type in the first *y* value of "0.094" and press the TAB key. This will move the active cursor over to the "ENTER *X*–*Y* DATA" button. Press ENTER on the keypad or click the button with the mouse and this will enter the first set of *x* and *y* values for plotting. This will also move the active cursor back to the *x* input data box ready for the next value to be added. Enter the second *x* data value of "1.2" from Table 6.3 and press the TAB key. Enter the next associated *y* data value of "0.137" into the *y* input data box and press the TAB key. Press the ENTER key to input the second set of *x* and *y* values into the program for plotting. Continue in this way for the remaining 5 sets of *x* and *y* values from Table 6.3.

6.8.3 Plotting the Data

After the last data point has been inputted into the graphing program, press the "PLOT *X*–*Y* DATA" button. After pressing the button the results of the plotting will be displayed in the form's graph box and should look like the page in Figure 6.12. If you are not seeing the same graph as that displayed in Figure 6.12, then an error may have taken place during the inputting of the data values. In this case, click the "CLEAR DATA" button and reenter the data values and try clicking the "PLOT *X*–*Y* DATA" button again. At any time the "CLEAR DATA" button can be clicked to reinitialize the graphing page and the data entry can be started again. As before, take care that each *x*–*y* data value needs to be typed into the data input boxes and the "ENTER *X*–*Y* DATA" button clicked for each *x*–*y* data set. If the graph continues to be different than that depicted in Figure 6.12, then keep resetting the page and practice inputting the data until the proper results are obtained.

6.8.4 Linear Regression of the Data

We can next perform regression analysis of the inputted *x* and *y* values by pressing the "LINEAR REGRESSION" button. After doing this, the results of the regression analysis should appear in the output boxes located under the graph. This is depicted in Figure 6.13. The first linear regression output box is the linear regression factor or commonly known as the "*R* squared" value that indicates the linearity of the data. A value of $R^2 = 1$ would

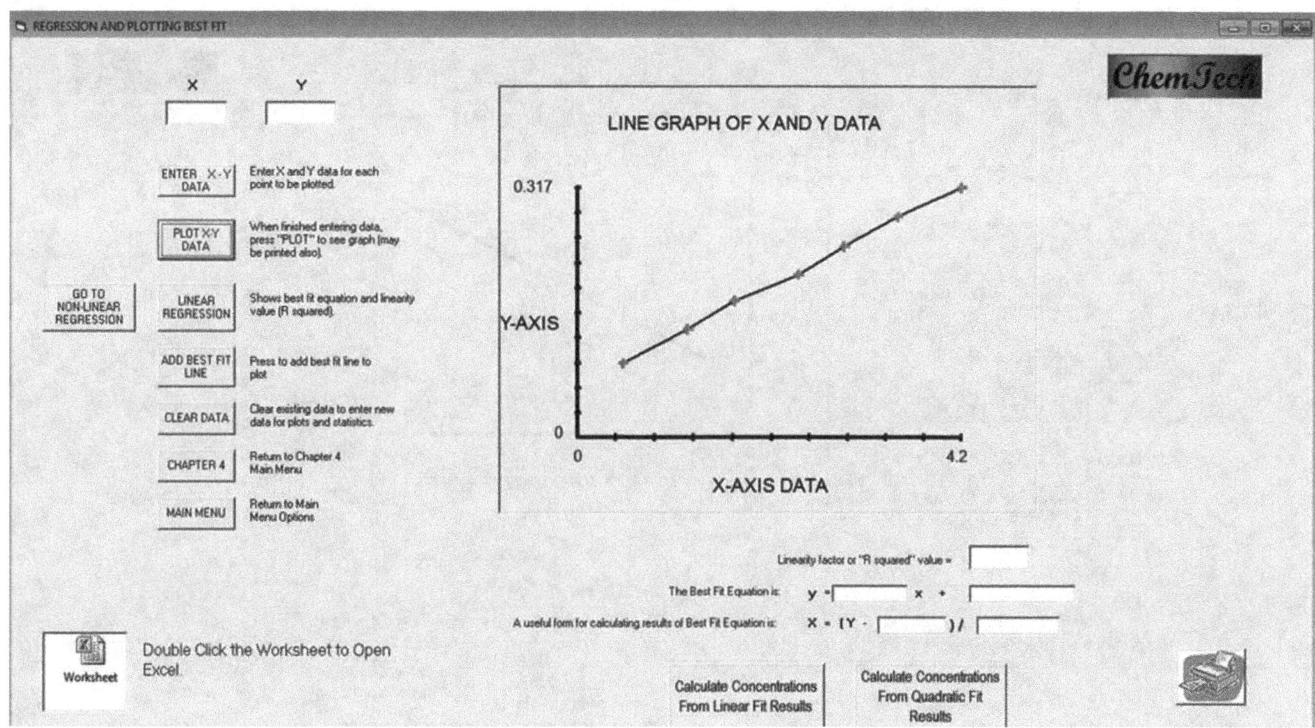

FIGURE 6.12 Result of clicking the "PLOT *X–Y* DATA" button from the data inputted from Table 6.3. The line graph should look like that illustrated in the page.

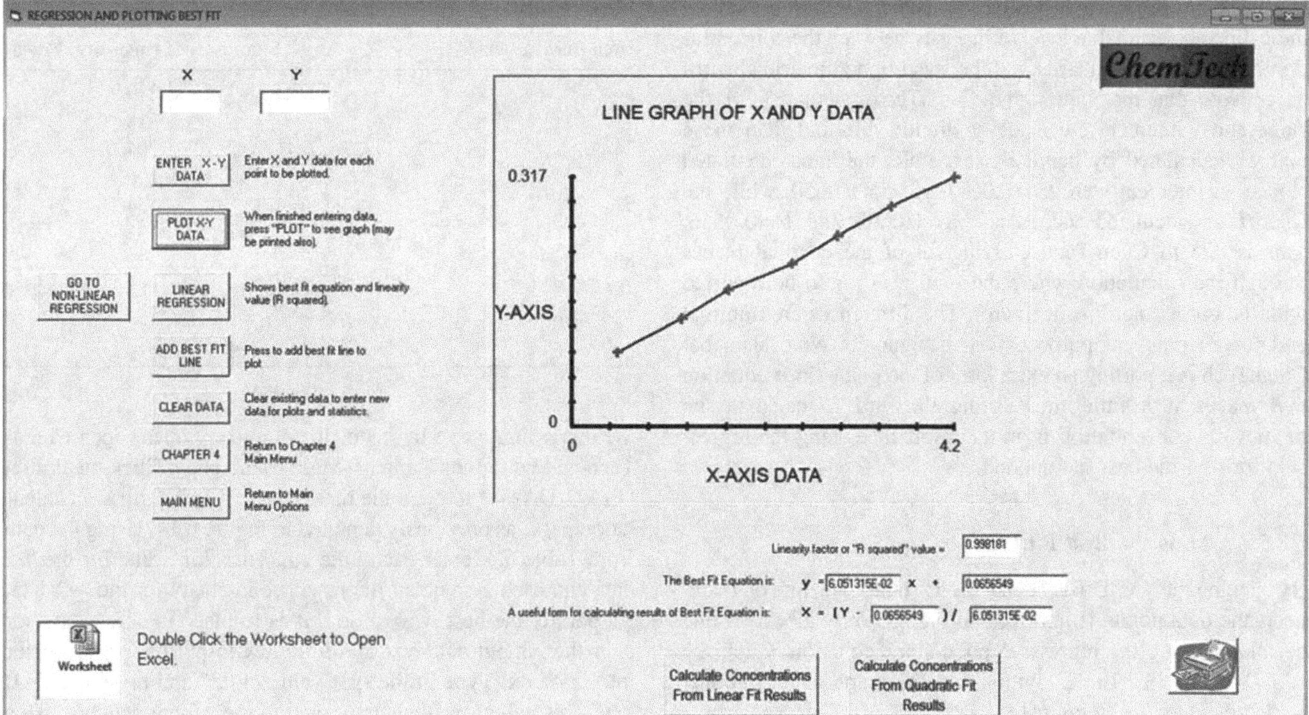

FIGURE 6.13 Result of clicking the "LINEAR REGRESSION" button from the data inputted from Table 6.3. The regression results should match those listed in the output boxes under the graph.

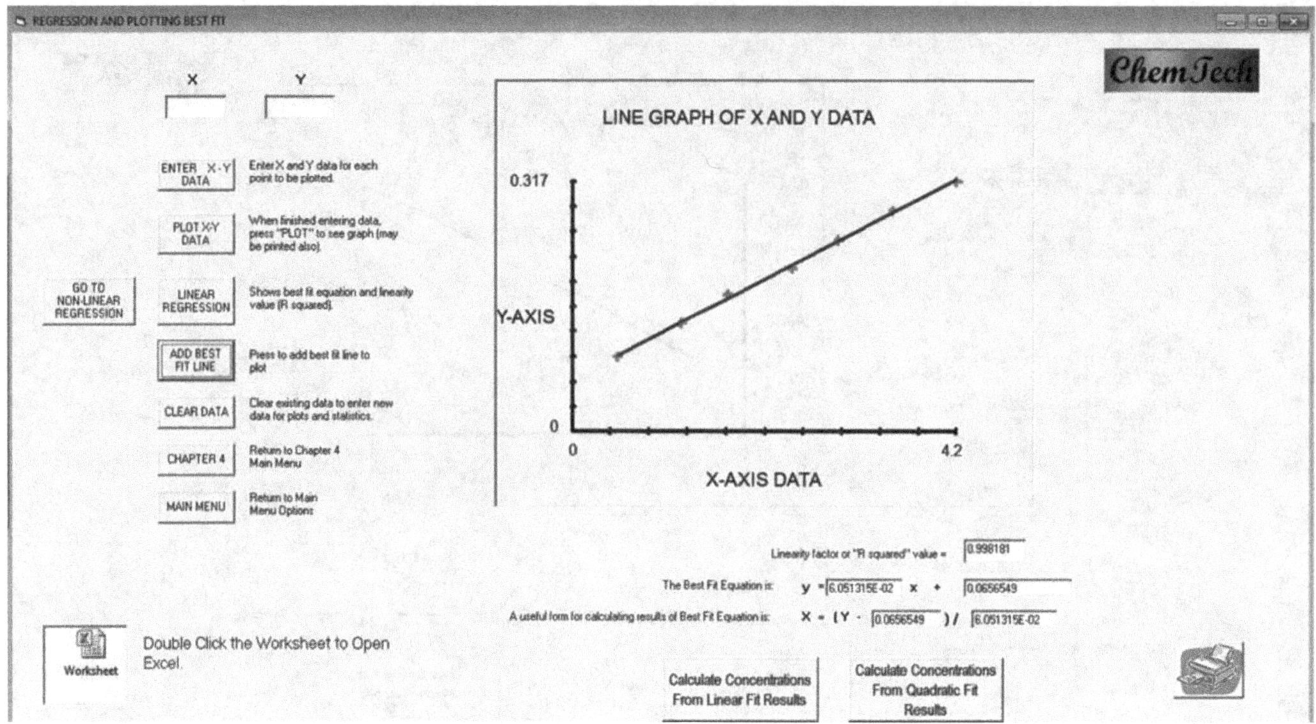

FIGURE 6.14 Result of clicking the "ADD BEST FIT LINE" button from the data inputted from Table 6.3. The line connecting each data point should be replaced with a "straight" best fit line that passes through the data points.

indicate a perfectly linear relationship between the x and y values. For the data listed in Table 6.3, an R^2 value of 0.998 was obtained indicating an acceptable level of linearity between the concentration of the Cu(II) standards and the spectrophotometric absorbance. Note that the values that ChemTech calculated for the slope and y-intercept are actually slightly different than those that we calculated by hand. For example, the hand-calculated slope and intercept were $m = 0.0607$ and $b = 0.0650$, while the ChemTech-calculated values are $m = 0.0605$ and $b = 0.0656$. This is due to ChemTech carrying all of the decimal places through the calculation where the final result is to be reported with the proper significant figures. The differences are minimal and for all practical purposes can be ignored. Note also that ChemTech is reporting an extra form of the point-slope equation that makes it a little more straightforward to calculate an unknown's concentration from its absorbance using the regression form of the best fit equation.

6.8.5 Adding the Best Fit Line

By clicking the "ADD BEST FIT LINE" button the line that connects the data points is removed and replaced by a best fit line calculated using the regression point-slope equation. Click the "ADD BEST FIT LINE" button and the graph should appear as that depicted in Figure 6.14.

6.8.6 Entering a Large Set of Data

Now let us look at an example of a set of x–y data points ($n = 10$) that would be a very time-consuming task to plot and calculate the regression curve by hand. If not already open, open ChemTech and go to the Chapter 6 Main Menu page. Click the button "PLOT DATA" to open the data graphing and regression calculation page, as previously depicted in Figure 6.14. Using the data from Table 6.4, enter each urea concentration value for the "x" data and each associated fluorescence intensity for the "y" data. As before, the best way to do this is to place the mouse cursor over the x input data box and left click to put the active cursor into that box. Type in the first value of "5" and press the TAB key. This will move the cursor over to the y input data box. This can also be obtained by using the mouse to move the cursor into the y input data box by clicking in the box. Type in the first y value of "229" and press the TAB key. This will move the active cursor over to the "ENTER X–Y DATA" button. Press ENTER on the keypad or click the button with the mouse and this will

TABLE 6.4 Linear Set of Data Collected of Concentration Versus Fluorescence.

Urea (mM, x data)	Fluorescence (Intensity, y data)
5	229
11	218
24	204
36	196
52	154
77	133
89	99
106	85
124	52
156	33

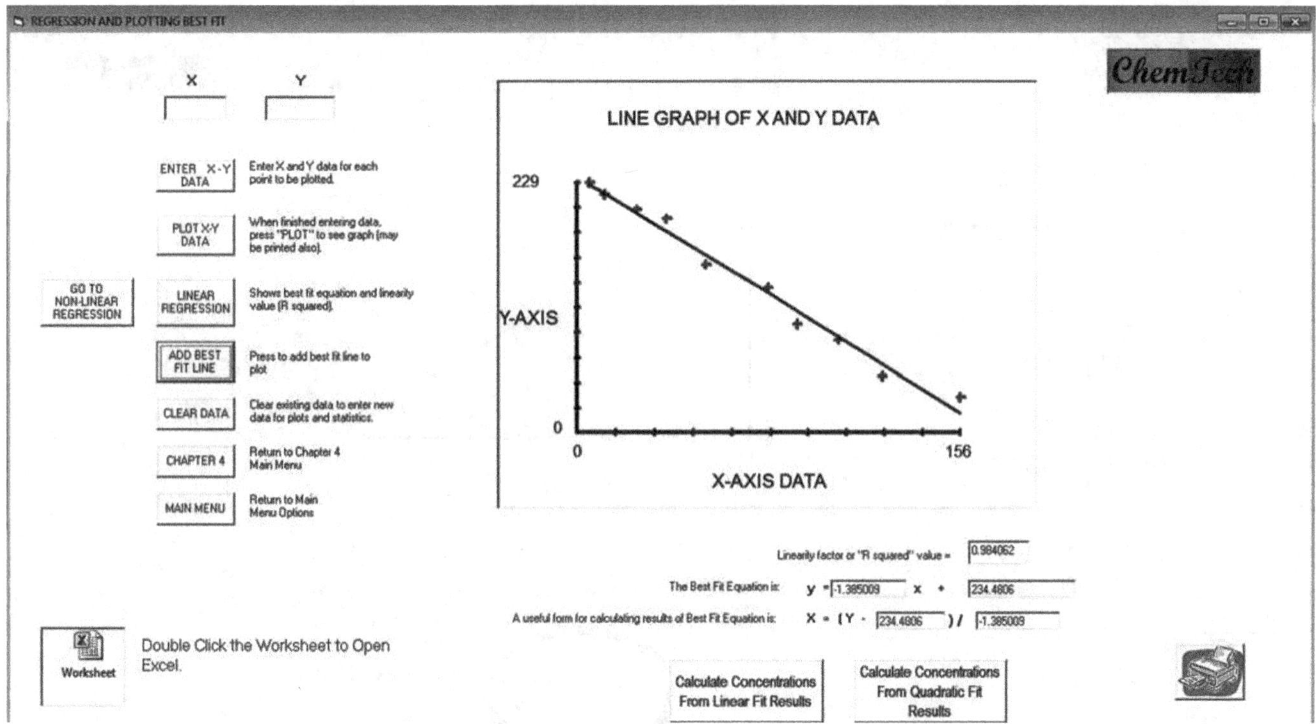

FIGURE 6.15 Final graphing and linear regression results from data entered from Table 6.4. Note the curve possesses a negative slope indicating an inverse relationship.

enter the first set of x and y values for plotting. This will also move the active cursor back to the x input data box ready for the next value to be added. Enter the second x data value of "11" from Table 6.4 and press the TAB key. Enter the next associated y data value of "218" into the y input data box and press the TAB key. Press the ENTER key to input the second set of x and y values into the program for plotting. Continue in this way for the remaining 8 sets of x and y values from Table 6.4. After the last data point has been inputted into the graphing program, press the "PLOT *X–Y* DATA" button. Next, perform regression analysis of the inputted x and y values by pressing the "LINEAR REGRESSION" button. Finally, click the "ADD BEST FIT LINE" button. The graph should appear as that depicted in Figure 6.15. For the data listed in Table 6.4, an R^2 value of 0.984 was obtained. The ChemTech calculated slope and y-intercept should be $m = -1.385$ and $b = 234.48$. Note that this particular relationship is what is known as an "inverse relationship" where the slope is negative. If you are not seeing the same graph as that displayed in Figure 6.15, then an error may have taken place during the inputting of the data values. In this case, click the "CLEAR DATA" button and reenter the data values and try clicking the "PLOT *X–Y* DATA" button again. At any time the "CLEAR DATA" button can be clicked to reinitialize the graphing page and the data entry can be started again. As before, take care that each x–y data value needs to be typed into the data input boxes and the "ENTER *X–Y* DATA" button clicked for each x–y data set. If the graph continues to be different than that depicted in Figure 6.15, then keep resetting the page and practice inputting the data until the proper results are obtained.

6.9 CALCULATING CONCENTRATIONS

The next step after curve fitting is to calculate the concentration of a sample from the measured value using the curve fit equation. In Table 6.4, the fluorescence values were plotted against the urea concentration in millimeter. The fitted curve was found to be $y = -1.385x + 234.48$. At the bottom of the page, depicted in Figure 6.15 is a button labeled Calculate Concentration. This is depicted in Figure 6.16. In ChemTech press the button to bring up a page as shown in Figure 6.17.

The ChemTech program uses the most recent curve that was fitted on the previous page to calculate a concentration. Type the value of 135 into the input box with the label "y =" and press the Calculate Concentration button. The page should display a result of 71.83 for the curve fit concentration "x =" as depicted in Figure 6.18.

6.10 NONLINEAR CURVE FITTING

Often in the analytical laboratory the chemist will be confronted with a concentration versus measurement system that deviates from Beer's law. Later in Chapter 13 we will take an in-depth look at Beer's law which states that the absorbance of a substance is proportional to the concentration of the substance, the path length that the absorbing light travels through the sample, and the probability that absorbance takes place with an individual part of the substance, such as each molecule or atom. Systems will deviate from Beer's law and produce a nonlinear relationship between the concentration and the measurement system such as a UV/Vis

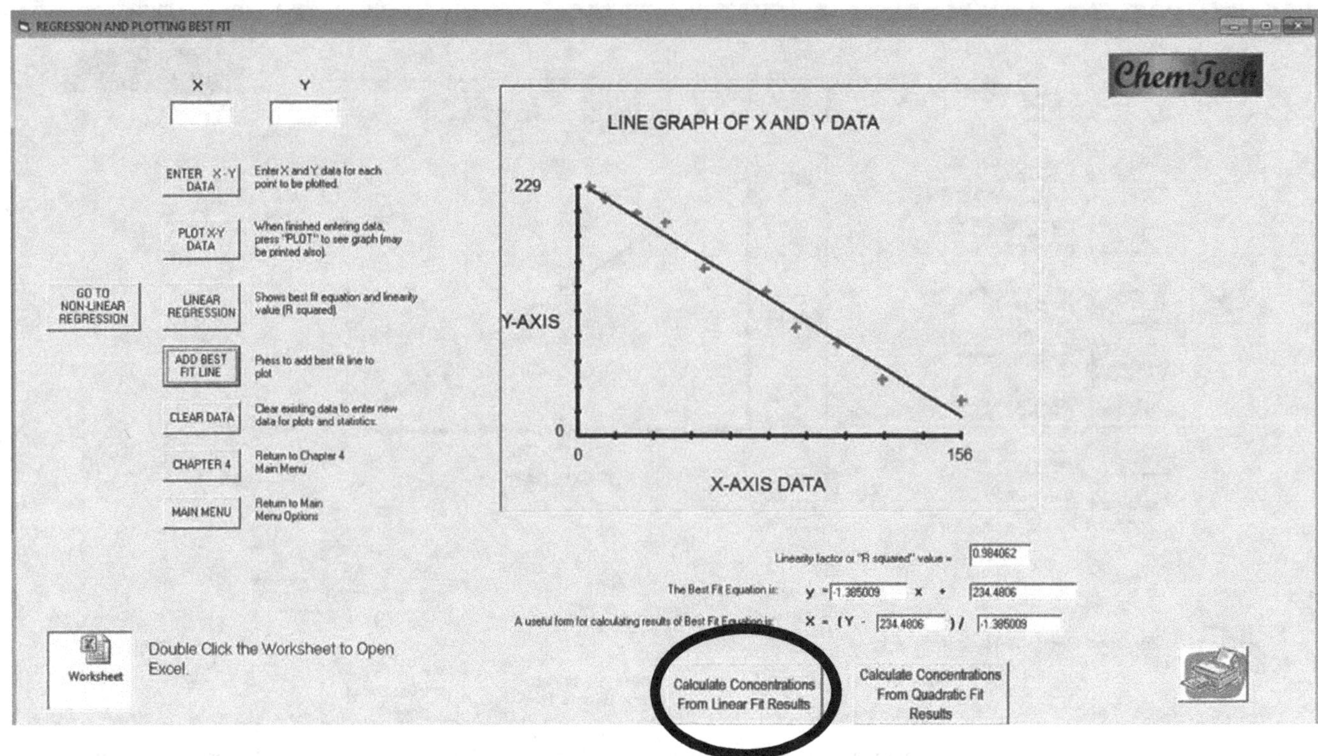

FIGURE 6.16 ChemTech page illustrating link button used to open page for calculating curve fit concentrations.

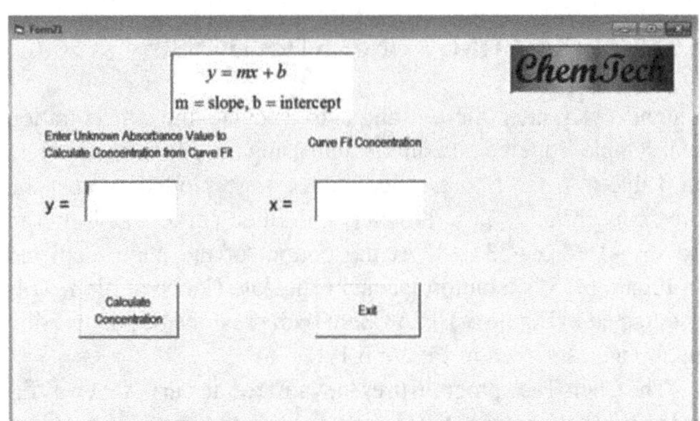

FIGURE 6.17 ChemTech page used to calculate curve fit concentrations.

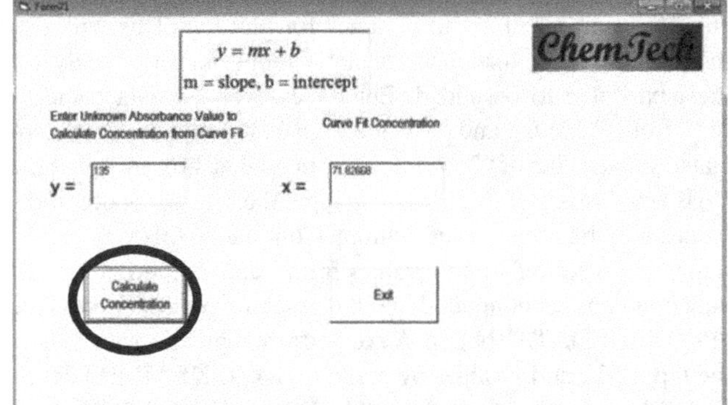

FIGURE 6.18 Using ChemTech to calculate a concentration from a curve fit.

detector or a fluorescence detector. When a relationship deviates from Beer's law it is usually due to the influence of another factor besides the concentration and response. This is often seen with detectors that are impact multiplier detectors, but can be observed in just about any system used.

The curve fitting used for nonlinear systems is more complicated than that used for linear systems. We just covered the use of a calculated slope (m) and y-intercept (b) to set up a general point-slope equation from curve data as $y = mx + b$ for a linear system. For nonlinear curve fitting, matrix notation is used and n simultaneous linear equations are used. Typically, computer programs are used for nonlinear curve fitting such as Microsoft Excel® or for more complex work a program such as OriginPro may be used.

If we return to our main graphing page in ChemTech, a link button to do nonlinear curve fitting is located on the left as depicted in Figure 6.19. Click on the link to open the page as depicted in Figure 6.20.

In this page, enter the data that are listed in Table 6.5. Mouse click into the x input box to activate that box ready to accept data. Begin entering the data that is listed in Table 6.5 by typing in the x value, hit the tab key, type in the associated y value, hit the tab key and then press enter. This will enter into the page the set of x and y values, and then bring the cursor back to the x data box. Go ahead and enter the rest of the data in Table 6.5 in the same manner. When completed, click on the Graph the Data button, as depicted in Figure 6.21. A graph should appear in the page the same as that

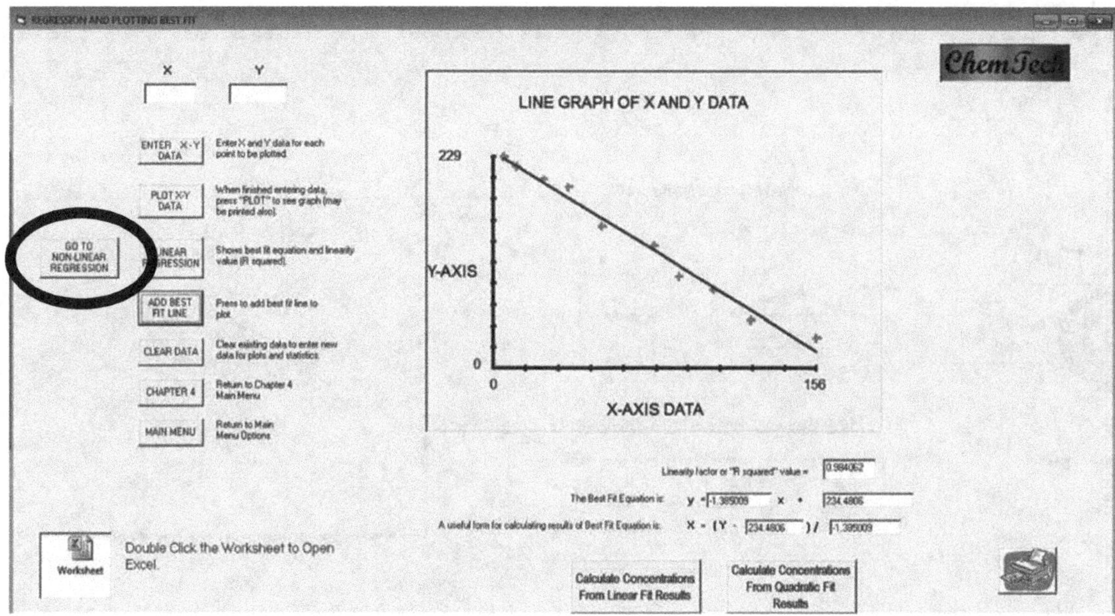

FIGURE 6.19 Link for opening the nonlinear curve fitting page.

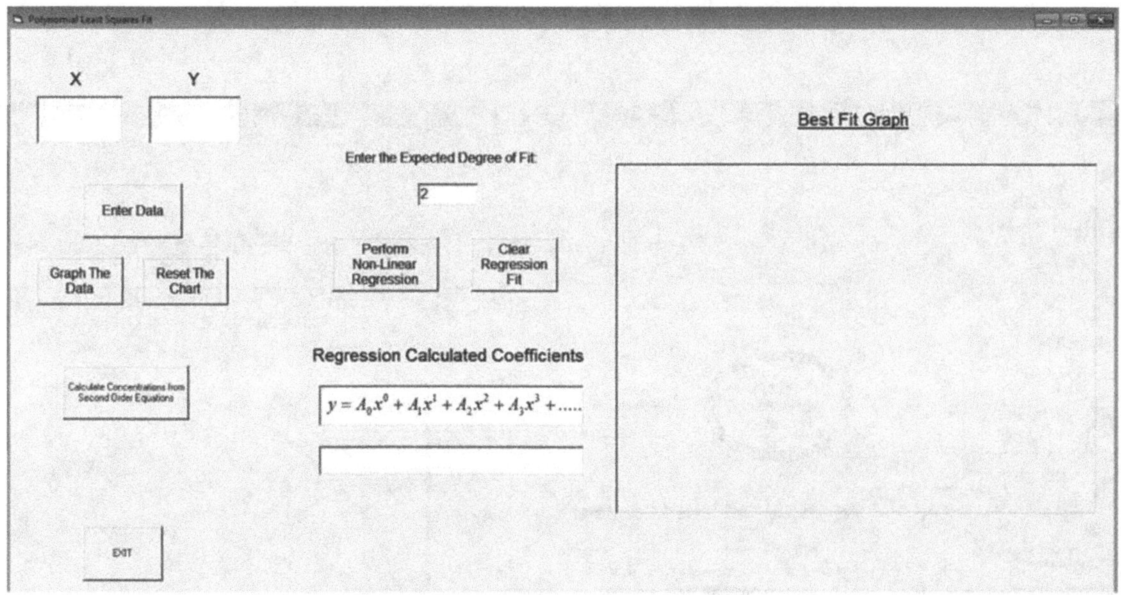

FIGURE 6.20 Nonlinear curve fitting page.

TABLE 6.5 Nonlinear Set of Data Collected of Concentration Versus Fluorescence.

Urea (mM, x data)	Fluorescence (Intensity, y data)
2	11
5	40
8	57
11	69
13	75
17	84
23	90
35	93

shown in Figure 6.21. As you can see, the plot of the data is not a straight line, but rather is curved. This is the plot of a nonlinear relationship.

Next, we want to get the coefficients of the second-order equation that best fits this line. This is nonlinear least-squares regression analysis and is somewhat complicated. To get the coefficients, press the Perform Non-Linear Regression button. The coefficients will be displayed in the output box beneath the example of higher order equations, as depicted in Figure 6.22. The coefficients, and subsequent second-order equation calculated is $-0.1522x^2 + 7.731x + 0.8686 = y$.

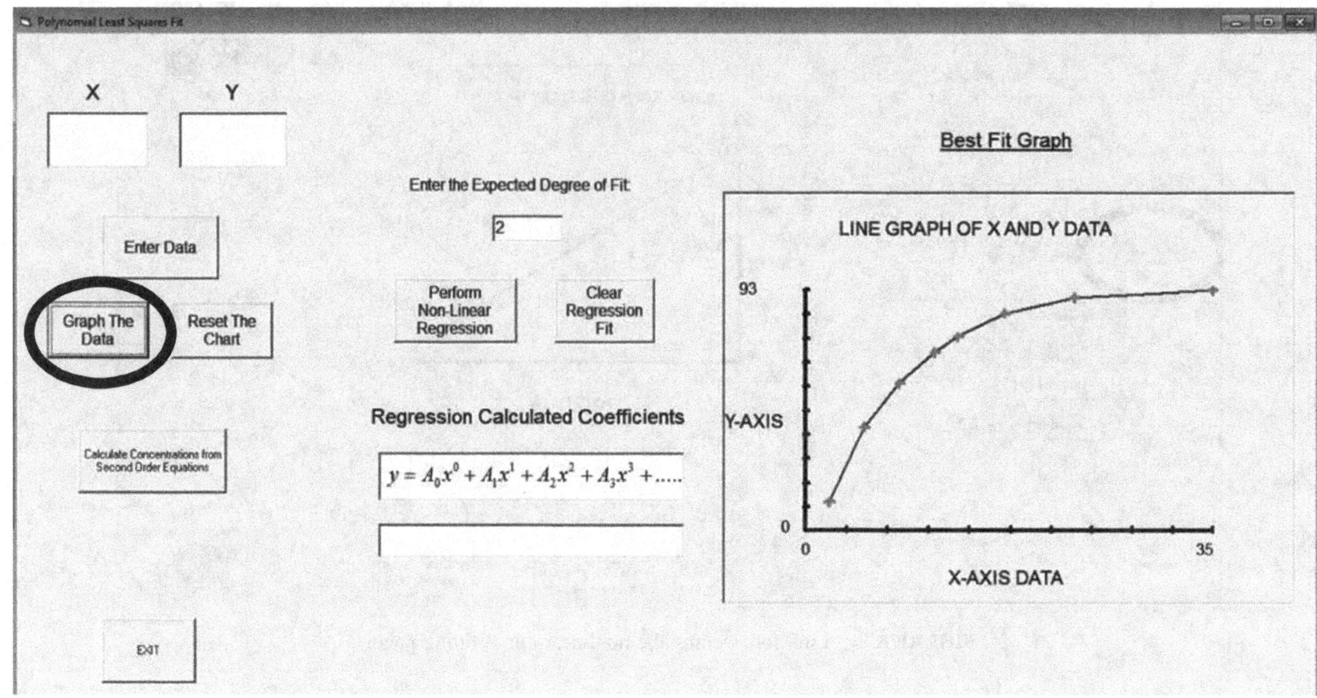

FIGURE 6.21 Plotting of the nonlinear data from Table 6.5.

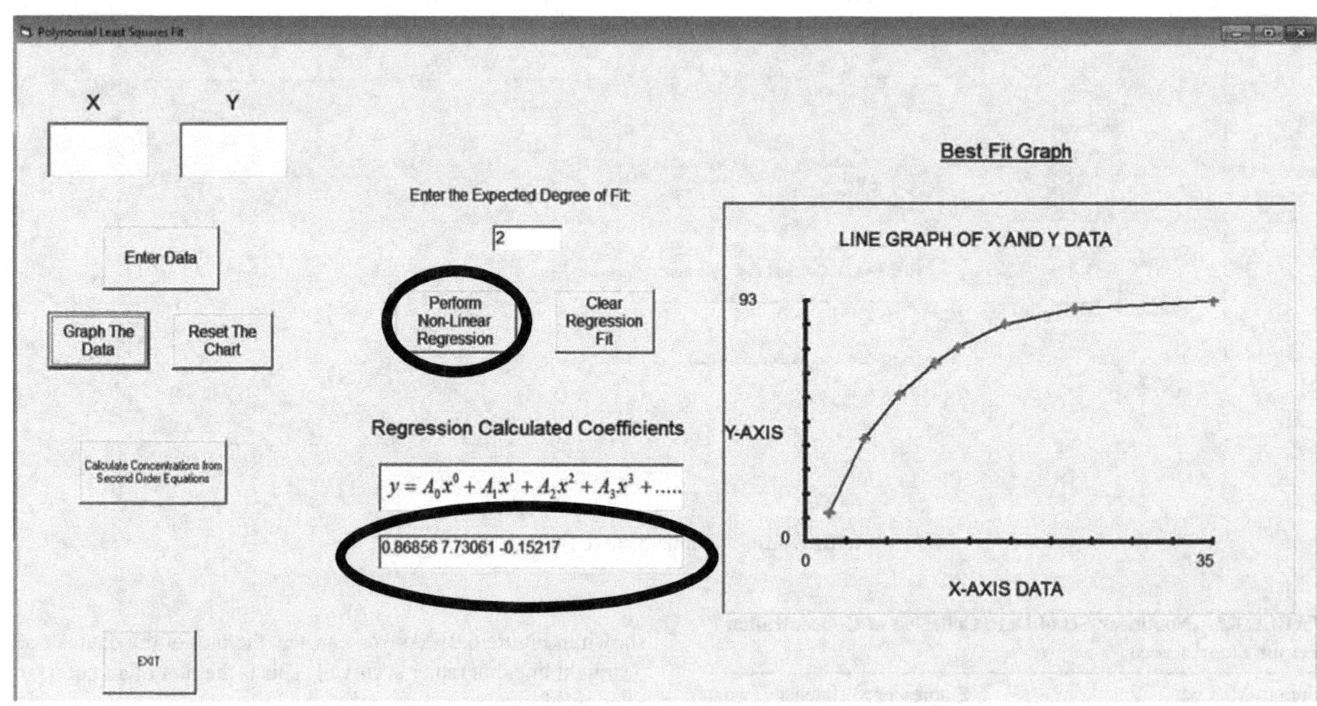

FIGURE 6.22 Calculation of the nonlinear regression coefficients.

The page also contains a link to a page where concentrations can be calculated from the nonlinear regression. Click the button labeled Calculate Concentrations from Second Order Equations on the left as depicted in Figure 6.23. This will bring up the page depicted in Figure 6.24.

The input page is in the form to accept data as $0 = ax^2 + bx + c$. The user has to determine the coefficients a, b, and c. The generalized forms of the second-order equation coefficients are generated from the previous page, and the second-order equation is in the form $y = ax^2 + bx + c$. If we want to calculate possible

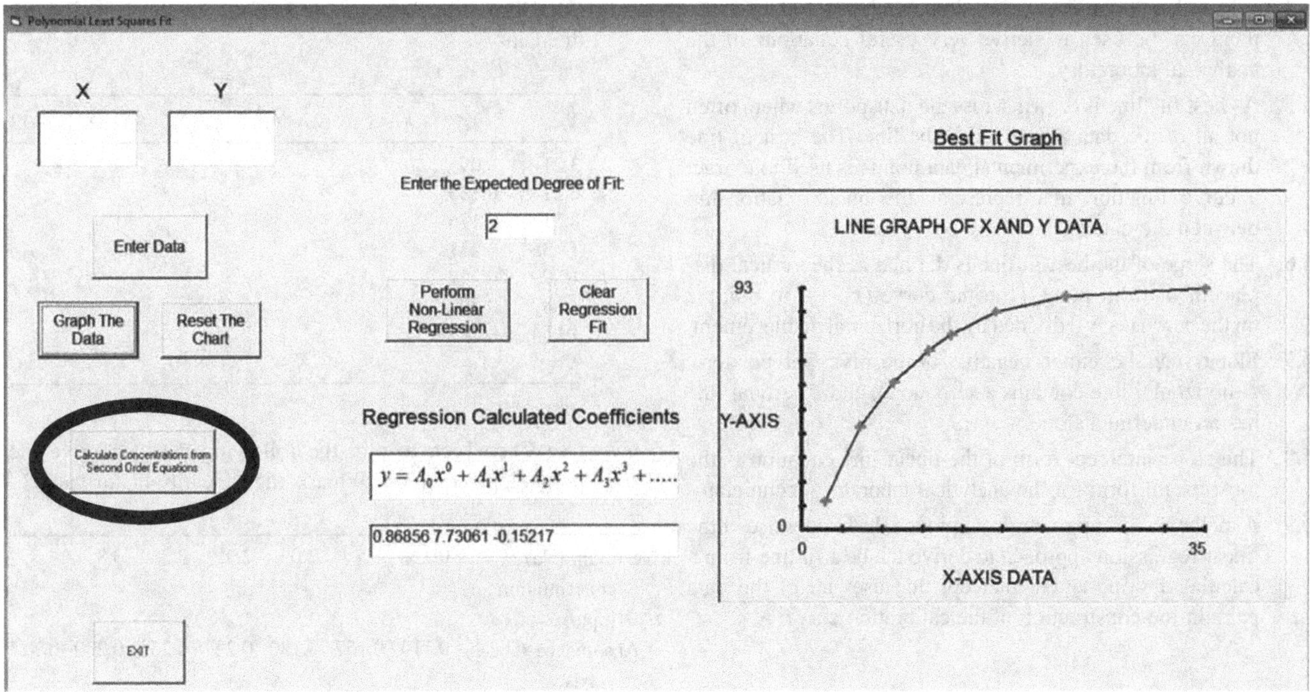

FIGURE 6.23 Button link labeled Calculate Concentrations from Second Order Equations to bring up concentration calculating page.

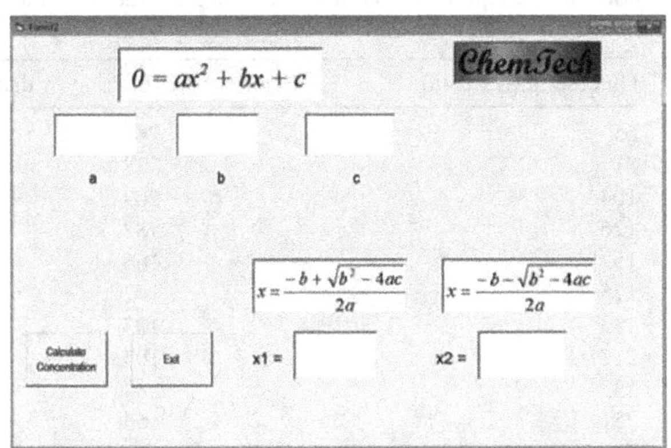

FIGURE 6.24 Page used to calculate second-order fit concentrations.

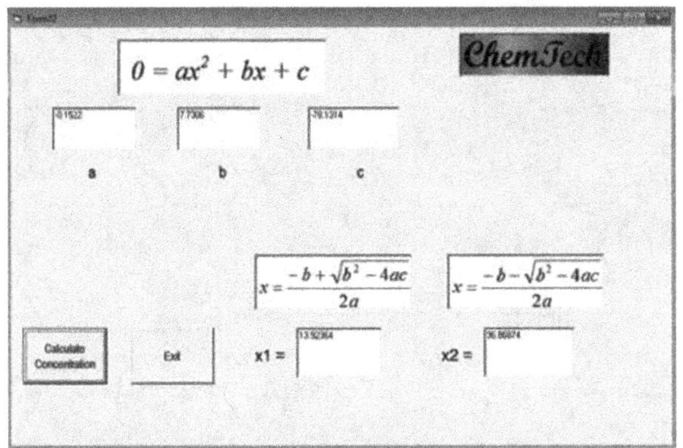

FIGURE 6.25 Results of calculating the concentration of an unknown urea sample.

roots for the second-order equation based upon a response value collected for an unknown concentration, then we need to substitute the unknown's response value for y and rearrange the equation into the form $0 = ax^2 + bx + c$. For example, suppose we have the generalized form of the second-order equation from the previous page as $y = -0.1522x^2 + 7.731x + 0.8686$ for our data from Table 6.5. Let us also suppose that we collected a value of 79 for an unknown urea sample's fluorescence. We need to plug 79 into the equation as y to get $0 = -0.1522x^2 + 7.7306x - 78.1314$. We are now ready to enter these coefficients into the page to calculate the two possible concentrations of the unknown (two roots). After entering the three coefficients and clicking the calculate concentrations button, the results and page should look like that in Figure 6.25. The two possible concentrations are 13.9 and 36.9 mm. The result of 13.9 mm falls on the curve and is the correct concentration for the unknown.

6.11 CHAPTER KEY CONCEPTS

6.1 Rectangular Cartesian coordinate system (named after Descartes) is a system to represent ordered pairs of numbers (x, y) in different quadrants of a set of axes laid in the plane of the page.

6.2 The fundamental design of a graph is based upon the rectangular Cartesian coordinate system where the plane of the paper is divided into four quadrants.

6.3 When a point in two-dimensional space within the predefined confines of the system is to be located, its distances from the x- and y-axes with respect to each other are used to designate the point's location.

6.4 When data points are plotted, the relationship between them can be used to derive very useful equations in the analytical laboratory.

6.5 A "best fit" line is drawn across the data points where often not all of the data points touch the line. The best fit line drawn from the experimental data points is used to extract a curve function that represents the linear relationship between the data.

6.6 The slope of the best fit line is defined as the vertical displacement from point 1 on the curve (x_1, y_1) to point 2 on the curve (x_2, y_2) divided by the horizontal displacement.

6.7 Slopes can be either negative or positive but nonzero. A horizontal line contains a zero slope, and a vertical line has an undefined slope.

6.8 The slope-intercept form of the linear line equation is the most useful form for the analytical laboratory technician.

6.9 A mathematical approach is employed that makes use of the linear regression approach to derive the best fit line from a calculated slope and y-intercept that uses all of the data pairs in the construction of the calibration curve.

6.12 CHAPTER PROBLEMS

6.1 Using graph paper plot the following sets of data (2, 5), (6, 10), (10, 15), (13, 19).

6.2 Draw a best fit line through the data points.

6.3 Estimate the y-intercept and the slope using rise over run.

6.4 From the slope and y-intercept, derive the slope-intercept form of the equation for the data in Problem 6.1.

6.5 Use Equation 6.13 to calculate the slope (m), and Equation 6.14 to calculate the intercept (b) for the data in Problem 6.1.

6.6 Using the following table, use Equation 6.13 to calculate the slope (m), and Equation 6.14 to calculate the intercept (b). Derive the slope-intercept form of the equation for the data.

X_i	Y_i	$X_i - \bar{X}$	$(X_i - \bar{X})^2$	$Y_i - \bar{Y}$	$(X_i - \bar{X})(Y_i - \bar{Y})$
3.21	0.095				
6.21	0.135				
8.81	0.176				
12.35	0.231				
13.99	0.257				
16.87	0.292				
19.54	0.325				
$\bar{X} =$	$\bar{Y} =$		$\Sigma =$		$\Sigma =$

6.7 Use ChemTech to plot the following data to derive the point-slope equation. What is the Linearity factor value?

Sodium (Na) concentration (ppm) x-axis	0.7	1.5	2.0	2.9	3.3	3.8	4.5
Absorbance (abu) y-axis	0.114	0.157	0.180	0.227	0.253	0.290	0.327

6.8 Use ChemTech to plot the following data to derive the point-slope equation. What is the Linearity factor value?

Glucose (μM, x data)	Fluorescence (Intensity, y data)
65	244
81	228
104	209
126	187
152	166
175	147
193	123
211	105
233	84
251	66

7

USING MICROSOFT EXCEL® IN THE LABORATORY

7.1 Introduction to Excel®
7.2 Opening Excel® in ChemTech
7.3 The Excel® Spreadsheet
 7.3.1 Spreadsheet Menus and Quick Access Toolbars
7.4 Graphing in Excel®
 7.4.1 Making Column Headings
 7.4.2 Entering Data into Columns
 7.4.3 Saving the Spreadsheet
 7.4.4 Constructing the Graph
 7.4.5 The Chart Wizard
 7.4.6 The Chart Source Data
 7.4.7 Chart Options
7.5 Charts in Excel® 2010
7.6 Complex Charting in Excel® 97-2003
 7.6.1 Calcium Atomic Absorption (AAS) Data
 7.6.2 Entering Ca Data into Spreadsheet
 7.6.3 Average and Standard Deviation
 7.6.4 Constructing the Calibration Curve
 7.6.5 Entering the Chart Options
 7.6.6 Error Bars
 7.6.7 Trendline
7.7 Complex Charting in Excel® 2010
 7.7.1 Entering the Data
 7.7.2 Using the Formula Search Function
 7.7.3 Inserting the Chart
 7.7.4 Formatting the Chart
7.8 Statistical Analysis Using Excel®
 7.8.1 Open and Save Excel® StatExp.xls
 7.8.2 Sulfate Data
 7.8.3 Excel® Confidence Function
 7.8.4 Excel® Student's *t* Test
 7.8.5 Excel® Tools Data Analysis

7.1 INTRODUCTION TO EXCEL®

Up to this point ChemTech has taught us in a simple and straightforward manner how to do simple statistics upon a given set of data, such as calculating averages and standard deviations, and also how to plot simple linear graphs. The laboratory technician however will often have instances where a relationship between a set of data may not be linear, or a more advanced statistical treatment of data is needed. We will now move on to using Microsoft Excel®, which will allow us to do more extensive statistical treatment of data and graphing including both linear and nonlinear relationships. Whether the student at this point is well versed in the use of Excel®, or this is the first introduction to it, we will begin at the very basic description of the use of Excel®, and then move into more advanced features of the spreadsheet software.

7.2 OPENING EXCEL® IN ChemTech

If not already open, open ChemTech and go to the Chapter 7 Main Menu page. Double click the button "Microsoft Excel®" to open a standard workbook. The workbook "Book1" is the basic "spreadsheet" that ChemTech uses and opens when we start Microsoft Excel® from ChemTech. The blank form of Book1 is depicted in Figure 7.1. This is a spreadsheet in Excel® 1997-2003 format. Figure 7.2 depicts the Excel® 2010 format. The two formats are basically the same, but Excel® 2010 has placed more emphasis on what they are calling the "Quick Access Toolbar," which is just an enhanced toolbar. We will take a moment to go over the layout and basic features of the workbook so that we may become more familiar with its construction and use.

7.3 THE EXCEL® SPREADSHEET

A workbook in Excel® is a workspace made up of spreadsheets. The workbook in Figure 7.1 is opened to "Sheet 1," as you can see as the active sheet (the sheet tabs are located at the bottom left of the sheet window and in this case contains one sheet). The spreadsheets are made up of rows and columns where a particular location of row-column is a cell. The first cell in the furthest most upper left-hand corner is the A1 cell for column A and row 1. This particular version of Excel® is the 2000 version which

Analytical Chemistry: A Chemist and Laboratory Technician's Toolkit, First Edition. Bryan M. Ham and Aihui MaHam.

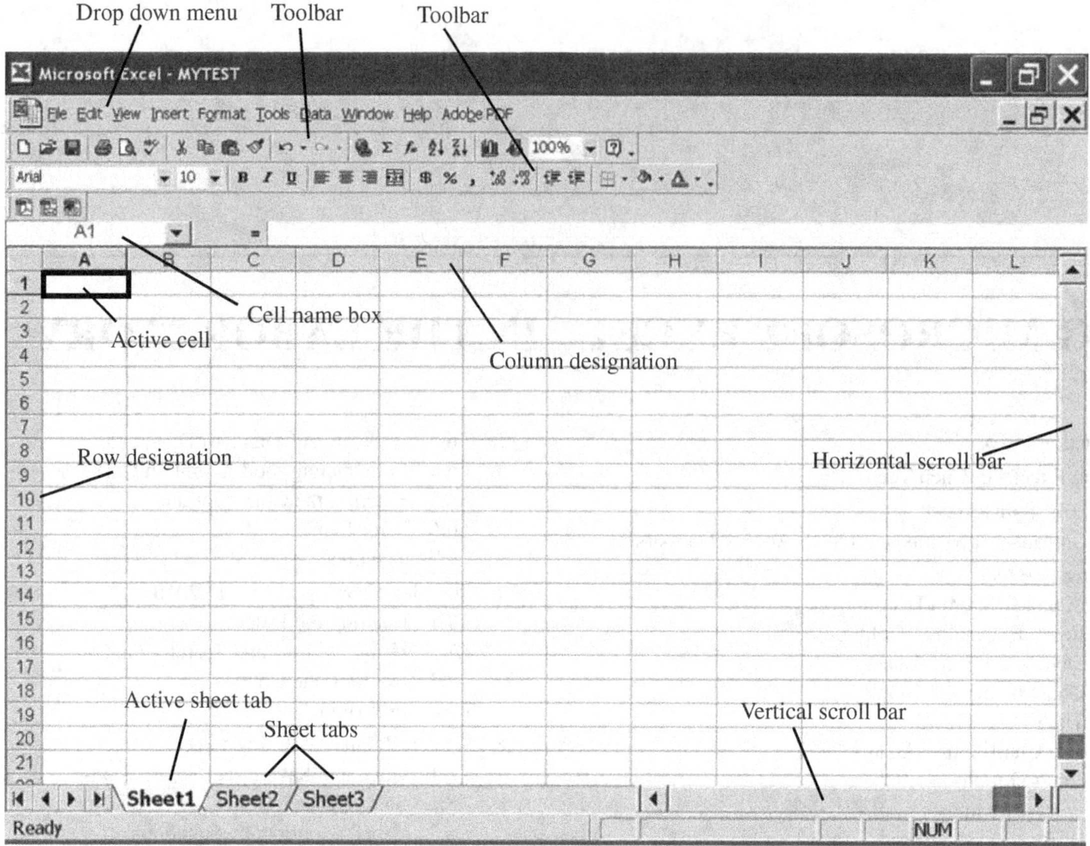

FIGURE 7.1 Basic example of the Microsoft Excel® workbook containing sheets (spreadsheets, Excel® 97-2003). The page contains menus and toolbars at the top, the spreadsheet cells where data are entered, and horizontal and vertical scroll bars. Note that the active cell is designated by a heavy lined box (cell A1 in this example). Used with permission from Microsoft.

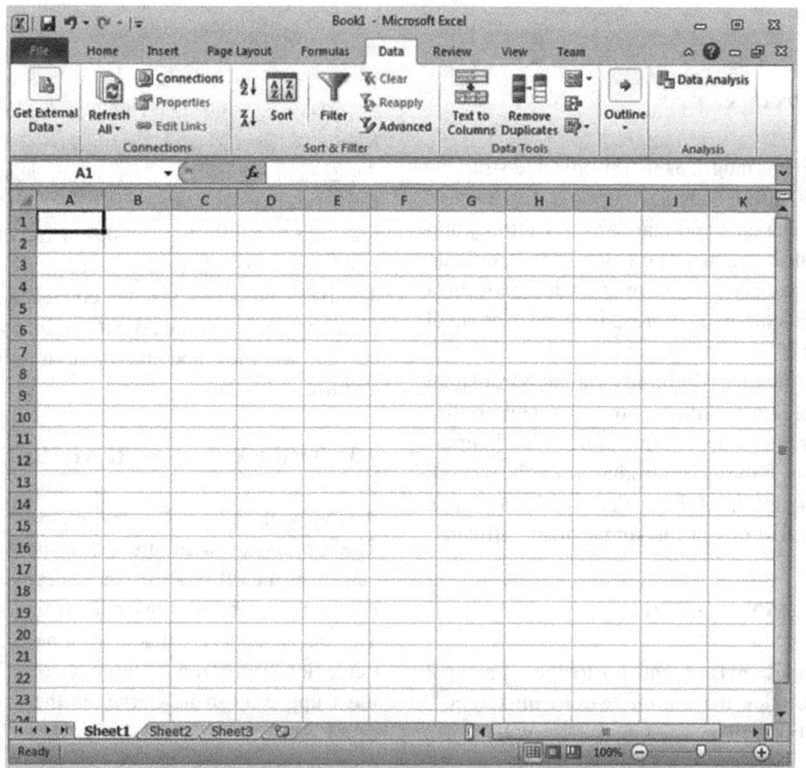

FIGURE 7.2 Basic example of the Microsoft Excel® workbook containing sheets (spreadsheets, Excel® 2010). Excel® 2010 has placed more emphasis on what they are calling the "Quick Access Toolbar," which is just an enhanced toolbar.

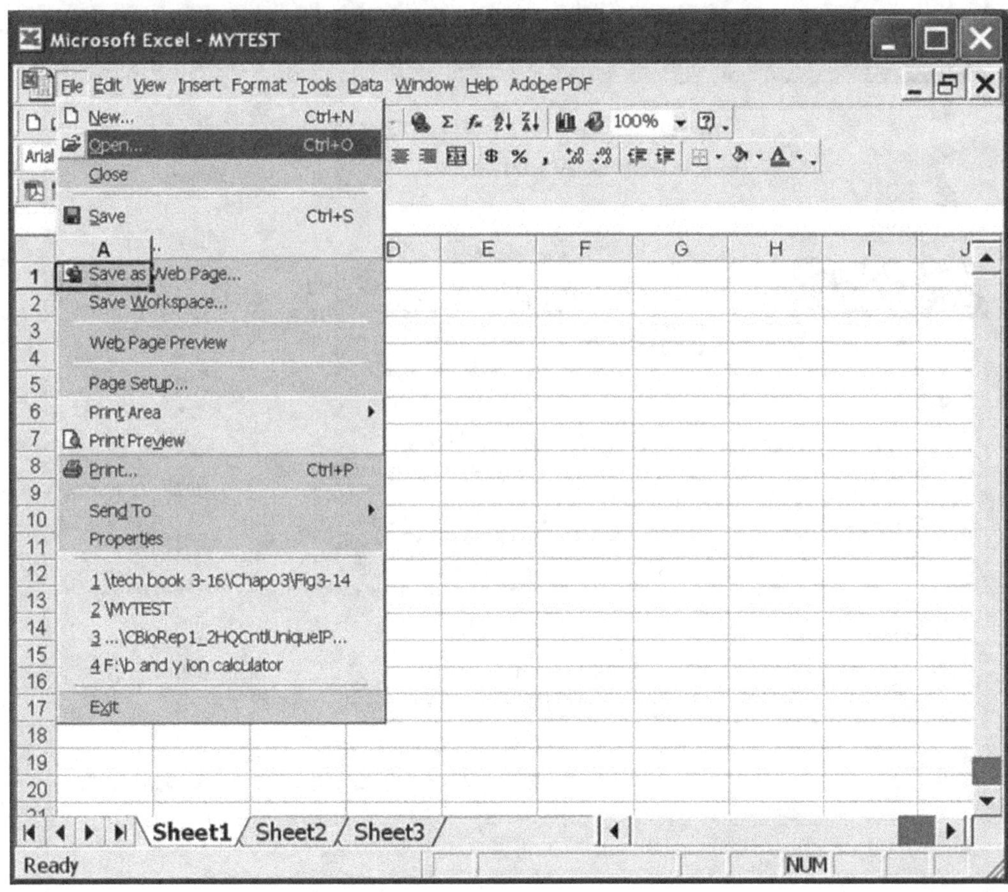

FIGURE 7.3 Example of the Excel® 97-2003 drop-down "File" menu where operations such as file Open, Close, Save, and Print may be selected and executed.

contains 65,536 rows (1 to 65,536) and 217 columns (A to IV) equating to a total of 14,221,312 cells. While the sizes of the spreadsheets and some features change with each new version, the basic construction and fundamental uses remain almost entirely the same. For example, the 95 version contained 16,385 rows (1 to 16,385) and 217 columns (A to IV) equating to a total of 3,555,545 cells, however the data entry, statistical treatment, and graphing are still quite similar and it is easy to adjust from one version to the next (i.e., if your computer is using a newer version than the 2000 version, the following tutorial will still be applicable, e.g., the 2003 version contains the same number of rows and columns as the 2000 version).

7.3.1 Spreadsheet Menus and Quick Access Toolbars

Referring back to the spreadsheet depicted in Figure 7.1 we see that the top of the page contains menus that are similar to those used in Microsoft Word, which many may already be familiar with. For example, clicking on the File drop-down menu results in the choices for file manipulation such as Open, Save, Close, Print, Exit, and so on as depicted in Figure 7.3. As we go through a few examples and exercises, we will become more familiar with the use and functionality of Excel®. Figure 7.4 depicts file manipulation in Excel® 2010.

7.4 GRAPHING IN EXCEL®

A major use of Microsoft Excel® is the graphing ability. Originally, Excel® was developed primarily as a business tool but quickly found use in just about every other discipline. It is relatively easy and straightforward to use and has been utilized by chemist in the chemistry laboratory.

7.4.1 Making Column Headings

Let us begin with a simple graphing exercise that will require the typing in of data into the spreadsheet and the subsequent construction of a graph. From Table 7.1, in our spreadsheet type into the A1 cell the label for our *x* data "Urea (mm)." Do this by placing the mouse pointer over the A1 cell and left click to make that cell the active cell. An alternative way to move around the spreadsheet is to use the arrow keys on the keypad. The arrow keys will move the active cell cursor to different places within the spreadsheet according to the arrow used. Take a moment and try both of these movements, the mouse cursor and left clicking, and the arrow keys to get a feel for moving about the spreadsheet in a small, local area. Again, make the A1 cell the active cell and type "Urea (mm)." The typed phrase can be put into the cell by either moving the mouse cursor away to another cell and left clicking,

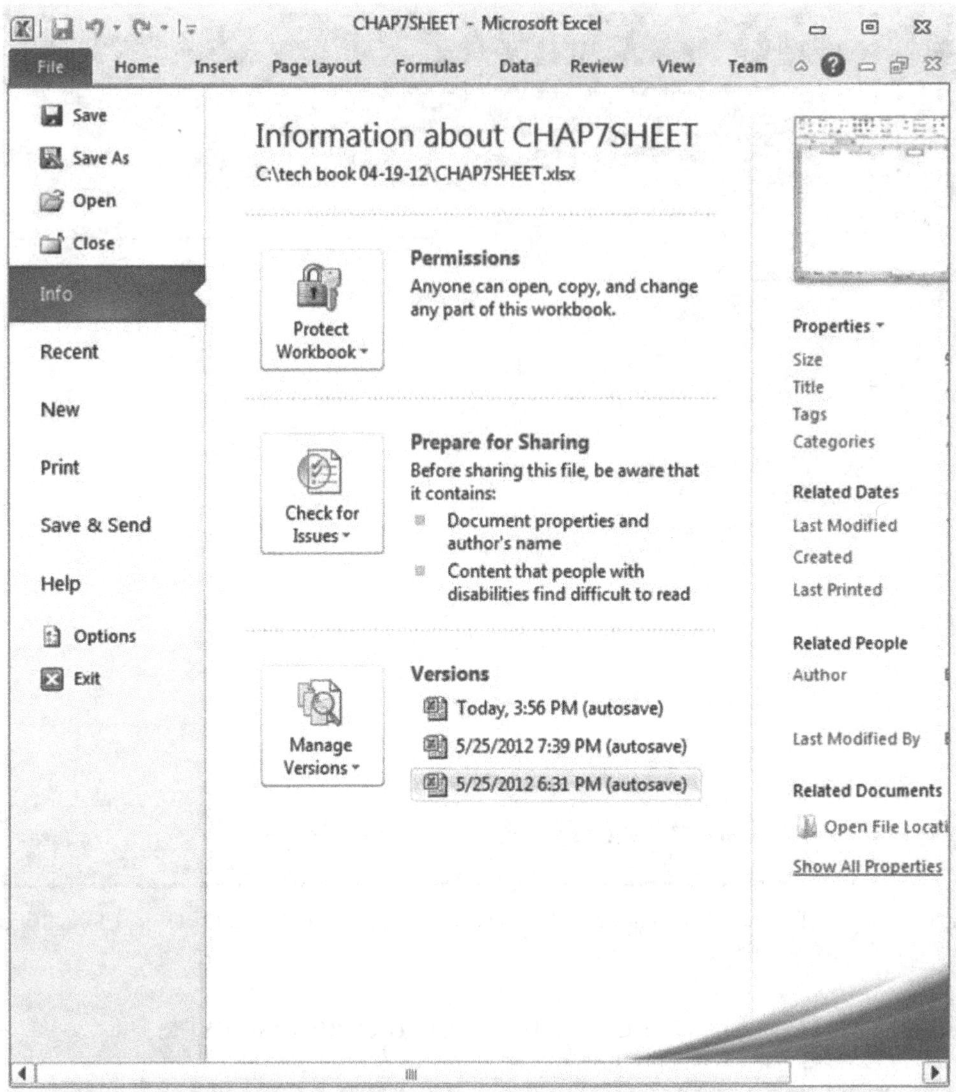

FIGURE 7.4 Example of the Excel® 2010 drop-down "File" menu where operations such as file Open, Close, Save, and Print may be selected and executed.

TABLE 7.1 Experimental data.

Urea (mM, x data)	Fluorescence (Intensity, y data)
5	229
11	218
24	204
36	196
52	154
77	133
89	99
106	85
124	52
156	33
167	27
179	19
190	12

or by pressing the Enter key which will move the cursor to the cell directly below and make that the active cell. This can be helpful if a large amount of data is needed to be input where the value is typed, the Enter key pressed, the next data is typed and the Enter key pressed, and so on. Now, into cell B1 our y data label "Fluorescence (Intensity)" is typed in and press the Enter key. In a spreadsheet, data can be listed in either rows or columns, it is up to the choice of the user. In this example, we will input data into columns.

7.4.2 Entering Data into Columns

Move the mouse cursor (or arrow key over to) to the A2 cell and click to make it the active cell. Now we are ready to begin entering our data into the spreadsheet in preparation for graphing. Type in the first Urea mM x data "5" and press Enter. The value of "5" has been entered into the A2 cell and the cursor has been moved to the A3 cell making it the active cell. Continue in this manner entering the rest of the *x* data values from Table 7.1 by typing in the next value (e.g., "11" would be the next *x* data value) and pressing Enter. When you have finished entering in the *x* data values move the mouse cursor over to the B2 cell and left click to make it the active cell. Type into the cell the "229" *y* data value, press the Enter key and continue until all of the *y* data values have been entered. The spreadsheet should look like the one depicted in Figure 7.5a.

7.4.3 Saving the Spreadsheet

Note however that the name of the workbook is "Book1" which is our base Excel® workbook (see red ellipse in figure). At this point we need to rename the workbook as "UreaFluor" (or any name wanted but must be remembered). Click on the drop-down menu "File" and select "Save As," type in "UreaFluor," and finally press the "Save" button to complete the saving process. The workbook should now have the heading file name as that depicted by the ellipse in Figure 7.5b.

7.4.4 Constructing the Graph

Let us move on now to graphing the set of data from Table 7.1 that we have just entered into our spreadsheet. To select the cells that we want to graph move the mouse cursor over the A1 cell, hold down the mouse cursor with the left mouse button and drag over until the B14 cell. This has highlighted the cells that are wanted to graph. This should look like the spreadsheet in Figure 7.5 where the cells from A1 to B14 are highlighted. An alternative method for selecting cells in preparation for plotting or data manipulation is to left click the A1 cell, hold down the "Shift" key and left click the A14 cell. This approach will also highlight the section of cells as depicted in Figure 7.6. This approach can be useful if there is a large area of the spreadsheet that must be included in the plotting (as we will see and is typical of Excel®, there are also other ways of doing the same operation such as selecting cells for plotting).

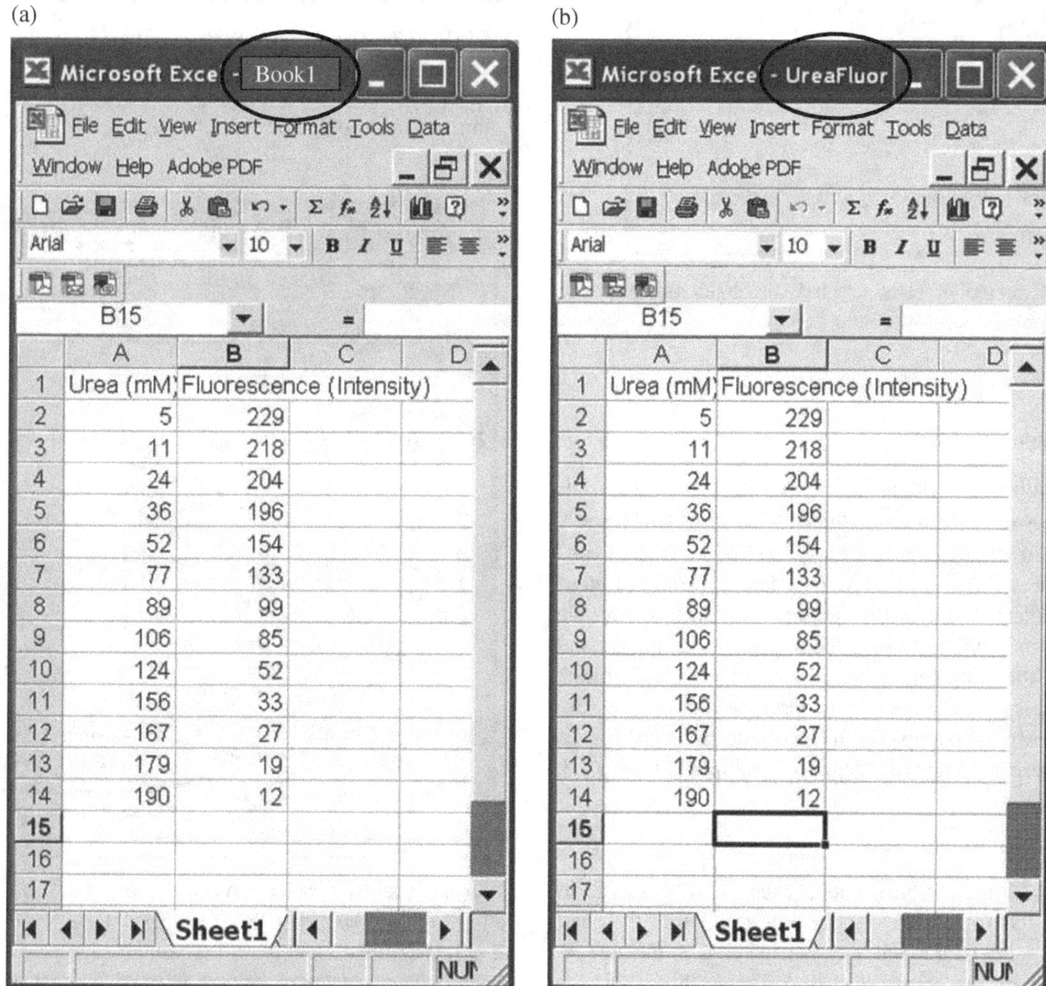

FIGURE 7.5 (a) Example of the spreadsheet after adding the data column headings and entering the associated data from Table 7.1. The current name of the workbook is "Book1" as encircled. (b) Same spreadsheet but with the workbook renamed to "UreaFluor."

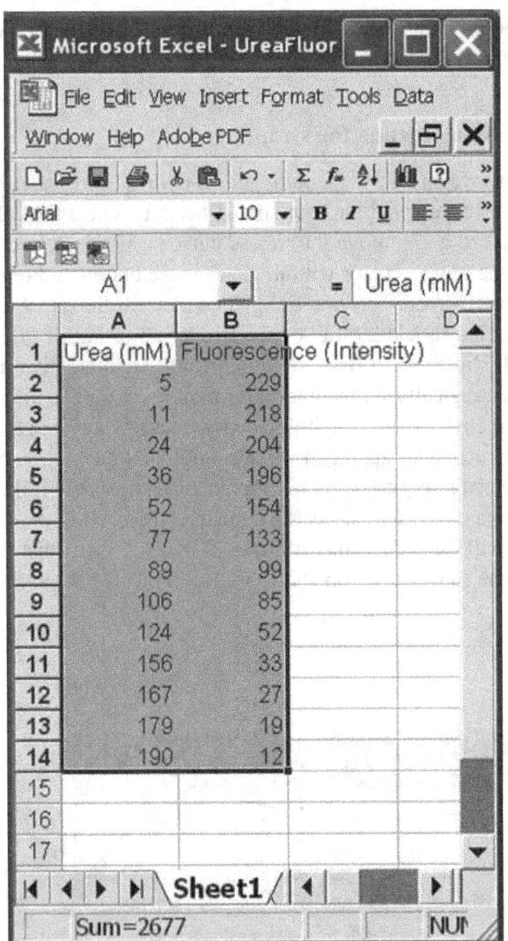

FIGURE 7.6 Example of cells selected (highlighted) within the spreadsheet in preparation for plotting.

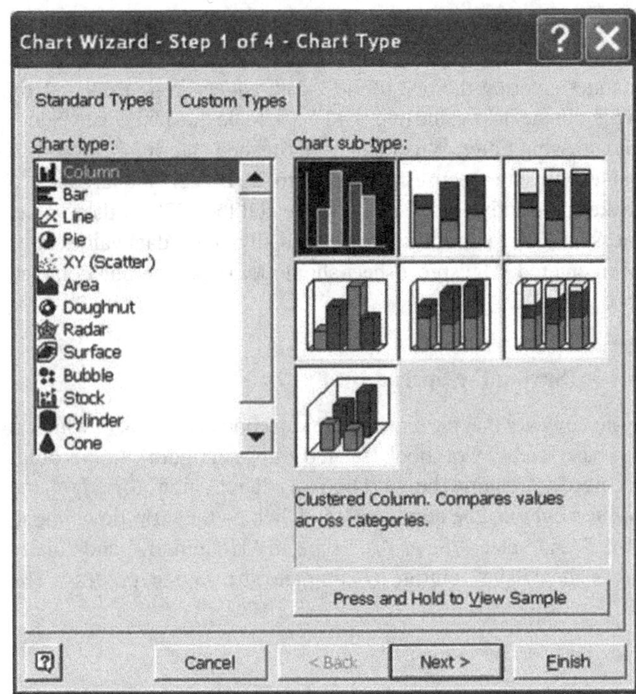

FIGURE 7.7 The "Chart Wizard—Step 1 of 4—Chart Type" selection window listing a variety of chart types that can be selected and used to present data. The current window has the column chart subtype selected.

7.4.5 The Chart Wizard

Next, we can either click on the "Chart Wizard" icon in the toolbar ( ) or choose in the drop-down menu "Insert" then "Chart." Both will bring up the "Chart Wizard—Step 1 of 4—Chart Type" selection window shown in Figure 7.7. Here, a variety of chart types can be selected and used to present data. In this example the "*XY* (Scatter)" type is to be selected along with the second subtype as "Scatter with data points connected by smooth lines." Note that there is a button under the chart types that by pressing and holding down an example of the selected graph type can be previewed. These three steps are depicted by ellipse in Figure 7.8.

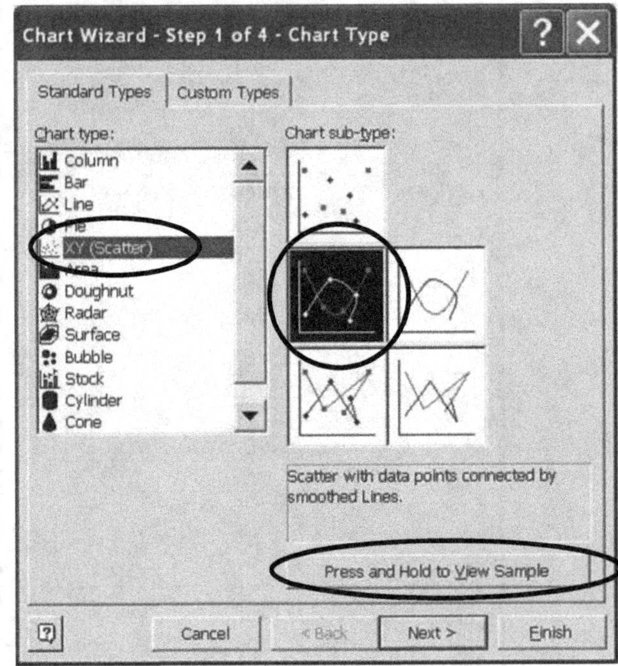

FIGURE 7.8 The "Chart Wizard—Step 1 of 4—Chart Type" selection window. The "XY (Scatter)" type, to the left in red circle, is selected along with the second subtype as "Scatter with data points connected by smooth lines." Note that by pressing and holding down the button under the chart types, an example of the selected graph type can be previewed.

7.4.6 The Chart Source Data

Click the "Next" button to bring about "Step 2 of 4" of the Chart Wizard, as depicted in Figure 7.9, for a description of the chart source data. For the data range, the selected cells in the spreadsheet are listed by the spreadsheet designation "Sheet1!$A$1:$B$14," and we have the series in columns. At this point the data range can also be selected for the graph by clicking the small

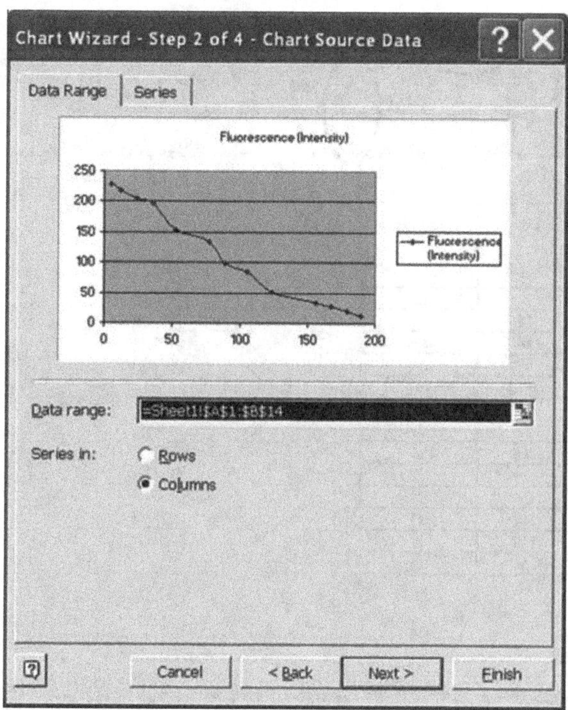

FIGURE 7.9 The "Chart Wizard—Step 2 of 4—Chart Source Data" page with Data Range tab selected.

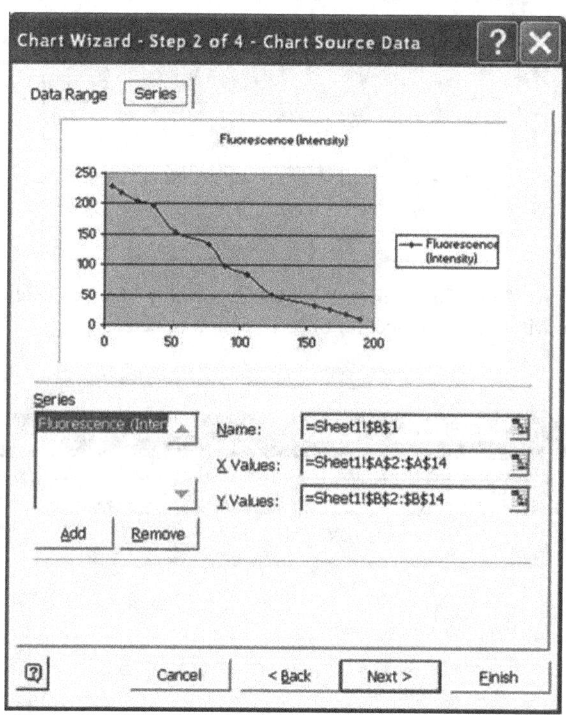

FIGURE 7.10 The "Chart Wizard—Step 2 of 4—Chart Source Data" page with Series tab selected.

spreadsheet button to the right of the Data range box. Clicking this button brings the user to the spreadsheet where a data range selection can be made for plotting. To return to the Chart Wizard, click on the small button or close the input box. Clicking the second tab "Series" brings up a window that is depicted in Figure 7.10. This page allows the user to add and remove data series contained within the spreadsheet. At this point we will not be adding or removing data series in our example (this is a more advanced application of plotting that we will use in a later example).

7.4.7 Chart Options

Clicking the Next button will bring up the "Chart Wizard—Step 3 of 4—Chart Options" page where a variety of formatting of the graph can be done. The first "Titles" tab allows the entering of a Chart title and labels for the *x* and *y* axes. For the chart title, type in "Urea Concentration versus Fluorescence." For the Value (*X*) axis type in "Urea (mM)," for the Value (*Y*) axis type in "Fluorescence (Intensity)." This is depicted in Figure 7.11. Click the second "Axes" tab. The Value (*X*) axis and Value (*Y*) axis boxes should be checked. Click the third "Gridlines" tab. If there is a check mark in any of the gridline selection boxes, remove them by clicking on the box. Click the fourth "Legend" tab. If the Show legend box is checked, remove the legend by clicking on the box. A legend is not necessary in this example because there is only one data series. Finally, click on the fifth "Data Labels" tab and ensure that none of the data label boxes are checked. Click the Next button and this will bring up the "Chart Wizard—Step

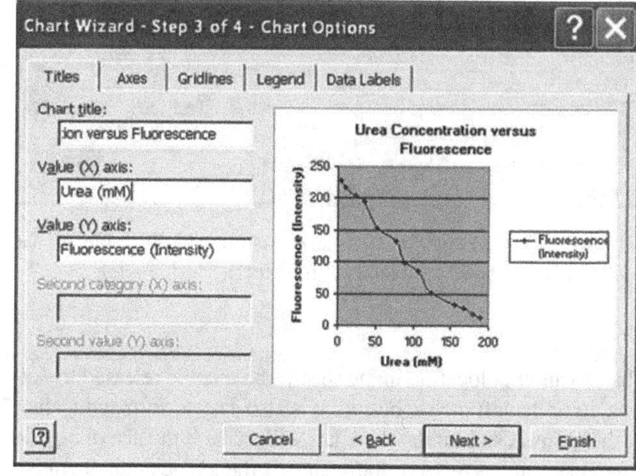

FIGURE 7.11 The "Chart Wizard—Step 3 of 4—Chart Options" page with Titles tab selected. Note that the Chart title and *X* and *Y* axes titles have been typed into the Chart Options window.

4 of 4—Chart Location" page as shown in Figure 7.12. Select the "As object in: Sheet1" option to place the graph within our currently opened spreadsheet. Click the Finish button to complete the construction of the chart. The chart has now been placed into the spreadsheet and should look something similar to that depicted in Figure 7.13. The chart can now be edited and moved around the spreadsheet if needed. To move the chart place the mouse cursor over the chart, click the left button and drag the

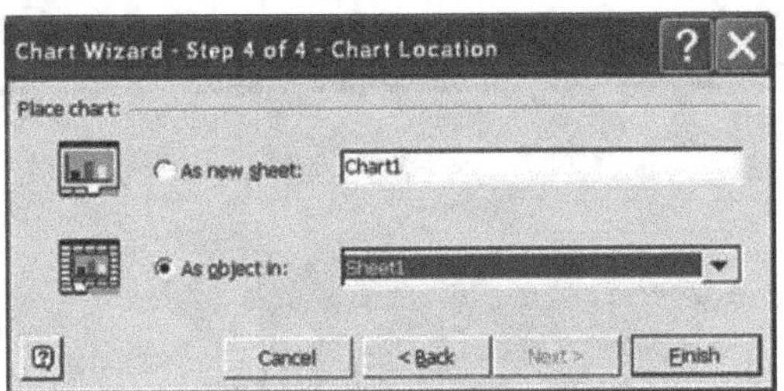

FIGURE 7.12 The "Chart Wizard—Step 4 of 4—Chart Location" page with the "As object in: Sheet1" option selected.

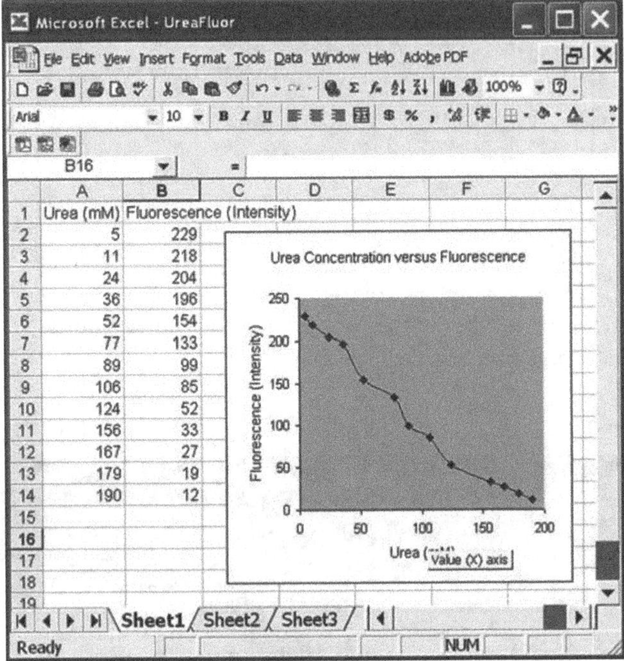

	A	B
1	Urea (mM)	Fluorescence (Intensity)
2	5	229
3	11	218
4	24	204
5	36	196
6	52	154
7	77	133
8	89	99
9	106	85
10	124	52
11	156	33
12	167	27
13	179	19
14	190	12

FIGURE 7.13 The chart that was just constructed from Table 7.1 has been finalized and is now located in Sheet1.

chart to another location in the sheet. The chart is released by letting go of the left mouse button. It is also easy to format the chart at this point. Clicking on the chart will make it the active object. The chart can be resized, the scales of the axis can be changed, boarders, shading, and gridlines can be added or removed, and the font sizes can be changed. Finally, as depicted in Figure 7.13, all of the steps that have just been covered can be accessed of the chart by right clicking the chart. This will bring up the menu as shown in Figure 7.13. There is also a Chart option on the drop-down main menu that may be used when the chart has been selected to access the various options of the chart.

7.5 CHARTS IN EXCEL® 2010

The graphing function in Excel® 2010 is a little different as compared to what was just covered. We will now take a brief look at graphing in Excel® 2010. Using the same set of data, click

	A	B
1	Urea	Fluorescence
2	(mM, x data)	(Intensity, y data)
3	5	229
4	11	218
5	24	204
6	36	196
7	52	154
8	77	133
9	89	99
10	106	85
11	124	52
12	156	33
13	167	27
14	179	19
15	190	12

FIGURE 7.14 ToolBar ribbon for inserting a chart in Excel® 2010.

the insert tab on the ToolBar to get the ribbon shown in Figure 7.14. Select the Charts shortcut to open up the Chart Types. Next click on Scatter and then All Chart Types… to bring up a chart page as shown in Figure 7.15. At this point we have a blank Chart object as depicted in Figure 7.16. The next step is to assign the data range by clicking on the Select Data in the ToolBar ribbon. This is depicted in Figure 7.17. Clicking OK will result in the Chart depicted in Figure 7.18. Excel® 2010 makes it a little easier and more straightforward to make charts using the ToolBar ribbon selections. The Quick Layout can be used to configure the chart with a number of saved styles. Also, by selecting the data table first, and then inserting the chart, Excel® will automatically assign the chart data and construct the chart that is depicted in Figure 7.18. Go ahead and try it out.

7.6 COMPLEX CHARTING IN EXCEL® 97-2003

7.6.1 Calcium Atomic Absorption (AAS) Data

Let us now take a look at a slightly more complex example of plotting data in Microsoft Excel®. In this example, a number of replicate analyses have been gathered that will allow us to plot an average and construct error bars based upon the standard deviation of the measurements. Table 7.2 contains a series of data points ($n = 3$) that have been collected for a set of calcium standards that were measured on an atomic absorption spectrophotometer (AAS). The concentrations of the set of calcium standards are listed in the first column of the table in parts per million (ppm).

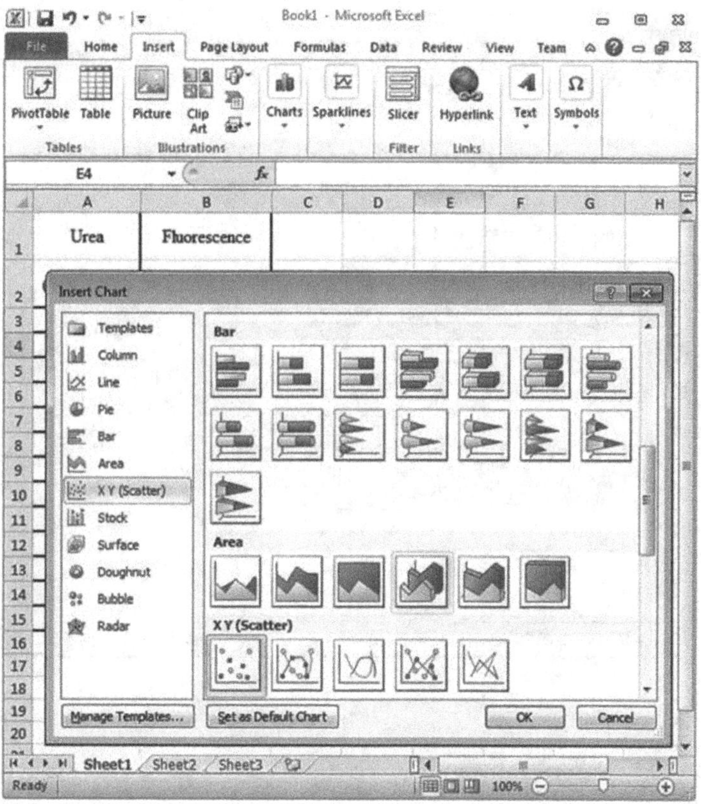

FIGURE 7.15 Scatter and then All Chart Types… to bring up a chart page in Excel® 2010.

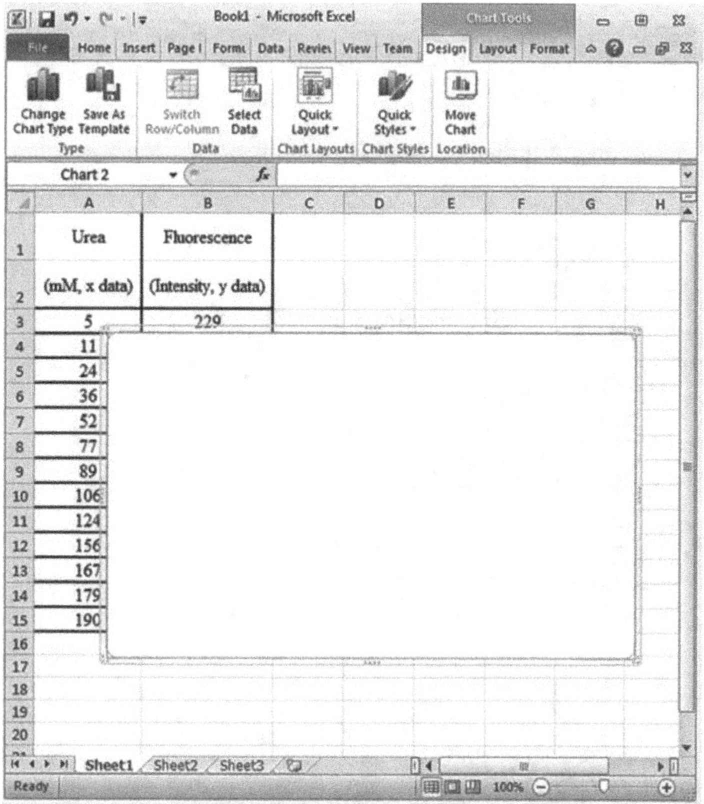

FIGURE 7.16 Blank Chart object.

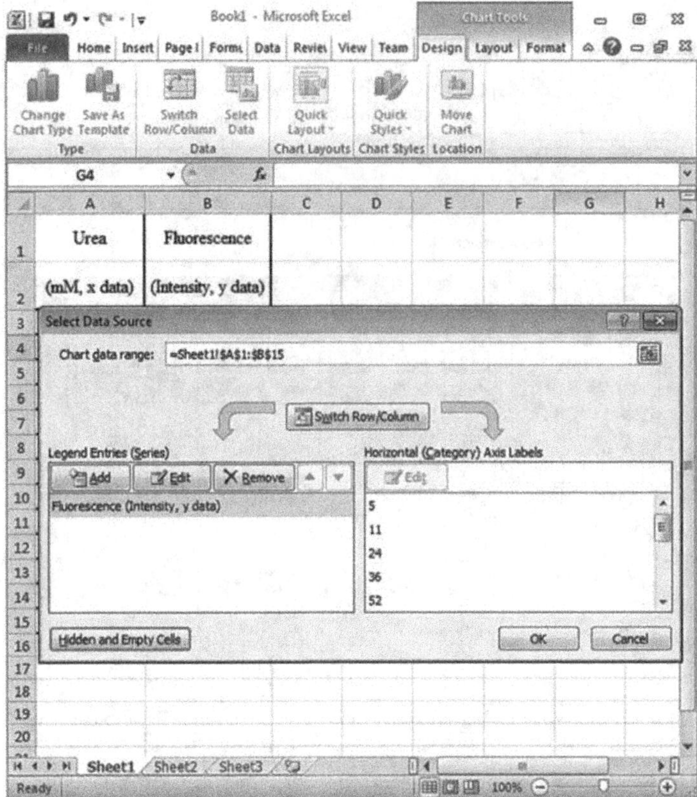

FIGURE 7.17 Assigning the data range for the Chart.

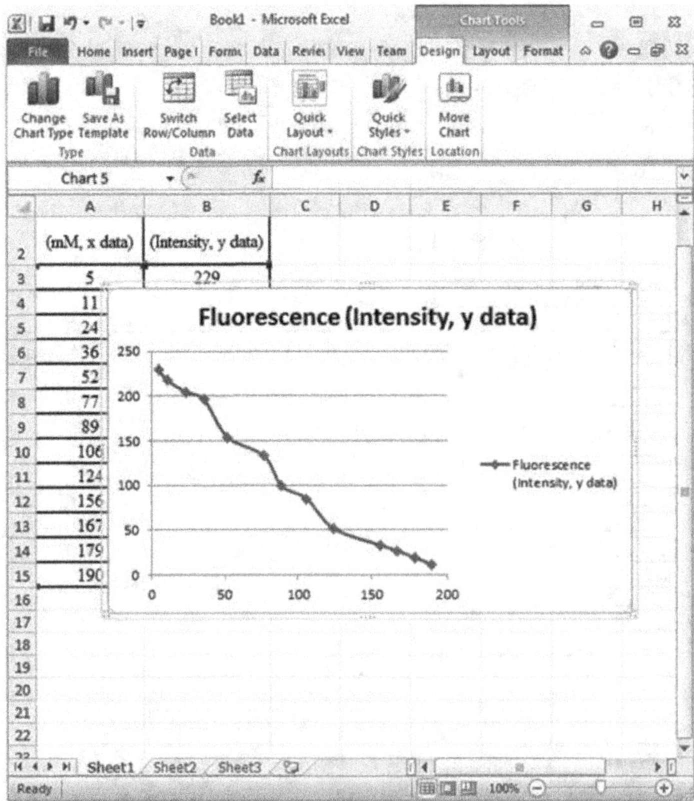

FIGURE 7.18 Constructed chart.

TABLE 7.2 Experimental data.

Calcium concentration (ppm)	Absorbance (abu)		
	Run 1	Run 2	Run 3
0.3	1109	1069	1155
1.2	1225	1168	1233
2.1	1308	1241	1276
3.3	1472	1319	1389
4.2	1497	1523	1472
5.1	1569	1599	1579
5.9	1688	1647	1625
7.2	1785	1765	1742
8.0	1833	1898	1849
10.1	2066	2012	2051

7.6.2 Entering Ca Data into Spreadsheet

If not already open, open ChemTech and go to the Chapter 7 Main Menu page. Click the button "Microsoft Excel®" to open the standard workbook "Book1." The workbook "Book1" is the basic "spreadsheet" that ChemTech uses and opens when we start Microsoft Excel® from ChemTech. Construct a spreadsheet based upon Table 7.2. In the blank cell A1 type in "Ca (ppm)," in cell B1 type in "Absorbance," in cell B2 type "Run 1," in cell C2 type "Run 2," and in cell D3 type "Run 3." Type in the calcium standard concentrations from 0.3 to 10.1 ppm in cells A3 to A12. Next, type in the absorbance data for Runs 1 through 3 in the same manner. Rename the workbook as "CalciumAAS" (or any name wanted but must be remembered). Click on the drop-down menu "File" and select "Save As," type in "CalciumAAS," and finally press the "Save" button to complete the saving process. An example of how the workbook and spreadsheet should look is depicted in Figure 7.19.

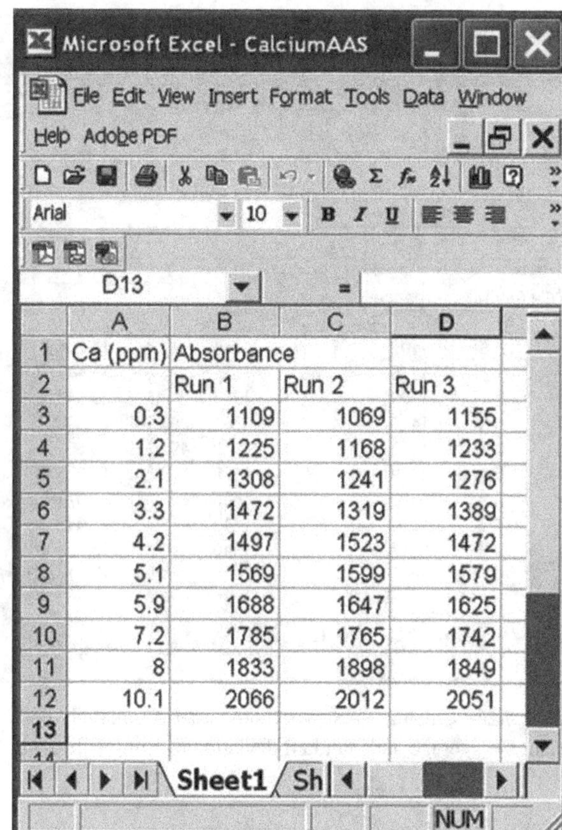

FIGURE 7.19 CalciumAAS workbook.

7.6.3 Average and Standard Deviation

When a series of replicate data has been collected such as that in Table 7.2, an average can be calculated and plotted using functions within the spreadsheet. This is generally a more accurate representation of the response of the standards with increasing concentration in instrumental analysis as compared to any single run. This is also typically how a calibration curve is constructed for a particular analytical methodology. In cell E2 of the spreadsheet type in "Average." In cell E3 of the spreadsheet type in "=AVERAGE(B3:D3)" and press Enter. This will calculate the average of the three replicate analyses (Run 1, Run 2, and Run 3) of the 0.3 ppm calcium standard and place the value of "1111" into cell E3. Click on the cell E3, go to the drop-down menu and press "Edit" and then "Copy." Click on the cell E4, hold down the left mouse button and drag down to cell E12 to select the cell range from E4 to E12. Go to the drop-down menu and select "Edit" and then "Paste" to paste the average calculation into these cells. In cell F2 of the spreadsheet type in "StdDev." Click on cell F3 and type in "=STDEV (B3:D3)" to calculate the standard deviation of the three replicate analyses (Run 1, Run 2, and Run 3) of the 0.3 ppm calcium standard and place the value of "43.03487" (Note: the exact number will depend upon the number format of the cell) into cell F3. Click on the cell F3, go to the drop-down menu, press "Edit," and then "Copy." Click on the cell F4, hold down the left mouse button and drag down to cell F12 to select the cell range from F4 to F12. Go to the drop-down menu and select "Edit" and then "Paste" to paste the average calculation into these cells. The spreadsheet should now look like that depicted in Figure 7.20.

7.6.4 Constructing the Calibration Curve

We are now ready to construct our calibration curve of calcium concentration (ppm) as measured by atomic absorption spectrophotometry. As often with Microsoft Excel®, there are a number of ways of plotting these data. The first approach we will take here is one of the simplest and will introduce a way to remove unwanted series of data as needed. First, click on the A3 cell, hold down the left mouse button, and drag until the E12 cell to select the data range. Click on the "Chart Wizard" icon in the toolbar ( ) to open up the Chart Wizard. Select the "XY (Scatter)" type of graph along with the second sub type as "Scatter with data points connected by smooth lines" and press the Next button. This chart however will contain multiple series due to the data range that was initially selected. This is shown in Figure 7.21. Click on the "Series" tab to view the series that have been included in the data range plotted as depicted in Figure 7.22. In the series

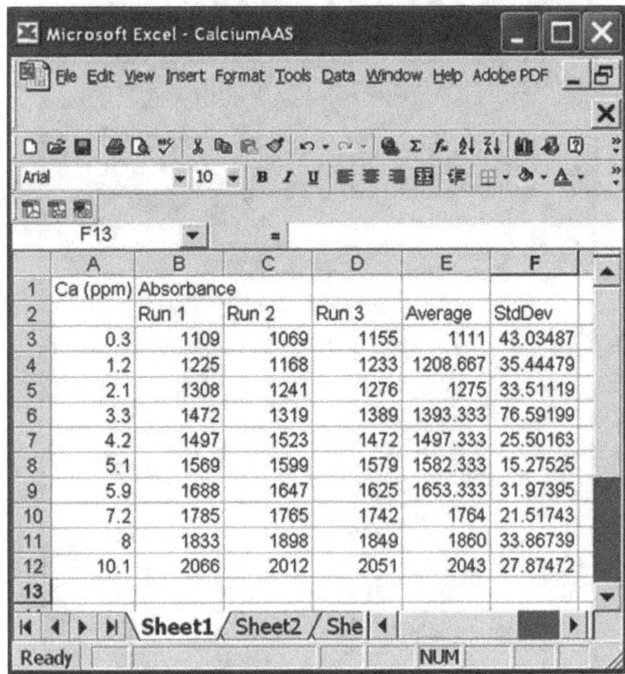

FIGURE 7.20 CalciumAAS workbook after formatting cells to calculate the average and standard deviation of Runs 1 through 3.

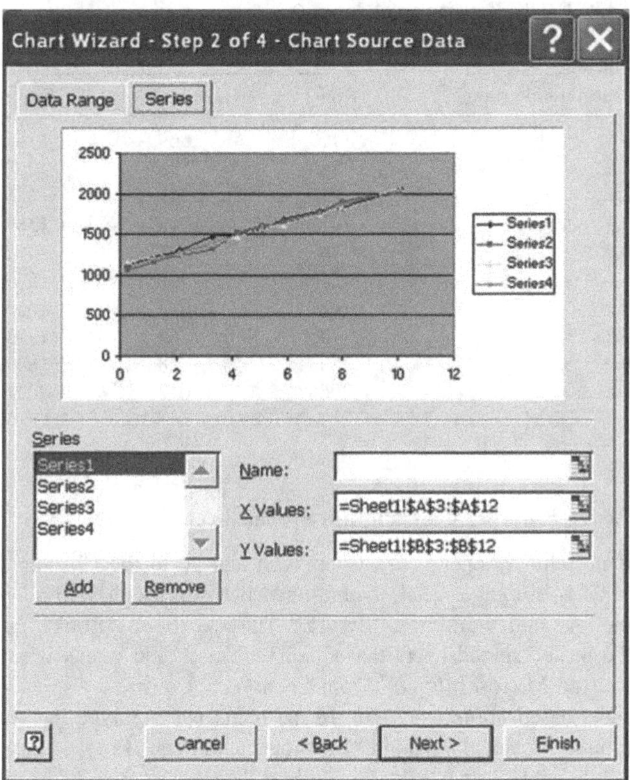

FIGURE 7.22 The "Chart Wizard—Step 2 of 4—Chart Source Data" page with Series tab selected.

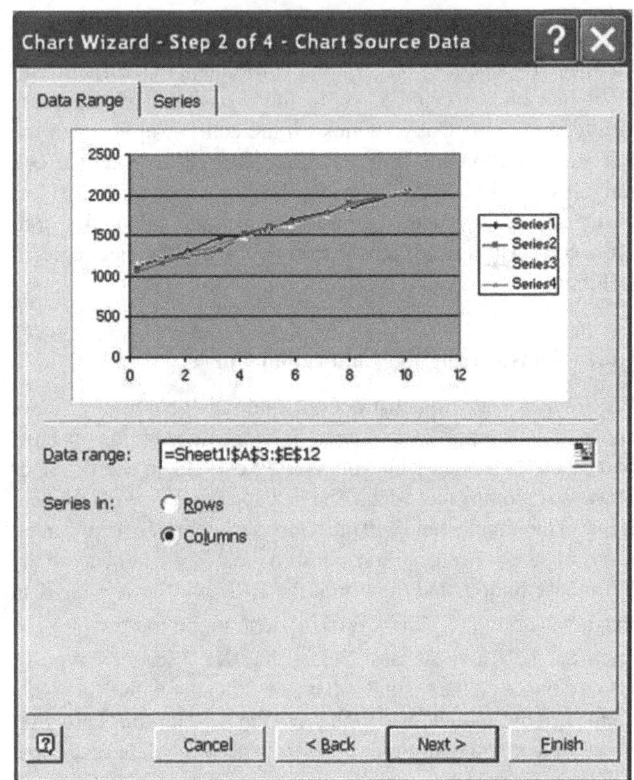

FIGURE 7.21 The "Chart Wizard—Step 2 of 4—Chart Source Data" page with Data Range tab selected.

window in the bottom left there are four series listed as Series1, Series2, Series3, and Series4. Click on Series1 to display the location of the series. To the right will be the locations as *X* Values: "=Sheet1!$A$3:$A$12" and *Y* Values: "=Sheet1!$B$3:$B$12." This is not the series that we want to plot; this is the data values for Run 1. Clicking on the second Series2 will show the series location as *X* Values: "=Sheet1!$A$3:$A$12" and *Y* Values: "=Sheet1!$C$3:$C$12." Note that the location of the *X* Values is the same, and this is correct. The furthest most column of data selected in the data range will be automatically designated as the *X* Values. However, the Series2 is still not the data that we want to plot. As might be guessed, the Series4 is the data that we want to plot as *X* Values: "=Sheet1!$A$3:$A$12" and *Y* Values: "=Sheet1!$E$3:$E$12." We can remove these unwanted series by clicking on each one and pressing the "Remove" button located below the list of series. Go ahead and remove Series1, Series2, and Series3. This will leave the Series4 in the plot, which is the average of the three runs in Table 7.2.

7.6.5 Entering the Chart Options

Clicking the Next button will bring up the "Chart Wizard—Step 3 of 4—Chart Options" page where again a variety of formatting of the graph can be done. The first "Titles" tab allows the entering of a Chart title and labels for the *x* and *y* axes. For the chart title type in "Calcium Concentration versus Absorbance." For the Value (*X*) axis type in "Calcium (ppm)," for the Value (*Y*) axis type in "Absorbance (abu)." Click the second "Axes" tab. The Value

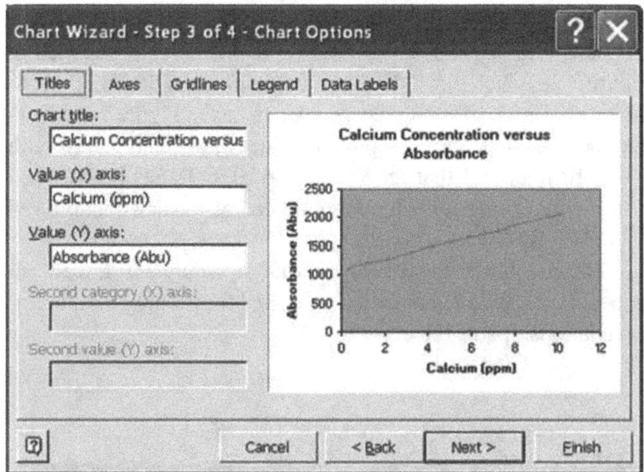

FIGURE 7.23 The "Chart Wizard—Step 3 of 4—Chart Options" page.

(*X*) axis and Value (*Y*) axis boxes should be checked. Click the third "Gridlines" tab. If there is a check mark in any of the gridline selection boxes, remove them by clicking on the box. Click the fourth "Legend" tab. If the Show legend box is checked, remove the legend by clicking on the box. A legend is not necessary in this example because there is only one data series. Finally, click on the fifth "Data Labels" tab and ensure that none of the data label boxes are checked. The page should now look like that shown in Figure 7.23 for the "Chart Wizard—Step 3 of 4—Chart Options" page. Click the Next button and this will bring up the "Chart Wizard—Step 4 of 4—Chart Location" page. Select the "As object in: Sheet1" option to place the graph within our currently opened spreadsheet. Click the Finish button to complete the construction of the chart. The chart has now been placed into the spreadsheet and should look like that depicted in Figure 7.24. Clicking on the chart will make it the active object. The chart can be resized, the scales of the axis can be changed, boarders, shading, and gridlines can be added or removed, and the font sizes can be changed. Left click on the *Y* axis to place it into focus, now right click to open a drop-down menu. Choose the "Format Axis" option. Click on the second tab "Scale" and type into the Minimum box "1000" and type into the Maximum box "2100." Left click on the *X* axis to place it into focus, now right click to open a drop-down menu. Choose the "Format Axis" option. Click on the second tab "Scale" and type into the Minimum box "0" and type into the Maximum box "12." This will rescale the *X* axis and should look like the chart in Figure 7.25.

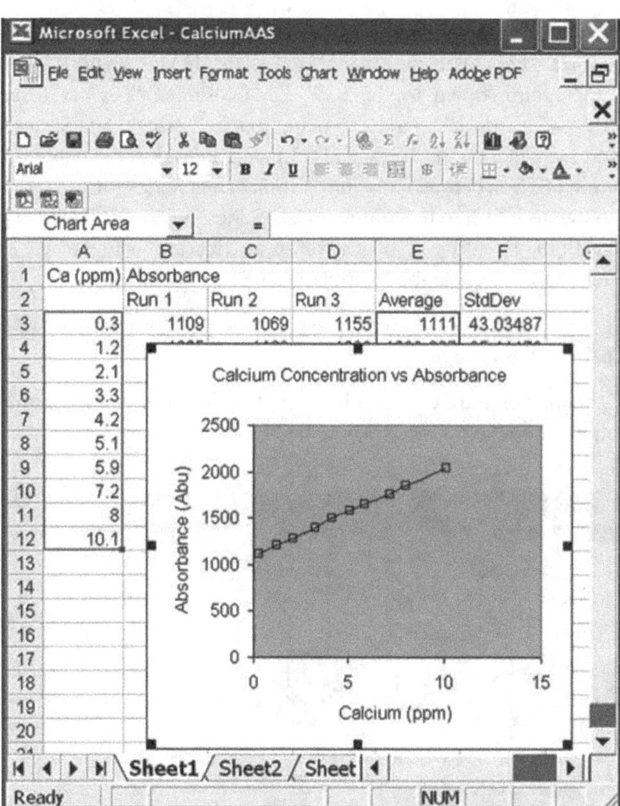

FIGURE 7.24 Calcium concentration versus absorbance chart.

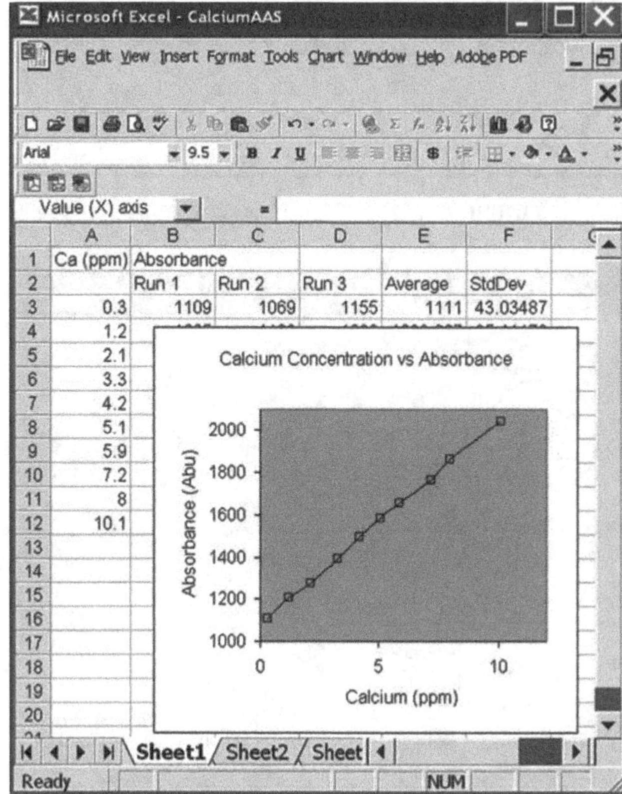

FIGURE 7.25 Calcium concentration versus absorbance chart with rescaled *X* and *Y* axes.

7.6.6 Error Bars

We are now ready to add our *Y* error bars to the plot. These *Y* error bars give us an idea about the scatter in each data point measurement. A large standard deviation value will result in a wide error bar indicating the amount of variance in the measurement of that data point. Place the mouse cursor over one of the plotted data points and left click to activate the data points. Right click to open up a menu and select "Format Data Series." Under the "Patterns" tab for "Line" select "None." Under the "*Y* Error Bars" tab select "Custom" at the bottom and click on the "+" spreadsheet icon that is to the right of the input box. This is depicted in Figure 7.26. In the spreadsheet select the range F3 to F12 by clicking on the F3 cell and dragging down to the F12 cell. Close the "Format Data Series—Custom +" window to return to the "Format Data Series" window. Click on the "-" spreadsheet icon and select the range F3 through F12 as before. Close the "Format Data Series—Custom -" window to return to the "Format Data Series" window. The "Display Both" should now be blackened as depicted in Figure 7.27. Click "OK" to close the "Format Data Series" window.

7.6.7 Trendline

Right click on the data series in the plot again and select the menu option "Add Trendline." Select the "Linear" type as depicted in Figure 7.28. Click the "Options" tab and select "Display equation on chart" and "Display R-squared value on chart" as depicted in Figure 7.28. Click the "OK" button to implement the additions to the chart. The chart should now look like that in Figure 7.29. Note that the line connecting the plotted values has been replaced with a best-fit line, and that an R^2 value of 0.9991 has been added to the chart along with the slope-intercept linear equation $y = 95.503x + 1086.1$.

This concludes the basic tutorial of Microsoft Excel® in respect to the fundamental skills required for plotting simple linear relationships in Excel® 97-2003.

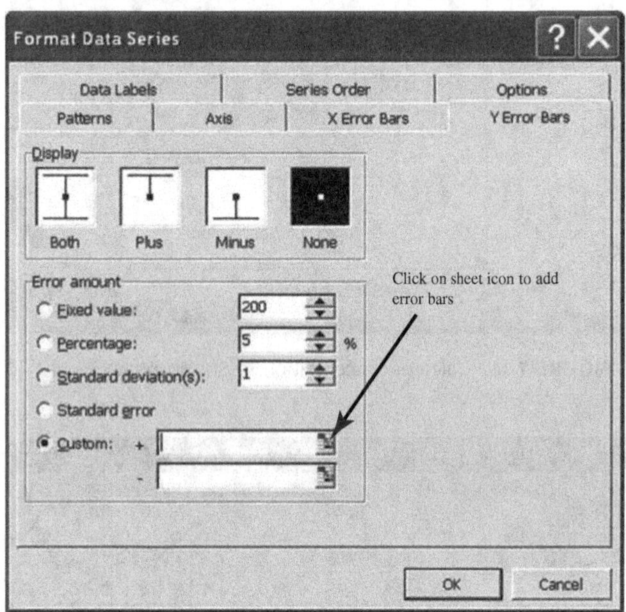

FIGURE 7.26 The "Format Data series" page.

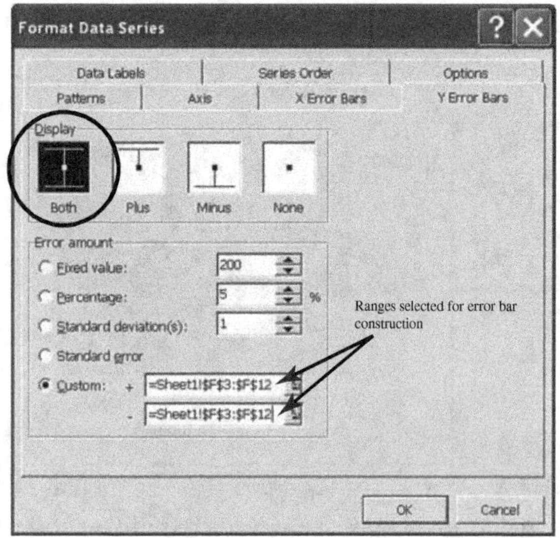

FIGURE 7.27 The "Format Data series" page of the "Y Error Bars" tab.

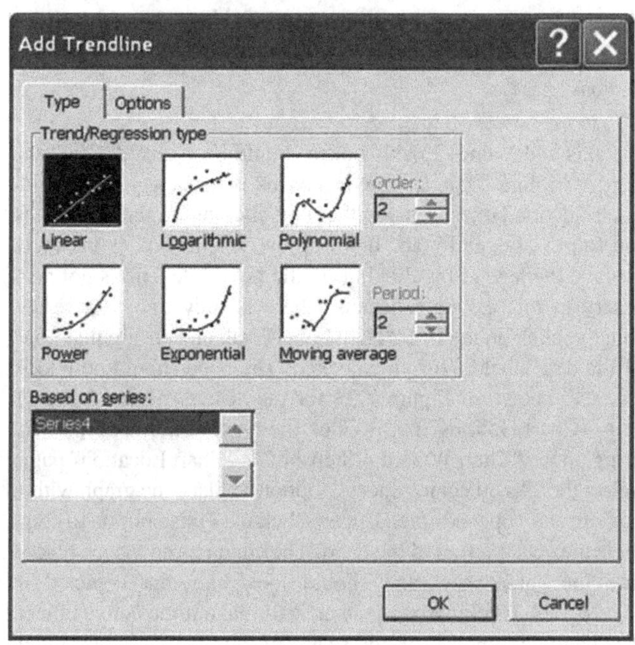

FIGURE 7.28 The "Add Trendline" page.

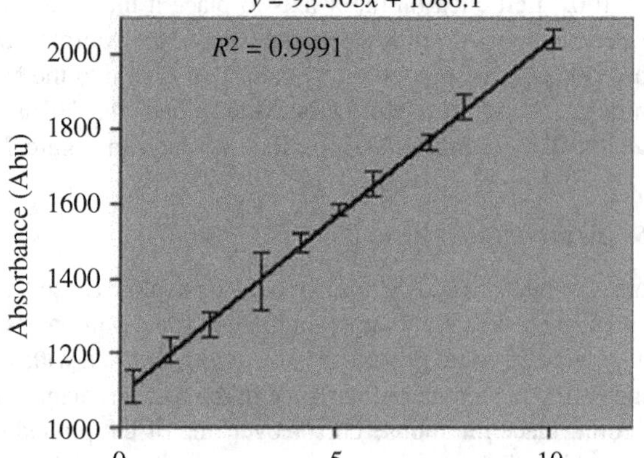

FIGURE 7.29 Calcium concentration versus absorbance chart with best-fit line, R^2 value, and slope-intercept equation.

7.7 COMPLEX CHARTING IN EXCEL® 2010

7.7.1 Entering the Data

Constructing charts in Excel® 2010 is very similar to that of Excel® 97-2003, both use the basic construct of cells in a spreadsheet. For Excel® 2010, open a new spreadsheet and type into the cells we covered earlier for the headings and data of Table 7.2. Next type in a new heading into cell E2 as "Average" and F2 as "StdDev" for the standard deviation.

7.7.2 Using the Formula Search Function

Place the cursor arrow over the E3 cell and click to make it the active cell. Next, click on the Formula tab and then click on the Insert Function shortcut in the ToolBar ribbon. A window will open that allows the selection of a function. Type into the Search "Average" and click the GO button. This will bring up the window shown in Figure 7.30. Click the OK button which will bring up the functions argument window as shown in Figure 7.31. In the Number1 box put the data range "B3:D3" for the average. Click the OK button which will place the average into the E3 cell. Next, highlight the F3 cell and click it to make it active. Click the Insert Function shortcut to open the functions window. Search the functions for Standard Deviation and select the STDEV function and click OK. This will open up the standard deviation function window as shown in Figure 7.32. In the Number1 box type in the data range "B3:D3" for the standard deviation. Click the OK button which will place the standard deviation into cell F3 as depicted in Figure 7.33. Next highlight cells E3 and F3, right click and press copy to copy these two functions onto the clipboard. Now highlight cells E4 through F12, right click, and select paste. This will paste the functions into these cells and also perform the calculations. The spreadsheet should now look like the one depicted in Figure 7.34.

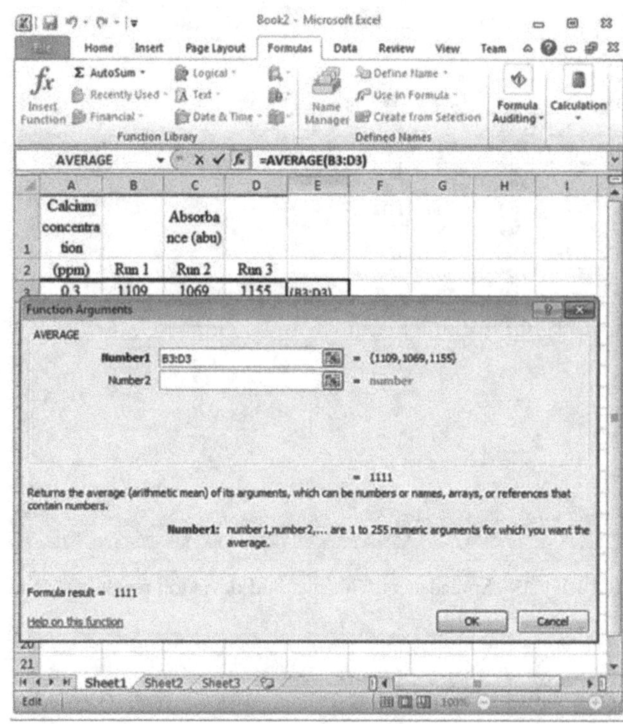

FIGURE 7.31 Functions argument window.

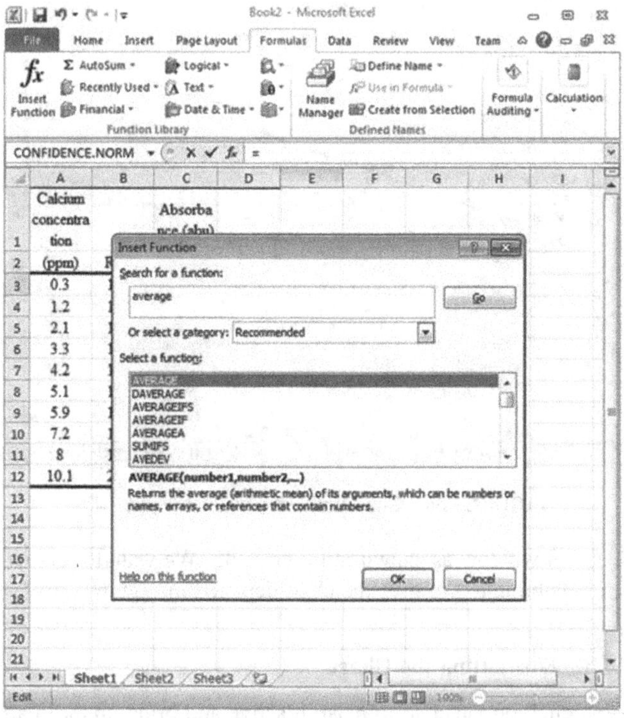

FIGURE 7.30 Inserting the Average function into the spreadsheet.

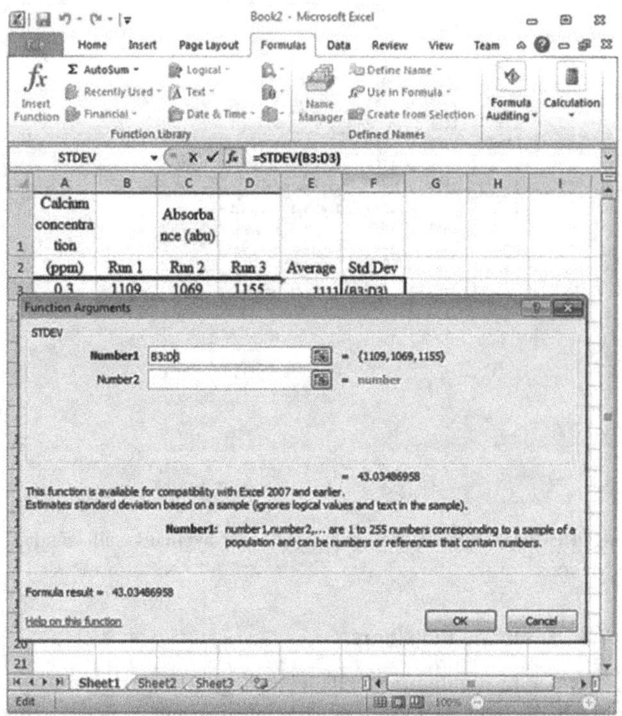

FIGURE 7.32 Standard Deviation function window.

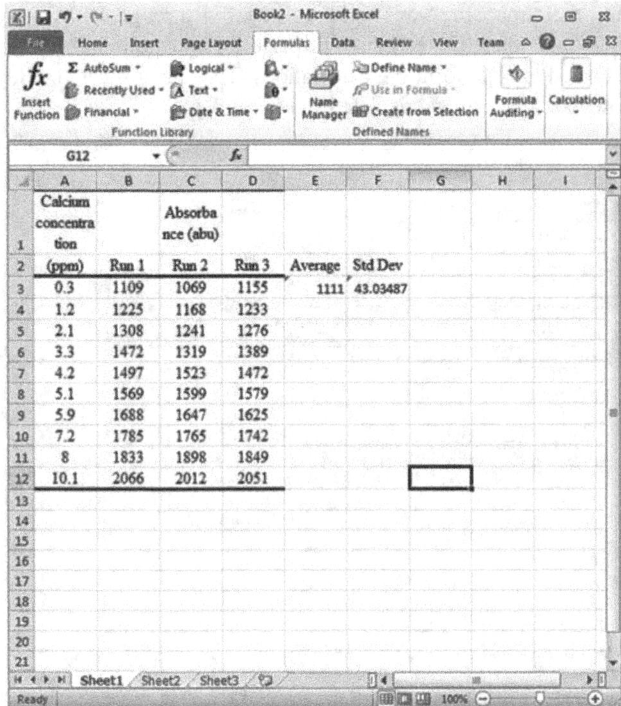

FIGURE 7.33 Spreadsheet with standard deviation result inserted.

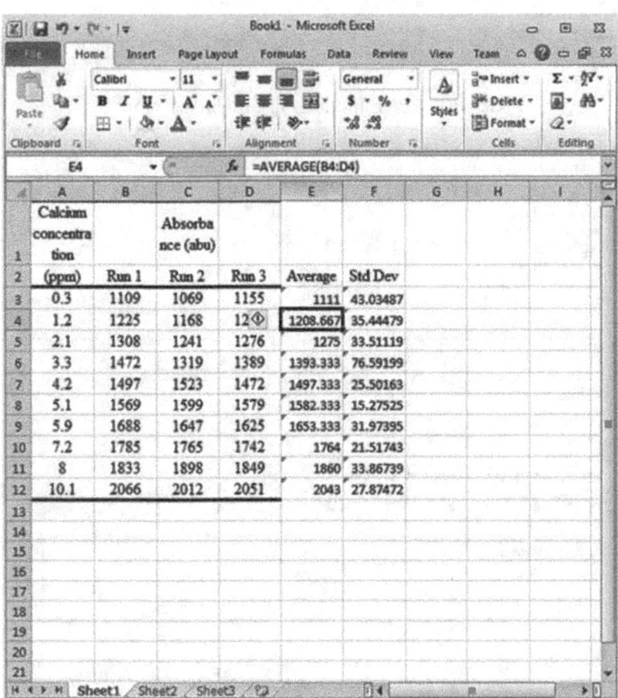

FIGURE 7.34 Spreadsheet populated with averages and standard deviations.

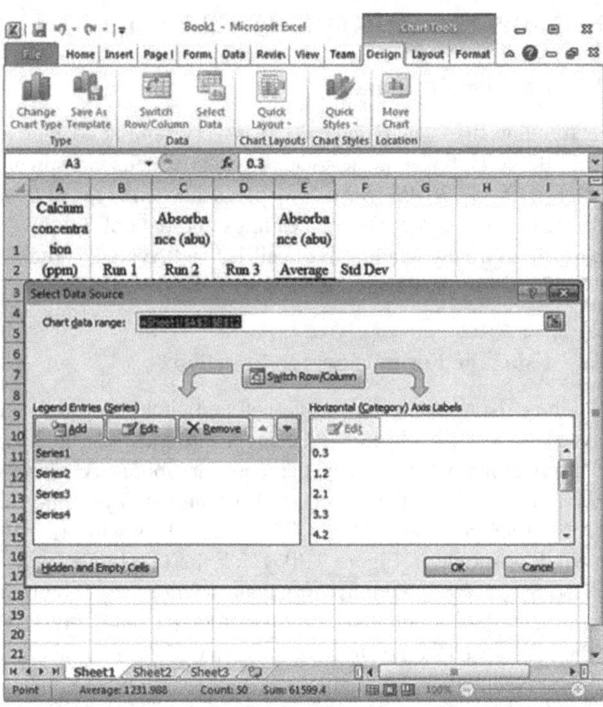

FIGURE 7.35 Spreadsheet with data selection window open.

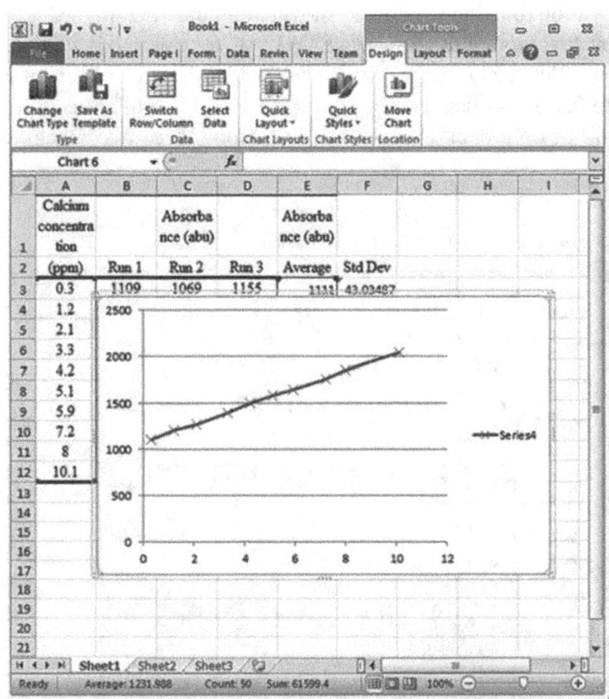

FIGURE 7.36 Chart with proper dataset plotted.

7.7.3 Inserting the Chart

Let us start by highlighting cells A3 through E12 and select Insert and Charts and select the with straight lines and markers chart. Next click the Select Data shortcut. This will open up the data selection window as shown in Figure 7.35. We want to remove series 1 through 3 to give the chart depicted in Figure 7.36.

7.7.4 Formatting the Chart

Next, click on the chart to make it active and bring up the chart formatting Toolbars. Click on the Layout tab and then click the

Chart Title selection and select Above Chart. This will insert a text box above the chart. Click inside the chart title text box and type in Calcium Concentration vs Absorbance. Next select axis titles and place a title box for the x and y axes. You can call

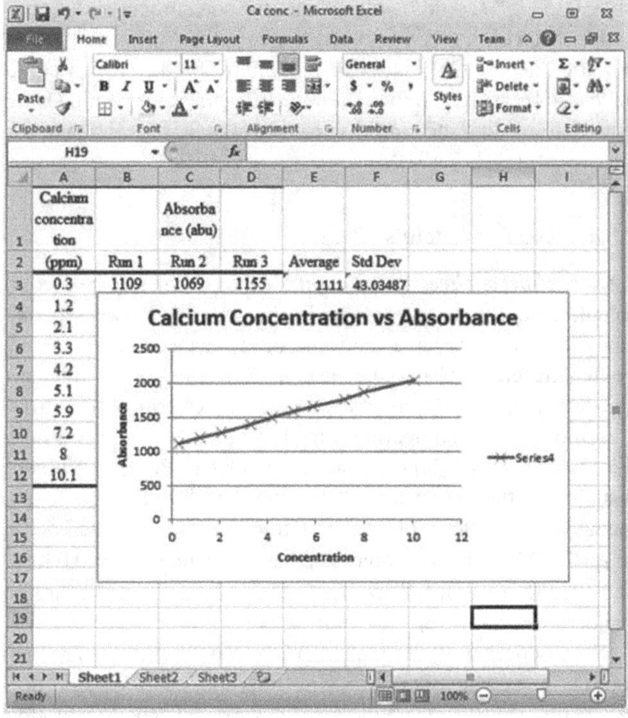

FIGURE 7.37 Chart with labels.

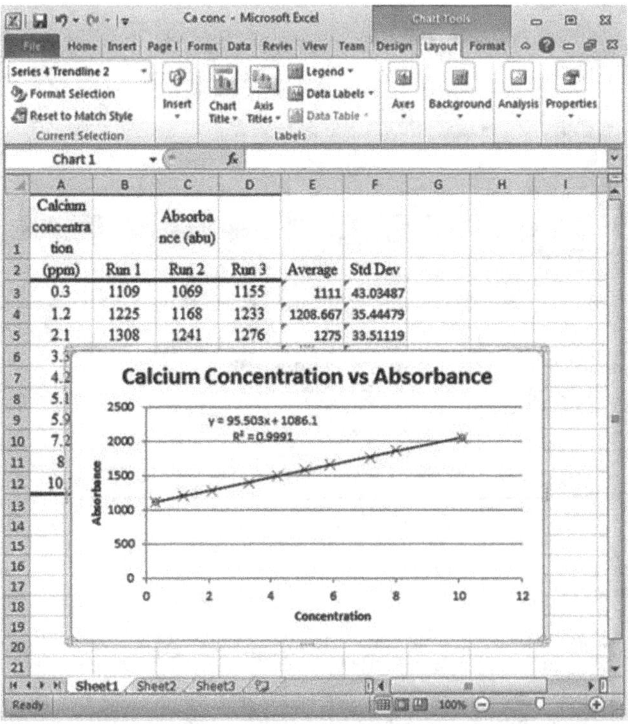

FIGURE 7.38 Chart with Trendline.

the y axis Absorbance and the x axis Concentration. The chart should now look like the one depicted in Figure 7.37. Next click layout and legend and then select None. Next click the data line to highlight the data markers. Right click and select Format Data Series. Click on Line Color and select No Line. Next click on Analysis and select Trendline. Next select More Trendline Options. Select Linear and click on Display equation on chart and Display R^2 on chart. The chart should now look like that depicted in Figure 7.38.

This concludes the basic tutorial of Microsoft Excel® in respect to the fundamental skills required for plotting simple linear relationships in Excel® 2010. We will next cover the use of spreadsheets for simple statistical calculations.

7.8 STATISTICAL ANALYSIS USING EXCEL®

7.8.1 Open and Save Excel® StatExp.xls

In plotting the data of the calcium analysis by atomic absorption spectroscopy in Table 7.2, we used the CalciumAAS workbook spreadsheet to calculate an average of the three replicate analyses by typing into the spreadsheet the function "=AVERAGE(B3:D3)." The average values were then used in the chart as the dependent Y axis values. We also calculated a standard deviation of the replicate analyses by typing into the spreadsheet "=STDEV(B3:D3)" that was used for constructing the Y error bars in the chart. Excel® contains both a wide variety of functions such as AVERAGE and STDEV that can be inserted into the spreadsheet. If not already open, open ChemTech and go to the Chapter 7 Main Menu page. Click the button "Microsoft Excel®" to open the standard workbook "Book1." The workbook "Book1" is the basic "spreadsheet" that ChemTech uses and opens when we start Microsoft Excel® from ChemTech. Go ahead and save the blank spreadsheet as "C:\StatExp.xls" for our statistical example work.

7.8.2 Sulfate Data

Table 7.3 lists the experimental data for the gravimetric measurement ($n = 10$) of sulfur content as molarity (M) sulfate [SO_4^{2-}].

The average sulfate concentration is $\bar{X} = 2.106$ M with a sample standard deviation $s = 0.0358$ M. If we choose a factor of risk level at $\alpha = 0.05$ (i.e., confidence of a 95% probability that μ would fall within our limits) then the degrees of freedom is

TABLE 7.3 Experimental data.

X_i
2.125
2.103
2.110
2.131
2.108
2.129
2.107
2.111
2.009
2.130

$df = 9$ and the Student's t value is 2.262. Using Equation 5.12, the confidence limits were calculated as

$$\bar{X} \pm \frac{ts}{\sqrt{n}} = 2.106\,\text{M} \pm \frac{(2.262)(0.0358\,\text{M})}{\sqrt{10}} \\ = 2.106\,\text{M} \pm 0.026\,\text{M} \tag{5.12}$$

From these calculations it can be determined that with a probability of 95%, μ will fall within the sulfate concentration $[SO_4^{2-}]$ range of 2.080–2.132 M.

7.8.3 Excel® Confidence Function

Excel® can also be used for calculations such as this by either typing the function into the cell, or alternatively by using the "Insert" "Function" capability. Go to the main drop-down menu and click on "Insert" and choose "Function" if using Excel® 97-2000. If using Excel® 2010 click on "Formulas" tab and then click on "Insert Functions." This will open up the "Insert Function" window. Select the category "Statistical" and scroll down and select "CONFIDENCE." There will be a brief description of the function under the scroll down window such as "CONFIDENCE.NORM(alpha,standard_dev,size) Returns the confidence interval for a population mean, as depicted in Figure 7.39. Pressing the "OK" button will bring up a "Function Arguments" page. Type into the input boxes "0.025" for "Alpha," "0.0358" for "Standard_dev," and "10" for "Size." An example of this is depicted in Figure 7.40. Press the "OK" button and the value of "0.025375" should be placed into the spreadsheet at the original location of the active cell. Also note in the "f_x function" input box the line "=CONFIDENCE(0.025,0.0358,10)," which could have also been typed into the cell to obtain the same result. Using this result we find the confidence interval as 2.106 ± 0.025 M for a range of 2.081–2.131 M.

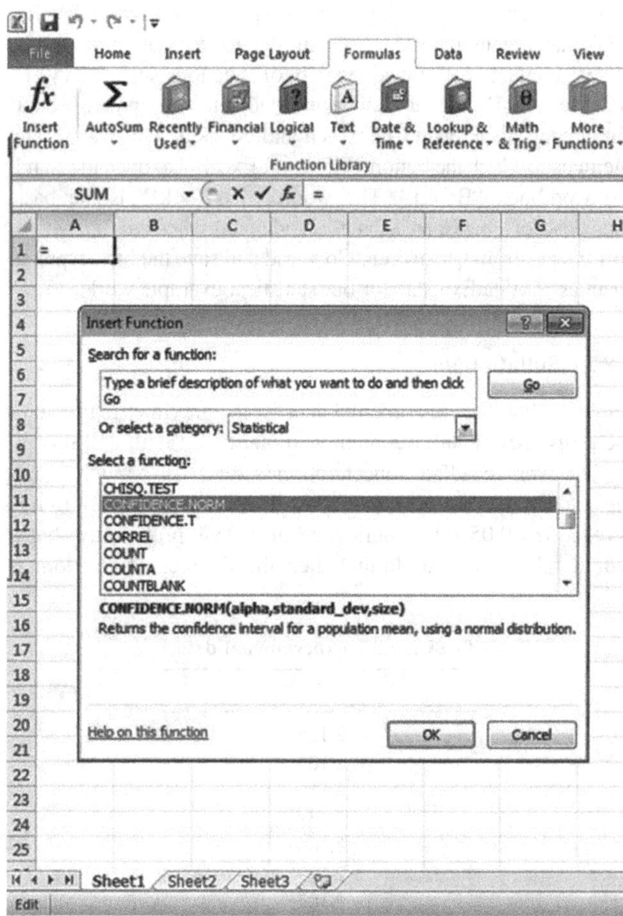

FIGURE 7.39 Microsoft Excel® Function window. Insert Function window illustrating the selection of a category and a function. Note the function's description at the bottom of the window.

7.8.4 Excel® Student's t Test

7.8.4.1 Spreadsheet Calculation I Excel® can also be used to calculate the Student's t test to determine whether two separate analytical methods have the same mean in the absence of any systematic error. Using the data in Example 5.8 for the gas chromatography (GC) measurement of cholesterol content from a corn oil extract comparing a method recently developed on a new high-pressure liquid chromatography (HPLC) instrument installed in the laboratory. Using the data listed in Table 7.4 the average cholesterol concentration for the GC method is $\bar{X}_{GC} = 3.50\%$ with a sample standard deviation $s_{GC} = 0.030\%$, and for the new HPLC method $\bar{X}_{HPLC} = 3.52\%$ with a sample standard deviation $s_{HPLC} = 0.037\%$.

When comparing the means of two sets of data we are performing a statistical hypothesis in order to answer the question of whether the means are the same or different. The assumption

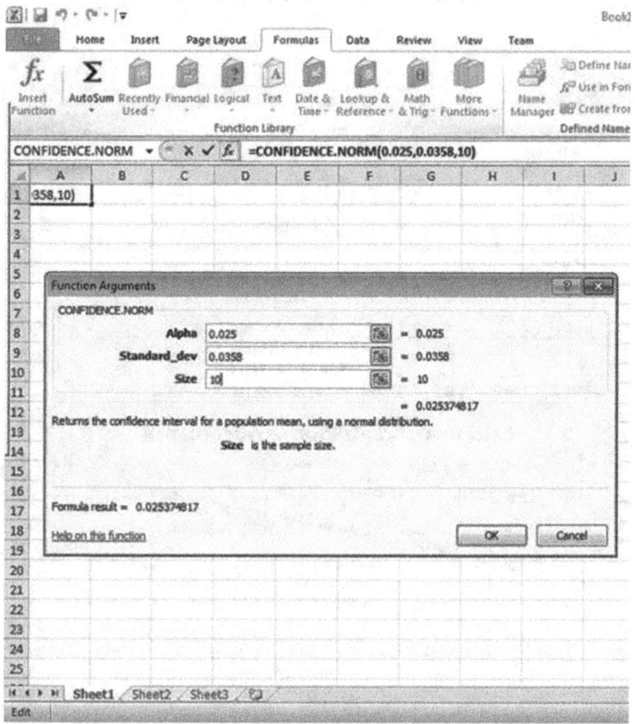

FIGURE 7.40 "Function Arguments" page. Type into the input boxes "0.025" for "Alpha," "0.0358" for "Standard_dev," and "10" for "Size."

TABLE 7.4 Experimental data.

X_i (GC, %)	X_i (HPLC, %)
3.51	3.52
3.54	3.56
3.49	3.48
3.47	3.56
3.52	3.46
3.46	3.51
3.53	3.52

that the means are the same is called the *null hypothesis* and is illustrated as H_o, where $\mu_1 - \mu_2 = 0$. The case where the means are not equal is called the alternative hypothesis and is illustrated as H_1, where $\mu_1 - \mu_2 \neq 0$. If we choose a factor of risk level at $\alpha =$ 0.05 (i.e., confidence of a 95% probability that μ would fall within our limits), then the degrees of freedom is taken from the sample set with the lowest number of data. In this problem, the $df = 6$ and the associated Student's t value is 2.015. Using Equation 5.16 and setting $\mu_1 - \mu_2 = 0$, the calculated t value for the comparison was

$$t = \frac{(3.50-3.52)}{\sqrt{\frac{0.030^2}{7}+\frac{0.037^2}{7}}} \quad (5.16)$$

$$t = -1.111$$

The absolute value of the calculated t value (1.111) is less than the Student's t test value (2.015) thus indicating that there is not a significant difference in the means obtained from the two methodology.

*7.8.4.2 **Spreadsheet Calculation II*** We essentially calculated this by hand using Equation 5.16. If the data in Table 7.4 are entered into a spreadsheet, then the t value can be calculated using the functions in the spreadsheet. Set up the "StatExp" spreadsheet as depicted in Figure 7.41. In the B9 cell we have entered the function "=COUNT(B2:B8)," in cell B10 "=AVERAGE(B2:B8)," and in cell B11 "=STDEV(B2:B8)." In the C9 cell we have entered the function "=COUNT(C2:C8)," in cell C10 "=AVERAGE(C2:C8)," and in cell C11 "=STDEV(C2:C8)." On the basis of these functions in the spreadsheet in cell B13 we can now enter a formula based upon Equation 5.16 to calculate the t statistic:

$$t = \frac{(\bar{x}_1-\bar{x}_2)}{\sqrt{\frac{s_1^2}{n_1}+\frac{s_2^2}{n_2}}} \quad (5.16)$$

$$t = \frac{(\text{B10}-\text{C10})}{\sqrt{\frac{\text{B11}^2}{\text{B9}}+\frac{\text{C11}^2}{\text{C9}}}} \quad (7.1)$$

What is actually typed into cell B13 is "=((B10-C10)/SQRT((B11^2/B9)+(C11^2/C9)))." This returns the value "–0.70638" for the t statistic.

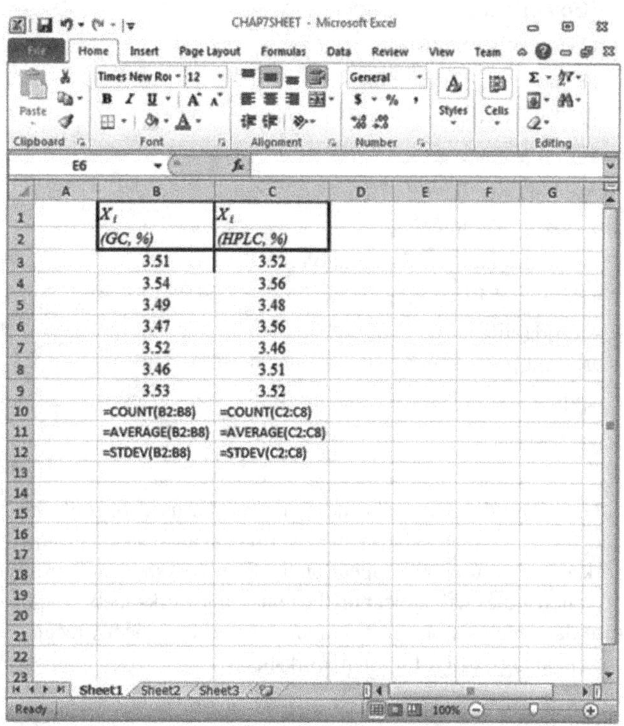

FIGURE 7.41 Set up the "StatExp" spreadsheet including in the B9 cell the function "=COUNT(B2:B8)," in cell B10 "=AVERAGE(B2:B8)," and in cell B11 "=STDEV(B2:B8)." In the C9 cell the function "=COUNT(C2:C8)," in cell C10 "=AVERAGE(C2:C8)," and in cell C11 "=STDEV(C2:C8)."

Alternatively, the "Insert" "Function" window can be used to calculate and return the result of the t-Test. However, the t statistic is not returned by a probability value representing the significance of the difference. In general, if this is used, a probability value of "$p < 0.05$" indicates a significant difference in the means of the two sample populations. This is a useful tool if a quick test is desired for comparing means for a difference that is statistically significant.

7.8.5 Excel® Tools Data Analysis

*7.8.5.1 **Analysis ToolPak*** Finally, a more advanced way of calculating the t statistic in the spreadsheet is to utilize the "Tools" "Data Analysis" menu selection. This is from the Analysis ToolPak add-in that can be installed during Excel® setup, or by clicking the main drop-down menu "Tools" and selecting "Templates and add-ins" to install the Analysis ToolPak. For Excel® 2010, click the File tab, click Options, and then click the Add-Ins category. In the Manage box, select Excel® Add-ins and then click Go. In the Add-Ins available box, select the Analysis ToolPak check box, and then click OK. Tip: If Analysis ToolPak is not listed in the Add-Ins available box, click Browse to locate it. If you are prompted that the Analysis ToolPak is not currently installed on your computer, click Yes to install it.

*7.8.5.2 **ToolPak Functions*** The following is a list of the functions available in the ToolPak.

+Anova

+Correlation

+Covariance
+Descriptive Statistics
+Exponential Smoothing
+*F*-Test Two-Sample for Variances
+Fourier Analysis
+Histogram
+Moving Average
+Random Number Generation
+Rank and Percentile
+Regression
+Sampling
+*t*-Test
+*z*-Test

Note The data analysis functions can be used on only one worksheet at a time. When you perform data analysis on grouped worksheets, results will appear on the first worksheet and empty formatted tables will appear on the remaining worksheets. To perform data analysis on the remainder of the worksheets, recalculate the analysis tool for each worksheet.

7.8.5.3 Data Analysis* t*-Test: Two-Sample Assuming Unequal Variances Go to the "Tools" main drop-down menu and select "Data Analysis" to open the Data Analysis window as shown in Figure 7.42. Scroll down the choices and click on "*t*-Test: Two-Sample Assuming Unequal Variances" and click "OK." This will bring up a second input window where ranges and parameters for the *t* statistic calculation are entered. Set up the ranges and parameters as depicted in Figure 7.43 and press the "OK" button. This will place into the spreadsheet the results of the *t*-Test calculations, and should resemble that of Figure 7.44.

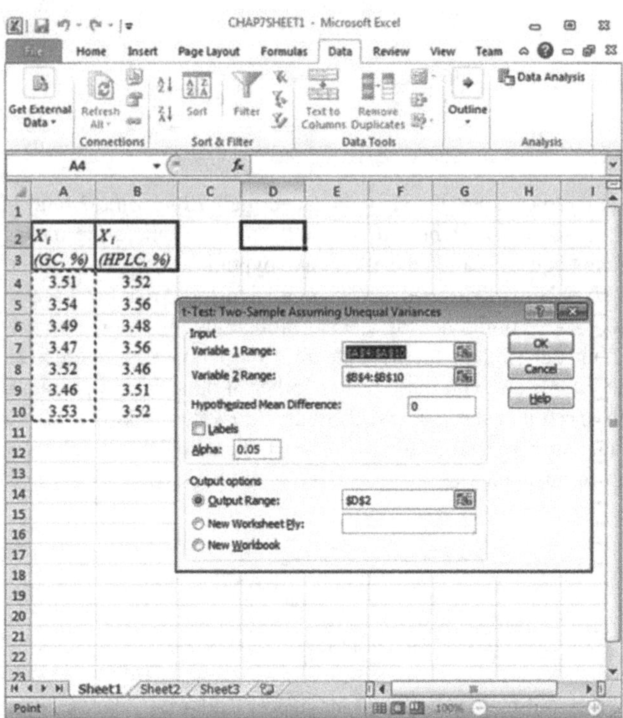

FIGURE 7.43 Set up of the ranges and parameters.

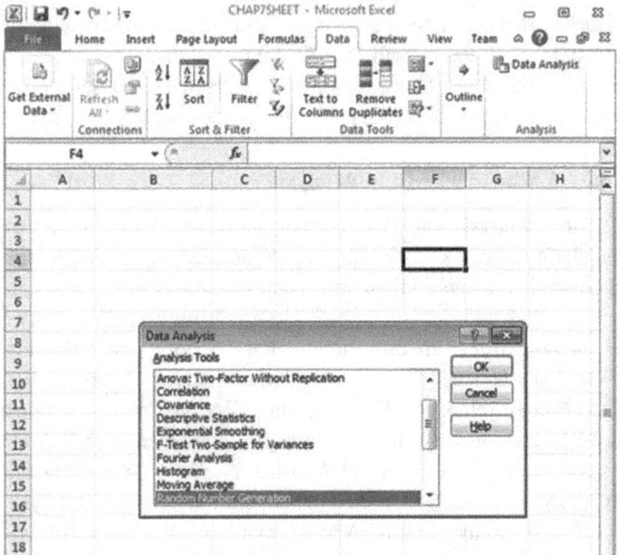

FIGURE 7.42 Data Analysis window.

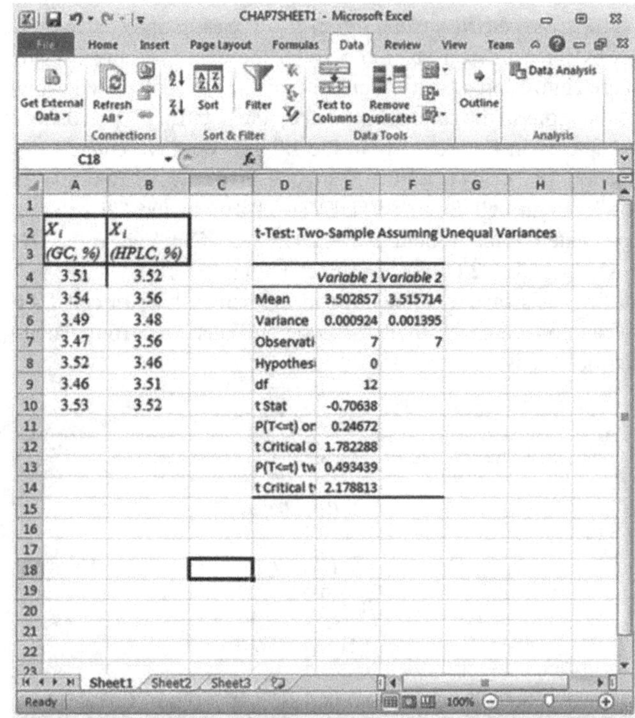

FIGURE 7.44 Spreadsheet results of the *t*-Test calculations.

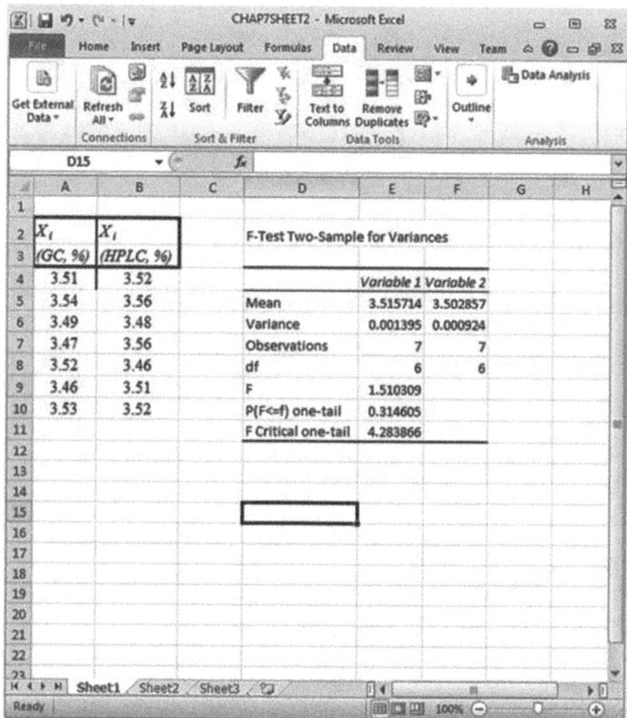

FIGURE 7.45 Spreadsheet results of the *F*-Test calculations.

	A	B	C	D	E	F	G
1							
2	X_i (GC, %)	X_i (HPLC, %)			XI (GC, %)	XI (HPLC, %)	
3	3.51	3.52					
4	3.54	3.56		Mean	3.502857	Mean	3.515714
5	3.49	3.48		Standard Error	0.011488	Standard E	0.014118
6	3.47	3.56		Median	3.51	Median	3.52
7	3.52	3.46		Mode	#N/A	Mode	3.52
8	3.46	3.51		Standard Deviatio	0.030394	Standard I	0.037353
9	3.53	3.52		Sample Variance	0.000924	Sample Va	0.001395
10				Kurtosis	-1.516803	Kurtosis	-0.85562
11				Skewness	-0.329688	Skewness	-0.20065
12				Range	0.08	Range	0.1
13				Minimum	3.46	Minimum	3.46
14				Maximum	3.54	Maximum	3.56
15				Sum	24.52	Sum	24.61
16				Count	7	Count	7

FIGURE 7.46 Spreadsheet results of the Descriptive Statistics calculations.

A small table of results is generated using the "*t*-Test: Two-Sample Assuming Unequal Variances" tool listing the mean of the two variable data sets, the variances (standard deviation squared), the null hypothesis of equal means, the degrees of freedom (df), and the *t* Stat calculated at −0.70638. Notice that this is the same value as that obtained in cell B13. The output table also contains a "*t* Critical one-tail" and a "*t* Critical Two-tail" value extracted from *t*-Test critical tables (Table 5.6) for statistical difference determination. The *t* Stat result −0.70638 calculated using the spreadsheet Analysis ToolPak is less than the *t* Critical values agreeing with the previous *t*-test calculations that the means of the two methods are not statistically different.

7.8.5.4 Analysis ToolPak* F*-test Another statistical test from the Analysis ToolPak that can be applied to the comparison of GC and HPLC methods listed in Table 7.4 is the *F*-test. The *F*-test is a ratio of the variances $\left(F = s_1^2/s_2^2\right)$ that have been arranged to give a number greater than 1. A value closer to 1 indicates that the variances are similar for the two methods. While the *F*-test ratio can easily be calculated within the spreadsheet, the use of the Analysis ToolPak also gives a results output table that includes a critical value for comparison that would otherwise require lookup in an *F*-test critical value table. If the resultant *F*-test value is less than 1, then the order of the inputted variables needs to be reversed. Go to the "Tools" main drop-down menu and select "Data Analysis" to open the Data Analysis window as shown in Figure 7.42. Scroll down the choices and click on "*F*-Test Two-Sample for Variances" and click "OK." This will bring up a second input window where ranges and parameters for the *F* statistic calculation are entered. Set up the "Variable 1 Range:" as C1 to C8, and the "Variable 2 Range:" as B1 to B8, click the labels box, put the "Output Range:" as D2, and press the "OK" button. This will place into the spreadsheet the results of the *F*-Test calculations, and should resemble that of Figure 7.45. The *F* value is 1.51 which is much smaller than the "*F* Critical one-tail" value of 4.28 indicating that the precisions of the two methods are not significantly different.

7.8.5.5 Analysis ToolPak Statistical Summary The Analysis ToolPak can also be used to calculate and give an output table of a statistical summary of data including the mean, standard error, standard deviation, maximum, minimum, and count. Go to the "Tools" main drop-down menu and select "Data Analysis" to open the Data Analysis window as shown in Figure 7.42. Scroll down the choices and click on "Descriptive Statistics" and click "OK." This will bring up a second input window where ranges and parameters for the descriptive statistic calculations are entered. Set up the input range as "$A$2:$B$9," put Grouped By: as Columns, click the "Labels in first row" box, put the Output Range as "$H$31," click the "Summary statistics" box, and press the "OK" button. This will place into the spreadsheet the results of the descriptive statistics calculations, and should resemble that of Figure 7.46.

There are numerous other calculations that can be performed in Excel® using functions, the Analysis ToolPak, macros, and visual basic for applications (VBA). Some of these approaches will be used in subsequent chapters.

8

MAKING LABORATORY SOLUTIONS

8.1 Introduction
8.2 Laboratory Reagent Fundamentals
8.3 The Periodic Table
 8.3.1 Periodic Table Descriptive Windows
8.4 Calculating Formula Weights
8.5 Calculating the Mole
8.6 Molecular Weight Calculator
8.7 Expressing Concentration
 8.7.1 Formal (*F*) Solutions
 8.7.2 Molal (*m*) Solutions
 8.7.3 Molar (*M*) Solutions
 8.7.4 Normal (*N*) Solutions
8.8 The Parts per (PP) Notation
8.9 Computer-Based Solution Calculations
 8.9.1 Computer-Based Concentration Calculation—Molarity I
 8.9.2 Computer-Based Concentration Calculation—Molarity II
 8.9.3 Computer-Based Concentration Calculation—Normality I
 8.9.4 Computer-Based Concentration Calculation—Normality II
8.10 Reactions in Solution
8.11 Chapter Key Concepts
8.12 Chapter Problems

8.1 INTRODUCTION

The preparation of laboratory solutions is a very important step in many analyses and calibrations. The proper makeup of solutions will help to ensure that the testing done and the methods used will result in the most accurate and precise measurements being done. Care needs to be taken in selecting the right reagent, the proper grade of the reagent, the weighing or taking an aliquot of the reagent, and the final volume of the prepared solution.

8.2 LABORATORY REAGENT FUNDAMENTALS

There are some basic, fundamental analytical aspects that we need to consider about reagents before calculating the makeup of the various laboratory solutions that will be needed to perform a variety of analytical methodologies and instrumental calibrations. The analyst in the laboratory needs to become familiar with the chemist's fundamental quantity known as the mole. A mole is designated by Avogadro's number of 6.023×10^{23} molecules of any given substance (6.023×10^{23} mol^{-1}). One mole of a substance is the gram formula weight obtained by adding the standard atomic weights of the elements that the substance's formula is composed of. The standard atomic weights of the elements are listed in the periodic table in Appendix IV, or alternatively in ChemTech found as a link in the Main Menu options. If not already open, open ChemTech and go to the Main Menu page. Click on the "PERIODIC TABLE OF THE ELEMENTS" button to open up the periodic table (or alternatively if needed go to the periodic table in Appendix IV).

8.3 THE PERIODIC TABLE

The periodic table that the laboratory analyst works with lists the average weighted mass for the various elements that are listed in the periodic table. The elements in the periodic table have relative weights based on the carbon isotope of mass number 12 (^{12}C). However, carbon in the periodic table is listed as having a standard atomic weight of 12.0107 g mol^{-1} while by definition the carbon 12 isotope has a unified atomic mass unit (u) of 12.000000 u. The abundances of the natural isotopes of carbon are ^{12}C (12.000000 u by definition) at 98.93(8)% and ^{13}C (13.003354826 (17) u) at 1.07(8)%. Therefore, the weighted average of the two carbon isotopes gives an atomic weight of 12.0107 g mol^{-1} as is found in the periodic table. This is true of all the elements where the average atomic weight values in the periodic table represent the sum of the mass of the isotopes for that particular element according to their

Analytical Chemistry: A Chemist and Laboratory Technician's Toolkit, First Edition. Bryan M. Ham and Aihui MaHam.

natural abundances (see Appendix I for a listing of the naturally occurring isotopes). For example, chlorine has two isotopes at ^{35}Cl (34.968852721(69 u at 75.78(4)% natural abundance) and ^{37}Cl (36.96590262(11) u at 24.22(4)% natural abundance) equating to standard atomic weight of 35.453(2) g mol^{-1}. In calculating grams and moles in the laboratory, we will always use the standard atomic weights that are listed in the periodic table (later in Chapter 12 we will see when it is necessary to use the unified atomic mass units in place of the standard atomic weights of the elements).

8.3.1 Periodic Table Descriptive Windows

With the periodic table open in ChemTech, click on the carbon element symbol. This will open up a descriptive page of carbon listing some elemental facts and also a picture of a common form of the element. This is an interactive periodic table that contains elemental facts and pictures of each of the elements. For 15 of the elements there is also included a rotatable three-dimensional movie of the element (see listing of elements with movies at bottom of the periodic table window). Close the carbon elemental facts window and click on the element zinc. In the zinc elemental facts page, double click on the zinc.mov link to open up a three-dimensional movie of the element. Using the left button of the mouse, the zinc metal block can be rotated 360°. Go ahead and try it. These movies are included for informational purposes only and are not in fact used in our specific laboratory work. Close the movie to return back to the zinc elemental facts page. The periodic table can be used to obtain an atomic weight of an element or for any of the descriptive information contained within the elemental facts page if needed.

8.4 CALCULATING FORMULA WEIGHTS

To calculate the formula weight of a substance, the atomic weights of all of the elements contained within the formula are added. For example, water has the formula H_2O, which consists of two hydrogen atoms and one oxygen atom. The formula weight (FW) would therefore be calculated as

$$\text{FW } H_2O = 2 \times 1.00794 \text{ g mol}^{-1} + 1 \times 15.9994 \text{ g mol}^{-1}$$
$$\text{FW } H_2O = 18.01528 \text{ g mol}^{-1}$$

This is the same for ions, inorganic compounds, and organic substances:

$$\text{FW } Cl^- = 1 \times 35.453 \text{ g mol}^{-1}$$
$$\text{FW } Cl^- = 35.453 \text{ g mol}^{-1}$$

8.5 CALCULATING THE MOLE

The mole of a substance is the formula weight of that substance expressed in grams. Thus, for water, 1 mole H_2O = 18.01528 g. To calculate the number of moles of a substance, the number of grams is divided by the formula weight as

$$\text{Moles of substance} = \frac{\text{grams}}{\text{FW}} \tag{8.1}$$

Here, the formula weight (FW) can be an ionic weight such as that of phosphate (PO_4^{3-}, 94.97136 g mol^{-1}), a molecular weight such as for sucrose ($C_{12}H_{22}O_{11}$, 342.29648 g mol^{-1}), or an atomic weight such as for sodium metal (Na, 22.98977 g mol^{-1}). Thus, 3.56 g of sucrose equals

$$\text{Moles of sucrose} = \frac{3.56 \text{g}}{342.29648 \text{g mol}^{-1}} \tag{8.2}$$
$$\text{Moles of sucrose} = 0.0104 \text{ mol}$$

8.6 MOLECULAR WEIGHT CALCULATOR

Learning to do this by hand helps the analyst to understand mole calculations, but it is not always necessary in today's lab where computer-based programs can assist the analyst with calculations of this kind. ChemTech contains one such program, Molecular Weight Calculator, for the analyst's use in the laboratory that is quite diverse in its calculations. If not already open, open ChemTech and go to the Main Menu. Click on the "CALCULATE FORMULA WEIGHTS" button to open the molecular weight calculator. The window for the molecular weight calculator should look like that in Figure 8.1. The program will allow the calculation of formula weights in a wide variety of representations. At this point, we will be using the approach of typing in the formula such as H_2O, and then press the "Calculate" button. Be sure to also have the "Average" choice selected for the element mode of calculation. Go ahead and type in H_2O and press "Calculate." The displayed result should be "18.01528."

8.7 EXPRESSING CONCENTRATION

There are many ways that the analyst in the laboratory can use to express the concentration of a solution. Table 8.1 lists the most

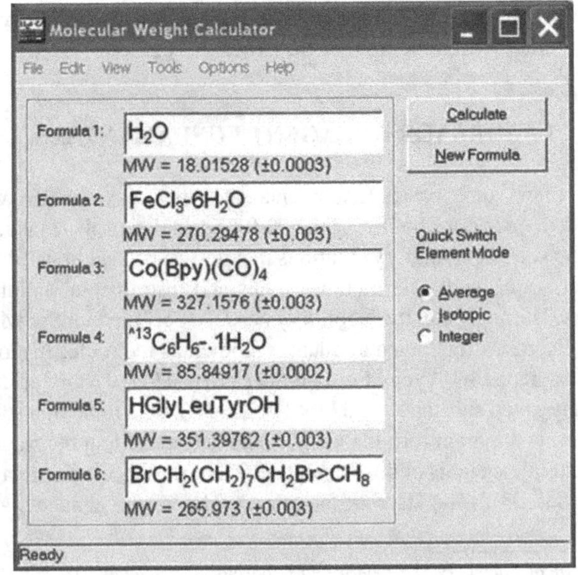

FIGURE 8.1 Main view of molecular weight calculator program.

TABLE 8.1 Forms of Expressing Concentration and Their Calculation.

Concentration	Label	Calculation
Formal	F	$\frac{\text{gram-formula-weight of solute}}{\text{liters of solution}}$
Molal	M	$\frac{\text{moles of solute}}{\text{kilograms of solvent}}$
Mole fraction	$\mathcal{N}$	$\frac{\text{moles of solute}}{\text{moles of solute + moles of solvent}}$
Molar	M	$\frac{\text{moles of solute}}{\text{liters of solution}}$
Normal	N	$\frac{\text{equivalents of solute}}{\text{liters of solution}}$
Volume percent	vol %	$\frac{\text{liters of solute}}{\text{liters of solution}} \times 100$
Weight percent	wt %	$\frac{\text{grams of solute}}{\text{grams of solute + grams of solvent}} \times 100$
Parts per million	Ppm	$\frac{\text{milligrams of solute}}{\text{kilograms of solution}}$ or $\frac{\text{milligrams of solute}}{\text{liters of solution}}$
Parts per billion	Ppb	$\frac{\text{micrograms of solute}}{\text{liters of solution}}$
Parts per trillion	Ppt	$\frac{\text{nanograms of solute}}{\text{liters of solution}}$
Grams per volume	–	$\frac{\text{grams of solute}}{\text{liters of solution}}$

commonly used forms of concentration designation. Though the molar, normal, and parts per million (ppm) are the most widely used concentration expressions, we will take a look at the calculation of a number of the types of concentration expressions listed in Table 8.1. When the term "solute" is used in Table 8.1, it refers to the substance that is being dissolved in the solution. For example, in an aqueous salt solution consisting of 20 ppm sodium chloride (NaCl), the sodium chloride would be the solute while the aqueous (water, H_2O) would be the solvent making up the solution.

8.7.1 Formal (F) Solutions

8.7.1.1 Formal (F) Solution Example What is the formal concentration of a solution made up of 3.165 g of Na_2HPO_4 in 829 ml of water? The formal calculation is based on the formula weight of the original form of the solute irrespective of whether it dissociates into other species.

$$F = \frac{\text{gram-formula-weight of solute}}{\text{liters of solution}} \tag{8.3}$$

$$F = \frac{\frac{3.165\,\text{g}\,Na_2HPO_4}{141.95884\,\text{g}\,\text{mol}^{-1}}}{0.8291} \tag{8.4}$$

$$F = 0.02689$$

8.7.2 Molal (m) Solutions

8.7.2.1 Molal (m) Solution—Simple Example What is the molal concentration (molality) of a solution made up of 1.524 g of pure formic acid (CH_2O_2) in 0.627 kg of methanol (CH_4O)? The molality calculation is based on the moles of solute and kilograms of solvent.

$$m = \frac{\text{moles of solute}}{\text{kilograms of solvent}} \tag{8.5}$$

$$m = \frac{\frac{1.524\,\text{g}\,CH_2O_2}{46.02538\,\text{g}\,\text{mol}^{-1}}}{0.627\,\text{kg}} \tag{8.6}$$

$$m = 0.05281$$

Often in the laboratory the analyst is working with reagents that are in different percentage purities, or may already be in solution in a predetermined concentration. Let us look at a similar example as 5.2 but in a more complex situation where the analyst must take into account a number of physical factors before the final concentration calculation.

8.7.2.2 Molal (m) Solution—Complex Example What is the molal concentration of a solution made up of 1.524 ml of concentrated (90.5% purity) formic acid (CH_2O_2, $d_4^{20} = 1.20$ g ml^{-1}) in 627 ml of methanol (CH_4O, $d_4^{20} = 0.7915$ g ml^{-1})?

Our first step is to convert milliliters of formic acid into moles of formic acid taking into account that the formic acid solution is 90.5% pure, and that its density is 1.20 g ml^{-1}:

$$\text{Moles}\,CH_2O_2 = \frac{1.524\,\text{ml} \times 1.20\,\text{g}\,\text{ml}^{-1} \times 0.905}{46.02538\,\text{g}\,\text{mol}^{-1}} \tag{8.7}$$

$$\text{Moles}\,CH_2O_2 = 0.03596$$

Next, we need to convert milliliters of methanol into kilograms:

$$\text{kg}\,CH_4O = 627\,\text{ml} \times 0.7915\,\text{g}\,\text{ml}^{-1} \times 1\,\text{kg} \times 1000\,\text{g}^{-1}$$

$$\text{kg}\,CH_4O = 0.4963$$

Finally, the molality of the solution can be calculated.

$$m = \frac{0.03596\,\text{mol}\,CH_2O_2}{0.4963\,\text{kg}\,CH_4O} \tag{8.8}$$

$$m = 0.07246$$

As can be seen from Examples 8.2 and 8.3 though the problems started out with the same numeric values, the final concentration calculations are quite different. As the analyst begins to learn, there are numerous tables that are very useful in the analytical chemistry laboratory. One such table is the listing of some common acids and bases that are often used in the laboratory along with their associated concentrations as illustrated in Table 8.2. The reagents are usually purchased from vendors and typically contain the same concentrations. For example,

TABLE 8.2 Listing of Some Common Laboratory Acids and Bases Including Physical Properties Necessary for Solution Preparations.

Reagent	Formula and Formula Weight	Purity % by Weight	Density g ml^{-1} (25 °C)	Molarity
Acetic acid glacial	CH_3COOH 60.052	99.8	1.049	17.43
Ammonium hydroxide	NH_4OH 35.046	57	0.90	14.80
Ammonia solution[a]	NH_3 17.030	28	0.90	14.80
Formic acid	HCOOH 46.025	96.0	1.22	25.45
Hydrobromic acid	HBr 80.912	48	1.49	8.84
Hydrochloric acid	HCl 36.461	37	1.2	12.18
Hydrofluoric acid	HF 20.006	51–55	1.15	29.32–31.62
Hydroiodic acid	HI 127.912	57	1.701	7.58
Nitric acid	HNO_3 63.013	70	1.400	15.55
Oxalic acid in H_2O	HO_2CCO_2H	–	0.99	0.1 N
Perchloric acid	$HClO_4$ 100.458	70	1.664	11.59
Phosphoric acid	H_3PO_4 97.995	85	1.685	14.62
Potassium hydroxide	KOH 56.106	45	1.456	11.68
Sodium hydroxide	NaOH 39.997	40	1.327	10.0
Sulfuric acid	H_2SO_4 98.079	97	1.840	18.20

[a] Same solution as ammonium hydroxide expressed as ammonia.
Note: Need to reference Sigma-Aldrich and Merck Index.

the concentrated hydrochloric acid (HCl) reagent used in the laboratory is not a 100% pure solution but rather an approximate 37% purity. Pure hydrochloric acid is in fact a gas at room temperature and not a solution. The concentrated hydrochloric acid reagent the analyst uses in the laboratory is a saturated solution prepared by bubbling pure HCl gas through water. A concentrated (saturated) solution of HCl can be prepared in the laboratory by bubbling pure HCl gas through water (in fact, this is how the reagent was prepared in the laboratory before commercial availability quite a while ago), but this is not practical or convenient in the volumes that concentrated HCl are typically used in the laboratory, nor very safe due to the high toxicity of the pure HCl gas.

8.7.3 Molar (*M*) Solutions

8.7.3.1 Molar (M) Solution Example What is the molar concentration (molarity) of a solution made up of 4.357 g of pure sodium chloride (NaCl) in 637 ml of H_2O? The molar calculation is based on the moles of solute and liters of final solution. In the makeup of molar solutions, typically the salt is weighed into a volumetric flask, a portion of solvent is added to dissolve the salt, and then the volume is brought to the final calibrated mark of the flask.

$$M = \frac{\text{moles of solute}}{\text{liters of solution}} \tag{8.9}$$

$$M = \frac{\dfrac{4.357\,\text{g NaCl}}{58.44247\,\text{g mol}^{-1}}}{0.6371} \qquad M = 0.1170 \tag{8.10}$$

An important point to note here is that the molar solution is different from the formal solution in that the formal solution is based on the FW of the solute irrespective of dissolution and any subsequent solute dissociation, while the molar solution can represent the original solute or any dissociated species derived from dissolution.

To illustrate this, the sodium chloride solution in Example 8.4 has a concentration that can be represented as both 0.1170 *F* and 0.1170 *M* NaCl. However, because of the fact that sodium chloride dissociates ~100% in solution, it is also 0.1170 *M* Na^+ and 0.1170 *M* Cl^-. This can be shown by way of the use of what is known as "gravimetric factors." Gravimetric factors are multiplication factors used to convert the amount of one substance into the amount of another related substance using the following relationship:

$$\text{Gravimetric factor} = \frac{m \times \text{formula weight (substance needed)}}{\text{n} \times \text{formula weight (second substance)}} \tag{8.11}$$

For example, to convert grams NaCl into grams Na^+, the following relationship is used:

$$\text{grams Na}^+ = \text{grams NaCl} \times \frac{1 \times \text{formula weight Na}^+}{1 \times \text{formula weight NaCl}} \tag{8.12}$$

$$\text{grams Na}^+ = 4.357\,\text{grams} \times \frac{1 \times 22.98977}{1 \times 58.44247} \qquad \text{grams Na}^+ = 1.714\text{ g} \tag{8.13}$$

Thus the molarity of the sodium is

$$M = \frac{\dfrac{1.714\,\text{g Na}^+}{22.98977\,\text{g mol}^{-1}}}{0.6371} \qquad M = 0.1170 \tag{8.14}$$

And in a similar manner, to convert grams NaCl into grams Cl^-, the following relationship is used:

$$\text{grams Cl}^- = \text{grams NaCl} \times \frac{1 \times \text{formula weight Cl}^-}{1 \times \text{formula weight NaCl}} \tag{8.15}$$

$$\text{grams Cl}^- = 4.357 \text{ grams} \times \frac{1 \times 35.4527}{1 \times 58.44247} \quad (8.16)$$

$$\text{grams Cl}^- = 2.643 \text{ g}$$

Thus the molarity of the chloride is

$$M = \frac{\dfrac{2.643 \text{ g Cl}^-}{35.4527 \text{ g mol}^{-1}}}{0.6371} \quad (8.17)$$

$$M = 0.1170$$

In this particular case, making a sodium chloride (NaCl) salt solution that is 0.1170 M, a solution that is also 0.1170 M in Na^+ and Cl^- is also produced due to ~100% dissociation in aqueous solution.

8.7.3.2 Molar (M) Solution of K_2CO_3 A solution is made up of 3.891 g of pure potassium carbonate (K_2CO_3) in 862 ml of H_2O? Assuming 100% disassociation, what is the molar concentration (molarity) of the potassium carbonate, the potassium ion (K^+), and the carbonate ion (CO_3^{-2})?

Molarity of potassium carbonate:

$$M = \frac{\dfrac{3.891 \text{ g K}_2\text{CO}_3}{138.2055 \text{ g mol}^{-1}}}{0.8621} \quad (8.18)$$

$$M = 0.03266$$

Molarity of potassium ion:

$$\text{grams K}^+ = \text{grams K}_2\text{CO}_3 \times \frac{2 \times \text{formula weight K}^+}{1 \times \text{formula weight K}_2\text{CO}_3} \quad (8.19)$$

$$\text{grams K}^+ = 3.891 \text{ grams} \times \frac{2 \times 39.0983}{1 \times 138.2055} \quad (8.20)$$

$$\text{grams K}^+ = 2.2015 \text{ g}$$

$$M = \frac{\dfrac{2.2015 \text{ g K}^+}{39.0983 \text{ g mol}^{-1}}}{0.8621} \quad (8.21)$$

$$M = 0.06532$$

Molarity of carbonate ion:

$$\text{grams CO}_3^{2-} = \text{grams K}_2\text{CO}_3 \times \frac{1 \times \text{formula weight CO}_3^{2-}}{1 \times \text{formula weight K}_2\text{CO}_3} \quad (8.22)$$

$$\text{grams CO}_3^{2-} = 3.891 \text{ grams} \times \frac{1 \times 60.0089}{1 \times 138.2055} \quad (8.23)$$

$$\text{grams CO}_3^{2-} = 1.689 \text{ g}$$

$$M = \frac{\dfrac{1.689 \text{ g CO}_3^{2-}}{60.0089 \text{ g mol}^{-1}}}{0.8621} \quad (8.24)$$

$$M = 0.03266$$

From the above exercise, we see that the molarity of the potassium ion is twice that of the original potassium carbonate and also the carbonate ion. This is because the formula for potassium carbonate, K_2CO_3, contains 2 moles of potassium and 1 mole of carbonate for every mole of potassium carbonate.

Sometimes the laboratory analyst finds it more convenient to work in millimoles. This allows volumetric calculations to be based on milligrams and milliliters versus grams and liters. A millimole (mmoles) is the expression of the moles of a substance as milligrams divided by the formula weight.

$$\text{mmoles} = \frac{\text{mg}}{\text{formula weight}} \quad (8.25)$$

$$\text{Molarity} = \frac{\text{mmoles}}{\text{ml}} \quad (8.26)$$

Some useful expressions involving molar calculations are as follows:

$$(\text{liters})(M) = \text{moles} \quad (8.27)$$

$$(\text{moles})(\text{formula weight}) = \text{grams} \quad (8.28)$$

$$(\text{liters})(M)(\text{formula weight}) = \text{grams} \quad (8.29)$$

$$(\text{ml})(M) = \text{mmoles} \quad (8.30)$$

$$(\text{mmoles})(\text{formula weight}) = \text{mg} \quad (8.31)$$

$$(\text{ml})(M)(\text{formula weight}) = \text{mg} \quad (8.32)$$

8.7.4 Normal (N) Solutions

When working with normal solutions, the "equivalents" of solute need to be taken into account. The equivalents are based on the moles of a reactant that the solute is able to contribute to some type of chemical reaction. For acid and base reactions, an equivalent (eq) of acid is the quantity of acid that supplies 1 mole of H^+, while an equivalent of base is the quantity of the base reacting with 1 mole of H^+. The use of normality effectively "normalizes" the concentrations to take into account the reacting ratio of the species in solution. This allows a direct use of the normality of the solution when performing titration calculations (will be covered in Chapter 10). Normality is expressed as

$$N = \frac{\text{equivalents of solute}}{\text{liters of solution}}, \quad (8.33)$$

where

$$\text{equivalents of solute} = \frac{\text{g solute}}{\text{equivalent weight solute}}. \quad (8.34)$$

The equivalent is related to the formula weight as follows:

$$1 \text{ eq of HCl} = 1 \text{ mole of HCl}$$
$$= 36.461 \text{ g of HCl (formula weight of HCl)}.$$

This is due to the fact that HCl supplies 1 mole of H^+ (formula is H_1Cl_1); thus, 1 eq of HCl equals the formula weight of HCl. Let us look at sulfuric acid (H_2SO_4):

$$1 \text{ eq of } H_2SO_4 = ½ \text{ mole } H_2SO_4$$
$$= 49.0397 \text{ g of } H_2SO_4 (½ \text{ formula weight of } H_2SO_4).$$

This is due to sulfuric acid supplying 2 moles of H^+ for every mole of H_2SO_4. When dealing with bases, it is the number of base substance (OH^-) that is supplied to neutralize the acid (H^+). For example, the weight of an equivalent of sodium hydroxide NaOH is

$$1 \text{ eq of NaOH} = 1 \text{ mole NaOH}$$
$$= 39.997 \text{ g of NaOH (formula weight of NaOH)}.$$

Sodium hydroxide in solution supplies 1 mole of OH^- for every mole of NaOH; thus, 1 eq of NaOH equals the formula weight of NaOH. The same for bases also applies when more than 1 mole of base is supplied. For example, for calcium hydroxide $Ca(OH)_2$ (slightly soluble in water):

$$1 \text{ eq of } Ca(OH)_2 = ½ \text{ mole } Ca(OH)_2$$
$$= 37.0463 \text{ g of } Ca(OH)_2$$
$$\times (½ \text{ formula weight of } Ca(OH)_2).$$

This is due to calcium hydroxide supplying 2 moles of base (OH^-), which requires 2 moles of acid (H^+) to neutralize.

Therefore, we can set up a general expression for the equivalent weight of a substance as

$$\text{equivalent weight} = \frac{\text{formula weight}}{\text{number of } H^+}, \quad (8.35)$$

where the "number of H^+" represents either the number of H^+ supplied by an acid (e.g., 2 H^+ supplied by H_2SO_4) or the number of H^+ required to neutralize the number of base (hydroxyl OH^-) molecules supplied by the base (e.g., NaOH supplies 1 OH^- requiring 1 H^+ to neutralize it). Thus, the equivalents of solute is expressed as

$$\text{equivalents of solute} = \frac{\text{g solute}}{\text{equivalent weight}}. \quad (8.36)$$

The complete expression of normality is

$$N = \frac{\frac{\text{g solute}}{\text{equivalent weight}}}{\text{liters of solution}} \quad (8.37)$$

In abbreviated notation, this becomes

$$N = \frac{\frac{\text{g}}{\text{eq wt}}}{1} \quad (8.38)$$

$$N = \frac{\text{eq}}{1} \quad (8.39)$$

Normality can also be expressed in milligrams and milliliters using milliequivalents (meq):

$$\text{meq} = \frac{\text{mg}}{\text{eq wt}} \quad (8.40)$$

$$N = \frac{\text{meq}}{\text{ml}} \quad (8.41)$$

8.7.4.1 Normal (N) Solution Calculation Example What is the normality of a solution made up of the addition of 5 ml of concentrated sulfuric acid to water that is brought to a final volume of 350 ml. Note: the addition of concentrated sulfuric acid to water is a very exothermic reaction that generates substantial heat. When mixing acids with water a rule of thumb is ALWAYS to add acid to water, never add water to acid. Also, when preparing an acidic solution from concentrated sulfuric acid, the water contained within the Erlenmeyer flask or beaker for the solution is first immersed in an ice bath and allowed to cool thoroughly. The sulfuric acid is then added very slowly to the chilled water and stirred often to dissipate the generated heat into the ice bath. Adding the concentrated acid too quickly will generate a dangerous amount of heat and also produces sputtering out of the solution of the acid which is also very dangerous if contacted to skin or eyes. Always wear the proper protective clothing and eye protection when working in the laboratory. As covered in Chapter 2 Laboratory Safety always flush exposed skin for 5 min with water and splashes to the eye should be flushed for 15 min.

We must first convert the milliliters of sulfuric acid into grams. From Table 8.2, we obtain the purity and density of the sulfuric acid from which the amount can be obtained:

$$\text{weight } H_2SO_4 = 5 \text{ ml} \times \frac{1.840 \text{ g}}{\text{ml}} \times 0.97$$
$$\text{weight } H_2SO_4 = 8.924 \text{ g} \quad (8.42)$$

The eq wt of H_2SO_4 is 49.0397 g (½ the formula weight due to 2 H^+ available) thus the normality is

$$N = \frac{\frac{8.924 \text{ g}}{49.0397 \text{ g}}}{0.3501}.$$
$$N = 0.5199 \quad (8.43)$$

An alternative approach would be to use the stated molarity of the sulfuric acid of 18.20 *M*, where the relationship between molarity and normality is

$$N = M \times nH^+$$
$$N = 18.20 \times 2 \quad (8.44)$$
$$N = 36.40$$

$$N_I V_I = N_F V_F$$
$$(18.20 \text{ N})(5 \text{ ml}) = N_F(350 \text{ ml}) \quad (8.45)$$

$$N_F = \frac{(18.20 \text{ N})(5 \text{ ml})}{(350 \text{ ml})}.$$
$$N_F = 0.52 \quad (8.46)$$

The analyst should note that normality does not represent a general expression of concentration but is substance specific according to reacting ratios. It has become more common in the laboratory to list the molarity of a solution as representing its general use concentration, and then calculating the normality of the specific analytes in the solution according to the use of the solution (e.g., if used for titrations). In light of this, The National Institute of Standards and Testing (NIST) has stated that the concentration designation of normality is obsolete and has recommended that it should eventually be discontinued.

8.8 THE PARTS PER (PP) NOTATION

An often convenient representation of a concentration is to use the "parts per" notation such as "ppm," which denotes that there is one part of something for a million parts of something else. The concentration unit of ppm is often used in trace analysis of analytes of interest from for example trace metals analysis (e.g., ≤1.5 ppm mercury, Hg), pesticide analysis (e.g., 3.64 ppm malation), and mycotoxin analysis (e.g., 1.55 ppm total afla toxin). Interestingly, the ppm designation was not recently introduced but stems back from a British beer poisoning epidemic in 1900. A Royal Commission headed by Lord Kelvin (William Thomson) initiated an arsenic tolerance based on the ppm concentration designation. PPM can be a weight/weight (wt/wt) relationship such as mg/kg, or a weight/volume (wt/v) such as mg/l. If the numeric value for ppm is 10,000 or greater, it is converted to % (i.e., 10,000 ppm = 1%).

$$\text{ppm} = \frac{\text{milligrams of solute}}{\text{kilograms of solution}} \text{ or } \frac{\text{milligrams of solute}}{\text{liters of solution}} \tag{8.47}$$

The parts per billion (ppb) and parts per trillion (ppt) are subsequent 10^{-3} factors and calculated as follows:

$$\text{ppb} = \frac{\text{micrograms of solute}}{\text{liters of solution}} \tag{8.48}$$

$$\text{ppt} = \frac{\text{nanograms of solute}}{\text{liters of solution}} \tag{8.49}$$

8.9 COMPUTER-BASED SOLUTION CALCULATIONS

Now that we have gone through the calculating and making of laboratory solutions by hand, let us now turn to computer-based solution calculations. ChemTech has the ability to help in simple solution calculations for both the preparation of molar and normal solutions. Open ChemTech and click on "Chapter 8: Laboratory Solutions and Titrations" to open up the Chapter 8 main menu shown in Figure 8.2. Click on the "Making Laboratory Solutions" button to open up the main menu shown in Figure 8.3. This main menu contains links to making molar solutions, normal solutions, and buffer solutions at different pH values. Click on the first button "Making Molar Solutions I" to open up the molar solution solver page as shown in Figure 8.4.

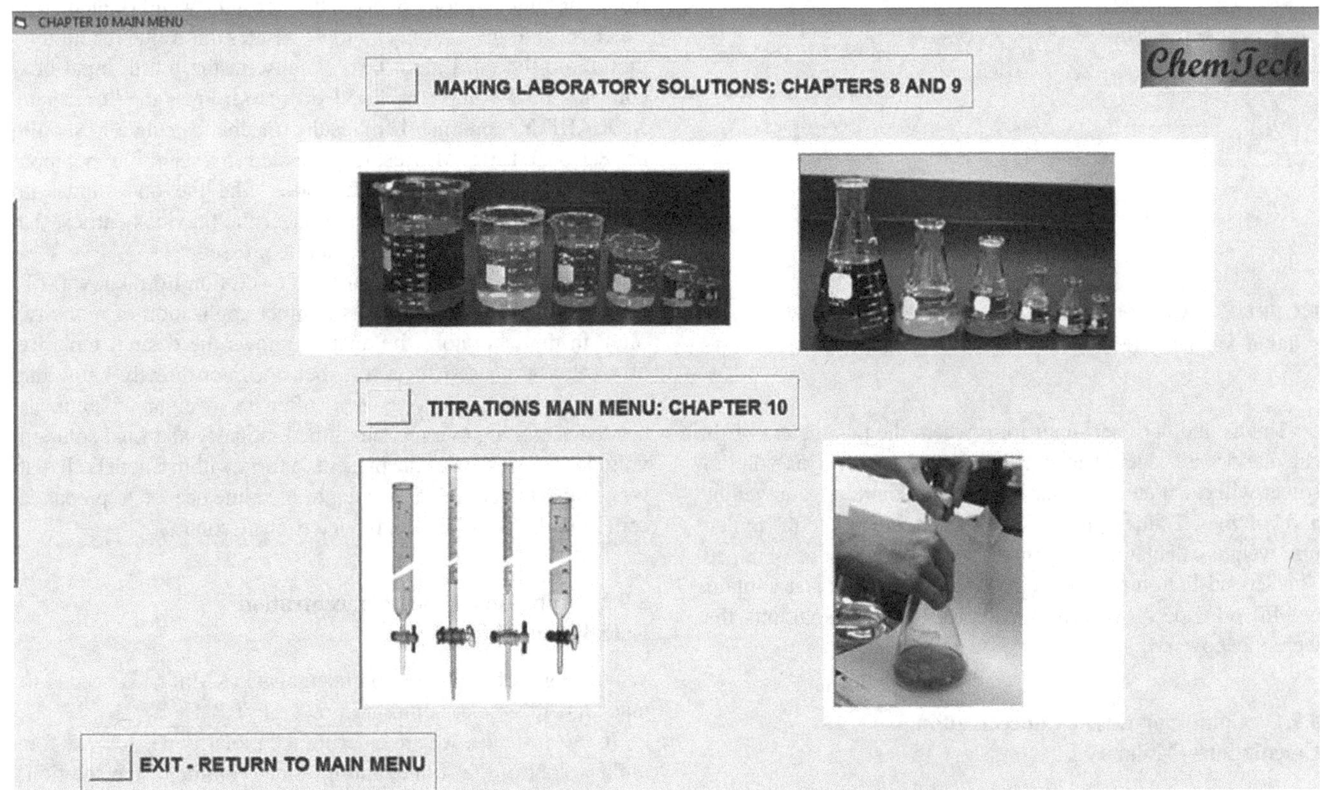

FIGURE 8.2 Chapter 8 Laboratory Solutions and Titrations main menu.

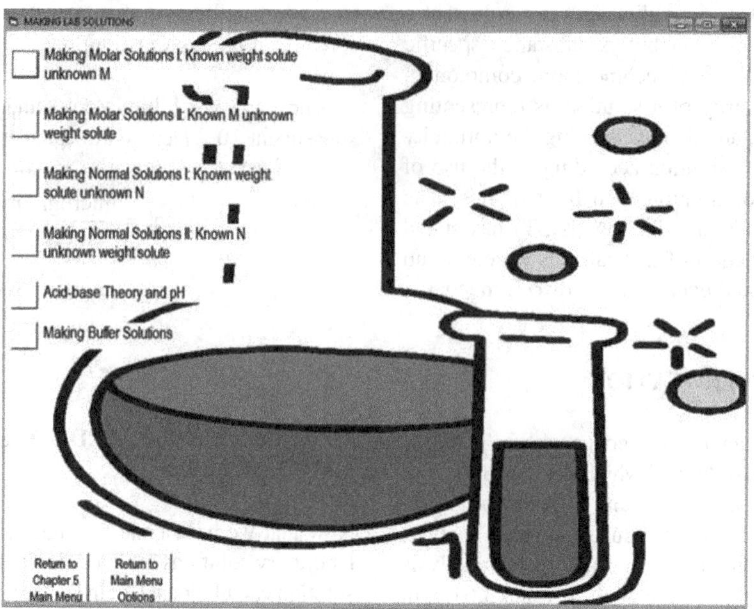

FIGURE 8.3 Chapter 8 Making Laboratory Solutions main menu.

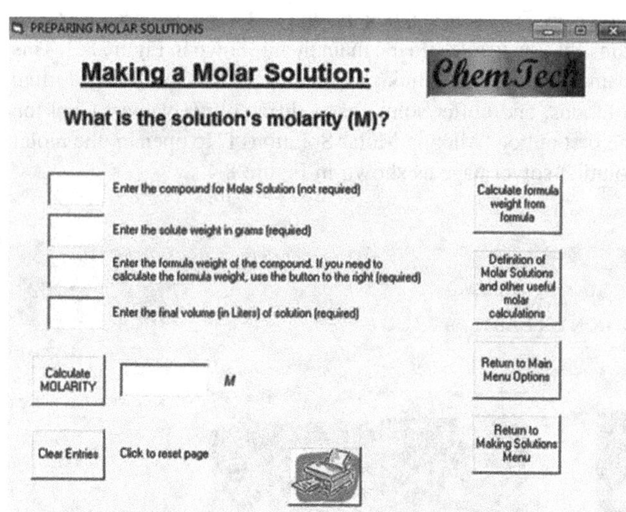

FIGURE 8.4 Chapter 8 Making a Molar Solutions I: What is the solution's molarity (M)?

This is another user input form where the parameters of the calculation are entered to get the needed result. This solution solver will calculate the molarity (M) of a solution from weights and volumes. To the right of the page there is a link to the molecular weight calculator if a formula weight needs to be obtained. The second button on the right brings up a page that contains helpful reminders concerning calculating molar solutions that were just covered.

8.9.1 Computer-Based Concentration Calculation—Molarity I

What is the sodium chloride molarity (M) of a solution made up of 6.883 g NaCl dissolved in 1.452 l H_2O?

In the Chapter 8 option Making Molar Solutions I: Known weight solute unknown M (What is the solution's molarity (M)?) page as shown in Figure 8.4, enter the compound in the first input box as "NaCl" (this is actually not required but can be useful if the page is printed out with the parameters and results). In the second input box, the solute's weight is entered (in grams) as 6.883. Next, click the "Calculate formula weight from formula" button to open the formula weight calculator. Enter NaCl to get 58.44247 g mol^{-1}. Enter this value (excluding units) into the third input box. Finally, in the fourth input box enter the final volume as 1.452 (in liters). Press the "Calculate MOLARITY" button. The results of the calculation should be 8.111153E-02 M (result presented in scientific notation when ≤ 0.09). The page should look like the one shown in Figure 8.5. To reset the page and clear the previous entries, the bottom "Clear Entries" button can be pressed.

While calculating the molarity of a solution from known solute weight and final volume is a direct application of molarity, often in the laboratory the analyst knows the desired molarity of a solution needed for a test, but does not directly know the weight of solute needed. Thus, often the weight of solute is needed according to a predetermined molarity and final solution volume. The next example presents a form within ChemTech that we can use to determine the weight of solute needed to prepare a certain molar solution at a predetermined volume.

8.9.2 Computer-Based Concentration Calculation—Molarity II

What weight of potassium permanganate ($KMnO_4$) is needed to make a 0.155 molar solution in 695 ml of H_2O?

In this instance, we are entering the molarity desired, the formula weight of the solute, and the final volume of the solution. The page also contains a link to the formula weight calculator and a page outlining molar calculations.

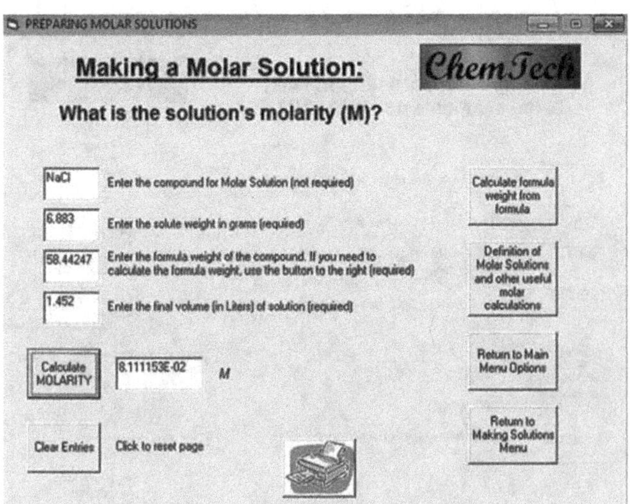

FIGURE 8.5 Chapter 8 option Making a Molar Solutions I: Known weight solute unknown M (What is the solution's molarity (M)?) results page with input from Example 8.7.

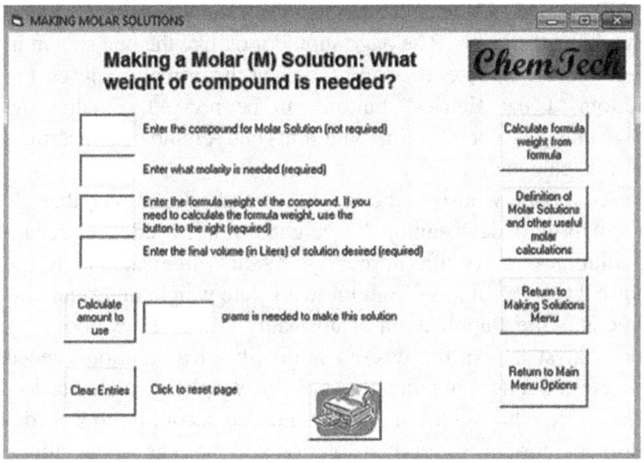

FIGURE 8.6 Chapter 8 option Making Molar Solutions II: Known M unknown weight solute (What weight of compound is needed?)

In the Chapter 8 option Making Molar Solutions II: Known M unknown weight solute (What weight of compound is needed?) page as shown in Figure 8.6, enter the compound in the first input box as "$KMnO_4$" (this is actually not required but can be useful if the page is printed out with the parameters and results). In the second input box, the molarity of the solution is entered as 0.155. Next, click the "Calculate formula weight from formula" button to open the formula weight calculator. Enter $KMnO_4$ to get 158.03395 g mol^{-1}. Enter this value (excluding units) into the third input box. Finally, in the fourth input box enter the final volume as 0.695 (in liters). Press the "Calculate amount to use" button. The results of the calculation should be 17.0242 g. The page should look like the one shown in Figure 8.7. To reset the page and clear the previous entries the bottom "Clear Entries" button can be pressed.

The calculations involved with normal solutions are also included in ChemTech Chapter 8. Open Chapter 8 option Making

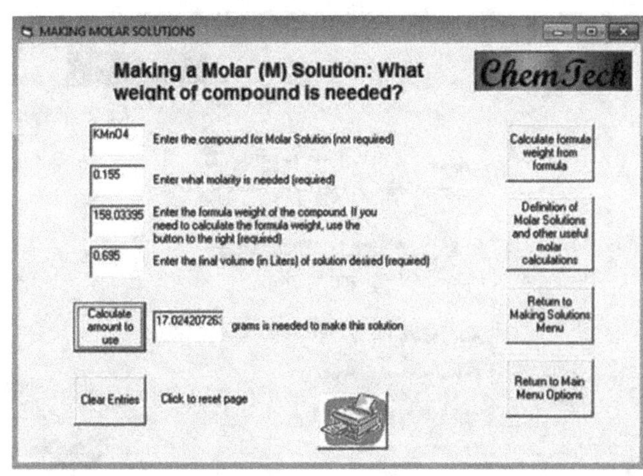

FIGURE 8.7 Chapter 8 option Making Molar Solutions II: Known M unknown weight solute (What weight of compound is needed?) page illustrated with input from Example 8.8 and subsequent result.

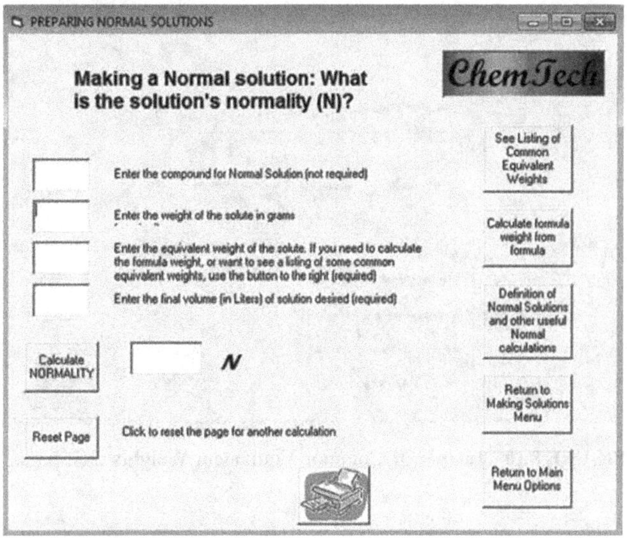

FIGURE 8.8 Chapter 8 option Making Normal Solutions I: What is the solution's Normality?

Laboratory Solutions main menu as depicted in Figure 8.3. Below the options for making molar solutions I and II that was just covered are similar buttons for procedures to make normal solutions. Click on the button for Making Normal Solutions I: Known weight solute unknown N to open up a calculation page as shown in Figure 8.8. This page is for calculating the normality of a solution when the weight of solute and final volume are known. The page also contains links to some useful pages needed when calculation the normality of solutions. Let us look at a specific example.

8.9.3 Computer-Based Concentration Calculation—Normality I

What is the hydrochloric acid normality (N) of a solution made up of 3.55 g HCl dissolved in 739 ml H_2O?

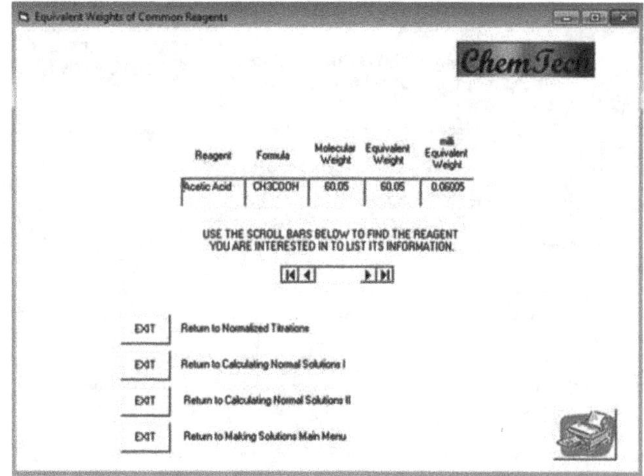

FIGURE 8.9 Listing of Common Equivalent Weights page.

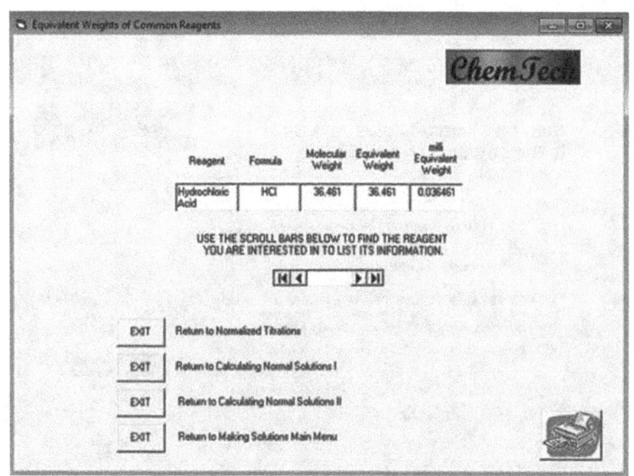

FIGURE 8.10 Listing of Common Equivalent Weights HCl page.

In the Chapter 8 option Making Normal Solutions I: Known weight solute unknown N page as shown in Figure 8.8, enter the compound in the first input box as "HCl" (this is actually not required but can be useful if the page is printed out with the parameters and results). In the second input box, the solute's weight is entered (in grams) as 3.55. While the page does contain a link for calculating formula weights from the formula, the page also contains a link to a listing of some common equivalent weights. Click the "See Listing of Common Equivalent Weights" button to open the page that is shown in Figure 8.9. The page is listing the associated data for acetic acid. Use the horizontal data scroll bar located under the data boxes to move to hydrochloric acid. When there, the page should look like that in Figure 8.10. The value needed can either be written down or the page can be printed using the printer button at the bottom of the page. Click the "Return to Calculating Normal Solutions I: button to return to our previous page. Enter the equivalent weight of 36.461 g mol^{-1} (excluding units) into the third input box. Finally, in the fourth input box enter the final volume as 0.739 (in liters). Press the

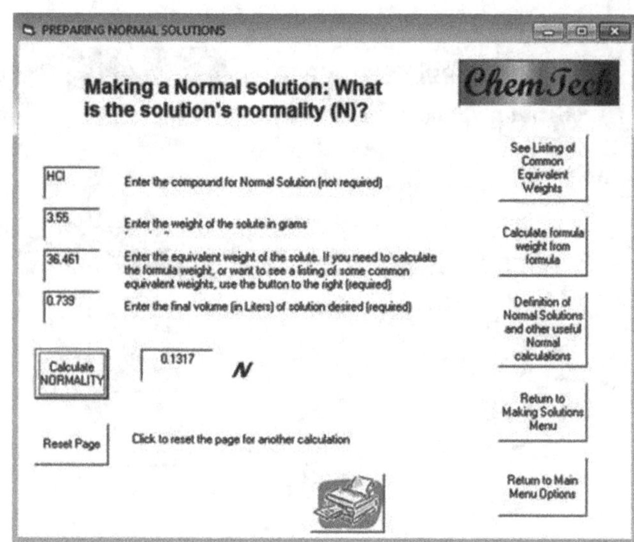

FIGURE 8.11 Results of Normality calculation.

"Calculate NORMALITY" button. The results of the calculation should be 0.1317 N. The page should look like the one shown in Figure 8.11. To reset the page and clear the previous entries, the bottom "Clear Entries" button can be pressed. Finally, the page includes a helpful link with hints and reminders concerning normal calculations.

As was previously the case with molar calculations, there is also a form for determining the weight of solute needed to prepare a solution of a specific normality. Again, while calculating the normality of a solution from known solute weight and final volume is a direct application of normality, often in the laboratory the analyst knows the desired normality of a solution needed for a test, but does not directly know the weight of solute needed. Thus, often the weight of solute is needed according to a predetermined normality and final solution volume. The next example presents a form within ChemTech that we can use to determine the weight of solute needed to prepare a certain normal solution at a predetermined volume.

8.9.4 Computer-Based Concentration Calculation—Normality II

What weight of sulfuric acid (H_2SO_4) is needed to make a 0.515 normal solution in 825 ml of H_2O?

In this instance, we are entering the normality desired, the equivalent weight of the solute, and the final volume of the solution. The page also contains a link to common equivalent weights, the formula weight calculator, and a page outlining normal calculations.

In the Chapter 8 option Making Normal Solutions II: Known N unknown weight solute page as shown in Figure 8.12, enter the compound in the first input box as "H_2SO_4" (this is actually not required but can be useful if the page is printed out with the parameters and results). In the second input box, the normality of the solution is entered as 0.515. This page also contains a link for calculating formula weights from the formula, and a link to

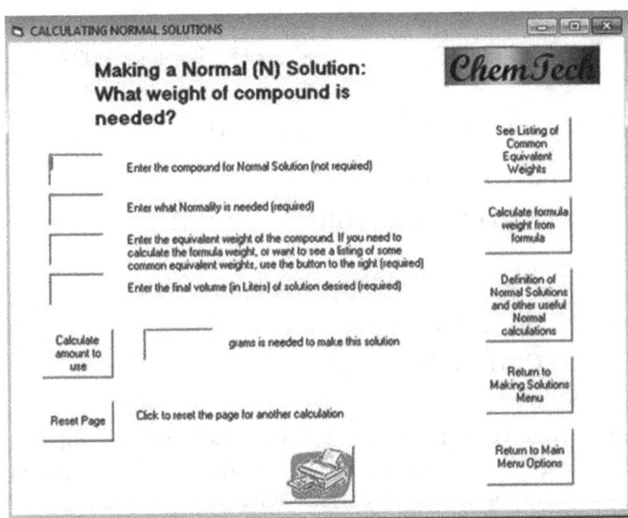

FIGURE 8.12 Making Normal Solutions II: Known N unknown weight solute page.

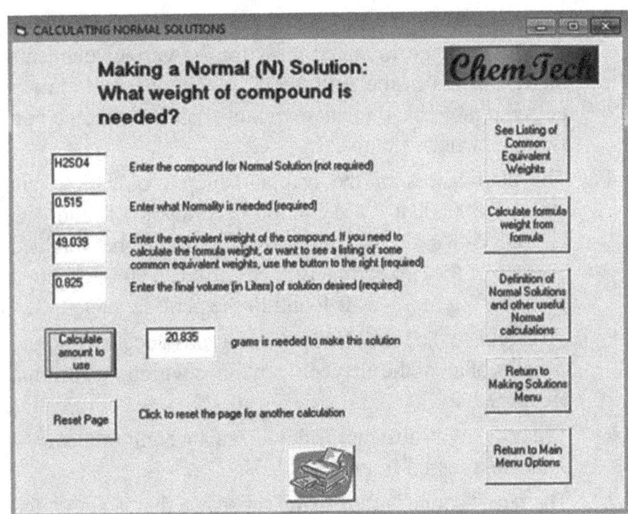

FIGURE 8.14 Result of Normality calculation.

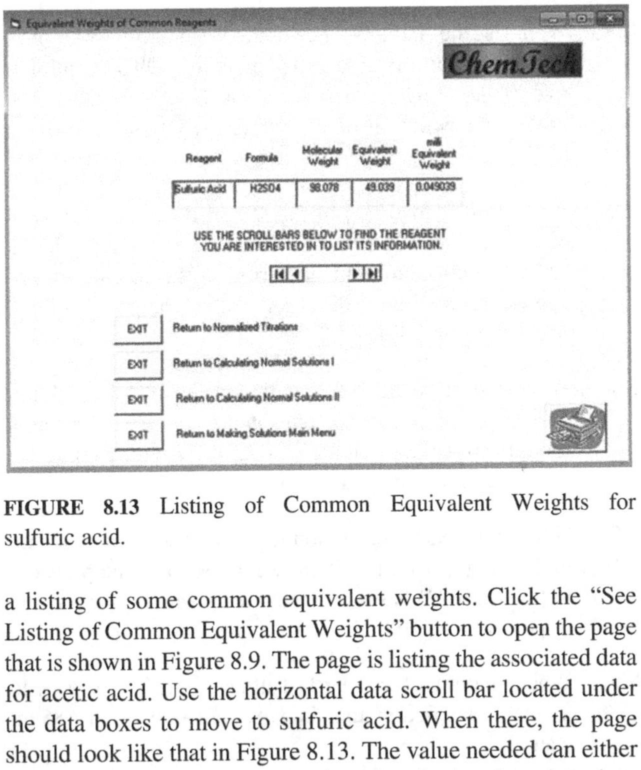

FIGURE 8.13 Listing of Common Equivalent Weights for sulfuric acid.

a listing of some common equivalent weights. Click the "See Listing of Common Equivalent Weights" button to open the page that is shown in Figure 8.9. The page is listing the associated data for acetic acid. Use the horizontal data scroll bar located under the data boxes to move to sulfuric acid. When there, the page should look like that in Figure 8.13. The value needed can either be written down or the page can be printed using the printer button at the bottom of the page. Click the "Return to Calculating Normal Solutions II: button to return to our previous page. Enter the equivalent weight of 49.039 g mol^{-1} (excluding units) into the third input box. Finally, in the fourth input box enter the final volume as 0.825 (in liters). Press the "Calculate amount to use" button. The results of the calculation should be 20.835 g. The page should look like the one shown in Figure 8.14. To reset the page and clear the previous entries, the bottom "Clear Entries" button can be pressed. Finally, the page includes a link to helpful hints and reminders concerning normal calculations.

8.10 REACTIONS IN SOLUTION

In the previous sections, we have been considering the concentrations of solutes in solution. This has included processes in solution such as the dissolving and solvation of a salt in solution such as dissolving sodium chloride NaCl in water where the salt is dissociated into sodium ions Na^+ and chloride ions Cl^-. We have also looked at the addition of acids and bases to solutions such as the base sodium hydroxide, which dissociates into a sodium ion Na^+ and a hydroxyl ion OH^-, and also simple acids such as hydrochloric acid HCl, which when added to water dissociates into a hydrogen proton H^+ and a chloride ion Cl^-. These are simple reactions in solution that we have represented as completely dissociating (100%) in solution into various components of the compound. We can note at this time that most of the work done in the analytical laboratory is in aqueous solutions. Though sometimes it is possible that the technician will also do analyses in other solvents, such as organic solvents (e.g., acetonitrile, acetone, and chloroform), the theory and procedures presented here can be translated to other solution matrices directly without the specific consideration of a solvent's properties (some special considerations will be presented, such as pH determinations). Often it is the case that if analyses are done in solvents other than water, a specific method will be available and followed. If not, the general approach presented here can be applied to other solvents.

8.11 CHAPTER KEY CONCEPTS

8.1 The preparation of laboratory solutions is a very important step in many analysis and calibrations.

8.2 A mole is designated by Avogadro's number of 6.023×10^{23} molecules of any given substance (6.023×10^{23} mol^{-1}). One mole of a substance is the gram formula weight obtained by adding the standard atomic weights of the elements that the substance's formula is composed of.

8.3 The periodic table that the laboratory analyst works with lists the average weighted mass for the various elements that are listed in the periodic table. The elements in the periodic table have relative weights based on the carbon isotope of mass number 12 (^{12}C).

8.4 The abundances of the natural isotopes of carbon are ^{12}C (12.000000 u by definition) at 98.93(8)% and ^{13}C (13.003354826(17) u) at 1.07(8)%. Therefore, the weighted average of the two carbon isotopes gives an atomic weight of 12.0107 g mol^{-1} as is found in the periodic table.

8.5 To calculate the formula weight of a substance, the atomic weights of all of the elements contained within the formula are added.

8.6 The mole of a substance is the formula weight of that substance expressed in grams.

8.7 The term "solute" refers to the substance that is being dissolved in the solution, while the solvent is the substance that is dissolving the solute.

8.8 Gravimetric factors are multiplication factors used to convert the amount of one substance into the amount of another related substance.

8.9 Some useful expressions involving concentration calculations are as follows:

a. (liters) (M) = moles
b. (moles) (formula weight) = grams
c. (liters) (M) (formula weight) = grams
d. (ml) (M) = mmoles
e. (mmoles) (formula weight) = mg
f. (ml) (M) (formula weight) = mg
g. $\text{mmoles} = \dfrac{\text{mg}}{\text{formula weight}}$
h. $\text{Molarity} = \dfrac{\text{mmoles}}{\text{ml}}$
i. $N = M \times nH^+$
j. $N_I V_I = N_F V_F$

8.10 The equivalent is related to the formula weight as follows: eq of compound = 1 mole of compound = formula weight of compound.

8.12 CHAPTER PROBLEMS

8.1 Calculate the formula weight of each of the following listing the contribution from each element.

a. $\text{FW } Cu_2SO_4 = 2 \times 63.546\ \text{g mol}^{-1} + 1 \times 32.064\ \text{g mol}^{-1} + 4 \times 15.9994\ \text{g mol}^{-1}$

b. $\text{FW } C_6H_6O_2 = 6 \times 12.0107\ \text{g mol}^{-1} + 6 \times 1.00794\ \text{g mol}^{-1} + 2 \times 15.9994\ \text{g mol}^{-1}$

c. $\text{FW } C_6H_{16}NPO_4 = 6 \times 12.0107\ \text{g mol}^{-1} + 16 \times 1.00794\ \text{g mol}^{-1} + 1 \times 14.00674\ \text{g mol}^{-1} + 1 \times 30.973761\ \text{g mol}^{-1} + 4 \times 15.9994\ \text{g mol}^{-1}$

d. $\text{FW } Ca_3Si_2O_4(OH)_6 = 3 \times 40.078 + 2 \times 28.0855 + 10 \times 15.9994 + 6 \times 1.00794$

8.2 Calculate the formula weight of each of the following.

a. CaMg(AsO4)(OH)
b. $NaMg_3(Si_3Al)O_{10}(OH)$
c. C_4H_8
d. $Cr(CO)_6$
e. $Hg(C_6H_5)_2$
f. $C_{10}H_{16}N_2O_8$

8.3 What is the molal concentration (molality) of a solution made up of 0.987 g of pure sulfuric acid (H_2SO_4) in 0.400 kg of ethanol (CH_3CH_2OH)?

8.4 What is the molal concentration of a solution made up of 1.321 ml of concentrated (92.8% purity) sulfuric acid (H_2SO_4, $d_4^{20} = 1.84$ g ml^{-1}) in 550 ml of ethanol (CH_3CH_2OH, $d_4^{20} = 0.789$ g ml^{-1})?

8.5 What is the molar concentration (molarity, M) of a solution made up of 6.239 g of pure (>99.9%) lithium hydroxide (LiOH) in 857 ml of H_2O? What is the molar concentration of the lithium (Li^+) and hydroxide (OH^-)?

8.6 A solution is made up of 2.597 g of pure (Na_2HPO_4) in 728 ml of H_2O? Assuming 100% disassociation, what is the molar concentration (molarity) of the sodium hydrogen phosphate, the sodium ion (Na^+), and the hydrogen phosphate ion ($HPO_4{}^{-2}$)?

8.7 What is the normality of a solution made up of the addition of 6.5 ml of concentrated phosphoric acid to water that is brought to a final volume of 475 ml? Assume complete dissociation of the acid.

8.8 What is the parts per million (ppm) of a solution made up of 12.3 mg lactose in 750 ml water? What is the parts per billion (ppb)? What is the lactose concentration expressed in percentage (%)?

8.9 What amount of copper(I) sulfate (Cu_2SO_4) is needed to make a solution that is 108 ppm copper in 685 ml water?

8.10 What is the potassium nitrate (KNO_3) molarity (M) of a solution made up of 5.968 g KNO_3 dissolved in 1.233 l H_2O?

8.11 What weight of potassium permanganate $KMnO_4$ is needed to make a 0.15 molar solution in 425 ml of H_2O?

8.12 What is the nitric acid normality (N) of a solution made up of 2.21 g HNO_3 dissolved in 987 ml H_2O?

8.13 What weight of sulfuric acid (H_2SO_4) is needed to make a 0.125 normal solution in 850 ml of H_2O?

9

ACID–BASE THEORY AND BUFFER SOLUTIONS

9.1 Introduction
9.2 Acids and Bases in Everyday Life
9.3 The Litmus Test
9.4 Early Acid–Base Descriptions
9.5 Brønsted–Lowry Definition
9.6 The Equilibrium Constant
9.7 The Acid Ionization Constant
9.8 Calculating the Hydrogen Ion Concentration
9.9 The Base Ionization Constant
9.9.1 OH^- Ion Concentration Example
9.9.2 Percent Ionization Example
9.10 Ion Product for Water
9.11 The Solubility Product Constant (K_{sp})
9.11.1 Solubility of Silver(I) Thiocyanate
9.11.2 Solubility of Lithium Carbonate
9.12 The pH of a Solution
9.13 Measuring the pH
9.13.1 The Glass Electrode
9.14 Buffered Solutions—Description and Preparing
9.14.1 Le Chatelier's Principle
9.14.2 Titration Curve of a Buffer
9.14.3 Natural Buffer Solutions
9.14.4 Calculating Buffer pH
9.14.5 Buffer pH Calculation I
9.15 ChemTech Buffer Solution Calculator
9.16 Chapter Key Concepts
9.17 Chapter Problems

9.1 INTRODUCTION

The uses of acids and bases in the analytical laboratory are quite numerous: there are volumetric methods to determine the percentage of an acid or base in a solution, which we will cover later in this chapter, the analyst measures the acidity or basicity of a solution as pH, the analyst uses acid and base solutions to adjust the pH of solutions, an acid or base may be added to neutralize a solution, and buffer solution are made from acids, bases, and their conjugate salts.

9.2 ACIDS AND BASES IN EVERYDAY LIFE

The student is in fact quite aware of acids and bases in everyday life if you stop to think about it. For example, citric acid ($C_6H_8O_7$, the formula does not do this weak organic acid justice as it contains three organic acid moieties, see Figure 9.1), an organic acid, is found in citrus fruits such as oranges, grapefruits, lemons, and limes; hydrochloric acid (HCl) is the major acid in stomach digestive juices; carbonic acid (H_2CO_3) is what gives soft drinks their fizz and zip; and acetic acid ($C_2H_4O_2$, see Figure 9.1) is the one that gives vinegar its sour taste and recognizable smell. The sour taste is one of the general characteristics of an acidic water solution. The word *acid* is derived from the Latin *acidus*, which means sour or sharp tasting. The tastes of basic solutions in water are described to be bitter. Some common examples of basic solutions are the ammonia (NH_3) solutions that are used as household cleaners and *lye* solutions that are derived from sodium hydroxide (NaOH). Basic solutions also will feel slippery like soap when rubbed between the fingers.

9.3 THE LITMUS TEST

Another general characteristic of acidic and basic solutions is the well-known "litmus" test. Litmus paper is an easy and straightforward way of telling if something is acidic or basic. If dipped in solution and the litmus paper turns red the solution is acidic, if turned blue the solution is basic. Litmus was originally the name of the dye erythrolitmin that is extracted from small plants found in the Netherlands. Today we often hear of a litmus test

Analytical Chemistry: A Chemist and Laboratory Technician's Toolkit, First Edition. Bryan M. Ham and Aihui MaHam.

as pertaining to a simple answer one way or the other. For the technician, litmus paper as it comes in many forms (often most conveniently in strips) and pH ranges from manufacturers is a very useful tool in the laboratory (see Figure 9.2). We will take a closer look at the definition of pH and the measurement of pH in the analytical laboratory in a later section.

9.4 EARLY ACID–BASE DESCRIPTIONS

The description of acids and bases, and their reaction behavior in aqueous solution has a bit of a long history and stems back to 1776 where the French chemist Antoine Lavoisier first described acids as substances containing oxygen, or oxyacids, such as nitric acid (HNO_3) or sulfuric acid (H_2SO_4). (Miessler, L. M., Tar, D. A., (1991) p166—Table of discoveries attributes Antoine Lavoisier as the first to posit a scientific theory in relation to oxyacids.) This of course is a limited view as it does not take into account many other acids such as iodic acid (HI) or hydrochloric acid (HCl). This was followed by Justus von Liebig in 1838 who was working primarily with organic acids. (Miessler, L. M., Tar, D. A., (1991) p166—Table of discoveries attributes Justus von Liebig's publication as 1838.) In this work, von Liebig described the relationship of acids containing metal-replaceable hydrogen. Next there is the description by Svante Arrhenius around 1884 where he described the dissociation of acids in solution to form the hydronium ion (an oxonium) (H_3O^+) ion, and bases which when added to aqueous solution formed the hydroxide (OH^-) ion. (Miessler, L. M., Tar, D. A., (1991) p165) Coming from this description was the general rule that the neutralization reaction of an acid and base produced a salt and water.

$$\text{Acid} + \text{base} \rightarrow \text{salt} + \text{water} \tag{9.1}$$

$$HCl + NaOH \rightarrow NaCl + H_2O \tag{9.2}$$

Finally, we arrive at the Brønsted–Lowry definition, which is now the generally recognized description of simple acid–base reactions in solution. The Brønsted–Lowry description was independently developed in 1923 by both Johannes Nicolas Brønsted in Denmark and Martin Lowry in England where the idea that acids are compounds that can donate protons (H^+) and bases are compounds that can accept protons. (Miessler, L. M., Tar, D. A., (1991), p167–169—According to this page, the original definition was that "acids have a tendency to lose a proton.") In this definition, the neutralization reaction of an acid and base in solution produces the "conjugate" of the original acid or base.

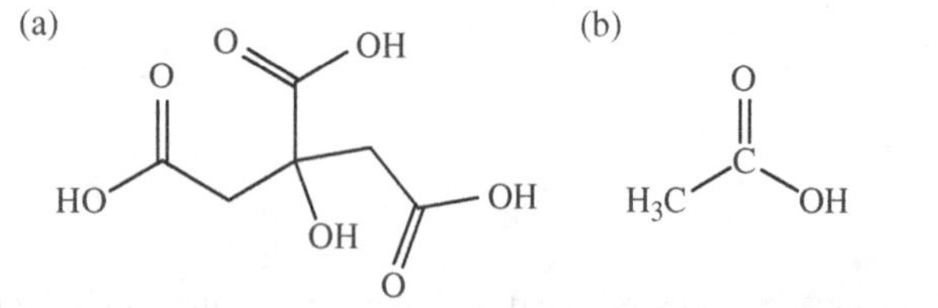

FIGURE 9.1 (a) Structure of citric acid that contains three organic acid moieties (–C=OOH). (b) Structure of acetic acid that contains one organic acid moiety (–C=OOH).

9.5 BRØNSTED–LOWRY DEFINITION

According to the Brønsted–Lowry definition, when a strong acid is dissolved in water it dissociates completely into a solvated hydrogen proton (hydronium ion) and its conjugate base. The general representation of this is depicted in Figure 9.3, where the double arrow "$\rightleftharpoons$" represents a system in a state of equilibrium (the forward reaction is in equilibrium with the reverse reaction). In general, acid_1 (AH) will produce a base_1 (A^-), while base_2 (H_2O here) will produce acid_2 (hydronium ion).

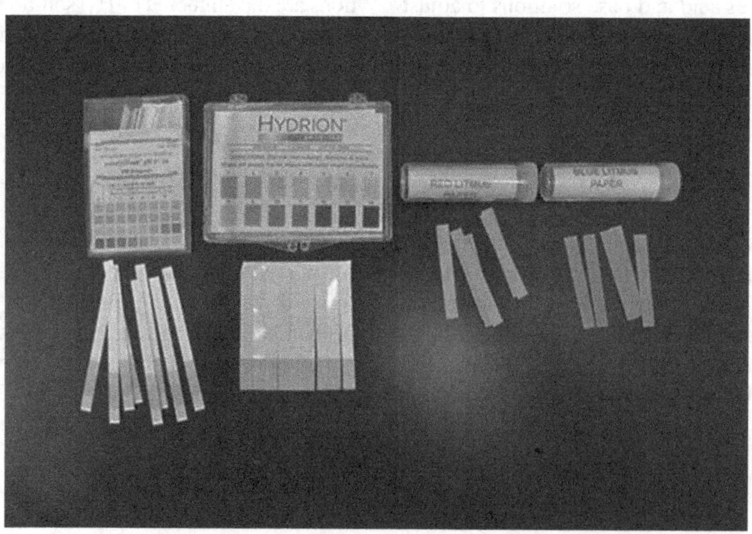

FIGURE 9.2 Examples of commercially available pH test strips and pH indicator papers. Some test strips cover the entire pH range from 1 to 14.

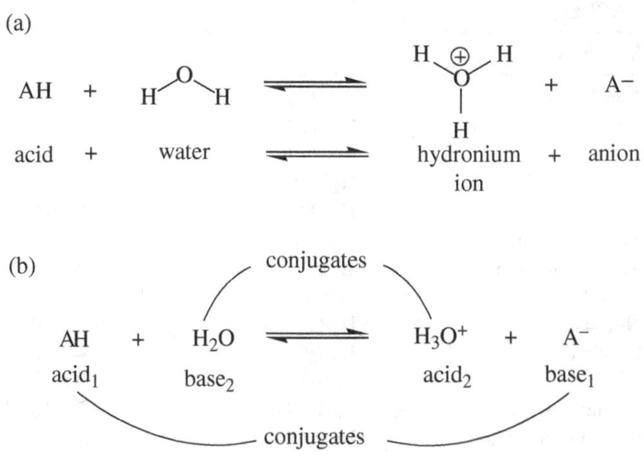

(a)

$$\mathrm{AH} + \mathrm{H_2O} \rightleftharpoons \mathrm{H_3O^+} + \mathrm{A^-}$$

acid + water ⇌ hydronium ion + anion

(b)

$$\underset{\text{acid}_1}{\mathrm{AH}} + \underset{\text{base}_2}{\mathrm{H_2O}} \rightleftharpoons \underset{\text{acid}_2}{\mathrm{H_3O^+}} + \underset{\text{base}_1}{\mathrm{A^-}}$$

conjugates

conjugates

FIGURE 9.3 (a) Production of the hydronium ion upon addition of an acid to water. (b) Relationship of acid and base conjugates.

Using hydrochloric acid (HCl) as an example, we have

$$\mathrm{HCl\,(acid_1) + H_2O(base_2) \rightarrow H_3O^+\,(acid_2) + Cl^-\,(base_1)}. \tag{9.3}$$

Using sulfuric acid (H_2SO_4) as an example, we have

$$\mathrm{H_2SO_4(acid_1) + H_2O\,(base_2) \rightarrow H_3O^+\,(acid_2) + HSO_4^-\,(base_1)}. \tag{9.4}$$

The principle for the solvation of a base in aqueous solution is similar:

$$\mathrm{B\,(base_2) + H_2O\,(acid_1) \rightarrow BH^+\,(acid_2) + OH^-\,(base_1)}. \tag{9.5}$$

Using sodium hydroxide (NaOH) as an example, we have

$$\begin{aligned}\mathrm{NaOH\,(base_2) + H_2O\,(acid_1)} &\rightarrow \mathrm{Na^+\,(spectator\ ion)}\\ &+ \mathrm{OH^-\,(base_1) + H_2O\,(acid_2)}\end{aligned} \tag{9.6}$$

Using ammonia (NH_3) as an example, we have

$$\mathrm{NH_3(base_2) + H_2O(acid_1) \rightarrow NH_4^+\,(acid_2) + OH^-\,(base_1)} \tag{9.7}$$

In the simplest of terms, when an acid and base are mixed together, there is a neutralization reaction that takes place:

$$\mathrm{H_3O^+\,(acid) + OH^-\,(base) \rightarrow 2H_2O}. \tag{9.8}$$

9.6 THE EQUILIBRIUM CONSTANT

In the laboratory, the technician will frequently use solutions that may be either "acidic" or "basic." An acidic solution of course contains a dissolved acid, such as 0.1 M HCL, while a basic solution contains a dissolved base, such as 0.1 M NaOH. The expression used in the analytical laboratory of the acidity or basicity of a solution is the pH of the solution. The pH of a solution is the representation of the molar hydrogen ion concentration [H^+], (*M*), in solution. Let us spend a moment to consider a little deeper the behavior of acids and bases in solution as systems at equilibrium that will lead us to the definition and calculation of pH.

For the dissociation of an acid in aqueous solution, we have the general Equation 9.9:

$$\underset{\text{acid}}{\mathrm{AH}} + \mathrm{H_2O} \rightleftharpoons \underset{\substack{\text{hydronium}\\ \text{ion}}}{\mathrm{H_3O^+}} + \underset{\text{base}}{\mathrm{B^-}} \tag{9.9}$$

Reactions such as the dissolution of an acid in aqueous solution can be expressed in equation notation as the molar concentration of the products divided by the reactants. This form of expression is equal to the *equilibrium constant* of the reaction (K_c) and written as such:

$$K_c = \frac{[\mathrm{H_3O^+}][\mathrm{A^-}]}{[\mathrm{AH}][\mathrm{H_2O}]} \tag{9.10}$$

The molar concentration of water, [H_2O], in solutions that are dilute acids or bases is always 55.51 *M*, so the equilibrium equation can be written as

$$K_c = \frac{[\mathrm{H_3O^+}][\mathrm{A^-}]}{[\mathrm{AH}](55.51)}. \tag{9.11}$$

9.7 THE ACID IONIZATION CONSTANT

We can now derive what is known as the *acid ionization constant*, K_a, for a general acid in aqueous solution (after substituting [H^+] in the place of [H_3O^+]):

$$K_c \times 55.51 = \frac{[\mathrm{H^+}][\mathrm{A^-}]}{[\mathrm{AH}]} = K_a. \tag{9.12}$$

Since the molar concentration of water has been incorporated into the ionization constant, the dissolution of an acid in aqueous solution is usually written without H_2O as

$$\mathrm{AH} \rightleftharpoons \mathrm{H^+} + \mathrm{A^-}. \tag{9.13}$$

For the dissolution of hydrochloric acid in aqueous solution:

$$\mathrm{HCl} \rightleftharpoons \mathrm{H^+} + \mathrm{Cl^-}. \tag{9.14}$$

The expression for the acid ionization constant is

$$K_a = \frac{[\mathrm{H^+}][\mathrm{Cl^-}]}{[\mathrm{HCl}]}. \tag{9.15}$$

Table 9.1 lists some common acid ionization constants (K_a). As a general guideline, if K_a is greater than 1×10^3 then the acid is very strong; if K_a is between 1×10^3 and 1×10^{-2} then the acid

TABLE 9.1 Acid Ionization Constants (K_a) in Aqueous Solution at 25 °C.

Substance	Formula	K_a
Acetic acid	$HC_2H_3O_2$	1.8×10^{-5}
Acrylic acid	$HC_3H_3O_2$	5.5×10^{-5}
Aluminum 3+ ion	$Al^{3+}(aq)$	1.4×10^{-5}
Ammonium ion	NH_4^+	5.6×10^{-10}
Anilinium ion	$C_6H_5NH_3^+$	1.4×10^{-5}
Arsenic acid	H_3AsO_4	6.0×10^{-3}
	$H_2AsO_4^-$	1.0×10^{-7}
	$HAsO_4^{2-}$	3.2×10^{-12}
Arsenous acid	H_3AsO_3	6.6×10^{-10}
Ascorbic acid	$H_2C_6H_6O_6$	6.8×10^{-5}
	$HC_6H_6O_6^-$	2.8×10^{-12}
Bromoacetic acid	$HC_2H_2BrO_2$	1.3×10^{-3}
Benzoic acid	$HC_7H_5O_2$	6.3×10^{-5}
Beryllium 2$^+$ ion	$Be^{2+}(aq)$	3×10^{-7}
Boric acid	H_3BO_3	5.9×10^{-10}
Butyric acid	$HC_4H_7O_2$	1.5×10^{-5}
Carbonic acid	H_2CO_3	4.4×10^{-7}
	HCO_3^-	4.7×10^{-11}
Chloroacetic acid	$HC_2H_2ClO_2$	1.4×10^{-3}
Chlorous acid	$HClO_2$	1.1×10^{-2}
Chromic acid	H_2CrO_4	1.5×10^{-1}
	$HCrO_4^-$	3.2×10^{-7}
Chromium 3$^+$ ion	$Cr^{3+}(aq)$	6.6×10^{-4}
Citric acid	$H_3C_6H_5O_7$	7.4×10^{-4}
	$H_2C_6H_5O_7^-$	1.7×10^{-5}
	$HC_6H_5O_7^{2-}$	4.0×10^{-7}
Cobalt 2$^+$ ion	$Co^{2+}(aq)$	1.3×10^{-9}
Codeine ammonium ion	$HC_{18}H_{21}O_3\ N^+$	1.1×10^{-8}
Cyanic acid	$HOCN$	3.5×10^{-4}
Dichloroacetic acid	$HC_2HCl_2O_2$	5.5×10^{-2}
Diethylammonium ion	$(C_2H_5)_2NH_2^+$	1.4×10^{-11}
Dimethylammonium ion	$(CH_3)_2NH_2^+$	1.7×10^{-11}
Ethylammonium ion	$C_2H_5NH_3^+$	2.3×10^{-11}
Ethylenediammonium ion	$NH_2CH_2CH_2NH_3^+$	1.9×10^{-11}
Fluoroacetic acid	$HC_2H_2FO_2$	2.6×10^{-3}
Formic acid	$HCHO_2$	1.8×10^{-4}
Hydrazinium ion	$N_2H_5^+$	1.2×10^{-8}
Hydrazoic acid	HN_3	1.9×10^{-5}
Hydrocyanic acid	HCN	6.2×10^{-10}
Hydrofluoric acid	HF	6.6×10^{-4}
Hydrogen peroxide	H_2O_2	2.2×10^{-12}
Hydrogen selenate ion	$HSeO_4^-$	2.2×10^{-2}
Hydrogen sulfate ion	HSO_4^-	1.1×10^{-2}
Hydroselenic acid	H_2Se	1.3×10^{-4}
	HSe^-	1×10^{-11}
Hydrosulfuric acid	H_2S	1.0×10^{-7}
	HS^-	1×10^{-19}
Hydrotelluric acid	H_2Te	2.3×10^{-3}
	HTe^-	1.6×10^{-11}
Hydroxylammonium ion	$HONH_3^+$	1.1×10^{-6}
Hypobromous acid	$HOBr$	2.5×10^{-9}
Hypochlorous acid	$HClO$	2.9×10^{-8}
Hypoiodous acid	HOI	2.3×10^{-11}
Hyponitrous acid	$H_2N_2O_2$ (HON=NOH)	8.9×10^{-8}
	$HN_2O_2^-$ ($HON{=}NO^-$)	4×10^{-12}
Iodic acid	HIO_3	1.6×10^{-1}
Iodoacetic acid	$HC_2H_2IO_2$	6.7×10^{-4}
Iron 2$^+$ ion	$Fe^{2+}(aq)$	3.2×10^{-10}
Iron 3$^+$ ion	$Fe^{3+}(aq)$	6.3×10^{-3}
Isoquinolinium ion	$HC_9H_7N^+$	4.0×10^{-6}
Lactic acid	$HC_3H_5O_3$	1.3×10^{-4}
Malonic acid	$H_2C_3H_2O_4$	1.5×10^{-3}
	$HC_3H_2O_4^-$	2.0×10^{-6}
Methylammonium ion	$CH_3NH_3^+$	2.4×10^{-11}
Morphinium ion	$HC_{17}H_{19}O_3\ N^+$	1.4×10^{-8}
Nickel 2+ ion	$Ni^{2+}(aq)$	2.5×10^{-11}
Nitrous acid	HNO_2	7.2×10^{-4}
Oxalic acid	$H_2C_2O_4$	5.4×10^{-2}
	$HC_2O_4^-$	5.3×10^{-5}
Phenol	C_6H_5OH	1.0×10^{-10}
Phenylacetic acid	$HC_8H_7O_2$	4.9×10^{-5}
Phosphoric acid	H_3PO_4	7.1×10^{-3}
	$H_2PO_4^-$	6.3×10^{-8}
	HPO_4^{2-}	4.2×10^{-13}
Phosphorous acid	H_3PO_3	3.7×10^{-2}
	$H_2PO_3^-$	2.1×10^{-7}
Piperidinium ion	$HC_5H_{11}N^+$	7.7×10^{-12}
Propionic acid	$HC_3H_5O_2$	1.3×10^{-5}
Pyridinium ion	$C_5H_5NH^+$	6.7×10^{-6}
Pyrophosphoric acid	$H_4P_2O_7$	3.0×10^{-2}
	$H_3P_2O_7^-$	4.4×10^{-3}
	$H_2P_2O_7^{2-}$	2.5×10^{-7}
	$HP_2O_7^{3-}$	5.6×10^{-10}
Pyruvic acid	$HC_3H_3O_3$	1.4×10^{-4}
Quinolinium ion	$HC_9H_7N^+$	1.6×10^{-5}
Selenous acid	H_2SeO_3	2.3×10^{-3}
	$HSeO_3^-$	5.4×10^{-9}
Succinic acid	$H_2C_4H_4O_4$	6.2×10^{-5}
	$HC_4H_4O_4^-$	2.3×10^{-6}
Sulfurous acid	H_2SO_3	1.3×10^{-2}
	HSO_3^-	6.2×10^{-8}
Thiophenol	HSC_6H_5	3.2×10^{-7}
Trichloroacetic acid	$HC_2Cl_3O_2$	3.0×10^{-1}
Triethanolammonium ion	$HC_6H_{15}O_3\ N^+$	1.7×10^{-8}
Triethylammonium ion	$(C_2H_5)_3\ NH^+$	1.9×10^{-11}
Trimethylammonium ion	$(CH_3)_3NH^+$	1.5×10^{-10}
Urea hydrogen ion	$NH_2CONH_3^+$	6.7×10^{-1}
Zinc 2+ ion	$Zn^{2+}(aq)$	2.5×10^{-10}

is strong; if K_a is between 1×10^{-2} and 1×10^{-7} the acid is weak; and if the K_a is less than 1×10^{-7} the acid is very weak.

9.8 CALCULATING THE HYDROGEN ION CONCENTRATION

What the acid ionization constant is illustrating is that acids ionize and release into solution different concentrations of the hydrogen proton [H^+]. The analytical chemist can use the ionization constants to calculate the concentration of the hydrogen ion in solution giving a relative value as to the "acidity" of the solution. We should note here that unless the ionization constant is large, such as 10^{-2}, the amount of original species can be used in the denominator. For example, let us compare the hydrogen ion concentration in aqueous solution of a 0.1 *M* hydrofluoric acid (HF) solution to that of a 0.1 *M* hypochlorous acid (HClO) solution. The 0.1 M HF solution contains

$$HF \rightleftharpoons H^+ + F^- \quad (9.16)$$

$$K_a = \frac{[H^+][F^-]}{[HF]} \quad (9.17)$$

$$6.6 \times 10^{-4} = \frac{[H^+][F^-]}{0.1} \quad (9.18)$$

Because the same amount of F^- is dissolved as that of H^+, we can express the equation as

$$6.6 \times 10^{-4} = \frac{(x)(x)}{0.1} \quad (9.19)$$

$$x^2 = (6.6 \times 10^{-4})(0.1) \quad (9.20)$$

$$x \approx \sqrt{6.6 \times 10^{-5}} \quad (9.21)$$

$$x \approx 8.1 \times 10^{-3} = [H^+] \quad (9.22)$$

The 0.1 M HClO solution contains

$$HClO \rightleftharpoons H^+ + CLO^- \quad (9.23)$$

$$K_a = \frac{[H^+][CLO^-]}{[HClO]} \quad (9.24)$$

$$3.0 \times 10^{-8} = \frac{[H^+][ClO^-]}{0.1} \quad (9.25)$$

Because the same amount of ClO^- is dissolved as that of H^+, we can express the equation as

$$3.0 \times 10^{-8} = \frac{(x)(x)}{0.1} \quad (9.26)$$

$$x^2 = (3.0 \times 10^{-8})(0.1) \quad (9.27)$$

$$x = \sqrt{3.0 \times 10^{-9}} \quad (9.28)$$

$$x = 5.5 \times 10^{-5} = [H^+] \quad (9.29)$$

There is ~150 times more $[H^+]$ in the HF solution as that in the HClO solution; therefore, the HF solution is more acidic than the HClO solution.

9.9 THE BASE IONIZATION CONSTANT

Analogous to this, the base ionization constant (K_b) is represented in a similar manner as follows:

$$B + H_2O \rightleftharpoons BH^+ + OH^- \quad (9.30)$$

$$K_c = \frac{[BH^+][OH^-]}{[B][H_2O]} \quad (9.31)$$

We can now derive what is known as the *base ionization constant*, K_b, for a general base in aqueous solution (after substituting 55.51 *M* for $[H_2O]$):

$$K_c \times 55.51 = \frac{[BH^+][OH^-]}{[B]} = K_b \quad (9.32)$$

For the reaction of ammonia in water, the base ionization constant is

$$NH_3 + H_2O \rightleftharpoons NH_4^+ + OH^- \quad (9.33)$$

$$K_b = \frac{[NH_4^+][OH^-]}{[NH_3]} \quad (9.34)$$

Table 9.2 lists some common base ionization constants (K_b). As a general guideline, if K_b is greater than 0.1, then the base is very strong; if K_b is between 0.1 and 1×10^{-2}, then the base is strong; if K_b is between 1×10^{-2} and 1×10^{-7}, then the base is weak; and if the K_b is less than 1×10^{-7}, then the base is very weak.

9.9.1 OH^- Ion Concentration Example

Calculate the OH^- ion concentration of a solution that is made up of 0.50 M sodium cyanide, where K_b of the cyanide ion, CN^-, is 1.6×10^{-5}.

Sodium is a spectator ion, so we write the equilibrium equation as

$$CN^- + H_2O \leftrightarrow HCN + OH^-. \quad (9.35)$$

The concentration of the species at the start is 0.50 M for CN^-, and 0 M for both HCN and OH^-. At equilibrium, the concentrations are $0.50 - x$ for CN^-, HCN $= x$, and $OH^- = x$.

$$K_b = 1.6 \times 10^{-5} = \frac{[HCN][OH^-]}{[CN^-]} \quad (9.36)$$

TABLE 9.2 Base Ionization Constants (K_b) in Aqueous Solution at 25 °C.

Substance	Formula	K_b
Ammonia	NH_3	1.8×10^{-5}
Aniline	$C_6H_5NH_2$	4.0×10^{-10}
Methylamine	CH_3NH_2	4.4×10^{-4}
Ethylamine	$CH_3CH_2NH_2$	5.6×10^{-4}
Dimethylamine	$(CH_3)_2NH$	5.9×10^{-4}
Trimethylamine	$(CH_3)_3$ N	6.3×10^{-5}
Hydrazine	N_2H_4	9.8×10^{-7}
		1.3×10^{-15}
Ethylenediamine	$H_2NCH_2CH_2NH_2$	3.6×10^{-4}
		5.4×10^{-7}
Hydroxylamine	$HONH_2$	9.1×10^{-9}
Calcium hydroxide	$Ca(OH)_2$	3.7×10^{-3}
Lithium hydroxide	LiOH	2.3
Potassium hydroxide	KOH	0.3
Sodium hydroxide	NaOH	0.6

$$1.6\times 10^{-5}=\frac{(x)(x)}{0.50-x} \quad (9.37)$$

$$x^2=8.0\times 10^{-6}-1.6\times 10^{-5}x \quad (9.38)$$

$$x^2+1.6\times 10^{-5}x-8.0\times 10^{-6}=0 \quad (9.39)$$

$$x=2.8\times 10^{-3}\text{M}=[\text{OH}^-] \quad (9.40)$$

In this calculation, we included the amount of loss of the CN^- ion, as 0.50 – x, but if we calculate using just the original amount of 0.50 M, the same result of 2.8×10^{-3} M is obtained. This is illustrated in the next example.

9.9.2 Percent Ionization Example

What is the percent ionization of a 0.10 M solution of methylamine?

From Table 9.2 we see that $K_b=4.4\times 10^{-4}$.

The ionization reaction is

$$CH_3NH_2+H_2O\leftrightarrow CH_3NH_3^+ +OH^- \quad (9.41)$$

$$K_b=\frac{[CH_3NH_3^+][OH^-]}{[CH_3NH_2]} \quad (9.42)$$

$$4.4\times 10^{-4}=\frac{(x)(x)}{0.10} \quad (9.43)$$

$$x^2=4.4\times 10^{-5} \quad (9.44)$$

$$x=\sqrt{4.4\times 10^{-5}} \quad (9.45)$$

$$x=6.6\times 10^{-3}\text{M}=[\text{OH}^-] \quad (9.46)$$

The percent ionization equals

$$\text{Percent ionization}=\frac{[OH^-]}{[CH_3NH_2]}\times 100 \quad (9.47)$$

$$\text{Percent ionization}=\frac{6.6\times 10^{-3}}{0.1}\times 100 \quad (9.48)$$

$$\text{Percent ionization}=6.6\%$$

9.10 ION PRODUCT FOR WATER

To simplify the relative comparison of the acidity of solutions, chemist devised a general way of expressing the hydrogen ion concentration as pH. Water itself ionizes to a small extent according to the following equation:

$$2H_2O\rightleftharpoons H_3O^+ +OH^-. \quad (9.49)$$

The equilibrium equation for the ionization of water is written as

$$K_c=\frac{[H_3O^+][OH^-]}{[H_2O]^2}. \quad (9.50)$$

The ionization constant, K_w, for water is thus derived as

$$K_c=\frac{[H_3O^+][OH^-]}{(55)^2} \quad (9.51)$$

$$K_c\times (55)^2=[H_3O^+]\,[OH^-] \quad (9.52)$$

$$K_w=[H^+]\,[OH^-]. \quad (9.53)$$

Measurements have shown that the molar concentrations of H^+ and OH^- in solution are both 1.0×10^{-7} mol/l. Thus the *ion product for water*, K_w, is

$$K_w=\left(1.0\times 10^{-7}\right)\left(1.0\times 10^{-7}\right) \quad (9.54)$$

$$K_w=1.0\times 10^{-14} \quad (9.55)$$

The relationship between the ion product for water, K_w, the acid ionization constant, K_a, and the base ionization constant, K_b, is $K_w=K_a\times K_b$.

9.11 THE SOLUBILITY PRODUCT CONSTANT (K_{SP})

When a solution contains a precipitate such as a metal–anion complex, denoted MA, there will be a small amount of the dissociated metal cation (M^+) and the anion (A^-) in solution. While for acid–base dissociation, the amount of the acid or base determines the concentration of the conjugate acid and bases, the concentrations of the metal cation (M^+) and the anion (A^-) in solution is not dependent upon the amount of the precipitated MA. The equilibrium constant for a precipitate is written as

$$M_aA_b(s)=aM+bA \quad (9.56)$$

$$K=\frac{[M]^a[A]^b}{1}$$

$$K_{sp}=[M]^a[A]^b$$

The solubility product constant is dependent upon the temperature of the given solution. The K_{sp} values listed in Table 9.1 give us an indication of the solubility of the compounds at 25 °C. For example, a small value for the K_{sp} indicates an insoluble substance, while a large value indicates a more soluble substance. Table 9.3 lists the product solubility constants (K_{sp}) of many compounds encountered in the laboratory.

9.11.1 Solubility of Silver(I) Thiocyanate

What is the solubility of silver(I) thiocyanate (AgSCN) at 25 °C with $K_{sp}=1.03\times 10^{-12}$?

$$AgSCN(s)=Ag^+ +SCN^-$$

$$K_{spAgSCN}=[Ag^+][SCN^-]$$

Each mole of AgSCN that dissolves in water produces one mole of $[Ag^+]$ and one mole of $[SCN^-]$. Thus, we can set the solubility as $[Ag^+]=[SCN^-]$.

TABLE 9.3 Product Solubility Constants (K_{sp}).

Compound	Formula	K_{sp} (25 °C)
Aluminum hydroxide	$Al(OH)_3$	3×10^{-34}
Aluminum phosphate	$AlPO_4$	9.84×10^{-21}
Barium bromate	$Ba(BrO_3)_2$	2.43×10^{-4}
Barium carbonate	$BaCO_3$	2.58×10^{-9}
Barium chromate	$BaCrO_4$	1.17×10^{-10}
Barium fluoride	BaF_2	1.84×10^{-7}
Barium hydroxide octahydrate	$Ba(OH)_2 \times 8H_2O$	2.55×10^{-4}
Barium iodate	$Ba(IO_3)_2$	4.01×10^{-9}
Barium iodate monohydrate	$Ba(IO_3)_2 \times H_2O$	1.67×10^{-9}
Barium molybdate	$BaMoO_4$	3.54×10^{-8}
Barium nitrate	$Ba(NO_3)_2$	4.64×10^{-3}
Barium selenite	$BaSeO_4$	3.40×10^{-8}
Barium sulfate	$BaSO_4$	1.08×10^{-10}
Barium sulfite	$BaSO_3$	5.0×10^{-10}
Beryllium hydroxide	$Be(OH)_2$	6.92×10^{-22}
Bismuth arsenate	$BiAsO_4$	4.43×10^{-10}
Bismuth iodide	BiI	7.71×10^{-19}
Cadmium arsenate	$Cd_3(AsO_4)_2$	2.2×10^{-33}
Cadmium carbonate	$CdCO_3$	1.0×10^{-12}
Cadmium fluoride	CdF_2	6.44×10^{-3}
Cadmium hydroxide	$Cd(OH)_2$	7.2×10^{-15}
Cadmium iodate	$Cd(IO_3)_2$	2.5×10^{-8}
Cadmium oxalate trihydrate	$CdC_2O_4 \times 3H_2O$	1.42×10^{-8}
Cadmium phosphate	$Cd_3(PO_4)_2$	2.53×10^{-33}
Cadmium sulfide	CdS	1×10^{-27}
Caesium perchlorate	$CsClO_4$	3.95×10^{-3}
Caesium periodate	$CsIO_4$	5.16×10^{-6}
Calcium carbonate (aragonite)	$CaCO_3$	6.0×10^{-9}
Calcium carbonate (calcite)	$CaCO_3$	3.36×10^{-9}
Calcium fluoride	CaF_2	3.45×10^{-11}
Calcium hydroxide	$Ca(OH)_2$	5.02×10^{-6}
Calcium iodate	$Ca(IO_3)_2$	6.47×10^{-6}
Calcium iodate hexahydrate	$Ca(IO_3)_2 \times 6H_2O$	7.10×10^{-7}
Calcium molybdate	$CaMoO$	1.46×10^{-8}
Calcium oxalate monohydrate	$CaC_2O_4 \times H_2O$	2.32×10^{-9}
Calcium phosphate	$Ca_3(PO_4)_2$	2.07×10^{-33}
Calcium sulfate	$CaSO_4$	4.93×10^{-5}
Calcium sulfate dihydrate	$CaSO_4 \times 2H_2O$	3.14×10^{-5}
Calcium sulfate hemihydrate	$CaSO_4 \times 0.5H_2O$	3.1×10^{-7}
Cobalt(II) arsenate	$Co_3(AsO_4)_2$	6.80×10^{-29}
Cobalt(II) carbonate	$CoCO_3$	1.0×10^{-10}
Cobalt(II) hydroxide (blue)	$Co(OH)_2$	5.92×10^{-15}
Cobalt(II) iodate dihydrate	$Co(IO_3)_2 \times 2H_2O$	1.21×10^{-2}
Cobalt(II) phosphate	$Co_3(PO_4)_2$	2.05×10^{-35}
Cobalt(II) sulfide (alpha)	CoS	5×10^{-22}
Cobalt(II) sulfide (beta)	CoS	3×10^{-26}
Copper(I) bromide	$CuBr$	6.27×10^{-9}
Copper(I) chloride	$CuCl$	1.72×10^{-7}
Copper(I) cyanide	$CuCN$	3.47×10^{-20}
Copper(I) oxide	Cu_2O	2×10^{-15}
Copper(I) iodide	CuI	1.27×10^{-12}
Copper(I) thiocyanate	$CuSCN$	1.77×10^{-13}
Copper(II) arsenate	$Cu_3(AsO_4)_2$	7.95×10^{-36}
Copper(II) hydroxide	$Cu(OH)_2$	4.8×10^{-20}
Copper(II) iodate monohydrate	$Cu(IO_3)_2 \times H_2O$	6.94×10^{-8}
Copper(II) oxalate	CuC_2O_4	4.43×10^{-10}
Copper(II) phosphate	$Cu_3(PO_4)_2$	1.40×10^{-37}
Copper(II) sulfide	CuS	8×10^{-37}

TABLE 9.3 *(Continued)*

Compound	Formula	K_{sp} (25 °C)
Europium(III) hydroxide	$Eu(OH)_3$	9.38×10^{-27}
Gallium(III) hydroxide	$Ga(OH)_3$	7.28×10^{-36}
Iron(II) carbonate	$FeCO_3$	3.13×10^{-11}
Iron(II) fluoride	FeF_2	2.36×10^{-6}
Iron(II) hydroxide	$Fe(OH)_2$	4.87×10^{-17}
Iron(II) sulfide	FeS	8×10^{-19}
Iron(III) hydroxide	$Fe(OH)_3$	2.79×10^{-39}
Iron(III) phosphate dihydrate	$FePO_4 \times 2H_2O$	9.91×10^{-16}
Lanthanum iodate	$La(IO_3)_3$	7.50×10^{-12}
Lead(II) bromide	$PbBr_2$	6.60×10^{-6}
Lead(II) carbonate	$PbCO_3$	7.40×10^{-14}
Lead(II) chloride	$PbCl_2$	1.70×10^{-5}
Lead(II) chromate	$PbCrO_4$	3×10^{-13}
Lead(II) fluoride	PbF_2	3.3×10^{-8}
Lead(II) hydroxide	$Pb(OH)_2$	1.43×10^{-20}
Lead(II) iodate	$Pb(IO_3)_2$	3.69×10^{-13}
Lead(II) iodide	PbI_2	9.8×10^{-9}
Lead(II) oxalate	PbC_2O_4	8.5×10^{-9}
Lead(II) selenate	$PbSeO_4$	1.37×10^{-7}
Lead(II) sulfate	$PbSO_4$	2.53×10^{-8}
Lead(II) sulfide	PbS	3×10^{-28}
Lithium carbonate	Li_2CO_3	8.15×10^{-4}
Lithium fluoride	LiF	1.84×10^{-3}
Lithium phosphate	Li_3PO_4	2.37×10^{-4}
Magnesium ammonium phosphate	$MgNH_4PO_4$	3×10^{-13}
Magnesium carbonate	$MgCO_3$	6.82×10^{-6}
Magnesium carbonate pentahydrate	$MgCO_3 \times 5H_2O$	3.79×10^{-6}
Magnesium carbonate trihydrate	$MgCO_3 \times 3H_2O$	2.38×10^{-6}
Magnesium fluoride	MgF_2	5.16×10^{-11}
Magnesium hydroxide	$Mg(OH)_2$	5.61×10^{-12}
Magnesium oxalate dihydrate	$MgC_2O_4 \times 2H_2O$	4.83×10^{-6}
Magnesium phosphate	$Mg_3(PO_4)_2$	1.04×10^{-24}
Manganese(II) carbonate	$MnCO_3$	2.24×10^{-11}
Manganese(II) hydroxide	$Mn(OH)_2$	2×10^{-13}
Manganese(II) iodate	$Mn(IO_3)_2$	4.37×10^{-7}
Manganese(II) oxalate dihydrate	$MnC_2O_4 \times 2H_2O$	1.70×10^{-7}
Manganese(II) sulfide (green)	MnS	3×10^{-14}
Manganese(II) sulfide (pink)	MnS	3×10^{-11}
Mercury(I) bromide	Hg_2Br_2	6.40×10^{-23}
Mercury(I) carbonate	Hg_2CO_3	3.6×10^{-17}
Mercury(I) chloride	Hg_2Cl_2	1.43×10^{-18}
Mercury(I) fluoride	Hg_2F_2	3.10×10^{-6}
Mercury(I) iodide	Hg_2I_2	5.2×10^{-29}
Mercury(I) oxalate	$Hg_2C_2O_4$	1.75×10^{-13}
Mercury(I) sulfate	Hg_2SO_4	6.5×10^{-7}
Mercury(I) thiocyanate	$Hg_2(SCN)_2$	3.2×10^{-20}
Mercury(II) bromide	$HgBr_2$	6.2×10^{-20}
Mercury(II) hydroxide	HgO	3.6×10^{-26}
Mercury(II) iodide	HgI_2	2.9×10^{-29}
Mercury(II) sulfide (black)	HgS	2×10^{-53}
Mercury(II) sulfide (red)	HgS	2×10^{-54}
Neodymium carbonate	$Nd_2(CO_3)_3$	1.08×10^{-33}
Nickel(II) carbonate	$NiCO_3$	1.42×10^{-7}
Nickel(II) hydroxide	$Ni(OH)_2$	5.48×10^{-16}
Nickel(II) iodate	$Ni(IO_3)_2$	4.71×10^{-5}
Nickel(II) phosphate	$Ni_3(PO_4)_2$	4.74×10^{-32}
Nickel(II) sulfide (alpha)	NiS	4×10^{-20}

(continued)

TABLE 9.3 *(Continued)*

Compound	Formula	K_{sp} (25 °C)
Nickel(II) sulfide (beta)	NiS	1.3×10^{-25}
Palladium(II) thiocyanate	$Pd(SCN)_2$	4.39×10^{-23}
Potassium hexachloroplatinate	K_2PtCl_6	7.48×10^{-6}
Potassium perchlorate	$KClO_4$	1.05×10^{-2}
Potassium periodate	KIO_4	3.71×10^{-4}
Praseodymium hydroxide	$Pr(OH)_3$	3.39×10^{-24}
Radium iodate	$Ra(IO_3)_2$	1.16×10^{-9}
Radium sulfate	$RaSO_4$	3.66×10^{-11}
Rubidium perchlorate	$RuClO_4$	3.00×10^{-3}
Scandium fluoride	ScF_3	5.81×10^{-24}
Scandium hydroxide	$Sc(OH)_3$	2.22×10^{-31}
Silver(I) acetate	$AgCH_3COO$	1.94×10^{-3}
Silver(I) arsenate	Ag_3AsO_4	1.03×10^{-22}
Silver(I) bromate	$AgBrO_3$	5.38×10^{-5}
Silver(I) bromide	$AgBr$	5.35×10^{-13}
Silver(I) carbonate	Ag_2CO_3	8.46×10^{-12}
Silver(I) chloride	$AgCl$	1.77×10^{-10}
Silver(I) chromate	Ag_2CrO_4	1.12×10^{-12}
Silver(I) cyanide	$AgCN$	5.97×10^{-17}
Silver(I) iodate	$AgIO_3$	3.17×10^{-8}
Silver(I) iodide	AgI	8.52×10^{-17}
Silver(I) oxalate	$Ag_2C_2O_4$	5.40×10^{-12}
Silver(I) phosphate	Ag_3PO_4	8.89×10^{-17}
Silver(I) sulfate	Ag_2SO_4	1.20×10^{-5}
Silver(I) sulfide	Ag_2S	8×10^{-51}
Silver(I) sulfite	Ag_2SO_3	1.50×10^{-14}
Silver(I) thiocyanate	$AgSCN$	1.03×10^{-12}
Strontium arsenate	$Sr_3(AsO_4)_2$	4.29×10^{-19}
Strontium carbonate	$SrCO_3$	5.60×10^{-10}
Strontium fluoride	SrF_2	4.33×10^{-9}
Strontium iodate	$Sr(IO_3)_2$	1.14×10^{-7}
Strontium iodate hexahydrate	$Sr(IO_3)_2 \times 6H_2O$	4.55×10^{-7}
Strontium iodate monohydrate	$Sr(IO_3)_2 \times H_2O$	3.77×10^{-7}
Strontium oxalate	SrC_2O_4	5×10^{-8}
Strontium sulfate	$SrSO_4$	3.44×10^{-7}
Thallium(I) bromate	$TlBrO_3$	1.10×10^{-4}
Thallium(I) bromide	$TlBr$	3.71×10^{-6}
Thallium(I) chloride	$TlCl$	1.86×10^{-4}
Thallium(I) chromate	Tl_2CrO_4	8.67×10^{-13}
Thallium(I) hydroxide	$Tl(OH)_3$	1.68×10^{-44}
Thallium(I) iodate	$TlIO_3$	3.12×10^{-6}
Thallium(I) iodide	TlI	5.54×10^{-8}
Thallium(I) sulfide	Tl_2S	6×10^{-22}
Thallium(I) thiocyanate	$TlSCN$	1.57×10^{-4}
Tin(II) hydroxide	$Sn(OH)_2$	5.45×10^{-27}
Yttrium carbonate	$Y_2(CO_3)_3$	1.03×10^{-31}
Yttrium fluoride	YF_3	8.62×10^{-21}
Yttrium hydroxide	$Y(OH)_3$	1.00×10^{-22}
Yttrium iodate	$Y(IO_3)_3$	1.12×10^{-10}
Zinc arsenate	$Zn_3(AsO_4)_2$	2.8×10^{-28}
Zinc carbonate	$ZnCO_3$	1.46×10^{-10}
Zinc carbonate monohydrate	$ZnCO_3 \times H_2O$	5.42×10^{-11}
Zinc fluoride	ZnF	3.04×10^{-2}
Zinc hydroxide	$Zn(OH)_2$	3×10^{-17}
Zinc iodate dihydrate	$Zn(IO_3)_2 \times 2H_2O$	4.1×10^{-6}
Zinc oxalate dihydrate	$ZnC_2O_4 \times 2H_2O$	1.38×10^{-9}
Zinc selenide	$ZnSe$	3.6×10^{-26}
Zinc selenite monohydrate	$ZnSe \times H_2O$	1.59×10^{-7}
Zinc sulfide (alpha)	ZnS	2×10^{-25}
Zinc sulfide (beta)	ZnS	3×10^{-23}

$$K_{spAgSCN} = [Ag^+][SCN^-]$$
$$K_{spAgSCN} = [Ag^+]^2$$
$$1.03 \times 10^{-12} = [Ag^+]^2$$
$$[Ag^+] = \sqrt{1.03 \times 10^{-12}}$$
$$[Ag^+] = 1.01 \times 10^{-6}\text{M}$$

$$\text{Solubility of AgSCN} = \text{form wt}_{AgSCN} \times [Ag^+]$$
$$\text{Solubility of AgSCN} = 165.95164 \times 1.01 \times \times 10^{-6}\text{M}$$
$$\text{Solubility of AgSCN} = 1.68 \times 10^{-4}\text{M}$$

9.11.2 Solubility of Lithium Carbonate

What is the solubility of lithium carbonate (Li_2CO_3) at 25 °C with $K_{sp} = 8.15 \times 10^{-4}$?

$$Li_2CO_3 = 2Li + CO_3^{2-}$$
$$K_{spLi2CO3} = [Li^+]^2 [CO_3^{2-}]$$

For each mole of lithium carbonate that dissolves there are 2 moles of lithium ions to 1 mole of carbonate ions. From this we see that the solubility of lithium carbonate is equal to the concentration of the carbonate $[CO_3^{2-}]$. We also see that the concentration of the lithium ion is twice that of the carbonate ion.

We set x as the number of moles per liter of lithium carbonate that dissolve by

$$[CO_3^{2-}] = x$$
$$[Li^+] = 2[CO_3^{2-}] = 2x$$

We now substitute into the solubility product equation and get

$$8.15 \times 10^{-4} = (2x)^2(x)$$
$$8.15 \times 10^{-4} = 4x^3$$
$$x = \sqrt[3]{\frac{8.15 \times 10^{-4}}{4}}$$
$$X = 0.0588\text{ M}$$

9.12 THE pH OF A SOLUTION

Assigning pure water as a neutral pH solution gives a hydrogen ion concentration of 1.0×10^{-7} M. On the basis of this, chemists define the pH of a solution as

$$\text{pH} = \log\frac{1}{[H^+]} \quad \text{or} \quad \text{pH} = -\log[H^+] \tag{9.57}$$

The pH of pure water is therefore

$$\text{pH} = -\log[H^+] \tag{9.58}$$
$$\text{pH} = -\log(1 \times 10^{-7}) \tag{9.59}$$
$$\text{pH} = 7.00 \tag{9.60}$$

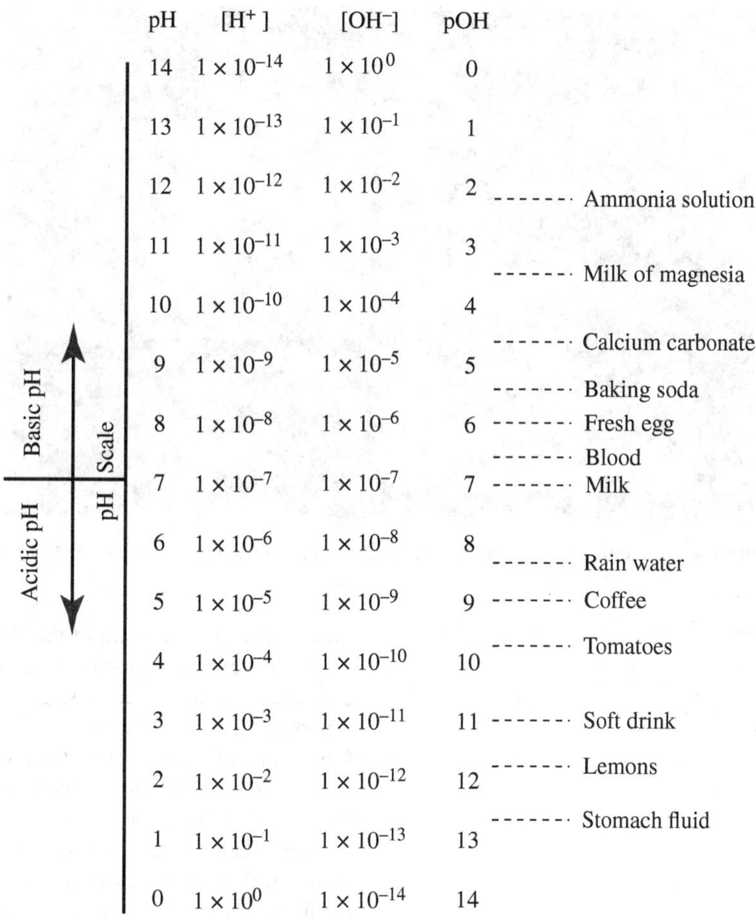

FIGURE 9.4 The pH scale, the hydrogen ion concentration [H^+], the hydroxide ion concentration [OH^-], the pOH, and the pH of some common solutions and substances.

The pH scale, the hydrogen ion concentration [H^+], the hydroxide ion concentration [OH^-], the pOH, and the pH of some common solutions and substances are illustrated in Figure 9.4.

The pH scale in general is defined as

If the solution pH is 7.0, the solution is neutral.

If the solution pH is <7.0, the solution is acidic.

If the solution pH is >7.0, the solution is basic.

The pOH in Figure 9.4 is the negative log of the molar [OH^-] concentration.

$$\mathrm{pOH} = -\log[\mathrm{OH}^-] \tag{9.61}$$

A useful relationship exists between pH, pOH, and the *ion product for water*, K_w:

$$\mathrm{pH} + \mathrm{pOH} = \mathrm{p}K_w \tag{9.62}$$

If we know the pH of a solution, then we can calculate the pOH from the relationship in Equation 9.12. Suppose a solution has a pH of 5.9. The value of pK_w is $-\log(1 \times 10^{-14}) = 14$, thus:

$$5.9 + \mathrm{pOH} = 14 \tag{9.63}$$

$$\mathrm{pOH} = 8.1 \tag{9.64}$$

9.13 MEASURING THE pH

The laboratory technician and chemical analyst will often measure the pH of a solution or of a substance dissolved in water. The pH is not usually calculated from a determination of the hydrogen ion [H^+] molar concentration by any wet chemical analyses such as titration, but typically by a pH meter, or sometimes as an approximate estimation by pH paper or strips as depicted in Figure 9.2. An example of a typical pH meter is depicted in Figure 9.5. The pH meter system consists of a pH sensing electrode and a readout screen.

9.13.1 The Glass Electrode

The heart of the pH meter is the glass electrode depicted in Figure 9.6.

The pH meter determines the [H^+] concentration in an unknown solution by measuring the difference in potential of two internal electrodes. The outer reference electrode and its solution do have contact with the solution to be measured through a junction or salt bridge as depicted in Figure 9.6. The inner electrode is completely isolated from the test solution and is associated with a glass sensing membrane. The glass sensing membrane

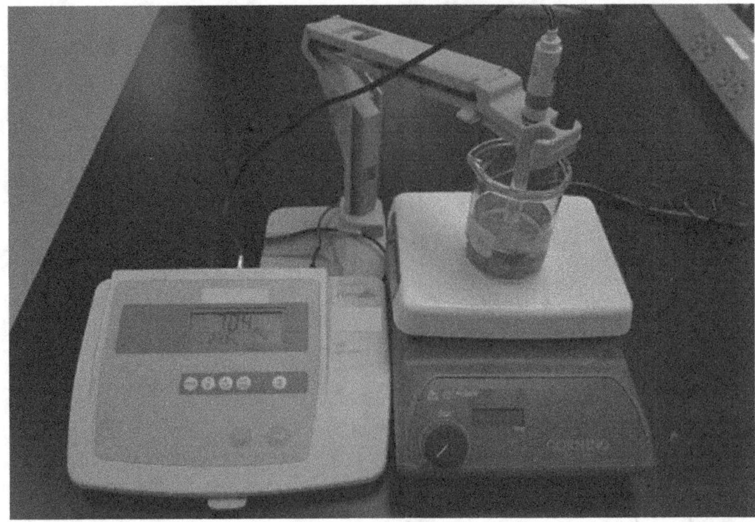

FIGURE 9.5 Example of a bench top pH meter including LCD readout and pH electrode.

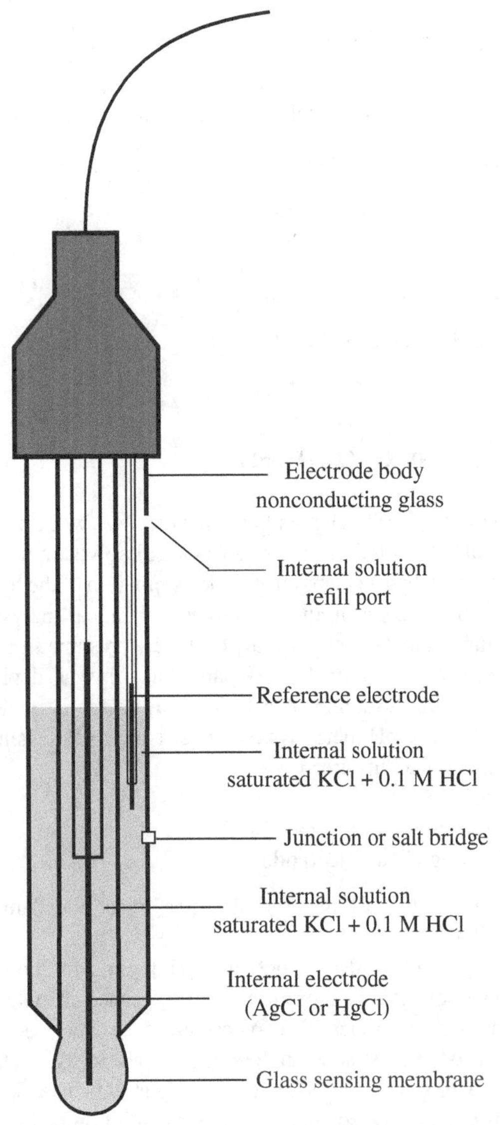

FIGURE 9.6 pH meter glass electrode.

bulb located at the bottom of the electrode is composed of three layers. One classic example of a glass pH sensing electrode consists of an inner hydrated glass gel (usually silicic acid, H_4SiO_4), a middle dry glass layer, and an outer hydrated glass gel. A potential difference across the membrane develops due to differences in $[H^+]$ concentrations within the electrode and the test solution. The dry glass boundary experiences an ion exchange between alkali metal ions in the glass and hydrogen ions, which results in the conductance of current through the dry glass by alkali metal movement. The internal solution that is contained with the reference electrode (outer electrode sleeve) needs to be periodically added through the refill port to replenish solution lost through evaporation. The pH electrode is calibrated using external solutions that are made of buffers of known pH (the technician needs to follow the specific instructions pertaining to the pH meter model that is being used in the laboratory for calibrations and pH solution measurements). The meters often have thermocouples to adjust for temperature, and give direct readouts of pH with ±0.01 pH units.

9.14 BUFFERED SOLUTIONS—DESCRIPTION AND PREPARING

Buffer solutions are solutions that contain a weak acid and the salt of the acid, or they contain a weak base and the salt of the base, or an acid salt such as potassium acid phthalate. These types of solutions have the ability to react with strong acids and strong bases, when added in small amounts, while maintaining the buffered solutions original pH (or a small change in pH). As an example of this, suppose we have made a buffered solution out of phosphoric acid (H_3PO_4) and monobasic sodium monophosphate (NaH_2PO_4) one of its salts. If we add some hydrochloric acid to the buffered solution, the basic sodium monophosphate reacts with the acid to produce more phosphoric acid molecules and the pH does not change appreciably:

$$H^+ + H_2PO_4^- \rightarrow H_3PO_4 \tag{9.65}$$

Also in the other direction if we add some sodium hydroxide (NaOH) to the buffered solution, the added base hydroxide ion will react with and remove from the solution hydrogen ions:

$$OH^- + H^+ \rightarrow H_2O \tag{9.66}$$

9.14.1 Le Chatelier's Principle

In response to this (according to Le Chatelier's principle), the phosphoric acid will ionize to form more hydrogen ions:

$$H_3PO_4 \rightarrow H^+ + H_2PO_4^- \tag{9.67}$$

Le Chatelier's principle states that if a chemical system at equilibrium experiences a change in concentration, temperature, volume, or partial pressure, then the equilibrium shifts to counteract the imposed change, and a new equilibrium is established. This principle also applies to a buffered solution composed of a base and its salt. Suppose we have a buffered solution of ammonia and its salt ammonium chloride. If we add hydrochloric acid to the buffered solution, hydroxide ions will be converted to water thus removing the hydroxide ions from solution. The ammonia in the solution will in response ionize to form more hydroxide ions:

$$NH_3 + H_2O \rightarrow NH_4^+ + OH^- \tag{9.68}$$

On the other hand, if some sodium hydroxide is added to the buffered solution the acidic ammonium ion will react with the hydroxide ion and form more of the ammonia in solution:

$$OH^- + NH_4^+ \rightarrow H_2O + NH_3 \tag{9.69}$$

9.14.2 Titration Curve of a Buffer

A titration curve can help illustrate the changes in pH according to an acid or base being added to the buffer system. Such a curve of the addition of either acid or base to a buffer solution is depicted in Figure 9.7. The bottom scale of the graph is the addition of an acid (H^+) to the buffer solution, while the top scale is the addition of a base (OH^-). The middle portion of the curve illustrates the buffer region of the solution. As can be seen, with the addition of either an acid or base, the pH changes (left scale) very little. The buffer capacity is the amount of acid or base that the solution can take until the pH changes by 1 unit. The greatest buffer capacity of the solution is at the midpoint of the curve where the conjugate acid and base of the buffer are in solution at equal molar concentrations.

9.14.3 Natural Buffer Solutions

Buffered solutions are important in many areas of analytical chemistry, biology, biochemistry, bacteriology, microbiology, and physiology. Both synthetic organic reactions are often controlled by pH as well as physiological reactions involving biosynthesis and the action of enzymes. The major intracellular buffer in living organisms is the conjugate acid base pair of

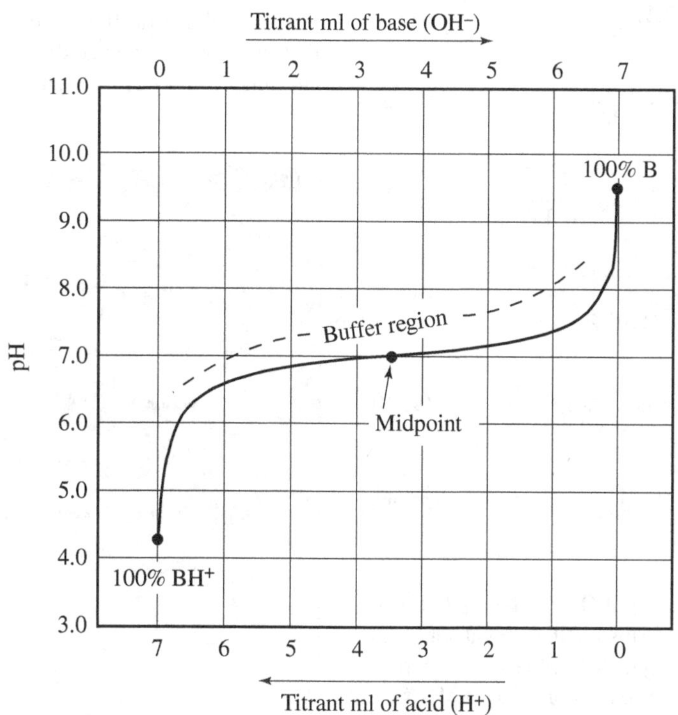

FIGURE 9.7 Titration curve of a buffered solution.

dihydrogenphosphate–monohydrogenphosphate ($H_2PO_4^-$–HPO_4^{2-}), while the major extracellular buffer is the conjugate acid–base pair of carbonic acid–bicarbonate (H_2CO_3–HCO_3^-). The carbonic acid–bicarbonate conjugate acid base pair is the buffer system responsible for the maintenance of an approximate 7.4 pH in blood.

9.14.4 Calculating Buffer pH

A very useful relationship for calculating the pH of buffers, or the amounts of the conjugate acid base pair needed is derived as follows:

$$K_a = \frac{[H^+][A^-]}{[HA]} \quad (9.70)$$

$$[H^+] = K_a \times \frac{[HA]}{[A^-]} \quad (9.71)$$

$$-\log[H^+] = -\log K_a - \log\frac{[HA]}{[A^-]} \quad (9.72)$$

$$pH = pK_a - \log\frac{[HA]}{[A^-]} \quad (9.73)$$

$$pH = pK_a + \log\frac{[A^-]}{[HA]} \quad (9.74)$$

9.14.5 Buffer pH Calculation I

What is the ratio of $[HPO_4^{2-}]$:$[H_2PO_4^-]$ required to keep an intracellular compartment at a pH of 7.8?

$$pH = pK_a + \log\frac{[A^-]}{[HA]} \quad (9.75)$$

$$7.8 = -\log(6.2\times10^{-8}) + \log\frac{[HPO_4^{2-}]}{[H_2PO_4^-]} \quad (9.76)$$

$$7.8 = 7.2 + \log\frac{[HPO_4^{2-}]}{[H_2PO_4^-]} \quad (9.77)$$

$$\frac{[HPO_4^{2-}]}{[H_2PO_4^-]} = \text{antilog}(0.6) \quad (9.78)$$

$$\frac{[HPO_4^{2-}]}{[H_2PO_4^-]} = 4 \quad (9.79)$$

From this we see that the ratio of $[HPO_4^{2-}]$:$[H_2PO_4^-]$ needed to keep the pH at 7.8 would be 4:1. This is a theoretical approach to calculating the ratio of acid to conjugate base needed to prepare a buffer solution. This is only an approximation, and the pH of the buffer solution should be checked with a pH meter and the pH adjusted with strong acid or base to the desired pH.

9.15 ChemTech BUFFER SOLUTION CALCULATOR

The ChemTech Toolkit program contains a page that can be used to prepare six different buffer solutions that encompass the entire pH range. If not already open, open ChemTech, go to the Main Menu and select the Chapter 9 Main Menu. In the Chapter 9 Main Menu select the Making Solutions link. Then select the Making Buffer Solutions link. This will bring up a set of options as depicted in Figure 9.8. As can be seen by each of the six options, there is an acetate buffer for pH range 3.6–5.6, two sets of phosphate buffers with ranges 2.0–3.2, and 5.6–8.0. The citrate buffer range is 3.0–7.0, and this is followed by two buffers in the basic pH range. These buffers cover the typical ranges that are often needed for a laboratory buffer. However, there are numerous buffers that can be required by specific methodology where the buffer described in the method should be prepared and used. This buffer solution preparation page in ChemTech will be used as a guideline for preparing several types of common buffers. For convenience, there are many buffer solutions and dry forms that are commercially available to the laboratory analyst for purchase either used as is or mixing the powder in a certain volume of water.

In the buffer preparation page let us select "Option 1" Preparation procedure to make an acetate buffer in the pH range of 3.6–5.6. Selecting Option 1 will bring up the page shown in Figure 9.9.

This is a calculator page where the analyst may enter into the "Enter pH desired" input textbox a certain pH that he would like to prepare a buffer solution for. Upon clicking the "Enter pH desired" button, the page will display the volumes of Solution A needed to mix with Solution B. At the top of the page there is the recipe for preparing the two solutions A and B. As we have learned, the buffer solution is composed of the acid and its salt. For preparing the acetate buffer, Solution A (0.1 *M* acetic acid) is made by adding ~500 ml of distilled and deionized (DDI) water to a 1 L volumetric flask. Next, 5.8 ml (6.0 g) of glacial acetic acid (water-free acetic acid) is added to the 1 l volumetric flask. The final volume is brought to 1 l with the DDI water and the

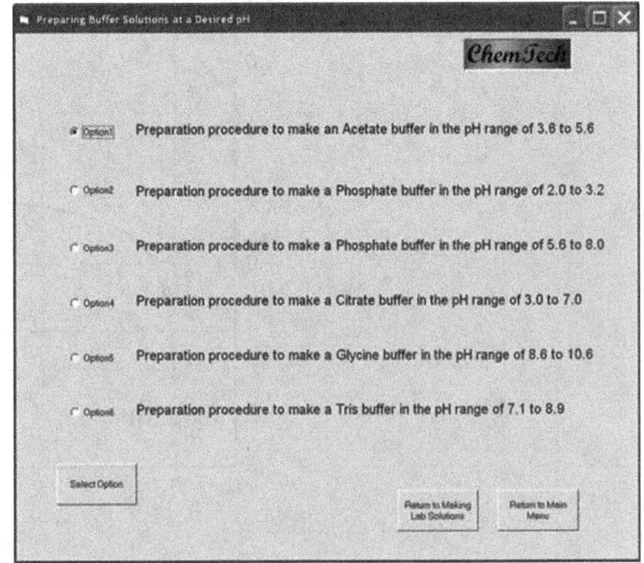

FIGURE 9.8 Buffer preparation page in ChemTech.

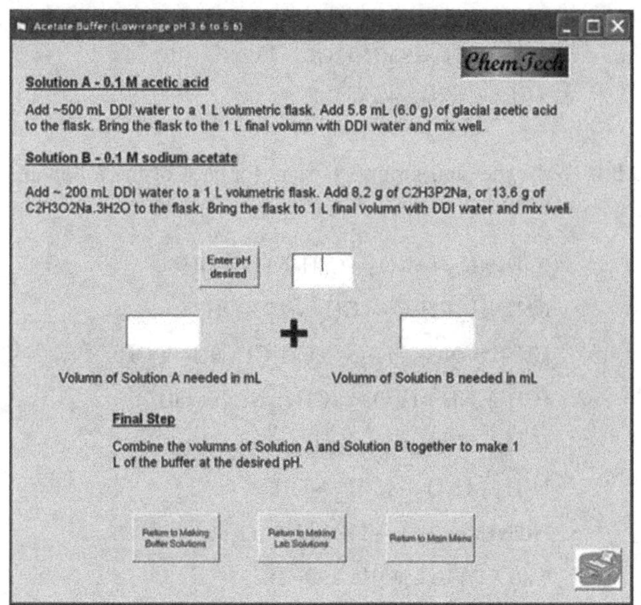

FIGURE 9.9 Acetate buffer preparation page in pH range of 3.6–5.6.

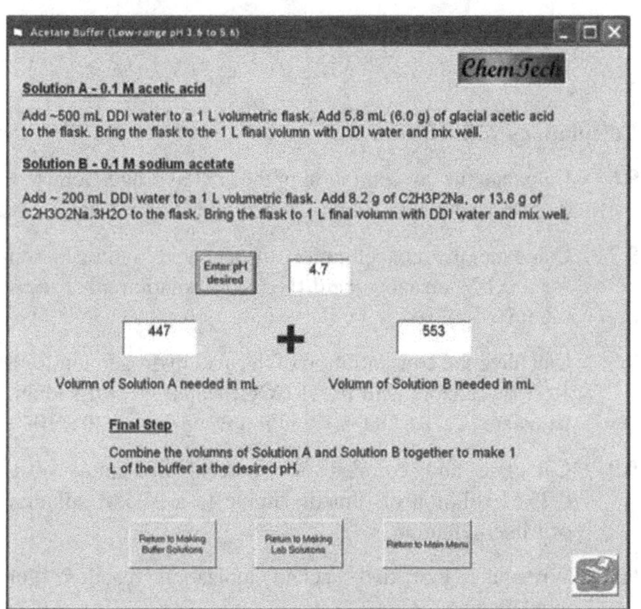

FIGURE 9.10 Output for the preparation of an acetate buffer at pH 4.7.

solution is mixed well. For the preparation of Solution B (0.1 *M* sodium acetate), ~200 ml DDI water is added to a volumetric flask. Then, 8.2 g of sodium acetate ($C_2H_3P_2Na$), or alternatively 13.6 g of $C_2H_3P_2Na \cdot 3H_2O$ (sodium acetate with three waters of hydration), is added to the flask and mixed well. Finally, the volume is brought to 1 l with DDI water and the solution is mixed well.

Suppose we would like to prepare an acetate buffer at a pH of 4.7 for some specific application. By entering 4.7 into the desired pH entry box and clicking the Enter button, the page displays the result as shown in Figure 9.10.

The result shown in Figure 9.10 instructs that 447 ml of Solution A are to be mixed with 553 ml of Solution B to prepare 1 l of the pH 4.7 acetate buffer. This is illustrated in the "Final Step" instruction at the bottom of the page. Note in the bottom right-hand corner a printer button that can be pressed to print the instructions for preparing the buffer solutions A and B and the volumes needed for the desired buffer pH. In an analogous manner, any of the six options depicted in Figure 9.8 can be chosen to prepare various buffer solutions. A note of caution, the buffer calculators can only be used for the stated pH ranges. If a pH is entered that is outside of the particular selected buffer range, then ChemTech will ask for a pH to be entered within the buffer range stated. Secondly, the calculator gives a starting point for preparing the buffer at a desired pH. After preparing the buffer solution, if a very accurate pH is needed, then the buffer should be checked with a pH meter and adjusted with either a strong acid or base until the specific pH is obtained.

9.16 CHAPTER KEY CONCEPTS

9.1 The word *acid* is derived from the Latin *acidus*, which means sour or sharp tasting.

9.2 The tastes of basic solutions in water are described to be bitter.

9.3 Litmus paper is an easy and straightforward way of telling if something is acidic or basic. If dipped in solution and the litmus paper turns red the solution is acidic, if turned blue the solution is basic.

9.4 The neutralization reaction of an acid and base produces a salt and water.

9.5 Acids are compounds that can donate protons (H^+) and bases are compounds that can accept protons. The neutralization reaction of an acid and base in solution produces the "conjugate" of the original acid or base.

9.6 The pH of a solution is the representation of the molar hydrogen ion concentration [H^+], (*M*), in solution.

9.7 The equilibrium constant (K_c) of reactions such as the dissolution of an acid in aqueous solution can be expressed in equation notation as the molar concentration of the products divided by the reactants.

9.8 The molar concentration of water, [H_2O], in solutions that are dilute acids or bases is always 55.51 *M*.

9.9 The base ionization constant, K_b, for a general base in aqueous solution is equal to $K_c \times 55.51$.

9.10 The acid ionization constant, K_a, for a general acid in aqueous solution is equal to $K_c \times 55.51$.

9.11 The ion product for water, K_w, is 1.0×10^{-14}.

9.12 The relationship between the ion product for water, K_w; the acid ionization constant, K_a; and the base ionization constant, K_b, is $K_w = K_a \times K_b$.

9.13 The pH scale in general is defined as

a. If the solution pH is 7.0, the solution is neutral.

b. If the solution pH is <7.0, the solution is acidic.

c. If the solution pH is >7.0, the solution is basic.

9.14 The pH of a solution

a. $pH = -\log [H^+]$
b. $pOH = -\log[OH^-]$
c. $pH + pOH = pK_w$

9.15 Buffer solutions are solutions that contain a weak acid and the salt of the acid, or they contain a weak base and the salt of the base, or an acid salt such as potassium acid phthalate.

9.16 Le Chatelier's principle states that if a chemical system at equilibrium experiences a change in concentration, temperature, volume, or partial pressure, then the equilibrium shifts to counteract the imposed change, and a new equilibrium is established.

9.17 CHAPTER PROBLEMS

Ionization Reactions and Constants

9.1 Write the conjugate base formula(s) for each of the following acids.

a. Hydrochloric acid (HCl)
b. Sulfuric acid (H_2SO_4)
c. Fumaric acid ($C_2H_2(COOH)_2$)
d. Acetic acid (CH_3COOH)
e. Formic acid (HCOOH)

9.2 Write the reaction between the acids in Problem 9.1 and ammonia (NH_3).

a. Hydrochloric acid (HCl)
b. Sulfuric acid (H_2SO_4)
c. Fumaric acid ($C_2H_2(COOH)_2$)
d. Acetic acid (CH_3COOH)
e. Formic acid (HCOOH)

9.3 Write the equilibrium constant(s) for the reactions in Problem 9.2.

a. Hydrochloric acid (HCl)
b. Sulfuric acid (H_2SO_4)
c. Fumaric acid ($C_2H_2(COOH)_2$)
d. Acetic acid (CH_3COOH)
e. Formic acid (HCOOH)

9.4 Write the equilibrium constant for each of the following reactions.

a. $H_2CO_3 = H^+ + HCO_3^-$
b. $Ni^{2+} + 4CN^- = Ni(CN)_4^{2-}$
c. $HCO_3 = H^+ + CO_3^{2-}$
d. $Ag + 2NH_3 = Ag(NH_3)_2^+$
e. $H_2CO_3 = 2H^+ + CO_3^{2-}$

9.5 Write the equilibrium constant for each of the following reactions.

$$C_6H_5NH_2 + H_2O \leftrightarrow C_6H_5NH_3^+ + OH^-$$
$$CH_3NH_2 + H_2O \leftrightarrow CH_3NH_3^+ + OH^-$$
$$CH_3CH_2NH_2 + H_2O \leftrightarrow CH_3CH_2NH_3^+ + OH^-$$
$$(CH_3)_2NH + H_2O \leftrightarrow (CH_3)_2NH_2^+ + OH^-$$
$$(CH_3)_3N + H_2O \leftrightarrow (CH_3)_3NH^+ + OH^-$$
$$N_2H_4 + H_2O \leftrightarrow N_2H_5^+ + OH^-$$
$$HONH_3 + H_2O \leftrightarrow HONH_4^+ + OH^-$$
$$Ca(OH)_2 \leftrightarrow CaOH^+ + OH^-$$
$$LiOH \leftrightarrow Li^+ + OH^-$$
$$KOH \leftrightarrow K^+ + OH^-$$
$$NaOH \leftrightarrow Na^+ + OH^-$$

Calculations with K_a and K_b

9.6 Calculate the concentration of the acid hydrogen ion for a 0.05 M acetic acid solution when $K_a = 1.74 \times 10^{-5}$.

9.7 Calculate the concentration of the acid hydrogen ion for a 0.1 M arsenous acid (H_3AsO_3) solution when $K_a = 6.6 \times 10^{-10}$.

9.8 Calculate the concentration of the acid hydrogen ion for a 1.65 g/l benzoic acid (C_6H_5COOH) and 2.97 g/l sodium benzoate (C_6H_5COONa) solution when $K_a = 6.30 \times 10^{-5}$.

9.9 Calculate and compare the percent ionization of a 0.25 M solution of dimethylamine to a 0.25 M solution of trimethylamine.

9.10 Write the first and second ionization reactions for ethylenediamine.

9.11 Write expressions for the first and second ionization constants for ethylenediamine.

9.12 What is the percent ionization for the first and second ionization reactions for a 0.5 M ethylenediamine solution?

Solubility Product K_{sp} Calculations

9.13 Write the dissociation equation for each.

a. Aluminum hydroxide ($Al(OH)_3$)
b. Barium carbonate ($BaCO_3$)
c. Cadmium arsenate ($Cd_3(AsO_4)_2$)
d. Calcium oxalate monohydrate ($CaC_2O_4 \times H_2O$)

e. Calcium phosphate ($Ca_3(PO_4)_2$)
f. Iron(II) carbonate ($FeCO_3$)
g. Iron(II) hydroxide ($Fe(OH)_2$)

9.14 Write the solubility product constant expression for each of the following compounds.

a. Lead(II) bromide ($PbBr_2$)
b. Lithium carbonate (Li_2CO_3)
c. Lithium fluoride (LiF)
d. Lithium phosphate (Li_3PO_4)
e. Magnesium oxalate dehydrate ($MgC_2O_4 \times 2H_2O$)
f. Palladium(II) thiocyanate ($Pd(SCN)_2$)
g. Silver(I) sulfite (Ag_2SO_3)
h. Zinc carbonate monohydrate ($ZnCO_3 \times H_2O$)

9.15 What is the solubility of manganese(II) carbonate ($MnCO_3$) at 25 °C?

9.16 Calculate the concentration of the potassium ion in a solution that is in equilibrium with potassium hexachloroplatinate (K_2PtCl_6) while the solution contains sodium hexachloroplatinate so that $[PtCl_6]^{2-} = 0.01$ M.

9.17 What is the solubility of cadmium phosphate ($Cd_3(PO_4)_2$) at 25 °C with $K_{sp} = 2.53 \times 10^{-33}$?

Calculations Involving pH

9.18 Calculate the pH of the following solutions.

a. 0.025 M solution of nitric acid (HNO_3).
b. 0.005 M sulfuric acid (H_2SO_4)
c. 0.002 M hydrochloric acid (HCl)
d. 0.01 M iodic acid (HIO_3)

9.19 Calculate the pH of the following solutions.

a. 0.0147 M solution of sodium hydroxide (NaOH).
b. 0.001 M lithium hydroxide (LiOH)
c. 0.005 M calcium hydroxide ($Ca(OH)_2$)
d. 0.01 M potassium hydroxide (KOH)

9.20 Calculate the pH of a 0.25 M cyanic acid (HOCN) solution.

9.21 Calculate the pH of a 0.15 M solution of ethylamine ($CH_3CH_2NH_2$).

9.22 What is the pH of a solution made up of 0.30 M acetic acid ($HC_2H_3O_2$), which has 0.15 M of sodium acetate ($NaC_2H_3O_2$) added to it?

Buffer Solution Calculations

9.23 A solution contains 0.015 M pyruvic acid ($HC_3H_3O_3$, $K_a = 1.4 \times 10^{-4}$) and 0.015 M sodium pyruvate ($NaC_3H_3O_3$).

a. What is the pH of the solution?
b. What is the pH when 1.5 ml of 0.10 M hydrochloric acid is added to 30 ml of the buffer solution?
c. What is the pH when 1.5 ml of 0.10 M sodium hydroxide is added to 30 ml of the buffer solution?

9.24 What molar ratio of bicarbonate ion (HCO_3^-) to carbonic acid (H_2CO_3) is needed to maintain a pH of 6.9?

9.25 What is the pH change when 2.0 ml of a 1.0 M HCl solution is added to 100 ml of a buffered solution that is 0.5 M in nitrous acid (HNO_2) and 0.5 M in sodium nitrite.

9.26 What volume of a 0.1 M acetic acid solution is needed to be mixed with a 0.1 M sodium acetate solution to make 1 liter of a pH 4.9 buffer?

10

TITRATION—A VOLUMETRIC METHOD OF ANALYSIS

10.1 Introduction
10.2 Reacting Ratios
10.3 The Equivalence Point
10.4 Useful Relationships for Calculations
10.5 Deriving the Titration Equation
  10.5.1 Titration Calculation Example
10.6 Titrations in ChemTech
  10.6.1 Acid/Base Titrations Using Molar Solutions
  10.6.2 Titration Calculation Example
10.7 Acid/Base Titration Endpoint (Equivalence Point)
10.8 Acid/Base Titration Midpoint
10.9 Acid/Base Titration Indicators
  10.9.1 The Ideal Indicator
10.10 Titrations Using Normal Solutions
  10.10.1 Normal Solution Titration Example
10.11 Polyprotic Acid Titration
10.12 ChemTech Calculation of Normal Titrations
10.13 Performing a Titration
  10.13.1 Titration Glassware
  10.13.2 Titration Steps
10.14 Primary Standards
10.15 Standardization of Sodium Hydroxide
  10.15.1 NaOH Titrant Standardization Example
10.16 Conductometric Titrations (Nonaqueous Solutions)
10.17 Precipitation Titration (Mohr Method for Halides)
  10.17.1 Basic Steps in Titration
  10.17.2 Important Considerations
10.18 Complex Formation with Back Titration (Volhard Method for Anions)
  10.18.1 Iron(III) as Indicator
  10.18.2 Chloride Titration
  10.18.3 The General Calculation
  10.18.4 Chloride Titration
10.19 Complex Formation Titration with EDTA for Cations
  10.19.1 EDTA–Metal Ion Complex Formation
  10.19.2 The Stability Constant
  10.19.3 Metal Ions Titrated
  10.19.4 Influence of pH
  10.19.5 Buffer and Hydroxide Complexation
  10.19.6 Visual Indicators
10.20 Chapter Key Concepts
10.21 Chapter Problems

10.1 INTRODUCTION

A very useful and widely employed technique in the analytical laboratory is the titrimetric (volumetric) method of analysis. Titrimetric analysis is the process of measuring a substance of unknown concentration in a solution of interest via reaction with a standard that we have made that contains a known substance concentration. If we take a known weight or volume of our solid or solution of interest, we can calculate the concentration of the unknown from the measured use of our known concentration solution. Chemists call the standard solution the *titrant* solution.

10.2 REACTING RATIOS

In a titration analysis, the reaction taking place is known with a predetermined stoichiometry. The analyst knows what compound is reacting with what compound and the moles of each. Suppose we have a compound "A" that is the titrant and a compound "B" that is our known substance of interest that we want to determine the concentration of. A simplified representation of the reaction is

$$a\text{A} + b\text{B} \rightarrow \text{products.} \tag{10.1}$$

Analytical Chemistry: A Chemist and Laboratory Technician's Toolkit, First Edition. Bryan M. Ham and Aihui MaHam.

In the above reaction, a and b are the moles of reactants, and are necessary to know to calculate concentrations. For example, the reaction of a titrant of dissolved silver (Ag^+) and chloride (Cl^-) in our test solution to produce the insoluble compound silver chloride (AgCl) is represented as

$$Ag^+(aq) + Cl^-(aq) \rightarrow AgCl(s), \quad (10.2)$$

where $a = 1$, $b = 1$, aq = aqueous (meaning the substance is in solution), and s = solid (meaning the substance is not in solution). For the determination of sodium hydroxide with the titrant sulfuric acid, the reaction is

$$H_2SO_4 + 2NaOH \rightarrow Na_2SO_4 + 2H_2O. \quad (10.3)$$

In this case, $a = 1$ and $b = 2$.

10.3 THE EQUIVALENCE POINT

When we are performing titrations, it is most optimal if the reaction basically goes to completion in a very short amount of time. For most titrations that we will be looking at this is the case. Next, the analyst needs a way to determine that the reaction between the two compounds, the titrant and the sample, is complete and a stoichiometric amount of the titrant has been added. The point at which the stoichiometry of the reaction has been reached is known as the *equivalence point*. This can be accomplished with an indicator or with some type of potentiometric measurement (often a pH meter). The determination of this point is called the *endpoint* of the titration and ideally is equal to the *equivalence point*. However, due to various reasons, this sometimes is not the case (we will assume for our purposes that the endpoint is equal to the equivalence point).

Let us consider our simplified reaction equation again, this time listing the reactants according to a titration:

$$a\text{A(titrant)} + b\text{B(substance titrated, titrand)} \rightarrow \text{products}. \quad (10.4)$$

10.4 USEFUL RELATIONSHIPS FOR CALCULATIONS

In the titration reaction aA again is the titrant, the solution of known concentration, and B is the substance titrated, the titrand containing the compound of interest with unknown concentration. In the reaction, "a" is the moles of A that react with "b" moles of B such as an acid–base neutralization reaction. Depending on whether the analysis in the laboratory is working in grams or milligrams, there are some useful relationships that we will go over as applied to titrations:

$$(\text{ml})(M) = \text{mmoles} \quad (10.5)$$

$$(\text{mmoles})(\text{milli form wt}) = \text{g}, \quad \text{alternatively}(\text{mmoles})(\text{form wt}) = \text{mg} \quad (10.6)$$

$$(\text{ml})(M)(\text{milli form wt}) = \text{g}, \quad \text{alternatively}(\text{ml})(M)(\text{form wt}) = \text{mg} \quad (10.7)$$

10.5 DERIVING THE TITRATION EQUATION

An important consideration in our titrations is the reacting ratio between the titrant and the substance titrated. This allows us to convert from the amount of titrant consumed in the titration reaction to the amount of substance titrated:

$$(\text{mmoles}_A)\left(\frac{b}{a}\right) = \text{mmoles}_B. \quad (10.8)$$

In this way, if we combine in our calculation the volume of titrant consumed in the titration reaction with the known molarity of the titrant, the analyst can calculate the amount of substance titrated as

$$(\text{ml}_A)(\text{M}_A)\left(\frac{b}{a}\right)(\text{milli form wt}_B) = \text{g}_B. \quad (10.9)$$

If we know the starting material sample weight that was titrated, then the percentage B in the sample can be calculated by dividing both sides of the equation by the sample weight and multiplying by 100:

$$\frac{(\text{ml}_A)(M_A)\left(\frac{b}{a}\right)(\text{milli form wt}_B)\ (100)}{\text{g}_S} = \frac{\text{g}_B}{\text{g}_S} \times 100 \quad (10.10)$$

$$\%\text{B} = \frac{(\text{ml}_A)(M_A)\left(\frac{b}{a}\right)(\text{milli form wt}_B)(100)}{\text{g}_S}. \quad (10.11)$$

10.5.1 Titration Calculation Example

Let us look at a specific example. Suppose a powder sample is thought to be pure sodium carbonate. A 1.068 g portion of the sample that was dissolved in 50 ml distilled deionized (DDI) water required 25.86 ml of a 0.5 M HCl titrant to neutralize the base present. The base present is sodium carbonate (Na_2CO_3), which has a milli form weight of 0.103 g/mmol, and reacts according to the following equation:

$$2HCl + Na_2CO_3 \rightarrow 2NaCl + H_2CO_3. \quad (10.12)$$

In this case, $a = 2$ and $b = 1$.

This can be directly entered into Equation 10.7 as follows:

$$\%Na_2CO_3 = \frac{(\text{ml}_{HCl})(M_{HCl})\left(\frac{1}{2}\right)(\text{milli form wt}_{Na_2CO_3})(100)}{\text{g}_S} \quad (10.13)$$

$$\%Na_2CO_3 = \frac{(25.86)(0.5)\left(\frac{1}{2}\right)(0.103)(100)}{1.068} \quad (10.14)$$

$$\%Na_2CO_3 = 62.3$$

The sample is obviously not pure sodium carbonate.

10.6 TITRATIONS IN ChemTech

We have now gone over the fundamentals for titrations using molar solutions. The Equations 10.1–10.4 can be used in many different instances. ChemTech can also be used in the laboratory for calculating the results of titrations using molar solutions. The section in ChemTech can be used as a tool in the laboratory for calculating titrations, and also can be used as a source of reference and reminder in the laboratory for the analyst. If not already open, open ChemTech and go to the Main Menu options. In the Main Menu options, click on the Chapter 10 link to pull up the Main Menu for Chapter 10. In the Chapter 10 Main Menu, choose the link for the Titrations Main Menu, which will open a page as depicted in Figure 10.1.

10.6.1 Acid/Base Titrations Using Molar Solutions

Click on the first link "Acid/Base Titrations Using Molar Solutions" to open up the titration calculation page depicted in Figure 10.2. The calculation page contains input boxes used to calculate the results of a titration, and also links to other tools and useful information. At the bottom right-hand side of the page, there is also a button that can be used to print the page if needed. Let us look at using the page to calculate the results of a specific titration.

10.6.2 Titration Calculation Example

Suppose a submitted aqueous sample is calling for an acidity measurement as sulfuric acid (H_2SO_4). A 50 ml aliquot (this will be the sample weight in grams as the density of water is 1 g/ml) of the sample required 11.26 ml of a 0.015 *M* NaOH titrant to neutralize the acid present. The acid present is being represented as sulfuric acid (H_2SO_4). The milli form wt (g/mmol) of sulfuric acid can be determined in two ways using the titration calculation page. First, there is a link to a listing of some of the more common analytes. Click on the third link, that is, "See listing of common milli form wt" button, to pull up a page as shown in Figure 10.3. Use the bottom scroll bar to thumb through the listing until sulfuric acid is found.

Record the value of 0.098078 g/mmol as the milli formula weight. If the analyte of interest is not found to be included in the list, an alternative way to obtain the milli formula weight is to use the molecular weight calculator link. Press the exit key of the page shown in Figure 10.3 to return to the molar titration calculation page. To the right, there is a button that links to "Calculate formula weight from formula." Press the button to bring up the formula weight calculator. Enter "H_2SO_4" into the top Formula 1 input box and press the Calculate button. Below

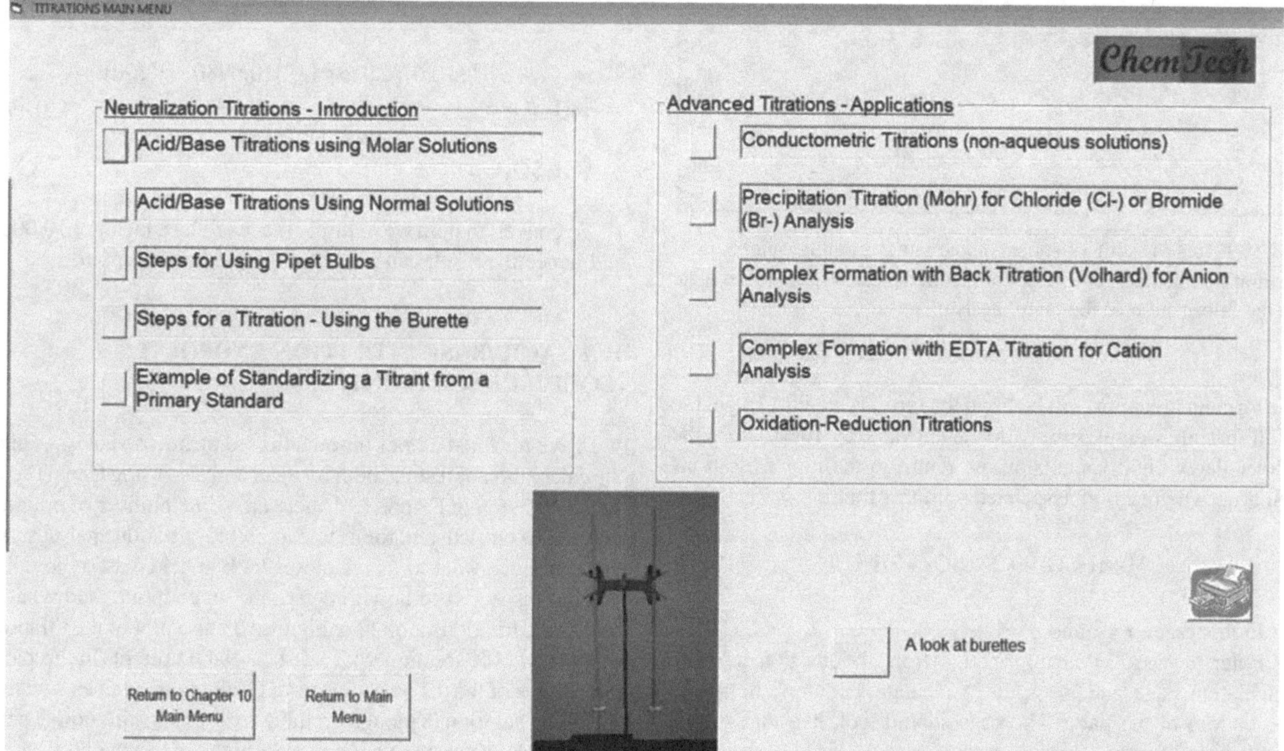

FIGURE 10.1 Titrimetric (volumetric) main menu.

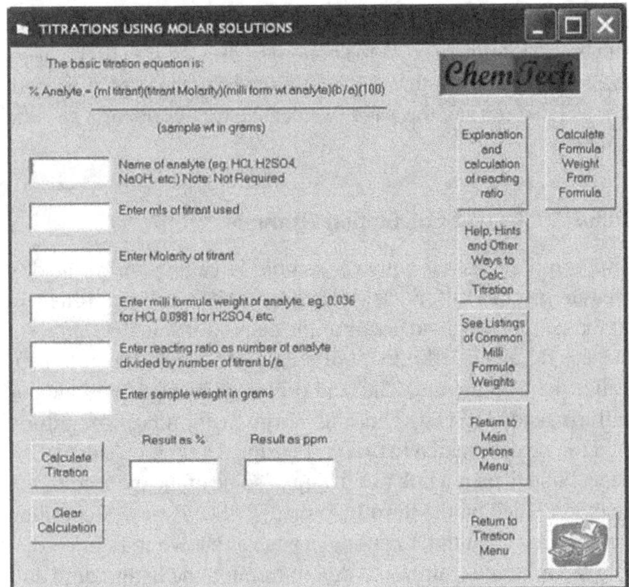

FIGURE 10.2 Titrations using molar solutions calculation page with included links.

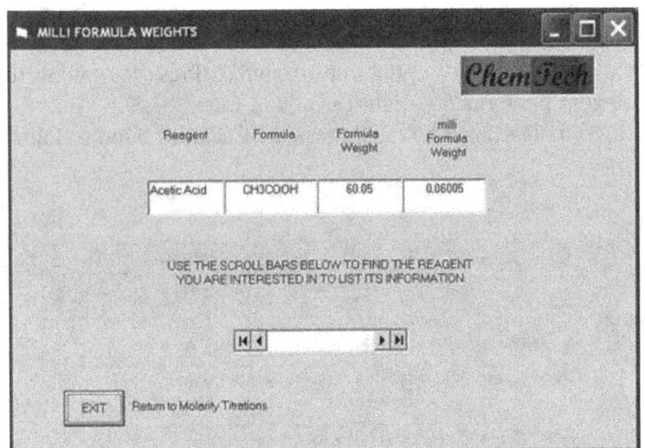

FIGURE 10.3 Listing of some common milli formula weights used in titration calculations. The bottom scroll bar can be used to navigate to the information of the target analyte.

the formula input box will be listed the formula weight. To get the milli formula weight, simply divide the value by 1000. Close the Molecular Weight Calculator and return to the molar titrations window. The reaction associated with this titration is

$$2NaOH + H_2SO_4 \rightarrow 2NaSO_4 + 2H_2O. \tag{10.15}$$

In this case, $a = 2$ and $b = 1$.

Enter the associated information from the titration into the calculation page and press the "Calculate Titration" button at the bottom of the page. The page should look like that shown in Figure 10.4.

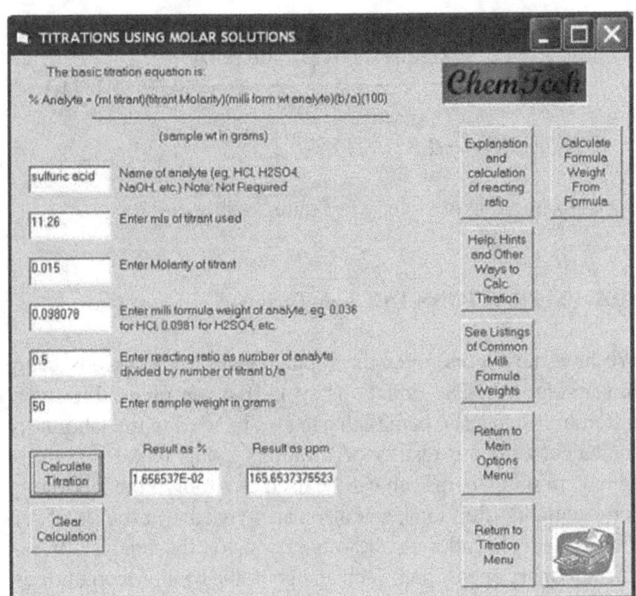

FIGURE 10.4 Results of titration calculation in Example 10.2.

Note the result of 165 ppm. Let us use Equation 10.11 to calculate this result by hand. The information from Example 10.2 can be directly entered into Equation 10.11 as follows:

$$\%H_2SO_4 = \frac{(mL_{NaOH})(M_{NaOH})\left(\frac{1}{2}\right)(\text{milli form wt}_{H_2SO_4})(100)}{g_S} \tag{10.16}$$

$$\%H_2SO_4 = \frac{(11.26)(0.015)\left(\frac{1}{2}\right)(0.098078)(100)}{50} \tag{10.17}$$

$$\%H_2SO_4 = 0.0166$$

To convert to ppm we multiply the result by a factor of 10,000, and hence the result can also be expressed as 166 ppm.

10.7 ACID/BASE TITRATION ENDPOINT (EQUIVALENCE POINT)

We have now had a brief introduction to titrimetric analyses and the calculation of results, both by hand and also using ChemTech. As can be seen in Figure 10.1 there are still a number of options that can be chosen, and for us to explore, for titration analyses. At this point, we want to take a closer look at acid–base type titrations in respect as to how to determine an endpoint, and what a plot of the titrations look like. Figure 10.5 is a plot of the titration of a 0.1 *M* HCl solution with a 0.1 *M* NaOH titrant. In the plot, the volume of the 0.1 M NaOH titrant is the *x*-axis values, and the pH of the solution is plotted on the y-axis. In this titration experiment, a pH meter was used to measure the pH as the titrant was

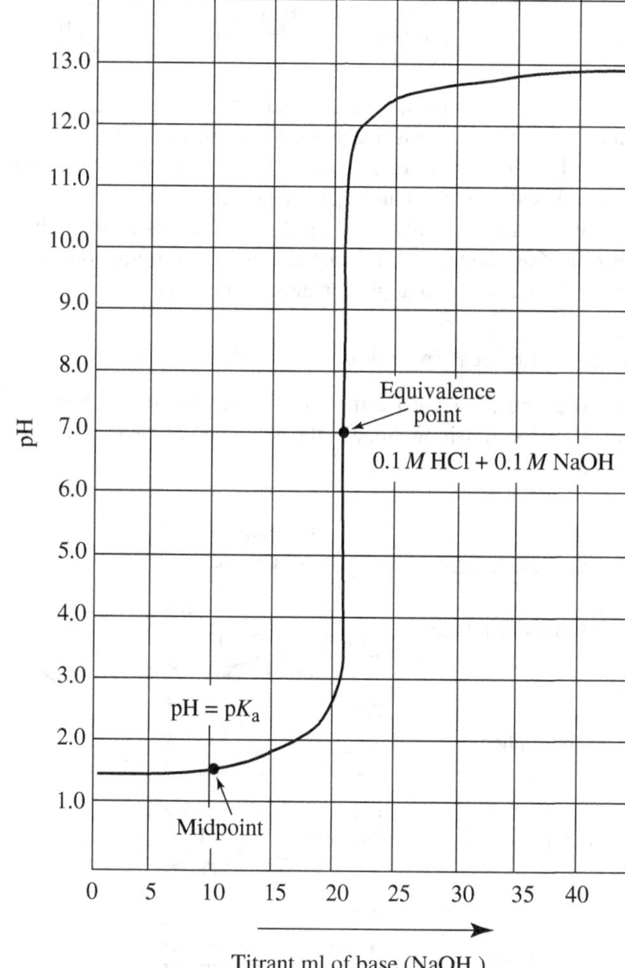

FIGURE 10.5 Titration curve of the measurement of 0.1 M HCl plotting pH of the solution versus volume of titrant (0.1 M NaOH) added.

added. Note at the beginning of the titration the pH starts out at approximately 1.5 pH. With the addition of the NaOH titrant, the pH will change slowly as illustrated with the first part of the titration curve. The curve illustrates that as the titration approaches the equivalence point, the pH will change very quickly. In the initial rise of the curve, there is a midpoint that can be used to describe and predict how a titration curve will look. At the midpoint, there is a direct relationship between the pH of the titration and the pK_a value of the analyte being measured. If we refer back to the acid ionization constant equation, the pH of the solution can be obtained with the following relationship:

$$K_a = \frac{[H^+][A^-]}{[HA]} \tag{10.18}$$

$$[H^+] = K_a \times \frac{[HA]}{[A^-]} \tag{10.19}$$

$$-\log[H^+] = -\log K_a - \log\frac{[HA]}{[A^-]} \tag{10.20}$$

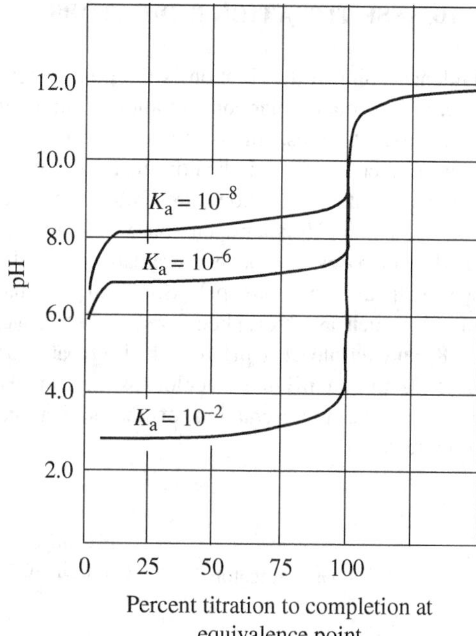

FIGURE 10.6 Titration behavior at the equivalence point depending upon the magnitude of the ionization constant K_a. The weaker the acid (smaller K_a) the less dramatic in change of pH at the equivalence point.

$$\mathrm{pH} = \mathrm{p}K_a - \log\frac{[HA]}{[A^-]} \tag{10.21}$$

$$\mathrm{pH} = \mathrm{p}K_a + \log\frac{[A^-]}{[HA]} \tag{10.22}$$

10.8 ACID/BASE TITRATION MIDPOINT

However, at the midpoint of the titration as shown in Figure 10.5, the concentration of the protonated acid [HA] is equal to the concentration of the deprotonated acid [A^-], thus the equation becomes

$$\mathrm{pH} = \mathrm{p}K_a + \log 1.0 \tag{10.23}$$

$$\mathrm{pH} = \mathrm{p}K_a + 0 \tag{10.24}$$

$$\mathrm{pH} = \mathrm{p}K_a \tag{10.25}$$

A consequence of this is that a when strong acid is titrated with a strong base, such as our example in Figure 10.5 of HCl titrated with NaOH, the rise at the equivalence point of the pH value will be very sharp. In contrast to this, a weak acid titrated with a strong base will have a much less dramatic rise in the pH at the equivalence point. From Table 9.1, the K_a value for hypochlorous acid (HClO) is small at 3.0×10^{-8} indicating a very weak acid. Its titration curve will only have a slight pH change of approximately 2–3 units when titrated with a strong base such as NaOH. The comparison of three different titration curves for three acids with decreasing ionization constants is depicted in Figure 10.6. This can make it difficult sometimes to catch the equivalence point.

10.9 ACID/BASE TITRATION INDICATORS

The equivalence point in the titration is the point where in the solution there is an equal molar concentration of titrant and analyte. In acid–base titrations, this is the point of neutralization. There are a number of visual indicators that are used to show the equivalence point of a titration, often more convenient than doing titrations with a pH meter. Acid–base indicators are weak organic acids and bases that have a dramatic color change in accordance with the solutions pH. Some of the indicators change in color such as chlorophenol red, which is yellow at a pH of 4.8, and purple at a pH of 6.4. There are some like phenolphthalein that at pH 8.2 is colorless, and at pH 9.8 is red/violet. A simple representation of the ionization of an acid–base indicator is

$$\underset{\substack{\text{acidic form}\\ \text{color 1}}}{\text{HInd}} \rightleftharpoons \text{H}^+ + \underset{\substack{\text{basic form}\\ \text{color 2}}}{\text{Ind}^-} . \tag{10.26}$$

Since the indicator is, for acidic indicators, a weak acid at low concentration, the stronger analyte acid will titrate first. When the analyte has been neutralized by the titrant base, the weak indicator will react next and then change color. Figure 10.7 lists a number of commonly used acid–base indicators with their pH transition ranges and associated colors. The transition range for the indicators is in approximately 2 pH units.

10.9.1 The Ideal Indicator

The ideal indicator for a certain acid–base titration is chosen to have its color transition range in the pH region of the *equivalence*

Acid–base indicator	pH range		color change	
Brilliant green	0.0	2.6	Yellow	Green
Methyl green	0.1	2.3	Yellow	Blue
Thymol blue	1.2	2.8	Red	Yellow
Quinaldine red	1.4	3.2	Colorless	Pink
2,4-Dinitro phenol	2.8	4.7	Colorless	Yellow
Bromophenol blue	3.0	4.6	Yellow	Blue/violet
Congo red	3.0	5.2	Blue	Yellow/orange
Methyl orange	3.1	4.2	Red	Yellow/orange
Bromocresol green	3.8	5.4	Yellow	Blue
Methyl red	4.4	6.2	Red	Yellow/orange
Litmus	5.0	8.0	Red	Blue
Bromophenol red	5.2	6.8	Orange/yellow	Purple
4-Nitrophenol	5.4	7.5	Colorless	Yellow
Bromothymol blue	6.0	7.6	Yellow	Blue
Phenol red	6.4	8.2	Yellow	Red/violet
Cresol red	7.0	8.8	Orange	Purple
Thymol blue	8.0	9.6	Yellow	Blue
Phenolphthalein	8.2	9.8	Colorless	Red/violet
Thymolphthalein	9.3	10.5	Colorless	Blue
Alkali blue	9.4	14.0	Violet	Pink
Indigo carmine	11.5	13.0	Blue	Yellow
Epsilon blue	11.6	13.0	Orange	Violet

FIGURE 10.7 Common acid–base indicators including pH range and example of color change.

point of the titration. The *endpoint* of the titration is when the color changes for the indicator. The equivalence point and the endpoint are essentially taken as the same for acid–base indicator titrations. As depicted in Figure 10.6, the equivalence point can be different for different acid–base titrations. For example, the acid with the K_a at 10^{-2} has its equivalence point within the pH change region of 6–8, thus a good indicator for this titration might be bromothymol blue or phenol red. For the acid with the K_a at 10^{-6} and equivalence point within the pH change region of 8–10, a good indicator might be thymol blue or phenolphthalein. Finally, for the acid with the K_a at 10^{-8} and equivalence point within the pH change region of 9–11, a good indicator might be thymolphthalein or alkali blue. Often, when a routine titration is being set up in the analytical laboratory, a pH meter in conjunction with a visual indicator will be used. This is done to determine precisely where the equivalence point occurs (by observing the very rapid change in pH with a small addition of titrant) and the associated color of the indicator. A reference solution of the indicator's color at the equivalence point pH can then be kept and used to obtain the equivalence point of the routine titrations by comparing the color of the samples during the titration.

10.10 TITRATIONS USING NORMAL SOLUTIONS

Titrations are also performed using normal solutions where equivalent weights are used in place of formula weights. If needed, refer back to Chapter 8, Section 8.7.4 Normal (*N*) Solutions, for the calculation of normal solutions. For the simple titration reaction equation, the normality of the titrant is used along with the milli equivalent weight of the analyte. Let us look at an example of calculating a titration using normality.

$$a\text{A (titrant)} + b\text{B (substance titrated)} \rightarrow \text{products} \qquad (10.27)$$

$$\%\text{B} = \frac{(\text{ml}_\text{A})(N_\text{A})(\text{milli eq wt}_\text{B})(100)}{\text{g}_\text{S}} \qquad (10.28)$$

10.10.1 Normal Solution Titration Example

Suppose the laboratory has received a submitted sample of a solution that is acidic. The sample's submitter would like the ppm acidity as phosphoric acid (H_3PO_4) determined. The following curve (Fig. 10.8) was generated using a pH meter and a 50 ml burette (see volumetric glassware) for the titration of a 20 ml aliquote of the sample using 0.2 *N* NaOH as the titrant. As can be seen in the curve, there are two equivalence points that are being observed in the titration. The first equivalence point is occurring at a pH of approximately 5, while the second is occurring at a pH of ~10. We know from Table 9.1 that there are in fact three acid ionization constants associated with phosphoric acid along with three midpoints.

In the next section, we are going to take a closer look at polyprotic acid titrations, such as phosphoric acid using normal solutions.

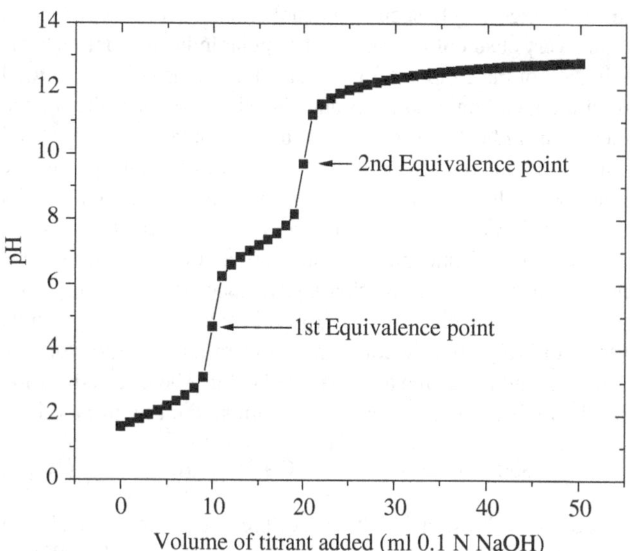

FIGURE 10.8 Titration results of submitted sample for % acidity as H_3PO_4.

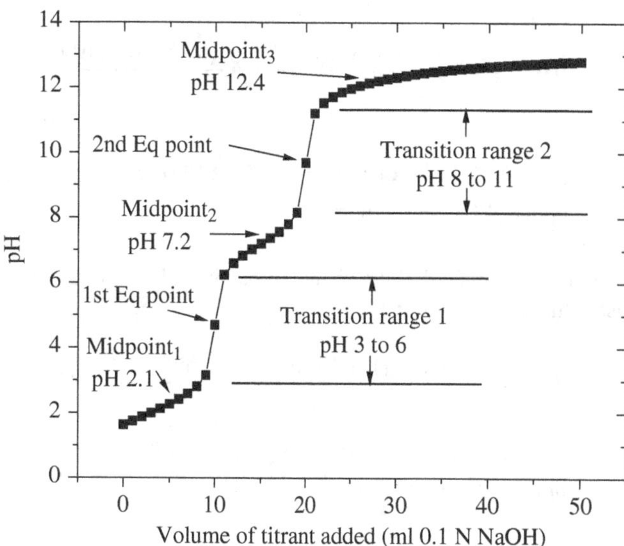

FIGURE 10.9 Phosphoric acid titration curve denoting the midpoints, the equivalence points, and the pH transition ranges.

10.11 POLYPROTIC ACID TITRATION

Phosphoric acid is a polyprotic acid and thus possesses two distinguishable equivalence points, and thus also two pH transition ranges. Figure 10.9 depicts the titration curve denoting the midpoints, the equivalence points, and the pH transition ranges. With using a pH meter we are able to detect and measure these two distinct transition ranges. This brings us to an important aspect of visual pH indicators as listed in Figure 10.7. The analyst, knowing that phosphoric acid is a polyprotic acid, would need to select an indicator that would have a transition range appropriate for the measurement desired. The measurement for ppm acidity as phosphoric acid would require an indicator with a transition range in the pH range 8–11, such as phenolphthalein. Phenolphthalein, with a very clear color change, is a popular indicator for the determination of the endpoint for weak acid titrations such as the total titration of carbonic, sulfuric, and phosphoric acids. If the analyst picked an indicator in the first transition range, say methyl red with a pH transition range of pH 4.4–6.2, the titration equivalence point would have been taken at the first equivalence point and a concentration would have been reported that is much less than the total amount of phosphoric acid present. Let us consider this numerically. Suppose the first equivalence point (actually endpoint) was determined at requiring 11.5 ml of the 0.2 N NaOH titrant to bring about a color change from red to yellow/orange using methyl red as the acid–base indicator. The overall reaction for the titration of phosphoric acid with sodium hydroxide is

$$3NaOH + H_3PO_4 \rightarrow 3H_2O + Na_3PO_4. \quad (10.29)$$

For phosphoric acid, there are three equivalents of H^+ for each mole of H_3PO_4; thus, the equivalent weight of H_3PO_4 is the formula weight divided by 3 = 97.995/3 = 32.665. If the analyst were simply using the equivalent weight of phosphoric acid and the methyl red endpoint, his/her calculated result for the titration would be

$$\%H_3PO_4 = \frac{(11.5\,\text{ml}_{NaOH})(0.2N_{NaOH})(0.032665)(100)}{20\,g_S} \quad (10.30)$$

$$\begin{aligned} \text{ppm}\,H_3PO_4 &= \% \times 10{,}000 = 0.3756 \times 10{,}000 \\ \text{ppm}\,H_3PO_4 &= 3756\ \text{ppm} \end{aligned} \quad (10.31)$$

On the other hand, if the phenolphthalein endpoint of 23 ml is used, the result would be

$$\%H_3PO_4 = \frac{(23\,\text{ml}_{NaOH})(0.2N_{NaOH})(0.032665)(100)}{20\,g_S} \quad (10.32)$$

$$\begin{aligned} \text{ppm}\,H_3PO_4 &= \% \times 10{,}000 = 0.7513 \times 10{,}000 \\ \text{ppm}\,H_3PO_4 &= 7513\ \text{ppm} \end{aligned} \quad (10.33)$$

However, phosphoric acid is in fact triprotic and the final equivalence endpoint is actually at 30 ml of titrant. If we look back at the titration curve in Figure 10.9, we know that the third midpoint is at a pH of 12.4. If we use 30 ml of titrant in our calculation, the final result is

$$\%H_3PO_4 = \frac{(30\,\text{ml}_{NaOH})(0.2N_{NaOH})(0.032665)(100)}{20\,g_S} \quad (10.34)$$

$$\begin{aligned} \text{ppm}\,H_3PO_4 &= \% \times 10{,}000 = 0.9799 \times 10{,}000 \\ \text{ppm}\,H_3PO_4 &= 9799\ \text{ppm} \end{aligned} \quad (10.35)$$

Thus, the titration of total phosphoric acid from a phenolphthalein endpoint is an estimation (in fact, an underestimation) of the phosphoric acid present. The third dissociation for phosphoric acid is so weak ($K_a = 4.4 \times 10^{-13}$) that competition is present with the autoprotolysis of water ($K_w = 1 \times 10^{-14}$). Scheme 10.1 depicts the three dissociation reactions of phosphoric acid, the K_a values, and the midpoint pH values.

10.12 ChemTech CALCULATION OF NORMAL TITRATIONS

The ChemTech program may also be used to calculate the results of titrations using normal solutions. If not already open, open ChemTech and go to the Chapter 10 Main Menu. Click on the Titrations Main Menu button, and then click on the second button "Acid/Base Titrations Using Normal Solutions" to bring up a titration calculation page as shown in Figure 10.10. This is a

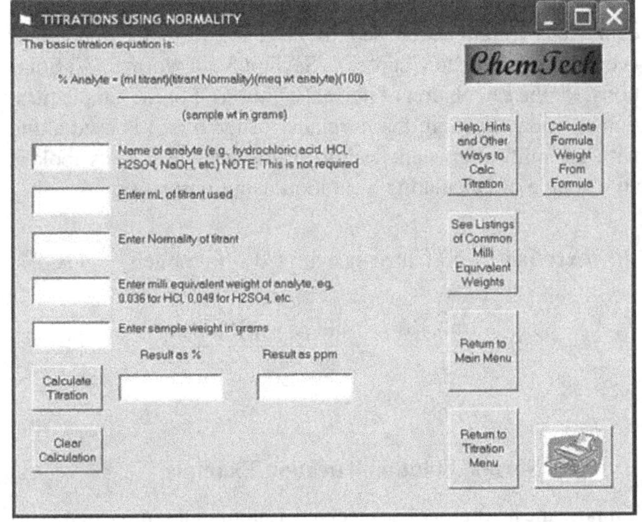

FIGURE 10.10 Titrations using normal solutions calculation page.

					K_a	Midpoint pH pH = –log K_a
H_3PO_4	$\rightleftharpoons$	H^+	+	$H_2PO_4^-$	7.5×10^{-3}	2.1
$H_2PO_4^-$	$\rightleftharpoons$	H^+	+	HPO_4^{2-}	6.2×10^{-8}	7.2
HPO_4^{2-}	$\rightleftharpoons$	H^+	+	PO_4^{3-}	4.4×10^{-13}	12.4

SCHEME 10.1 The three dissociation reactions of phosphoric acid, the K_a values, and the midpoint pH values.

similar input page that we have now become quite familiar with. Type in the data from Example 10.10.1 to become familiar with using the calculator.

This page also contains a link to view help, hints and reference for normal titrations, a link to look up some common milli equivalent weights, the formula weight calculator, and a printer button at the bottom.

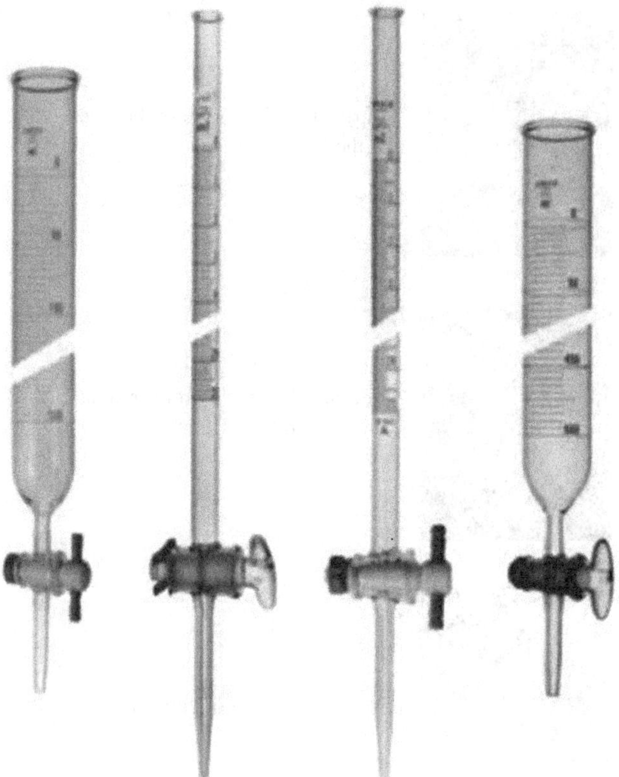

FIGURE 10.11 Burettes used to transfer solutions and for titrations.

10.13 PERFORMING A TITRATION

Now that we have covered the theory behind titrations and the calculations involved, we will look at the steps "in the laboratory" for performing a titration. Titrations involve volumetric analysis, which means that exact volumes of solutions are used within a small limit of error. Some of the glassware used in titrations includes volumetric flasks, which are of ranging sizes anywhere from 5up to 2000 ml or more (examples of all common glassware used in the analytical laboratory may be found in Chapter 2).

10.13.1 Titration Glassware

The next important piece of glassware used in titrations is the burette. Burettes contain graduations that are used to measure the amount of titrant added to reach an endpoint to an accuracy of ±0.01 ml. Figure 10.11 depicts some examples of commercially available burettes. Often burettes are controlled by a stopcock at the bottom that is now usually made out of Teflon®. Figure 10.12 is a close-up view of the burette stopcock and the handling of the stopcock.

10.13.2 Titration Steps

The general steps in a titration are as follows:

1. The burette is placed into a titration clamp stand and filled with the titrant usually past the 0 ml mark at the top. The excess titrant is then allowed to slowly drain into a waste beaker bringing the volume back down to the 0 ml mark. A filled burette in a titration clamp stand is depicted in Figure 10.13.
2. The tip of the burette is then placed down into the beaker or flask holding the sample to be titrated, but not into the liquid but kept above. The positioning of the burette tip is depicted in Figure 10.14(a). The tip of the burette is not held above the flask as some of the titrant may miss the solution. This is a quantitative reaction where all of the

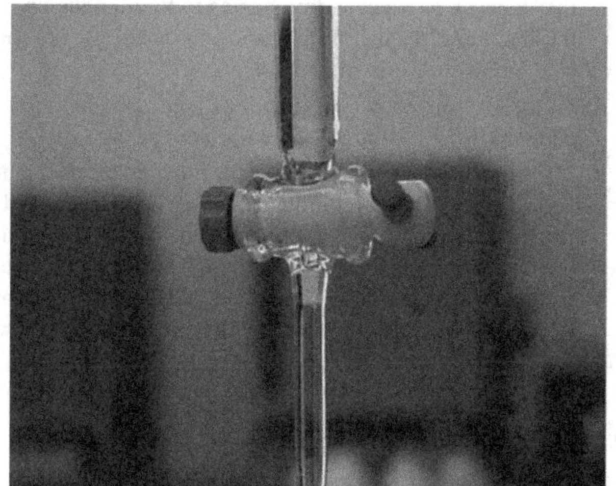

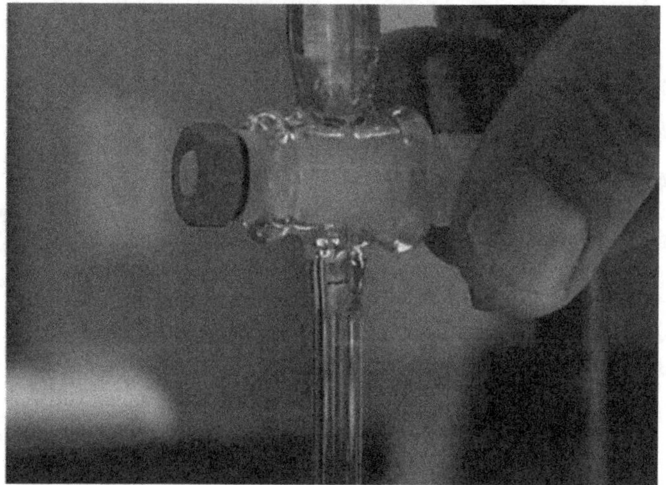

FIGURE 10.12 Titration burette stopcock.

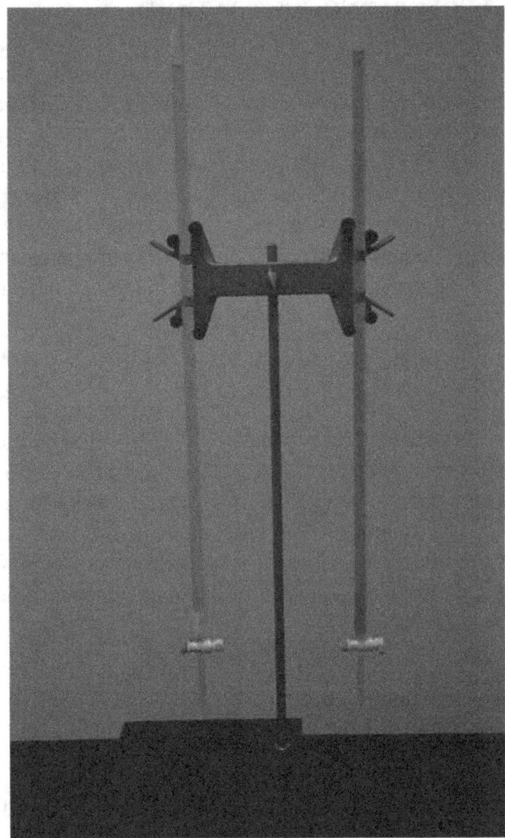

FIGURE 10.13 Filled burette in titration clamp-stand.

titrant used in the calculation must have gone into the titrand solution. A piece of white paper can also be placed under the flask to facilitate in detecting color changes (see Fig. 10.14(a)).

3. Before starting the titration, record the initial reading on the burette. This is especially important if starting at a reading other than 0 ml. To start the titration, grasp the burette stopcock with the left hand and the flask with the right hand. This is best illustrated in Figure 10.14(d).
4. With the right hand, moderately swirl the flask solution in a circular motion while slowly opening the stopcock to allow flow of the titrant into the solution.
5. As the titrant is added, you will notice a brief swirl of color, which quickly disappears. As the endpoint nears, the swirl of color will continue to increase in duration. In the case of phenolphthalein, the color change is from colorless to red/violet (see Fig. 10.7).
6. As depicted in Figure 10.14(b), periodically (perhaps every 5 ml of titrant addition) stop the addition and rinse the sides of the flask with DDI water to ensure an accurate endpoint determination (washes splashed material back into solution).
7. Near the endpoint, the color swirl will last much longer and the endpoint is reached when the color change is persistent. Often, with practice the analyst will come to a point in the titration just before the endpoint that requires less than a drop of titrant. A "half" drop can be added by just opening the stopcock to where a drop begins to form on the tip of the burette. The "half-drop" can be washed off into the solution as shown in Figure 10.14(c).

(a)

Place the tip of the buret inside the flask.

(b)

During the titration rinse the sides of the flask.

(c)

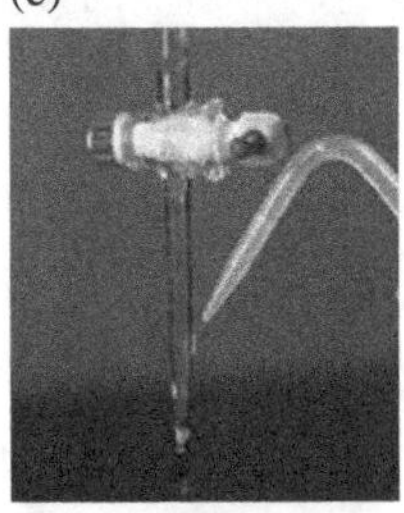

Adding a "half" drop of titrant near the endpoint.

(d)

A red/violet phenolphthalein endpoint.

FIGURE 10.14 Steps involved in a titration.

8. Figure 10.14(d) shows the reaching of the phenolphthalein endpoint where the red/violet color now persists.
9. The burette volume is recorded for the amount of titrant added. A 50 ml burette will start at the top as 0 ml and graduate to the bottom at 50 ml. If the burette is started at 0 ml, the final reading is the milliliter titrant added. If the burette reading is started at a reading >0 ml, the amount of titrant added is found by subtracting the initial reading from the final reading.

10.14 PRIMARY STANDARDS

Typically, the titrants that the analyst is using in the laboratory are "secondary standards" that have been titrated with a "primary

TABLE 10.1 Common Primary Acids and Bases Used for Titrant Standardization.

	Formula	Formula weight
Primary acids		
Potassium hydrogen *o*-phthalate (KHP)	$o\text{-}C_6H_4(COOK)(COOH)$	204.2212
Sulfamic acid	NH_2SO_3H	97.09476
Oxalic acid dihydrate	$HOOC\text{–}COOH{\cdot}2H_2O$	126.06544
Benzoic acid	C_6H_5COOH	122.12134
Primary bases		
Sodium carbonate	Na_2CO_3	105.98844
Tris(hydroxymethyl)aminomethane (Tris base)	$NH_2C(CH_2OH)_3$	121.13508
4-Aminopyridine	$C_5H_4N(NH_2)$	94.11462

standard" to obtain its exact concentration. The laboratory reagents that are used to make titrants, such as an HCl solution or sodium hydroxide, do not have an exact concentration "HCl reagent is approximately 37% pure" or are not in a pure form (solid NaOH reagent contains some water). Thus, when preparing a titrant in the laboratory for routine use, the analyst will first make a carefully prepared solution from a primary standard and use this to obtain an accurate concentration of the routine titrant. It should be noted though that in today's analytical laboratory there are many commercially available secondary standards that may be purchased and used directly within the manufacturer's stated expiration period of the titrant. However, the laboratory analyst should be familiar with the preparation of primary and secondary titration solutions. Table 10.1 is a listing of some common primary acids and bases that are used in the analytical laboratory for standardizing secondary titrants.

10.15 STANDARDIZATION OF SODIUM HYDROXIDE

We will look at the preparation of a specific example of standardizing a sodium hydroxide titrant using the primary acid potassium hydrogen *o*-phthalate (KHP). This example is also located in ChemTech for future review and reference in the laboratory, found in the Chapter 10 Titrations Main Menu under "Example of Standardizing a Titrant from a Primary Standard." The ChemTech review is in normality, but in the case of KHP and NaOH, normality and molarity are interchangeable. This is not always the case; one must always remember to take into account the reacting ratio $\frac{(b)}{(a)}$ of the particular titrant/titrand reaction.

10.15.1 NaOH Titrant Standardization Example

Sodium hydroxide (NaOH) is a very common titrant to use in the analytical laboratory as a strong base. It is a strong monoprotic base; thus, its molarity and normality are equal. Sodium hydroxide in the solid pellet form often contains a small amount of both water and sodium carbonate, approximately 97%–98% pure. The sodium carbonate comes from the reaction of the sodium hydroxide with carbon dioxide in the ambient atmosphere. For the titration of weak acids, the preparation of sodium hydroxide solution free of sodium carbonate is essential as the carbonate ion present can act as a buffer and distort the titration endpoint. For the titration of strong acids, this is less of a problem. The first step in preparing the sodium hydroxide titrant is to purify the sodium hydroxide. Distilled water is first boiled to remove any dissolved carbon dioxide. A solution of nearly saturated sodium hydroxide is then prepared. Sodium hydroxide has a solubility of 1 g per 0.9 ml H_2O. A saturated solution is approximately 27.8 *M* NaOH. Upon sitting, the saturated NaOH solution will clear (no cloudiness) and any sodium carbonate will precipitate out of solution and settle to the bottom. The purified sodium hydroxide in solution can then be carefully decanted from the top of the saturated solution. To make the working titrant to be standardized, 3.6 ml of the saturated solution is then brought to 1 l in a volumetric flask with freshly boiled distilled water. The concentration of the sodium hydroxide titrant is now approximately 0.1 *M* NaOH.

O
||
C–O⁻K⁺ / C–OH
||
O

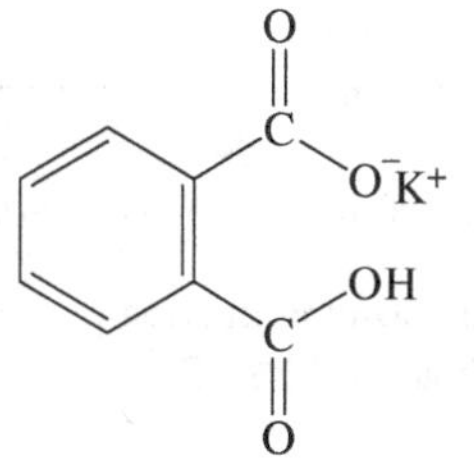

FIGURE 10.15 Structure of potassium hydrogen *o*-phthalate.

A convenient and often used primary acid standard for standardizing sodium hydroxide titrants is potassium hydrogen orthophthalate (KHP, see Table 10.1). KHP is a moderately weak organic (structure shown in Figure 10.15) acid ($K_a = 3.9 \times 10^{-6}$, $pK_a = 5.4$) that gives a good endpoint using phenolphthalein as indicator.

To standardize the sodium hydroxide titrant, exactly 999.9 mg of the primary acid standard potassium hydrogen *o*-phthalate was weighed and dissolved in 100 ml of DDI water producing a 0.049 *M* KHP solution. A 35 ml aliquot of the 0.049 *M* KHP solution was taken and titrated with the approximately 0.1 *M* NaOH solution to a phenolphthalein endpoint. The titration of the KHP titrand pH versus milliliter NaOH added is depicted in Figure 10.16. For illustrative purposes, the midpoint, endpoint, and the phenolphthalein indicator pH transition range are included in the titration plot.

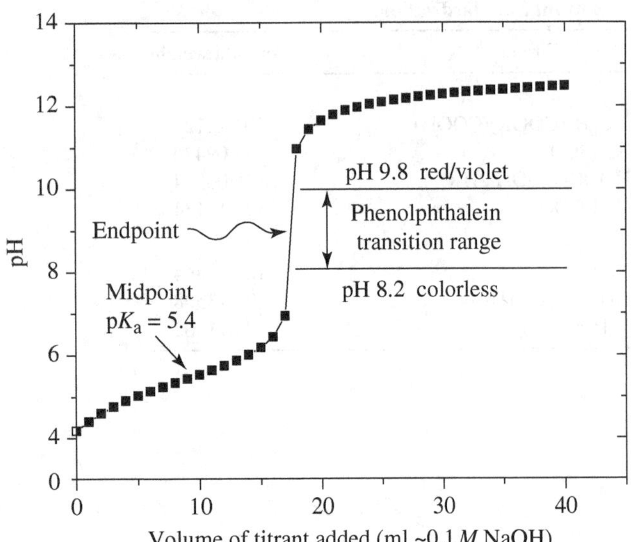

FIGURE 10.16 Standardization titration plot of approximately 0.1 *M* NaOH with the primary acid KHP.

The amount of the approximately 0.1 *M* NaOH solution required to reach the endpoint was 17.7 ml. The neutralization reaction taking place in solution is

$$NaOH + C_6H_4(COOK)(COOH) \rightarrow H_2O + C_6H_4(COOH)_2. \tag{10.36}$$

We had taken a 35 ml aliquot of the original KHP solution, so the amount of KHP titrated was

$$\begin{aligned} \text{mg KHP} &= \frac{999.9 \text{ mg}}{100 \text{ ml}} \times 35 \text{ ml} \\ \text{mg KHP} &= 350.0 \text{ mg} \end{aligned} \tag{10.37}$$

From our above titration relationships, we have

$$\text{mg KHP} = (\text{ml}_{\text{NaOH}})(M_{\text{NaOH}})\left(\frac{b}{a}\right)(\text{form wt}_{\text{KHP}}), \tag{10.38}$$

$$\text{where in this reaction } \frac{b}{a} = \frac{1}{1} = 1. \tag{10.39}$$

Rearranging the above equation, we have

$$M_{\text{NaOH}} = \frac{\text{mg}_{\text{KHP}}}{(\text{ml}_{\text{NaOH}})(\text{form wt}_{\text{KHP}})} \tag{10.40}$$

$$\begin{aligned} M_{\text{NaOH}} &= \frac{350.0 \text{ mg}}{(17.7 \text{ ml})(204.2212 \text{ g mol}^{-1})} \\ M_{\text{NaOH}} &= 0.0968 \end{aligned} \tag{10.41}$$

As can be seen from the titration, the original solution consisting of 3.6 ml saturated NaOH solution diluted to 1 l to make an approximately 0.1 *M* NaOH titrant was in fact an approximate, and the exact concentration is 0.0968 *M* NaOH. We now have a secondary titrant standard that may be used for titrations of both strongly and weakly acidic solutions.

10.16 CONDUCTOMETRIC TITRATIONS (NONAQUEOUS SOLUTIONS)

In conductometric titrations, the conductance in the solution is measured during the addition of the titrant. Conductometric titrations can be useful when the equivalence point is difficult to discern using an indicator or when other problems make the titration difficult, such as a nonaqueous solution. Figure 10.17 depicts the basic components of a conductometric titration including a reaction vessel where the titration takes place, a magnetic stir bar for constant mixing, and two platinum electrodes where the solution's conductance is measured. The device typically used for the measurement in the change in conductance of the solution during the titration is a system of resistors known as a Wheatstone bridge. (J.G. Dick In: Analytical Chemistry: McGraw-Hill, New York, NY 1973, p. 560.) The equivalent conductance ($\Lambda°$) for a given solution of a dissolved salt is the addition of the ionic species present:

$$\Lambda^{\circ} = \lambda_{+}^{\circ} + \lambda_{-}^{\circ}. \tag{10.42}$$

Table 10.2 lists some values of equivalent ionic conductance at 25 °C and infinite dilution.

To illustrate the principle behind conductometric titrations, we will look at an acid–base titration where a solution of HCl is being titrated with the strong base, NaOH. Figure 10.18 depicts the results of the titration where the conductance of the solution is recorded with each addition of titrant. At the beginning of the titration, the equivalent ionic conductance (λ_i°) is due entirely to the hydrogen ion H^+ present. As the sodium hydroxide titrant is added, the hydroxide OH^- added is neutralizing the hydrogen ion H^+ present to produce water, and the effective result is that the conductance in the solution by H^+ is being replaced by the Na^+ added. The equivalent ionic conductance of H^+ is $\lambda_{H+}^{\circ} = 349.82$, and for Na^+ is $\lambda_{Na+}^{\circ} = 50.11$, thus the conductance in the solution is decreasing as illustrated by the negative slope in Figure 10.18 (square data point plot line). During the first part of the titration, the concentration of the chloride ion $[Cl^-]$ is staying constant $(\lambda_{Cl-}^{\circ} = 76.34)$ and not contributing to the "change" in solution conductance. At the equivalence point [HCl] = [NaOH], and the contribution to the conductance is from the $[Na^+]$ and $[Cl^-]$ present. As the OH^- concentration increases, the conductance of the solution also begins to increase as illustrated by the positive slope of the line in Figure 10.18 (circle data point plot line). The equivalent ionic conductance of OH^- is $\lambda_{OH-}^{\circ} = 198.0$. Because the equivalent ionic conductance of OH^- (198.0) is less than that of H^+ (349.82), the slope of the second curve is lower in magnitude as compared to the first. By plotting the conductance against the titrant added in milliliter, we can extrapolate both lines to a point where they intersect to obtain the equivalence point. This value would be the milliliter

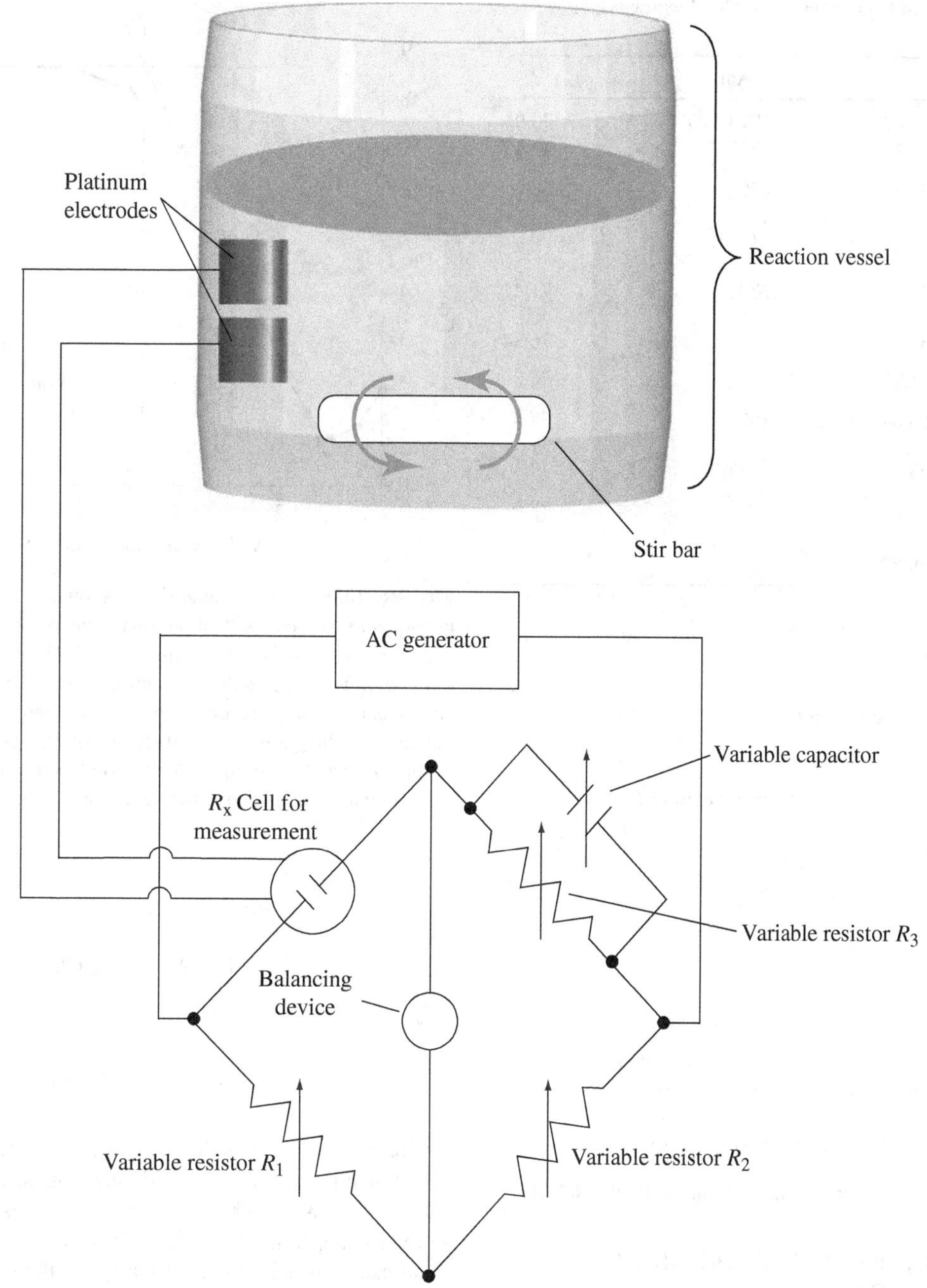

FIGURE 10.17 Apparatus used in a conductometric titration including a reaction vessel and a Wheatstone bridge.

titrant added that is used in the titration equation to obtain the amount of HCl present.

This method is found most useful in nonaqueous-type titrations where an appropriate indicator is not readily available to be used. An example of this is the titration of acetic acid (a weak acid) with ammonia (a weak base) in a nonaqueous solvent. In the nonaqueous solvent, the weak acid and weak base are essentially neutral with very little solution conductance. The salt produced by the combination of the acetic acid and ammonia is ammonium acetate (CH_3COONH_4), producing the acetate ion $\left(\lambda^{o}_{CH_3COO^-} = 40.9\right)$ and the ammonium ion $\left(\lambda^{o}_{NH_4^+} = 73.4\right)$. The conductometric titration curve of acetic acid in nonaqueous solvent with ammonia is depicted in Figure 10.19. The salt produced is ammonium acetate and the rising linear curve from the titration is due to the increase of the ions of this salt produced in solution during the titration. When the equivalence point is reached, the neutral ammonia in excess does not cause a change in the conductance of the solution (straight line at top of curve in Fig. 10.19). The equivalence point can be obtained by extrapolation of the upward curve and the top, level curve to a point of intersection.

TABLE 10.2 Values of Equivalent Ionic Conductance (λ_i^o) at 25 °C and Infinite Dilution.

Cation	λ_+^o	Anion	λ_-^o
Li^+	38.69	CH_3COO^-	40.9
Na^+	50.11	HCO_3^-	44.48
$\frac{1}{2}Mg^{2+}$	53.06	ClO_4^-	68.0
$\frac{1}{2}Cu^{2+}$	54	$\frac{1}{2}C_2O_4^{2-}$	70
$\frac{1}{2}Sr^{2+}$	59.46	NO_3^-	71.44
$\frac{1}{2}Ca^{2+}$	59.50	Cl^-	76.34
Ag^+	61.92	I^-	76.8
$\frac{1}{2}Ba^{2+}$	63.64	Br^-	78.4
$\frac{1}{2}Pb^{2+}$	73	$\frac{1}{2}SO_4^{2-}$	79.8
NH_4^+	73.4	OH^-	198.0
K^+	73.52		
H^+	349.82		

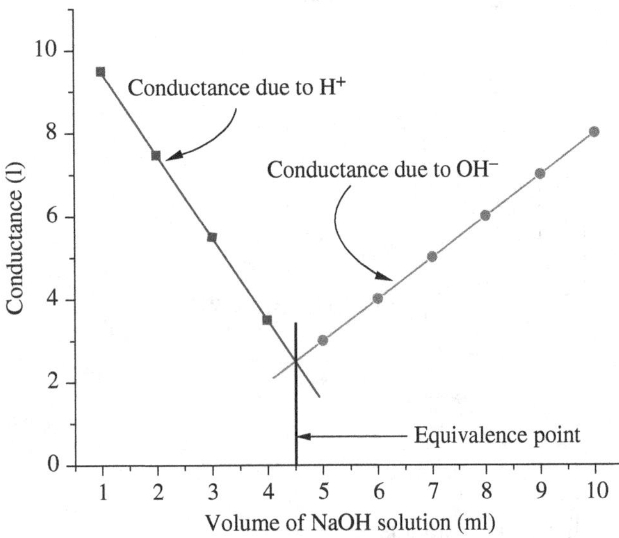

FIGURE 10.18 Conductometric titration of HCl with NaOH.

10.17 PRECIPITATION TITRATION (MOHR METHOD FOR HALIDES)

A common example of a precipitation titration method is the Mohr method used for the determination of the halides, bromide (Br^-) and chloride (Cl^-). An important consideration of this approach is that the method should only be used with a titrand pH ranging from 7 to 10. The Mohr method is also susceptible to interference by carbonates. A standardized silver nitrate ($AgNO_3$) solution is used to titrate the halides present in a titrand forming white insoluble salts of silver bromide (AgBr) or silver chloride (AgCl). A soluble chromate salt such as potassium chromate (K_2CrO_4) is used as the indicator. When the equivalence point is reached, the excess silver will form a red insoluble silver chromate (Ag_2CrO_4) salt precipitate:

$$Br^- + Ag^+ \rightleftharpoons AgBr(s) \tag{10.43}$$

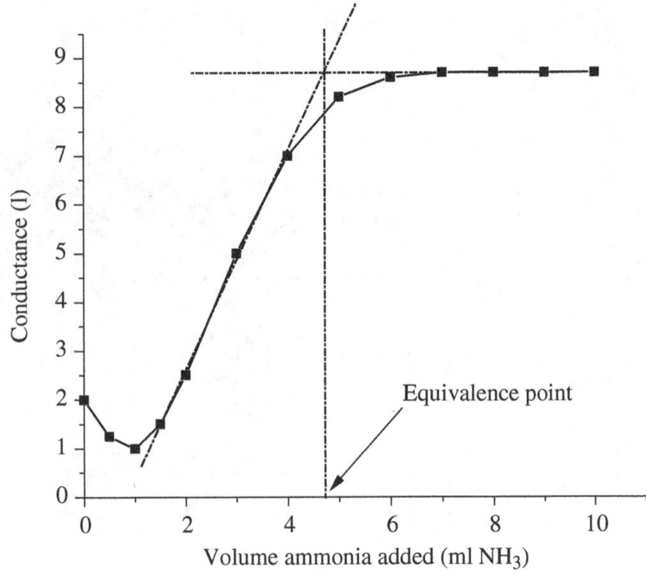

FIGURE 10.19 Conductometric titration curve of acetic acid in nonaqueous solvent with ammonia. The rising linear curve from the titration is due to the increase of the acetate and ammonium ions produced in solution during the titration. When the equivalence point is reached, the neutral ammonia in excess does not cause a change in the conductance of the solution (straight line at top of curve). The equivalence point obtained by extrapolation of the upward curve and the top, level curve to a point of intersection.

or

$$Cl^- + Ag^+ \rightleftharpoons AgCl(s) \tag{10.44}$$

and

$$2Ag^+ + CrO_4^{2-} \rightleftharpoons Ag_2CrO_4(s) \tag{10.45}$$

The titration is simple and straightforward except for the visual determination of the endpoint. The silver halide precipitate is white and the chromate indicator makes the solution have a yellow appearance; thus, it is often necessary to add a small excess of silver ion to clearly see the red silver chromate at the endpoint. This often necessitates the analysis of an indicator blank to subtract the excess silver required to see the endpoint.

10.17.1 Basic Steps in Titration

The following summary of the Mohr method can be found in ChemTech under the Chapter 10 Titrations Main Menu:

1. A standardized silver nitrate ($AgNO_3$) solution is used to titrate the halides present in a titrand forming white insoluble salts of silver bromide (AgBr) or silver chloride (AgCl).
2. A soluble chromate salt, such as potassium chromate (K_2CrO_4), is used as the indicator.

3. When the equivalence point is reached, the excess silver will form a red insoluble silver chromate (Ag_2CrO_4) salt precipitate:

$$Br^- + Ag^+ \rightleftharpoons AgBr(s) \quad (10.43)$$

or

$$Cl^- + Ag^+ \rightleftharpoons AgCl(s) \quad (10.44)$$

and

$$2Ag^+ + CrO_4^{2-} \rightleftharpoons AgCrO_4(s) \quad (10.45)$$

10.17.2 Important Considerations

1. Method should only be used with a titrand pH ranging from 7 to 10.
2. Method is susceptible to interference by carbonates.
3. Visual determination of the endpoint is difficult.
4. The silver halide precipitate is white and the chromate indicator makes the solution have a yellow appearance.
5. Often necessary to add a small excess of silver ion to clearly see the red colored chromate at the endpoint.
6. Analysis of an indicator blank is needed to subtract the excess silver required to see the endpoint.

10.18 COMPLEX FORMATION WITH BACK TITRATION (VOLHARD METHOD FOR ANIONS)

Back titrations can be applied to a number of types of titrations that we have covered. They can be effective for a titration reaction that is slow or requires energy to be input for the reaction to go to a stoichiometric equivalence point. We will look at a specific type of complex formation reaction that uses a visual indicator in conjunction with a back titration. This titration is based on the Volhard method, which was initially developed for the measurement of silver(I) (Ag^+) ions in solution, but the basic premises of the titration has been adapted to the indirect measurement of the halides Br^-, Cl^-, and I^-, which has a more general usefulness. The Volhard method titrates the silver ion with the thiocyante ion (SCN^-) that produces the mostly insoluble silver thiocyanate salt (AgSCN).

$$Ag^+ + SCN^- \rightleftharpoons AgSCN(s) \quad (10.46)$$

10.18.1 Iron(III) as Indicator

The visual indicator used is the iron(III) "ferric ion" (Fe^{3+}), which by itself is colorless in solution, but complexed with the thiocyanate titrant forms an intensely red iron thiocyanate $Fe(SCN)^{2+}$ soluble complex.

$$Fe^{3+} + SCN^- \rightleftharpoons Fe(SCN)^{2+}. \quad (10.47)$$

10.18.2 Chloride Titration

In contrast to the Mohr method which is conducted in a pH range of 7–10, the Volhard method is in a highly acidic solution that is approximately 1×10^{-3} M HNO_3 at the equivalence point (~pH 3). In alkaline solutions, the carbonates can be slightly soluble interfering with halide titration analyses. This problem is avoided with the Volhard method. The basic premise of the method is to add an excess of silver nitrate ($AgNO_3$) to the titrand solution containing the halide to be measured (chloride Cl^- in our example). The insoluble silver chloride (AgCl) salt formed will precipitate out of solution. Because the silver nitrate was added in excess of the chloride, the solution will still contain silver ions (Ag^+). The remaining silver ions are then titrated with the thiocyanate (potassium thiocyanate (KSCN) is generally used to make the titrant solution) producing the insoluble white silver thiocyanate (AgSCN) salt. When all of the silver has reacted with the thiocyanate, the weaker iron(III) ion will then react with the thiocyanate producing the soluble red iron thiocyanate complex $Fe(SCN)^{2+}$. The three steps are illustrated below:

$$Ag^+ + Cl^- \rightleftharpoons AgCl(s) \quad (10.48)$$

$$Ag^+ + SCN^- \rightleftharpoons AgSCN(s) \quad (10.49)$$

$$Fe^{3+} + SCN^- \rightleftharpoons Fe(SCN)^{2+} \quad (10.50)$$

The silver halide precipitates of bromide (AgBr) and iodide (AgI) are less soluble than silver thiocyanate and do not pose a problem with this titration. However, silver chloride is in fact more soluble than silver thiocyanate and will reconvert back from silver chloride to silver thiocyanate during the titration. This has been addressed by filtering out the silver chloride as it forms but is not very practical. A second approach that has been very successful is the addition of nitrobenzene to the titrand solution. With vigorous shaking, the immiscible nitrobenzene (which is also denser than water and goes to the bottom of the flask) will coat the silver chloride precipitate and prevent it from going back into solution. The addition of nitrobenzene (2–5 ml nitrobenzene per 50–100 ml titrand) is standard procedure for chloride analysis using the Volhard method.

10.18.3 The General Calculation

The general calculation to use here is to determine the actual amount of titrant (T) that was used to combine with the halide (X^-).

$$a\text{T}_{\text{titrant}} + b\text{X}^-_{\text{halide}} \rightarrow \text{products} + \text{excess T}_{\text{titrant}} \quad (10.51)$$

$$c\text{C}_{\text{backtitrant}} + d\text{T}_{\text{excesstitrant}} \rightarrow \text{products} \quad (10.52)$$

$$\text{T}_{\text{net}} = \text{T}_{\text{added}} - \text{T}_{\text{excess}} \quad (10.53)$$

$$\text{mmoles}_T = [(\text{ml}_T)(M_T)] - \left[\text{ml}_C(M_C)\left(\frac{d}{c}\right)\right] \quad (10.54)$$

$$\text{mg}_{X^-} = (\text{mmoles}_T)\left(\frac{b}{a}\right)(\text{form wt}_{X^-}) \quad (10.55)$$

10.18.4 Chloride Titration

10.18.4.1 Volhard Chloride Analysis Example Let us look at a specific example of a chloride analysis. Suppose 20 ml of a 0.11 M silver nitrate ($AgNO_3$) titrant was added to a 25 ml titrand solution that also contained 2.5 ml of nitrobenzene. It required 5.32 ml of a 0.105 M thiocyanate (SCN^-) solution to titrate to a red endpoint. How many milligrams of chloride are present in the titrand?

Using our relationships above:

$$\text{mmoles}_T = [(20\,\text{ml}_T)(0.110\,M_T)] - \left[5.32\,\text{ml}_C(0.105\,M_C)\left(\frac{1}{1}\right)\right] \quad (10.56)$$

$$\text{mmoles}_T = 2.2\,\text{mmoles} - 0.559\,\text{mmoles} \quad (10.57)$$

$$\text{mmoles}_T = 1.64\,\text{mmoles} \quad (10.58)$$

$$\text{mg}_{Cl^-} = (1.64\,\text{mmoles})\left(\frac{1}{1}\right)(35.4527) \quad (10.59)$$

$$\text{mg}_{Cl^-} = 58.1\,\text{mg} \quad (10.60)$$

The Volhard method can be used to determine the concentration of a number of anions, including sulfide (S^{2-}), chromate (CrO_4^{2-}), cyanide (CN^-), thiocyanate (SCN^-), carbonate (CO_3^{2-}), and oxalate ($C_2O_4^{2-}$).

10.18.4.2 The Titration Steps The following summary of the Volhard method can be found in ChemTech under the Chapter 10 Titrations Main Menu:

1. Add an excess of silver nitrate ($AgNO_3$) to the titrand solution containing the halide to be measured.
2. The insoluble silver chloride (AgCl) salt formed will precipitate out of solution.
3. Because the silver nitrate was added in excess of the chloride, the solution will still contain silver ions (Ag^+).
4. The remaining silver ions are then titrated with the standardized thiocyanate (potassium thiocyanate, KSCN) producing the insoluble white silver thiocyanate (AgSCN) salt.
5. When all of the silver has reacted with the thiocyanate, the weaker iron(III) ion will then react with the thiocyanate producing the soluble red iron thiocyanate complex $Fe(SCN)^{2+}$.

The three titration reactions are

$$Ag^+ + Cl^- \rightleftharpoons AgCl(s) \quad (10.48)$$

$$Ag^+ + SCN^- \rightleftharpoons AgSCN(s) \quad (10.49)$$

$$Fe^{3+} + SCN^- \rightleftharpoons Fe(SCN)^{2+} \quad (10.50)$$

<u>NOTES</u>

1. The method is in a highly acidic solution that is approximately 1×10^{-3} M HNO_3 at the equivalence point (~pH 3).
2. The addition of nitrobenzene to the titrand solution for chloride analysis (2–5 ml nitrobenzene per 50–100 ml titrand).

10.19 COMPLEX FORMATION TITRATION WITH EDTA FOR CATIONS

Many of the di- and trivalent cationic metals can be titrated using a complex formation technique. For monovalent cationic species (e.g., alkali metals Li^+, Na^+, and K^+, Ag^+, and NH_4^+), ion-selective electrodes (see Chapter 9), spectrophotometric (see Chapter 13), or atomic (XRF, ICP-MS, AAS, etc., see Chapters 17, 18, 19, and 20) methodologies may be employed. Polydentate ligands (also known as multidentate ligands) are used to react with the metal ions to form water-soluble complexes in a 1:1 molar ratio. The polydentate ligands form coordination complexes with the metal ions by donating an unshared electron pair to the metal ion forming a metal–ligand bond. The complex ions that are formed between the polydentate ligand and metal ion are called chelates. The chelates that are formed are usually non-linear, but form what is known as a chelate ring structure. The most commonly used polydentate ligand is EDTA (ethylenediaminetetraacetic acid). Figure 10.20 shows the structure of EDTA

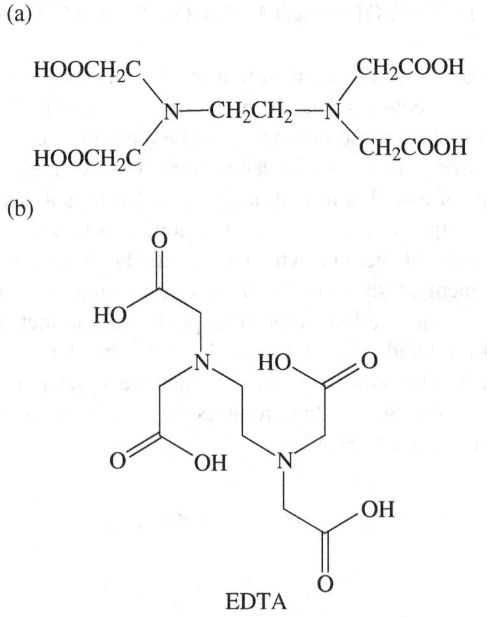

FIGURE 10.20 Structure of EDTA (ethylenediaminetetraacetic acid) in the fully protonated acid form H_4Y. (a) Typically drawn linear form of the structure illustrating the four carboxylic acid groups. (b) EDTA structure in its lowest energy state.

in the fully protonated acid form, H_4Y. The structure in Figure 10.20(a) is the typically drawn linear form of the structure while that in (b) is a representation of the structure in its lowest energy state. The fully protonated acid form, and the monosodium salt, however are not generally used due to a low solubility in water. The disodium dihydrate salt $Na_2H_2Y{\cdot}2H_2O$ is the more commonly used form of EDTA in complex titrations.

10.19.1 EDTA–Metal Ion Complex Formation

The complex formed between EDTA and metals is always of a 1:1 ratio, thus the direct relationship between titrant used (EDTA) and metal ion present in the titrand can be used to quantitate the titration. EDTA is an acid and the complexation reaction between EDTA and metal ions releases hydrogen protons (H^+) and increases the acidity of the solution. If we represent the metal ion as M^{n+}, and the dibasic form of EDTA as H_2Y^{2-}, the production of the acid proton with metal ion complexation is represented as

$$M^{n+} + H_2Y^{2-} + 2H_2O \rightleftharpoons M(Y)^{(n-4)} + 2H_3O^+. \qquad (10.61)$$

Some specific examples of metal ion–EDTA complexes producing H^+ include

$$Ca^{2+} + H_2Y^{2-} \rightleftharpoons CaY^{2-} + 2H^+ \qquad (10.62)$$

$$Fe^{3+} + H_2Y^{2-} \rightleftharpoons FeY^- + 2H^+ \qquad (10.63)$$

$$Th^{4+} + H_2Y^{2-} \rightleftharpoons ThY + 2H^+ \qquad (10.64)$$

During the titration, the production of H_3O^+ will lower the pH of the solution. This has an effect on the quantitative conversion of the metal ion M^{n+} to $M(Y)^{(n-4)}$ at the equivalence point due to a change in the stability (or also called the formation) constant (K_{MY}) of the metal–ligand complex. A general reaction for the metal–EDTA complex is

$$M^{n+} + Y^{4-} \rightleftharpoons M(Y)^{(n-4)} \qquad (10.65)$$

10.19.2 The Stability Constant

Note that H_2O is not included, and that EDTA is in the fully deprotonated state as Y^{4-}. The stability constant (K_{MY}) is defined as

$$K_{MY} = \frac{[MY]}{[M][Y]}. \qquad (10.66)$$

The larger the magnitude of K_{MY} the larger stretch in the "observed" endpoint will be seen in the titration equating to a more accurate determination (estimation) of the equivalence point. This is depicted in Figure 10.21 along with the effect that pH has upon the EDTA metal ion magnesium (Mg^{2+}) complex formation titration used here as an example. Analogous to the K_{MY} "observed" endpoint stretch, the pH of the system set at an optimal range will increase the accuracy of the estimated determination of the equivalence point. As can be seen in the plot, the titration of the magnesium ion (Mg^{2+}) is optimal at a pH of 10.

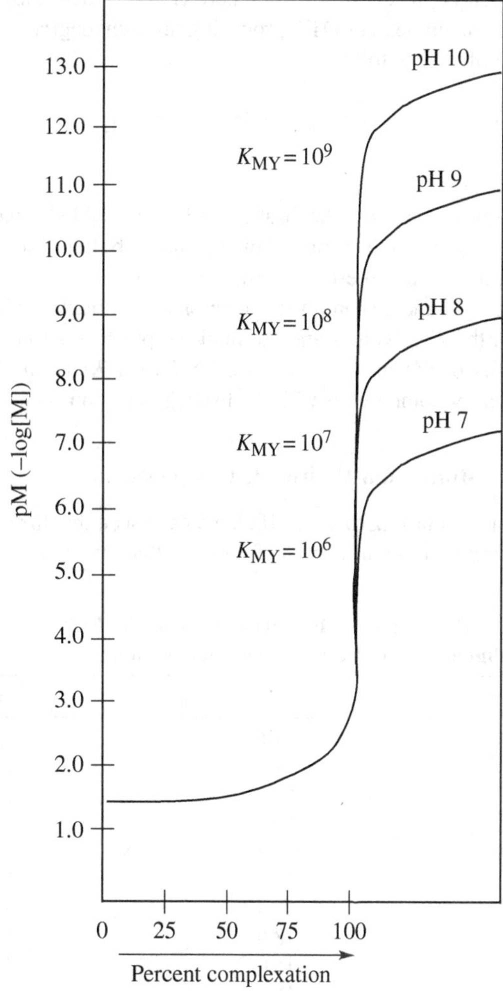

FIGURE 10.21 Effect upon the observed equivalence points of the stability constant (K_{MY}), and the pH during an EDTA metal ion complex formation titration. The plot also illustrates the effect pH has upon the titration curve for the EDTA magnesium ion (Mg^{2+}) complex formation titration.

Table 10.3 lists a number of the most commonly EDTA-titrated metal ions along with the minimal pH needed for the titration solution. Also included in Table 10.3 are the associated log K_{MY} values for each metal ion.

10.19.3 Metal Ions Titrated

The complexing of metal ions with EDTA produces quite complicated structures. For example, the complex formed between EDTA and the magnesium ion is a heptadentate (coordination number CN = 7) complex where bonds are formed with magnesium between the four carboxyl groups, the two nitrogen atoms, and one water molecule. The structure for the EDTA magnesium heptadentate complex is depicted in Figure 10.22.

10.19.4 Influence of pH

The pH of the solution has an effect upon the titration due to the interaction of the hydrogen proton (H^+) with the uncomplexed

EDTA that is present as Y^{4-}. As more H^+ are added, they react with the uncomplexed EDTA producing different degrees of protonation of Y^{4-} as follows:

$$nH^+ + Y^{4-} \rightleftharpoons HY^{3-} + H_2Y^{2-} + H_3Y^- + H_4Y, \quad (10.67)$$

where $n = 1, 2, 3$, or 4.

To control the pH of the titration, buffers are added usually to both the titrant and the titrand. Some common buffer systems for EDTA titrations are presented in the next section. As can be seen in Table 10.3, there is an inverse relationship (though not linear) between the K_{MY} values and the minimal pH. The stronger the binding is of the metal ions to EDTA (larger K_{MY}) the lesser the effect by competition with the hydrogen proton binding.

10.19.5 Buffer and Hydroxide Complexation

Therefore, according to Table 10.3, we have seen that the acidity of the titration is an influencing factor and that the titration of the barium ion (Ba^{2+}) or the magnesium ion (Mg^{2+}) with EDTA would be carried out at a pH value much higher than say the copper ion (Cu^{2+}). Thus, various buffers are added to maintain the desired pH of the system. Metal ions in solution however can also complex with the hydroxide ion (OH^-) of a high pH system, and also with complexing buffers such as ammonia (NH_3):

$$M^{2+} + OH^- \rightarrow M(OH)^+ \quad (10.68)$$

$$M^{2+} + 4NH_3 \rightarrow M(NH_3)_4^{2+} \quad (10.69)$$

The complexing of the metal ion with the hydroxide ion or a complexing buffer will effectively reduce the amount of available metal ion to complex with EDTA, resulting in an underestimation of the concentration. In some methods that the analyst may use, these aspects are in fact employed to stabilize a metal in solution by making a soluble complex with ammonia versus an insoluble complex with hydroxide, or to precipitate out one metal as an insoluble hydroxide complex while titrating another species present. For example, at pH 10, both calcium (Ca^{2+}) and magnesium (Mg^{2+}) are titrated with EDTA (water hardness analysis), but raising the pH to 12 or higher with the base hydroxide (OH^-) will precipitate out the magnesium as magnesium hydroxide ($Mg(OH)_2$) and leave the calcium to be titrated and measured. Another method for the analysis of lead (Pb^{2+}) in the presence of nickel (Ni^{2+}) incorporates the addition of the cyanide ion (CN^-) to complex with nickel, forming $Ni(CN)_4^{2-}$ allowing the titration and measurement of the lead only. Finally, at a pH of approximately 9, the copper ion (Cu^{2+}) will form a complex with ammonia that is both pH and ammonia concentration dependent:

$$Cu(NH_3)_4^{2+} + HY^{3-} + H_2O \rightleftharpoons Cu(Y)^{2-} + 4NH_3 + H_2O. \quad (10.70)$$

Note the form of the EDTA at this pH (HY^{3-}). This is the case of many of the metal ions listed in Table 10.3 and illustrates the complexity of these titrations, which are pH dependent and complexing species (hydroxide and buffer) dependent. Even though ammonia may form complexes itself with metal at certain pH, the ammonia (NH_3)–ammonium chloride (NH_4Cl) buffer salt system is often used in order to avoid the insoluble precipitation complexes most of the metals from with the hydroxide ion.

TABLE 10.3 Common Metal Ions Titrated with EDTA Including the Minimum pH Needed of the Titration Solution.

Metal ion	Log K_{MY}	Minimum pH
Fe^{3+}	25.1	1.5
Th^{4+}	23.2	1.7
Cr^{3+}	23	1.8
Bi^{3+}	22.8	1.8
Hg^{2+}	22.2	2.2
VO^{2+}	18.8	3.0
Cu^{2+}	18.8	3.2
Ni^{2+}	18.6	3.2
Pb^{2+}	18.0	3.3
Al^{3+}	16.1	3.9
Ce^{3+}	16.0	3.9
Cd^{2+}	16.5	4.0
Co^{2+}	16.3	4.1
Zn^{2+}	16.5	4.1
La^{3+}	15.4	4.2
Mn^{2+}	14.0	4.9
Fe^{2+}	14.0	5.1
Ca^{2+}	10.7	7.3
Mg^{2+}	8.7	10.0
Sr^{2+}	8.6	10.0
Ba^{2+}	7.8	11.2

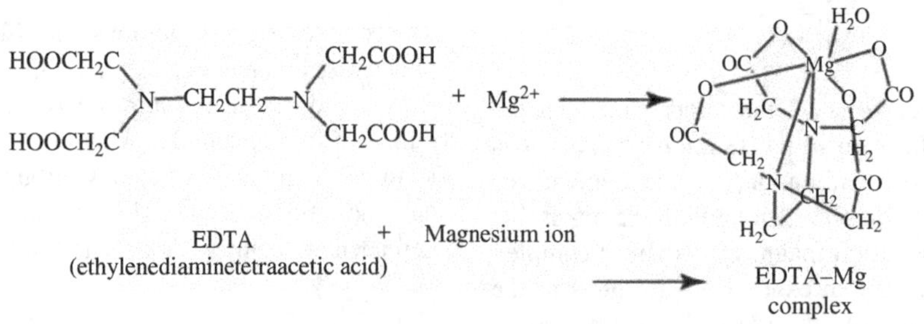

FIGURE 10.22 The complexing of the magnesium ion (Mg^{2+}) with EDTA producing a heptadentate (coordination number CN = 7) complex where bonds are formed with magnesium between the four carboxyl groups, between the two nitrogen atoms, and one water molecule.

10.19.6 Visual Indicators

There are a number of available visual indicators that are used for the EDTA–metal ion complex formation titrations. These are highly colored organic compounds often called *metallochromic indicators* that upon binding with metals will form intensely colored complexes. These indicators are often also pH dependent and will shift in color according to states of protonation/deprotonation. For example, a widely used visual indicator in EDTA complexation titrations is Eriochrome black T. The IUPAC name for Eriochrome black T (EBT) as a neutral salt is sodium (4Z)-4-[(1-hydroxynaphthalen-2-yl-hydrazinylidene]-7-nitro-3-oxo y-naphthalene-1-sulfonate, molecular formula is $C_{20}H_{12}N_3O_7SNa$, and molecular weight is 461.381 g/mol. The structure of the neutral sodium salt is depicted in Figure 10.23.

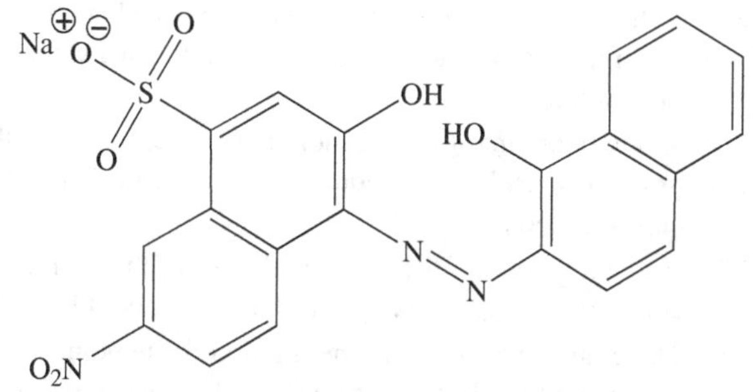

$HOC_{10}H_6N{=}NC_{10}H_4(OH)(NO_2)SO_3Na$

Eriochrome black T (neutral sodium salt)

FIGURE 10.23 EDTA visual indicator Eriochrome Black T.

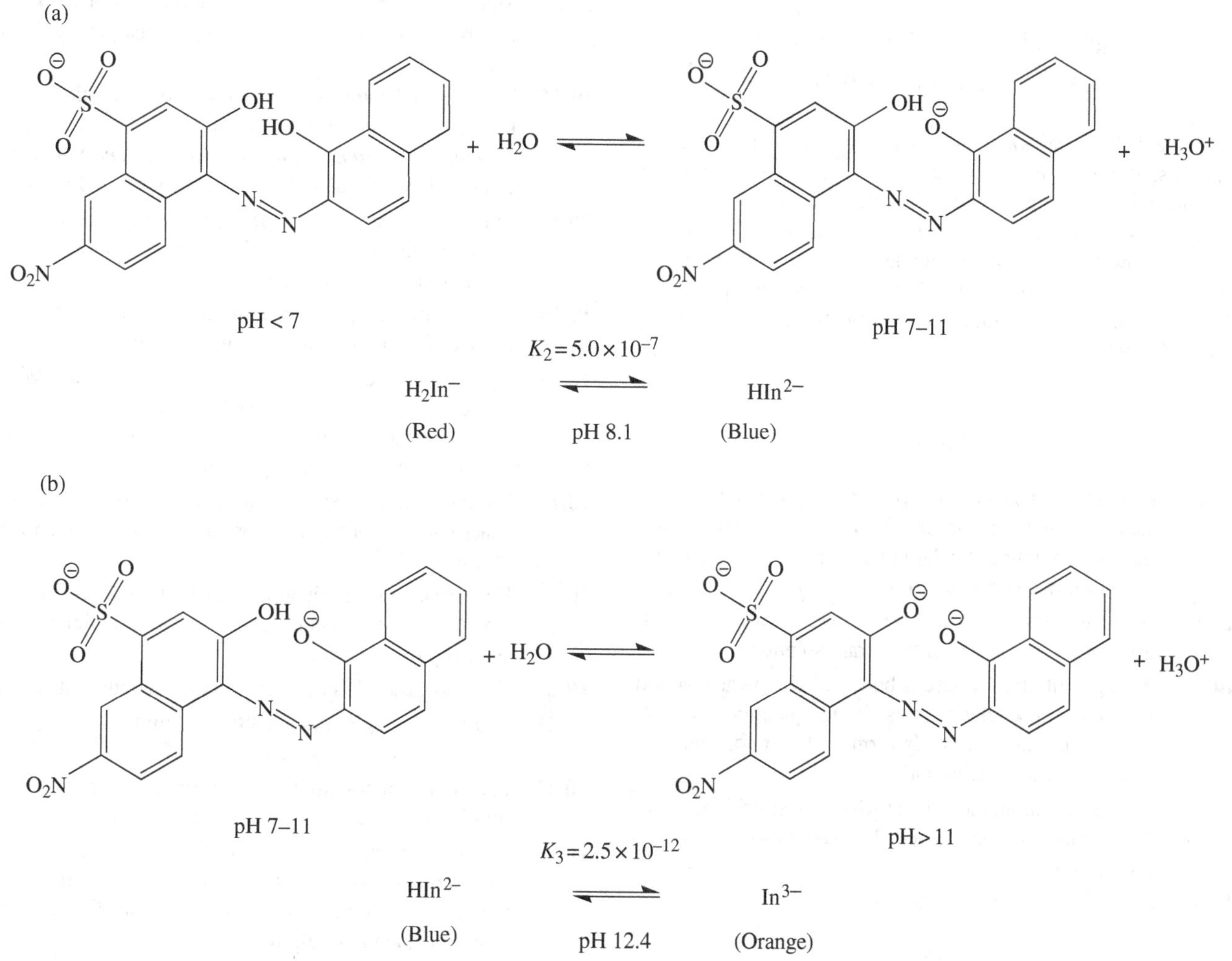

Eriochrome black T is also a pH-dependent visual indicator, and its transitions from red to blue to orange in increasing pH are depicted in Figure 10.24. At a pH < 7, EBT exists as H_2In^- and has a red appearance. In the pH range of 7–11, EBT exists as HIn^{2-} and has a blue appearance. At a pH above 11, EBT exists as In^{3-} and possesses an orange appearance.

EBT is a visual indicator used under conditions that transition in alkaline or basic pH ranges. When EBT complexes with a metal ion in solution at a pH of 10, it complexes in the form MIn^-, which is red in color. When the metal ion has been titrated to the equivalence point with EDTA, the metal ions are removed from the indicator complex and the indicator is in the blue form

(a)

$+ H_2O \rightleftharpoons$ $+ H_3O^+$

pH < 7 pH 7–11

$$H_2In^- \underset{\text{pH } 8.1}{\overset{K_2 = 5.0 \times 10^{-7}}{\rightleftharpoons}} HIn^{2-}$$

(Red) (Blue)

(b)

$+ H_2O \rightleftharpoons$ $+ H_3O^+$

pH 7–11 pH > 11

$$HIn^{2-} \underset{\text{pH } 12.4}{\overset{K_3 = 2.5 \times 10^{-12}}{\rightleftharpoons}} In^{3-}$$

(Blue) (Orange)

FIGURE 10.24 Eriochrome black T pH dependent color transitions from red to blue to orange in increasing pH. (a) At a pH < 7 EBT exists as H_2In^- and has a red color. In the pH range between 7 and 11 EBT exists as HIn^{2-} and has a blue color. (b) At a pH above 11 EBT exists as In^{3-} and possesses an orange color.

of HIn^{2-}. Thus, the visual indicator change for the endpoint of the titration is a transition in color of the solution from red to blue. Let us look at the specific sequence of events for the titration of magnesium ion (Mg^{2+}) with EDTA at a pH of 10 and EBT as indicator. Initially, we have a solution containing magnesium ions as Mg^{2+}. The EBT visual indicator is added to the solution and we now have a solution of free Mg^{2+} ions and Mg^{2+} ions complexed to EBT:

$$Mg^{2+} + HIn^{2-} + H_2O \rightleftharpoons MgIn^-(\text{red}) + H_3O^+. \quad (10.71)$$

The EDTA titrant is added and the reaction between the Mg^{2+} ions and the EDTA takes place until all of the free Mg^{2+} in solution is complexed with the EDTA:

$$Mg^{2+} + H_2Y^{2-}(\text{EDTA}) + 2H_2O \rightleftharpoons MgY^{2-} + 2H_3O^+. \quad (10.72)$$

The weak red complex between the magnesium ion and EBT will be broken by EDTA and the magnesium ion will become complexed to the EDTA leaving the EBT in its blue form of HIn^{2-} at pH 10:

$$\begin{aligned} &MgIn^-(\text{red}) + H_2Y^{2-}(\text{EDTA}) + H_2O \rightleftharpoons \\ &MgY^{2-} + HIn^{2-}(\text{blue}) + H_3O^+. \end{aligned} \quad (10.73)$$

While EBT is an ideal indicator at alkaline pH, there are other visual indicators that are used for metal analysis at lower pH often with as acetate–acetic acid buffer system. These include NAS, xylenol orange, and arsenazo I. For example, NAS forms a pale-yellow to pale-orange complex with most of the M(II) species. NAS in solutions of pH <3.5 will be red-violet, and above pH 3.5 is red-orange. It is typically used in the pH range of 3–9.

10.20 CHAPTER KEY CONCEPTS

10.1 Titrimetric analysis is the process of measuring a substance of unknown concentration in a solution of interest via reaction with a standard that we have made that contains a known substance concentration.

10.2 The solution being titrated is the titrand, while the standard solution is known as the titrant solution.

10.3 During a titration where a balanced chemical reaction takes place, the relationships among quantities of reactants and products typically form a ratio of positive integers called the reacting ratio.

10.4 During a titration, the point at which the stoichiometry of the reaction has been reached is known as the *equivalence point.*

10.5 Fundamental titration equation:

$$\%B = \frac{(\text{ml}_A)(M_A)\left(\frac{b}{a}\right)(\text{milli form wt}_B)(100)}{g_S}$$

10.6 ChemTech can be used in the laboratory for calculating the results of titrations. In the Main Menu options click on the Chapter 10 link to pull up the Main Menu for Chapter 10.

10.7 ChemTech can be used for titrations using either molar solutions or normal solutions.

10.8 At the midpoint, there is a direct relationship between the pH of the titration and the pK_a value of the analyte being measured.

10.9 At the midpoint of the titration, the concentration of the protonated acid [HA] is equal to the concentration of the deprotonated acid [A^-].

10.10 At the midpoint of the titration, the pH of the solution is equal to the pK_a of the species being titrated, $pH = pK_a$.

10.11 The equivalence point in the titration is the point where in the solution there is an equal molar concentration of titrant and analyte. In acid–base titrations, this is the point of neutralization.

10.12 Acid–base indicators are weak organic acids and bases that have a dramatic color change in accordance with the solutions pH.

10.13 Since the indicator is, for acidic indicators, a weak acid at low concentration, the stronger analyte acid will titrate first.

10.14 The ideal indicator for a certain acid–base titration is chosen to have its color transition range in the pH region of the *equivalence point* of the titration. The *endpoint* of the titration is when the color changes for the indicator.

10.15 Polyprotic acids, such as phosphoric acid, possess multiple distinguishable equivalence points, and thus also multiple pH transition ranges.

10.16 The ChemTech program may be used to calculate the results of titrations using normal solutions.

10.17 Open ChemTech and go to the Chapter 10 Main Menu. Click on the Titrations Main Menu button, and then click on the second button, "Acid/Base Titrations Using Normal Solutions" to bring up a titration calculation page.

10.18 Titrations involve volumetric analysis, which means that exact volumes of solutions are used within a small limit of error.

10.19 Burettes contain graduations that are used to measure the amount of titrant added to reach an endpoint with an accuracy of ±0.01 ml.

10.20 Titrants in the laboratory are usually secondary standards.

10.21 A primary standard is used to determine the concentration of the secondary standard.

10.22 In conductometric titrations, the conductance in the solution is measured during the addition of the titrant.

10.23 Conductometric titrations can be useful when the equivalence point is difficult to discern using an indicator or when other problems make the titration difficult such as a nonaqueous solution.

10.24 An example of a useful conductometric titration is the titration of the weak acid acetic acid with the weak base ammonia in an organic solvent.

10.25 A common example of a precipitation titration method is the Mohr method used for the determination of the halides bromide (Br^-) and chloride (Cl^-).

10.26 A standardized silver nitrate ($AgNO_3$) solution is used to titrate the halides present in a titrand forming white insoluble salts of silver bromide (AgBr) or silver chloride (AgCl).

10.27 A soluble chromate salt, such as potassium chromate (K_2CrO_4), is used as the indicator. When the equivalence point is reached, the excess silver will form a red insoluble silver chromate (Ag_2CrO_4) salt precipitate.

10.28 Back titrations can be effective for a titration reaction that is slow or requires energy to be input for the reaction to go to a stoichiometric equivalence point.

10.29 The Volhard method titrates the silver ion with the thiocyante ion (SCN^-) that produces the mostly insoluble silver thiocyanate salt (AgSCN).

10.30 Many of the di- and trivalent cationic metals can be titrated using a complex formation technique with the polydentate ligand EDTA (ethylenediaminetetraacetic acid).

10.21 CHAPTER PROBLEMS

Preparing Titration Solutions

10.1 What is the amount of substance (in milligram) in each of the following?

a. 150 ml of 0.5 M HCl.
b. 0.510 l of 0.25 N H_2SO_4.
c. 750 ml of 0.125 M NaOH.
d. 0.450 l of 0.55 M AgCl.
e. 665 ml of 0.100 N Na_2SO_4.

10.2 What volume of 18 N sulfuric acid solution is needed to make 500 ml of a 0.25 N solution? What volume for a 0.25 M solution?

10.3 Some common acid reagents in the laboratory come in the form of solutions. For the following acids, determine the molarity and normality of the reagents.

Acid	Purity (%)	Density (g/cm^3)
Sulfuric acid (H_2SO_4)	95	1.84
Phosphoric acid (H_3PO_4)	85	1.69
Hydrochloric acid (HCl)	37	1.18
Nitric acid (HNO_3)	70	1.51

10.4 A sodium hydroxide solution in the lab has a concentration of 50%. What is the molarity of the solution? What volume is needed to make 650 ml of a 0.525 M solution?

10.5 What weight of benzoic acid ($C_7H_6O_2$) is needed to make 2.5 l of a 0.05 M solution?

Calculations with Titrations

10.6 A KHP standard was weighed at 0.5092 g and dissolved in water. It took 22.15 ml of a sodium hydroxide solution to titrate to endpoint. What is the molarity of the sodium hydroxide solution?

10.7 An HCl solution was standardized at 0.1159 M. What dilution is needed to make a 500 ml of a 0.050 M solution?

10.8 A 0.095 g primary standard of sodium carbonate dissolved in 100 ml DDI water required 15.36 ml of an HCl solution to titrate to endpoint. What is the molarity of the HCl solution?

$$2HCl + Na_2CO_3 \rightarrow 2NaCl + H_2CO_3$$

10.9 A 7.5 ml sample required 28.5 ml of a 0.15 M sodium hydroxide solution to titrate to endpoint.

a. What is the weight of acid in the sample as acetic acid (CH_3COOH)?
b. What is the concentration of acetic acid in the original sample?

Titration Concentration Calculations

10.10 A 0.650 g sample titrated for basicity as sodium hydroxide required 31.53 ml of a 0.1025 M HCl titrant to titrate to the endpoint. What percentage concentration is the sodium hydroxide?

10.11 An industrial waste stream sample is taken and brought to the lab. The lab technician measures the volume of the sample at 1.521 l. For titration, a 10 ml aliquot is taken from the sample and mixed with 50 ml of DI water. The sample requires 12.36 ml of a 0.1256 M sodium hydroxide titrant. What is the sulfuric acid concentration in parts per million (ppm), molarity (M), and normality (N)?

10.12 A powder sample is brought to the lab for analysis of the amount of sodium dihydrogen phosphate. A 2.545 g sample is dissolved in water and requires 38.24 mL of a 0.09892 M sodium hydroxide solution to titrate to endpoint. What is the concentration of the sodium dihydrogen phosphate?

10.13 A 2.547 g sample was titrated for phosphoric acid (H_3PO_4) content using a standardized 0.1589 M calcium hydroxide ($Ca(OH)_2$) solution requiring 32.19 ml of the titrant. The following two reactions represent the neutralization. Decide which reaction is most likely observed and calculate the phosphoric acid content of the sample.

$$H_3PO_4 + Ca(OH)2 \rightarrow CaHPO_4 + 2H_2O$$

$$2H_3PO_4 + 3Ca(OH)2 \rightarrow Ca3(PO_4)2 + 6H_2O$$

10.14 If 4.435 g of oxalic acid ($H_2C_2O_4$) required 48.93 ml of a 2.0136 N sodium hydroxide (NaOH), what is the equivalent weight and molecular weight of the oxalic acid?

10.15 If a solution is made up of 16.83 g of phosphoric acid (H_3PO_4) dissolved in 250 ml of DI water, what volume of a 0.255 M potassium hydroxide (KOH) solution is needed to completely neutralize the phosphoric acid?

$$3KOH + H_3PO_4 \rightarrow K_3PO_4 + 3H_2O$$

10.16 A basic solution is made by adding 38.69 g of pure sodium oxide (Na_2O) to 500 ml of DI water. What volume of a 2.175 M solution of sulfuric acid is needed to completely neutralize the sodium hydroxide that is produced?

10.17 25 ml of a 0.255 M silver nitrate ($AgNO_3$) titrant was added to a 50 ml titrand solution that contains the chromate ion (CrO_4^{2-}). It required 12.39 ml of a 0.1998 M thiocyanate (SCN^-) solution to titrate to a red endpoint. How many milligrams of chromate ion are present in the titrand? What is the concentration of the chromate ion in ppm and molarity?

$$2Ag^+ + CrO_4^{2-} \rightarrow Ag_2CrO_4$$

10.18 A 0.502 g sample is dissolved in 50 ml DI water. 50 ml of a 0.0512 M solution of EDTA was added to complex with aluminum(III) in a 1:1 ratio. The excess EDTA was titrated with a 0.0408 M zinc(II) titrant (1:1) that required 16.73 ml. What is the percentage of aluminum(III) in the sample?

10.19 A 5.326 g sample is dissolved in 500 ml DI water. The calcium present is precipitated as calcium oxalate. The precipitate is filtered, isolated, and dissolved in acid. The liberated calcium is analyzed by titrating the oxalate acid with 26.74 ml of a 0.0505 M potassium permanganate solution. What is the percentage of calcium in the sample?

10.20 Write equilibrium equations between the metals Cr^{3+}, Co^{2+}, La^{3+}, and Th^{4+} and EDTA (H_2Y^{2-}).

11

OXIDATION–REDUCTION (REDOX) REACTIONS

11.1 Introduction
11.2 Oxidation and Reduction
11.3 The Volt
11.4 The Electrochemical Cell
11.5 Redox Reaction Conventions
 11.5.1 Electrode Potential Tables
 11.5.2 The Standard Hydrogen Electrode (SHE)
 11.5.3 The SHE Half-Reaction
 11.5.4 Writing the Standard Electrode Potentials
 11.5.5 Drawing a Galvanic Cell
 11.5.6 Calculating the Cell Potential
11.6 The Nernst Equation
 11.6.1 Nernst Equation Example I
 11.6.2 Nernst Equation Example II
 11.6.3 Nernst Equation Example III
11.7 Determining Redox Titration Endpoints
11.8 Potentiometric Titrations
 11.8.1 Detailed Potentiometer
 11.8.2 Half-Reactions
 11.8.3 The Nernst Equation
 11.8.4 Assumed Reaction Completion
 11.8.5 Calculated Potentials of Ce^{4+}
11.9 Visual Indicators Used in Redox Titrations
11.10 Pretitration Oxidation–Reduction
 11.10.1 Reducing Agents
 11.10.2 Oxidizing Agents
11.11 Ion-Selective Electrodes
11.12 Chapter Key Concepts
11.13 Chapter Problems

11.1 INTRODUCTION

In the Brønsted–Lowery definition of acid–base neutralization reactions, an acid was a species that donates a hydrogen proton (H^+) while a base is a species that accepts a hydrogen proton (e.g., OH^-) (see Chapter 9, Section 9.5). The analyst is able to use this generally straightforward concept to perform a number of titrations such as acidity as sulfuric acid (H_2SO_4) or alkalinity as potassium hydroxide (KOH). Neutralization reactions are essentially instantaneous; thus, a consideration of reaction completeness was not involved (we did learn the concept of back titration in Section 10.4 to address this problem). We have also found in the preceding chapter that there are numerous other titration methods that the analyst may use in the course of sample analysis in the analytical laboratory including precipitation (Section 10.9) and conductometric (Section 10.8). Here we come to yet another titration methodology to add to our list of volumetric analyses called oxidation–reduction titrations. Even though this chapter may appear to cover aspects of chemistry that are somewhat advanced at times, oxidation–reduction reactions are very important in the analytical laboratory and need to be presented to a certain degree. This will help the analyst to better understand some basic principles of chemical reactions that are routinely conducted in the analytical laboratory. To this effect, we will begin will a very fundamental treatment of oxidation–reduction reactions and move to its application to titrations and finally to ion-selective electrodes.

11.2 OXIDATION AND REDUCTION

Analogous to acid–base reactions, oxidation–reduction reactions have a definition for each half as well: *oxidation* is the process where an atom, ion, or molecule has lost one or more electrons, while *reduction* is the process where an atom, ion, or molecule has gained one or more electrons. Oxidation–reduction reactions (sometimes shortened to *redox* reactions) are taking place in solution (as far as we are concerned here) where there are no free electrons; thus, one must take place with the other. In other words, if something is oxidized then something else must be reduced. An *oxidizing agent* is a species that is able to accept electrons and thus becomes reduced:

$$A_{ox} + ne^- \rightarrow A_{red}. \tag{11.1}$$

Analytical Chemistry: A Chemist and Laboratory Technician's Toolkit, First Edition. Bryan M. Ham and Aihui MaHam.

A reducing agent is a species that is able to donate electrons and thus becomes oxidized:

$$B_{red} \rightarrow B_{ox} + ne^-. \tag{11.2}$$

The oxidation reaction and the reduction reaction are each called half-reactions of the entire redox process. The sum of the two half-reactions represents the redox reaction taking place:

$$A_{ox} + ne^- \rightarrow A_{red} \tag{11.1}$$

$$B_{red} \rightarrow B_{ox} + ne^- \tag{11.2}$$

$$A_{ox} + B_{red} \rightarrow A_{red} + B_{ox} \tag{11.3}$$

A specific example of the reduction of iron(III) (Fe^{3+}) to iron(II) (Fe^{2+}) and the oxidation of copper (Cu) to Cu(II) (Cu^{2+}) in a solution together would be

$$2Fe^{3+} + 2e^- \rightarrow 2Fe^{2+} \tag{11.4}$$

$$+ Cu^0(s) \rightarrow Cu^{2+} + 2e^- \tag{11.5}$$

$$2Fe^{3+} + Cu^0(s) \rightarrow 2Fe^{2+} + Cu^{2+} \tag{11.6}$$

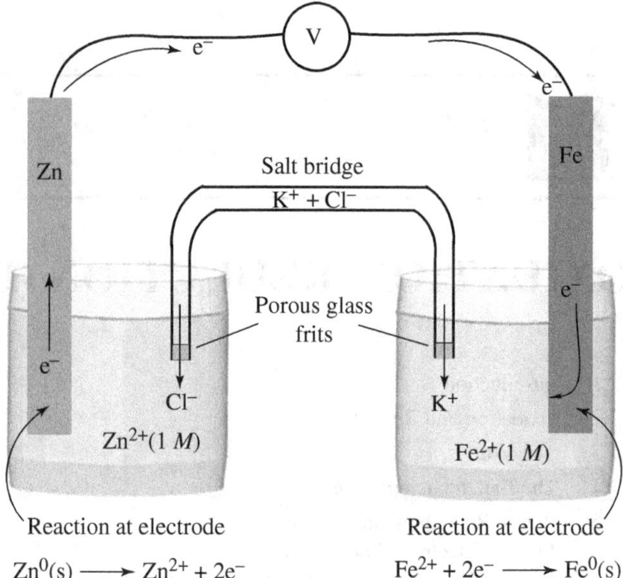

FIGURE 11.1 An electrochemical cell base upon the oxidation and reduction reaction of zinc and iron. In this galvanic cell zinc is being oxidized to zinc(II) (Zn^{2+}), and iron(II) (Fe^{2+}) is being reduced to iron (Fe).

11.3 THE VOLT

We now have the basic premise of what a redox reaction is and how one species is oxidized while a second species is reduced. The idea of the transfer of electrons from one species to another in the redox reaction may bring about the thought of the "flow" of electrons such as what is associated with "electricity" (electrons flowing through a wire). This is in fact the case with redox reactions where the flow of electrons does produce a current that can be measured with a galvanometer (voltage measuring device) when the redox reactions are set up as a circuit. The equilibrium state of redox reactions (redox equilibria) has been defined through galvanic cell electromotive forces (emfs). Chemists define a galvanic cell as a system containing a spontaneous chemical reaction that releases electrical energy that may perform work. The potential or voltage of the cell is the emf having units of volts (V). The definition of a volt is the required emf to give one joule (J) of energy to a one coulomb (C) electrical charge:

$$1\,V = \frac{1\,J}{C}. \tag{11.7}$$

11.4 THE ELECTROCHEMICAL CELL

An electrochemical cell is an example of a galvanic cell that is composed of a two-electrode system where oxidation is taking place at one electrode and reduction is taking place at the second electrode. The electrodes for the two solutions are connected by a wire allowing the flow of electrons, and the circuit is completed through the use of a salt bridge. Let us look at a specific example of an electrochemical cell through the oxidation of zinc (Zn) at the left electrode and the reduction of iron(II) (Fe^{2+}) at the right electrode. A schematic of the galvanic cell is depicted in Figure 11.1. In the left single potential electrode cell, there is a pure metal strip of zinc that is suspended in a 1 *M* solution of the zinc metal salt (Zn^{2+}). In the right single potential electrode cell, there is a pure metal strip of iron that is suspended in a 1 *M* solution of the iron metal salt (Fe^{2+}). To complete the circuit, a salt bridge containing potassium chloride (KCl) is immersed into both solutions (may be either a tube filled with KCl solution that is inserted into the solutions, or more usually a tube containing a saturated KCl solution with agar plugs at each end allowing transfer (diffusion) of K^+ or Cl^- into solution). The salt bridge allows the two solutions to stay electrically neutral as the redox reaction takes place at each electrode. Because of the difference in potential of each of the two connected solutions, the redox reaction will take place. In the left cell, the zinc will go into solution as the zinc(II) cation (Zn^{2+}) leaving behind two extra electrons on the zinc metal electrode. In the right cell, the iron(II) in solution will plate onto the iron metal electrode and accept two electrons thus reducing to neutral iron (Fe). The potential difference in the two cells will produce a spontaneous redox reaction where the electrons will flow from the zinc electrode to the iron electrode thus producing a measurable current. The current is measured via an ammeter or voltmeter as shown in Figure 11.1 between the two cells.

11.5 REDOX REACTION CONVENTIONS

11.5.1 Electrode Potential Tables

As depicted in Figure 11.1, the potential difference between the two cells is measured, and for this particular electrochemical cell

TABLE 11.1 Standard Reduction Potentials.

Half-reaction			
Oxidant	$\rightleftharpoons$	Reductant	E° (V)[a]
$Ca^+ + e^-$	$\rightleftharpoons$	Ca	−3.80
$Li^+ + e^-$	$\rightleftharpoons$	Li(s)	−3.0401
$K^+ + e^-$	$\rightleftharpoons$	K(s)	−2.931
$Ba^{2+} + 2e^-$	$\rightleftharpoons$	Ba(s)	−2.912
$Ca^{2+} + 2e^-$	$\rightleftharpoons$	Ca(s)	−2.868
$Na^+ + e^-$	$\rightleftharpoons$	Na(s)	−2.71
$Mg^+ + e^-$	$\rightleftharpoons$	Mg	−2.70
$Mg^{2+} + 2e^-$	$\rightleftharpoons$	Mg(s)	−2.372
$H_2 + 2e^-$	$\rightleftharpoons$	$2H^-$	−2.23
$Mn^{2+} + 2e^-$	$\rightleftharpoons$	Mn(s)	−1.185
$2H_2O + 2e^-$	$\rightleftharpoons$	$H_2(g) + 2OH^-$	−0.8277
$Zn^{2+} + 2e^-$	$\rightleftharpoons$	Zn(Hg)	−0.7628
$Zn^{2+} + 2e^-$	$\rightleftharpoons$	Zn(s)	−0.7618
$Fe^{2+} + 2e^-$	$\rightleftharpoons$	Fe(s)	−0.44
$PbSO_4(s) + 2e^-$	$\rightleftharpoons$	$Pb(s) + SO_4^{2-}$	−0.3588
$Pb^{2+} + 2e^-$	$\rightleftharpoons$	Pb(s)	−0.13
$Fe^{3+} + 3e^-$	$\rightleftharpoons$	Fe(s)	−0.04
$2H^+ + 2e^-$	$\rightleftharpoons$	$H_2(g)$	0.0000
$S(s) + 2H^+ + 2e^-$	$\rightleftharpoons$	$H_2S(g)$	+0.14
$Sn^{4+} + 2e^-$	$\rightleftharpoons$	Sn^{2+}	+0.15
$Cu^{2+} + e^-$	$\rightleftharpoons$	Cu^+	+0.159
$Fc^+ + e^-$	$\rightleftharpoons$	Fc(s)	+0.641
$O_2(g) + 2H^+ + 2e^-$	$\rightleftharpoons$	$H_2O_2(aq)$	+0.70
$Fe^{3+} + e^-$	$\rightleftharpoons$	Fe^{2+}	+0.77
$O_2(g) + 4H^+ + 4e^-$	$\rightleftharpoons$	$2H_2O$	+1.229
$Pb^{4+} + 2e^-$	$\rightleftharpoons$	Pb^{2+}	+1.69
$MnO_4^- + 4H^+ + 3e^-$	$\rightleftharpoons$	$MnO_2(s) + 2H_2O$	+1.70
$Ag^{2+} + e^-$	$\rightleftharpoons$	Ag^+	+1.98
$O_3(g) + 2H^+ + 2e^-$	$\rightleftharpoons$	$O_2(g) + H_2O$	+2.075
$HMnO_4^- + 3H^+ + 2e^-$	$\rightleftharpoons$	$MnO_2(s) + 2H_2O$	+2.09
$FeO_4^{2-} + 3e^- + 8H^+$	$\rightleftharpoons$	$Fe^{3+} + 4H_2O$	+2.20
$F_2(g) + 2e^-$	$\rightleftharpoons$	$2F^-$	+2.87
$F_2(g) + 2H^+ + 2e^-$	$\rightleftharpoons$	$2HF(aq)$	+3.05

[a] http://creativecommons.org/licenses/by-sa/3.0/

the potential difference is +0.323 V. The potential difference of this setup can also be calculated (instead of measured) using electrode potential tables. A tabulation of a select number of standard reduction potential half-reactions is listed in Table 11.1. A more extensive listing of potentials and half-reactions are found in Appendix IV.

11.5.2 The Standard Hydrogen Electrode (SHE)

The potential of a half-reaction, such as the zinc electrode in Figure 11.1, or the iron electrode, cannot be measured by itself (for all practical purposes of the analysts' laboratory work). Instead, a reference electrode is set up with the electrode of interest to measure the potential.

The reference electrode used is the standard hydrogen electrode (SHE) that has been assigned a value of exactly zero for the standard potential ($E^\circ = 0.00$ V) for a hydrogen gas (H_2) pressure of 1 atm and a hydrogen ion (H^+) activity in solution at unity (all standard potentials are measured and assigned at a concentration of 1 *M*).

11.5.3 The SHE Half-Reaction

The half-reaction for the SHE is then

$$2H^+ + 2e^- \rightleftharpoons H_2(g) \quad E^\circ = 0.00 \text{ V}. \tag{11.8}$$

If we couple our zinc electrode to the SHE, then we can measure the potential difference. The measured potential difference is then assigned to the standard zinc electrode. When the SHE is coupled to the zinc electrode (Zn–Zn^{2+}), the potential difference measured is 0.763 V. The polarity (which can be determined from the galvanic cell setup) of the hydrogen electrode is found to be positive, while the polarity of the zinc electrode is negative. Thus, the standard potential assigned to the zinc electrode (E°) is −0.763 V. When the SHE is coupled to the iron electrode (Fe–Fe^{2+}), the potential difference measured is 0.440 V. The polarity (which can be determined from the galvanic cell setup) of the hydrogen electrode is found to be positive, while the polarity of the iron electrode is negative. Thus, the standard potential assigned to the iron electrode (E°) is −0.440 V.

11.5.4 Writing the Standard Electrode Potentials

The internationally agreed form of writing the standard electrode potentials is to write the electrode reactions, from left to right, as reductions:

$$2H^+ + 2e^- \rightleftharpoons H_2(g) \quad E^\circ = 0.00\text{V} \tag{11.9}$$

$$Zn^{2+} + 2e^- \rightleftharpoons Zn \quad E^\circ = -0.763\text{V} \tag{11.10}$$

$$Fe^{2+} + 2e^- \rightleftharpoons Fe \quad E^\circ = -0.440\text{V} \tag{11.11}$$

11.5.5 Drawing a Galvanic Cell

There are a number of conventions that are used when drawing a galvanic cell such as that shown in Figure 11.1. The flow of electrons is drawn from left to right where the reaction taking place at the left electrode is oxidation, and the electrode on the right is reduction. The electrode on the left has a negative polarity and is called the anode, while the electrode on the right has a positive polarity and is called the cathode. Figure 11.1 is the schematic representation of the galvanic cell. This is usually replaced by the following written form, first in the general form followed by the example form of iron and zinc:

$$A_{ox} + ne^- \rightarrow A_{red} \tag{11.1}$$

$$B_{red} \rightarrow B_{ox} + ne^- \tag{11.2}$$

$$A_{ox} + B_{red} \rightarrow A_{red} + B_{ox} \tag{11.3}$$

$$B_{red\,E_l}|B_{ox}(1\,M)|\;|A_{ox}(1\,M)|_{E_r}A_{red} \tag{11.12}$$

$$Zn_{E_l}|Zn^{2+}(1\,M)|\;|Fe^{2+}(1\,M)|_{E_r}Fe \tag{11.13}$$

11.5.6 Calculating the Cell Potential

The potential of the cell is then calculated as the difference of the right cell potential from the left cell potential:

$$E^{o}_{cell} = E_r - E_l. \qquad (11.14)$$

In the convention of writing the half-reactions illustrated in Equations 11.1 and 11.2, the oxidation reaction in Equation 11.2 is reversed of that listed in the reduction half-reaction tables. When this is done, the E^{o}_{cell} sign is also reversed, so Equation 11.14 is written as

$$E^{o}_{cell} = E_{ox} + E_{red}. \qquad (11.15)$$

The standard potentials for the cell half-reactions are listed in Table 11.1. These are used in calculating the potential of any particular galvanic cells.

11.5.6.1 ***Iron and Zinc Cell*** Let us consider again the cell depicted in Figure 11.1. The overall reaction is obtained by writing the right reaction as a reduction, then subtracting the left reaction (as a reduction) from the right. This includes the standard potentials obtained from Table 11.1. The equations sometimes require balancing the exchanged electrons by multiplying one of the reactions. In this case, it is not needed. When done however, the potential value is not changed.

$$Fe^{2+} + 2e^- \rightleftharpoons Fe \quad E^{o}_{r} = -0.440\,V \qquad (11.16)$$

$$Zn^{2+} + 2e^- \rightleftharpoons Zn \quad E^{o}_{l} = -0.763\,V \qquad (11.17)$$

$$Fe^{2+} + Zn \rightleftharpoons Fe + Zn^{2+} \quad E^{o}_{cell} = 0.323\,V \qquad (11.18)$$

The resultant emf value (E^{o}_{cell}) gives the overall potential of the cell and indicates the polarity of the right-hand cell. If the E^{o}_{cell} value is positive, then the reaction is a spontaneous flow of electrons from the left to the right, where the left electrode has a negative polarity and is the anode, and the right electrode has a positive polarity and is the cathode. If the two cells are reversed, the E^{o}_{cell} value will be negative but the polarities and reactions will still be the same.

$$Fe^{2+} + 2e^- \rightleftharpoons Fe \quad E^{o}_{r} = -0.440\,V \qquad (11.16)$$

$$Zn \rightleftharpoons Zn^{2+} + 2e^- \quad E^{o}_{l} = +0.763\,V \qquad (11.17)$$

$$Fe^{2+} + Zn \rightleftharpoons Fe + Zn^{2+} \quad E^{o}_{cell} = 0.323\,V \qquad (11.18)$$

11.5.6.2 ***Nickel and Silver Cell*** When given two sets of reactions to be made into a galvanic cell, the reaction with the lowest potential is reversed (the sign of the potential is also reversed) and is designated as the anode. For example, if we have a galvanic cell consisting of a nickel electrode in contact with a solution of Ni^{2+} ions, and a silver electrode in contact with a solution of Ag^+ ions, the reactions (from Table 11.1), potentials, and standard cell are

$$Ni^{2+}(aq) + 2e^- \rightarrow Ni(s)(E^{o} = -0.26V)$$
$$Ag^{+}(aq) + e^- \rightarrow Ag(s)(E^{o} = 0.80V)$$

$$\begin{array}{lll} \text{Anode:} & Ni(s) \rightarrow Ni^{2+}(aq) + 2e^- & E^{o} = 0.26\,V \\ \text{Cathode:} & Ag^{+}(aq) + e^- \rightarrow Ag(s) & \underline{E^{o} = 0.80\,V} \\ & & E^{o} = 1.06\,V \end{array}$$

$$Ni(s) + 2Ag^{+}(aq) \rightarrow Ni^{2+}(aq) + 2Ag(s)$$

$$Ni(s)_{E_l}|Ni^{2+}(1\,M)|\;\;|Ag^{+}(1\,M)|_{E_r}Ag(s)$$

The cell has a potential of E^{o} = 1.06 V indicating that the reaction is spontaneous.

11.6 THE NERNST EQUATION

The *Nernst* equation describes the relationship between the potential of a galvanic cell and the activities (or as an approximate the molar *M* concentrations) of the redox species. For our galvanic cell reaction involving a left and right electrode coupled with a salt bridge, the redox reaction is

$$aA + bB \rightleftharpoons cC + dD. \qquad (11.19)$$

The potential of the galvanic cell as dependent upon the concentrations is written as

$$E = E^{o} - \frac{2.303RT}{nF}\log\frac{[C]^c[D]^d}{[A]^a[B]^b}, \qquad (11.20)$$

where *E* is the potential in volts (V), E^{o} is the standard electrode potential in volts (V), *R* is the gas constant (8.314 joules per degree-mole), T is the absolute temperature (273 K + degrees C), *n* is the number of electrons in the reaction, and *F* is the Faraday constant (96,487 coulombs per Eq. 11.20).

To simplify the *Nernst* equation, the redox reaction is normally assumed to be performed at room temperature of 25 °C (298 K); thus, the term 2.303 *RT/F* becomes 0.05915. The *Nernst* equation then becomes

$$E = E^{o} - \frac{0.059}{n}\log\frac{[C]^c[D]^d}{[A]^a[B]^b}. \qquad (11.21)$$

For the redox half-reaction:

$$A_{ox} + ne^- = A_{red}. \qquad (11.1)$$

The *Nernst* equation can be written as

$$E = E^{o} + \frac{0.059}{n}\log\frac{[A_{ox}]}{[A_{red}]}. \qquad (11.22)$$

This form of the equation can be used to calculate different aspects of a test electrode versus a standard reference electrode of known potential, such as the SHE.

11.6.1 Nernst Equation Example I

A platinum electrode is immersed into a solution containing 0.0256 M of Hg_2^{2+}, and is connected by a salt bridge to a reference electrode. What is the potential of the platinum electrode?

Using Table 11.1, the half-reaction and standard electrode potential are

$$Hg_2^{2+} + 2e^- \rightleftharpoons 2Hg^o(s) \quad E^o = 0.789\,V. \tag{11.23}$$

The *Nernst* equation then becomes

$$E = 0.789 + \frac{0.059}{2}\log\frac{[0.0256]}{1} \tag{11.24}$$

$$E = 0.742\,V\,\infty. \tag{11.25}$$

Note that Hg^o(s) is in the standard state and has a value of unity. The *Nernst* equation is effected only by the species in solution.

If the test electrode potential is measured, then the concentration of a species in solution can also be measured.

11.6.2 Nernst Equation Example II

A platinum electrode is placed into a solution containing silver(I) (Ag^+) ions along with a reference electrode. The potential measured on the platinum electrode is 0.765 V. What is the molar concentration of the silver in solution?

$$Ag^+ + e^- \rightleftharpoons Ag^o(s) \quad E^o = 0.800\,V \tag{11.26}$$

$$0.765\,V = 0.800\,V + 0.059\log[Ag^+] \tag{11.27}$$

$$[Ag^+] = \text{antilog}\left(\frac{0.765 - 0.800}{0.059}\right) \tag{11.28}$$

$$[Ag^+] = 0.255M \tag{11.29}$$

This example illustrates that with a calibrated test electrode (such as a silver ion sensing electrode) and known reference electrode, the concentration of different species in solution can be determined. These are called ion-selective electrodes and will be further discussed later.

Returning back to our galvanic cells (as depicted in Figure 11.1), let us look at the application of the *Nernst* equation to a complete redox reaction.

11.6.3 Nernst Equation Example III

The following voltaic cell, at a particular point in time, has the following concentrations as determined in the laboratory:

$$\text{Pt}|Cr^{2+}(0.15\,M) + Cr^{3+}(0.024\,M)|_{E_l} \; |Sn^{4+}(0.10\,M) + Sn^{2+}(0.065\,M)|_{E_r}\text{Pt}. \tag{11.30}$$

From Table 11.1 the following half-reactions describe the cell:

$$Sn^{4+} + 2e^- \rightleftharpoons Sn^{2+} \quad E_r^o = +0.15\,V \tag{11.31}$$

$$Cr^{3+} + e^- \rightleftharpoons Cr^{2+} \quad E_l^o = -0.41\,V \tag{11.32}$$

In order to balance the number of electrons in each half-reaction, we will have to multiply the left electrode reduction by "2." The next step is to obtain the potential of the cell (E_{cell}^o) by subtracting the left electrode potential from the right as follows:

$$Sn^{4+} + 2e^- \rightleftharpoons Sn^{2+} \quad E_r^o = +0.15\,V \tag{11.33}$$

$$-\left(2Cr^{3+} + 2e^- \rightleftharpoons 2Cr^{2+}\right) \quad -\left(E_l^o = -0.41\,V\right) \tag{11.34}$$

$$Sn^{4+} + 2Cr^{2+} \rightleftharpoons Sn^{2+} + 2Cr^{3+} \quad E_{cell}^o = 0.56\,V \tag{11.35}$$

The standard potential value $E_{cell}^o = 0.56\,V$ can now be used to calculate the potential of the cell due to the effect of the concentrations of the species present using the *Nernst* equation:

$$E_{cell} = E_{cell}^o - \frac{0.059}{n}\log\frac{[Sn^{2+}][Cr^{3+}]^2}{[Sn^{4+}][Cr^{2+}]^2} \tag{11.36}$$

$$E_{cell} = 0.56 - \frac{0.059}{2}\log\frac{[0.065][0.024]^2}{[0.10][0.15]^2} \tag{11.37}$$

$$E_{cell} = 0.56 - (-0.052) \tag{11.38}$$

$$E_{cell} = 0.61\,V \tag{11.39}$$

The voltaic cell potential value $E = 0.61$ V tells us that the reaction will be spontaneous with the flow of electrons from the left to the right. It also tells us that the left electrode has a negative polarity and is the anode, while the right electrode has a positive polarity and is the cathode.

We can also calculate the equilibrium constant (K) for the reaction couple from the two electrode standard potentials. At equilibrium, the potential of the cell will equal zero, $E_{cell} = 0$ V. Thus, we have the following relationship:

$$E_{cell} = E_{cell}^o - \frac{0.059}{n}\log\frac{[Sn^{2+}][Cr^{3+}]^2}{[Sn^{4+}][Cr^{2+}]^2} \tag{11.40}$$

$$0 = E_{cell}^o - \frac{0.059}{n}\log K \tag{11.41}$$

$$K = \text{antilog}\left(\frac{nE_{cell}^o}{0.059}\right) \tag{11.42}$$

$$K = \text{antilog}\left(\frac{2(0.56)}{0.059}\right) \tag{11.43}$$

$$K = 9.6 \times 10^{18} \tag{11.44}$$

The large equilibrium constant value indicates that the reaction has a great tendency to go to the right. If the standard potentials are known for the two redox couples, the equilibrium constant can be determined and used as an indication whether the reactions under consideration will go to completion thus being suitable candidates for a redox titration or not.

11.7 DETERMINING REDOX TITRATION ENDPOINTS

There are two primary ways of determining a redox titration endpoint in the analytical laboratory by the analyst that include the use of a potentiometer where the data gathered are plotted and using a visual indicator. Often when setting up a routine analysis, a potentiometric titration analysis may be performed first in order to select and use the appropriate visual indicator. In the following sections, we will first look at the determination of endpoints in redox titrations, then cover the reagents that are typically involved in redox titrations followed by some specific examples of applying redox titration methodology for analyte measurement. The following sections will present these topics however in a rather general aspect to give the analyst an introduction to, and basic understanding of, redox titration methodology. Unless the method is being developed from the bottom up, specific step-by-step instructions are usually available and followed in the specific laboratory that the technician is working in. This outline, though, is sufficient as a starting point for developing a redox titration method if needed.

11.8 POTENTIOMETRIC TITRATIONS

A rudimentary schematic for a potentiometric titration is depicted in Figure 11.2. The galvanic cell setup includes a reference electrode (right cell) and an indicator electrode (left cell). The reference electrode is the SHE that has arbitrarily been given a potential of 0.000 V (see discussion in Section 11.5 Redox Reaction Conventions) for a hydrogen gas (H_2) pressure of 1 atm and a hydrogen ion (H^+) activity in solution at unity. A more convenient reference electrode that is often employed is the saturated calomel reference electrode (SCE) that does not require the use of hydrogen gas (an extremely flammable gas). For demonstration purposes, we are using the SHE electrode in Figure 11.2. The indicator electrode in the left cell is an inert platinum electrode. Inert electrodes such as platinum (Pt) or gold (Au) do not participate directly in the redox reactions but have the role of transferring the electrons associated with the reactions. The potential that is developed is directly dependent upon the molar concentrations of the oxidized and reduced redox couples in the solution. In the potentiometric titration, we are titrating iron(II) (Fe^{2+}) to the higher iron(III) (Fe^{3+}) oxidation state with cerium(IV) (Ce^{4+}) that subsequently reduces to the lower cerium(III) (Ce^{3+}) oxidation state.

11.8.1 Detailed Potentiometer

The previous electrochemical cell depicted in Figure 11.1 simply showed a voltmeter in connection between the two half-cells for measuring the potential of the cell. In Figure 11.2, a more detailed design of a potentiometer is shown in connection to the two half-cells. To measure the emf of the cell, the draw of current from the electrodes must be at a very small value to keep the electrodes at their equilibrium potentials. An ordinary voltmeter will draw a heavy amount and thus is not feasible to use. Instead, potentiometers like the one shown are used. In this type of measurement, a variable voltage is used to oppose the cell voltage of the system being measured. A null-point detector (galvanometer, G) that is very sensitive gives a reading to indicate when the opposing potential equals the potential of the cell. The battery is used to supply the opposing emf. A standard cell is used to set the potentiometer. When the test cell is used on the potentiometer, the difference measured is equal to the test cell's potential. These systems are very sensitive and do not affect the potential of the test cell to any appreciable amount.

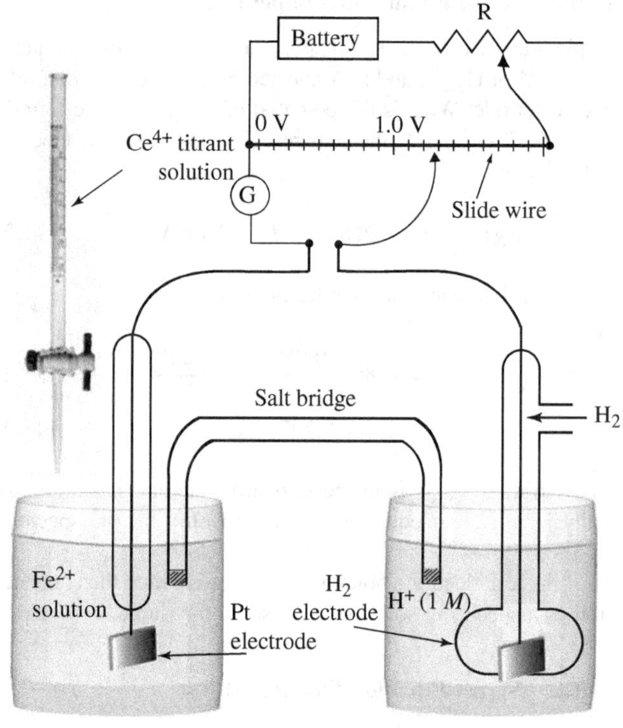

FIGURE 11.2 Potentiometric titration setup for the titration of iron(II) (Fe^{2+} titrand) with cerium(IV) (Ce^{4+} titrant). The reference electrode used is the standard hydrogen electrode (SHE). Also illustrated is a simple design of a potentiometer.

11.8.2 Half-Reactions

The solution to be titrated was made by dissolving 10 mmol of iron(II) chloride in 75 ml of 1 *N* aqueous H_2SO_4 solution. The titrant used for the titration consists of 0.155 *M* cerium(IV) (Ce^{4+}) in 1 *N* aqueous H_2SO_4 solution. Looking at Table 11.1 for the standard potential, we note that there are actually four listed for the cerium(IV) half-reaction:

$$Ce^{4+} + e^- \rightleftharpoons Ce^{3+} \tag{11.45}$$

$$E^o = 1.28\ \text{V in } 1N\ \text{HCl} \tag{11.46}$$

$$E^o = 1.44\ \text{V in } 1N\ H_2SO_4 \tag{11.47}$$

$$E^o = 1.61\ \text{V in } 1M\ HNO_3 \tag{11.48}$$

$$E^o = 1.70\ \text{V in } 1M\ HClO_4. \tag{11.49}$$

The titrant and titrand are both in 1 N H_2SO_4, so the standard potential value used is $E^o = 1.44$ V. The redox couple also consists of (obtained from Table 11.1)

$$Fe^{3+} + e^- \rightleftharpoons Fe^{2+} \quad E^o = 0.771\,V. \tag{11.50}$$

It is standard convention to subtract the half-reaction with the lowest potential from the half-reaction with the greater potential:

$$Ce^{4+} + e^- \rightleftharpoons Ce^{3+} \quad E^o = 1.44\,V \tag{11.51}$$

$$-\left(Fe^{3+} + e^- \rightleftharpoons Fe^{2+}\right) \quad -(E^o = 0.771\,V) \tag{11.52}$$

$$Ce^{4+} + Fe^{2+} \rightleftharpoons Ce^{3+} + Fe^{3+} \quad E^o_{cell} = 0.669\,V \tag{11.53}$$

This gives a cell standard potential of $E^o_{cell} = 0.669\,V$. Initially, the potential in the cell will be due to the Fe^{3+}/Fe^{2+} potential before the addition of any titrant. At this point, the concentration of the Fe^{3+} in the solution it is not known, but we can approximate that 99.9% of the iron in the solution if from the iron(II) chloride reagent that was used in preparing the solution. Thus, 0.1% of the solution would be composed of Fe^{3+}. The initial concentrations are therefore

$$\left[Fe^{2+}\right] = \frac{0.010\,\text{mol}}{0.0751} \times 0.999 = 0.133M \tag{11.54}$$

$$\left[Fe^{3+}\right] = \frac{0.010\,\text{mol}}{0.0751} \times 0.001 = 1.33 \times 10^{-4} M \tag{11.55}$$

11.8.3 The Nernst Equation

Using the Nernst equation, we can calculate the potential of the cell before titration, and then after the addition of the titrant. A plot of the potential in volts against the milliliter titrant added gives a titration curve similar to the ones that were constructed for the other types of titrations that have been covered (e.g., neutralization, conductometric, precipitation, and complexometric).

$$E_{cell} = E^o_{cell} - \frac{0.059}{n}\log\frac{[Fe^{2+}]}{[Fe^{3+}]} \tag{11.56}$$

$$E_{cell} = 0.771\,V - \frac{0.059}{1}\log\frac{0.133}{1.33 \times 10^{-4}} \tag{11.57}$$

$$E_{cell} = 0.594\,V. \tag{11.58}$$

Upon addition of 10 ml of titrant, the potential of the system can be calculated from either redox couple potential as the system is in equilibrium:

$$E_{cell} = E^o_{cell} - \frac{0.059}{n}\log\frac{[Fe^{2+}]}{[Fe^{3+}]} \tag{11.59}$$

$$E_{cell} = E^o_{cell} - \frac{0.059}{n}\log\frac{[Ce^{3+}]}{[Ce^{4+}]} \tag{11.60}$$

11.8.4 Assumed Reaction Completion

The amount of Ce^{4+} that has been added to the system is 0.00155 mol and it is assumed that approximately 100% of the cerium(IV) has been reduced to cerium(III). In actuality, there is a very small amount of cerium(IV) left in solution and could be obtained from the equilibrium constant (this would be necessary if the potential of the cell is being calculated from the Ce^{4+}/Ce^{3+} couple). The amount of iron(II) that has been oxidized to iron(III) will equal the original amount minus the amount of cerium(IV) added. The amount of iron(III) produced will equal the amount of cerium(IV) added. For simplicity, note that the volume of the solution in the left-hand cell is now 75 ml + 10 ml = 85 ml, but is the same for all of the species involved and thus can be canceled out of the calculation.

$$Ce^{4+} = \sim \text{very small amount} \tag{11.61}$$

$$Ce^{3+} = 0.00155\,\text{mol} \tag{11.62}$$

$$Fe^{2+} = 0.010 - 0.00155 = 0.00845\,\text{mol} \tag{11.63}$$

$$Fe^{3+} = 0.00155\,\text{mol} \tag{11.64}$$

It is easier to use the iron redox couple since we directly know the amounts involved:

$$E_{cell} = E^o_{cell} - \frac{0.059}{n}\log\frac{[Fe^{2+}]}{[Fe^{3+}]} \tag{11.65}$$

$$E_{cell} = 0.771 - \frac{0.059}{1}\log\frac{0.00845}{0.00155} \tag{11.66}$$

$$E_{cell} = 0.728\,V \tag{11.67}$$

Thus, at 10 ml addition of the Ce^{4+} titrant to the test solution containing the iron(II), the potential measured by the potentiometer for the galvanic cell depicted in Figure 11.2 is 0.728 V. Using Equation 11.6, we can calculate the potential at each addition of titrant:

$$E_{cell} = 0.771 - 0.059 \times \log\frac{\left(0.01\,\text{mol} - \dfrac{\text{ml}\,Ce^{4+}}{1000} \times 0.155\,M\right)}{\left(\dfrac{\text{ml}\,Ce^{4+}}{1000} \times 0.155\,M\right)} \tag{11.68}$$

The potentials calculated with the application of Equation 11.6 from 10 ml of titrant added up to 64.51 ml are listed in Table 11.1. At the equivalence point, we know that there will be the following equal molar relationships:

$$\left[Fe^{2+}\right] = \left[Ce^{4+}\right] \tag{11.69}$$

$$\left[Fe^{3+}\right] = \left[Ce^{3+}\right]. \tag{11.70}$$

If we add the two cell half-reaction potential equations:

$$E_{cell} = 0.771 - \frac{0.059}{1}\log\frac{[Fe^{2+}]}{[Fe^{3+}]} + E_{cell} = 1.44 - \frac{0.059}{1}\log\frac{[Ce^{3+}]}{[Ce^{4+}]}, \tag{11.71}$$

we obtain

$$2E_{\text{cell}} = 2.211 - 0.059\log\frac{[Fe^{2+}][Ce^{3+}]}{[Fe^{3+}][Ce^{4+}]}, \quad (11.72)$$

resulting in

$$2E_{\text{cell}} = 2.211 - 0.059\log(1) \quad (11.73)$$

$$2E_{\text{cell}} = 2.211 \quad (11.74)$$

$$E_{\text{cell}} = 1.106\,\text{V} \;\; (\text{at equivalence point}). \quad (11.75)$$

Thus, if the transfer of electrons for the two half-reactions is the same (in this case $n = 1$), then the following relationship at the equivalence point holds:

$$E_{\text{eqpt}} = \frac{E_1^{\circ} + E_2^{\circ}}{2} \quad (11.76)$$

We can thus take Equation 11.7 above and set it up to solve for the milliliter of Ce^{4+} titrant needed at the equivalence point (1.106 V):

$$E_{\text{cell}} = 0.771 - 0.059 \times \log\frac{\left(0.01\,\text{mol} - \frac{\text{ml Ce}^{4+}}{1000} \times 0.155\,M\right)}{\left(\frac{\text{ml Ce}^{4+}}{1000} \times 0.155\,M\right)} \quad (11.77)$$

$$E_{\text{cell}} = 0.771 - 0.059 \times \log\frac{\left(0.01\,\text{mol} - \frac{x}{1000} \times 0.155\,M\right)}{\left(\frac{x}{1000} \times 0.155\,M\right)} \quad (11.78)$$

$$1.106 = 0.771 - 0.059 \times \log\frac{(0.01\,\text{mol} - 1.55 \times 10^{-4}x)}{1.55 \times 10^{-4}x} \quad (11.79)$$

$$x = 64.51599\,\text{ml} \quad (11.80)$$

After the equivalence point, it will be required to calculate the concentrations of the Fe^{2+}/Fe^{3+} redox couple using the equilibration constant. As before, it is easier to now use the Ce^{4+}/Ce^{3+} redox couple and assume that the concentrations of $[Fe^{2+}]$ and $[Fe^{3+}]$ are negligible

$$E_{\text{cell}} = 01.44 - 0.059 \times \log\frac{(0.01\,\text{mol})}{\left(\frac{\text{ml Ce}^{4+}}{1000} \times 0.155\,M\right)}. \quad (11.81)$$

11.8.5 Calculated Potentials of Ce^{4+}

The potential versus milliliter Ce^{4+} titrant from 64.52 to 100 ml are listed in Table 11.2 using the relationship illustrated in Equation 11.81. We are now in a position to plot the calculated potential of the galvanic cell in Figure 11.2 against the milliliter of Ce^{4+} titrant added. The graph constructed from the data in Table 11.2 is depicted in Figure 11.3. Now that we have had an introduction to the specific measurement (theoretical) and plotting of a redox titration, let us look at some other aspects involved with redox titrations, such as the appropriate visual indicators.

TABLE 11.2 Calculated Potentials for the Addition of Ce^{4+} Titrant.

Milliliter Ce^{4+} Added	Half-Cell Reaction	Cell Potential (E, V)
10	$[Fe^{2+}]/[Fe^{3+}]$	0.728
20	$[Fe^{2+}]/[Fe^{3+}]$	0.750
30	$[Fe^{2+}]/[Fe^{3+}]$	0.767
40	$[Fe^{2+}]/[Fe^{3+}]$	0.784
50	$[Fe^{2+}]/[Fe^{3+}]$	0.803
60	$[Fe^{2+}]/[Fe^{3+}]$	0.837
61	$[Fe^{2+}]/[Fe^{3+}]$	0.844
62	$[Fe^{2+}]/[Fe^{3+}]$	0.853
63	$[Fe^{2+}]/[Fe^{3+}]$	0.866
64	$[Fe^{2+}]/[Fe^{3+}]$	0.894
64.5	$[Fe^{2+}]/[Fe^{3+}]$	0.984
64.51	$[Fe^{2+}]/[Fe^{3+}]$	1.008
64.51599 (eq pt)	$[Fe^{2+}]/[Fe^{3+}]$	1.106
64.52	$[Ce^{4+}]/[Ce^{3+}]$	1.440
64.6	$[Ce^{4+}]/[Ce^{3+}]$	1.440
70	$[Ce^{4+}]/[Ce^{3+}]$	1.442
80	$[Ce^{4+}]/[Ce^{3+}]$	1.446
90	$[Ce^{4+}]/[Ce^{3+}]$	1.448
100	$[Ce^{4+}]/[Ce^{3+}]$	1.451

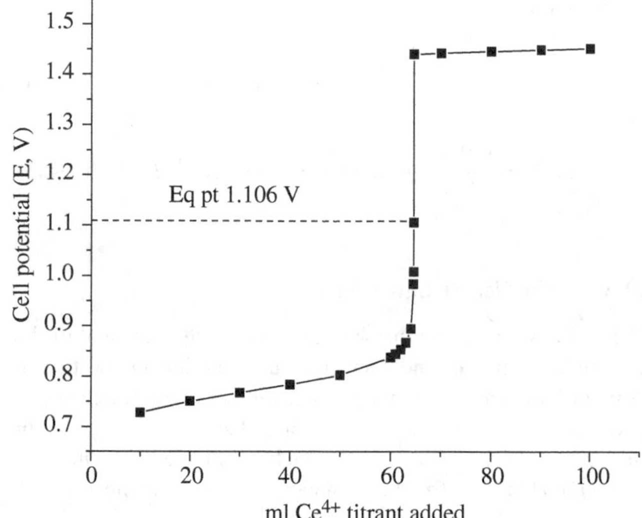

FIGURE 11.3 Potentiometric titration curve of Fe^{2+} with Ce^{4+}.

11.9 VISUAL INDICATORS USED IN REDOX TITRATIONS

Similar to the visual indicators used in acid–base neutralization titrations, the visual indicators used in redox titrations need to possess the attributes of a clear color change at a transition potential within the range of the titration equivalence point. Potassium permanganate ($KMnO_4$), a strong oxidizing agent titrant used in some redox titrations, on one hand can act as its own visual indicator due to its deep blue/purple coloring. In this case, the solution being titrated is clear, and the titrant in the oxidized form of manganese (permanganate ion MnO_4^-) is deeply colored, while the reduced form (Mn^{2+}) is colorless. During the titration, the colored permanganate titrant is reduced and the solution stays clear until the endpoint where a slight excess of the permanganate ion will turn the titrand solution into a faint to slightly strong pink color. The titration of iron(II) (Fe^{2+}) is a common redox titration that employs the permanganate ion as the titrant. The reduction reaction of the permanganate ion is as follows:

$$MnO_4^- + 8H^+ + 5e^- \rightleftharpoons Mn^{2+} + 4H_2O \quad E^o = 1.51\,V. \qquad (11.82)$$

We shall look at the permanganate titrant a little closer in the Redox Titration Reagents and Applications section below.

There are a number of visual redox indicators that undergo redox reactions in potential ranges anywhere from 0.15 to 1.25 V. One of the most commonly used indicators is Tris(1,10-phenanthroline) iron(II) sulfate also known as "ferroin," often associated with Ce(IV) titrations. Also used are diphenylamine and diphenylamine-sulfonate when using dichromate as the titrant. The redox couple reaction for an indicator in general is represented as

$$Ind^+_{(oxidized,\ color\ 1)} + e^- \rightleftharpoons Ind_{(reduced,\ color\ 2)}. \qquad (11.83)$$

The potential equation is thus

$$E = E^o_{Ind} - 0.059 \log \frac{[Ind]}{[Ind^+]}. \qquad (11.84)$$

While calculations using the potentials can be used to decide upon the proper indicator to use, in most instances the visual indicator is outlined in the method, or if not, the choice can almost always be made from the known transition of the titrant/titrand redox couple. For example, from above we know that the equivalence point for the titration of iron(II) with cerium(IV) is at a potential of 1.106 V. From references, we know that the transition of the ferroin visual indicator is at a potential of 1.147 V making it a suitable indicator for the titration.

11.10 PRETITRATION OXIDATION–REDUCTION

In most cases, the species to be titrated may not exist entirely in a single oxidation state. For example, a solution to be titrated for iron content usually consists of a mixture of iron in both the iron(II) (Fe^{2+}) and iron(III) (Fe^{3+}) oxidation states. A way to alleviate this is to either oxidize or reduce the entire iron content to a single oxidation state before the redox titration.

11.10.1 Reducing Agents

For pretitration reduction, this can be done using a metal reductor column where the most commonly used is the Jones reductor, a column that has been packed with granular zinc that has been coated with zinc amalgam, a mixture of zinc and mercury. Zinc amalgam is produced by mixing zinc metal with mercury(II) chloride, where the mercury is reduced to elemental mercury forming an amalgam with the zinc. The reason for coating the zinc with zinc amalgam (Zn(Hg)) is to prevent the reaction of the zinc metal with acids that will produce hydrogen gas (H_2) and release the oxidized metal into the solution. The solution containing the species to be reduced (the titrand) is passed slowly over the Jones reductor column and collected. The column is then washed with a dilute acid solution and the wash is combined with the solution and then titrated.

TABLE 11.3 Some Reduction Products of Metal Species Treated with Either the Silver of Jones Reductors.

	Reduction Product	
Metal species	Silver reductor Ag(HCl)	Jones reductor Zn(Hg)
Vanadium(V) V^{5+}	V^{4+}	V^{2+}
Molybdenum(VI) Mo^{6+}	Mo^{5+}	Mo^{3+}
Copper(II) Cu^{2+}	Cu^{1+}	Cu^0
Chromium(III) Cr^{3+}	Not reduced	Cr^{2+}
Titanium(IV) Ti^{4+}	Not reduced	Ti^{3+}
Iron(III) Fe^{3+}	Fe^{2+}	Fe^{2+}
Uranium(VI) U^{6+}	U^{4+}	$U^{3+} + U^{4+}$

An alternative column used for reducing species before redox titration is the silver reductor, sometimes used in place of the Jones reductor. A comparison of some metal species and their oxidation/reduction states using either the Jones or silver reductor columns is listed in Table 11.3.

When using these columns, blanks should often be run to exclude excess metal present or interferences from the reductor columns. This is usually done by passing an acidic solution, approximately 100–200 ml of either 1 M H_2SO_4 or 1 M HCl, and titrating the acidic wash solution. The acidic wash solution should not require more than 0.05 ml of titrant to change the respective indicator used.

There are alternatives to the preparation and use of reductor columns for lowering the oxidation state of a metal before titration. For example, stannous chloride (tin(II) chloride, $SnCl_2$) is sometimes used for reducing iron(III) to iron(II) in acidic solution. Gases such as hydrogen sulfide (H_2S) and sulfur dioxide are also sometimes employed to reduce metals by bubbling the gas through an acidic solution containing the metal species of interest for reduction.

11.10.2 Oxidizing Agents

There are a number of oxidizing agents that are used to increase the oxidation state of a particular metal species in preparation for redox titration. The most commonly used are listed in Table 11.4 along with the metal species and oxidation products. Most are used in acidic solution with the exception of oxidation of manganese and chromium with hydrogen peroxide, which is performed in a basic solution (alkaline pH).

TABLE 11.4 Some Oxidizing Agents and the Metal Species Oxidation State Change with Treatment for Pretitration Oxidation (in Acidic Solution Unless Otherwise Denoted).

Oxidizing Agent	Metal Species	Oxidation Product
Potassium periodate (KIO_4)	Manganese(II) Mn^{2+}	Mn^{7+} (as MnO_4^-)
Peroxodisulfates ($S_2O_8^{2-}$) (trace Ag^+ present)	Iron(II) Fe^{2+}	Fe^{3+}
	Cerium(III) Ce^{3+}	Ce^{4+}
	Manganese(II) Mn^{2+}	Mn^{7+} (as MnO_4^-)
	Chromium(II) Cr^{2+}	Cr^{6+} (as $Cr_2O_7^{2-}$)
Sodium bismuthate ($NaBiO_3$)	Cerium(III) Ce^{3+}	Ce^{4+}
	Manganese(II) Mn^{2+}	Mn^{7+} (as MnO_4^-)
	Chromium(III) Cr^{3+}	Cr^{6+} (as $Cr_2O_7^{2-}$)
Hydrogen peroxide (H_2O_2)	Iron(II) Fe^{2+}	Fe^{3+}
	Tin(II) Sn^{2+}	Sn^{5+}
	Manganese(II) Mn^{2+} (basic)	Mn^{4+} (as MnO_2)
	Chromium(III) Cr^{3+} (basic)	Cr^{6+} (as CrO_4^{2-})
Bromine (Br_2)	Manganese(II) Mn^{2+} (NH_3 solution)	Mn^{4+} (as MnO_2)

11.11 ION-SELECTIVE ELECTRODES

The potentiometric titration methodology that we just studied measures the potential of a cell with changing concentrations of the redox couples present, and is therefore a titration method. A second method used that measures the potential of a cell at one concentration point is known as *direct potentiometry*. The effective calibration of the indicator electrode allows the direct measurement of the concentration of the species of interest. An example of this is the pH measurements using a pH electrode depicted in Figure 11.4.

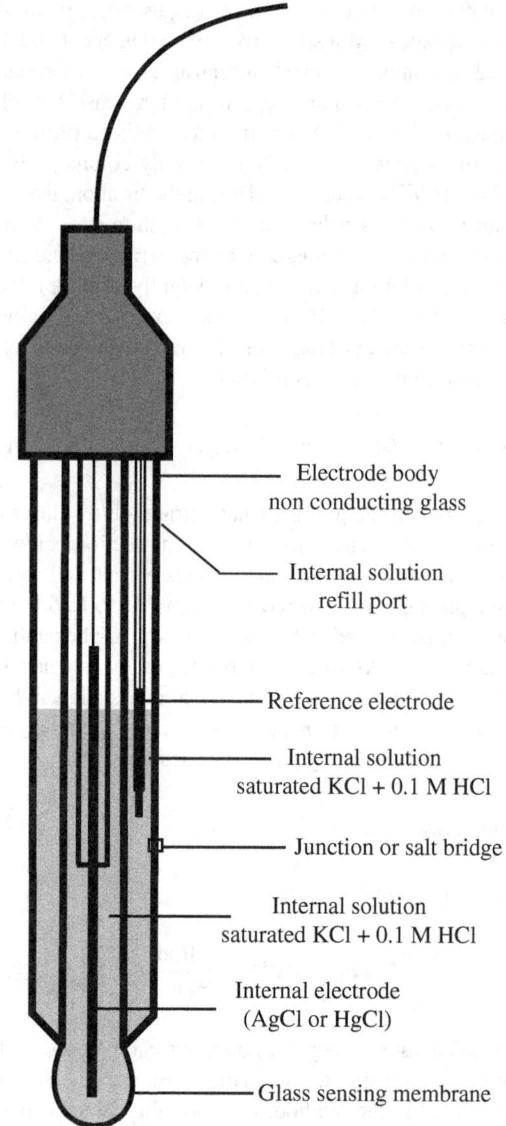

FIGURE 11.4 pH meter glass electrode as an example of an ion-selective electrode.

11.12 CHAPTER KEY CONCEPTS

11.1 Oxidation–reduction reactions have a definition for each half: *oxidation* is the process where an atom, ion, or molecule has lost one or more electrons, while *reduction* is the process where an atom, ion, or molecule has gained one or more electrons.

11.2 Oxidation–reduction reactions (sometimes shortened to *redox* reactions) are taking place in solution (as far as we are concerned here) where there are no free electrons; thus, one must take place with the other.

11.3 An *oxidizing agent* is a species that is able to accept electrons and thus becomes reduced.

11.4 A reducing agent is a species that is able to donate electrons and thus becomes oxidized.

11.5 The oxidation reaction and the reduction reaction are each called half-reactions of the entire redox process. The sum of the two half-reactions represents the redox reaction taking place.

$$A_{ox} + ne^- \rightarrow A_{red}$$

$$B_{red} \rightarrow B_{ox} + ne^-$$

$$A_{ox} + B_{red} \rightarrow A_{red} + B_{ox}$$

11.6 Redox reactions result in a flow of electrons that produces a current which can be measured with a galvanometer (voltage measuring device) when the redox reactions are set up as a circuit.

11.7 A galvanic cell is a system containing a spontaneous chemical reaction that releases electrical energy to perform work.

11.8 A volt is the required emf to give one joule (J) of energy to a one coulomb (C) electrical charge.

11.9 An electrochemical cell is an example of a galvanic cell that is composed of a two-electrode system where oxidation is taking place at one electrode and reduction is taking place at the second electrode.

11.10 The voltage as the potential difference between the two cells is measured. The potential difference of the cell can also be calculated (instead of measured) using electrode potential tables.

11.11 The potential of a half-reaction cannot be measured by itself. Instead, a reference electrode is set up with the electrode of interest to measure the potential.

11.12 The reference electrode used is the SHE that has been assigned a value of exactly zero for the standard potential ($E^o = 0.00$ V) for a hydrogen gas (H_2) pressure of 1 atm and a hydrogen ion (H^+) activity in solution at unity (all standard potentials are measured and assigned at a concentration of 1 *M*).

11.13 A number of conventions are used when drawing a galvanic cell such as that shown in Figure 11.1. The flow of electrons is drawn from left to right where the reaction taking place at the left electrode is oxidation, and the electrode on the right is reduction.

11.14 The electrode on the left has a negative polarity and is called the anode, while the electrode on the right has a positive polarity and is called the cathode.

11.15 The potential of the cell is calculated as the difference of the right cell potential from the left cell potential.

$$B_{ox\,E_l}|B_{red}(1\,M)|\;|A_{ox}(1\,M)|_{E_r}A_{red}.$$

11.16 The *Nernst* equation describes the relationship of the potential of a galvanic cell upon the activities (or as an approximate the molar *M* concentrations) of the redox species.

$$E = E^o + \frac{0.059}{n}\log\frac{[A_{ox}]}{[A_{red}]}.$$

11.17 The potential of the cell is then calculated as the difference of the right cell potential from the left cell potential.

$$E_{cell} = E_r - E_l.$$

11.18 We can also calculate the equilibrium constant (*K*) for the reaction couple from the two electrode standard potentials. At equilibrium, the potential of the cell will equal zero, $E_{cell} = 0$ V. Thus, we have the following relationship:

$$K = \text{antilog}\left(\frac{nE^o_{cell}}{0.059}\right).$$

11.19 There are two primary ways of determining a redox titration endpoint in the analytical laboratory by the analyst that includes the use of a potentiometer where the data gathered are plotted, or using a visual indicator.

11.13 CHAPTER PROBLEMS

11.1 Write the standard voltaic cell representation for each of the following shown as "anode, cathode."

a. K^+/K, $Br_2(aq)/Br^-$
b. Co^{2+}/Co, Cu^{2+}/Cu
c. Na^+/Na, NO_2^-/N_2O_4
d. Mg^{2+}/Mg, Au^{3+}/Au
e. I_2/I^-, Zn^{2+}/Zn

11.2 Write the half-reaction and cell reaction for the standard cells in Problem 11.2.

11.3 Calculate the cell voltage for the cell reactions in Problem 11.2. Which are spontaneous?

11.4 Balance the following reactions and write the electrochemical cell for each of the reactions below.

a. $Ni + Ag^+ \rightarrow Ni^{2+} + Ag$
b. $Ca + Cl_2 \rightarrow Ca^{2+} + 2Cl^-$
c. $Au^{3+} + Ag \rightarrow Au + Ag^+$
d. $Pb^{2+} + Fe^{2+} \rightarrow Pb + Fe^{3+}$
e. $Sn^{4+} + Br^- \rightarrow Sn^{2+} + Br_2$

11.5 Write the half-reactions for the cells in Problem 11.4.

11.6 Designate the anode and cathode reactions in Problem 11.5, and calculate the cell voltages for each cell reaction? Are they spontaneous?

11.7 For each of the following electrodes, calculate the potential.

a. $Ni^0(s)/Ni^{2+}$ (0.15 M)
b. $Ca^0(s)/Ca^{2+}$ (0.04 M)
c. $Au^0(s)/Au^{3+}$ (5.5×10^{-5} M)
d. $Pb^0(s)/Pb^{2+}$ (0.250 M)
e. $Sn^0(s)/Sn^{2+}$ (4.25×10^{-3} M)

11.8 For each of the following, calculate the molar concentration of the ionic species.

a. $Ag^+ + e^- \rightarrow Ag^o(s)$ ($E^o = 0.7996$ V, $E = 0.765$ V)
b. $Tl^{3+} + 3e^- \rightarrow Tl(s)$ ($E_0 = 0.72$, $E = 0.67$)
c. $Pd^{2+} + 2e^- \rightarrow Pd(s)$ ($E_0 = 0.915$, $E = 0.884$)
d. $Fe^{3+} + e^- \rightarrow$
Fe^{2+} ($E_0 = 0.77$, $E = 0.71$, $[Fe^{2+}] = 1.6 \times 10^{-4}$M
e. $[AuI_2]^- + e^- \rightarrow$
$Au(s) + 2I^-$ ($E_0 = 0.58$, $E = 0.53$, $[I^-] = 0.0015$

11.9 Calculate the potential of the cell due to the effect of the concentrations of the species present according to the following cell. Is the concentration effect significant?

$$\mathrm{Pt}|\mathrm{Cu}^+\ (0.125\,M) + \mathrm{Cu}^{2+}\,(0.050\,M)|_{\mathrm{E_l}}$$
$$|\mathrm{Fe}^{3+}\,(0.11\,M) + \mathrm{Fe}^{2+}\,(0.055\,M)|_{\mathrm{E_r}}\ \mathrm{Pt}$$

11.10 Calculate the potential of the cell due to the effect of the concentrations of the species present according to the following cell. Is the concentration effect significant?

$$\mathrm{Pt}|\mathrm{Cr}^{2+}\ (0.0035\,M) + \mathrm{Cr}^{3+}\,(0.150\,M)|_{\mathrm{E_l}}$$
$$|\mathrm{Hg}_2^{3+}\,(0.025\,M) + \mathrm{Hg}\,(0.055\,M)|_{\mathrm{E_r}}\ \mathrm{Pt}$$

11.11 Calculate the potential of the cell due to the effect of the concentrations of the species present according to the following cell. Is the concentration effect significant?

$$\mathrm{Pt}|\mathrm{Ag}^{2+}\ (0.115\,M) + \mathrm{HCN}\,(0.075\,M) + \mathrm{H}^+\ (0.088\,\mathrm{M})|_{\mathrm{E_l}}$$
$$|\mathrm{Ag}^+\ (0.066\,\mathrm{M}) + \mathrm{Fe(CN)}_6^{4-}\,(0.045\,M)|_{\mathrm{E_r}}\ \mathrm{Pt}$$

11.12 Calculate the equilibrium constant, *K*, for the cells in Problems (a) 9, (b) 10, and (c) 11.

11.13 An elemental zinc strip is placed into a solution of copper sulfate. When this is done, a reaction will take place. At equilibrium, what will be the relative concentrations of the Zn^{2+} and the Cu^{2+}?

$$\mathrm{Zn} + \mathrm{Cu}^{2+} \rightarrow \mathrm{Zn}^{2+} + \mathrm{Cu}$$

11.14 For the following reaction:

$$2\mathrm{Au}^{3+} + 3\mathrm{Ni} \rightarrow 2\mathrm{Au} + 3\mathrm{Ni}^{2+}$$

a. Calculate the equilibrium constant K.
b. What is the expression used for K?
c. What is the concentration of Ni^{2+} at equilibrium if the concentration of Au^{3+} is 0.55 M?

11.15 The following voltaic cell, at a particular point in time, has the following concentrations as determined in the laboratory:

$$\mathrm{Pt}|\mathrm{V}^{2+}\ (0.15\,M) + \mathrm{V}^{3+}\,(0.025\,M)|_{\mathrm{E_l}}$$
$$|\mathrm{Hg}^{2+}\,(0.25\,M) + \mathrm{Hg}_2^{2+}\,(0.055\,M)|_{\mathrm{E_r}}\ \mathrm{Pt}$$

a. Calculate the cell potential.
b. Calculate the equilibrium constant K.

11.16 A 0.5113 g sample of an iron metal standard, dissolved in acid and reduced (as Fe^{2+}), is titrated with a laboratory solution of permanganate (MnO_4^-). If the titration requires 35.2 ml of the permanganate titrant, what is the normality of the permanganate solution? The titration reaction is as follows:

$$\mathrm{MnO}_4^- + 5\mathrm{Fe}^{2+} + 8\mathrm{H}^+ \rightarrow 5\mathrm{Fe}^{3+} + \mathrm{Mn}^{2+} + 4\mathrm{H_2O}$$

11.17 A 0.3251 g sample of an iodine standard is titrated with a laboratory solution of arsenous acid (H_3AsO_3). If the titration requires 25.9 ml of the arsenous acid titrant, (a) what is the molarity of the arsenous acid solution? (b) What are the half-reactions and potentials? (c) What is the change in oxidation number for the arsenic? (d) Why at a pH <4 is the reaction reversible?

$$\mathrm{H_3AsO_3} + \mathrm{I_2} + \mathrm{H_2O} \rightarrow \mathrm{H_3AsO_4} + 2\mathrm{I}^- + 2\mathrm{H}^+$$

11.18 A titrant solution is made by weighing 2.005 g of a primary potassium dichromate ($K_2Cr_2O_7$) standard into a 100 ml volumetric flask and brought to volume with DI water. An iron ore solid sample containing iron (siderite, $FeCO_3$, as ferrous, Fe^{2+}) was taken and 2.901 g was acid-digested and brought to volume with DI water in a 100 ml volumetric flask. A 10 ml aliquot was pipetted into a 250 ml Erlenmeyer flask. 10 ml of 1 M H_2SO_4, 10 ml of 1 M H_3PO_4, and 5 drops of sodium diphenylamine sulfonate solution (indicator) were added. The solution required 4.3 ml of the standard potassium dichromate to endpoint titration. The overall reaction for the titration is as follows:

$$\mathrm{K_2Cr_2O_7} + 6\mathrm{Fe(NH_4)_2(SO_4)_2} + 7\mathrm{H_2SO_4} \rightarrow$$
$$3\mathrm{Fe_2(SO_4)_3} + \mathrm{Cr_2(SO_4)_3} + \mathrm{K_2SO_4}$$
$$+ 6\mathrm{(NH_4)_2SO_4} + 7\mathrm{H_2O}$$

a. What are the half-redox reactions, and what is the combined reaction?
b. What is the molarity of the potassium dichromate ($K_2Cr_2O_7$) titrant?
c. What is the molarity of the unknown iron (as ferrous, Fe^{2+}) solution?
d. What is the percentage of iron (as ferrous, Fe^{2+}) in the original iron ore sample?
e. What is the percentage of siderite ($FeCO_3$) in the ore sample?

12

LABORATORY INFORMATION MANAGEMENT SYSTEM (LIMS)

12.1 Introduction
12.2 LIMS Main Menu
12.3 Logging in Samples
12.4 Entering Test Results
12.5 Add or Delete Tests
12.6 Calculations and Curves
12.7 Search Wizards
12.7.1 Searching Archived Samples
12.7.2 General Search
12.7.3 Viewing Current Open Samples
12.8 Approving Samples
12.9 Printing Sample Reports

12.1 INTRODUCTION

This chapter covers a program that is utilized in most laboratories today, the Laboratory Information Management System (or Software) abbreviated as "LIMS." LIMS are used to input sample data as they are collected in the laboratory, track sample progress, record sample data such as company, type, and tests needed, and so on. We will now be introduced to a LIMS example that we will use to log in samples, input data, search samples, approve samples, and print reports and certificates of analysis (C of A). Also becoming more common in laboratories are electronic laboratory notebooks that are often coupled with LIMS, and we will take a brief look at using them.

12.2 LIMS MAIN MENU

Let us start by opening the LIMS Main Menu. Go to the Chem-Tech Main Menu and click on the Chapter 12 button. This will bring up the ACME Labs LIMS Main Menu as depicted in Figure 12.1. The Main Menu contains a set of buttons as links to Log in Samples, Enter Test Results, Calculations and Curves, Search Wizards, Approve Samples, and Reports and C of A's. We will spend time on each of these sections to learn to get familiar with working with a LIMS system. Even though there are many types of LIMS systems used in laboratories, getting used to working with one will help prepare the chemist or technician for any particular LIMS used in the Lab.

12.3 LOGGING IN SAMPLES

Logging in samples that have come into the laboratory is a common use of LIMS. Click on the "Log In Samples" button to open the page shown in Figure 12.2.

For the Lab No. enter the value 101 and press the Tab key to move to the next field, Sample Date. For the sample date, click on the calendar to the right of the field and choose today's date and press the Tab key to move to the next field, Company. For company, enter ACME and press the Tab key. For commodity, enter corn and press the Tab key. For testing code, enter "1234" and press the Tab key. Finally, enter your or any name into the "Sampled By" field. The Login form should now look something like that in Figure 12.3 (except for the date and name). Click the "Enter Current Sample" button to place this new sample into the sample database. Next click the "Return To Main Menu" button to return to the main menu.

12.4 ENTERING TEST RESULTS

LIMS systems are used in the laboratory to transcribe data that have been collected while working on samples. In the main menu,

Analytical Chemistry: A Chemist and Laboratory Technician's Toolkit, First Edition. Bryan M. Ham and Aihui MaHam.

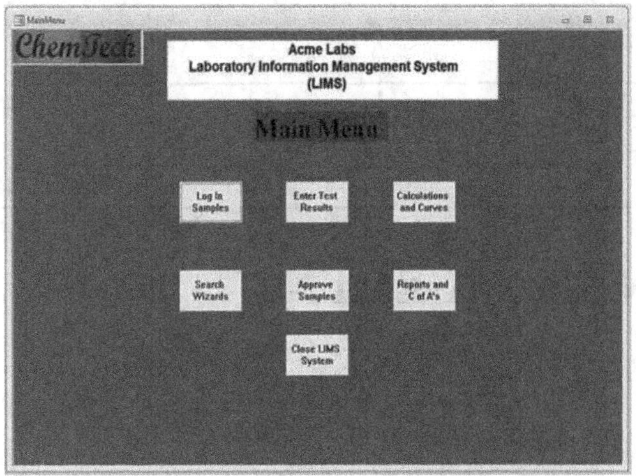

FIGURE 12.1 ACME Labs LIMS Main Menu.

click the "Enter Test Results" button to open up a page as shown in Figure 12.4. Your actual page may not look exactly as the one shown in Figure 12.4, but it will contain a number of samples including the last sample which we just logged in (Lab No. 101). In this page, if we click on the laboratory number the form to the right will accept values for the sample tests. For the first input field, Flash Point there is listed to the right a minimum and a typical value. If an out-of-range value is entered, a warning message is given. Go ahead and enter 200 into the Flash Point field and press Enter. A message box appears instructing the analyst that the sample is out of specifications and supervision should be consulted. The message box is depicted in Figure 12.5. Press OK to close the message box and change the Flash Point value to 250. Fill in the rest of the test result boxes. The input page should look like that shown in Figure 12.6. Let us close the input page by clicking the exit (close door) box at the bottom of the form. This will return us to the Main Menu page.

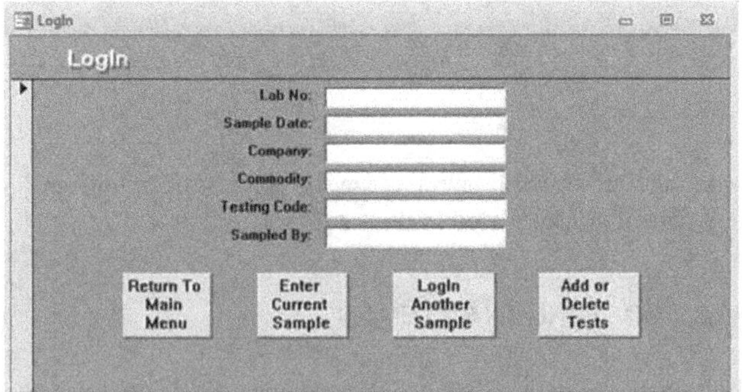

FIGURE 12.2 Sample log in page.

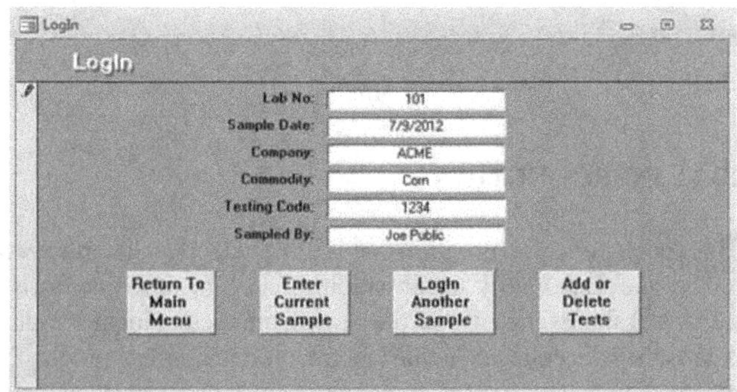

FIGURE 12.3 The Sample Log page filled in.

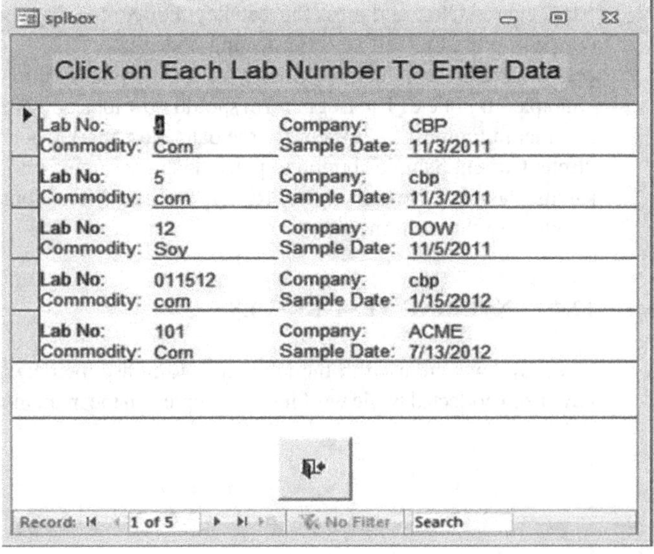

inputform

	Value	Min	Max	Typ
Flash Point	250	250		275
Halphen Test	Neg			Neg
Sap Value	194	188	194	191
Unsap	1.3		1.3	
FFA, as Oleic	2.300		2.3	
Moist and Vol	0.5		0.5	
Insol Impur	0.3		0.3	
Color	2.5		2.5	
Linolenic Acid	1		1.0	
Moist and Imp	2	0		
Peroxide Valu	3	0		
Linoleic Acid	4	0		
PUFA	5	0		

FIGURE 12.4 Enter Test Results page.

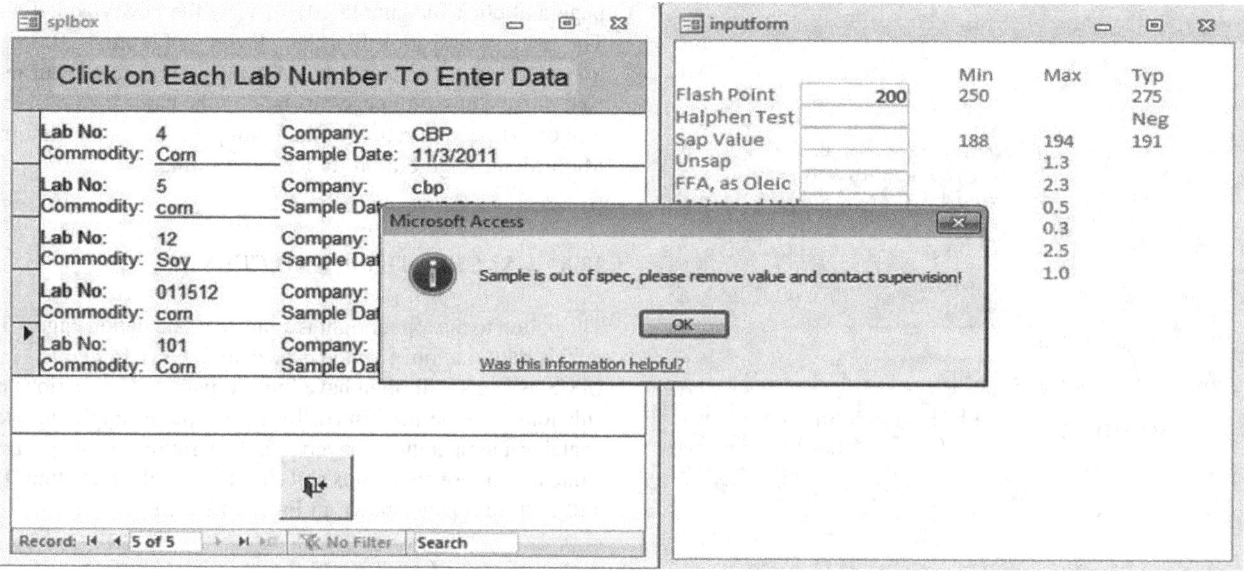

FIGURE 12.5 Out of specification (Spec) warning window.

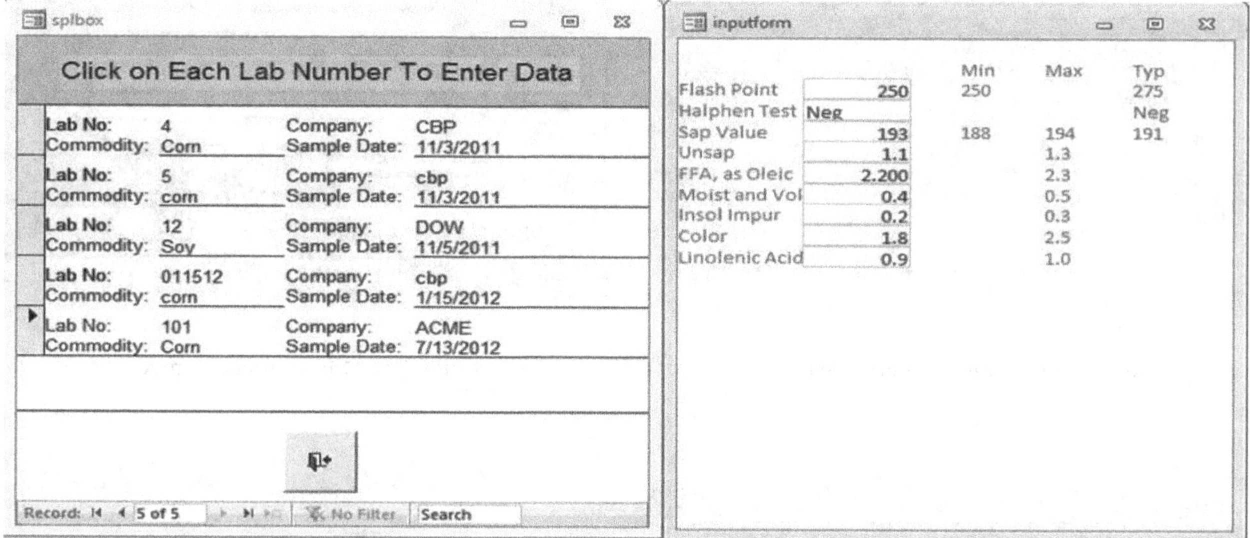

FIGURE 12.6 Enter Test Results page filled out.

12.5 ADD OR DELETE TESTS

In the Main Menu, click the Log In Samples button to open up the page that is shown in Figure 12.2. At the bottom right is the "Add or Delete Tests" button. We can use this to both add tests to and delete tests from samples that have been logged in. Press the button to open up the page shown in Figure 12.7. The page contains 13 tests that can be either added or deleted. The 13 tests can be viewed one by one using the Record scroll bar located at the bottom of the form. Type the laboratory number 101 into the Lab No. field. The test that is illustrated as the first test is the Flash Point as seen in the Tests field. This test already exists for sample 101, so if we try to add it we get the message box depicted in Figure 12.8. There are nine tests that have been automatically assigned to sample 101 when it was logged in ending with Linolenic Acid. Scroll the tests over using the bottom Record scroll bar to test number 10 which is "Moist and Impur," which stands for moisture and impurities. Click the "Add Test" button to add this test to the sample. Scroll over to the next test 11 "Peroxide Value" and click the Add Test button. Do the same for test 12 "Linoleic Acid" and test 13 "PUFA," which stands for polyunsaturated fatty acids. Press the Return To Login button and then press the Return To Main Menu button. Let us go back to the Enter Test Results page and click on sample 101 to view the newly added tests. The page should look like that shown in Figure 12.9. Let us go ahead and fill in test results for the newly added four tests. Exit the test entering page by clicking the close box at the bottom of the page to return to the login page. Press the Return to Main Menu Page button to close the form.

12.6 CALCULATIONS AND CURVES

The button to the top far right is a link to a calculation page. Click on this button to open up the page that is shown in Figure 12.10. LIMS systems will often have links to pages where various calculations may be performed. In this simple example, the page contains a temperature converter and a titration calculator. Enter 1 into the choose option box and click the calculation button. This will pull up a page shown in Figure 12.11. Enter 78.6 into the

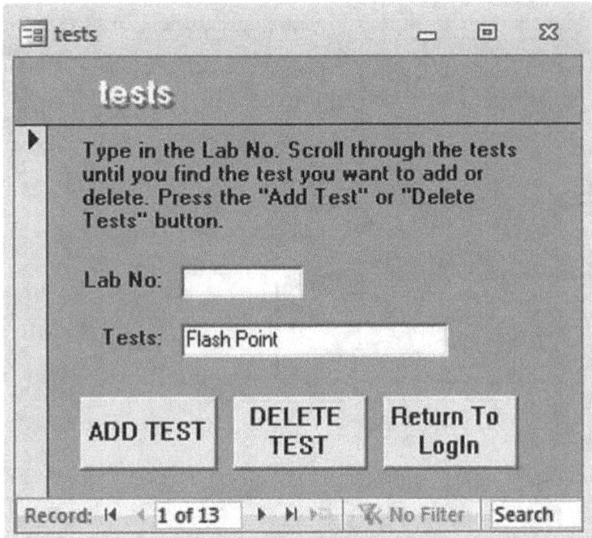

FIGURE 12.7 Add or Delete Tests page.

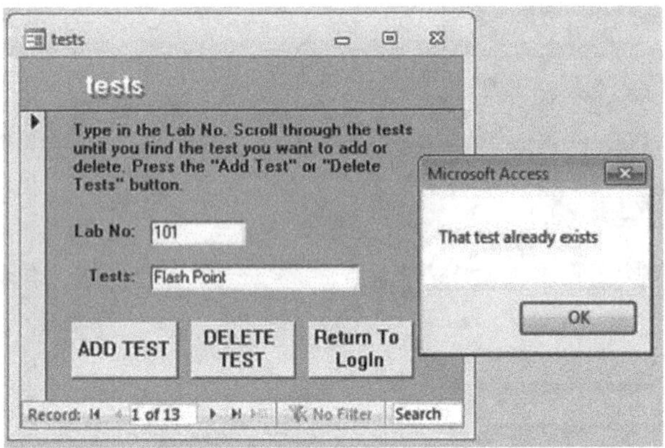

FIGURE 12.8 Existing test message box.

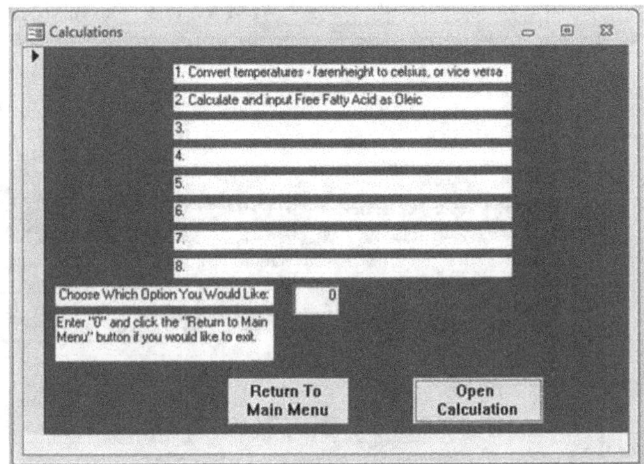

FIGURE 12.10 Calculation page.

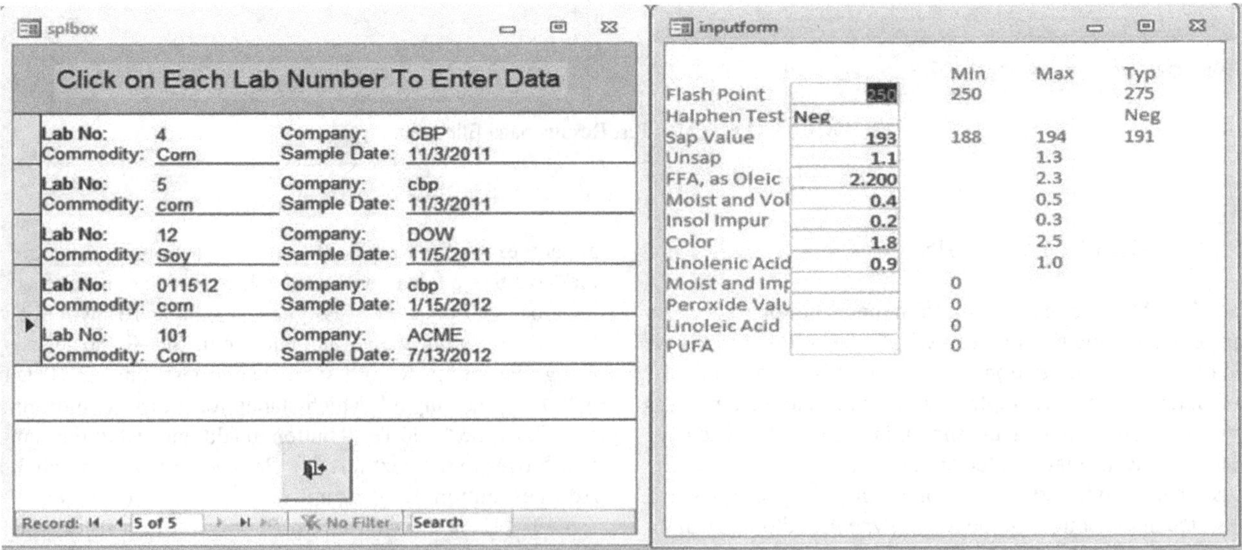

FIGURE 12.9 Sample 101 with newly added tests.

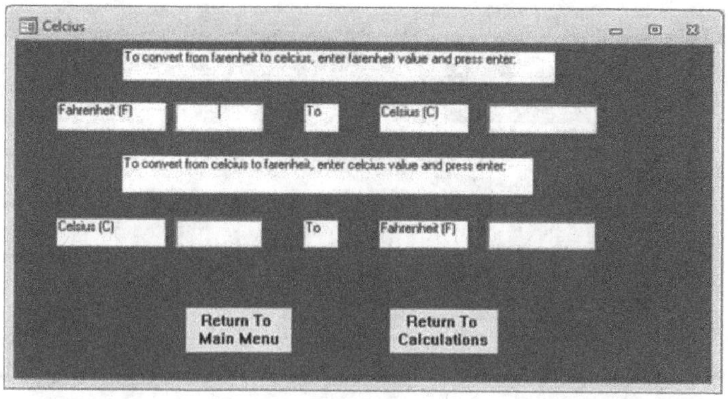

FIGURE 12.11 Temperature converter calculation page.

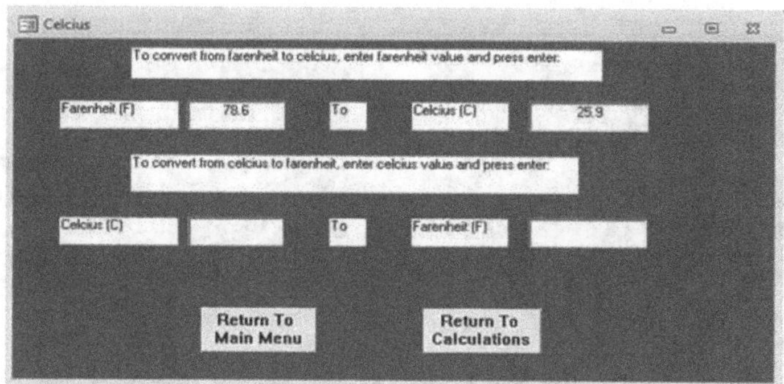

FIGURE 12.12 Celsius as 25.9 in the Celsius result box.

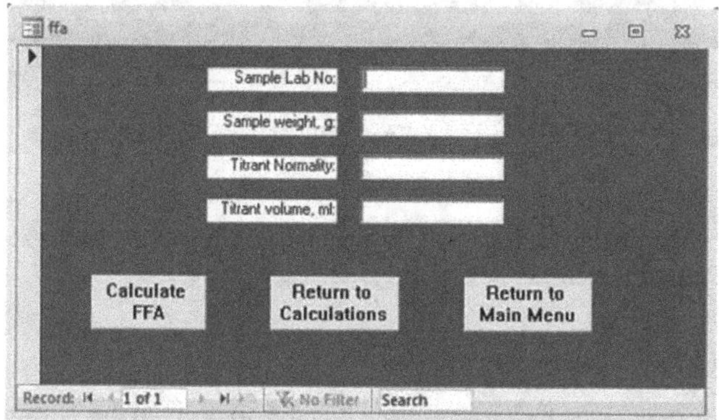

FIGURE 12.13 Free Fatty Acid as oleic acid titration data input page.

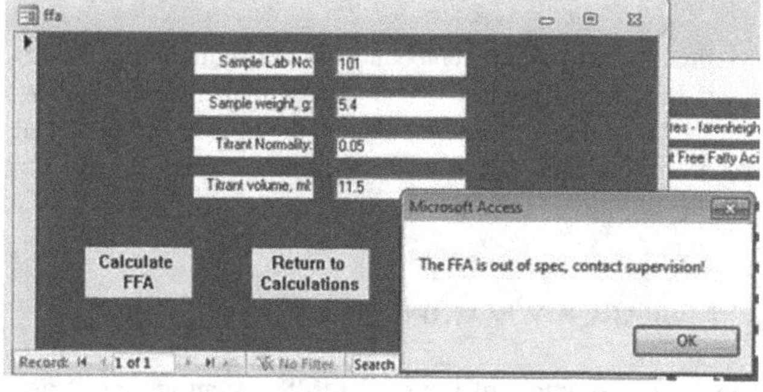

FIGURE 12.14 Titration free fatty acid out of specification message box page.

Fahrenheit input box and press enter. The value for Celsius as 25.9 should be in the Celsius result box, as depicted in Figure 12.12. Press the Return to Calculations page, enter 2 into the option box and click the calculation button. A page will open up like the one depicted in Figure 12.13. This is a page that will take data obtained from a titration and will calculate a sample's free fatty acid (FFA) as oleic acid result. For sample number, type 101 into the input box. For sample weight, 5.4; for titrant normality, 0.05; and for titrant volume, 11.5. Press the Calculate FFA button. A warning page should pop up instructing that the FFA value is out of specification. The form and message box should look like that depicted in Figure 12.14. Press the OK button to close the message box, remove the 11.5 titrant volume amount and replace with 5.5. Now press the Calculate FFA button to get the next message box depicted in Figure 12.15. Now a message box opens with the calculated FFA value of 1.44% and a message asking if the analyst would like to add that test result to the test list for sample number 101. Click the OK button

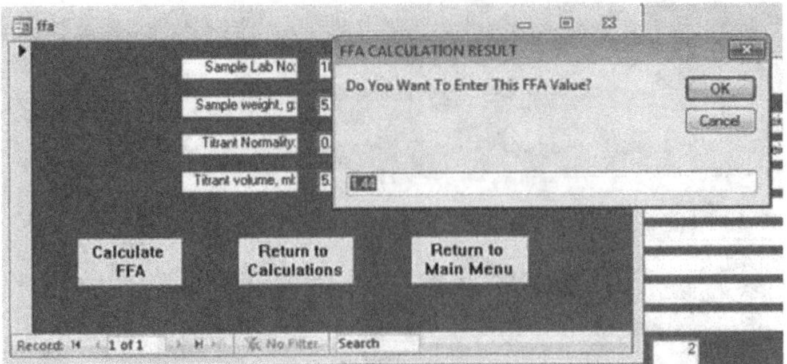

FIGURE 12.15 Message box opens the calculated FFA value of 1.44% and a message asking if the analyst would like to add that test result to the test list for sample number 101.

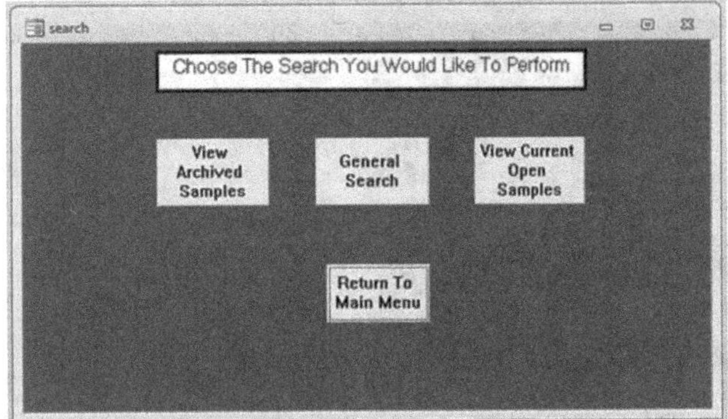

FIGURE 12.16 The Search Wizards page.

and the Return to Main Menu button. In the Main Menu, click the Enter Tests Results button and click on sample number 101. We can see that 1.44 has been entered into the FFA test result. LIMS can be used for calculations and also imputing data automatically into the sample test results.

12.7 SEARCH WIZARDS

Next in our example of a typical LIMS system are the Search Wizards. We can use them to look at open samples and samples that have been closed. Click on the Search Wizards button to open up the page that is depicted in Figure 12.16. In this page, there are three choices for searching: there is searching archived samples, a general search, and then viewing the current open samples.

12.7.1 Searching Archived Samples

Let us start with the archive sample search. Archived samples are samples that have been completed and placed into an archive database. Press the View Archived Samples button to open up the page shown in Figure 12.17. Type into the Lab No. input box the sample number "4" and click the Find Sample button. The Search Result page is shown in Figure 12.18. The page

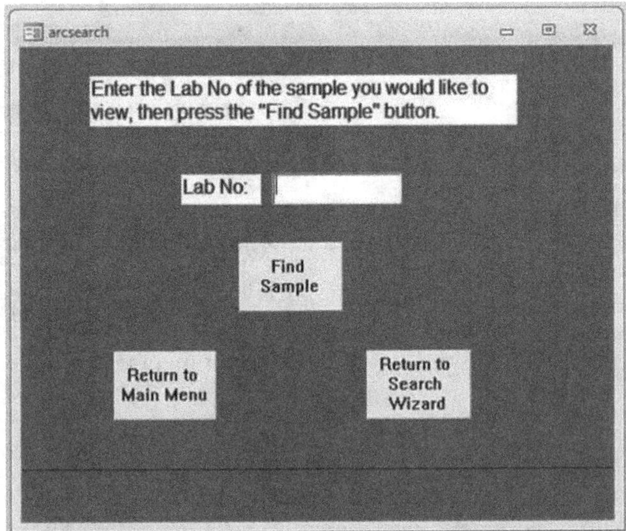

FIGURE 12.17 View Archived Samples page.

can be printed for reference if wanted. Click the Return to Search Wizard button to return to the search main page.

12.7.2 General Search

Next, click on the General Search button to open up the general search page shown in Figure 12.19. In this search page, we have a number of options for searching the database for samples. The search can be performed with a laboratory number, a sample date, or sample descriptions such as the commodity (e.g., corn). Let us go ahead and enter the search term "corn" and press the Perform Search button. A page similar to that depicted in Figure 12.20 should be displayed. All of the samples associated with corn are listed in the table to the right. Click the form button on the left for View Test Data. A sample Lab No input page will open as depicted in Figure 12.21. Enter sample number 102 into the Lab No input box and click the Find Sample button. A sample display page will be brought up as similar to the one depicted in Figure 12.22. At this point, we can print the search results or return to one of the other menus.

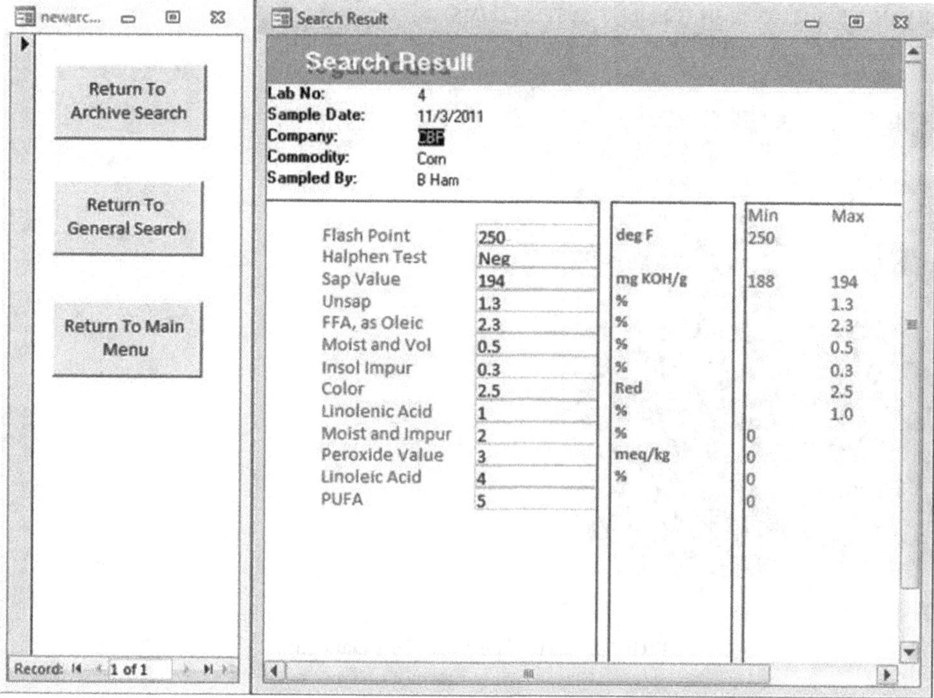

FIGURE 12.18 Archived sample search results page.

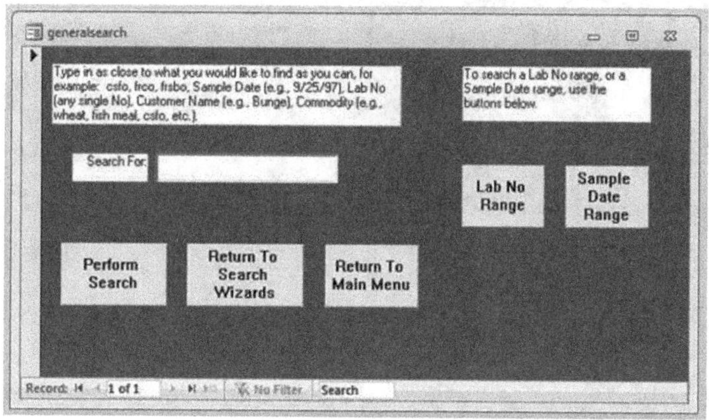

FIGURE 12.19 General Search Page.

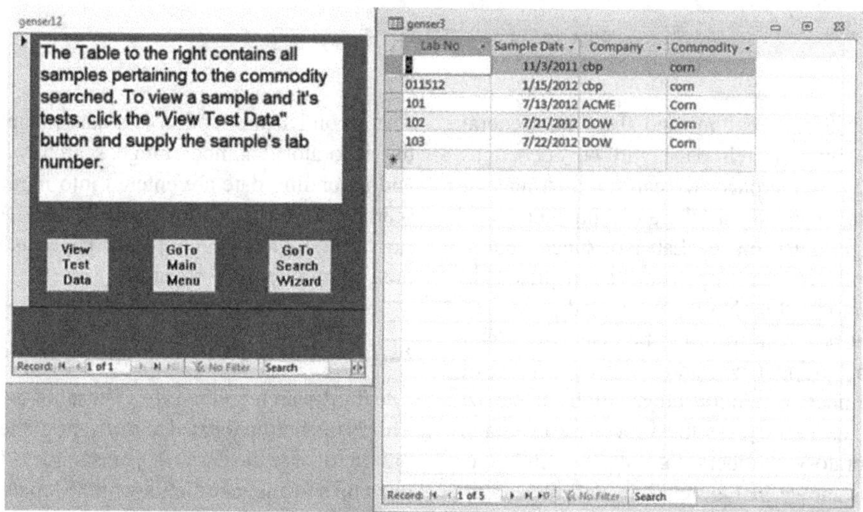

FIGURE 12.20 General Search Result page for searching "corn."

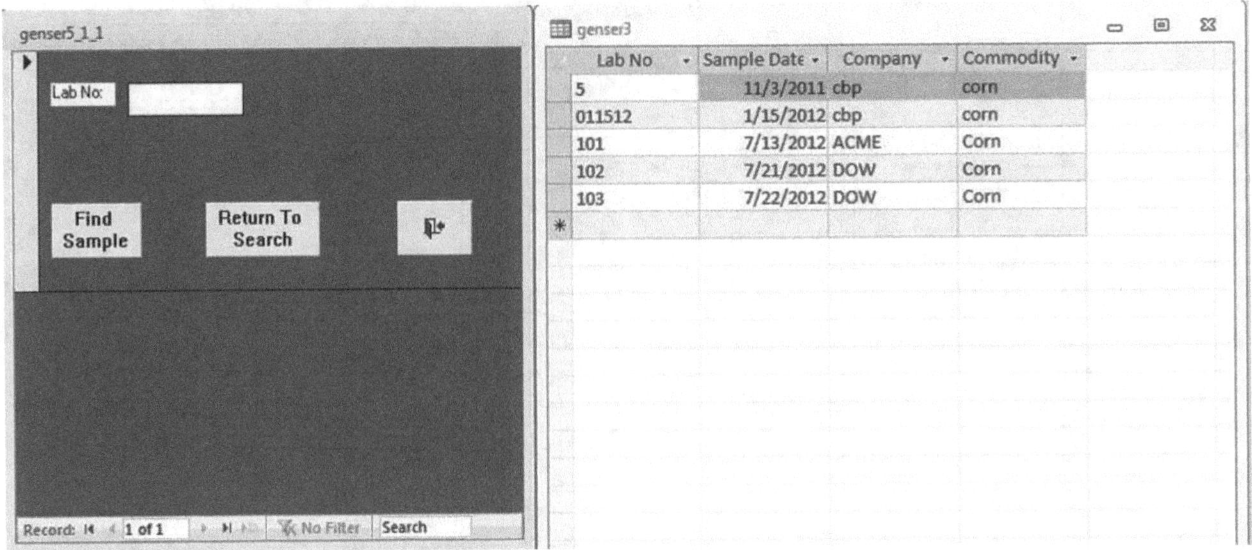

FIGURE 12.21 The View Test Data page.

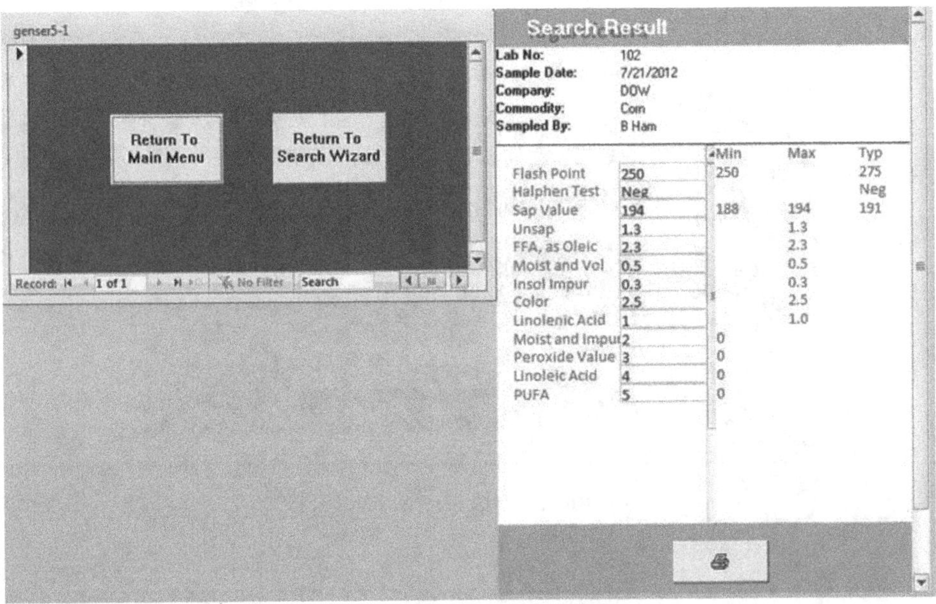

FIGURE 12.22 Sample display page.

Press the Return To Search button, and then the General Search button. The General Search page will be opened as depicted in Figure 12.23. In the page, we can see a number of search options including searching a laboratory number range and a sample date range. Click on the Lab No Range button. A laboratory no input box will open up as depicted in Figure 12.24. Enter 1 into the Lab No input box and press OK. A second Lab No input box will open and be displayed. Enter 500 into the Final Lab No input box and press OK. A list of the found laboratory numbers within this range will be displayed as shown in Figure 12.25. Next, press the View Test Data and enter one of the laboratory numbers into the search page. A search result page as shown in Figure 12.26 will be shown.

The second search option is a date range search. It is similar to the laboratory number range search where a beginning date and an ending date are entered into input boxes. Go ahead and search a date range from January 1,2010 to a current date and run through the different results pages as we just did for the laboratory number range search.

12.7.3 Viewing Current Open Samples

The third search option is to view all of the current open samples. This is the button located to the right of the search page. Click the button to view the current open samples. The search page should open up a result page similar to that depicted in Figure 12.27. The

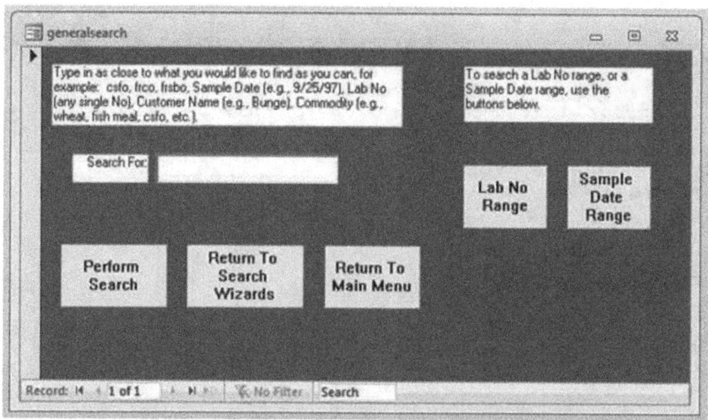

FIGURE 12.23 General Search page.

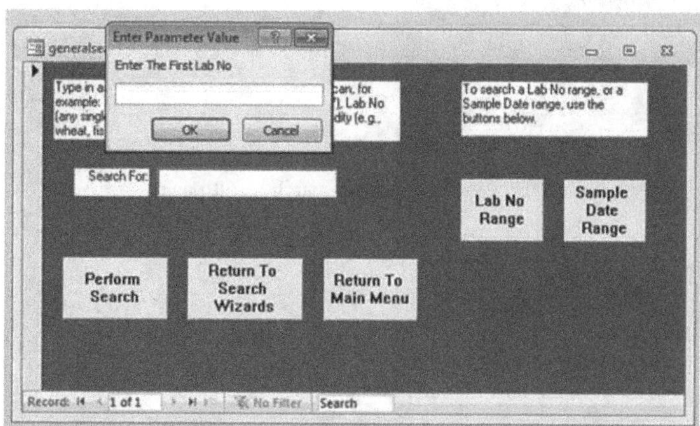

FIGURE 12.24 Input box for Lab Number search range.

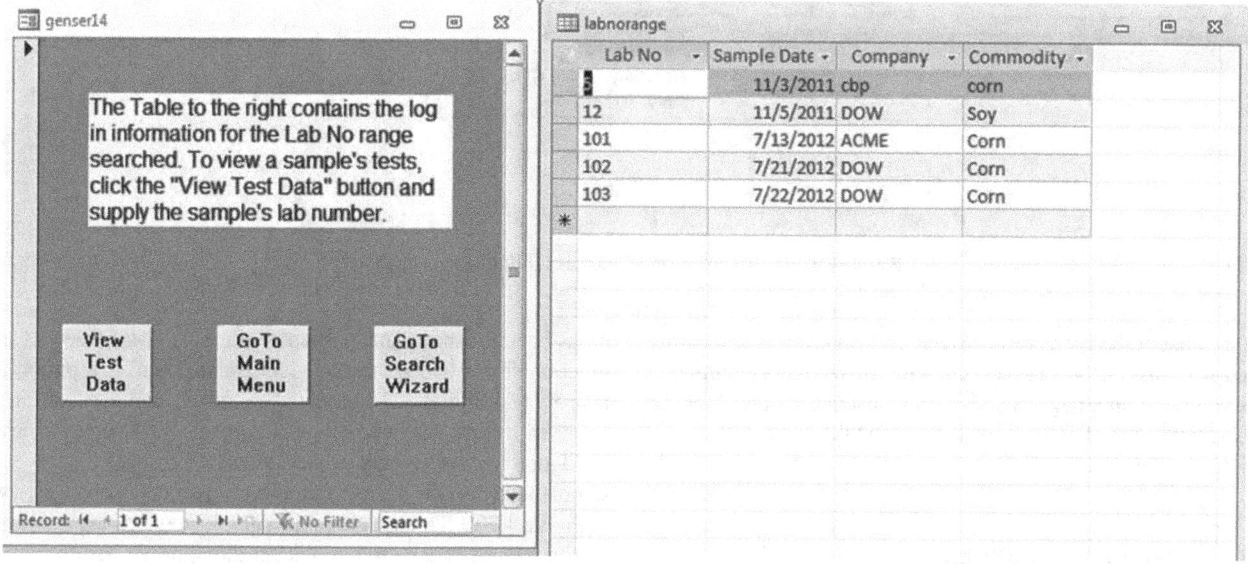

FIGURE 12.25 Table of laboratory numbers found within the search range.

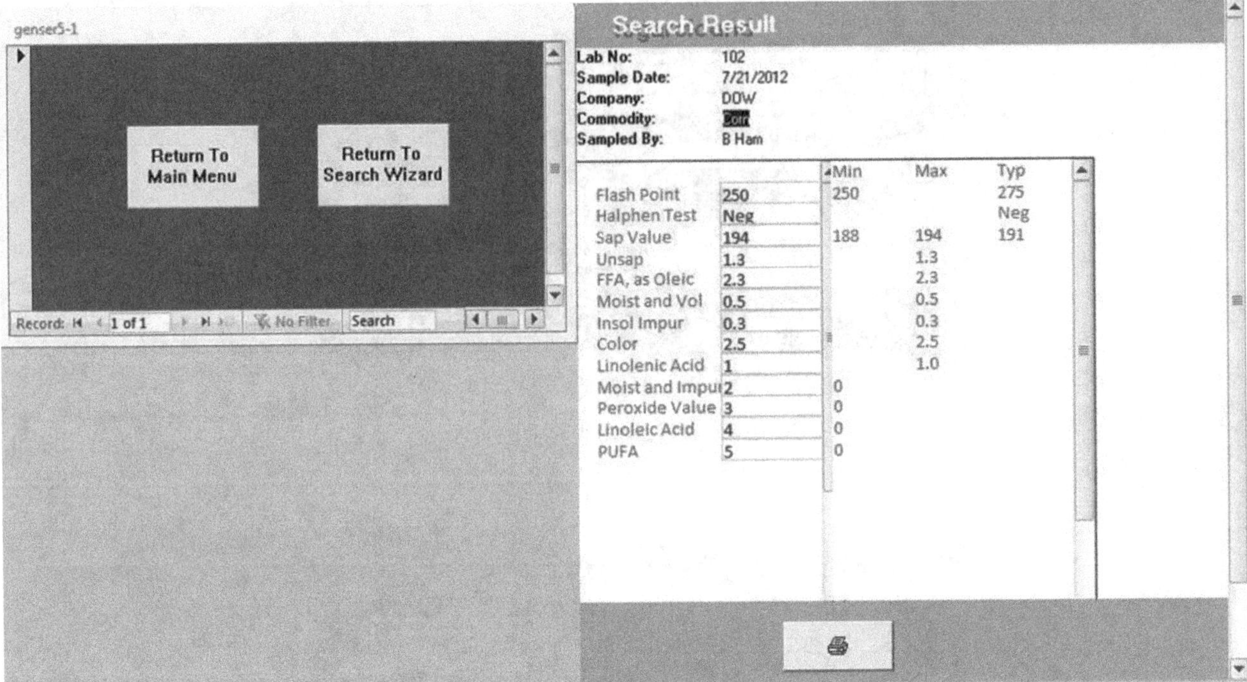

FIGURE 12.26 Sample search result page.

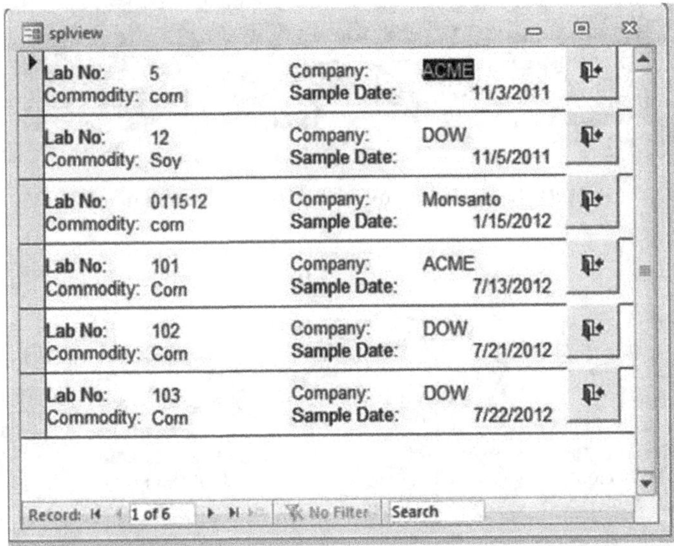

FIGURE 12.27 Current open samples initial search page.

page is a list of the current open samples. To view the data associated with one of the samples, click on the laboratory number. This will open up a page that shows the test data of the sample. A schematic of the list of open samples and the test data page is depicted in Figure 12.28.

12.8 APPROVING SAMPLES

In the tracking and lifetime of samples, another function and use of the LIMS system is to approve and close the sample out after it has been finished. This particular example of a LIMS has this functionality. If we go to the Main Menu, there is a button for approving samples. Click on the Approve Samples button to open up the list of current open samples. This page is shown in Figure 12.29. To look at the associated test results for each sample, click on the sample number to open up a page as shown in Figure 12.30. To the right of the sample numbers is an approve button and an exit button. Click on the approve button next to sample number 102 or 104, or any that may be in your table. Next, a question box will open up asking if you want to print a report as depicted in Figure 12.31. Click on the cancel button.

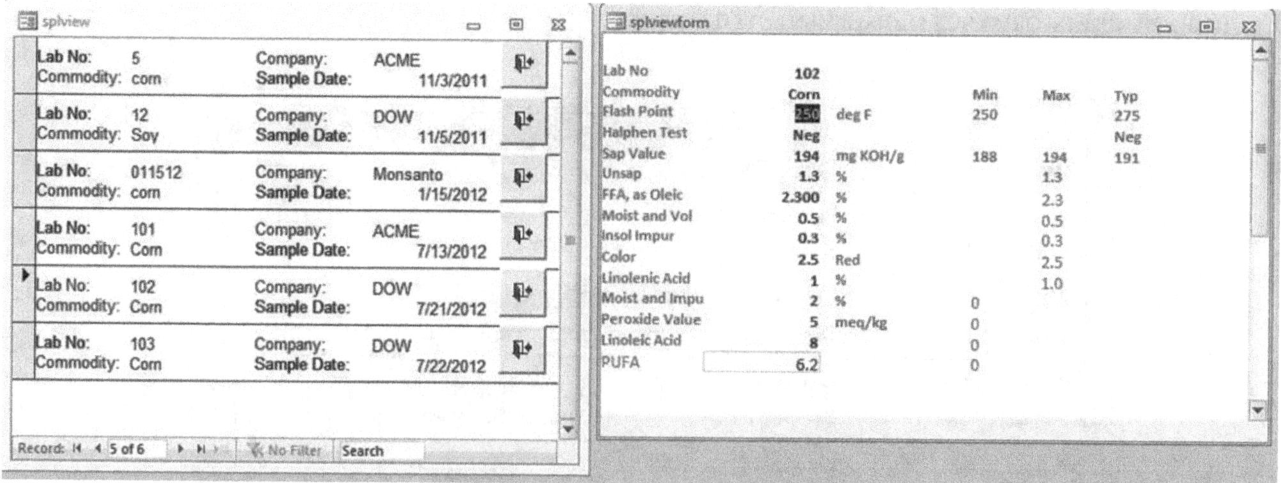

FIGURE 12.28 List of open samples and test data page.

oilbox

Lab No:	5	Company:	ACME	Approve Sample	
Commodity:	corn	Sample Date:	11/3/2011		
Lab No:	12	Company:	DOW	Approve Sample	
Commodity:	Soy	Sample Date:	11/5/2011		
Lab No:	011512	Company:	Monsanto	Approve Sample	
Commodity:	corn	Sample Date:	1/15/2012		
Lab No:	101	Company:	ACME	Approve Sample	
Commodity:	Corn	Sample Date:	7/13/2012		
Lab No:	102	Company:	DOW	Approve Sample	
Commodity:	Corn	Sample Date:	7/21/2012		
Lab No:	103	Company:	DOW	Approve Sample	
Commodity:	Corn	Sample Date:	7/22/2012		

FIGURE 12.29 The approve sample current open sample list.

oilbox

Lab No:	5	Company:	ACME	Approve Sample
Commodity:	corn	Sample Date:	11/3/2011	
Lab No:	12	Company:	DOW	Approve Sample
Commodity:	Soy	Sample Date:	11/5/2011	
Lab No:	011512	Company:	Monsanto	Approve Sample
Commodity:	corn	Sample Date:	1/15/2012	
Lab No:	101	Company:	ACME	Approve Sample
Commodity:	Corn	Sample Date:	7/13/2012	
Lab No:	102	Company:	DOW	Approve Sample
Commodity:	Corn	Sample Date:	7/21/2012	
Lab No:	103	Company:	DOW	Approve Sample
Commodity:	Corn	Sample Date:	7/22/2012	

approveform

Lab No	y			
Commodity	Corn	Min	Max	Typ
Flash Point	250	250		275
Halphen Test	Neg			Neg
Sap Value	194	188	194	191
Unsap	1.3		1.3	
FFA, as Oleic	2.300		2.3	
Moist and Vol	0.5		0.5	
Insol Impur	0.3		0.3	
Color	2.5		2.5	
Linolenic Acid	1		1.0	
Moist and Imp	2	0		
Peroxide Valu	5	0		
Linoleic Acid	8	0		
PUFA	6.2	0		

FIGURE 12.30 Open sample list for approval including test data.

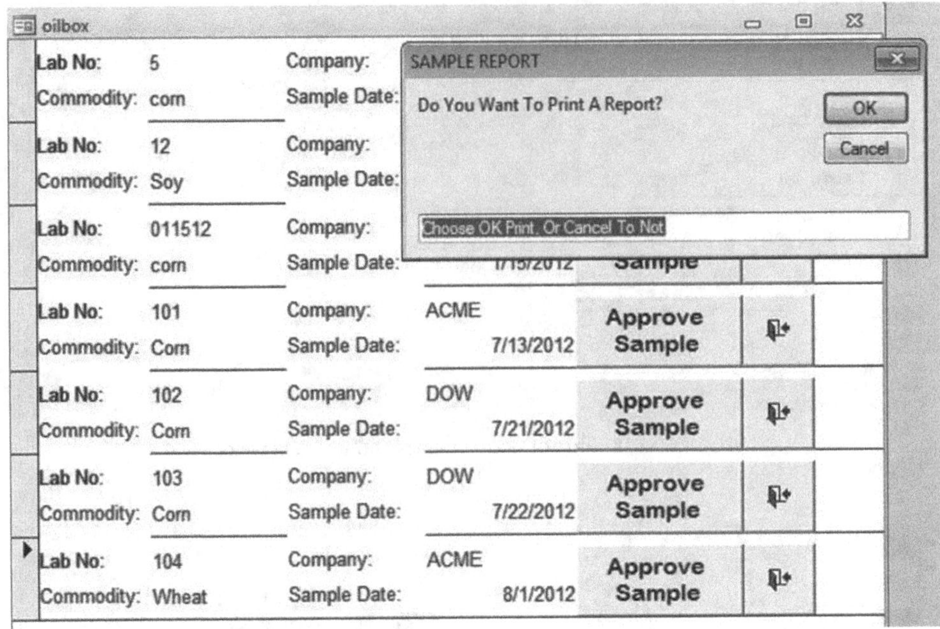

FIGURE 12.31 Last step in approving a sample with the option to print a report.

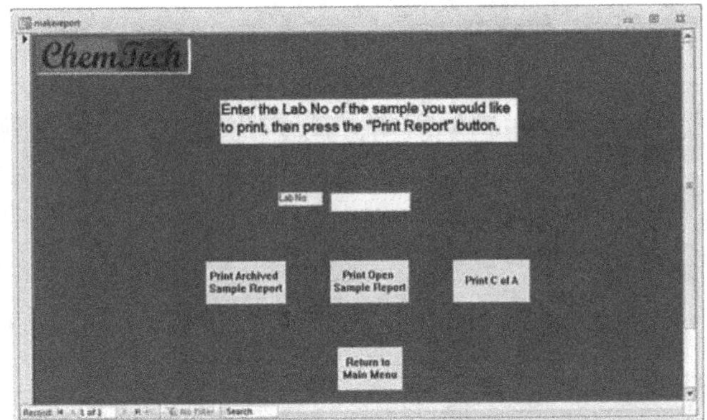

FIGURE 12.32 Printing reports page.

The sample will be removed from the open sample table and placed into the archives. We can now see this sample in the archive search.

12.9 PRINTING SAMPLE REPORTS

The next functionality of the example LIMS system is printing reports of samples. To the bottom right of the Main Menu page is a button for printing sample reports. Click the Reports and C of A's (certificates of analysis) button to open up the page as shown in Figure 12.32. Both archived samples and current open samples can have a report printed. For example, we can enter 102 for the sample number and print the report by pressing the Print Open Sample Report. A report will be displayed as shown in Figure 12.33.

FIGURE 12.33 Open sample report.

13

ULTRAVIOLET AND VISIBLE (UV/VIS) SPECTROSCOPY

13.1 Introduction to Spectroscopy in the Analytical Laboratory
13.2 The Electromagnetic Spectrum
13.3 Ultraviolet/Visible (UV/Vis) Spectroscopy
13.3.1 Wave and Particle Theory of Light
13.3.2 Light Absorption Transitions
13.3.3 The Color Wheel
13.3.4 Pigments
13.3.5 Inorganic Elemental Analysis
13.3.6 The Azo Dyes
13.3.7 UV-Visible Absorption Spectra
13.3.8 Beer's Law
13.4 UV/Visible Spectrophotometers
13.5 Special Topic (Example)—Spectrophotometric Study of Dye Compounds
13.5.1 Introduction
13.5.2 Experimental Setup for Special Topic Discussion
13.5.3 UV/Vis Study of the Compounds and Complexes
13.6 Chapter Key Concepts
13.7 Chapter Problems

13.1 INTRODUCTION TO SPECTROSCOPY IN THE ANALYTICAL LABORATORY

The analytical laboratory utilizes the phenomenon of the electromagnetic spectrum for an untold number of analyses. Chemists and technicians in the analytical laboratory often make use of the special interaction of molecules with electromagnetic radiation, with the assistance of analytical instrumentation, such as ultraviolet/visible (UV/Vis) spectrophotometers, fluorometers, and Fourier transform infrared spectrometers (FTIR) to measure, identify, and even quantitate compounds of interest. Sometimes, another type of analysis is done called colorimetric analysis, where the eye is used to distinguish differences in color. Even though colorimetric analysis is still used, we will focus upon spectrophotometric analysis. As we will see both qualitative and quantitative analyses are routinely performed by the chemist and technician in the analytical laboratory with the employment of electromagnetic radiation. Most often, UV/Vis and fluorimetry analysis is done with compounds in solution where the analyte of interest will either be measured directly according to its spectroscopic properties, or it may be combined with a chromophore or fluorophore to allow its measurement. Let us start by looking at the basics of electromagnetic radiation, specifically the electromagnetic spectrum, and how we utilize this phenomenon all around us to measure compounds in the analytical laboratory.

13.2 THE ELECTROMAGNETIC SPECTRUM

The visible spectrum, what most of us are usually familiar with constitutes but a small part of the total electromagnetic radiation spectrum. Almost all of the radiation that surrounds us cannot be seen, but is detected by special spectrophotometric instruments. The entire electromagnetic spectrum comprises very short wavelengths (including gamma and X-rays) to very long wavelengths (including microwaves and broadcast radio waves). Let us look at the chart in Figure 13.1 that displays the regions of the spectrum, the source that produces the radiation, and some conversion factors.

13.3 ULTRAVIOLET/VISIBLE (UV/VIS) SPECTROSCOPY

Our world fortunately is not in black and white like an old television set, but rather is filled with many different shades of color. The reason we see something as red is due to a specific wavelength portion of the visible spectrum being absorbed by the object while the other colors will form the color that we see, red in this instance. We can think of the human eye functioning as a spectrometer that is analyzing the light observed from the object that appears red. The sunlight (or white light) that we

Analytical Chemistry: A Chemist and Laboratory Technician's Toolkit, First Edition. Bryan M. Ham and Aihui MaHam.

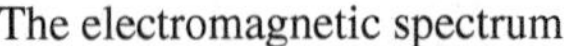

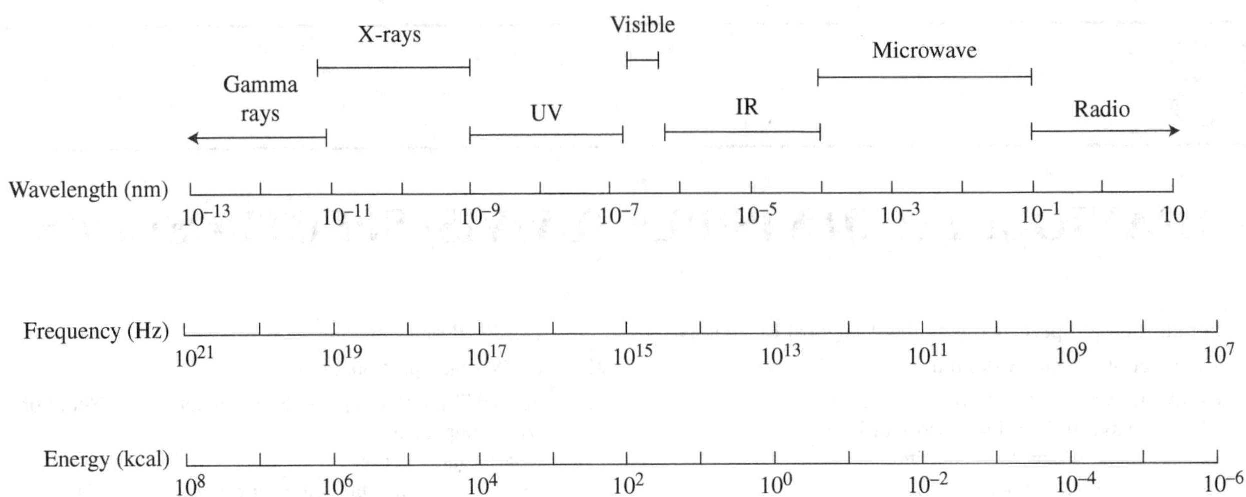

FIGURE 13.1 The electromagnetic spectrum in wavelength, frequency, and energy.

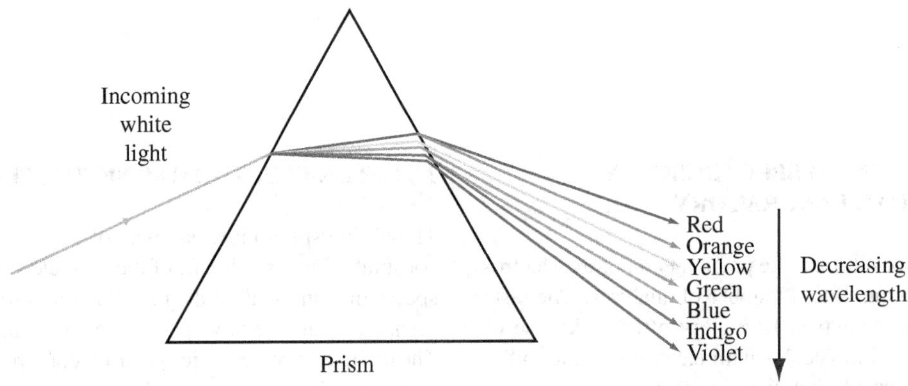

FIGURE 13.2 Separating the visible portion of the spectrum from 400 to 800 nm using a prism.

see as uniform or homogeneous in color is composed of a broad range of radiation wavelengths in the ultraviolet (UV), visible, and infrared (IR) portions of the spectrum.

The visible portion of the spectrum, roughly from 400 to 800 nm, can be separated into the component colors that we see by passing sunlight through a prism. When done, the prism acts to bend the light in differing degrees according to the wavelength. This is depicted in Figure 13.2. The colors of the spectrum that we see from sunlight, in order of decreasing wavelength, are often remembered by the mnemonic: *ROY G BIV*.

The wavelengths seen as particular colors in the visible portion of the spectrum are listed in Figure 13.3.

Violet:	400–420 nm
Indigo:	420–440 nm
Blue:	440–490 nm
Green:	490–570 nm
Yellow:	570–585 nm
Orange:	585–620 nm
Red:	620–780 nm

FIGURE 13.3 Wavelength ranges for the visible spectrum.

13.3.1 Wave and Particle Theory of Light

Visible light electromagnetic radiation is described in a dual manner as a wave phenomenon (wave theory), characterized by a wavelength or frequency which is required to describe optical effects such as diffraction and refraction. Light is also described by a particle theory, which is required to describe the absorption and emission of radiant energy. The "wavelength" is the distance between adjacent peaks (or troughs), and may be designated in meters, centimeters, or nanometers (10^{-9} m). The "frequency" is the number of wave cycles that travel past a fixed point per unit of time, and is usually given in cycles per second, or hertz (Hz). A simple schematic of wavelength is depicted in Figure 13.4. Red is the longest visible wavelength and violet is the shortest.

In the entire spectrum depicted in Figure 13.1, which includes the visible region, the energy carried by a photon of a certain wavelength is proportional to its frequency. The following equations describe these relationships:

$$\nu = \frac{c}{\lambda}, \tag{13.1}$$

where

ν = frequency
λ = wavelength
c = velocity of light ($c = 3 \times 10^{10}$ cm/s).

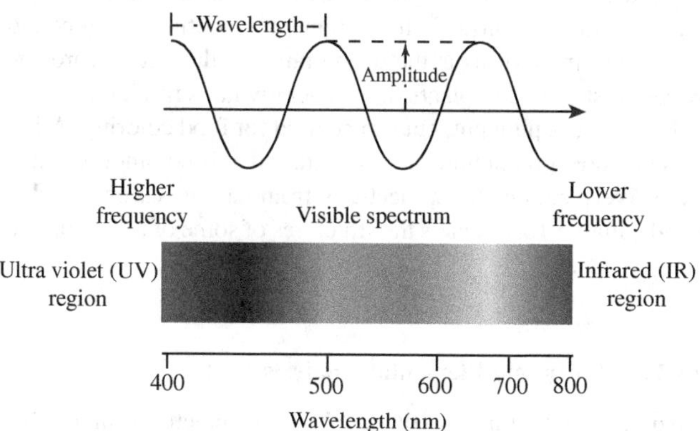

FIGURE 13.4 Visible light electromagnetic radiation wavelength and amplitude.

$$\Delta E = h\nu, \tag{13.2}$$

where

E = energy
ν = frequency
h = Planck's constant ($h = 6.6 \times 10^{-27}$ erg s).

We can combine these two expressions by substituting Equation 13.1 into Equation 13.2 to get the following Equation 13.3, a useful relationship that relates the energy and the wavelength.

$$\Delta E = h\frac{c}{\lambda}. \tag{13.3}$$

13.3.2 Light Absorption Transitions

When a compound absorbs light, it produces a transition of an outer valence shell electron from its ground energy state to a higher energy level called an excited state. These transitions are depicted in Figure 13.5. The highest energy transition involves the absorption of ultraviolet light (190–400 nm) to the second electronic excited state. The next transition is the absorption of visible light (400–800 nm) to the first electronic excited state. Finally, there is the absorption in the infrared region (1000–1500 nm), which is the lowest energy transition. According

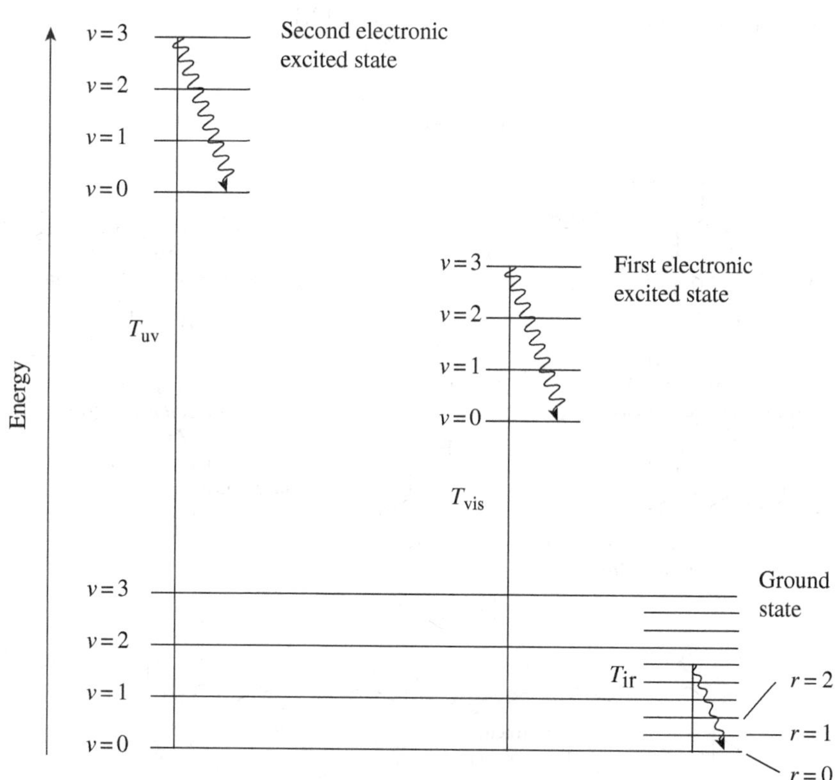

FIGURE 13.5 Three transitions involving the absorption of light. These include T_{uv} for absorption in the ultraviolet region, T_{vis} for absorption in the visible region, and T_{ir} for the infrared region.

to Figure 13.1, energy in the ultraviolet region has shorter wavelengths and higher frequencies and energies. Each electronic excited state also has a number of vibrational levels ($\nu = 0$, 1, 2, 3). When the electron relaxes, it will drop down to the lowest vibrational level.

13.3.3 The Color Wheel

The color wheel shown in Figure 13.6 is used to predict what color will be seen when a substance absorbs energy at different specific wavelengths. If something absorbs energy in a wavelength range, then the complimentary color, which is directly opposite in the color wheel, is observed. For example, suppose something absorbs from 580 to 600 nm light, then the complimentary color observed for the absorbing substance will be blue. The example of red earlier would mean that the substance or object is absorbing somewhere in the green region from 490 to 560 nm.

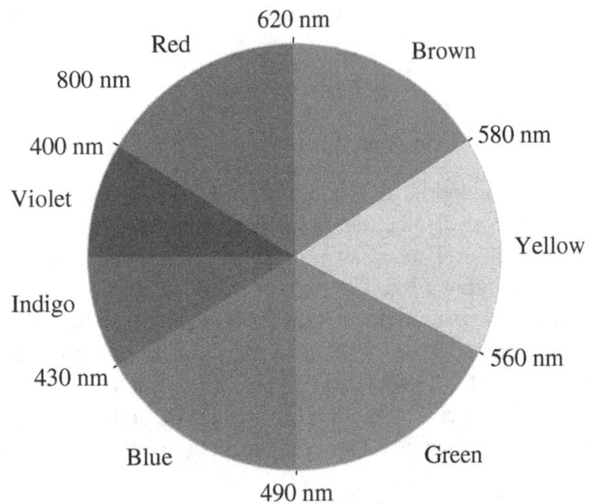

FIGURE 13.6 The color wheel.

13.3.4 Pigments

There are many examples of natural compounds, often referred to as pigments or organic dyes that exhibit bold colors. Included in these are the crimson pigment, kermesic acid, the blue dye, indigo, and the yellow saffron pigment, crocetin. A compound familiar to most of us is the deep-orange hydrocarbon carotene widely distributed in plants. It is generally not sufficiently stable to be used as a pigment, but can be used for food coloring. All of these colored compounds have a feature that is commonly called extensively conjugated π electrons from carbon–carbon double bonds. Figure 13.7 depicts the structures of some of these colored compounds.

13.3.5 Inorganic Elemental Analysis

Also related to this are the nonbonding electrons of double bonds involving inorganic element such as Mn=O, which also demonstrates color. There is a number of chromophore reagents used to detect elements. This is done by reacting the element with the chromophore reagent forming a characteristic color (Table 13.1).

Some natural organic pigments

Kermesic acid
(Carminic acid)
from the insect *Coccus cacti*

Z=H
Indigo
from *Isatis tinctoria* (woad)
Z=Br
Punicin or tyrian purple
from mollusks of the genus *Murex*

Crocetin
from saffron

β-Carotene
from carrots

FIGURE 13.7 Examples of natural organic pigments.

TABLE 13.1 Elements Analyzed, Reagents, and Colors Formed for Inorganic Elemental Analysis.

Element to Detect	Reagent Used	Color Formed at λ nm
Al	8-Hydroxyquinoline	Yellow—295
Bi	Thiourea	Yellow—295
Ca	Calcein	Yellow/green—520
Cl as Cl_2	*o*-Tolidine	Yellow—295
Co	Ammonium thiocyanate	Blue—620
Cr	Diphenylcarbazide	Red/violet—540
Cu	FerroZine	Brown—470
F as F^-	Cerous alizarin complexone	Wine red—512
Fe	1,10-Phenanthroline	Red—512
Fe	FerroZine	Red—562
Mg	*o,o*′-Dihydroxyazobenzene	Orange—485
Mn	Periodate	Purple—520
Mn	Thiothenoyltrifluoroacetone	Purple—450
Mo	Thiolactic acid	Yellow/brown—450
P as PO_4^{3-}	Molybdate, hydrazine	Blue—830
Pb	Dithiozone	Pink
S as SO_3^{2-}	Iodine	I_3
Ti	Hydrogen peroxide	Yellow—295
U	Arsenazo I or III	Violet/blue—640
Zn	Dithizone	Pink

13.3.6 The Azo Dyes

Another type of colored compound is the azo dyes; however, they are not natural but have been prepared synthetically. The azo dyes get their name from the azo group, –N=N–, which is part of their structure. The synthetic azo dyes make up approximately 70% of the dyes that are used in both the textiles industry and food manufacture. The aromatic side groups stabilize the internal N=N group making it part of an extended delocalized conjugated system. Figure 13.8 depicts a few of the azo dyes commonly in use today as coloring agents for food. The azo dyes are relatively inexpensive to produce, nontoxic, and more stable than the natural dyes.

Sources:

- http://en.wikipedia.org
- http://www.chm.bris.ac.uk/webprojects2002/price/azo.htm
- http://mst.dk/udgiv/publications/1999/87-7909-548-8/html/kap05_eng.htm
- Fennema, O.R.: Food Chemistry, 3rd edition, 1996
- Bateman et al; The effects of a double blind, placebo controlled, artificial food colourings and benzoate preservative challenge on hyperactivity in a general population sample of preschool children. *Archives of Disease in Childhood* 2004;

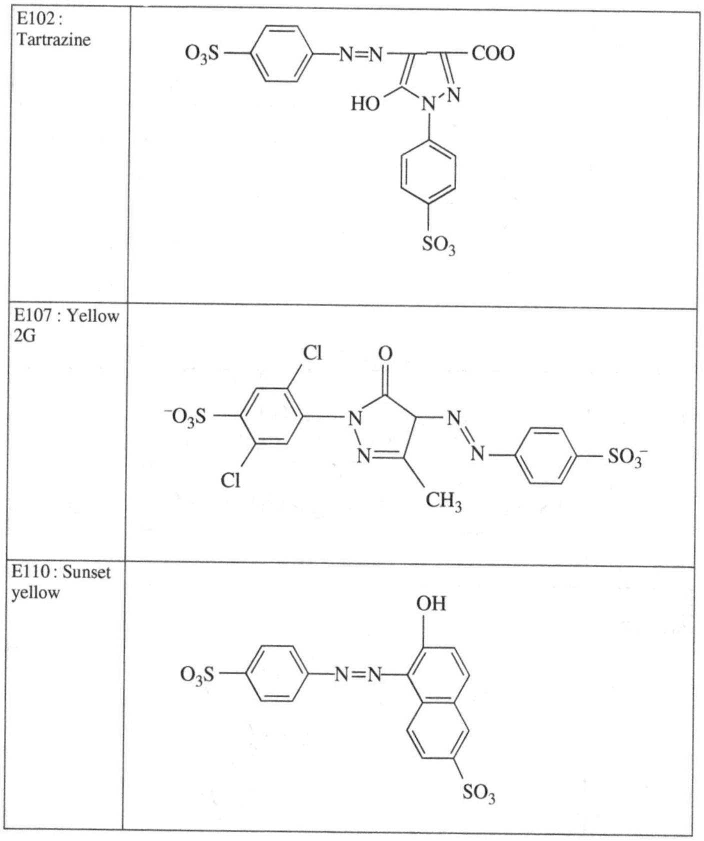

FIGURE 13.8 Examples of the azo dyes.

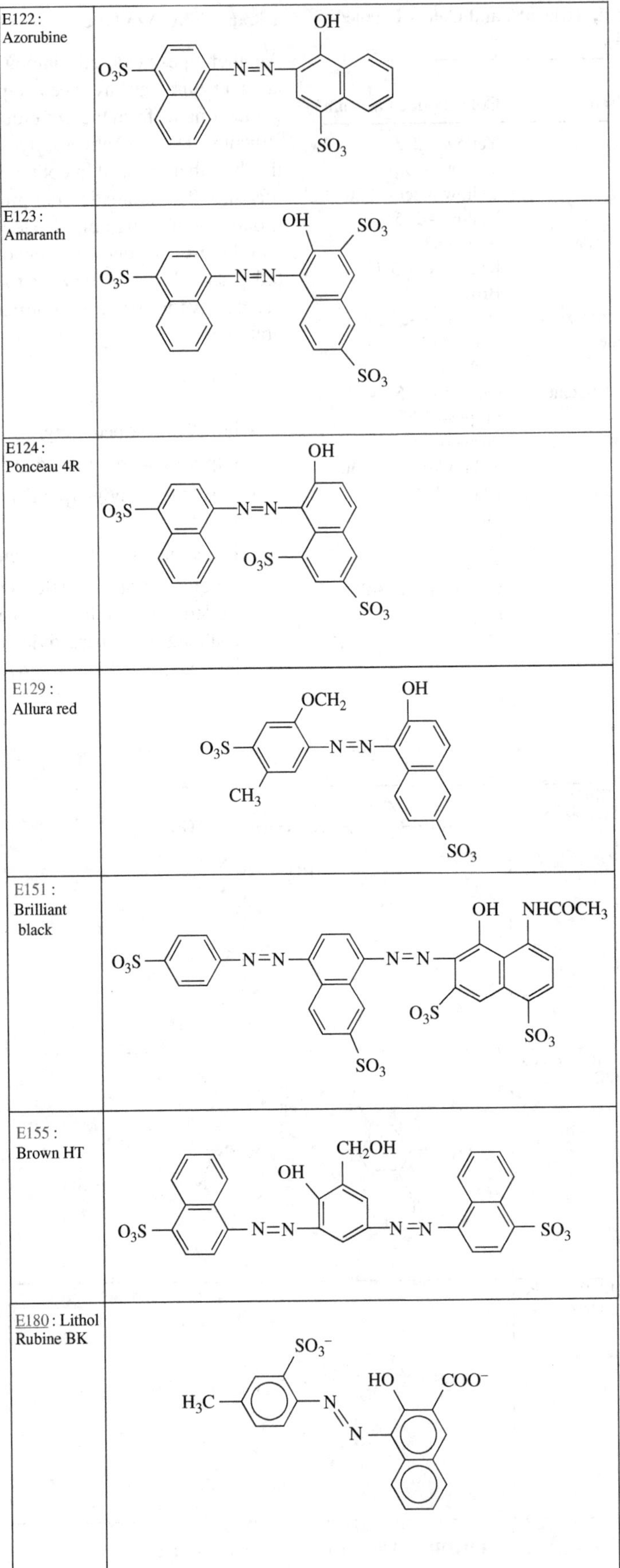

FIGURE 13.8 (Continued)

89:506–511 plus reactions : **Eigenmann PA**, Haengelli CA. Food colourings and preservatives—allergy and hyperactivity. *Lancet* 2004; **364**:823–4 and Stevenson et al., Rejoinder to Eigenmann PA, Haengelli CA, Food colourings and preservatives—allergy and hyperactivity (*Lancet* 2004; 364:823–4) and an erratum, *Archives of Disease in Childhood* 2005; **90**:875.

Table 13.2 contains a list of FDA-approved food color additives.

TABLE 13.2 FDA-Approved Food Color Additives.

Color Additives Approved for Use in Human Food Part 73, Subpart A: Color additives exempt from batch certification				
21 CFR Section	Straight Color	EEC#	Year Approved	Uses and Restrictions
§73.30	Annatto extract	E160b	1963	Foods generally
§73.40	Dehydrated beets (beet powder)	E162	1967	Foods generally
§73.75	Canthaxanthin	E161g	1969	Foods generally, NTE 30 mg/lb of solid or semisolid food or per pint of liquid food; May also be used in broiler chicken feed
§73.85	Caramel	E150a–d	1963	Foods generally
§73.90	β-Apo-8′-carotenal	E160e	1963	Foods generally, NTE: 15 mg/lb solid, 15 mg/pt liquid
§73.95	β-Carotene	E160a	1964	Foods generally
§73.100	Cochineal extract	E120	1969	Foods generally
	Carmine		1967	
§73.125	Sodium copper chlorophyllin	E141	2002	Citrus-based dry beverage mixes NTE 0.2% in dry mix; extracted from alfalfa
§73.140	Toasted partially defatted cooked cottonseed flour	–	1964	Foods generally
§73.160	Ferrous gluconate	–	1967	Ripe olives
§73.165	Ferrous lactate	–	1996	Ripe olives
§73.169	Grape color extract	E163?	1981	Nonbeverage food
§73.170	Grape skin extract (enocianina)	E163?	1966	Still & carbonated drinks & ades; beverage bases; alcoholic beverages (restrict. 27 CFR Parts 4 & 5)
§73.200	Synthetic iron oxide	E172	1994	Sausage casings NTE 0.1% (by wt)
§73.250	Fruit juice	–	1966	Foods generally
			1995	Dried color additive
§73.260	Vegetable juice	–	1966	Foods generally
			1995	Dried color additive, water infusion
§73.300	Carrot oil	–	1967	Foods generally
§73.340	Paprika	E160c	1966	Foods generally
§73.345	Paprika oleoresin	E160c	1966	Foods generally
§73.350	Mica-based pearlescent pigments	–	2006	Cereals, confections and frostings, gelatin desserts, hard and soft candies (including lozenges), nutritional supplement tablets and gelatin capsules, and chewing gum
§73.450	Riboflavin	E101	1967	Foods generally
§73.500	Saffron	E164	1966	Foods generally
§73.575	Titanium dioxide	E171	1966	Foods generally; NTE 1% (by wt)
§73.585	Tomato lycopene extract; tomato lycopene concentrate	E160	2006	Foods generally
§73.600	Turmeric	E100	1966	Foods generally
§73.615	Turmeric oleoresin	E100	1966	Foods generally
Color Additives Approved for Use in Human Food Part 74, Subpart A: Color additives subject to batch certification				
21 CFR Section	Straight Color	EEC#	Year Approved	Uses and Restrictions
§74.101	FD&C Blue No. 1	E133	1969	Foods generally
			1993	Added Mn spec
§74.102	FD&C Blue No. 2	E132	1987	Foods generally
§74.203	FD&C Green No. 3	–	1982	Foods generally
§74.250	Orange B	–	1966	Casings or surfaces of frankfurters and sausages; NTE 150 ppm (by wt)
§74.302	Citrus Red No. 2	–	1963	Skins of oranges not intended or used for processing; NTE 2.0 ppm (by wt)
§74.303	FD&C Red No. 3	E127	1969	Foods generally
§74.340	FD&C Red No. 40	E129	1971	Foods generally
§74.705	FD&C Yellow No. 5	E102	1969	Foods generally
§74.706	FD&C Yellow No. 6	E110	1986	Foods generally

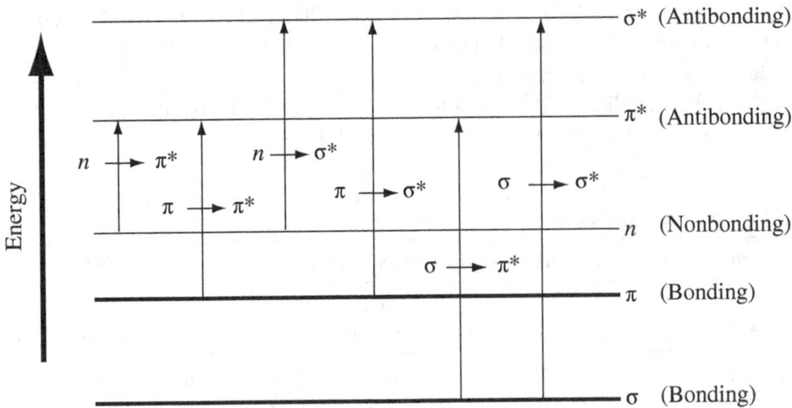

FIGURE 13.9 Color producing electronic energy changes.

13.3.7 UV-Visible Absorption Spectra

We have looked at a number of organic molecules that absorb light and are observed to have colored solutions. As we saw earlier by the color wheel, the color of a compound, usually in water or an organic solvent depends on the wavelength of absorbed light. Let us take a deeper look at how atoms and molecules absorb light to produce different colors. In this respect, when we refer to light, we are talking about white light that contains wavelength ranges from 200 to 800 nm. When atoms or molecules of our sample that we are testing are exposed to light having an energy that matches a possible electronic transition within the molecule, some of the light energy will be absorbed as the electron (π-bonding or n-bonding) is promoted to a higher energy orbital. Always, energetically favored electron promotion will be from the "highest occupied molecular orbital" (HOMO) to the "lowest unoccupied molecular orbital" (LUMO), producing what is called an excited state.

Let us start by looking at a simple gaseous atom electronic transition example. Suppose a ground-state magnesium atom (Mg_0) absorbs a photon of 590 nm wavelength and promotes to the first excited state (Mg_1). The electron has transitioned from the 3s orbital to the 3p orbital. Suppose the magnesium atom absorbs a photon of 335 nm wavelength and promotes to the second excited state (Mg_2). An electron has now transitioned from the 3s orbital to the 4p orbital. The electronic configuration of a ground-state magnesium atom is $1s^22s^22p^63s^23p^0$, or in short notation $[Ne]3s^23p^0$. The electronic configuration of the first excited state magnesium atom is $[Ne]3s^13p^1$, and for the second excited state magnesium atom $[Ne]3s^13p^04p^1$. The absorption of light for gaseous atoms generally has very sharp absorption versus wavelength spectra.

The light absorption behavior for molecules is quite different than the simple gaseous magnesium atom just discussed. There are three types of energy changes when molecules absorb light energy: *electronic*, which is the change in the energy of electrons in the molecule; *vibrational*, which is the change in the distance of atoms in the molecule; and *rotational*, which is the change in energy of the rotation of the molecule around a center. The types of electronic energy changes that produce color are associated with two transitions that are depicted in Figure 13.9. These are the left two transitions involving π-bonding and n nonbonding transitioning to an excited state π∗ antibonding. The energy changes involving vibrational and rotational changes are lower energy changes in the infrared energy, while the electronic changes involve higher energies in the ultraviolet and visible spectra. Molecular absorption of UV/Vis light involves many transitions including sublevel transitions that generally result in a much broader spectrum, depicted in Figure 13.10 for trans-β-carotene.

13.3.8 Beer's Law

In Figure 13.10, we can see that the absorbance of a substance can increase and decrease with different wavelengths. Often, we use what are called absorbance maximum wavelengths for spectrophotometric analysis. It may also be obvious that as the concentration of a substance increases in solution, then the associated absorbance will also increase due to a greater number of species being present. This brings us to an important relationship between concentration and absorbance used in the analytical laboratory called Beer–Lambert law (or for short usually referred to as Beer's law). Beer's law states that the absorbance of a substance is proportional to the concentration of the substance, the path length that the absorbing light travels through the sample, and the probability that absorbance takes place with an individual part of the substance, such as each molecule or atom.

The passage of light through a container holding a substance onto a detector is depicted in Figure 13.11. If the incident light is designated as P_0 and the light that has passed through to the detector as P, then we can express the absorbance (A) as the difference in the logarithms of the incident and detected light:

$$\text{Absorbance}\,(A) = \log P_0 - \log P = \log\frac{P_0}{P}. \qquad (13.4)$$

A spectrophotometer is used to measure the different wavelengths that are being absorbed by the analyte, and the magnitude of the absorbance.

Aldrich Cat. No.:	85,555-3
Sigma Prod. No.:	C9750
CAS No.:	*[7235-40-7]*
C.I. No.:	40800
Mol. Form.:	$C_{40}H_{56}$
F.W.:	536.89
m.p.:	178–179°C
Appearance:	Red-orange powder
λ_{max}**:**	450(478) nm in hexane
Solubility:	H_2O 0.6 mg/ml EGME 3 mg/ml EtOH 2 mg/ml

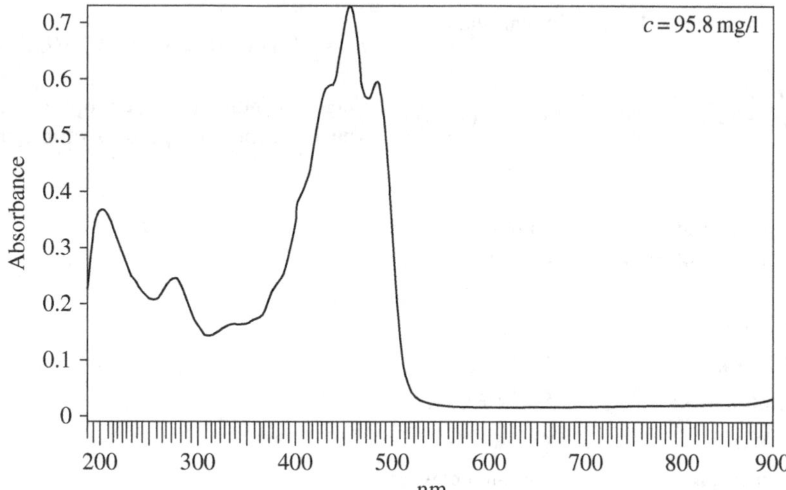

FIGURE 13.10 Beta-carotene UV/Vis spectrum. A large maxima is observed at approximately 450 nm due to the extensive double bond conjugation contained within the structure of beta-carotene. (Reproduced with permission of Sigma-Aldrich Co. LLC from The Sigma-Aldrich Handbook of Stains, Dyes and Indicators by Floyd J. Green, 1990, Aldrich Chemical Company, Inc., p. 194.)

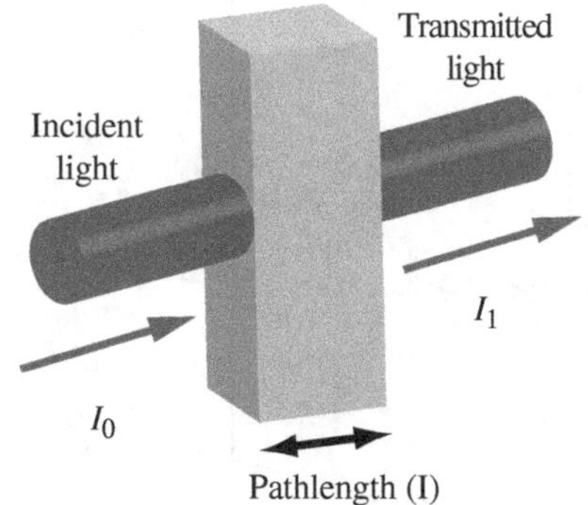

FIGURE 13.11 Incident and transmitted light through a sample.

Beer's law is generally referred to as the fundamental law of spectroscopy written in the following form:

$$\log\frac{P_0}{P} = A = abc, \qquad (13.5)$$

where

a = absorptivity, a characteristic of a substance that is constant at a given wavelength

b = the path length that the incident light travels through the analyte solution, in centimeters

c = the concentration of the analyte in solution, expressed in mg/l (ppm), g/100 ml, and so on.

The units of a will depend on the concentration units, and a is also specific to a single wavelength (thus monochromatic).

When the concentration of the absorbing species is expressed in molarity (M), the molar absorptivity (ε) is used and Beer's law is expressed in the form:

$$A = \varepsilon bc, \qquad (13.6)$$

where

ε = molar absorptivity, a characteristic of a substance that is constant at a given wavelength

b = the path length that the incident light travels through the analyte solution, in centimeters

c = the concentration of the analyte in solution, expressed in molarity, M, with units mole/L

ε is expressed in l/mole cm, and it is generally designated at the wavelength of maximum absorbance.

There is a direct relationship between the absorbance and the transmittance of the incident light energy passing through the analyte solution. The transmittance T is equal to the ratio of the transmitted light P to that of the incident beam radiant power P_0.

$$T = \frac{P}{P_0}. \tag{13.7}$$

The relationship of transmittance to Beer's law can thus be expressed as

$$-\log \frac{P}{P_0} = -\log T = \varepsilon bc. \tag{13.8}$$

A UV/Vis spectrum obtained from a spectrophotometer can be illustrated in two manners as either absorbance (A) versus wavelength (λ, nm) or percent transmittance (%T). In quantitative analysis, the absorbance is used and usually ranges from 0 to 2. Percent transmittance ranges from 0 to 100. Figure 13.12 depicts the UV/Vis spectrum of carminic acid as absorbance versus wavelength. In the spectrum, there is a small amount of absorbance at approximately 500 nm due to short segments of conjugated double bonding. The major absorbance occurs in the lower UV region from approximately 280 to 200 nm.

The absorbance and %T can be obtained interchangeably from the following relationship:

$$A = -\log T = -\log(\%T/100). \tag{13.9}$$

13.4 UV/VISIBLE SPECTROPHOTOMETERS

There are numerous spectrophotometers on the market today. The majorities of spectrophotometers are computer controlled

Aldrich Cat. No.:	22,925-3
Sigma Prod. No.:	C3522
CAS No.:	*[1260-17-9]*
C.I. No.:	75470
Mol. Form.:	$C_{22}H_{20}O_{13}$
F.W.:	492.40
Appearance:	Dark-red powder
λ_{max}**:**	495nm in methanol
Solubility:	H_2O 4 mg/ml EGME 4 mg/ml EtOH 2 mg/ml

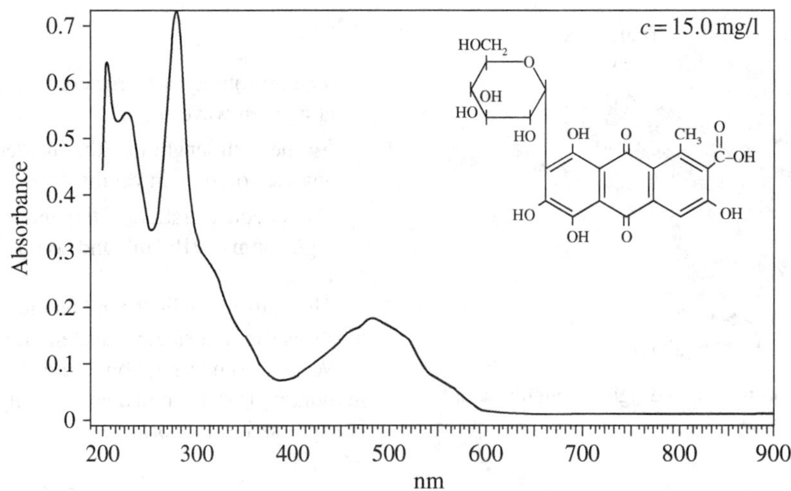

FIGURE 13.12 The UV/Vis spectrum of carminic acid as absorbance versus wavelength. Small amount of absorbance at approximately 500 nm due to short segments of conjugated double bonding. The major absorbance occurs in the lower UV region from approximately 280 to 200 nm. (Reproduced with permission of Sigma-Aldrich Co. LLC from The Sigma-Aldrich Handbook of Stains, Dyes and Indicators by Floyd J. Green, 1990, Aldrich Chemical Company, Inc., p. 193.)

FIGURE 13.13 Scale of absorbance and transmittance on a spectrophotometer.

FIGURE 13.14 GENESYS∗ 20 Visible Spectrophotometer as an example of an instrument with readout and keypad. (Used by permission from Thermo Fisher Scientific, the copyright owner.)

and offer software that allows the user many functions toward UV/Vis studies. The computer control may consist of an LCD readout screen with a keypad right on the instrument, or a personal computer connected to the instrument. The earliest spectrophotometric instruments used had very simple readouts, where a scale was presented with absorbance (A) on one side and percent transmittance ($\%T$) on the other. A needle pointer that spanned the scale gave the operator a location to read the values, as depicted in Figure 13.13.

A basic design of the spectrophotometer is depicted in Figure 13.14 for the GENESYS∗ 20 Visible Spectrophotometer, where this instrument has a readout and keypad located to the left of the top face. On the right is the sample compartment where in this model a test tube containing the sample is inserted into a holder in the compartment.

A similar designed spectrophotometer is depicted in Figure 13.15 of the Beckman Coulter DU 730. Here, we can see the sample compartment has been opened to the right ready for sample insertion.

An example of a computer-controlled spectrophotometer using a computer screen visual output is depicted in Figure 13.16 for the Beckman Coulter DU 800 series instrument.

The internal optics of these spectrophotometers can be quite complex. Figure 13.17 is the Varian Cary 300 series spectrophotometer that also is computer controlled with full-screen output. The internal optics for the Varian Cary 300 is depicted in Figure 13.18, where we can see that the light paths actually pass through a number of processes and redirections.

Cuvettes or cells are what are generally used to hold samples for introduction into the spectrophotometer for measurement. The cells can be made of quartz or plastic depending upon what wavelength range is to be measured. If in the visible range from 380 to 800 nm, analytical labs will often use plastic disposable cells for measurement. If the UV range is being measured from 190 to 380 nm, it is usually best to use a quartz cell. The cuvettes or cells come in a variety of sizes and volumes with different path

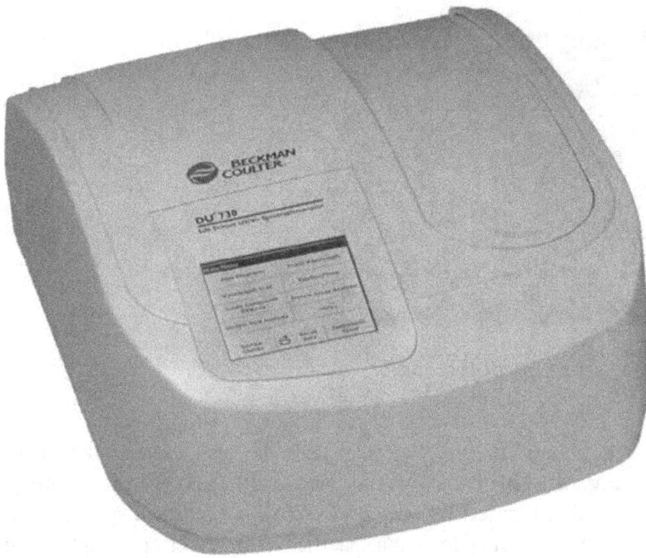

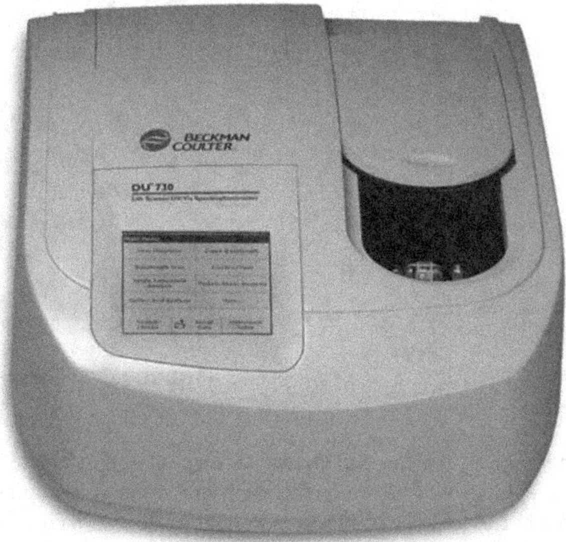

FIGURE 13.15 Beckman Coulter DU 730 spectrophotometer. (Photo courtesy of Beckman Coulter, Inc.)

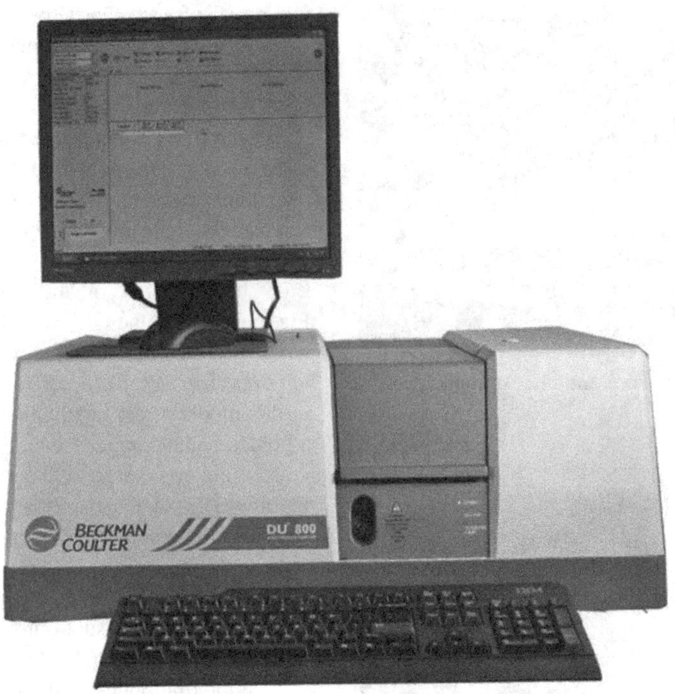

FIGURE 13.16 Beckman Coulter DU 800 series spectrophotometer. (Photo courtesy of Beckman Coulter, Inc.)

FIGURE 13.17 Varian Cary 300 series spectrophotometer.

length. The typical rectangular cuvette cell as depicted in Figure 13.19 will have a path length of 1 cm and a volume of 3.5 ml. The sizes available include large cells with path lengths up to 10 cm and volume of 35 ml. Microcells are also available that may have a 1 cm path length with a volume of 1.4 ml, as depicted in Figure 13.20. Cylindrical cells are also used as depicted in Figure 13.21 that allow long path lengths and even flow through. Finally, Figure 13.22 depicts examples of plastic disposable cells.

We will now look at a specific example of a spectrophotometric study of colored dye compounds.

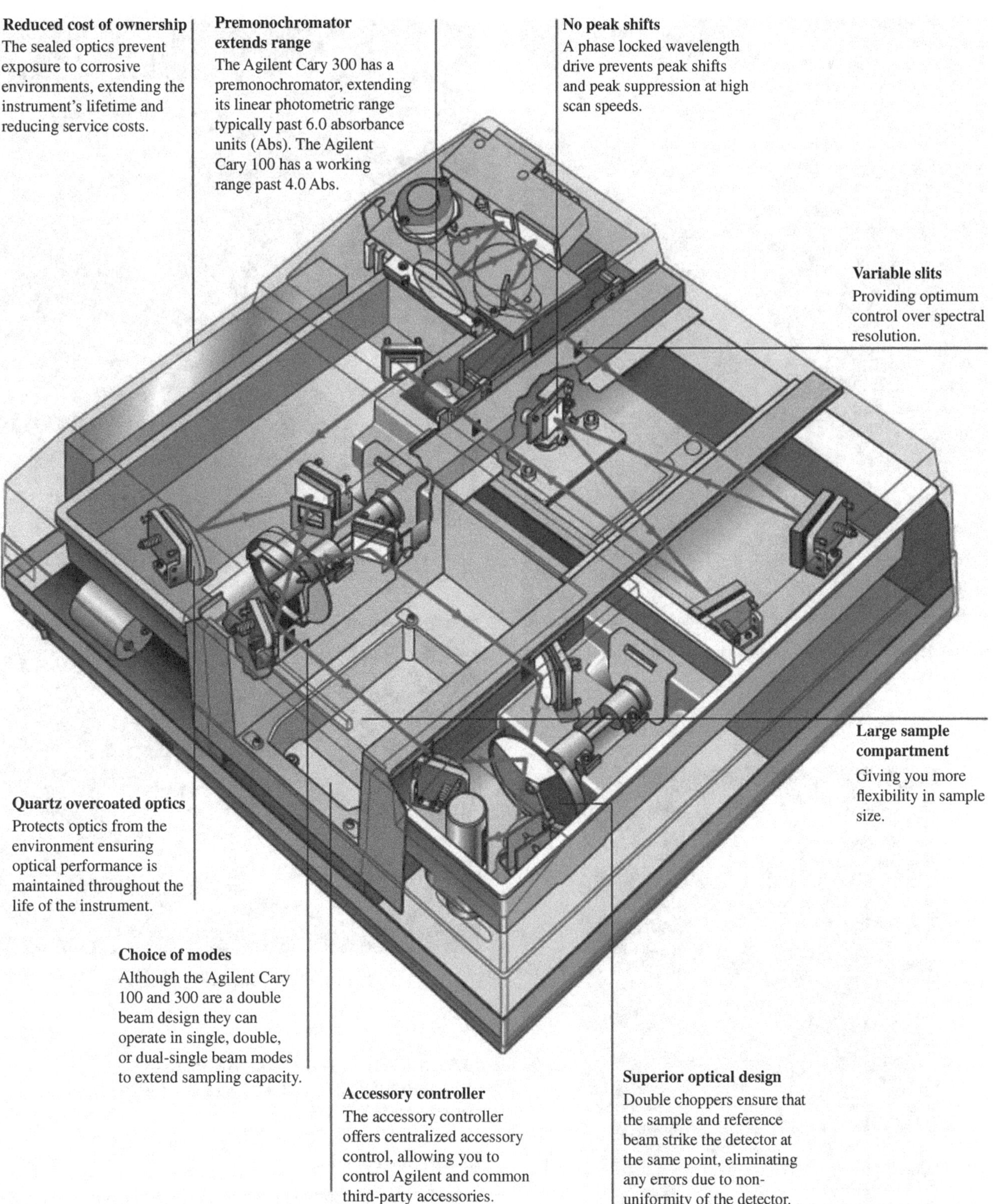

FIGURE 13.18 Internal optics of the Varian Cary 300 series spectrophotometer.

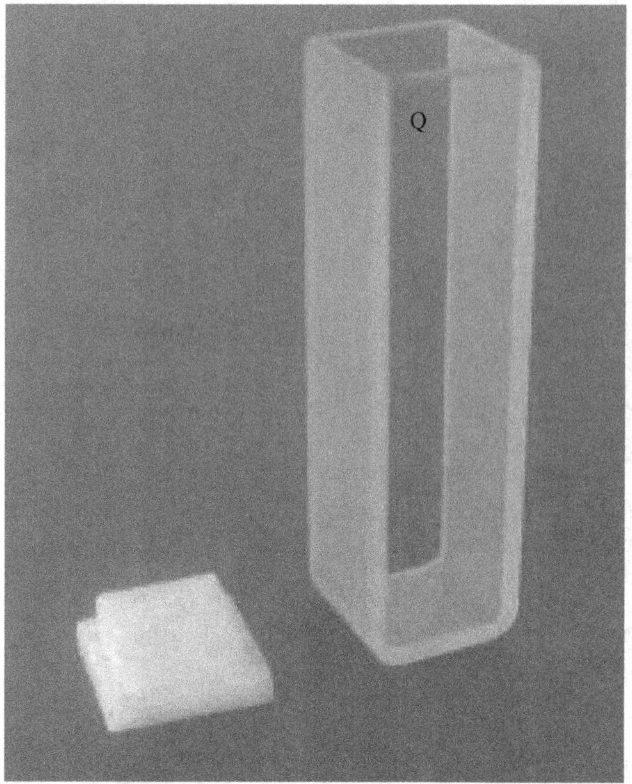

FIGURE 13.19 Rectangular quartz cuvette cell with 1 cm path length and a volume of 3.5 ml.

FIGURE 13.20 Micro quartz cuvette cell with 1 cm path length and a volume of 1.4 ml.

FIGURE 13.21 Cylindrical quartz cells.

FIGURE 13.22 Disposable plastic cells.

13.5 SPECIAL TOPIC (EXAMPLE)—SPECTROPHOTOMETRIC STUDY OF DYE COMPOUNDS

13.5.1 Introduction

One of the major problems with anticancer therapy is the effective delivery of the anticancer agent to the tumor site. Skin permeability studies have demonstrated that the coupling of the anticancer agent A007 with a carrier dye molecule (methylene green, MEG; methylene blue, MEB; and toluidine blue, TB; see Figure 13.23 for structures) increases A007's therapeutic activity, with a rating from most effective, to least, as MEG > MEB > TB. Electrospray ionization tandem mass spectrometry (ESI-MS/MS) has been used to study the characteristics and

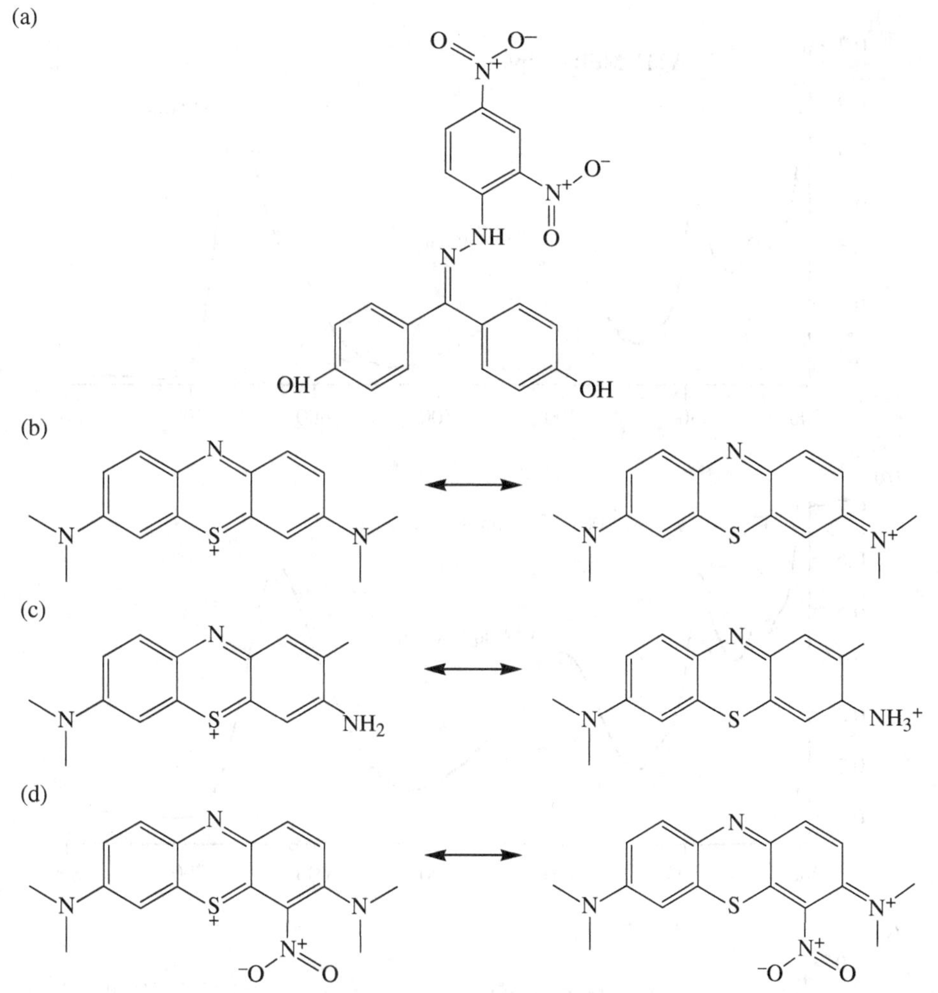

FIGURE 13.23 Structures of (a) 4,4′-dihydroxybenzophenone-2,4-dinitrophenylhydrazone (A007), (b) methylene blue (MEB) double salt, (c) toluidine blue (TB) double salt, and (d) methylene green (MEG) double salt.

relative binding strengths of the three dyes, MEG, MEB, and TB, complexed with A007. The gas-phase binding strengths of the three complexes, determined by ESI-MS/MS, were ranked in the following order: MEG > MEB > TB, which indicates that a stronger binding between the carrier and anticancer agent directly correlates to a greater efficacy. A high-performance liquid chromatography (HPLC) method has been developed for the determination of A007 in plasma, but not for the complexes that are formed between A007 and carrier compounds. An HPLC method is being developed in order to rank the binding strengths of carrier:A007 complexes. This will effectively gave a "solution-phase" ranking of binding strengths. The question to answer is whether the solution phase-bound A007:dye complexes have unique absorbance maxima that will allow measurement using HPLC coupled with a UV/Vis detector.

13.5.2 Experimental Setup for Special Topic Discussion

The UV/VIS experiments were performed with a PC-controlled CARY 500 Scan UV–Vis–NIR Spectrophotometer (Varian Australia Pty. Ltd.). The fluorescence experiments (we will look at the results of this study in the next Fluoroscopy Chapter 14) were performed with a PC-controlled Photon Technology International Fluorometer (PTI, Canada).

To mimic physiological conditions, all solutions are at a pH of 7.4, but adjusted due to the change in H^+ activity for a 50% methanol and 50% water mixture.

13.5.3 UV/Vis Study of the Compounds and Complexes

Figure 13.24 depicts UV/Vis spectra of the complexes formed between the three dyes and A007. What is observed from the spectra is that each compound contains a unique absorbance peak which can be used to detect and monitor the compounds and complexes without interference. The three dyes used in the study, methylene green, methylene blue, and toluidine blue, have unique absorbance maxima at 656, 662, and 630 nm, respectively. The anticancer agent A007 has an absorbance maximum between 398 and 405 nm. Both dye and A007 peaks are found in the complexes, and the A007 max at 398 nm can be used to detect the complex for HPLC retention times. The question set forth has been answered.

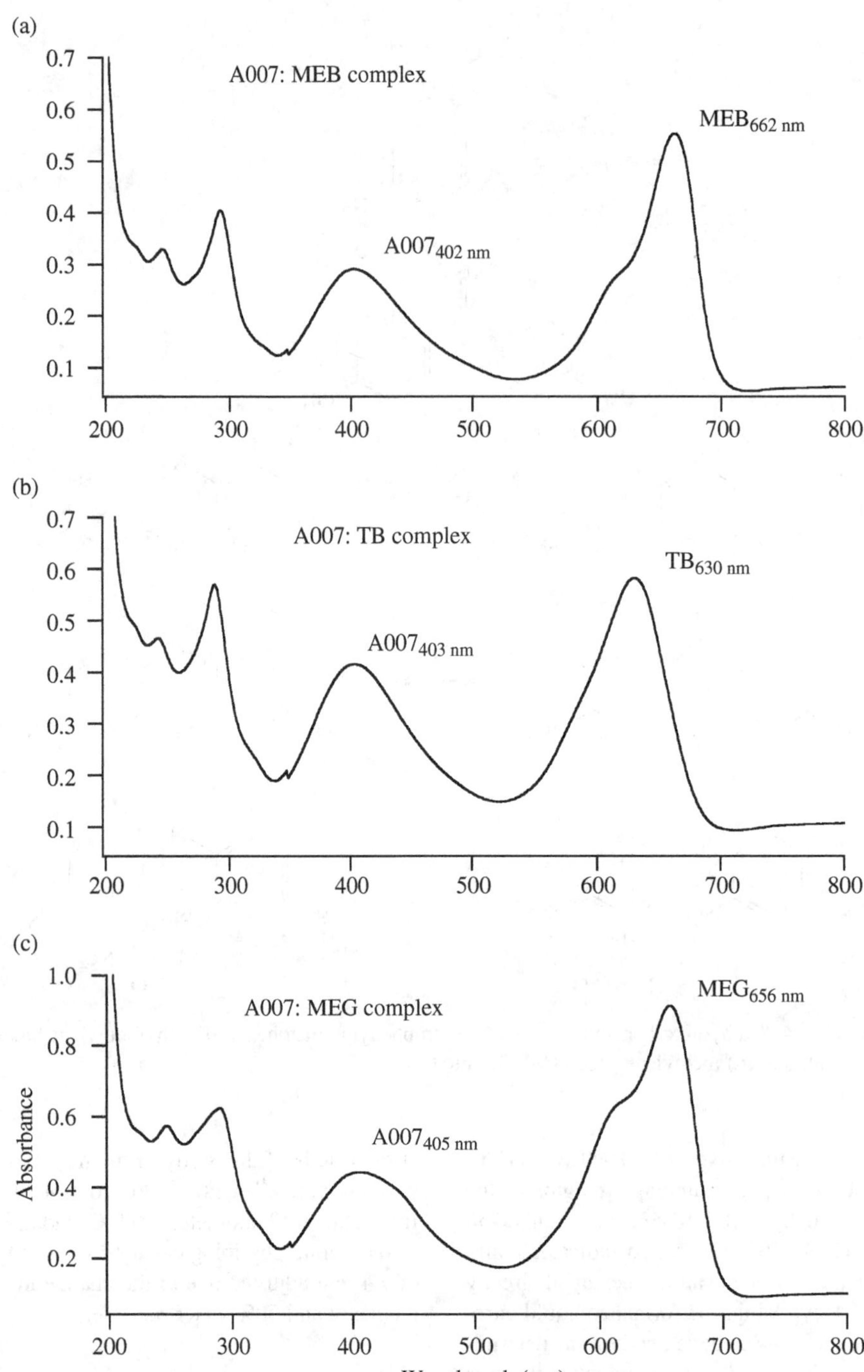

FIGURE 13.24 UV/Vis absorbance spectrums of A007:Dye complexes illustrating absorbance maximums of (a) 405 n + m for A007, 662 nm for MEB, (b) 630 nm for TB, and (c) 656 nm for MEG.

13.6 CHAPTER KEY CONCEPTS

13.1 The analytical laboratory utilizes the phenomenon of the electromagnetic spectrum for analyses.

13.2 Chemists and technicians make use of the special interaction of molecules with electromagnetic radiation, with the assistance of analytical instrumentation, such as UV/Vis spectrophotometers, fluorometers, and FTIR to measure, identify, and even quantitate compounds of interest.

13.3 UV/Vis and fluorimetry analysis is done with compounds in solution where the analyte of interest will either be measured directly according to its spectroscopic properties, or it may be combined with a chromophore or fluorophore to allow its measurement.

13.4 The entire electromagnetic spectrum comprises very short wavelengths (including gamma and X-rays) to very long wavelengths (including microwaves and broadcast radio waves).

13.5 The sunlight (or white light) that we see as uniform or homogeneous in color is composed of a broad range of radiation wavelengths in the ultraviolet (UV), visible, and infrared (IR) portions of the spectrum.

13.6 Visible light electromagnetic radiation is described in a dual manner as a wave phenomenon (wave theory), and also by a particle theory which is required to describe the absorption and emission of radiant energy.

13.7 The "wavelength" is the distance between adjacent peaks (or troughs), and may be designated in meters, centimeters or nanometers (10^{-9} m).

13.8 The "frequency" is the number of wave cycles that travel past a fixed point per unit of time, and is usually given in cycles per second, or hertz (Hz).

13.9 The energy carried by a photon of a certain wavelength is proportional to its frequency.

$$\nu = \frac{c}{\lambda}$$

$$\Delta E = h\nu$$

$$\Delta E = h\frac{c}{\lambda}$$

13.10 When a compound absorbs light, it produces a transition of an outer valence shell electron from its ground energy state to a higher energy level called an excited state.

13.11 The highest energy transition is the absorption of ultraviolet light (190–400 nm) to the second electronic excited state. The next transition is the absorption of visible light (400–800 nm) to the first electronic excited state. Finally, there is the absorption in the infrared region (1000–1500 nm), which is the lowest energy transition.

13.12 The color wheel is used to predict what color will be seen when a substance absorbs energy at different specific wavelengths.

13.13 If something absorbs energy in a wavelength range, then the complimentary color, which is directly opposite in the color wheel, is observed.

13.14 The nonbonding electrons of double bonds involving inorganic element such as Mn=O also demonstrates color.

13.15 Another type of colored compound is the azo dyes; however, they are not natural but have been prepared synthetically. The azo dyes get their name from the azo group, –N=N–, which is part of their structure.

13.16 Energetically favored electron promotion will be from the HOMO to the LUMO, producing what is called an excited state.

13.17 There are three types of energy changes when molecules absorb light energy: *electronic* which is the change in the energy of electrons in the molecule; *vibrational*, which is the change in the distance of atoms in the molecule; and *rotational*, which is the change in energy of the rotation of the molecule around a center.

13.18 Beer's law states that the absorbance of a substance is proportional to the concentration of the substance, the path length that the absorbing light travels through the sample, and the probability that absorbance takes place with an individual part of the substance, such as each molecule or atom.

$$\text{Absorbance } (A) = \log P_0 - \log P = \log\frac{P_0}{P}.$$

13.19 A spectrophotometer is used to measure the different wavelengths that are being absorbed by the analyte, and the magnitude of the absorbance.

13.20 Beer's law is generally referred to as the fundamental law of spectroscopy written in the following form:

$$\log\frac{P_0}{P} = A = abc.$$

13.21 There is a direct relationship between the absorbance and the transmittance of the incident light energy passing through the analyte solution.

$$A = -\log T = -\log(\%T/100).$$

13.22 The majorities of spectrophotometers are computer controlled and offer software that allows the user many functions toward UV/Vis studies.

13.23 Cuvettes or cells are what are usually used to hold samples for introduction into the spectrophotometer for measurement.

13.7 CHAPTER PROBLEMS

13.1 What is the absorbance of a solution with a %T of 73% at 350 nm?

13.2 Determine the absorbance of each solution.

a. Transmittance of 0.63.
b. Transmittance of 0.97.
c. Transmittance of 0.32.
d. Transmittance of 0.71.

13.3 Determine the absorbance of each solution.

a. Percent transmittance of 0.55%.
b. Percent transmittance of 68.9%.
c. Percent transmittance of 82.63%.
d. Percent transmittance of 95%.

13.4 If a system has a desired absorbance range of 0.095–0.155, what is the associated percent transmission (%*T*) range?

13.5 What is the absorptivity, a, of a copper solution that is 3.2×10^{-2} M and has an absorbance of 0.299 at 266 nm in a 1.50 cm cell?

13.6 What is the concentration of a solution that has an absorbance of 0.325 at 460 nm in a 2.0 cm cell, and a molar absorptivity, ε, of 12.5 l/mole cm?

13.7 A solution in the lab has an absorbance of 0.663 in a 1.0 cm cell, too concentrated for an accurate measurement. The sample is diluted from 1 to 5 ml and then measured in a 2.0 cm cell. If the measurement is linear, what is the expected absorbance?

13.8 If the molar absorptivity of the analyte in Problem 13.7 is $\varepsilon = 8.32$ l/mole cm, what is the concentration of the solution?

13.9 The absorbance of a 0.855 M standard of a complex between magnesium and ASN (mg-ASN) was collected from 320 to 490 nm in 10 nm steps. Plot the data and determine the best wavelength to collect them.

Wavelength (nm)	Absorbance	Wavelength (nm)	Absorbance
320	0.125	410	0.267
330	0.158	420	0.255
340	0.174	430	0.231
350	0.199	440	0.217
360	0.221	450	0.189
370	0.244	460	0.163
380	0.275	470	0.146
390	0.288	480	0.131
400	0.281	490	0.114

13.10 What is the molar absorptivity, ε, of the mg-ASN complex if the solution in Problem 13.9 was 0.855 M in a 1.0 cm cell?

13.11 What is the concentration of an mg-ASN solution that has an absorbance of 0.198 at 288 nm in a 0.5 cm cell?

13.12 If we were to measure the solution described in Problem 13.11 using a 1.0 cm cell, would the absorbance be doubled?

13.13 A colored calcium complex was measured with the following standards and tabulated. Using ChemTech or Excel®, plot the data.

Standard	Ca concentration (ppm)	Absorbance
1	1.1	0.110
2	3.2	0.215
3	5.1	0.313
4	7.2	0.435
5	10.0	0.588

13.14 From the slope of the best fit equation, determine the absorptivity in l/mg cm, and the molar absorptivity in l/mole cm. Assume that the cell path length is 1 cm.

13.15 If an unknown sample has an absorbance of 0.339, calculate the concentration of the sample in ppm and in molarity (M).

13.16 The chromophore retinal absorbance at 280 nm was measured for a set of standards and recorded. Using ChemTech or Excel®, plot the data.

Standard	Retinal concentration (%)	Absorbance
1	0.15	0.048
2	0.25	0.093
3	0.38	0.136
4	0.50	0.171
5	0.75	0.208

13.17 If an unknown sample has an absorbance of 0.198, using the ChemTech-ToolKit calculate the concentration of the sample in % and in molarity. The molar mass of retinal is 284.44 g/mol.

13.18 If an unknown sample has an absorbance of 0.210, using the ChemTech-ToolKit calculate the concentration of the sample in % and in molarity.

14

FLUORESCENCE OPTICAL EMISSION SPECTROSCOPY

14.1 Introduction to Fluorescence
14.2 Fluorescence and Phosphorescence Theory
14.2.1 Radiant Energy Absorption
14.2.2 Fluorescence Principle—Jabłoński Diagram
14.2.3 Excitation and Electron Spin States
14.3 Phosphorescence
14.4 Excitation and Emission Spectra
14.5 Rate Constants
14.5.1 Emission Times
14.5.2 Relative Rate Constants (k)
14.6 Quantum Yield Rate Constants
14.7 Decay Lifetimes
14.8 Factors Affecting Fluorescence
14.8.1 Excitation Wavelength (Instrumental)
14.8.2 Light Source (Instrumental)
14.8.3 Filters, Optics, and Detectors (Instrumental)
14.8.4 Cuvettes and Cells (Instrumental)
14.8.5 Structure (Sample)
14.9 Quantitative Analysis and Beer–Lambert Law
14.10 Quenching of Fluorescence
14.11 Fluorometric Instrumentation
14.11.1 Spectrofluorometer
14.11.2 Multidetection Microplate Reader
14.11.3 Digital Fluorescence Microscopy
14.12 Special Topic—Flourescence Study of Dye-A007 Complexes
14.13 Chapter Key Concepts
14.14 Chapter Problems

14.1 INTRODUCTION TO FLUORESCENCE

In Chapter 13, Ultraviolet/Visible (UV/Vis) Absorption Spectroscopy, we studied the theory and application of UV/Vis spectroscopy where the absorbance of radiant energy was used to detect and quantitate molecules. In molecular absorbance photospectroscopy, the absorbance of the incident radiant energy is proportional to the concentration of the analyte according to Beer's law:

$$\text{Absorbance (A)} = \log P_0 - \log P = \log \frac{P_0}{P} \quad (14.1)$$

$$\log \frac{P_0}{P} = A = abc, \quad (14.2)$$

where

a = absorptivity, a characteristic of a substance that is constant at a given wavelength

b = the path length that the incident light travels through the analyte solution, in centimeters

c = the concentration of the analyte in solution, expressed in mg l^{-1} (ppm), g/100 ml, and so on.

Fluorescence and the closely related phosphorescence are processes that include absorbance of radiant energy of infrared, visible, or ultraviolet light, which is followed by the release of quanta of energy that are subsequently detected and measured. Collectively, fluorescence, phosphorescence, and chemiluminescence are known as molecular luminescence. In chemiluminescense, an excited species is formed through a chemical reaction that results in emission of energy. There are in fact many types of other luminescence including radioluminescence, electroluminescence, electroluminescence, and bioluminescence. Incandescence is the emission of radiation from a hot body system due to its condition of high temperature. Luminescence, on the other hand includes all other types of light emission. Most organic molecules and inorganic compounds will luminescence or emit photons when excited to higher electronic states. In proteins, the aromatic amino acids tyrosine, tryptophan, and phenylalanine fluoresce. The nucleotides cytosine, thymine, guanine, uracil, and adenine in DNA and RNA fluoresce. Aromatic compounds makeup the largest group of luminescing compounds. Fluorescence and phosphorescence are photoluminescent processes, where it is incident energy that excites the species resulting in emission of photons. Fluorescence is a faster process as compared to phosphorescence, where the release of energy is almost instantaneous.

Analytical Chemistry: A Chemist and Laboratory Technician's Toolkit, First Edition. Bryan M. Ham and Aihui MaHam.

Phosphorescence is emission that is accompanied with a change in electron spin that results in irradiation that lasts up to a few seconds.

14.2 FLUORESCENCE AND PHOSPHORESCENCE THEORY

14.2.1 Radiant Energy Absorption

Radiant energy is absorbed by atoms or molecules in definite units, which are called quanta. One quantum of energy (E) is directly proportional to its frequency of oscillations represented as

$$E = h\nu = \frac{hc}{\lambda}, \quad (14.3)$$

where

ν = frequency (s^{-1})
λ = wavelength (either nm or cm)
h = Plank's constant (6.626×10^{-34} J s)
c = speed of light (3.0×10^{8} m)

In physics, the energy value is coupled to the work using units of Joules J. Often in spectroscopy the unit used to represent energy is the inverse centimeter (cm^{-1}) where $hc \sim 2 \times 10^{-23}$.

The quantity used in quantum efficiency calculations is the einstein based on the number of single molecules in a gram molecule as N quanta, where $N = 6.023 \times 10^{23}$. In this way, the absorption of one einstein is the energy needed for the reaction of one gram mole.

Excited species are formed with the absorption of incident radiation (light), which imparts energy to the molecules. The excited molecules then release some of this energy in the form of light emission. The efficiency in which the light is absorbed and then released is called the quantum efficiency and is described as

$$\Phi E = \frac{\text{einsteins emitted}}{\text{einsteins absorbed}} = \frac{\text{quanta emitted}}{\text{quanta absorbed}}. \quad (14.4)$$

Note that in the perfect sense where the quanta emitted is equal to the quanta absorbed, the quantum efficiency would be unit or equal to 1 ($\Phi E = 1$). Thus, since the quanta emitted can never be greater than the quanta absorbed, the quantum efficiency will always be less than or equal to 1 ($\Phi E \leq 1$).

14.2.2 Fluorescence Principle—Jabłoński Diagram

Jabłoński diagrams describe light absorption and emission and demonstrate different molecular processes happening in the excited states during fluorescence. Figure 14.1 is a schematic of a typical Jabłoński diagram. In the Jabłoński diagram, S_0 shows a singlet ground state, and S_1 and S_2 show first and second excited states, respectively. At each different state, the notations 0, 1, and 2 mean different vibrational energy levels. T_1 means the first triplet state. When a fluorophore absorbs light, it will excite from the singlet ground state to the first or second excited states. Internal conversion usually occurs when the excited fluorophore molecules relax from the second excited state to the first excited state. This process normally happens before the emission starts, so generally we see the absorption and emission spectra are the mirror images of each other because the electrons excited do not change the nuclear geometry. Intersystem crossing happens when the electrons in the first excited state S_1 transit to the triplet state T_1. This process will emit phosphorescence which is lower in energy compared to fluorescence.

14.2.3 Excitation and Electron Spin States

Excitation of atomic and molecular species can come in many forms. For example, gamma rays produce nuclear excitation reactions, X-rays produce transitions of inner atomic electrons, ultraviolet and visible lights produce transitions of outer atomic electrons, infrared produces molecular vibrations, far infrared produces molecular rotations, and long radio and radar waves produce oscillations of mobile or free electrons. All of these

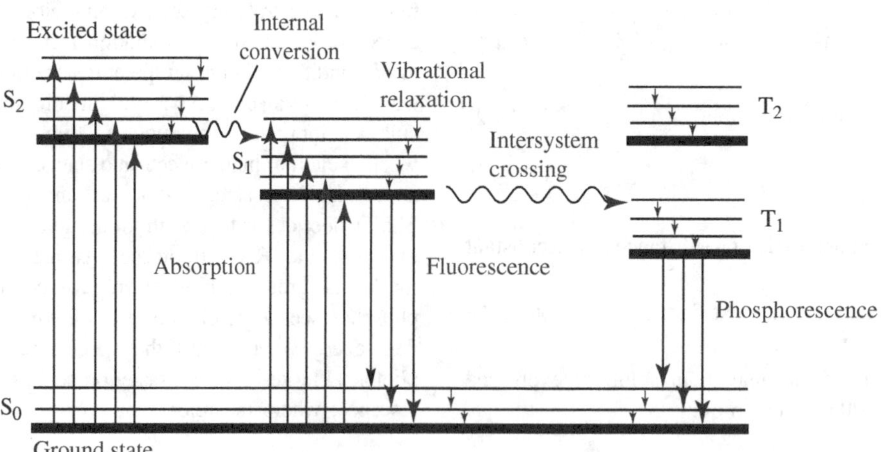

FIGURE 14.1 Jabłoński diagram.

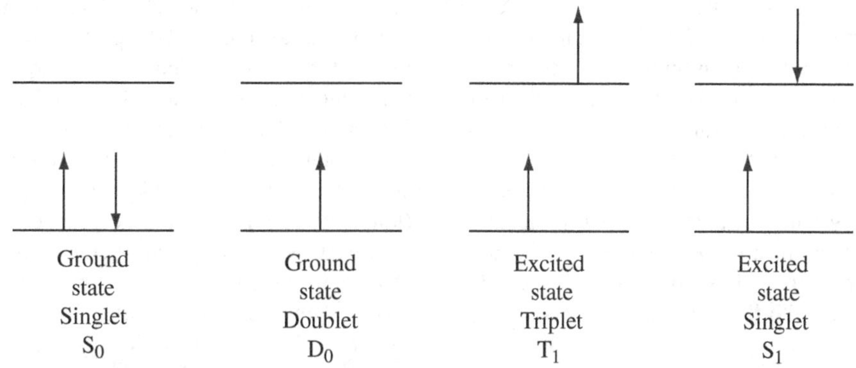

FIGURE 14.2 The spins for singlet ground state (S_0), doublet ground state (D_0), triplet excited state (T_1), and singlet excited state (S_1).

are examples of excitation produced from the absorbance of incident energy. In fluorescence and phosphorescence, the processes involve the outer atomic electrons. Take for instance the absorption of energy of gaseous atomic calcium at 422.7 nm exciting a valence shell electron from the 4s level to the 3d level. After a few seconds, the atom will release energy at a wavelength of 422.7 nm in all directions returning to the ground state. For atoms, the emitted wavelength is often at a value that is the same as the absorbed wavelength because atoms do not have the same vibrational energy levels as molecules. The type of fluorescence with the release of energy without a change in frequency is known as resonance fluorescence and is common with excited atoms. Most organic molecules will release energy at longer wavelength than the excitation due to a shift known as the Stokes shift.

14.2.3.1 Quantum Numbers Electrons in atoms possess four quantum numbers (*n*, *l*, *ml*, and *ms*). The definitions of the quantum numbers are

- *n* principle (1 s, 3p...)
- *l* angular momentum ($l = 0 = $ s, $l = 1 = $ p...)
- *s* spin ($s = +1/2$ or $-1/2$), and
- *m* magnetic.

The Pauli exclusion principle states that two electrons cannot have the same four quantum numbers and that only two electrons can occupy each orbital and their spins must be opposite. Spin pairing results in what is known as a diamagnetic state where the molecule does not exhibit a magnetic field and are not repelled or attracted by a static electric field. When a molecule has an unpaired electron, it will possess a magnetic moment, will be attracted to a magnetic field, and is called paramagnetic. Free radicals are an example of paramagnetic molecules. If two electrons are in the same orbital, they will have the same n, l, and m quantum numbers, but different spins.

14.2.3.2 Electron Spin States We know that two electrons that are located in the same orbital must have different spins:

$$s = +\frac{1}{2} \text{ or } -\frac{1}{2} \tag{14.5}$$

$$S = \sum |s_i|, \tag{14.6}$$

where +1/2 is a spin-up while −1/2 is a spin-down. The state of the electron is determined from its multiplicity:

$$2S + 1 \text{ (either 1, 2, 3...).} \tag{14.7}$$

The spins for singlet ground state (S_0), doublet ground state (D_0), triplet excited state (T_1), and singlet excited state (S_1) are depicted in Figure 14.2. The singlet, doublet, and triplet states are derived for the multiplicity. The simplest state is the rest ground state of the singlet S_0, where the two electrons are in the lowest level and have opposite spins. The multiplicity is $2S + 1 = 2(+1/2 - 1/2) + 1 = 1$ for the singlet state. For the excited singlet state, the multiplicity is still $2S + 1 = 2(+1/2 - 1/2) + 1 = 1$ for the singlet state. For the doublet state, the multiplicity is $2(+1/2) + 1 = 2$. For the excited triplet state, the multiplicity is $2(+1/2 + 1/2) + 1 = 3$.

The transition from the ground singlet state to the excited singlet state has a greater probability of taking place as compared to the transition to the excited triplet state. This has been born out through experimentation where the absorption band for the singlet to triplet state is much weaker than that of the singlet to singlet state. The transition to the triplet state requires a change in electron spin or electronic state. Under normal conditions at room temperature, most molecules will exist in the ground singlet state.

14.3 PHOSPHORESCENCE

Phosphorescence differs from fluorescence where in phosphorescence there is an electron spin change resulting in a long excited-state lifetime on the order of seconds to minutes. In comparison, the fluorescence process (predominantly singlet state) does not involve an electron spin change (triplet state) resulting in much shorter excited-state lifetimes on the order of less than 10^{-5} s. Even though quantum theory does predict the existence of a triplet excited state, the direct transition from the ground singlet state to the excited triplet state is forbidden. The process that produces the triplet state is generally derived from an excited singlet state that possesses a lowest vibrational level that has the same energy as an upper vibrational level of the triplet state. The conversion from the excited singlet state to the excited triplet state is known as intersystem crossing and is depicted in Figure 14.1. The emission of light with phosphorescence is delayed as compared to fluorescence. In fluorescence, the transition from the excited singlet state to the ground state with the emission of light takes place rapidly from 10^{-6} to 10^{-9} s. In phosphorescence, the transition from the excited triplet state to the ground state is much slower on the order of 10^{-4} sup to 1–2 min. After the intersystem crossing, the excited triplet state possesses lower energy than the associated excited singlet state. Thus, we observe the emission in phosphorescence to take place at higher wavelengths than with fluorescence.

14.4 EXCITATION AND EMISSION SPECTRA

As we have seen in the Jabloński diagram, when a system absorbs incident radiant energy elevating an electron to an excited state, the fluorescent mode of relaxation is to release energy. Figure 14.3(a) depicts a UV spectrum from 200 to 700 nm wavelength for the absorption and subsequent emission. At lower wavelength, there are four bands of absorbance at 240, 290, 340, and 410 nm. The bands representing the energy emission are at longer wavelengths of 500, 580, 625, and 680 nm. Emission or fluorescence bands are primarily at lower frequency (longer wavelength) for most organic compounds. This is known as the Stokes shift named after the Irish physicist George G. Stokes. In the simplest sense, the shift will be approximately 10 nm. However, this is often not the case as observed in Figure 14.3(a). Usually, when a molecule absorbs energy and the electron is elevated to a higher energy state, relaxation processes take place that releases a small amount of the initial absorbed energy. Often the relaxation process is in the form of vibrational relaxation and internal conversion. Following this, the compound emits energy as fluorescence but at a lower energy

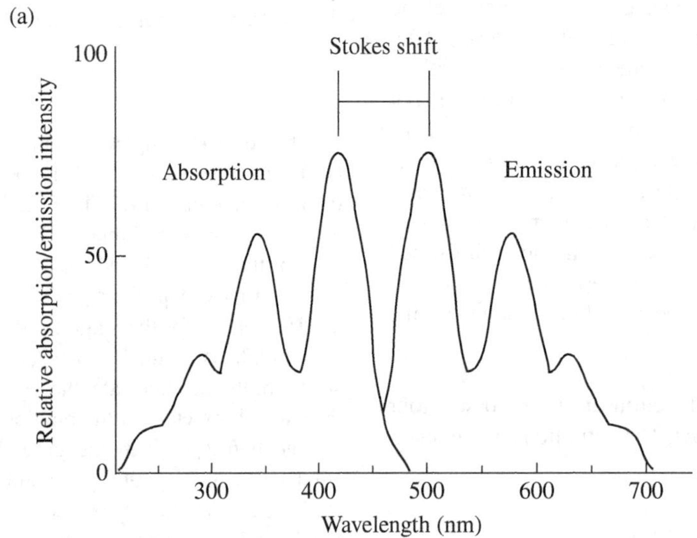

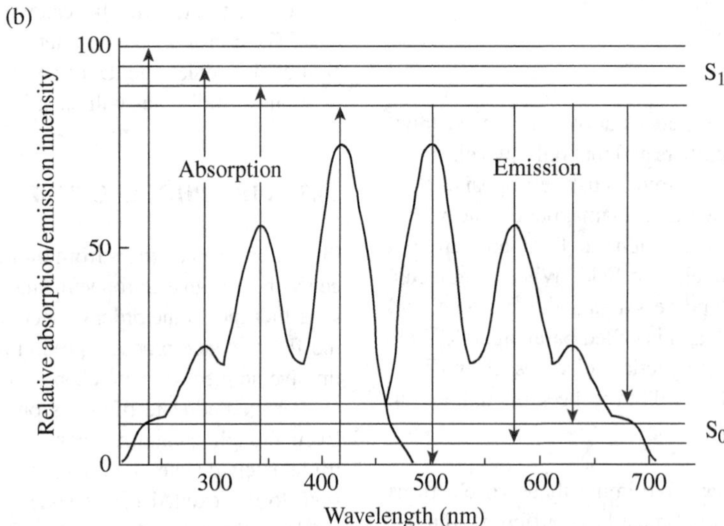

FIGURE 14.3 Absorption and emission spectra for a typical organic compound. (a) Emission or fluorescence bands are at lower frequency (longer wavelength) as compared to the absorption bands known as the Stokes shift named after the Irish physicist George G. Stokes. (b) Overlay of the absorption and emission lines associated with the energy bands that are illustrated in Jabloński diagrams.

level, which we see as the Stokes shift (also called a red shift). Figure 14.3(b) shows an overlay of the absorption and emission lines that were illustrated in the Jabłoński diagram of Figure 14.1. This gives us a better idea of the absorption and emission spectra and how they relate to the energy processes that are shown in the Jabłoński diagram.

14.5 RATE CONSTANTS

14.5.1 Emission Times

The absorption of the incident radiant energy by atoms and molecules is very fast for the elevation of the electron from the singlet ground state to the excited singlet state $S_0 \rightarrow S_1$ on the order of 10^{-13} to 10^{-15} s. Resonant emission, which is common in atoms for the relaxation of the singlet excited-state electron to the singlet ground state $S_1 \rightarrow S_0$, is on the order of 10^{-5} to 10^{-9} s. The same is for nonresonant emission, which is common in molecules is also on the order of 10^{-5} to 10^{-9} s. As mentioned previously, the transitions involving the triplet state are theoretically forbidden, which means in practice that they have a low probability of taking place. This, then equates to a much slower emission rate. The nonresonant emission from the triplet excited state to the singlet ground state $T_1 \rightarrow S_0$ is slow on the order of 10^{-5} to 10 s or slower.

14.5.2 Relative Rate Constants (k)

Rate parameters are used to describe the emission process of excited species and the probability or likelihood that fluorescence will take place. As was presented earlier, the quantum yield or quantum efficiency (ΦE) is a quantity used to determine whether a molecule will fluoresce or phosphoresce, where $\Phi E = 0.0$ to 1.0.

$$\Phi E = \frac{\text{einsteins emitted}}{\text{einsteins absorbed}} = \frac{\text{quanta emitted}}{\text{quanta absorbed}}. \quad (14.4)$$

The quantum efficiency can be expressed in terms of the relative rate constants (k) of the deactivation processes that return the excited state to the ground state. Figure 14.4 depicts the various deactivation processes and the associated rate constant designations.

14.6 QUANTUM YIELD RATE CONSTANTS

The general rate equation for the quantum yield of both fluorescence and phosphorescence is expressed as

$$\Phi_{\text{fluor}} = \frac{k_{\text{fluor}}}{k_{\text{fluor}} + k_{\text{int con}} + k_{\text{ext con}} + k_{\text{ISC}} + k_{\text{pre dis}} + k_{\text{dis}}}, \quad (14.8)$$

where

k_{fluor} = fluorescence	The release of energy as emission not involving a spin change such as $S \rightarrow S$.
$k_{\text{int con}}$ = internal conversion	The transition to a lower state when vibrational energy levels match without the release of radiation.
$k_{\text{ext con}}$ = external conversion	The transition to a lower state by collisional deactivation without the release of radiation.
k_{ICS} = intersystem crossing	A transition with spin change such as $S \rightarrow T$.
$k_{\text{pre dis}}$ = predissociation	Relaxation to ground state with enough energy break bonds and cause dissociation.
k_{dis} = dissociation	Excitation to a vibrational state with enough energy to break bonds and cause dissociation.

The quantum yield constants for the associated transitions are depicted in Figure 14.4. This give a better view of what each of these rate constant transitions stand for. For fluorescence, the quantum yield for the $S_1 \rightarrow S_0$ transition equation is

$$\Phi_{\text{fluor1}} = \frac{k_{FM}}{k_{FM} + k_{TM} + k_{GM} + k_{\text{pre dis}} + k_{\text{dis}}}. \quad (14.9)$$

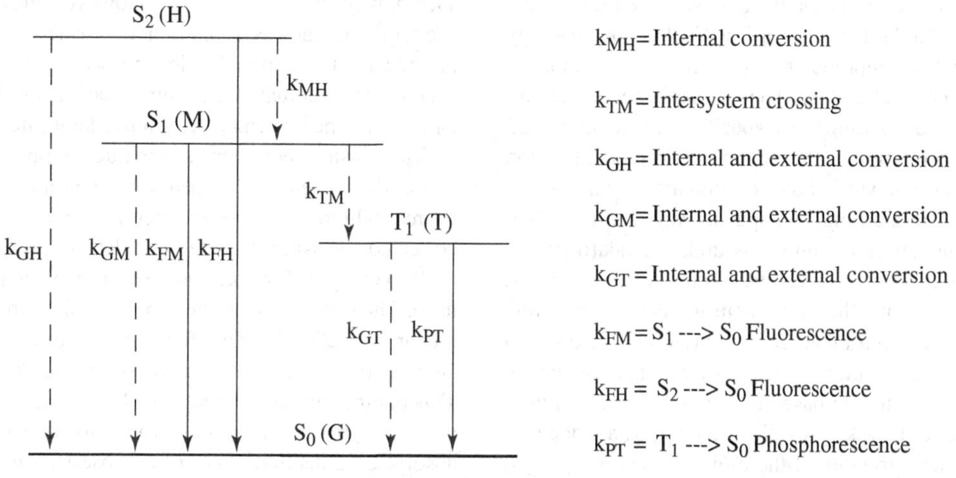

k_{MH}= Internal conversion

k_{TM}= Intersystem crossing

k_{GH}= Internal and external conversion

k_{GM}= Internal and external conversion

k_{GT} = Internal and external conversion

$k_{FM} = S_1$ ---> S_0 Fluorescence

$k_{FH} = S_2$ ---> S_0 Fluorescence

$k_{PT} = T_1$ ---> S_0 Phosphorescence

FIGURE 14.4 Deactivation processes and the associated rate constant designations.

For this transition, the k_{MH} and k_{TM} are considered very small and not significant in the quantum yield calculation. The fluorescence quantum yield for the $S_2 \rightarrow S_0$ transition equation is

$$\Phi_{fluor2} = \frac{k_{FH}}{k_{FH} + k_{GH} + k_{MH} + k_{pre\ dis} + k_{dis}}. \quad (14.10)$$

The phosphorescence quantum yield for the $T_1 \rightarrow S_0$ transition equation is

$$\Phi_{phos} = \frac{k_{PT}}{k_{PT} + k_{GT} + k_{pre\ dis} + k_{dis}}. \quad (14.11)$$

14.7 DECAY LIFETIMES

The lifetimes of the decay processes are also based on the rate constants. For the $S_1 \rightarrow S_0$ transition, the lifetime equation is

$$\tau_M = \frac{1}{k_{FM} + k_{TM} + k_{GM} + k_{pre\ dis} + k_{dis}}. \quad (14.12)$$

The lifetime for the $S_2 \rightarrow S_0$ transition is

$$\tau_H = \frac{1}{k_{FH} + k_{GH} + k_{MH} + k_{pre\ dis} + k_{dis}}. \quad (14.13)$$

The phosphorescence lifetime for the $T_1 \rightarrow S_0$ transition is

$$\tau_T = \frac{1}{k_{PT} + k_{GT} + k_{pre\ dis} + k_{dis}}. \quad (14.14)$$

14.8 FACTORS AFFECTING FLUORESCENCE

Fluorescence and phosphorescence are similar to UV/Visible analysis at low concentrations where in both there is a direct relationship between concentration and the intensity of emission. We are able to collect standard spectra in UV/Visible spectroscopy that are for the most part reproducible from instrument to instrument, and laboratory to laboratory. The standard spectra are collected under standard conditions of specific wavelengths, cell pathways, and concentrations, often reported as molar extinction coefficients (ϵ). The analyst is able to compare spectra which have been collected into a library to spectra collected in his/her laboratory, and calculate concentrations under standard conditions using the extinction coefficients. With fluorescence and phosphorescence, this is not the case. There are factors which distort and influence the spectra that require corrections in order to obtain true (instrument independent) spectra. The distorting influences are both instrument based and sample based. In general, the magnitudes of k_{FM}, k_{FH}, k_{PT}, k_{dis}, and $k_{pre\ dis}$ are dependent upon the chemical structure of the molecule, and k_{GH}, k_{GM}, and k_{GT} upon the environment.

14.8.1 Excitation Wavelength (Instrumental)

The excitation wavelength (λ, nm) can have a negative impact on decays by bringing about bond breakage designated as either dissociation or predissociation. The larger the values for k_{dis} and $k_{pre\ dis}$, the lower the quantum yield will be. Both constants are located in the denominator of the quantum yield equations as illustrated in Equation 14.8. Shorter wavelengths from 220 to 190 nm are high in energy and can induce bond breakage. The absorption of ultraviolet radiation with wavelengths shorter than 250 nm usually does not result in fluorescence. This is due to the energetic state high enough to deactivate the excited electron through dissociation.

14.8.2 Light Source (Instrumental)

There is also observed variation in the light source intensity and wavelength, an instrumental influence. The variation occurs from experiment to experiment and also takes place during a single experimental run. Numerous light sources have been incorporated into fluorescence instrumentation, including lasers, light-emitting diodes (LEDs), xenon arc, tungsten-halogen, and mercury-vapor lamps. These lamps have either a continuum of energy over a wide range or a series of discrete lines. The tungsten-halogen lamp is an example of a continuum over a wide range, and the mercury-vapor lamp is of discrete lines. The mercury-vapor lamp is one of the most common discrete line light sources used in instrumentation. The mercury-vapor lamp is a gas discharge lamp using an electric arc through vaporized mercury producing very close to white light (light covering the entire visible spectrum). The strongest peaks of the mercury-vapor emission line spectrum include 184 nm (ultraviolet), 254 nm (ultraviolet), 365 nm (ultraviolet), 405 nm (violet), 436 nm (blue), 546 nm (green), and 578 nm (yellow-orange). Figure 14.5 depicts (a) the spectral distribution of a commercially available mercury lamp, and (b) a xenon arc lamp spectrum showing a continuum of energy from approximately 250 to 700 nm. The figure contains the major wavelengths listed above but also includes a few more. Sometimes, the lamps are also coated with fluorophores to add extra wavelengths, and doped with other metals or gasses to increase the wavelength range coverage. Mercury arc discharge lamps are ideal wavelength sources due to their coverage of the spectrum from 184 to 578 nm. However, they do possess some shortcomings such as a significantly greater amount of fluctuation in intensity as compared to incandescent lamps, LEDs, and laser sources. The fluctuations are attributable to the deterioration of the lamp components producing shorter lamp lifetimes.

Most instruments correct for fluctuations in the light source of the fluorometer using beam splitters that direct some of the source light to a reference detector. The primary detector is thus corrected for using the reference detector.

Wavelengths for excitation lower than 250 nm are usually not used. The energy transmitted to the analyte from wavelengths in the range of 200–250 nm often will transmit enough energy to the analyte to bring about deactivation of the fluorescence process. This in turn increases the k_{dis} and $k_{pre\ dis}$ values. The highly energetic $\sigma \rightarrow \sigma*$ (sigma bond orbital to sigma antibond orbital) absorption transition is not often observed in fluorescence excitation using wavelengths greater than 250 nm. The transitions

(a)

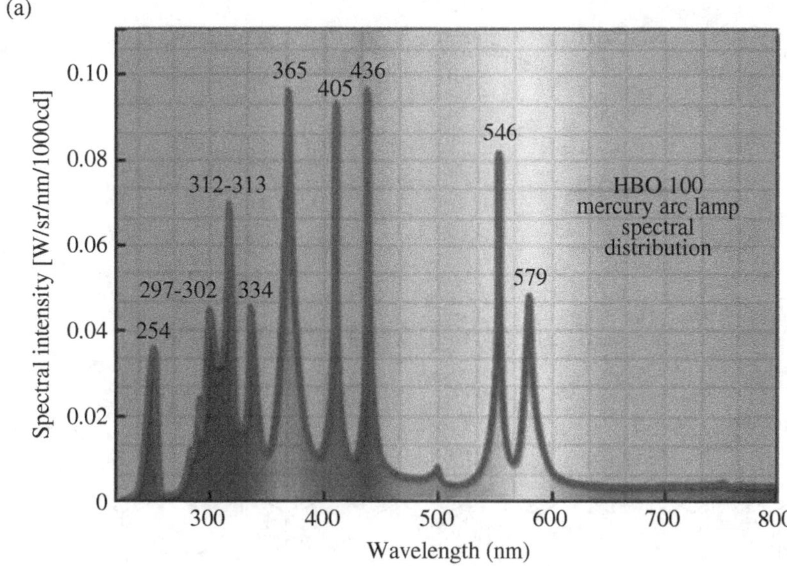

(b)

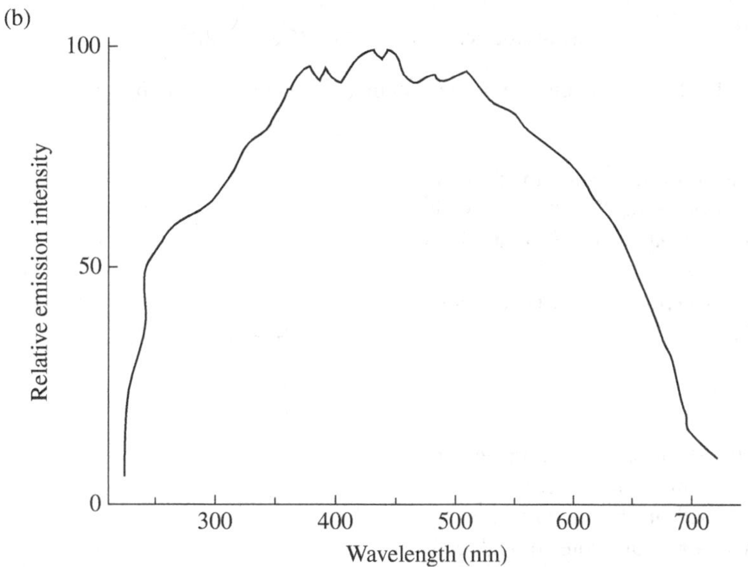

FIGURE 14.5 (a) Spectral distribution of mercury lamp with emission line spectrum including 184 nm (ultraviolet), 254 nm (ultraviolet), 365 nm (ultraviolet), 405 nm (violet), 436 nm (blue), 546 nm (green), and 578 nm (yellow-orange), and (b) a xenon-arc lamp spectrum showing a continuum of energy from approximately 250 nm to 700 nm. (Credit and/or copyright notice: www.zeiss.com/campus, © Mike Davidson, FSU, Tallahassee.)

usually observed for the fluorescence radiation are the $\pi* \rightarrow \pi$ (pi antibond orbital to pi bond orbital) transitions. More on these orbital transitions will be discussed in the consideration of structures below.

14.8.3 Filters, Optics, and Detectors (Instrumental)

The fluorometric instrumentation contains numerous filters, slits, and monochromators involved in isolating and transferring various wavelengths for excitation, and intensity collection. With time, these change in transmission efficiency and also specificity. Detectors also will deteriorate with time resulting in changing the detector quantum efficiency and thus decreasing in sensitivity due to a decrease in the number or percentage of photons collected. There is also variation in response from detector to detector. All of these processes have a distinct effect upon the measurement of fluorescence making it difficult to standardize across all instruments. Later in this chapter we shall take a closer look at the inner components of fluorometer instrumentation.

14.8.4 Cuvettes and Cells (Instrumental)

As was also the case with UV/Visible spectrophotometric analyses, the sample holder needs to be composed of an inert substance that does not absorb in a useful wavelength range. Quartz has been the most used material for cuvettes in UV, visible, and near-infrared (NIR) spectroscopy measurements due to its ideal transmittance from 200 to 2500 nm. The crystalline structure of quartz is made up exclusively of pure silicon dioxide (SiO_2) while other glass tends to have impurities. Glass cuvettes

(a)

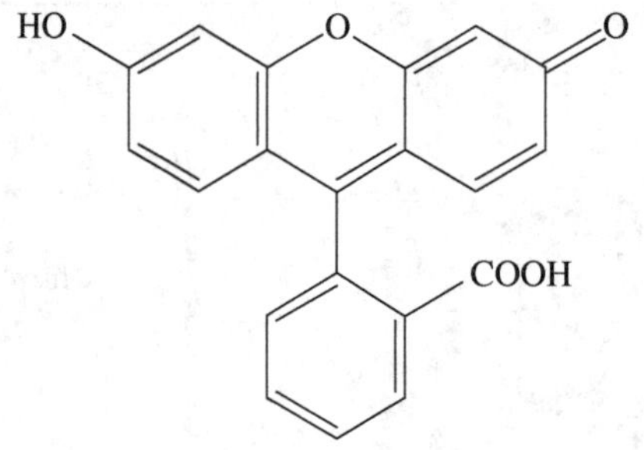

Fluorescein, $C_{20}H_{12}O_5$, MW 332.31 g mol^{-1}

(b)

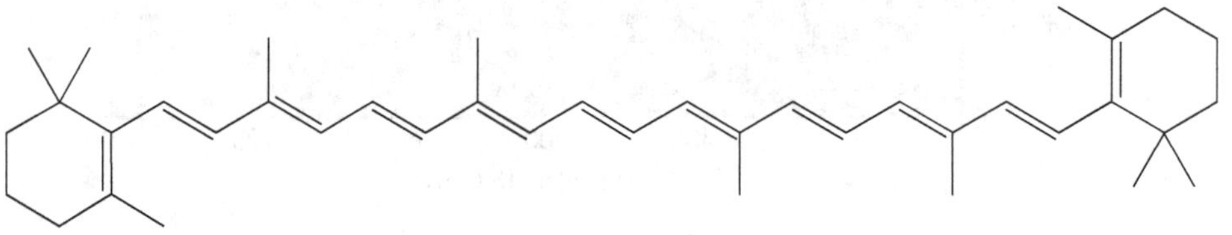

Beta-carotene, $C_{40}H_{56}$, MW 536.87 g mol^{-1}

FIGURE 14.6 (a) Structure of fluorescein. (b) Structure of beta-(β)-carotene.

available have a transmittance range from 320/360 to 2500 nm, Plastic cuvettes available cover a range of 220–900 nm, while polystyrene cover 350–900 nm. See Chapter 13, Section 13.3.8 UV/Visible Spectrophotometers, and Figures 13.19, 13.20, 13.21, and 13.22 for examples of cuvettes used in UV, visible, NIR, and fluorescence spectroscopy.

14.8.5 Structure (Sample)

14.8.5.1 Fluorescein and Beta-(β)-Carotene Aromatic containing compounds are the most common fluorescent structure that the analyst is likely to work with in the laboratory. These compounds contain low-energy $\pi \rightarrow \pi*$ transitions through conjugated pi bonds. Fluorescein, a commonly used aromatic fluorescent tracer is depicted in Figure 14.6(a) as an example of an organic fluorescent compound. Fluorescein is a dark orange/red synthetic organic powder compound. It has an absorption maximum at 494 nm and an emission maximum at 521 nm. Other nonaromatic compounds that fluoresce include highly conjugated double-bond structures (alternating single and double bonds) such as beta-(β)-carotene, which has an excitation wavelength of 280 nm and a fluorescence maximum at 320–340 nm. β-carotene is a terpenoid and is a strongly colored red-orange pigment found abundantly in plants and fruits. Figure 14.6(b) depicts the repeating conjugated double-bond structure of β-carotene. The double-bond system contains pi (π) bonding.

14.8.5.2 Diatomic Oxygen Molecular Orbital Diagram The molecular orbital diagram for the diatomic oxygen molecule (O_2) is illustrative of the possible $\sigma \rightarrow \sigma*$ transition and the $\pi* \rightarrow \pi$ transition due to its double-bond characteristic. The molecular orbital diagram for the diatomic oxygen molecule (O_2) is depicted in Figure 14.7. As shown in the figure, the high energy

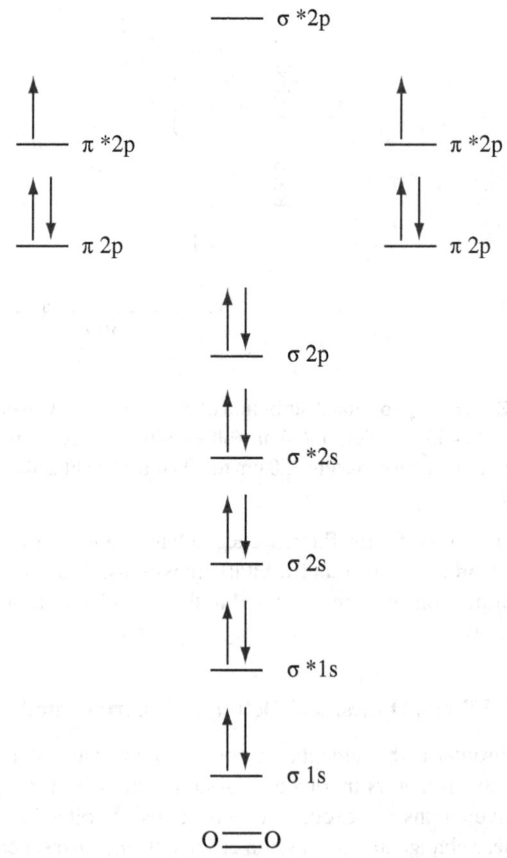

FIGURE 14.7 Molecular orbital diagram for the diatomic oxygen molecule (O_2).

transition of an electron from the $\sigma 2p \rightarrow \sigma{*}2p$ is not likely to take place as compared to the energy required for the transition $\pi 2p \rightarrow \pi{*}2p$. This is only a visual example as molecular orbitals for organic compounds are much more complex than this with overlap from neighboring orbitals.

In general, there are few conjugated aliphatic compounds that fluorescence while there are numerous aromatic compounds that do. For efficient fluorescence, the compound should have a short lifetime for the emission of energy from the $\pi{*} \rightarrow \pi$ transition, usually on the order of 10^{-9} to 10^{-7} s.

14.8.5.3 Examples of Nonfluorescent and Fluorescent Compounds Fluorescence increases with the number of rings in the structure, the amount of condensation, and sometimes the substitution. However, many simple heterocyclic compounds do not fluoresce, such as pyridine (nitrogen substitution), furan (oxygen substitution), thiophene (sulfur substitution), and pyrrole (nitrogen substitution). Examples of nonfluorescent structures are depicted in Figure 14.8(a). The lowest energy transition for these simple heterocyclic compounds involves the $n \rightarrow \pi{*}$ transition, which prevents fluorescence by rapidly converting to the triplet state. When simple heterocyclic compounds are fused with a benzene ring, fluorescence is produced. Two examples of fluorescent, fused ring heterocyclic compounds, quinolone and indole, are depicted in Figure 14.8(b). Both compounds contain fused aromatic rings, and a heterocyclic nitrogen substitution. The fusion with an aromatic ring allows the low-energy $\pi \rightarrow \pi{*}$ transitions resulting in fluorescence. Finally, examples of fluorescent, aromatic fused ring systems are depicted in Figure 14.8(c), along with their associated excitation and emission band maximum wavelengths as naphthalene, anthracene, biphenyl, and fluorene.

14.8.5.4 Other Structural Influences

14.8.5.4.1 Rigidity and Substitution Rigidity plays a role in fluorescence with an increase in fluorescence with an increase in rigidity. Nonrigid compounds have an increased probability of internal conversion and can also undergo low-frequency vibrations, which help to dissipate the internal energy. Oxygenation as a compound substitution also increases fluorescence, such as

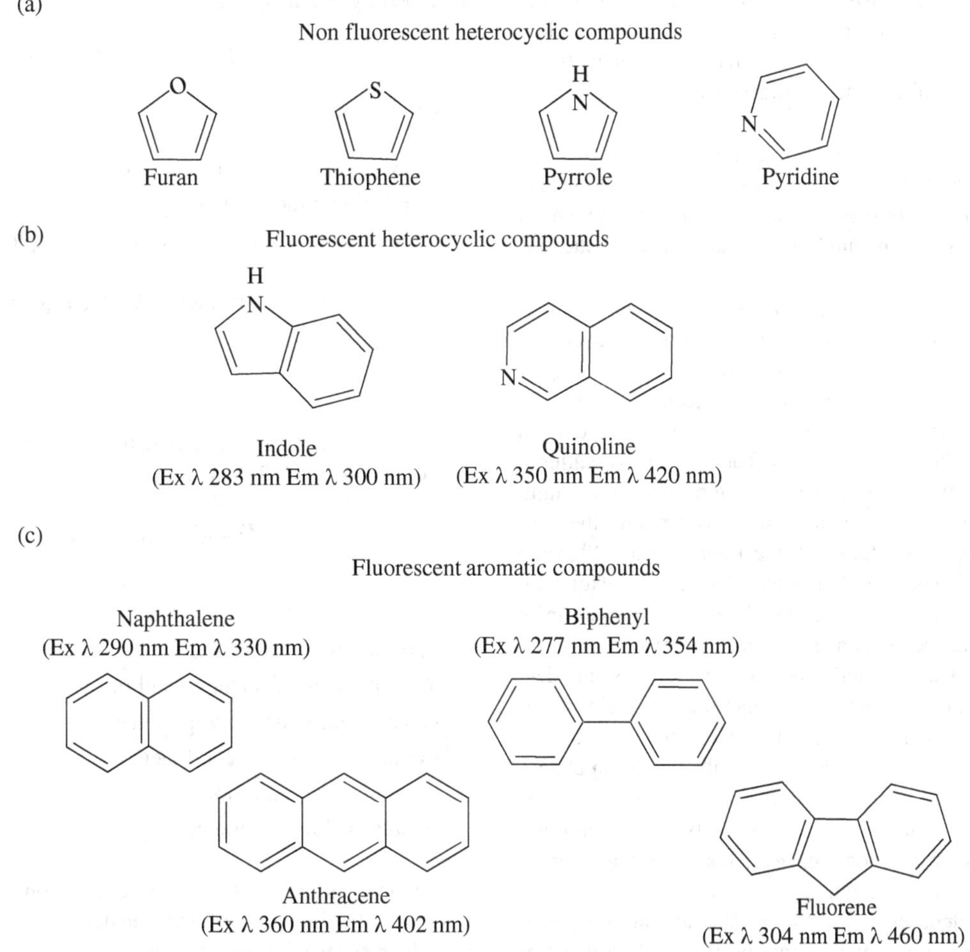

FIGURE 14.8 Examples of nonfluorescent and fluorescent compounds. (a) Simple nonfluorescent heterocyclic compounds as pyridine (nitrogen substitution), furan (oxygen substitution), thiophene (sulfur substitution), and pyrrole (nitrogen substitution). (b) Fluorescent, fused ring heterocyclic compounds of quinolone and indole. (c) Fluorescent, aromatic fused ring systems along with their associated excitation and emission band maximum wavelengths as naphthalene, anthracene, biphenyl, and fluorene.

phenol, the phenolate ion, and anisole. While halogenated aromatic substances substituted with fluorine, chlorine, or bromine do fluoresce, the intensity of the fluorescence decreases with increasing atomic weight of the halogen. Iodine-substituted compounds tend to undergo predissociation where the ruptured bonds absorb the excitation energy and undergo internal conversion. Organic carboxylic acids containing a carbonyl on an aromatic ring (e.g., benzoic acid) usually have low fluorescent quantum yields because the energy of the $n \rightarrow \pi*$ transition is less than that of the $\pi \rightarrow \pi*$ transition, but the $n \rightarrow \pi*$ transition has a very low yield.

14.8.5.4.2 Temperature, pH, and Solvent Effects Temperature, pH, and the solvent all have an effect upon the fluorescence of a system. Typically, fluorescence decreases with increasing temperature due to the increase in the frequency of collisions, which increases the deactivation of the excited molecule through external conversion. Solvent blanks need to be analyzed to determine whether they contain impurities that will fluoresce. Also, viscous solvent can have an effect upon fluorescence. Finally, the pH of the system's solvent can also have an influence upon fluorescence. Aniline is a good example of this where the neutral solution species exists in three resonance forms giving the excited aniline molecule a greater stability. In acidic solution, the protonated anilinium ion does not have resonance forms resulting in the quenching of the fluorescence.

14.8.5.5 Scattering (Sample) While not something that can be observed by the analyst, scattering of the incident light can have a major effect upon fluorescent measurements. The two most influential types of scattering are Rayleigh and Raman scattering.

14.8.5.5.1 Rayleigh–Tyndall Scattering When light is scattered, it is scattered in all directions, thus it cannot be excluded from the fluorescence detection system. Rayleigh scattering is a process that is derived from the analyte molecules themselves in solution while Tyndall scattering is a process derived from small particles in colloidal suspensions. The light that is scattered by Rayleigh–Tyndall scattering possesses the same wavelength as that of the original incident light. When this happens, the scattered light is collected together with the fluorescence. Often the fluorescence emission band is close in wavelength to the excitation band, and thus the scattering interferes with the analysis as it can be difficult to separate the scattering influence because it is in all directions. To remove the interference of the excitation incident wavelength band, the detector is often placed at an angle of 90° to the cell. This is possible because the fluorescence process is in all directions. However, the scattering is also in all directions.

14.8.5.5.2 Raman Scattering The other type of scattering process taking place with fluorescence analysis is Raman scattering. An artifact of the Rayleigh scattering is the conversion of some of the incident energy into vibrational and rotational energy. This results in a scattered energy of slightly lower frequency, lower energy, and subsequent longer wavelength. This is often observed as a weak emission band that can be confused with the actual fluorescence emission band, or it can interfere with the fluorescence band. The amount of energy that is transferred to either the analyte or the solvent is always constant, and is particular to the compound. The Raman band always appears separated from the incident wavelength by the same frequency difference. There are tables that demonstrate that all solvents containing hydrogen linked either to carbon or oxygen have Raman bands. If the Raman band is interfering with the fluorescence measurement, then a slight change in the incident excitation wavelength will correspond to a change in the wavelength of the Raman band. This is often used to eliminate interference from solvent Raman bands. While the Raman scattering is usually quite weak, analysts often will run a blank solvent fluorescent measurement to inspect for any possible Raman band influence on analyte measurement.

14.9 QUANTITATIVE ANALYSIS AND BEER–LAMBERT LAW

In both fluorescence and phosphorescence, the luminescent power P_L is proportional to the number of molecules in the excited state. We can relate this to the radiant power that is absorbed by the analyte by

$$P_L = \Phi_L(P_0 - P), \tag{14.15}$$

where

Φ_L = quantum efficiency yield of fluorescence
P_0 = incident radiant power
P = radiant power emitting from the sample.

In the form of Beer–Lambert law, Equation 14.15 can be written as

$$P_L = \Phi_L P_0 \varepsilon b C. \tag{14.16}$$

Equation 14.16 can be rewritten in the form of intensities for dilute solutions (<0.05) as

$$F_I = I_0 \Phi_f (2.303 \varepsilon C l), \tag{14.17}$$

where

F_I = fluorescence intensity
Φ_L = quantum efficiency yield of fluorescence
I_0 = incident radiant energy intensity
ε = molar extinction coefficient
C = molar concentration
l = solid volume of the beam.

The intensity of the fluorescence emission is directly proportional to the intensity of the incident radiation. Because of this fluorescence, sensitivity can reach levels as low as 10^{-12} moles, as compared to 10^{-8} moles for UV/Vis measurements. For quantitative analysis, a series of standards are measured and the fluorescence intensity is plotted against the standard concentrations. The dynamic linear range is determined and then used to calculate concentrations of test samples. This approach is very similar to what was presented in Chapter 13 for UV/Visible calibration curves.

14.10 QUENCHING OF FLUORESCENCE

There are other processes happening in solution that may adversely affect the fluorescent measurements. Quenching of the fluorescence is a phenomenon by which the excited-state energy of the analyte is lost due to collisions with either other analyte species or impurities in the solution. The energy that would have been observed as fluorescence is dissipated through nonradiative energy transfer from the excited analyte to the quenching agent. Quenching due to collisions is called dynamic quenching. Dissolved oxygen in solution can bring about dynamic quenching. Self-absorption can be another problem where the emission band wavelength is close enough to the excitation band wavelength that the emitted light is reabsorbed by the analyte. To alleviate this, the sample is often diluted before analysis. Diluting also helps with quenching from impurities in the solvents by reducing their concentrations to a point of insignificance.

14.11 FLUOROMETRIC INSTRUMENTATION

14.11.1 Spectrofluorometer

The spectrofluorometer is one of the major instruments used in fluorescence applications. The light source of the fluorometer transmits the selected wavelength to the sample, which then emits fluorescence. The emitted fluorescence intensity is directly proportional to the sample concentration. The fluorometer spectra are presented as intensity versus wavelength. For a set of known standards, the maximum fluorescence intensity at a given wavelength is regressed with the known concentration. The resultant linear relationship is used for samples where the fluorescence intensity obtained from the spectra will correspond directly to the sample's concentration.

In our study of fluorescence, we will look at two spectrofluorometers: the PTI model Quant Master 400 spectrofluorometer (PTI, QuantaMaster™, ON, Canada) and an LS 55 Luminescence spectrometer (PerkinElmer, USA) which is relatively new. Figure 14.9 is an illustration of one of the layouts used for the PTI spectrofluorometer where the system is designed as modular components. Figure 14.10 depicts the light paths and internal optics used in the fluorometer (also called fluorimeter). Compared to the PTI model spectrofluorometer, the LS 55 spectrometer is smaller and more compact. A picture of the LS 55 fluorimeter is depicted in Figure 14.11. All of the optical components are under the same cover in contrast to the PTI, which is multicomponent. The monochromatic slits are automatic and do not need to be changed manually. This important aspect can help to increase the accuracy of the instrument. The instrument can change from a liquid sample to solid sample holder easily, which increases the versatility and application ability of the spectrometer. However, even though the spectrometer's response and accuracy are improved, the major components that make up a spectrometer do not fundamentally change. The major components include a light source, monochromators, and a photomultiplier tube (PMT) detector.

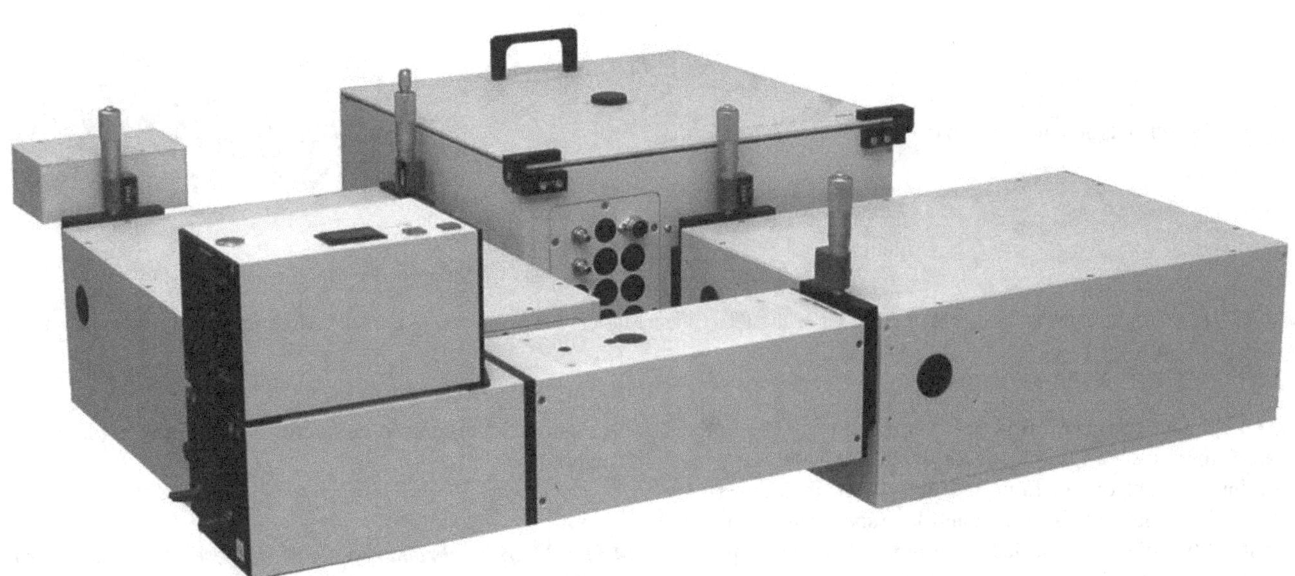

FIGURE 14.9 Configuration used for the PTI QuantaMaster™ 400 spectrofluorometer where the system is designed as modular components. (Reprinted with permission from HORIBA Scientific.)

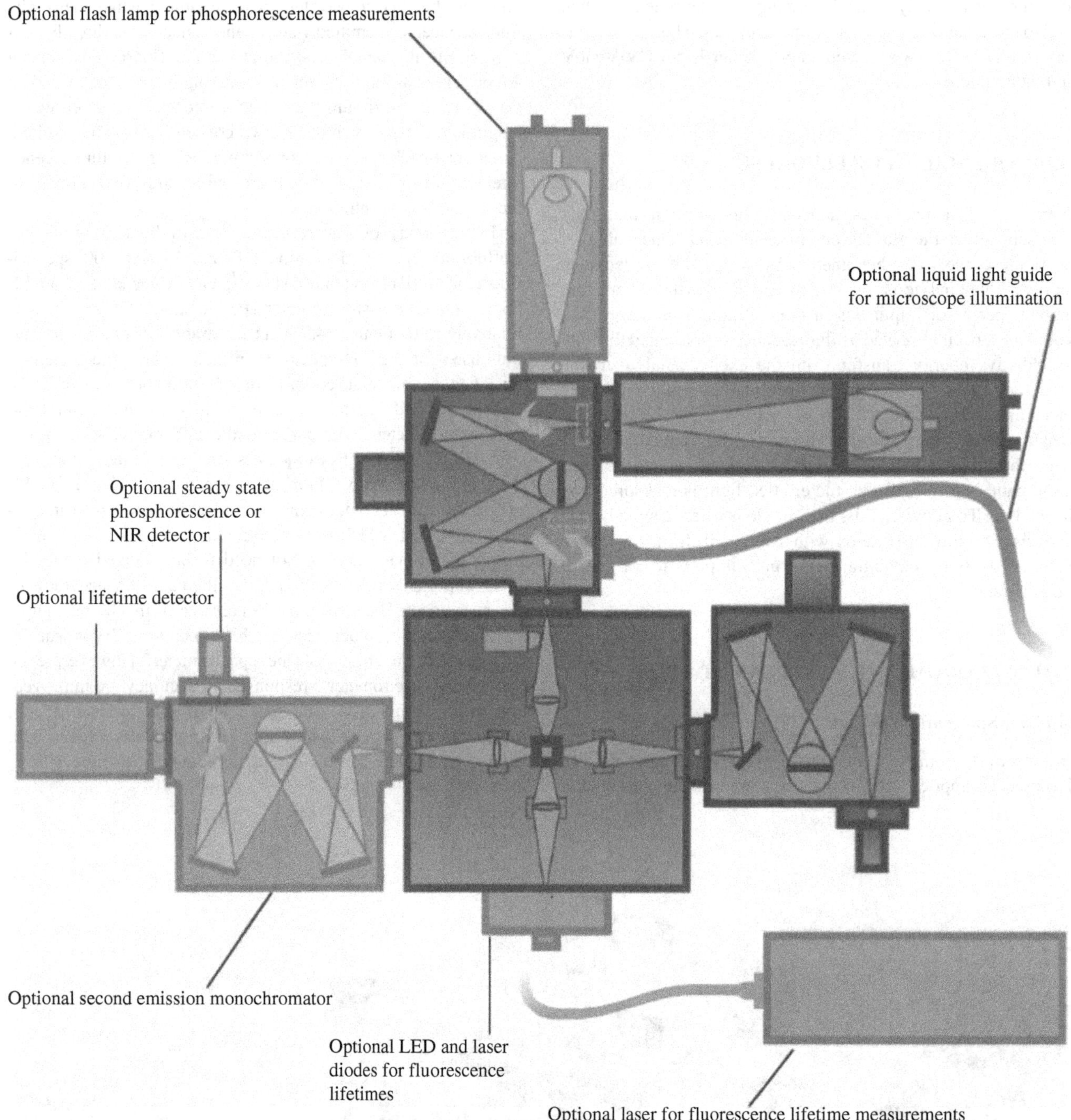

FIGURE 14.10 Light paths and internal optics used in the PTI QuantaMaster^{TM} 400 spectrofluorometer. (Reprinted with permission from HORIBA Scientific.)

14.11.1.1 Light Source Both spectrometers, the PTI and the Perkin Elmer, use a 75 W xenon (Xe) lamp. The lamp supplies a continuous light output from 250 to 700 nm, thus covering both the ultraviolet and visible regions of the spectrum. The lamp consists of two electrodes sealed under high pressure in a quartz bulb with Xe gas. When the power is turned on, a high-voltage pulse is generated between the two electrodes, which will induce collisions between the Xe gas. The collisions ionize the Xe atoms by removing electrons. The recombination of the removed electrons with the ionized Xe atoms will result in a continuum of light.

14.11.1.2 Monochromators The purpose of the monochromator is to disperse white light into various colors of wavelengths. The spectrofluorometer has two monochromators; one is to select an excitation wavelength while the other is to select an emission wavelength.

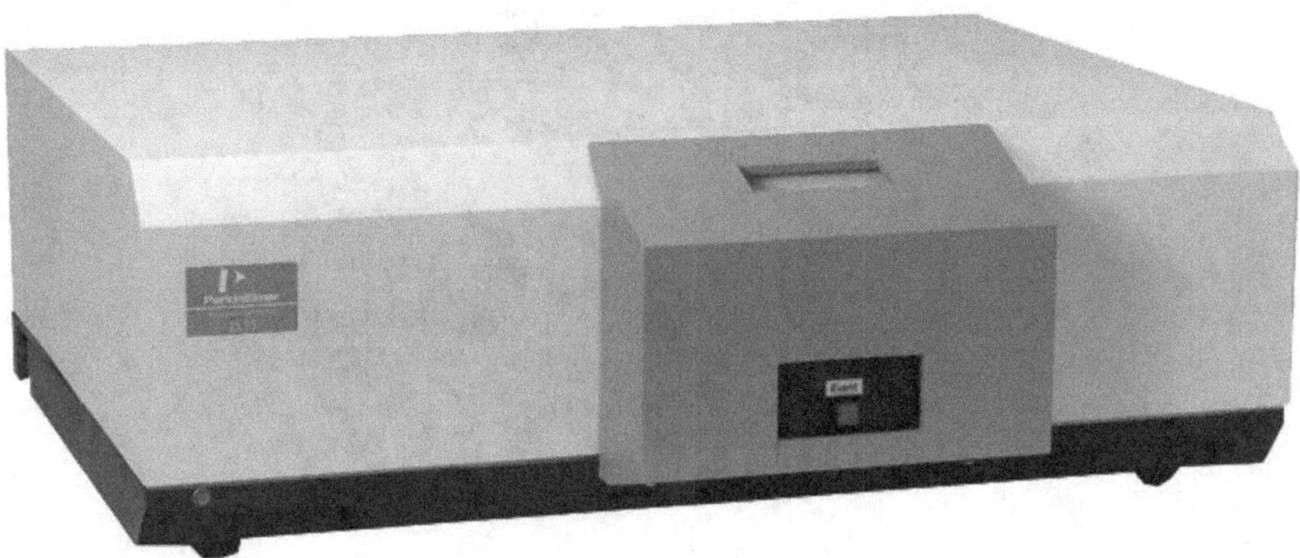

FIGURE 14.11 LS 55 luminescence spectrometer. (Reprinted with permission from PerkinElmer).

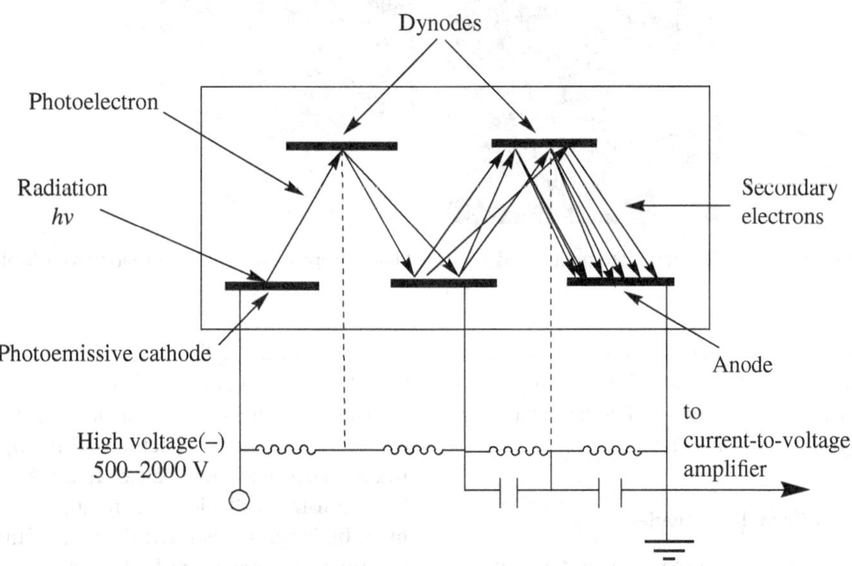

FIGURE 14.12 Schematic of a photomultiplyer tube.

14.11.1.3 ***Photomultiplier tube (PMT)*** The purpose of the PMT is to amplify electrons, derived from incident photons entering the photomultiplier, to an extremely large number of electrons, thus increasing the analyte response many times fold. PMTs are used as the detector for most types of spectrofluorometers. A PMT output is taken as a current source by the instrument and the light intensity emitted by the analyte is proportional to the current, which is used to correlate concentration to fluorescence intensity.

Figure 14.12 depicts the principle of a PMT. Within the PMT vacuum area is a photocathode and a series of dynodes. Incident photons hit the surface of the photocathode where an electron will be ejected. The potential difference between the photocathode and dynode will accelerate the ejected electrons to the first dynode. Several additional electrons are ejected by the differential potential. This process continues along the dynode chain, where more electrons are ejected and collected. When a new current pulse arrives at the cathode, a new cycle of this process is started again. By this process, amplification of the electrons is generated that represents amplification of the incident signal.

The PTI spectrofluorometer specifically utilizes digital photon counting detector with a discriminator and high-voltage power supply. At constant high voltage, the PMT is very sensitive. The measurement is performed when each photon hits the photocathode of the PMT. Individual photon results in a count at the anode which can be detected. The light hitting on the

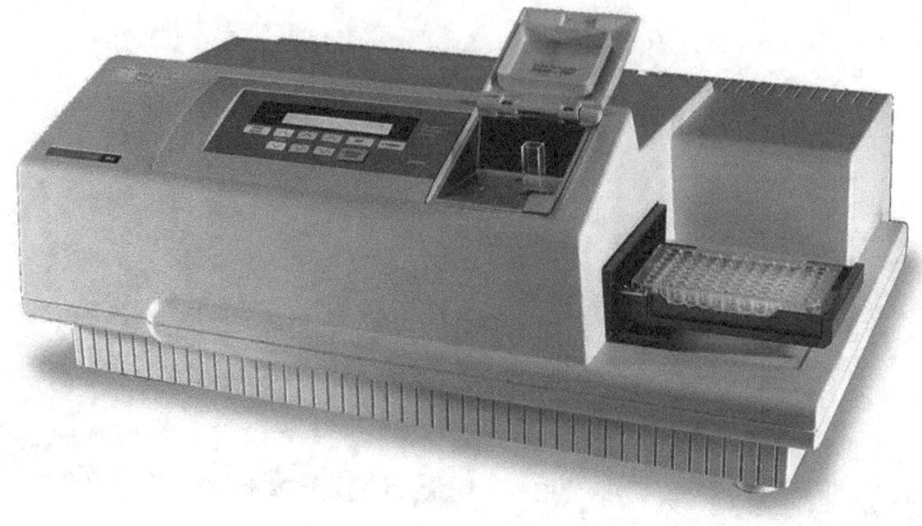

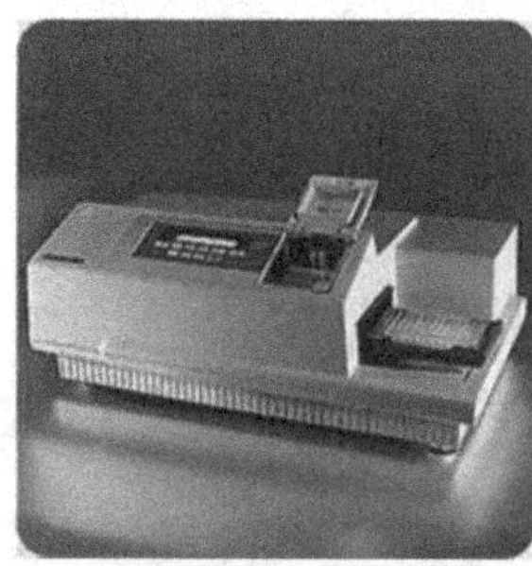

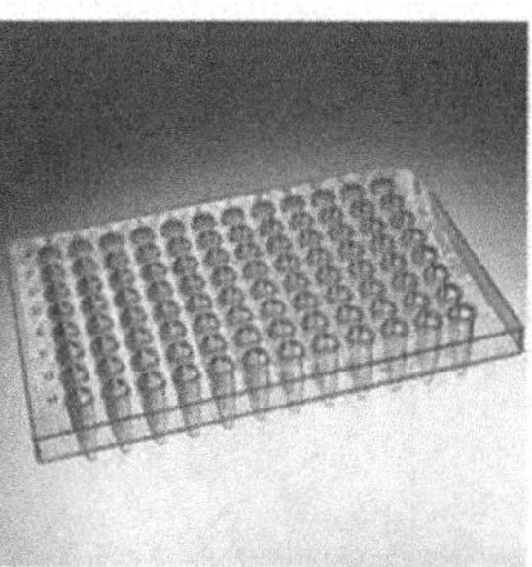

FIGURE 14.13 The SpectraMax M2 microplate reader and 96 well plate. (Reprinted with permission from Molecular Devices, Inc.)

photomultiplier detector is proportional to the count rate, which is the number of counts per second. The detector is usually operated with a discriminator to discriminate a low-level noise signal from a higher level signal from the incident photons.

14.11.2 Multidetection Microplate Reader

Another useful way to do fluorescence measurement is to use a microplate reader. This is a fluorimeter that accepts a 96 well microplate for fluorescence measurements. This can often be advantageous when a large number of samples need to be analyzed at one time, and also when small volumes of sample are being used (~100–400 µl). The design and function of the SpectraMax M2 microplate reader (Molecular Devices, Inc.) performs similarly to a spetrofluorometer, and is depicted in Figure 14.13. However, this is the only system that can provide both dual-mode measurement for a cuvette port and 6-384 microplate reading. The major components of a multidetection microplate reader are similar to the spectrofluorometer, such as the light source, the monochromator, and the photodetector. There is slight difference between the two such as the light source for the microplate reader uses a 50 W xenon flash lamp (versus a 75 W for the spectrofluorometer). It has very similar functions for the monochromator. The dual monochromators of the microplate reader are flexible to select any absorbance wavelength range between 200 and 1000 nm, any excitation wavelength between 250 and 850 nm and any emission wavelength from 360 to 850 nm.

The multidetection microplate reader can measure and obtain endpoint and kinetic spectra, and multipoint well scanning for fluorescence and absorbance. It can be applied to the field of biochemistry, cell biology, immunology, nuclear biology, and microbiology. A schematic diagram of the multidetection microplate reader is shown in Figure 14.14.

14.11.3 Digital Fluorescence Microscopy

Digital fluorescence imaging microscopy system is an extremely sensitive technique capable of single molecule observations, and is a highly specialized tool for fluorescent measurement. It can distribute to a single molecule measurement and visualize specific fluorescent molecules in intracellular locations. Primarily used instrument in the research work is inverted fluorescence microscopy (Olympus IX-70). Figure 14.15 depicts a typical digital fluorescence microscope used. The major components in a microscope are the light source, filter cubes, objectives and grating, and a charge-coupled device (CCD), which are individually described next.

14.11.3.1 Light Source The fluorescence microscope is a very sensitive instrument that requires a bright, white light

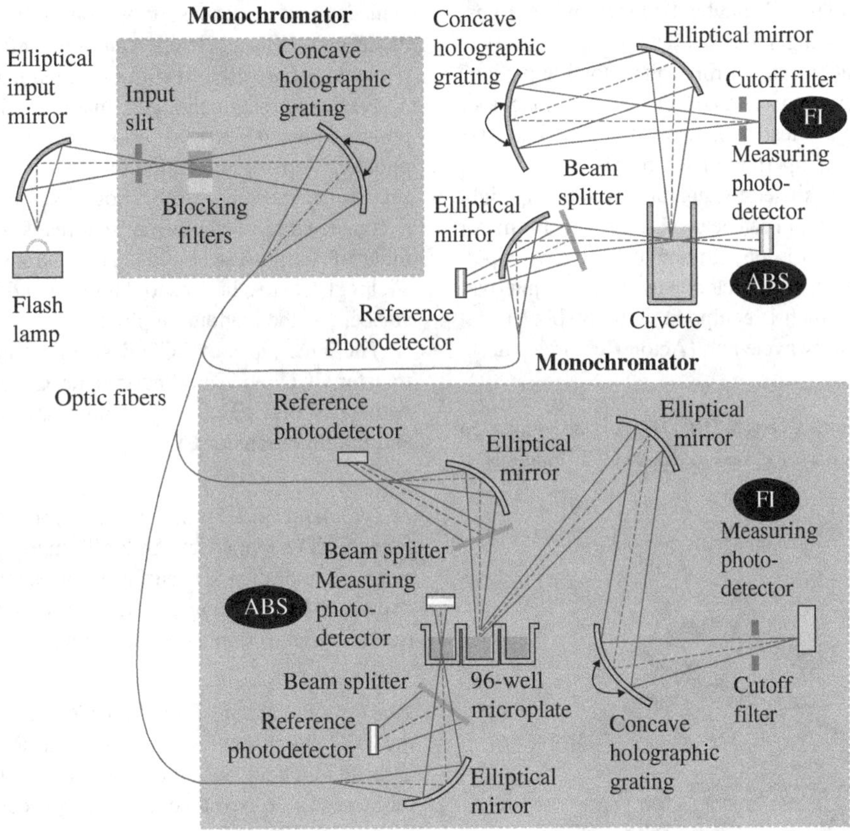

FIGURE 14.14 Diagram of the spectramax microplate reader. (Reprinted with permission from Molecular Devices, Inc.)

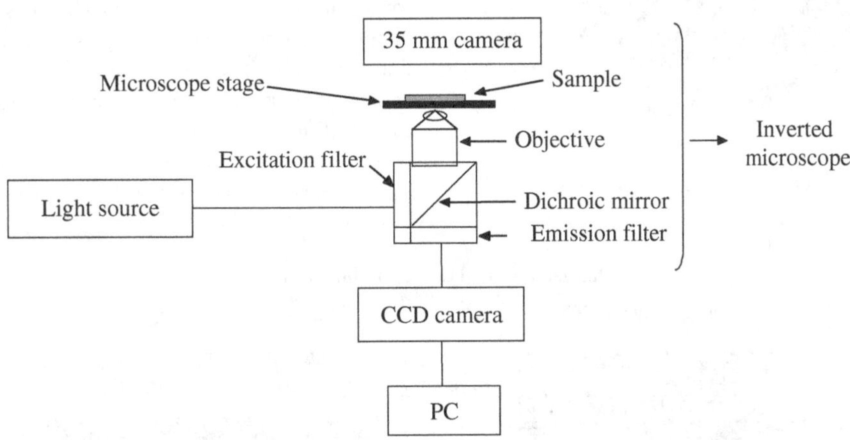

FIGURE 14.15 Diagram of a typical digital fluorescence microscope.

source. A 100 W mercury arc lamp is the most commonly used light source. This type of lamp provides a bright, continuous emission across the visible range from 400 to 750 nm, plus a UV range of 200 to 399 nm. Mercury lamps also have very distinct and sharp emission lines that are very importantly used to characterize the mercury arc lamp and calibrate the spectra.

14.11.3.2 Filter Cube To deliver the light to the specimen from the lamp, then collect the emitting fluorescence and form a fluorescence image, the proper filter cube need to be selected. A filter cube is typically made up of an excitation filter, a dichromic mirror, and an emission filter. The excitation filter is used to excite the specimen by selectively transmitting a narrow band of wavelengths. The dichromic mirror is used to reflect the excited light to the specimen and to transport the collected emission to the CCD detector. The emission filter is used to transmit the emission fluorescence from the specimen and block the residual excited light. Figure 14.16 depicts the filter cube components.

14.11.3.3 Objectives and Grating The objective is often considered the most important part of the microscope because the image qualities are produced by it. The objective is positioned between the specimen and the filter cube. Its function is to transmit the fluorescence-inducing light (excitation wavelength) to the specimen from the dichromic mirror while allowing passage of the emitted fluorescence to the CCD camera for images or spectra. Analytical laboratories often use 10×, 20×, and 40× objectives with a numerical aperture of 0.5 or 0.9.

The diffraction grating is used to separate the mercury light into individual wavelengths and can be used as a monochromator and as a spectrograph in microscopy. In one of our microscopes, we use a triple grating to achieve efficiency light throughput over a broad spectral region, which is equipped with 150 blz (blaze) and 300 blz at an optimum wavelength (Acton Research, Inc.).

14.11.3.4 ***Charged-Coupled Device (CCD)*** A charge-coupled device (CCD) is a photon detector used in digital CCD cameras. It is made up of thousands, or millions, of pixels, which are silicone diode photosensors. Pixels can store information from incident photons that are used to comprise the microscope image.

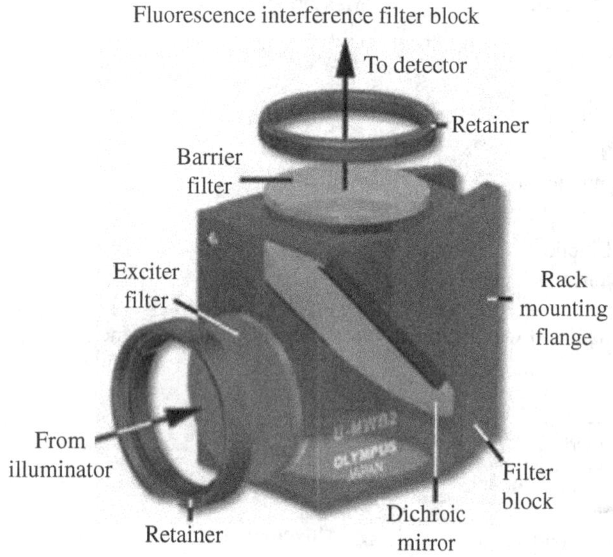

FIGURE 14.16 Filter cube and its components.

Pixels are semiconductor materials that can trap and hold photon-induced electrons (photoelectrons) derived from incident photons. A pixel is coupled to a charge storage region that will accumulate and store the photoelectrons. This storage region is connected to one amplifier that reads out the amount of accumulated charge. The stored charge is transferred through the parallel registers to a linear serial register and then to an output mode adjacent to the readout amplifier.

The three types of CCD designs are full-frame CCD, frame-transfer CCD, and interline-transfer CCD. Figure 14.17 depicts two types of CCD architectures, namely frame-transfer CCD and interline-transfer CCD.

14.11.3.4.1 Full-Frame CCD One microscope uses full-frame CCD supplied by Andor Technology, Inc. In this design, every pixel of the CCD surfaces corresponds to the image being collected. During image collection, exposures are usually controlled by an electrochemical shutter.

14.11.3.4.2 Frame-Transfer CCD In this design, one half of the CCD chip is masked and used as a storage space. After exposure, all of the pixels in the image side are transferred to pixels on the storage side. No camera shutter is needed because transferring time for the image is only a fraction of the exposure time.

14.11.3.4.3 Interline-Transfer CCD In this design, imaging rows and masked storage transfer rows are parallel pixels of columns. Camcorders and video cameras typically use interline-transfer CCDs because they provide high-quality images and can be read out at video rate. This type of CCD can be used for dim fluorescent specimens because of the low camera read noise and improved camera electronics. Interline-transfer CCDs can be very fast and do not require a shutter to control the

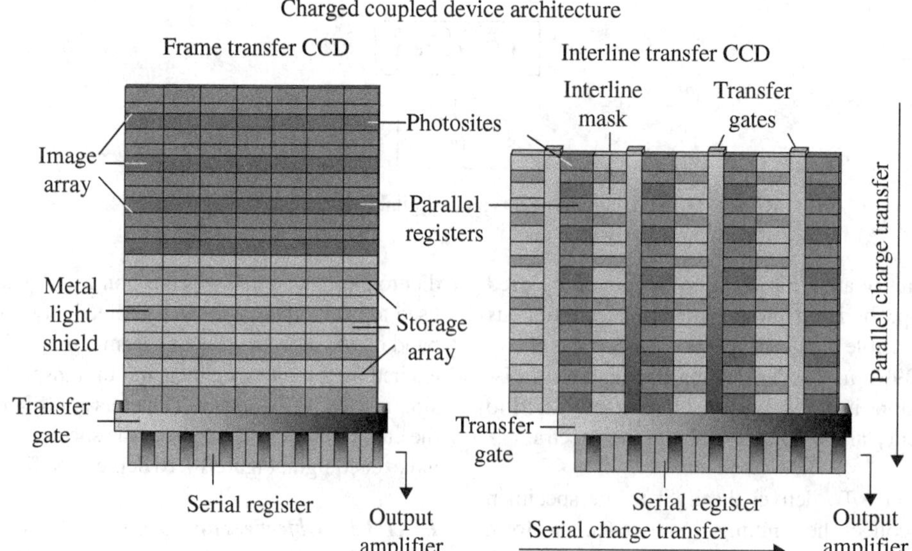

FIGURE 14.17 Frame transfer CCD and interline transfer CCD.

exposure. With current technologies, the spatial resolution and light-collecting efficiency can reach those of a full-frame CCD.

14.12 SPECIAL TOPIC—FLOURESCENCE STUDY OF DYE-A007 COMPLEXES

The in-solution binding behavior of the anticancer agent A007 and three dye compounds was studied using fluorescence. The structures of the compounds are depicted in Figure 14.18. Density Functional Theory (DFT) was used to study the gas-phase binding structure of the anticancer A007 and methylene blue (MEB). The most stable configuration is depicted in Figure 14.19 for a one-to-one binding.

The dyes used in the study naturally fluoresce at specific wavelengths. It was initially felt that the complex between the dyes and the anticancer agent A007 enhanced this fluorescence. Using the method of continuous variation, a fluorescence maximum was obtained for each dye-A007 complex versus mole fraction. Figure 14.20 depicts the stoichiometries determined

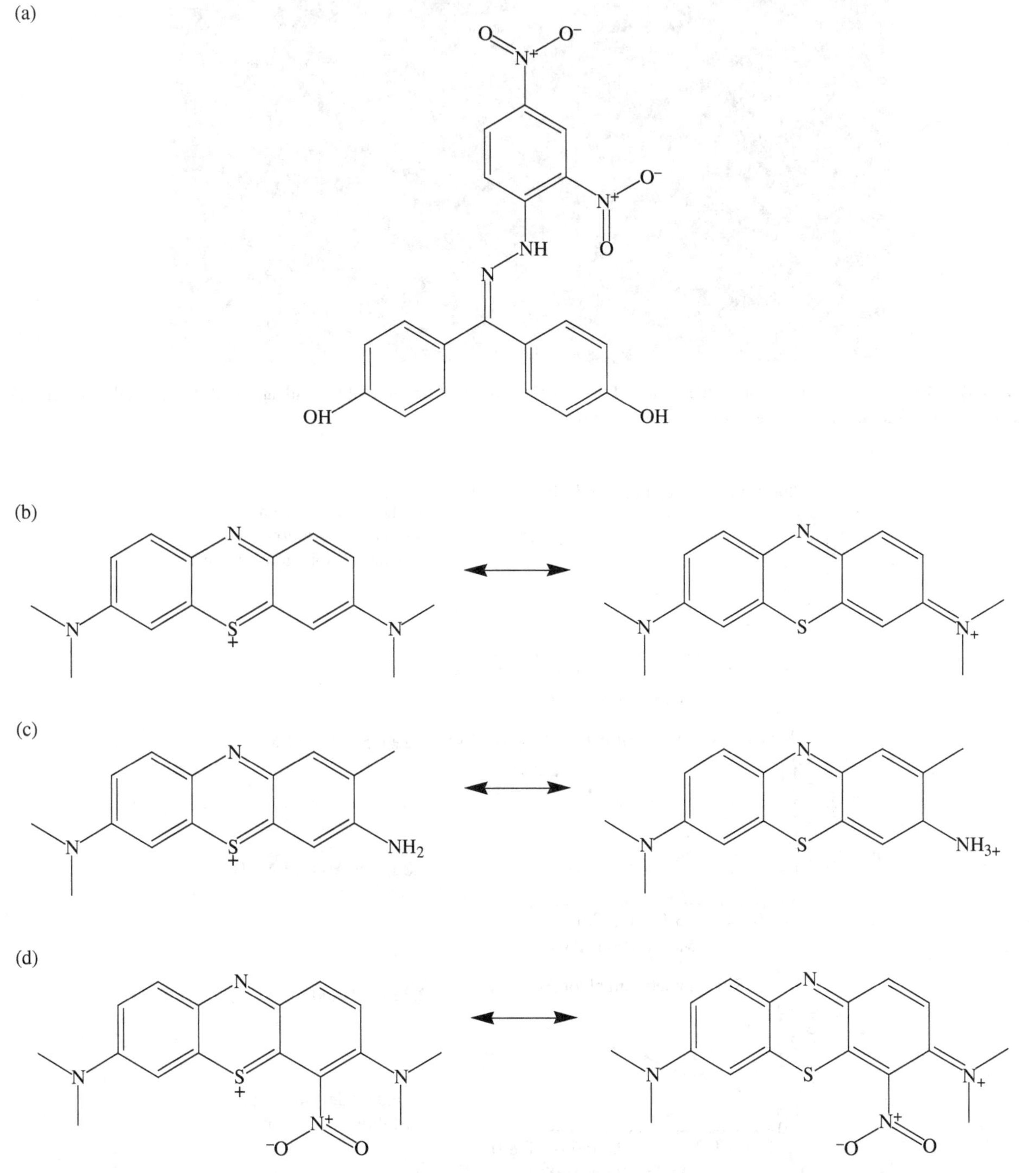

(a) (b) (c) (d)

FIGURE 14.18 Structures of (a) 4,4′-Dihydroxybenzophenone-2,4-dinitrophenylhydrazone (A007)[1], (b) methylene blue (MEB) double salt, (c) toluidine blue (TB) double salt, and (d) methylene green (MEG) double salt.

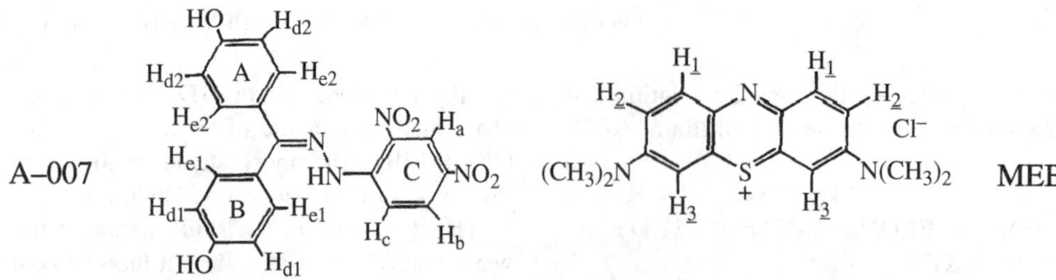

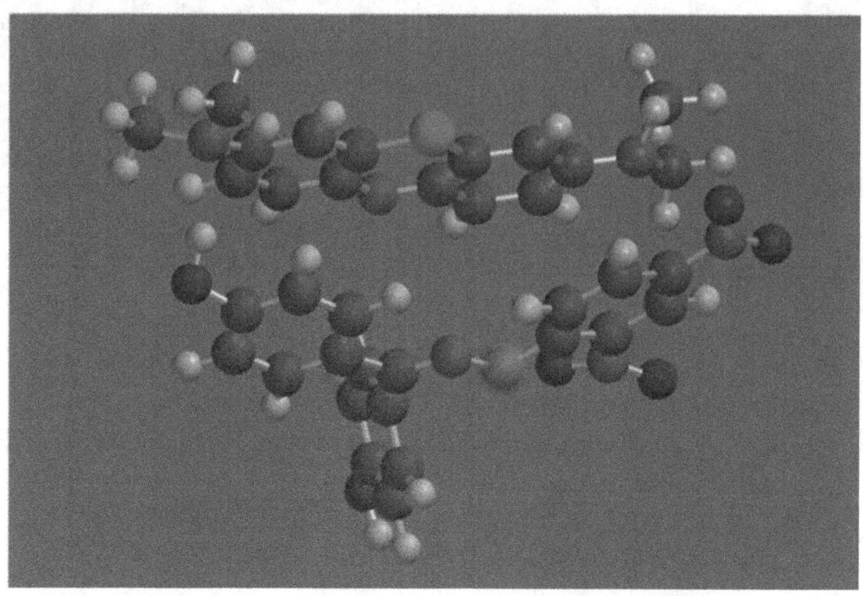

FIGURE 14.19 Density Functional Theory (DFT) study of the gas phase binding structure of the anticancer A007 and methylene blue (MEB). The most stable configuration for a one-to-one binding is shown.

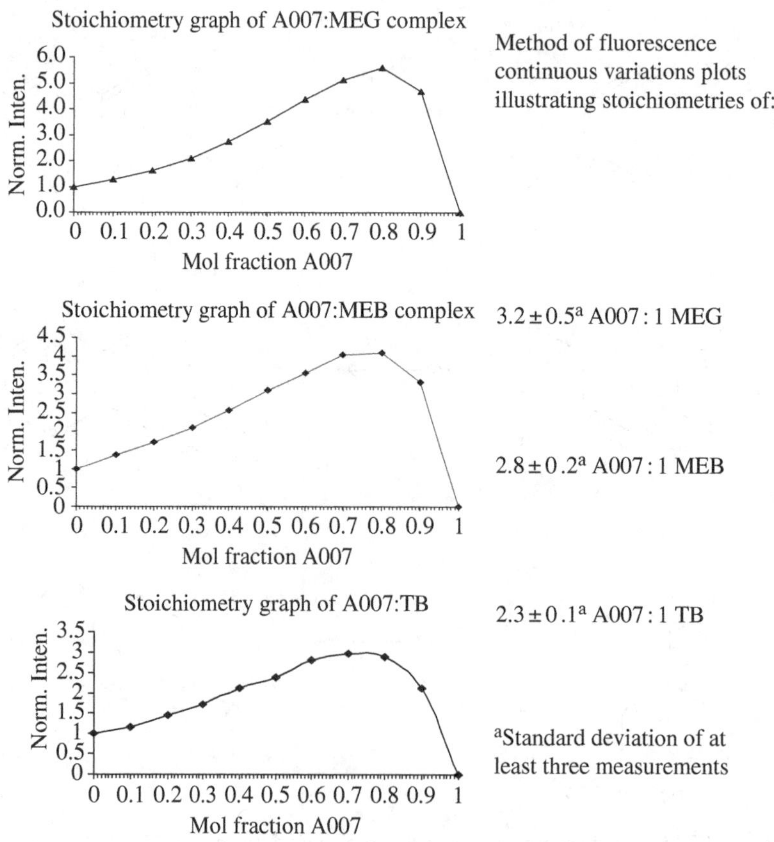

FIGURE 14.20 Stoichiometries determined for the three complexes.

for the three complexes. These dyes however are known to produce dimers in the liquid phase, where the production of the dimer will quench the fluorescence. Further spectroscopic experiments showed that only one species of complex was being formed in solution between A007 and the dyes, but the enhancement in fluorescence was attributable to the decrease in dye dimer only, and not to the formation of the A007:dye complex.

14.13 CHAPTER KEY CONCEPTS

14.1 In molecular absorbance photospectroscopy, the absorbance of the incident radiant energy is proportional to the concentration of the analyte according to Beer's law.

14.2 Fluorescence and the closely related phosphorescence are processes that include absorbance of radiant energy of infrared, visible, or ultraviolet light which is followed by the release of quanta of energy that is subsequently detected and measured.

14.3 Collectively, fluorescence, phosphorescence, and chemiluminescence are known as molecular luminescence.

14.4 Most organic molecules and inorganic compounds will luminescence or emit photons when excited to higher electronic states.

14.5 Fluorescence and phosphorescence are photoluminescent processes where it is incident energy that excites the species resulting in emission of photons. Fluorescence is a faster process as compared to phosphorescence where the release of energy is almost instantaneous.

14.6 Phosphorescence is emission that is accompanied with a change in electron spin that results in irradiation that lasts up to a few seconds.

14.7 Radiant energy is absorbed by atoms or molecules in definite units, which are called quanta.

14.8 Excited species are formed with the absorption of incident radiation (light), which imparts energy to the molecules. The excited molecules then release some of this energy in the form of light emission.

14.9 The efficiency in which the light is absorbed and then released is called the quantum efficiency and is described as

$$\Phi E = \frac{\text{einsteins emitted}}{\text{einsteins absorbed}} = \frac{\text{quanta emitted}}{\text{quanta absorbed}}.$$

14.10 Since the quanta emitted can never be greater than the quanta absorbed, the quantum efficiency will always be less than or equal to 1 ($\Phi E \leq 1$).

14.11 Jabłoński diagrams describe light absorption and emission and demonstrate different molecular processes happening in the excited states during fluorescence.

14.12 Excitation of atomic and molecular species comes in many forms including: gamma rays produce nuclear excitation reactions, X-rays produce transitions of inner atomic electrons, ultraviolet and visible lights produce transitions of outer atomic electrons, infrared produces molecular vibrations, far infrared produces molecular rotations, and long radio and radar waves produce oscillations of mobile or free electrons.

14.13 In fluorescence and phosphorescence, the processes involve the outer atomic electrons.

14.14 The type of fluorescence with the release of energy without a change in frequency is known as resonance fluorescence and is common with excited atoms. Most organic molecules will release energy at longer wavelength than the excitation due to a shift known as the Stokes shift.

14.15 Electrons in atoms possess four quantum numbers (n, l, ml, and ms).

14.16 The Pauli exclusion principle states that two electrons cannot have the same four quantum numbers and that only two electrons can occupy each orbital and their spins must be opposite.

14.17 Phosphorescence differs from fluorescence where in phosphorescence there is an electron spin change resulting in a long excited-state lifetime on the order of seconds to minutes.

14.18 Emission or fluorescence bands are primarily at lower frequency (longer wavelength) for most organic compounds. This is known as the Stokes shift named after the Irish physicist George G. Stokes.

14.19 The absorption of the incident radiant energy by atoms and molecules is very fast for the elevation of the electron from the singlet ground state to the excited singlet state $S_0 \rightarrow S_1$ on the order of 10^{-13} to 10^{-15} s.

14.20 Rate parameters are used to describe the emission process of excited species and the probability or likelihood that fluorescence will take place.

14.21 The lifetimes of the decay processes are also based on the rate constants.

14.22 There are factors which distort and influence the spectra that require corrections in order to obtain true (instrument independent) spectra. The distorting influences are both instrument based and sample based.

14.23 The excitation wavelength (λ, nm) can have a negative impact on decays by bringing about bond breakage designated as either dissociation or predissociation.

14.24 There is also observed variation in the light source intensity and wavelength, an instrumental influence.

14.25 The fluorometric instrumentation contains numerous filters, slits, and monochromators involved in isolating and transferring various wavelengths for excitation, and intensity collection. With time, these change in transmission efficiency and also specificity.

14.26 As was also the case with UV/Visible spectrophotometric analyses, the sample holder needs to be composed of an inert substance that does not absorb in a useful wavelength range. Quartz has been the most used material for cuvettes in UV, visible, and NIR spectroscopy measurements due to its ideal transmittance from 200 to 2500 nm.

14.27 Aromatic containing compounds are the most common fluorescent structures.

14.28 In general, there are few conjugated aliphatic compounds that fluorescence while there are numerous aromatic compounds that do. For efficient fluorescence, the compound should have a short lifetime for the emission of energy from the $\pi* \rightarrow \pi$ transition, usually on the order of 10^{-9} to 10^{-7} s.

14.29 Fluorescence increases with the number of rings in the structure, the amount of condensation, and sometimes the substitution.

14.30 Rigidity plays a role in fluorescence with an increase in fluorescence with an increase in rigidity.

14.31 Temperature, pH, and the solvent all have an effect upon the fluorescence of a system.

14.32 While not something that can be observed by the analyst, scattering of the incident light can have a major effect upon fluorescent measurements. The two most influential types of scattering are Rayleigh and Raman scattering.

14.33 In both fluorescence and phosphorescence, the luminescent power P_L is proportional to the number of molecules in the excited state.

14.34 For quantitative analysis, a series of standards are measured and the fluorescence intensity is plotted against the standard concentrations. The dynamic linear range is determined and then used to calculate concentrations of test samples.

14.35 Quenching of the fluorescence is a phenomenon by which the excited-state energy of the analyte is lost due to collisions with either other analyte species or impurities in the solution.

14.36 The spectrofluorometer is one of the major instruments used in fluorescence applications.

14.14 CHAPTER PROBLEMS

14.1 What is the energy of a wavelength of 395 nm (395×10^{-9} m)?

14.2 What is the wavelength of a system with energy of 9.50×10^{-19} J?

14.3 What is the quantum efficiency of a system that was measured to have 1.344 einstein absorbed and 0.992 einstein emitted?

14.4 What are the quanta emitted if a system has a quantum efficiency of 0.65 and the quanta absorbed are 289.0?

14.5 Using the chapter as a reference, define the following in your own words: (a) fluorescence, (b) resonance fluorescence, (c) dissociation, (d) predissociation, (e) singlet state, (f) triplet state, (g) internal conversion, (h) external conversion, (i) vibrational relaxation, (j) intersystem crossing, (k) quantum yield, phosphorescence.

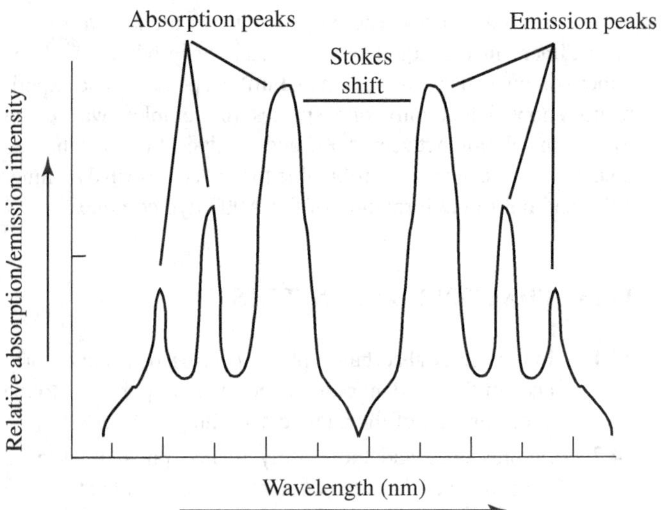

FIGURE 14.21 Unknown fluorescence excitation and emission spectrum.

14.6 Often fluorometric analyses are more sensitive as compared to UV/Visible analyses. Explain why this is so.

14.7 Roughly draw the following spectrum and label the absorption peaks, the emission peaks, and the Stokes shift (Fig. 14.21).

14.8 Explain in your own words the Stokes shift.

14.9 Explain how fluorescence is different from phosphorescence.

14.10 If a compound has a known quantum yield (ΦE) of 0.755, and rate constants consisting of k_{diss} and $k_{prediss} = 0$, $k_{int\ con} = 2.36 \times 10^{-12}$, $k_{ext\ con} = 6.21 \times 10^{-14}$, and $k_{ISC} = 1.19 \times 10^{-16}$, at a certain state, can the rate constant for fluorescence (k_{fluor}) be determined?

14.11 Using the constants given in Problem 14.10, what affect upon the quantum yield be observed if the wavelength has been reduced to a lower value resulting in a $k_{diss} = 3.39 \times 10^{-12}$, and a $k_{prediss} = 2.47 \times 10^{-13}$?

14.12 Using Figure 14.4 as a guide and the rate constants in Problems 14.10 and 14.11, what would be the lifetime (τ) for the $S_1 \rightarrow S_0$ transition?

14.13 If the lifetime value calculated in Problem 14.12 has units of nanoseconds, what is the lifetime in seconds and minutes? Does this appear to be fluorescence or phosphorescence?

14.14 For the lifetime determined in Problem 14.12, if a higher wavelength was used that resulted in no dissociations (i.e., k_{diss} and $k_{predis} = 0$), what affect upon the lifetime would this have?

14.15 The fluorescence intensity of a system is dependent upon both the incident radiant intensity (I_0) and the quantum yield (Φ_L). What influence does changing either of these have on the fluorescence intensity?

14.16 What is the molar concentration (C) of a quinine sample with the following fluorescence-related parameters?

$F_I = 355$
$\Phi_L = 0.669$
$I_0 = 1054$
$\varepsilon = 2.056\ M^{-1}\ cm^{-1}$
$l = 0.989$ cm

14.17 Under standard conditions, the following parameters were determined. Calculate the molar extinction coefficient (ϵ).

$F_I = 990$
$\Phi_L = 0.472$
$I_0 = 824$
$C = 0.0523$ M
$l = 1.326$ cm

14.18 A solution in the laboratory has a fluorescence intensity of 50,000, too concentrated for an accurate measurement. The sample is diluted one to ten (1:10) and then measured again under the same conditions. If the measurement is linear, what is the expected fluorescence intensity of the diluted sample?

14.19 If the molar extinction coefficient of the analyte in Problem 14.18 is $\epsilon = 5.25 \times 10^{-2}\ M^{-1} \cdot cm^{-1}$, the quantum yield is 0.939, the incident intensity is 10,500, and the beam volume is 2.55, what is the concentration of the solution?

14.20 The absorbance from 220 to 580 nm of a 0.1 M rhodamine derivative was collected in 20 nm steps. Plot the data and determine the best wavelength to excite the fluorophore for analysis.

Wavelength (nm)	Absorbance	Wavelength (nm)	Absorbance
220	0.125	400	0.140
240	0.158	420	0.168
260	0.174	440	0.211
280	0.162	460	0.225
300	0.130	480	0.236
320	0.145	500	0.208
340	0.169	520	0.177
360	0.188	540	0.149
380	0.157	580	0.131

14.21 The fluorescence emission of the rhodamine derivative fluorophore was collected from 540to 880 nm in 20 nm steps. Plot the data and determine the best wavelength to collect the emission of the fluorophore for analysis.

Wavelength (nm)	Emission	Wavelength (nm)	Emission
540	1310	720	1570
560	1490	740	1880
580	1770	760	1690
600	2080	780	1450
620	2360	800	1300
640	2250	820	1620
660	2110	840	1740
680	1680	860	1580
700	1400	880	1250

14.22 A series of standards of the rhodamine derivative were measured with an excitation wavelength of 480 nm (λ_{ex}) and emission wavelength of 620 nm (λ_{em}). Using ChemTech-TookKit or Excel®, plot the data.

Standard	Rhodamine derivative concentration (ppm)	Emission intensity
1	1.7	1100
2	3.7	2150
3	5.6	3130
4	7.8	4350
5	10.6	5880

14.23 From the slope of the best-fit equation, determine the extinction coefficient in $1\ mg^{-1}\ cm^{-1}$, and the molar extinction coefficient in $1\ mole^{-1}\ cm^{-1}$. The rhodamine derivative has a molar mass of $480\ g\ mol^{-1}$.

14.24 If an unknown sample has a fluorescence emission intensity of 3899, calculate the concentration of the sample in ppm and in molarity.

14.25 The fluorophore anthracene at 360 nm excitation and 402 nm emission was measured for a set of standards and recorded. Using ChemTech-ToolKit or Excel®, plot the data.

Standard	Anthracene concentration (ppm)	Fluorescence Intensity
1	50	520
2	100	949
3	200	1550
4	350	2100
5	500	2207

14.26 If an unknown sample has a fluorescence emission of 1789 using the above conditions, using the ChemTech-

ToolKit calculate the concentration of the sample in ppm and in molarity. The molar mass of anthracene is 178.23 g mol^{-1}.

14.27 The concentration determined in Problem 14.25 is for a solution that is directly measured. If, for another sample, the original sample makeup consists of 10.994 g dissolved in 500 ml, and an aliquot of 50 ml was diluted to 200 ml, and the fluorescence intensity was measured at 1355, what is the concentration of anthracene in the original sample?

14.28 If an unknown sample has an emission of 2200, using the ChemTech-ToolKit calculate the concentration of the sample in percentage and in molarity.

15

FOURIER TRANSFORM INFRARED (FTIR) SPECTROSCOPY

15.1 Introduction
15.2 Basic IR Instrument Design
15.3 The Infrared Spectrum and Molecular Assignment
15.4 FTIR Table Band Assignments
15.5 FTIR Spectrum Example I
15.6 FTIR Spectrum Example II
15.7 FTIR Inorganic Compound Analysis
15.8 Chapter Key Concepts
15.9 Chapter Problems

15.1 INTRODUCTION

Infrared (IR) analysis is another analytical laboratory instrumental analysis with relation to the UV/Vis instrumental analysis just covered. In UV/Vis analysis, the wavelengths used are approximately 190–400 nm for the ultraviolet (UV) region, 400–800 nm for the visible (Vis), and finally 1000–15,000 nm for the infrared region (IR). The name infra comes from "inferior" as the infrared region is next to but less in energy than the red region of the spectrum. In Figures 13.11 and 14.1, we saw two types of energy diagrams that depict the absorption of electromagnetic radiation producing the processes that we use in UV/Vis analyses and fluorescence analyses. Figure 15.1 depicts a comparison of the UV absorption transition process, the Vis and the infrared transition absorption processes. With the infrared absorption process, energy of longer wavelengths is being absorbed with translation to processes involving molecular vibrations. There are two types of molecular vibrations that take place: stretching and bending. Stretching is the movement of atoms back and forth along the bond axis. Bending may involve the change in bond angles between two atoms that are each attached to a third. Bending can also take place as a movement of atoms in respect to the whole molecule. This type of bending is referred to as rocking, scissoring, wagging, and twisting. The spectral region of 1000–15,000 nm is usually expressed in micrometers, as 1–15 μm, and was primarily dictated by the optic crystals being used. The internal optics is not made of glass or quartz, because both of these absorb energy greatly in the infrared region. In order to overcome this, inorganic salts are used for internal transmission optics and sample holders. Examples of inorganic salts used and their spectral regions are listed in Table 15.1.

The spectral regions used in Table 15.1 are in values of inverse centimeters (cm^{-1}). Both wavelength in micrometer and wavenumber in cm^{-1} are used in presenting infrared spectra, though wavenumber has become the most commonly used scale when reporting and discussing IR spectra. The relationship between wavelength and wavenumber is as follows:

$$\text{Wavenumber, cm}^{-1} = \frac{1}{\text{wavelength, }\mu\text{m}} \times 10^4. \qquad (15.1)$$

15.2 BASIC IR INSTRUMENT DESIGN

The major components of an infrared spectrometer include the infrared radiation source, the beam path for sample and reference, the monocromators, and the detector as depicted in Figure 15.2. The instrument layout depicted in Figure 15.3 is a simple design of an IR instrument including the IR source, interferometer, beam splitter, sample compartment, and detector (illustration by Thermo-Nicolet). The infrared radiation sources used are either a *Nernst glower* or a *globar*. The Nernst glower comprises a small rod of zirconium and yttrium oxides that is heated to approximately 2000 °C. The globar consists of sintered silicon carbide that is heated to approximately 1400 °C. Both rods are heated electrically. At these elevated temperatures, the IR sources emit IR radiation in a wavelength range that is used for IR molecular analyses. For the monochromator, a dispersing diffraction grating

Analytical Chemistry: A Chemist and Laboratory Technician's Toolkit, First Edition. Bryan M. Ham and Aihui MaHam.

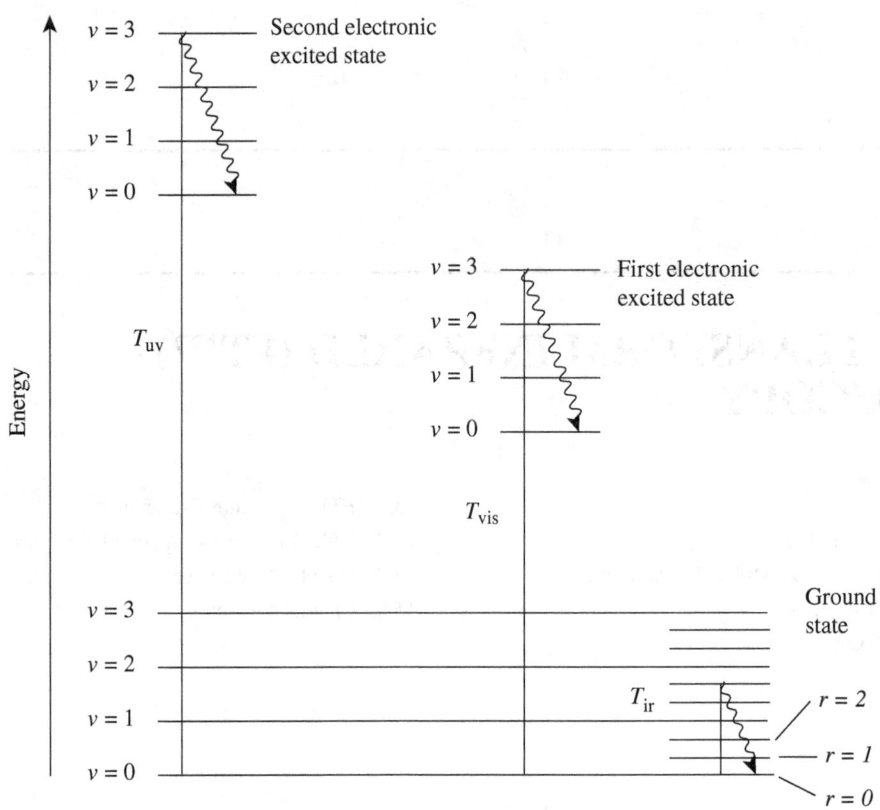

FIGURE 15.1 Absorption electronic transitions for T_{uv}, T_{vis}, and T_{ir}.

TABLE 15.1 Inorganic Salts Used for IR Optics.

Inorganic Salt	Spectral Range (cm^{-1} Practical)	Properties
AgCl	4000–400	Water insoluble
BaF_2	4000–750	
CaF_2	4000–1100	Water insoluble
CsBr	4000–250	Hygroscopic
CsI	4000–200	
KBr	4000–400	Hygroscopic
KCl	4000–500	Hygroscopic
MgF_2	4000–1400	
NaCl	4000–600	Hygroscopic
ZnS	4000–715	Water insoluble
ZnSe	4000–550	

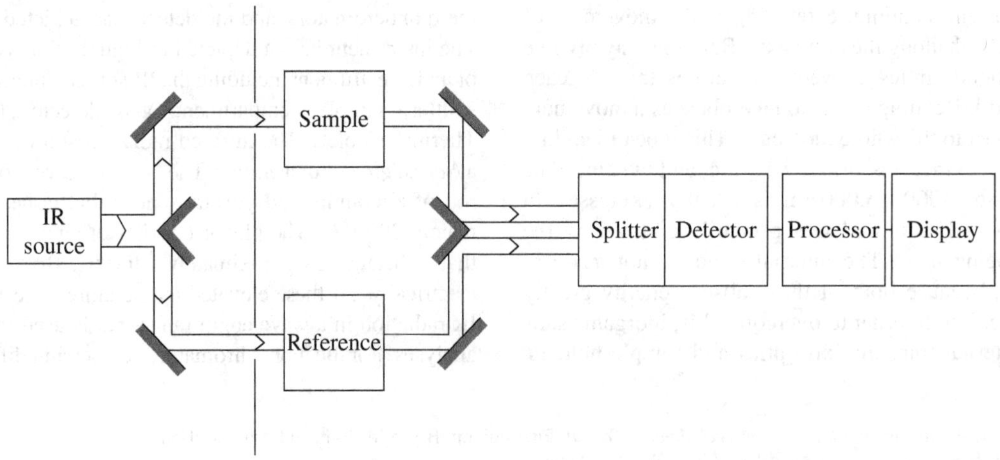

FIGURE 15.2 Basic components of an infrared spectrophotometer. (Reprinted with permission under the Wiki Creative Commons Attribution-Sharealike 3.0 Unported License.)

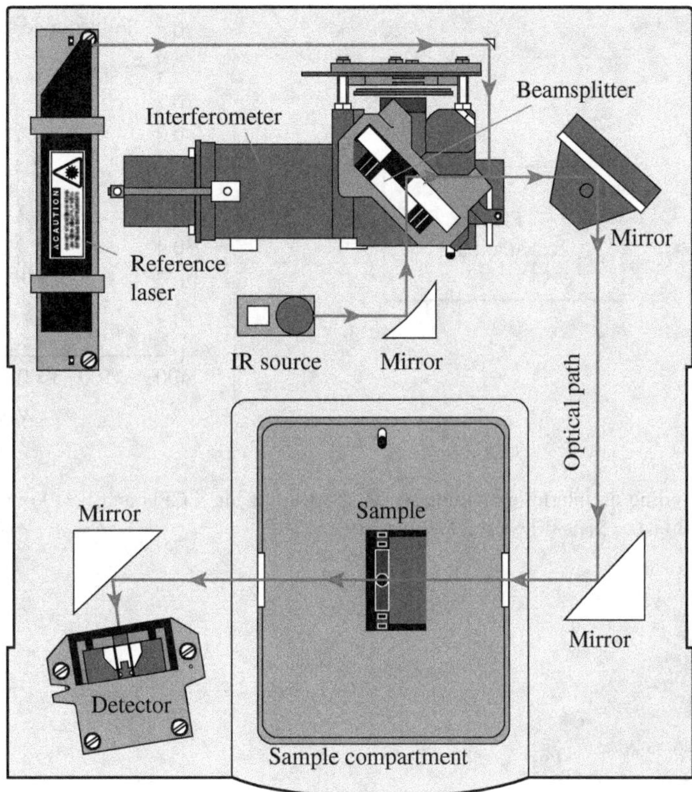

FIGURE 15.3 A simple infrared spectrophotometer including the IR source, interferometer, beam splitter, sample compartment, and detector. (Used by permission from Thermo Fisher Scientific, the copyright owner.)

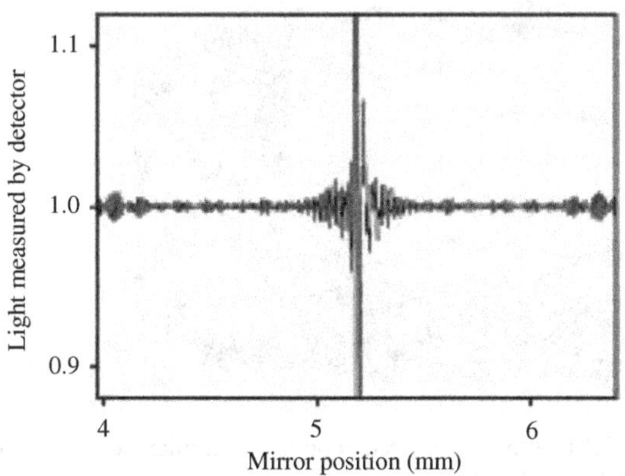

FIGURE 15.4 Interferogram obtained from an FTIR measurement. The horizontal axis is the position of the mirror, and the vertical axis is the amount of light detected. This "raw data" can be Fourier transformed to get the actual IR spectrum. (Reprinted with permission under the Wiki Creative Commons Attribution-Sharealike 3.0 Unported License.)

is typically used. The Fourier transform infrared (FTIR) spectrometers will have an interferometer after the IR source and before the beam splitting. The advantage of the FTIR is that the infrared spectrum is scanned very quickly using a moving mirror system. The interferogram scans are added together reducing the background noise significantly. The IR instruments used today in analytical laboratories are solely FTIR spectrometers. The typical configuration in an FTIR instrument is the Michelson interferometer, which uses a beam splitter and two mirrors. An interferogram is produced by the Michelson interferometer such as the one depicted in Figure 15.4. Fourier transform mathematical algorithm is used to convert the interferogram in Figure 15.4 into an IR spectrum as depicted in Figure 15.5. In order to get a proper spectrum, the IR source beam must be sampled also as a reference to adjust for instrument characteristics influencing the IR beam measurements, such as the IR source and the detector. This is illustrated in the double-beam layout shown in Figure 15.2. For raw spectral data processing, the sample transmission spectrum is divided by the reference transmission spectrum.

An example of a commercial-grade FTIR instrument is depicted in Figure 15.6 for the Agilent Cary 620 spectrochemical imaging system where the FTIR instrument has been coupled to a microscope. This configuration enables single point and spatial mapping with IR spectral imaging. Figure 15.7 depicts the internal optics of the microscope.

15.3 THE INFRARED SPECTRUM AND MOLECULAR ASSIGNMENT

A very powerful and valuable aspect of FTIR spectrometers is molecular assignments for the IR spectra. Figure 15.8 depicts an FTIR spectrum of the ambient background gases in the FTIR sample compartment. This is usually collected in order to subtract out from a sample spectrum. Note that in this particular

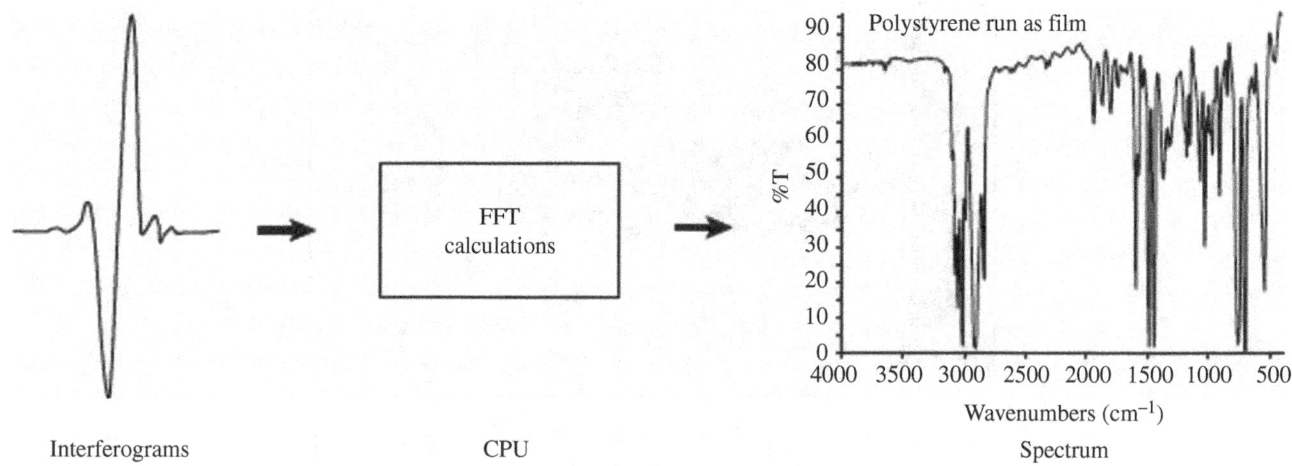

FIGURE 15.5 Illustration of converting an interferogram into an IR spectrum using a CPU and Fast Fourier transform (FFT) calculations. (Used by permission from Thermo Fisher Scientific, the copyright owner.)

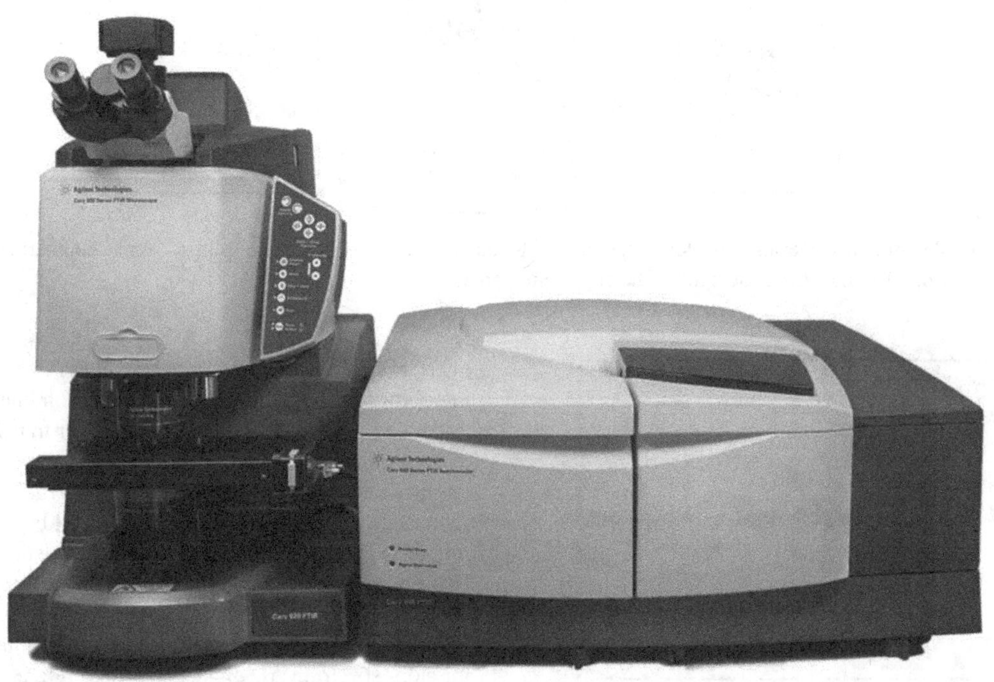

FIGURE 15.6 Agilent Cary 620 spectrochemical imaging system where the FTIR instrument has been coupled to a microscope. (© Agilent Technologies, Inc. 2014, Reproduced with Permission, Courtesy of Agilent Technologies, Inc.)

spectrum, the *x*-axis is in wavenumbers (cm^{-1}) and the *y*-axis is percent transmittance. This is often a convenient form used in functional group assignment to a sample. When quantitative analysis is being performed, the preferred output is to use absorbance. The major wavenumbers displayed in the background spectrum are for water vapor (H_2O) at approximately 1500 and 3800 cm^{-1}, and CO_2 at approximately 675 and 2300 cm^{-1}. The background ambient atmosphere in the sample compartment has these bands that can interfere in sample analysis if they are not removed. Different bonds, most associated with functional groups in an organic compound will absorb infrared radiation at different wavenumbers. We can utilize this aspect to assign structures to both known compounds and also to elucidate the structure and identity of an unknown compound or sample.

15.4 FTIR TABLE BAND ASSIGNMENTS

Table 15.2 depicts a listing of absorption peaks for the different types of bonds making up organic structures and also listing the relative intensities displayed on the FTIR spectra as weak, medium, or strong. Table 15.2 is a relatively simple table for FTIR spectral band assignments for the most commonly observed organic functional groups. A more complex assignment table is presented in Figure 15.9, which includes both organic and

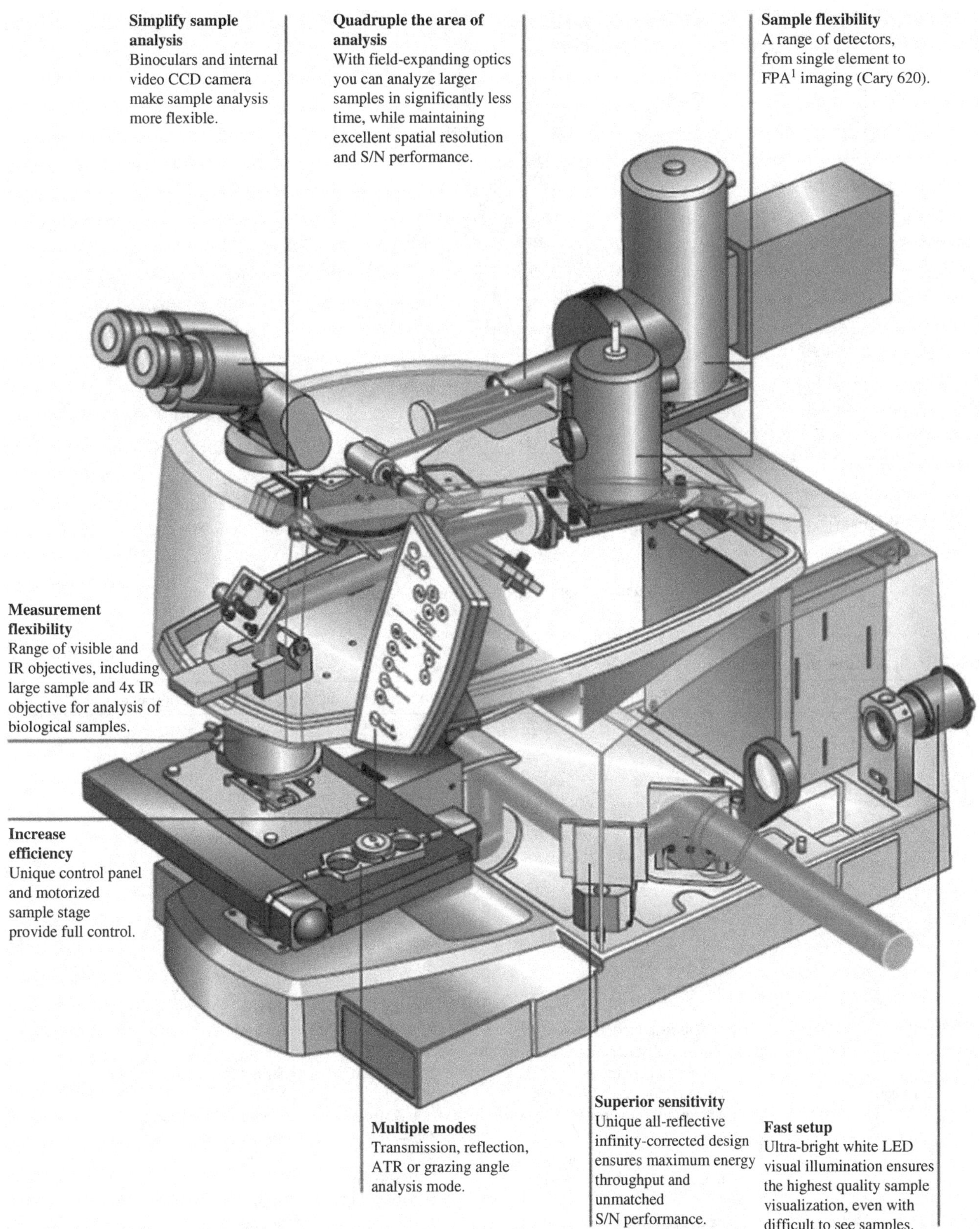

FIGURE 15.7 Internal optics of the 620 FTIR Microscope. (© Agilent Technologies, Inc. 2014, Reproduced with Permission, Courtesy of Agilent Technologies, Inc.)

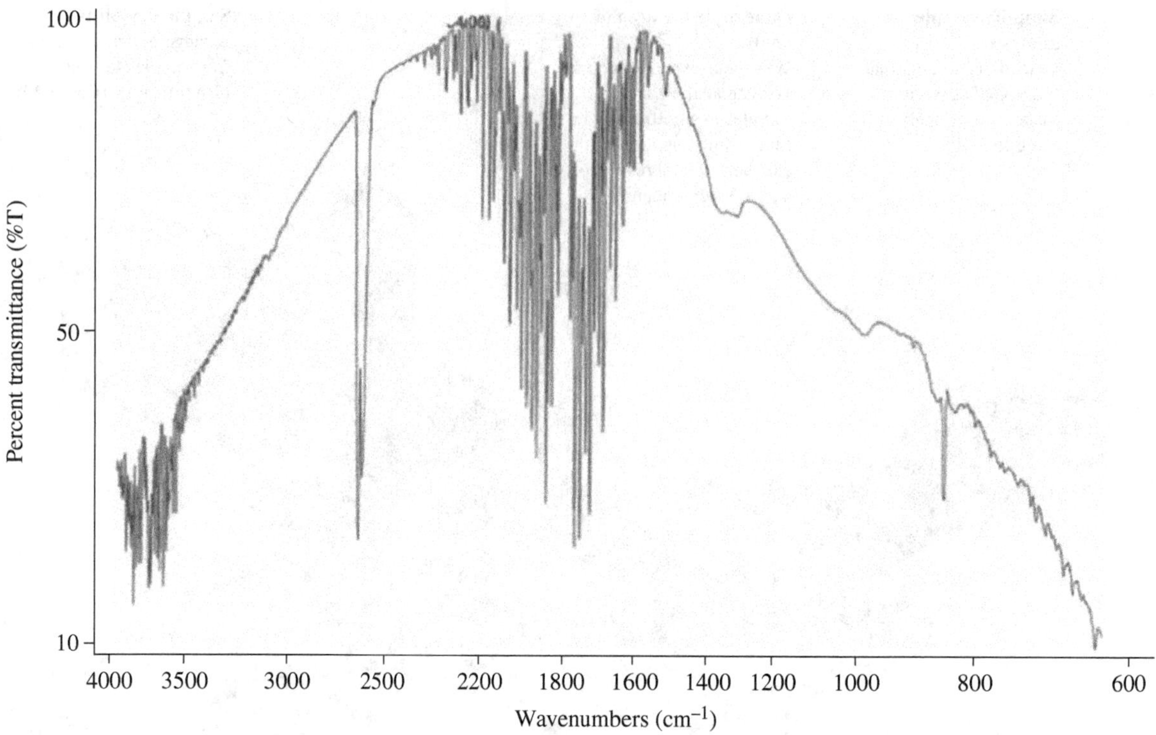

FIGURE 15.8 FTIR background spectrum.

TABLE 15.2 Infrared Spectroscopy Correlation Table[a].

Bond	Type of bond	Specific type of bond	Absorption peak	Appearance
C—H	Alkyl	Methyl	$1260\ cm^{-1}$	Strong
			$1380\ cm^{-1}$	Weak
			$2870\ cm^{-1}$	Medium to strong
			$2960\ cm^{-1}$	Medium to strong
		Methylene	$1470\ cm^{-1}$	Strong
			$2850\ cm^{-1}$	Medium to strong
			$2925\ cm^{-1}$	Medium to strong
		Methane	$2890\ cm^{-1}$	Weak
	Vinyl	$C{=}CH_2$	$900\ cm^{-1}$	Strong
			$2975\ cm^{-1}$	Medium
			$3080\ cm^{-1}$	Medium
		C=CH	$3020\ cm^{-1}$	Medium
		Monosubstituted alkenes	$900\ cm^{-1}$	Strong
			$990\ cm^{-1}$	Strong
		cis-Disubstituted alkenes	$670–700\ cm^{-1}$	Strong
		trans-Disubstituted alkenes	$965\ cm^{-1}$	Strong
		Trisubstituted alkenes	$800–840\ cm^{-1}$	Strong to medium
	Aromatic	Benzene/substituted benzene	$3070\ cm^{-1}$	Weak
		Monosubstituted benzene	$700–750\ cm^{-1}$	Strong
			$690–710\ cm^{-1}$	Strong
		ortho-Disubstituted benzene	$750\ cm^{-1}$	Strong
		meta-Disubstituted benzene	$750–800\ cm^{-1}$	Strong
			$860–900\ cm^{-1}$	Strong
		para-Disubstituted benzene	$800–860\ cm^{-1}$	Strong
	Alkynes	Any	$3300\ cm^{-1}$	Medium
	Aldehydes	Any	$2720\ cm^{-1}$	Medium
			$2820\ cm^{-1}$	
C—C	Acyclic C—C	Monosubstituted alkenes	$1645\ cm^{-1}$	Medium
		1,1-Disubstituted alkenes	$1655\ cm^{-1}$	Medium
		cis-1,2-Disubstituted alkenes	$1660\ cm^{-1}$	Medium
		trans-1,2-disubstituted alkenes	$1675\ cm^{-1}$	medium
		Trisubstituted, tetrasubstituted Alkenes	$1670\ cm^{-1}$	Weak

TABLE 15.2 (*Continued*)

Bond	Type of bond	Specific type of bond	Absorption peak	Appearance
	Conjugated C—C	Dienes	1600 cm^{-1}	Strong
			1650 cm^{-1}	Strong
	With benzene ring		1625 cm^{-1}	Strong
	With C=O		1600 cm^{-1}	Strong
	C=C (both sp^2)	Any	1640–1680 cm^{-1}	Medium
	Aromatic C=C	Any	1450 cm^{-1}	Weak to strong (usually 3 or 4)
			1500 cm^{-1}	
			1580 cm^{-1}	
			1600 cm^{-1}	
	C≡C	Terminal alkynes	2100–2140 cm^{-1}	Weak
		Disubstituted alkynes	2190–2260 cm^{-1}	Very weak (often indistinguishable)
C=O	Aldehyde/ketone	Saturated aliphatic/cyclic 6-membered	1720 cm^{-1}	
		α,β-Unsaturated	1685 cm^{-1}	
		Aromatic ketones	1685 cm^{-1}	
		Cyclic 5-membered	1750 cm^{-1}	
		Cyclic 4-membered	1775 cm^{-1}	
		Aldehydes	1725 cm^{-1}	Influence of conjugation (as with ketones)
	Carboxylic acids/derivates	Saturated carboxylic acids	1710 cm^{-1}	
		Unsaturated/aromatic carboxylic acids	1680–1690 cm^{-1}	
		Esters and lactones	1735 cm^{-1}	Influenced by conjugation and ring size (as with ketones)
		Anhydrides	1760 cm^{-1}	
			1820 cm^{-1}	
		Acyl halides	1800 cm^{-1}	
		Amides	1650 cm^{-1}	Associated amides
		Carboxylates (salts)	1550–1610 cm^{-1}	
		Amino acid zwitterions	1550–1610 cm^{-1}	
O—H	Alcohols, phenols	Low concentration	3610–3670 cm^{-1}	
		High concentration	3200–3400 cm^{-1}	Broad
	Carboxylic acids	Low concentration	3500–3560 cm^{-1}	
		High concentration	3000 cm^{-1}	Broad
N—H	Primary amines	Any	3400–3500 cm^{-1}	Strong
			1560–1640 cm^{-1}	Strong
	Secondary amines	Any	>3000 cm^{-1}	Weak to medium
	Ammonium ions	Any	2400–3200 cm^{-1}	Multiple broad peaks
C—O	Alcohols	Primary	1040–1060 cm^{-1}	Strong, broad
		secondary	~1100 cm^{-1}	Strong
		Tertiary	1150–1200 cm^{-1}	Medium
	Phenols	Any	1200 cm^{-1}	
	Ethers	Aliphatic	1120 cm^{-1}	
		Aromatic	1220–1260 cm^{-1}	
	Carboxylic acids	Any	1250–1300 cm^{-1}	
	Esters	Any	1100–1300 cm^{-1}	Two bands (distinct from ketones, which do not possess a C—O bond)
C—N	Aliphatic amines	Any	1020–1220 cm^{-1}	Often overlapped
	C=N	Any	1615–1700 cm^{-1}	Similar conjugation effects to C=O
	C≡N (nitriles)	Unconjugated	2250 cm^{-1}	Medium
		Conjugated	2230 cm^{-1}	Medium
	R—N—C (isocyanides)	Any	2165–2110 cm^{-1}	
	R—N=C=S	Any	2140–1990 cm^{-1}	
C—X	Fluoroalkanes	Ordinary	1000–1100 cm^{-1}	
		Trifluromethyl	1100–1200 cm^{-1}	Two strong, broad bands
	Chloroalkanes	Any	540–760 cm^{-1}	Weak to medium
	Bromoalkanes	Any	500–600 cm^{-1}	Medium to strong
	Iodoalkanes	Any	500 cm^{-1}	Medium to strong
N—O	Nitro compounds	Aliphatic	1540 cm^{-1}	Stronger
			1380 cm^{-1}	Weaker
		Aromatic	1520, 1350 cm^{-1}	Lower if conjugated

[a] Reprinted with permission under the Wiki Creative Commons Attribution-Sharealike 3.0 Unported License.

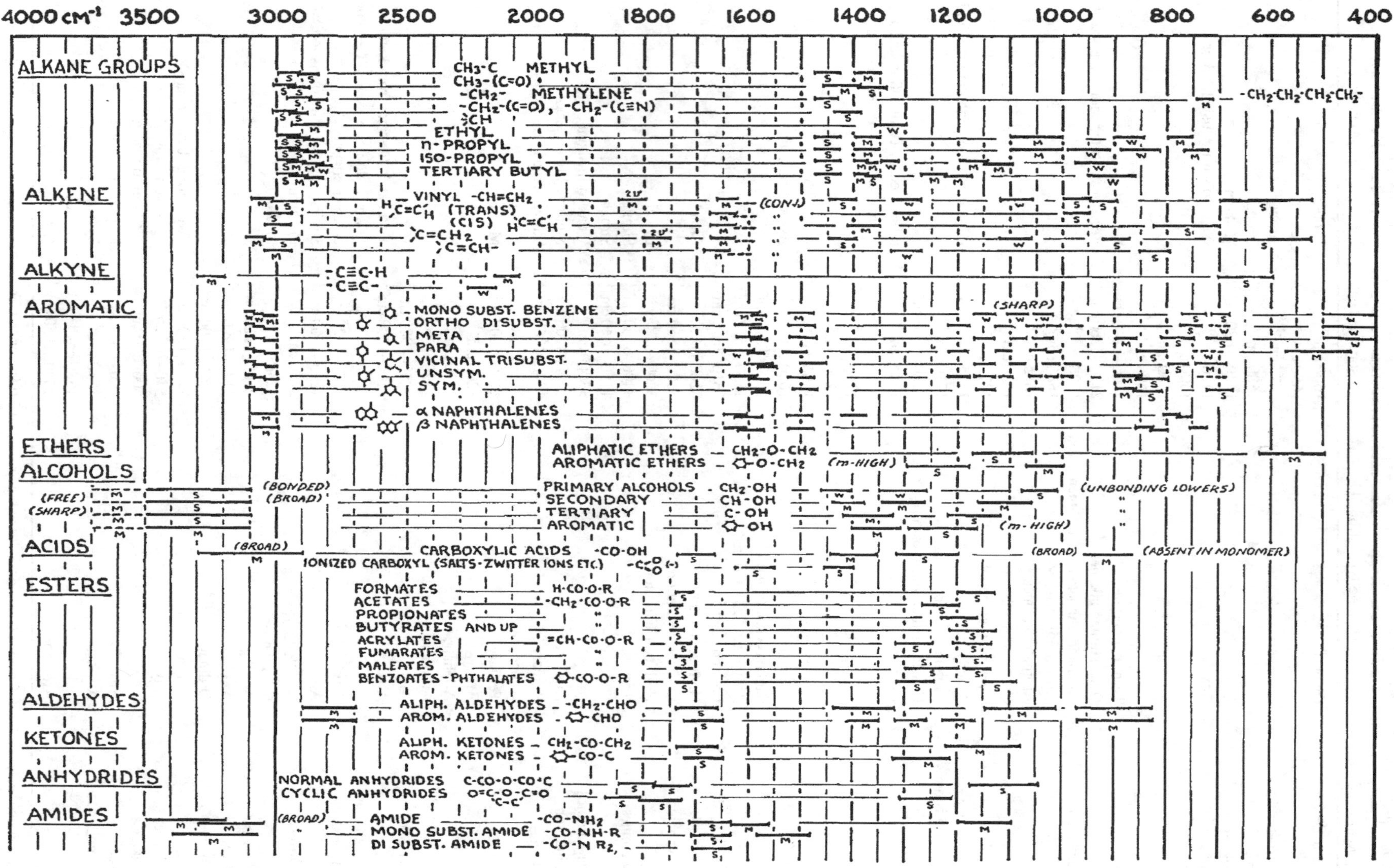

FIGURE 15.9 Spectra–Structure Correlations in the Infrared Region. (N.B. Colthup Spectra-Structure Correlations in the Infrared Region J. Opt. Soc. Am. 40 (6), 1950, 397–400. Reprinted with permission from the Optical Society of America, 2014.)

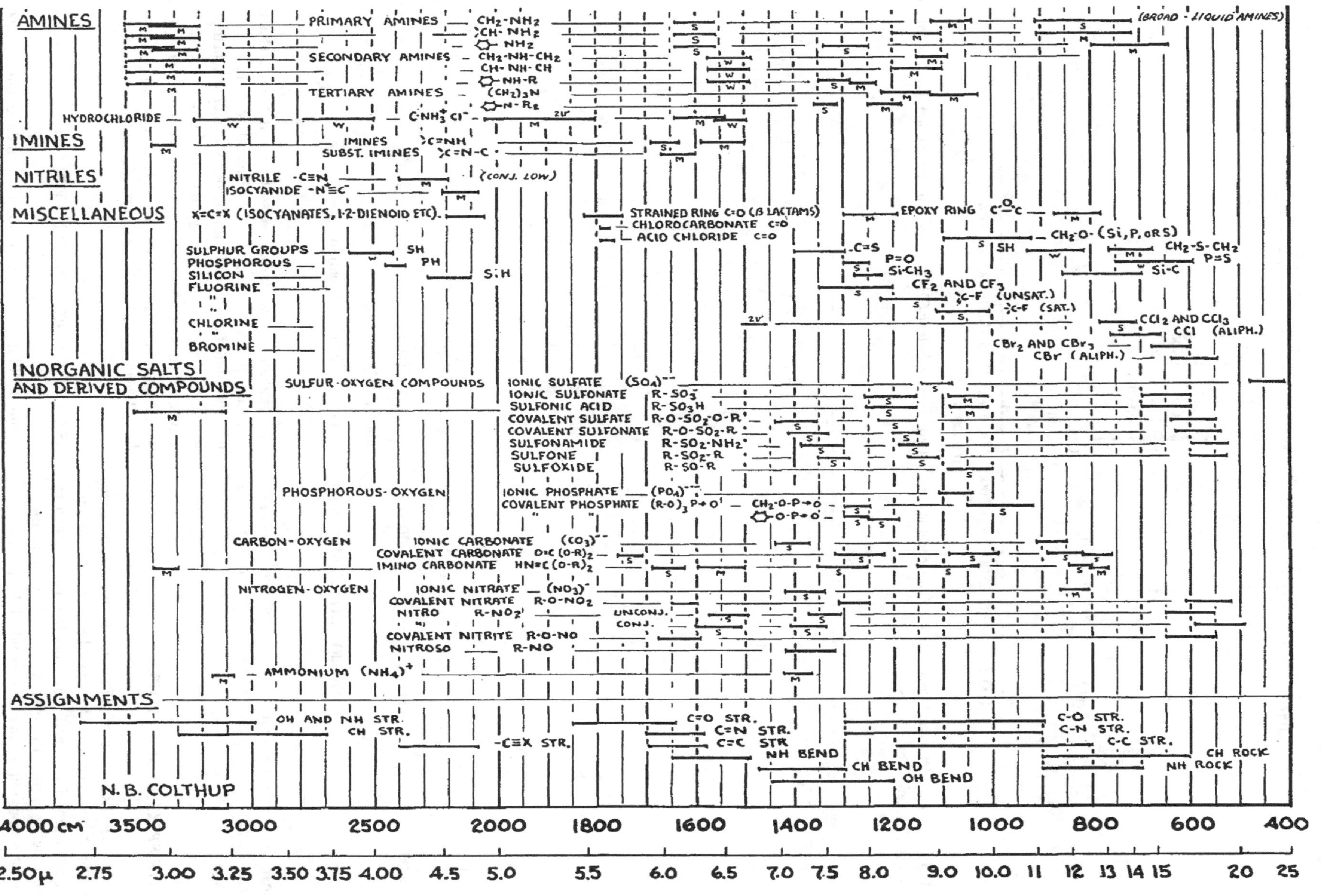

FIGURE 15.9 (Continued)

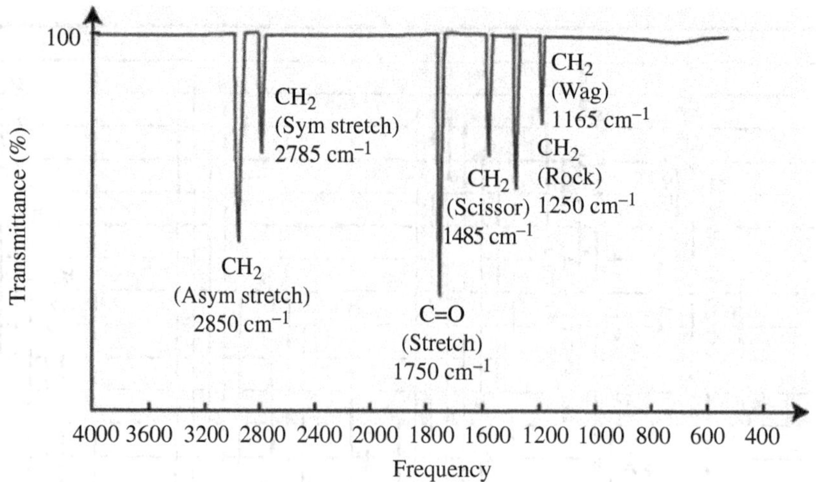

FIGURE 15.10 Characteristic bands associated with the organic carbon-hydrogen methylene CH_2 bond, and for the carbonyl functional group C=O stretch at 1750 cm^{-1} for ketones.

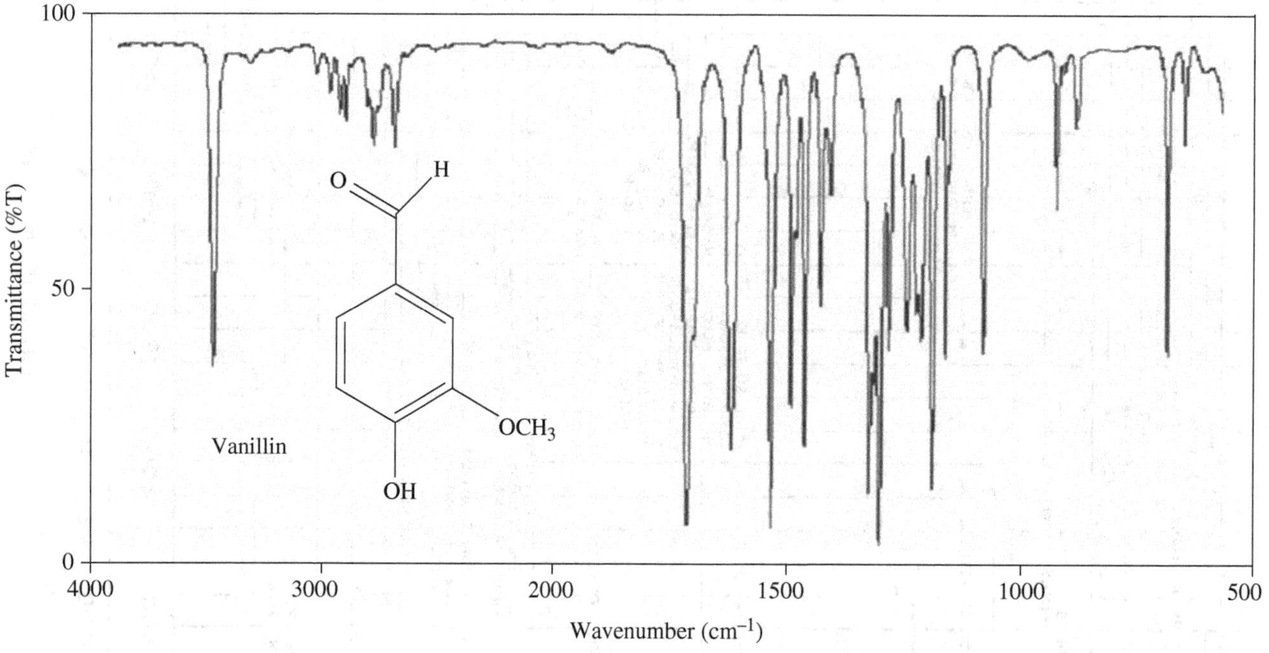

FIGURE 15.11 FTIR spectrum of the organic compound vanillin.

inorganic band assignments. Figure 15.9 is a well-known and still quite useful correlation table for FTIR spectral band functional group assignments.

15.5 FTIR SPECTRUM EXAMPLE I

If we look at the FTIR spectrum in Figure 15.10, we can see some characteristic bands associated with the organic carbon–hydrogen methylene CH_2 bond. There are five bands listed in the figure ranging from 1165 to 2850 cm^{-1}. It is common to observe most of these in organic compound FTIR spectra. Note the descriptions to the vibrations associated to the absorbance of IR radiation as wag, rock, scissor, and stretching. This is the common terminology used for the different vibrations and structure deformities associated with the IR. Also included in the spectrum is the band for the carbonyl functional group C=O stretch at 1750 cm^{-1} for ketones, an important spectral range area associated with the carbonyl group. From Table 15.2, we also see 1725 cm^{-1} for aldehydes, 1710 cm^{-1} for carboxylic acids, and so on.

15.6 FTIR SPECTRUM EXAMPLE II

Figure 15.11 depicts an FTIR spectrum of the organic compound vanillin. As can be seen, the spectrum contains numerous bands

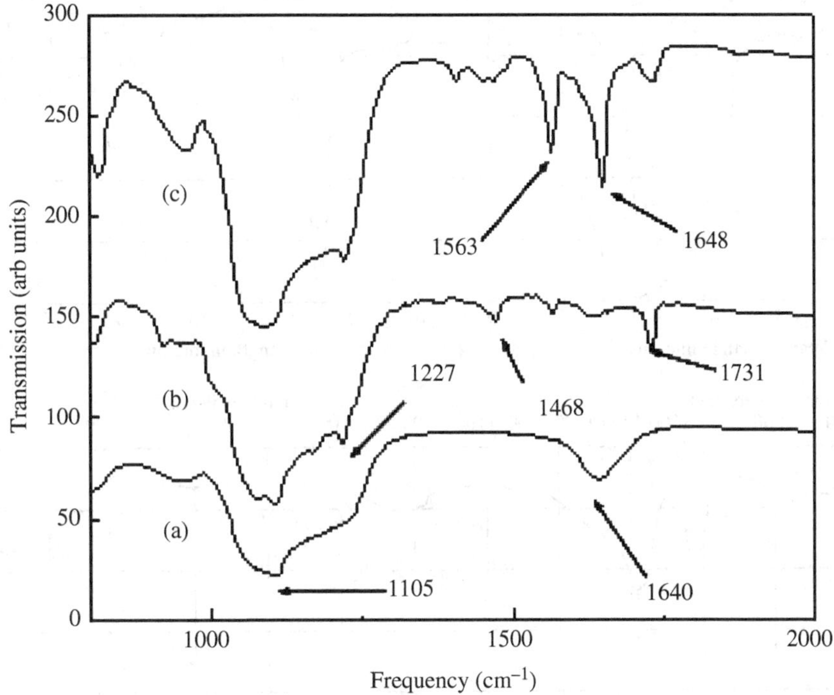

FIGURE 15.12 FTIR spectra monitoring different steps of the lipobeads synthesis: (a) bare silica beads, (b) silica lipobeads prior to acylation, and (c) silica lipobeads following acylation with sebacoyl chloride.

that at first glance appears to be nondescriptive; however, many of the characteristic functional group absorbance bands are represented. Starting at the higher wavenumbers, there is a band at approximately 3500 cm^{-1} for the hydroxyl OH group, there are aromatic C–H bands at 3000–3200 cm^{-1}, CH_2 bands at 2700–2850 cm^{-1}, aromatic aldehyde at 1700 cm^{-1}, and aromatic ether at 1220–1260 cm^{-1}. We can readily see that the FTIR spectrum does represent the different groups contained within the vanillin structure (Figure 15.12).

15.7 FTIR INORGANIC COMPOUND ANALYSIS

Inorganic compound analysis is also performed using FTIR spectroscopy. Many inorganic compounds have FTIR spectra detailed enough to identify the compound. Figure 15.13 depicts the IR spectra for four selected inorganic compounds including: sodium carbonate (Na_2CO_3), potassium cyanide (KCN) ($KHCO_3$, K_2CO_3 impurities), ammonium phosphate, dibasic ($(NH_4)_2HPO_4$), and copper sulfate ($CuSO_4$). In the spectra, a star (∗) denotes a spectral band arising from the Nujol oil that is used to disperse and hold the inorganic compounds for spectral analysis. Most laboratories now use attenuated total reflectance (ATR) apparatus that allows the measurement of powdered compounds directly without the need for the use of Nujol oil. Spectra like these can be searched against known libraries to identify the compounds when there are enough discerning features. Table 15.3 contains the characteristic IR bands in the spectra of Figure 15.13 for the selected inorganic compounds. For the spectral intensity, the following abbreviations are used: m = medium, w = weak, vw = very weak, and s = strong to describe the expected observed bands.

15.8 CHAPTER KEY CONCEPTS

15.1 In UV/Vis analysis, the wavelengths used are approximately 190–400 nm for the ultraviolet (UV) region, 400–800 nm for the visible (Vis), and finally 1000–15,000 nm for the infrared region (IR).

15.2 The name infra comes from “inferior” as the infrared region is next to but less in energy than the red region of the spectrum.

15.3 With the infrared absorption process, energy of longer wavelengths is being absorbed with translation to processes involving molecular vibrations.

15.4 There are two types of molecular vibrations that take place: stretching and bending. Stretching is the movement of atoms back and forth along the bond axis. Bending may involve the change in bond angles between two atoms that are each attached to a third.

15.5 Bending also takes place as a movement of atoms in respect to the whole molecule referred to as rocking, scissoring, wagging, and twisting.

15.6 Glass and quartz both absorb energy greatly in the infrared region.

15.7 Inorganic salts are used for internal transmission optics and sample holders.

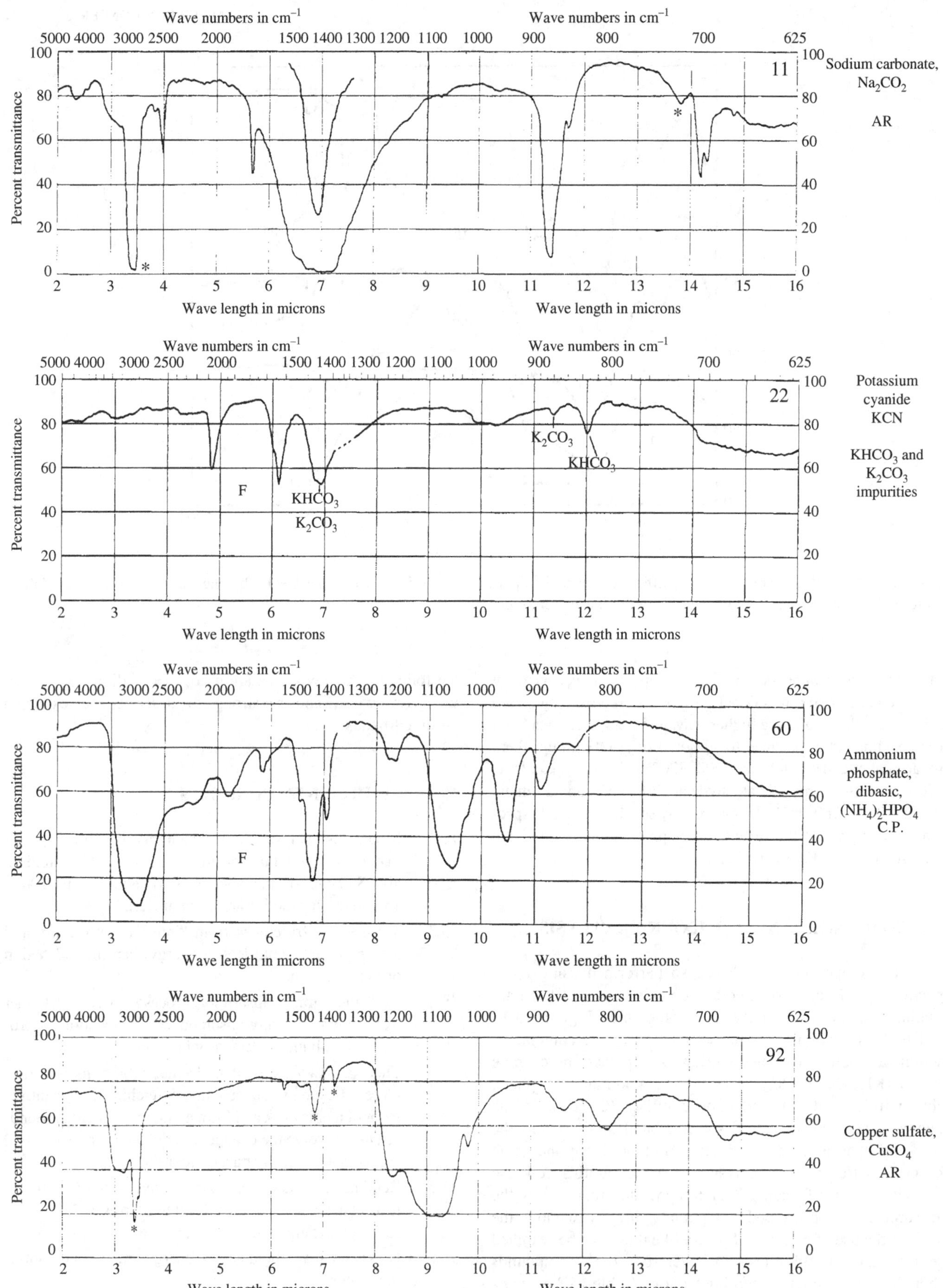

FIGURE 15.13 FTIR spectra of sodium carbonate Na_2CO_3, potassium cyanide KCN ($KHCO_3$, K_2CO_3 impurities), ammonium phosphate, dibasic $(NH_4)_2HPO_4$, and copper sulfate $CuSO_4$. (Reprinted with permission from Analytical Chemistry, F.A. Miller, C.H. Wilkins Infrared Spectra and Characteristic Frequencies of Inorganic Ions 1952, pp. 1253–1294. Copyright 1952 American Chemical Society.)

TABLE 15.3 Characteristic IR Bands for the Selected Inorganic Compounds. For Intensity, m = medium, w = weak, vw = very weak, s = strong, vs = very strong[a].

Sodium carbonate (Na_2CO_3)

Wavenumber (cm^{-1})	Micrometers (μm)	Intensity
700	14.3	m
705	14.2	m
855	11.7	vw
878	11.4	s
1440	6.95	vs
1755	5.7	m
2500	4.0	m
2620	3.82	vw
3000	3.3	m

Potassium cyanide (KCN) ($KHCO_3$, K_2CO_3 impurities)

Wavenumber (cm^{-1})	Micrometers (μm)	Intensity
833	12.0	m
882	11.35	vw
1440	6.95	s
1635	6.12	s
2070	4.83	s

Ammonium phosphate, dibasic ($(NH_4)_2HPO_4$)

Wavenumber (cm^{-1})	Micrometers (μm)	Intensity
850	11.75	vw
900	11.1	m
953	10.5	s
1060	9.45	vs
1195	8.37	m
1220	8.22	m
1415	7.05	m
1470	6.8	s
1530	6.55	vw
1710	5.85	w
1920	5.2	m
2200	4.55	w
2860	3.5	vs

Copper sulfate ($CuSO_4$)

Wavenumber (cm^{-1})	Micrometers (μm)	Intensity
680	14.7	m
805	12.45	m
860	11.6	m
1020	9.8	w
1090	9.2	vs
1200	8.35	s
1600	6.25	w
3300	3.15	s

[a] Reprinted with permission from Analytical Chemistry, F.A. Miller, C.H. Wilkins Infrared Spectra and Characteristic Frequencies of Inorganic Ions 1952, pp. 1253–1294.

15.8 Wavenumber is the most commonly used scale when reporting and discussing IR spectra. The relationship between wavelength and wavenumber is as follows:

$$\text{Wavenumber, cm}^{-1} = \frac{1}{\text{wavelength, μm}} \times 10^4.$$

15.9 The major components of an infrared spectrometer include the infrared radiation source, the beam path for sample and reference, the monochromator, and the detector.

15.10 The infrared radiation sources used are either a *Nernst glower* or a *globar*. The Nernst glower comprises a small rod of zirconium and yttrium oxides that is heated to approximately 2000 °C. The globar consists of sintered silicon carbide that is heated to approximately 1400 °C.

15.11 For the monochromator, a dispersing diffraction grating is typically used.

15.12 The FTIR spectrometers will have an interferometer after the IR source and before the beam splitting.

15.13 The advantage of the FTIR is that the infrared spectrum is scanned very quickly using a moving mirror system. The interferogram scans are added together reducing the background noise significantly.

15.14 The IR instruments used today in analytical laboratories are solely FTIR spectrometers.

15.15 A very powerful and valuable aspect of FTIR spectrometers is molecular assignments for the IR spectra.

15.16 Different bonds, most associated with functional groups in an organic compound will absorb infrared radiation at different wavenumbers. We can utilize this aspect to assign structures to both known compounds and also to elucidate the structure and identity of an unknown compound or sample.

15.17 A spectrum of the ambient background gases in the FTIR sample compartment is usually collected in order to subtract out from a sample spectrum.

15.18 Absorption peaks for the different types of bonds making up organic structures list the relative intensities as weak, medium, or strong.

15.19 Wag, rock, scissor, and stretching are common terminologies used for the different vibrations and structure deformities associated with the IR.

15.20 Inorganic compound analysis is also performed using FTIR spectroscopy. Many inorganic compounds have FTIR spectra detailed enough to identify the compound.

15.21 Most laboratories now use ATR apparatus that allows the measurement of powdered compounds directly without the need for the use of Nujol oil.

15.22 Spectra can be searched against known libraries to identify the compounds when there are enough discerning features.

15.9 CHAPTER PROBLEMS

15.1 The following spectrum is a background FTIR spectrum of ambient air. What are the peaks in the spectrum attributed to?

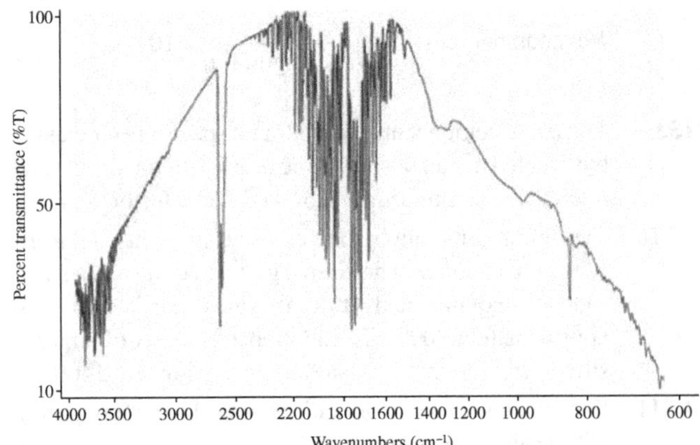

15.2 Without a background spectrum, spectra of samples can contain interference peaks from carbon dioxide and water. Why are these two compounds IR active?

15.3 List whether the following are expected to be IR active or IR inactive for the compound acetylene.

HC≡CH

Acetylene

(a)

← C≡C →

Stretch

(b)

HC≡C—H →

Stretch

(c)

← H—C≡C—H →

Stretch

15.4 List whether the following are expected to be IR active or IR inactive for the compound ethene.

CH_2=CH_2

Ethene

(a)

← C—C →

Stretch

(b)

H, H / C=C / H, H (with arrows)

Stretch

(c)

H, H / C=C / H, H (with arrows)

Stretch

(d)

C—C →

Stretch

15.5 Vegetable seeds were extracted with petroleum ether. The ether was removed on a steam bath leaving behind an oily substance. The oily substance was smeared onto a KBr window, and an FTIR spectrum was collected.

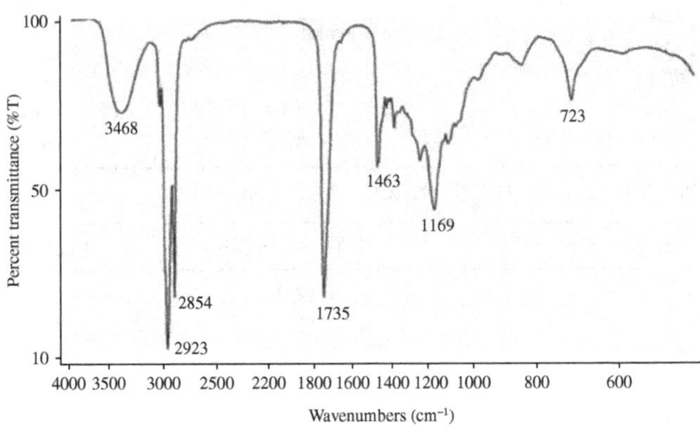

a. List the major transmittance peak wavenumbers in the spectrum.
b. Using Table 15.2 and Figure 15.9, assign functional groups to the FTIR spectral peaks.
c. From the extraction information and the spectrum, what is the material?

15.6 The following FTIR spectrum is of mineral oil, a substance often used as a carrier for other compounds used for collecting spectra. Nujol is a brand of mineral oil by Plough Inc., CAS number 8012-95-1, and density 0.838 g/ml at 25 °C, used in infrared spectroscopy.

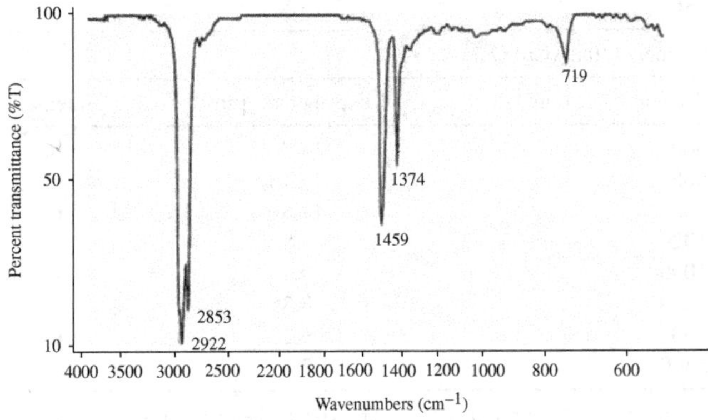

a. List the major peaks and make assignments.
b. Compare this spectrum to the spectrum of Problem 15.1. How are they similar and/or different?

15.7 The following FTIR spectrum was obtained by crushing oyster shells, mixing some of the powder with mineral oil, and collecting the spectrum.

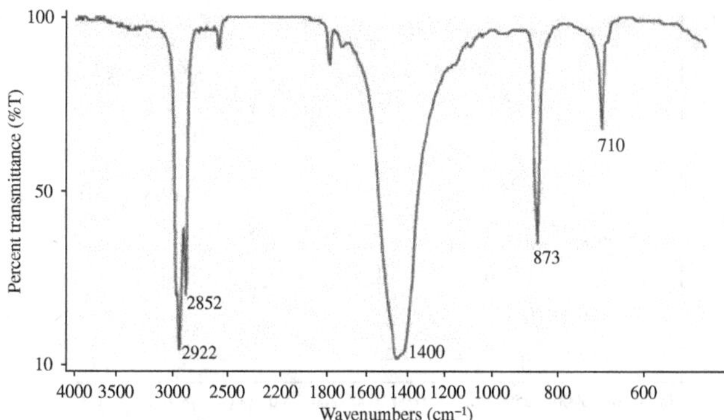

a. List the major transmittance peak wavenumbers in the spectrum.
b. Using Table 15.2 and Figure 15.9, assign functional groups to the FTIR spectral peaks.
c. From the extraction information and the spectrum, what is the material?

15.8 The following FTIR spectrum is that of potassium phosphate.

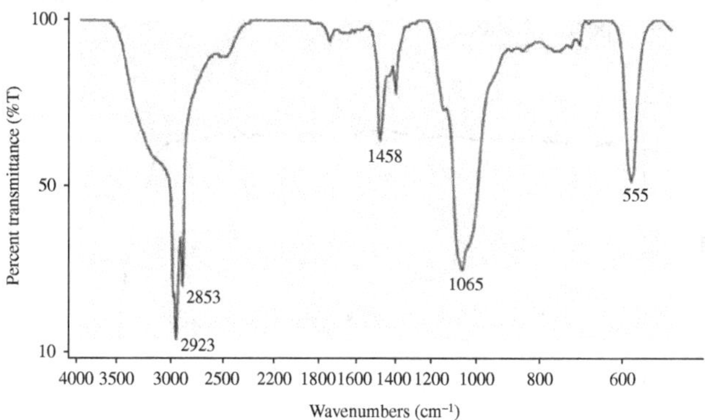

a. List the major transmittance peak wavenumbers in the spectrum.
b. Using Table 15.2 and Figure 15.9, assign functional groups to the FTIR spectral peaks.

15.9 The following FTIR spectrum is that of the aromatic compound benzene.

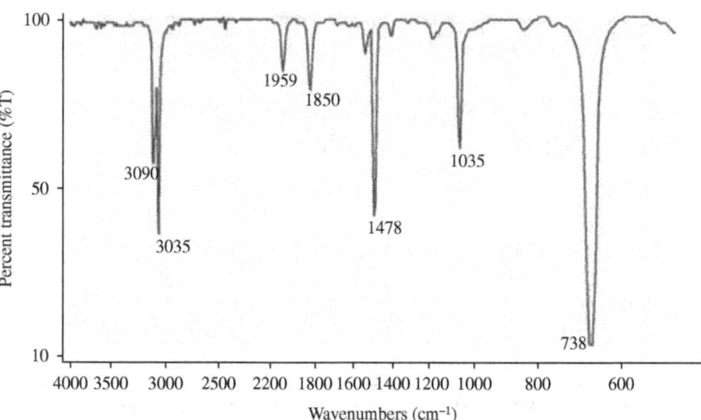

a. List the major transmittance peak wavenumbers in the spectrum.
b. Using Table 15.2 and Figure 15.9, assign functional groups to the FTIR spectral peaks.

15.10 The following FTIR spectrum is that of a simple derivative of the aromatic compound benzene. The compound is phenol.

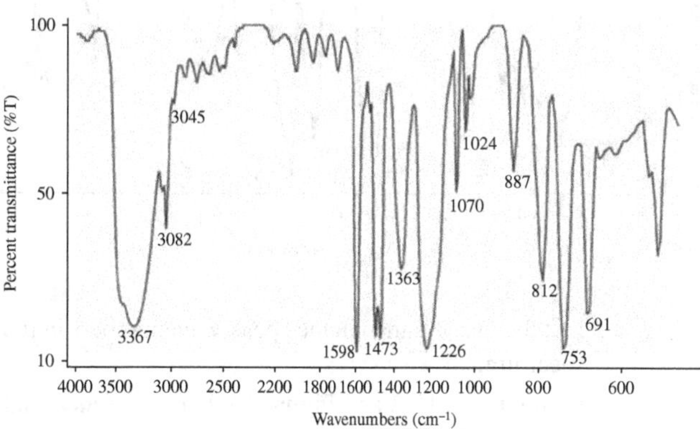

a. List the major transmittance peak wavenumbers in the spectrum.
b. Using Table 15.2 and Figure 15.9, assign functional groups to the FTIR spectral peaks.
c. Why is the spectrum for phenol more complicated than the spectrum in Problem 9 for benzene? What differentiates the two?

15.11 The following FTIR spectrum is that of a simple derivative of the aromatic compound benzene. The compound is benzoic acid.

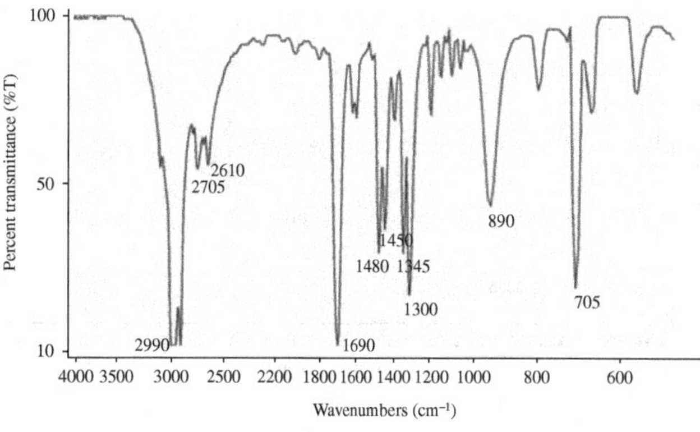

a. List the major transmittance peak wavenumbers in the spectrum.
b. Using Table 15.2 and Figure 15.9, assign functional groups to the FTIR spectral peaks.
c. Why is the spectrum for benzoic acid more complicated than the spectrum in Problem 15.9 for benzene, and different from that of phenol? What differentiates the spectrum?

15.12 The following FTIR spectrum is that of a second simple organic carboxylic acid compound acetic acid (ethanoic acid).

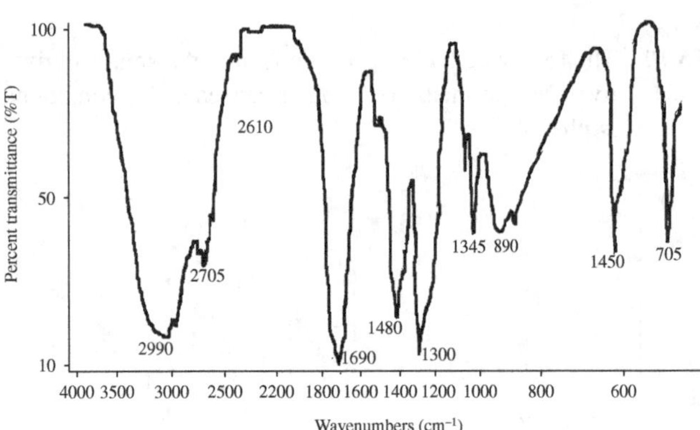

a. List the major transmittance peak wavenumbers in the spectrum.
b. Using Table 15.2 and Figure 15.9, assign functional groups to the FTIR spectral peaks.
c. What are the similarities and differences between the FTIR spectra for acetic acid and benzoic acid?

15.13 The following FTIR spectrum is that of starch. Starch is a carbohydrate consisting of a large number of glucose units joined by glycosidic bonds. This polysaccharide is produced by most green plants as an energy store. It is the most common carbohydrate in human diets and is contained in large amounts in such staple foods as potatoes, wheat, maize (corn), rice, and cassava.

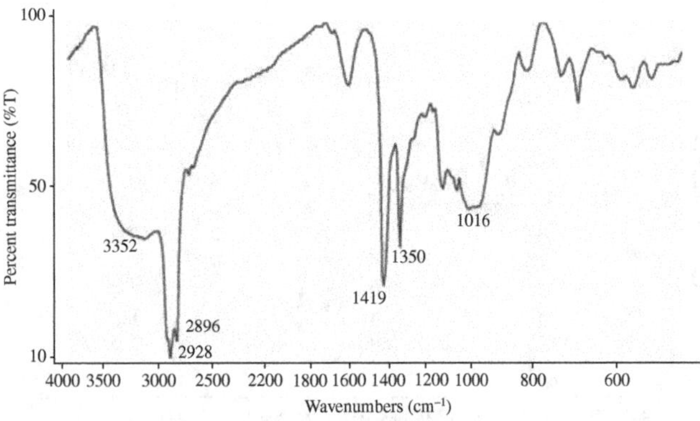

a. List the major transmittance peak wavenumbers in the spectrum.
b. Using Table 15.2 and Figure 15.9 and the structure of starch, assign functional groups to the FTIR spectral peaks.

15.14 Use the assignments that have been presented in the preceding problems to decipher the following FTIR spectrum.

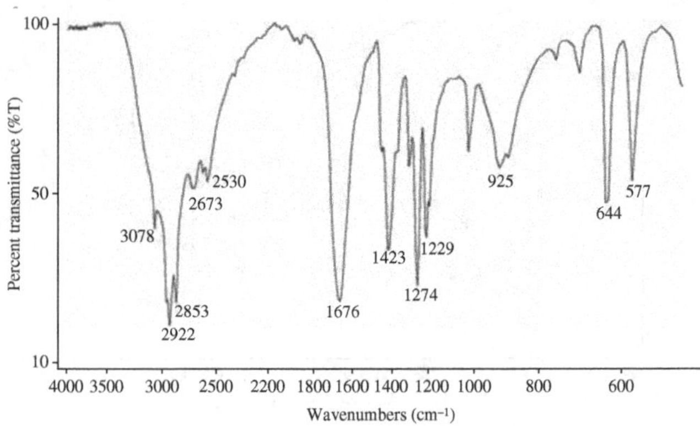

a. List the major transmittance peak wavenumbers in the spectrum.
b. Using Table 15.2 and Figure 15.9, assign functional groups to the FTIR spectral peaks.
c. What are some possible structures or identifications for the compound?

15.15 The FTIR spectrum is that of anhydrous sodium chloride (NaCl). Explain why the spectrum is void of any IR spectral bands.

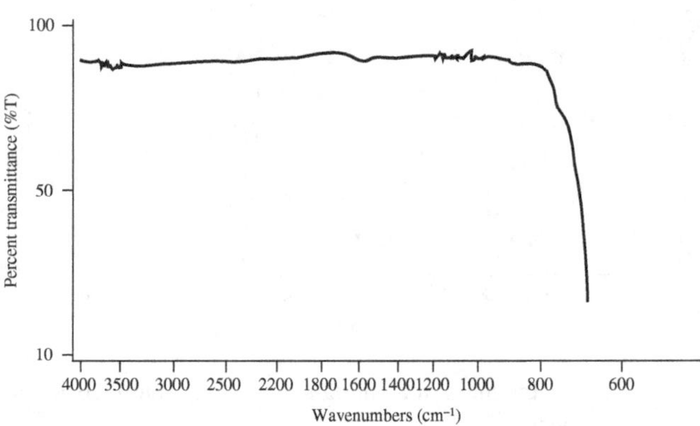

15.16 The FTIR spectrum is that of magnesium chloride hexahydrate ($MgCl_2{\cdot}6H_2O$). Explain why the spectrum possesses IR spectral bands. Identify the functional groups that are responsible for the spectral bands.

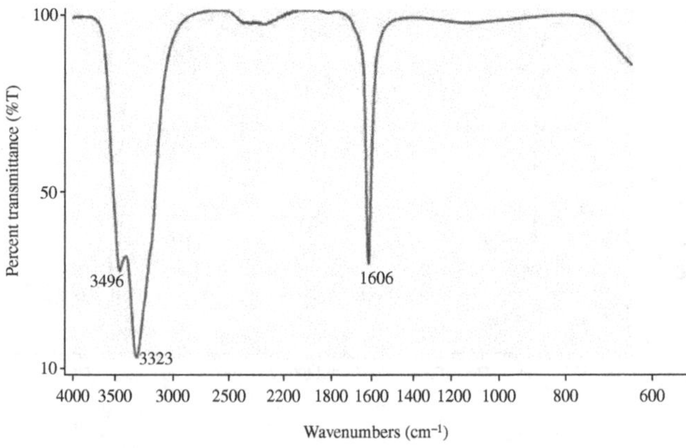

16

NUCLEAR MAGNETIC RESONANCE (NMR) SPECTROSCOPY

16.1 Introduction
16.2 Frequency and Magnetic Field Strength
16.3 Continuous-Wave NMR
16.4 The NMR Sample Probe
16.5 Pulsed Field Fourier Transform NMR
16.6 Proton NMR Spectra Environmental Effects
16.6.1 Chemical Shift
16.6.2 Spin–Spin Splitting (Coupling)
16.6.3 Interpretation of NMR Spectra
16.7 Carbon-13 NMR
16.7.1 Introduction
16.7.2 Carbon-13 Chemical Shift
16.7.3 Carbon-13 Splitting
16.7.4 Finding the Number of Carbons
16.7.5 Carbon-13 NMR Examples
16.8 Special Topic—NMR Characterization of Cholesteryl Phosphate
16.8.1 Synthesis of Cholesteryl Phosphate
16.8.2 Single-Stage and High-Resolution Mass Spectrometry
16.8.3 Proton Nuclear Magnetic Resonance (^{1}H-NMR)
16.8.4 Theoretical NMR Spectroscopy
16.8.5 Structure Elucidation
16.9 Chapter Key Concepts
16.10 Chapter Problems
References

16.1 INTRODUCTION

The analytical instrumental analysis technique of nuclear magnetic resonance (NMR) is another very useful tool for compound structural information. Unlike UV/Vis and FTIR spectroscopy where energy absorption involves electrons and bonds, the NMR process is a nuclear interaction brought about in the presence of a strong magnetic field. The four most common nuclei that are used in NMR studies are ^{1}H, ^{13}C, ^{19}F, and ^{31}P. Each of the nuclei have a spin quantum number of ½, and each nuclei have two spin states: $I = +1/2$ and $I = -1/2$. The related magnetic quantum numbers are $m = +1/2$ and $m = -1/2$. Similar to when electricity flows through a coil of wire a magnetic field is produced, so to a charged nucleus spinning creates a magnetic field. The physical property of the spinning nucleus is described by the magnetic moment μ that is orientated along the axis of spin and is proportional to the angular momentum p. The magnetic moment is expressed as

$$\mu = \gamma p, \tag{16.1}$$

where γ is a proportionality constant called the magnetogyric ratio (radian·T^{-1} s^{-1}). The ratio is specific for each nuclei. The magnetogyric ratio for ^{1}H is 2.6752×10^8 radian·T^{-1} s^{-1}, for ^{13}C is 6.7283×10^7 radian·T^{-1} s^{-1}, for ^{19}F is 2.5181×10^8 radian·T^{-1} s^{-1}, and for ^{31}P is 1.0841×10^8 radian·T^{-1} s^{-1}. Not all nuclei have a magnetic moment, only those with an odd sum of protons and neutrons as illustrated in Table 16.1.

16.2 FREQUENCY AND MAGNETIC FIELD STRENGTH

When an organic compound containing hydrogen nuclei is placed into a magnetic field, the magnetic moments (m) of each nuclei will align either with the magnetic field (low energy state, where $m = +1/2$) or against the external magnetic field (high energy state, where $m = -1/2$). The low energy state can be utilized to absorb energy at a certain frequency (radio frequency range) in order to raise it to the high energy state allowing measurement of the transition. Illustrative example of the energy levels of the two magnetic moments of the nucleus in an applied magnetic field is shown in Figure 16.1.

Analytical Chemistry: A Chemist and Laboratory Technician's Toolkit, First Edition. Bryan M. Ham and Aihui MaHam.

TABLE 16.1 Magnetic Properties of Selective Elements Used in NMR Studies and for Comparison.

Nucleus	Magnetogyric Ratio (radian·T^{-1}·s^{-1})	Spin Number	Isotopic Abundance (%)	Absorption Frequency (MHz)
^{1}H	2.6752×10^{8}	½	99.98	600[a]
^{2}H (deuterium)	4.1066×10^{7}	1	0.016	91.5
^{12}C		0	98.9	
^{13}C	6.7283×10^{7}	½	1.11	150
^{19}F	2.5181×10^{8}	½	100.0	561
^{31}P	1.0841×10^{8}	½	100.0	242

[a] All calculated with a 14 T magnet.

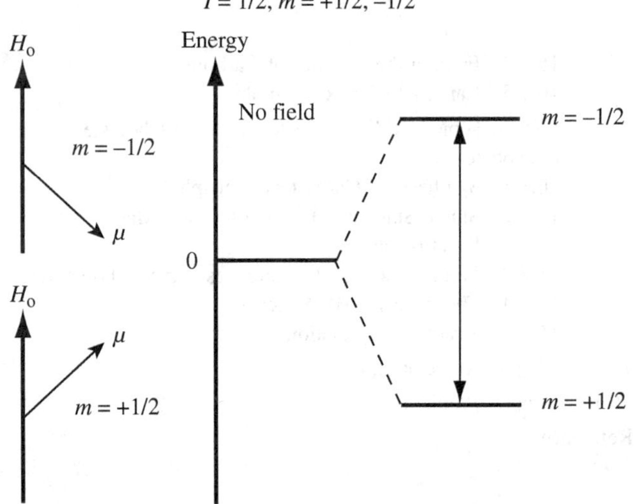

FIGURE 16.1 Magnetic moments in the presence of a strong applied magnetic field.

The potential energy of the two states can be derived from Planck's equation as follows:

$$E = -\frac{\gamma h}{4\pi}. \tag{16.2}$$

For the lower energy state, $m = +1/2$:

$$E_{+1/2} = -\frac{\gamma h}{4\pi} B_0. \tag{16.3}$$

For the higher energy state, $m = -1/2$:

$$E_{-1/2} = -\frac{\gamma h}{4\pi} B_0. \tag{16.4}$$

The difference between the two energy states is represented by ΔE:

$$\Delta E = \frac{\gamma h}{4\pi} B_0 - \left(-\frac{\gamma h}{4\pi} B_0\right)$$
$$\Delta E = \frac{\gamma h}{2\pi} B_0 \tag{16.5}$$

The frequency of the radiation required to bring about the transition can be obtained by substituting into Equation 16.5 the relationship $\Delta E = h\nu_0$ giving the expression:

$$\nu_0 = \frac{\gamma B_0}{2\pi}. \tag{16.6}$$

We can see that there is a direct relationship between the frequency of the transition and the magnetic field strength.

The frequency absorbed in the strong external magnetic field is characteristic of the isotope. NMR instruments used for proton resonance are often referred to as proton magnetic resonance (PMR) instruments, and are usually designated by the frequency for the proton transition instead of the magnet strength (also referred to as H-NMR). Using the magnetogyric ratio and Equation 16.6, if the frequency is known then the magnet strength can be calculated, and vice versa. For example, a 900 MHz (megahertz or million hertz) PMR will have a magnetic field strength of

$$B_0 = \frac{\nu_0 2\pi}{\gamma} = \frac{(9.0 \times 10^{8})2\pi}{2.6752 \times 10^{8}} = 21T. \tag{16.7}$$

In this case, the 21 T magnet is referred to as a 900 MHz magnet.

16.3 CONTINUOUS-WAVE NMR

Older NMR spectrometers are generally called continuous-wave (CW) NMR spectrometers due to the use of a steady, nonchanging magnetic field. Nuclei can absorb energy by either holding the magnetic field constant while changing the radio frequency, or holding the radio frequency oscillator constant and varying the magnetic field. These were often equipped with permanent magnets that have field strengths of 0.7 T (30 MHz proton), 1.4 T (60 MHz proton), and 2.1 T (90 MHz proton). Permanent magnets are very sensitive to heat and also require extensive shielding. Both sensitivity and resolution increase with increasing field strength, thus stronger magnets than these are desirable. The permanent magnets used in the CW-NMR were also unstable and tended to drift with time making extended period experiments difficult. The instruments were designed where the magnetic field is induced around the sample, and unlike optical spectroscopy that uses the electrical field of electromagnetic radiation to interact with absorbing species, NMR uses the magnetic field of the radiation to excite the absorbing species. If the magnetic vector's rotational frequency of the induced radiation is the same as the precessional frequency of the nucleus, then absorption with flipping will occur. An oscillator coil is mounted in the NMR instrument acting as the source radiation for absorbance at 90° to the direction of the fixed magnetic field. In order to alter the applied magnetic field over a small range, two field sweep generators made of a pair of coils are located parallel to the permanent magnet faces. A PMR instrument may have a fixed 60 MHz radio frequency and a small magnetic sweep range of 16.7 ppm. Both the magnetic field strength and the radio frequency are proportional and can be plotted against the NMR spectrum. The source frequency can be swept also where when the frequency resonates with the nuclei and produces a transition, the frequency when the absorbance takes place can be stored and plotted to produce an NMR spectrum. The radio frequency source consists of a coil around the sample tube. The same is for the radio frequency receiver. A simple component representation of the configuration of a CW NMR is depicted in Figure 16.2. An example of NMR spectrum is depicted in Figure 16.3 for a compound with

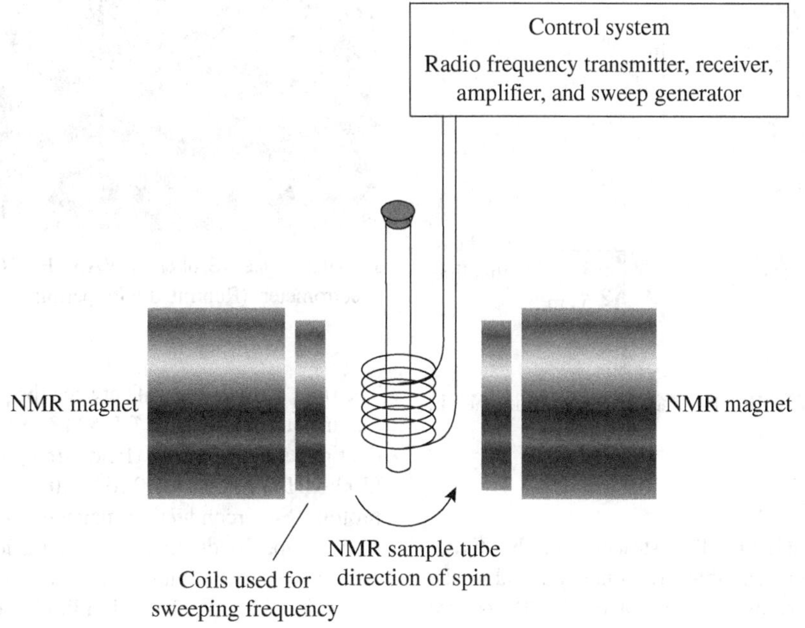

FIGURE 16.2 Design of a continuous wave NMR apparatus.

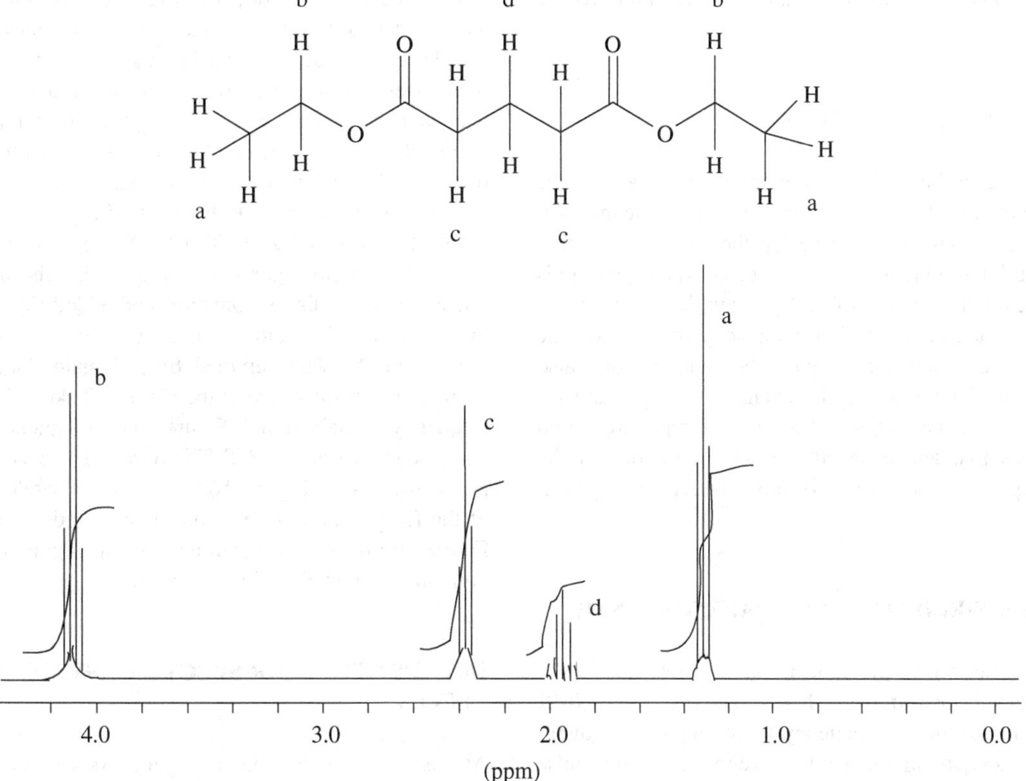

FIGURE 16.3 Simple proton NMR spectrum of 2-amino-3-methyl-pentanoic acid.

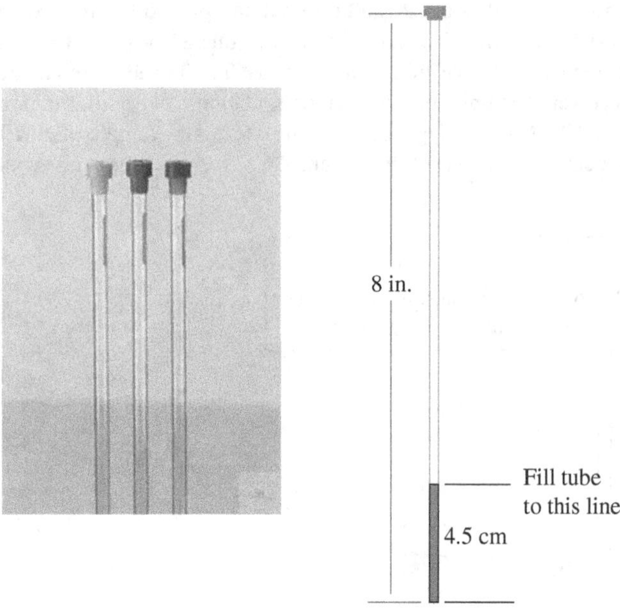

FIGURE 16.4 Examples of NMR tubes and the proper filling height with sample.

molecular formula of $C_9H_{16}O_4$. The structure in the figure illustrates the hydrogen protons that are contained within the compound. Also shown are assignments of the NMR spectral peaks to the associated structural protons, in the form of a, b, c, and d. The spectrum also contains integration lines giving the ratio of protons. Later, we will take a closer look at some example NMR spectra and the assignment of structural protons to the NMR peaks.

16.4 THE NMR SAMPLE PROBE

The NMR sample probe is also an important component of the NMR instrumentation. The probe compartment is where the sample sits within the magnetic field. Usually, the sample is spinning anywhere from 30 to 60 revolutions per second. The spinning is achieved using an air-driven turbine. Spinning the sample helps to greatly reduce the effects of inhomogeneity in the magnetic field, the cell walls, and the sample. The sample probe also contains the coils for producing the excitation energy, and for the detection. The actual tubes that hold the sample are 5-mm o.d. glass tubes that are designed to hold 0.5 ml of sample. Figure 16.4 depicts examples of NMR tubes, and the proper sample height.

16.5 PULSED FIELD FOURIER TRANSFORM NMR

Pulsed field Fourier transform nuclear magnetic resonance (FT-NMR) is an NMR measurement that differs from CW NMR where instead of scanning a frequency wave range, all frequencies (spectral lines) are simultaneously excited for a brief pulse period (τ, 1–100 μs), then the pulse is repeated a number of times.

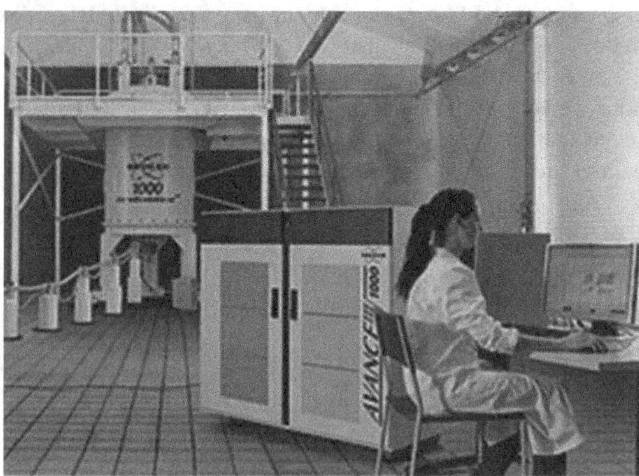

FIGURE 16.5 Bruker AVANCE 1000 MHz (23.3 T) NMR spectrometer. (Reprinted with permission from Bruker BioSpin.)

The time T of the interval between the pulses is usually 1–3 s. The magnets used in current FT-NMR instrumentation are superconducting magnets with field strengths that range from 7 T (300 MHz proton), 14 T (600 MHz proton), to 21 T (900 MHz proton). Superconducting magnets produce their magnetic field from a superconducting solenoid made of either coiled niobium and tin wire or niobium and titanium wire that is supercooled. The solenoid coil is bathed in liquid helium that keeps the solenoid at 4 K. The liquid helium bath is surrounded by a liquid nitrogen bath. The cost of filling the liquid nitrogen Dewar (required about every 10–30 days), and the liquid helium Dewar (about every 70–130 days) is well offset by the advantages of using the superconducting magnet. Advantages include much simpler configurations requiring a smaller size, greater stability over extended intervals, low maintenance and operation costs, and extensively stronger field strengths. The increased sensitivity obtained by using superconducting magnets with pulsed Fourier transform NMR has allowed the measurement of less-sensitive nuclei such as ^{13}C and ^{31}P. Figure 16.5 is a picture of the Bruker AVANCE 1000 MHz (23.3 T) NMR spectrometer.

The FID decay signals following each pulse are digitized by a fast analog-to-digital converter and added together increasing the signal-to-noise ratio of the measurement and thus increasing the sensitivity. The summed time domain data are then fast Fourier transformed (using the Cooley–Tukey algorithm) into a frequency domain signal. Figure 16.6(a) depicts an input signal pulse sequence of τ and T. The duration of τ is 1–10 μs, and T is usually 1–3 s. Figure 16.6(b) is an expanded view of one of the RF pulses. Figure 16.6(c) is a time domain signal, and a Fourier transform of the time domain signal to a frequency domain is depicted in Figure 16.6(d).

16.6 PROTON NMR SPECTRA ENVIRONMENTAL EFFECTS

At this point, we are able to begin to look at the final output of an NMR experiment, such as a PMR spectrum of an organic

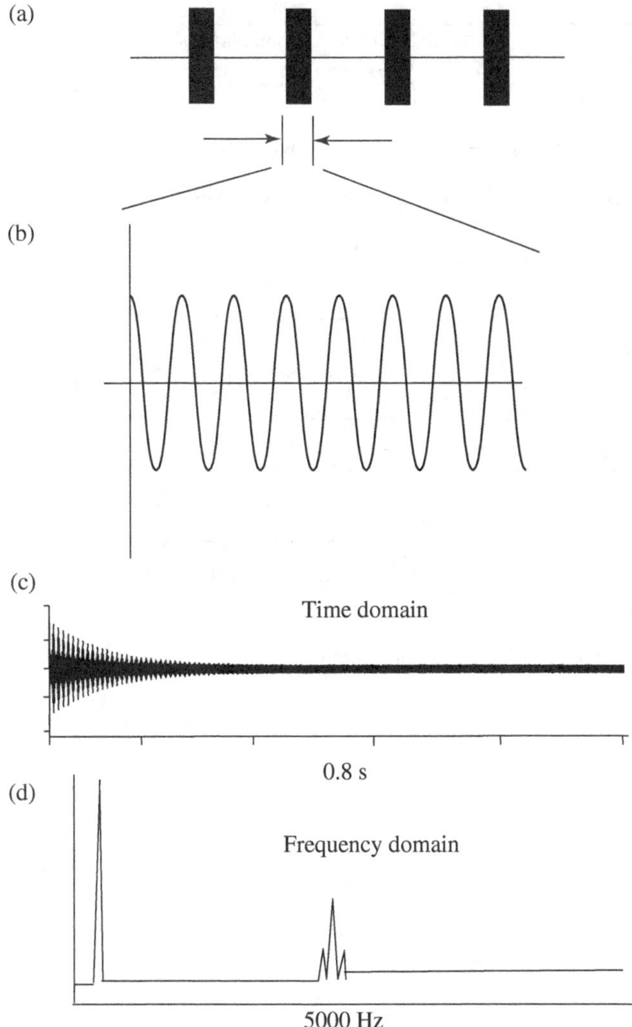

FIGURE 16.6 (a) An input signal pulse sequence of τ and T. The duration of τ is from 1 to 10 μs, and T is usually 1–3 s. (b) Expanded view of one of the RF pulses. (c) Time domain signal. (d) Fourier transform of the time domain signal to a frequency.

compound. There are two processes observed in NMR that are used in obtaining structural information of a compound called *chemical shift* and *spin–spin splitting (coupling)*.

16.6.1 Chemical Shift

Protons in a molecule do not absorb radiation at the same frequency in a given magnetic field strength. The absorbing protons are influenced by the protons that surround them. Each proton has a magnetic field induced by its electrons as they circulate around the nuclei that tends to shield the proton from the external magnetic field of the NMR instrument. This effect is clearly seen with covalent bonds, and the structure and composition of the organic compound. The local magnetic field around the nucleus will oppose the external field thus decreasing the magnitude of the field affecting the nucleus during the NMR experiment. Therefore, the external field will have to be increased in strength to cause nuclear resonance. The magnitude of the shielding that takes place depends directly on the electron density around the proton. The greater the electronegativity of the protons environment the less shielding will take place. An electronegative environment will draw the electrons away from the nucleus thus decreasing the shielding effect. For example, halogens will decrease the shielding effect. To give a type of universal magnitude to the shielding effect, a difference in the chemical shift, expressed as parts per million (ppm), from a reference standard is used. A typical reference standard is tetramethylsilane (TMS, $(CH_3))_4Si$) that contains 12 protons that are identical in their environment with a large degree of shielding taking place from the central electropositive silicon atom. An electropositive environment will not draw the electrons away from the proton nucleus thus allowing the shielding to take place. TMS was chosen because it is chemically inert, soluble in most nonpolar solvents, and has a single sharp resonance line at a high magnetic applied field that does not interfere with most organic compounds. The chemical shift is expressed as a delta (δ) in ppm using the following expression:

$$\text{Chemical shift } (\delta), \text{ ppm} = \frac{(RF_S - RF_{TMS})}{RF_{TMS}} \times 10^6, \qquad (16.8)$$

where

RF_S = resonance frequency of sample, in Hz
RF_{TMS} = resonance frequency of TMS reference, in Hz.

In general, for most NMR instrumentation the chemical shift for protons is from 1 to 13 ppm. We can try to correlate the chemical shift with observed in the NMR spectrum with structural characteristics of the organic compound. Figure 16.7 depicts a chart that roughly shows correlations between chemical shifts and the chemical makeup adjacent to the proton.

An example of chemical shift is depicted in Figure 16.8 for ethanol with TMS as an internal standard. In the figure, we can see a good example of the influence of shielding and deshielding and the chemical shift. The hydroxyl proton is experiencing the greatest amount of deshielding due to the electronegative oxygen, and thus is shifted far to the left. By convention, the x-axis is plotted with the magnetic field increasing from left to right. The internal reference TMS is situated to the far right and designated as 0 δ, with increasing chemical shift from right to left (increasing δ from 0 to 13 ppm). The methylene group ($-CH_2$) next to the hydroxyl group is experiencing less deshielding, and then finally the terminal methyl ($-CH_3$) group is influenced the least in comparison with the greatest amount of shielding taking place.

16.6.2 Spin–Spin Splitting (Coupling)

Another type of nuclear interaction that is manifested onto the NMR spectrum is the splitting of the NMR peaks. There is a small interaction or coupling between the protons on one group and the protons that are adjacent. This interaction has been shown to be between the magnetic field of the proton and the bonding electrons and not an influence through the open space between the protons. The magnetic moment of the nucleus that is being observed in the NMR spectrum is interacting with the magnetic moment of

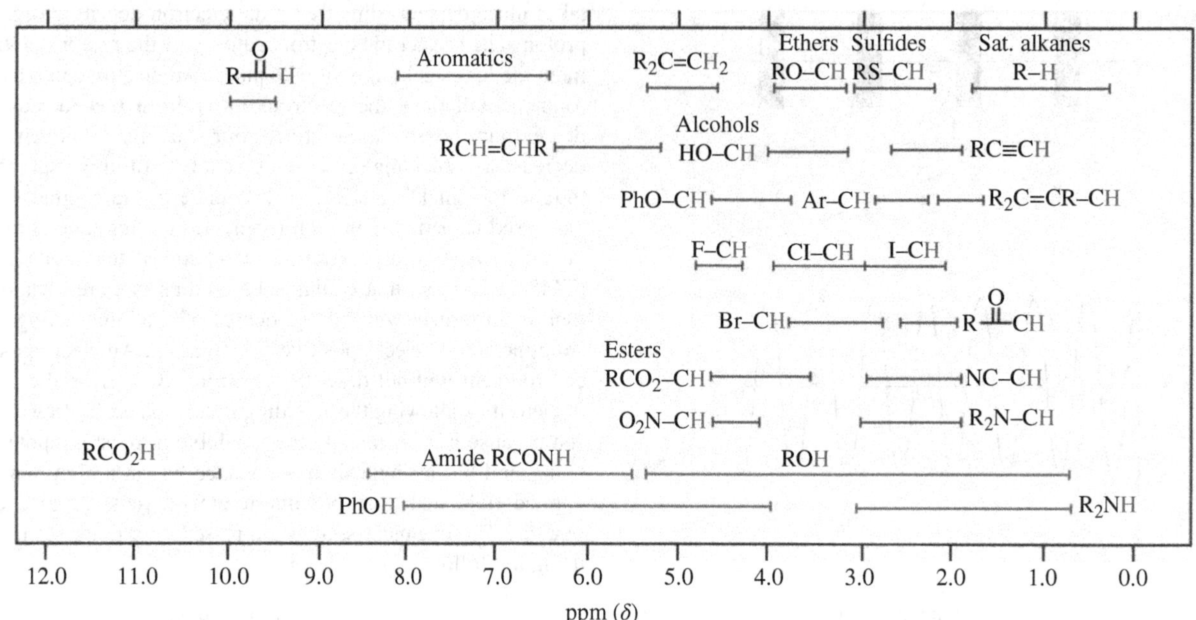

FIGURE 16.7 Proton NMR chemical shifts correlated with function group.

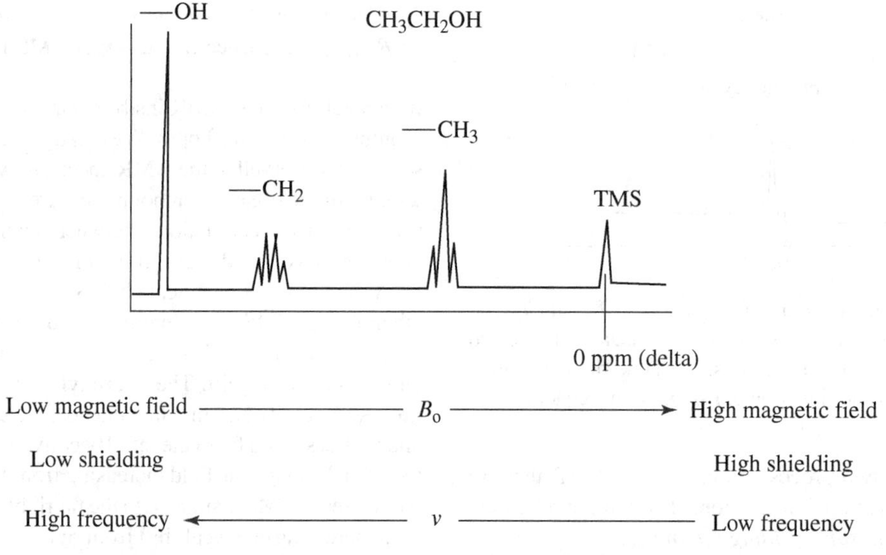

FIGURE 16.8 Example of chemical shift for ethanol with TMS as an internal standard.

adjacent nuclei. The energy levels of the transitions are being split, which results in multiple peaks on the spectrum. In Figure 16.9, instead of one peak for the methylene group ($-CH_2$) and the terminal methyl ($-CH_3$) group, like what is seen with the proton on the hydroxyl group, the methylene group ($-CH_2$) is split into four peaks and the terminal methyl ($-CH_3$) group into three. The rule for splitting is the number of adjacent equivalent nucleus (n) plus one ($n + 1$) equals the number of spectral peaks. The number of spectral peaks ($n + 1$) is referred to as the *multiplicity*. The distance between the peaks in Hertz (Hz) is called coupling constant (J) and is the same for the interacting nuclei. This is illustrated for our ethanol spectrum in Figure 16.8, where the coupling constant for the four methylene group ($-CH_2$) nuclei and the three terminal methyl ($-CH_3$) nuclei are equal at ~7 Hz. The relative areas under the peaks are also predictable and are in integer ratios. For example, the ratio for the terminal methyl ($-CH_3$) group split into three is 1:2:1. The ratio for the methylene group ($-CH_2$) split into four peaks is 1:3:3:1. The splitting of the peaks and the ratios are due to the possible combinations of the spin states of the nuclei. Note that the splitting of the peaks also involves slight chemical shifts of the different spin states. The relative intensities of the first-order (spectra where the chemical shift is large compared to the coupling constant) multiplets are presented in the following Pascal's triangle in Table 16.2.

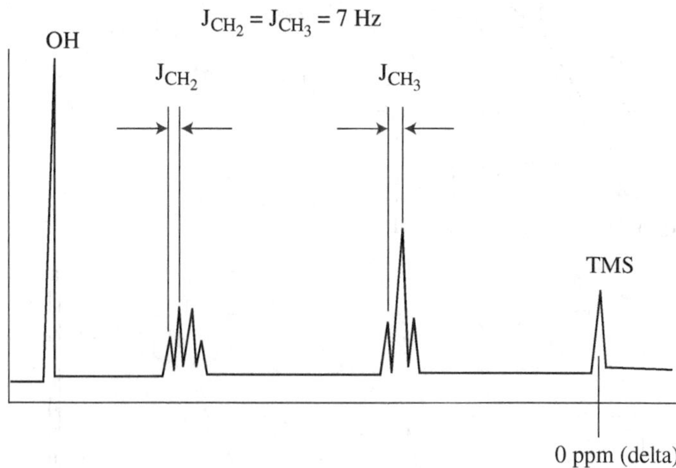

FIGURE 16.9 Example of J coupling constants.

TABLE 16.2 Spin–Spin Coupling Patterns of First-Order Multiplets.

Equivalent hydrogen protons (n)	Number of peaks multiplicity (n + 1)	Name	Splitting pattern and ratios
0	1	Singlet	1
1	2	Doublet	1:1
2	3	Triplet	1:2:1
3	4	Quartet	1:3:3:1
4	5	Quintet	1:4:6:4:1
5	6	Sextet	1:5:10:10:5:1
6	7	Septet	1:6:15:20:15:6:1

16.6.3 Interpretation of NMR Spectra

16.6.3.1 2-Amino-3-Methyl-Pentanoic Acid The first example NMR spectrum we will look at is depicted in Figure 16.10, which is of the organic compound 2-amino-3-methyl-pentanoic acid. The structure of 2-amino-3-methyl-pentanoic acid is included in the figure with letters denoting the hydrogens that are giving rise to the peaks in the NMR spectrum. The lower <1 ppm group of peaks is due to the two methyl groups in the structure labeled as "a." The two small series of peaks labeled as "b" are associated with the $-CH_2$ methylene group next to the alkane $-CH_3$ methyl groups. Next are the "c" and "d" hydrogens, all with different ppm shifts due to their local environment. Finally, there is a peak at approximately 4.8 ppm, which is associated with the protonated form of the deuterium oxide solvent, HDO versus D_2O. The deuterium oxide undergoes exchanges with both the $-NH_2$ amino group and the –OH hydroxyl group of the carboxylic acid.

16.6.3.2 Unknown I An unknown sample was analyzed using NMR to give the following spectrum depicted in Figure 16.11. FTIR analysis showed that the compound has a major peak at 1710 cm^{-1}, and a 1% solution of the unknown material was found to change litmus paper to red. The compound was found to have an empirical formula of $C_5H_{10}O_2$.

Upon inspection of the NMR spectrum and the added FTIR and litmus information above, the most likely assignment of the broad singlet with shift of 11.4 ppm would be for the hydroxyl proton of a carboxylic acid moiety containing compound. This agrees with the FTIR peak at 1720 cm^{-1} indicative of a carboxyl acid group, and the red (acidic) litmus paper test. We can conclude from this that the unknown compound contains an organic acid group. The triplet located at approximately 2.5 ppm is derived by the splitting of a $-CH_2-$ group by two protons. The triplet is also shifted downfield most likely due to the carboxyl group. The structure is indicated as HO–COO–CH_2–CH_2– now. The pentuplet signal at ~1.7 ppm is split by two $-CH_2-$ groups on each side, $(2+2)+1=5$ peaks (pentuplet). This now indicates the structure as HO–COO–CH_2–CH_2–CH_2–. The sextuplet signal at ~1.35 ppm is split by a $-CH_2-$ group on one side, and a $-CH_3$ methyl group on the other side, $(2+3)+1=6$ peaks (sextuplet). This is also supported by the triplet at ~0.35 ppm, which corresponds to a $-CH_3$ methyl group that is split by one $-CH_2$ methylene group. This now indicates the final structure as HO–COO–CH_2–CH_2–CH_2–CH_3. Figure 16.12 depicts the assignment of the different groups of the proposed structure to the corresponding NMR peaks, labeled as a, b, c, d, and e.

16.7 CARBON-13 NMR

16.7.1 Introduction

As we saw in the introduction, NMR Spectroscopy is not just limited to the activity of protons. The four most common nuclei that are used in NMR studies are ^1H, ^{13}C, ^{19}F, and ^{31}P. Carbon-13 comprises 1.1% of the naturally occurring carbon (as opposed to carbon-12 at 98.9% abundance) found in nature. At 1.1%, carbon-13 is abundant enough to be useful in NMR studies. It has found direct applications because carbon is the central atom of all organic compounds. The nuclear spin of carbon is ½ (I = ½) allowing its measurement by NMR.

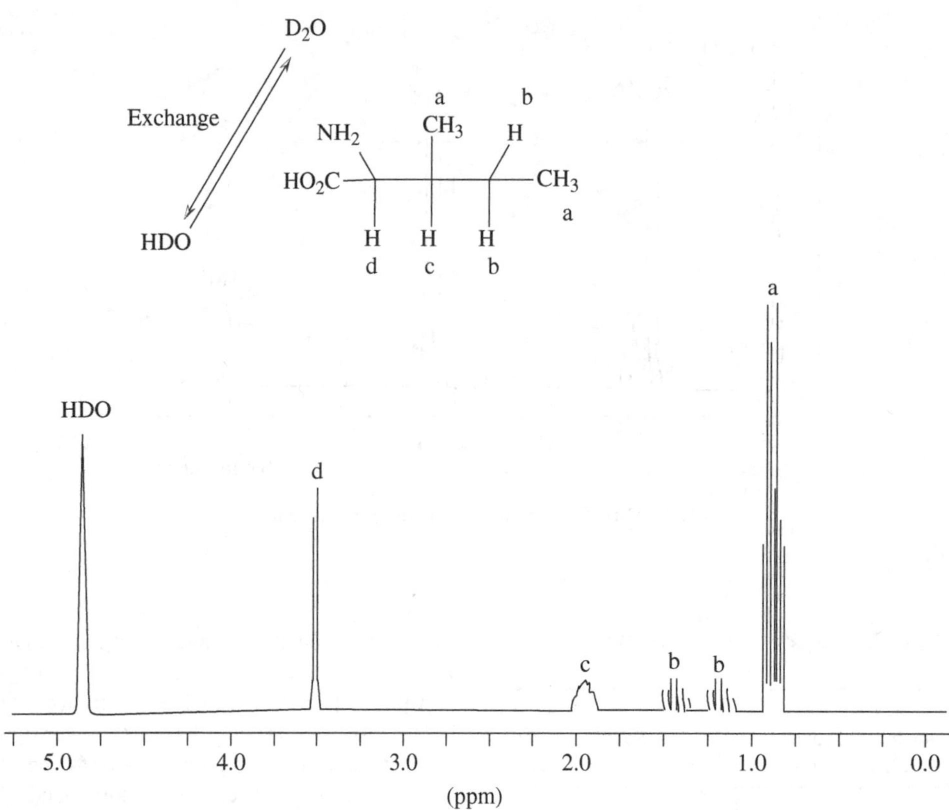

FIGURE 16.10 Proton NMR spectrum of 2-Amino-3-Methyl-Pentanoic Acid.

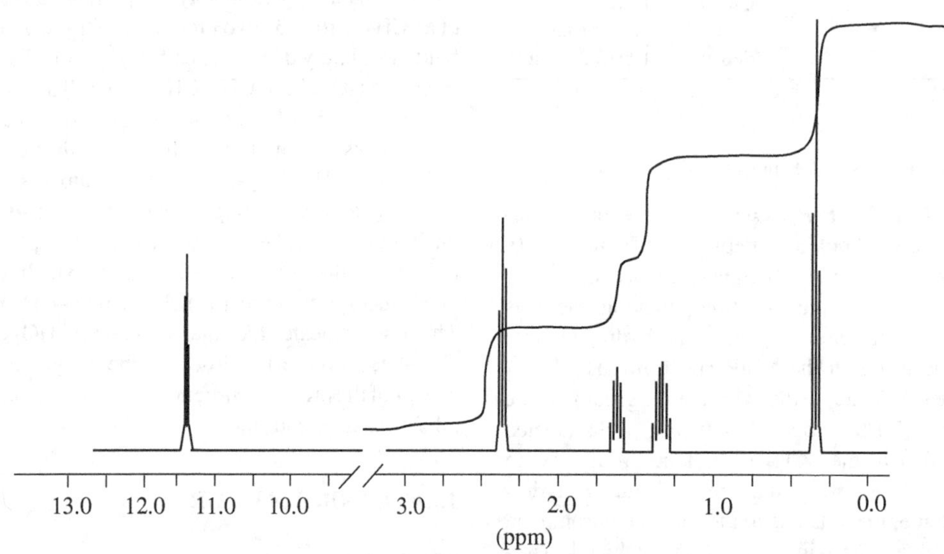

FIGURE 16.11 Proton NMR spectrum of Unknown I.

16.7.2 Carbon-13 Chemical Shift

The effect of chemical shift is observed also with other different nuclei that are measured with NMR such as the carbon-13 chemical shifts depicted in Figure 16.13. Observe though that a different scale is used for the *x*-axis for the carbon shifts depicted in Figure 16.13. While proton NMR has shifts typically in the range of 1–10 ppm, the range observed in carbon-13 NMR runs from 0 to 200 ppm.

Carbons that have a different environment respond to the magnetic field and resonance differently. Electrons shield the nucleus and subsequently reduce the effective magnetic field. This in turn

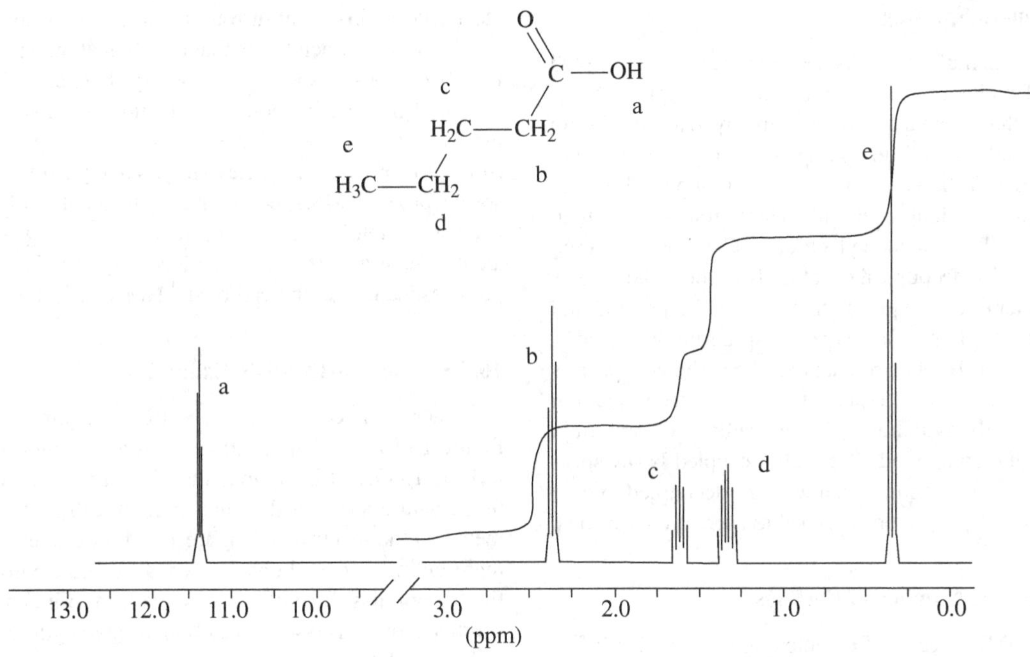

FIGURE 16.12 Proton NMR spectrum of Unknown I with assignments of protons to the corresponding spectral shift peaks as, a, b, c, d, and e.

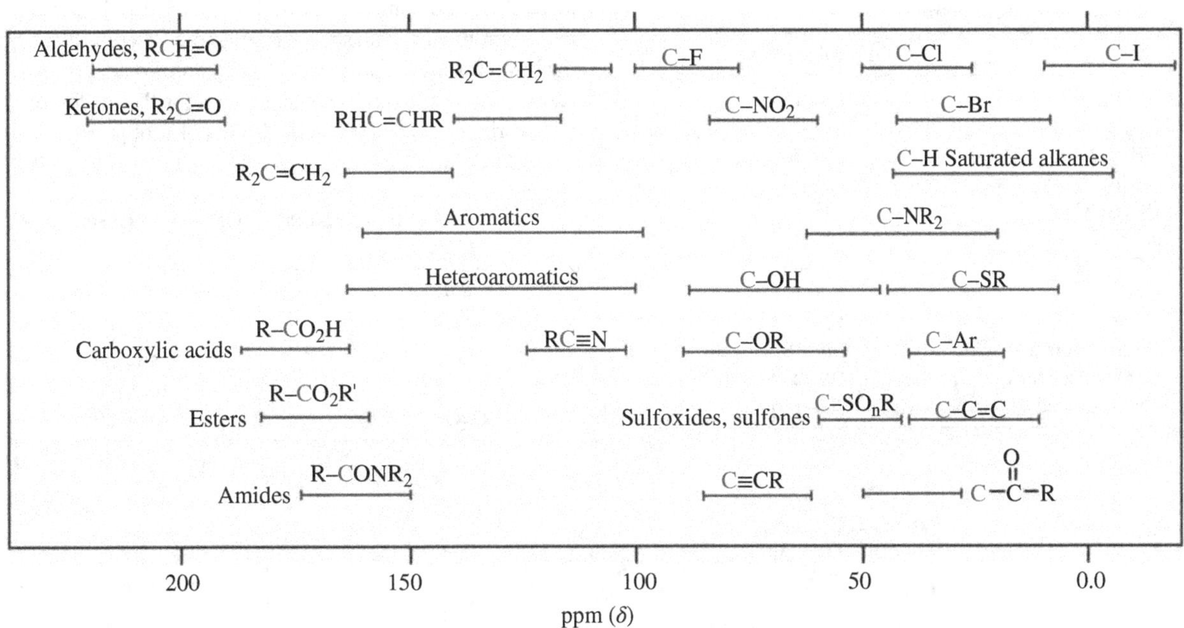

FIGURE 16.13 Carbon-13 chemical shifts.

requires energy of lower frequency in order to cause the magnetic resonance. This shows up as a lower ppm chemical shift on the NMR spectrum. In contrast to this, an environment that is not electron dense requires a higher frequency for resonance. These two environments are associated with electron-withdrawing groups and electron-donating groups. Electron-withdrawing groups are those that remove the electron density from around the carbon of interest. In Figure 16.13, some examples of electron-withdrawing groups are the amides, esters, carboxylic acids, ketones, and aldehydes, all found at 150 ppm or greater. Halogens on the other hand are electron donating and are found at much lower chemical shifts in Figure 16.13, such as the carbons covalently attached to chloride or bromide, both with a chemical shift <50 ppm. Table 16.3 is a numerical listing of typical chemical shifts in carbon-13 NMR spectra. Table 16.4 is a simplification of Table 16.3.

16.7.3 Carbon-13 Splitting

In the NMR spectra, multiple peaks are observed if not decoupled and removed. This is because the carbon atoms couple with the hydrogen atoms that are attached to them directly. If a carbon atom is a quaternary carbon with no hydrogens, then the carbon will appear as a singlet. If the group is a methane group (–CH), then the carbon appears as a doublet, if a methylene group ($–CH_2$) then a triplet, and finally as a methyl group ($–CH_3$) as a quartet. A technique is used to decouple the protons from the carbon atoms by irradiating them at a frequency that excites the protons and interrupts their coupling. This type of spectrum is called a *proton-decoupled* NMR spectrum and all the carbons appear as singlets. The proton-decoupled spectrum allows the determination of the number of carbons and/or equivalent carbons are present in the structure of the compound. The proton-coupled NMR spectrum can also be used in conjunction with the decoupled to give the number of hydrogen protons attached to each of the carbons.

16.7.4 Finding the Number of Carbons

The carbon-13 NMR spectra of organic compounds give information about: (i) the number of different types of carbon atoms that are contained within the compound; (ii) the description of the electronic environment surrounding the carbon atom; and (iii) the number of direct neighbors that a carbon atom has through splitting. In carbon-13 NMR spectra, every chemically distinct carbon or shared groups of carbons have a unique resonance. We also see that carbon-13 carbons have a very broad range of resonances, from 1 to 250 ppm; thus, they are less likely to overlap as is often seen in proton NMR spectra. The end result of all of this is that the analyst can tell how many different carbons or groups of equivalent carbons are present in a molecule by counting the number of peak resonances in the carbon-13 NMR spectrum.

TABLE 16.3 Typical Chemical Shifts in Carbon-13 NMR Spectra.

Carbon	Chemical shift (ppm)
RCH_3	10–15
R_2CH_2	15–25
$CH_3CO–$	20–30
R_3CH	25–35
RCH_2NH_2	35–45
RCH_2Cl	40–45
RCH_2OH	50–65
C=C	115–140
C—aromatic	125–150
C=O—esters and acids	170–185
C=O—aldehydes	190–200
C=O—ketones	200–220

16.7.5 Carbon-13 NMR Examples

A rather simple carbon-13 NMR spectrum is depicted in Figure 16.14 for ethanol. Ethanol has a very limited structure as $CH_3–CH_2–OH$. It is obvious that there are only two carbons in the structure and that they are in chemically different environments. In the spectrum of Figure 16.14, there are only two peaks representing two carbons in two different environments. The first carbon has three hydrogens attached to it and is bonded to another carbon. The second carbon has two hydrogens, a carbon, and a bond to oxygen. If we refer back to Figure 16.14 and Tables 16.4 and 16.5, we can figure out which peak is associated with which carbon. Looking at Figure 16.13, we can see that C–H saturated alkanes have resonances <50 ppm, and that the C–OH of alcohols is >50 ppm. Tables 16.4 and 16.5 also list these two carbon environments with similar resonances. Therefore, we attribute the lower resonance peak as the C–H saturated alkane methyl CH_3 carbon and the higher resonance peak for that of the methylene CH_2 carbon attached to the OH alcohol group. The CH_3 group

TABLE 16.4 Simplified Carbon-13 Chemical Shifts.

Carbon	Chemical shift (ppm)
C–C	0–50
C–O	50–100
C=C	100–150
C=O	150–220

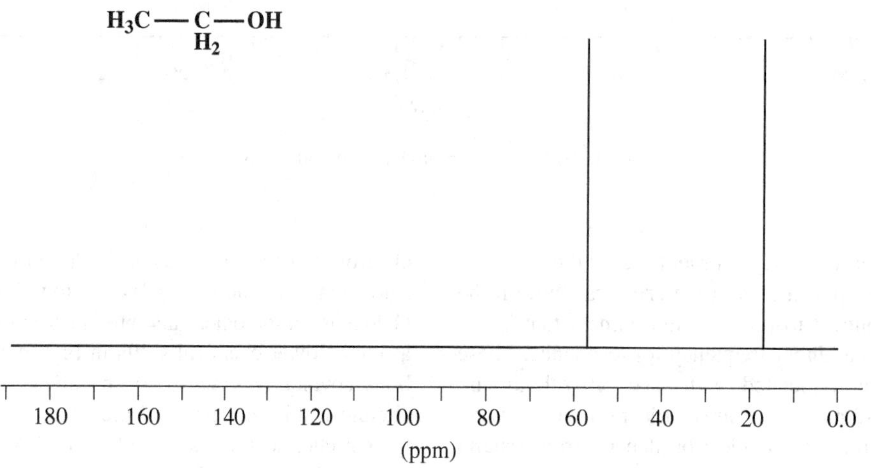

FIGURE 16.14 Carbon-13 NMR spectrum for ethanol.

is at about 18 ppm while the CH_2 group is at about 60 ppm. Can you explain why these two carbons are at the resonance values they are? The RCH_3 group is at 18 ppm instead of 10–15 ppm range due to the electronegativity of the alcohol hydroxyl group. The electronegative oxygen pulls electrons away from the carbon nucleus. This in turn leaves the carbon nucleus more exposed to the external magnetic field. As was stated earlier, the carbon will need a smaller external magnetic field to bring it into resonance as compared to if it was in an environment that is less electronegative. The end result is that the smaller the magnetic field needed for resonance, the higher the chemical shift will be.

TABLE 16.5 Proton NMR Spectral Results.

Node	Theoretical Shift (ppm)	Experimental Shift (ppm)
CH3(18)		0.68–0.73 (0.69)[27]
CH3(26), CH3(27)		0.88–0.89 (0.86–0.87)[27]
CH3(21)		0.92 (0.91)[27]
CH3	1.01	1.01 (1.0)[27]
CH3	1.01	1.02
CH3	1.06	1.08
CH3	1.16	1.15–1.18
CH2	1.25	
CH2	1.25	
CH2	1.25	
CH3	1.26	1.275
CH2	1.29	1.29
CH2	1.36	1.36
CH	1.40	1.40
CH2	1.40	1.40
CH2	1.44	
CH	1.44	
CH	1.45	
CH2	1.47	1.46
CH2	1.47	1.47
CH	1.47	1.48
CH	1.64	1.64
CH	1.83	1.83
CH2	1.92	1.90
OH	2.0	2.01
CH2	2.11	2.11
CH	3.25	
C=C–H	5.37	5.46 (5.3)[27]

16.8 SPECIAL TOPIC—NMR CHARACTERIZATION OF CHOLESTERYL PHOSPHATE

In metabolomics studies, often an unknown biomolecule is observed in a system under study by mass spectrometry. The structural elucidation and subsequent identification of an unknown small biomolecule is a process that often involves a number of steps. Figure 16.15 is a single-stage ESI Q-TOF mass spectrum of a biological extract containing unknown biomolecules in a solution of 1:1 chloroform/methanol with 10 mM LiCl. The spectrum contains primarily lithium adducts of the lipid species in the biological extract as $[M + Li]^+$. The predominant biomolecule observed in the biological extract at *m/z* 473.5 (also observed at *m/z* 489 as the sodiated species) will be the subject of study concerning its identification in the following example of identifying an unknown biomolecule.

Figure 16.16 is an ESI Q-TOF single-stage negative ion mode mass spectrum of the same biological extract as depicted in Figure 16.15 in a 1:1 chloroform/methanol solution containing 1 mM ammonium acetate. The spectrum is composed of deprotonated biomolecule species as $[M - H]^-$. In the lower molecular weight region of the spectrum, there are four peaks observed representing the free fatty acids: myristic acid at *m/z* 227.3; palmitic acid at *m/z* 255.4; oleic acid at *m/z* 281.4; and stearic acid at *m/z* 283.4. The peak at *m/z* 465.5 is suspected to be the same biomolecule as that of *m/z* 473.5 observed in positive ion mode.

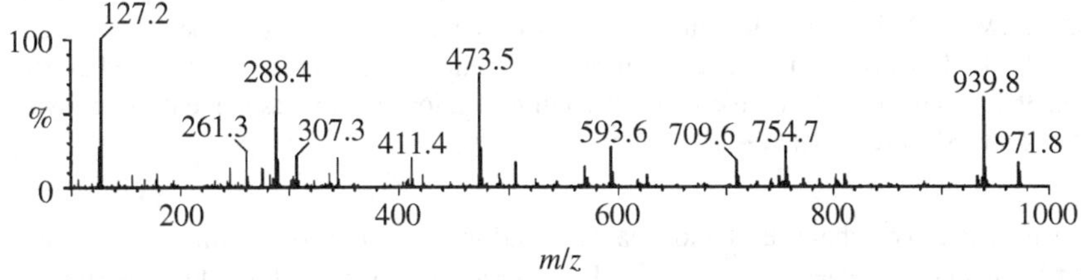

FIGURE 16.15 Single stage ESI Q-TOF mass spectrum of a biological extract in a solution of 1:1 chloroform/methanol with 10 mM LiCl containing primarily lithium adducts as $[M + Li]^+$. *m/z* 473.5 (also observed at *m/z* 489 as the sodiated species) is the subject of study concerning its identification.

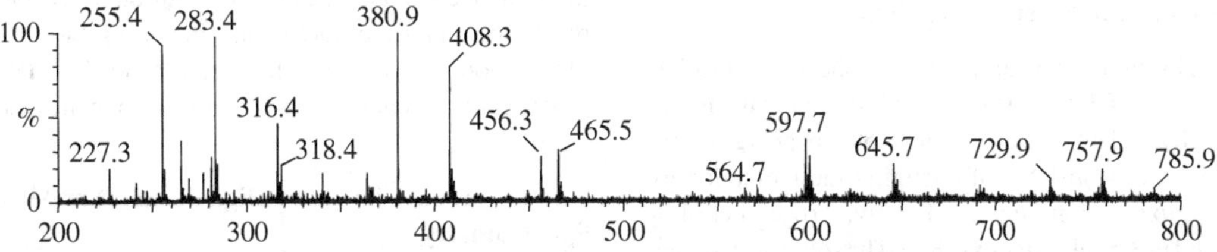

FIGURE 16.16 ESI/Q-TOF single stage mass spectrum of the biological extract collected in negative ion mode in a 1:1 chloroform/methanol solution containing 1 mM ammonium acetate. Spectrum is comprised of deprotonated lipid species as $[M–H]^-$ illustrating the free fatty acids myristic acid at *m/z* 227.3, palmitic acid at *m/z* 255.4, oleic acid at *m/z* 281.4, and stearic acid at *m/z* 283.4, and an unknown species at *m/z* 465.5.

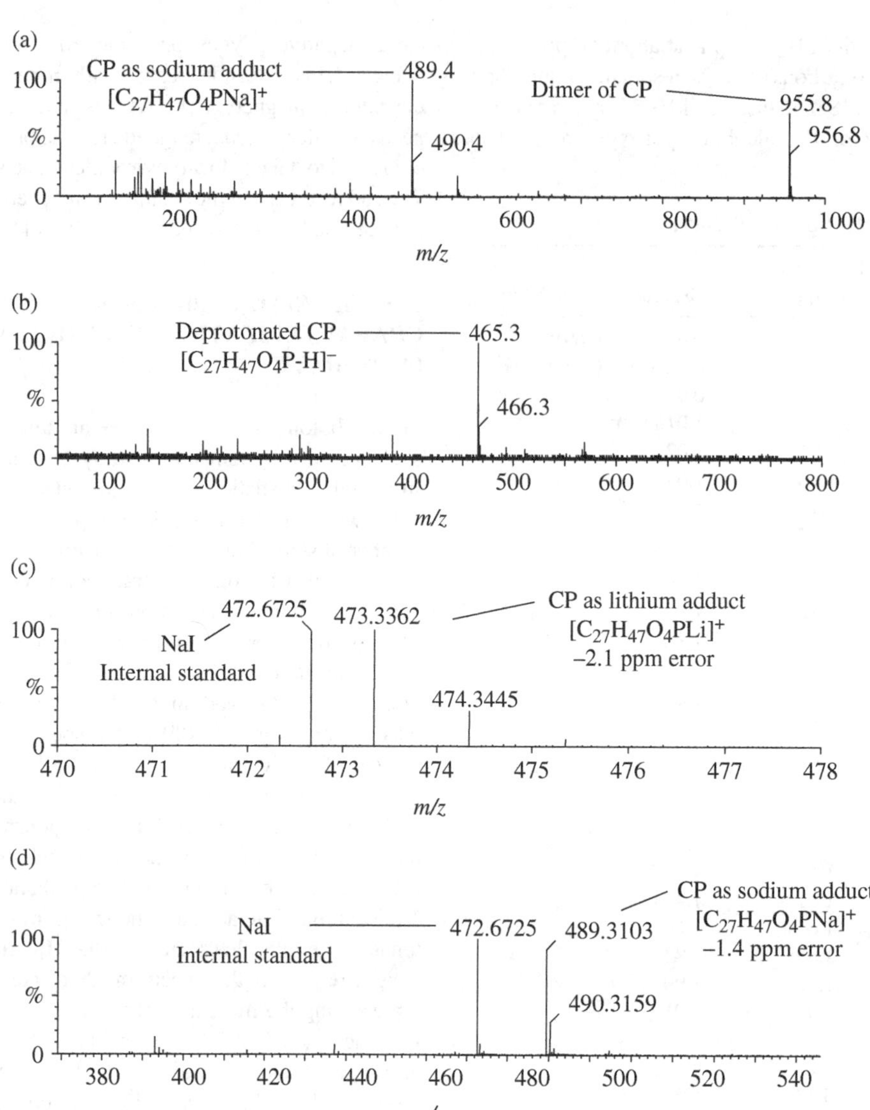

FIGURE 16.17 (a) Single stage mass spectrum in positive ion mode of the synthesized cholesteryl phosphate in an acidic solution containing NaI (1% acetic acid in 1:1 $CHCl_3$/MeOH). (b) Single stage mass spectrum in negative ion mode of the synthesized cholesteryl phosphate in a 10 mM ammonium hydroxide 1:1 $CHCl_3$/MeOH solution. (c) High resolution mass measurement of the synthesized CP as the lithium adduct $[C_{27}H_{47}O_4PLi]^+$ at *m/z* 473.3362 (calculated −2.1 ppm error). (d) High resolution mass measurement of the synthesized CP as the sodium adduct $[C_{27}H_{47}O_4PNa]^+$ at *m/z* 489.3103 (calculated −1.4 ppm error).

The unknown species was suspected to be cholesteryl phosphate. To verify this, the cholesteryl phosphate species was synthesized, characterized, and compared to the unknown for confirmation of its identification.

16.8.1 Synthesis of Cholesteryl Phosphate

The first step was to synthesize the cholesteryl phosphate species. Cholesteryl phosphate (CP) was synthesized following the methodology outlined by Sedaghat et al. [1] and Gotoh et al., [2] where the major modification consisted of changing the methodology from a gram scale to a milligram scale. Briefly, 0.67 g of cholesterol was dissolved in 3.5 ml of dry pyridine. The solution was then cooled in ice water, and phosphorus oxychloride (0.175 ml dissolved in 3.5 ml acetone) was added slowly with stirring. A precipitate was formed immediately and the solution was allowed to cool for 10 min in the ice water. The precipitate (cholesteryl phosphochloridate) was filtered and washed with 15 ml cold dry acetone. The solid was dissolved in tetrahydrofuran (THF) and refluxed for 2 h after adding aqueous NaOH (2.1 equivalent). The resulting CP precipitate was filtered and washed with dry cold acetone and dried in a Vacufuge™ (Eppendorf). High-resolution mass measurements (including both the sodium and lithium adducts) of the synthesized CP and H-NMR results are presented for verification of its identification in the next two sections.

16.8.2 Single-Stage and High-Resolution Mass Spectrometry

The synthesized CP was next characterized using high-resolution mass spectrometry. Figure 16.17a is a single-stage mass spectrum in positive ion mode acquired on the Q-TOF mass spectrometer of the synthesized CP in an acidic solution containing NaI (1% acetic acid in 1:1 $CHCl_3$/MeOH). The two major peaks in the spectrum are at *m/z* 489.4 for CP as the sodium adduct, and *m/z* 955.8 for a CP dimer. Figure 16.17b is a single-stage mass spectrum in negative ion mode acquired on the Q-TOF mass spectrometer of the synthesized CP in a 10 mM ammonium hydroxide 1:1 $CHCl_3$/MeOH solution. The major peak in the spectrum is at *m/z* 465.3 for the deprotonated form of CP as $[C_{27}H_{47}O_4P{-}H]^-$. Figure 16.17c is the high-resolution mass measurement of the synthesized CP as the lithium adduct $[C_{27}H_{47}O_4PLi]^+$ at *m/z* 473.3362 (*m/z* 472.6725 is the internal standard NaI peak used for the exact mass measurement). The theoretical mass of CP as the lithium adduct is *m/z* 473.3372 equating to a calculated −2.1 ppm error. Figure 16.12d is the high-resolution mass measurement of the synthesized CP as the sodium adduct $[C_{27}H_{47}O_4PNa]^+$ at *m/z* 489.3103 (*m/z* 472.6725 is the internal standard NaI peak used for the exact mass measurement). The theoretical mass of CP as the sodium adduct is *m/z* 489.3110 equating to a calculated −1.4 ppm error.

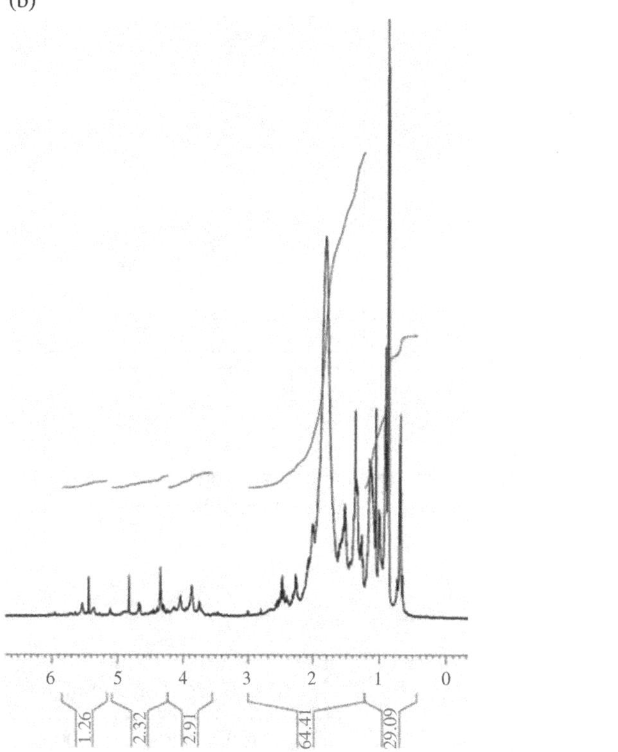

FIGURE 16.18 (a) Structure of cholesteryl phosphate with the carbon atoms numbered. (b) Experimental proton NMR results of the synthesized cholesteryl phosphate.

16.8.3 Proton Nuclear Magnetic Resonance (^{1}H-NMR)

The next step was to characterize the synthesized CP using NMR to confirm its synthesis. Single-stage proton (^{1}H-NMR) spectra were obtained on a Bruker DRX 800 (800 MHz) NMR spectrometer (Bruker, Bremen, Germany). The spectra of the synthesized CP were obtained at 800 MHz (shifts in ppm) with deuterated chloroform ($CDCl_3$) used as solvent. Figure 16.18 contains the structure of CP with the carbon atoms numbered and the experimental proton NMR results of the synthesized CP.

16.8.4 Theoretical NMR Spectroscopy

The NMR proton shifts were calculated using ChemDraw (Cambridge Soft Corporation, Cambridge, MA). The theoretical proton shifts for the structure are depicted in Figure 16.19, where (a) represents the structure of CP with the proton shifts labeled and (b) is the theoretical NMR spectrum generated by ChemDraw using the ChemNMR feature. Table 16.5 illustrates a comparison of the theoretically generated proton NMR shifts versus the experimentally obtained proton NMR shifts. Good agreement was observed between the theoretical and the experimental proton NMR shifts.

16.8.5 Structure Elucidation

Product ion mass spectra were collected of the unknown biomolecule to compare to the product ion spectra of the synthesized CP. Figure 16.20 is a product ion spectrum of the *m/z* 473.5 biomolecule, as $[M{+}Li]^+$, in the biological extract collected on a Q-TOF mass spectrometer in a solution of 1:1 chloroform/methanol with 10 mM LiCl. The major products produced and the associated fragmentation pathways are depicted in Figure 16.21. The *m/z* 255.3 product ion, $[C_{11}H_{21}O_4PLi]^+$, is formed through cleavage of the cholesteryl backbone as depicted in Figure 16.16, $[C_{27}H_{47}O_4P + Li{-}C_{16}H_{26}]^+$. The *m/z* 237.2 product ion is derived through neutral water loss from the *m/z* 255.3 product ion, $[C_{11}H_{21}O_4PLi{-}H_2O]^+$. The *m/z* 293.3 and *m/z* 311.3 product ions, which differ by 18 amu (H_2O), are formed in an analogous manner with their respective structures depicted in Figure 16.21. The *m/z* 311.3 product ion, $[C_{15}H_{29}O_4PLi]^+$, is formed through cleavage of the cholesteryl backbone as $[C_{27}H_{47}O_4P + Li{-}C_{12}H_{18}]^+$. The *m/z* 293.3 product ion is derived through neutral water loss from the *m/z* 311.3 product ion, $[C_{15}H_{29}O_4PLi{-}H_2O]^+$. The *m/z* 473.4 lipid species has been identified as the lithium adduct of CP with the ionic formula

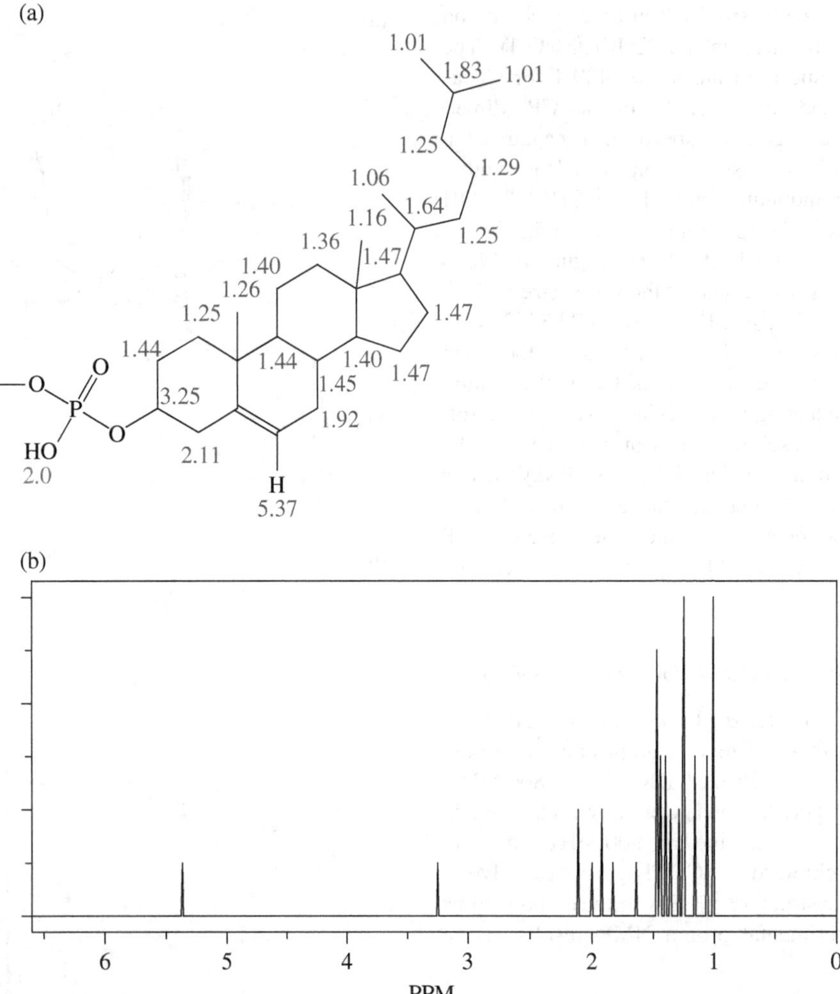

FIGURE 16.19 (a) Structure of cholesteryl phosphate with the proton shifts labeled. (b) Theoretical NMR spectrum of cholesteryl phosphate generated by ChemDraw using the ChemNMR feature.

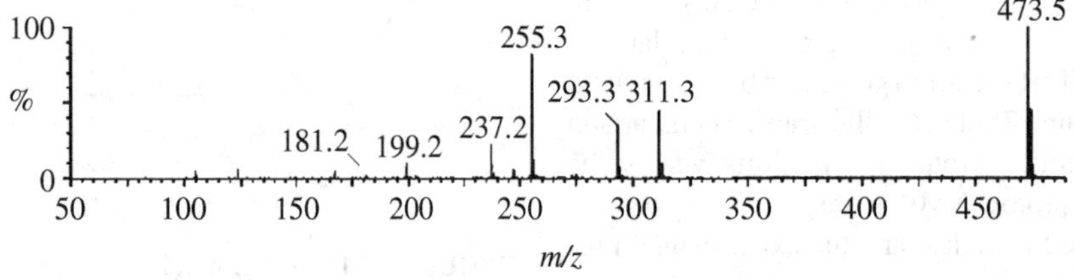

FIGURE 16.20 Product ion spectrum of *m/z* 473.7 identified as cholesteryl phosphate. Product ion spectrum of synthesized cholesteryl phosphate as the lithium adduct $[C_{27}H_{47}O_4P + Li]^+$ at *m/z* 473.4. The spectrum is identical to the product ion spectrum of the unknown *m/z* 473.5 species observed in biological extract.

of $[C_{27}H_{47}O_4P + Li]^+$ and a theoretical mass of 473.3372 Da. High-resolution mass measurements of the *m/z* 473.4 lipid species in meibum was determined at *m/z* 473.3361 equating to a mass error of −2.3 ppm. The *m/z* 489.3 lipid species has been identified as the sodium adduct of CP with the ionic formula of $[C_{27}H_{47}O_4P + Na]^+$ and a theoretical mass of 489.3110 Da. High-resolution mass measurements of the *m/z* 489.3 lipid species in meibum was determined at *m/z* 489.3099 equating to a mass error of −2.2 ppm.

Figure 16.22 depicts the product ion spectrum of the *m/z* 465.5 species observed in the biological extract obtained using the Q-TOF mass spectrometer with an electrospray ionization source

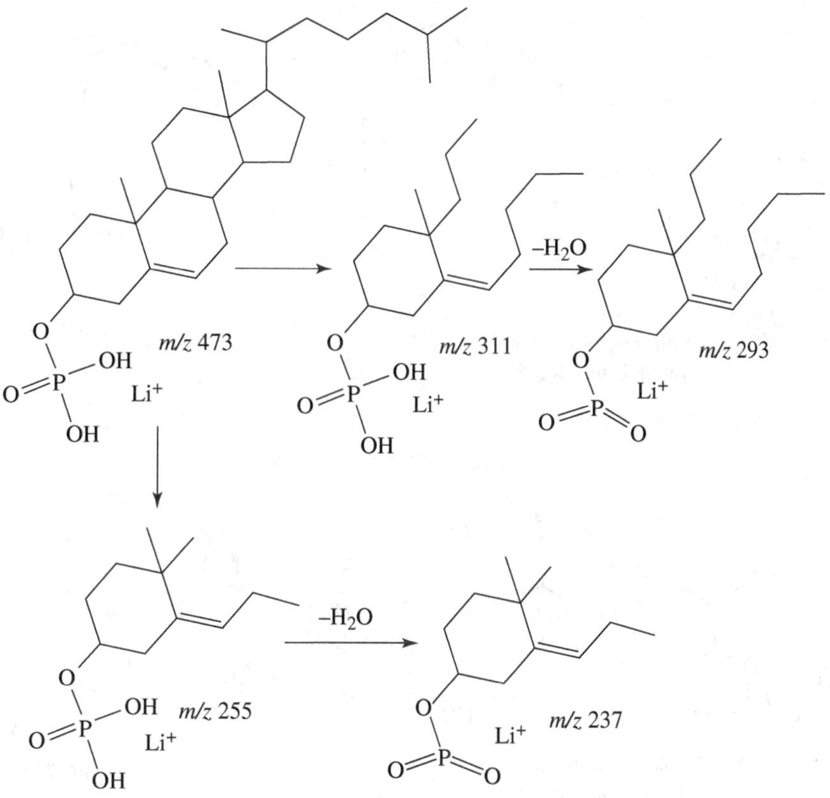

FIGURE 16.21 Structure of cholesteryl phosphate and fragmentation pathways from product ion spectrum in Figure 8.29.

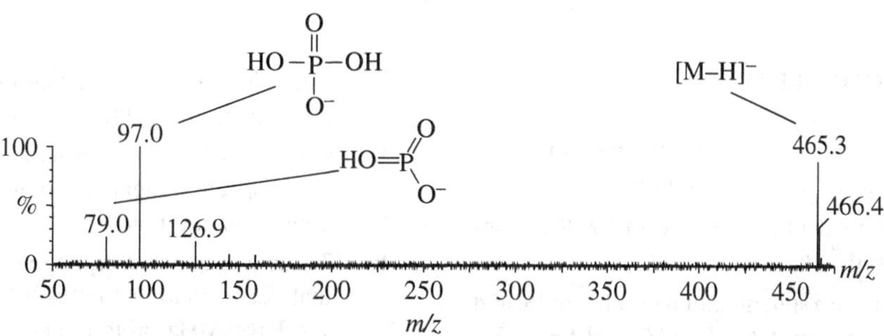

FIGURE 16.22 Q-TOF MS negative ion mode product ion spectrum of the *m/z* 465.5 species in the biological extract identified as the deprotonated form of cholesteryl phosphate with the ionic, deprotonated formula $[M–H]^-$ of $[C_{27}H_{47}O_4P–H]^-$.

collected in negative ion mode. In Figure 16.22, there are two major product ion peaks observed at *m/z* 127.0 and 97.0. The *m/z* 465.5 lipid species has been identified as the deprotonated form of CP with the ionic formula $[M–H]^-$ of $[C_{27}H_{47}O_4P–H]^-$ and a theoretical mass of 465.3134 Da. High-resolution mass measurements of the *m/z* 465.5 lipid species in the biological extract was determined at *m/z* 465.3140 equating to a mass error of 1.3 ppm. The structure of the *m/z* 465.5 lipid species is depicted in Figure 16.23 along with proposed fragmentation pathways that describe the production of the *m/z* 97.0 and *m/z* 79.0 product ions. Both the *m/z* 97.0 and 79.0 product ions represent the H_2PO_4 phosphate portion of the headgroup. This same H_2PO_4 phosphate product ion at *m/z* 97.0 is also observed in the product ion spectra of the phosphorylated lipid standards 1-palmitoyl-2-oleoyl-*sn*-glycero-3-phosphate (16:0–18:1 PA or POPA, spectrum not shown), and 1-palmitoyl-2-oleoyl-*sn*-glycero-3-[phospho-*rac*-(1-glycerol)] (16:0–18:1 PG or POPG, spectrum not shown). The product ion spectrum of the synthesized CP is identical to the *m/z* 465.5 species in the biological extract and confirms its identification as CP.

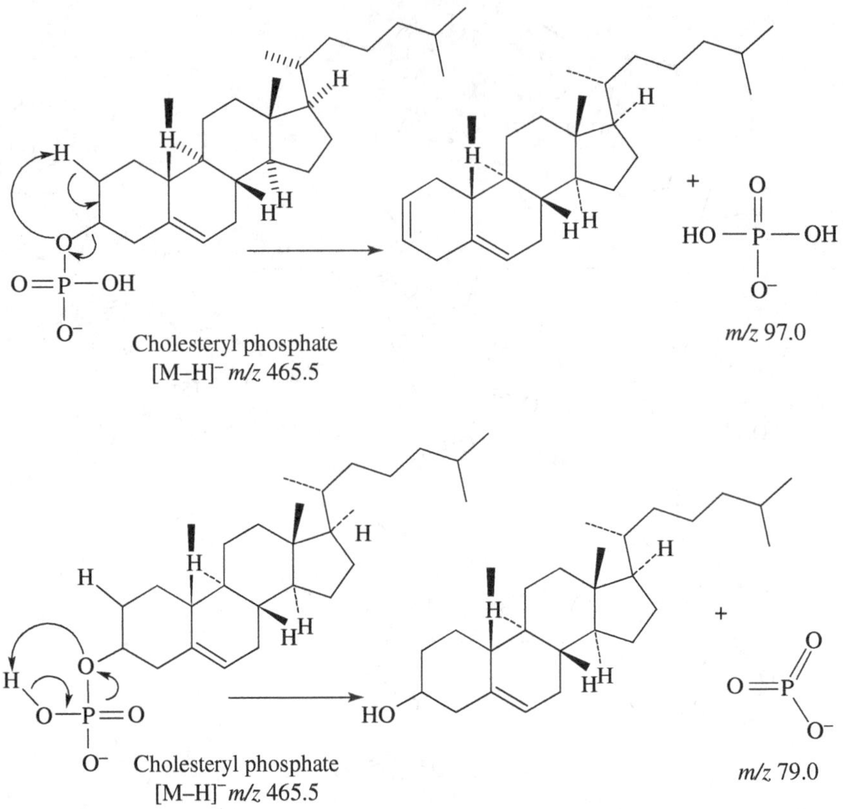

FIGURE 16.23 Structure of the *m/z* 465.5 lipid species and fragmentation pathways.

16.9 CHAPTER KEY CONCEPTS

16.1 The NMR process is a nuclear interaction brought about in the presence of a strong magnetic field.

16.2 The most common nuclei that are used in NMR studies are ^{1}H, ^{13}C, ^{19}F, and ^{31}P.

16.3 Each of the nuclei have a spin quantum number of ½, and each nuclei have two spin states $I = +1/2$ and $I = -1/2$.

16.4 The physical property of the spinning nucleus is described by the magnetic moment μ that is orientated along the axis of spin and is proportional to the angular momentum p. The magnetic moment is expressed as $\mu = \gamma p$, where γ is a proportionality constant called the magnetogyric ratio (radian·T^{-1} s^{-1}).

16.5 Not all nuclei have a magnetic moment, only those with an odd sum of protons and neutrons.

16.6 When an organic compound containing hydrogen nuclei are placed into a magnetic field, the magnetic moments (m) of each nuclei will align either with the magnetic field (low energy state, where $m = +1/2$) or against the external magnetic field (high energy state, where $m = -1/2$).

16.7 There is a direct relationship between the frequency of the transition and the magnetic field strength.

16.8 Older NMR spectrometers are generally called CW NMR spectrometers due to the use of a steady, nonchanging magnetic field.

16.9 These were often equipped with permanent magnets that have field strengths of 0.7 T (30 MHz proton), 1.4 T (60 MHz proton), and 2.1 T (90 MHz proton).

16.10 Permanent magnets are very sensitive to heat and also require extensive shielding. Both sensitivity and resolution increase with increasing field strength, thus stronger magnets than these are desirable.

16.11 The NMR sample probe is also an important component of the NMR instrumentation.

16.12 The sample is spinning anywhere from 30 to 60 revolutions per second. The spinning is achieved using an air-driven turbine. Spinning the sample helps to greatly reduce the effects of inhomogeneity in the magnetic field, the cell walls, and the sample.

16.13 Pulsed field FT-NMR is an NMR measurement that differs from CW NMR, where instead of scanning a frequency wave range all frequencies (spectral lines) are simultaneously excited for a brief pulse period (τ, 1–100 μs), then the pulse is repeated a number of times.

16.14 The magnets used in current FT-NMR instrumentation are superconducting magnets with field strengths that range from 7 T (300 MHz proton), 14 T (600 MHz proton), to 21 T (900 MHz proton).

16.15 Superconducting magnets produce their magnetic field from a superconducting solenoid made of either coiled niobium and tin wire or niobium and titanium wire that is supercooled.

16.16 The solenoid coil is bathed in liquid helium that keeps the solenoid at 4 K. The liquid helium bath is surrounded by a liquid nitrogen bath.

16.17 Advantages include much simpler configurations requiring a smaller size, greater stability over extended intervals, low maintenance and operation cost, and extensively stronger field strengths.

16.18 During the time interval T between RF pulses, the excited nuclei will undergo relaxation where a signal is emitted. This is a time domain radio frequency signal called the *free induction decay (FID) signal* and is the signal collected from the FT-NMR analysis.

16.19 There are two processes observed in NMR that are used in obtaining structural information of a compound called *chemical shift* and *spin–spin splitting (coupling).*

16.20 Each proton has a magnetic field induced by its electrons as they circulate around the nuclei that tends to shield the proton from the external magnetic field of the NMR instrument.

16.21 The local magnetic field around the nucleus will oppose the external field thus decreasing the magnitude of the field affecting the nucleus during the NMR experiment.

16.22 The magnitude of the shielding that is taking place depends directly on the electron density directly around the proton. The greater the electronegativity of the protons environment the less shielding will take place.

16.23 To give a type of universal magnitude to the shielding effect, a difference in the chemical shift, expressed as ppm, from a reference standard is used. A typical reference standard is tetramethylsilane (TMS, $(CH_3))_4Si$) that contains 12 protons that are identical in their environment with a large degree of shielding taking place from the central electropositive silicon atom.

16.24 In general, for most NMR instrumentation the chemical shift for protons is from 1 to 13 ppm.

16.25 Another type of nuclear interaction that is manifested onto the NMR spectrum is the splitting of the NMR peaks.

16.26 There is a small interaction or coupling between the protons on one group and the protons that are adjacent.

16.27 The rule for splitting is the number of adjacent equivalent nucleus (n) plus one ($n + 1$) equals the number of spectral peaks. The number of spectral peaks ($n + 1$) is referred to as the *multiplicity*.

16.28 The distance between the peaks in Hertz (Hz) are called coupling constants (J) and are the same for the interacting nuclei.

16.10 CHAPTER PROBLEMS

16.1 The following proton NMR spectrum is that of a simple compound where only one peak is observed at 7.4 ppm (Fig. 16.24). The peak at 0.0 ppm is for hydrogen in the

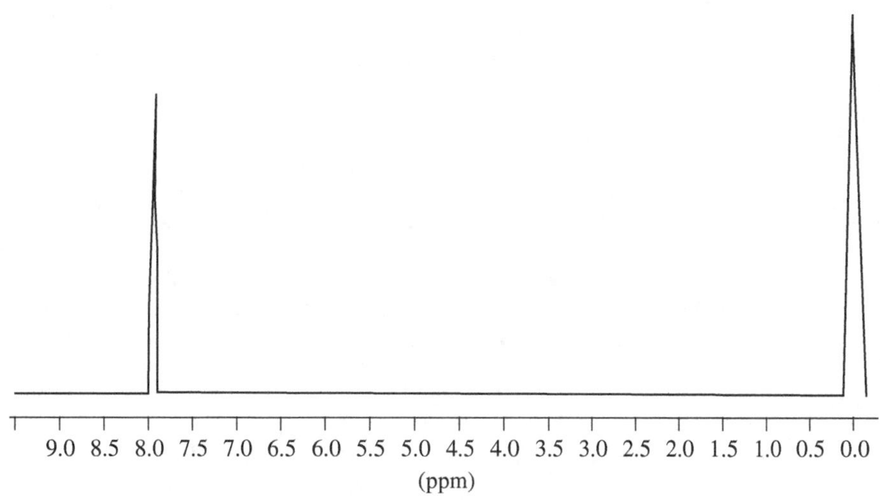

FIGURE 16.24 Unknown NMR spectrum.

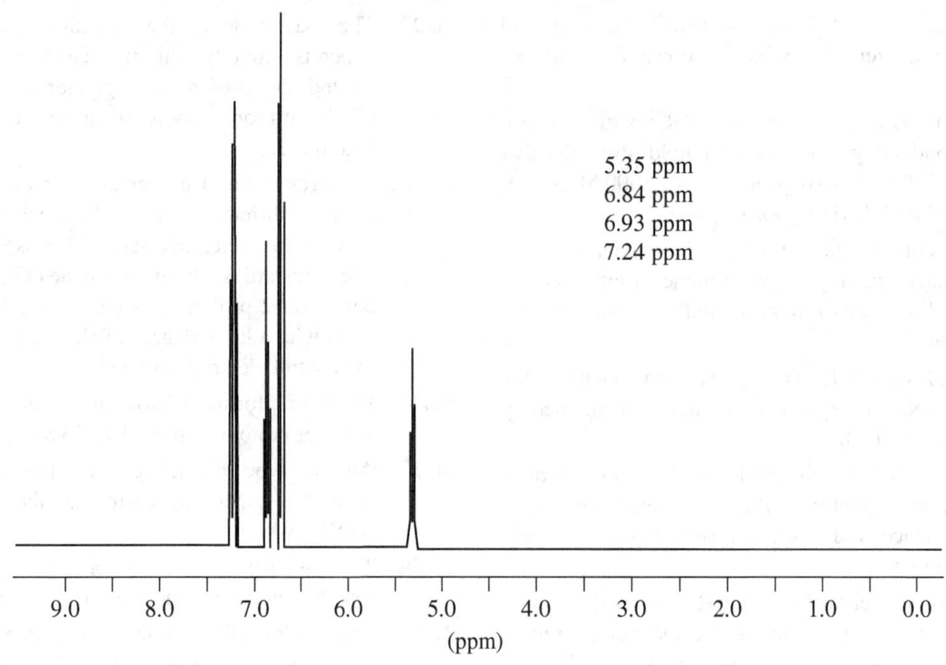

FIGURE 16.25 Unknown NMR spectrum.

TMS. Using the assignments in Figure 7, identify what is most likely the compound?

16.2 The following NMR spectrum is that of phenol. Identify the protons responsible for the spectral peaks. Explain the multiple peaks observed around 7 ppm (Fig. 16.25).

REFERENCES

1. Sedaghat, S.; Désaubry, L.; Streiff, S.; Ribeiro, N.; Michels, B.; Nakatani, Y.; Ourisson, G. *Chem. Biodiver.* 2004, **1**, 124–128.
2. Gotoh, M.; Ribeiro, N.; Michels, B.; Elhabiri, M.; Albrecht-Gary, A. M.; Yamashita, J.; Hato, M.; Ouisson, G.; Nakatani, Y. *Chem. Biodiver.* 2006, **3**, 198–209.

17

ATOMIC ABSORPTION SPECTROSCOPY (AAS)

17.1 Introduction
17.2 Atomic Absorption and Emission Process
17.3 Atomic Absorption and Emission Source
17.4 Source Gases and Flames
17.5 Block Diagram of AAS Instrumentation
17.6 The Light Source
17.7 Interferences in AAS
17.8 Electrothermal Atomization—Graphite Furnace
17.9 Instrumentation
17.10 Flame Atomic Absorption Analytical Methods

17.1 INTRODUCTION

Analysis of metals in the analytical laboratory is a very useful technique that is used in multiple disciplines, including food testing and safety analyses, environmental analyses, water analysis, toy safety (lead in toys), additives in lubricating oils and greases (Ca, Na, Ba, Li, Zn, Mg), petrochemical products, animal feeds (Cu, Cr, Mn, Fe, Zn), and contract labs to name a few. Some examples include calcium analysis in animal feeds, metal profile of grains, metal content of soils for heavy metal contamination (e.g., mercury, lead, and cadmium), lead content in toys, metals and elements used in some pesticides, herbicides, and fumigants such as arsenic and phosphorus. The sample prep approaches to isolate and analyze the metals are almost as numerous as the elements themselves. The clinical laboratory also does metal analysis of blood samples for Fe, Na, Mg, Ca, Li, and K. Often, sample prep is optimized for a certain matrix such as grains, soil, waste water, and so on. There are general approaches that may be used to first attempt to isolate the metals from a sample.

17.2 ATOMIC ABSORPTION AND EMISSION PROCESS

The process of atomic absorption involves the absorption of radiant energy at specific wavelengths by the elements in a gaseous state.

$$\text{Absorption} = -\log(I_t/I_0). \tag{17.1}$$

The process of atomic emission involves the release of radiant energy at specific wavelengths by the elements in a gaseous state.

$$\text{Emission} = -\log(I_0/I_t). \tag{17.2}$$

The absorption follows the Beer–Lambert law where the amount of incident radiant energy is directly proportional to the concentration of the absorbing element. The Beer–Lambert law is expressed as

$$I_t = I_0\left(10^{-abc}\right), \tag{17.3}$$

where I_0 is the intensity of the source radiation available for absorption by the element, I_t is the transmitted energy after absorption, a is the absorptivity of the element, b is the path length, and c is the concentration of the element.

The absorption of the radiant energy is not a nuclear process, but involves the valence electrons where the absorbed radiant energy elevates an electron to the next higher orbital energy level. Each element will absorb at a wavelength that is particular to that element, and then release the energy at a wavelength that is also specific to that element. These two processes, atomic absorption and atomic emission, are utilized for elemental analysis.

Figure 17.1(a) depicts the absorption and emission processes. With the absorption of radiant energy, an electron is elevated from the ground state to an excited state. The energy is then emitted and the electron returns to the ground state. The figure in 17.1(b) depicts the possible levels of absorption and emission as an example.

Figure 17.2(a) is a continuous spectrum of white light where smooth transition is seen from wavelength to wavelength. Figure 17.2(b) depicts a spectrum where absorption is observed. There are sharp black lines where discrete wavelengths of energy have been absorbed by an element. An emission spectrum is shown in Figure 17.2(c) where sharp colored bands are seen

Analytical Chemistry: A Chemist and Laboratory Technician's Toolkit, First Edition. Bryan M. Ham and Aihui MaHam.

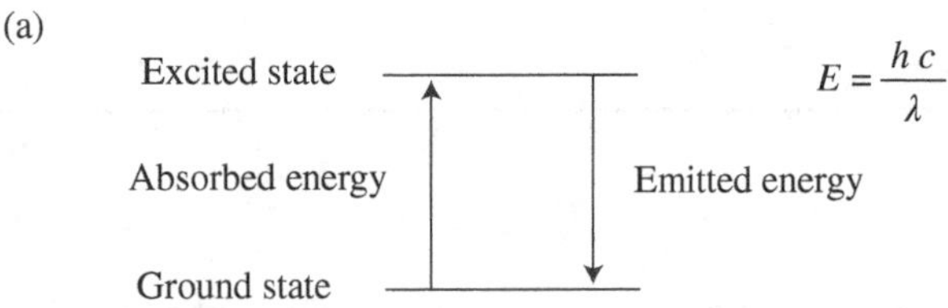

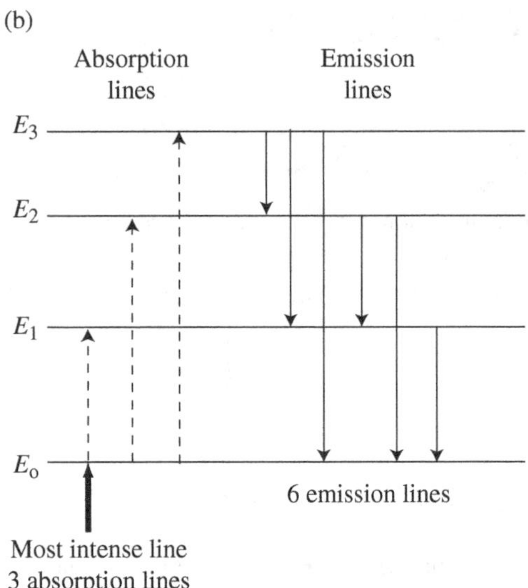

FIGURE 17.1 (a) The absorption and emission process. (b) The possible levels of absorption and emission as an example.

FIGURE 17.2 Spectra representing (a) a continuous spectrum, (b) atomic absorption spectrum, and (c) atomic emission spectrum.

against a black background representing discrete energy released from the excited electrons as they fall back down to their original energy orbital levels. The bands are very sharp as compared to molecular absorption bands. The element spectral bands are around 0.003 nm while molecular bands are ≥25 nm. If solvent or matrix is present in the flame, the bands can overlap resulting in interference. Each element has its own spectrum, for example sodium has a unique band at 589 nm. Both of these processes, atomic absorption and emission, are used to measure metals and some nonmetals usually in solution.

17.3 ATOMIC ABSORPTION AND EMISSION SOURCE

In most cases, samples that have been prepared for the analysis of metals are in solution. To do the analysis, a source is needed to transfer the analytes in solution to an essentially gaseous state while also removing the solvent. Transferring the metals in solution into a gaseous state with the solvent removed will allow the metal atoms to pass through a beam of radiant energy. While passing through the beam, the metals of interest will absorb energy thus allowing their measurement. Figure 17.3 depicts the basic makeup of an atomic absorption spectroscopy sample introduction system and source. The important steps are to introduce the sample in a liquid state into the source. This can be done by positive suction from the nebulizer gases, or with a peristaltic pump. The liquid will pass through a nebulizer that transforms the sample solution into a mist. The mist is then carried into the flame where the solvent is removed and the metals are converted into a gaseous state. Only about 10%–15% of the sample solution is converted to the finest mist spray droplet size making it to the flame. Most of the sample condenses out into large droplets and flow out the drain. Figure 17.4 shows an actual source and burner for an atomic absorption instrument. In current instruments, the sources are often interchangeable. This allows removing one source and installing another allowing different source approaches to analyzing samples.

17.4 SOURCE GASES AND FLAMES

The two main gas mixtures used for the source flames include air–acetylene and nitrous oxide–acetylene. The air–acetylene is used for elements that are not prone to refractory conditions. Refractory conditions exist where the element exists as an oxide that is not converted to the gaseous element in the flame. A hotter flame is required to reduce the element to the nonionized state. Table 17.1 lists the fuel, oxidant, and temperatures of some flame mixtures used. Figure 17.5 depicts a flame for a burner source.

17.5 BLOCK DIAGRAM OF AAS INSTRUMENTATION

At this point, let us look at a block diagram of the major components of atomic absorption and emission spectrometry instrumentation. Figure 17.6 depicts a simple block diagram of the atomic absorption spectrometer (AAS) instrument. As we have already covered, the sample is introduced into the flame through a source, or as listed in Figure 17.6 the atomizer. The sample is transported and nebulized in the atomizer chamber. The aerosol spray of fine droplets produced is mixed with the gases and then passed into the flame where desolvation and vaporization take place. The atoms are now in a gaseous state shown as a cloud above the atomizer. The radiation source is located to the left. We will look more closely at the radiation source in the next section. The wavelength being used will be selected using a monochromator. This is followed by signal detection, amplification, and finally, signal processing with a PC.

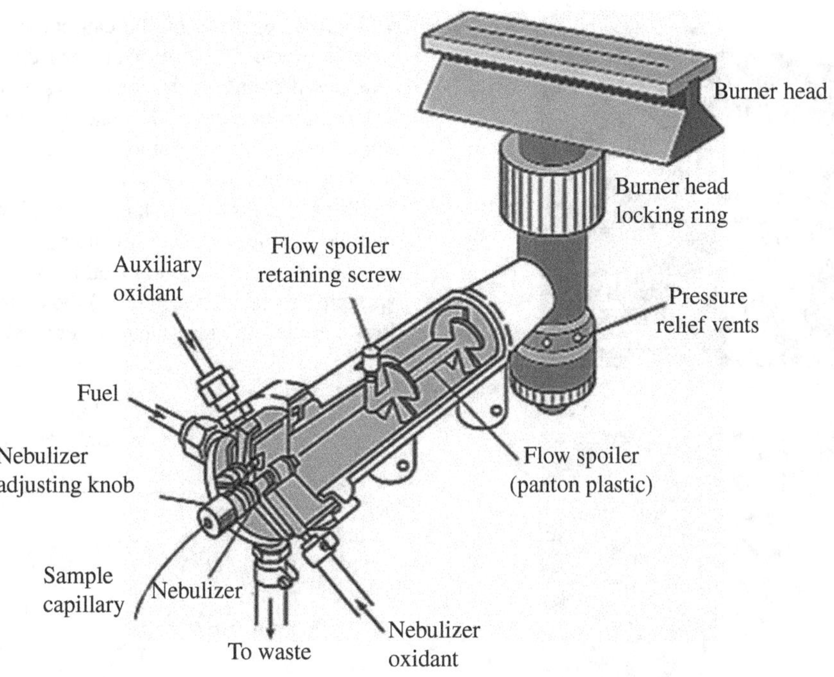

FIGURE 17.3 Basic make up of a source and burner for an atomic absorption spectroscopy instrument.

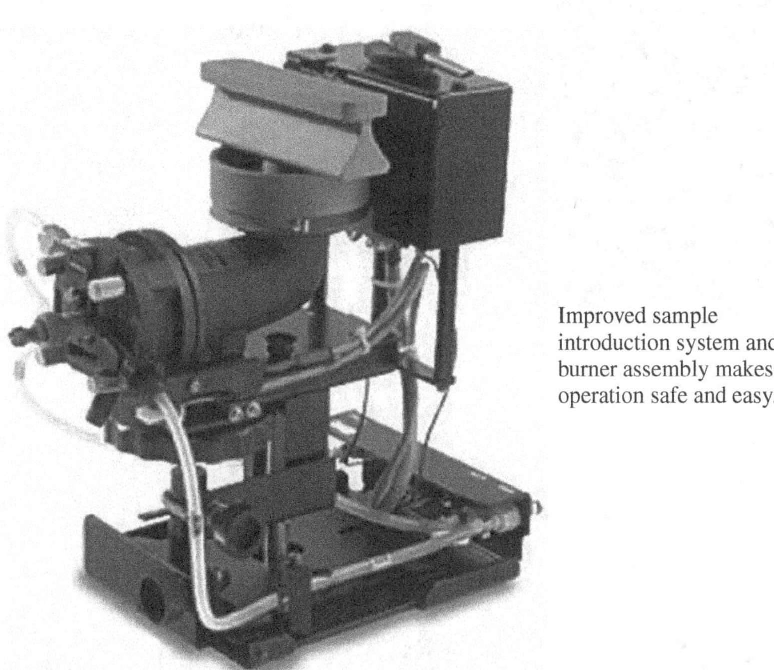

Improved sample
introduction system and
burner assembly makes
operation safe and easy.

FIGURE 17.4 Example of a flame coming out of an atomic absorption spectroscopy instrument source.

TABLE 17.1 Temperature of Atomic Absorption Flames.

Fuel	Oxidant	Temperature (°C)
Natural gas	Air	1700–1900
Hydrogen	Air	2000–2100
Acetylene	Air	2100–2400
Hydrogen	Oxygen	2550–2700
Natural gas	Oxygen	2700–2800
Acetylene	Oxygen	3050–3150
Acetylene	Nitrous oxide	2600–2800

17.6 THE LIGHT SOURCE

The hollow cathode lamp is the most common light source used in AAAs. An example of a hollow cathode lamp is depicted in Figure 17.7. The lamp consists of a cathode that is either made of the metal of interest or supports a small layer of that metal. There is a tungsten anode in parallel with the cathode, both sealed in a glass tube that is filled at 1–5 torr of either neon or argon. A potential is applied across the cathode and anode that accelerates ionized gas atoms toward the cathode. The gaseous atoms

(a)

(b)

FIGURE 17.5 (a) Example of an air-acetylene flame for a source. (b) Example of a nitrpus-oxide flame.

will strike the metal on the cathode and sputter (knock off and release) metal atoms into the gas cloud. The sputtered metal atoms will collide with the rare gas atoms causing excitation of valence electrons to higher energy orbitals. When the electrons relax back to their ground state, a quantum of energy will be released as emission of light.

The reactions that are taking place in the hollow cathode lamp include: (i) the ionization of the rare gas neon or argon, (ii) the sputtering of the cathode metal atoms, (iii) the excitation of the sputtered metal atoms, and (iv) the emission of light. These hollow cathode lamp reactions are depicted in Figure 17.8. The light

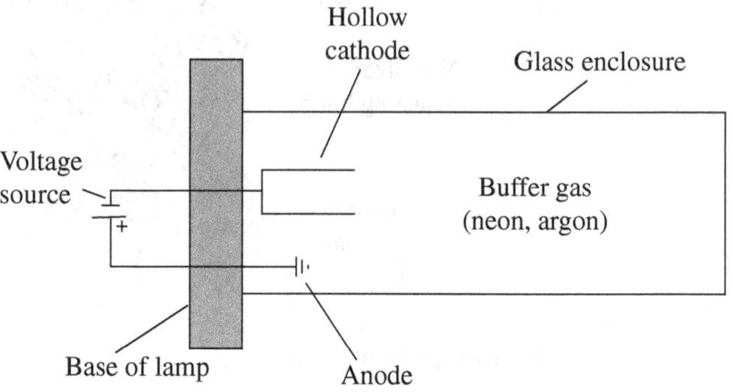

FIGURE 17.7 A hollow cathode lamp light source.

(1) $Ar + e^-$	$\longrightarrow Ar^+ + 2e^-$	ionization of rare gas
(2) $M(s) + Ar^+$	$\longrightarrow M(g) + Ar$	sputtering of cathode atoms
(3) $M(g) + Ar^+$	$\longrightarrow M^*(g) + Ar$	excitation of metal atoms
(4) $M(g) + Ar^+$	$\longrightarrow M(g) + h\nu$	emission of light

FIGURE 17.8 Reactions taking place in the hollow cathode lamp including (1) the ionization of the rare gas neon or argon, (2) the sputtering of the cathode metal atoms, (3) the excitation of the sputtered metal atoms, and (4) the emission of light.

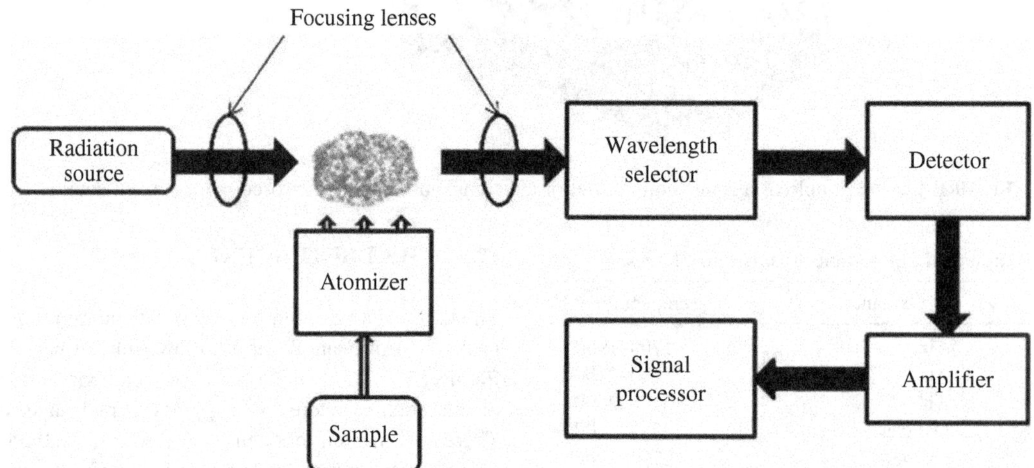

FIGURE 17.6 Block diagram of the major components that make up an atomic absorption spectrometry instrument.

that is emitted is the same as that absorbed by the same element. In other words, the hollow cathode lamp used to measure sodium in a sample will have sodium as the metal for the cathode. Manufacturers offer both single-element lamps and multielement lamps. Multielement lamps can save some effort by the analyst having to measure many elements. Using all single-element lamps requires changing the lamp every time a new element is to be measured. Many AAS instruments have lamp turrets that can hold four to five lamps and can be rotated to each different lamp.

17.7 INTERFERENCES IN AAS

There are two types of interferences that are encountered in atomic absorption spectroscopy: chemical interferences (including ionization) and spectral interferences. When stable or refractory compounds are formed, chemical interference takes place. When refractory compounds form, they are not completely atomized in the flame. The temperature of the flame may be too low. An example is the analysis of calcium in the presence of phosphate where the stable compound calcium phosphate is formed and does not decompose in an air–acetylene flame (2100–2400 °C).

$$3Ca^{2+} + 2PO_4^{3-} \rightarrow Ca_3(PO_4)_2.$$

The solution is to use the hotter nitrous oxide–acetylene flame (2600–2800 °C), which breaks down the refractory compound and releases the calcium as gaseous atoms. Other approaches are to add a chelating agent, such as EDTA:

$$Ca_3(PO_4)_2 + 3EDTA \rightarrow 3Ca(EDTA) + 2PO_4^{3-}$$

or the addition of a release agent, such as lanthanum chloride:

$$Ca_3(PO_4)_2 + 2LaCl_3 \rightarrow 3CaCl_2 + 2LaPO_4.$$

Ionization interference takes place when the element of interest ionizes in the flame. The ionized form will have a different spectrum than that of the neutral gaseous atom. This is especially a problem with the alkali metal which has the lowest ionization energies. This is usually removed by adding an element that is more easily ionized than the elements of interest. The added element will ionize in the flame and there will be an excess of electrons in the flame suppressing the ionization of the other elements present. An example is the addition of 1000 ppm cesium chloride (CsCl) when analyzing potassium (K) or sodium (Na).

17.8 ELECTROTHERMAL ATOMIZATION—GRAPHITE FURNACE

Another approach to elemental analysis is electrothermal atomization using graphite furnaces. In this approach, a small sample volume (3–5 μl) is deposited onto a platform in the center of a small graphite tube. The tube is then electrically heated, which in turn atomizes the sample. The lamp is aligned to pass through the center of the tube for the atomic absorption. Figure 17.9(a) depicts examples of graphite tubes used in the graphite furnace. Note the hole in the middle of the tube. This is where the sample is introduced. Figure 17.9(b) depicts a graphite tube with a platform for sample addition. Figure 17.10 shows a graphite furnace sample introduction source where a robotic arm with a sample tube will insert the tube in the middle of the graphite tube and deposit the sample. Once the sample is introduced into the graphite tube there are three stages that the graphite tube goes through. First, the temperature of the tube is slightly elevated for the sample drying stage at 125 °C for 20 s. Second, the temperature is increased for ashing of the organic matter present. For sodium (Na), the ashing temperature is 900 °C for 60 s. Molecular species have broad absorption bands and will interfere with the narrow

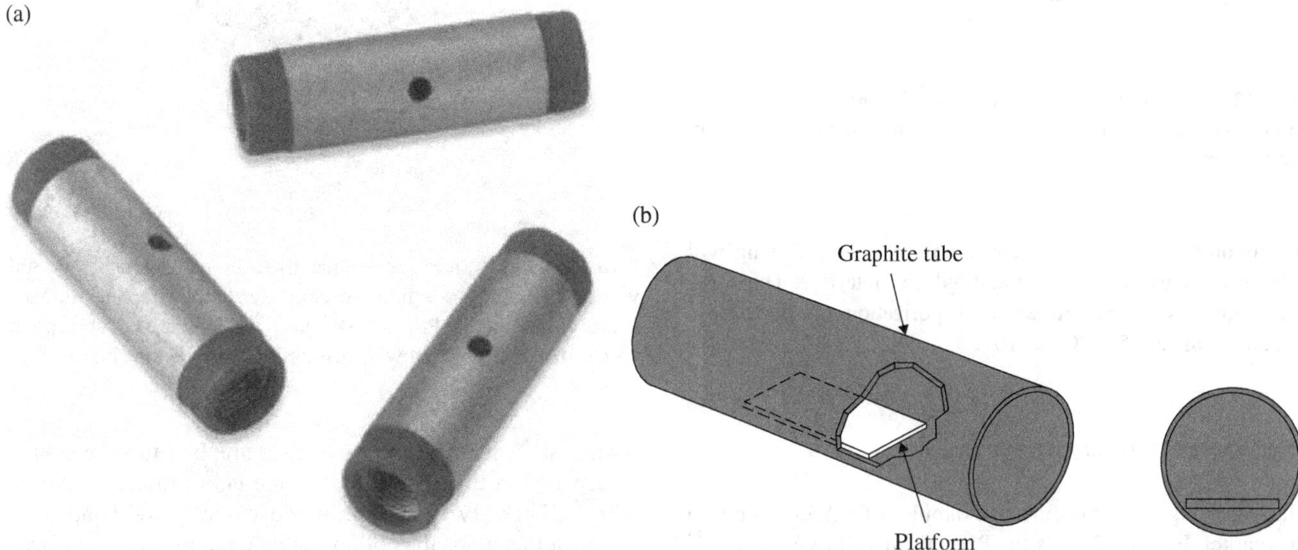

FIGURE 17.9 (a) Examples of graphite tubes used in the graphite furnace including a hole for sample introduction. (b) Illustration of the sample platform that is inside of the graphite tube.

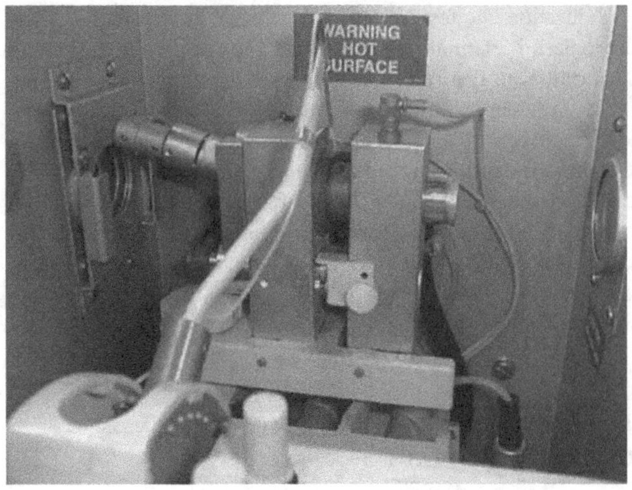

FIGURE 17.10 Graphite furnace sample introduction system. The robotic arm holds a sample introduction tube that is inserted in the middle hole of the graphite tube where the sample is introduced.

FIGURE 17.11 PerkinElmer PinAAcle™ 900 version atomic absorption spectrometer. (Reprinted with permission from PerkinElmer).

absorption bands of the elements being measured. Ashing will help ensure they are removed and will not interfere. Third, the temperature is raised further for vaporization of the analyte sodium atoms at 1500 °C for 10 s.

(a)

(b)

FIGURE 17.12 (a) Flame source used in the AAnalyst™ 800 version atomic absorption spectrometer. (b) Graphite furnace source used in the PinAAcle™ 900 version atomic absorption spectrometer. (Reprinted with permission from PerkinElmer).

17.9 INSTRUMENTATION

The following are some current examples of AAAs offered in the market. Figure 17.11 is the PerkinElmer PinAAcle 900™ version AAA. Some of the models allow changing the source from a flame source as depicted in Figure 17.12(a) of the AAnalyst™ 800 version AAA, to a graphite furnace source as depicted in Figure 17.12(b) of the PerkinElmer PinAAcle 900™. Figure 17.13 shows the PinAAcle™ 900 lamp compartment that holds the hollow cathode lamps for metals analysis. Figure 17.14 depicts the Agilent 280Z AAS. The compartment on the right holds an eight lamp turret.

FIGURE 17.13 The PinAAcle™ 900 lamp compartment that holds the hollow cathode lamps for metals analysis. (Reprinted with permission from PerkinElmer).

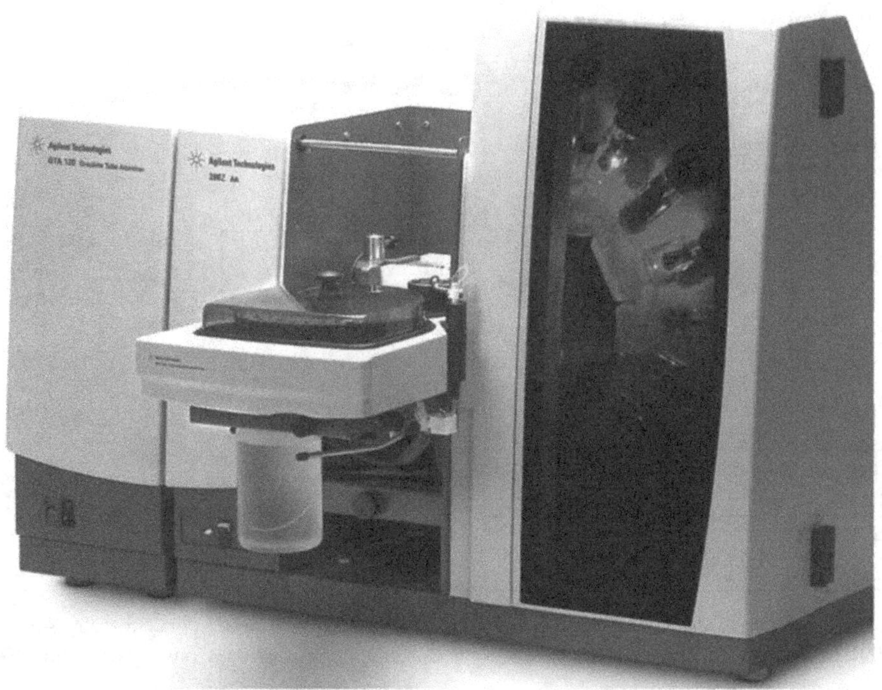

FIGURE 17.14 The Agilent 280Z atomic absorption spectrometer (AAS). (© Agilent Technologies, Inc. 2014, Reproduced with Permission, Courtesy of Agilent Technologies, Inc.)

17.10 FLAME ATOMIC ABSORPTION ANALYTICAL METHODS

An excellent reference is the Agilent Flame Atomic Absorption Spectrometry Analytical Methods (manual part number 8510000900) which is a listing of conditions and methods for the comprehensive analysis of metals and some nonmetals. The listing includes sample and standard preparations, discussions on ionization, and the instrument apparatus including parameters. As an example, an excerpt of silver (Ag) from the Agilent Flame Atomic Absorption Spectrometry Analytical Methods is depicted in Figure 17.15.

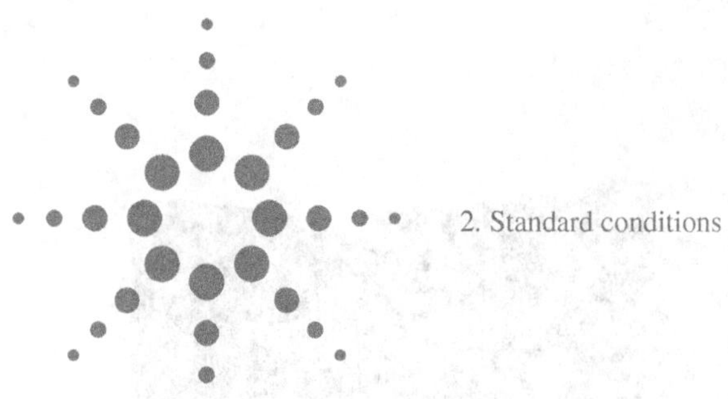

2. Standard conditions

Ag (Silver)

A.W.107.9

Preparation of standard solutions

Recommended standard materials

Silver metal strip or wire	99.99%
Silver nitrate (AgNOa)	99.99%

Solution technique

Dissolve 1.000 g of silver in 20 ml of 1:1 nitric acid and dilute quantitatively to 1 liter to give 1000 μg/ml Ag.

Recommended instrument parameters

Atomic absorption

Working conditions (Fixed)

Lamp current	4 mA
Fuel	acetylene
Support	air
Flame stoichiometry	oxidizing

Working conditions (variable)

Wavelength (nm)	Slit width (nm)	Optimum working range (μg/ml)
328.1	0.5	0.02–10
338.3	0.5	0.06–20

Flame emission

Wavelength	328.1 nm
Slit width	0.1 nm
Fuel	acetylene
Support	Nitrous oxide

Interferences

No chemical interferences have been observed in air-acetylene flames

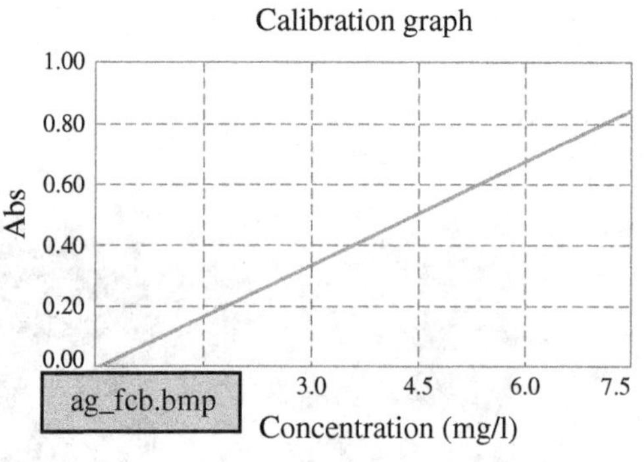

FIGURE 17.15 An excerpt of silver (Ag) parameters from the Agilent Flame Atomic Absorption Spectrometry Analytical Methods (manual part number 8510000900). (© Agilent Technologies, Inc. 2014, Reproduced with Permission, Courtesy of Agilent Technologies, Inc.)

18

ATOMIC EMISSION SPECTROSCOPY

18.1 Introduction
18.2 Elements in Periodic Table
18.3 The Plasma Torch
18.4 Sample Types
18.5 Sample Introduction
18.6 ICP-OES Instrumentation
18.6.1 Radially Viewed System
18.6.2 Axially Viewed System
18.6.3 Ergonomic Sample Introduction System
18.6.4 Innovative Optical Design
18.6.5 Advanced CID Camera Technology
18.7 ICP-OES Environmental Application Example

18.1 INTRODUCTION

The next atomic spectroscopy technique we will look at is optical atomic emission spectroscopy from excited atoms in inductively coupled plasma. This technique is known as inductively coupled plasma optical emission spectroscopy (ICP-OES). ICP-OES has a number of advantages over the flame atomic absorption spectroscopy covered in the previous chapter including the ability to measure 75 elements of the periodic table, this in comparison to approximately 62 elements analyzed by AAS. With ICP-OES, there are a substantial more amount of elements that can also be analyzed simultaneously, instead of one-by-one as with AAS. Also, since ICP-OES is an emission-based atomic measurement, there is no need for all of the lamps that are required with atomic absorption spectroscopy.

18.2 ELEMENTS IN PERIODIC TABLE

Another advancement with ICP-OES is the extension of elements that can be measured as compared with AAS (75 elements vs 62). The ICP-OES methodology can also be more sensitive, especially for some of the more difficult elements. Figure 18.1 depicts periodic tables that are associated with each technique. The periodic table in Figure 18.1(a) is for the elements that can be measured by AAS. Also included are the types of flame used for the element, and some alternative techniques such as the hydride vapor generator (HVG) or a mercury vapor unit (MVU). The elements analyzed by these techniques include arsenic (As), selenium (Se), and antimony (Sb), as well as mercury (Hg). Figure 18.1(b) has a periodic table with detection limits for use with ICP-OES. Note the inclusion as compared with Figure 18.1 (a) of carbon (C), nitrogen (N), sulfur (S), chlorine (Cl), bromine (Br), and iodine (I).

Table 18.1 lists some elements that are analyzed with ICP-OES, including sensitivities (listed as μg/l, or ppb) according to whether the ICP source is arranged axial or radial. In the next section, we will look at the plasma sources including axial and radial.

18.3 THE PLASMA TORCH

The plasma is an electrical conducting gaseous mixture that has a large amount of cations and electrons. Argon is typically used for the plasma gas where it is argon ions and electrons that make up the conductive gas. The argon flows through a radio frequency induction coil that has produced a strong fluctuating magnetic field. The argon ions are initiated through a spark. The produced argon ions will revolve around in a circular path within the induction coil magnetic field. Resistance to the motion by the argon ions produces ohmic heating into what is known as the argon plasma torch. The argon plasma torch that is produced in the radio frequency field is a high-temperature torch reaching temperatures as high as 10,000 K. Figure 18.2 depicts an argon plasma torch source that is in the radial position. Liquid samples are atomized into the torch where desolvation takes place liberating gaseous metal atoms. Within the plasma torch, collisional excitation takes place to excite the metal atoms and ions to higher excited states. The elevated electrons then relax emitting distinct wavelengths of energy. Figure 18.3 depicts an actual argon plasma torch.

Analytical Chemistry: A Chemist and Laboratory Technician's Toolkit, First Edition. Bryan M. Ham and Aihui MaHam.

(a)

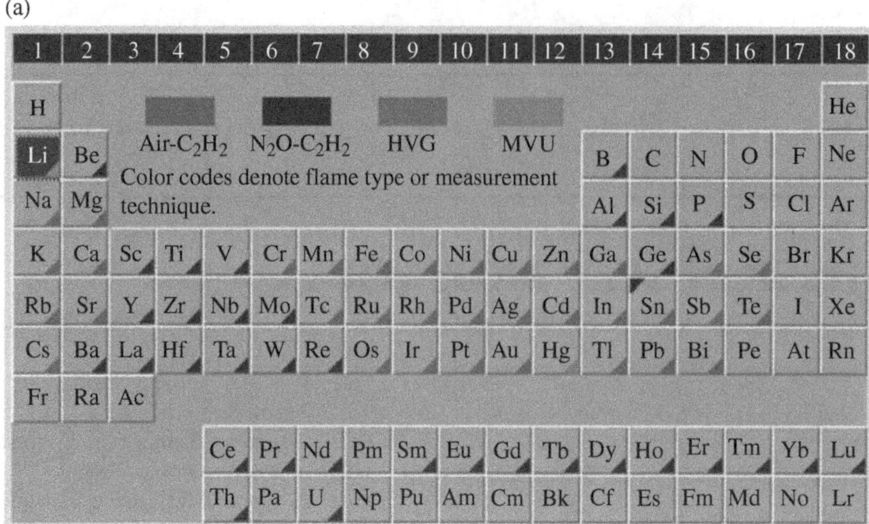

(b)

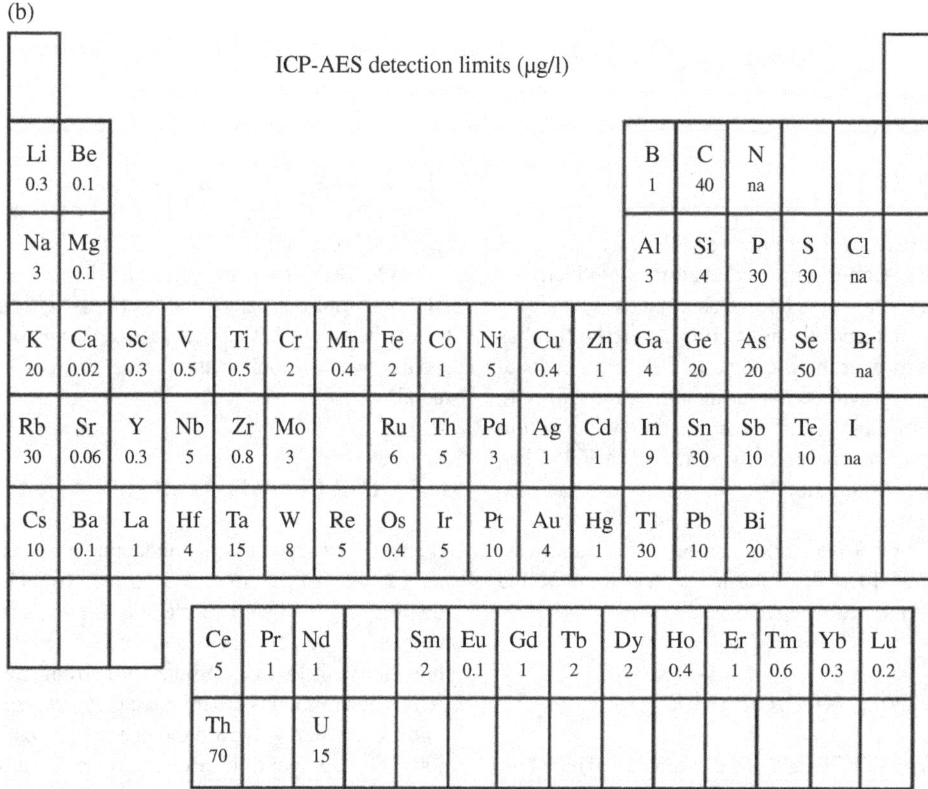

ICP-AES detection limits (μg/l)

Li 0.3	Be 0.1											B 1	C 40	N na			
Na 3	Mg 0.1											Al 3	Si 4	P 30	S 30	Cl na	
K 20	Ca 0.02	Sc 0.3	V 0.5	Ti 0.5	Cr 2	Mn 0.4	Fe 2	Co 1	Ni 5	Cu 0.4	Zn 1	Ga 4	Ge 20	As 20	Se 50	Br na	
Rb 30	Sr 0.06	Y 0.3	Nb 5	Zr 0.8	Mo 3		Ru 6	Th 5	Pd 3	Ag 1	Cd 1	In 9	Sn 30	Sb 10	Te 10	I na	
Cs 10	Ba 0.1	La 1	Hf 4	Ta 15	W 8	Re 5	Os 0.4	Ir 5	Pt 10	Au 4	Hg 1	Tl 30	Pb 10	Bi 20			

Ce 5	Pr 1	Nd 1		Sm 2	Eu 0.1	Gd 1	Tb 2	Dy 2	Ho 0.4	Er 1	Tm 0.6	Yb 0.3	Lu 0.2
Th 70		U 15											

FIGURE 18.1 (a) Elements that can be measured by AAS. Also included are the types of flame used for the element, and some alternative techniques such as the hydride vapor generator (HVG) or a mercury vapor unit (MVU). The elements analyzed by these techniques include arsenic (As), selenium (Se), and antimony (Sb), as well as mercury (Hg). (b) Periodic table with detection limits for use with ICP-OES. Notice the inclusion of carbon (C), nitrogen (N), sulfur (S), chlorine (Cl), bromine (Br), and iodine (I).

To compare the axial torch position to the radial torch position, Figure 18.4 depicts these orientations with a couple of different views.

18.4 SAMPLE TYPES

The methodology of ICP-OES is applied to numerous samples and matrices in a variety of analytical laboratories. Some general types of samples include agricultural and food, biological and clinical, geological, environmental and water.

18.5 SAMPLE INTRODUCTION

The sample is introduced into the argon plasma torch through a cyclonic spray chamber and nebulizer/atomizer. A peristaltic pump is used to move the sample from the sample vial into the

TABLE 18.1 Sensitivities of Some Elements with Plasma Torch Radial or Axial[a].

Element	Wavelength (nm)	Radial (μg/l)	Axial (μg/l)	Element	Wavelength (nm)	Radial (μg/l)	Axial (μg/l)
Ag	328.068	1	0.3	Mg	279.553	0.04	0.01
Al	167.019	0.9	0.1	Mn	257.61	0.08	0.03
As	188.98	5	1	Mo	202.032	1.5	0.5
Au	242.794	2.5	1	Na	589.592	2	0.15
B	249.772	0.6	0.1	Ni	231.604	1.4	0.3
Ba	455.403	0.15	0.03	P	177.434	5	1.5
Be	313.042	0.04	0.01	Pb	220.353	5	0.8
Bi	223.061	6	2	S	181.972	9	3
Ca	396.847	0.06	0.01	Sb	206.834	5	2
Cd	214.439	0.6	0.05	Se	196.026	6	2
Ce	418.659	7	2	Si	251.611	2.5	1
Co	238.892	1	0.2	Sn	189.925	7	1
Cr	267.716	0.9	0.15	Sr	407.771	0.05	0.01
Cu	327.395	1	0.3	Ti	334.941	0.25	0.1
Fe	238.204	0.8	0.1	Tl	190.794	6	1.5
Hg	184.887	2	0.8	V	292.401	0.7	0.2
K	766.491	4	0.3	Zn	213.857	0.5	0.2
Li	670.783	1	0.06	Zr	343.823	0.9	0.3

[a] © Agilent Technologies, Inc. 2014, Reproduced with Permission, Courtesy of Agilent Technologies, Inc.

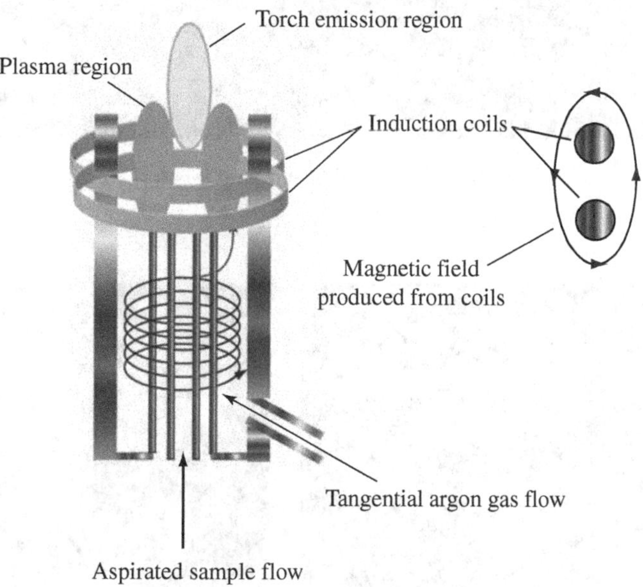

FIGURE 18.2 Argon plasma torch source illustrating the radial position.

spray chamber. Figure 18.5 depicts a peristaltic pump. The tubing that draws the sample runs through the pump. Rollers in the pump turn in a circle and roll across the tubing causing a steady flow of the sample through the tubing. Peristaltic pumps, due to the rotating rollers, roll across the sample tubing causing the flow almost pulseless. The sample flows into the spray chamber through a nebulizer. The nebulizer produces an aerosol spray of fine droplets into the spray chamber. The spray chamber acts to select the finest mist of droplets to further flow into the plasma torch. The spray chamber contains a drain at the bottom to allow the unused spray to condense and flow out of the spray chamber. Figure 18.6(a) and (b) depicts a nebulizer and a spray chamber, respectively. The sample fluid flows from the peristaltic pump through the nebulizer that is coupled with the spray chamber. Figure 18.7 shows a picture of the nebulizer coupled with the spray chamber.

18.6 ICP-OES INSTRUMENTATION

There is numerous ICP-OES instrumentation that is available from various vendors. All of these instruments are doing the same analysis based off of the argon plasma torch, but differ in their configurations and associated software. Figure 18.8 depicts the Agilent 5100 ICP-OES. The instrument comes in two orientations for the plasma torch.

FIGURE 18.3 Argon plasma torch.

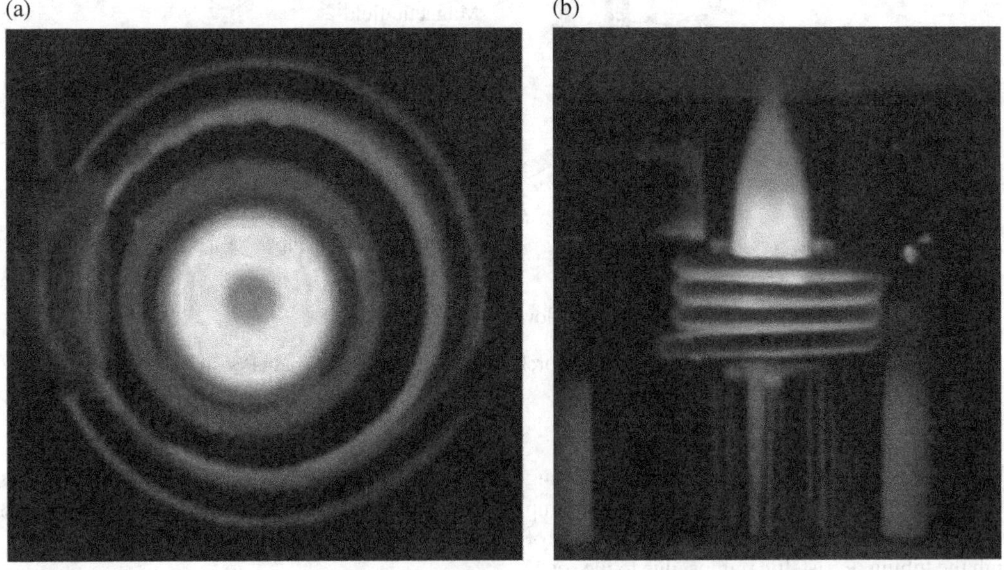

FIGURE 18.4 Axial (a) and radial (b) views of the plasma torch. (Reprinted with permission from Shimadzu).

18.6.1 Radially Viewed System

Vertically oriented, radially viewed plasma is ideal for the most difficult applications, including the analysis of oils and organic solvents, geological/metal digests, and high TDS solutions, for example, brines. Includes full PC control of plasma viewing height from 0 to 20 mm and horizontal adjustment of ±3 mm to optimize sensitivity and minimize interferences. Viewing height may be adjusted under PC control for each emission line of interest.

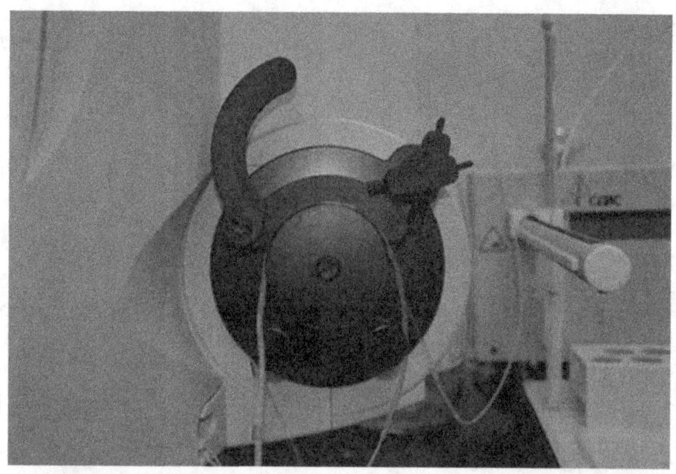

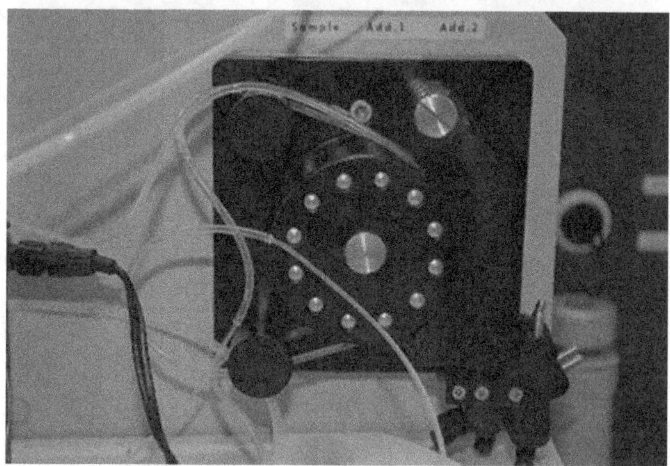

FIGURE 18.5 A peristaltic pump used to move sample from the sample vial into the spray chamber.

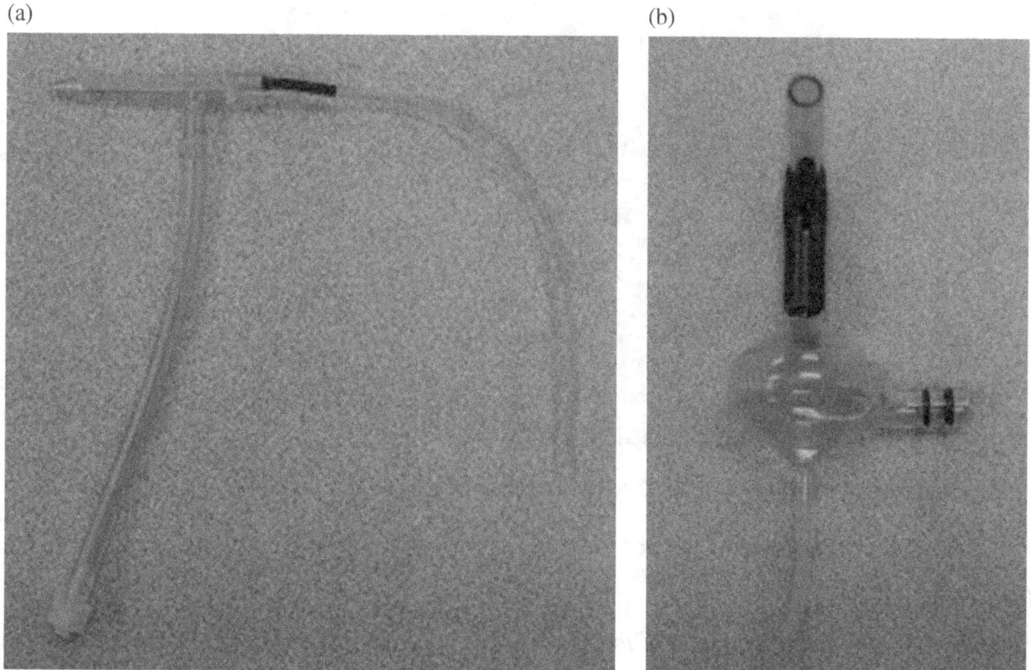

FIGURE 18.6 (a) ICP nebulizer. (b) Spray chamber.

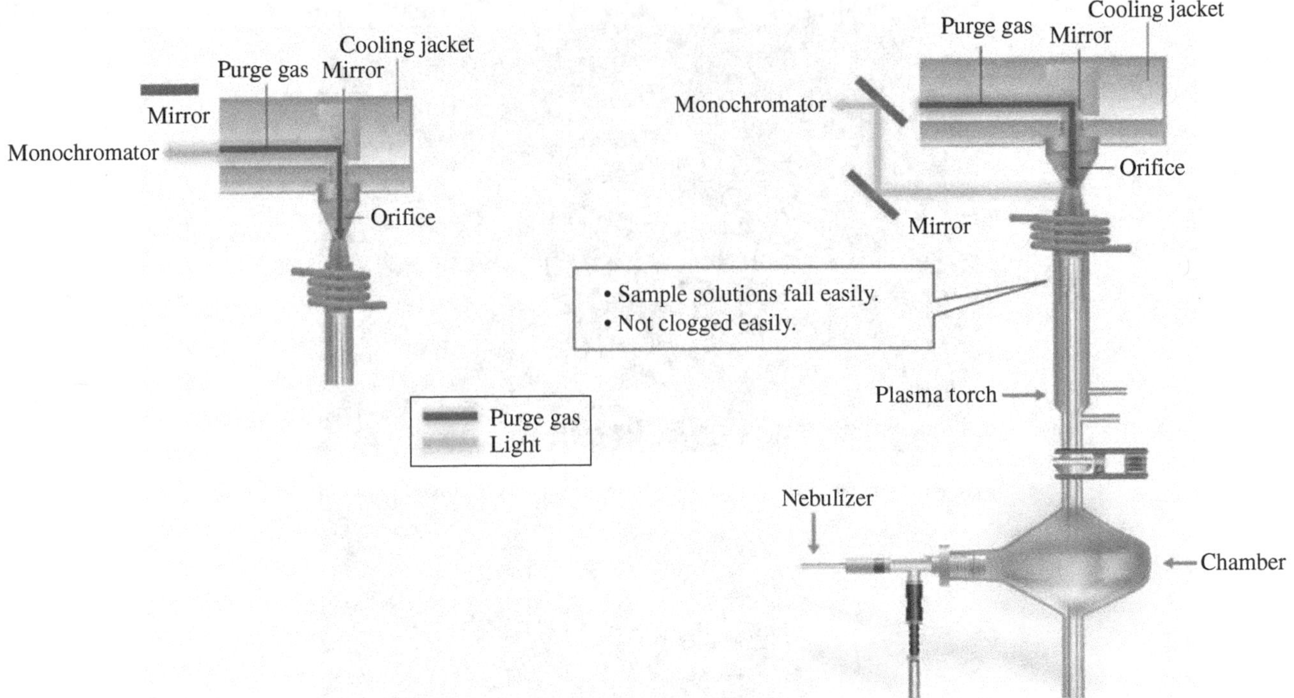

FIGURE 18.7 Picture of the coupling of the nebulizer with the spray chamber.

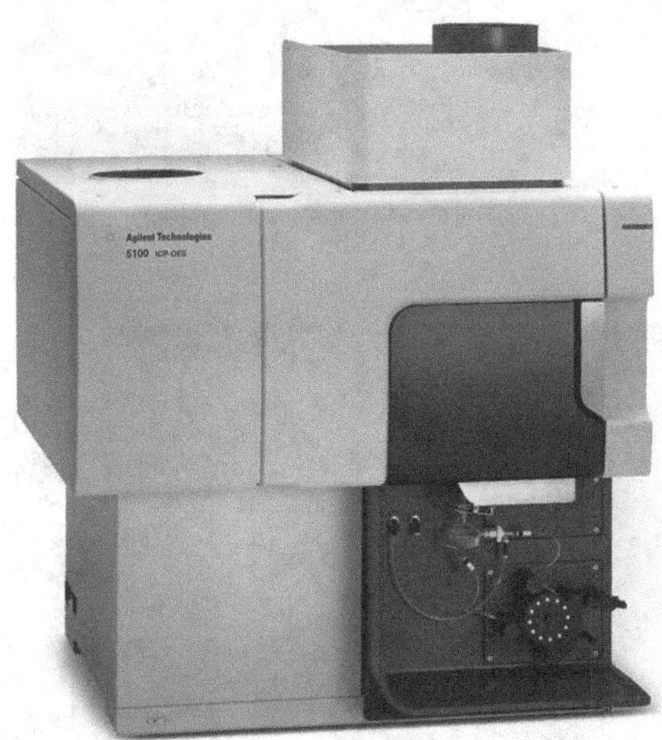

FIGURE 18.8 The Agilent 5100 ICP-OES instrument. (© Agilent Technologies, Inc. 2014, Reproduced with Permission, Courtesy of Agilent Technologies, Inc.)

18.6.2 Axially Viewed System

Horizontally oriented, axially viewed plasma is ideal for high-sensitivity analyses. Provides a 3- to 12-fold improvement in detection limits compared to radial viewing. The axially viewed plasma system features a unique cooled cone interface (CCI) to prevent the cooler plasma tail from being viewed by the optics. This reduces interferences, improves the system's tolerance to high dissolved solids, and extends the linear dynamic range compared to conventional axial systems. The CCI is a superior plasma interface with lower running costs compared to shear gas systems. Includes full X, Y adjustment of plasma viewing position under PC control.

A second ICP-OES instrument illustrated is the Thermo Scientific iCAP 6000 Series ICP Emissions Spectrometer, shown in Figure 18.9. Some highlights of the instrument are listed below.

18.6.3 Ergonomic Sample Introduction System

An open architecture sample introduction system enables easy access to the peristaltic pump, nebulizer, and spray chamber and torch configuration in the iCAP 6000 Series. A unique drain sensor is integrated within the sample introduction system to ensure the plasma is extinguished safely and the liquid flow is

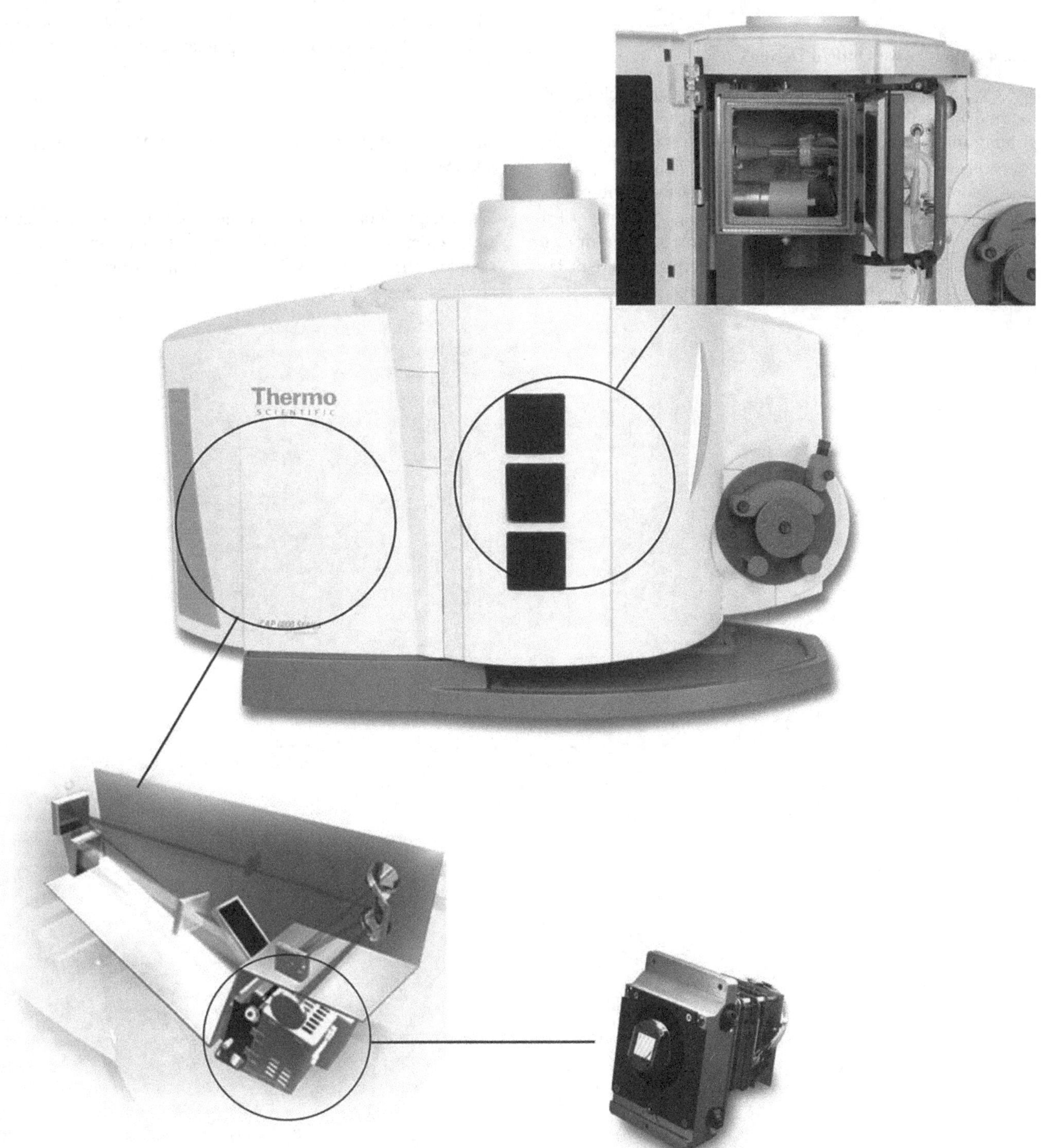

FIGURE 18.9 Thermo iCAP 6000 Series ICP Emissions Spectrometer. (Used with permission from Thermo Fisher Scientific, the copyright owner.)

controlled in the event of a blockage or leak. The self-aligning enhanced matrix tolerance torch employs an integral orientation lock to establish reliable plasma gas connections automatically and is optimized to operate with almost 20% lower argon usage than typical ICPs. A screw-threaded center tube enables rapid disassembly from the torch body without removing the torch from the torch box and facilitates fast and efficient maintenance without switching off the plasma.

18.6.4 Innovative Optical Design

The elegant fore-optic and polychromator design employs only four moving optical components to enable exceptional analytical stability and sensitivity across the entire wavelength range. Compact echelle-based spectrometer design with unique all-spherical mirror configuration achieves exceptional analytical resolution over the entire area of the detector chip. The polychromator features a highly efficient gas distribution system to purge air from the optical tank and plasma interface, ensuring maximum UV wavelength transmission while reducing purge gas costs.

18.6.5 Advanced CID Camera Technology

The fourth-generation charge injection device (CID) complements the optical design and enables access to over 50,000 analytical wavelengths. Inherently nonblooming nondestructive readout capability delivers increased signal to background ratios to achieve exceptional sensitivity and analyte detection limits, while enabling wide dynamic range. Powerful simultaneous data acquisition capability enables the display of a two-dimensional CID image of the entire spectrum for "live" or postrun processing and fast, efficient generation of qualitative and semiquantitative data for any element.

18.7 ICP-OES ENVIRONMENTAL APPLICATION EXAMPLE

We will now look at a complete method including setup, results, and validation for an ICP-OES application for environmental samples. The following is an application note from Agilent Technologies (Fig. 18.10).

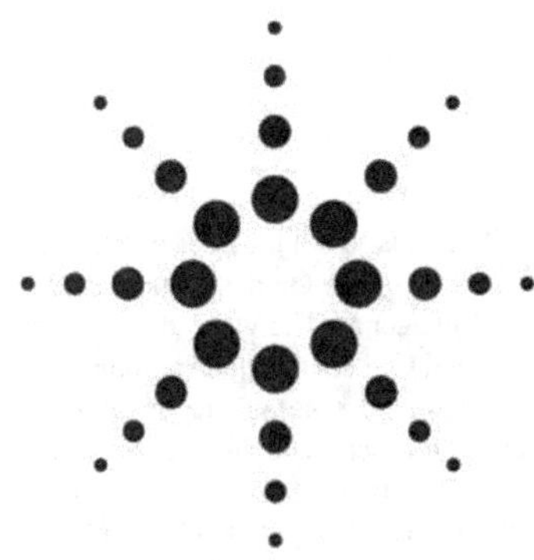

A Complete Method for Environmental Samples by Simultaneous Axially Viewed ICP-OES following US EPA Guidelines

Application Note

Inductively Coupled Plasma-Optical Emission Spectrometers

Authors

Scott Bridger
Mike Knowles

Introduction

With the growing demand for elemental analysis of environmental samples and the financial pressures being applied to the modern laboratory, development of a universal method for a wide range of sample types is needed. The Agilent Vista ICP-OES, with simultaneous measurement of the entire elemental spectrum facilitates such universal methods.

The Vista instrument has a number of distinct advantages over similar ICP-OES systems. Firstly, the VistaChip is the only single Charge Coupled Device (CCD) that allows full coverage of the spectrum from 165-785nm, with a pixel processing speed of 1 MHz and exceptional anti-blooming properties. These features allow both trace level analytes and major analytes to be determined in the same measurement. Secondly, the RF robustness of the Vista ICP-OES permits the analysis of difficult samples, up to 5% total dissolved solids, using an axially viewed plasma. Finally, the Cooled Cone Interface (CCI) of the axially viewed Vista eliminates the cooler tail of the plasma, reducing Easily Ionizable Element (EIE) interferences and maximizing linear dynamic range. The CCI consists of a cooled nickel cone with a large orifice at its tip, positioned to view the optimum region of the axial plasma.

The greatest challenge in creating one method for all analytes is achieving the dynamic range coverage from low parts-per-billion for the toxic elements to high parts-per-million for the Group I and II elements. With the Vista this is further facilitated by software features such as MultiCal, which allows multiple wavelengths to be used simultaneously for the same element to provide complete coverage of the linear dynamic range. MultiCal allows the user to assign the valid linear dynamic range to each wavelength used.

The user enters the allowable minimum and maximum concentration for each wavelength so that the software can then automatically assign sample results to the appropriate wavelengths. The software preferences can then be set to only display concentrations that fall within this valid range. By combining multiple wavelengths

FIGURE 18.10 Agilent Application Note ICPES-29. (© Agilent Technologies, Inc. 2014, Reproduced with Permission, Courtesy of Agilent Technologies, Inc.)

in this way, the Vista manages the full dynamic range capabilities of the VistaChip from sub ppb levels to low percentage levels. Another software capability, Adaptive Integration, automatically assigns the integration time for each wavelength in real time, to achieve the optimum signal to noise ratio. For example, a high level signal for a matrix element such as Na, might be assigned multiple, shorter integration times, ensuring that this signal is within range and also improving precision statistics through the multiple readings. Simultaneously, a low level analyte of interest such as Pb, might be assigned the full integration time requested by the user, thereby ensuring optimum signal to noise ratio and detection limits. With Adaptive Integration these two measurement sequences can be conducted simultaneously, whereas conventional systems have to sequence these different integration times with the resultant longer analysis times.

As a result the Vista simultaneous ICP-OES is able to measure all required elements in a single environmental analysis using axial viewing. Alternative techniques such as dual viewed plasmas, require the samples to be analyzed first with axial viewing and then with radial viewing to accommodate the linear dynamic range of the target elements.

The use of the dual viewed plasma will therefore significantly lengthen the analysis time. Direct analysis using the Vista provides a significant saving, in analysis time and running costs particularly argon consumption.

In this work, the steps to develop a universal method for the analysis of waters and wastewaters are reviewed. As a measure of success, the US EPA guidelines for data quality control for these sample types have been used. The primary guiding documents for this analysis type from the US EPA are CLP ILMO 4.0 [1] and ILM05.0 [2] and Methods 200.7 [3] and 6010B [4]. These protocols describe strict rules for establishing calibration validity, linear dynamic range and management of interferences, thus ensuring data quality. It should be noted that these protocols are 'living documents' which undergo a process of continual development. For example, ILM04.0 is currently undergoing revision to ILM05.0 [2].

In this work, terminology from the ILM04.0 and ILM05.0 documentation is used, however a table of analysis sequence is offered which translates the protocols into the language of the different source documents. The method developed here has been applied to typical water and waste water samples.

The data Quality Control Protocols (QCP) provided as standard with the Vista software have been used in this work to meet the US EPA data validation guidelines. The Vista QCP package consists of a series of automated tests designed around these guidelines however these are adaptable to any other protocol by the use of a simple programmable language and user definable tests. The QCP software allows the user to specify the corrective action that will occur on a QCP solution test failure with options such as Recalibrate and Repeat With Samples, Flag and Continue and Stop. With the addition of the Varian autosampler and diluter which provides on-line over range dilution, the Vista ICP requires minimal supervision during the analysis, resulting in further resource savings.

Instrument Set up

An Agilent Vista simultaneous ICP spectrometer with an axially viewed plasma was used for this analysis. The instrument was fitted with the mass flow controller option on the nebulizer gas and with the 3 channel peristaltic pump option. The operating conditions for the instrument were obtained by following the criteria documented in the SOW (Statement of Works) for Methods 200.7, 6010B and ILM 04.0 and 05.0. Parameters were then optimised to obtain the best performance from the ICP-OES system.

The final instrument operating conditions are given in Table 1. Particular attention was paid to the Method Detection Limits (MDL) in the final acceptance of the operating conditions. All test solutions and calibrants were from Inorganic Ventures (Lakewood, NJ, USA) using their US EPA 200.7 kit.

Table 1. Instrument Operating Conditions

Power	1.40 kW
Plasma gas flow	15.0 L/min
Auxiliary gas flow	0.75 L/min
Nebuliser type	SeaSpray Glass Concentric (Glass Expansion, Melbourne Australia).
Nebuliser gas flow	0.75 L/min
Pump speed	15 rpm
Sample tubing	White/White
Internal standard tubing	Orange/White
Sample delay	40 sec
Rinse time	40 sec between each sample
Replicate time	30 sec
Stabilisation time	10 sec
Replicates	2
Background correction	Left and right off peak
Autosampler	Agilent SPS-5 5

FIGURE 18.10 (Continued)

An ionization buffer consisting of up to 1% $CsCl_2$ (Merck, Germany) with 10 mg/L yttrium (EM Science Gibbstown, NJ, USA) as internal standard and 0.1% Triton X100 (LabChem, Auburn, Australia), was connected to the sample flow via a post-pump T-piece (1/16" diameter, Cole Palmer, Illinois, USA part number 6365-77). The $CsCl_2$ ionization buffer is used to suppress the ionization effects of EIE, resulting in improved calibration linearity. This approach has been approved by the US EPA in one region of the USA [5], and so it is expected that written approval in other regions for this approach should be reasonably obtained. The yttrium is added as internal standard and the addition of the Triton X100 provides improved spraychamber wetting [6] for optimum precision.

Results - Method Detection Limits (MDL) and Linear Dynamic Range (LDR)

Having optimized the instrument conditions the MDL's were measured in accordance with USEPA documentation for a range of replicate read times. The definitions of Instrument Detection Limits (IDL) versus Method Detection Limits (MDL) and indeed Contract Required Detection Limit (CRDL) in the EPA literature are often confused.

In some documents the IDL is taken to mean an instrument detection limit achieved under manufacturer's recommended conditions in a dilute acid matrix. In this work the definition of Instrument Detection Limit (IDL) was taken from Exhibit E-10 of the ILM 04.0 Statement of Work [1].

To paraphrase this definition "the IDL shall be determined as 3xStandard Deviation of seven consecutive measurements of a standard solution at a concentration of 3-5 x the manufacturer's suggested IDL on three non-consecutive days". In other documentation this technique is described as an MDL [4] -this is probably a more appropriate designation in distinguishing between the ultimate IDL obtainable at any time and the more representative MDL obtained over several days. The results of determination of the IDLs by the ILM0.40/05.0 method are shown in the Table 2. These detection limits were obtained by averaging a pool of results from four separate Vista instruments around the world [8]. Due to the inherent uncertainty in detection limit measurements the results have then been rounded to only one significant figure. The IDLs obtained in this way must meet the levels specified in Exhibit C of the ILM04.0/05.0 Exhibit C is the table of Contract Required Detection Limits (CRDL). Table 2 shows that the CRDLs are met with a replicate read time of 30 sec.

Table 2.

Element	CRDL ILM 04.0 [1] (ug/L)	CRDL ILM 05.0 [2] (ug/L)	IDL 60sec (µg/L)	IDL 30sec (µg/L)
Ag 328.068	10	5	0.5	0.7
Al 236.705	200	200	10	12
Al 308.215	200	200	1	1
As 188.980	10	5	2	3
Ba 233.527	200	20	0.2	1
Ba 585.367	200	20	0.5	3
Be 234.861	5	1	0.1	0.2
Be 249.473	5	1	1	2
Be 313.042	5	1	0.2	0.5
Ca 370.602	5000	5000	200	300
Ca 315.887	5000	5000	1	2
Cd 226.502	5	2	0.2	0.3
Co 228.615	50	5	0.3	0.6
Co 238.892	50	5	1	2
Cr 267.716	10	5	0.2	0.5
Cu 327.395	25	5	0.6	1
Fe 259.940	100	100	0.5	1
Fe 258.588	100	100	1	2
K 404.721	5000	5000	1000	2000
K 766.491	5000	5000	2	2
Mg 383.829	5000	5000	5	10
Mn 257.610	15	10	0.2	0.5
Mn 261.020	15	10	3	5
Na 330.237	5000	5000	70	300
Na 589.592	5000	5000	1	2
Ni 231.604	40	20	1	1
Pb 220.353	3	3	2	2
Sb 206.834	60	5	2	4
Se 196.026	5	5	3	4
Tl 190.794	10	5	2	3
V 292.401	50	10	0.5	1
Zn 206.20	20	10	0.5	0.6

* IDLs calculated over 3 non-consecutive days [1,8] and rounded to one significant figure

FIGURE 18.10 (Continued)

Linear Range Analysis (LRA)

According to ILM04.0/05.0, a linear range verification check standard must be analyzed and reported quarterly for each analyte. The concentrations of the analytes in the LRA standard define the upper limit of the ICP linear range beyond which results cannot be reported without dilution. The analytes in the LRA standard must be recovered to within ± 5% of their true values. It is in the interest of every laboratory therefore to formulate an LRA standard with acceptable recoveries at the highest possible concentrations for each element. In most cases, high concentrations are generally only expected for the major elements such as Fe, K, Ca, Na, Mg and possibly Al. Table 3 shows the results of the LRA obtained during this work. It should be noted that silver is particularly prone to precipitation from solution at high concentrations. The US EPA recommends adding an excess of hydrochloric acid to avoid this precipitation and limiting the maximum concentration of Ag to 2 mg/L in solution [1]. In this work it was found that Ag calibrations became curved at concentrations of 5 mg/L or higher, and so the Ag calibration range was restricted to 2 mg/L to obtain good linearity.

Using the Vista's MultiCal feature a second wavelength was added for the elements Fe, K, Na, Ca and Al as shown in Table 3.

During the analysis Vista automatically assigns sample results to the wavelength that has the appropriate user defined linear dynamic range (LDR). In the same way, the automatic data QCP tests and actions are only applied to those wavelengths for which the results fall within the specified LDR. For example, referring to Table 3, an iron result of 70 ppm would be automatically measured and QC-assessed against the 258.258 nm wavelength, not the 259.940 nm wavelength.

The LRA results in Table 3 include some later work in which useful alternate wavelengths were found for a number of elements. These alternate wavelengths are indicated in the table. Note that these wavelengths were found to be suitable for analysis from the detection limit to the LDR limit, but with the MultiCal feature it is possible to restrict the lower concentration limit of the calibration to a non-zero value as mentioned above for iron 259.940 nm.

It should also be noted that the on-line overrange dilution capability of the Vista can be used in conjunction with MultiCal to ensure complete compliance with the US EPA regulations with unattended operation.

Table 3. Linear Range Analysis for Recommended Wavelengths for the 22 US EPA Elements. Note That Some Additional Elements Studied In This Work Have Been Included Such As Boron.

Element	Curve type	Minimum concentration per line (mg/L)	Maximum concentration per line (mg/L)
Ag 328.068	Linear	0	2
Al 236.705	Linear	200	2000
Al 308.215	Linear	0	200
As 188.980	Linear	0	100
B 249.772	Linear	0	100
Ba 585.367	Linear	0	100
Be 313.042 (alternates 234.861 and 249.473 nm)	Linear	0	10
Ca 370.602	Linear	0	2000
Ca 315.887	Linear	0	200
Cd 226.502	Linear	0	10
Co 228.615 (alternate 238.892 nm)	Linear	0	100
Cr 267.716	Linear	0	100
Cu 327.395	Linear	0	100
Fe 259.940	Linear	100	2000
Fe 258.258 (alternate 258.588)	Linear	0	100
K 404.721 (alternate 693.876 nm)	Linear	100	2000
K 766.491	Linear	0	100
Mg 383.829	Linear	0	2000
Mn 261.020	Linear	0	1000
Na 330.237	Linear	50	2000
Na 589.592	Linear	0	100
Ni 231.604	Linear	0	100
Pb 220.353	Linear	0	50
Pb 283.305	Linear	0	100
Sb 206.834	Linear	0	10
Se 196.026	Linear	0	10
Tl 190.794	Linear	0	10
V 311.837 (alternate 292.401 nm)	Linear	0	100
Zn 334.502	Linear	0	100

4

FIGURE 18.10 (Continued)

The recommended background correction points for each wavelength are shown in Table 4.

Table 4. Background Correction Points for Recommended Wavelengths (n.u. Indicates "not used")

Element	Wavelength (nm)	BC point left (nm)	BC point right (nm)
Ag	328.068	0.020	n.u.
Al	308.215	0.020	n.u.
Al	236.705	0.020	n.u.
As	188.890	0.020	n.u.
Ba	585.367	0.062	n.u.
Be	234.861	0.018	n.u.
Be	249.473	n.u.	0.020
Ca	370.602	0.024	n.u.
Ca	315.887	0.024	n.u.
Cd	226.502	0.020	n.u.
Co	238.892	0.020	n.u.
Co	228.615	0.016	n.u.
Cr	267.716	0.020	n.u.
Cu	327.395	0.020	n.u.
Fe	259.940	0.020	n.u.
Fe	258.588	0.020	n.u.
K	766.491	0.113	n.u.
K	693.876	n.u.	0.087
K	404.721	0.020	n.u.
Mg	383.829	0.036	n.u.
Mn	261.02	0.020	n.u.
Na	589.592	0.080	n.u.
Na	330.237	0.028	n.u.
Ni	231.604	0.020	n.u.
Pb	283.305	0.020	n.u.
Pb	220.353	0.012	n.u.
Sb	206.834	0.020	n.u.
Se	196.026	0.012	n.u.
Tl	190.794	0.011	n.u.
V	292.401	0.024	n.u.
Zn	334.502	0.022	n.u.

Initial and Continuing Calibration Verification (ICV, CCV) and Analytical Samples

The QC tests outlined in the various SOWs are designed to ensure the accuracy and precision of the results produced. The results shown in Figures 1-4 are in accordance with the specification detailed in CLP ILM 04.0/ILM 05.0 SOW [1]. The Initial Calibration Verification (ICV) test is conducted immediately after instrument calibration. The ICV solution is a check standard either obtained from the EPA or from a secondary source, other than that used to prepare the calibration standards. All analytes in the ICV must be recovered within ±10% of the certified value.

The Continuing Calibration Verification test is used to ensure the validity of the calibration throughout the analysis run and is carried out at a frequency of 10% (every 10 analytical samples) or every 2 hours, whichever is more frequent. The definition of an analytical sample is best given by exception - the Glossary of ILM0.40 defines an Analytical Sample as "any solution ... on which analysis is performed excluding instrument calibration, ICV, ICB, CCV and CCB". This means that if a sample is automatically diluted, the frequency counter must be incremented by the number of dilutions - this is done automatically by the Vista software.

The CCV is also measured at the beginning (but not before the ICV) and end of the analysis run. The CCV must be recovered between 90% and 110% (ILM04.0/05.0) of the true value. Method 200.7 features an Instrument Performance Check (IPC) solution and requires that the first time the IPC is analyzed (ie: equivalent to the ICV) it must be recovered within ± 5% and the precision of each measurement of the IPC must be less than 3%. Method 6010 B requires a recovery of ± 10% for both the ICV and CCV with a precision of less than 5%.

If the CCV test fails, the problem must be corrected, the instrument recalibrated and all samples since the last successful ICV (Initial Calibration Verification), CCV or check standard must be reanalyzed. Figure 1, shows the trends for

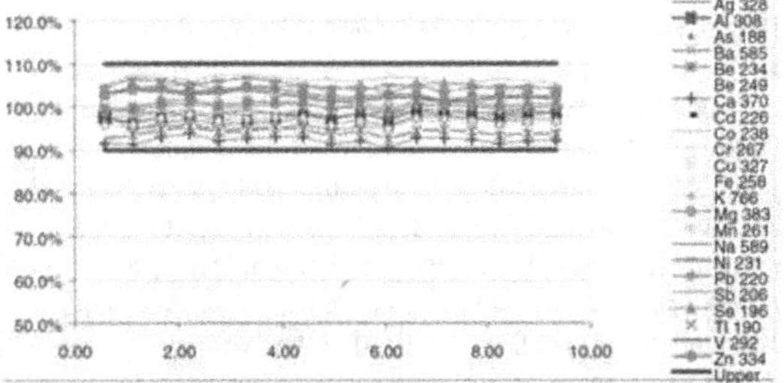

Figure 1. Percentage recoveries of Continuing Calibration Verification standards over a 10 hour period with no internal standard correction for all US EPA 22 elements at all wavelengths used. All recoveries were within the ± 10% limits.

the CCV results over a 10 hour period of continuous analysis of water samples without the use of internal standard. All CCV's fall within the US EPA acceptance criteria and the largest RSD for any element was 2%.

5

FIGURE 18.10 (Continued)

Interference Check Solutions (ICSA and ICSAB) and Inter-Element Corrections (IEC)

Interference Check Solutions (ICS) are used to confirm that interfering elements likely to be encountered in environmental samples do not cause incorrect measurements of analyte concentrations.

According to the US EPA criteria, inter-element interference corrections are achieved by using Inter Element Correction (IEC) factors.

These are calculated by observing the effect of known amounts of interferents on the analyte wavelengths. A table of IEC factors is generated and applied to the sample results during the analysis [4]. In ILM04.0/05.0, two solutions are analyzed - Interference Check Samples A and AB, where A contains 4 interferents only (Al, Ca, Fe and Mg) and AB contains the interferents plus 16 analytes. In the ICSA, the analytes must be within ± 2xCRDL (for elements with CRDLs ≧ 10 ug/L) and in ICSAB the analytes should be recovered within ± 20%. If the recoveries are outside these limits the "Recalibrate and Repeat With Samples" action should be conducted. For elements with CRDLs > 10 ug/L the results of the ICSA are simply reported with no test applied. The ICSA and ICSAB tests are applied at the beginning and end of the run and at a frequency of not greater than 20 samples. In Method 200.7, 17 single element, Spectral Interference Check (SIC) solutions are prepared. Concentration results at analyte wavelengths are then compared to the IDL or 3 sigma control limits of the calibration blank.

Only those failing these criteria need be tested daily, otherwise SIC testing can be conducted weekly. The results are then compared to a concentration range about the calibration blank, to determine whether SIC factors need updating. The Vista software automates the measurement, calculation and tabulation of the IEC factors. It is important to note that for accurate IEC calculations, when an internal standard correction is applied it should be applied to both the interferents and analytes. The IEC factors used in this work are not reproduced here because the factors will vary according to the analytical conditions used.

For example, use of different background correction techniques or locations will alter the appropriate IEC factor. For non EPA methods other techniques can be used to account for spectral interferences such as spectral deconvolution or the selection of alternative wavelengths.

The results for the ICSA solution analyzed at this frequency over a 10 hour period are shown in Figure 2. It can be seen that the results for Pb and Se, for example, with CRDLs of 3 and 5 μg/L respectively, are well within the allowable range for the 10 hour period.

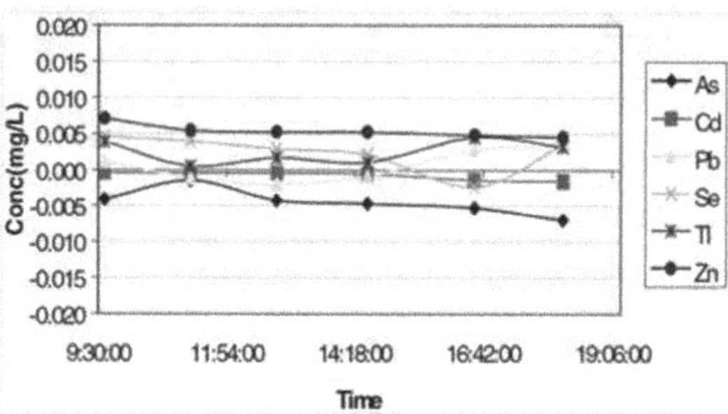

Figure 2. Concentrations of selected elements in the ICSA interferents only solution, showing that the analytes are present within the ± 2x CRDL limits for elements with CRDL < 10 ug/L over 10 hours.

Figure 3 plots the percentage recovery for the ICSAB over a 10 hour analysis period at the required frequency. Of particular interest are the results for aluminium. The Al concentration in the ICSA and ICSAB was 500 mg/L but it can be seen from Figure 3 that excellent recoveries of between 105–110% were obtained over the entire analysis period. These results indicate the excellent stability of the Vista ICP-OES at high concentrations with the axially viewed plasma.

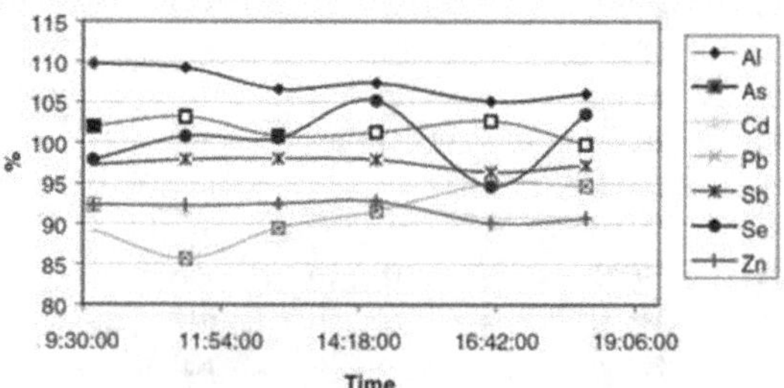

Figure 3. Percentage recoveries of selected elements in the ICSAB solution in the presence of the interferents over 10 hours are within the ± 20% limits. Note the recovery of the Al interferent which has a concentration of 500 mg/L.

6

FIGURE 18.10 (Continued)

Contract Required Detection Limit Test for ICP (CRI)

According to ILM04.0/05.0, to 'verify linearity near the CRDL' the laboratory is required to analyze a standard at a concentration of 'two times the CRDL or IDL whichever is greater' at the beginning and end of each sample analysis run of up to 20 samples [1]. The CRI is measured after the ICV but before the ICS. The limits to be applied to this test are not specified in the SOW [1], however the results for the analytes of interest are plotted over the 10 hour analysis period in Figure 4.

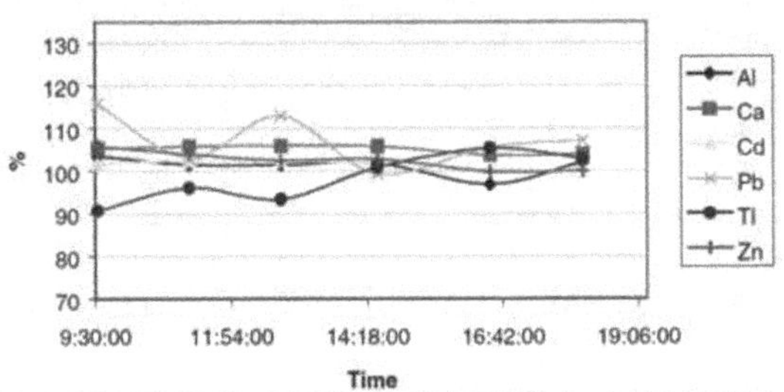

Figure 4. Percentage recoveries of the contract required detection limit test solution for ICP (CRI) for selected elements over 10 hours.

Duplicates and Spike Sample Analysis

In addition to the above tests, Duplicates and Spike Sample Analyses are required. A duplicate pair is created by processing two aliquots of the same sample through the sample preparation procedures. The Relative Percent Difference (RPD) between the original and duplicate sample is then calculated as:

$$RPD = \frac{|\text{Sample - Duplicate Conc}| \times 100}{(\text{Sample + Duplicate Conc})/2}$$

One duplicate pair must be analyzed for each batch of 20 samples. A control limit of 20% RPD is applied to the duplicate pair for concentrations greater than 5 × CRDL. For concentrations less than this value the limit of ± CRDL is applied to the difference in the concentrations. If one result of the pair is below the 5 x CRDL limit, the ± CRDL limit is applied. If both results are less than the IDL, the RPD is not reported. The RPD of the duplicate experiment should indicate any problems due to analyte losses or contamination during the sample preparation process. Although the US EPA does not recommend an action under ILM04.0/05.0, it is generally accepted that the analysis needs to be stopped and the problem corrected before proceeding. As a result, the duplicate pair is usually analyzed at the beginning of the sample batch rather than at the end. See Table 5 for some typical duplicate results using an NIST reference as a sample.

A Spiked Sample Analysis (SSA) is performed to assess the effect of the sample matrix on analyte recoveries.

A known amount of analytes is spiked into the sample prior to digestion and the spiked sample (sometimes known as a Matrix Spike or under Method 200.7 a Lab Fortified Matrix (LFM)) is then processed through the sample preparation procedures. One SSA is required per Sample Delivery Group (batch of up to 20 samples) according to ILM04.0/05.0 and a LFM is required at a frequency of 10% according to Method 200.7. The percent recovery of the analytes is calculated and compared to the control limits 75–125% (ILM04.0/05.0) or 70–130% (Method 200.7).

If analytes are outside these limits the sample results in the SDG must be flagged and under some circumstances a post digestion spike of the affected samples may be required, see ILM04.0/05.0 Exhibit E-6.

See Table 5 for some typical SSA results.

FIGURE 18.10 (Continued)

Table 5. NIST Water Sample 1643d

Element	1643d (mg/L)	Duplicate (mg/L)	QC spike (mg/L)	1643d certified (mg/L)	% Recovery 1643d (LCS)	RPD duplicate) %	QC spike concentration	% spike recovery
Al 308.215	0.141	0.128	2.074	0.1276	111.3	9.9	2	96.6
As 188.980	0.047	0.048	2.032	0.056	96.1	2.6	2	99.2
B 249.772	0.134	0.134	0.128	0.144	103.6	0.5		
Ba 585.367	0.504	0.501	2.659	0.506	110.1	0.5	2	107.8
Be 313.042	0.011	0.011	0.062	0.0125	100.0	0.1	0.05	101.3
Ca 315.887	28.896	28.311		31.04	101.3	2.0		
Cd 226.502	0.006	0.005	0.056	0.00647	94.3	0.7	0.05	101.5
Co 228.615	0.023	0.023	0.537	0.025	103.1	0.1	0.5	102.7
Cu 327.395	0.019	0.019	0.285	0.0205	104.3	2.9	0.25	106.6
Fe 259.940	0.096	0.092	1.196	0.0912	112.1	4.7	1	110.0
4Mg 383.829	7.337	7.356		7.9889	102.3	0.3		
Mn 261.020	0.030	0.028	0.543	0.037	84.1	6.9	0.5	102.6
Mo 202.032	0.104	0.105	1.131	0.1129	103.4	1.2	1	102.7
Na 589.592	23.386	23.338	-	22.07	117.5	0.2		
Ni 231.604	0.053	0.054	0.551	0.058	103.2	0.8	0.5	99.6
Pb 220.353	0.016	0.017	0.533	0.0181	102.0	4.9	0.5	103.3
Sb 206.834	0.038	0.038	0.036	0.054	78.7	0.8		
Si 251.611	2.782	2.784	2.816	2.7	114.6	0.1		
Sr 430.544	0.249	0.249	0.243	0.294	94.2	0.0		
Tl 190.794	0.008	0.009	2.039	0.00728	136.8	7.9	2	101.5
V 311.837	0.035	0.035	0.577	0.035	112.6	0.5	0.5	108.4
Zn 206.200	0.063	0.063	0.560	0.072	97.2	0.1	0.5	99.4

Laboratory Control Sample (LCS)

The LCS is an ILM04.0/05.0 required check standard obtained from the US EPA which is processed through the same preparation procedures as the samples1. If an LCS is not available from the EPA, an ICV may be used (ILM04.0 page E-25). One LCS must be analyzed in each sample delivery group. The analytes must be recovered within limits of 80–120% of the true value for the analytes or the analysis must be terminated and the samples in that SDG redigested and reanalyzed.

In this work, a NIST Water Sample 1643d was used for the Duplicate, SSA and LCS tests. The reported values obtained in Table 5 for 1643d and the duplicate were obtained after a dilution of 9ml of sample to a final volume of 10 mL. The LCS percentage recovery, RPD of the duplicate and the percentage recovery of the SSA on NIST 1643d were all within specification.

Other Protocol Tests

The US EPA protocols also demand other tests that have not been discussed here in detail. These include Serial Dilutions, Preparation Blanks and Initial and Continuing Calibration Blanks. According to ILM0.40 [1] a serial dilution experiment consists of a five fold dilution of each sample type (water, soil). The diluted result is then compared to the original undiluted result and after allowing for dilution the results must agree within ± 10%. The analytes tested in the sample must be at concentrations of 50 times the IDL (Instrument Detection Limit) or higher. If the serial dilution is out of control for an analyte(s), the sample results associated with that serial dilution must be flagged in the laboratory reports. This may indicate that chemical or physical interferences are occuring for the sample type. A Preparation Blank is an aliquot of deionized, distilled water which has been processed through the sample preparation and analysis process [1]. The presence of contaminants in Preparation Blank is therefore indicative of contamination problems in the sample preparation process. One Preparation Blank (PBLK) is required per sample delivery group. If the absolute found concentration of the PBLK is less than or equal to the CRDL for any analyte no action is required. If any analyte in the PBLK is found at a concentration greater than the CRDL or less than the negative CRDL the lowest concentration of any sample in the SDG for that analyte must be 10 times the CRDL. Otherwise the samples must be redigested and reanalyzed and the sample concentration is not corrected for the blank value.

A typical Vista worksheet for this analysis is shown in Figure 5.

FIGURE 18.10 (Continued)

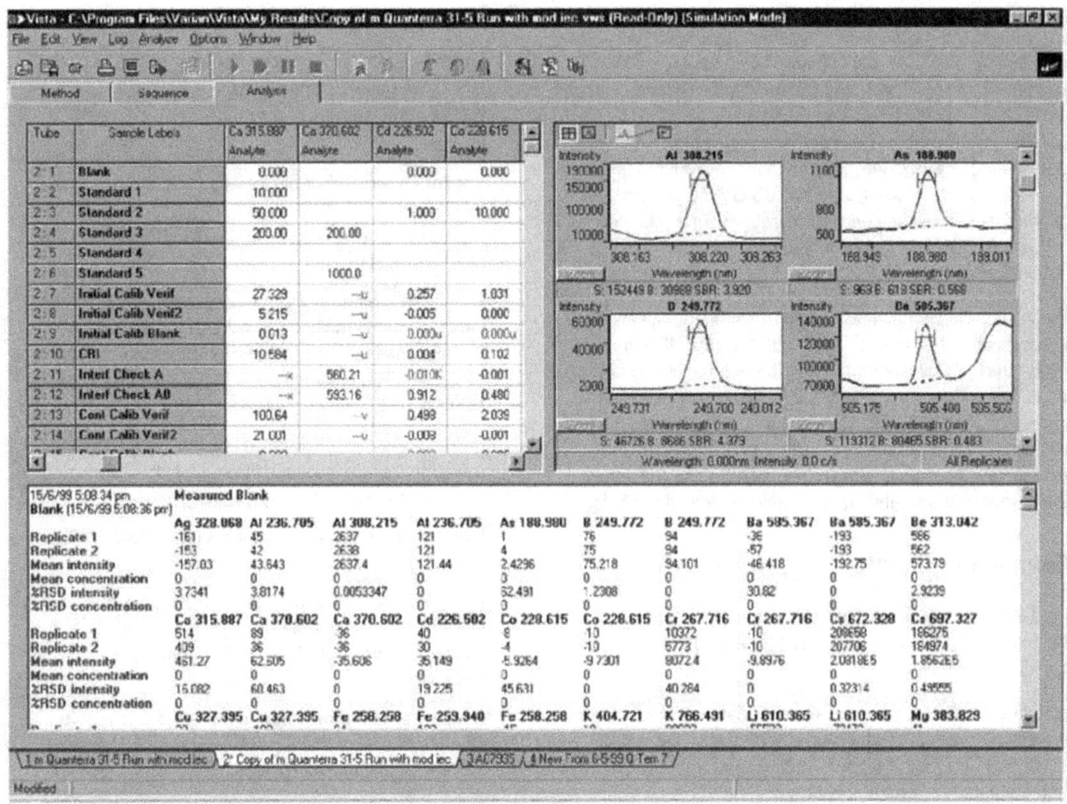

Figure 5. A typical analysis worksheet for the analysis of waters and waste waters showing two wavelengths for Ca spanning 0–1000 mg/L and the basic EPA start up sequence - calibration, ICV, ICB, CRI.

Speed of Analysis

Meeting the requirements of the US EPA protocols is time consuming due to the high overhead of quality controls required to analyze a batch of samples and the large number of variables that must be monitored to ensure compliance. The Vista simultaneous ICP-OES has been shown to meet all the rigorous US EPA specifications with a sample analysis time of 2.5 mins/sample, including a rinse time of 40 seconds per sample, two 30 second integrations plus sample delay and stabilization times. The use of dilute nitric acid and Triton X-100 in the rinse solution was found to be useful in aiding the rapid rinse out of the system.

Conclusion

In this work, a universal simultaneous ICP-OES method meeting US EPA environmental regulations has been developed for waters and waste-waters. The Agilent Vista ICP-OES provides the unique advantage of being able to achieve this task in a single analysis from an axially viewed plasma system. This avoids the time delay and costs related to repeating analyses via either other techniques or by using dual viewed ICP-OES systems.

Long term stability over 10 hours was extremely good with the general long term precision of 1% for most elements and a maximum of 2%. This was also established by noting that the CCV recoveries over the 10 hour period were all within specified limits.

FIGURE 18.10 (Continued)

Vista's MultiCal provides the benefit of combining wavelengths to cover a wider linear dynamic range from low parts per billion to high parts per million. The suitability of this approach was proved with the CRI and Linear Dynamic Range tests. The ability to choose any wavelength from the VistaChip CCD facilitates use of this extended linear dynamic range and also allows flexibility to choose wavelengths to avoid interferences. Using standard US EPA conditions, it was shown that successful compliance with the Interference Correction Standards tests could be achieved over the 10 hour period.

The Vista software provides complete automation of all the protocols identified in this article with the capability to customize these QC protocols to meet the requirements of other regulatory bodies other than the US EPA. In this work, full automatic compliance with the required US EPA protocols was possible without need for further customization. The Vista ICP-OES has been shown to meet all of the regulatory requirements in a single, fast and fully automated analysis.

Acknowledgments

The authors would like to thank Quanterra Environmental Services staff (CA, CO and TN USA) for their early insight into the US EPA protocols.

In addition to Reference 8, thanks to Christine Rivera, Patrick Simmons, Steven Eckroat, Douglas Shrader, Michelle Cree and John Dabritz all of Agilent Technologies, Inc. USA for their input and reviews of this article.

References

1. US EPA Contract Laboratory Program, Statement of Works for Inorganics, Multi-Media, Multi-Concentration, Document Number ILMO 4.0.
2. ILM05.0 documentation was placed at the following website for review. This website was current at the time of publishing: http://epa.gov/oamsrpod/pollard/hq9915 909/ilm050c.pdf.
3. US EPA Washington, DC. Determination of Metal and Trace Elements in Water and Wastes by Inductively Coupled Plasma-Atomic Emission Spectrometry- Method 200.7, US EPA, Environmental Monitoring and Support Laboratory, Cincinnati, OH, May 1994.
4. SW-846- Inductively Coupled Plasma- Atomic Emission Spectrometry- Method 6010B, U.S.E.P.A. Washington, DC December 1996.
5. Private communication to Michael Knowles, October 1999.
6. A. P. M de Win, How to Improve Wettability in Cloud Chambers for Use in the Inductively Coupled Plasma, Journal of Analytical Atomic Spectrometry, Vol. 3, April 1988, 487.
7. D. Johnson, Determination of Metals in a 3% sodium chloride Matrix by Axially-viewed ICP-AES, Varian at Work #19, December 1996.
8. Data supplied to Michael Knowles courtesy of Douglas Shrader, North American Atomic Spectroscopy Marketing Manager (Wood Dale, IL), John Dabritz Atomic Spectroscopy Product Specialist Western Region (Walnut Creek CA) and Michelle Cree Atomic Spectroscopy Specialist North East Region, all of Agilent Technologies Inc, USA.

10

FIGURE 18.10 (Continued)

ILM04.0/05.0	Method 200.7	Method 6010B	Brief definition of function according to ILM04.0/05.0
Analytical Sample	Sample - this method does not clearly indicate counting rules	Sample - this method does not clearly indicate counting rules	Everything other than the ICV, ICB, CCV, CCB, calibration standards and calibration blank.
Sample Delivery Group (SDG)	Group of 20 samples	Not clearly stated	A unit within a sample case that is used to identify a group of samples for delivery. An SDG is a group of 20 or fewer real samples within a case. Note that since sample deliveries will not include Preparation Blanks, ICSA and ICSAB and other solutions that are counted as Analytical Samples, the number of Analytical Samples for an SDG may be higher than 20.
Contract Required Detection Limit (CRDL)	No equivalent	Estimated IDLs are provided but not mandated.	The detection limits for each of the 22 elements that must be met to comply with the Statement of Work or Method.
Instrument Detection Limit (IDL)	Method detection limits (MDL) - 7 replicates of fortified reagent water at 2-3 x instrument detection limit	Method detection limits (MDL) - at 3-5 times anticipated detection limits - 7 replicates on 3 non-consecutive days 'for additional confirmation'.	The detection limit calculated by multiplying by 3 the average of the standard deviations obtained on 3 nonconsecutive days from a standard solution at a concentration of 3-5 times the instrument manufacturers suggested IDL. Seven consecutive measurements are taken to define the standard deviation under the same conditions as the proposed analytical method.
Initial Calibration Verification (ICV)	See IPC below.	ICV	Used to initially verify the validity of the calibration by measuring the recovery of analytes in this standard immediately after calibration. A second source standard.
Initial Calibration Blank (ICB)	Calibration blank	Calibration blank	A "calibration blank" that immediately follows the ICV. Usually compared to the Instrument Detection Limit (IDL).
Continuing Calibration Verification (CCV)	Initial performance check (IPC) standard - combines both ICV and CCV - first time limits are + 5%	CCV - accompanied by calibration blank - can use ICV instead - ± 10% limits plus < 5 %RSD	Used to verify the validity of the calibration on an on-going basis. This standard is measured at a frequency of every 10 'analytical samples'. Recovery limits + 10%
Continuing Calibration Blank (CCB)	Calibration blank	Calibration blank	A 'calibration blank' that immediately follows the CCV. Usually compared to the Instrument Detection Limit (IDL).
Contract Required Detection Limit Test for ICP (CRI)	No equivalent	No equivalent	A standard containing the analytes at 2 times the CRDL or IDL whichever is greater. No control limits yet applicable. Not required for Al, Ba, Ca, Fe, Mg, Na and K. Analyzed at the beginning and end of the SDG.
Interference Check Solution (ICSA)	Spectral interference check (SIC) solutions - up to 17 interferent solutions tested - test for 10% error in baseline	Interference check sample - analytes at 0.5-1.0 mg/L - interferents at 100 mg/L of Al, Ca, Cr, Cu, Fe, Mg, Mn, Ni, Ti and V are shown in Table 2 of the method. - test for 20% error in baseline	Used to check the effect of interferents on the determination of the analytes. ICSA is the interferents only solution and contains Al, Ca and Mg at 500 mg/L and Fe at 200 mg/L. Analytes should be within 0 ± 2 x CRDL. Analyzed at the beginning and end of the SDG. Note the 20% error limit in the 6010B method probably refers to ± 10% since otherwise the method is exactly the same as 200.7.
Interference Check Solution (ICSAB)	See above.	See above.	Used to check the effect of interferents on the determination of the analytes. ICSAB contains both interferents and 16 selected analytes. Analytes must be recovered within ± 20%. Analyzed at the beginning and end of the SDG.
Preparation Blank (Prep Blk)	Lab reagent blank (LRB)	Method blank	Also known as a Reagent Blank. A volume of deionized distilled water processed through the sample preparation procedure. Analytes are then monitored versus the CRDL. In ILM04.0/05.0, no preparationblank correctionof samples is done. One per SDG.
No equivalent	Lab fortified blank (LFB) - an aliquot of LRB which has been spiked with analytes - one per SDG	No equivalent	
Lab Control Sample (LCS)	Quality control sample (QCS) - 3 analyses to within + 5% recovery	No equivalent	A control standard sourced from the EPA or another independent source. The LCS is put through the sample preparation process and then the recovery of the analyes is calculated and compared to ± 20% limits. One per SDG.

Continued on next page.

11

FIGURE 18.10 (Continued)

ILM04.0/05.0	Method 200.7	Method 6010B	Brief definition of function according to ILM04.0/05.0
Duplicate (Dup)	Laboratory duplicates (LD1 and LD2)	Matrix spiked duplicate Samples - measure 2 duplicates of a spiked sample - RPD to 20% - and spike recovery to ± 25%	A duplicate aliquot of a sample, put through the sample preparation procedure. Acts as a monitor for contamination and losses during sample preparation. The Relative Percent Difference between the duplicate and the sample is calculated and compared to ± 20% limits or CRDL limits. One per SDG. Note the Matrix Spiked Duplicate of 6010B tries to combine both the Matrix Spike and Duplicate tests into one but most ICP software is currently not designed to handle this combination.
Serial Dilution (Ser)	Dilution test	Dilution test - do for new or unusual matrices	Conduct a five fold dilution on one sample from each SDG. The calculated result after correction for dilution must agree within ± 10% of the undiluted sample result.
Spiked Sample Analysis (SSA) or Matrix Spike (MS)	Lab fortified matrix (LFM) - spike every 10 % of samples - recover to ± 30%	No equivalent (see PDSA)	A spike is added to a sample prior to the digestion or sample preparation procedures. The recovery of the spike is calculated and compared to ± 25% limits. A matrix spike is required for each SDG.
Linear Range Analysis (LRA)	Linear dynamic range (LDR) - verified annually - top standard recovered to - 10% limit - dilute all samples that are more than 90% of the LDR	LDR - as per 200.7	A standard which is analyzed quarterly to confirm the linearity of analytical calibrations. A high level standard must be recovered within 5% of the true value. This defines the upper limit of the linear dynamic range.
Post Digestion Spike - only required if the SSA fails	Analyte addition test - spike at 20-100 times the MDL - recover to ± 15% limits	Post digestion spike addition (PDSA) - do for new or unusual matrices - spike at 10-100 times the MDL - recover to ± 25% limits	Post digestion spikes are often used to assess whether the Method of Standard Additions is required. This is done separately from pre-digestion spikes because the pre¬digestion spike might be indicative of contamination picked up during the sample preparation procedure.

12

FIGURE 18.10 (Continued)

Typical Analysis Orders For US EPA Methods

#	Analytical sample count ILM04.0/05.0	ILM04.0/05.0	Method 200.7	6010B
2	0	Samples 20	Samples 20	Samples
1	0	Calibration blank	Calibration blank	Calibration blank
4	0	Standards	Standards*	Standards
5	0	ICV	Initial performance check (IPC)	ICV
6	0	ICB	Calibration blank	Calibration blank
7	1	CRI	Lab reagent blank method	Blank
8	2	ICSA	Lab fortified	Blank CCV
9	3	ICSAB	S1(Lab duplicate 1)	Calibration blank
10	0	CCV	S1(Lab duplicate 2)	Sample 1
11	0	CCB	S1(Lab fortified matrix)	S1 D1 (Matrix spike duplicate 1)
12	1	Prep	Blank S1 (Dilution test)	S1 D2 (Matrix spike duplicate 2)
13	2	LCS	S1 (Analyte addition test)	S1 Post digestion spike
14	3	Sample 1	Sample 2	S1 Dilution test
15	4	S1 DUPLICATE	Sample 3	Sample 2
16	5	S1 SPIKE	Sample 4	Sample 3
17	6	S1 DILUTION	IPC	Sample 4
18	7	Sample 2	Cal Blk	Sample 5
19	8	Sample 3	Sample 5	Sample 6
20	9	Sample 4	Sample 6	CCV
21	10	Sample 5	Sample 7	CCB
22	0	CCV	Sample 8	Sample 7
23	0	CCB	Sample 9	Sample 8
24	1	Sample 6	Sample 10	Sample 9
25	2	Sample 7	Sample 11	Sample 10
26	3	Sample 8	Sample 12	Sample 11
27	4	Sample 9	Sample 13	Sample 12
28	5	Sample 10	Sample 14	Sample 13
29	6	Sample 11	IPC	Sample 14
30	7	Sample 12	Cal Blk	Sample 15
31	8	Sample 13	Sample 15	Sample 16
32	9	Sample 14	S 15 (LFM)	CCV
33	10	Sample 15	Sample 16	CCB
34	0	CCV	Sample 17	Sample 17
35	0	CCB S	ample 18	Sample 18
36	1	Sample 16	Sample 19	Sample 19
37	2	Sample 17	Sample 20	Sample 20
38	3	Sample 18	IPC	CCV
39	4	Sample 19	Cal Blk	CCB
40	5	Sample 20		
41	6	CRI		
42	7	ICSA		
43	8	ICSAB		
44	0	CCV		
45	0	CCB		
	Efficiency	44.4%	51.3%	51.3%

* *It is assumed that the Initial Demostration of Performance, including annual demonstration of LDR and MDLs has been done prior to analysis*

* *It is assumed that the 17 Spectral Interference Check (SIC) standards have been analysed prior to the instrument calibration. Only those correction factors exceeding certain criteria need be tested daily*

* *Note that a QCS (Quality Control Sample) is required by 200.7 on a quarterly basis - but since this is not a daily operation it is not indicated in the analysis list above*

13

FIGURE 18.10 (Continued)

For More Information

For more information on our products and services, visit our Web site at www.agilent.com/chem

www.agilent.com/chem

Printed in the USA
November 1, 2010
ICPES-29

FIGURE 18.10 (Continued)

19

ATOMIC MASS SPECTROMETRY

19.1 Introduction
19.2 Low-Resolution ICP-MS
 19.2.1 The PerkinElmer NexION® 350 ICP-MS
 19.2.2 Interface and Quadrupole Ion Deflector (QID)
 19.2.3 The Collision/Reaction Cell
 19.2.4 Quadrupole Mass Filter
19.3 High-Resolution ICP-MS

19.1 INTRODUCTION

Another instrumental application using the inductively coupled plasma source is the addition of a mass spectrometer as the detector. This type of instrument is called an inductively coupled plasma mass spectrometer (ICP-MS). In this instrument, the plasma torch is used to ionize the elements allowing their gas-phase separation and measurement using mass analyzers. Later in Chapters 27–29, we will be taking a detailed look at mass spectrometry. For a mass spectrometer to measure something such as a molecule or element, the species must be in an ionized state. This is where the plasma torch comes in; it is hot enough to ionize elements. This allows the measurement of the elements. The mass spectrometer measures the elements according to their mass-to-charge ratio (*m/z*). The elements ionize primarily in the plus one (+1) ionization state; thus, the mass-to-charge ratio (*m/z*) measured is usually the atomic mass of the element. There are two types of ICP-MS instruments that are typically encountered in today's laboratories, low resolution and high resolution (Table 19.1).

19.2 LOW-RESOLUTION ICP-MS

19.2.1 The PerkinElmer NexION® 350 ICP-MS

The ICP-MS sample introduction using the peristaltic pump to move the sample solution into the nebulizer producing the fine droplets, followed by the spray chamber, is basically the same as the ICP–OES instrumentation. The plasma torch used in ICP-MS instrumentation is also the same as that used with ICP–OES. The ion optics, mass analyzers, and detectors after the sample introduction into the instrument from the plasma torch are where the ICP-MS instrumentation differs from the ICP–OES. Let's take a look at the PerkinElmer NexION® 350 ICP-MS instrument, illustrated in Figure 19.1(a). The inner workings of the PerkinElmer NexION® 350 ICP-MS instrument are illustrated in Figure 19.1(b). Another example is the Agilent 7900 Series ICP-MS, which also uses a quadrupole mass analyzer, illustrated in Figure 19.2(a). The inner components that make up the instrument are illustrated in Figure 19.2(b).

19.2.2 Interface and Quadrupole Ion Deflector (QID)

The interface is the introduction area of the instrument from the plasma torch into the lower pressure inside of the ICP-MS. The interface in Figure 19.1(b) is the triple cone interface where the ions produced from the plasma torch at atmospheric pressure enter the instrument where the environment is under vacuum pumped down by vacuum pumps. Keeping the system under vacuum reduces interferences from ambient gasses and helps keep the system dry. The vacuum is controlled by two pumps, a mechanical rough pump and a turbomolecular pump. The ions produced in the torch and neutrals enter the instrument into the quadrupole ion deflector (QID). The QID functions to remove interferences and allow ions to continue on into the interior of the instrument. The QID is illustrated in Figure 19.3. The QID is a quadrupole that has been set at a right angle to the incoming ion beam. This alignment will bend the ion beam at a 90° angle. The neutrals and solvent will continue on their straight path and not enter into the instrument.

19.2.3 The Collision/Reaction Cell

Following the QID in the configuration of the NexION® 300 ICP-MS is the collision/reaction cell comprised of a single

Analytical Chemistry: A Chemist and Laboratory Technician's Toolkit, First Edition. Bryan M. Ham and Aihui MaHam.

TABLE 19.1 Comparison of ICP Techniques[a].

Technique	Metal	Sensitivity	Advantages	Disadvantages
ICP–OES	Most metals and nonmetals	ppt	Multiple elements measured simultaneously	Interferences
ICP-MS	Most metals and nonmetals	ppb to ppm	Multiple elements measured simultaneously	Interferences
GFAA	Most metals	ppt	Few interferences	One element at a time
Hydride AA	Elements that form hydrides (Sb, Te, As, Se, Pb, Tl)	ppt to ppb	Few interferences with high sensitivity	One element at a time, takes time
Cold vapor for mercury	Hg	ppt	Few interferences with high sensitivity	Only single element analyzed, takes time

[a] © Agilent Technologies, Inc. 2014, Reproduced with Permission, Courtesy of Agilent Technologies, Inc.

(a)

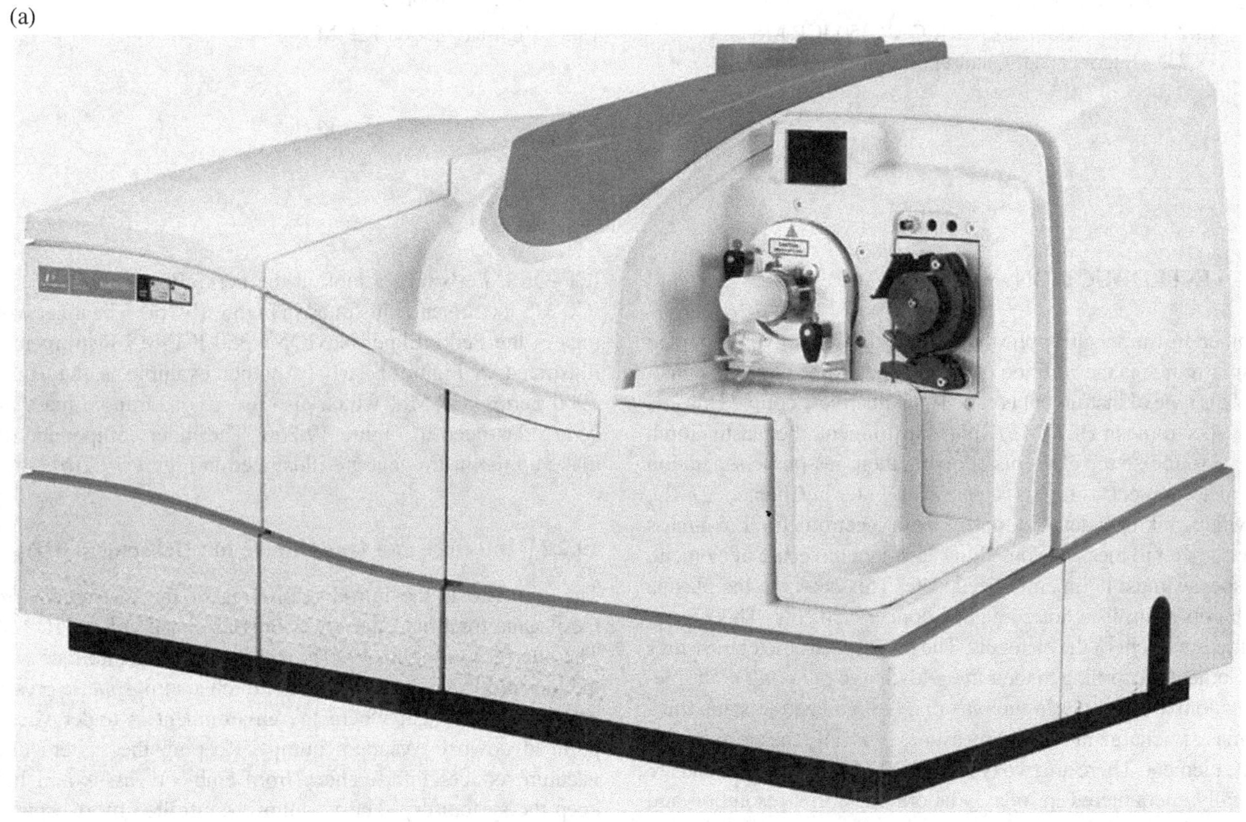

(b)

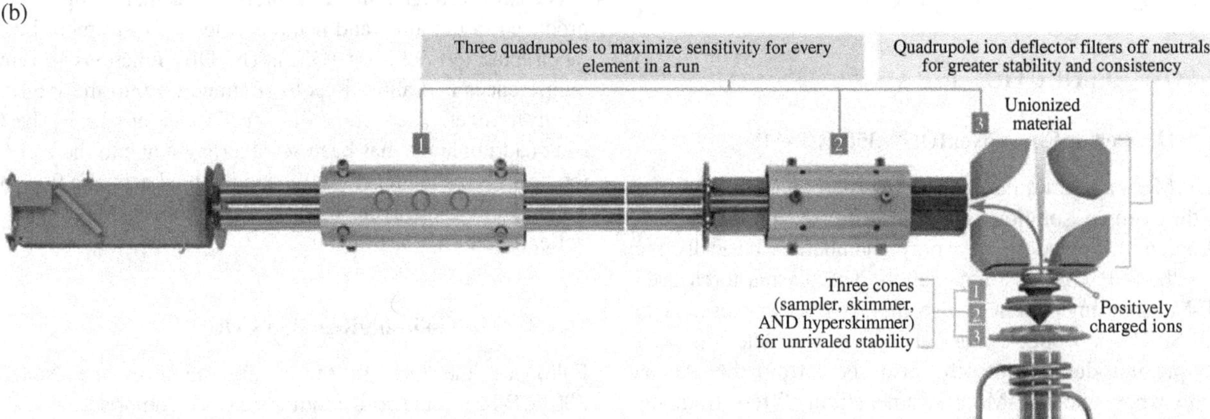

FIGURE 19.1 (a) The PerkinElmer NexION® 350 ICP-MS instrument. (b) The inner workings of the ICP-MS instrument. (Reprinted with Permission from PerkinElmer.)

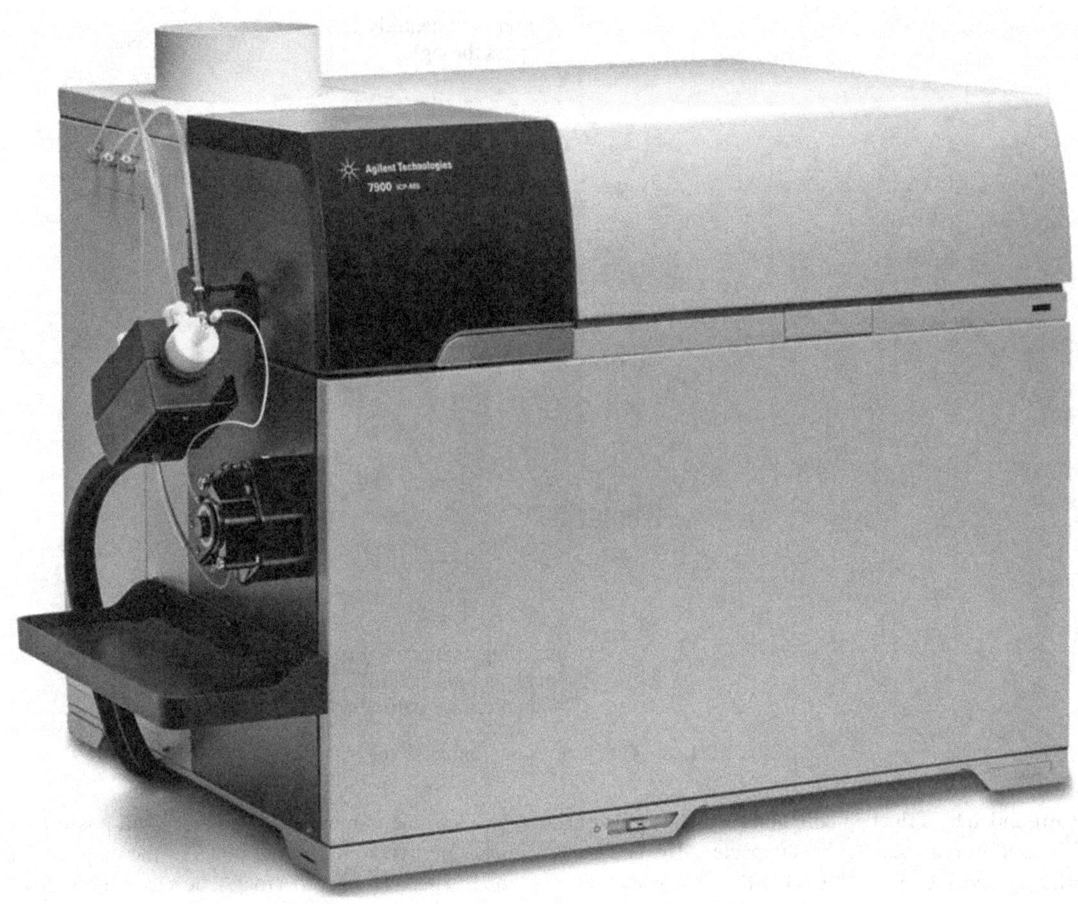

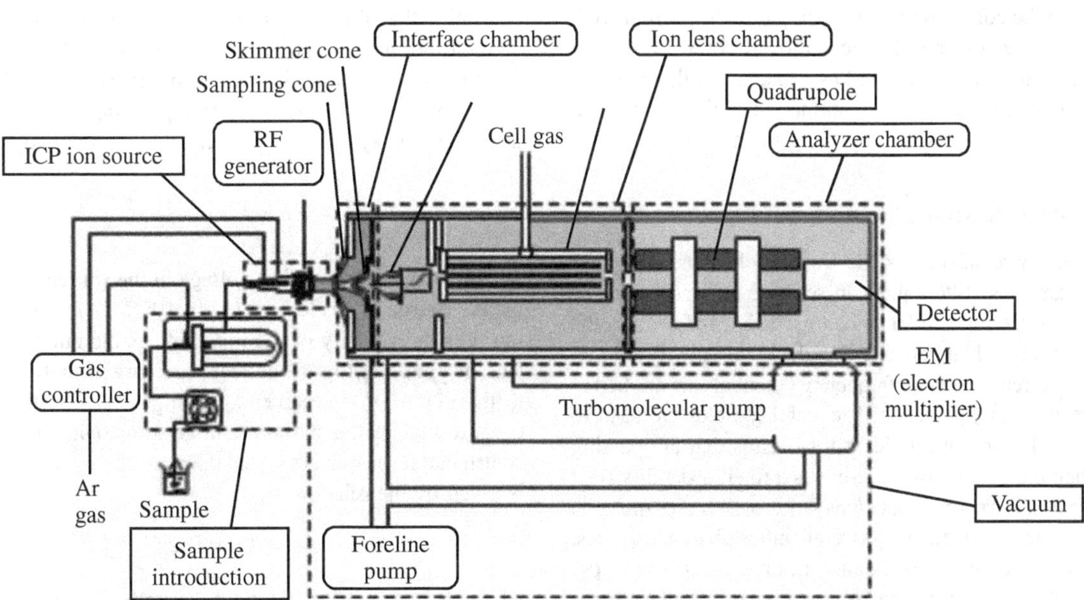

FIGURE 19.2 (a) The Agilent 7900 ICP-MS. (b) Inner components that make up the instrument including the quadrupole mass analyzer. (© Agilent Technologies, Inc. 2014, Reproduced with Permission, Courtesy of Agilent Technologies, Inc.)

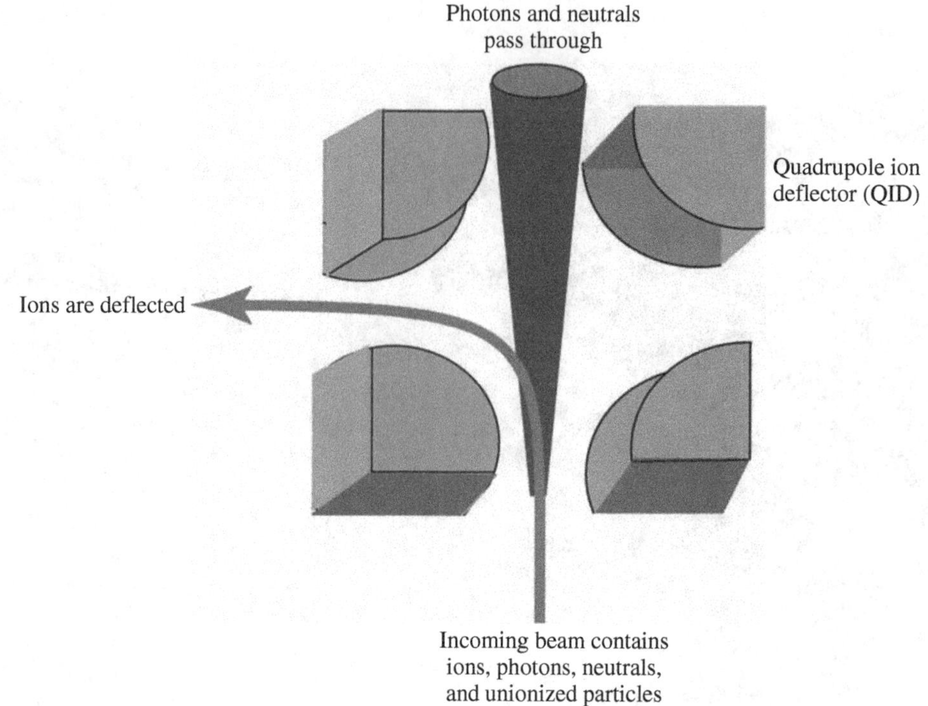

FIGURE 19.3 The quadrupole ion deflector (QID).

quadrupole surrounded by a floating collision cell. The collision/reaction cell is used to remove interfering species. The cell can remove interfering species in a collision cell mode where the cell is filled with a stationary collision gas. The interfering species are larger than the ionized elements and collide more frequently with the collision gas, thus being effectively removed. The interfering species can also be removed when the cell is acting as a reaction cell. When used as a reaction cell, a reactive gas is filled into the cell such as ammonia, which reacts and neutralizes the interfering species making them unstable.

19.2.4 Quadrupole Mass Filter

Following the universal cell is the quadrupole mass analyzer. Quadrupoles are mass filters that can separate ions according to their mass-to-charge ratio (*m/z*). When used as a mass filter, the ion beam is introduced tangentially into the quadruple. Combinations of direct current and radio frequency potentials are applied to the poles creating stability fields. The stability fields are used to filter masses and scan the ion beam for masses that are passing through. When the quadrupole is acting as a filter, a stability field is created that allows only one mass-to-charge ratio (*m/z*) to pass through effectively filtering out all other charged species. In ICP-MS, the quadrupole is scanned at rates of 5000 *m/z* per second. This allows the simultaneous measurement of all elements that are passing through the quadrupole mass analyzer.

19.3 HIGH-RESOLUTION ICP-MS

High-resolution ICP-MS instrumentation incorporates the use of an electric–magnetic sector mass spectrometer. When charged particles enter a magnetic field, they will possess a circular orbit that is perpendicular to the poles of the magnet. This phenomenon has been applied to a magnetic sector mass analyzer, which is a momentum separator. Figure 19.4 illustrates the basic design of the magnetic sector mass analyzer.

Notice that the slits are normally placed collinear with the apex of the magnet. The ions enter a flight tube (first field-free region) from a source through a source exit slit and travel into the magnetic field. The accelerating voltage in the source will determine the kinetic energy (KE) that is imparted to the ions:

$$KE = zeV = \frac{1}{2}mv^2 \tag{19.1}$$

where V is the accelerating voltage in the source, e is the fundamental charge of an electron (1.60×10^{-19} C), m is the mass of the ion, ν is the velocity of the ion, and z is the number of charges. The magnetic field will deflect the charged particles according to the radius of curvature of the flight path (r) that is directly proportional to the mass-to-charge ratio (*m/z*) of the ion. The centripetal force that is exerted upon the ion by the magnetic field is given by the relationship

$$F_M = Bzev \tag{19.2}$$

where B is the magnetic field strength. The centrifugal force acting upon the ion from the initial momentum is given by

$$F_c = \frac{mv^2}{r} \tag{19.3}$$

For the ion to reach the detector, the centripetal force acting upon the ion from the magnet (down-pushing force) must equal

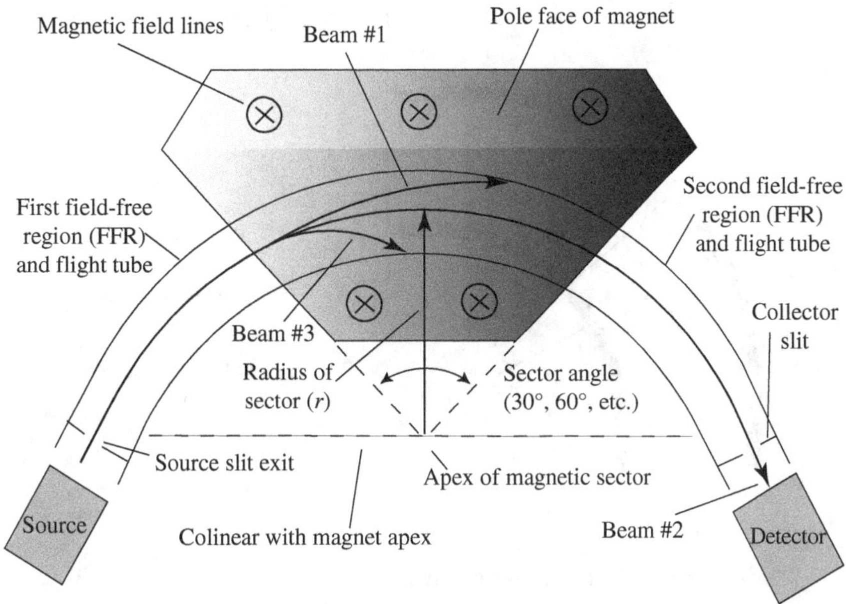

FIGURE 19.4 Basic components of a magnetic sector mass analyzer. The magnetic field lines are directed into the plane of the paper for negative ion mode which follows the right hand rule. For positive ion mode the polarity is switched and the magnetic field lines are directed out of the plane of the paper.

the centrifugal force acting upon the ion from the initial momentum of the charged ion (upwardly pushing force). This equality is represented by

$$F_M = F_C \tag{19.4}$$

$$Bzev = \frac{mv^2}{r} \tag{19.5}$$

where B is the magnetic field strength, e is the electron fundamental charge, m is the mass of the ion, v is the velocity of the ion, z is the number of charges, and r is the radius of the curvature of the sector. Solving Equation 19.5 for v and substituting into Equation 19.1, we can derive the relationship between the mass-to-charge ratio (m/z) of the ion and the force of the magnetic field strength

$$Bzev = \frac{mv^2}{r} \tag{19.5}$$

$$v = \frac{Bzer}{m} \tag{19.6}$$

$$KE = zeV = \frac{1}{2}mv^2 \tag{19.1}$$

$$zeV = \frac{1}{2}m\left(\frac{Bzer}{m}\right)^2 \tag{19.7}$$

$$zeV = \frac{1}{2}m\frac{B^2z^2e^2r^2}{m^2} \tag{19.8}$$

$$V = \frac{1}{2}\frac{B^2zer^2}{m} \tag{19.9}$$

$$\frac{m}{z} = \frac{B^2er^2}{2V} \tag{19.10}$$

This relationship is often referred to as the scan law. In normal experimental operation, the radius of the sector is fixed, and the accelerating voltage V is held constant, while the magnetic field strength B is scanned. From Equation 19.10, it can be seen that a higher magnetic field strength equates to a higher mass-to-charge ratio stability path. The magnetic field strength B is typically scanned from lower strength to higher strength for m/z scanning in approximately 5 s. In exact mass measurements and when calibrating the magnetic sector mass analyzer, the magnetic field strength B of the magnet is kept fixed, while the accelerating voltage is scanned. This is done to avoid the slight error that is inherent in the magnet known as historesis where after the magnet has scanned its original strength is slightly changed from its initial condition before the scanning was commenced. Scanning the accelerating voltage does not have this problem. The calibration of the magnetic sector mass analyzer is a nonlinear function; therefore, several points are often required for calibration. In calibrating, a standard mass is scanned by scanning the acceleration voltage and holding the magnetic field strength constant. When the mass is detected, the values are set for that set of acceleration voltage versus magnetic field strength. The resolution of the magnetic sector mass analyzer is directly proportional to the magnet radius and the slit width of the source exit slit and the slit width of the collector slit:

$$\text{Resolution} \propto \frac{r}{S_1 + S_2} \tag{19.11}$$

where:

r = radius of the magnet
S_1 = slit width of the source exit slit
S_2 = slit width of the collector slit

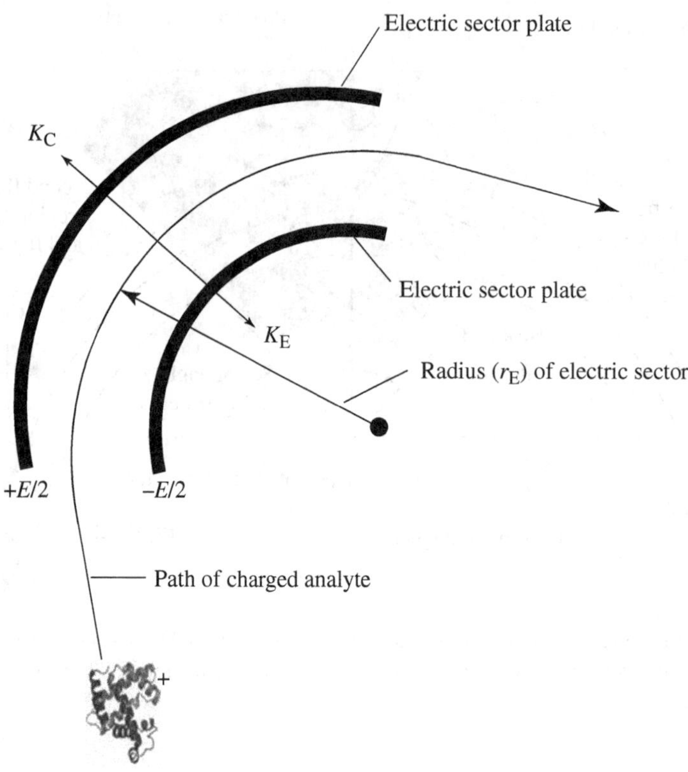

FIGURE 19.5 Path of charged particle through the electric sector energy filter. Forces acting upon the particle include the centrifugal force (K_C) and the deflecting force (K_E).

It can be seen from the relationship in Equation 19.11 that a larger magnet radius and smaller slit widths will equate to a larger resolution value. However, decreasing the slit widths causes a loss in sensitivity, as the amount of ions allowed to pass through will decrease. The typical resolution obtained under normal scan mode for a magnetic sector mass analyzer is ≤5000. This means that the magnetic sector mass analyzer can separate an m/z 5000 molecular ion from an m/z 5001 molecular ion, an m/z 500.0 molecular ion from an m/z 500.1 molecular ion, an m/z 50.00 molecular ion from an m/z 50.01 molecular ion, and so on. Two factors that limit the mass resolution of the magnetic sector mass analyzer are angular divergence of the ion beam and the kinetic energy spread of the ions as they leave the source. Decreasing the source exit slit width will reduce the angular divergence effect upon the resolution; however, this will also decrease the sensitivity of the mass analyzer. The spread in the kinetic energy of the ions as they leave the source is caused by slight differences in the initial velocity of the ions imparted to them by the source. In EI ionization, the initial kinetic energy spread is approximately 1 to 3 eV. To address the kinetic energy spread distribution in a magnetic sector mass analyzer, an electric sector that acts as a kinetic energy separator is used in combination with the magnetic sector. The electric sector shown in Figure 19.5 is used as an energy filter where only one energy of ions will pass through the electric field. The electric sector will only pass ions that have the exact energy match as that with the acceleration voltage in the source. In combination with slits, the electric sector can be called an energy focuser. The electric sector is constructed of two cylindrical plates with potentials of $+1/2E$ and $-1/2E$, where the total field $= eE = (+1/2E) + (-1/2E)$. For ions to pass through the electric sector, the deflecting force (K_E) must equal the centrifugal force (K_C). This equality is represented in the following Equation 19.12 as

$$eE = \frac{mv^2}{r_E} \tag{19.12}$$

where eE is the deflecting force (K_E) and mv^2/r_E is the centrifugal force (K_C). For the ions arriving into the electric sector, the kinetic energy (KE) equals

$$KE = zeV = \frac{1}{2}mv^2 \tag{19.1}$$

$$v^2 = \frac{2zeV}{m} \tag{19.13}$$

substituting Equation 19.13 into Equation 19.12, we get

$$eE = \frac{m\left(\frac{2zeV}{m}\right)}{r_E} \tag{19.14}$$

$$E = \frac{2zV}{r_E} \tag{19.15}$$

$$r_E = \frac{2zV}{E} \tag{19.16}$$

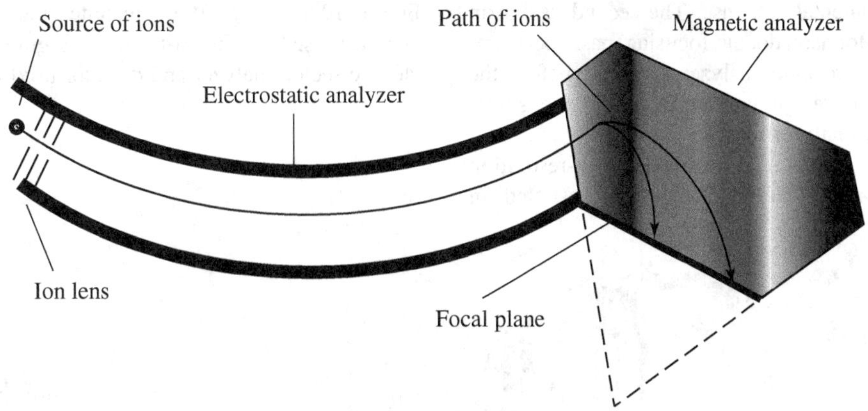

FIGURE 19.6 Mattauch-Herzog double-focusing geometry mass spectrometer.

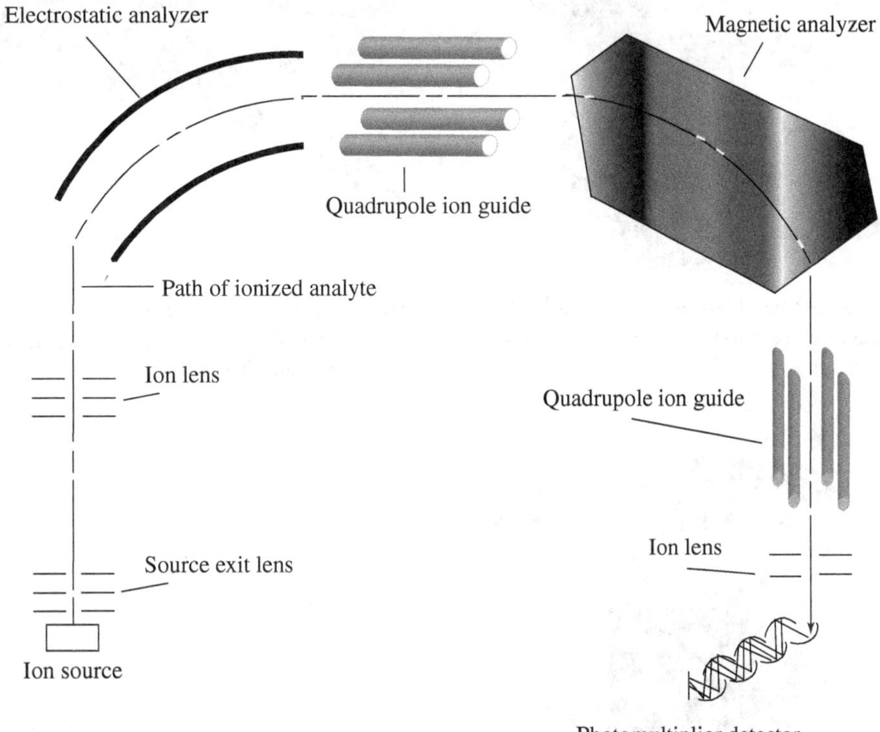

FIGURE 19.7 Nier-Johnson double-focusing geometry mass spectrometer.

From this relationship, we see that the radius of the curvature will equal two times the accelerating voltage divided by the total electric field strength. What ions pass through the electric sector does not depend upon the mass or charge of the ions but upon how well the ion's kinetic energy matches the field according to the acceleration voltage within the source. This relationship demonstrates that the electric sector is not a mass analyzer like the magnetic sector but rather is an energy filter.

By combining the magnetic sector with the electric sector, a double-focusing effect can be achieved where the magnetic sector does directional focusing, while the electric sector does energy focusing. Two types of double-focusing geometries are illustrated in Figures 19.6 and 19.7. The double-focusing geometry illustrated in Figure 19.6 is known as the Mattauch–Herzog double-focusing geometry. The electric sector (electrostatic analyzer) located before the magnetic analyzer in the instrumental design has effectively been added to gain in the resolution of the mass-analyzed ions. The Mattauch–Herzog double-focusing geometry results in a focal plane within the mass analyzer. This is a static analyzer plane where a photoplate or microchannel array detectors can be placed parallel to the focal plane. In this design, the different mass-to-charge ratios will be focused at different points on the microchannel plate. The detection of the separated mass-to-charge ratio ions will require no scanning increasing the sensitivity in the instrumental response. The second design in Figure 19.7 is the Nier–Johnson double-focusing mass spectrometer. Here, also the electrostatic analyzer is placed before the magnetic analyzer. In this design, ions are focused onto a single point, and the magnetic analyzer must be scanned.

An example of a commercial available high-resolution ICP-MS is the Thermo Scientific Element 2 illustrated in Figure 19.8. Figure 19.9 illustrates the magnetic sector mass analyzer used in the instrument. Figure 19.10 illustrates the electric sector analyzer and detector used in the instrument.

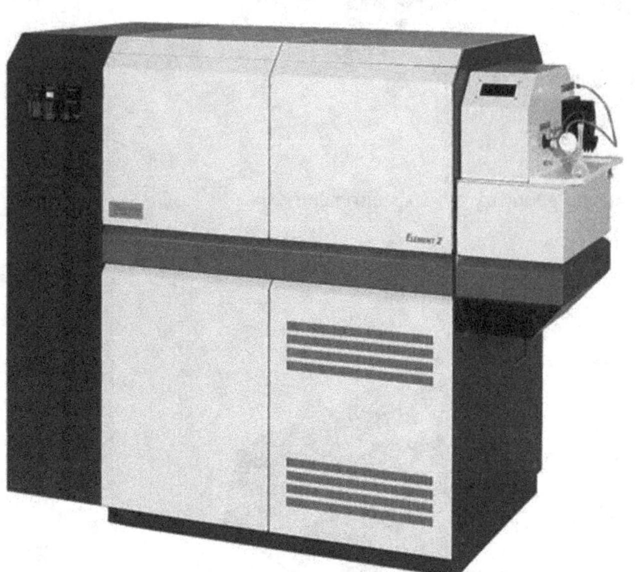

FIGURE 19.8 Thermo Scientific Element 2 available high resolution ICP-MS. (Used by permission from Thermo Fisher Scientific, the copyright owner.)

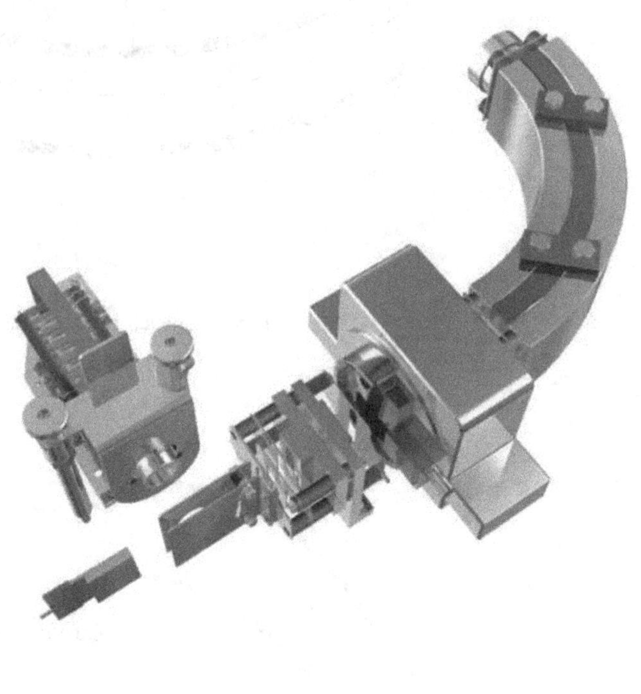

FIGURE 19.10 The electric sector analyzer and detector used in the Thermo Scientific Element 2 high resolution ICP-MS. (Used by permission from Thermo Fisher Scientific, the copyright owner.)

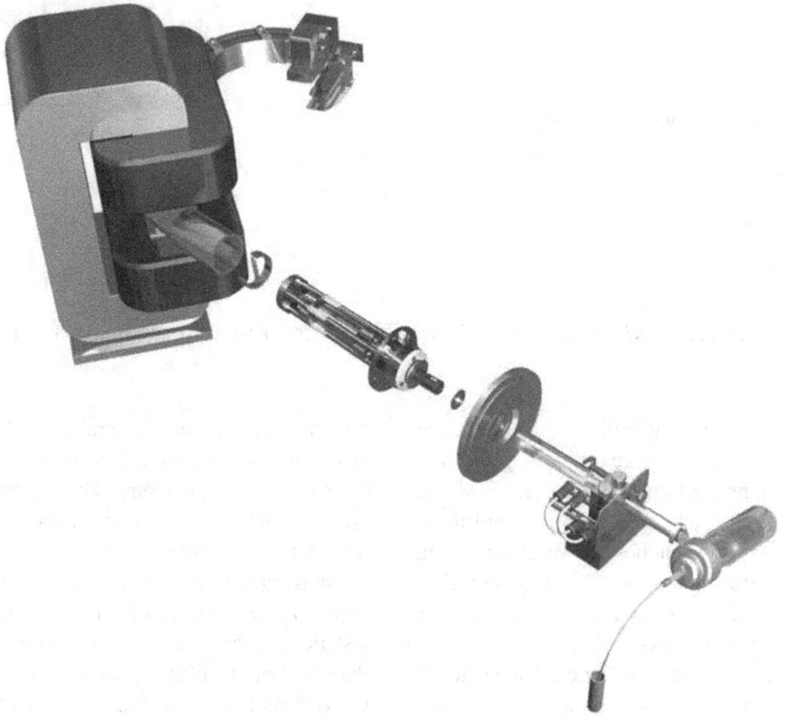

FIGURE 19.9 The magnetic sector mass analyzer used in the Thermo Scientific Element 2 high resolution ICP-MS. (Used by permission from Thermo Fisher Scientific, the copyright owner.)

20

X-RAY FLUORESCENCE (XRF) AND X-RAY DIFFRACTION (XRD)

20.1 X-Ray Fluorescence Introduction
20.2 X-Ray Fluorescence Theory
20.3 Energy-Dispersive X-Ray Fluorescence (EDXRF)
  20.3.1 EDXRF Instrumentation
  20.3.2 Commercial Instrumentation
20.4 Wavelength Dispersive X-Ray Fluorescence (WDXRF)
  20.4.1 Introduction
  20.4.2 WDXRF Instrumentation
20.5 Applications of XRF
20.6 X-ray Diffraction (XRD)
  20.6.1 Introduction
  20.6.2 X-Ray Crystallography
  20.6.3 Bragg's Law
  20.6.4 Diffraction Patterns
  20.6.5 The Goniometer
  20.6.6 XRD Spectra

20.1 X-RAY FLUORESCENCE INTRODUCTION

When a material has been bombarded with high-energy X-rays or gamma rays, an emission of characteristic secondary or fluorescent X-rays will take place. This phenomenon has been utilized in many different types of elemental analyses in numerous sample matrices including metals, glass, ceramics, ash, soil, and foodstuffs. The first X-ray fluorescence (XRF) technique was demonstrated in the 1960s, and the first commercially available instrumentation was available by the 1970s. Today, XRF instrumentation is found in many types of laboratories analyzing samples from environmental, food, petroleum, and building materials.

20.2 X-RAY FLUORESCENCE THEORY

To produce the XRF, samples are subjected to high-energy short-wavelength irradiation by X-rays or gamma rays. The irradiation results in the absorption of the energy producing ionization of the material in the form of an inner electron ejection. Upon ionization, the electronic structure of the atom will become unstable. To stabilize the system, an electron from a higher orbital will drop down to fill the inner orbital missing electron space. With this, transition energy will be released in the form of a photon. The energy released will be equal to the energy involved in the energy level transition of the electron in the atom's two orbitals involved. Due to this, the energy emitted is characteristic of the particular element involved. A sodium atom will have different energy transitions from that of gold and so forth for all of the different elements in the periodic table. The term fluorescence is used when a process absorbs energy at a lower wavelength–higher energy radiation and then releases the energy in the form of photons usually at higher wavelength–lower energy radiation.

This process of XRF is illustrated in Figure 20.1 where (a) is the process of an inner shell electron absorbing radiant energy. In Figure 20.1(b), the electron has gained enough energy to be ejected (i.e., energy greater than its ionization energy), producing an ionized atom with a vacant electron shell. The picture in Figure 20.1(c) illustrates an outer shell electron dropping down to the inner shell vacant orbital and in the process emitting a photon of energy. The number of photons emitted per unit time is proportional to the concentration of the element.

When the process of XRF takes place, there are only a limited number of transitions that can take place where the electron from the outer shell drops into the vacant inner shell. The main transitions are given names: an $L \rightarrow K$ transition is traditionally called Kα, an $M \rightarrow K$ transition is called Kβ, and an $M \rightarrow L$ transition is called Lα. Each transition yields a fluorescent photon with a characteristic energy. The difference in energy of the initial and final orbital is the energy that is released. Using Planck's law, the wavelength of this fluorescent radiation can be calculated:

$$\lambda = h\frac{c}{E} \quad (20.1)$$

Analytical Chemistry: A Chemist and Laboratory Technician's Toolkit, First Edition. Bryan M. Ham and Aihui MaHam.

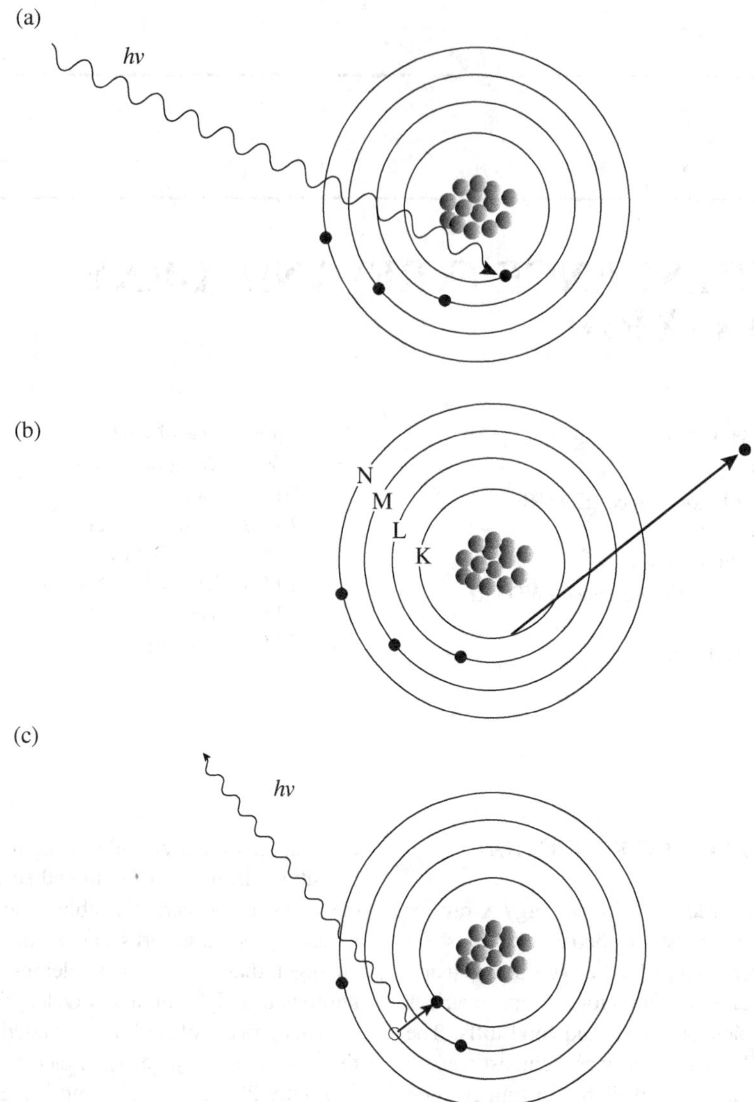

FIGURE 20.1 (a) Process of an inner shell electron absorbing radiant energy. (b) Electron has gained enough energy to be ejected (i.e., energy greater than its ionization energy) producing an ionized atom with a vacant electron shell. (c) Outer shell electron dropping down to the inner shell vacant orbital and in the process emitting a photon of energy.

There are two approaches for measuring the fluorescent radiation, either by sorting the energies of the photons (energy-dispersive analysis) or by separating the wavelengths of the radiation (wavelength-dispersive analysis). With both approaches, the intensity of each characteristic radiation is directly proportional to the amount of each element in the material. This allows the application of this methodology to analytical chemistry.

20.3 ENERGY-DISPERSIVE X-RAY FLUORESCENCE (EDXRF)

20.3.1 EDXRF Instrumentation

20.3.1.1 Basic Components Energy-dispersive X-ray fluorescence (EDXRF) uses detectors to resolve spectral peaks due to the X-rays with different energy emitting from the samples being analyzed. EDXRF instrumentation utilizes high-resolution detectors and computers in order to resolve the spectral peaks. The EDXRF instrumentation is more simple and inexpensive as compared to other types of XRF spectrometers. The basic components consist of an X-ray source and a detector.

20.3.1.2 X-Ray Sources To excite the atoms and ionize the atoms in a sample, a radiation source is required that has sufficient energy to expel the tightly held inner electrons. In most instrumentation, the X-ray source consists of a 50 to 60 kV 50–300 W X-ray tube. Handheld EDXRF instruments use radioisotopes such as Fe-55, Cd-109, Cm-244, Am-241, or Co-57 as the X-ray source. Some simply use a small X-ray tube. Figure 20.2 illustrates a few examples of commercially available X-ray sources used. The inner workings of an X-ray source tube

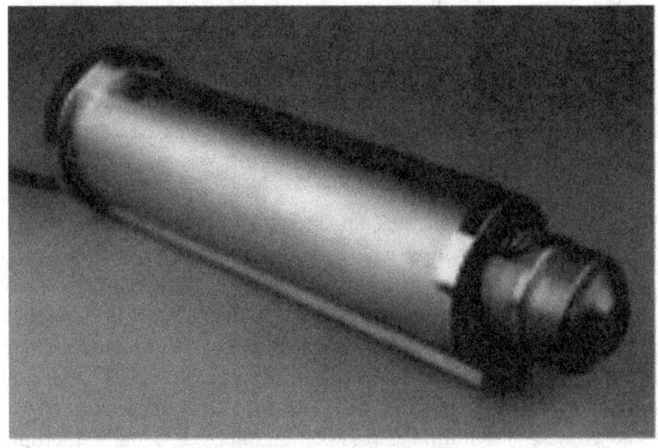

FIGURE 20.2 Examples of X-ray source tubes. (Image courtesy of Varian Medical Systems, Inc. All rights reserved.)

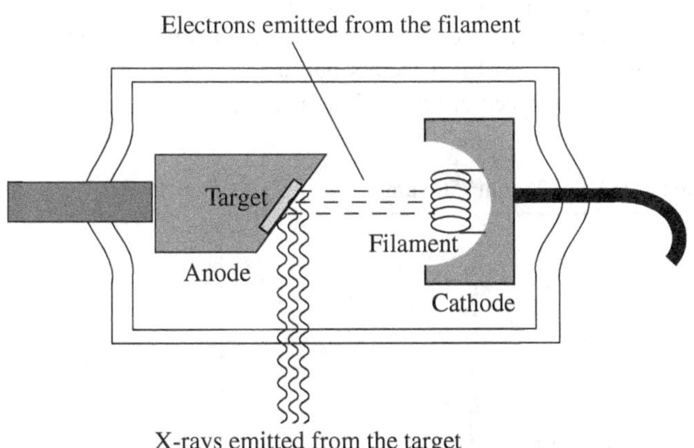

FIGURE 20.3 Inner workings of an X-ray source tube including the filament at the cathode and the target material located at the anode.

are illustrated in Figure 20.3. In this simple type of X-ray source, the source parts are housed within a glass tube under vacuum. Within the source is a filament that is heated by passing a current through it. As the filament heats up, it emits electrons. At the filament end, there is a high negative potential called the cathode. Across from the filament is an anode at high positive potential where a high voltage is placed across the cathode and the anode. The high voltage up to 150 kV produces as beam of electrons from the heated filament to the anode. Thus, a current is produced. The accelerated electrons strike a target material that is placed at the anode, as illustrated in Figure 20.3. The electrons emitted from the filament at the cathode collide with the anode material usually made up of tungsten, molybdenum, or copper. The colliding electrons ionize the anode target material. About 1% of the energy released by the ionized anode target material is X-rays. This X-ray photon-generating effect is called the Bremsstrahlung effect from the German words *bremsen* for braking and *strahlung* for radiation.

20.3.1.3 Detectors The detector used in EDXRF instrumentation produces electrical pulses that coincide with the energy of the incident X-rays produced from the samples under irradiation. Most detectors use Peltier-cooled Si(Li) detectors or Peltier-cooled silicon drift detectors (SDD). An example of an SDD is illustrated in Figure 20.4. The inner working of the SDD is illustrated in Figure 20.5. In the figure, we see a beryllium window that allows the passage of the X-rays. Under the window is the detector that measures the X-rays. Under the detector is a Peltier cooling

FIGURE 20.4 Example of a Peltier-cooled silicon drift detector (SDD). (Reprinted with permission from Amptek Inc.)

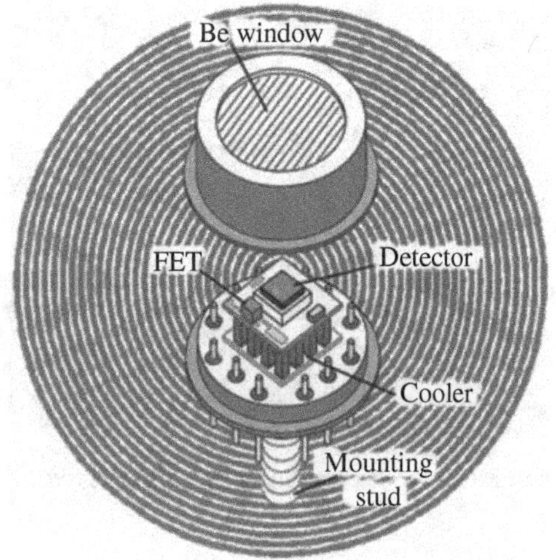

FIGURE 20.5 Inner workings of the SDD including a beryllium window that allows the passage of the X-rays, a detector that measures the X-rays, and a Peltier cooling chip. (Reprinted with permission from Amptek Inc.)

chip. Figure 20.6 illustrates a spectrum of the measurement of iron (^{55}Fe) at 5.89 keV.

When a photon is released from the sample, the energy-dispersive spectrometers measure the energy with the detector. The two most predominant detectors have been based upon silicon semiconductors such as lithium-drifted silicon crystals and high-purity silicon wafers.

Figure 20.7 illustrates an energy-dispersive spectrum of a sample of paint chips where multiple elements are measured simultaneously. The spectrum was produced by simultaneously scanning a broad range of energies that can detect the presence of multiple elements at the same time. In the spectrum, there are peaks for silicon (Si), sulfur (S), titanium (Ti), and lead (Pb). Often with EDXRF instruments, the entire spectrum can be viewed in real time as it is being collected. Usually, multiple spectra are collected and summed together to reduce background noise and to increase sensitivity.

20.3.1.3.1 Si(Li) Detectors The Si(Li) detectors consist of a 3–5 mm thick silicon junction type p–i–n diode (same as PIN diode) with a bias of −1000 V across it. In the center, the lithium-drifted part forms the nonconducting i-layer. The layer p-type Li compensates the residual acceptors. A voltage pulse is formed when a photon from the emitted X-ray passes through the detector that causes electron hole pairs to form. Low conductivity is required for the proper function of the detector where cooling with liquid nitrogen is the most efficient. However, often Peltier cooling chips are used but with a slight loss in resolution.

20.3.1.3.2 Wafer Detectors High-purity silicon wafers with low conductivity are also used that are cooled by Peltier chips. These detectors are cheaper and easier to maintain than the liquid nitrogen-cooled Si(Li) detectors but have a lower resolution to distinguish between different photon energies.

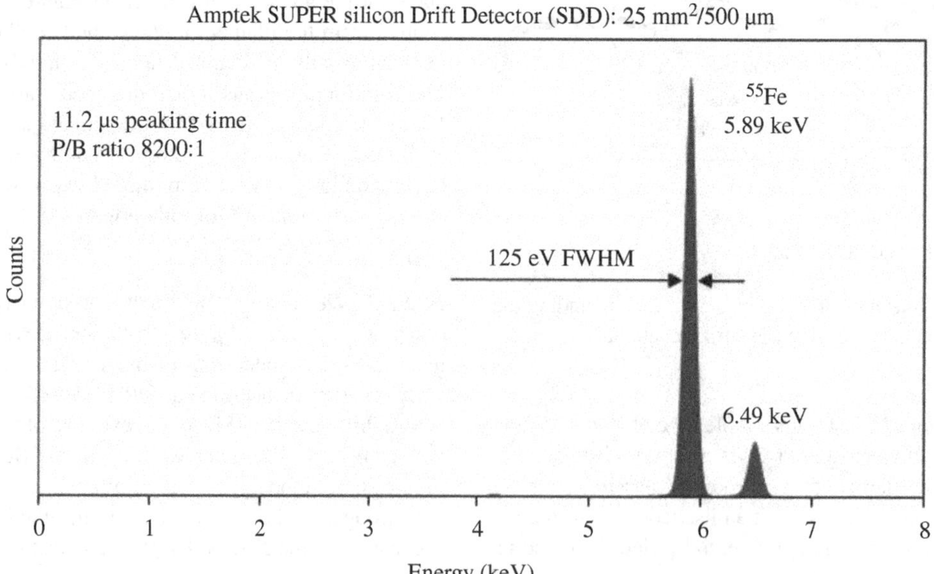

FIGURE 20.6 Spectrum of the measurement of iron (^{55}Fe) at 5.89 keV. (Reprinted with permission from Amptek Inc.)

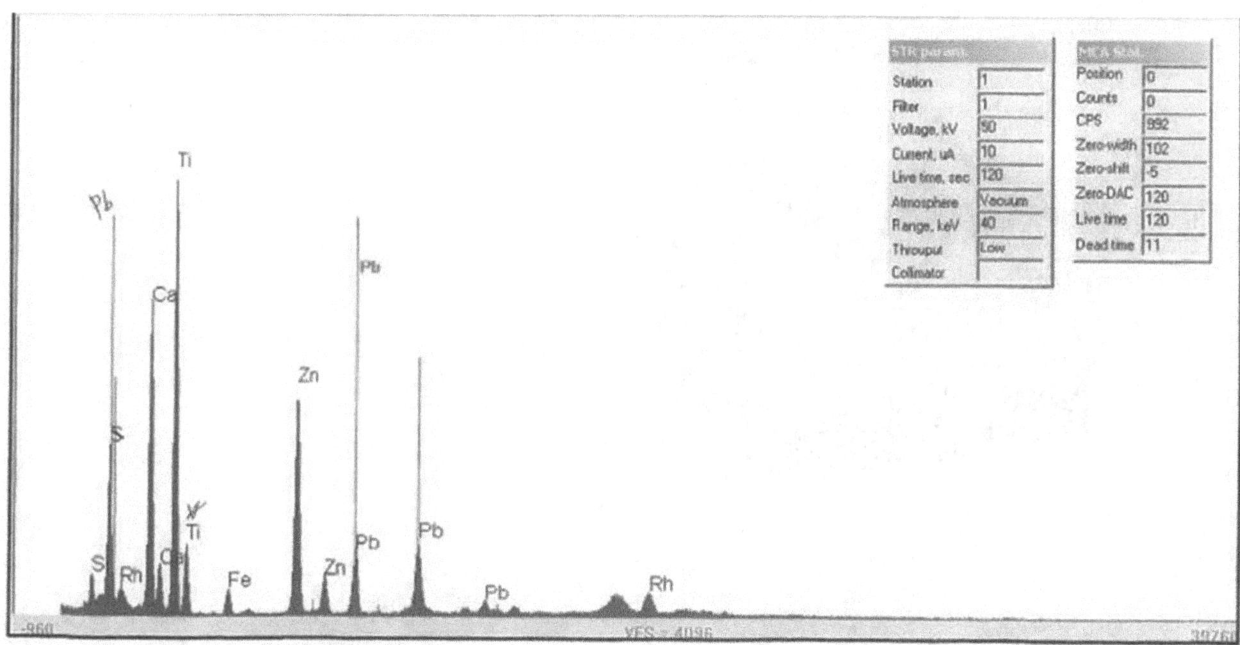

FIGURE 20.7 Energy-dispersive spectrum of paint chips where multiple elements are measured simultaneously including silicon (Si), sulfur (S), titanium (Ti), and lead (Pb).

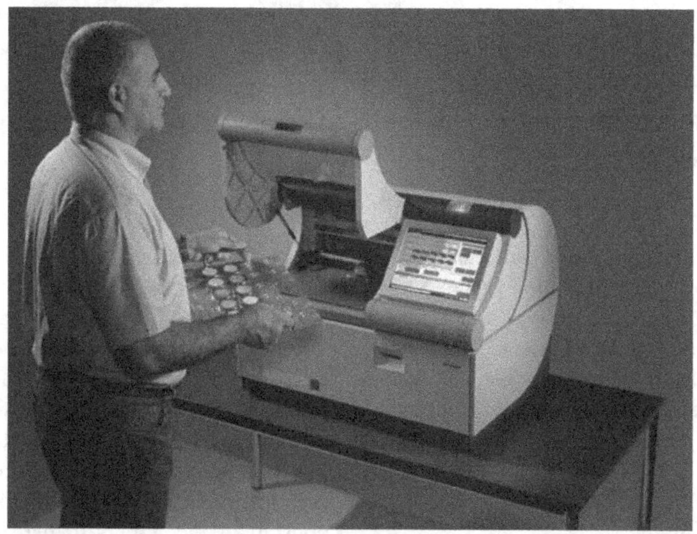

FIGURE 20.8 Bruker S2 Ranger EDXRF spectrometer. (Reprinted with permission from Bruker AXS Inc.)

20.3.2 Commercial Instrumentation

There are a number of commercially available EDXRF spectrometers from large floor models to midsize benchtop models and to small handheld models. Figure 20.8 illustrates a Bruker S2 Ranger EDXRF spectrometer. The picture illustrates the loading of samples in a sample loading tray. Figure 20.9(a) illustrates an Epsilon 3^{XLE} EDXRF benchtop spectrometer from PANalytical. A diagram of the EDXRF is illustrated in Figure 20.9(b). Finally, Figure 20.10 illustrates a handheld X-MET7000 series EDXRF from Oxford Instruments.

20.4 WAVELENGTH DISPERSIVE X-RAY FLUORESCENCE (WDXRF)

20.4.1 Introduction

Wavelength dispersive X-ray fluorescence (WDXRF) spectrometers use diffractive optics to separate the photons on a single crystal for detection instead of high-resolution solid-state detectors. This makes them a little easier to maintain as they do not require the use of liquid nitrogen for cooling the detector. They are usually tuned to measure wavelengths of the emission lines

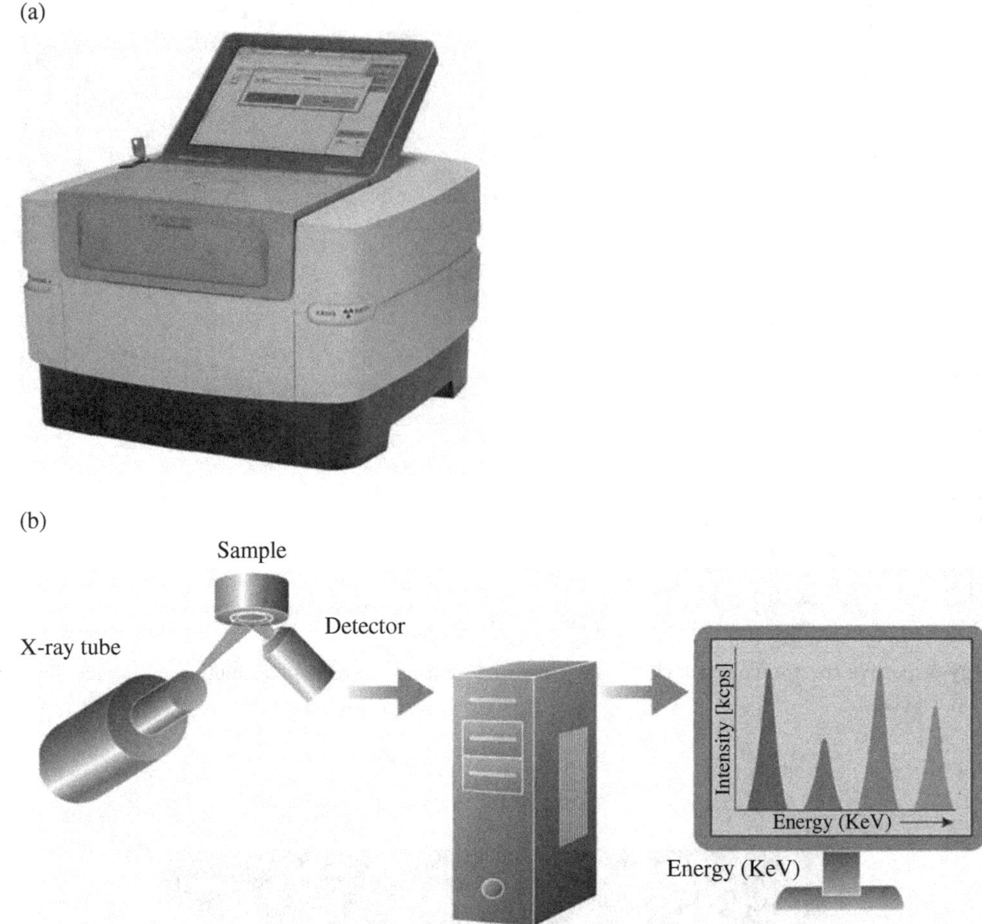

FIGURE 20.9 (a) Epsilon 3^{XLE} EDXRF Benchtop Spectrometer from PANalytical. (b) Diagram of an energy-dispersive X-ray fluorescence (EDXRF) spectrometer. (Image courtesy of PANalytical)

of specific elements of interest, although they can scan a wide range of wavelengths to produce a spectrum plot. Figure 20.11 illustrates a WDXRF spectrum. The spectrum was produced by scanning separate wavelengths one by one to produce the resultant multielement spectrum. In the figure, the light element scans are shown along the top, and the heavy elements are shown in the bottom spectrum.

20.4.2 WDXRF Instrumentation

For excitation, WDXRF instruments also use X-ray tube sources as those illustrated in Figures 20.2 and 20.3. Usually, higher energy X-ray tubes are used though in WDXRF instruments due to lower efficiency in the fluorescence system, somewhere in the 1–4 kW range. Crystals are used as diffraction devices to diffract X-rays emitted from the sample onto the detector for measurement. The diffracted X-rays for measurement are the ones that have the relationship of the Bragg equation

$$\eta \cdot \lambda = 2d \cdot \sin(\theta) \qquad (20.2)$$

where d is the atomic spacing within the crystal, n is an integer, and theta (θ) is the angle between the sample and the detector. Other wavelengths that do not satisfy this relationship are scattered and not registered by the detector. Another component used in the system is collimators that effectively limit the angular spread of the X-rays and increasing the resolution. In WDXRF instruments, the resolution is obtained by the components prior to the detector so relatively low-resolution detectors can be used such as proportional counters. Figure 20.12 illustrates the relationship of the X-rays to the crystal according to the Bragg Equation 20.2.

20.4.2.1 Simultaneous WDXRF Instrumentation The components of the WDXRF instrument can be fixed to make a single channel for analyzing a single element. In simultaneous WDXRF instruments, there are a number of fixed single channels that are arranged in a circle surrounding the sample with an X-ray source tube in the middle facing upward. Each fixed single channel has a fixed geometry crystal monochromator, a detector, and the associated electronics for processing the signals. This arrangement

FIGURE 20.10 A hand held X-MET8000 series EDXRF from Oxford Instruments. Examples of field applications using the portable XRF. (Reprinted with permission from Oxford Instruments.)

allows the simultaneous measurement of a number of elements with no moving parts. Usually, only 15 to 20 elements can be set up for analysis simultaneously due to a limit in the area that the channels can be set up around the sample fluorescing. Figure 20.13 shows an example of a commercially available simultaneous WDXRF spectrometer from PANalytical called the Axios FAST simultaneous WDXRF spectrometer. Figure 20.14 shows the inner workings of the simultaneous WDXRF spectrometer where multiple elements can be measured simultaneously with multiple channels. Let's take a closer look at one of the channels in the WDXRF spectrometer as illustrated in Figure 20.15. The X-ray source produces primary X-rays that are focused onto the sample. Within the sample, atoms are ionized, producing fluorescent X-rays. The fluorescent X-rays are passed through a

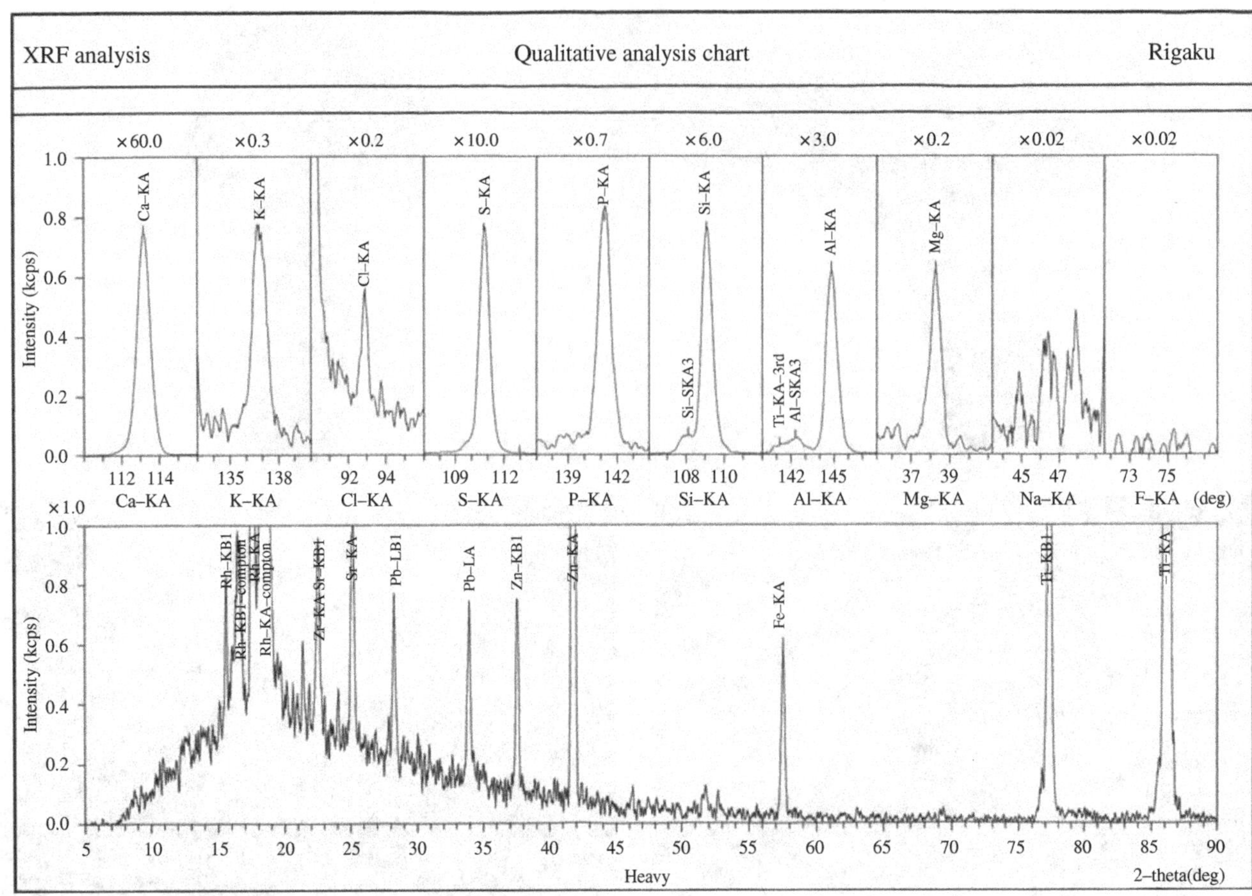

FIGURE 20.11 WDXRF spectrum produced by scanning separate wavelengths one by one to produce a resultant multi-element spectrum.

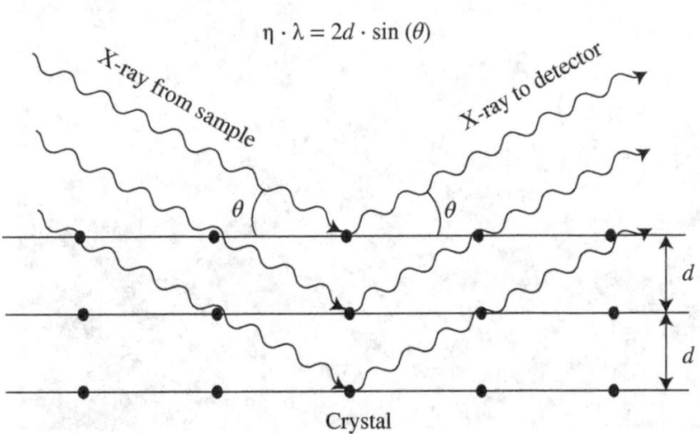

FIGURE 20.12 X-ray diffraction by the crystal according to the Bragg equation.

collimator allowing focused X-rays to pass through. The focused X-rays are directed to the analyzing crystal that diffracts X-rays according to Bragg's law, which are further refined through a collimator located right before the detector where the X-rays are measured.

20.4.2.2 Sequential WDXRF Instrumentation In sequential WDXRF instruments, the system contains a set single monochromator, detector, and electronics. The system moves through a

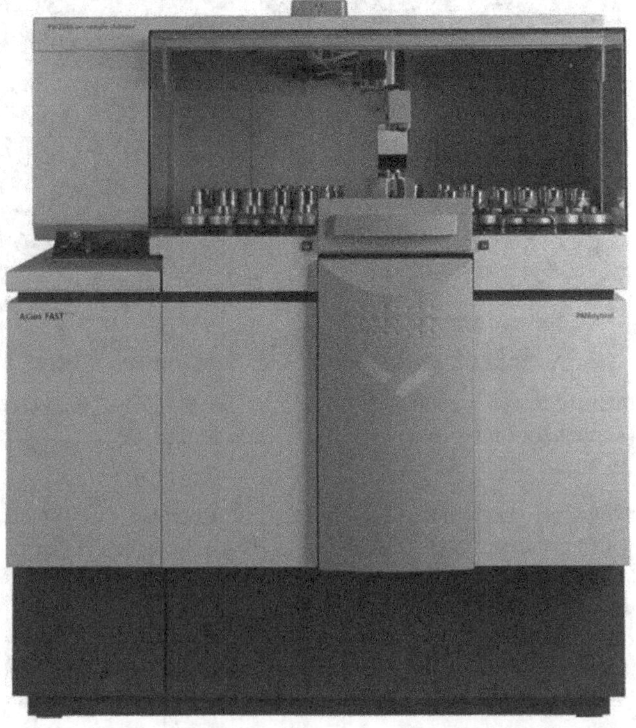

FIGURE 20.13 The PANalytical Axios FAST simultaneous WDXRF spectrometer. (Image courtesy of PANalytical.)

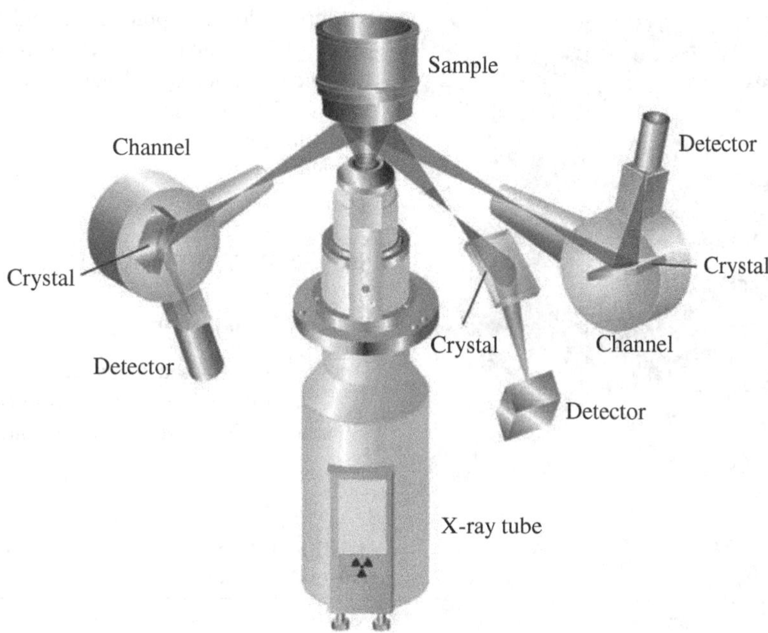

FIGURE 20.14 Inner workings of the simultaneous WDXRF spectrometer where multiple elements can be measured simultaneously with multiple channels. (Image courtesy of PANalytical.)

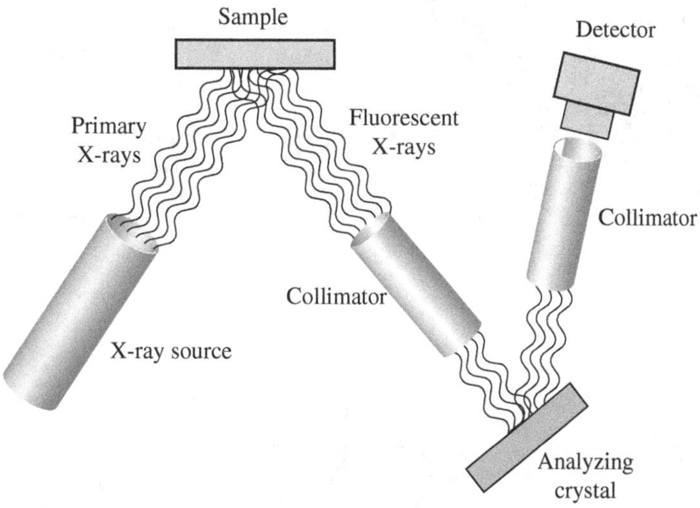

FIGURE 20.15 A closer look at a single channel within the simultaneous WDXRF spectrometer.

sequence of wavelengths where the appropriate X-ray tube power, the appropriate crystal, and the detector settings are selected for each wavelength. The distances between the X-ray source, the sample, and the crystal detector system can be kept short, resulting in a minimal loss in signal detection. The complexity of the system has been a challenge, and the measurement is a little longer than that of simultaneous configurations.

Figure 20.16 demonstrates an example of a commercially available sequential WDXRF spectrometer from PANalytical called the Axios^{mAX} sequential WDXRF spectrometer. The Axios XRF is reported to be able to simultaneously measure up to 14 elements at a time. Figure 20.17 shows the inner workings of the sequential WDXRF spectrometer where multiple elements can be measured simultaneously. The sequential WDXRF uses two moving parts including the crystal mill and the goniometer. Table 20.1 lists some common crystals used in XRF, their elemental makeup, and the associated Bragg parameters.

20.5 APPLICATIONS OF XRF

The technique of XRF, whether using WDXRF or EDXRF, can be used for a large variety of elemental analysis applications. The method can be used to span almost the entire periodic table of elements. Figure 20.18 illustrates a periodic table listing the energies and spectral lines associated with XRF analysis from beryllium (Be) to americium (Am). XRF has been used for both qualitative analysis and quantitative analysis with good

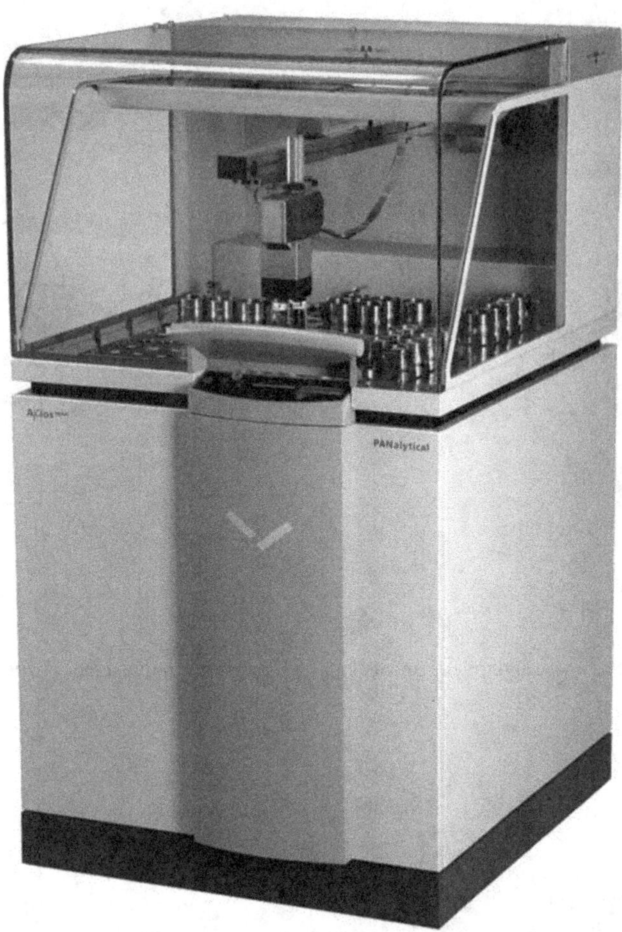

FIGURE 20.16 The PANalytical AXIOS^{mAX} sequential WDXRF spectrometer. (Image courtesy of PANalytical.)

sensitivity down to ppm levels and good dynamic range up to 100% of the sample. The XRF technique is used in many industries including geological and mineralogy, environmental, textile, and the food industry. Table 20.2 lists elements measured by XRF along with the fluorescent electron shell line and associated wavelength.

20.6 X-RAY DIFFRACTION (XRD)

20.6.1 Introduction

Another X-ray-based analytical technique used in laboratories is X-ray diffraction (XRD) spectroscopy. In most analytical laboratories, XRD in its simplest and most direct application is

TABLE 20.1 Some Common Crystal Elemental Makeup and Bragg Parameters.

Material	Plane	d (nm)	Min λ (nm)	Max λ (nm)
LiF	200	0.2014	0.053	0.379
LiF	220	0.1424	0.037	0.268
LiF	420	0.0901	0.024	0.169
ADP	101	0.5320	0.139	1.000
Ge	111	0.3266	0.085	0.614
Graphite	001	0.3354	0.088	0.630
InSb	111	0.3740	0.098	0.703
PE	002	0.4371	0.114	0.821
KAP	1010	1.325	0.346	2.490
RbAP	1010	1.305	0.341	2.453
Si	111	0.3135	0.082	0.589
TlAP	1010	1.295	0.338	2.434

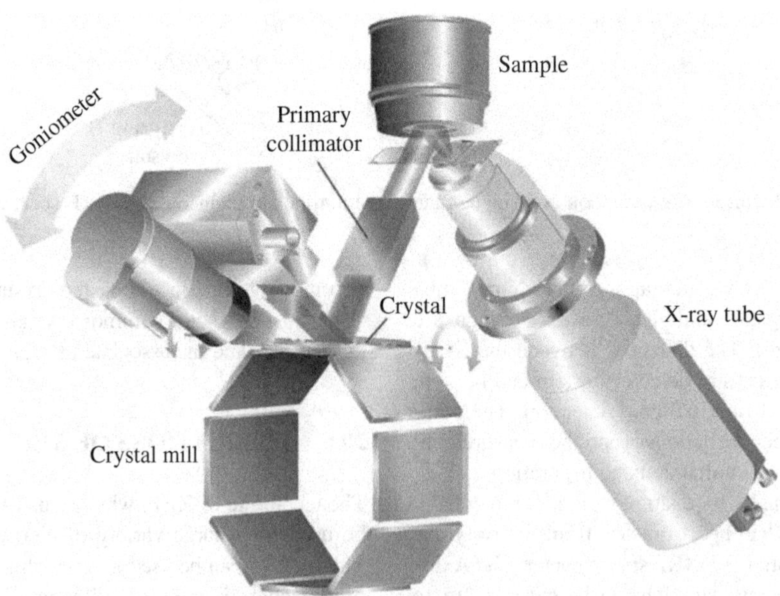

FIGURE 20.17 Inner workings of the sequential WDXRF spectrometer where multiple elements can be measured simultaneously with the crystal mill and the goniometer. (Image courtesy of PANalytical.)

Periodic table of elements and X-ray energies

www.bruker.com/hhxrf

1 H 1.01 0.0007 Hydrogen																	2 He 4.00 0.0002 Helium
3 Li 6.94 0.53 Lithium	4 Be 9.01 1.85 Beryllium Kα 0.108											5 B 10.81 2.34 Boron Kα 0.183	6 C 12.01 2.27 Carbon Kα 0.277	7 N 14.01 0.001 Nitrogen Kα 0.392	8 O 16.00 0.001 Oxygen Kα 0.525	9 F 19.00 0.001 Fluorine Kα 0.677	10 Ne 20.18 0.0009 Neon Kα 0.849
11 Na 22.99 0.97 Sodium Kα 1.040	12 Mg 24.31 1.74 Magnesium Kα 1.254											13 Al 26.98 2.70 Aluminium Kα 1.486	14 Si 28.09 2.33 Silicon Kα 1.740	15 P 30.97 1.82 Phosphorus Kα 2.010	16 S 32.07 2.07 Sulfur Kα 2.309	17 Cl 35.45 0.003 Chlorine Kα 2.622	18 Ar 39.95 0.002 Argon Kα 2.958
19 K 39.10 0.86 Potassium Kα 3.314	20 Ca 40.08 1.54 Calcium Kα 3.692 Lα 0.341	21 Sc 44.96 2.99 Scandium Kα 4.093 Lα 0.395	22 Ti 47.87 4.54 Titanium Kα 4.512 Lα 0.452	23 V 50.94 6.11 Vanadium Kα 4.953 Lα 0.510	24 Cr 52.00 7.15 Chromium Kα 5.415 Lα 0.572	25 Mn 54.94 7.44 Manganese Kα 5.900 Lα 0.637	26 Fe 55.85 7.87 Iron Kα 6.405 Lα 0.705	27 Co 58.93 8.86 Cobalt Kα 6.931 Lα 0.775	28 Ni 58.69 8.91 Nickel Kα 7.480 Lα 0.849	29 Cu 63.55 8.93 Copper Kα 8.046 Lα 0.928	30 Zn 65.38 7.13 Zinc Kα 8.637 Lα 1.012	31 Ga 69.72 5.91 Gallium Kα 9.251 Lα 1.098	32 Ge 72.64 5.32 Germanium Kα 9.886 Lα 1.188	33 As 74.92 5.78 Arsenic Kα 10.543 Lα 1.282	34 Se 78.96 4.81 Selenium Kα 11.224 Lα 1.379	35 Br 79.90 3.12 Bromine Kα 11.924 Lα 1.481	36 Kr 83.80 0.004 Krypton Kα 12.648 Lα 1.585
37 Rb 85.47 1.53 Rubidium Kα 13.396 Lα 1.692	38 Sr 87.62 2.64 Strontium Kα 14.165 Lα 1.806	39 Y 88.91 4.47 Yttrium Kα 14.958 Lα 1.924	40 Zr 91.22 6.51 Zirconium Kα 15.775 Lα 2.044	41 Nb 92.91 8.57 Niobium Kα 16.615 Lα 2.169	42 Mo 95.94 10.22 Molybdenum Kα 17.480 Lα 2.292	43 Tc (98) 11.50 Technetium Kα 18.367 Lα 2.423	44 Ru 101.07 12.37 Ruthenium Kα 19.279 Lα 2.558	45 Rh 102.91 12.41 Rhodium Kα 20.216 Lα 2.697	46 Pd 106.42 12.02 Palladium Kα 21.177 Lα 2.838	47 Ag 107.87 10.50 Silver Kα 22.163 Lα 2.983	48 Cd 112.41 8.69 Cadmium Kα 23.173 Lα 3.133	49 In 114.82 7.31 Indium Kα 24.210 Lα 3.286	50 Sn 118.71 7.29 Tin Kα 25.271 Lα 3.444	51 Sb 121.76 6.69 Antimony Kα 26.359 Lα 3.604	52 Te 127.60 6.23 Tellurium Kα 27.473 Lα 3.768	53 I 126.90 4.93 Iodine Kα 28.612 Lα 3.938	54 Xe 131.29 0.006 Xenon Kα 29.775 Lα 4.110
55 Cs 132.91 1.87 Cesium Kα 30.973 Lα 4.285	56 Ba 137.33 3.59 Barium Kα 32.194 Lα 4.466	57 La 138.91 6.15 Lanthanum Kα 33.442 Lα 4.647	72 Hf 178.49 13.31 Hafnium Lα 7.899 Mα 1.646	73 Ta 180.95 16.65 Tantalum Lα 8.146 Mα 1.712	74 W 183.84 19.25 Tungsten Lα 8.398 Mα 1.775	75 Re 186.21 21.02 Rhenium Lα 8.652 Mα 1.843	76 Os 190.23 22.61 Osmium Lα 8.911 Mα 1.907	77 Ir 192.22 22.65 Iridium Lα 9.175 Mα 1.980	78 Pt 195.08 21.46 Platinum Lα 9.442 Mα 2.050	79 Au 196.97 19.28 Gold Lα 9.713 Mα 2.123	80 Hg 200.59 13.53 Mercury Lα 9.989 Mα 2.195	81 Tl 204.37 11.85 Thallium Lα 10.269 Mα 2.271	82 Pb 207.20 11.34 Lead Lα 10.551 Mα 2.342	83 Bi 208.98 9.81 Bismuth Lα 10.839 Mα 2.423	84 Po (209) 9.32 Polonium Lα 11.131 Mα 2.499	85 At (210) 7.00 Astatine Lα 11.427 Mα 2.577	86 Rn (222) 0.01 Radon Lα 11.727 Mα 2.654
87 Fr (223) 1.87 Francium Lα 12.031 Mα 2.732	88 Ra (226) 5.50 Radium Lα 12.339 Mα 2.806	89 Ac (227) 10.07 Actinium Lα 12.652 Mα 2.900															
				58 Ce 140.12 6.77 Cerium Lα 4.839 Mα 0.884	59 Pr 140.91 6.77 Praseodymium Lα 5.035 Mα 0.927	60 Nd 144.24 7.01 Neodymium Lα 5.228 Mα 0.979	61 Pm (145) 7.26 Promethium Lα 5.432 Mα 1.023	62 Sm 150.36 7.52 Samarium Lα 5.633 Mα 1.078	63 Eu 151.96 5.24 Europium Lα 5.849 Mα 1.131	64 Gd 157.25 7.90 Gadolinium Lα 6.053 Mα 1.181	65 Tb 158.93 8.23 Terbium Lα 6.273 Mα 1.240	66 Dy 162.50 8.55 Dysprosium Lα 6.498 Mα 1.293	67 Ho 164.93 8.80 Holmium Lα 6.720 Mα 1.348	68 Er 167.26 9.07 Erbium Lα 6.949 Mα 1.404	69 Tm 168.93 9.32 Thulium Lα 7.180 Mα 1.462	70 Yb 173.04 6.97 Ytterbium Lα 7.416 Mα 1.526	71 Lu 174.47 9.84 Lutetium Lα 7.655 Mα 1.580
				90 Th 232.04 11.72 Thorium Lα 12.968 Mα 2.996	91 Pa 231.04 15.37 Protactinium Lα 13.291 Mα 3.082	92 U 238.03 18.95 Uranium Lα 13.614 Mα 3.171	93 Np (237) 20.45 Neptunium Lα 13.946 Mα 3.250	94 Pu (244) 19.84 Plutonium Lα 14.282 Mα 3.339	95 Am (243) 13.69 Americium Lα 14.620 Mα 3.438	96 Cm (247) 13.51 Curium	97 Bk (247) 14.79 Berkelium	98 Cf (251) 15.1 Californium	99 Es (252) 13.5 Einsteinium	100 Fm (257) Fermium	101 Md (258) Mendelevium	102 No (259) Nobelium	103 Lr (262) Lawrencium

Key:
Atomic number — 35
Atomic weight — 79.90
Density (g/cm³) — 3.12
Symbol — Br
Element name — Bromine
Energy (keV) — Kα 11.924, Lα 1.481
Spectral line — Kα, Lα

FIGURE 20.18 Periodic table of the elements listing the energies and spectral lines associated with X-ray fluorescence analysis from beryllium (Be) to americium (Am).

TABLE 20.2 Elements Measured by XRF, Electron Shell Lines Responsible for Fluorescence, and the Associated Wavelengths.

Element	Line	Wavelength (nm)
Li	Kα	22.8
Be	Kα	11.4
B	Kα	6.76
C	Kα	4.47
N	Kα	3.16
O	Kα	2.362
F	K$\alpha_{1,2}$	1.832
Ne	K$\alpha_{1,2}$	1.461
Na	K$\alpha_{1,2}$	1.191
Mg	K$\alpha_{1,2}$	0.989
Al	K$\alpha_{1,2}$	0.834
Si	K$\alpha_{1,2}$	0.7126
P	K$\alpha_{1,2}$	0.6158
S	K$\alpha_{1,2}$	0.5373
Cl	K$\alpha_{1,2}$	0.4729
Ar	K$\alpha_{1,2}$	0.4193
K	K$\alpha_{1,2}$	0.3742
Ca	K$\alpha_{1,2}$	0.3359
Sc	K$\alpha_{1,2}$	0.3032
Ti	K$\alpha_{1,2}$	0.2749
V	Kα_1	0.2504
Cr	Kα_1	0.2290
Mn	Kα_1	0.2102
Fe	Kα_1	0.1936
Co	Kα_1	0.1789
Ni	Kα_1	0.1658
Cu	Kα_1	0.1541
Zn	Kα_1	0.1435
Ga	Kα_1	0.1340
Ge	Kα_1	0.1254
As	Kα_1	0.1176
Se	Kα_1	0.1105
Br	Kα_1	0.1040
Kr	Kα_1	0.09801
Rb	Kα_1	0.09256
Sr	Kα_1	0.08753
Y	Kα_1	0.08288
Zr	Kα_1	0.07859
Nb	Kα_1	0.07462
Mo	Kα_1	0.07094
Tc	Kα_1	0.06751
Ru	Kα_1	0.06433
Rh	Kα_1	0.06136
Pd	Kα_1	0.05859
Ag	Kα_1	0.05599
Cd	Kα_1	0.05357
In	Lα_1	0.3772
Sn	Lα_1	0.3600
Sb	Lα_1	0.3439
Te	Lα_1	0.3289
I	Lα_1	0.3149
Xe	Lα_1	0.3016
Cs	Lα_1	0.2892
Ba	Lα_1	0.2776
La	Lα_1	0.2666
Ce	Lα_1	0.2562
Pr	Lα_1	0.2463
Nd	Lα_1	0.2370
Pm	Lα_1	0.2282
Sm	Lα_1	0.2200
Eu	Lα_1	0.2121
Gd	Lα_1	0.2047
Tb	Lα_1	0.1977
Dy	Lα_1	0.1909
Ho	Lα_1	0.1845
Er	Lα_1	0.1784
Tm	Lα_1	0.1727
Yb	Lα_1	0.1672
Lu	Lα_1	0.1620
Hf	Lα_1	0.1570
Ta	Lα_1	0.1522
W	Lα_1	0.1476
Re	Lα_1	0.1433
Os	Lα_1	0.1391
Ir	Lα_1	0.1351
Pt	Lα_1	0.1313
Au	Lα_1	0.1276
Hg	Lα_1	0.1241
Tl	Lα_1	0.1207
Pb	Lα_1	0.1175
Bi	Lα_1	0.1144
Po	Lα_1	0.1114
At	Lα_1	0.1085
Rn	Lα_1	0.1057
Fr	Lα_1	0.1031
Ra	Lα_1	0.1005
Ac	Lα_1	0.0980
Th	Lα_1	0.0956
Pa	Lα_1	0.0933
U	Lα_1	0.0911
Np	Lα_1	0.0888
Pu	Lα_1	0.0868
Am	Lα_1	0.0847
Cm	Lα_1	0.0828
Bk	Lα_1	0.0809
Cf	Lα_1	0.0791
Es	Lα_1	0.0773
Fm	Lα_1	0.0756
Md	Lα_1	0.0740
No	Lα_1	0.0724

used to identify the crystalline structure and compositional makeup of powdered samples, for example, determining if a sample is simply glass (SiO_2) or perhaps a laboratory glassware known as Pyrex, which is a borosilicate that is approximately 80% silica, 13% boric oxide, 4% sodium oxide, and 2%–3% aluminum oxide.

20.6.2 X-Ray Crystallography

The XRD technique used in laboratories has its beginnings from X-ray crystallography, which is a method that can determine the atomic and molecular structure of crystals. When X-rays are directed toward the crystal sample, the crystalline atoms will cause the incoming X-rays to diffract into many specific directions. The angles and energies of the diffracted X-rays can be interpreted into a three-dimensional structure of the density of the electrons within the crystal. The positions of the atoms in the crystal can then be determined along with their bonding.

20.6.3 Bragg's Law

We had previously had an introduction to Bragg's law in Section 20.4.2 where crystals are used as diffraction devices to diffract X-rays emitted from the sample onto the detector for measurement. The diffracted X-rays for measurement are the ones that have the relationship of the Bragg equation

$$\eta \cdot \lambda = 2d \cdot \sin(\theta) \quad (20.2)$$

where d is the atomic spacing within the crystal, n is an integer, and theta (θ) is the angle between the sample and the detector. This relationship was derived by the English physicists Sir W.H. Bragg and his son Sir W.L. Bragg in 1913. The relationships of Equation 20.2 are an example of X-ray wave interference that is called XRD. The relationships in Equation 20.2 are direct evidence for the periodic atomic structure of crystals. In 1915, the Braggs were awarded the Nobel Prize in Physics in 1915 for their work in describing the crystalline structure of inorganic compounds such as NaCl, ZnS, and diamond.

The structures of crystals are regular arrays of atoms in orderly spaced rows. When X-rays are projected onto the crystal, the atoms' electrons that make up the crystal will scatter a portion of the X-rays. Approximately 96% to 98% of the incident X-ray beam is transmitted through the crystal, and the remaining portion (2% to 4%) is scattered. The electron scatterer produces a secondary spherical wave through elastic scattering of the incident X-rays. The atoms that make up the crystal are in an array pattern, so the X-rays are scattered in an ordered array. The scattered X-rays undergo either destructive interference or constructive interference depending upon whether the wave periods are in sync or not. These two types of scattered X-rays are illustrated in Figure 20.19. On the left is an example of constructive interference where the waves are in phase and the amplitudes are additive. On the right is an example of destructive interference where the amplitudes are subtracted. At this point, we can refer back to Figure 20.12 where it is the X-rays that satisfy the Bragg relationship that is constructive. The other waves cancel one another out that do not satisfy the Bragg relationship in other directions through the destructive interference.

20.6.4 Diffraction Patterns

When the incident beam of X-rays is directed upon the sample crystal, the X-rays will diffract according to the atom spacing in the crystalline lattice. One way to detect the diffraction pattern is to place a photographic plate after the sample crystal where the diffracted X-rays will leave a black spot. The basic design of this setup is illustrated in Figure 20.20. The majority of the incident X-ray is transmitted through the sample, while a small portion is diffracted onto the photographic plate. X-rays are able to undergo scattering because their wavelengths are similar to that of the spacing between the planes in the crystal on the order of 1 to 100 Å. Figure 20.21 illustrates a photographic film that has collected a diffraction pattern. The diffracted X-rays emanate out from the sample crystal in cone patterns according to the relationships between the crystal lattice spacing (d) and θ of Equation 20.2.

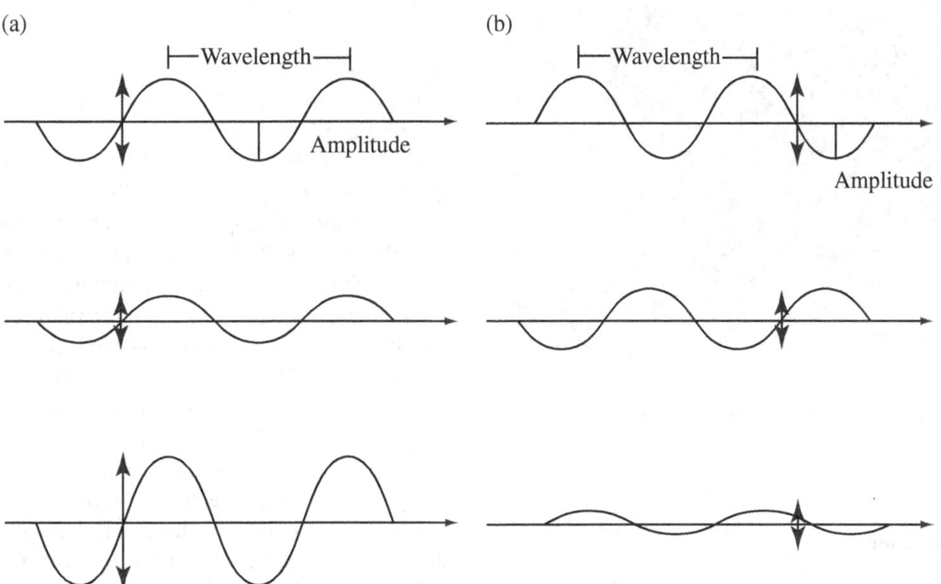

FIGURE 20.19 Examples of (a) waves experiencing constructive interference and (b) waves experiencing destructive interference.

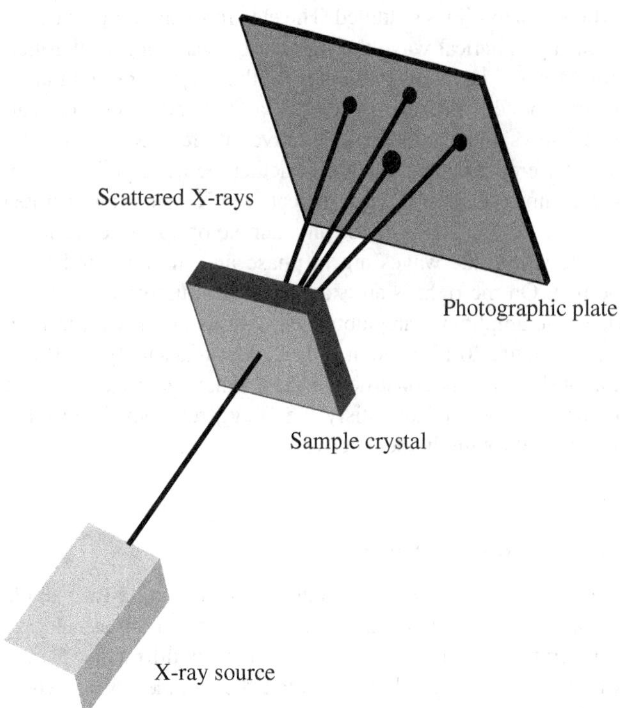

FIGURE 20.20 Scattered X-rays being detected by a photographic plate.

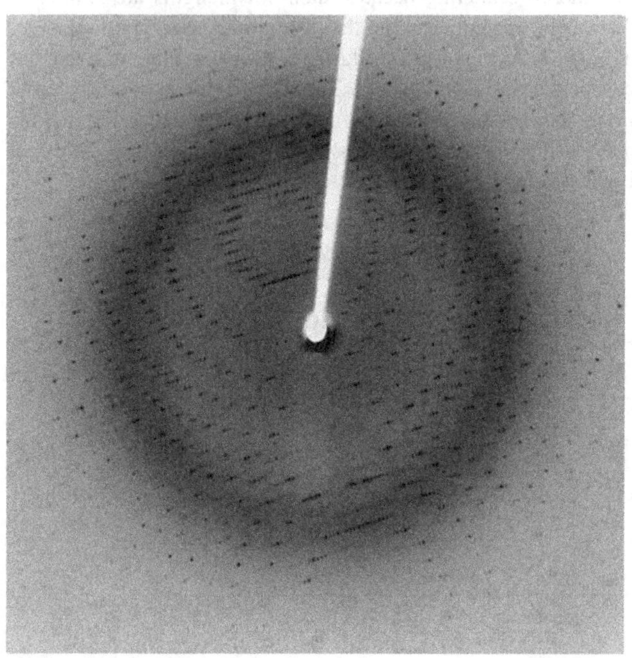

FIGURE 20.21 Photographic film after collecting a diffraction pattern.

20.6.5 The Goniometer

There are numerous XRD instruments commercially available for laboratory analysis. Figure 20.22 illustrates the Thermo Scientific ARL X'TRA powder diffractometer. Thermo Scientific ARL X'TRA powder diffractometer features X-ray optics on the goniometer to provide resolution in the low-angle region where peaks can be observed down to 0.5°. The ARL X'TRA instrument for powder XRD features a goniometer and a Peltier-cooled Si(Li) solid-state detector providing a high-performance instrument at an affordable price. Combining 25 years of XRD experience and engineering excellence, ARL X'TRA offers a cost-effective solution for high-quality powder diffraction applications in the areas of academic research, chemicals, pharmaceuticals, polymers, semiconductors, thin films, metals, and minerals. Figure 20.23 illustrates the goniometer part of the instrument including the X-ray source, the sample stage, and the detector. In the goniometer, the X-ray source and the detector are able to rotate through the angles associated with the diffraction measurement. The sample stage does not move; thus, loose powder samples are stable and do not shift around. Figure 20.24 illustrates the movement of the goniometer. This design is known as the vertical θ–θ Bragg–Brentano setup, which replaces the traditional vertical or horizontal θ–2θ geometry.

In an XRD measurement, a crystal is mounted on a goniometer and gradually rotated while being bombarded with X-rays, producing a diffraction pattern of regularly spaced spots known as *reflections*. The two-dimensional images taken at different rotations are converted into a three-dimensional model of the density of electrons within the crystal using the mathematical method of Fourier transforms, combined with chemical data known for the sample. Poor resolution (fuzziness) or even errors may result if the crystals are too small, or not uniform enough in their internal makeup.

20.6.6 XRD Spectra

The XRD spectra that are generated by most software processing the data collected from XRD instrumentation experiments are not in the form of diffraction patterns but in the form of peaks on an x- and *y*-axis scale. The *x*-axis is multiples of the diffraction angle 2θ, and the *y*-axis is the intensity. The spectrum illustrated in Figure 20.25 is the XRD spectrum of silicon oxide (SiO or SiO_2), or more specifically quartz. Quartz is the second most abundant mineral in the Earth's continental crust, after feldspar. It is made up of a continuous framework of SiO_4 silicon–oxygen tetrahedra, with each oxygen being shared between two tetrahedra, giving an overall formula SiO_2. The top spectrum is the actual spectrum collected with the XRD, while the bottom spectra are the results of searching the spectrum against a library of spectra for identification. In this way, a sample is analyzed and then searched against a library for possible identifications. The XRD spectrum for silicon oxide as quartz is a relatively simple spectrum at approximately 27 2θ. The next XRD spectrum illustrated in Figure 20.26 for calcium sulfate ($CaSO_4$) is more complex with a number of peaks being observed. The spectra in the bottom of the figure are the matching results from searching a database of spectra in a spectral library. The best match was to a spectrum of calcium sulfate hydrate. The next spectrum illustrated in Figure 20.27 is that of zirconium yttrium oxide. The spectrum is well characterized as illustrated by the matching of the peaks to the library search results.

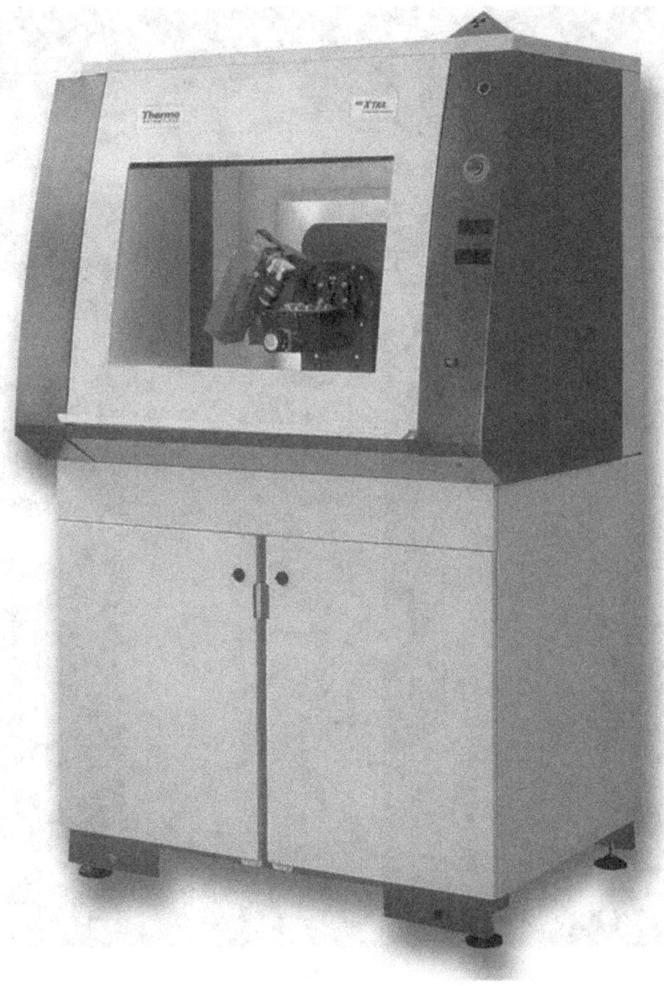

FIGURE 20.22 Thermo Scientific ARL X'TRA Powder Diffractometer. (Used by permission from Thermo Fisher Scientific, the copyright owner.)

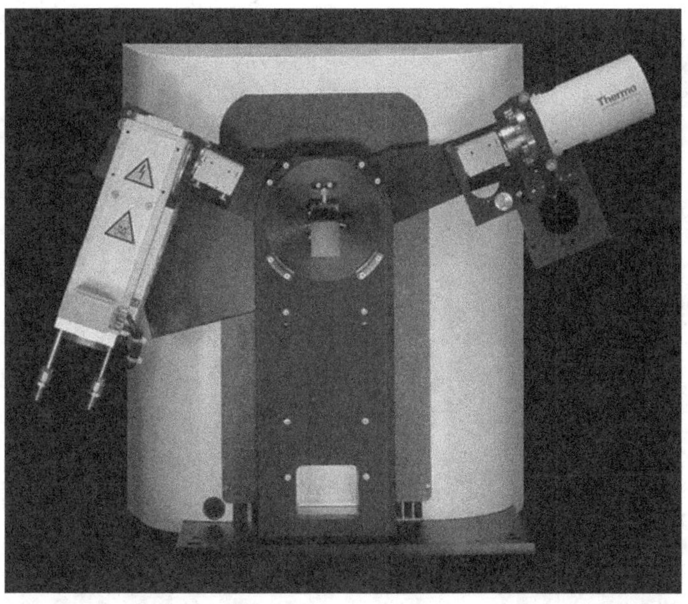

FIGURE 20.23 The Goniometer of the XRD instrument including the X-ray source, the sample stage, and the detector. (Used by permission from Thermo Fisher Scientific, the copyright owner)

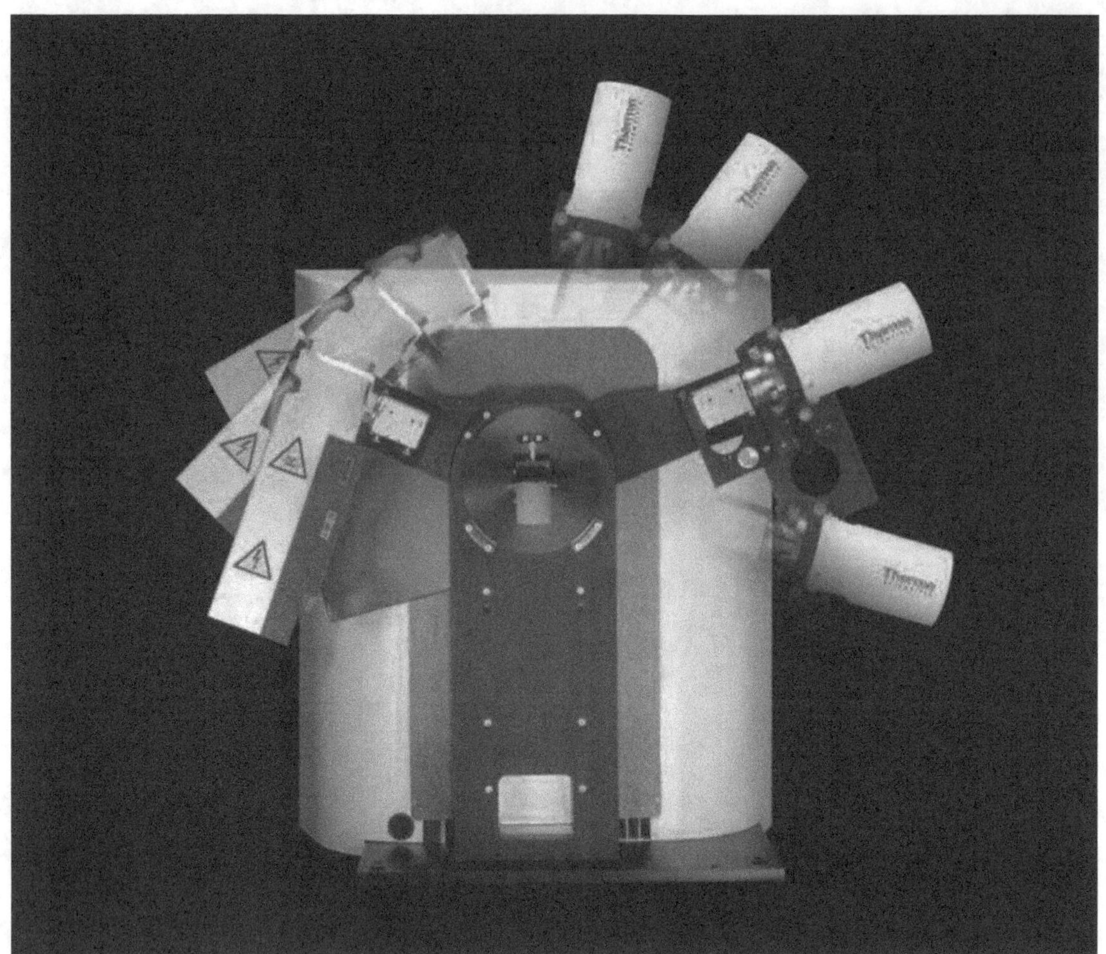

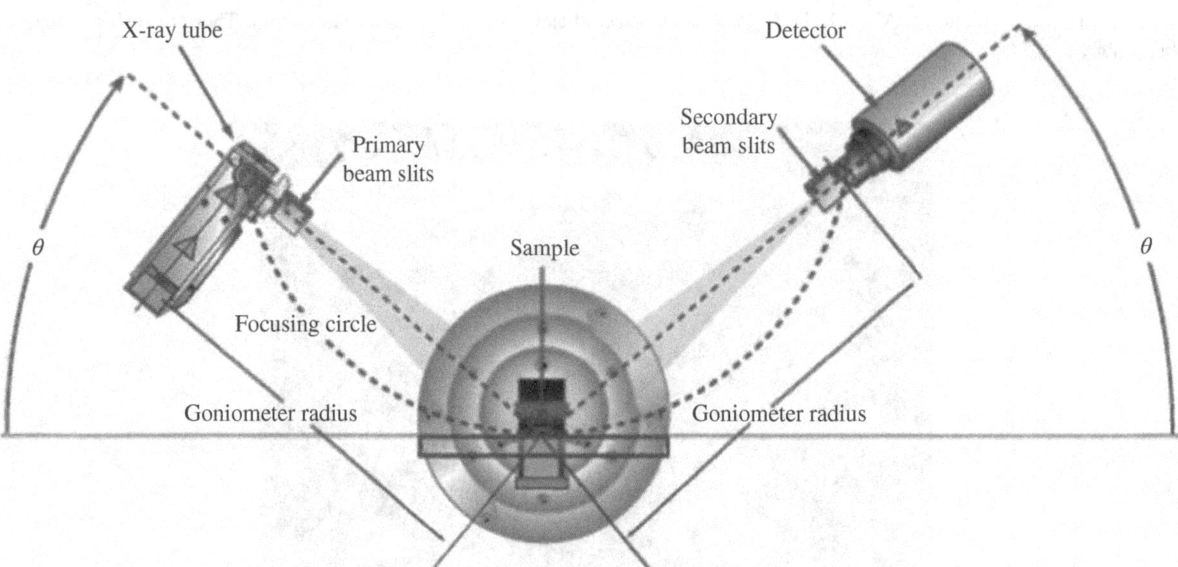

FIGURE 20.24 Movement of the Goniometer known as the vertical θ–θ Bragg–Brentano setup which replaces the traditional vertical or horizontal θ–2θ geometry. (Used by permission from Thermo Fisher Scientific, the copyright owner)

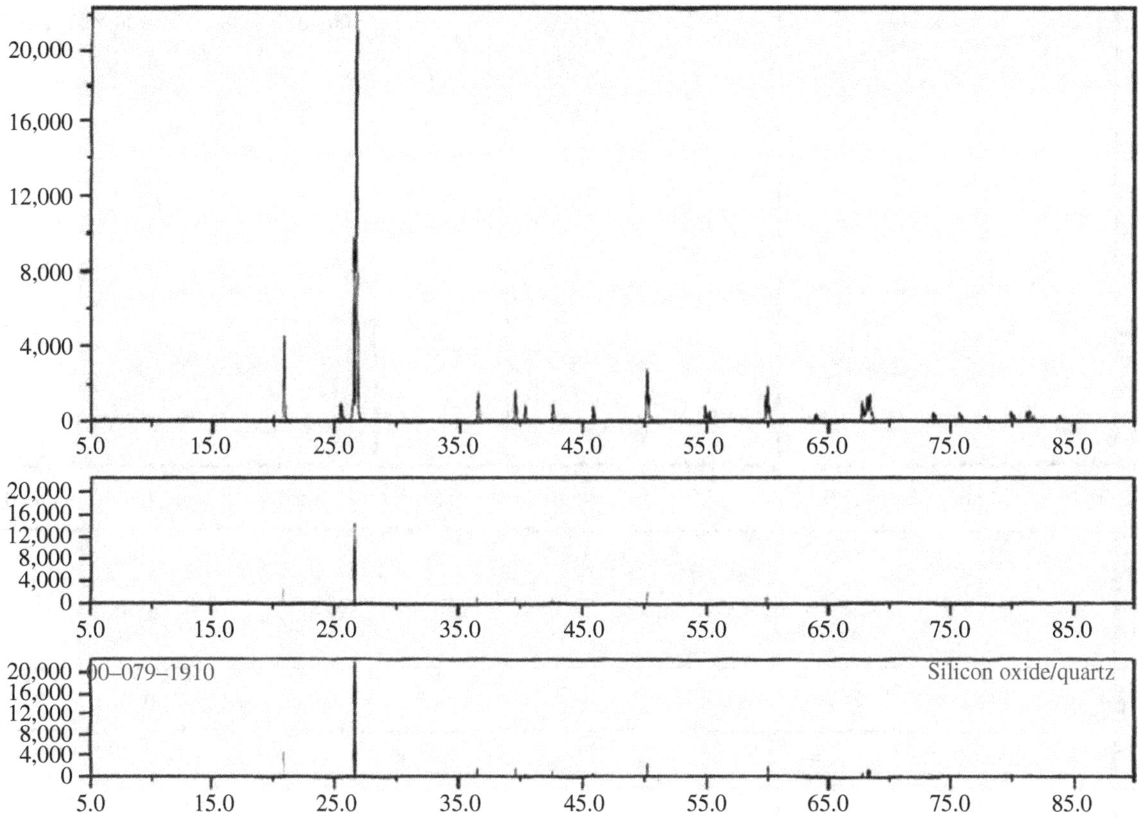

FIGURE 20.25 XRD spectrum of silicon oxide (SiO or SiO_2), or more specifically quartz.

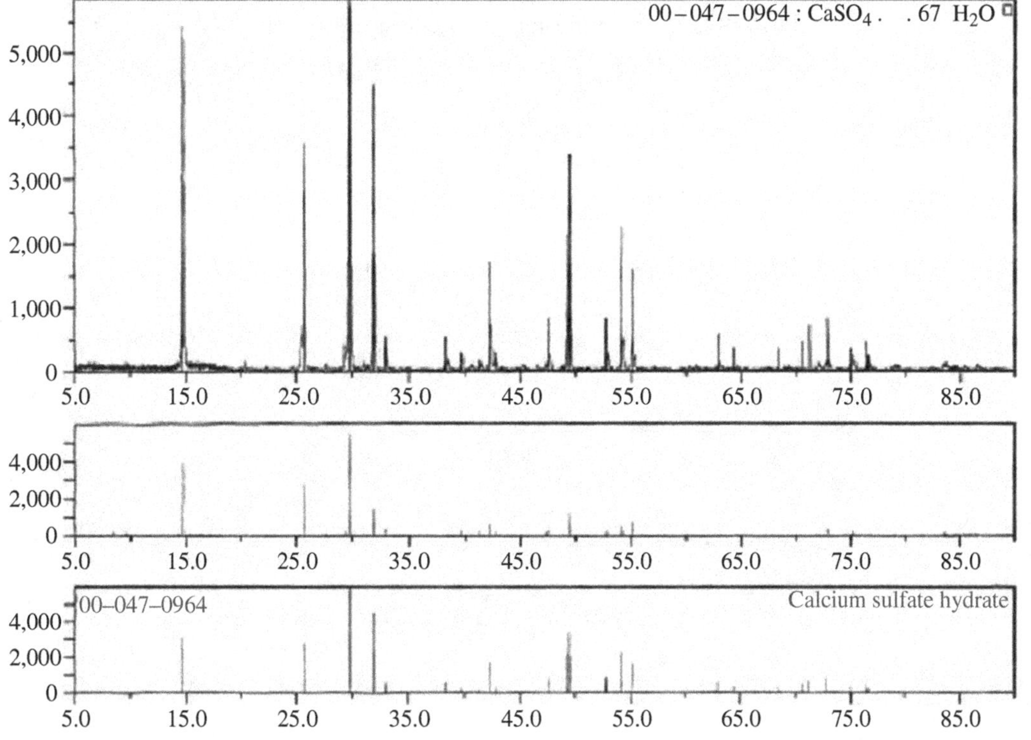

FIGURE 20.26 XRD spectrum of calcium sulfate ($CaSO_4$).

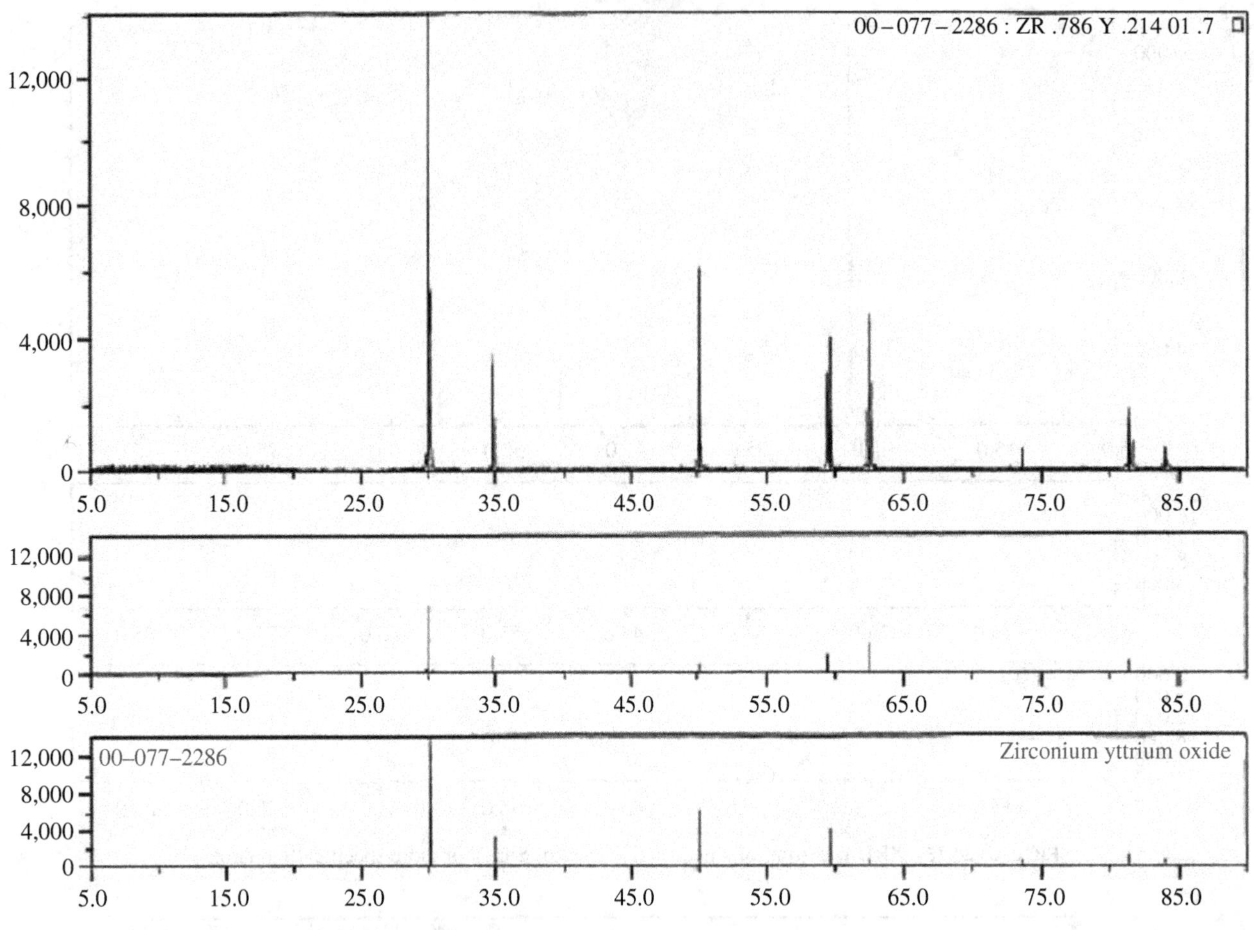

FIGURE 20.27 XRD spectrum of zirconium yttrium oxide.

21

CHROMATOGRAPHY—INTRODUCTION AND THEORY

21.1 Preface
21.2 Introduction to Chromatography
21.3 Theory of Chromatography
21.4 The Theoretical Plate Number N
21.5 Resolution R_S
21.6 Rate Theory Versus Plate Theory
21.6.1 Multiple Flow Paths or Eddy Diffusion (A Coefficient)
21.6.2 Longitudinal (Molecular) Diffusion (B Coefficient)
21.6.3 Mass Transfer Resistance between Phases (C_S and C_M Coefficients)
21.7 Retention Factor k'
References

21.1 PREFACE

This chapter is designed in a way that the student is introduced to the basics of the instrumental analysis technique under consideration followed by examples using the technique. The instrumental technique covered in this chapter involves chromatography, an extremely useful and quite common technique found in most analytical laboratories that the chemist or technician may find himself working in. The chromatographic techniques covered in this chapter and Chapters 22–27 include column liquid chromatography (LC), high-performance liquid chromatography (HPLC), solid-phase extraction (SPE), thin-layer chromatography (TLC), and gas-liquid chromatography (GLC). The chapter starts with the basic theory behind chromatography and then looks in detail at the instrumental techniques listed above that include the important components of the chromatography instrumentation. There are illustrative examples throughout the chapters to help the technician in mastering each section followed by a set of problems to be worked at the end of the chapter.

21.2 INTRODUCTION TO CHROMATOGRAPHY

Chromatography is the separation of analyte species using a combination of a mobile phase and a stationary phase. The analytes will spend different amounts of time in each of the two phases according to their affinities for each phase. For example, if analyte A has a greater affinity for the stationary phase as compared to analyte B, then it will move slower through the chromatography column. This is because analyte B will spend more time in the mobile phase as compared to analyte A and thus will move through the column quicker. Chromatography has its roots going back to a Russian botanist named Mikhail Tswett who in 1901 devised a method for separating plant pigments. Tswett reported this work in 1906 [1] where he described using a petrol ether/ethanol mobile phase to separate carotenoids and chlorophylls with calcium carbonate stationary phase in column chromatography. He named the methodology after the Greek words *chroma* (color) and *graphein* (to write) to form chromatography. Today, chromatography has taken on many forms including liquid column chromatography (LC), high-performance/pressure liquid chromatography (HPLC), flash chromatography (FC), paper chromatography, TLC, supercritical fluid chromatography (SFC), and gas chromatography (GC), all having their own unique advantages for separating analytes of interest.

21.3 THEORY OF CHROMATOGRAPHY

All of the chromatographic methods listed previously separate compounds based upon the same partitioning of the analytes between the mobile phase and the stationary phase. This is a relatively simple relationship to describe as a state of equilibrium (though not actually realized due to continuous mobile-phase movement) for the amount of solute A at any one point in time partitioned in the mobile phase, A_{mobile}, versus partitioned in the stationary phase, $A_{stationary}$:

$$A_{mobile} \rightleftharpoons A_{stationary} \tag{21.1}$$

Analytical Chemistry: A Chemist and Laboratory Technician's Toolkit, First Edition. Bryan M. Ham and Aihui MaHam.

The molar distribution (c_A) ratio for analyte A (also known as the partition ratio or partition coefficient) is written with the molar concentration of A in the stationary phase (c_S) always in the numerator:

$$c_A = \frac{\text{molar concentration of solute in stationary phase}}{\text{molar concentration of solute in mobile phase}} \tag{21.2}$$

$$c_A = \frac{c_S}{c_M} \tag{21.3}$$

As the compounds move through the chromatographic column, they partition themselves into like groups or "bands" that if colored can be seen moving down the column. If not colored, the bands can be rendered visible by shining a UV light source onto the column. Let's look at the principle of chromatography by looking at a simple column chromatography experiment. In Figure 21.1(a), a solution containing three compounds is being applied to the top of a column packed with chromatography stationary phase. Compound A is represented by a red band

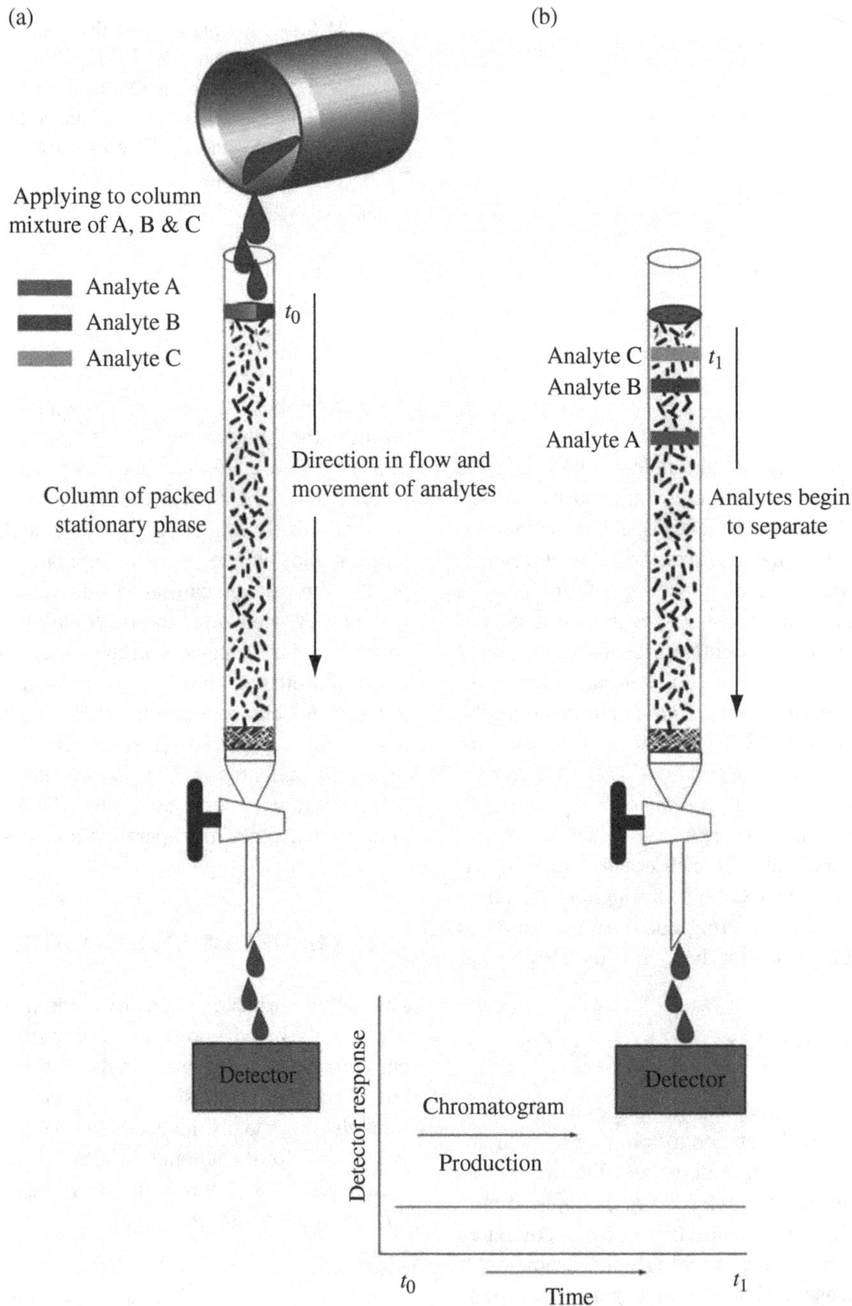

FIGURE 21.1 Illustration of column chromatography. (a) Sample has been placed onto the top of the column (t_0) where the three compounds are mixed. (b) Mobile phase has been continuously flowing through the column (t_1) resulting in the separation of the three compounds. Chromatogram shows no registering of a compound elution and subsequent detector response.

(■), compound B by a blue band (■), and compound C by a yellow band (■). Upon first applying the solution mixture to the column (time zero, t_0), the three compounds are mixed together at the head of the column. As mobile phase is added to the column, the compounds begin to move through the column spending part of their time associated with either the mobile phase or the stationary phase. This is illustrated in Figure 21.1 (b) where the three compounds have begun to separate into distinct bands moving down the column (time 1, t_1).

At the bottom of Figure 21.1, a detector is registering the elution of the compounds from the column. At this point, no compounds have made it through the column, and the "chromatogram" is simply registering a baseline that represents the mobile phase that has been zeroed for no detector response. After a set amount of time of mobile-phase flow (t_2), the three compounds have moved through the column and are close to eluting from the column, as illustrated in Figure 21.2(a). After more mobile phase has flowed through the column (t_3), we have breakthrough

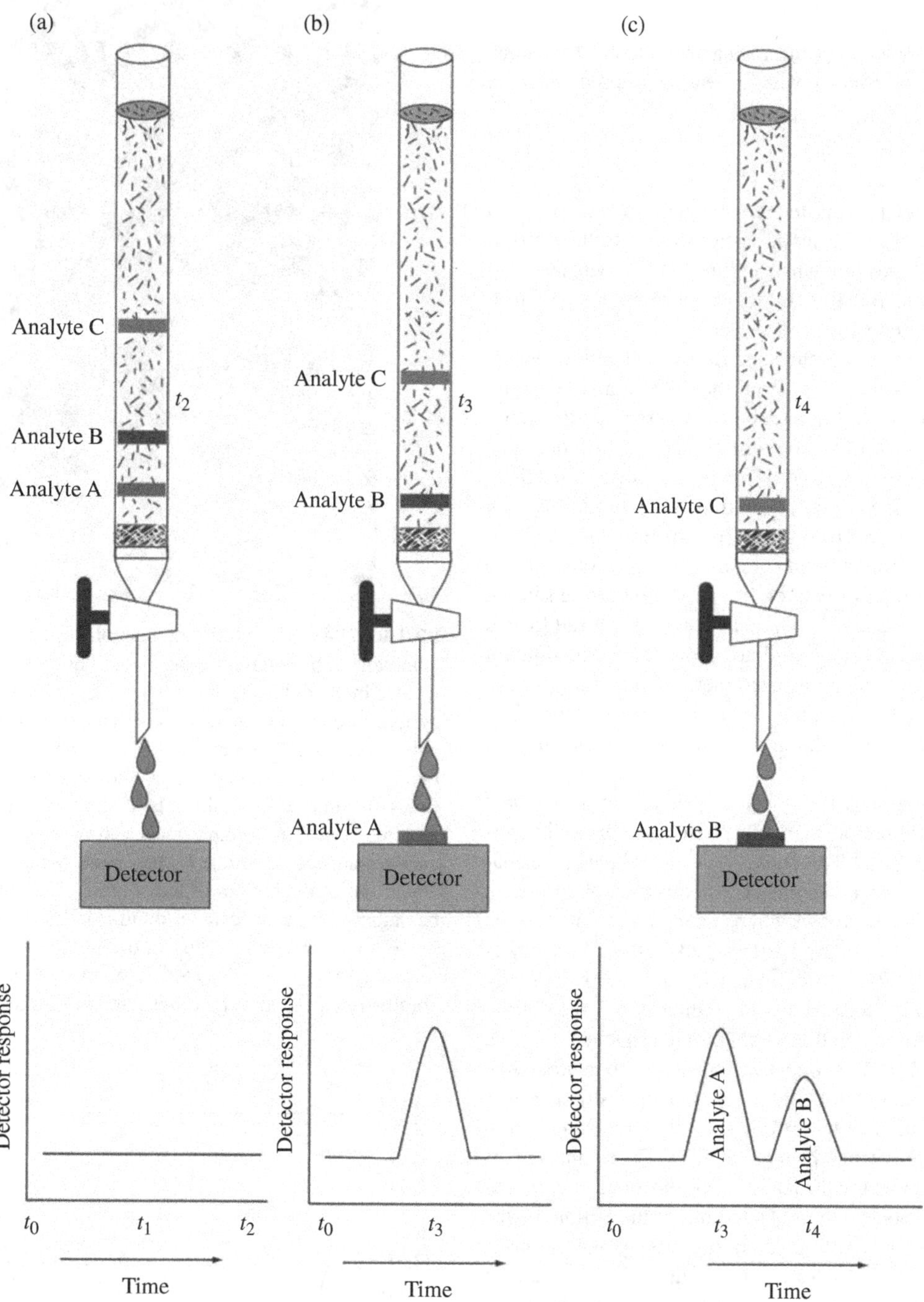

FIGURE 21.2 Elution profiles of compounds A and B: (a) at time t_2 the three compounds have moved through the column and are close to eluting from the column, (b) at time t_3 there is breakthrough of compound A and it is registered by the detector resulting in a response in the form of a Gaussian peak shape, (c) at time t_4 there is breakthrough of compound B that is also registered by the detector producing a chromatogram with two peaks.

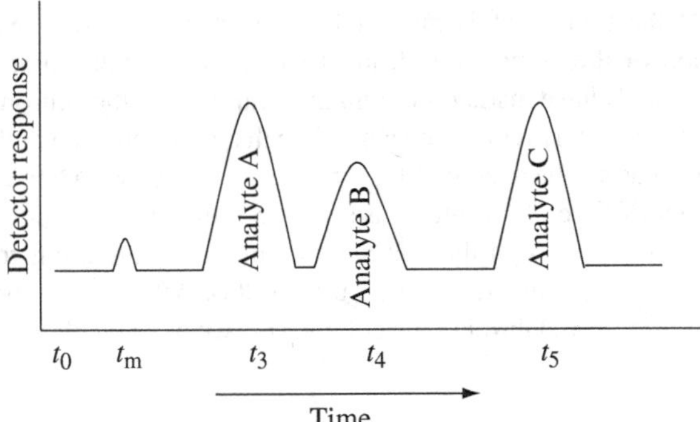

FIGURE 21.3 Chromatogram illustrating the elution of all three compounds from the column with increasing elapsed retention times ($t_5 > t_4 > t_3$).

of compound A, and it is registered by the detector resulting in a response in the form of a Gaussian peak shape as illustrated in Figure 21.2(b). At time t_4 shown in Figure 21.2(c), there is breakthrough of compound B that is also registered by the detector producing a chromatogram with two peaks.

After a certain amount of time and flow of the mobile phase, all three compounds will elute from the column and be registered by the detector producing a chromatogram as illustrated in Figure 21.3. Notice in the figure that there is included a t_m peak just right after time zero (t_0). This peak represents the dead time of the column sometimes referred to as a nonretained peak or an injection peak. This peak represents the time for a mobile-phase molecule (or nonretained compound) to pass through the column from the time of injection or application of the sample till the detector is reached. The t_m peak can be produced from a change in refractive index or baseline absorption (depending on the type of detector), or the measurement of something in the injected sample that is nonretained in the mobile-phase flow stream eluting from the column from the solvent making up the sample.

Better chromatography in the form of peak shapes is obtained by reconstituting the sample in mobile phase sometimes eliminating the nonretained peak. To obtain a nonretained peak, the analyst may sometimes need to inject a compound dissolved in mobile phase that he/she knows will be nonretained. An example of a column chromatography system used in the laboratory is illustrated in Figure 21.4. The column for the chromatography is the white column located in the right side of the figure. A sample is introduced into the system through an injection valve and sample loop. The injection valve and sample loop are in line with the column through a liquid pump, the top instrument box shown in the figure. The pump flows the injected sample onto the top of the column using the mobile phase. The sample passes through the column where the analytes are separated due to their partitioning coefficients and are eluted out of the bottom of the column. The bottom of the column has tubing that connects the liquid flow to a UV/Vis detector that is the bottom instrument box in the figure. The separated analytes move through the detector, are measured, and result in a chromatogram such as that illustrated in Figure 21.3.

The retention time of a compound, t_R, is the time that has elapsed from the injection or application of the sample onto

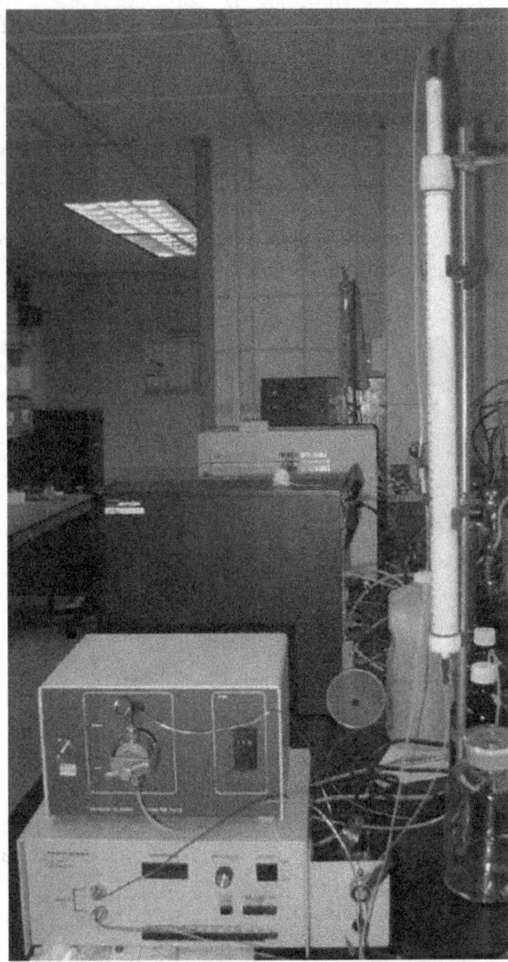

FIGURE 21.4 Column chromatography system example used in the laboratory. The column for the chromatography is the white column located in the right side of the figure. A sample is introduced into the system through an injection valve and sample loop. The injection valve and sample loop are in line with the column through a liquid pump, the top instrument box shown in the figure. The pump flows the injected sample onto the top of the column using the mobile phase. The sample passes through the column where the analytes are separated due to their partitioning coefficients, and are eluted out of the bottom of the column. The bottom of the column has tubing that connects the liquid flow to a UV/Vis detector that is the bottom instrument box in the figure. The separated analytes move through the detector, are measured, and result in a chromatogram such as that illustrated in Figure 21.3.

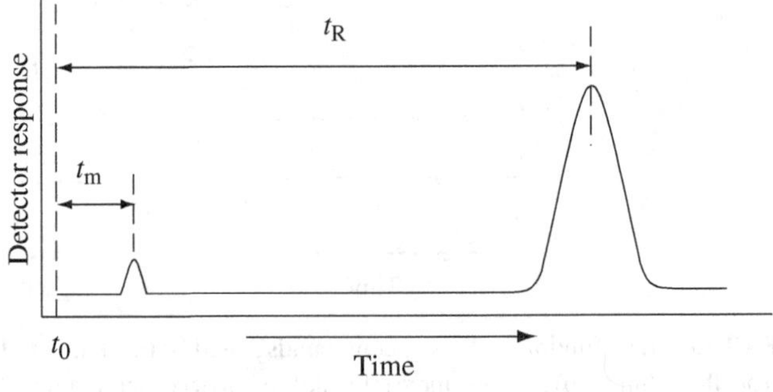

FIGURE 21.5 Retention times representing sample introduction at t_0, nonretained peak at t_m, and analyte retention time at t_R.

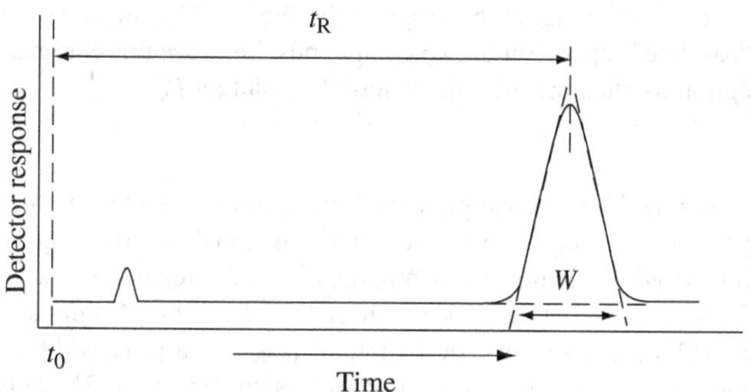

FIGURE 21.6 Chromatographic determination of number of theoretical plates N from peak width and retention time.

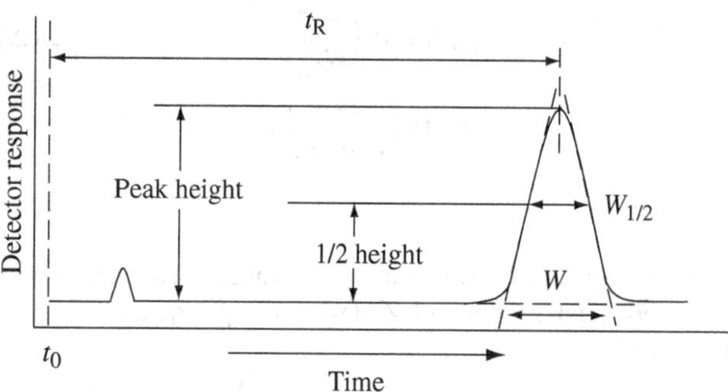

FIGURE 21.7 Chromatographic determination of number of theoretical plates N from peak width and retention time, or peak width at ½ peak height and retention time.

the column till the detection of the compound after elution from the column. The dead time, t_m, is again the time that a nonretained compound (or injection peak) is measured after the start of the chromatography. These two times are illustrated in Figure 21.5 along with the initial sample introduction as time zero, t_0. Sometimes, methods report an "adjusted" or "normalized" retention time that is obtained by subtracting the dead time t_m from the compound retention time t_R.

21.4 THE THEORETICAL PLATE NUMBER N

A term often used to describe the effectiveness of a column to separate analytes is the *height equivalent of a theoretical plate* (*HETP* or *H*) of the equilibrium state of the partitioning of the analyte between the stationary phase and the mobile phase. This terminology is the same used in distillation. The *theoretical plate number*, N, of a column (dimensionless) is defined as the length of the column (L) in cm divided by the *height equivalent of a theoretical plate* (H) in cm of a single plate:

$$N = \frac{L}{H} \tag{21.4}$$

The more theoretical plates that a column has, the greater its ability will be to separate compounds, especially analytes that have similar distribution coefficients. If it is assumed that the peak shape of an analyte will be in a Gaussian shape, then a single plate height is related to the variance in the measurement of the peak at the end of the column $\left(H = \frac{\sigma^2}{L}\right)$. It is estimated that a single plate contains approximately 34% of the analyte. Due to these relationships, we can calculate the number of theoretical plates N from a chromatogram. To obtain N from a chromatogram, we use the following relationship:

$$N = 16\left(\frac{t_R}{W}\right)^2 \tag{21.5}$$

where t_R is the retention time of the analyte and W is the width of the peak at its base. This is illustrated in the chromatogram in Figure 21.6 where tangents are drawn along the peak sides dropping down to the baseline to properly determine the peak width.

An alternative method for calculating N is to use the peak width at ½ the height of the peak, which is often viewed as a more accurate representation of N. Using the peak width at ½, the height of the peak will negate any ambiguity at determining the beginning and ending of the peak at the baseline, which may be difficult due to interference from overlapping peaks or baseline noise. Drawing a tangent by hand though is usually quite accurate. Software included in today's analytical laboratories that control the chromatographic instrumentation and process the data have functions that allow the determination of the number of theoretical plates automatically, thus relieving the analyst from performing this task by hand. Figure 21.7 illustrates a combination of peak width at the baseline and peak width at ½ peak height:

$$N = 5.54\left(\frac{t_R}{W_{1/2}}\right)^2 \tag{21.6}$$

The theoretical plate number is commonly used by column manufacturers as a measure of column efficiency and ability. When purchasing a chromatography column, the manufacturer will include a value for N for a particular analyte. To reproduce this value and thus check your column, the same compound under similar conditions should be used.

Let's use the chromatogram in Figure 21.7 to look at an illustration of calculating the theoretical plate number for this column and compound.

Example 21.1 A caffeine standard was loaded onto the chromatography column resulting in the chromatogram illustrated in Figure 21.7. The retention time of the caffeine was found to be $t_R = 7.45$ min. Dropping the left tangent of the peak to the baseline resulted in an intercept at 7.21 min. Dropping the right tangent of the peak to the baseline resulted in an intercept at 7.69 min. The peak width at baseline is then 7.69 − 7.21 min = 0.48 min. The theoretical number of plates N using Equation 21.5 is therefore:

$$N = 16\left(\frac{t_R}{W}\right)^2 \quad (21.5)$$

$$N = 16\left(\frac{7.45\,\text{min}}{0.48\,\text{min}}\right)^2 \quad (21.7)$$

$$N = 3854 \quad (21.8)$$

The peak width at ½ the peak height was found to be 0.28 min. The theoretical number of plates N using Equation 21.6 is therefore:

$$N = 5.54\left(\frac{t_R}{W_{1/2}}\right)^2 \quad (21.6)$$

$$N = 5.54\left(\frac{7.45\,\text{min}}{0.28\,\text{min}}\right)^2 \quad (21.9)$$

$$N = 3922 \quad (21.10)$$

The theoretical number of plates using the two equations gives very similar results in this case (a calculated 2% difference in the value of N). Notice that the units of min cancel out of the calculation and that N is dimensionless.

21.5 RESOLUTION R_S

A second important term that can be obtained from an experimental chromatogram is called *resolution* R_s. The resolution of a column is the quantitative measurement of the column's ability to separate two compounds. The resolution, like the theoretical plate number, can be obtained from an experimental chromatogram using the following relationship:

$$R_s = \frac{2\left[(t_R)_B - (t_R)_A\right]}{W_A + W_B} \quad (21.11)$$

The following chromatogram in Figure 21.8 illustrates a "baseline" separation of two compounds. Let's use this chromatogram to illustrate the calculation of resolution R_s.

Example 21.2 A sample containing two closely related compounds was analyzed on a column resulting in the chromatogram illustrated in Figure 21.8. Compound A had a retention time of 12.6 min $(t_R)_A$ and a peak width of W_A of 0.2 min. Compound B had a retention time of 13.1 min $(t_R)_B$ and a peak width of W_B of 0.4 min. The resolution R_s using Equation 21.11 is therefore

$$R_s = \frac{2\left[(t_R)_B - (t_R)_A\right]}{W_A + W_B} \quad (21.11)$$

$$R_s = \frac{2[13.1\,\text{min} - 12.6\,\text{min}]}{0.2\,\text{min} + 0.4\,\text{min}} \quad (21.12)$$

$$R_s = 1.7 \quad (21.13)$$

Notice here too that the units of min cancel out of the calculation and that R_s is also dimensionless.

What happens to the calculation of resolution when two compounds do not baseline separate as that found in Figure 21.8? This is the case of the chromatogram illustrated in Figure 21.9(a). Here, the peaks representing the elution of the two compounds are actually overlapping and the baseline is not visible. The chromatogram in Figure 21.9(b) shows the setup for calculating the resolution from the chromatogram. Notice that each peak actually contains a portion of the other peak. In the case of compounds eluting from the column at the same time, this is known as "coelution" and the calculation of resolution is not possible (other factors are a problem in coelution also such as peak identification and quantitation).

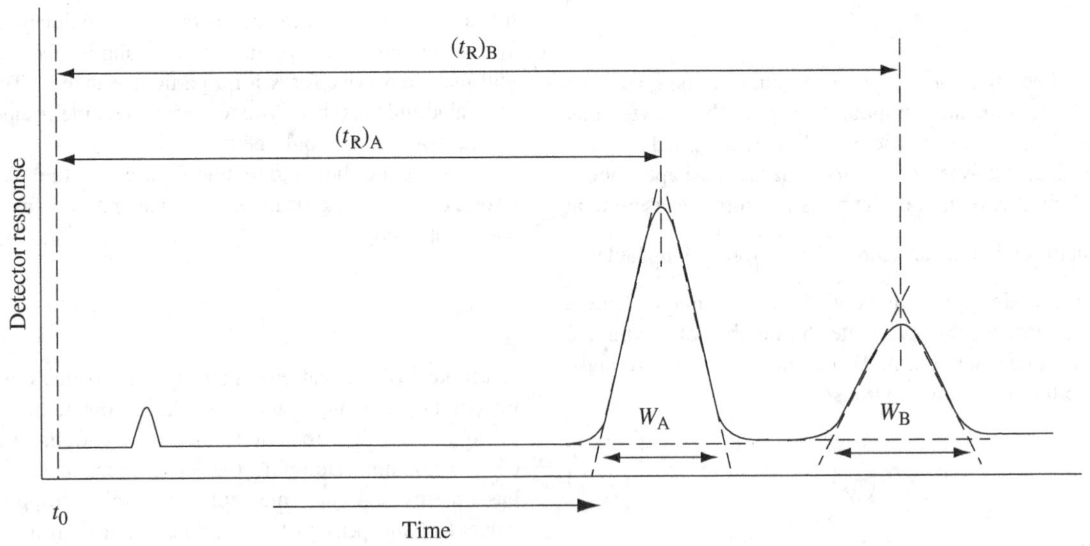

FIGURE 21.8 Chromatographic determination of the quatitative value of resolution R_s for the separation of two compounds.

(a)

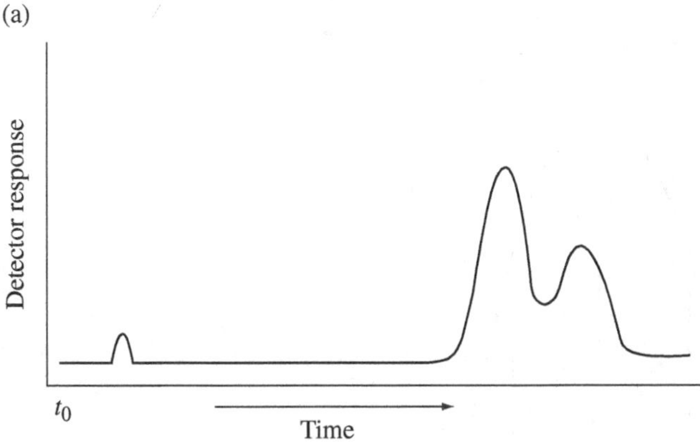

(b)

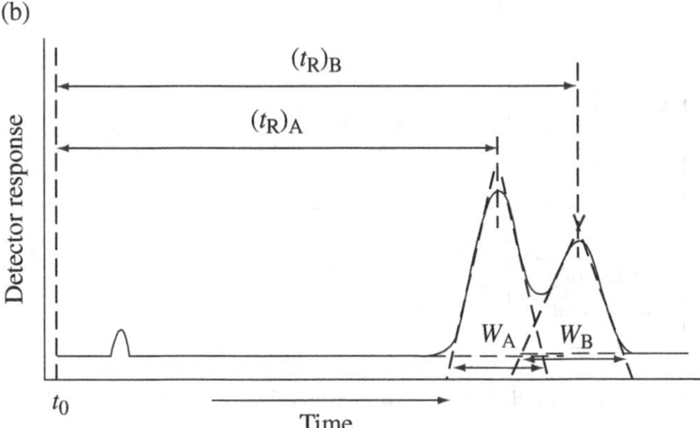

FIGURE 21.9 Chromatographic determination of the quantitative value of resolution R_s for the separation of two compounds that are not baseline resolved.

Example 21.3 A sample containing two closely related compounds was analyzed on a column resulting in the chromatogram illustrated in Figure 21.9 where the two compounds are not baseline resolved. Compound A had a retention time of 12.6 min $(t_R)_A$ and a peak width of W_A of 0.2 min. Compound B had a retention time of 12.8 min $(t_R)_B$ and a peak width of W_B of 0.4 min. The resolution R_s using Equation 21.11 is therefore

$$R_s = \frac{2\left[(t_R)_B - (t_R)_A\right]}{W_A + W_B} \tag{21.11}$$

$$R_s = \frac{2[12.8\,\text{min} - 12.6\,\text{min}]}{0.2\,\text{min} + 0.4\,\text{min}} \tag{21.14}$$

$$R_s = 0.7 \tag{21.15}$$

Typically, it is desirable to set up a chromatography method where the resolution between two neighboring compounds is 1 or greater.

21.6 RATE THEORY VERSUS PLATE THEORY

As presented in Section 21.4, "The Theoretical Plate Number *N*," the term used to describe the effectiveness of a column to separate analytes is the *height equivalent of a theoretical plate* (*HETP* or *H*) of the equilibrium state of the partitioning of the analyte between the stationary phase and the mobile phase. This terminology is the same as that used in distillation where the fractionating column's number of theoretical plates describes the column's ability to separate compounds according to their differences in boiling points. The *theoretical plate number*, *N*, of a column (dimensionless) is defined as the length of the column (*L*) in cm divided by the *height equivalent of a theoretical plate* (*H*) in cm of a single plate:

$$N = \frac{L}{H} \tag{21.4}$$

The more theoretical plates that a column has, the greater its ability will be to separate compounds, especially analytes that have similar distribution coefficients. From the relationship of the plate number "*N*" to the plate height "*H*" in Equation 21.4, we see that *N* increases with increasing length of the column "*L*" and with decreasing the plate height "*H*." The *efficiency* of a column can be thought of in terms of the plate height "*H*" where the plate height is the distance that must be traveled through the column for a separation to take place that is the same as an equilibrium separation of compounds between the mobile and stationary phases. Equation 21.4 can be rearranged into the expression relating plate height *H* to column length *L* and plate number *N*:

$$H = \frac{L}{N} \tag{21.16}$$

A more efficient column in separating compounds will have a lower plate height value than a less efficient column. A column with a plate height of $H = 0.1$ mm will have greater efficiency in separating compounds than a column with a plate height of $H =$ 1.0 mm. By increasing the column length, we can increase the number of theoretical plates, which in turn will increase the efficiency of the column. However, this will also increase the time required for the separation to take place. With longer separation times, a peak broadening effect is observed that in turn has a negative effect upon the efficiency of the separation.

In Section 21.3, we also described that the peak shape of an analyte will be in a Gaussian shape and that a single plate height is related to the variance in the measurement of the peak at the end of the column where *L* is the length of the column:

$$H = \frac{\sigma^2}{L} \tag{21.17}$$

This relationship is illustrated in Figure 21.10. For a given length of column *L*, a narrower peak will have a smaller value for σ equating to a smaller value for the plate height and thus greater efficiency. Conversely, a wider peak will have a greater value for σ equating to a larger value for the plate height and thus a reduced efficiency.

At this point, we begin to ask, what are the factors that affect the shape of the Gaussian peak illustrated in Figure 21.10? Or more importantly, what are the parameters of our separation that can be optimized to obtain a narrow peak shape for a compound

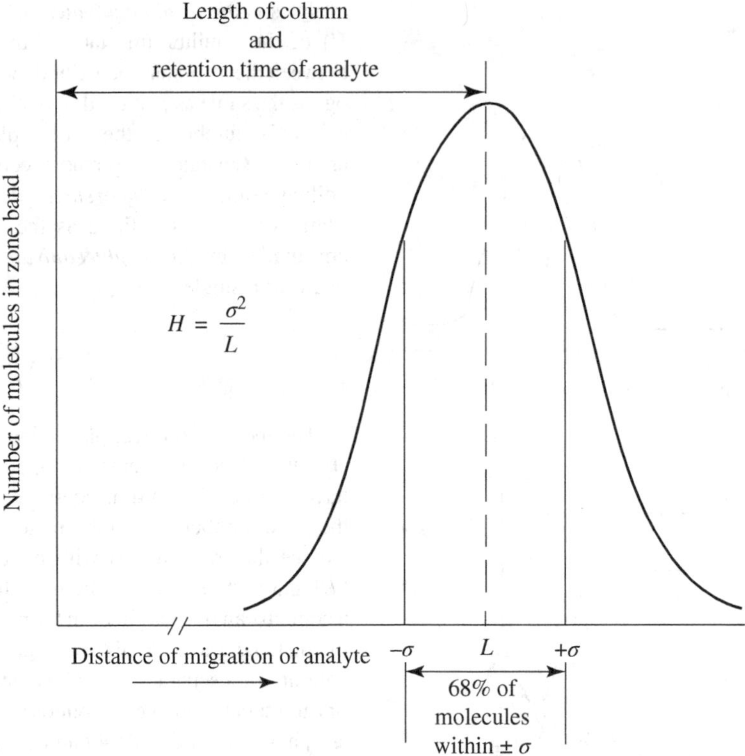

FIGURE 21.10 Illustration of the relationship between the plate height H, the column length L, and the variation σ in the Gaussian peak shape of the eluting compound.

giving us a more efficient separation? Equation 21.17 tells us that with a longer column, we should get a smaller value for H; however, experimental analysis shows us that in fact we may obtain a much broader peak for the eluting compound resulting in a decreased efficiency. The relationship in Equation 21.17 describes the plate height in terms of an equilibrium theory. To better understand the efficiency of the separation, and thus the parameters affecting the efficiency, a more accurate description was developed and reported in 1956 by van Deemter et al. [2] where the chromatographic column is described according to a rate theory in terms of partitioning dynamics versus what before was described as equilibrium conditions. This is in fact the case as the mobile phase is always moving; thus, a true equilibrium state cannot be obtained. In simplified form, the rate theory describes a relationship of column parameters according to the linear gas flow rate (F) through the column as cm of column length per sec (cm/s). The *van Deemter equation* is expressed in the simplified form as

$$H = A + \frac{B}{F} + (C_S + C_M)F \tag{21.18}$$

where H is the plate height in centimeters, A is the coefficient for *multiple flow paths* (*eddy diffusion*), B is the coefficient for *longitudinal diffusion*, C is the coefficient for *mass transfer between phases* (C_S for stationary and C_M for mobile), and F is the mobile-phase linear velocity in centimeters per second (cm/s). The van Deemter equation describes the parameters that attribute to zone broadening. The term "zone broadening" is referring to the widening of a zone or band as it moves through the chromatographic column. Another way to think about this is to consider zone broadening as peak broadening. Let's take a closer look at the three coefficients in the van Deemter equation.

A graphical representation of the three terms (A, the coefficient for *multiple flow paths* (*eddy diffusion*); B, the coefficient for *longitudinal diffusion*; and C, the coefficient for *mass transfer between phases* (C_S for stationary and C_M for mobile) and their behavior with increasing F, the mobile-phase linear velocity in centimeters per second (cm/s)) for the van Deemter equation is illustrated in Figure 21.11. As illustrated in the figure, the A coefficient term has a constant effect upon the efficiency with increasing F, while the $(C_S + C_M)F$ coefficient term has an increasing effect upon the efficiency with increasing F.

21.6.1 Multiple Flow Paths or Eddy Diffusion (A Coefficient)

When a band of analytes are passing through a chromatographic column packed with stationary-phase particles, the individual molecules making up the band will encounter slightly different paths to follow. Some paths will be either a little longer or a little shorter as compared to each other. The end result will be an elongation or stretching out of the band as the molecules that

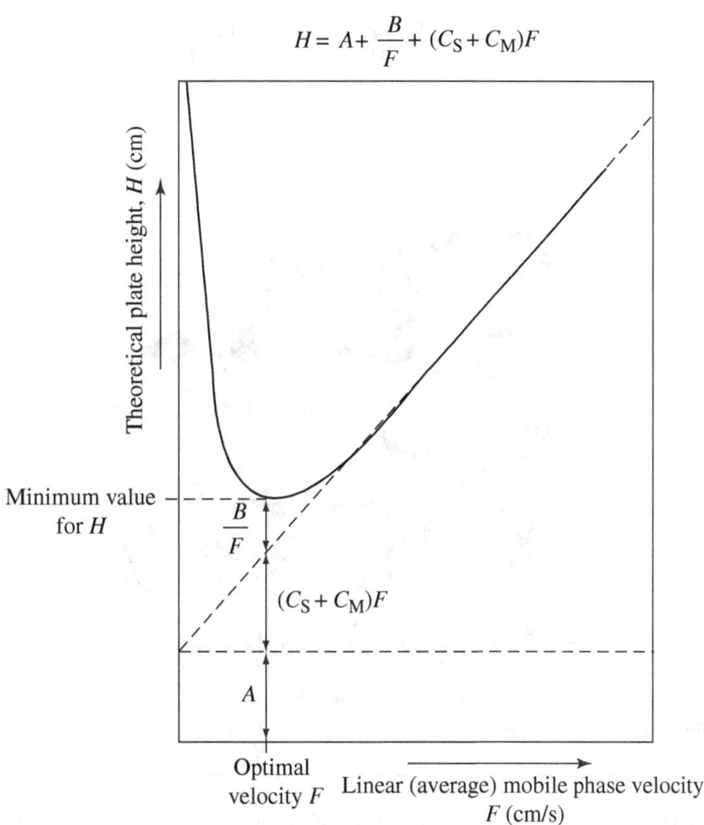

FIGURE 21.11 Theoretical plate height (H) versus Linear (Average) Mobile Phase Velocity (F) van Deemter plot. The influence upon the shape of the curve by A the coefficient for *multiple flow paths (eddy diffusion)*, B the coefficient for *longitudinal diffusion*, and C the coefficient for *mass transfer between phases* (C_S for stationary and C_M for mobile) are included in the plot. The curve minimum gives the minimum value for H in conjunction with the most optimal velocity F.

on average traveled a set of shorter paths will elute from the column quicker than the molecules that on average traveled a set of longer paths thus eluting slightly slower. The deleterious effect of multiple flow paths can be alleviated through the incorporation of smaller particle sizes that are of a very uniform shape and that have been efficiently packed. If we take a closer look at this, it is easier to see how the stationary-phase particle size will have an effect upon the paths traveled by the analyte molecules. In Figure 21.12, the column has been packed with particles that possess a wide variety of shapes and sizes. The packing of the particles, due to the shapes, will not be in a very uniform fashion. This will in turn create a variety of alternative paths that the analyte molecules may transverse.

In the figure, the molecule that has traveled path A has actually traveled a farther distance than the molecule that has moved along path B. Path A will take more time to move through resulting in a slower elution time. In contrast, the molecule in path B will elute faster. This results in the spreading out or broadening of the zone (peak), decreased resolution, and lower column efficiency. However, if we pack our column with small, uniformly shaped particles as those illustrated in Figure 21.13, the resulting paths A and B will be very similar in distance and time. This results in a much narrower zone eluting from the column, higher resolution, and therefore increased column efficiency. Lastly, as we observed before in Figure 21.11, the A term has a constant effect upon the resolution.

21.6.2 Longitudinal (Molecular) Diffusion (B Coefficient)

The zone broadening produced by longitudinal or molecular diffusion is the diffusion of the molecules in the forward and reverse direction from the concentrated center of the zone to the more dilute outer regions of the zone. The longitudinal effect upon resolution is decreased with the use of small, uniformly shaped particles as those illustrated in Figure 21.13. As shown in Equation 21.18, the longitudinal diffusion is inversely proportional to the linear velocity (F) of the mobile phase. Thus, as the mobile-phase velocity is increased, the B/F term in Equation 21.18 decreases, and in turn, a smaller value for H is obtained. This effect is related to the residence time of the analytes in the column; with a higher mobile-phase velocity, the analytes spend less time in the column, and thus, diffusion is decreased. This is illustrated in Figure 21.11 in the left portion of the curve in the form of a negative slope. Notice however that there is an optimal flow rate and that above this the resolution will again decrease. In gas chromatography, longitudinal diffusion is often viewed as band broadening caused by the diffusion of the

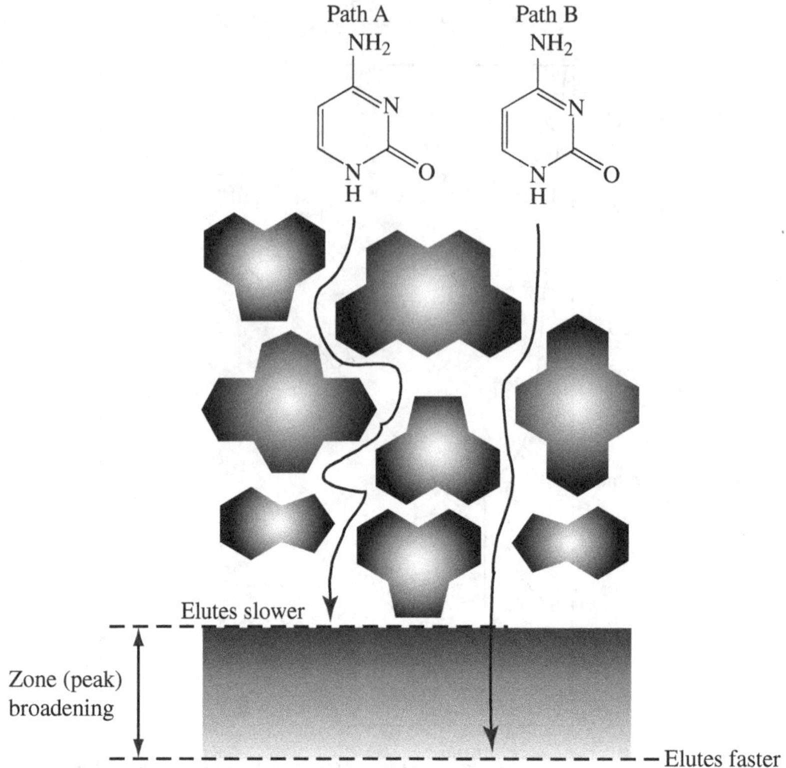

FIGURE 21.12 Effect stationary phase particle size has upon the paths of the analyte molecules during a chromatographic separation. The molecule that experiences Path A will travel a farther distance than the molecule in Path B, resulting in different elution times and subsequent zone broadening.

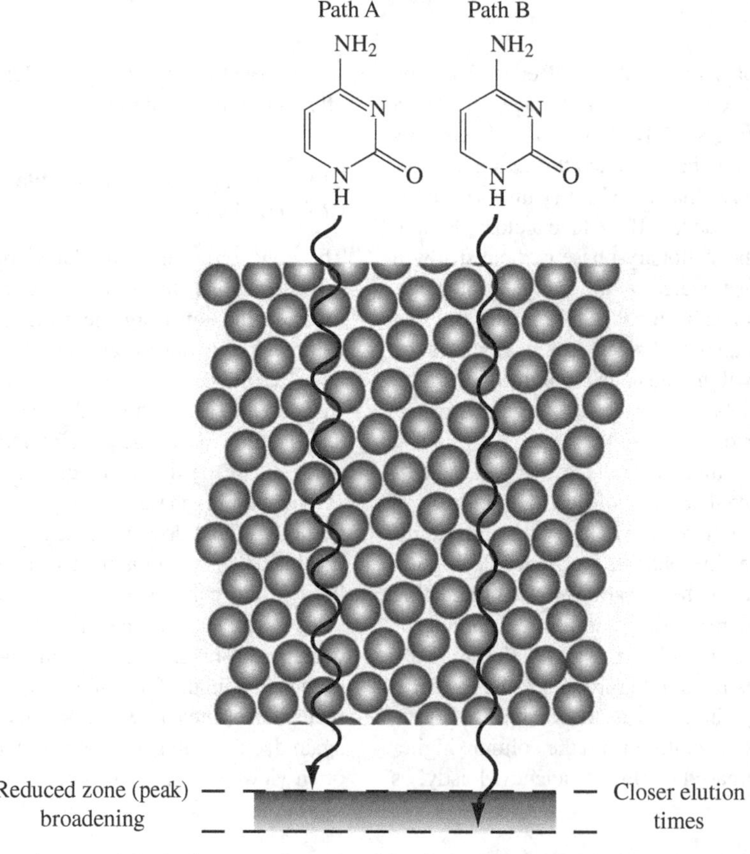

FIGURE 21.13 Effect stationary phase particle size has upon the paths of the analyte molecules during a chromatographic separation. The paths taken in A and B are very similar resulting in a narrower zone eluting from the column, higher resolution, and increased column efficiency.

analyte in the forward direction of the center of the band in the carrier gas mobile phase.

21.6.3 Mass Transfer Resistance between Phases (C_S and C_M Coefficients)

Along the length of the column, equilibrium in concentration of the analytes between the two phases (mobile and stationary) as we stated before is not actually achieved. If our chromatography system were in fact a static system, then equilibrium of the distribution of the analyte within the phases would take place. However, this is a dynamic system where the mobile phase is constantly sweeping analytes forward. There is also a time lag taking place with the movement of the analyte from within each phase to the interface of the phases where exchange takes place. Thus, analytes at the front of the band in the mobile phase, due to nonequilibrium conditions, are slightly swept forward by the movement of the mobile phase broadening the band in the forward direction. Analytes at the back of the band in the stationary phase, due to nonequilibrium conditions, lag slightly behind broadening the band in the backward direction. As shown in Equation 21.18, the mass transfer effect is directly proportional to the linear velocity of the mobile phase. As would be guessed with increasing mobile-phase velocity, the approach to equilibrium is even more reduced, leading to a lowering of the efficiency. A thinner stationary phase can reduce the lag time in the analytes moving from the interior of the stationary phase to the phase interface region.

In a more complex representation, the van Deemter equation can be expanded to include very specific parameters of the chromatography system and thus more precise information on the causes of efficiency loss and possible approaches for optimization [3]:

$$H = 2\lambda d_p + \frac{2GD_m}{\mu} + \frac{\omega\mu(d_p \text{ or } d_c)^2}{D_m} + \frac{Rd_f^2\mu}{D_s} \qquad (21.19)$$

where:

H is the plate height.
λ is the packing particle shape.
μ is the linear velocity (flow rate) of the mobile phase.
d_p is the particle diameter.
G, ω, and R are the constants.
D_m is the mobile-phase diffusion coefficient.
d_c is the capillary diameter.
d_f is the film thickness.
D_s is the stationary-phase diffusion coefficient.

We can see in greater detail from the relationships in Equation 21.19 that the three parameters just discussed that affect the plate height and therefore the zone broadening (efficiency) of the column are in fact (i) eddy diffusion, independent of the flow rate (μ) and directly dependent upon particle packing shape (λ) and diameter (d_p), ($2\lambda d_p$); (ii) longitudinal diffusion, proportionally dependent on the flow rate (μ), $\left(\frac{2GD_m}{\mu}\right)$; and (iii) mass transfer resistance, inversely proportional to the flow rate (μ), the mobile-phase diffusion coefficient (D_m), and the stationary-phase diffusion coefficient (D_s) and directly proportional to the film thickness (d_f), $\left(\frac{\omega\mu(d_p \text{ or } d_c)^2}{D_m} + \frac{Rd_f^2\mu}{D_s}\right)$.

In the following sections, we will be looking closely at analytical chromatography techniques, both instrumental and benchtop, that the technician will encounter in most analytical laboratories: high-performance liquid chromatography (HPLC), SPE, TLC, and gas-liquid chromatography (GLC). The column efficiency (plate height H) in the HPLC and GLC techniques can be accurately described with the van Deemter Equation 21.18. The plots of plate height H versus linear mobile-phase velocity for the two techniques have the basic structure of that illustrated in Figure 21.11 but are slightly different from each other due to a couple of factors. The plots illustrated in Figure 21.14 show experimental fits of the van Deemter equation (modified for B and C) for the chromatographic techniques (Fig. 21.14: (A) liquid chromatography (LC), (B) SFC, and (C) solvating gas chromatography (SGC) [4]).

We can see from the figure that the mobile-phase velocities used in liquid chromatography (A) are lower than those used in gas chromatography (B). This is in fact almost always the case where, per length of column, analytes elute much faster in gas chromatography than in liquid chromatography. A second comparison between these two chromatography techniques is that plate heights are generally about 10 times lower in liquid chromatography than in gas chromatography (although in this particular study, they are more similar than that due to other factors). One advantage that can be utilized in gas chromatography to help improve separation efficiencies is the length of the column. With gas chromatography, it is possible to construct columns that are very long on the order of anywhere from 15 m up to 200 m. This in turn tremendously increases the number of theoretical plate resulting in very high separation efficiencies. Due to constraints such as high pressure drops, LC columns are usually not constructed in lengths greater than 50 cm in general applications and uses. Obviously, 200 m compared to 50 cm is quite a difference in length and results in often far superior separation in gas chromatography as compared to liquid chromatography. As we will see later though, gas chromatography is somewhat limited to analytes that are either semivolatile or volatile or are derivatized to be volatile, giving huge and diverse application to LC when analytes are not conducive to GC analyses.

21.7 RETENTION FACTOR k'

In liquid chromatography and gas chromatography, the retention factor (also previously called capacity factor) k' is used to describe the retention behavior of a compound on a particular column. The retention factor uses the compound's retention time t_R and the nonretained retention time t_m and is calculated using Equation 21.20:

$$k' = \frac{t_R - t_m}{t_m} \qquad (21.20)$$

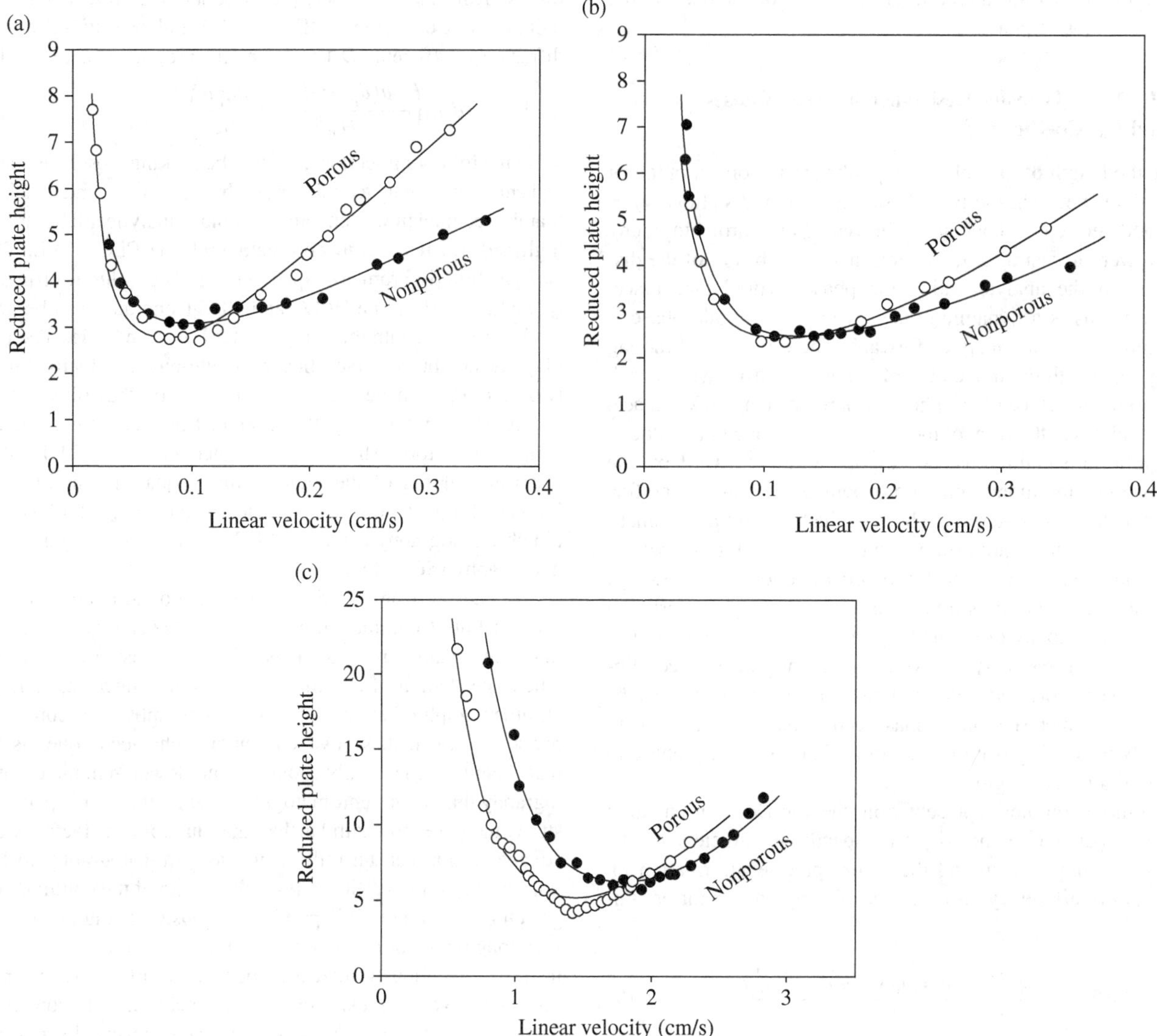

FIGURE 21.14 Van Deemter curves for packed capillary LC, SFC, and SGC. Conditions: (a) LC, 55 cm × 250 μm i.d. fused-silica capillary columns packed with 5-ím porous (120 Å) and nonporous silica particles, nitromethane test solute, 25 °C, UV detector (214 nm); (b) SFC, 75 cm × 250 μm i.d. fused-silica capillary columns packed with 5-μm porous (120 Å) silica deactivated with poly(methylhydrosiloxane) and cross-linked with SE-54 (10% w/w) and nonporous silica deactivated with poly(methylhydrosiloxane) and cross-linked with SE-54 (1.5% w/w), 45 °C, 230 atm, carbon dioxide mobile phase, methane test solute, FID; (c) SGC, 120 °C; other conditions are the same as for SFC. Solid lines represent regression analysis. (Reprinted with permission from Naijun Wu, Qinglin Tang, Yi Shen, Milton L. Lee, Porous and nonporous particles in packed capillary column solvating gas chromatography. *Anal. Chem.* 1999, **71**, 5084–5092. Copyright 1999 American Chemical Society.)

In general, under typical analytical measurement conditions, a good range for the retention factor k' of an analyte is between 2 and 10. If the k' value is much greater than 10, then the retention time of the compound becomes quite long, and peak broadening begins to take place, lowering the resolution. If the k' value is lower than 1, then the compound is eluting too quickly, and problems associated with reproducibly measuring the retention time and peak shape for area determination begin to arise.

Now that we understand the basic theory of chromatography (particularly liquid chromatography at this time), let's move on to an analytical technique that is one of the most common instrumental analyses that the technician will most likely encounter in the analytical laboratory, high-performance/pressure liquid chromatography, HPLC.

REFERENCES

1. Tswett, M. Adsorption analysis and chromatographic methods. Application to the chemistry of chlorophylls. *Ber. Deut. Botan. Ges.* 1906, **24**, 384. Translated to English by H.H. Strain and J. Sherma, *J. Chem. Ed.* 1967, 44, 238.
2. van Deemter, J.J.; Zuiderweg, E.J.; Klinkenberg, A. Longitudinal diffusion and resistance to mass transfer as causes of non ideality in chromatography. *Chem. Eng. Sci.* 1956, **5**, 271–289.
3. van Deemter, J.J.; Zuiderweg, E.J.; Klinkenberg, A. Longitudinal diffusion and resistance to mass transfer as causes of non ideality in chromatography. *Chem. Eng. Sci.* 1956, **5**, 271–289.
4. Wu, N.; Tang, Q.; Shen, Y.; Lee, M.L. Porous and nonporous particles in packed capillary column solvating gas chromatography. *Anal. Chem.* 1999, **71**, 5084–5092.

22

HIGH PERFORMANCE LIQUID CHROMATOGRAPHY (HPLC)

22.1 HPLC Background
22.2 Design and Components of HPLC
22.2.1 HPLC Pump
22.2.2 HPLC Columns
22.2.3 HPLC Detectors
22.2.4 HPLC Fraction Collector
22.2.5 Current Commercially Available HPLC Systems
22.2.6 Example of HPLC Analyses

22.1 HPLC BACKGROUND

Liquid chromatography offered to analysts a way of separating biological compounds into discrete bands eluting off of a chromatography column, allowing their detection and collection (fraction collecting) in a pure state. However, the columns were usually of dimensions (5 cm × 100 cm) that required large particle sizes (5 μm) of the stationary phase that resulted in slow separations (sometimes many hours) that often were not efficient. Attempts at speeding up the separations by applying pressure to a mobile-phase reservoir on top of the column (called flash chromatography and is still widely used in LC preparatory work) actually resulted in lowering the plate heights of the column, thus lowering the column's ability to separate compounds. With experimentation, it was found that using smaller particle sizes for the stationary phase resulted in a more efficient separation of compounds. Smaller particle sizes for the stationary phase result in a greater surface area overall for the stationary phase that equates to larger plate heights (increase in the efficiency of separations). However, smaller particle sizes also results in a greater resistance to flow of the mobile phase through the column meaning even longer separation times. With time, chromatographers began to realize that a way to greatly improve on liquid chromatography was to combine high pressure to the system, which reduced the separation time with smaller stationary-phase particle sizes, which increase the column efficiencies. Improvements in liquid chromatography eventually produced the systems in use today known as high-performance/high-pressure liquid chromatography, HPLC. Most analysts prefer to use the term high-"performance" liquid chromatography or just simple "HPLC" for short. As just discussed, the major advancement in LC was the application of high pressure through the use of liquid pumps to increase the flow of the mobile phase through columns that now contained much smaller particle sizes for the stationary-phase packing material.

22.2 DESIGN AND COMPONENTS OF HPLC

There are a large number of manufacturers of HPLC systems that the analyst may come to use in the laboratory. Some of the most common vendors include Agilent Technologies, Beckman Coulter, Buck Scientific, Dionex, Hitachi, PerkinElmer Inc., Shimadzu Scientific Instruments, Thermo Fisher Scientific, Varian Inc., and Waters Corporation. Though there are differences in designs, the basic principle behind each vendor HPLC system is the same: a high-pressure pump to drive the mobile phase, an injection port to introduce the sample, a column for the chromatography, and some form of detector to register the separated components. Figure 22.1 illustrates the basic components of an HPLC system and how each component of the system is connected to each other. The solvent reservoirs, located in the top left-hand side of Figure 22.1 as four bottles, hold the different mobile phases being used in the HPLC methodologies. HPLC systems will always require at least one solvent reservoir. In this diagram, there is a regulated helium source that is plumbed into the solvent reservoir bottles ending in "spargers." This is used to bubble helium into the solvent reservoirs to degas the mobile phases. An alternative to helium spargers is in-line vacuum degassers where the mobile phases are carried through a vacuum box through a gas-permeable tubing that allows the removal of dissolved and entrained gasses from the mobile phases.

Analytical Chemistry: A Chemist and Laboratory Technician's Toolkit, First Edition. Bryan M. Ham and Aihui MaHam.

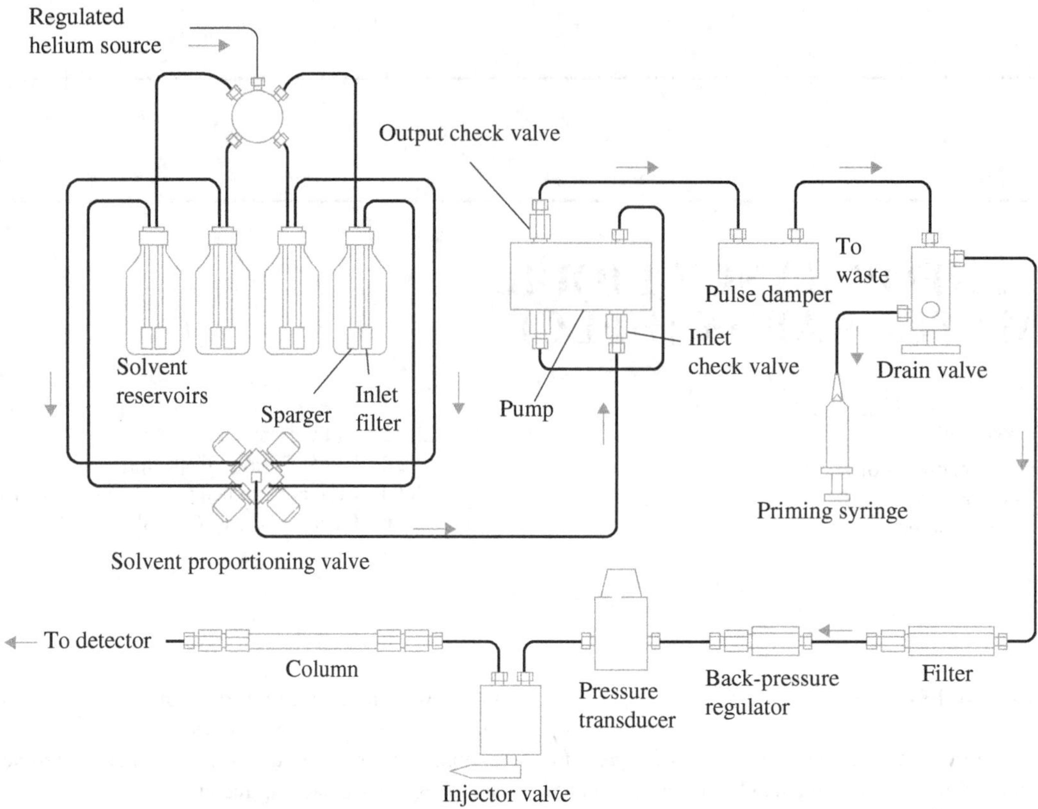

FIGURE 22.1 Diagram of the basic components of an HPLC system and their interconnectivity. (Reprinted with permission from PerkinElmer.)

Dissolved gasses in the mobile phases can come out of the mobile phase under high pressures and create bubbles in the system or make the mobile-phase flows unstable. Various HPLC systems have their own approach to mixing the mobile phases prior to reaching the injection valve and analytical column. The system in Figure 22.1 has a solvent proportioning valve located right beneath the solvent reservoirs where the prescribed percentages of the mobile phases are proportioned prior to the pump. Let's look at another system that closely illustrates the action of an HPLC pump in a simplified view.

On the right-hand side of Figure 22.2(a), the solvent reservoirs (mobile phases) are connected to a mobile-phase filling (mixing) chamber. In the illustration, it is mobile phase A, methanol, being introduced into the chamber along with mobile phase D, water. In this case, it is 30% mobile phase A that is being mixed with 70% mobile phase D. The mobile-phase mixing chamber is connected to the head of the pump through a transfer line (see Fig. 22.1). When the metering piston is forced into the mobile-phase mixing chamber as illustrated in Figure 22.2(b), the fluid in the mixing chamber is forced into the pump head and in the process is completely mixed together at a 30:70 ratio. The next action of the pump is to drive the delivery piston into the pump head, as illustrated in 22.2(a), thus forcing the mixed mobile phase through the system to the column. This pumping action is cycled back and forth through Figure 22.2(a) to (b), continuously delivering a mobile-phase composition controlled by the method being used.

If a constant ratio of mobile phases is being pumped through the system, such as the 30:70 mobile phase A to D as illustrated here, the HPLC run is called an "isocratic" run. If on the other hand the composition of the mobile-phase mixture is changing with time, then the run is called a "gradient" run. For example, if the HPLC analysis starts off at 100% mobile phase A but is then programmed to change in composition to 30% mobile phase A and 70% mobile phase D over a 20 min period, the mobile-phase composition changes in integral steps with time according to the gradient. Let's look at an actual HPLC system being used in the laboratory. Figure 22.3 illustrates a Waters HPLC system. The HPLC system is broken into three main components: (a) the pump for flowing the mobile phase through the system, (b) a set of detectors for measuring the separated analytes, and (c) a fraction collector for collecting the separated analytes after separation by the column and measurement by the detectors. Analytes in this way can be collected in a relatively pure form that may be used in further analyses such as FTIR or NMR spectroscopy. The system also contains a set of mobile phases that are actually located behind the system (out of view) in glass bottles. The tubing connecting the mobile phases to the rest of the HPLC system run through a vacuum degasser prior to the pump. The vacuum degasser is located between the pump and the detectors in Figure 22.3 (slightly visible in the figure but mainly is a box with the mobile phase running into and out of with tubing labeled A, B, C, and D). We will now take a closer look at each of the three main components of this HPLC system. Figure 22.4 is a closeup view of the pump whose purpose is to flow mobile phase through the HPLC system. The top part of the pump shown in Figure 22.3 (a) but not in Figure 22.4 is a programmable controller for

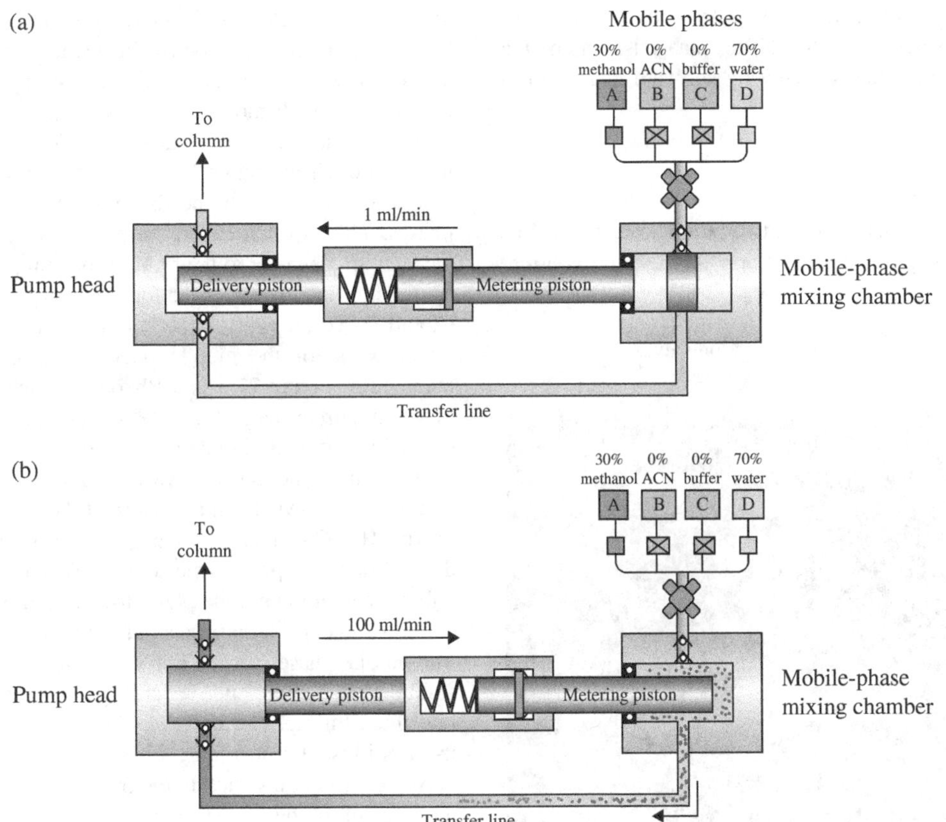

FIGURE 22.2 Diagram of the basic components of an HPLC pump. (Reprinted with permission from PerkinElmer.)

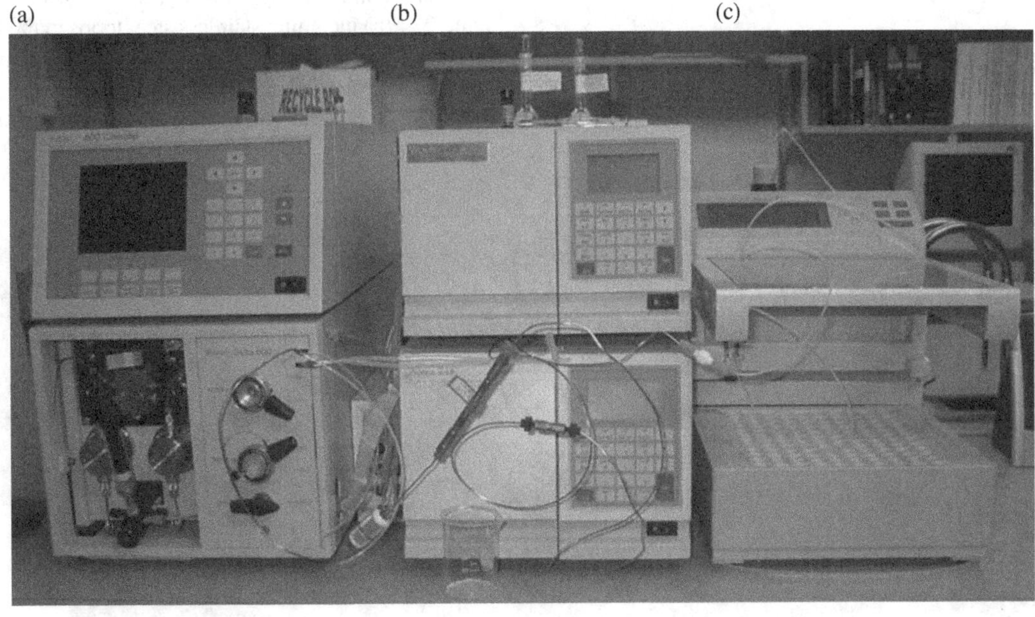

FIGURE 22.3 A Waters HPLC system being used in the laboratory. The HPLC system is broken into three main components: (a) the pump for flowing the mobile phase through the system, (b) a set of detectors for measuring the separated analytes, and (c) a fraction collector for collecting the separated analyte(s) after chromatographic separation by the column and subsequent measurement by the detectors.

the pump where settings such as flow rates and gradient compositions can be entered for the HPLC analysis. Often, the HPLC system is controlled remotely by software located on a connected PC.

22.2.1 HPLC Pump

In Figure 22.4, we see that there are two pump heads that house pistons for delivering the mobile phase through the column,

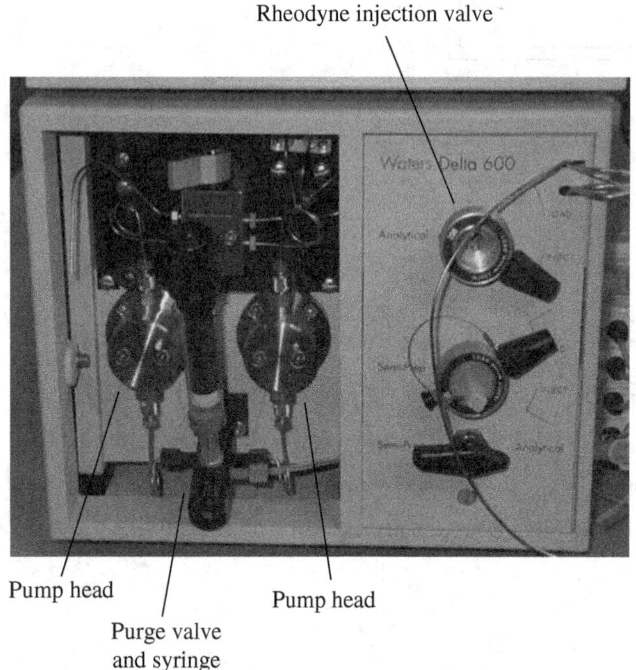

FIGURE 22.4 Close up view of the Waters HPLC pump. Included in the figure are two pump heads, a purge valve and purge syringe, and a Rheodyne injection valve. Notice also the vacuum degasser to the right of the pump.

through the detectors, and finally to the fraction collector (refer back to Figure 22.3 for a look at the inside workings of the pump). There is also a purge valve and purge syringe used to flush the pump heads with mobile phase. This is done by opening the purge valve and pulling on the syringe. This in turn pulls mobile phase through the pump heads effectively replacing any other mobile phase left in the heads with the desired current mobile phases. This is often called "priming" the pumps. Notice also the vacuum degasser to the right of the pump. What can be seen are the four inlet ports for the four mobile-phase reservoirs A, B, C, and D. To the right of the inlet ports (not seen) are identical outlet ports for the mobile phase after the degassing. Also included in Figure 22.4 is a Rheodyne injection valve. This is a very important part of the HPLC system, and we shall take a closer look at its design and its function.

The Rheodyne injection valve is a small 6-port/3-way valve system that allows the introduction of the sample to be analyzed into the HPLC system's mobile phase. The sample is injected into the mobile phase prior to the column. When injected, the sample will flow with the mobile phase to the head or beginning of the column and deposit there as a plug of nonseparated compounds. The mobile-phase composition can be then changed to begin to solubilize the compounds according to their affinities and separate them through the column. The Rheodyne injection valve itself is illustrated in Figure 22.5(a) as a cylindrical body. The valve consists of a switch in the front of the valve used to move the channels of the valve from the load position to the inject position by holding the arm of the valve and moving it in one direction or the other. In the rear of the valve is located what is known as an injection loop. An injection loop is a tube that will hold a certain known volume of sample or standard solution to be injected into the mobile-phase stream.

Different sizes and types of injection loops are illustrated in Figure 22.5(b) including stainless steel loops to the left and polyether ether ketone (PEEK, a polymer of repeating monomers of two ether and ketone groups making a heat- and solvent-resisting thermoplastic material) tubing loops on the right. Injection loop

FIGURE 22.5 (a) Close up view of the Rheodyne injection valve. Located in the back of the injection valve is a sample loop. (b) Examples of different volumes of sample injection loops.

volumes can range anywhere from 1–5 to 100–1000 μl. Typical injection loop volumes used in the laboratory range from 5 to 50 μl in most applications. The greater the volume of the injection loop, the larger amount of sample will be introduced onto the column. For example, let's consider the amount of sample introduced in the following Example 22.1.

Example 22.1 Suppose we have a standard consisting of 20.6 ppm caffeine. How much sample is introduced onto the HPLC column if we use a 5 μl injection loop? How much if a 20 μl injection loop is used?

First, remember that 20.6 ppm equals $\frac{20.6\ \text{mg}}{1}$ (wt/v) and that 5 μl = 5×10^{-6} l.

$$\begin{aligned}\text{Amount on column} &= \left(\frac{20.6\ \text{mg}}{1}\right)\left(5\times10^{-6}\ \text{l}\right)\\ &= 10.3\times10^{-5}\ \text{mg}\\ &= 10.3\times10^{-2}\ \mu\text{g}\\ &= 103\ \text{ng}\end{aligned} \tag{22.1}$$

Using the same approach, a 20 μl injection would equate to 412 ng of caffeine on column.

The internal valve system within the Rheodyne injection valve is relatively simple to understand with a visual example, as illustrated in Figure 22.6(a) and (b). The valve consists of a set of six ports and two channels that can connect flow between the ports. Figure 22.6(a) illustrates some of the specific designs for flow

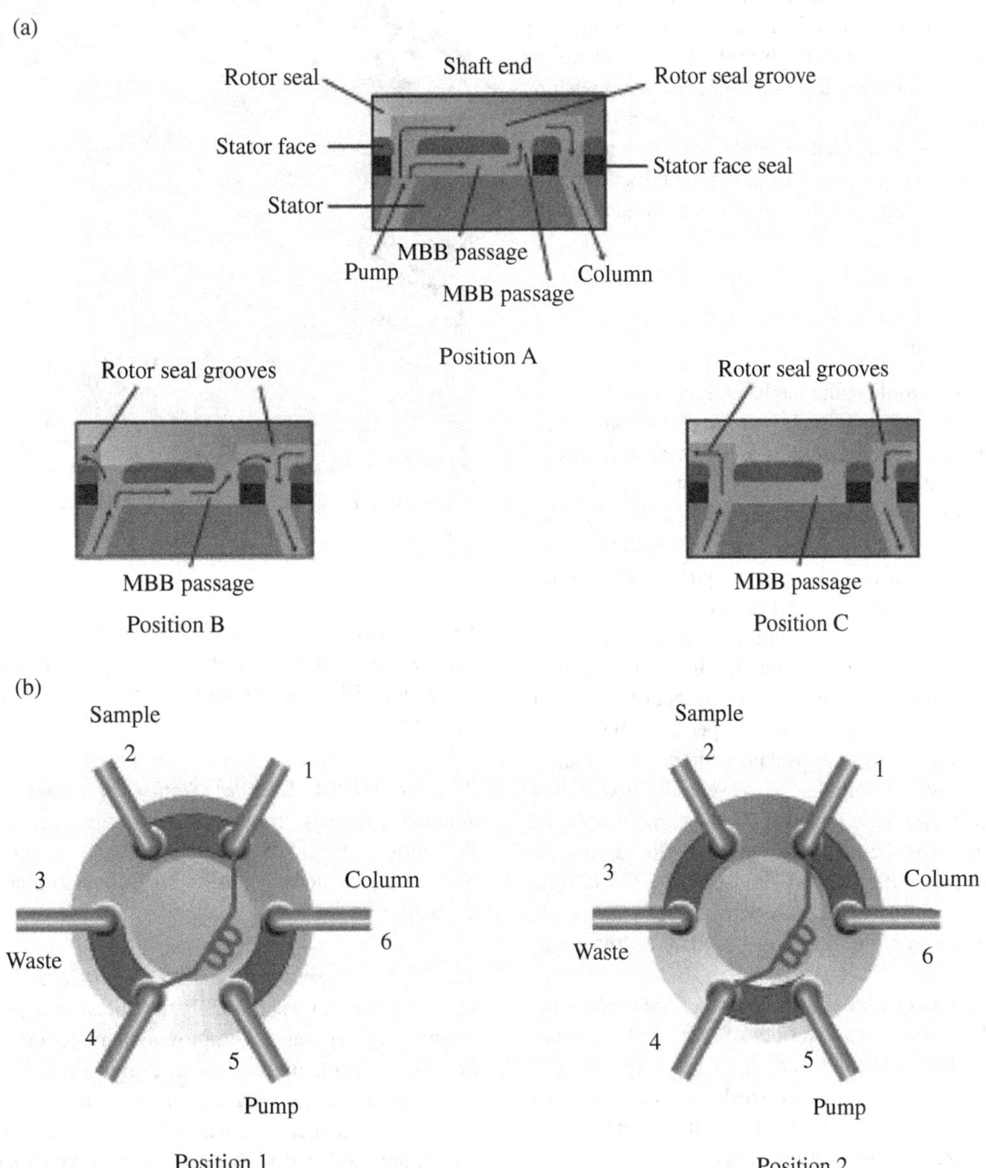

FIGURE 22.6 (a) Close up view of the flow paths within the internal channels of the Rheodyne injection valve. (b) Valve switching flow paths through the position A "load" and the position B "inject."

through the ports and channels. In this example though, let's consider the two positions listed in Figure 22.6(b). In position A (load), the mobile phase is flowing from the pump into port 2, which has a channel connecting it to port 3 where the column is connected. In this configuration, the mobile phase is flowing from the pump directly to the column. Port 4 has the needle port aligned with it and is connected to port 1 through the injection loop. Port 1 has a channel connecting it to port 6 that goes to a waste collection bottle. In this load position A, sample is forced through the injection loop using an injection syringe that is pushed into port 4, and the syringe plunger is pushed through the syringe barrel. This is done to wash the injection loop with sample and to load the injection loop with sample in preparation to introduction to the column. Excess sample being pushed through the loop is directed to port 6 to waste. The valve is then switched to the position B (inject) in Figure 22.6(b). This moves the channels connecting the ports from what is referred to as the "1–6" position (position A) to the "1–2" position (position B). In the inject position B, there is a channel now connecting the pump to the injection loop through ports 2 and 1 where the mobile phase runs through the injection loop, effectively pushing the loaded sample out of the injection loop to port 4. In this position, port 4 is now connected through a channel to port 3, the column. The sample is thus pushed out of the injection loop and onto the beginning of the column for the start of the chromatographic separation.

22.2.2 HPLC Columns

After injection of the sample using the Rheodyne injection valve into the HPLC system, we come to the second part of our system illustrated in Figure 22.3, (b) the column and detectors. A closeup view of the column and detectors is illustrated in Figure 22.7. The sample introduced into the mobile-phase flow will actually load onto what is known as a guard column. The guard column is usually a very short version of the principle analytical column made up of the same stationary phase. The purpose of the guard column is to filter out any impurities that may otherwise become permanently bound to the analytical column. In time, this impurity buildup can ruin an analytical column, so the guard column is sacrificed and changed out whenever it is suspected of being very dirty. Because the guard column is packed with the same stationary phase as the analytical column, it is essentially just a small extension of the column and will have the same chromatographic effect on the analytes as the analytical column. Guard columns are HPLC system parts often called "consumables," while the principle analytical column is not and with proper care can be used for many more sample injections as that of the guard column.

The HPLC column is where the separation of the compounds is taking place. There are a wide variety of sizes and stationary phases used for HPLC columns ranging from large (10 mm width × 50 cm length) for preparatory work, analytical size for flows in the range of 0.1–1 ml/min (4 mm diameter × 50 mm length), capillary size for flows in the range of 1.0–500 µl/min (500 µm diameter × 5 cm length), and nanosize for flows in the range of 10–500 nl/min (30 to 150 µm width × 15 to 150 cm length). The nanosize HPLC columns are however typically used when a nanoflow HPLC system is coupled to a mass analyzer and will be discussed in more detail in Chapter 28, "Mass Spectrometry."

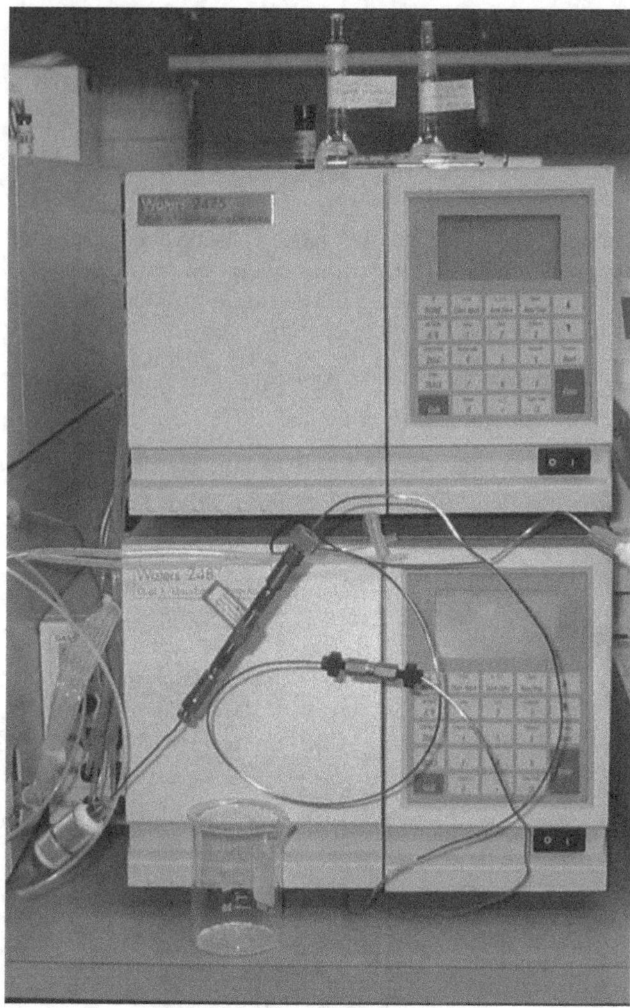

FIGURE 22.7 Close up view of the HPLC system's column and detectors.

22.2.2.1 HPLC Column Stationary Phases HPLC column stationary phases employ primarily three types of chemistries that target ranges of compound types according to affinities and are called normal phase (NP), reverse phase (RP), and ion exchange (IEC).

22.2.2.1.1 Normal-Phase HPLC (NP-HPLC) Normal-phase HPLC (NP-HPLC) is probably the oldest approach in liquid chromatography separations employing a number of different stationary-phase chemistries. The basic premise behind normal-phase chromatography is using a polar stationary phase and nonpolar mobile phases such as hexane, isooctane, and dichloromethane, which are modified with more polar solvents such as methanol and water. Most organic compounds have some degree of polarity due to various function groups. When the organic compound

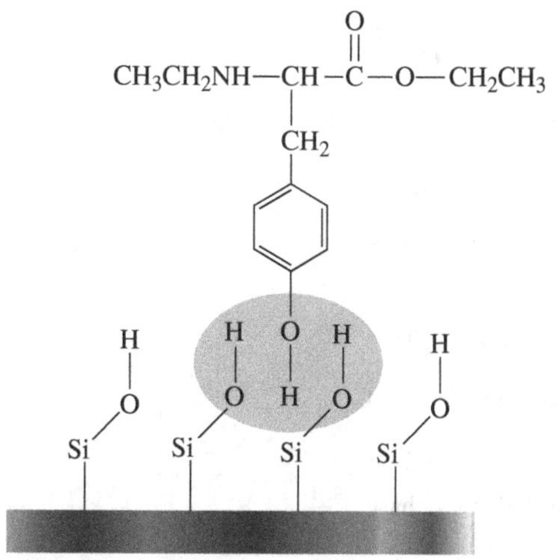

FIGURE 22.8 Normal phase (NP) silica column stationary phase illustrating the hydrogen bonding interaction (oval region) between the silanol (–Si–OH) groups of the stationary phase and the organic compound's hydroxyl (–OH) group.

is dissolved or solvated in a nonpolar solvent such as hexane, the sample can be loaded onto the column where the use of hydrophilic attractions (e.g., hydrogen bonding, polar–polar, pi–pi, dipole–dipole, and dipole–induced dipole) will enable an interaction between the organic compound and the stationary phase that exhibits a much higher affinity as compared to the nonpolar mobile phase. Let's look at this in respect to a normal-phase silica column where the stationary phase is comprised of silanol (–Si–OH) groups as illustrated in Figure 22.8. The hydrogen bonding that takes place with the stationary phase will be a stronger attraction (greater affinity) than the hydrophobic attraction between the organic compound and the nonpolar normal-hexane ($CH_3CH_2CH_2CH_2CH_2CH_3$) mobile phase. Usually, a more polar solvent such as methanol (CH_3–OH) will be added to the mobile phase, perhaps at a 60:40 normal-hexane/methanol percent composition to give some polar–polar interaction with the organic compound. This will allow the elution of the compound off of the column. This will also allow the chromatographic separation of the components in the sample according to their different affinities for the stationary phase versus mobile phase.

Other stationary phases used in NP-HPLC include cyanopropyl-bonded endcapped silica (LC-CN) for polar compounds such as pesticides, herbicides, steroids, antibiotics, phenols, and dyes; aminopropyl-bonded silica (LC-NH_2); and diol-bonded silica (LC-Diol), all four used for the chromatographic separation of polar compounds.

22.2.2.1.2 Reversed-Phase HPLC (RP-HPLC) Reversed-phase HPLC (RP-HPLC) is a newer approach in liquid chromatography separations employing a number of different stationary phases that are nonpolar and differ only in aliphatic chain length. The basic premise behind reversed-phase chromatography is using a nonpolar stationary phase and polar mobile phases, primarily mixtures of water, methanol, and acetonitrile, which are all three miscible in any compositional combination. This is in fact the opposite or "reverse" of normal-phase chromatography and hence the name reversed phase. In reversed-phase HPLC, the sample is usually dissolved in water, which is a very polar solvent. The stationary phase in reversed-phase chromatography is comprised of chemically modified silica where the hydrophilic silanol groups have been reacted with hydrophobic alkyl groups. Most organic compounds, besides having some degree of polarity due to various function groups, are also nonpolar. When the organic compound is dissolved or solvated in a polar solvent such as water, the sample can be loaded onto the column where the use of hydrophobic attractions (e.g., van der Waals forces or dispersion forces) will enable an interaction between the organic compound and the stationary phase that exhibits a much higher affinity as compared to the polar mobile phase. The chemistry of the reversed phase is derived from normal-phase silica where the stationary phase comprised of silanol (–Si–OH) groups has been derivatized with different alkyl group chain lengths to produce a hydrophobic stationary phase. The alkyl chain lengths are typically C-4, C-8, and C-18, as illustrated in Figure 22.9. Usually, a less polar solvent such as acetonitrile or methanol will be added to the mobile phase in a gradual increasing rate to give some nonpolar–nonpolar interaction with the organic compound.

This will allow the elution of the compound off of the column. The hydrophobic interaction between the organic compound and the stationary phase increases with increasing alkyl chain length. For example, proteins (which are very large macromolecules) are often separated with a C-4 or C-8 column, while peptides (much smaller and more water soluble) are separated using a C-18 column. Figure 22.10 gives a better idea of how these stationary-phase particles may look like. The backbone of the particle is made up of a spherical ball of silica. The outer surface of the silica particle is coated with the stationary phase.

22.2.2.1.3 Ion Exchange HPLC (IEX-HPLC) The principle of ion exchange chromatography is based upon a charge–charge coulomb interaction between the stationary phase and the analyte. Thus, in cation exchange chromatography (CEC), the stationary phase has a negative charge and the analytes have a positive charge. Opposite to this is anion exchange chromatography (AEC) where the stationary phase has a positive charge and the analytes possess a negative charge. The typical approach in ion exchange chromatography is to load the sample onto the column in a weakly ionic buffer where the analytes will be bound to the stationary phase due to charge–charge coulomb interactions. The analytes are then displaced from the stationary phase by introducing a stronger ionic buffer solution. Competition will take place between the charge-bound analyte and the ion being introduced into the system in the mobile phase. This effectively shifts the equilibrium of the bound analyte toward the mobile phase.

22.2.2.1.3.1 Cation Exchange Chromatography (CEC) For CEC, the equilibrium is represented as

$$\text{R-X}^-\text{C}^+ + \text{A} + \text{B}^- \rightleftharpoons \text{R-X}^-\text{A}^+ + \text{C}^+ + \text{B}^- \qquad (22.2)$$

where R-X^-C^+ is the stationary phase bound with a cation, A^+B^- is the analyte in a neutral form bound with a buffer base, R-X^-A^+

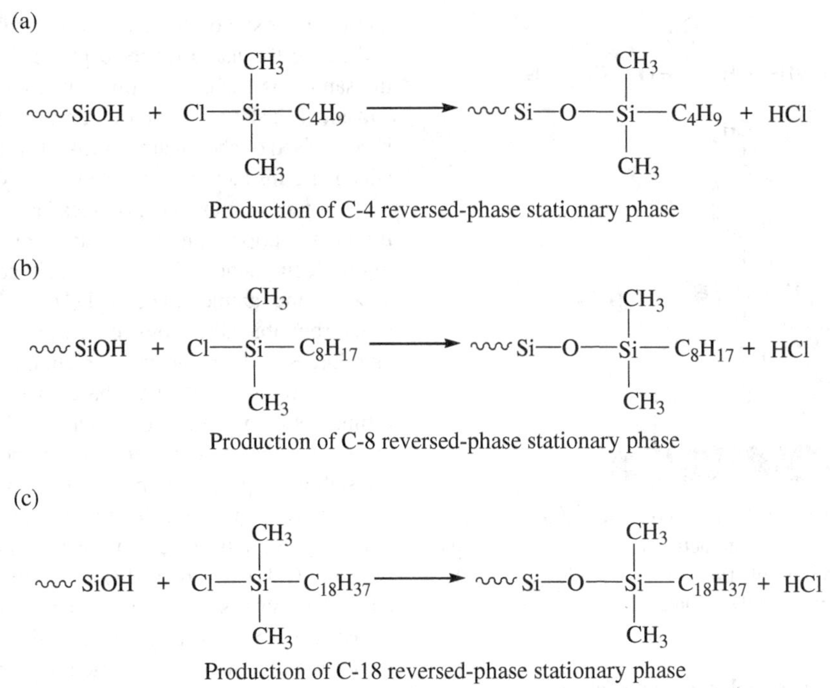

(a)

$$\sim\!\sim SiOH + Cl{-}Si(CH_3)_2{-}C_4H_9 \longrightarrow \sim\!\sim Si{-}O{-}Si(CH_3)_2{-}C_4H_9 + HCl$$

Production of C-4 reversed-phase stationary phase

(b)

$$\sim\!\sim SiOH + Cl{-}Si(CH_3)_2{-}C_8H_{17} \longrightarrow \sim\!\sim Si{-}O{-}Si(CH_3)_2{-}C_8H_{17} + HCl$$

Production of C-8 reversed-phase stationary phase

(c)

$$\sim\!\sim SiOH + Cl{-}Si(CH_3)_2{-}C_{18}H_{37} \longrightarrow \sim\!\sim Si{-}O{-}Si(CH_3)_2{-}C_{18}H_{37} + HCl$$

Production of C-18 reversed-phase stationary phase

FIGURE 22.9 Production of the reversed phase stationary phase is derived from normal phase silica where the stationary phase comprised of silanol (–Si–OH) groups has been derivatized with different alkyl group chain lengths to produce a hydrophobic stationary phase. The alkyl chain lengths are typically C-4, C-8, and C18 for increasing degrees of hydrophobic interaction.

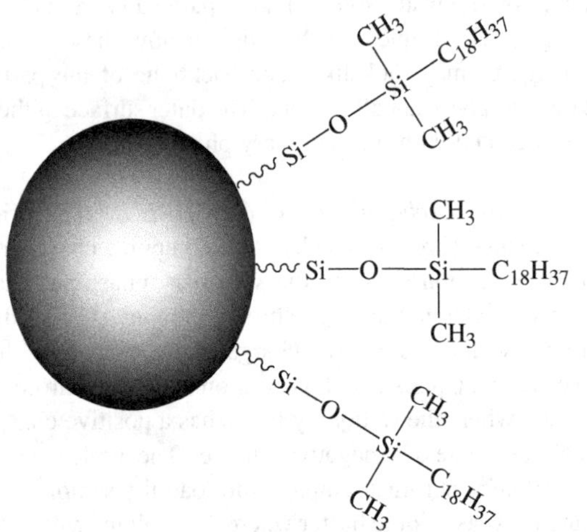

FIGURE 22.10 Stationary phase silica particle coated with reversed phase hydrophobic phase. The coating would actually be a complete coverage of the sphere in all directions.

is the analyte bound to the stationary phase, and C^+ and B^- are the ionic species involved in the equilibrium. If the system contains a cation (C^+) that has a weaker interaction with the stationary phase than the analyte A^+, the equilibrium will be shifted to the right, and the analyte will be bound to the stationary phase. As a cation (C^+) that has a stronger interaction with the stationary phase than the analyte (A^+) is introduced into the system, then the equilibrium will be shifted to the left, and the analyte will now be more predominantly in the mobile phase and thus will move down the

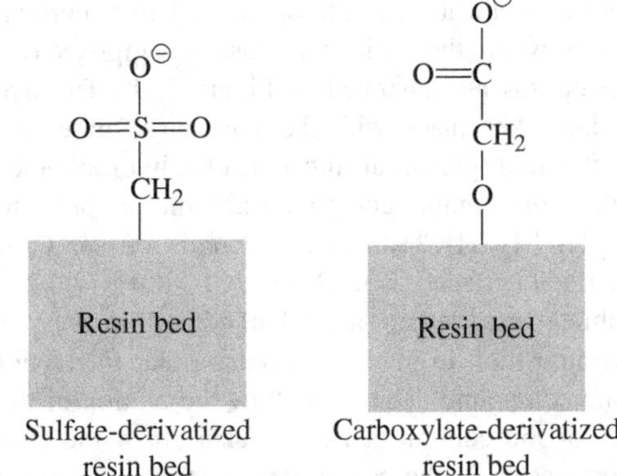

FIGURE 22.11 Sulfate and carboxylate derivatized resin beds used as the stationary phase in cation exchange chromatography (CEC).

column. Obviously, different analytes in the sample will have different coulombic interactions with the stationary phase and thus will be separated by the ion exchange chromatography. Two commonly used resin derivatives for the stationary phase in CEC are sulfate derivatives and carboxylate derivatives, as illustrated in Figure 22.11. The interaction between the protonated and positively charged amino acid lysine and a sulfate ion exchange-derivatized stationary-phase resin is illustrated in Figure 22.12.

The counterions are also included in the figure but are relatively dilute and are not appreciably participating in the interaction and are essentially spectator ions at this state of the equilibrium. Some

FIGURE 22.12 Cation exchange chromatography charge-charge interaction between the protonated and positively charged amino acid lysine and a sulfate ion exchange derivatized stationary phase resin. The counter ions (B^- and C^+) are dilute and are not participating in the interaction and are spectator ions at this state of the equilibrium.

TABLE 22.1 Buffers for Cation Exchange Chromatography[a].

Molecule	p*K*a	dp*K*a (°C)	Counterion
Maleic acid	2.00		Sodium
Malonic acid	2.88		Sodium
Citric acid	3.13	−0.0024	Sodium
Lactic acid	3.81		Sodium
Formic acid	3.75	0.0002	Sodium or lithium
Butanedioic acid	4.21	−0.0018	Sodium
Acetic acid	4.76	0.0002	Sodium or lithium
Malonic acid	5.68		Sodium or lithium
Phosphate	7.20	−0.0028	Sodium
HEPES	7.55	−0.0140	Sodium or lithium
BICINE	8.35	−0.0180	Sodium

[a] These values were taken from the Pharmacia Biotech, "Ion Exchange Chromatography, Principles and Methods," guidebook.

typical cation exchange buffers are listed in Table 22.1. An example of a cation exchange HPLC method conditions would consist of a mobile phase A that is 15 mM sodium acetate (pH 4.5). In this mobile phase, the concentration of the acetate ion is 15 mM, and the concentration of the sodium ion is also 15 mM. The sample is loaded onto the cation exchange column in mobile phase A where the analytes will bind to the stationary phase due to electrostatic interaction. Mobile phase B will consist of 15 mM sodium acetate with 1 M sodium chloride (NaCl) at a pH of 4.5. A gradient elution will be performed with increasing concentration of mobile phase B. As the concentration of mobile phase B increases, so will the concentration of the sodium ion. This will shift the equilibrium of Equation 22.2 to the left, displacing the analyte from the stationary phase and subsequently eluting it from the column. CEC is typically performed using buffers at pHs between 4 and 7 and using a gradient HPLC method from a mobile phase A containing a buffer listed in Table 22.1 to a mobile phase B solution containing the buffer with 1 M NaCl.

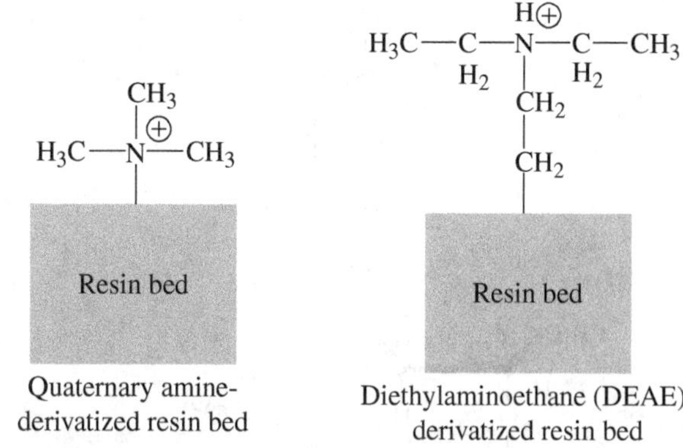

FIGURE 22.13 Simple quaternary amine derivative and a diethylaminoethane (DEAE) derivative resin beds used as the stationary phase in anion exchange chromatography (AEC).

22.2.2.1.3.2 Anion Exchange Chromatography (AEC) For AEC, the equilibrium is represented as

$$\text{R-X}^+\text{A}^- + \text{M}^-\text{C}^+ \rightleftharpoons \text{R-X}^+\text{M}^- + \text{C}^+ + \text{A}^- \qquad (22.3)$$

where $R\text{-}X^+A^-$ is the stationary phase bound with an anion, M^-C^+ is the analyte molecule (M^-) in a neutral form bound with a buffer cation, $R\text{-}X^+M^-$ is the analyte bound to the stationary phase, and C^+ and A^- are the ionic species involved in the equilibrium. If the system contains an anion (A^-) that has a weaker interaction with the stationary phase than the analyte M^-, the equilibrium will be shifted to the right, and the analyte will be bound to the stationary phase. As an anion (A^-) that has a stronger interaction with the stationary phase than the analyte (M^-) is introduced into the system, then the equilibrium will be shifted to the left, and the analyte will now be more predominantly in the mobile phase and thus will move down the column. Obviously, different analytes in the sample will have different coulomb interactions with the stationary phase and thus will be separated by the ion exchange chromatography. Two commonly used resin derivatives for the stationary phase in AEC are a simple quaternary amine derivative and a diethylaminoethane (DEAE) derivative, as illustrated in Figure 22.13. The interaction between thymine and a diethylaminoethane (DEAE) derivative ion exchange stationary-phase resin is illustrated in Figure 22.14.

The counterions are also included in the figure but are relatively dilute, are not appreciably participating in the interaction, and are essentially spectator ions at this state of the equilibrium. Some typical anion exchange buffers are listed in Table 22.2. An example of an anion exchange HPLC method conditions would consist of a mobile phase A that is 25 mM ethanolamine (pH 10.0). In this mobile phase, the concentration of the ethanolamine ion is 25 mM, and the concentration of the chloride ion (Cl^-) is also 25 mM. The sample is loaded onto the anion exchange column in mobile phase A where the analytes will bind

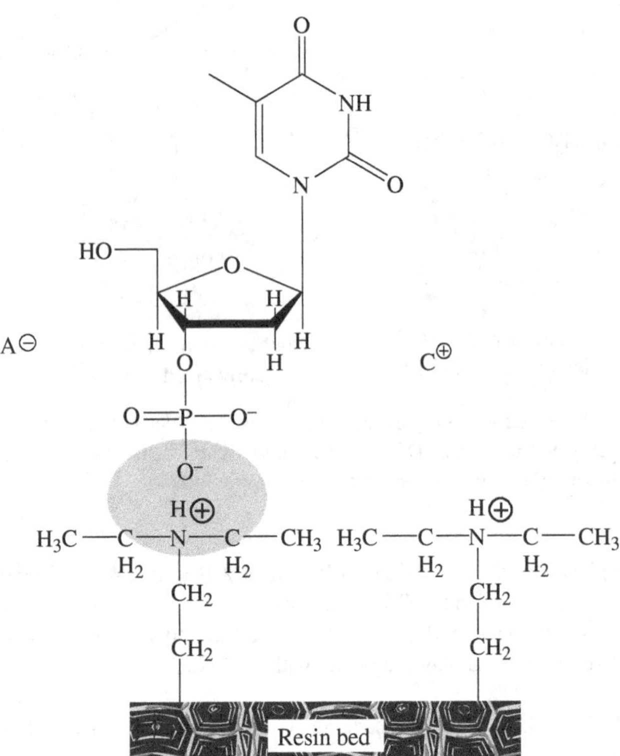

FIGURE 22.14 Anion exchange chromatography charge-charge interaction between thymine and a diethylaminoethane (DEAE) derivative resin bed. The counter ions (A^- and C^+) are dilute and are not participating in the interaction and are spectator ions at this state of the equilibrium.

TABLE 22.2 Buffers for Anion Exchange Chromatography[a].

Molecule	p*K*a	dp*K*a (°C)	Counterion
N-Methyl piperazine	4.75	−0.015	Chloride
Piperazine	5.68	−0.015	Chloride or formate
L-Histidine	5.96		Chloride
bis-Tris	6.46	−0.017	Chloride
bis-Tris propane	6.80		Chloride
Triethanolamine	7.76	−0.020	Chloride or acetate
Tris	8.06	−0.028	Chloride
N-Methyl-diethanolamine	8.52	−0.028	Chloride
Diethanolamine	8.88	−0.025	Chloride
1,3-Diaminopropane	8.64	−0.031	Chloride
Ethanolamine	9.50	−0.029	Chloride
Piperazine	9.73	−0.026	Chloride
1,3-Diaminopropane	10.47	−0.026	Chloride
Piperidine	11.12	−0.031	Chloride
Phosphate	12.33	−0.026	Chloride

[a] These values were taken from the Pharmacia Biotech, "Ion Exchange Chromatography, Principles and Methods," guidebook.

to the stationary phase due to electrostatic interaction. Mobile phase B will consist of 25 mM ethanolamine with 1 M sodium chloride (NaCl) at a pH of 10.0. A gradient elution will be performed with increasing concentration of mobile phase B. As the concentration of mobile phase B increases, so will the concentration of the chloride ion. This will shift the equilibrium of

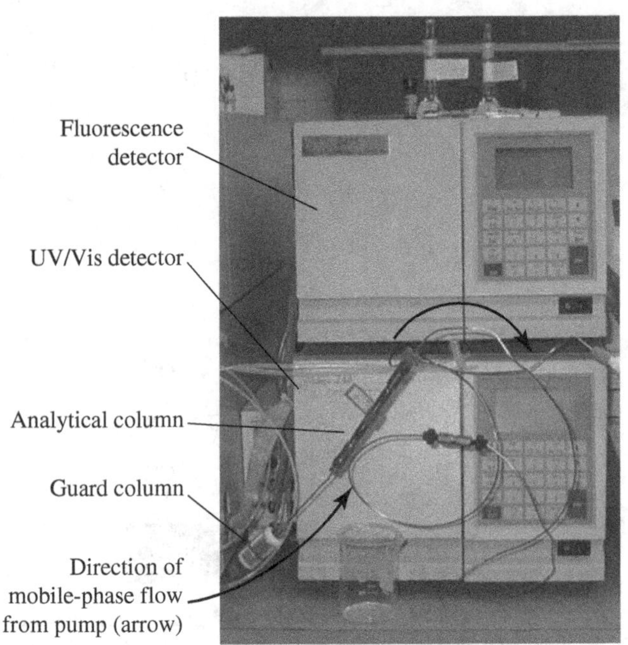

FIGURE 22.15 (a) Close up view of the HPLC system's mobile phase flow direction, guard column, analytical column, and UV/Vis (bottom) and fluorescence (top) detectors.

Equation 22.3 to the left, displacing the analyte from the stationary phase and subsequently eluting it from the column. AEC is typically performed using buffers at pHs between 7 and 10 and using a gradient HPLC method from a mobile phase A containing a buffer listed in Table 22.2 to a mobile phase B solution containing the buffer with 1 M NaCl.

22.2.3 HPLC Detectors

As the analytes are separated and eluted from the analytical column, they will flow with the mobile phase through the tubing (see stainless steel tubing in the figure) into the detectors for analyte measurement. The bottom detector in Figure 22.15 is a UV/Vis detector, which is a very common detector used in HPLC systems. The UV/Vis detector is programmable to measure at one particular wavelength (e.g., $\lambda = 230$ nm) and multiple wavelengths simultaneously (e.g., $\lambda = 230$, 254, and 595 nm) or the entire UV/Vis spectrum ($\lambda = 190$–800 nm). The last example is done using a diode array detector (DAD) where the UV/Vis spectrum of the analyte may be obtained. This allows for qualitative, quantitative, and possible spectral identification of the analyte. The top detector is a fluorescence detector where the detector can be programmed to emit a certain wavelength to excite the analyte and subsequently measure a second wavelength from the fluorescence emitted by the analyte due to spectral excitation and subsequent relaxation.

The cells used in HPLC detectors are of a special design to allow a greater path length of the particular wavelength(s) being employed for analyte detection and measurement. This is obtained by routing the flow of the mobile phase through a detector cell in a "Z type" of configuration. This is illustrated

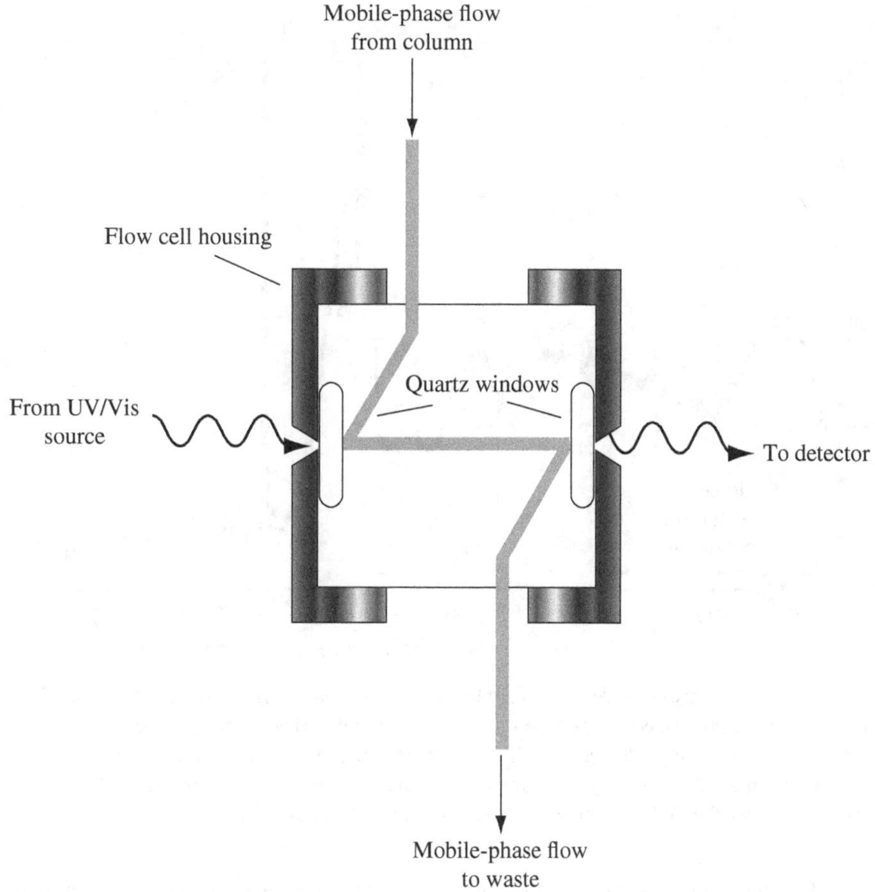

FIGURE 22.16 Z-type configuration of an HPLC detector cell. By diverting the path into a "Z" shape the path length is greatly increased thus increasing the sensitivity in the detection of compounds eluting from the HPLC column.

in Figure 22.16 where the "Z" in the tubing flow greatly increases the path length of the detector light.

Back in Chapter 13 (see Chapter 13 Section 13.3.7), we had learned about the processes of absorbance and the Beer's law relationship between absorbance (A), molar extinction coefficient (ε), molar concentration (c), and path length (b):

$$A = \varepsilon bc \tag{22.4}$$

From the Beer's law relationship in Equation 22.4, we can readily see that for a given concentration, c, an increase in path length, b, will result in a direct increase in absorbance, A, thus increasing the sensitivity. Most HPLC detector systems based upon an optical measurement incorporate the Z-shaped detector cell design. Volumes used in the Z-flow cells are up to 10 μl, and the path lengths around 10 cm.

Deuterium arc lamps are common ultraviolet (UV) wavelength range sources used in HPLC systems. Some examples of deuterium vacuum ultraviolet (VUV) source lamps are illustrated in Figure 22.17. The deuterium arc lamp emits a continuous spectrum in the UV region, a region that is very useful in detecting organic compounds by HPLC. The deuterium lamp is a low-pressure (vacuum) gas-discharge light source that uses a tungsten filament placed opposite to an anode. The glass

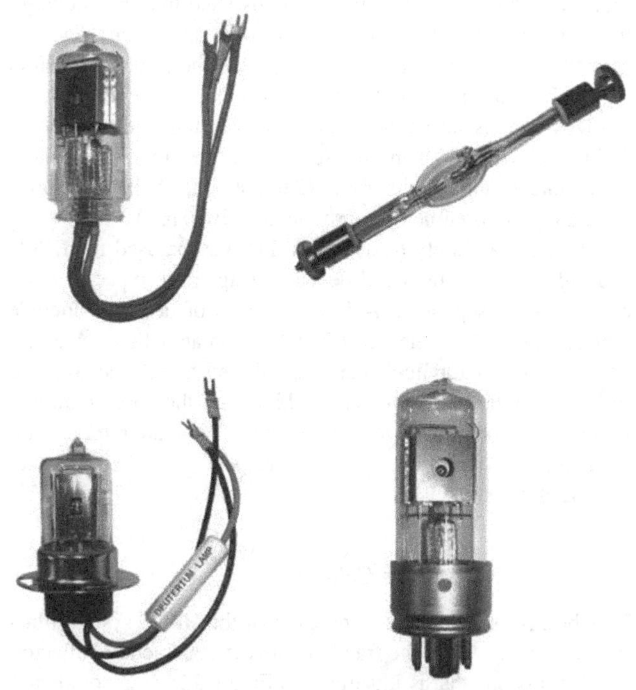

FIGURE 22.17 Examples of vacuum ultraviolet (VUV) light sources used in HPLC detector systems.

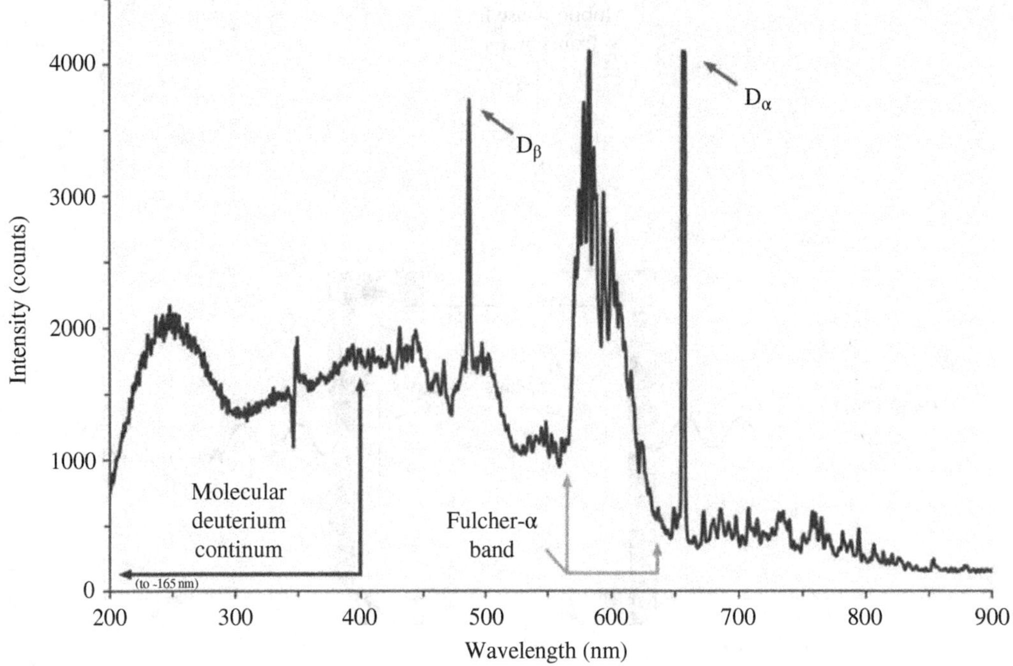

FIGURE 22.18 Emission spectrum of an ultraviolet deuterium arc lamp showing characteristic hydrogen Balmer lines (sharp peaks at 486 and 656 nm labeled D_β and D_α from left to right respectively), continuum emission in the ~160–400 nm region and Fulcher band emission between around 560 to 640 nm. The emission spectrum of deuterium differs slightly from that of protium due to the influence of hyperfine interactions, though these effects alter the wavelength of the lines by mere fractions of a nanometer and are too fine to be discerned by the spectrometer used here. (Reprinted with permission under the Wiki Creative Commons Attribution-Sharealike 3.0 Unported License.)

housing of the lamp is usually made out of fused quartz to withstand high temperatures and allow emission of UV light. To produce the continuous UV spectrum, an arc is produced between the tungsten filament and the anode. The lamp is also filled with molecular deuterium (D_2) in a gaseous state that gains energy from the arc transitioning the molecular deuterium from the ground state to an excited state. When the molecular deuterium relaxes back to the ground state, it emits electromagnetic energy (molecular emission) as a continuous spectrum primarily in the UV range of 180–370 nm (the spectrum in Figure 22.18 actually shows that the lamp emits from 112 to 900 nm). Deuterium lamps have a typical lifetime of approximately 2000 h.

There are a variety of detectors that may be used for HPLC analyses, usually employed depending upon the types of compounds being separated. A few examples of detectors include diode array UV/Vis detectors (DAD), evaporative light scattering detector (useful for lipid analyses), fluorescence detector, and refractive index detector. Table 22.3 lists the most common HPLC detectors employed in the analytical laboratory, their commercial availability, and their respective limits of detection (LOD).

TABLE 22.3 Performances of LC Detectors[a].

LC Detector	Commercially Available	Mass LOD (Commercial Detector)[b]	Mass LOD (State of the Art)[c]
Absorbance	Yes[d]	100 pg–1 ng	1 pg
Fluorescence	Yes[d]	1–10 pg	10 fg
Electrochemical	Yes[d]	10 pg–1 ng	100 fg
Refractive index	Yes	100 ng–1 μg	10 ng
Conductivity	Yes	500 pg–1 ng	500 pg
Mass spectrometry	Yes[e]	100 pg–1 ng	1 pg
FTIR	Yes[e]	1 μg	100 ng
Light scattering[f]	Yes	10 μg	500 ng
Optical activity	No	–	1 ng
Element selective	No	–	10 ng
Photoionization	No	–	1 pg–1 ng

[a] Reprinted with permission from Edward S. Yeung, Robert E. Synovec Detectors for liquid chromatography. Anal Chem 1986, 58(12), 1237A–1256A.
[b] Mass LOD is calculated for injected mass that yields a signal equal to five times the σ noise, using a mol wt of 200 g/mol. 10 μl injected for conventional or 1 μl injected for microbore LC.
[c] Same definition as b above, but the injected volume is generally smaller.
[d] Commercially available for microbore LC also.
[e] Commercially available, yet costly.
[f] Including low-angle light scattering and nephelometry.

22.2.4 HPLC Fraction Collector

This brings us to the last component of the HPLC system illustrated in Figure 22.3, the fraction collector. A fraction collector, such as this one that is illustrated in Figure 22.19, is not always part of the analytical laboratory HPLC system, but can be a very useful tool. The fraction collector does just as its name implies, it collects a fraction of the eluent coming off of the HPLC analytical column. The fraction can be either a collection of compounds coming off of the column during a predetermined time range, or it can be the collecting of a single compound eluting,

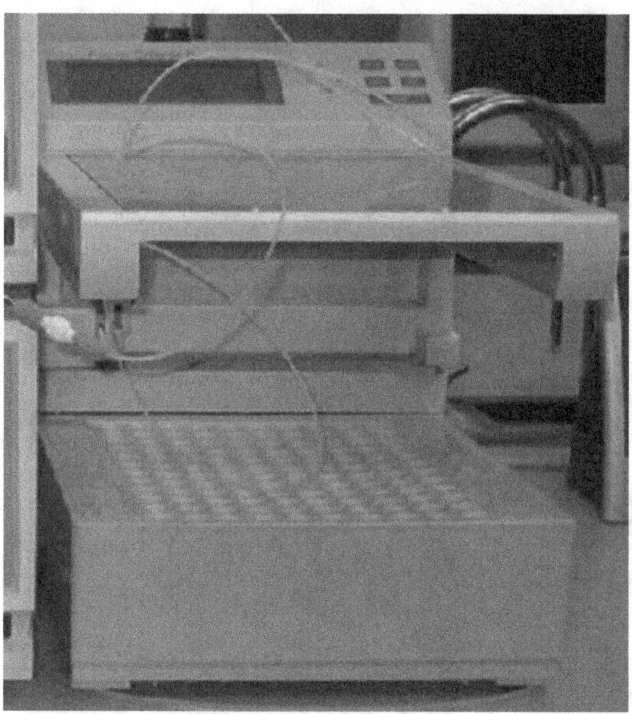

FIGURE 22.19 The fraction collector collects a fraction of the eluent coming off of the HPLC analytical column. The fraction can be either a collection of compounds coming off of the column during a predetermined time range, or it can be the collecting of a single compound eluting thus collecting it in a purified state.

thus collecting it in a purified state. In proteomics, a strong cation exchange (SCX) column is used to reduce the complexity of the peptides contained within a proteomic digestion. The HPLC SCX system coupled to a fraction collector is used to catch perhaps 10 fractions of the SCX-separated peptides. Each fraction is then loaded onto a reverse phase (RP) C18 column where the peptides are separated and measured using a mass spectrometer as a detector (see Chapter 32, "Metabolomics and Proteomics for a More Extensive Treatment of Proteomics").

22.2.5 Current Commercially Available HPLC Systems

There are numerous HPLC manufacturers providing a number of platforms. This can be a slight issue for the analytical laboratory technician as each one is configured differently in its design. Also, each system comes with the vendor's own software for controlling, operating, and processing the data obtained. This in turn requires the technician to learn the particular HPLC system being used in his or her laboratory, and sometimes, more than one system is being used in the same laboratory in which the technician needs to become familiar with. There are a couple of ways that the analytical laboratories approach this issue in order to help train the technician in the use of the HPLC system. Most if not all of the HPLC system manufacturers offer training courses in the upkeep, preventive maintenance, and general use of their systems, though this can be to some laboratories a costly approach as the technician is sent to the vendor location for training. In some, if not most laboratories, the personnel with prior experience with the system will in-house train the new technician in the use of the HPLC system. While this at first may seem a daunting task to the technician to have to learn perhaps so many different HPLC systems, the plus side of this is that they are actually all based upon the same principle and general design. Once you have become familiar with one particular system, it is easier to learn another and so forth. It is also just an aspect of time as the more the technician works with the system, the more familiar it becomes. A very important attitude for the technician to exhibit in the laboratory is the eagerness and willingness to learn new systems, instrumentation, software, and analytical techniques. This in turn makes the technician a more valuable asset to the laboratory and ultimately to himself. It is always difficult in the beginning to learn new instrumentation, but with a little time, it becomes much easier. There is no better way to learn a new system than to just begin to use it and work out the particulars as you encounter and experience them.

A very general list of HPLC manufacturers commonly encountered in the analytical laboratory includes Agilent Technologies; Beckman Coulter; Cole-Parmer; Hitachi; Perkin Elmer, Inc.; Shimadzu Scientific Instruments; Thermo Fisher Scientific; Varian, Inc.; and Waters Corporation. Manufacturers who supply HPLC columns, parts, and consumables includes Agilent Technologies; AkzoNobel; Beckman Coulter; Cole-Parmer; Hitachi; Merck KGaA; Perkin Elmer, Inc.; Phenomenex; SGE Analytical Science; Shimadzu Scientific Instruments; Shodex; Sigma-Aldrich; Thermo Fisher Scientific; Tosoh Corporation; Upchurch Scientific; Varian, Inc.; Waters Corporation; and W. R. Grace and Company.

Let's briefly take a look at some of these commercially available HPLC systems that the analytical laboratory technician may encounter in the laboratory. In Figure 22.20, a couple of Agilent Technologies 1200 Infinity Series HPLC systems are being illustrated. The 1200 Series is a modular system where each part of the HPLC system, the autosampler, the pump, the detector, and the vacuum degasser is separate modules that are usually stacked upon each other. A PC is used for instrument control, and also a "gameboy" type of handheld controller is also used. Figure 22.21 is the PerkinElmer Flexar™ FX-15 UHPLC system, which is also a modular system of stacked components. Figure 22.22 shows the Shimadzu Scientific Instruments Prominence HPLC systems: (a) the basic modules of the system and (b), (c), and (d) the increased complexity of the system by adding various modules with different functionalities such as the addition of a PC workstation and autosamplers. Finally, Figure 22.23 illustrates an example of the Thermo Scientific Ultimate 3000 Standard LC System. The Ultimate 3000 system is an example of more recent HPLC system design where the column packing material particles are much smaller, approximately 2 μm or less, creating high backpressure. These HPLC systems are able to deliver flows with very high pressure. The column bed volumes are much smaller with these particle sizes; thus, separation is optimized and can be performed much quicker.

22.2.6 Example of HPLC Analyses

22.2.6.1 HPLC Analysis of Acidic Pesticides Acidic herbicides are used extensively to control agricultural and small-water

FIGURE 22.20 Examples of Agilent Technologies Infinity 1200 Series HPLC systems. The 1200 Series is a modular system where each part of the HPLC system, the autosampler, the pump, the detector, and the vacuum degasser are each separate modules that are stacked upon each other. A PC is used for instrument control, and also a "gameboy" type of hand held controller that is also used. (© Agilent Technologies, Inc. 2014, Reproduced with Permission, Courtesy of Agilent Technologies, Inc.)

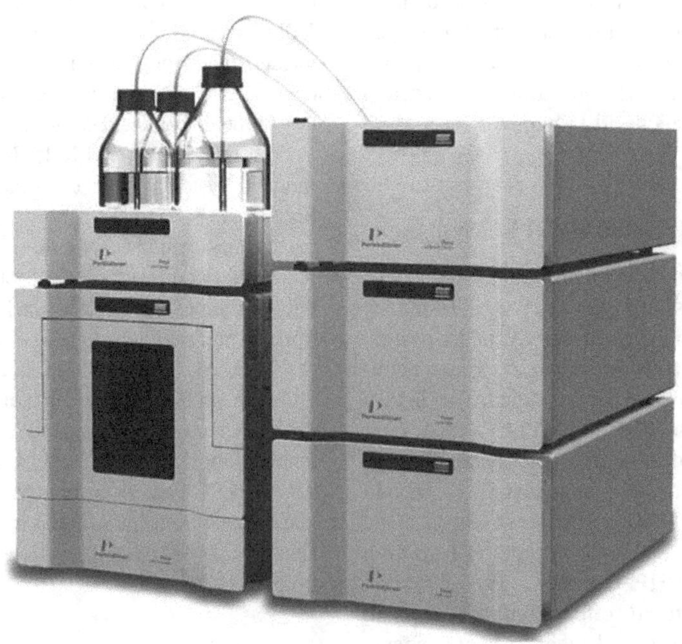

FIGURE 22.21 Example of a PerkinElmer Flexar™ FX-15 UHPLC system. (Reprinted with permission from PerkinElmer.)

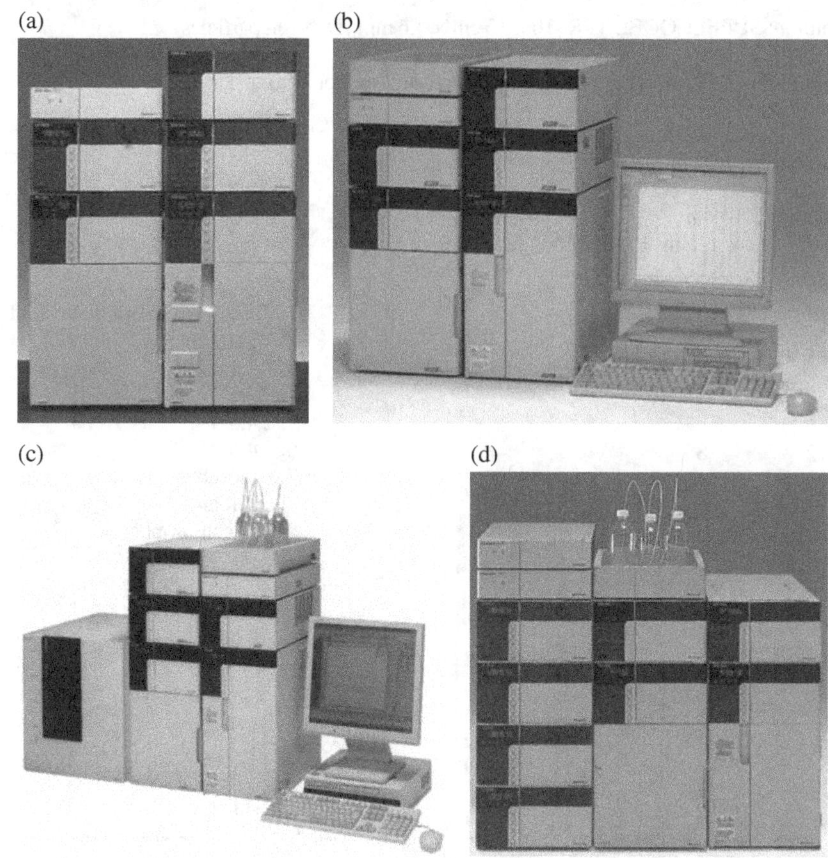

FIGURE 22.22 Examples of the Shimadzu Scientific Instruments Prominence HPLC systems: (a) the basic modules of the system, (b), (c), and (d) illustrate increasing complexity of the system by adding various modules with different functionalities. (Reprinted with permission from Shimadzu.)

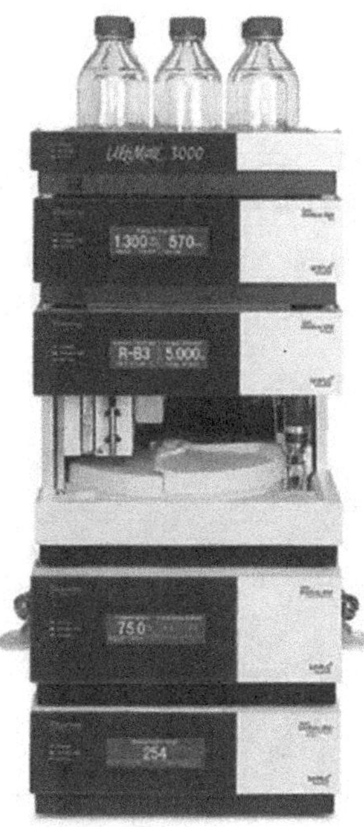

FIGURE 22.23 Example of the Thermo-Scientific Ultimate 3000 Standard LC System. (Used by permission from Thermo Fisher Scientific, the copyright owner.)

weeds. Concerns about the residues of pesticides in agricultural food products have led scientists to develop many analytical techniques. Some agencies routinely test samples taken from wells and other sources of drinking water. Most samples are concentrated, using methods such as solid-phase extraction and liquid–liquid extraction. Once concentrated, the samples are analyzed using any of a variety of detection, separation, and identification techniques, including HPLC or GC. Some analyses of herbicides and pesticides in fruits, vegetables, and drinking water have been conducted using solid-phase extraction and HPLC, with a silica-based HPLC column. The resin-based SUPELCOGEL™ TPR-100 column is an excellent choice for analysis of pesticides. The TPR-100 column exhibits high efficiency and reduced hydrophobicity, relative to silica-based columns, and is compatible with most mobile phases. It also alters selectivity relative to other reversed-phase resins. The TPR-100 column uses a patented templated polymerization process, starting with a 5 μm porous silica bead and ending with an exact copy of the silica in resin (the pores of the silica become the skeleton of the resin). Figure 22.24 illustrates the conditions used to separate a water sample containing nine acidic pesticides and the results obtained. The TPR-100 column completely separated these analytes and provided symmetrical peak shapes in less than ten minutes. Chemical names and structures of these pesticides are shown in Table 22.4.

Column: SUPELCOGEL TPR–100, 15 cm × 4.6 mm ID, 5-μm particles
Cat. No.: 59154
Mobile Phase: A = 0.025 M potassium phosphate (monobasic), pH 2.3
B = acetonitrile
gradient program shown on figure
Flow rate: 1 ml/min
Temp.: 45°C
Det.: UV, 215 nm
Inj.: 10 μl containing 100 μg/ml of each analyte

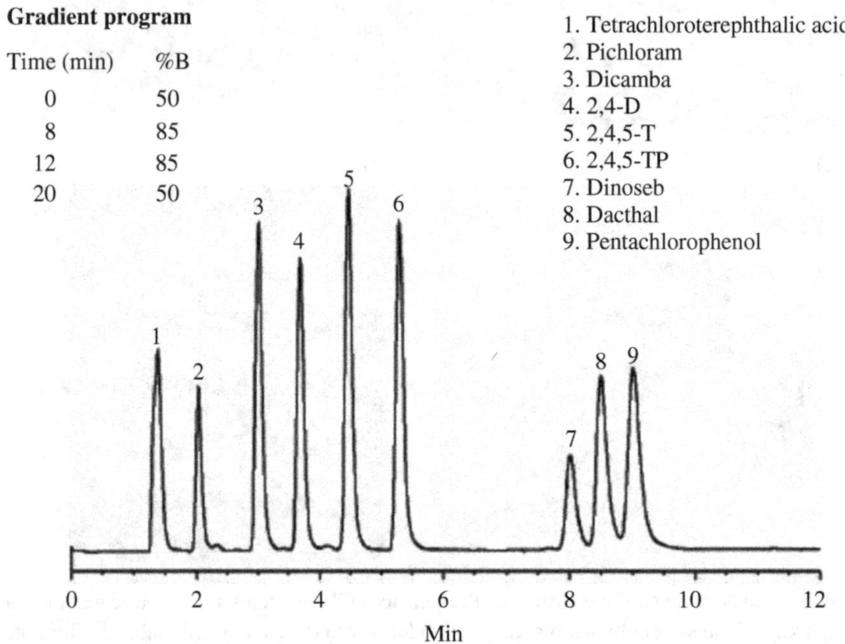

FIGURE 22.24 The conditions used and results obtained in the separation of a water sample containing nine acidic pesticides.

TABLE 22.4 Chemical Names and Structures of Acidic Pesticides.

Peak Number	Common Name	Chemical Name	Structure
1	Tetrachlorotere-phthalic acid	Tetrachlorotere-phthalic acid	
2	Pichloram	4-Amino-3,5,6-trichloropicolinic acid	
3	Dicamba	2-Methoxy-3,6-di-chlorobenzoic acid	

TABLE 22.4 (*Continued*)

Peak Number	Common Name	Chemical Name	Structure
4	2,4-D	2,4-Dichlorophenoxy acetic acid	Cl, OCH_2OOH, Cl
5	2,4,5-T	2,4,5-Trichlorophenoxy acetic acid	OCH_2COOH, Cl, Cl, Cl
6	2,4,5-TP	2-(2,4,5-Trichlorophenoxy) propionic acid	Cl, CH_3, OCHCOOH, Cl, Cl
7	Dinoseb	2-(sec-Butyl)-4,6-dinitrophenol	OH, O_2N, $CHCH_2CH_3$, CH_3, NO_2
8	Dacthal	Dimethyltetrachloro-terephthalate	$COOCH_3$, Cl, Cl, Cl, Cl, $COOCH_3$
9	Pentachlorophenol (PCP)	Pentachlorophenol	OH, Cl, Cl, Cl, Cl, Cl

23

SOLID-PHASE EXTRACTION

23.1 Introduction
23.2 Disposable SPE Columns
23.3 SPE Vacuum Manifold
23.4 SPE Procedural Bulletin

23.1 INTRODUCTION

The chromatographic technique utilizing solid-phase extraction (SPE) has become a widely used and quite common methodology in the analytical laboratory. Often, SPE is used to desalt or "cleanup" a sample prior to instrumental analyses. The principle behind SPE is the same as was covered in the beginning of this chapter for chromatography.

23.2 DISPOSABLE SPE COLUMNS

The most common SPE used in the laboratory utilizes small plastic disposable columns that are packed with a stationary phase that has been selected for a particular purpose, for example, the use of C-18 packing for reversed-phase extraction of nonpolar to moderately polar compounds, such as antibiotics, barbiturates, benzodiazepines, caffeine, drugs, dyes, essential oils, fat-soluble vitamins, fungicides, herbicides, pesticides, hydrocarbons, parabens, phenols, phthalate esters, steroids, surfactants, theophylline, and water-soluble vitamins, or C–NH_2 aminopropyl-bonded silica for normal-phase extraction of polar compounds, weak anion exchange for carbohydrates, weak anions, and organic acids. Some examples of SPE cartridges are illustrated in Figure 23.1 including small plastic disposable packed columns and also inline cartridges. The SPE tubes come in different loading volume capacities and different bed volume capacities. For example, the three SPE tubes in the top row and to the right are obviously different in the packing bed volume, increasing in packing material moving from left to right. It is also apparent that a larger volume of sample can be introduced at one time for the SPE tube on the top far left as compared to the other three. When specifying SPE cartridges, the volume size of the tube, the bed packing volume, and the loading capacity (maximum amount the packing material can handle of analyte before saturating the packing and premature breakthrough takes place) need to be considered.

23.3 SPE VACUUM MANIFOLD

The SPE tubes can be used individually for a single sample, or more commonly, a number of tubes for many samples are processed simultaneously. To do this, SPE tubes are placed onto an SPE vacuum manifold where multiple samples can be prepared at the same time. The vacuum manifold is also used to help aid in the flow of solvents through the tube during sample treatment steps such as loading, washing, and eluting. Figure 23.2 illustrates an SPE vacuum manifold where six SPE tubes are positioned on the manifold. The flow of the tubes is controlled by small hand turn valves under the tubes, and either the flow-through wash/waste or the eluting compounds are collected in glass test tubes placed inside the manifold and under the SPE tubes. At the bottom right is located an inlet fitting for connection of a vacuum source usually also consisting of a control valve and a pressure gauge.

23.4 SPE PROCEDURAL BULLETIN

Let's now take a look at a procedural bulletin (910) that is adapted from a vendor manual (reproduced with permission from Sigma-Aldrich Co. LLC) concerning SPE.

Analytical Chemistry: A Chemist and Laboratory Technician's Toolkit, First Edition. Bryan M. Ham and Aihui MaHam.

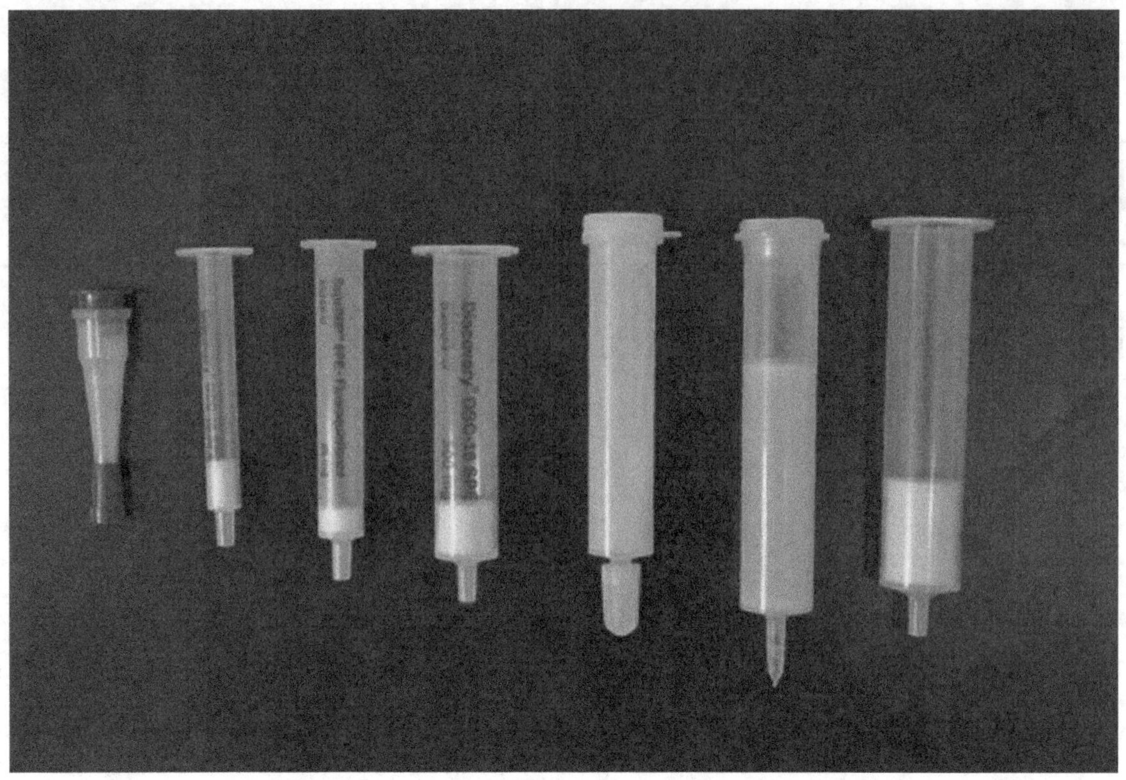

FIGURE 23.1 Examples of solid phase extraction (SPE) cartridges used in the analytical laboratory.

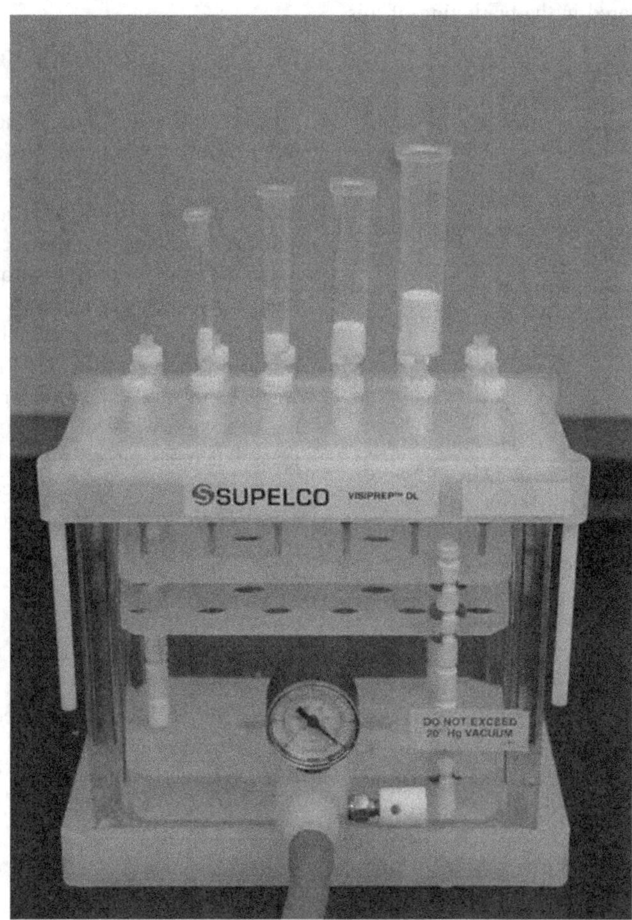

FIGURE 23.2 Solid phase extraction (SPE) vacuum manifold. Six SPE tubes are positioned on the manifold. SPE tube flow controlled by small hand turn valves under the tubes and either the flow-through wash/waste or the eluting compounds are collected in glass test tubes placed inside the manifold and under the SPE tubes. At the bottom right is located an inlet fitting for connection of a vacuum source usually also consisting of a control valve and a pressure gauge.

Bulletin 910

Guide to Solid Phase Extraction

	Page
Introduction	1
Phase Types	2
Reversed phase packings	
Normal phase packings	
Ion exchange packings	
Adsorption packings	
SPE Theory	3
How compounds are retained by the sorbent	
Reversed phase SPE	
Normal phase SPE	
Ion exchange SPE	
Secondary interactions	
The role of pH in SPE	
How to Use SPE	6
Selecting the proper extraction scheme	
The five-step SPE method development process	
Sample pretreatment options - Liquid samples - Solid samples	
SPE hardware and accessories for processing samples	

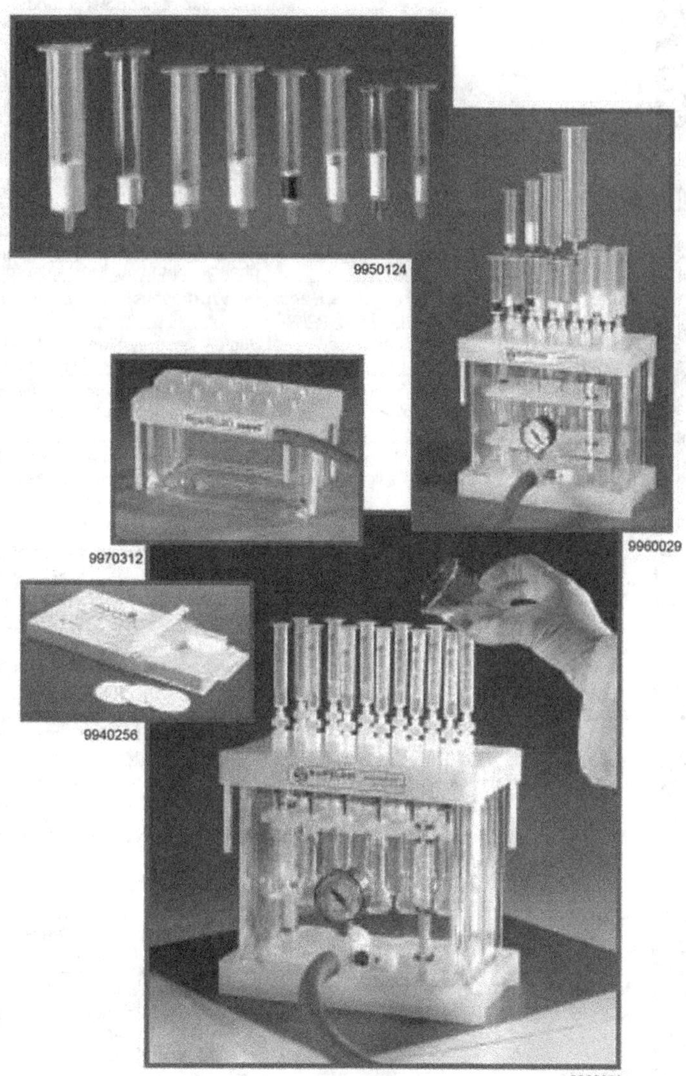

9950124 9970312 9960029 9940256 9960028

Introduction

Solid phase extraction (SPE) is an increasingly useful sample preparation technique. With SPE, many of the problems associated with liquid/liquid extraction can be prevented, such as incomplete phase separations, less-than-quantitative recoveries, use of expensive, breakable specialty glassware, and disposal of large quantities of organic solvents. SPE is more efficient than liquid/liquid extraction, yields quantitative extractions that are easy to perform, is rapid, and can be automated. Solvent use and lab time are reduced.

SPE is used most often to prepare liquid samples and extract semivolatile or nonvolatile analytes, but also can be used with solids that are pre-extracted into solvents. SPE products are excellent for sample extraction, concentration, and cleanup. They are available in a wide variety of chemistries, adsorbents, and sizes. Selecting the most suitable product for each application and sample is important.

SPE Phase Types

Mode	Phase	Packing	Applications
	Silica-Based Packing – 40µm particles, 60Å pores (unless otherwise noted).		
Reversed Phase	**LC-18**	octadecyl bonded, endcapped silica	For reversed phase extraction of nonpolar to moderately polar compounds, such as antibiotics, barbiturates, benzodiazepines, caffeine, drugs, dyes, essential oils, fat soluble vitamins, fungicides, herbicides, pesticides, hydrocarbons, parabens, phenols, phthalate esters, steroids, surfactants, theophylline, and water soluble vitamins.
Reversed Phase	**ENVI™-18**	octadecyl bonded, endcapped silica	Higher phase coverage and carbon content than LC-18, greater resistance to extreme pH conditions, and slightly higher capacity for nonpolar compounds. For reversed phase extraction of nonpolar to moderately polar compounds, such as antibiotics, caffeine, drugs, dyes, essential oils, fat soluble vitamins, fungicides, herbicides, pesticides, PNAs, hydrocarbons, parabens, phenols, phthalate esters, steroids, surfactants, water soluble vitamins. Also available in disk format.
Reversed Phase	**LC-8**	octyl bonded, endcapped silica	For reversed phase extraction of nonpolar to moderately polar compounds, such as antibiotics, barbiturates, benzodiazepines, caffeine, drugs, dyes, essential oils, fat soluble vitamins, fungicides, herbicides, pesticides, hydrocarbons, parabens, phenols, phthalate esters, steroids, surfactants, theophylline, and water soluble vitamins. Also available in disk format.
Reversed Phase	**ENVI-8**	octyl bonded, endcapped silica	Higher phase coverage and carbon content than LC-8, greater resistance to extreme pH conditions, and slightly higher capacity for nonpolar compounds. For reversed phase extraction of barbiturates, benzodiazepines, caffeine, drugs, dyes, essential oils, fat soluble vitamins, fungicides, herbicides, pesticides, PNAs, hydrocarbons, parabens, phenols, phthalate esters, steroids, surfactants, theophylline, water soluble vitamins.
Reversed Phase	**LC-4**	butyldimethyl bonded, endcapped silica (500Å pores)	Less hydrophobic than LC-8 or LC-18. For extraction of peptides and proteins.
Reversed Phase	**LC-Ph**	phenyl bonded silica	Slightly less retention than LC-18 or LC-8 material. For reversed phase extraction of nonpolar to moderately polar compounds, especially aromatic compounds.
Reversed Phase	**Hisep™**	hydrophobic surface enclaved by a hydrophilic network	For exclusion of proteins in biological samples; retains small molecules such as drugs under reversed phase conditions.
Normal Phase	**LC-CN**	cyanopropyl bonded, endcapped silica	For reversed phase extraction of moderately polar compounds, normal phase extraction of polar compounds, such as aflatoxins, antibiotics, dyes, herbicides, pesticides, phenols, steroids. Weak cation exchange for carbohydrates and cationic compounds.
Normal Phase	**LC-Diol**	diol bonded silica	For normal phase extraction of polar compounds.
Normal Phase	**LC-NH_2**	aminopropyl bonded silica	For normal phase extraction of polar compounds, weak anion exchange for carbohydrates, weak anions, and organic acids.
Ion Exchange	**LC-SAX**	quaternary amine bonded silica with Cl^- counterion	For strong anion exchange for anions, organic acids, nucleic acids, nucleotides, and surfactants. Capacity: 0.2meq/g.
Ion Exchange	**LC-SCX**	sulfonic acid bonded silica with Na^+ counterion	For strong cation exchange for cations, antibiotics, drugs, organic bases, amino acids, catecholamines, herbicides, nucleic acid bases, nucleosides, and surfactants. Capacity: 0.2meq/g.
Ion Exchange	**LC-WCX**	carboxylic acid bonded silica with Na^+ counterion	For weak cation exchange of cations, amines, antibiotics, drugs, amino acids, catecholamines, nucleic acid bases, nucleosides, and surfactants.
Adsorption	**LC-Si**	silica gel with no bonded phase	For extraction of polar compounds, such as alcohols, aldehydes, amines, drugs, dyes, herbicides, pesticides, ketones, nitro compounds, organic acids, phenols, and steroids.
	Alumina-Based Packing – Crystalline, chromatographic grade alumina, irregular particles, 60/325 mesh.		
Adsorption	**LC-Alumina-A**	acidic pH ~5	For anion exchange and adsorption extraction of polar compounds, such as vitamins.
Adsorption	**LC-Alumina-B**	basic pH ~8.5	For adsorption extraction of polar compounds, and cation exchange.
Adsorption	**LC-Alumina-N**	neutral pH ~6.5	For adsorption extraction of polar compounds. With pH adjustment, cation or anion exchange. For extraction of vitamins, antibiotics, essential oils, enzymes, glycosides, and hormones.
	Florisil®-Based Packing – Magnesium silicate, 100/120 mesh particles.		
Adsorption	**LC-Florisil**		For adsorption extraction of polar compounds, such as alcohols, aldehydes, amines, drugs, dyes, herbicides, pesticides, PCBs, ketones, nitro compounds, organic acids, phenols, and steroids.
Adsorption	**ENVI-Florisil▲**		For adsorption extraction of polar compounds, such as alcohols, aldehydes, amines, drugs, dyes, herbicides, pesticides, PCBs, ketones, nitro compounds, organic acids, phenols, and steroids.
	Graphitized Carbon-Based Packing – Nonbonded carbon phase.		
Adsorption	**ENVI-Carb**	nonporous, surface area 100m²/g, 120/400 mesh	For adsorption extraction of polar and nonpolar compounds.
Adsorption	**ENVI-Carb C**	nonporous, surface area 10m²/g, 80/100 mesh	For adsorption extraction of polar and nonpolar compounds.
	Resin-Based Packing – 80-160µm spherical particles.		
Adsorption	**ENVI-Chrom P▲▲**		For extraction of polar aromatic compounds such as phenols from aqueous samples. Also for adsorption extraction of nonpolar to midpolar aromatic compounds.

▲ SPE tubes that are packed with this material contain stainless steel or Teflon® frits, required by US Environmental Protection Agency Contract Laboratory Program (CLP) pesticide methods.

▲▲ Highly crosslinked, neutral, specially cleaned styrene-divinylbenzene resin. Very high surface area, mean pore size 110-175Å.

SPE Theory

How Compounds Are Retained by the Sorbent

Reversed Phase
(polar liquid phase, nonpolar modified solid phase)
Hydrophobic interactions
- nonpolar-nonpolar interactions
- van der Waals or dispersion forces

Normal Phase
(nonpolar liquid phase, polar modified solid phase)
Hydrophilic interactions
- polar-polar interactions
- hydrogen bonding
- pi-pi interactions
- dipole-dipole interactions
- dipole-induced dipole interactions

Ion Exchange
Electrostatic attraction of charged group on compound to a charged group on the sorbent's surface

Adsorption
(interactions of compounds with unmodified materials)
Hydrophobic and hydrophilic interactions may apply
Depends on which solid phase is used

Reversed Phase SPE

Reversed phase separations involve a polar (usually aqueous; see Table A on page 8) or moderately polar sample matrix (mobile phase) and a nonpolar stationary phase. The analyte of interest is typically mid- to nonpolar. Several SPE materials, such as the alkyl- or aryl-bonded silicas (**LC-18, ENVI-18, LC-8, ENVI-8, LC-4,** and **LC-Ph**) are in the reversed phase category. Here, the hydrophilic silanol groups at the surface of the raw silica packing (typically 60Å pore size, 40μm particle size) have been chemically modified with hydrophobic alkyl or aryl functional groups by reaction with the corresponding silanes.

$$\sim\sim\sim \text{Si-OH} + \text{Cl-}\underset{\text{CH}_3}{\overset{\text{CH}_3}{\text{Si}}}\text{-C}_{18}\text{H}_{37} \rightarrow \sim\sim\sim \text{Si-O-}\underset{\text{CH}_3}{\overset{\text{CH}_3}{\text{Si}}}\text{-C}_{18}\text{H}_{37} + \text{HCl}$$

Retention of organic analytes from polar solutions (e.g. water) onto these SPE materials is due primarily to the attractive forces between the carbon-hydrogen bonds in the analyte and the functional groups on the silica surface. These nonpolar-nonpolar attractive forces are commonly called van der Waals forces, or dispersion forces. To elute an adsorbed compound from a reversed phase SPE tube or disk, use a nonpolar solvent to disrupt the forces that bind the compound to the packing. **LC-18** and **LC-8** are standard, monomerically bonded silicas. Polymerically bonded materials, such as **ENVI-18** and **ENVI-8,** result in a more complete coverage of the silica surface and higher carbon loading. Polymeric bonding is more resistant to pH extremes, and thus is more suitable for environmental applications for trapping organic compounds from acidified aqueous samples. All silica-based bonded phases have some percentage of residual unreacted silanols that act as secondary interaction sites. These secondary interactions may be useful in the extraction or retention of highly polar analytes or contaminants, but may also irreversibly bind analytes of interest (see *Secondary Interactions* on page 5).

The following materials also are used under reversed phase conditions: **ENVI-Carb** (carbon-based), **ENVI-Chrom P** (polymer-based), and **Hisep** (polymer-coated and bonded silica).

Carbonaceous adsorption media, such as the **ENVI-Carb** materials, consist of graphitic, nonporous carbon that has a high attraction for organic polar and nonpolar compounds from both polar and nonpolar matrices. The carbon surface is comprised of atoms in hexagonal ring structures, interconnected and layered in graphitic sheets. The hexagonal ring structure demonstrates a strong selectivity for planar aromatic or hexagonal ring-shaped molecules and hydrocarbon chains with potential for multiple surface contact points. Retention of analytes is based primarily on the analyte's structure (size and shape), rather than on interactions of functional groups on the analyte with the sorbent surface. Elution is performed with mid- to nonpolar solvents. The unique structure and selectivity of ENVI-Carb materials, compared to bonded alkyl-silicas, makes them an excellent alternative when the bonded silicas will not work for an application.

Polymeric adsorption media such as the **ENVI-Chrom P** material also is used in reversed phase fashion. ENVI-Chrom P is a styrene/divinylbenzene material that is used for retaining hydrophobic compounds which contain some hydrophilic functionality, especially aromatics. Phenols are sometimes difficult to retain on C18-modified silica under reversed phase conditions, mainly due to their greater solubility in water than in organic matrices. The ENVI-Chrom P material has been shown to retain phenols well under reversed phase conditions. Elution steps can be done with mid- to nonpolar solvents, because the polymeric packing is stable in almost all matrices.

Hisep is a hydrophobic (C18-like) bonded silica that is coated with a hydrophilic polymer and is typically used under reversed phase conditions. The porous polymer coating prevents the adsorption of large, unwanted molecules onto the silica surface. The pores in the polymer allow small, hydrophobic organic compounds of interest (such as drugs) to reach the bonded silica surface, while large interfering compounds (such as proteins) are shielded from the bonded silica by the polymer and are flushed through the SPE tube. SPE procedures on Hisep material are similar to those on LC-18.

Normal Phase SPE

Normal phase SPE procedures typically involve a polar analyte, a mid- to nonpolar matrix (e.g. acetone, chlorinated solvents, and hexane), and a polar stationary phase. Polar-functionalized bonded silicas (e.g. **LC-CN, LC-NH$_2$,** and **LC-Diol**), and polar adsorption media (**LC-Si, LC-Florisil, ENVI-Florisil,** and **LC-Alumina**) typically are used under normal phase conditions. Retention of an analyte under normal phase conditions is primarily due to interactions between polar functional groups of the analyte and polar groups on the sorbent surface. These include hydrogen bonding, pi-pi interactions, dipole-dipole interactions, and dipole-induced dipole interactions, among others. A compound adsorbed by these mechanisms is eluted by passing a solvent that disrupts the binding mechanism—usually a solvent that is more polar than the sample's original matrix.

The bonded silicas—**LC-CN, LC-NH$_2$,** and **LC-Diol**—have short alkyl chains with polar functional groups bonded to the surface. These silicas, because of their polar functional groups, are much more hydrophilic relative to the bonded reversed phase silicas. As with typical normal phase silicas, these packings can be used to adsorb polar compounds from nonpolar matrices. Such SPE tubes have been used to adsorb and selectively elute compounds of very similar structure (e.g. isomers), or complex mixtures or classes of compounds such as drugs and lipids. These materials also can be used under reversed phase conditions (with aqueous samples), to exploit the hydrophobic properties of the small alkyl chains in the bonded functional groups.

The **LC-Si** material is underivatized silica commonly used as the backbone of all of the bonded phases. This silica is extremely hydrophilic, and must be kept dry. All samples used with this material must be relatively water-free. The functional groups that are involved in the adsorption of compounds from nonpolar matrices are the free hydroxyl groups on the surface of the silica particles. LC-Si may be used to adsorb polar compounds from nonpolar matrices with subsequent elution of the compounds in an organic solvent that is more polar than the original sample matrix. In most cases, LC-Si is used as an adsorption media, where an organic extract is applied to the silica bed, the analyte of interest passes through unretained, and the unwanted compounds adsorb onto the silica and are discarded. This procedure is usually called *sample cleanup*.

LC-Florisil and **ENVI-Florisil** SPE tubes are packed with a magnesium silicate that is used typically for sample cleanup of organic extracts. This highly polar material strongly adsorbs polar compounds from nonpolar matrices. The ENVI-Florisil SPE tubes are made with either Teflon® or stainless steel frits, a configuration necessary for environmental procedures specified in US EPA methods. ENVI-Florisil is specifically tested for low backgrounds via GC analysis.

LC-Alumina SPE tubes are also used in adsorption/sample cleanup-type procedures. The aluminum oxide materials can either be of acidic (Alumina-A, pH ~5), basic (Alumina-B, pH ~8.5), or neutral (Alumina-N, pH ~6.5) pH, and are classified as having Brockmann Activities of I. The activity level of the alumina may be altered from grade I through grade IV with the controlled addition of water, prior to or after packing this material into tubes.

Ion Exchange SPE

Ion exchange SPE can be used for compounds that are charged when in a solution (usually aqueous, but sometimes organic). Anionic (negatively charged) compounds can be isolated on **LC-SAX** or **LC-NH$_2$** bonded silica cartridges. Cationic (positively charged) compounds are isolated by using **LC-SCX** or **LC-WCX** bonded silica cartridges. The primary retention mechanism of the compound is based mainly on the electrostatic attraction of the charged functional group on the compound to the charged group that is bonded to the silica surface. In order for a compound to retain by ion exchange from an aqueous solution, the pH of the sample matrix must be one at which both the compound of interest and the functional group on the bonded silica are charged. Also, there should be few, if any, other species of the same charge as the compound in the matrix that may interfere with the adsorption of the compound of interest. A solution having a pH that neutralizes either the compound's functional group or the functional group on the sorbent surface is used to elute the compound of interest. When one of these functional groups is neutralized, the electrostatic force that binds the two together is disrupted and the compound is eluted. Alternatively, a solution that has a high ionic strength, or that contains an ionic species that displaces the adsorbed compound, is used to elute the compound.

Anion Exchange SPE

The **LC-SAX** material is comprised of an aliphatic quaternary amine group that is bonded to the silica surface. A quaternary amine is a strong base and exists as a positively-charged cation that exchanges or attracts anionic species in the contacting solution — thus the term strong anion exchanger (SAX). The pKa of a quaternary amine is very high (greater than 14), which makes the bonded functional group charged at all pHs when in an aqueous solution. As a result, LC-SAX is used to isolate strong anionic (very low pKa, <1) or weak anionic (moderately low pKa, >2) compounds, as long as the pH of the sample is one at which the compound of interest is charged. For an anionic (acidic) compound of interest, the pH of the matrix must be 2 pH units above its pKa for it to be charged. In most cases, the compounds of interest are strong or weak acids.

Because it binds so strongly, LC-SAX is used to extract strong anions only when recovery or elution of the strong anion is not desired (the compound is isolated and discarded). Weak anions can be isolated and eluted from LC-SAX because they can be either displaced by an alternative anion or eluted with an acidic solution at a pH that neutralizes the weak anion (2 pH units below its pKa). If recovery of a strongly anionic species is desired, use LC-NH$_2$.

The **LC-NH$_2$** SPE material that is used for normal phase separations is also considered to be a weak anion exchanger (WAX) when used with aqueous solutions. The LC-NH$_2$ material has an aliphatic aminopropyl group bonded to the silica surface. The pKa of this primary amine functional group is around 9.8. For it to be used as an anion exchanger, the sample must be applied at a pH at least 2 units below 9.8. The pH must also be at a value where the anionic compound of interest is also charged (2 pH units above its own pKa). LC-NH$_2$ is used to isolate and recover both strong and weak anions because the amine functional group on the silica surface can be neutralized (2 pH units above its pKa) in order to elute the strong or weak anion. Weak anions also can be eluted from LC-NH$_2$ with a solution that neutralizes the adsorbed anion (2 pH units below its pKa), or by adding a different anion that displaces the analyte.

Cation Exchange

The **LC-SCX** material contains silica with aliphatic sulfonic acid groups that are bonded to the surface. The sulfonic acid group is strongly acidic (pKa <1), and attracts or exchanges cationic species in a contacting solution – thus the term strong cation exchanger (SCX). The bonded functional group is charged over the whole pH range, and therefore can be used to isolate strong cationic (very high pKa, >14) or weak cationic (moderately high pKa, <12) compounds, as long as the pH of the solution is one at which the compound of interest is charged. For a cationic (basic) compound of interest, the pH of the matrix must be 2 pH units below its pKa for it to be charged. In most cases, the compounds of interest are strong or weak bases.

LC-SCX SPE tubes should be used to isolate strong cations only when their recovery or elution is not desired. Weak cations can be isolated and eluted from LC-SCX; elution is done with a solution at 2 pH units above the cation's pKa (neutralizing the analyte), or by adding a different cation that displaces the analyte. If recovery of a strongly cationic species is desired, use LC-WCX.

The **LC-WCX** SPE material contains an aliphatic carboxylic acid group that is bonded to the silica surface. The carboxylic acid group is a weak anion, and is thus considered a weak cation exchanger (WCX). The carboxylic acid functional group in LC-WCX has a pKa of about 4.8, will be negatively charged in solutions of at least 2 pH units above this value, and will isolate cations if the pH is one at which they are both charged. LC-WCX can be used to isolate and recover both strong and weak cations because the carboxylic acid functional group on the silica surface can be neutralized (2 pH units below its pKa) in order to elute the strong or weak cation. Weak cations also can be eluted from LC-WCX with a solution that neutralizes the adsorbed cation (2 pH units above its pKa), or by adding a different cation that displaces the analyte.

In many cases, the analyte in ion exchange SPE is eluted in an aqueous solution. If you must use an acidic or basic solution to elute an analyte from an SPE tube, but the extracted sample must be analyzed in an organic solvent that is not miscible with water, try to elute the compound with acidic methanol (98% methanol/2% concentrated HCl) or basic methanol (98% methanol/2% NH_4OH). The methanol can be evaporated quickly, and the sample may be reconstituted in a different solvent. If you need a stronger (more nonpolar) solvent to elute the analyte from the SPE tube, add methylene chloride, hexane, or ethyl acetate to the acidic or basic methanol.

Secondary Interactions

The primary retention mechanisms for compounds on the SPE materials are described above. For the bonded silicas, it is possible that secondary interactions will occur.

For *reversed phase bonded silicas*, the primary retention mechanism involves nonpolar interactions. However, because of the silica particle backbone, some polar secondary interactions with residual silanols — such as those described for normal phase SPE – could occur. If a nonpolar solvent does not efficiently elute a compound from a reversed phase SPE packing, the addition of a more polar solvent (e.g. methanol) may be necessary to disrupt any polar interactions that retain the compound. In these cases, methanol can hydrogen-bond with the hydroxyl groups on the silica surface, thus breaking up any hydrogen bonding that the analyte may be incurring.

The silanol group at the surface of the silica, Si-OH, can also be acidic, and may exist as an $Si-O^-$ group above pH 4. As a result, the silica backbone may also have cation exchange secondary interactions, attracting cationic or basic analytes of interest. In this case, a pH adjustment of the elution solvent may be necessary to disrupt these interactions for elution (acidic to neutralize the silanol group, or basic to neutralize the basic analyte). This can be done by using acidic methanol (98% MeOH:2% concentrated HCl) or basic methanol (98% MeOH:2% concentrated NH_4OH), or by mixtures of these with a more nonpolar, methanol-miscible solvent.

Normal phase bonded silicas will exhibit primary polar retention mechanisms via the bonded functional group, but also can have some secondary nonpolar interactions of the analyte with the small alkyl chain that supports the functional group. In this case, a more nonpolar solvent, or a mix of polar and nonpolar solvents, may be needed for elution. As with the reversed phase silicas, secondary polar or cation exchange interactions of the adsorbed compound may occur with the silica backbone.

Ion exchange bonded silicas can provide secondary nonpolar interactions of analytes with the nonpolar portions of their functional groups, as well as polar and cation exchange interactions of the analyte with the silica backbone. A delicate balance of pH, ionic strength, and organic content may be necessary for elution of the analyte of interest from these packings.

The Role of pH in SPE

Solutions used in SPE procedures have a very broad pH range. Silica-based packings, such as those used in HPLC columns, usually have a stable pH range of 2 to 7.5. At pH levels above and below this range, the bonded phase can be hydrolyzed and cleaved off the silica surface, or the silica itself can dissolve. In SPE, however, the solutions usually are in contact with the sorbent for short periods of time. The fact that SPE cartridges are disposable, and are meant to be used only once, allows one to use any pH to optimize retention or elution of analytes. If stability of the SPE cartridge at an extreme pH is crucial, polymeric or carbon-based SPE materials such as ENVI-Chrom P or ENVI-Carb may be used. These materials are stable over the pH range of 1-14.

For **reversed phase** SPE procedures on bonded silicas, if trapping the analyte in the tube is desired, the pH of the conditioning solution and sample (if mostly or entirely aqueous) should be adjusted for optimum analyte retention. If the compound of interest is acidic or basic you should, in most cases, use a pH at which the compound is not charged. Retention of neutral compounds (no acidic or basic functional groups) usually is not affected by pH. Conversely, you can use a pH at which the unwanted compounds in the sample are retained on the SPE packing, but the analyte of interest passes through unretained. Secondary hydrophilic and cation exchange interactions of the analyte can be used for retention at a proper pH. (For more detail, see *Secondary Interactions*).

For **adsorption media** (e.g. ENVI-Carb and ENVI-Chrom P) that are used under reversed phase conditions, a pH should be chosen to maximize retention of analytes on the sorbent as with reversed phase bonded silicas. Elution is usually done with an organic solvent, so pH is usually not a factor at this point. Surprisingly, phenols retain better on ENVI-Chrom P when applied in solutions at a neutral pH, where phenols can be charged, than at an acidic pH levels where they are neutral. This shows that adsorption media may have different selectivities than the bonded silicas for certain compounds, and that a range of pH levels of the sample and conditioning solutions should be investigated when using these materials.

In **normal phase** SPE procedures on bonded silicas or adsorption media, pH is usually not an issue, because the solvents used in these processes are typically nonpolar organic solvents, rather than water.

Retention in **ion exchange** SPE procedures depends heavily on the pH of the sample and the conditioning solutions. For retention of the analyte, the pH of the sample must be one at which the analyte and the functional groups on the silica surface are charged oppositely. For further details, see *Ion Exchange* on page 4.

Typical SPE Tube and Disk

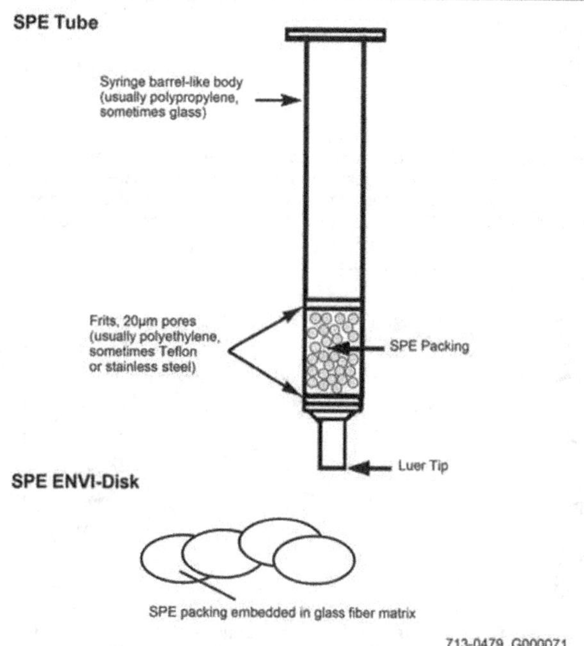

How to Use SPE

Solid phase extraction is used to separate compounds of interest from impurities in three ways. Choose the most appropriate scheme for your sample:

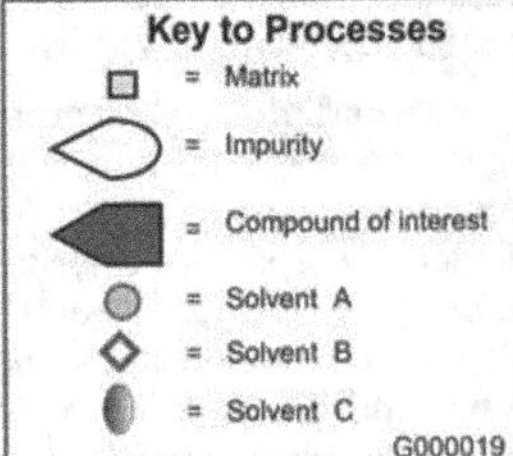

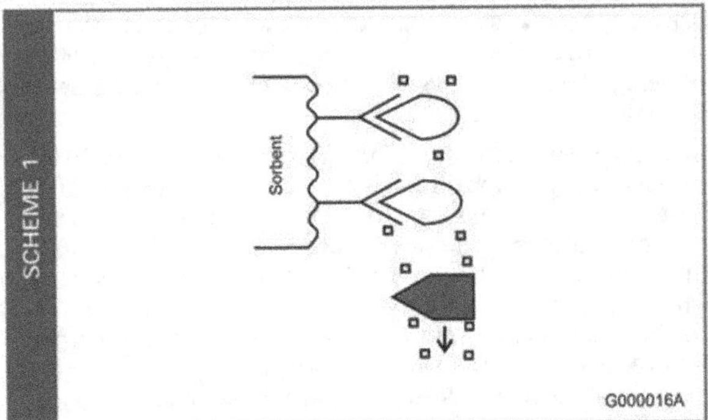

Selective Extraction. Select an SPE sorbent that will bind selected components of the sample — either the compounds of interest or the sample impurities. The selected components are retained when the sample passes through the SPE tube or disk (the effluent will contain the sample minus the adsorbed components). Then, either collect the adsorbed compounds of interest through elution, or discard the tube containing the extracted impurities.

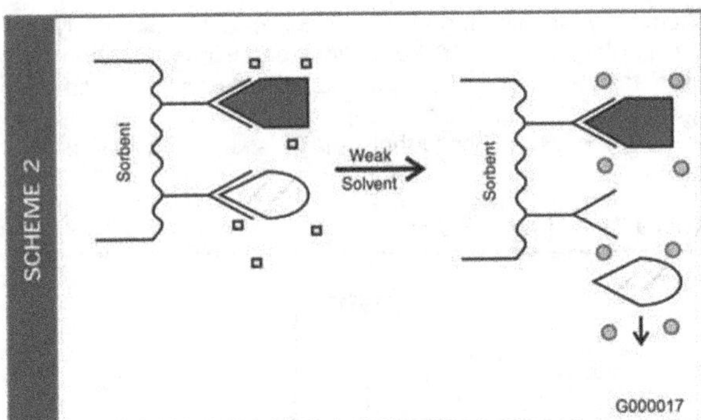

Selective Washing. The compounds of interest and the impurities are retained on the SPE packing when the sample passes through; the impurities are rinsed through with wash solutions that are strong enough to remove them, but weak enough to leave the compounds of interest behind.

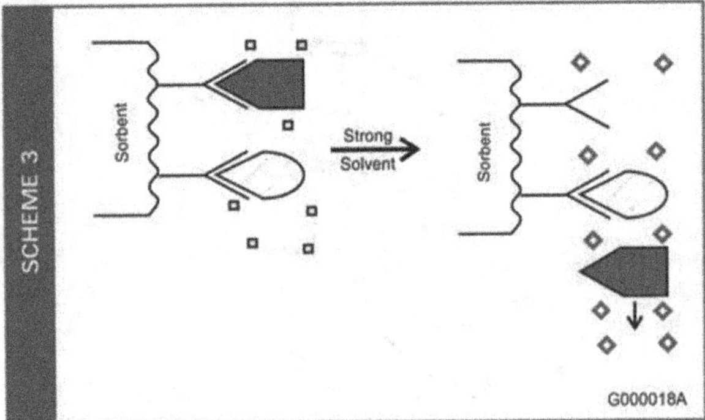

Selective Elution. The adsorbed compounds of interest are eluted in a solvent that leaves the strongly retained impurities behind.

SPE Is a Five-Step Process

The SPE process provides samples that are in solution, free of interfering matrix components, and concentrated enough for detection. This is done in five steps (summarized here and described on the next two pages).

- For reversed phase, normal phase, and ion exchange SPE procedures, all five steps typically are needed.
- For some sample cleanup procedures, only the first three steps may apply. Steps 1 and 2 are the same as shown. However, in step 3, the analyte is collected in the effluent as the sample passes through the tube; interfering impurities remain on the sorbent.

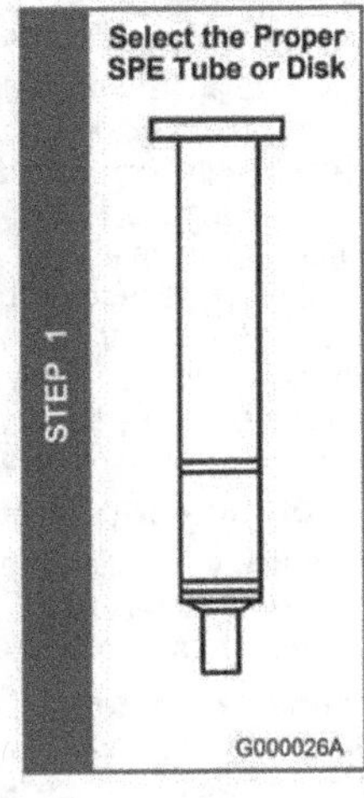

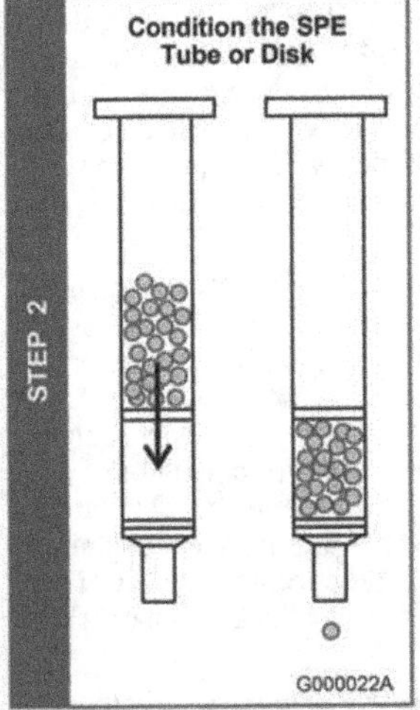

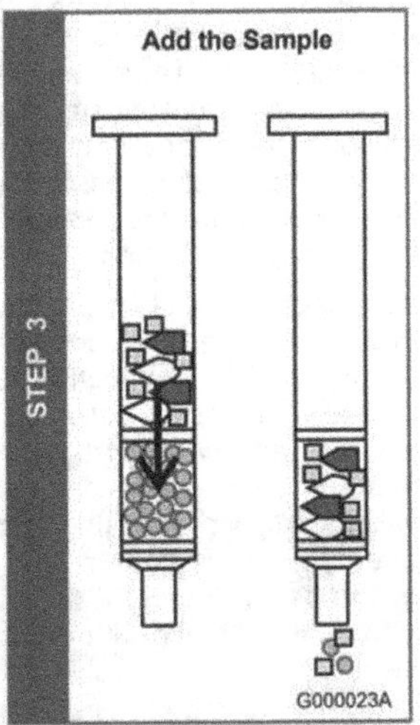

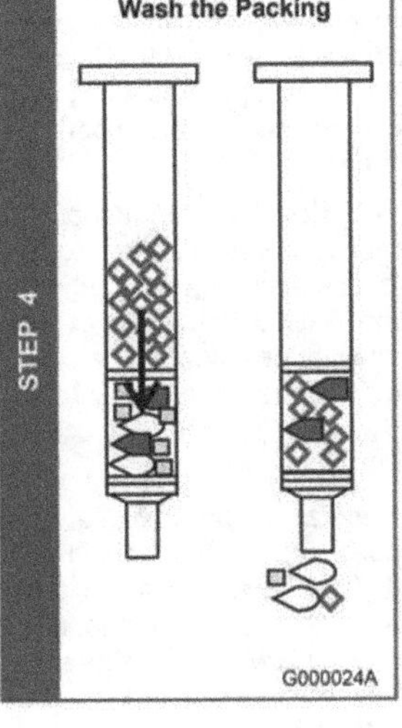

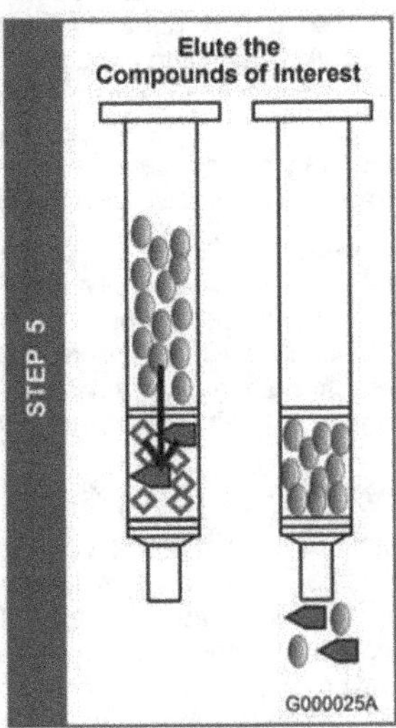

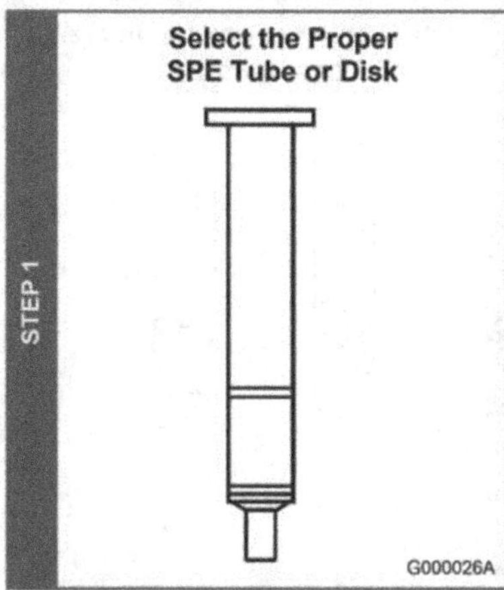

Note: An SPE disk is recommended for large volume samples, samples containing high amounts of particulates, or when a high flow rate is required during sampling.

Selecting an SPE Tube or Disk: Size

Selecting SPE Tube Size

If Your Sample Is . . .	Use Tube Size . . .
< 1mL	1mL
1mL to 250mL and the extraction speed is not critical	3mL
1mL to 250mL and a fast extraction procedure is required	6mL
10mL to 250mL and higher sample capacity is needed	12, 20, or 60mL
< 1 liter and extraction speed is not critical	12, 20, or 60mL

Selecting SPE Disk Size

If Your Sample Is . . .	Use Disk Size . . .
100mL to 1 liter	47mm
>1 liter and higher sample capacity is needed	90mm

Selecting an SPE Tube: Bed Weight

Reversed Phase, Normal Phase, and Adsorption-Type Procedures:

The mass of the compounds to be extracted should not be more than 5% of the mass of the packing in the tube.
In other words, if you are using a 100mg/1mL SPE tube, do not load more than 5mg of analytes.

Ion Exchange Procedures:

You must consider ion exchange capacity.

- LC-SAX and LC-SCX tubes have ~0.2meq/gram of sorbent capacity (1 meq = 1mmole of [+1] or [-1] charged species).
- LC-NH_2 and LC-WCX tubes: ion exchange capacities should be determined for your own application.

Selecting an SPE Tube: Sorbent Type

(Note: Refer to schematic on page 12.)

Is your sample matrix *aqueous* or *organic*?

If aqueous:

Is your analyte of interest more soluble in ***water*** or in ***organic solvents*** (e.g., hexane or dichloromethane)?

If more soluble in water, is your analyte charged or neutral?

Charged:

If weakly anionic (-) and acidic,
use an LC-SAX or LC-NH_2 tube.

If strongly anionic (-) and acidic:
- and you want to recover the extracted analyte,
  use an LC-NH_2 tube.
- and you do not want to recover the extracted analyte,
  use an LC-SAX tube.

If weakly cationic (+) and basic,
use an LC-SCX or LC-WCX tube.

If strongly cationic (+) and basic:
- and you want to recover the extracted analyte,
  use an LC-WCX tube.
- and you do not want to recover the extracted analyte,
  use an LC-SCX tube.

Neutral:
If analytes are difficult to extract using reversed phase packings (e.g. alcohols, sugars, glycols), try an ENVI-Carb or ENVI-Chrom P tube, or try to remove interferences by reversed phase extraction or by using an LC-SAX or LC-SCX tube.

If more soluble in organics, is your analyte charged or neutral?

Charged:
Try reversed phase or ion exchange extraction.

Neutral:
Try reversed phase extraction.

If organic:

Try any of the following.

Concentrate analyte by evaporation.

Evaporate to dryness and reconstitute with another solvent.

Use SPE.

Is the organic solvent ***polar*** and water-miscible (e.g. methanol or acetonitrile) or ***mid- to nonpolar*** and not water-miscible (e.g. dichloromethane or hexane)?
Refer to **Table A.**

If polar:
Dilute with water to <10% organic and follow the matrix scheme for aqueous analytes.

If mid- to nonpolar:
Use normal phase, or evaporate to dryness, reconstitute with water or a water-miscible solvent, then dilute with water as above and use the matrix scheme for aqueous analytes.

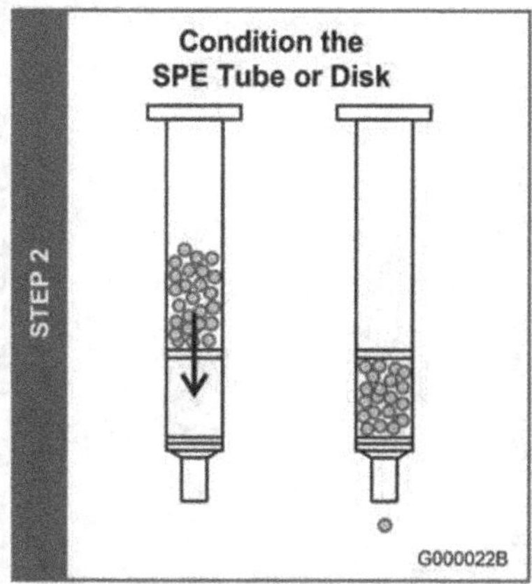

To condition the SPE tube packing, rinse it with up to one tube-full of solvent before extracting the sample. For disks, use a volume of 5-10mL.

Reversed phase type silicas and nonpolar adsorption media usually are conditioned with a water-miscible organic solvent such as methanol, followed by water or an aqueous buffer. Methanol wets the surface of the sorbent and penetrates bonded alkyl phases, allowing water to wet the silica surface efficiently.

Sometimes a *pre-conditioning solvent* is used before the methanol step. This solvent is usually the same as the elution solvent (see step 5), and is used to remove any impurities on the SPE tube that could interfere with the analysis, and may be soluble only in a strong elution solvent.

Normal phase type SPE silicas and polar adsorption media usually are conditioned in the organic solvent in which the sample exists.

Ion exchange packings that will be used for samples in nonpolar, organic solvents should be conditioned with the sample solvent. For samples in polar solvents, use a water-miscible organic solvent, then an aqueous solution with the proper pH, organic solvent content, and salt concentration.

To ensure that the SPE packing does not dry between conditioning and sample addition, allow about 1mm of the last conditioning solvent to remain above the top tube frit or above the surface of the disk. If the sample is to be introduced from a reservoir or filtration tube, add an additional 0.5mL of the final conditioning solution to a 1mL SPE tube, 2mL to a 3mL tube, 4mL to a 6mL tube, and so on. This prevents the tube from drying out before the sample actually reaches the tube. If the packing dries before the sample is added, repeat the conditioning procedure. Flush buffer salts from the tube with water before reintroducing organic solvents. If appropriate, attach the sample reservoir at this time using a tube adapter.

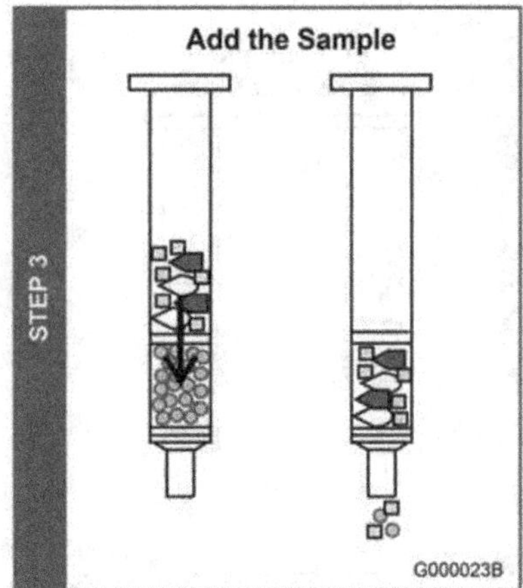

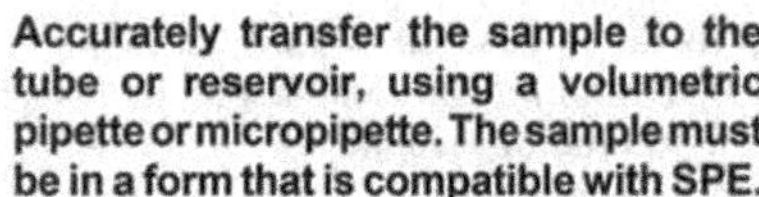

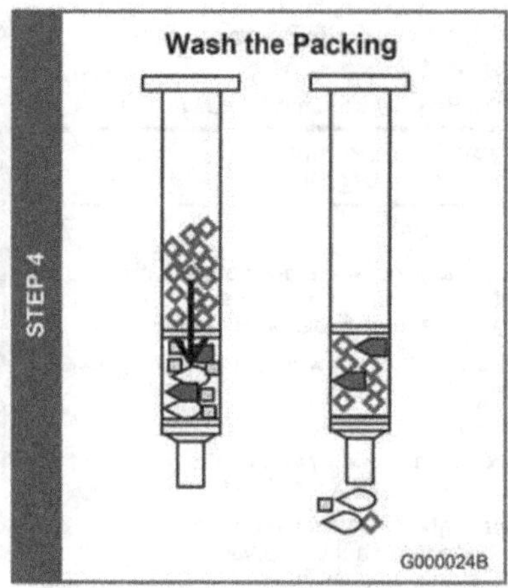

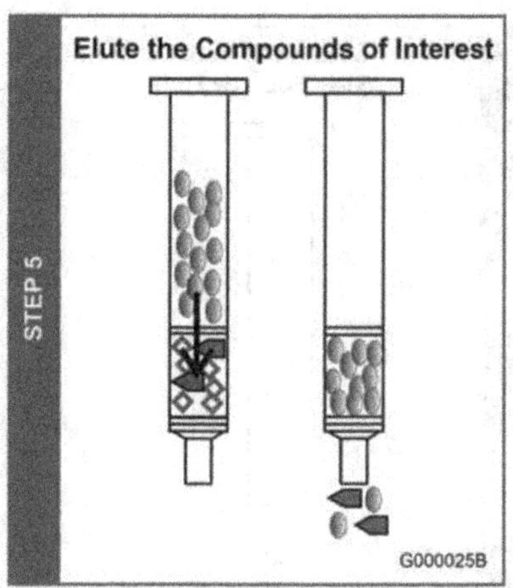

Accurately transfer the sample to the tube or reservoir, using a volumetric pipette or micropipette. The sample must be in a form that is compatible with SPE.

Total sample volume can range from microliters to liters (see step 1). When excessive volumes of aqueous solutions are extracted, reversed phase silica packings gradually lose the solvent layer acquired through the conditioning process. This reduces extraction efficiency and sample recovery. For samples >250mL, add small amounts of water-miscible solvents (up to 10%) to maintain proper wetting of reversed phase packings. Maximum sample capacity is specific to each application and the conditions used. If recoveries are low or irreproducible, test for analyte breakthrough using the following technique:

Attach two conditioned SPE tubes of the same packing together using an adapter. Pass the sample through both tubes. When finished, detach each tube and elute it separately. If the analyte is found in the extract of the bottom tube, the sample volume is too great or bed weight is too small, resulting in analyte breakthrough.

To enhance retention of appropriate compounds on the packing, and elution or precipitation of unwanted compounds, adjust the pH, salt concentration, and/or organic solvent content of the sample solution. To avoid clogging SPE tube frits or the SPE disk, pre-filter or centrifuge samples prior to extraction if possible.

Slowly pass the sample solution through the extraction device, using either vacuum or positive pressure. The flow rate can affect the retention of certain compounds. Generally, the flow rate should not exceed 2mL/min for ion exchange SPE tubes, 5mL/min for other SPE tubes, and may be up to 50mL/min for disks. Dropwise flow is best, when time is not a factor.

For some difficult sample matrices, additional pretreatment may be necessary. See Sample Pretreatment on the next page.

If compounds of interest are retained on the packing, wash off unwanted, unretained materials using the same solution in which the sample was dissolved, or another solution that will not remove the desired compounds. Usually no more than a tube volume of wash solution is needed, or 5-10mL for SPE disks.

To remove unwanted, weakly retained materials, wash the packing with solutions that are stronger than the sample matrix, but weaker than needed to remove compounds of interest. A typical solution may contain less organic or inorganic salt than the final eluant. It also may be adjusted to a different pH. Pure solvents or mixtures of solvents differing sufficiently in polarity from the final eluant may be useful wash solutions (see Table A).

If you are using a procedure by which compounds of interest are not retained on the packing, use about one tube volume of the sample solvent to remove any residual, desired components from the tube, or 5-10mL to remove the material from a disk. This rinse serves as the elution step to complete the extraction process in this case.

Rinse the packing with a small volume (typically 200μL to 2mL depending on the tube size, or 5-10mL depending on the disk size) of a solution that removes compounds of interest, but leaves behind any impurities not removed in the wash step. Collect the eluate and further prepare as appropriate.

Two small aliquots generally elute compounds of interest more efficiently than one larger aliquot. Recovery of analytes is best when each aliquot remains in contact with the tube packing or disk for 20 seconds to 1 minute. Slow or dropwise flow rates in this step are beneficial.

Strong and weak elution solvents for adsorbed compounds in SPE are described in Table A.

Table A. Characteristics of Solvents Commonly Used in SPE

Polarity			Solvent	Miscible in Water?
Nonpolar	Strong Reversed Phase	Weak Normal Phase	Hexane	No
			Isooctane	No
			Carbon tetrachloride	No
			Chloroform	No
↓	↑	↓	Methylene chloride (dichloromethane)	No
			Tetrahydrofuran	Yes
			Diethyl ether	No
			Ethyl acetate	Poorly
			Acetone	Yes
			Acetonitrile	Yes
			Isopropanol	Yes
			Methanol	Yes
			Water	Yes
Polar	Weak Reversed Phase	Strong Normal Phase	Acetic acid	Yes

Sample Pretreatment

In addition to ensuring proper pH of the sample (see *The Role of pH in SPE* on page 5), you should consider other sample pretreatment needs. The following section describes how some difficult sample matrices should be pretreated before being applied to the SPE device:

Liquids

Biological Matrices

Serum, plasma, and whole blood: Serum and plasma samples may not need to be pretreated for SPE. In many cases, however, analytes such as drugs may be protein-bound, which reduces SPE recoveries. To disrupt protein binding in these biological fluids, use one of the following methods for reversed phase or ion exchange SPE procedures:

- Shift pH of the sample to extremes (pH<3 or pH>9) with acids or bases in the concentration range of 0.1M or greater. Use the resulting supernatant as the sample for SPE.
- Precipitate the proteins using a polar solvent such as acetonitrile, methanol, or acetone (two parts solvent per one part biological fluid is typical). After mixing and centrifugation, remove the supernatant and dilute with water or an aqueous buffer for the SPE procedure.
- To precipitate proteins, treat the biological fluid with acids or inorganic salts, such as formic acid, perchloric acid, trichloroacetic acid, ammonium sulfate, sodium sulfate, or zinc sulfate. The pH of the resulting supernatant may be adjusted prior to use for the SPE procedure.
- Sonicate the biological fluid for 15 minutes, add water or buffer, centrifuge, and use the supernatant for the SPE procedure.

Urine: Urine samples may not require pretreatment for reversed phase or ion exchange SPE, but often is diluted with water or a buffer of the appropriate pH prior to sample addition. In some cases, acid hydrolysis (for basic compounds) or base hydrolysis (for acidic compounds) is used to ensure that the compounds of interest are freely solvated in the urine sample. Usually a strong acid (e.g. concentrated HCl) or base (e.g. 10M KOH) is added to the urine. The urine is heated for 15-20 minutes, then cooled and diluted with a buffer, and the pH adjusted appropriately for the SPE procedure. Enzymatic hydrolysis that frees bound compounds or drugs also may be used.

Cell Culture Media

Cell culture media may be used without pretreatment. Some methods may require dilution of the media with water or buffer at the proper pH to ensure that the analyte is freely solvated in the sample. If a particulate-laden cell culture medium is difficult to pass through the SPE device, it may need to be vortexed and centrifuged prior to SPE. Most SPE procedures for cell culture media are done using reversed phase or ion exchange methods.

Milk

Milk generally is processed under reversed phase or ion exchange SPE conditions. The sample may be diluted with water, or with mixtures of water and a polar solvent such as methanol (up to 50%). Some procedures may require precipitation of proteins by treatment with acid (typically HCl, H_2SO_4, or trichloroacetic acid). After precipitation, the sample is centrifuged and the supernatant is used for SPE.

Water Samples

Drinking water, groundwater, and wastewater samples may be extracted directly by SPE, as long as they are not heavily laden with solid particles. Groundwater and wastewater samples might need to be filtered prior to the SPE procedure. Filtering may reduce recoveries if compounds of interest are bound to the removed particles. If possible, do not filter the sample. Pass the unfiltered sample directly through the SPE device and, during elution, allow the solvent to pass through the particles on the adsorbent bed. This will improve recoveries, since particle-bound compounds of interest will be recovered using this process. In most cases, water samples are used with reversed phase or ion exchange SPE procedures.

Wine, Beer, and Aqueous Beverages

Aqueous and alcoholic beverages may be processed for SPE without pretreatment under reversed phase or ion exchange conditions. For reversed phase procedures, if alcohol content is high, dilution with water or buffer to <10% alcohol may be required. If necessary, solids in the sample can be removed by centrifugation or filtration prior to SPE.

Fruit Juices

Fruit juices typically are processed without pretreatment or are centrifuged for reversed phase or ion exchange SPE. If centrifuged, the resulting supernatant is used for the SPE procedure. Viscous juices may need to be diluted with water or buffer at the proper pH.

Liquid Pharmaceutical Preparations

Because liquid pharmaceuticals are mainly aqueous, these samples generally are processed by reversed phase or ion exchange SPE. If the preparation is viscous, dilution with water or an appropriate buffer may be necessary. Organic extracts of the preparation may be processed using normal phase SPE.

Oils

Hydrocarbon or fatty oils are commonly processed under normal phase conditions, because they cannot be diluted with water. The diluent is usually a mid-polar to nonpolar solvent such as hexane or a chlorinated solvent. The diluted sample is passed through a normal phase bonded silica or adsorption medium, and the sample is collected as it passes through. The compound of interest should pass through unretained, while impurities remain in the adsorbent. If the compound of interest is retained on the packing, successive washes of the SPE packing with increasingly polar solvents, or with mixtures of the diluent with a polar solvent, are performed until the analyte is recovered in one of the fractions. For collecting oil in water samples, reversed phase SPE is used.

Solids

Soil and Sediment

Soil and sediment samples typically are extracted with mid-polar to nonpolar solvents via Soxhlet extraction or sonication. The resulting extracts are then processed by normal phase SPE to remove interferences. The cleaned extracts then can be evaporated and reconstituted with another solvent for additional SPE (reversed phase, ion exchange, or normal phase) if necessary. If extraction efficiency of the compound of interest is pH-dependent, soil and sediment samples may need to be homogenized with water at the appropriate pH prior to extraction and SPE cleanup. In some cases, small amounts of soil or sediment can be homogenized with an appropriate solvent and then passed through the SPE device without pretreatment, as long as the particles do not clog the device. The analyte is then eluted with the appropriate solvent by passing it directly through any particles that rest on the SPE tube packing or disk.

Plant Tissues, Fruits, Vegetables, and Grains

Plant tissues, fruits, vegetables, and commodities such as animal feeds and grains are homogenized either in water, in a polar organic solvent (e.g. methanol or acetonitrile), or in mixtures of water with these solvents, for reversed phase or ion exchange cleanup procedures. After centrifugation or filtration to remove the precipitated proteins and solids, the pH of the sample may need to be adjusted. The analyte may adsorb onto the SPE packing or may simply pass through, free from interferences. The sample also may be homogenized with a mid-polar to nonpolar solvent for normal phase SPE procedures. Again, the sample may need to be centrifuged or filtered prior to SPE.

Meat, Fish, and Animal Tissues

Meat, fish, and other tissues can be processed in the same manner as described above for solid fruits and vegetables. In addition to homogenization with water, sample preparation for reversed phase and ion exchange SPE procedures may also involve hydrolysis or digestion of the meat or tissue with acid (typically HCl or trichloroacetic acid) or saponification with base (e.g. NaOH). Enzymatic hydrolysis also may be used. The sample can then be centrifuged and the supernatant used for the SPE procedure. Tissue extracts obtained with mid-polar to nonpolar solvents can be processed using normal phase procedures.

Tablets and Other Solid Pharmaceutical Preparations

Tablets and solid pharmaceutical preparations should be crushed into a fine powder, then extracted or homogenized with water or an appropriate buffer for reversed phase and ion exchange SPE procedures. A mid-polar to nonpolar solvent is used for normal phase cleanup procedures.

Hardware and Accessories for Processing Samples

SPE tubes can be processed individually using a single tube processor (Figure A) or with a syringe and an adapter (Figure B). The liquid sample is placed in the SPE tube, and the processor or syringe is used to provide positive pressure to force the liquid through the tube. Positive pressure from an air or nitrogen line also may be used to force the solutions through the tube.

Figure A. Single Tube Processor

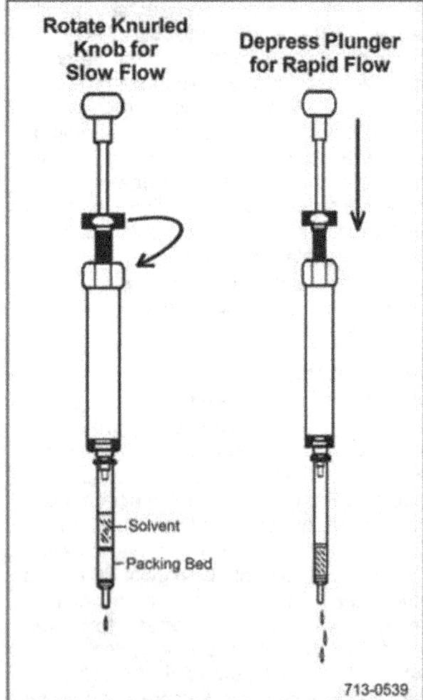

Figure B. Process Using Applied Pressure

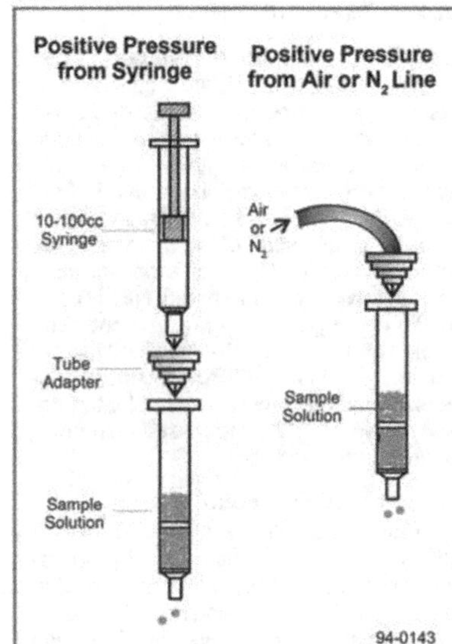

A solution also can be processed through a single SPE tube using a vacuum flask and rubber stopper (Figure C). The vacuum pulls a solution through the SPE tube. The solution then can be collected in a test tube located inside the flask.

Several SPE tubes can be processed using a centrifuge (Figure D). The solutions are placed in the SPE tubes and the centrifuge forces the solutions through the tubes into test tubes. Appropriate spin rates must be determined; they can vary depending on the type and mass of the packing in the tube and the volume of sample. For sample addition, follow the recommended flow rates discussed in step 3 (page 8).

Figure C. Configuration Using a Vacuum Flask

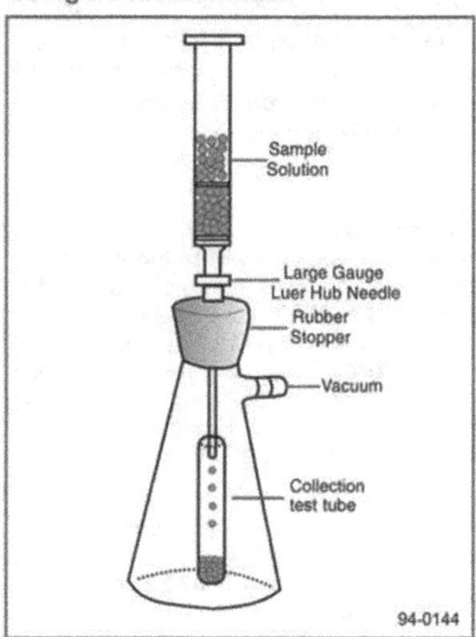

Figure D. Processing Several Tubes Using a Centrifuge

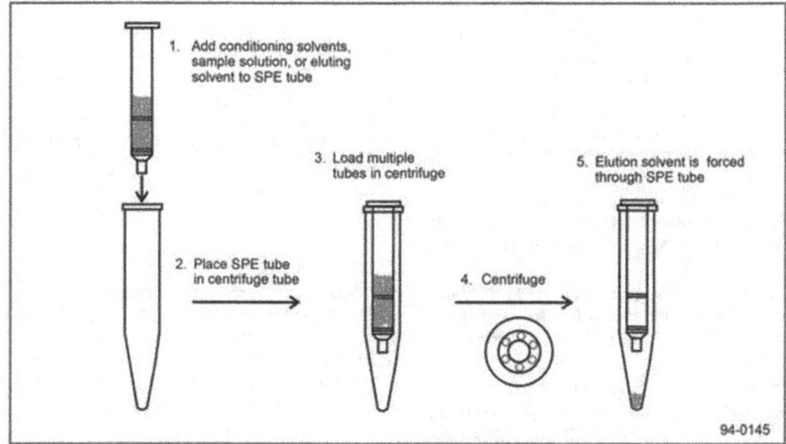

Multiple tubes can be processed simultaneously using a 12- or 24-port vacuum manifold. Supelco offers two types of Visiprep™ SPE vacuum manifolds – a standard lid version and a disposable liner version.

- Our *standard lid manifold* (Figure E) has unique flow control valves that allow easy control of flow through each SPE tube. Reusable stainless steel needle guides direct the sample into the glass basin below.
- Valves in the lid of the *DL manifold* (Figure F) contain a disposable Teflon® liner that directs the sample into the glass basin. The liner is conveniently disposable, is inert, and prevents cross-contamination in critical applications.

Both types of vacuum manifolds have a solvent resistant main vacuum gauge and valve used to monitor the vacuum and release the vacuum during processing.

Figure E. Visiprep Vacuum Manifold with Standard Lid

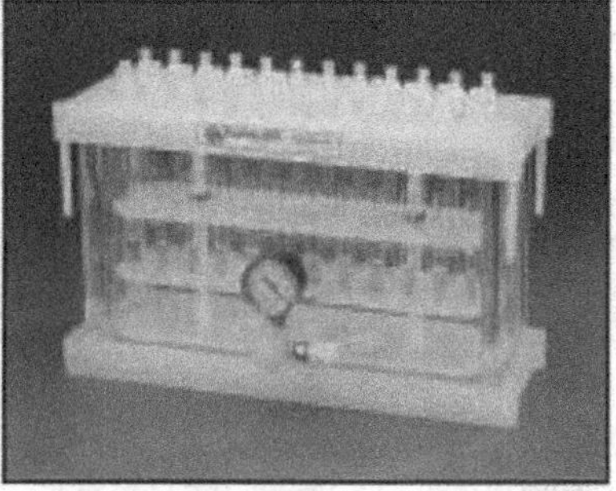

Figure F. Visiprep Vacuum Manifold with Disposable Liner

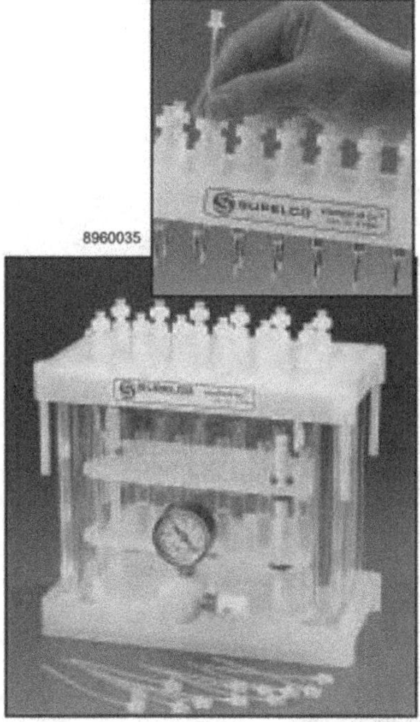

Unwanted solvents that collect in the bottom of the glass basin are pulled continuously through the gauge into the vacuum pump trap (Figure G) located between the pump and the manifold. This minimizes contamination by preventing buildup of unwanted waste solutions in the basin. The manifolds are equipped with an adjustable collection rack system that is placed inside the glass basin. The racks can be adjusted easily to accommodate many types and dimensions of collection vessels, such as small test tubes (10mm), large test tubes (16mm), volumetric flasks (1mL-10mL), and many types of autosampler vials.

Figure G. SPE Vacuum Pump Trap

The Preppy™ vacuum manifold (Figure H) is our simplest and most economical manifold. It too enables the analyst to simultaneously prepare up to 12 samples.

Figure H. Preppy Vacuum Manifold

The Preppy manifold consists of a chemical-resistant cover and gasket, a glass basin, a vacuum release vent, 12 individual flow control valves with knurled tops, and stainless steel solvent guide needles. Two optional collection vessel racks are available: one holds both 1mL and 4mL autosampler vials, and the other holds 15mL or 20mL vials. Adapters are available for the 1mL-4mL vessel rack to hold mini-centrifuge tubes. An optional vacuum gauge/bleed valve assembly can be installed to allow precise control of the vacuum used with the Preppy manifold.

SPE tubes can be processed individually or can be combined using an adapter to provide different selectivities. Small volumes are processed directly in the SPE tube. Larger volumes can be accommodated by using a reservoir with an adapter. For very large samples, a large volume sampler is available, which allows unattended sample processing (Figure I).

Figure I. Visiprep Large Volume Sampler

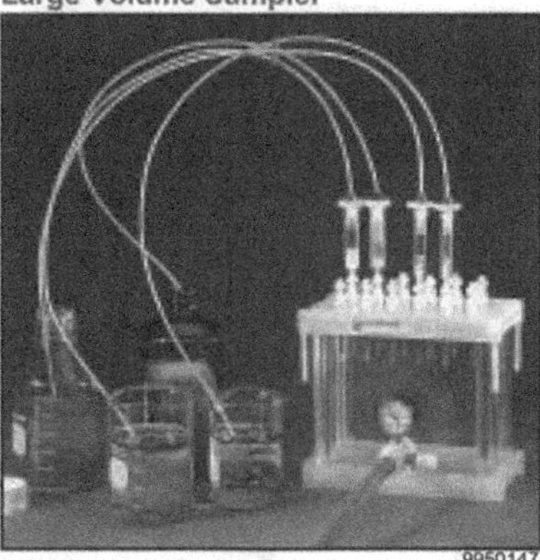

For more information on these accessories, please refer to Supelco's general catalog.

Visidry™ drying attachments (Figure J), available with 12 ports and 24 ports, can be used to dry the tubes or evaporate and concentrate collected samples during the SPE procedure.

Figure J. Visidry Drying Attachment

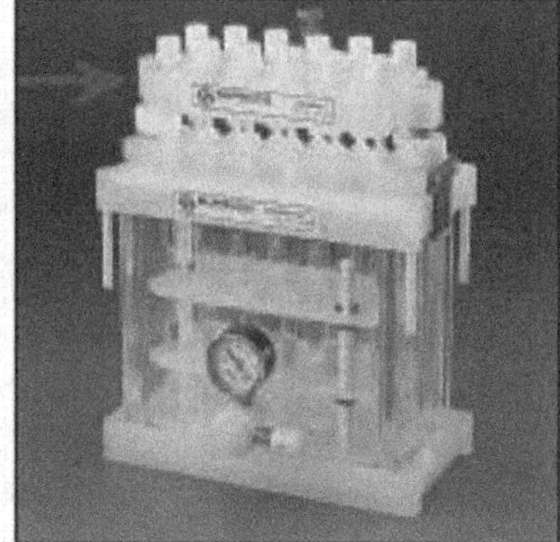

SPE disks can be processed on a vacuum filtration flask-type assembly (Figure K).

The ENVI-Disk Clamp (Figures L and M) is designed to eliminate potential leakage that is often observed with conventional flask clamps. The ENVI-Disk Holder (Figure N) is used when a standard filtration flask assembly is not available.

Figure K.

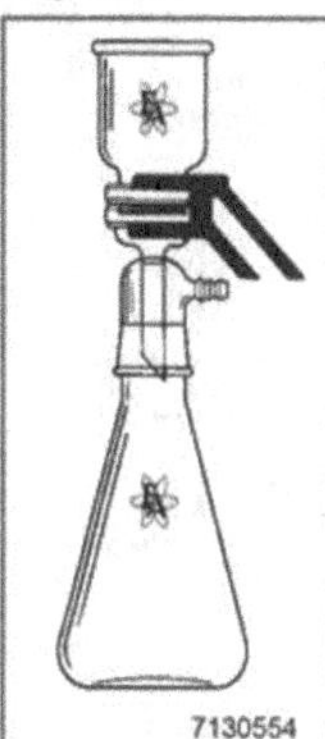

Figure L.

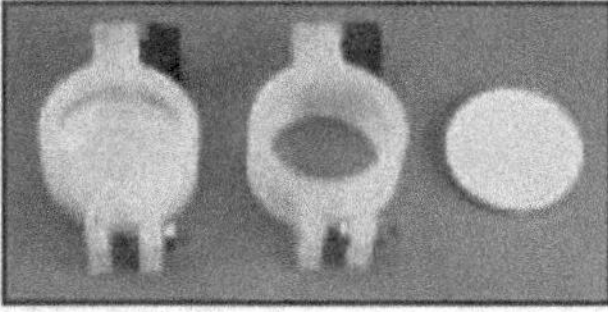

Figure M.

Figure N.

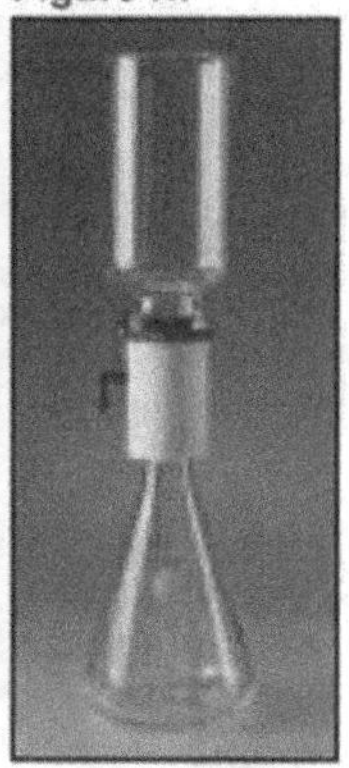

Sample Characteristics Determine Your SPE Procedure

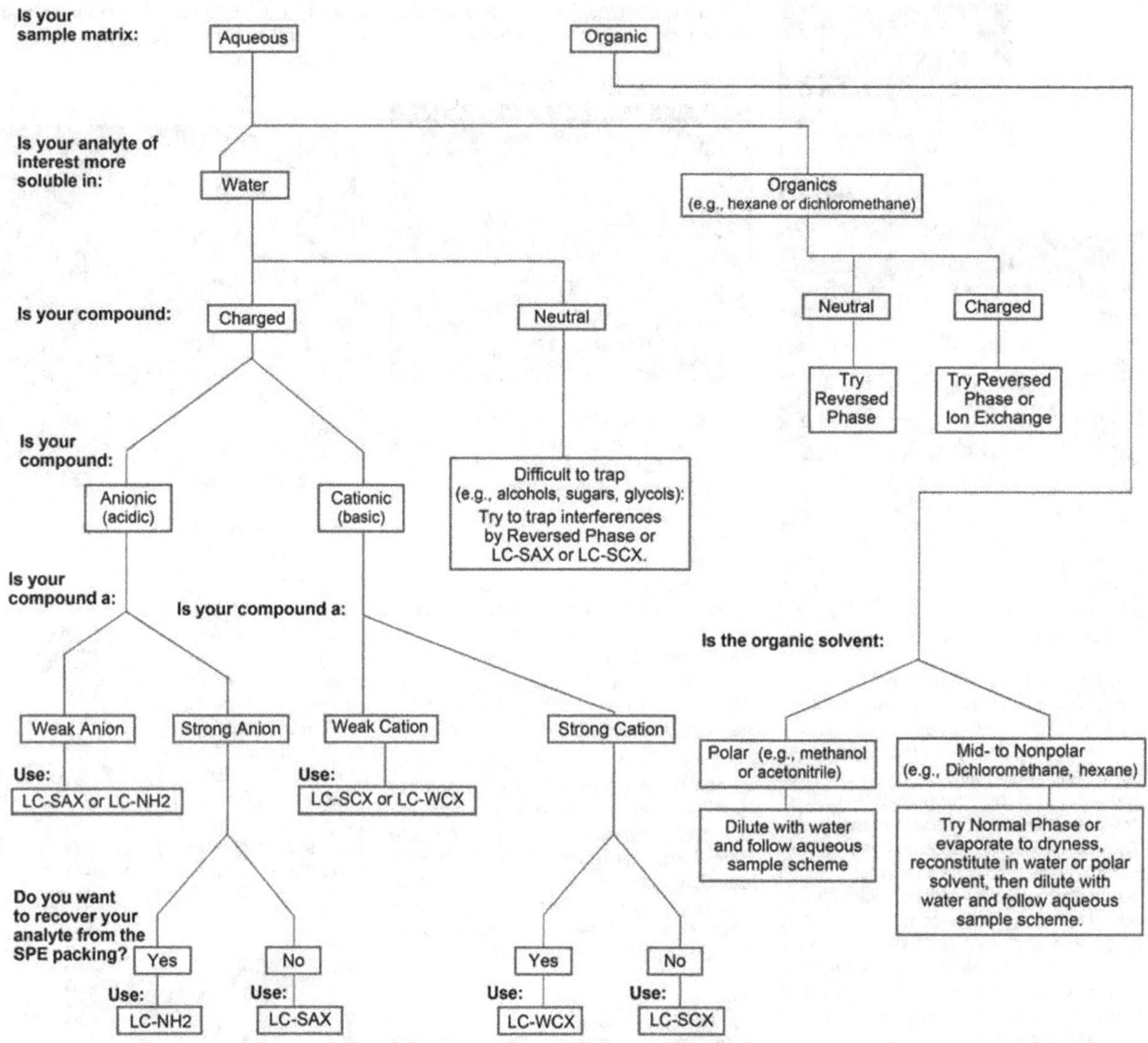

Choosing the proper SPE device for your application depends on:

- Sample volume
- Degree of contamination
- Complexity of sample matrix
- Quantity of compounds of interest
- Type and solvent strength of sample matrix

Patents
Visidry Drying Attachment – US patent 4,810,471; other patents pending.
Visiprep Vacuum Manifold – US patents D.289,861; 4,810,471; other patents pending.

Technical Service

If you need help in choosing the proper devices for your sample preparation applications, please contact our Technical Service chemists at 800-359-3041 or 814-359-3041.

Trademarks
ENVI, Hisep, Preppy, Supelclean, Visidry, Visiprep – Sigma-Aldrich Co.
Florisil – U.S. Silica Co.
Teflon – E.I. du Pont de Nemours & Co., Inc.

24

PLANE CHROMATOGRAPHY: PAPER AND THIN-LAYER CHROMATOGRAPHY

24.1 Plane Chromatography
24.2 Thin-Layer Chromatography
24.3 Retardation Factor (R_F) in TLC
24.3.1 Example I
24.3.2 Example II
24.4 Plate Heights (H) and Counts (N) in TLC
24.5 Retention Factor in TLC

24.1 PLANE CHROMATOGRAPHY

There is an alternative chromatography technique that may be encountered in the analytical laboratory called "plane chromatography." Just as the name suggests, plane chromatography utilizes a flat planar surface as the stationary phase in contrast to column chromatography that uses small spherical particles packed into a column. In plane chromatography, a sample is put onto the plane surface usually as a small droplet near the edge of the plane. The plane is then placed into a solvent, and the solvent begins to move over the plane due to capillary and wetting effects. As the solvent moves across the plane, the compounds in the sample will move with the solvent, again partitioning themselves between the stationary phase (the plane material) and the mobile phase (the solvent moving across the plane) according to chromatographic theory. A very simple example of this is paper chromatography using something as common in the laboratory as Whatman #1 filter paper cut into a rectangular shape. A sample can be spotted onto the bottom edge of the cut filter paper and then placed into a beaker containing a developing solvent. These steps are illustrated in Figure 24.1 wherein (a) a Whatman #1 filter paper is cut into a rectangular shape. This allows convenient placement of the paper into the beaker. In Figure 24.1(b), samples are spotted onto the edge of the filter paper using a pipette.

Next, the sample-spotted filter paper is placed into a beaker with about ¼ inch of solvent in the bottom. This is illustrated in Figure 24.2(a). Normally, a watch glass has been placed onto the top of the beaker in order to allow the saturation of the air within the beaker with the developing solvent. In Figure 24.2(b), the solvent has begun to move up the paper, and the chromatography begins to take place. This is the same process that we have previously learned where components will partition between the stationary phase and the mobile phase according to affinities. In Figure 24.2(c), the solvent has moved almost to the top of the paper, and the chromatography is now complete.

24.2 THIN-LAYER CHROMATOGRAPHY

While the process of paper chromatography gives a good example of plane chromatography, usually the technique of thin-layer chromatography (TLC) is mostly used in the analytical laboratory. In TLC, the process is the same as paper chromatography, but instead of using paper fiber as the stationary phase, usually silica or alumina is used. The name TLC is due to the thin layer of stationary phase that has been spread out upon a glass plate or microscope slide as a slurry that is allowed to dry producing the thin-layered stationary-phase plate. The TLC plate or slide is also spotted with sample like we just saw in paper chromatography, and the TLC plate is placed into a beaker (for TLC slides) or into a TLC developing tank for TLC plates. The silica gel used as the stationary phase contains –SiOH groups that form hydrogen bonds with polar groups such as alcohols, amines, and carboxylic acids. Alumina works well for aromatic hydrocarbons but may bond too strongly to carboxylic acids. Figure 24.3 illustrates some common types of TLC developing tanks and also a couple of TLC plates in the process of being developed.

Let's now take a closer look at the developed TLC plate from Figure 24.2 (the description is the same whether we are looking at paper chromatography or, say, a silica gel TLC plate, as we are now calling Figure 24.2). After developing the plate in the TLC chamber, the plate is removed and allowed to dry, such as illustrated in Figure 24.4. The TLC plate has a designated sample origin line, a solvent front, and the separated compounds.

Analytical Chemistry: A Chemist and Laboratory Technician's Toolkit, First Edition. Bryan M. Ham and Aihui MaHam.

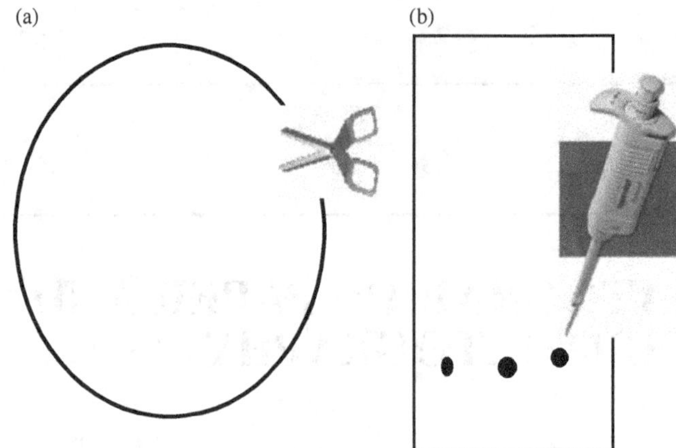

FIGURE 24.1 Initial steps in preparation for paper chromatography. (a) Whatman #1 filter paper is cut into a rectangular shape. (b) Samples are spotted onto the edge of the filter paper using a pipette.

(a) Step 1—paper with spots into beaker
(b) Step 2—solvent moves up paper
(c) Step 3—separated compounds
Cover watch glass for air solvent saturation
Beaker as developing chamber
Solvent front
Solvent front
Developing solvent (mobile phase)
Sample spotted paper (stationary phase)
Samples moving with mobile phase
Samples separated into individual components due to chromatography

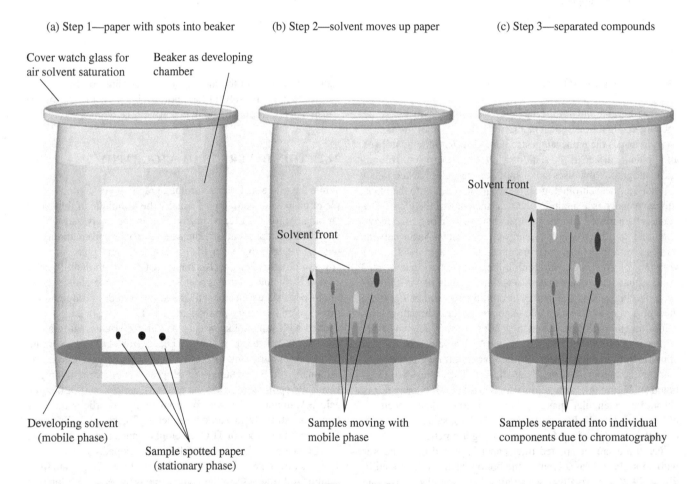

FIGURE 24.2 Steps in developing the paper chromatography. (a) Whatman #1 filter paper cut into a rectangular shape with sample spots placed into beaker. (b) Solvent moves up the paper with the beginning of the chromatography. (c) Chromatography complete with the solvent front near the top and sample components separated.

Often, the spots on a TLC plate will have become a little distorted and elongated. In this case, the part of the spot with the greatest intensity or absorbance, or visual concentrated region, is drawn through as the designated place for the compound. Also, spots often are not completely visible (or not visible at all) on the plate, and further processing of the plate is required such as spraying a fine mist of a sulfuric acid solution across the plate. The plate is then heated for a short period of time. The sulfuric acid will char

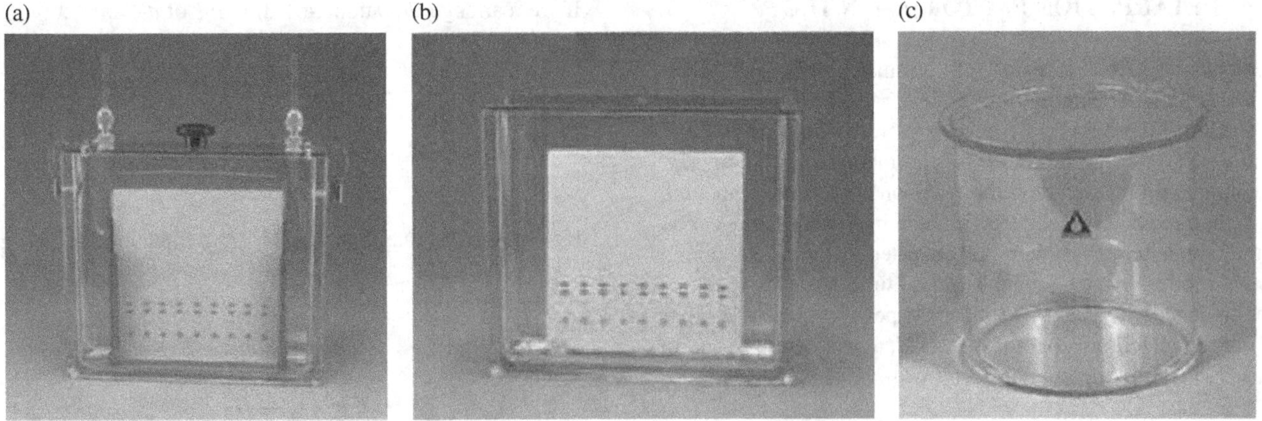

FIGURE 24.3 (a) and (b) Examples of TLC developing tanks and TLC plates. (c) Empty developing tank.

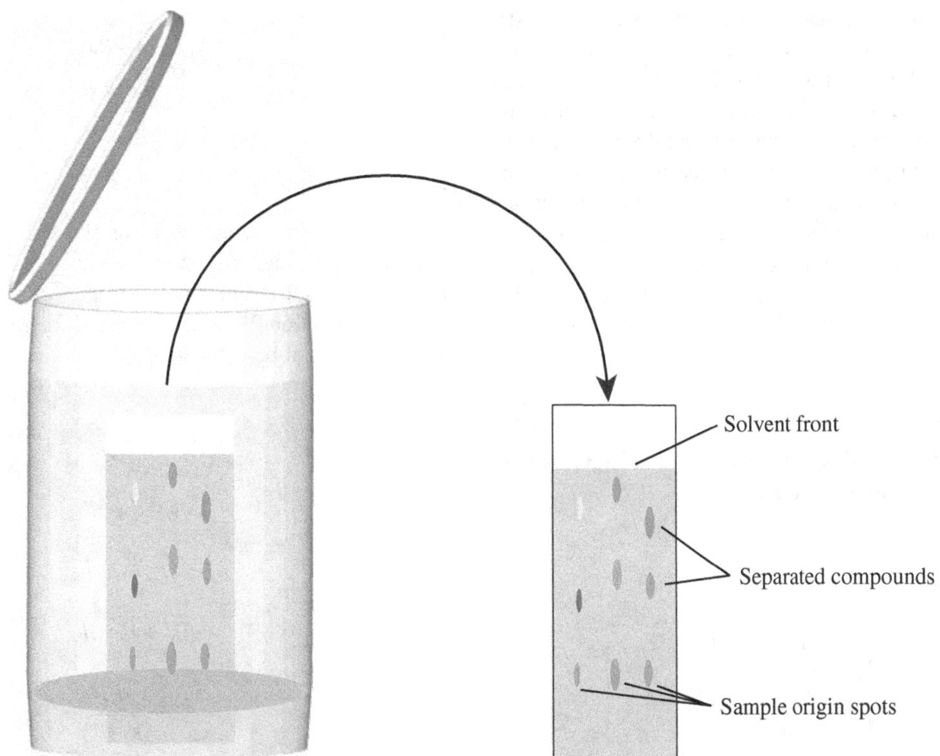

FIGURE 24.4 After developing the plate in the TLC chamber, the plate is removed and allowed to dry. The TLC plate has a designated sample origin spots, a solvent front, and the separated compounds.

any organic compounds present, and the separated components will then appear as black spots. This destroys the sample though, so some applications may call for spraying the plate with a fine mist of an iodine solution where upon heating a brown spot will appear, but the organic compound will not be charred. Other approaches for visualization of the TLC plate spots include the use of a fluorophore homogenously mixed with the chromatography media where the organic spot will block the background fluorescence. Sometimes, the analytes may contain conjugated double bonds, and this can be utilized to detect analytes on the plate by their fluorescence. Finally, holding a UV source over the plate in a dark chamber is probably the most common approach to nondestructively visualize plate spots. You may wonder why the need to avoid spot destruction such as sulfuric acid charring. The reason is to preserve the separated components that can be recovered by scrapping off the spot from the plate and then extracting the compound from the chromatography media with solvents. The extracted component can then be further analyzed such as by mass spectrometry or nuclear magnetic resonance (NMR) spectroscopy (see Chapter 16) for more specific identification and structural elucidation.

24.3 RETARDATION FACTOR (R_F) IN TLC

Many descriptive processes of chromatography that were reviewed in the beginning of this chapter can be translated over to TLC. In Figure 24.5, the concept of retention time (t_R) and dead time (t_m) were introduced for chromatography. To review, in column chromatography, the retention time of a compound, t_R, is the time that has elapsed from the injection or application of the sample onto the column till the detection of the compound after elution from the column. The dead time, t_m, is the time that a nonretained compound (or injection peak) is measured after the start of the chromatography. In TLC, a new descriptive term is used called the retardation factor (R_F):

$$R_F = \frac{d_R}{d_M} \tag{24.1}$$

where d_R is the distance that the analyte has traveled (migrated) from the origin spot and d_M is the distance that the solvent front has traveled. This is illustrated in Figure 24.5 for the TLC plate that was developed in Figure 24.4. Again, the horizontal lines used for linear measurement are drawn through the most intense (concentrated) portion of the spot, which is usually the estimated middle of the spot. From Equation 24.5, we can see that if an analyte were to migrate the same distance that the solvent front does, then the retardation factor would equal 1. On the other hand, if the analyte does not migrate at all and stays at the origin spot location, then the retardation factor will equal zero.

24.3.1 Example I

Rewrite the retention factor k' presented in Equation 24.3 in terms of the retardation factor R_F of Equation 24.1.

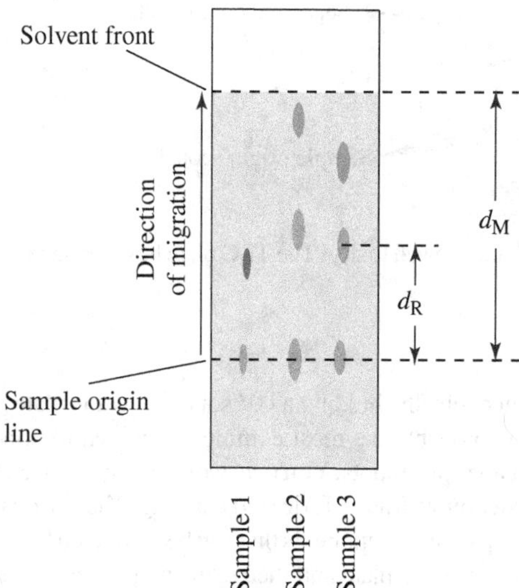

FIGURE 24.5 Linear measurements of the analytes after development of the TLC plate, where d_R is the distance that the analyte has traveled (migrated) from the origin spot, and d_M is the distance that the solvent front has traveled.

If we rearrange Equation 24.1 in terms of d_R, we get

$$R_F = \frac{d_R}{d_M} \tag{24.1}$$

$$d_R = R_F d_M \tag{24.2}$$

If we now substitute Equation 24.2 into Equation 24.1, we get

$$k' = \frac{d_M - d_R}{d_R} \tag{24.3}$$

$$k' = \frac{d_M - R_F d_M}{R_F d_M} \tag{24.4}$$

$$k' = \frac{(1 - R_F) d_M}{R_F d_M} \tag{24.5}$$

$$k' = \frac{(1 - R_F)}{R_F} \tag{24.6}$$

24.3.2 Example II

If we have a known retention factor of $k' = 2.64$ for compound A, what is the predicted migration distance d_R if the solvent migration distance is 6.78 cm?

Using Equation 24.3, the predicted migration distance d_R of compound A can be estimated:

$$k' = \frac{d_M - d_R}{d_R} \tag{24.3}$$

$$k' d_R = d_M - d_R \tag{24.7}$$

$$k' d_R + d_R = d_M \tag{24.8}$$

$$d_R(k' + 1) = d_M \tag{24.9}$$

$$d_R = \frac{d_M}{(k' + 1)} \tag{24.10}$$

$$d_R = \frac{6.78 \text{ cm}}{(2.64 + 1)} \tag{24.11}$$

$$d_R = 1.86 \text{ cm}$$

24.4 PLATE HEIGHTS (H) AND COUNTS (N) IN TLC

In Chapter 21 (Section 21.1.4 "Rate Theory versus Plate Theory"), the terminology used to describe the effectiveness of a column to separate analytes was introduced as the *height equivalent of a theoretical plate* (*HETP*, or *H*) of the equilibrium state of the partitioning of the analyte between the stationary phase and the mobile phase. This terminology can also be used in TLC. In

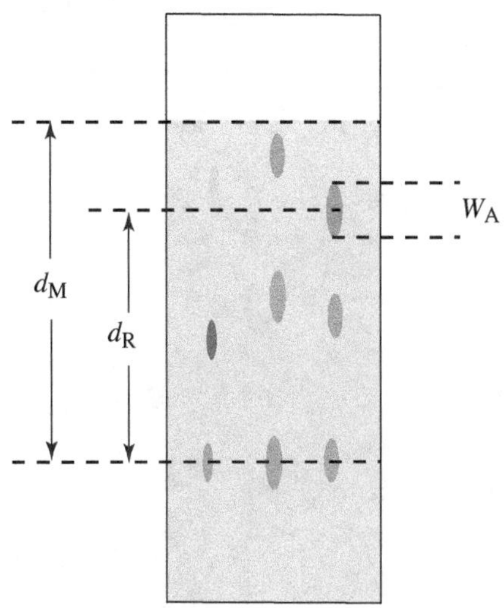

FIGURE 24.6 Measurement of the width (W_A) of the analyte after development of the TLC plate.

column chromatography, the *theoretical plate number*, N, of a column (dimensionless) is defined as the length of the column (L) in cm divided by the *height equivalent of a theoretical plate* (H) in cm of a single plate:

$$N = \frac{L}{H} \tag{24.12}$$

The more theoretical plates that a column has, the greater its ability will be to separate compounds, especially analytes that have similar distribution coefficients. To obtain N from a chromatogram, we use the following relationship:

$$N = 16\left(\frac{t_R}{W}\right)^2 \tag{24.13}$$

where t_R is the retention time of the analyte and W is the width of the peak at its base. This was illustrated in the chromatogram in Figure 21.6 where tangents are drawn along the peak sides dropping down to the baseline to properly determine the peak width. This same concept can be used in TLC where the substitution for the analyte retention time (t_R) is made with the analyte's distance traveled (migrated) from the origin spot (d_R):

$$N = 16\left(\frac{d_R}{W}\right)^2 \tag{24.14}$$

For the width (W) term in Equation 24.14, we have substituted the width of the spot, as illustrated in Figure 24.6. We can also express the *height equivalent of a theoretical plate* (*HETP*, or *H*) in terms of d_R as

$$H = \frac{d_R}{N} \tag{24.15}$$

24.5 RETENTION FACTOR IN TLC

In Section 24.3, the retention factor (also previously called capacity factor) k' was presented to describe the retention behavior of a compound on a particular column. The retention factor uses the compound's retention time t_R and the nonretained retention time t_m and is calculated using Equation 24.16:

$$k' = \frac{t_R - t_m}{t_m} \tag{24.16}$$

The retention factor k' can also be expressed in terms of TLC d_M and d_R values:

$$k' = \frac{d_M - d_R}{d_R} \tag{24.17}$$

In general, under typical analytical measurement conditions, a good range for the retention factor k' of an analyte is between 2 and 10. If the k' value is much greater than 10, then the retention time of the compound becomes quite long, and peak broadening begins to take place lowering the resolution. If the k' value is lower than 1, then the compound is eluting too quickly, and problems associated with reproducibly measuring the retention time and peak shape for area determination begin to arise.

25

GAS-LIQUID CHROMATOGRAPHY

25.1 Introduction
25.2 Theory and Principle of GC
25.3 Mobile-Phase Carrier Gasses in GC
25.4 Columns and Stationary Phases
25.5 Gas Chromatograph Injection Port
  25.5.1 Injection Port Septa
  25.5.2 Injection Port Sleeve (Liner)
  25.5.3 Injection Port Flows
  25.5.4 Packed Column Injection Port
  25.5.5 Capillary Column Split Injection Port
  25.5.6 Capillary Column Splitless Injection Port
25.6 The GC Oven
25.7 GC Programming and Control
25.8 GC Detectors
  25.8.1 Flame Ionization Detector (FID)
  25.8.2 Electron Capture Detector (ECD)
  25.8.3 Flame Photometric Detector (FPD)
  25.8.4 Nitrogen Phosphorus Detector (NPD)
  25.8.5 Thermal Conductivity Detector (TCD)

25.1 INTRODUCTION

Gas-liquid chromatography (GLC) is an important and widely used analytical tool in the laboratory. It is extremely conducive to the measurement of semivolatile and volatile organic compounds. The methodology is employed in many laboratory settings and sectors of industry including the petrochemical industry, the pharmaceutical industry, the cosmetics industry, the petroleum industry, flavor and fragrance industry, in clinical analyses, food and beverage industry, forensics, in environmental analyses, and contract laboratories. The main components of the gas chromatograph include an injection port where the sample is introduced, a column housing the stationary phase where the chromatography (separation) of the sample components takes place, a carrier gas supply used as the mobile phase and sometimes in the detector, a detector where the separated species are registered, and a PC for data collection and analyses. These main components of a gas chromatography (GC) system are illustrated in Figure 25.1. We shall in turn take a close look at each part of the gas chromatograph to become familiar with its principles, operations, and applications in the analytical laboratory.

25.2 THEORY AND PRINCIPLE OF GC

All of the expressions and descriptions previously presented in Chapter 21, "Theory of Chromatography," apply to GC. GC also separates compounds based upon the partitioning of the analytes between the mobile phase and the stationary phase. In GC, the simple relationship can again be described as a state of equilibrium (though not actually realized due to continuous mobile-phase movement) for the amount of solute A at any one point in time partitioned in the mobile phase, A_{mobile}, versus partitioned in the stationary phase, $A_{stationary}$:

$$A_{mobile} \rightleftharpoons A_{stationary} \tag{25.1}$$

In GC, the molar distribution (c_A) ratio for analyte A and plate theory in Chapter 20, "The Theoretical Plate Number *N*," also apply. As we saw before, the molar distribution (c_A) ratio for analyte A (also known as the partition ratio or partition coefficient) is written with the molar concentration of A in the stationary phase (c_S) always in the numerator:

$$c_A = \frac{\text{molar concentration of solute in stationary phase}}{\text{molar concentration of solute in mobile phase}} \tag{25.2}$$

$$c_A = \frac{c_S}{c_M} \tag{25.3}$$

We can calculate the molar distribution (c_A) ratio for analyte A from the analyte's retention time, t_R, and the dead time, t_m, for a nonretained compound. To refresh our understanding, the retention time of a compound, t_R, is the time that has elapsed from the

Analytical Chemistry: A Chemist and Laboratory Technician's Toolkit, First Edition. Bryan M. Ham and Aihui MaHam.

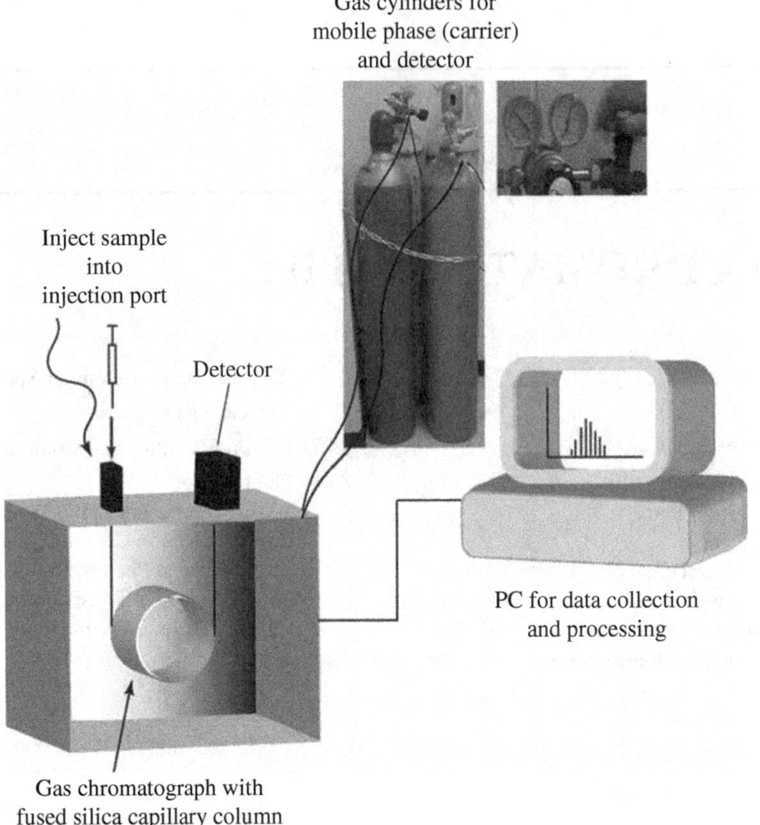

FIGURE 25.1 Main components of the gas chromatograph instrument (GC) includes the injection port where the sample is introduced, a column (Pyrex glass, stainless steel, or fused silica) housing the stationary phase where the chromatography (separation) of the sample components takes place, a carrier gas supply used as the mobile phase and sometimes in the detector, a detector where the separated species are registered, and a PC for data collection and analyses.

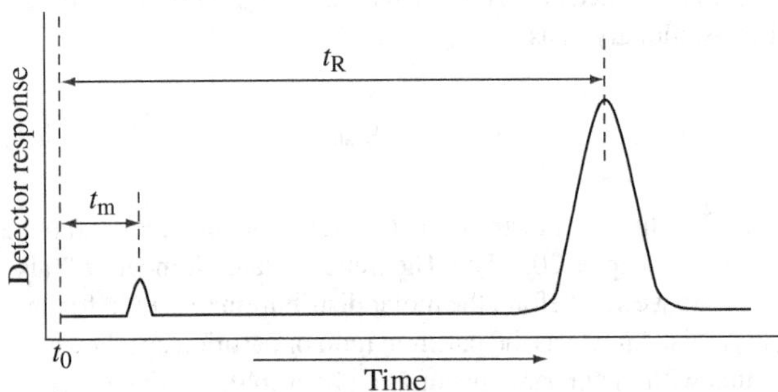

FIGURE 25.2 Retention times representing sample introduction at t_0, nonretained peak at t_m, and analyte retention time at t_R.

injection or application of the sample onto the column till the detection of the compound after elution from the column. The dead time, t_m, is again the time that a nonretained compound (or injection peak) is measured after the start of the chromatography. These two times are illustrated in Figure 25.2 along with the initial sample introduction as time zero, t_0. Sometimes, methods report an "adjusted" or "normalized" retention time ($t_R{}'$) that is obtained by subtracting the dead time t_m from the compound retention time t_R.

The relationship between the retention times and the molar distribution (c_A) ratio for analyte A is expressed in Equation 25.4:

$$t_R = t_m(c_A + 1) \tag{25.4}$$

If we measure t_R and t_m, we can calculate the molar distribution (c_A) ratio for analyte A using Equation 25.5. The adjusted retention time $t_R{}'$ is thus obtained from the following relationship:

$$t_R{}' = t_m c_A \tag{25.5}$$

In GC, the factors that affect c_A are strongly dependent upon the temperature of the system. As the temperature is increased, the partial pressure of the analyte increases, and the result is a shift in equilibrium toward the mobile gas phase versus the stationary phase. With increasing temperatures, analytes will elute quicker from the GC column. A general use column in GC is composed of methyl polysiloxane, separates analytes according to their boiling points, and is typically known as a boiling point column. Of course, modifications are made to the stationary phase (as listed in Table 21.2) such as the addition of the incorporation of 5% phenyl group, which will introduce a more nonaliphatic, aromatic influence on the separation of the species.

To summarize, all of the principles presented in Chapter 21 are applicable to GC directly without any alterations or new terms needed.

25.3 MOBILE-PHASE CARRIER GASSES IN GC

The separation of compounds in GC takes place in the heart of the system, the column, which is a partitioning affinity effect between a stationary phase and a mobile phase where in the technique of GC the mobile phase is a gas. The mobile phase possible for use can include a number of stable, nonreactive gasses such as hydrogen, helium, and nitrogen. According to the descriptive equations governing the separations in GC, the use of hydrogen gas will give the best performance in the separation of compounds (resolution). The *van Deemter equation* (Eq. 25.6) is also

$$H = A + \frac{B}{F} + (C_S + C_M)F \tag{25.6}$$

applicable to the efficiency description of gas chromatographic systems along with the effect of the equation parameters as we saw illustrated in Figure 21.11. A comparison of plate height versus linear velocity of the three-carrier gasses nitrogen (N_2), helium (He), and hydrogen (H_2) is illustrated in Figure 25.3. The nitrogen curve on the far left possesses the lowest plate height equating to the greatest separation efficiency between the three gasses. However, as illustrated with the x-axis, the optimal linear velocity range for nitrogen is narrow, thus allowing only a limited range of velocities. As illustrated in the figure, hydrogen has a similar plate height to helium, which is both just slightly greater than nitrogen and a quite broad optimal linear velocity range. This combination has given hydrogen the most readily found optimal conditions for efficient separations. Hydrogen gas however is highly flammable and is usually replaced by helium (nonflammable) as the carrier gas of choice in most analytical laboratories, though hydrogen can be used if care is taken. Lastly, in GC, the longitudinal diffusion term $\left(\frac{B}{F}\right)$ of the van Deemter equation (Eq. 25.6) has the greatest effect upon efficiency due to larger diffusion rates in gasses.

Carrier gasses are supplied to the GC instrumentation through two major means: compressed gas cylinders or gas generators. Compressed gas cylinders are the most common approach in analytical laboratories for supplying gas to the gas chromatograph. They must be regulated however in order to refresh with a full cylinder when the one in use is about to go empty. It is important to keep a steady, uninterrupted supply of the carrier gas to the GC column to keep it dry and well equilibrated. Gas cylinders are controlled using two-stage regulators that allow the setting of the exit flow and pressure from the cylinder to the instrumentation. Figure 25.4 (a) is an example of gas cylinders and (b) a two-stage regulator. Proper handling of compressed gas cylinders and regulators is very important in the analytical laboratory. A good source of safety information regarding compressed gas cylinders and their

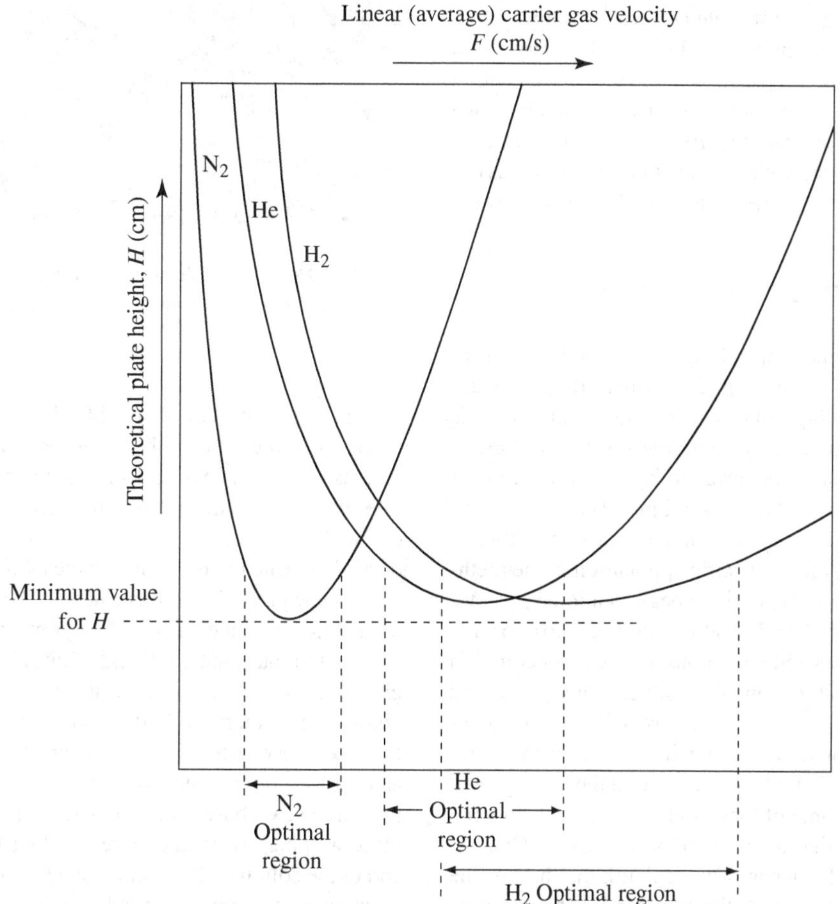

FIGURE 25.3 A comparison of plate height versus linear velocity of the three-carrier gases nitrogen (N_2), helium (He), and hydrogen (H_2).

(a)

(b)

FIGURE 25.4 (a) A set of gas cylinders used in the analytical laboratory. (b) A two-stage regulator.

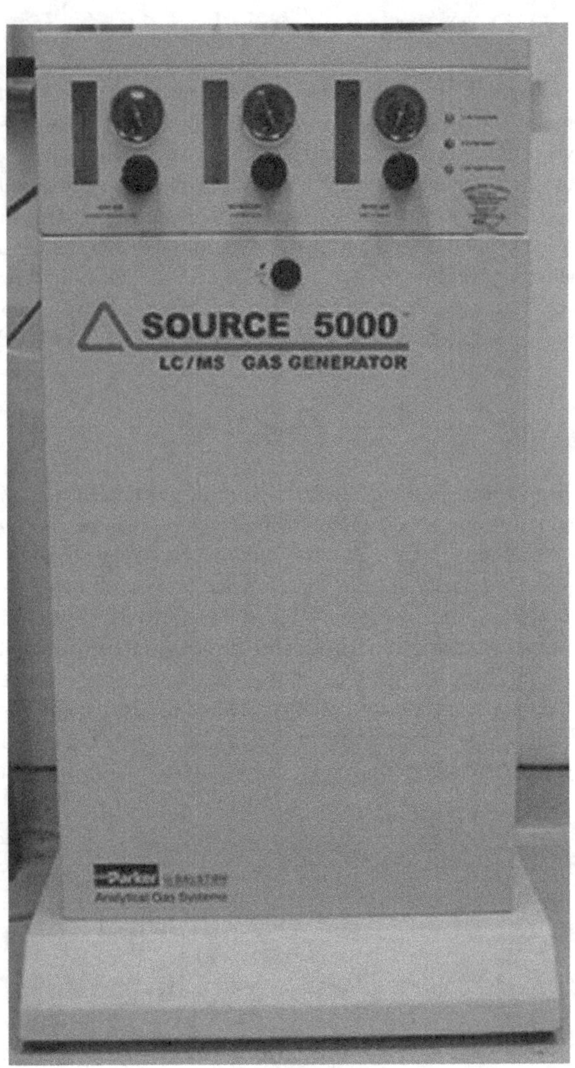

FIGURE 25.5 Example of an analytical laboratory gas generator. These can be used to supply air, O_2, He, H_2, N_2, and other gasses.

safe handling is the Department of Labor Occupational Safety and Health Administration (OSHA) website at www.osha.gov.

Another approach used in the analytical laboratory is to employ gas generators for supply to analytical instrumentation such as GCs and also mass spectrometers. The major advantage of using a gas generator is that they give a continuous, steady, uninterrupted supply of gas without encountering empty cylinders. Figure 25.5 illustrates a few examples of gas generators.

25.4 COLUMNS AND STATIONARY PHASES

GC falls into two main types depending upon the physical property of the stationary phase: there is GLC where the stationary phase is a very high boiling point, viscous liquid, and there is gas-solid chromatography (GSC) where the stationary phase is a solid material. In most laboratories today, the predominant type being employed is the GLC technique. There are many types of stationary phases that are commercially available to the analyst depending upon the specific application of the methodology. Table 25.1 lists many of the types of stationary phases that are available for use in GC. The table is broken into two parts including nonpolar phases and polar phases. The most common phases used are located at the top of the table. Many phases are very similar to each other being offered by different manufacturers and are contained within the same line, separated by a semicolon. A little later, we will look more closely at the physical and chemical properties of some of these stationary phases to get a better understanding behind the principles that govern GC.

The physical size and makeup of the column that houses the stationary phase have actually been through a series of substantial changes over the course of GC development. Older columns where constructed of either Pyrex glass or stainless steel with lengths of 1 m or less up to a maximum of perhaps 5 m or so. The diameters of these columns were anywhere between 3 and 6 mm. These were known as "packed" columns and were often packed by hand by the analyst in the laboratory. Figure 25.6 illustrates some packed columns where 25.6(a) is a stainless steel column, 25.6(b) is an example of a Pyrex glass column packed with a dark stationary material, and 25.6(c) is an example of a Pyrex glass column packed with a white stationary material. Stationary phases were bought in bottles from manufacturers and stored in the laboratory until a particular column type was needed. The analyst would then take an empty glass column and pack it with the stationary phase. One method for packing a column was to place a "plug" of temperature-resistant fiber glass wool at one end of the column with a fitting that kept the plug at the end while allowing flow through the column. The glass wool plugged end

TABLE 25.1 Common Stationary Phases Used in Gas Chromatography[a].

Nonpolar phases
• DB-5; HP-5; SE-54; SE-52; CP Sil 8 CB; RTX-5; BPX-5; Ultra-2; 5 % phenyl methyl siloxane; SPB-5; OV-3; PTE-5; NB-54; XTI-5; Mega 5MS; DB-5.625; RSL-200; MDN-5; OV-73; ZB-5; OV-5
• SE-30; DB-1; OV-101; OV-1; HP-1; methyl silicone; BP-1; SPB-1; Ultra-1; SP-2100; SF-96; RTX-1; E-301; DC-200; HP-101; RSL-150; Optima-1; ZB-1; 007-1; NB-30; LM-1; SPD-1
• Squalane
• Apiezon L
• CP Sil 5 CB
• DB-5MS; HP-5MS; RTX-5Sil MS
• Apolane
• Apiezon M
• Porapack Q
• CBP-1
• UCW-98
• PoraPLOT Q
• BP-5
• HP-PONA; Petrocol DH
• Cross-linked methyl silicone
• Apiezon
• PONA
• CP Sil 2
• Nonpolar
• LM-5
• JXR
• Vacuum Grease Oil (VM-4)
• MFE-73
• HT-5
• Silicon High Vacuum Grease (obsolete)
• RTX-5Sil
• PB-1
• SF96 + Igepal
• C103H208
• PS-255
• Tridecane
• Silicone oil
• Apiezon LH + KF
• PE-1HT
• DB5-30W
• DB-1MS
• SE-30 + Igepal
• Apiezon N
• PMS-1000
• CBP-5
• GP SP 2100 DB
• CP select for PCBs
• PoraPLOT
• OV101 (1% Carbowax 20M)
• RTX-1 PONA
• Triacontane
• HG-5
• Apiezon L + KF
• DB5-30N
• Permaphase DMS
• SimDist CB
• Optima 5
• Ultra-5
• Paraffin wax
• *n*-Dotriacontane
• PS-264

TABLE 25.1 (*Continued*)

Nonpolar phases
• Petrocol DH-100
• SE-30/SE-52
• C78, Branched paraffin
• SSP-1
• SE-33
• SE-52/54
• SPB-Sulfur
• MS5
• Polymethylsiloxane (PMS-20000)
• OV-101 + Igepal
Polar phases
• Carbowax 20M; DB-Wax; PEG-20M; BP-20; Innowax; CP-Wax 52CB; AT-Wax; HP-Wax; RTX-Wax; Carbowax; Supelcowax; HP-20M; Stabilwax
• FFAP; SP-1000; OV-351
• Supelcowax-10
• HP-Innowax
• DB-FFAP
• PEG-40M
• PEG 4000
• Thermon 600T
• TC-WAX FFS
• CP-WAX 57CB
• Carbowax 20M-TPA
• Carbowax 20M + Igepal (20:1)
• PEGA
• Polyethylene Glycol
• CBP-20
• ZB-Wax
• Carbowax 4000
• Carbowax 40M
• Supelcowax-20M
• PEG-2000
• Carbowax 6000 + Hyprose SP 80 (40: 60)
• Carbowax 400
• EPON 1001
• CP-Wax
• Stabilwax DA
• HP-FFAP
• Megawax
• Polyethylene glycol 4000
• Emulphor-Q (polyethylene glycol, ca. 40)
• EC-WAX

[a] Reprinted from the National Institute for Standards and Testing (NIST) webbook: http://webbook.nist.gov/chemistry/gc-ri/phases.html.

was then connected to a vacuum source, and the stationary phase was added to the other end of the column through a funnel. The vacuum helped to pull the stationary phase through the column packing it starting at the glass wool plug in the opposite end. To facilitate the packing, the analyst also would use a handheld engraver to gently touch the sides of the column while it was being packed using the engraver's rapid vibration to help move the packing through the column. This was a very technique-orientated method, and with practice, the analyst often became quite skillful in packing columns that gave good performance.

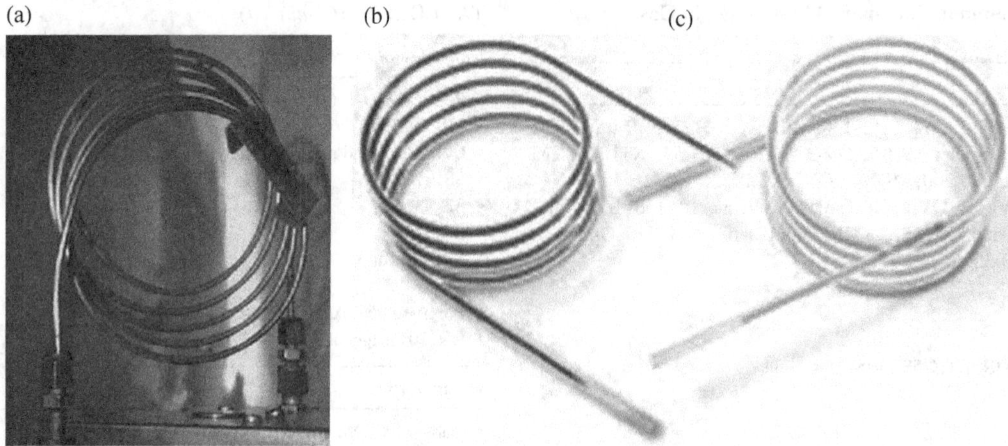

FIGURE 25.6 Examples of gas chromatography packed columns: (a) stainless steel column, (b) Pyrex glass column packed with a dark stationary phase, and (c) Pyrex glass column packed with a white stationary phase.

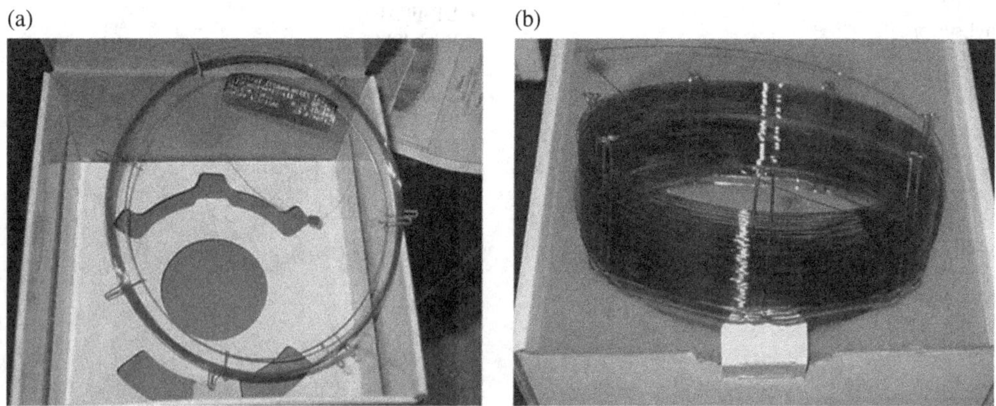

FIGURE 25.7 Examples of fused silica capillary columns: (a) a 15 m column, and (b) a 60 m column.

Packing columns is actually an art that has gone by the wayside in most analytical laboratories due to the more recently introduced fused silica (purified silica glass) capillary column. The fused silica column was a tremendous advancement over the older packed column in its ability to resolve compounds while requiring a much reduced sample size. The length of the columns could now be produced at 15, 30, 60, and even 100 m allowing a greatly enhanced chromatography. Two examples of fused silica capillary columns are illustrated in Figure 25.7 where (a) is a 15 m column and (b) is a 60 m column.

Column manufacturers and providers have spent the last 40 years developing the chemical properties of stationary phases, stationary phase thickness, column lengths, and column diameters specifically designed for established methodologies such as organic solvent analysis in environmental samples or pesticide analyses in foodstuffs. Many methods that the analyst will use in GC analyses will state what particular column and stationary phase was used to develop the method. The analyst will, unless he/she is developing a new method, either follow the guidelines of the method being used for column type or will consult the literature for examples of similar analyses. We will look at this in more detail momentarily, but let's first consider an introduction to the main components of a gas chromatograph instrument.

25.5 GAS CHROMATOGRAPH INJECTION PORT

In Figure 25.1, a simple presentation is given of the basic design of a gas chromatographic instrument. In this section, we will begin to take a closer look at the main components of a gas chromatograph. To start an analysis, the sample is introduced into the GC through the injection port (shown on the left of Fig. 25.1). The injection port is a heated component of the GC system. Figure 25.8(a) illustrates a very common GC used in analytical laboratories. This is a Hewlett Packard (HP) 5890 Series GC that is now a number of years old but is still quite commonly found in labs. We will be using the HP 5890 as a simple, basic model to study the main components of a GC. The GC illustrated below the HP 5890 in Figure 25.8(b) is the latest generation of this series of GC's called the Agilent (spun-off company from HP) 7890A Series GC (presented here as an example of a more recent GC system, though the basic components and functions are the same).

(a)

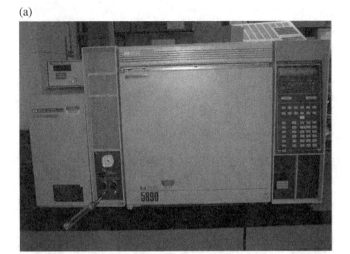

(b)

FIGURE 25.8 (a) A Hewlett Packard (HP) 5890 Series GC. (b) The latest generation of this series of GC's called the Agilent (spun-off company from HP) 7890A Series GC.

25.5.1 Injection Port Septa

On the top left side of the GC instrument is located the sample inlet injection port. This is where the sample is introduced into the GC. Figure 25.9 illustrates a close-up view of the injection port as seen looking at the top of the instrument. The sample to be analyzed by GC, usually contained within a 10 μl syringe, is injected into the inlet by piercing a rubber septum in the top center of the port and then rapidly and smoothly pressing in the syringe plunger, injecting the sample. Let's take a look at the injection port inlet from a cutaway side view. This is illustrated in Figure 25.10 where at the top of the inlet is the septum nut that we see in Figure 25.9. Under the septum nut is held a polymer septum, which holds the pressure within the injection port and gives a pierceable portal into the port for sample introduction. Figure 25.11(a) illustrates the changing of the septum under the septum nut. Because the injection port is heated to approximately 250–280 °C, it is too hot to directly touch by bare hand. A variety of septa are illustrated in Figure 25.11(b) where different colors and sizes are available according to instrument specifications.

25.5.1.1 *Merlin Microseal* A common septum found in the analytical laboratory is a septum called the Merlin Microseal, illustrated in Figure 25.12.

25.5.1.1.1 What It Is The Merlin Microseal septum and nut are patented, long-life replacements for the standard septum and septum nut in the capillary inlet system or the purged packed inlet system of a Hewlett Packard gas chromatograph. The Microseal septum incorporates two sequential seals to provide a much longer life between septum changes. It requires a 0.63 mm diameter blunt tip syringe. It is particularly useful with the HP 7673 autosampler, as it allows many more samples to be run unattended while reducing the risk of lost or compromised

FIGURE 25.9 Views of the GC injection port.

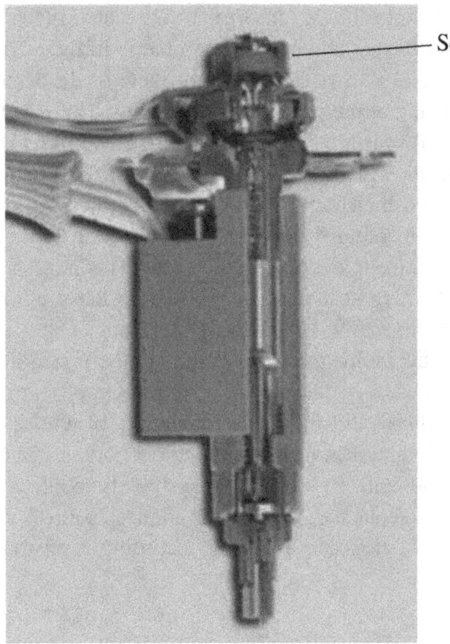

FIGURE 25.10 GC injection port.

data caused by septum leaks or by pieces of the septum falling into the injection port. Because the syringe insertion force is much lower, it also allows for much easier manual injections. While actual life will vary depending on operating conditions, the Microseal septum will typically sustain over 25,000 injections. This corresponds to an injection every 21 min around the clock for a year (this paragraph is from Microseal pdf – get permission to reprint).

25.5.1.1.2 How It Works A septum serves two functions in an injection port: sealing the injection port during the separation and sealing around the needle during the injection. The Merlin Microseal septum has two separate seals to accomplish these functions. The spring-assisted duckbill (see Fig. 25.12, at the bottom) seals effectively because pressure in the injection port helps to force the lips together. The duckbill is easily opened by the syringe needle as it passes through and exerts a very low drag force on the needle, reducing wear. The ring seal (see Fig. 25.12, at the top) contains a double O-ring, which forms a sliding seal around the syringe needle as it is inserted into the port. This seal is made before the duckbill seal is forced open, ensuring that the port remains sealed as the syringe needle penetrates into the port. After the sample is expelled from the syringe and as the needle is withdrawn, the duckbill reseals the port before the ring seal around the needle is opened. Because the two seals perform separate functions and are only slightly distorted in operation, rather than being pierced, the Microseal septum can be made from a high-temperature, very wear-resistant fluorocarbon elastomer. In combination with the blunt needle, this means the Microseal septum will not shed pieces into the injection port, even after thousands of injections. The Microseal nut replaces the standard septum nut. It provides support for the sealing flange at the top of the Microseal septum. This flange requires only a small,

(a)

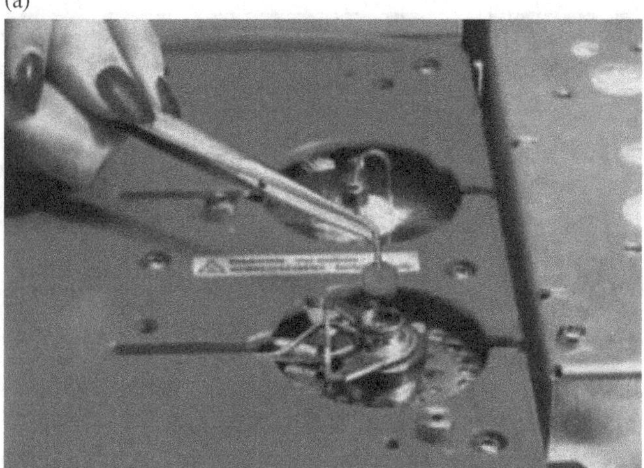

(b)

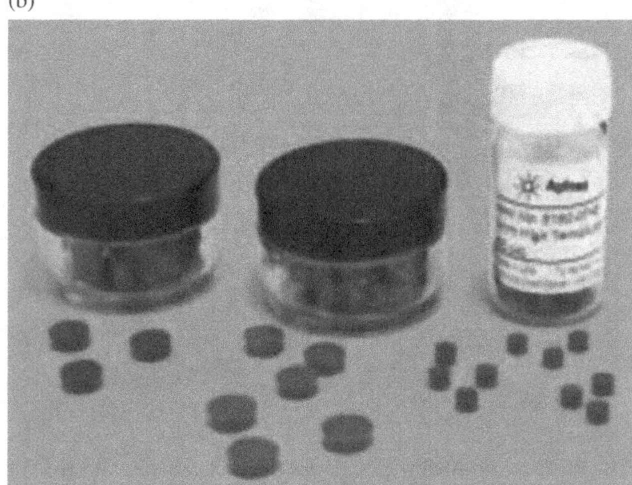

FIGURE 25.11 (a) Changing the GC injection port septum. (b) Examples of septa.

controlled amount of deformation to seal reliably on the rim of the septum pocket. The nut has a series of 12 guide marks at 30° intervals on its perimeter to aid in the tightening process. The nut is coated with titanium nitride to provide wear resistance. A PTFE liner in the nut prevents the Microseal septum from adhering to the Microseal nut during months of use (this paragraph is from Microseal pdf—get permission to reprint).

25.5.2 Injection Port Sleeve (Liner)

Below the septum is the injection port sleeve that is made of deactivated silica glass. In injecting the sample, the barrel of the needle pierces through the septum and comes into the middle of the glass sleeve. The injection port is heated, usually somewhere between 200 and 250 °C, allowing the vaporization of the sample being introduced. The glass sleeve is butted up against the septum making a seal. At the bottom of the glass sleeve is the beginning of the GC column. So, the sample is introduced and vaporized within the injection port glass sleeve and then swept into the GC column for separation. The primary purpose of the glass

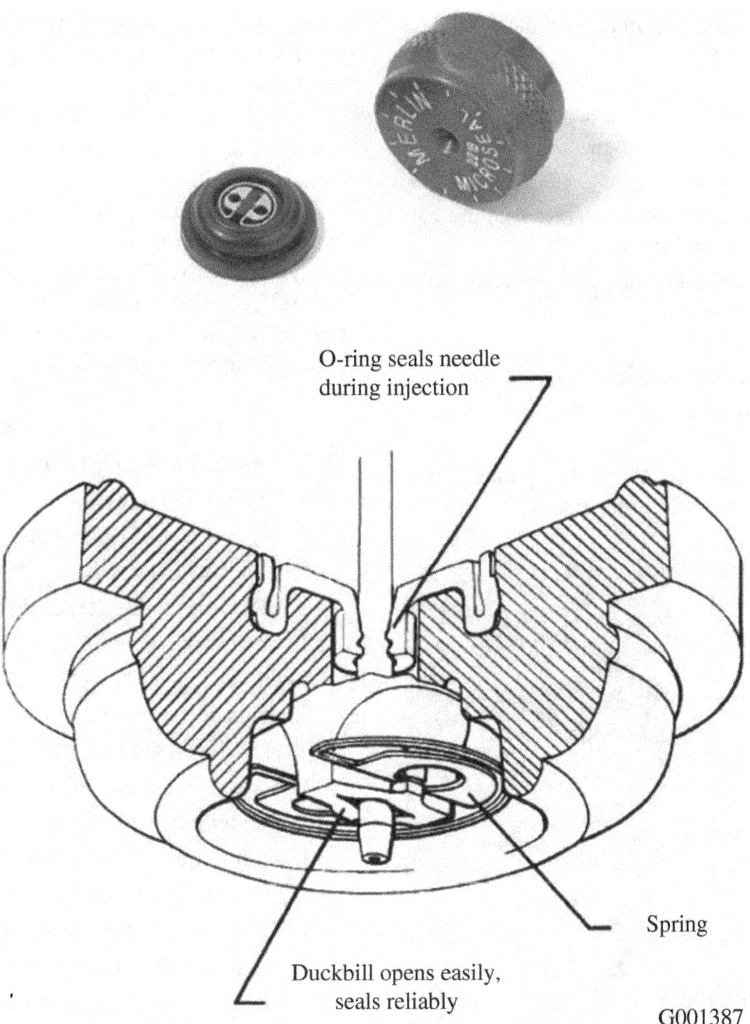

FIGURE 25.12 The Merlin Microseal septum.

injection sleeve is to provide an environment that is nonreactive to any compounds being introduced into the GC. The metal of the injection port tends to be quite active and can seriously affect sample analysis. Figure 25.13 illustrates the replacement on the injection port glass sleeve. Again, the injection port is hot and so is the sleeve. Because the septum and the sleeve have to be changed quite regularly (unless of course the Merlin Microseal is being used in place of a regular septum), the changing is usually done with the injection port at operating temperature. It should be strongly noted here that this, however, is usually done by an experienced operator who can change these disposable items in a safe and efficient way without injuring himself (e.g., getting skin burns from the hot port). The safest approach though is to turn off the oven and the injection port heaters and allow the system to cool prior to replacing the septum or injection liner. This approach is highly recommended for the inexperienced operator. In a high-throughput time-demanding laboratory environment, it can take quite a while to cool down the port and then reheat it back up after changing these disposable items. This can be viewed as a waste of a considerable amount of analysis time. Safety first must always be practiced, and the approach used to change these items needs to be chosen in accordance with safe laboratory practices and operator experience. Actually, in some laboratories where many injections are made, the septum often may require daily replacement. The sleeve on the other hand may only need to be replaced when it is suspected that it has become dirty or activated and is affecting sample analysis. This is usually anywhere from one to two weeks to months, really depending upon the amount of injections being done and the types of samples being introduced. For example, dilute standards in high-quality solvents will probably have little effect upon the glass sleeve, and the sleeve would not have to be replaced for quite some time. On the other hand, suppose very dirty samples or samples with very heavy, nonvolatile components are being injected such as small amounts of dissolved viscous oils. These conditions would favor the quick buildup of material on the glass sleeve that may deleteriously affect analysis. Sometimes, the analyst may try to clean the injection port sleeve by "baking it out" through raising the injection port temperature to 300–350 °C.

25.5.2.1 Attributes of a Proper Liner Sleeve design technology has actually produced quite a variety of glass sleeves designed to enhance analyses. Five important attributes of a good liner design include:

1. The liner design should minimize mass discrimination by ensuring complete vaporization of the sample before it reaches the column entrance.
2. The volume of the inlet liner must be larger than the volume of vaporized sample and solvent.
3. The liner must not react with the sample. This is especially important for polar solutes where the liner should be deactivated.
4. The addition of quartz wool increases the vaporization surface area for the sample and promotes efficient mixing of the sample and carrier gas.
5. The position of the quartz wool should be optimized corresponding to the needle depth in the liner.

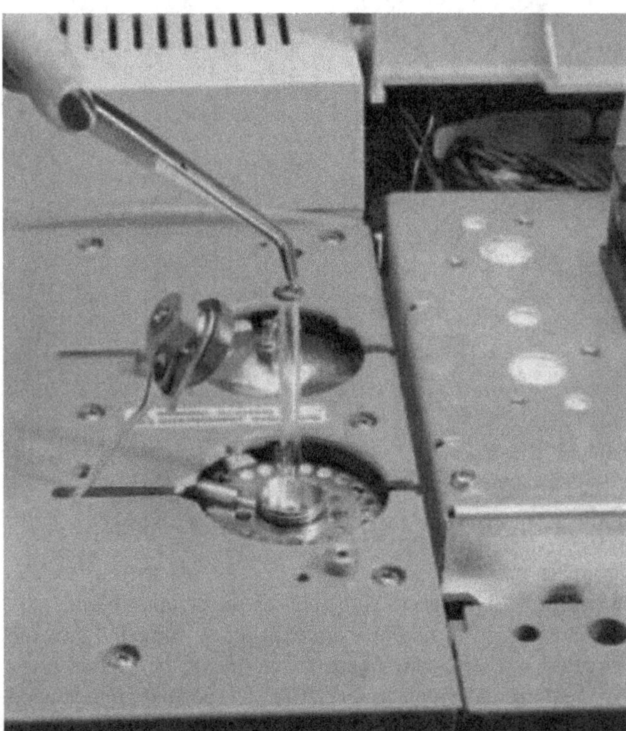

FIGURE 25.13 The changing of a GC injection port liner. The injection port is usually very hot thus direct handling of the liner is not recommended.

(Adapted from Speck & Burke Analytical Technical Note: ISSUES SURROUNDING THE GC INJECTION PORT, reprinted with permission.)

The injection port liner illustrated in Figure 25.14 has incorporated these different features such as a stationary glass wool plug and tapering.

Finally, at the bottom of the injection sleeve is the gold seal. The seal is made out of a thin layer of gold in order to be completely nonreactive. Examples of the gold seal are illustrated in Figure 25.15. The seal sometimes is also made out of stainless steel that is also nonreactive to organic compounds.

Item Description HP Part No.

1. Heater/sensor assembly 05890-61140
2. Heater, 70 W 19231-60620
3. Contact 1251-1679
4. PRT assembly 19231-60660
5. Contact 1251-5963
6. Flexible sleeving (ordered by the inch) 0890-0737
7. Shell weldment 19251-80570
8. Glass cloth 9300-0713
9. Insulation 0340-0686
10. Heat sink 18740-20940
11. Shaped insulation 19251-00020
12. Splitter tube 19251-80525
13. Front ferrule, 1/8-in. Brass 0100-0032
14. Back ferrule, 1/8-in. Brass 0100-0036
15. Tubing nut, 1/8-in. Brass 0100-0058
16. Retaining nut 19251-20620
17. Gold-plated seal 18740-20885 Stainless steel seal 18740-20880
18. Flat washer, stainless steel 2190-0701 (package of 12) 5061-5869
19. Reducing nut 18740-20800
20. Lower insulation 19243-00060 20A Insulation 19243-00065
21. Lower insulation cover 19243-00070
22. Insert assembly 19251-60575
23. Septa retainer (standard) 18740-60835
    Retainer nut for headspace 18740-60830
24. Insulation 19251-00120

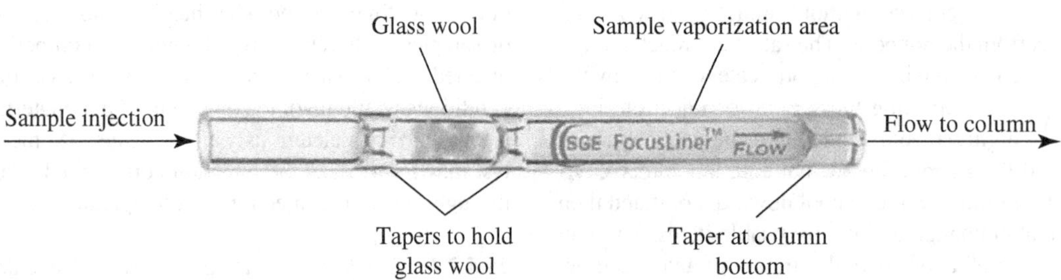

FIGURE 25.14 Close-up view of an injection port liner.

25. O-ring—High temp. (package of 12) 5180-4182 Graphite seal for split liner (Pkg of 12). 5180-4168 Graphite seal for splitless liner (Pkg of 12) 5180-4173.
26. Liner—Split/splitless (4 mm id) 19251-60540
    Split packed (4 mm id) 18740-60840
    Split unpacked (4 mm id) 18740-80190
    Splitless (2 ± 0.2 mm id) 18740-80220
    Direct (1.5 ± 0.2 mm id) 18740-80200
    Split/splitless with glass wool
    (4 mm id/deactivated) 5062-3587
    Splitless tapered one end (4 mm id/deact.) 5181-3316
    Splitless double tapered (4 mm id/deact.) 5181-3315
27. Septa, low bleed (25/pk) 5080-8894
28. Column seal/ferrule (see Consumables Catalog)
29. Column nut 5181-8830

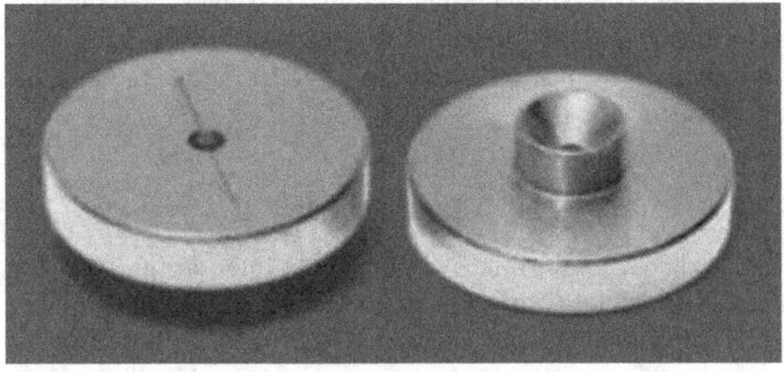

FIGURE 25.15 GC injection port gold seal.

The injection port is actually somewhat more complex than what we have described thus far. Figure 25.16 illustrates an exploded view of the injection port along with the name of each part. Notice also that the figure contains a number of examples of injection port liners in the top left portion of the figure. Exploded views such as that in Figure 25.16 are supplied by the manufacturers as sometimes the port needs to be repaired or extensively cleaned, both often performed by the laboratory analyst.

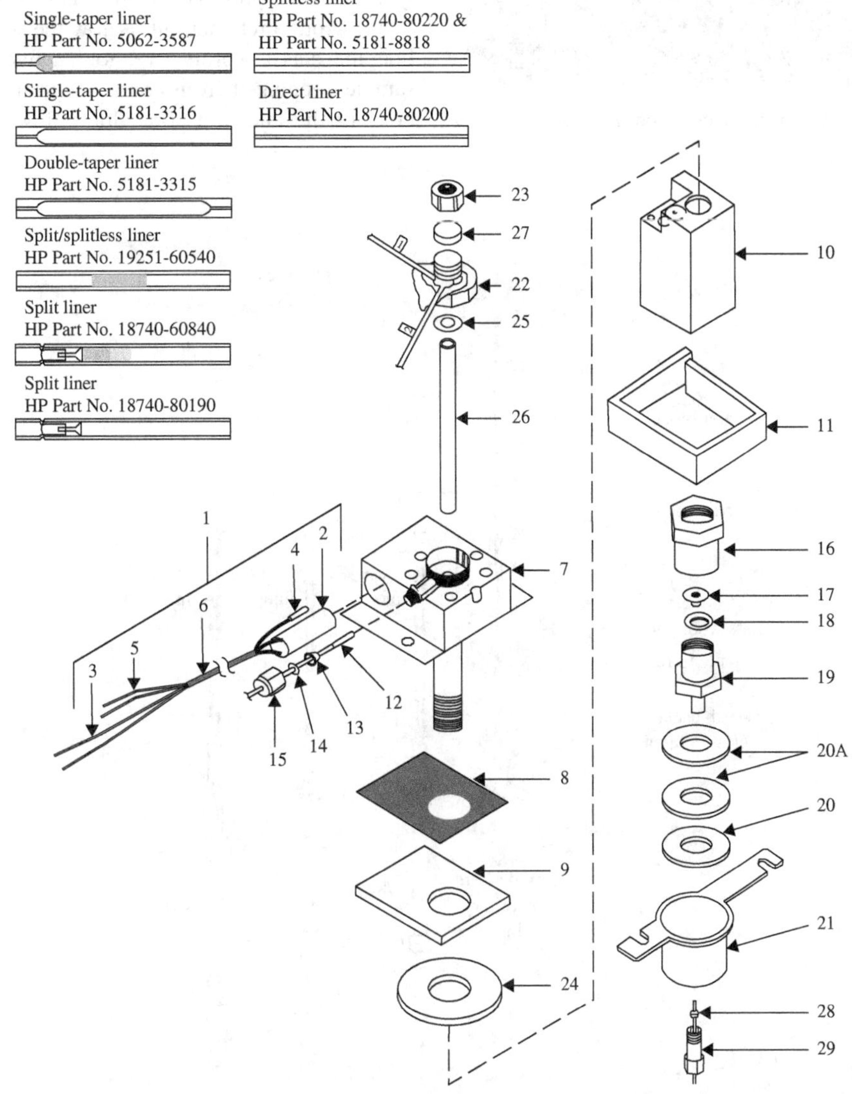

FIGURE 25.16 EPC Split/Splitless, Split-Only Capillary Inlet. (© Agilent Technologies, Inc. 2014, Reproduced with Permission, Courtesy of Agilent Technologies, Inc.)

25.5.3 Injection Port Flows

There are a number of flows that are coming into and through the injection port that are very important. There is a septum purge flow, a split flow, and the flow of carrier gas for the column. The primary purpose of the septum purge is to remove any residual vapor from the injection port after each injection. This will help to avoid contamination of the next injection and the observance of ghost peaks (peaks in the sample's chromatogram that is not actually a component of the current sample but are either from contamination or a previous sample injection). Often, the injection port flows are measured using a flowmeter such as the one illustrated in Figure 25.17, connected to the split flow outlet. The exhaust from this outlet is being passed through an activated carbon filter used to scrub out any organic compounds that would otherwise be expelled into the ambient environment.

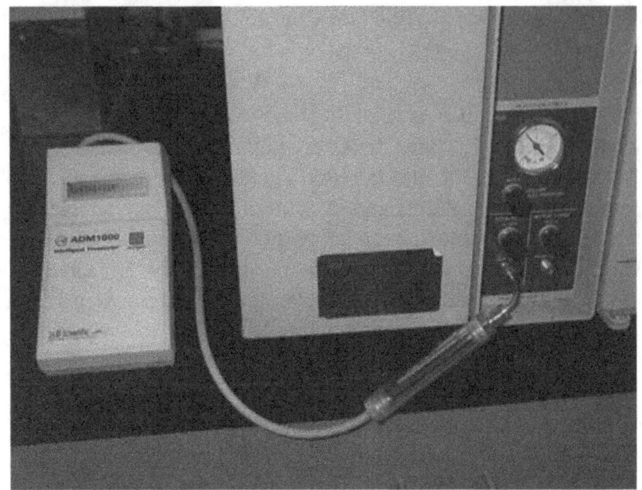

FIGURE 25.17 Measurement of injection port flows.

25.5.4 Packed Column Injection Port

The packed column inlet is the original inlet design used for both GLC and GSC. It is the simplest inlet system that requires little adjustment or optimization. Figure 25.18 illustrates the components of a packed column injection port inlet. As we were just introduced, the top of the injection port houses a septum used for a method of sealed injection of the sample to be analyzed into the port. The septum is held down tightly in place by the septum retainer nut. Below the septum is a simple glass liner that acts as a protective, nonreactive sleeve for the injected sample keeping it from contact with the metal surface of the body of the injection port. Going down the figure, there is a

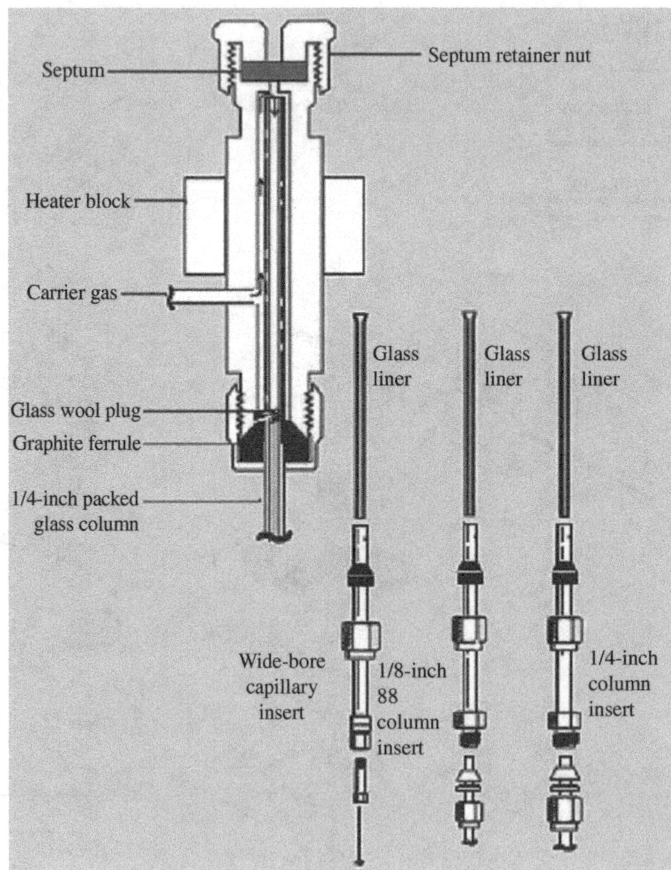

FIGURE 25.18 Packed column injection port inlet illustrating gas flow and glass liner inserts.

heater block that is temperature regulated to keep the whole injection port at an elevated temperature (~150–300 °C) allowing vaporization of the sample upon injection into the port. Progressing down the injection port body in the figure, we next see the carrier gas inlet into the port. The carrier gas enters the port and must travel up alongside the inner tube of the port thus effectively heating the gas prior to entering the glass liner at the top. The heated carrier gas will carry the vaporized sample into the column that is located at the bottom of the injection port. The figure also illustrates on the bottom right that the port can be used with three different sizes of columns. First is the wide-bore capillary column that typically is 0.5 mm in diameter with flows of 2 to 15 ml/min when using nitrogen (N_2), helium (He), or hydrogen (H_2) as carrier gas. Next, there is a setup for 1/8 in. in diameter stainless steel packed columns that typically incorporates flows of 20 to 40 ml/min when using nitrogen (N_2) as the carrier gas or 30 to 60 ml/min when using helium (He) or hydrogen (H_2) as carrier gas. Finally, there is the standard size ¼ in. diameter packed column that also can typically incorporate the same flows of 20 to 40 ml/min when using nitrogen (N_2) as the carrier gas or 30 to 60 ml/min when using helium (He) or hydrogen (H_2) as carrier gas. In the packed column inlet, there can be just one flow controlling the carrier gas that takes the entire sample into the column, or a septum purge flow can be added. Two different flow systems for packed column inlets are illustrated in Figure 25.19 where (a) shows the gas flow from a regulated carrier gas cylinder to the column for the simple packed column inlet system just described. The carrier gas flow system illustrated in Figure 25.19(b) includes a septum purge flow also.

Packed column GC still finds wide use when the higher resolutions that capillary columns give are not necessary, or the analysis of gasses are being performed when using GSC and a thermocouple detector (TCD).

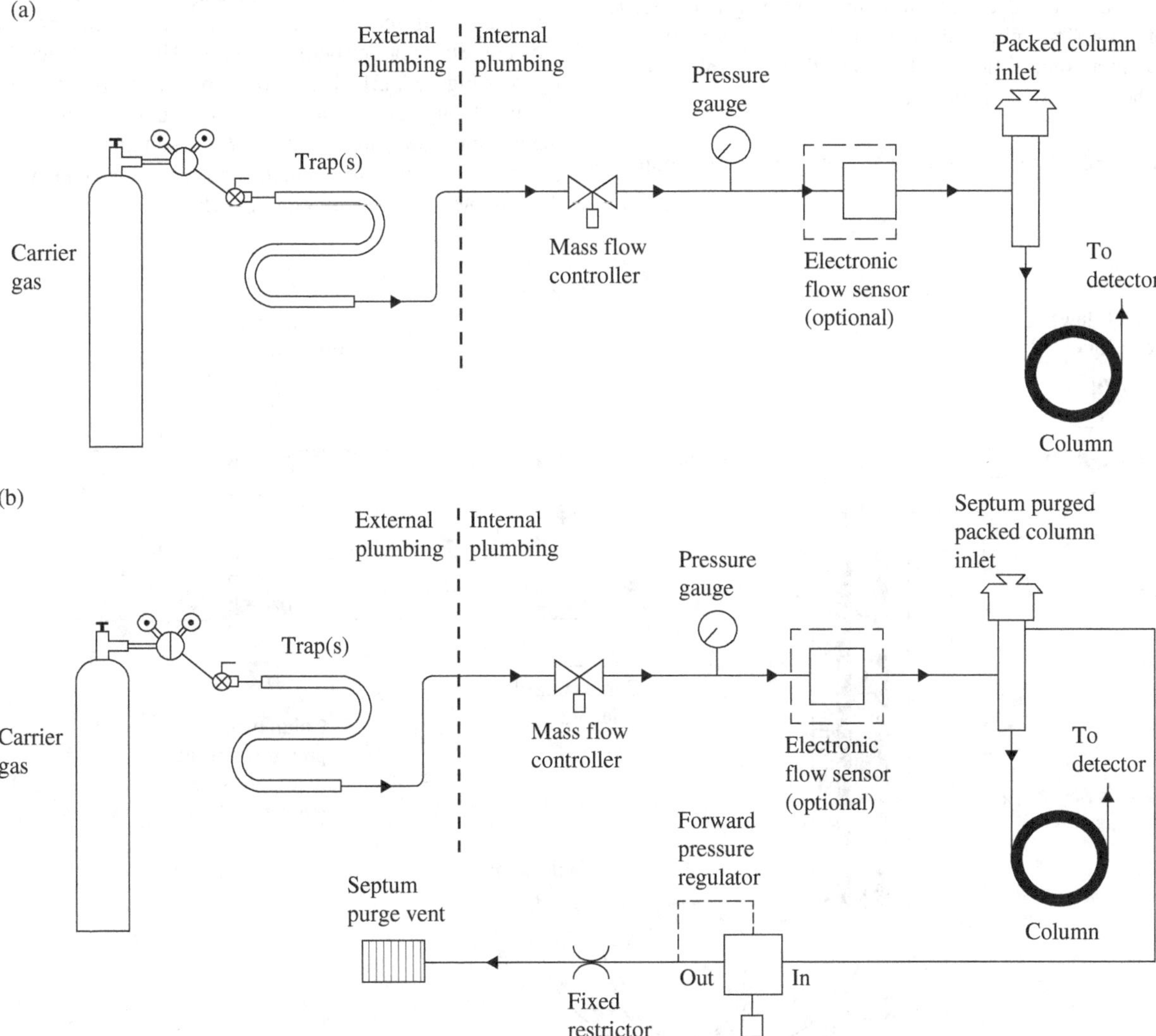

FIGURE 25.19 Packed column injection port inlets illustrating carrier gas flows for (a) a simple packed column injection port, and (b) a packed column injection port that also includes a septum purge flow.

25.5.5 Capillary Column Split Injection Port

With the advent of capillary columns that typically have diameters of <0.5 mm, injection port flows and sample introduction were needed for optimized analyses. Flows that are used for packed column GC were not applicable to capillary columns due to large back pressures at injection port flows of 20 to 60 ml/min. Another problem was the amount of sample being introduced onto the capillary column. It was found that often a much smaller amount of sample is needed for the capillary column as compared to the packed column with a diameter of ¼ in. The first design of an injection port inlet for capillary columns was called a split inlet where the injected sample is split and one fraction of the injected sample goes to the column and the other fraction is removed through a split exhaust port outlet. The inlet configuration for a split injection port is illustrated in Figure 25.20.

The flows illustrate and dictate the split ratio for the sample introduction into the column. To the top left of Figure 25.18, there is a total flow coming into the injection port of 104 ml/min, controlled by a total flow valve. There is a 3 ml/min septum purge flow that has reduced the flow entering the injection sleeve to 101 ml/min. The split vent flow here has been set to 100 ml/min leaving a column flow of 1 ml/min; thus, the split ratio is 1:100. The split ratio is defined as:

$$\text{Split ratio} = \text{column flow(ml/min)} : \text{split vent flow(ml/min)} \tag{25.7}$$

If we take a close look at the gold seal that is illustrated in Figure 25.15, two small slits can be seen on the face of the seal. This is where the split flow passes through under the injection port sleeve liner. The flows can either be computer controlled or controlled by hand. If the flows are computer controlled than usually, there are also electronic flow sensors that give a digital reading of the flows. If not, the flows can be measured and adjusted using a flowmeter such as the one illustrated in Figure 25.17. Typical flows are in the 1 ml/min range with split ratios between 1:25 and 1:500, depending upon sample conditions such as concentration of analytes. For example, a very dilute analyte solution may be injected with a split ratio of 1:25, while a very concentrated solution may be injected at 1:500.

25.5.6 Capillary Column Splitless Injection Port

The injection port that is illustrated in Figure 25.20 can also be used for a splitless injection technique into the capillary GC columns. The advantage and reason for using this approach is usually due to very low concentrations of analytes in the solution to be measured. The splitless injection is achieved by closing off the split valve exit tubing during the injection time and allowing about 1 minute of time for the sample to be carried onto the column. This is illustrated in the top figure of Figure 25.21 where all of the purge flow passes over the top and not through the injection port. The sample is being carried into the column slowly though at 1 ml/min. so the purge flow is kept rerouted in this way for a minute or two to allow enough time for the sample to enter the column. Upon completion of sample

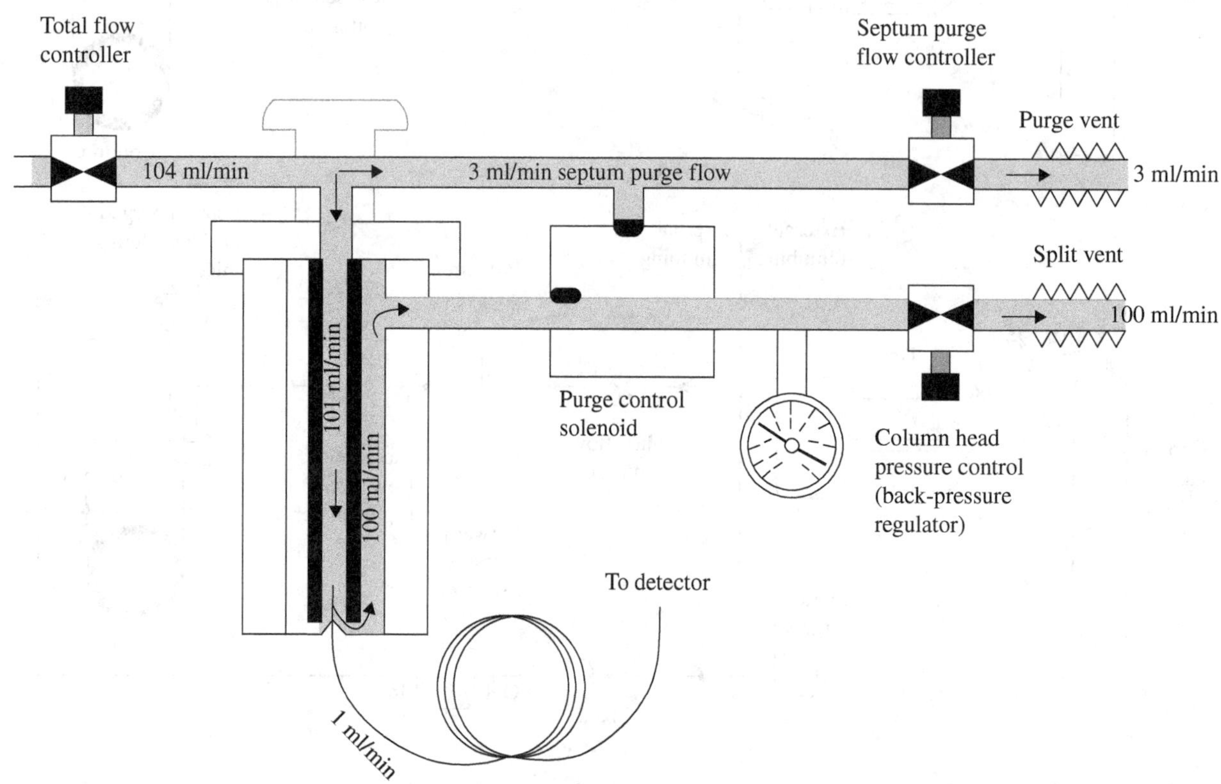

FIGURE 25.20 Split inlet injection port.

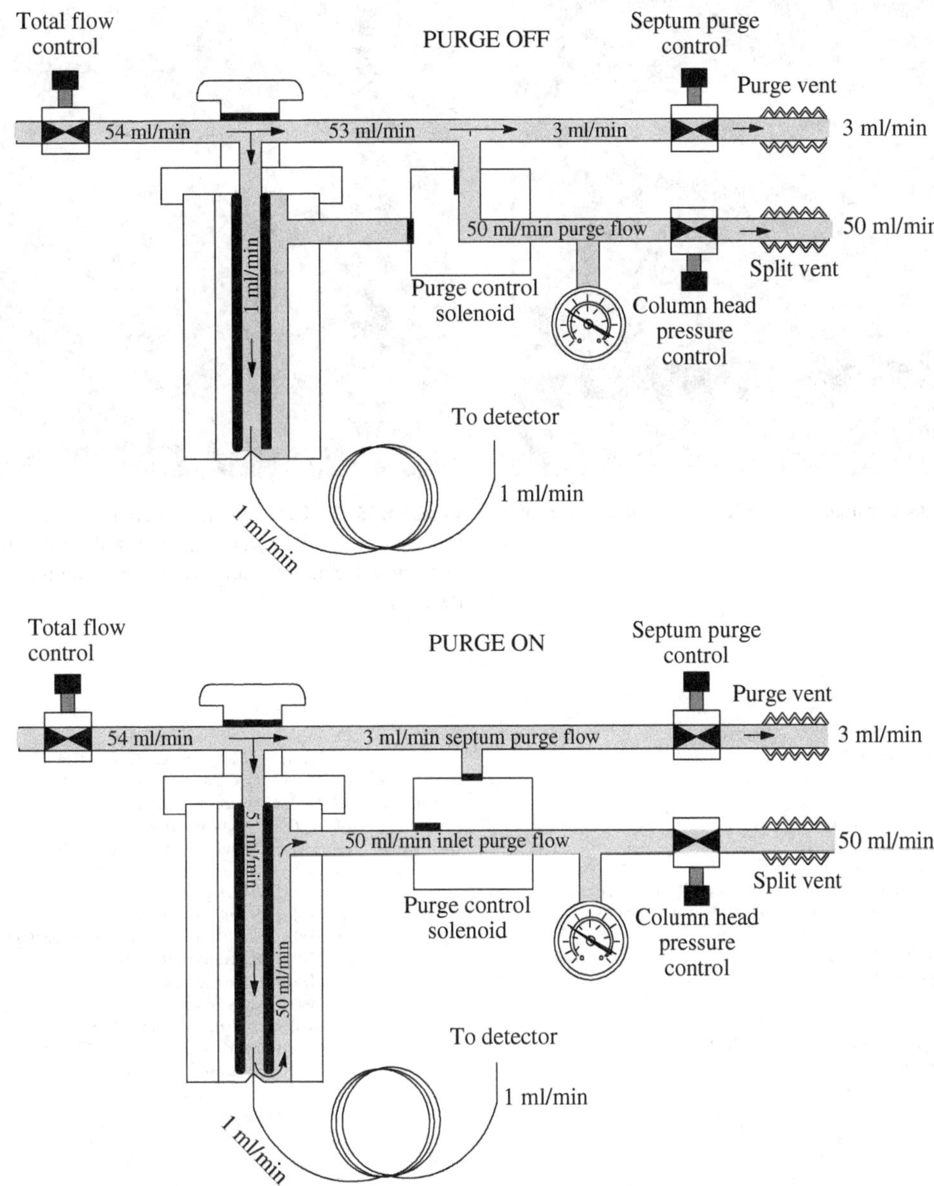

FIGURE 25.21 Splitless inlet. Top flow diagram is with "PURGE OFF" during splitless injection. Bottom flow diagram is with "PURGE ON" after sample introduction into the column to sweep out of the injection port any remaining sample residue.

introduction into the column, the split flow is redirected through the injection sleeve as illustrated in the bottom flow diagram of Figure 25.21. In the bottom figure, the flow is passing under the septum to purge the septum and is also flowing through the injection sleeve and out the split vent. This will clear out any extra sample remaining in the injection port after sample introduction eliminating any extra ghost peaks from being observed in the chromatogram.

25.6 THE GC OVEN

The next major component of the gas chromatograph is the GC oven. This is where the GC column is housed as illustrated in Figure 25.22 where the oven door has been opened showing the column inside. As the name implies, the GC oven can be programmed at elevated temperatures, usually the upper end being dictated by the column's heat maximum tolerance. Figure 25.23 is a close-up view of the column in the GC oven. To the top of the oven, we can see the column connected to the sample injection port. To the left of the column, we can see the column connected to the detector. The GC programmable oven is a very useful tool for use in optimizing the separation of analytes on the GC column.

Let's take a look at an example of programming the GC oven to separate a complex mixture of analytes. The top section of Figure 25.24 illustrates a chromatogram of separated fatty acid methyl esters (FAMEs) from a bacterial source, here called bacterial acid methyl esters (BAMEs). Often, samples such as this

FIGURE 25.22 The gas chromatograph GC oven housing the GC column.

FIGURE 25.23 Close up view of column in GC oven. Notice the two connections of the column including the top left into the injection port, and the middle left where the column is connected to the detector.

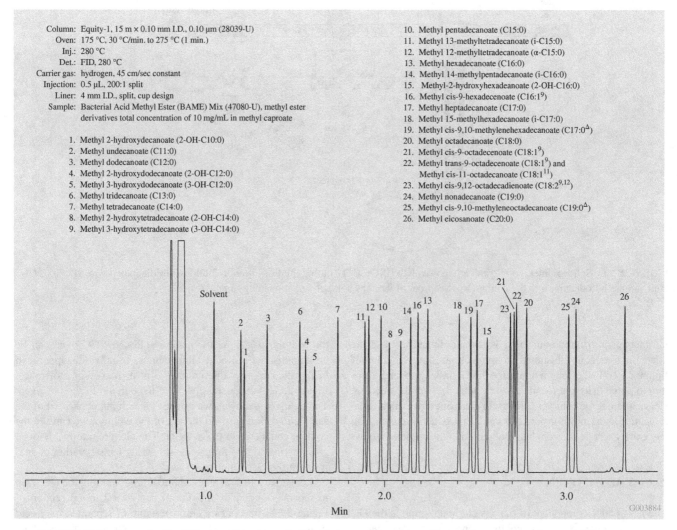

Column: Equity-1, 15 m × 0.10 mm I.D., 0.10 μm (28039-U)
Oven: 175 °C, 30 °C/min. to 275 °C (1 min.)
Inj.: 280 °C
Det.: FID, 280 °C
Carrier gas: hydrogen, 45 cm/sec constant
Injection: 0.5 μL, 200:1 split
Liner: 4 mm I.D., split, cup design
Sample: Bacterial Acid Methyl Ester (BAME) Mix (47080-U), methyl ester derivatives total concentration of 10 mg/mL in methyl caproate

1. Methyl 2-hydroxydecanoate (2-OH-C10:0)
2. Methyl undecanoate (C11:0)
3. Methyl dodecanoate (C12:0)
4. Methyl 2-hydroxydodecanoate (2-OH-C12:0)
5. Methyl 3-hydroxydodecanoate (3-OH-C12:0)
6. Methyl tridecanoate (C13:0)
7. Methyl tetradecanoate (C14:0)
8. Methyl 2-hydroxytetradecanoate (2-OH-C14:0)
9. Methyl 3-hydroxytetradecanoate (3-OH-C14:0)
10. Methyl pentadecanoate (C15:0)
11. Methyl 13-methyltetradecanoate (i-C15:0)
12. Methyl 12-methyltetradecanoate (α-C15:0)
13. Methyl hexadecanoate (C16:0)
14. Methyl 14-methylpentadecanoate (i-C16:0)
15. Methyl-2-hydroxyhexadeanoate (2-OH-C16:0)
16. Methyl cis-9-hexadecenoate ($C16:1^{9}$)
17. Methyl heptadecanoate (C17:0)
18. Methyl 15-methylhexadecanoate (i-C17:0)
19. Methyl cis-9,10-methylenehexadecanoate ($C17:0^{\Delta}$)
20. Methyl octadecanoate (C18:0)
21. Methyl cis-9-octadecenoate ($C18:1^{9}$)
22. Methyl trans-9-octadecenoate ($C18:1^{9}$) and Methyl cis-11-octadecanoate ($C18:1^{11}$)
23. Methyl cis-9,12-octadecadienoate ($C18:2^{9,12}$)
24. Methyl nonadecanoate (C19:0)
25. Methyl cis-9,10-methyleneoctadecanoate ($C19:0^{\Delta}$)
26. Methyl eicosanoate (C20:0)

FIGURE 25.24 Example of a GC oven temperature profile used to separate fatty acid methyl esters according to boiling points.

actually represent the total extractable acylglycerols present in the sample. The mono-, di-, and triacylglycerols extracted from the sample are chemically treated to cleave the fatty acid substituents from the glycerol backbone and then methylated (a transesterification) to make the fatty acid substituents more volatile as FAMEs. A chromatogram such as this is typically referred to as the sample's fatty acid profile and is used to characterize usually the triacylglycerols as these are the most abundant acylglycerols present. In this example, 26 BAMEs have been separated ranging from a hydroxylated 10-carbon fatty acyl chain (2-OH-C10:0) as methyl 2-hydroxydecanoate to a 20-carbon fatty acyl chain (C20:0) as methyl eicosanoate. For demonstrative purposes, we have added an oven profile below the chromatogram. The sample is injected into the injection port at time zero for the start of the run. The oven has an initial temperature of 175 °C that is held for 1 min. The oven is then ramped at 90 °C per minute to a final oven temperature of 275 °C and then held at 275 °C for 1 min. The BAMEs are separated according to boiling point where the boiling point increases with increasing fatty acyl chain length. We can see that a run like this is approximately 3 to 4 min for completion.

Columns routinely are often changed in the GC oven according to sample analyses needed. If the detector is a flame ionization detector (FID), the column can be switched out and replaced quite easily. If a mass selective detector (MSD, see Chapter 26) is being used, the MSD in older models will have to be vented to remove the detector vacuum prior to removal of the column. As a simple example, let's look at the changing of a column in a GC using an FID. Figure 25.25a illustrates a typical connection used for coupling a GC column to both the injection port and the detector consisting of a column nut and a graphite ferrule. It is best to follow the GC manufacturer's suggested procedures for installing the column into the GC oven. Briefly, for coupling the column into the injection port, the column is pushed through the column oven nut and the graphite ferrule. A couple of centimeters are then cut off of the end of the column in case any of the graphite ferrule is blocking the end of the column. Usually, a column depth position tool is used to correctly protrude approximately 3 to 5 cm of column past the ferrule. The column is then tightened into the injection port inside the left top of the oven. A similar measuring of protruding column is also done for the detector end of the column, and the column tightened into the detector similar to what is being shown in Figure 25.25b.

25.7 GC PROGRAMMING AND CONTROL

The two most common approaches to controlling the GC during analyses is the combination of an integrator and the GC control panel or using a PC with instrument controlling software. A picture of the GC control panel is illustrated in Figure 25.26. The use of an integrator is an older approach but can still be found quite common in many analytical chemistry

(a)

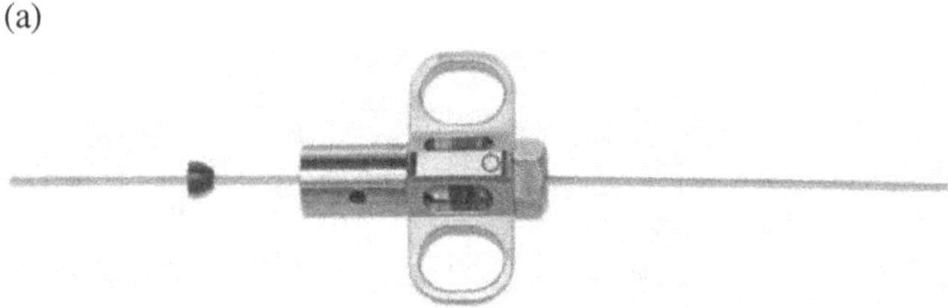

(b)

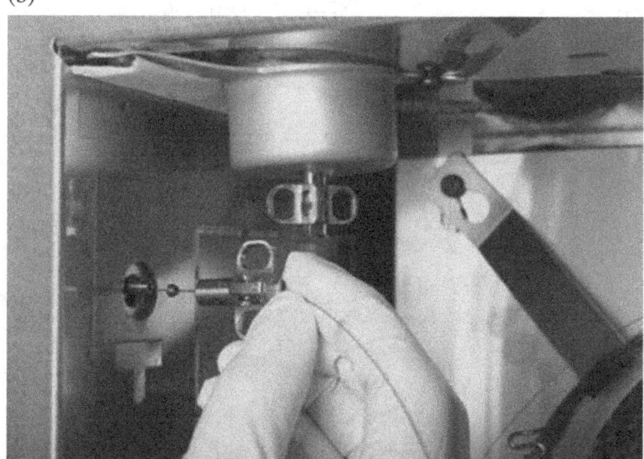

FIGURE 25.25 (a) Example of a self-tightening column nut and graphite ferrule used for connecting a GC column to both the injection port and the detector. (b) Changing of the column in the GC oven. (© Agilent Technologies, Inc. 2014, Reproduced with Permission, Courtesy of Agilent Technologies, Inc.)

FIGURE 25.26 The GC Control Panel.

FIGURE 25.27 Integrator GC Control.

labs. The use of a PC and manufacturer software however is the preferred control and is much more versatile in data processing and reporting. Figure 25.27 shows an integrator used to record data from a GC.

25.8 GC DETECTORS

There are a number of detectors that are available for use with GC. Some are general in application while others are more specific for targeting certain classes of compounds. Many current GCs are able to swap out one detector for another adding a higher level of versatility.

25.8.1 Flame Ionization Detector (FID)

The FID is the most commonly used detector found in analytical laboratories. It is robust and linear with a wide dynamic range. As long as a compound is combustible, it can be detected. The analyte as it elutes from the column will be transferred into the flame of the detector where decomposition takes place producing ionized species that are measured by the detector. Figure 25.28 is an illustration of an FID detector.

25.8.2 Electron Capture Detector (ECD)

The electron capture detector (ECD) is a more specific detector as compared to the FID. The ECD is used primarily for halogenated compounds such as pesticides and chlorophenols. The detector contains a beta emission source (electron emitter) of Ni^{63} foil. The Ni^{63} foil creates an electron cloud that produces a constant current within the source. As a compound with electron affinity passes through the detector, it captures electron and changes the charge within the detector creating a signal measuring the compound. Figure 25.29 illustrates an ECD.

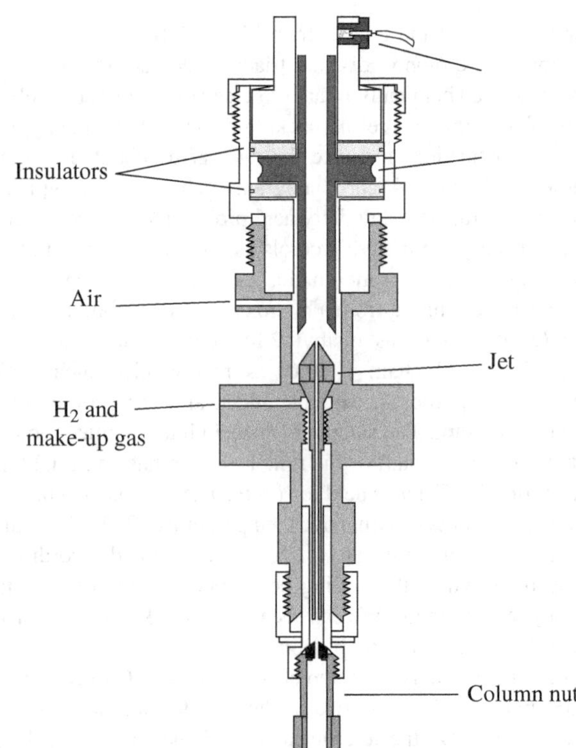

FIGURE 25.28 Flame ionization detector (FID). (© Agilent Technologies, Inc. 2014, Reproduced with Permission, Courtesy of Agilent Technologies, Inc.)

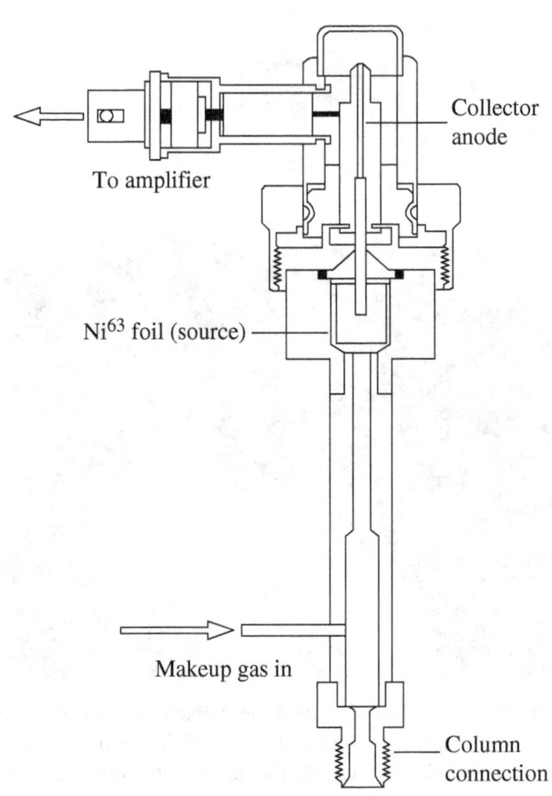

FIGURE 25.29 Electron capture detector (ECD). (Reprinted with permission from PerkinElmer.)

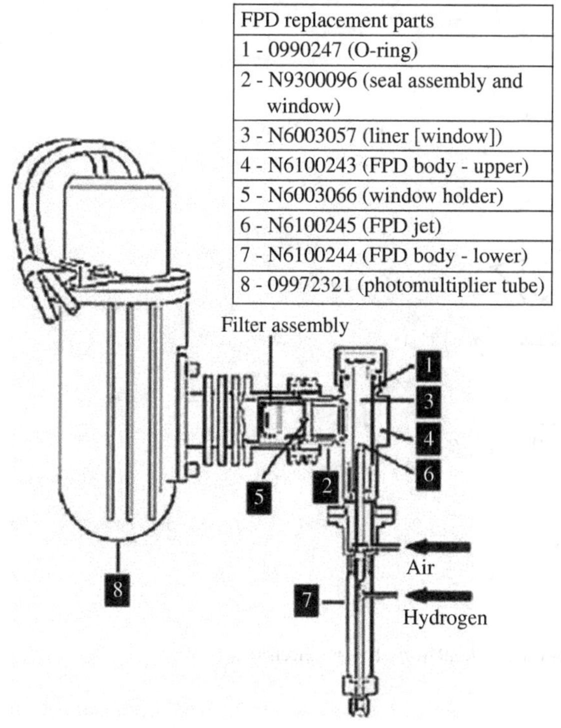

FIGURE 25.30 Flame photometric detector (FPD). (Reprinted with permission from PerkinElmer.)

25.8.3 Flame Photometric Detector (FPD)

The flame photometric detector (FPD) is used primarily to measure phosphorus- and sulfur-containing compounds. As the compound elutes from the column, it passes through a hydrogen/air flame which burns the compound liberating phosphorus as HPO and sulfur as S_2. The flame also excites the HPO and the S_2 bringing about optical emission at 510 and 526 nm for HPO and at 394 nm for S_2. Optical filters are used to isolate the signals for detection. The FPD has also been used to measure nitrogen, halogen, and some metals (Fig. 25.30).

25.8.4 Nitrogen Phosphorus Detector (NPD)

The nitrogen phosphorus detector (NPD) is a highly specific and sensitive detector for nitrogen and phosphorus. The detector contains a flame that also burns the compounds (similar to the FID) as they elute as seen with the previous detectors. However, the detector contains a rubidium or cesium silicate glass bead in a heater coil near the hydrogen flame. A schematic drawing of the NPD detector is illustrated in Figure 25.31. The bead which is heated emits electrons that are collected by an anode which provides a background current. The hydrogen flame is controlled to allow measurement of either phosphorus alone or the combination of phosphorus and nitrogen. When compounds elute from the column, they are combusted, and the partially decomposed

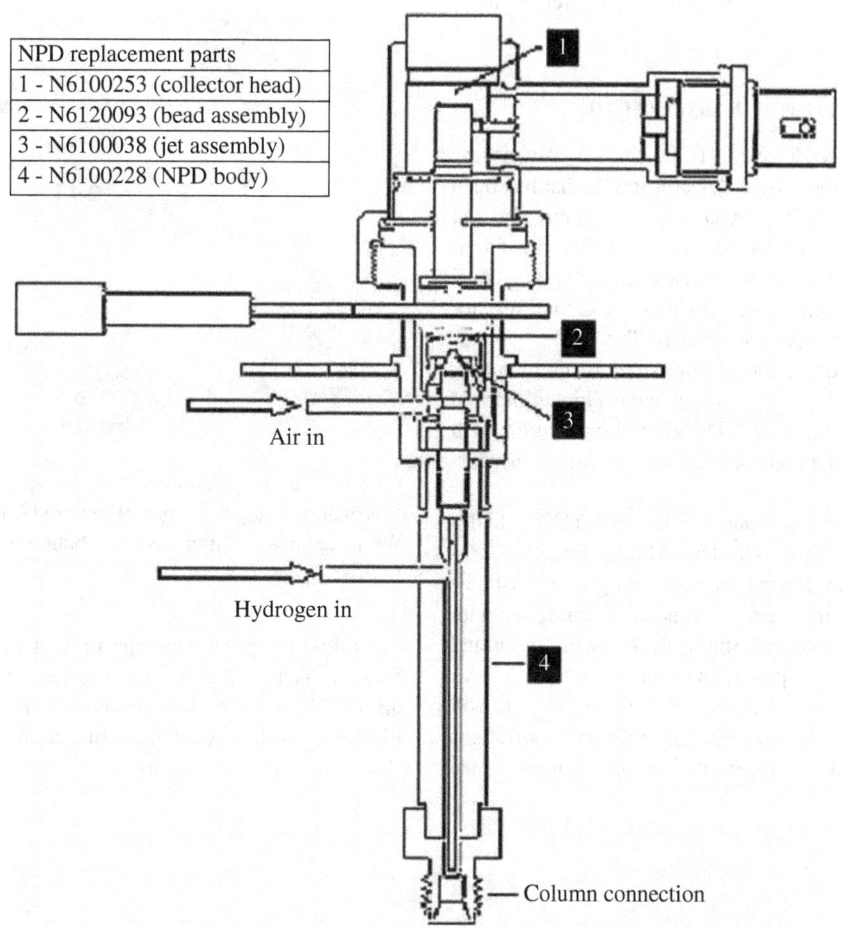

FIGURE 25.31 Schematic drawing of the nitrogen phosphorus detector (NPD). (Reprinted with permission from PerkinElmer.)

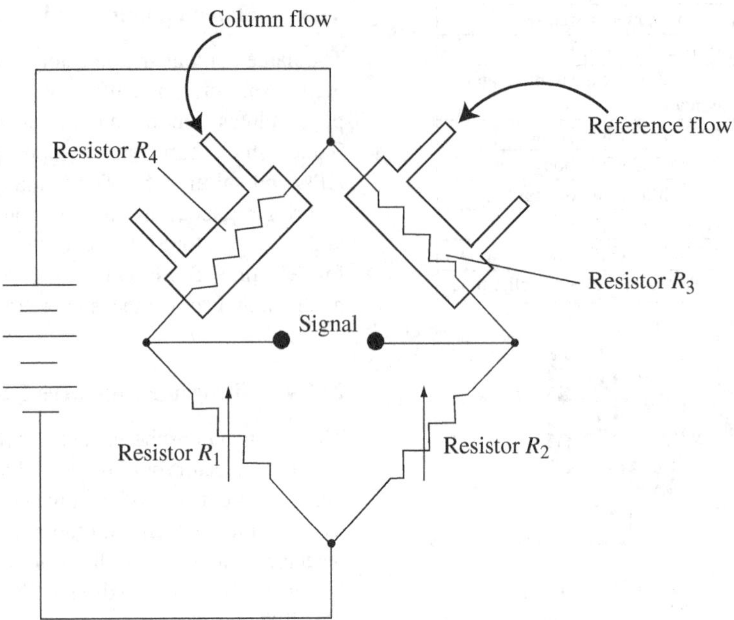

FIGURE 25.32 The resistor used in a TCD called a Wheatstone bridge circuit.

carbon–nitrogen and carbon–phosphorus are absorbed onto the silicate bead. The absorbed species induce a higher current by increasing the amount of electrons emitted from the bead. The detector is very sensitive but requires bead replacement periodically.

25.8.5 Thermal Conductivity Detector (TCD)

The thermal conductivity detector (TCD, known also as the katharometer) is a commonly used detector in GC that has been used for a number of years. The TCD is quite versatile and can detect a broad range of analytes due to its mechanism. Compounds eluting pass over a heated resistor changing the level of the resistor resulting in a measured signal. A second resistor has a reference carrier gas passing over it. The carrier gas has a different thermal conductivity as compared to compounds. Carrier gasses such as helium or hydrogen have a higher thermal conductivity compared to analytes. The eluting compounds will reduce the effluent thermal conductivity and produce a measurable signal.

Figure 25.32 illustrates the resistor used in a TCD called a Wheatstone bridge circuit. The Wheatstone bridge circuit is comprised of four resistors used to measure the changes in resistance across the resistors. The circuit cell is temperature controlled with a steady-state electrically heated filament. When a compound elutes from the column and passes over its filament as shown in the figure as Resistor 4, the filament will heat up and there will be a change in resistance. Resistor 3 has the reference carrier gas flow across it and is used to compensate in changes in resistance due to the temperature or the flow of the carrier gas. Unlike an FID, the TCD will detect any species that is eluting from the column. This includes noncombustible compounds. Figure 25.33 illustrates a few examples of commercially available TCD Wheatstone bridge circuits.

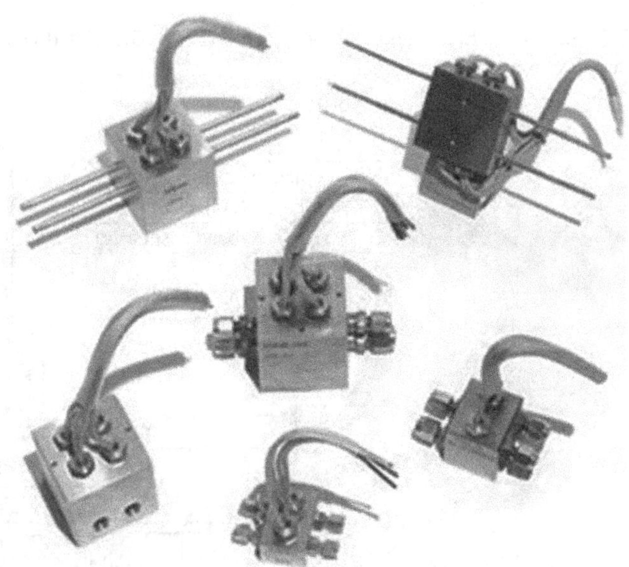

FIGURE 25.33 Examples of commercially available TCD Wheatstone bridge circuits. (Reprinted with permission under the Wiki Creative Commons Attribution-Sharealike 3.0 Unported License.)

26

GAS CHROMATOGRAPHY–MASS SPECTROMETRY (GC–MS)

26.1 Introduction
26.2 Electron Ionization (EI)
26.3 Electron Ionization (EI)/OE Processes
26.4 Oleamide Fragmentation Pathways: OE $M^{+\bullet}$ by Gas Chromatography/Electron Ionization Mass Spectrometry
26.5 Oleamide Fragmentation Pathways: EE $[M+H]^+$ by ESI/Ion Trap Mass Spectrometry
26.6 Quantitative Analysis by GC/EI–MS
26.7 Chapter Problems
References

26.1 INTRODUCTION

Gas chromatography–mass spectrometry is a useful tool for performing quantitative analysis of volatile metabolites that can be separated with a gas chromatograph. The response of the ion detectors coupled with the single quadrupole mass spectrometers is linear in mass regions, allowing quantitative analysis. We have covered gas chromatography extensively in the previous chapter. We will now look at using a mass spectrometer as a detector for the gas chromatography. The chapter starts with an overview of the ionization source, called electron ionization, that prepares the analytes for measurement in the mass spectrometer as they elute from the gas chromatography column.

26.2 ELECTRON IONIZATION (EI)

Electron ionization is the oldest ionization source that has been used in mass spectrometry and is still widely in use today. This design basically stems from the work of J.J. Thompson (1856–1940) who is considered the first mass spectrometrist to win the Nobel Prize. In 1906, he was awarded the Nobel Prize in Physics for his studies involving the discharge of electricity into gasses producing ionized species. Indeed, the recently introduced unit for the mass-to-charge (*m/z*) value in mass spectrometry, the Thompson (Th), has been proposed in his honor. Typically, electron ionization is used as an ionization source after analytes have been separated within a gas chromatograph. The most common mass analyzer used in this configuration is the single quadrupole mass filter, which will be covered in Chapter 30. Figure 26.1 illustrates the basic design of an electron ionization source. Incoming into the source from the left are the neutral analytes from some type of inlet system such as a gas chromatography column. The electron ionization source is not at atmospheric pressure (760 torr or 101.325 kPa) but at a reduced pressure of 10^{-3} Pa (10^{-5} torr), eliminating the ionization of ambient species by the electron beam and collisions with the gaseous analyte molecules. The neutral analytes in the gas phase pass through an electron beam that is generated by heating a filament that subsequently boils off electrons. Opposite the filament is a target collector for the boiled off electron beam at a potential of +5030 V while the filament is kept at a potential of +4930 V. This potential difference between the filament and the target is important and is kept constant from instrumentation to instrumentation. The potential difference between the filament and the target imparts to the electron beam energy of approximately 70 eV. By keeping this imparted energy constant at 70 eV from instrument to instrument, standardized spectra are obtained that are basically universal. This allows the searching of unknown EI spectra against EI spectral libraries due to reproducible spectra obtained from the standardized 70 eV electron beam impact-induced EI spectra. As the analyte passes through the electron beam, energy is transferred from the fast-moving electrons within the beam to the analyte. The excited analyte releases a portion of the absorbed energy through ejection of an electron, thus producing singly positive-charged molecular ions, $M^{+\bullet}$, that are odd electron (OE) radical ions:

$$M + e^-(70\text{ eV}) \rightarrow M^{+\bullet} + 2e^- \qquad (26.1)$$

Analytical Chemistry: A Chemist and Laboratory Technician's Toolkit, First Edition. Bryan M. Ham and Aihui MaHam.

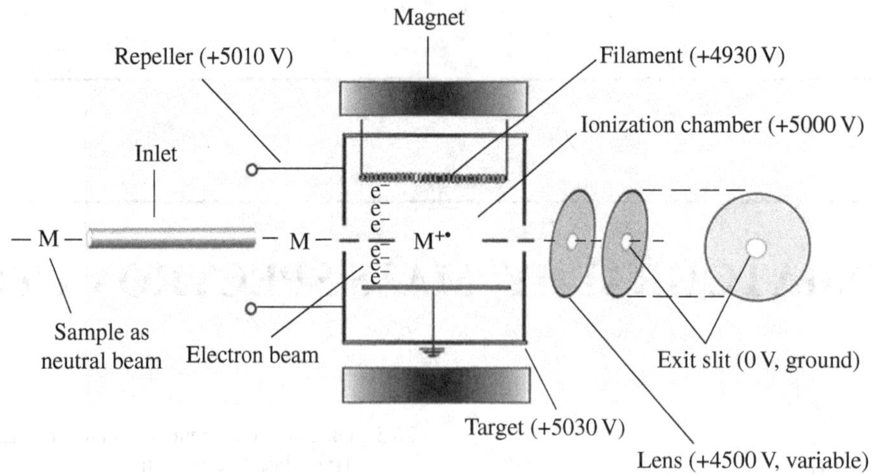

FIGURE 26.1 Basic design of an electron ionization (EI) source including potentials involved in moving the sample through the source. Movement of sample beam from neutral to ionized is from left to right. Sample progresses through the exit slit to the mass spectrometer.

Typical organic molecules have ionization potentials ranging from approximately 6 to 10 eV, which is the energy required to free an electron from the molecule. The excess energy remaining in the molecular ion produced from the electron ionization process generally causes fragmentation to take place, often severe enough to deplete the precursor molecular ion. The process of electron ionization is a "hard" ionization process resulting in considerable fragmentation of the molecular ion due to the excess energy imparted to the analyte ("∗∗" and "∗" represent excited states) from the electron beam. This process is illustrated in Equation 26.2 where the first step is the transfer of energy from the electron beam to the analyte, producing an excited state in the neutral analyte. The second step is the ejection of an electron from the excited analyte producing an OE molecular ion but still in an excited state due to excess energy imparted in the ionization process. The final step is fragmentation of the molecular ion producing both even electron (EE) and OE molecular ions:

$$\begin{array}{c} \text{ABCDE} + e^- (70\ \text{eV}) \rightarrow [\text{ABCDE}]{**} + e^- (<70\ \text{eV}) \\ [\text{ABCDE}]{**} \rightarrow [(\text{ABCDE})^{+\cdot}]{*} + e^- \end{array} \quad (26.2)$$

$$[(\text{ABCDE})^{+\cdot}]{*} \rightarrow \text{AB}^{+\cdot} + \text{CDE} (\text{OE radical product plus neutral})$$
$$\searrow \text{ABC}^+ + \cdot\text{DE} (\text{EE product plus radical neutral})$$

Located above and below the filament and the target trap used for the electron beam are the opposite poles of a magnet. The small magnetic field induces the electrons in the beam to follow a spiral trajectory from filament to target. This in effect causes the electrons to follow a longer path from filament to target producing a greater chance of ionizing the neutral analyte molecules passing through the electron beam. Behind the electron beam is a repeller that is kept at +5010 V used to give the newly formed molecular ions a push out of the ionization chamber, which is kept at +5000 V. The newly formed molecular ions then pass through a series of focusing lenses that are kept at +4500 V (but variation is allowed for ion beam focusing). Finally, the newly formed molecular ions pass through an exit slit that is held at 0 V (ground). The whole source has a potential gradient made up of a slope in voltages from the repeller to the exit slit that the molecular ion beam will follow leading out of the source and into the mass analyzer. Figure 26.2 illustrates a typical electron ionization mass spectrum where the molecular ion, $M^{+\bullet}$, is observed at *m/z* 296.2. The *y*-axis of the EI mass spectrum is in relative abundance (%). Typical of mass spectra, the *y*-axis has been normalized to the most abundant peak in the spectrum such as the *m/z* 54.8 peak in Figure 26.2. The *x*-axis is the mass-to-charge (*m/z*) ratio moving to increasing values from left to right. Notice in the EI mass spectrum the large degree of fragmentation taking place depleting the molecular ion at *m/z* 296.2.

26.3 ELECTRON IONIZATION (EI)/OE PROCESSES

Electron impact or electron ionization (EI) typically produces a charged species through the ejection of a single electron. The ejection of a single electron results in a molecular ion that is a positively charged radical, possesses an odd number of electrons, and is designated as an OE molecular ion $M^{+\bullet}$. The EI process is an energetic process that often results in significant fragmentation of the molecular ion producing a large number of product ions. The product ions produced in EI are comprised of both OE ions and EE ions. EE ions contain an even number of valence shell electrons and in positive ion mode are designated as $[M + nX]^{n+}$, for example, the protonated form of an EE ion is designated as $[M + H]^+$, the sodium adduct form as $[M + Na]^+$, and the ammonium adduct as $[M + NH_4]^+$. In negative ion mode, EE ions are designated as $[M + nX]^{n-}$, for example, chloride adducts $[M + Cl]^-$, and deprotonated species as $[M - nH]^{n-}$. EE ions are often derived by the softer ionization techniques such as electrospray ionization (ESI), fast atom bombardment (FAB), some types of chemical ionization (CI), and matrix-assisted laser desorption/ionization (MALDI).

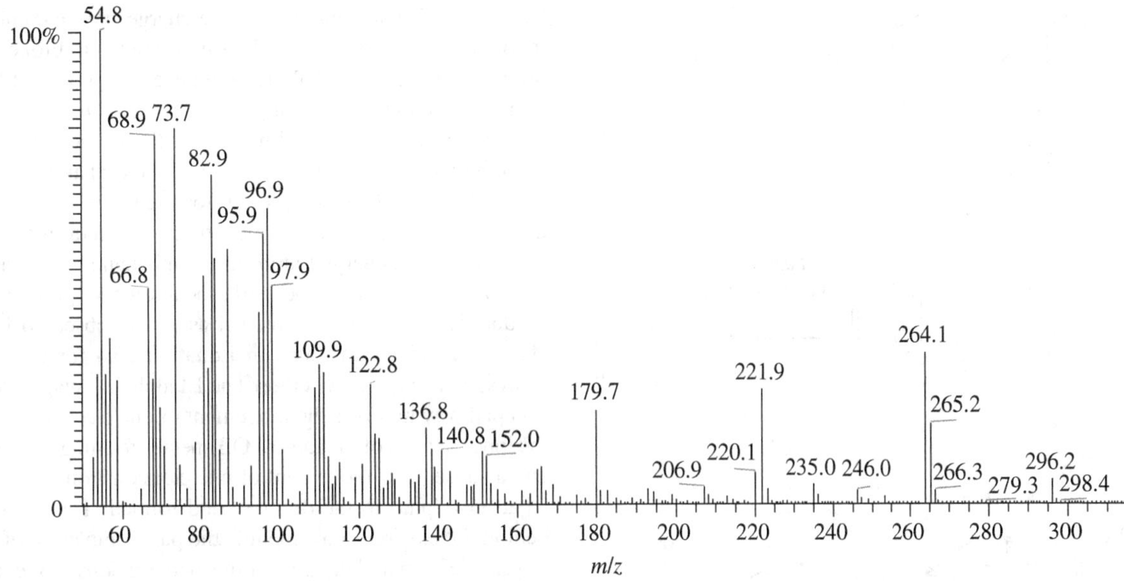

FIGURE 26.2 Odd electron product ion spectrum of methyl oleate ($C_{19}H_{36}O_2$) at *m/z* 296.2 as $M^{+\bullet}$ collected by electron impact ionization/single quadrupole mass spectrometry (EI–MS).

A simplified comparison between the production of OE molecular ions and EE ions is as follows:

Neutral pentanol —EI, $-e^-$→ OE, $M^{+\cdot}$ (26.1)

Neutral pentanol —ESI, H^+→ EE, $[M+H]^+$ (26.2)

where Equation 26.1 represents electron ionization (removal of one electron) of pentanol producing an OE molecular ion $M^{+\bullet}$ and Equation 26.2 represents the protonation of pentanol producing an EE ion $[M+H]^+$. The classification of ease of ionization by removal of an electron generally follows the trend of nonbonding > π > σ. An electron associated with a sigma bond (δ) requires the largest amount of energy to remove an electron and ionize the species. This is followed by pi bond (π) electrons associated with double and triple bonds and finally nonbonding electrons that generally require the least amount of energy for removal. The most common use of electron ionization is the coupling of a single quadrupole mass analyzer to gas chromatography instrumentation and still finds wide use for the analysis of volatile compounds. Due to the reproducibility of the molecular ion spectra obtained from the standardized 70 eV electron ionization beam used in EI sources, the generated molecular ion spectra are often searched against standard libraries for structural identification of the analyte of interest. However, it is useful to know the basics of some of the most common types of fragmentation pathways that take place using EI as the ionization technique (a comprehensive reference for OE spectral interpretation is given by McLafferty and Turecek [1]). With EI fragmentation, the dissociation is unimolecular where the activation of the fragmentation process has not involved collision-induced dissociation (CID) with a stationary target gas or CI of the ion/molecule type reaction. There are five basic types of bond cleavage that takes place during the fragmentation of a molecular ion when using EI: sigma cleavage (σ), inductive cleavage (*i*), alpha cleavage (α), retro-Diels–Alder, and hydrogen rearrangement (*r*H). When decomposition takes place with an OE molecular ion involving the breaking of a single bond, an EE ion is always produced along with a neutral radical. In breaking two bonds, an OE product ion is produced along with an EE neutral. The first type of cleavage mentioned, sigma cleavage (σ), is the predominant type of bond breakage that is observed

H_3C–CH(CH_3)–CH_2–CH_3 —$-e^-$→ [H_3C–C(CH_3)(H)$^{+\cdot}$–CH_2–CH_3] —σ→ H_3C–C^+H–CH_3 (EE product) + $\cdot CH_2$–CH_3 (OE neutral) (26.3)

Ether —$-e^-$→ molecular ion —Product *i*→ EE product + $\cdot$O–ethyl (OE neutral) (26.4)

Ether —$-e^-$→ molecular ion —α→ OE neutral + $^+$O= (EE product) (26.5)

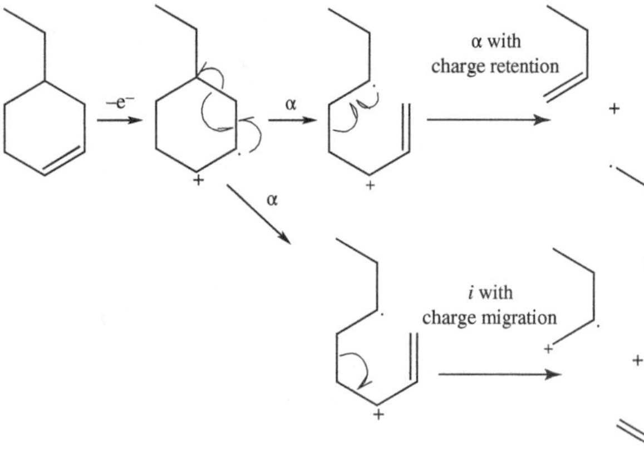

α with charge retention

α

i with charge migration

(26.6)

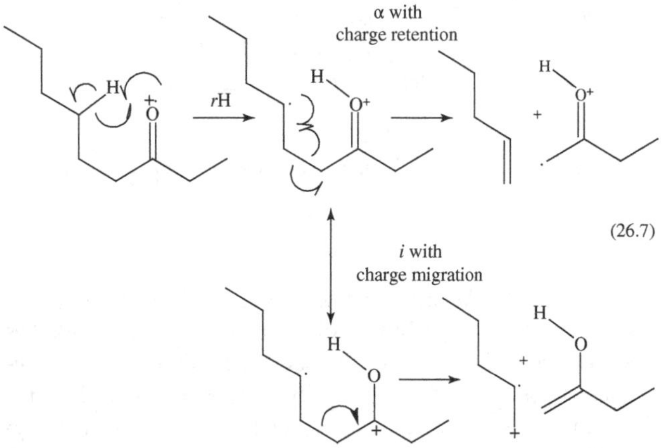

α with charge retention

rH

i with charge migration

(26.7)

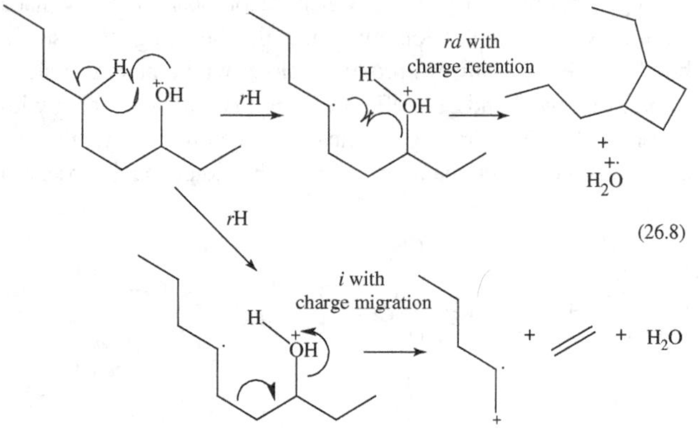

rd with charge retention

rH

i with charge migration

(26.8)

in EI spectra when saturated nonaromatic, aliphatic hydrocarbon species comprise the OE molecular ions. Sigma cleavage fragmentation of an OE molecular ion producing an EE product ion and radical OE neutral is illustrated in Equation 26.3 for the EI of isopentane. In the mechanistic equations used such as that found in Equations 26.4–26.8, a single fishhook arrow represents the movement of a single electron and represents a homolytic cleavage of the bond. An example of a single electron homolytic cleavage illustrated by a fishhook arrow can be found in the alpha-bond cleavage in Equation 26.5. When an electron pair is moved, such as in Equation 26.4 for the inductive cleavage, a full arrowhead is used. This type of cleavage represents a heterolytic cleavage that moves the charge site. In combinations involving these processes, when two bonds are broken, an OE product ion is produced along with the corresponding EE neutral fragment. Inductive cleavage (*i*), also known as charge site-initiated cleavage, is a unimolecular dissociation that involves the migration of a pair of electrons to the charged site of the molecular ion. The driving force for the movement of the electrons is greater as the electronegativity or electron-withdrawing tendency of the charged atom in the molecule is increased (i.e., Cl > O > N > C). This type of dissociation also produces an EE product ion and a neutral radical as is also observed for sigma cleavages and is illustrated in Equation 26.4 for the inductive cleavage of propylethyl ether. The EI alpha cleavage fragmentation pathway involves the movement of one electron to pair with the lone electron of the radical OE species forming a double bond and a subsequent movement of one electron from a bond that is adjacent (alpha) to the double bond formed. The driving force behind the electron movement is the pair completion of the lone radical electron and the greater the electron donation tendency of the radical atom in the molecule (i.e., N > O > C > Cl). Alpha cleavages also produce an EE product ion and a neutral radical. Equation 26.5 illustrates an alpha cleavage for propylethyl ether. In cyclic structures that contain an unsaturation moiety, the π-electrons often allow for the radical site and initial charge with EI. In the ensuing fragmentation pathways, an alpha cleavage takes place shifting the remaining π-electron to the alpha position creating a new double bond. This process is illustrated in the first step of the retro-Diels–Alder fragmentation process of Equation 26.6. A second alpha cleavage can take place resulting in a product ion where the charge has remained in its initial location, thus describing an alpha cleavage with charge retention (Equation 26.6 bottom). An alternative process can also take place following the initial alpha cleavage where the charge migrates within the structure, and an inductive cleavage is the fragmentation pathway. This is also illustrated in the latter part of Equation 26.6. Finally, a fifth type of fragmentation pathway often associated with the decomposition of an OE molecular ion produced by EI is gamma-hydrogen (γ-H) rearrangement accompanied by bond dissociation (known as the McLafferty rearrangement). The two types of bond dissociation that occur are β-cleavage when an unsaturated charge site is involved and adjacent cleavage when there is no unsaturation involved in the decomposition pathway. As was also the case with the retro-Diels–Alder decomposition pathway, if the second step in the γ-H rearrangement is alpha cleavage, charge retention will take place, while if the second step is an inductive cleavage, charge migration will take place. The γ-H rearrangement decomposition pathways are illustrated in the top half of Equation 26.7 for the singly unsaturated molecular ion of the aliphatic ketone nonanone resulting in the γ-H rearrangement and β-cleavage. The adjacent cleavage that takes place for a saturated molecular ion following γ-H rearrangement is illustrated in Equation 26.8 for 3-nonanol. This process also involves two mechanisms describing a product ion with charge retention and a product ion formed through charge migration. This completes our basic introduction to OE mass spectral processes observed using electron ionization as the source. We will next look at a specific example of spectral interpretation of an EI-generated mass spectrum.

26.4 OLEAMIDE FRAGMENTATION PATHWAYS: OE $M^{+•}$ BY GAS CHROMATOGRAPHY/ELECTRON IONIZATION MASS SPECTROMETRY

Fatty acid primary amides (FAPA) such as *cis*-9-octadecenamide (oleamide) [2] and the related *N*-acyl ethanolamines (NAEs) [3–7] are a special class of lipids that act as messengers or signaling molecules. Oleamide was first reported observed in a biological sample in a study by Arafat et al. [8] where oleamide was isolated from plasma. Six years later, it was reported that oleamide was observed in cerebrospinal fluid of mammals experiencing sleep deprivation by Cravatt et al. [9] Following Cravatt's work, others have also reported the apparent function of oleamide as an endogenous sleep-inducing lipid [10,11]. Figure 26.3 shows an EI spectrum of oleamide obtained by gas chromatography electron ionization single quadrupole mass spectrometry (GC/EI–MS) with the molecular ion $M^{+•}$ at *m/z* 281. The spectrum contains seven fragment ions that are illustrative of the diagnostic product ions observed in the EI fragmentation spectra of oleamide and the mechanistic pathways described previously for EI (Equations 26.3–26.9). The first predominant product ion observed in Figure 26.1, the *m/z* 264 product ion, is derived from ketene formation through neutral loss of ammonia NH_3 (−17 amu) $[M\text{-}NH_3]^{+•}$. This mechanism is illustrated in Equation 26.9 where the first step in the fragmentation pathway is a β-hydrogen shift to the amine group (α to the carbonyl and β to the amine) followed by inductive cleavage of ammonia.

$CH_3(CH_2)_7CH{=}CH(CH_2)_7C(=O)NH_2$ —($-e^-$)→ $CH_3(CH_2)_7CH{=}CH(CH_2)_6CH(H)C(=O)\overset{+\cdot}{N}H_2$ → $CH_3(CH_2)_7CH{=}CH(CH_2)_6CH(H)C(=O)\overset{+}{N}H_2(H)$ —(i)→ $CH_3(CH_2)_7CH{=}CH(CH_2)_6CH(H)C(=O)^{+}$ + NH_3 (*m/z* 264) (26.9)

The *m/z* 238 product ion is formed through σ-cleavage resulting in an aliphatic hydrocarbon loss $[M\text{-}CH_2CH_2CH_3]^+$ (Equation 26.10).

$CH_3(CH_2)_7CH{=}CH(CH_2)_7C(=O)NH_2$ —($-e^-$)→ $CH_3CH_2CH_2{+}CH_2(CH_2)_4CH{=}CH(CH_2)_7C(=O)NH_2$ —(σ)→ $CH_3CH_2CH_2\cdot$ + $^{+}CH_2(CH_2)_4CH{=}CH(CH_2)_7C(=O)NH_2$ (*m/z* 238) (26.10)

The *m/z* 59 and the *m/z* 222 product ions are formed through γ-hydrogen migration with subsequent inductive cleavages as illustrated in Equations 26.7 and 26.8 for 3-nonanone and 3-nonanol, respectively. The *m/z* 59 is a γ-hydrogen rearrangement mechanism accompanied by β-cleavage bond dissociation (McLafferty rearrangement). This is due to the unsaturated charge site being involved (carbonyl oxygen is ionized) with charge retainment producing $[CH_2COHNH_2]^{+•}$. The fragmentation pathway mechanism for the production of the *m/z* 59 product ion is illustrated in Equation 26.11. For the production of the *m/z* 222 product ion, the molecular ion is ionized at the amine nitrogen, which is a saturated moiety. This fragmentation pathway mechanism is γ-hydrogen migration with adjacent inductive cleavage involving charge migration $[M\text{-}CH_2CO\text{-}NH_3]^{+•}$. The fragmentation pathway mechanism for the production of the *m/z* 222 product ion is illustrated in Equation 26.12. Finally, the product ion at *m/z* 154 is derived from a radical site-initiated α-cleavage (allylic cleavage) between the C7 and the C8 carbons. This is not a sigma cleavage due to initial ionization of the sigma bond but is driven by the initial

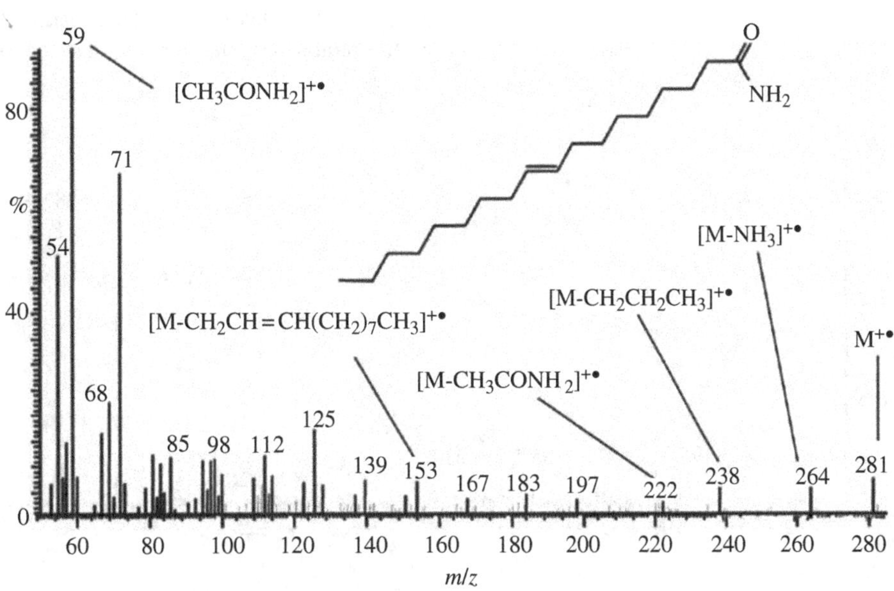

FIGURE 26.3 EI mass spectrum of *cis*-9-octadecenamide (oleamide).

$CH_3(CH_2)_7CH{=}CH(CH_2)_4$ ... NH_2 → $CH_3(CH_2)_7CH{=}CH(CH_2)_4$ ... O^+ ... NH_2 → $CH_3(CH_2)_7CH{=}CH(CH_2)_4$ + m/z 59 (26.11)

$CH_3(CH_2)_7CH{=}CH(CH_2)_4$ ... NH_2 → $CH_3(CH_2)_7CH{=}CH(CH_2)_4$ ... NH_2 → $CH_3(CH_2)_7CH{=}CH(CH_2)_4$ + m/z 222 + O + NH_3 (26.12)

removal of a π-electron from the unsaturation site between C9 and C10. Following the cleavage of the C7 and C8 sigma bond, there is a hydrogen migration from the C7 to the C8 carbon. A probable mechanism describing the radical site-initiated α-cleavage fragmentation pathway for the production of the m/z 154 product ion is illustrated in Equation 26.13.

$CH_3(CH_2)_7CH{=}CH(CH_2)_7C(O)NH_2$ $\xrightarrow{-e^-}$ $CH_3(CH_2)_7CH\overset{+\cdot}{-}CH{-}CH_2{-}CH_2(CH_2)_5C(O)NH_2$ → $CH_3(CH_2)_7\overset{+}{C}HCH{=}CH_2$ ... $C(CH_2)_5C(O)NH_2$ → $CH(CH_2)_5C(O)NH_2$ + $CH_3(CH_2)_7\overset{+}{C}H{=}CHCH_3$ m/z 154 (26.13)

26.5 OLEAMIDE FRAGMENTATION PATHWAYS: EE [M+H]$^+$ BY ESI/ION TRAP MASS SPECTROMETRY

We have now covered the electron ionization (EI)/OE mass spectrum of *cis*-9-octadecenamide (oleamide) including fragmentation pathways that describe the product ions produced. We will now compare the ESI mass spectrum of the EE protonated form of *cis*-9-octadecenamide (oleamide) as the precursor ion at m/z 282.6 [$C_{18}H_{35}NO + H]^+$. The protonated form of *cis*-9-octadecenamide was subjected to CID for structural information using a three-dimensional quadrupole ion trap mass spectrometer (ESI/IT/MS), and the product ion spectrum is illustrated in Figure 26.4. In ESI mass spectrometry, the precursor ions and the product ions are almost exclusively comprised of EE species. Often, analytes of interest are neutral; therefore, a cationizing agent (for positive ion mode analysis) is added to promote the gas-phase ionization of the analyte and to also help support the electrospray process. In positive ion mode, some typical cationizing agents added to the analyte solution prior to electrospray are volatile organic acids such as acetic acid or formic acid. These volatile organic acids will promote the gas-phase protonation of the analyte producing the EE precursor ion [M+H]$^+$. Metals are also often used as cationizing agents such as lithium, [M+Li]$^+$, or sodium, [M+Na]$^+$. Sometimes, it is also advantageous to use ammonium acetate as the solution additive prior to electrospray, which promotes the formation of ammonium adducts, [M+NH_4]$^+$. The ESI product ion spectrum of oleamide at m/z 282.6 illustrated in Figure 26.4 was obtained in an acidified (1% acetic acid) 1:1 chloroform/methanol solution.

The first major product ion observed in the ESI mass spectrum of Figure 26.4 is the m/z 265.6 peak. This is formed through neutral loss of ammonia (NH_3, −17 amu) from the EE precursor ion of protonated oleamide at m/z 282.6 [M+H–NH_3]$^+$. The protonation of the oleamide lipid species will take place as an association of the positive charge on the proton with the most basic site on the lipid. The most basic site is associated with the amide moiety of the lipid with a closer electrostatic sharing of the positive charge on the proton with the nonbonding electron pair of the amide

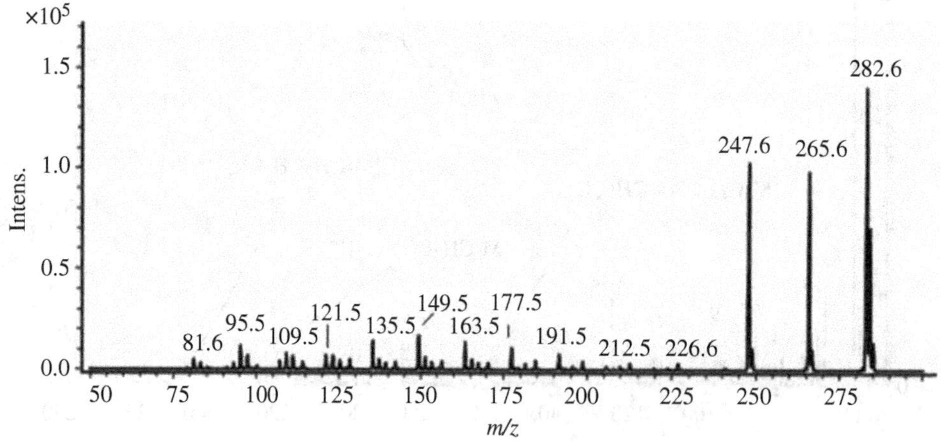

FIGURE 26.4 Electrospray ionization (ESI) mass spectrum of the even electron (EE) protonated form of *cis*-9-octadecenamide (oleamide) as the precursor ion at m/z 282.6 [$C_{18}H_{35}NO + H]^+$.

nitrogen. This association is illustrated in Equation 26.14 where the amide nitrogen is drawn as protonated. This same neutral loss of ammonia was also observed in the EI mass spectrum illustrated in Figure 26.3. The mechanism that described the production of the *m/z* 264 product ion in Figure 26.3 involved a two-step process, first of hydrogen transfer from the ß-carbon to the amide nitrogen followed by inductive cleavage to release the neutral NH_3 and subsequently produce an OE product ion as illustrated in Equation 26.9. In Equation 26.14, the production of the neutral loss of NH_3 from the protonated form of the precursor ion of

$$CH_3(CH_2)_7CH{=}CH(CH_2)_7C(=O)NH_2 \xrightarrow[ESI]{H^+} CH_3(CH_2)_7CH{=}CH(CH_2)_7C(=O)NH_3^+ \rightarrow CH_3(CH_2)_7CH{=}CH(CH_2)_7C^+{=}O \;(m/z\ 265.6) + NH_3 \quad (26.14)$$

oleamide is through a one-step inductive, heterolytic cleavage of the carbon–nitrogen bond producing an EE product ion. Fragmentation pathway mechanisms involved in the dissociation of EE precursor ions almost exclusively occur through an inductive, heterolytic cleavage of sigma bonds. A second predominant product ion observed in the ESI tandem mass spectrum of Figure 26.4 is the *m/z* 247.6 ion that on first inspection appears to be produced through neutral loss of ammonium hydroxide $[M+H–NH_4OH]^+$ (loss of 35 amu). The product ion spectrum illustrated in Figure 26.4 was obtained using an IT/MS. This type of mass analyzer allows the collection of multiple product ion spectra in the form of MS^n (where $n = 2$, 3, up to 9 maximum). This type of mass analysis involves the isolation of a precursor ion, product ion generation of the precursor, isolation of a first-generation product ion and subsequent production of second-generation product ions, and so on. Figure 26.5 illustrates the product ion spectrum of the *m/z* 265.6 product ion obtained from the *m/z* 282.6 precursor protonated ion of oleamide. This is an MS^3 product ion spectrum where the production of the *m/z* 282.6 precursor protonated ion of oleamide is the first mass analysis stage designated as MS, the production of the *m/z* 265.6 product ion is the second mass analysis stage designated as MS^2, and the product ion spectrum illustrated in Figure 26.5 is the third mass analysis stage designated as MS^3. The product ions observed in the MS^3 product ion spectrum illustrated in Figure 26.5 demonstrate that the production of the *m/z* 247.6 product ion is in two steps: the first step is neutral loss of ammonia from the *m/z* 282.6 precursor ion followed by a second step of neutral loss of H_2O, $[M+H–NH_3–H_2O]^+$. This is effectively loss of ammonium hydroxide (NH_4OH) in what requires a series of steps. The fragmentation pathway mechanism for the production of the *m/z* 247.6 product ion has similarities to the production of the *m/z* 265.6 product ion (Equation 26.14) in its first step of loss of ammonia (Equation 26.15). However, the *m/z* 265.6 product ion continues to decompose in a series of steps producing the *m/z* 247.6 product ion.

$$CH_3(CH_2)_7CH{=}CH(CH_2)_7C(=O)NH_2 \xrightarrow[ESI]{H^+} CH_3(CH_2)_7CH{=}CH(CH_2)_7C(=O)NH_3^+ \rightarrow NH_3 + CH_3(CH_2)_7CH{=}CH(CH_2)_6CH_2C^+{=}O \;(m/z\ 265.6) \rightarrow CH_3(CH_2)_7CH{=}CH(CH_2)_5CH_2CH{=}C{=}O^+H \rightarrow CH_3(CH_2)_7CH{=}CH(CH_2)_5CH_2C(H)(OH){=}C^+ \rightarrow CH_3(CH_2)_7CH{=}CH(CH_2)_5CH{=}C{=}\overset{+}{C}H + H_2O \;(m/z\ 247.6) \quad (26.15)$$

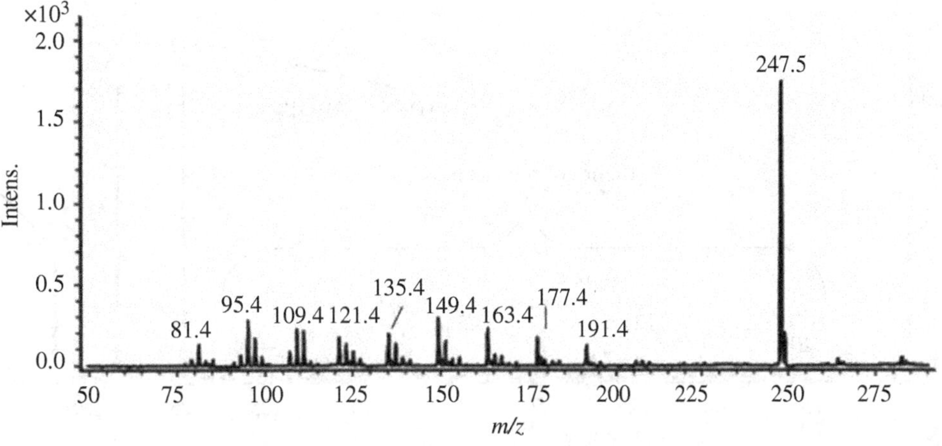

FIGURE 26.5 MS^3 product ion spectrum of the precursor protonated ion of oleamide at *m/z* 282.6, $[M+H]^+$. In the second mass analysis stage, the *m/z* 265.6 product ion was isolated and subjected to collision-induced dissociation.

Also observed in both Figures 26.4 and 26.5 is a series of peaks ranging from *m/z* 81.6 to *m/z* 226.6 that is often described as a "picket fence" series primarily from hydrocarbon chain decomposition (four series are included within *m/z* 81.6 to *m/z* 226.6). The predominant peaks in the picket fence series are odd m/z product ions that would appear to be derived from the fatty acyl chain hydrocarbon series. In OE spectra (e.g., electron impact (EI) spectra), hydrocarbon series are typically represented by the $C_nH^+_{2n+1}$ series[12]. Typically, for a straight-chain saturated aliphatic hydrocarbon series, the picket fence product ions will have a maximum around C_4 to C_5. Examples include C_3H_7 and C_5H_{11} for the $C_nH^+_{2n+1}$ series and C_3H_5 and C_5H_9 for the $C_nH^+_{2n+1}$ series. Notice that the picket fence series type of product ions produced through aliphatic hydrocarbon chain cleavage is EE product ions. Aliphatic hydrocarbon chain cleavage in OE spectra is produced by sigma bond ionization followed by carbon–carbon cleavage. Aliphatic hydrocarbon chain cleavage product ions are known to undergo rearrangements that are random resulting in cyclic formation and unsaturation. In the product ion spectra illustrated in Figures 26.4 and 26.5, there is also observed the production of aliphatic hydrocarbon chain cleavage. Unlike the case in OE/EI mass spectra where aliphatic hydrocarbon chain ionization has taken place (thus resulting in charge-driven fragmentation), the production of the *m/z* 81.6 to 226.6 product ions in Figures 26.4 and 26.5 is the result of charge-remote fragmentation processes, while the production of the *m/z* 265.6 and *m/z* 247.6 product ions is localized charge-driven fragmentation processes. Gross [12] has also observed charge-remote fragmentation for fatty acids as reported recently in an account of product ion mass spectra of lipids obtained using high-energy collisions. In the CID product ion spectrum of oleamide (Figs 26.4 and 26.5), there is also the observance of charge-remote fragmentation; however, these spectra were obtained at low-energy collisions with ES/IT/MS. The EE product ion spectrum of the protonated form of oleamide as the precursor ion is illustrated in Figure 26.6 listing the product ions discussed thus far including the picket fence series range from *m/z* 81.6 to 226.6.

The most predominant series as illustrated in Figures 26.4 and 26.5 is comprised of an aliphatic hydrocarbon product ion series of the form $C_nH_{2n-5}{}^+$. For example, the *m/z* 93.6 product ion has the product ion formula of $C_7H_9{}^+$. Using the rings plus double bonds formula results in a value of 3. This indicates that the series is comprised of EE product ions that contain three rings plus double bonds. As with OE/EI aliphatic hydrocarbon chain cleavages, the three rings plus double bonds value for the *m/z* 93.5 product ion indicate that cyclization, rearrangement, and isomerization are also taking place. The rest of the series includes *m/z* 107.5, 121.5, 135.5, 149.5, and so on. A second product ion in series observed in Figures 26.4 and 26.5 includes *m/z* 81.6, 95.5, 109.5, 123.5, 137.5, and 151.5. This series is represented by $C_nH^+_{2n-3}$ such as $C_6H_9^+$ for the *m/z* 81.6 product ion. An application of the rings plus double bonds equation results in a value of 2 indicating that this is an EE product ion series that contains two rings plus double bonds. In this product ion series from *m/z* 81.6 to 151.5, there is also cyclization, rearrangement, and isomerization taking place forming the product ions. Also included in the spectra is a series comprised of the product ions *m/z* 83.6, 97.5,

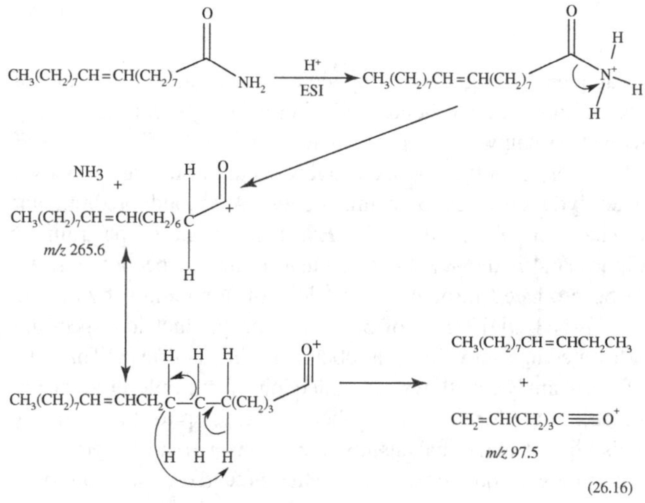

(26.16)

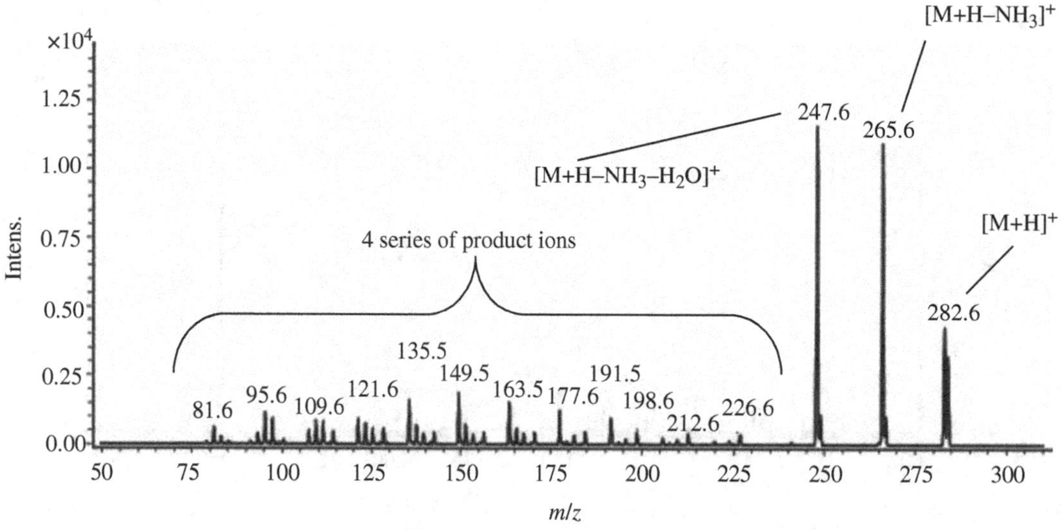

FIGURE 26.6 Product ion spectrum of the EE protonated form of oleamide as the precursor ion listing the production of the *m/z* 265.6 and 247.6 product ions and the *m/z* 81.6 to 226.6 picket fence product ion series.

111.5, 125.5, 139.5, and 153.5. This is an acylium product ion series observed in the form of $C_nH_{2n-3}O^+$ such as $C_6H_9O^+$ for the *m/z* 97.5 product ion. The fragmentation pathway mechanism for the production of this series of product ions would be derived through a two-step process involving the production of the *m/z* 265.6 product ion and then neutral loss of C_nH_{2n+2} from an initial acylium product ion (Equation 26.16). Lastly, also observed in the *m/z* 81.6 to *m/z* 226.6 series are even molecular weight product ions indicating the retainment of the amide moiety. These product ions are produced through single-step neutral losses of the C_nH_{2n} hydrocarbon series from the precursor ion at *m/z* 282.6, $[M+H–C_nH_{2n}]^+$. Examples include the *m/z* 226.6 product ion produced through neutral loss of C_4H_8, $[M+H–C_4H_8]^+$, the *m/z* 184.5 product ion produced through neutral loss of C_7H_{14}, $[M+H–C_7H_{14}]^+$, and the *m/z* 156.5 product ion produced through neutral loss of C_9H_{18}, $[M+H–C_9H_{18}]^+$. Equation 26.17 illustrates the production of the *m/z* 226.6 product ion through neutral loss of C_4H_8.

$$CH_3(CH_2)_7CH{=}CH(CH_2)_7C(O)NH_2 \xrightarrow[ESI]{H^+} CH_3CH_2CH_2CH_2CH_2(CH_2)_3CH{=}CH(CH_2)_7C(O)NH_3^+ \rightarrow CH_3CH_2CH{=}CH_2 + \underset{m/z\ 226.6}{CH_3(CH_2)_3CH{=}CH(CH_2)_7C(O)NH_3^+} \quad (26.17)$$

The method presented here is an internal standard (IS) method where a series of standards are analyzed, which contain a common IS that is also spiked into the samples. The response of the common IS is used to adjust the response of the current sample being measured. In this way, variations in sample injection amount and changes in the instrumental response can be adjusted for in a normalized way.

26.6 QUANTITATIVE ANALYSIS BY GC/EI–MS

The particular analysis presented here is the measurement of the fatty acid amides myristamide, palmitamide, oleamide, stearamide, and erucamide contained within a biological sample extract. The IS method presented here consists of dissolving or diluting the biological sample within a solvent solution containing an IS that is then quantitated against a series of fatty acid amide standards ratioed to the IS methyl oleate. The separation and quantitation of the fatty acid amides were performed on a gas chromatograph coupled to a single quadrupole mass spectrometer (GC–MS) with electron ionization (EI). Figure 26.7 illustrates the basic design and components of a GC/EI–MS system. Separations were obtained using a Restek Rtx-5MS Crossbond 5% diphenyl–95% PDMS 0.25 mm × 30 m, 0.25 μm df column. The GC oven temperature program consisted of an initial temperature of 50 °C for 2 min, ramp at 15 °C/min to 300 °C and hold for 10 min. The injection port temperature was 270 °C, and the transfer line was held at 280 °C. The injection volume was 2 μ l with a 1:5 split ratio. The electron ionization impact was at the standard 70 eV, and the quadrupole was scanned for a mass range of 50 to 470 amu.

Figure 26.8a is a total ion chromatogram (TIC) of the measurement of a multicomponent fatty acid amide standard consisting of 402 μM myristamide at 13.85 min, 298 μM palmitamide at 15.13 min, 290 μM oleamide at 16.19 min, 317 μM stearamide at 16.32 min, 290 μM erucamide at 18.31 min, and an IS spiked in all standards and samples at a constant concentration of 100 μM methyl oleate at 14.62 min. Calibration curves were constructed for a linear standard range spanning from approximately 400 to 90 μM fatty acid amides. The TICs for the serial dilutions used for

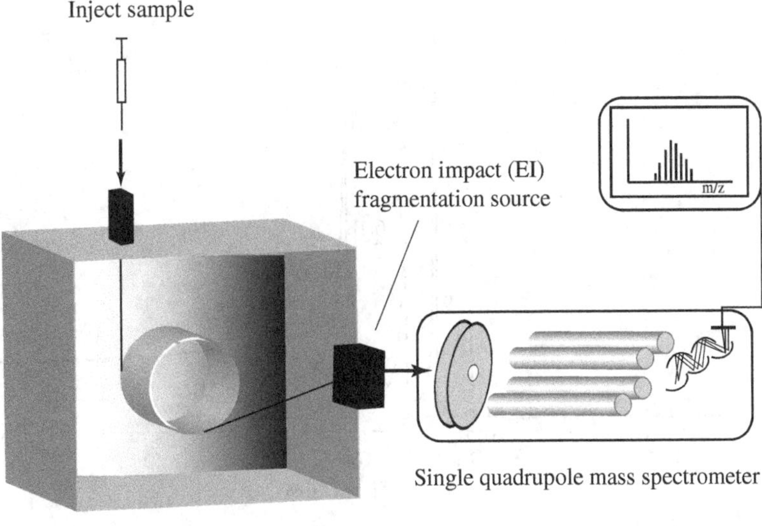

FIGURE 26.7 Basic design and components of a gas chromatograph coupled to a single quadrupole mass spectrometer (GC–MS) with electron ionization (EI).

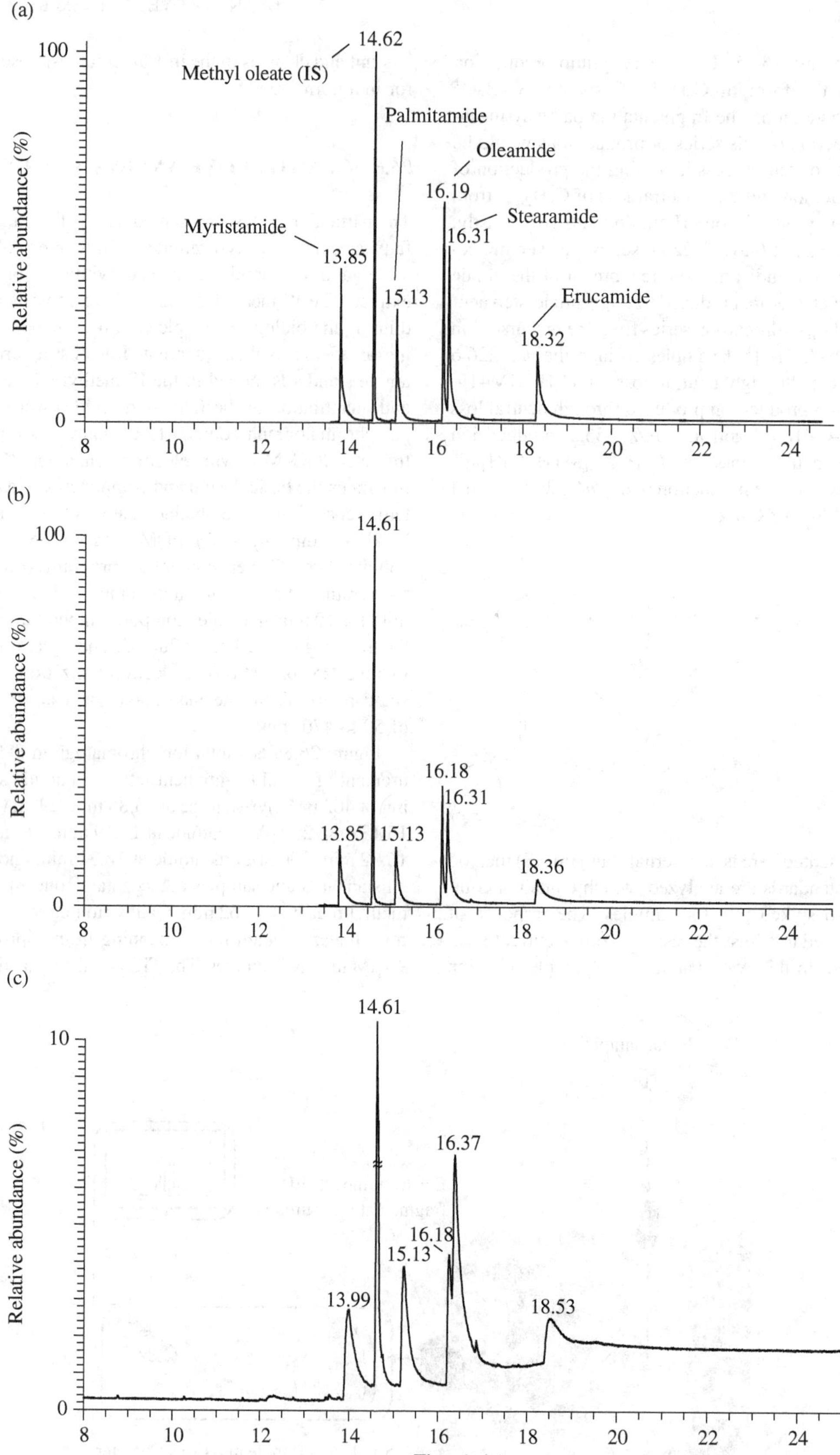

FIGURE 26.8 GC–MS TIC of (a) multicomponent fatty acid amide standard consisting of 4.02×10^{-4} M myristamide at 13.85 min, 2.98×10^{-4} M palmitamide at 15.13 min, 2.91×10^{-4} M oleamide at 16.19 min, 3.17×10^{-4} M stearamide at 16.32 min, 2.90×10^{-4} M erucamide at 18.31 min, and an internal standard spiked in all standards and samples at a constant concentration of 1×10^{-4} M methyl oleate at 14.62 min; (b) and (c) are serial dilutions of the multicomponent standard used for calibration curve construction.

construction of calibration curves are illustrated in Figures 26.8b and c. The TICs that are presented in Figures 26.9a and b illustrate the process of measuring a sample at different concentrations in order to target specific analytes within the sample. This is done to measure the species within the sample at responses that are within the linear response of the associated calibration curves. Figure 26.9a shows the TIC of a biological sample extract at a more diluted concentration, allowing the measurement of the most predominant fatty acid amide present at 16.20 min (oleamide). Notice the IS methyl oleate at 14.62 min, which was spiked into the sample prior to analysis. The TIC shown in Figure 26.9b represents the same biological extract sample as in Figure 26.9a but at a higher concentration. The biological extract sample illustrated in Figure 26.9b contains myristamide at 13.88 min, methyl oleate (IS) at 14.61 min, palmitamide at 15.12 min, oleamide at 16.22 min, stearamide at 16.32 min, and erucamide at 18.35 min. Table 26.1 lists the quantitative results for the fatty acid amides determined for five biological samples. As can be seen in the table, oleamide was consistently present with the largest quantity averaging at least four times greater quantity than the other fatty acid amide species present.

26.7 CHAPTER PROBLEMS

26.1 Methyl Oleate EI Mass Spectrum
Below is the odd electron product ion spectrum of methyl oleate ($C_{19}H_{36}O_2$) at *m/z* 296 as $M^{+\bullet}$ (Fig. 26.10) collected by electron impact ionization. Also included is the structure of methyl oleate along with its molecular formula and mass (Fig. 26.11). Try to use the example we just covered of *cis*-9-octadecenamide to

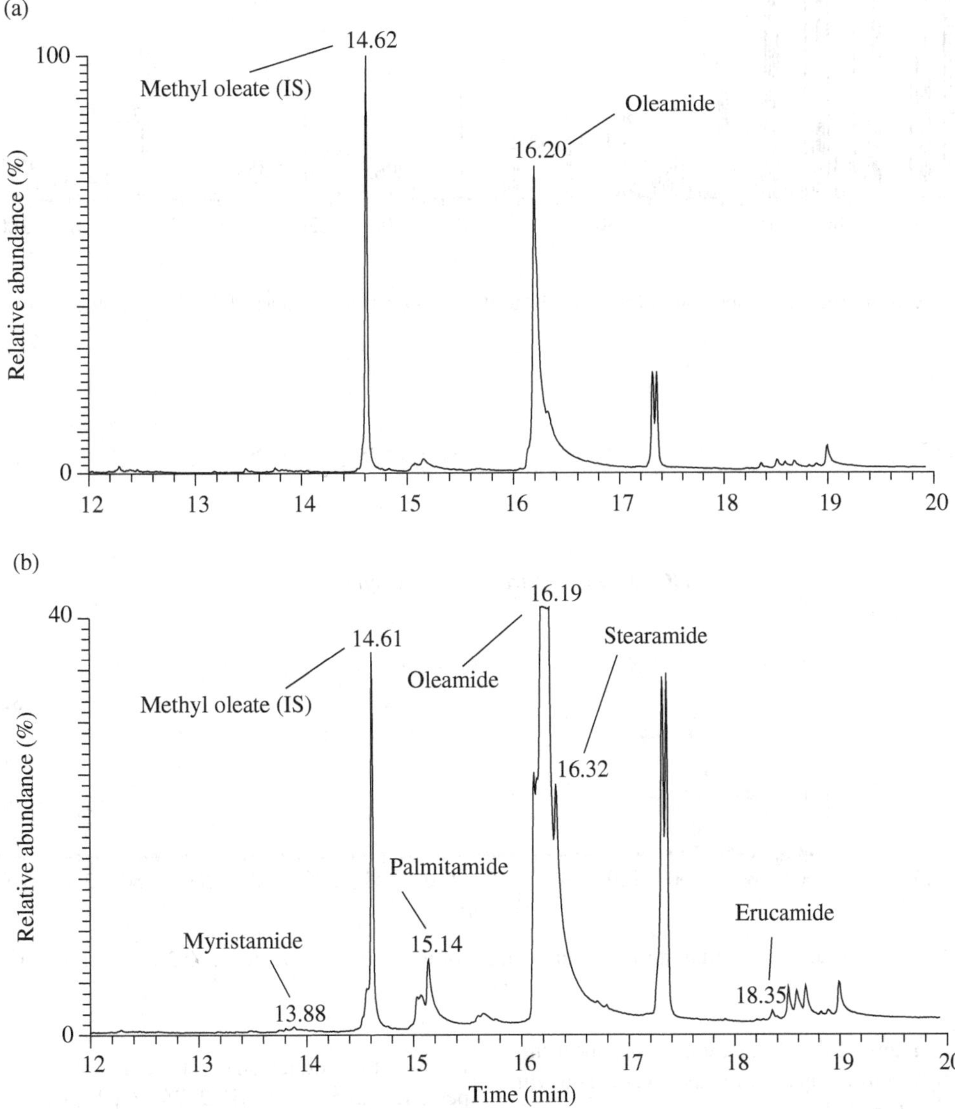

FIGURE 26.9 Total ion chromatograms of a biological sample extract at (a) more diluted concentration allowing the measurement of the most predominant fatty acid amide present at 16.20 min (oleamide) and (b) more concentrated sample containing myristamide at 13.88 min, methyl oleate (IS) at 14.61 min, palmitamide at 15.12 min, oleamide at 16.22 min, stearamide at 16.32 min, and erucamide at 18.35 min.

TABLE 26.1 Fatty Acid Amide Concentration in Biological Sample by GC–MS.

	Myristamide (μM)	Palmitamide (μM)	Oleamide (μM)	Stearamide (μM)	Erucamide (μM)
Sample 1	1.80	0.74	3.46	1.79	0.29
Sample 2	0.55	0.92	2.24	0.55	0.29
Sample 3	1.44	0.94	7.18	1.43	0.30
Sample 4	1.51	1.56	5.79	1.51	0.54
Sample 5	1.00	0.92	4.86	1.00	0.27
Average	1.26 ± 0.49	1.02 ± 0.31	4.71 ± 1.93	1.26 ± 0.49	0.34 ± 0.11

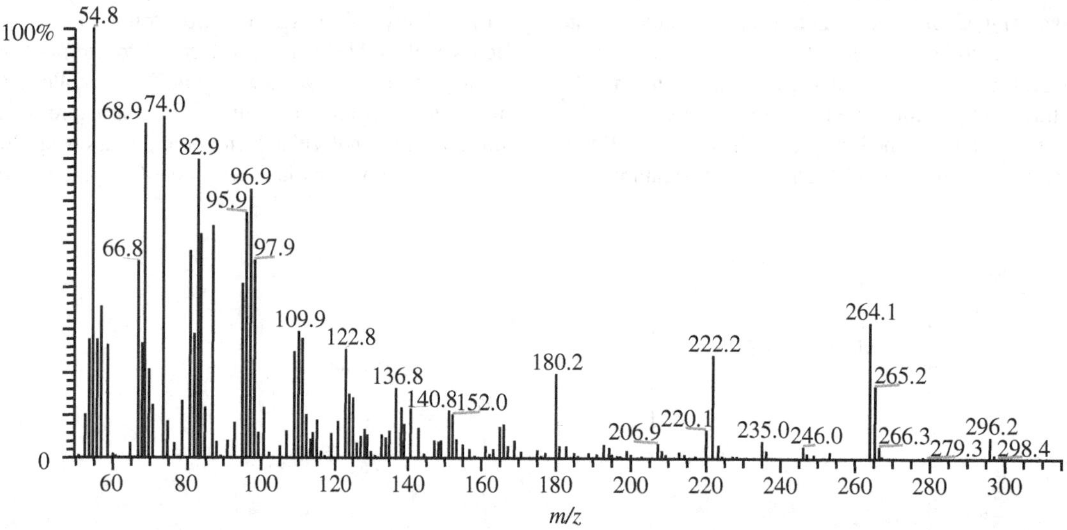

FIGURE 26.10 Odd electron product ion spectrum of methyl oleate ($C_{19}H_{36}O_2$) at m/z 296 as $M^{+\bullet}$ collected by electron impact ionization.

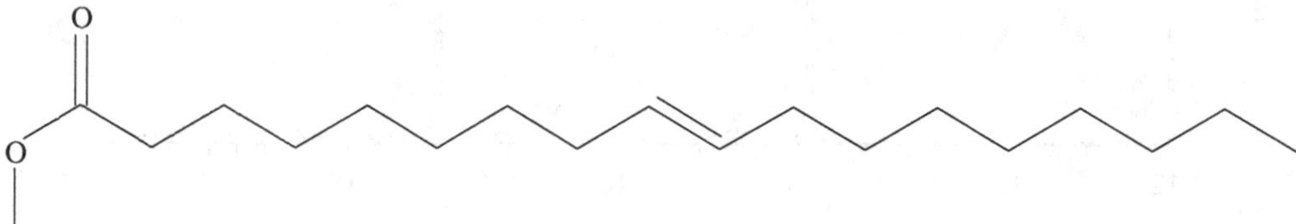

FIGURE 26.11 Structure of methyl oleate.

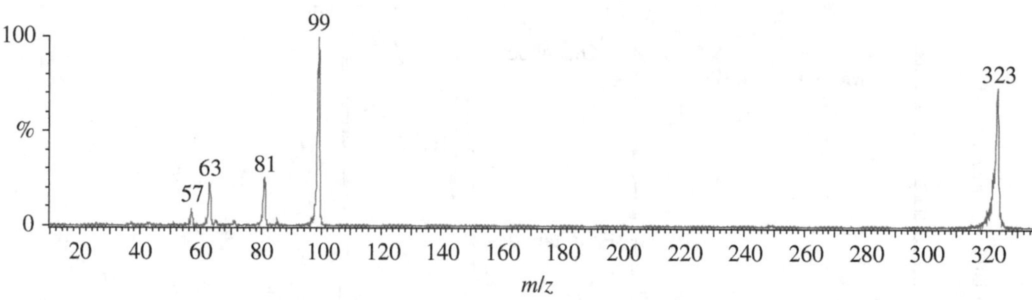

FIGURE 26.12 CID product ion spectrum for the lithium adduct of monopentadecanoin at m/z 323.2774, $[C_{15}H_{36}O_4 + Li]^+$.

assign structures to as many of the product ions in the spectrum that were generated by electron impact of methyl oleate. You will need to use Equations 26.3–26.8.

26.2 Problem: Lithiated Monopentadecanoin Product Ion Spectrum

Below is the product ion spectrum for the lithium adduct of monopentadecanoin at m/z 323.2774, $[C_{15}H_{36}O_4 + Li]^+$ (Fig. 26.12). Also included are the structure of monopentadecanoin and an exploded structure giving the major elements of the structure of monopentadecanoin and their masses (Fig. 26.13). The four

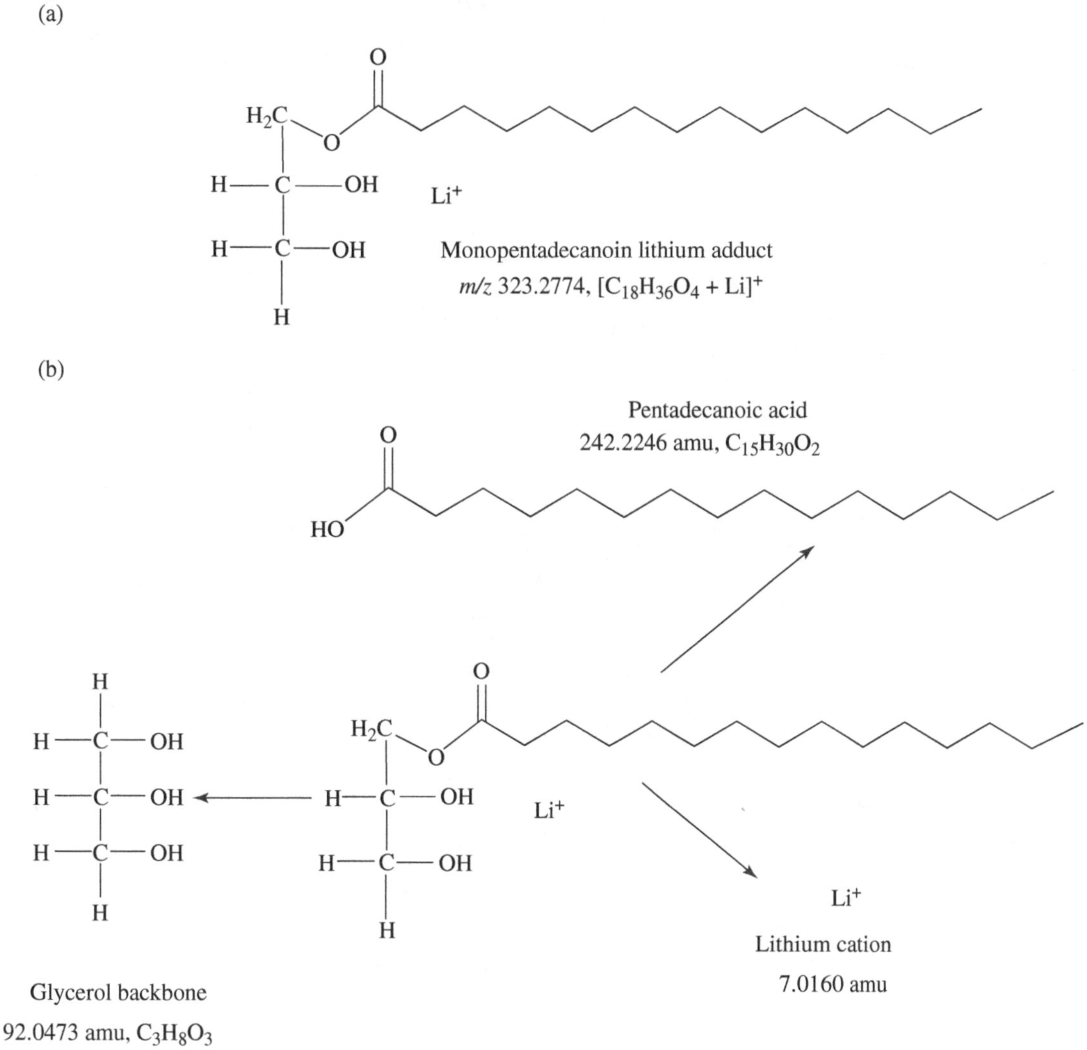

FIGURE 26.13 (a) Structure of monopentadecanoin lithium adduct *m/z* 323.2774, $[C_{18}H_{36}O_4 + Li]^+$. (b) Exploded structure of monopentadecanoin.

product ions are *m/z* 57, *m/z* 63, *m/z* 81, and *m/z* 99. Attempt to solve the product ion spectrum by following the steps below:

1. Subtract the masses of the product ions from the precursor ion and compare the masses to the exploded structure masses.
2. Try to draw reasonable fragmentation pathway mechanisms for the four product ions.

REFERENCES

1. McLafferty, F.W.; Turecek, F. *Interpretation of Mass Spectra.* 4th ed. Sausalito, CA: University Science Books; 1993.
2. Di Marzo V. *Biochim. Biophys. Acta* 1998, **1392**, 153–175.
3. Bachur, N.R.; Masek, K.; Melmon, K.L.; Udenfriend, S. *J. Biol. Chem.* 1965, **240**, 1019–1024.
4. Epps, D.E.; Natarajan, V.; Schmid, P.C.; Schmid, H.H. *Biochem. Biophys. Acta* 1980, **618**, 420–430.
5. Epps, D.E.; Palmer, J.W.; Schmid, H.H.; Pfieffer, D.R. *J. Biol. Chem.* 1982, **257**, 1383–1391.
6. Epps, D.E.; Schmid, P.C.; Natarajan, V.; Schmid, H.H. *Biochem. Biophys. Res. Commun.* 1979, **90**, 628–633.
7. Devane, W.A.; Hanus, L.; Breuer, A.; Pertwee, R.G.; Stevenson, L.A. *Science* 1992, **258**, 1946–1949.
8. Arafat, E.S.; Trimble, J.W.; Anderson, R.N.; Dass, C.; Desiderio, D.M. *Life Sci.* 1989, **45**, 1679–1687.
9. Cravatt, B.F.; Prospero-Garcia, O.; Siuzdak, G.; Gilula, N.B.; Henriksen, S.J.; Boger, D.L.; Lerner, R.A. *Science* 1995, **268**, 1506–1509.
10. Basile, A.S.; Hanus, L.; Mendelson, W.B. *NeuroReport* 1999, **10**, 947–951.
11. Stewart, J.M.; Boudreau, N.M.; Blakely, J.A.; Storey, K.B. *J. Therm. Biol.* 2002, **27**, 309–315.
12. Gross, M.L. *Int. J. Mass Spectrom.* 2000, **200**, 611–624.

27

SPECIAL TOPICS: STRONG CATION EXCHANGE CHROMATOGRAPHY AND CAPILLARY ELECTROPHORESIS

27.1 Introduction
 27.1.1 Overview and Comparison of HPLC and CZE
27.2 Strong Ion Exchange HPLC
27.3 CZE
 27.3.1 Electroosmotic Flow (EOF)
 27.3.2 Applications of CZE
27.4 Binding Constants by Cation Exchange and CZE
 27.4.1 Ranking of Binding Constants
 27.4.2 Experimental Setup
 27.4.3 UV/Vis Study of the Compounds and Complexes
 27.4.4 Fluorescence Study of the Dye/A007 Complexes
 27.4.5 Computer Modeling of the Complex
 27.4.6 Cation Exchange Liquid Chromatography Results
 27.4.7 Capillary Electrophoresis (CE)
27.5 Comparison of Methods
27.6 Conclusions
References

27.1 INTRODUCTION

27.1.1 Overview and Comparison of HPLC and CZE

High-performance liquid chromatography (HPLC) and capillary zone electrophoresis (CZE) are two analytical separation techniques that have been used as individual approaches for compound separations and as complimentary approaches. HPLC is a separation technique that utilizes a compound's affinities for either a stationary phase or a mobile phase, in order to perform a chromatographic separation. In reversed-phase HPLC (RP-HPLC), typically, the stationary phase is composed of a long-chain nonaromatic, aliphatic hydrocarbon (e.g., C18, for an 18-carbon chain) that is covalently bound to silica or polymer beads packed into a cylinder stainless steel tube. The mobile phase typically consists of mixtures of H_2O, methanol (MeOH), and acetonitrile (ACN) that are all three completely miscible in all molar fraction ratios. The sample is loaded onto the C18 column with a high H_2O ratio in the mobile phase where the affinity of organic compounds is greatest for the C18 stationary phase. The mobile phase is changed in a gradient fashion where the organic solvent (MeOH or ACN) is gradually increased. The affinity for the organic molecules now increases in the mobile phase, which helps to carry them along through the column. The separation of the individual organic constituents now takes place due to their different affinities for the stationary and mobile phases.

27.2 STRONG ION EXCHANGE HPLC

Strong ion exchange HPLC works upon similar principles as RP-HPLC, except that the compound of interest now has a strong ionic, electrostatic, noncovalent affinity for the stationary phase. This approach works well for compounds that possess a permanent charge such as an acid in basic solution or a base in acidic solution. Strong cation exchange high-performance liquid chromatography (SCEX-HPLC) uses an organically bound sulfonic acid group that has a high affinity for compounds that contain permanent positive charges. Due to the aromatic backbone of strong cation exchange stationary phases, there is also a slight reversed-phase interaction that can also take place. For the chromatographic elution of the compound of interest from the strong cation exchange column, typically, a salt gradient is used that displaces the analyte from the polysulfonic acid stationary phase.

27.3 CZE

CZE is another very useful separation technique that has traditionally been applied primarily to analytes of interest that also possess a permanent charge, just as is the case with strong ion exchange chromatography. Unlike liquid chromatography, CZE does not separate analytes according to their respective

Analytical Chemistry: A Chemist and Laboratory Technician's Toolkit, First Edition. Bryan M. Ham and Aihui MaHam.

affinities for mobile and stationary phases, rather the separation is achieved based off of analytes electrophoretic mobilities in an electric field. In positive mode electrophoresis, positively charged analytes move from the anodic (positive electrode) region of the separation phase to a cathodic (negative electrode) region of the separation phase, through a narrow bore, short fused silica tube. In general, small highly charged analytes have the greatest mobilities, followed by higher molecular weight similarly charged analytes; then by neutrals, which travel through the separation capillary due to a phenomenon known as electroosmotic flow (EOF); and lastly by oppositely charged ions. The forces acting upon the analytes are vectors; therefore, they are additive and result in the separation of analytes based upon their mass-to-charge ratios.

27.3.1 Electroosmotic Flow (EOF)

EOF results from bulk movement of the separation phase through the fused silica capillary due to the charged silanol groups of the fused silica wall. EOF can be influenced through pH changes (high pH produces deprotonated, charged silanol groups resulting in higher EOF, while low pH results in protonated silanol groups thus decreasing EOF), organic modifiers in the separation phase (a higher organic modifier such as MeOH reduces EOF), and also through neutral coating of the fused silica wall such as with polymers, which can reduce or eliminate EOF.

27.3.2 Applications of CZE

Since its early introduction [1–4], CZE is an analytical technique that has become an indispensable tool in many modern laboratories, especially in the pharmaceutical industry [5]. The application of capillary electrophoresis (CE) to the study of complexes, known as affinity capillary electrophoresis (ACE), has recently experienced a large growth. This methodology has been used to study the electrophoretic mobility and separation of various complexes [6], complex electrostatic interactions [7], complex binding and disassociation constants [8, 9], and complex stoichiometry [10]. In most studies, affinity CE ideally involves the measurement of a complex that stays bound throughout the electrophoretic separation [11]. The complexes that were evaluated by this study involve a purely electrostatic binding between the dye carrier compound and the anticancer agent A007, which have estimated high disassociation constants.

27.4 BINDING CONSTANTS BY CATION EXCHANGE AND CZE

Determination of relative binding constants between π-delocalized lymphangitic dyes and an anticancer agent by cation exchange liquid chromatography and affinity CZE will be looked at next.

27.4.1 Ranking of Binding Constants

One of the major problems with anticancer therapy is the effective delivery of the anticancer agent to the tumor site. Skin permeability studies have demonstrated that the coupling of the anticancer agent A007 with a carrier dye molecule (methylene green, MEG; methylene blue, MEB; and toluidine blue, TB; see Figure 27.1 for structures) increases A007's therapeutic activity [12], with a rating from most effective to least, as MEG>MEB>TB. Electrospray ionization tandem mass spectrometry [13–16] (ESI–MS/MS) has been used to study the characteristics and relative binding strengths of the three dyes MEG, MEB, and TB, complexed with A007. The gas-phase binding strengths of the three complexes, determined by ESI–MS/MS, were ranked in the following order: MEG>MEB>TB, which indicates that a stronger binding between the carrier and anticancer agent directly correlates to a greater efficacy. An HPLC method has been developed for the determination of A007 in plasma [17], but not for the complexes that are formed between A007 and carrier compounds. A cation exchange chromatography (CEC) method has been developed in order to rank the binding strengths of carrier/A007 complexes. The CEC method determined binding strengths with the same trend as was determined by ESI–MS/MS, namely, MEG>MEB>TB. This effectively gave a "solution-phase" ranking of binding strengths.

27.4.2 Experimental Setup

The UV/VIS experiments were performed with a PC-controlled Cary 500 Scan UV/Vis/NIR Spectrophotometer (Varian Australia Pty. Ltd.). The fluorescence experiments were performed with a PC-controlled Photon Technology International Fluorometer (PTI, Canada). The CE experiments were performed with a BioFocus 2000 Capillary Electrophoresis System (Bio-Rad, Hercules, California) with a UV/VIS variable wavelength detector. Uncoated fused silica capillaries were used with a total length of 56 cm, effective length of 50 cm, and 50 um i.d. The capillaries were maintained at a temperature of 25 °C. The injection time was 30 psi*sec pressure. All runs were made in a constant voltage mode with a running voltage of 30 kV. The polarity was + to – (positive polarity mode where the negative pole is at the capillary outlet). Between runs, the CE capillary was conditioned through a series of washes comprised of 13 min. 1 M NaOH, followed by 13 min. of separation phase. The separation phase consisted of (i) 50% MeOH/50% H_2O buffered with 25 mM mono- and dibasic hydrogen phosphate at a pH of 7.4 (adjusted) or (ii) 50% MeOH/50% H_2O buffered with 25 mM phosphate at a pH of 7.4 pH (adjusted) and spiked with either MEG, MEB, or TB.

The ACE method developed ranked the binding strengths of the carrier/A007 complexes, in the solution phase at an adjusted physiological pH (7.4), as MEG>MEB>TB. These results are in agreement with the ESI–MS/MS and CEC rankings that were previously determined.

The HPLC experiments were performed with a 1100 Series HPLC (Agilent Technologies, Palo Alto, CA) comprised of a vacuum degasser, a quaternary pump, and a diode array detector. The column used was a 150 × 1.0 mm, 5 μm, 300-A, PolySULFOETHYL A column (PolyLC, Columbia, MD). Instrumental parameters consisted of a 0.1 ml per min flow, 10 μl injection loop, and ambient column temperature. Mobile phases consisted of (i) 50% MeOH/50% H_2O, buffered at 7.4 pH (adjusted) with

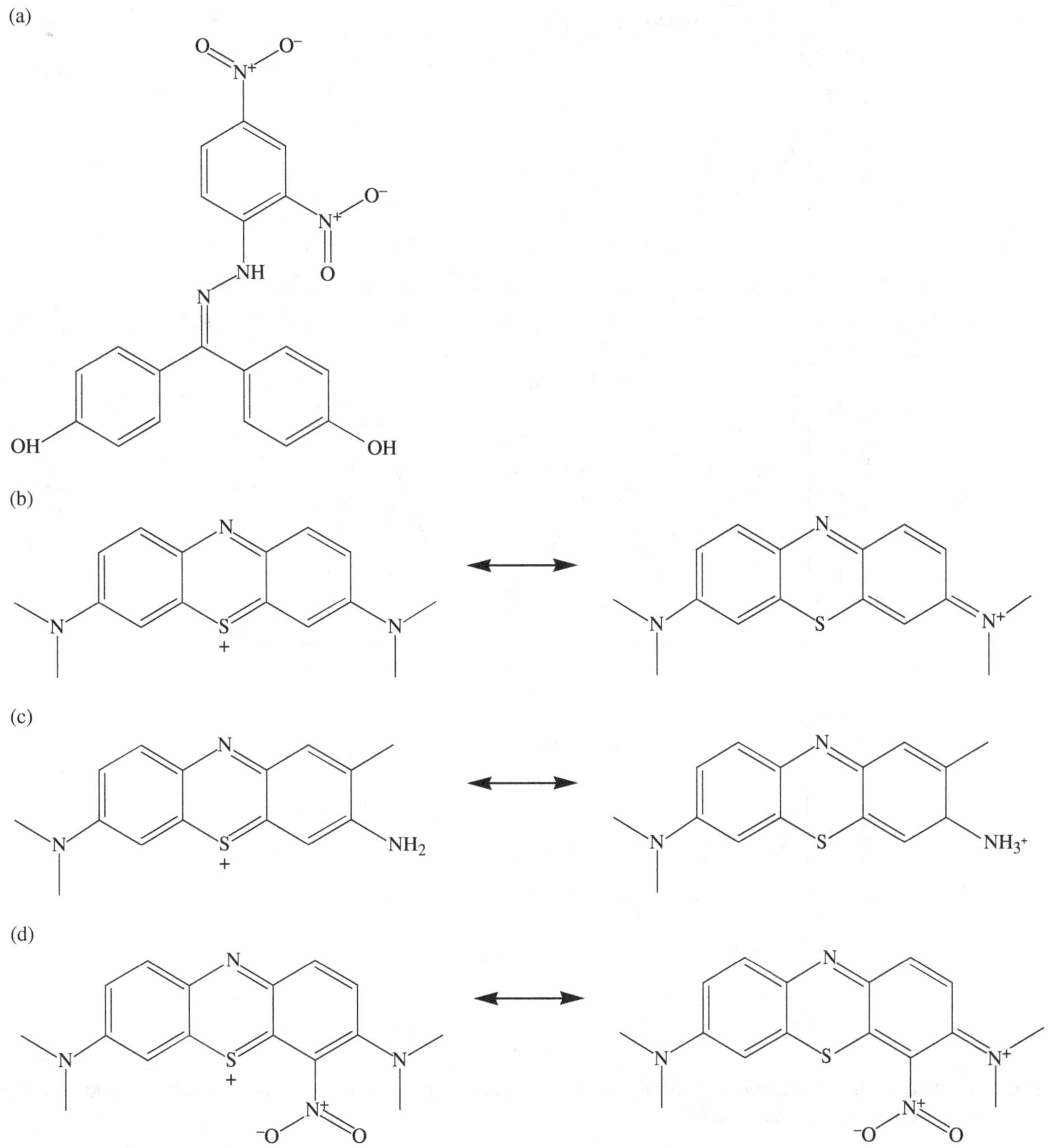

(a)

(b)

(c)

(d)

FIGURE 27.1 Structures of (a) 4,4′-dihydroxybenzophenone-2,4-dinitrophenylhydrazone (A007), (b) methylene blue (MEB) double salt, (c) toluidine blue (TB) double salt, and (d) methylene green (MEG) double salt.

15 mM triethylamine phosphate (TEAP); (ii) 50% MeOH/50% H_2O, buffered at 7.4 pH (adjusted) with 15 mM TEAP and spiked with either 5×10^{-5} M MEG, 5×10^{-5} M MEB, or 5×10^{-5} M TB; or (iii) 50% MeOH/50% H_2O, buffered at 7.4 pH (adjusted) with 15 mM TEAP and 0.5 M NaCl.

To mimic physiological conditions, all solutions are at a pH of 7.4 but adjusted due to the change in H^+ activity for a 50% methanol and 50% water mixture [18, 19].

27.4.3 UV/Vis Study of the Compounds and Complexes

Figure 27.2 illustrates UV/Vis spectra of the complexes formed between the three dyes and A007. What's observed from the spectra is that each compound contains a unique absorbance peak that can be used to detect and monitor the compounds and complexes without interference. The three dyes used in the study, MEG, MEB, and TB, have unique absorbance maximums at 656, 662,

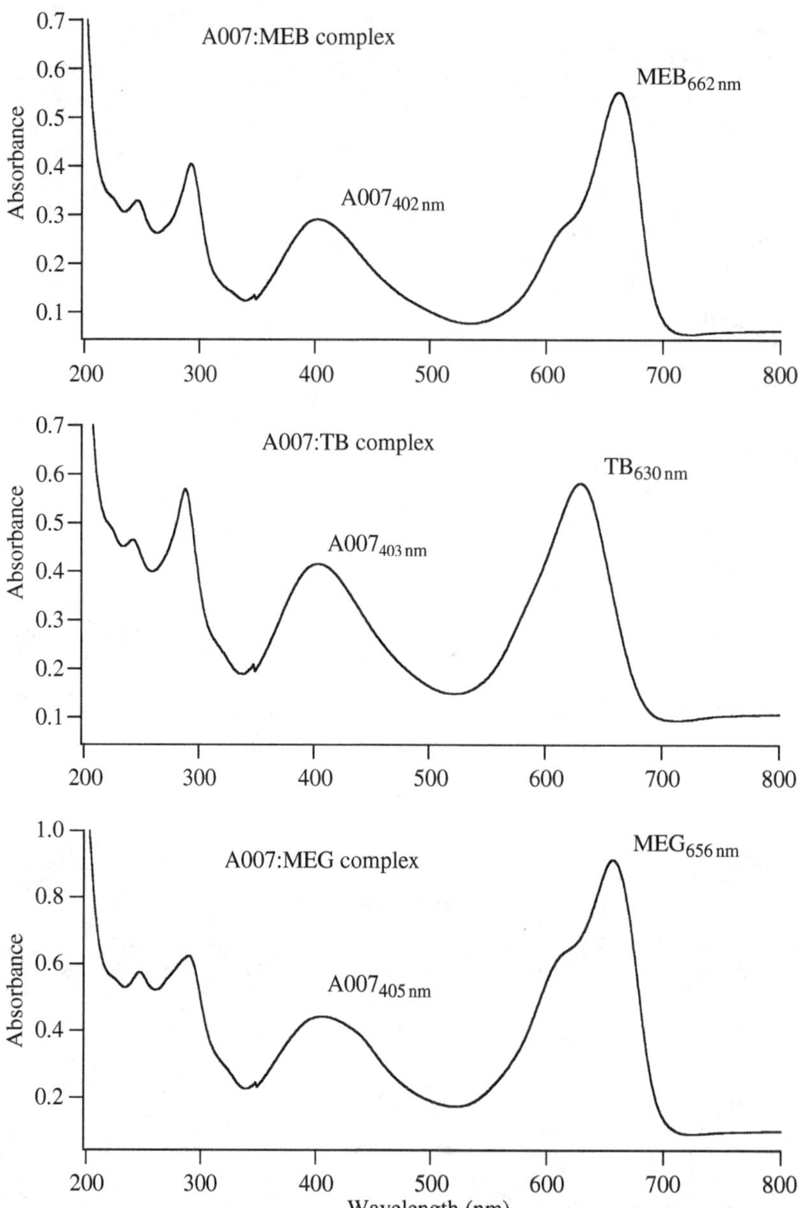

FIGURE 27.2 UV/Vis absorbance spectrums of A007/dye complexes illustrating absorbance maximums of (a) 405 nm for A007, 662 nm for MEB, (b) 630 nm for TB, and (c) 656 nm for MEG.

and 630 nm, respectively. The anticancer agent A007 has an absorbance maximum between 398 and 405 nm. Both dye and A007 peaks are found in the complexes, and the A007 max at 398 nm was used to detect the complex for mobility calculations and migration times (ACE) and retention times (CELC).

27.4.4 Fluorescence Study of the Dye/A007 Complexes

The dyes used in the study naturally fluoresce at specific wavelengths. It was initially felt that the complex between the dyes and the anticancer agent A007 enhanced this fluorescence. Using the method of continuous variation [20, 21], a fluorescence maximum was obtained for each dye/A007 complex versus mole fraction. Figure 27.3 illustrates the stoichiometries determined for the three complexes. These dyes however are known to produce dimers in the liquid phase [22], where the production of the dimer will quench the fluorescence. Further spectroscopic experiments showed that only one species of complex was being formed in solution between A007 and the dyes, but the enhancement in fluorescence was attributable to the decrease in dye dimer only and not to the formation of the A007/dye complex.

27.4.5 Computer Modeling of the Complex

Computer modeling studies were performed for the gas-phase electrostatic interaction between A007 and MEB. The lowest energy conformation is illustrated in Figure 27.4. Studies like this

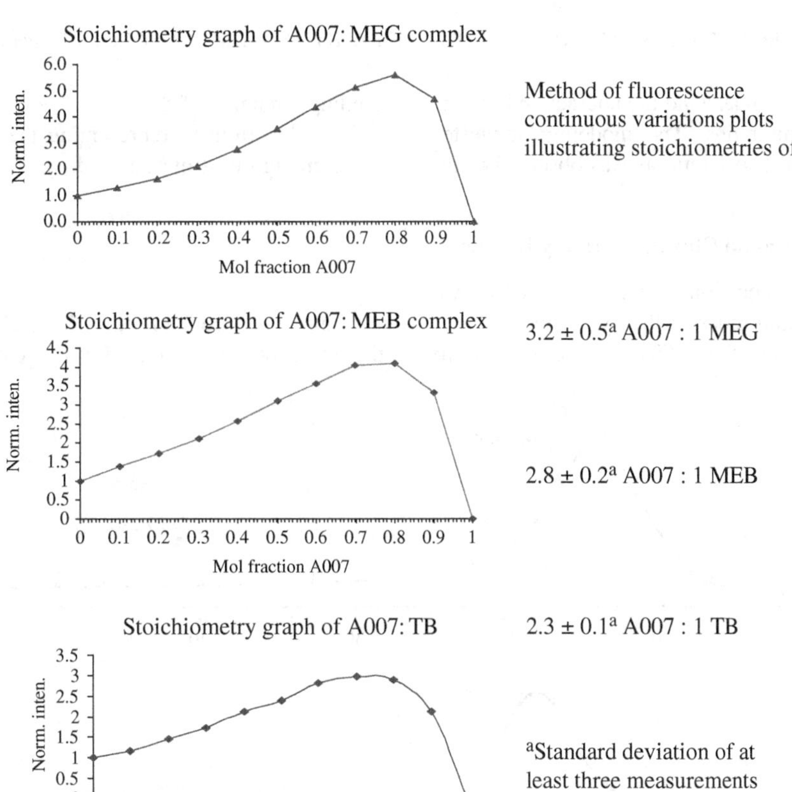

FIGURE 27.3 Stoichiometries determined for the three complexes.

A-007 MEB

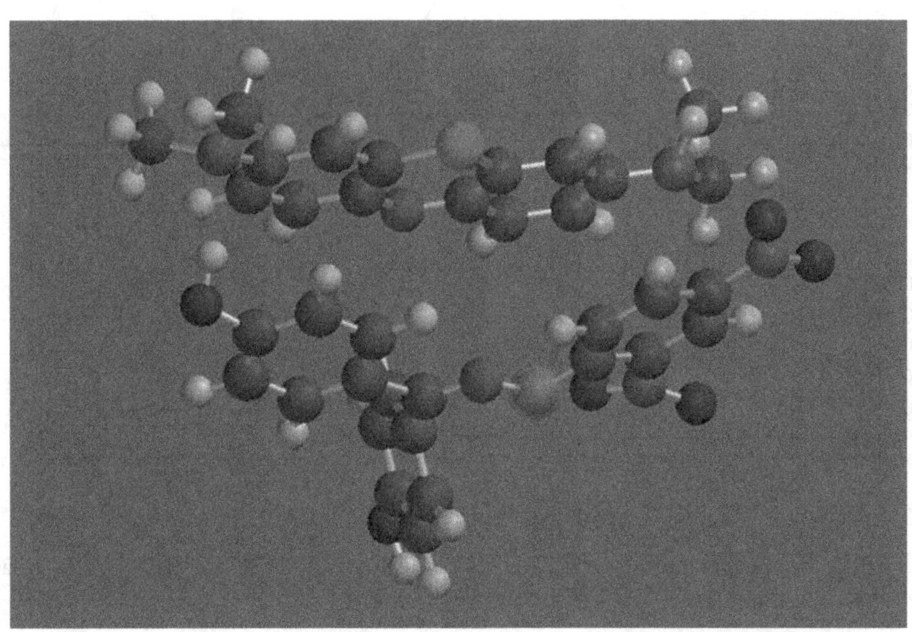

FIGURE 27.4 Computer modeling results of the gas-phase interaction of A007 and MEB.

are often performed to better understand the interaction between substances such as binding ratios. The modeling suggested that the interaction was in a 1 to 1 ratio as was observed in the fluorescence studies.

27.4.6 Cation Exchange Liquid Chromatography Results

A binding coefficient (BC) equation (see Equation 27.1) was derived based on the retention times of the dyes, the A007/dye complexes, and the anticancer agent A007, in order to rank the binding strengths of the A007/dye complexes. The scale spans a BC value from 0, representing the weakest binding, to 1.0, representing the strongest binding:

$$BC = \frac{\left(t_{\text{drug(carrier)}} - t_{\text{drug}}\right)}{\left(t_{\text{carrier}} - t_{\text{drug}}\right)} \tag{27.1}$$

Figure 27.5 contains the CELC chromatograms illustrating the retention times used for the calculation of the binding

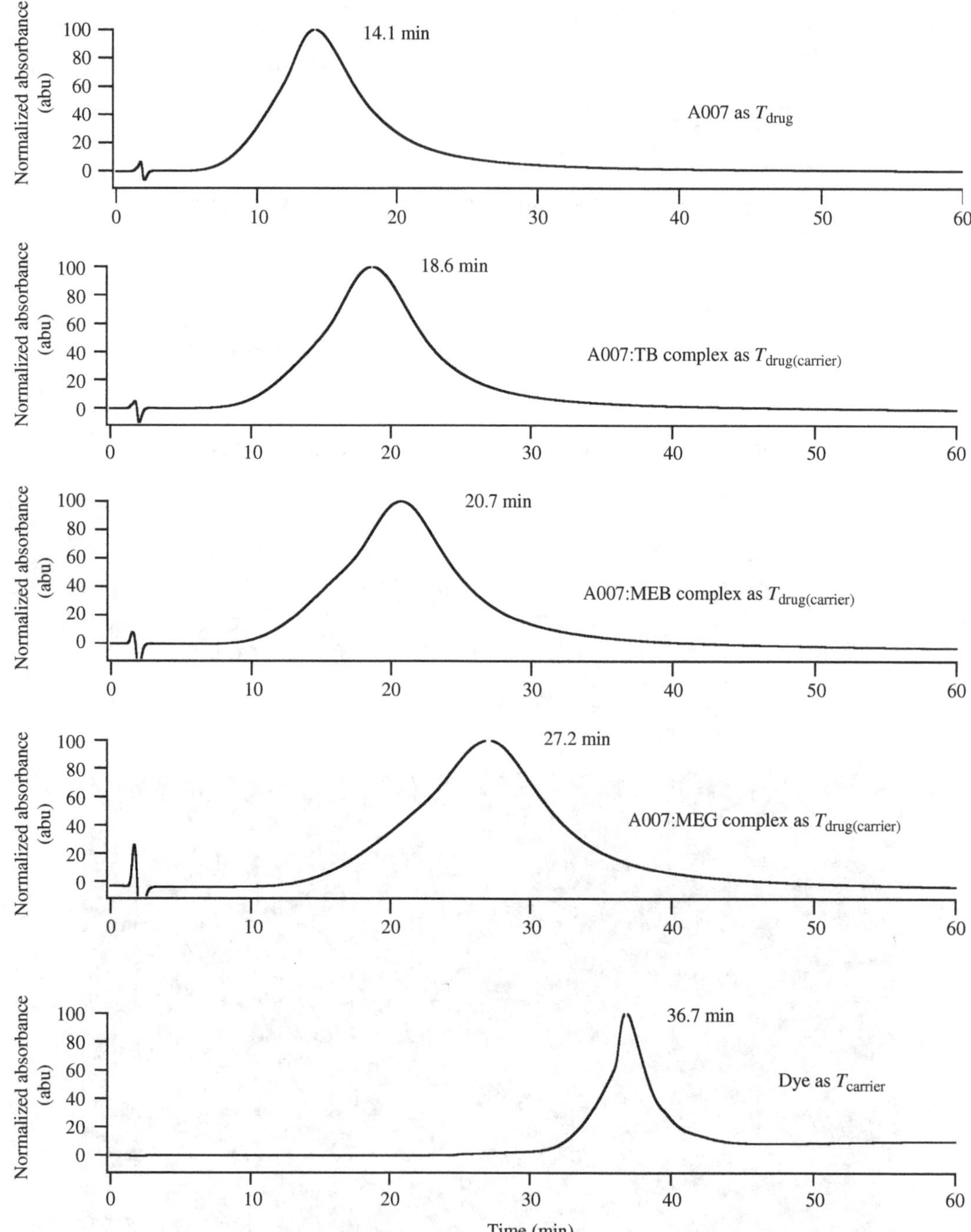

FIGURE 27.5 Typical CELC chromatograms of A007 (T_{drug}) at 5×10^{-5} M A007, A007/TB complex ($T_{drug(carrier)}$) at 5×10^{-5} M TB and A007, A007/MEB complex ($T_{drug(carrier)}$) at 5×10^{-5} M MEB and A007, A007/MEG complex ($T_{drug(carrier)}$) at 5×10^{-5} M MEG and A007, and dye ($T_{carrier}$).

constants. It is felt that this approach gives a direct measurement of the binding strength between the dyes and A007. The binding method is achieved through a kind of pseudostationary phase that is produced in the cation exchange column between the dye in the mobile phase and the sulfoethyl A stationary phase. With a stronger binding between the dye and A007, A007 will have a longer elution time, as is observed in the chromatograms contained in Figure 27.5. The binding strength scale of the three dyes was determined to be MEG > MEB > TB, with calculated BCs of 0.54 ± 0.05, 0.24 ± 0.02, and 0.18 ± 0.02, respectively (BC values are the average plus or minus one standard deviation of at least three determinations).

27.4.6.1 Description of HPLC Pseudophase The condensed-phase binding strength scale of the three dyes was determined to be MEG > MEB > TB using the HPLC method. This was in agreement with the gas-phase studies that were performed using mass spectrometry. In the HPLC experiments, the mobile phase contained the dye at concentrations of 5×10^{-5} M under testing (TB, MEB, and MEG), and A007 was injected onto the column. This will saturate the column with the dye, thus creating a pseudostationary phase in the column consisting of immobilized dye bound to the stationary phase, which is comprised of sulfoethyl groups. Figure 27.6 illustrates the pseudophase that has been set up with the strong cation exchange column stationary phase. As the A007 compound passes through the column, it will interact with the immobilized dye as the pseudostationary phase and the dye in the mobile phase. The A007 compound will thus elute faster with a lower binding strength as compared to a stronger binding strength due to a weaker interaction with the pseudostationary phase. This is what is observed in Figure 27.5 where TB has the weakest binding and elutes the fastest and MEG has the strongest binding and elutes the slowest.

27.4.7 Capillary Electrophoresis (CE)

27.4.7.1 Introduction CE is an analytical chemistry technique used to separate ionic species according to their charge and frictional forces and hydrodynamic radius. CE is also known as CZE. In CE, analytes that possess a charge move in a buffer or conductive liquid phase in the influence of an electric field.

27.4.7.2 CE Instrumentation There are numerous CE manufacturers that offer a wide variety of CE instrumentation and versions; however, the basic components are similar. Figure 27.7 illustrates the basic components that make up CE instrumentation. The separation medium is a conductive buffer system with an anode (positive polarity) at one end and a cathode (negative polarity) at the other. The two buffer reservoirs are connected by a buffer-filled capillary. A high voltage is placed across the buffer system. Within the electric field, charged species will all flow toward the cathode due to an action called EOF. This is a net flow of the buffer solution through the capillary that carries all compounds with it. Besides the movement of EOF, the compounds will be separated according to their electrophoretic mobility through the electric field. The charged species will migrate toward the pole of opposite charge and separate according to their migration but will be carried to the cathode by the EOF. In this simple system, a light path is set up to pass through the capillary to measure migrating species producing what is called an electropherogram. These systems are usually computer controlled with vendor software. The sample is introduced at the anode end often by a small positive pressure.

27.4.7.3 Theory of CE Separation Compounds are separated in CE according to their electrophoretic migration in an electric field. This is described by the following equation:

$$\mu = \frac{v}{E} = \frac{Q}{6\pi r\eta} \tag{27.2}$$

where Q is the effective charge of the ion, r is the total radius of the ion, and η is the viscosity of the separation phase.

Compounds will migrate and pass through the detector in the order of positive first, followed by neutrals, and then lastly negatively charged compounds.

27.4.7.4 Results of CE Binding Analysis of Dyes and A007

27.4.7.4.1 Mobility Change Titration Study Figure 27.8 illustrates the results of the change in mobilities of A007 versus increasing concentrations of the dye ligands titrated into the separation phase. There are initial similarities in the effective mobilities of A007 observed between the three systems at a dye ligand concentration of 5×10^{-5} M (MEG dye A007 mobility of 14.9 $cm^2\,kV^{-1}\,min^{-1}$, MEB dye A007 mobility of 14.3 $cm^2\,kV^{-1}\,min^{-1}$, and TB dye A007 mobility of 14.9 $cm^2\,kV^{-1}\,min^{-1}$). Following this, there is a subsequent decreasing mobility trend as the dye concentration is increased to 8×10^{-4} M. This suggests that the change in mobility could be attributed to a decrease in the EOF with increasing background electrolyte [23–25]. This type of effect is also observed with increasing percentages of organic solvent in the separation phase [26–30]. Salomon et al. [23] derived an expression that describes the change in electroosmotic mobility as a function of buffer (electrolyte) concentration:

$$\mu_{eo} = \left(\frac{Q_0}{n(1 + K_{wall}[M^+])}\right)\left(d_0 + \frac{1}{K'([M^+])^{1/2}}\right) \tag{27.3}$$

The μ_{eo} is inversely proportional to both the ionic strength of the solution $[M^+]$ and the association of the cations to the silica wall (K_{wall}). An increase in either of these leads to a decrease in the mobility. The studies reported by Salomon measured the change in electroosmotic mobilities for buffer concentrations ranging from approximately 2 to 40 mM. The mobility change experiments performed on the A007/dye complexes reported here ranged from 25.05 to 25.8 mM ionic. This is a much lower change in ionic strength of the separation phases suggesting perhaps a negligible contribution. The anticancer agent A007 at a pH of 7.4 is neutral; therefore, its mobility in buffer alone would represent the EOF of the system before titrating the separation phase with the dye ligand. A007's mobility in buffer alone was measured at 13.5 $cm^2\,kV^{-1}\,min^{-1}$. This result is very similar to the mobilities measured for A007 when 5×10^{-5} M dye (first data

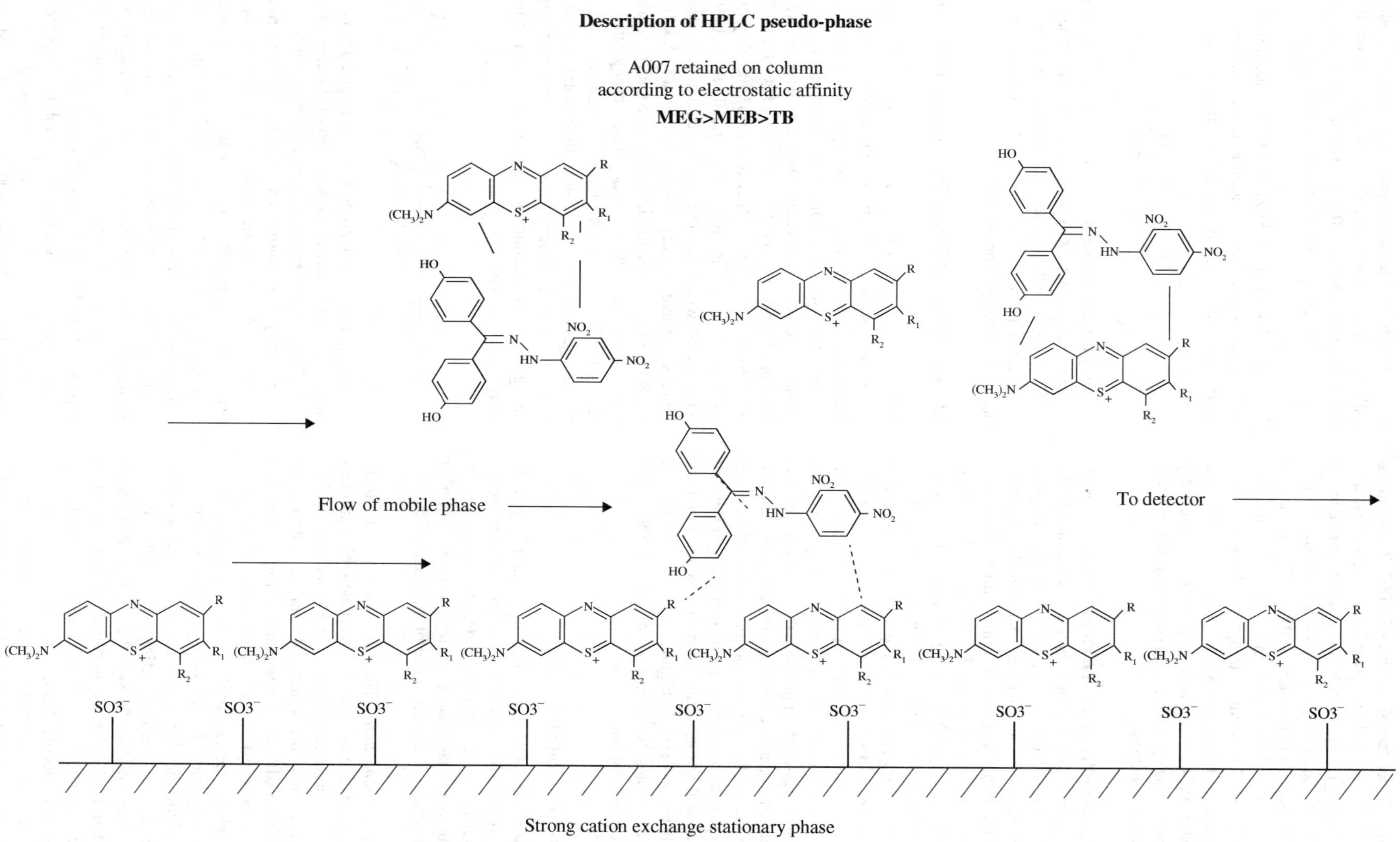

Description of HPLC pseudo-phase

A007 retained on column
according to electrostatic affinity
MEG>MEB>TB

Flow of mobile phase

To detector

Strong cation exchange stationary phase

FIGURE 27.6 Description of the pseudostationary phase that is set up from the interaction of the dye in the mobile phase and the strong cation exchange sulfoethyl stationary phase.

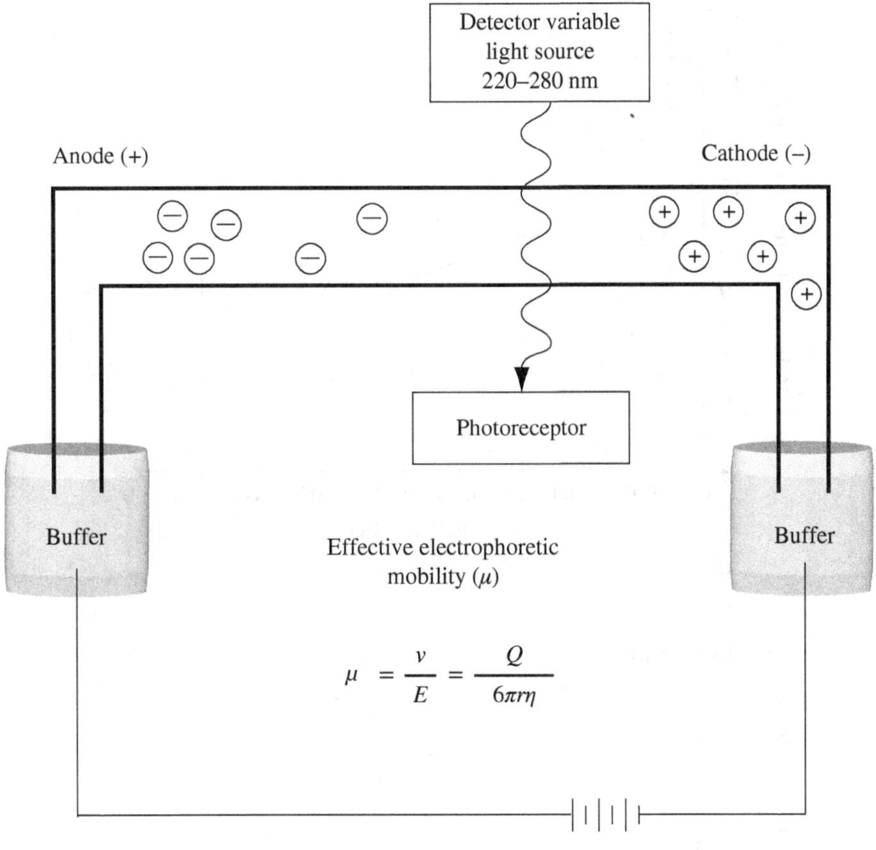

FIGURE 27.7 Basic design of capillary electrophoresis instrumentation.

point, Figure 27.8) had been titrated into the separation phases (MEG dye A007 mobility of 14.9 cm^2 kV^{-1} min^{-1}, MEB dye A007 mobility of 14.3 cm^2 kV^{-1} min^{-1}, and TB dye A007 mobility of 14.9 cm^2 kV^{-1} min^{-1}). Notice the slight increase in mobility of A007 in separation phases containing 5×10^{-5} M dye versus the mobility of A007 in buffer alone. The three dyes MEG, MEB, and TB are very similar in structure (see Figure 27.1), and all possess a permanent positive charge. Therefore, the change in EOF due to increasing concentrations of the dye ligands in the separation phase would be expected to be of similar magnitude for the three systems, as was observed with the mobility of A007 with the titration of 5×10^{-5} M dye into the separation phases. As higher dye concentrations are titrated into the separation phase, the mobilities of A007 are observed to decrease. This is attributed to the higher background electrolyte effect described by Equation 27.3. However, as the dye ligand is titrated into the separation phase, the A007/dye electrostatic interaction is strong enough to produce a difference in the mobility of A007 when comparing the three systems (e.g., A007 mobilities at 8×10^{-4} M dye ligand: 4.03 ± 0.24 cm^2 kV^{-1} min^{-1} for TB, 8.30 ± 0.76 cm^2 kV^{-1} min^{-1} for MEB, and 9.04 ± 0.78 cm^2 kV^{-1} min^{-1} for MEG). This illustrates that the dye ligand is playing an influential role in the change in A007 mobilities through a direct electrostatic interaction in addition to EOF changes due to high electrolyte background (see Figure 27.10). It was at the 8×10^{-4} M concentration of the dye ligand in the separation phase that the ranking of the relative binding strength coefficients of the A007/dye complexes was calculated.

The above discussion is illustrated pictorially in Figures 27.9 and 27.10. In Figure 27.9, the effect of the dye forming a cation rigid and diffuse double layer at the silica wall of the separation capillary influences the EOF of the system. Here, we see that the A007 mobility increases according to the electrostatic affinities of A007 with the dyes, ranked as MEG > MEB > TB. Figure 27.10 further illustrates the two processes being observed as described above affecting the mobility of A007 in the presence of the dyes. On the left of the figure, we saw with Equation 27.3 the influence upon EOF due to increasing electrolyte concentrations. On the right of the figure is a depiction of $v_{net}^+ > v_{eo}$ due to v^+ influence from dye in separation phase on A007.

27.4.7.4.2 Derivation of Binding Strength Equation A binding strength coefficient (BC) (Equation 27.4) was derived based on the ratio of the electrophoretic mobility of A007 in the presence of ligand to that of the mobility of the free A007. A BC of 0

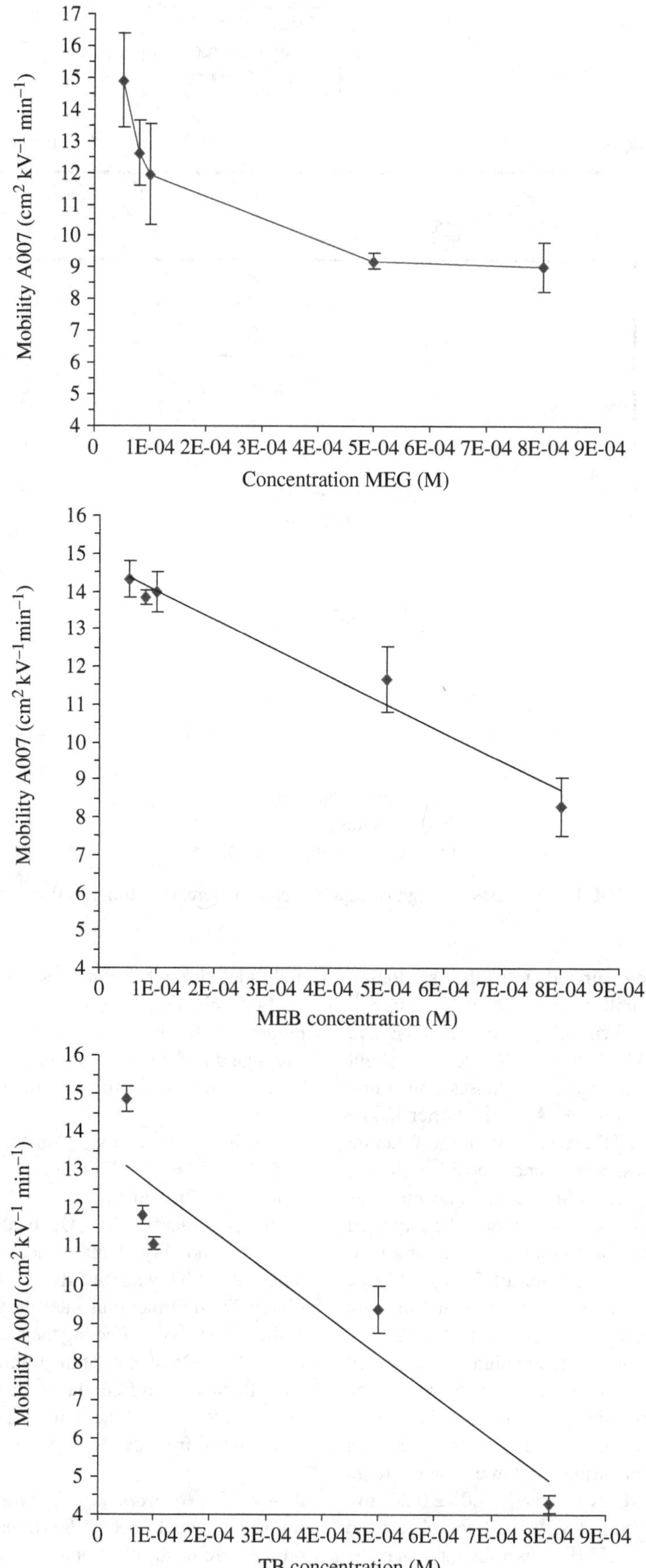

FIGURE 27.8 Plots of the mobility of A007 in increasing concentrations of methylene green, methylene blue, and toluidine blue.

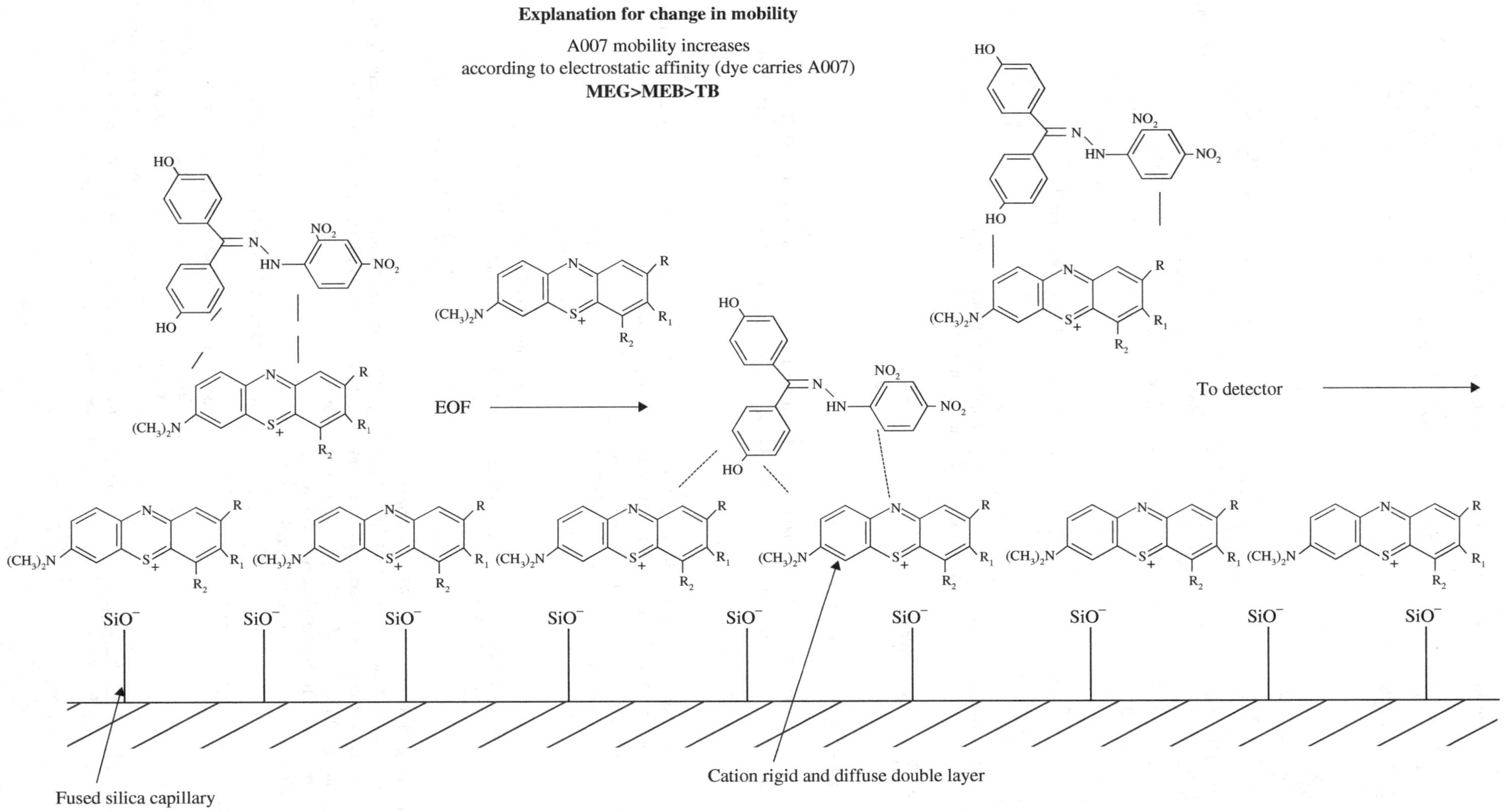
Explanation for change in mobility
A007 mobility increases
according to electrostatic affinity (dye carries A007)
MEG>MEB>TB
EOF
To detector
HO
NO2
HN
N
(CH3)2N
S+
R
R1
R2
SiO−
Fused silica capillary
Cation rigid and diffuse double layer

FIGURE 27.9 Dye in the separation phase producing a cation rigid and diffuse double layer.

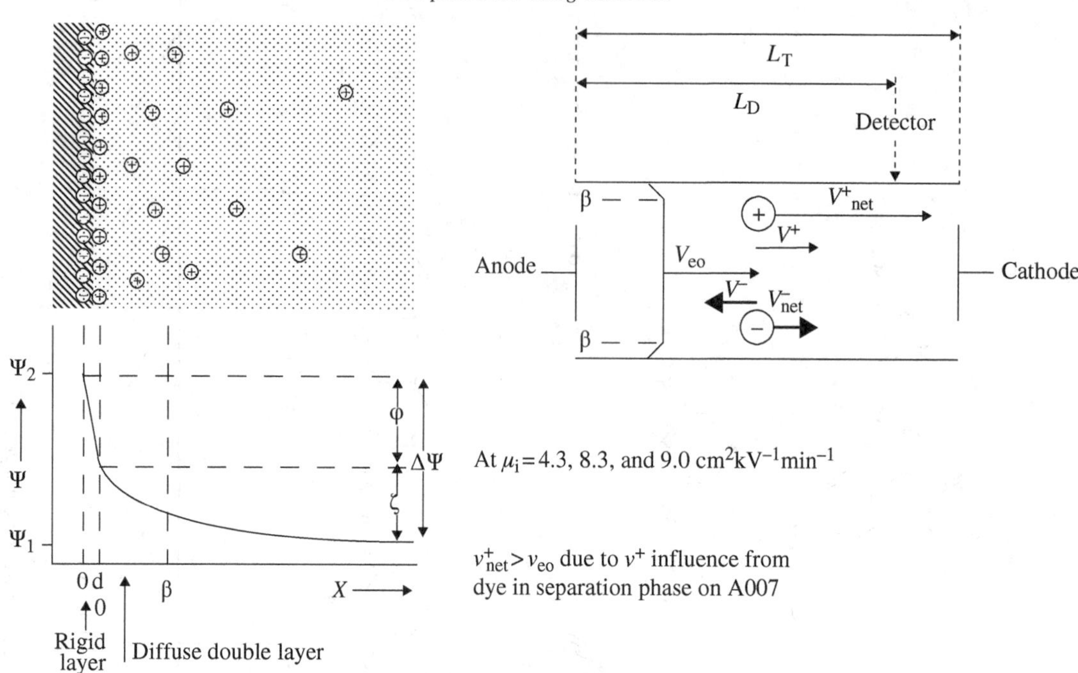

FIGURE 27.10 Two processes observed affecting the mobility of A007 in the presence of the dyes. Left, the influence upon electroosmotic flow due to increasing electrolyte concentrations. Right, depiction of $v^+_{net} > v_{eo}$ due to v^+ influence from dye in separation phase on A007.

would indicate the weakest binding, while a BC of 1 would indicate the strongest binding. A similar ratio relationship was derived by Srinivasan et al. [31] to directly calculate disassociation constants for noncovalent complexes between cyclodextrin and barbiturates. The relationship derived and used for the BC ranking is as follows:

$$BC = \frac{\mu_{complex}}{\mu_{drug}} \qquad (27.4)$$

27.4.7.5 *Electropherograms of Dye/A007 Complexes*

Figure 27.11 shows a series of electropherograms illustrating the migration times of the dyes and A007 as sample in separation phases containing the dye ligands MEG, MEB, and TB. The effective electrophoretic mobility, in the presence of EOF, of component i was calculated using the following equation [32]:

$$\mu_i = \frac{v_i}{E} = v_i \frac{L_t}{V} \left[\text{cm}^2\ \text{kV}^{-1}\ \text{min}^{-1}\right] \qquad (27.5)$$

where

$$v_i = v_{i(net)} - v_{eo} = \frac{L_D}{t_i} - \frac{L_D}{t_{eo}} \left[\text{cm}\,\text{min}^{-1}\right] \qquad (27.6)$$

Combining Equations 27.5 and 27.6 gives the standard mobility equation:

$$\mu_i = \left(\frac{L_D}{t_i} - \frac{L_D}{t_{eo}}\right) \times \frac{L_T}{V} \left[\text{cm}^2\ \text{kV}^{-1}\text{min}^{-1}\right] \qquad (27.7)$$

Equation (27.6) was further simplified, due to the absence of a measurable EOF, to

$$\mu_i = \frac{L_D L_T}{t_i V} \left[\text{cm}^2\ \text{kV}^{-1}\ \text{min}^{-1}\right] \qquad (27.8)$$

The calculated binding strength coefficients by ACE are listed in Table 27.1 along with the BC values calculated from the CELC results.

27.5 COMPARISON OF METHODS

As illustrated in Table 27.1, the ACE results show that MEG and MEB have very similar binding coefficients, while TB was well differentiated. The ranking by CELC shows binding coefficients that are more differentiated between the three, probably due to a much more direct electrostatic binding strength determination than what is measured with the ACE approach.

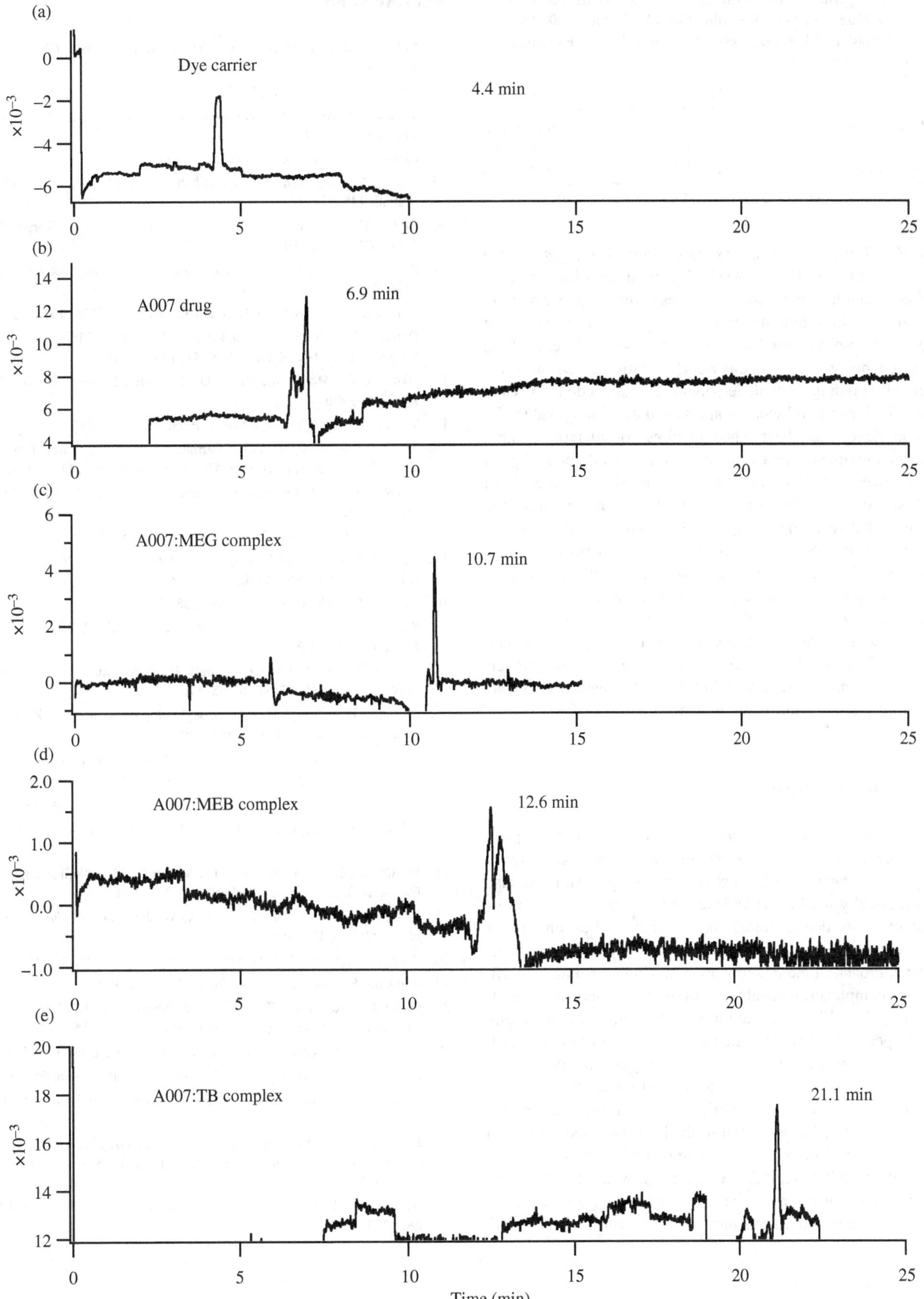

FIGURE 27.11 Electropherograms of (a) the three dyes (comigration), (b) the anticancer agent A007, (c) A007/MEG complex, (d) A007/MEB complex, and (e) A007/TB complex.

TABLE 27.1 Binding Constant Values for the Methylene Green, Methylene Blue, and Toluidine Blue Complexes with A007 by Affinity Capillary Electrophoresis (ACE) and Cation Exchange Chromatography (CEC).

Compound	ACE BC	CEC BC
Methylene green	0.67 ± 0.06	0.54 ± 0.05
Methylene blue	0.62 ± 0.06	0.24 ± 0.02
Toluidine blue	0.32 ± 0.02	0.18 ± 0.02

The ACE method does have some general advantages over the other two methods. The ES–MS/MS method describes a relative binding strength between the complexes in the gas phase only. In general, gas-phase binding strengths vary greatly in agreement with solution-phase binding strengths, always dependent on the system being studied. In this work, a solution-phase analytical method for binding strength determination was needed. The CEC approach gives a good separation between the ranking values for the complexes, but the method requires considerable column equilibration time with the changing of each mobile-phase ligand (approximately 2–3 h). In comparison, the ACE method is fast (analysis completed between 10 and 30 min) and requires little treatment between the changing of ligand separation phases (13 min caustic wash, followed by a 13 min separation-phase wash). A further advantage is that typical ACE methods generally require samples in the nanomolar to micromolar concentration ranges, as compared to millimolar levels as used in chromatographic methods such as the CELC method [33]. In general use, the ACE method would be utilized as a rapid screening determination of binding strengths, while the CELC method would act as a more precise, confirmation method.

27.6 CONCLUSIONS

ACE has become a versatile tool in the study of many descriptive parameters of complexes. Though this methodological approach has generally been used for high-affinity complexes, it can also be successfully used for complexes of low affinity, or high disassociation, as demonstrated through the application to this dye–anticancer agent complex. The ACE method offers an alternative approach in the determination of binding strength coefficients of complexes, with advantages over the other two methods studied, ES–MS/MS and cation exchange liquid chromatography, respectively. The ACE method requires less time for analysis and also smaller sample sizes as compared to ES–MS/MS and CELC. The cost for ES–MS/MS instrumentation can also be prohibitive for some laboratories. The ACE method is robust and can be applied as a general method for the determination of binding strengths of noncovalent complexes. In its specific use for the dye/A007 system, the ACE method would be utilized as a rapid screening determination of binding strengths, while the CELC method would act as a more precise, confirmation method.

REFERENCES

1. Mikkers, F.E.P.; Everaerts, F.M.; Verheggen, Th.P.E.M. *J. Chromatogr.*, 1979, **169**, 11.
2. Jorgenson, J.W.; Lukacs, K. D. *Anal. Chem.* 1981, **53**, 1298–1302.
3. Terabe, S.; Otsuka, K.; Ichikawa, K.; Tsuchiya, A.; Ando, T. *Anal. Chem.* 1984, **56**, 111.
4. Hjerten, S. *J. Chromatogr.*, 1985, **347**, 191.
5. Hadley, M.; Gilges, M.; Senior, J.; Shah, A.; Camilleri, P. *J. Chromatogr. B* 2000, **745**, 177–188.
6. Chu, Y.H.; Dunayevskiy, Y.M.; Kirby, D.P.; Vouros, P.; Karger, B.L. *J. Am. Chem. Soc.* 1996, **118**, 7827–7835.
7. Rao, J.; Colton, I.J.; Whitesides, G.M. *J. Am. Chem. Soc.* 1997, **119**, 9336–9340.
8. Rundlett, K.L.; Armstrong, D.W. *J. Chromatogr. A* 1996, **721**, 173–186.
9. Dunayevskiy, Y.M.; Lyubarskaya, Y.V.; Chu, Y.H.; Vouros, P.; Karger, B.L. *J. Med. Chem.* 1998, **41**, 1201–1204.
10. Okun, V.M.; Moser, R.; Blaas, D.; Kenndler, E. *Anal. Chem.* 2001, **73**, 3900–3906.
11. Wan, Q.H.; Le, X.C. *Anal. Chem.* 1999, **71**, 4183–4189.
12. Morgan, L.R.; Rodgers, A.H.; LeBlanc, B.W.; Boue, S.M.; Yang, Y.; Jursic, B.S.; Cole, R.B. *Bioorg. Med. Chem. Lett.* 2001, **11**, 2193–2195.
13. Whitehouse, C.M.; Dryer, R.N.; Yamashita, M.; Fenn, J.B. *Anal. Chem.* 1985, **57**, 675.
14. Fenn, J.B. *J. Am. Soc. Mass. Spectrom.* 1993, **4**, 524.
15. Kebarle, P., Ho, Y. In *Electrospray Ionization Mass Spectrometry;* Cole, R.B., Ed.; Wiley: New York, 1997; p. 17.
16. Cole, R.B. *J. Mass Spectrom.* 2000, **35**, 763–772.
17. Rodgers, A.H.; Subramanian, S.; Morgan Jr., L.R. *J. Chromatogr. B* 1995, **670**, 365–368.
18. Mussini, T.; Covington, A.K.; Longhi, P.; Rondinini, S. *Int. Union Pure Appl. Chem.*, 1985, **57** No. 6 865–876.
19. Mussini, T.; Covington, A.K.; Longhi, P.; Rondinini, S.; Zou, Z.Y. *Biochemica Acta*, 1983, **11**, 1593–1598.
20. Vosburge, W.C.; Cooper, G.R. *J. Am. Chem. Soc.* 1941, **63**, 437–442.
21. Stapelfeldt, H.; Skibsted, L.H. *J. Dairy Res.* 1999, **66**, 545–558.
22. Lee, C.; Sung, Y.W.; Park, J.W. *J. Phys. Chem. B* 1999, **103**, 893–898.
23. Salomon, K.; Burgi, D.S.; Helmer, J.C. *J. Chromatogr.* 1991, **559**, 69–80.
24. VanOrman, B.B.; Liversidge, G.G.; McIntire, G.L.; Olefirowicz, T.M.; Ewing, A.G. *J. Microcol. Sep.* 1990, **2**, 176–180.
25. Issaq, H.J.; Atamna, I.Z.; Muschik, G.M.; Janini, G.M. *Chromatographia* 1991, **32**, 155–161.
26. Wright, P.B.; Lister, A.S.; Dorsey, J.G. *Anal. Chem.* 1997, **69**, 3251–3259.
27. Fujiwara, S.; Honda, S. *Anal. Chem.*, 1987, **59**, 487–490.
28. Gorse, J.; Balchunas, A.T.; Swaile, D.F.; Sepaniak, M.J. *J. High Resolut. Chromatogr. Chromatogr. Commun.*, 1988, **11**, 554–559.
29. Bushey, M.M.; Jorgenson, J.W., *J. Microcol. Sep.*, 1989, **1**, 125–130.
30. Liu, J.; Cobb, K.; Novotny, M. *J. Chromatogr.*, 1988, **468**, 55–65.
31. Srinivasan, K.; Bartlett, M.G. *Rapid Commun. Mass Spectrom.* 2000, **14**, 624–632.
32. Kuhn, R.; Hoffstetter-Kuhn, S.; *Capillary Electrophoresis: Principles and Practice;* Springer-Verlag: Berlin Heidelberg, Germany, 1993; Chapter 2.
33. Ahmed, A.; Ibrahim, H.; Pastore, F.; Lloyd, D.K. *Anal. Chem.* 1996, **68**, 3270–3273.

28

MASS SPECTROMETRY

28.1 Definition and Description of Mass Spectrometry
28.2 Basic Design of Mass Analyzer Instrumentation
28.3 Mass Spectrometry of Protein, Metabolite, and Lipid Biomolecules
 28.3.1 Proteomics
 28.3.2 Metabolomics
 28.3.3 Lipidomics
28.4 Fundamental Studies of Biological Compound Interactions
28.5 Mass-to-Charge (*m/z*) Ratio: How the Mass Spectrometer Separates Ions
28.6 Exact Mass Versus Nominal Mass
28.7 Mass Accuracy and Resolution
28.8 High-Resolution Mass Measurements
28.9 Rings Plus Double Bonds ($r + db$)
28.10 The Nitrogen Rule in Mass Spectrometry
28.11 Chapter Problems
References

28.1 DEFINITION AND DESCRIPTION OF MASS SPECTROMETRY

During the past decade, mass spectrometry has experienced a tremendously large growth in its uses for extensive applications involved with complex biological sample analysis. Mass spectrometry is basically the science of the measurement of the mass-to-charge (*m/z*) ratio of ions in the gas phase. Mass spectrometers are generally comprised of three components: (i) an ionization source that ionizes the analyte of interest and effectively transfers it into the gas phase, (ii) a mass analyzer that separates positively or negatively charged ionic species according to their mass-to-charge (*m/z*) ratio, and (iii) a detector used to measure the subsequently separated gas-phase ions. Mass spectrometers are computer controlled, which allows the collection of large amounts of data and the ability to perform various and complex experiments with the mass spectral instruments. Applications of mass spectrometry include unknown compound identification, known compound quantitation, structural determination of molecules, gas-phase thermochemistry studies, ion–ion and ion molecule studies, and molecule chemical property studies. Mass spectrometry is routinely used to determine elements such as Li^+, Na^+, Cl^-, and Mg^{2+}; inorganic compounds such as $Li^+(H_2O)_x$ or $(TiO_2)_x^+$; and organic compounds including lipids, proteins, peptides, carbohydrates, polymers, and oligonucleotides (DNA/RNA).

28.2 BASIC DESIGN OF MASS ANALYZER INSTRUMENTATION

Typical mass spectrometric instrumentations that are used in laboratories and research institutions are comprised of six components: (i) an inlet, (ii) an ionization source, (iii) a mass analyzer, (iv) a detector, (v) a data processing system, and (vi) a vacuum system. Figure 28.1 illustrates the interrelationship of the six components that make up the fundamental construction of a mass spectrometer. The *inlet* is used to introduce a sample into the mass spectrometer and can be a solid probe, a manual syringe or syringe pump system, a gas chromatograph, or a liquid chromatograph. The inlet system can be either at atmospheric pressure as is shown in Figure 28.1, or it can be at a reduced pressure under vacuum. The *ionization source* functions to convert neutral molecules into charged analyte ions thus enabling their mass analysis. The ionization source can also be part of the inlet system. A typical inlet system and ionization source that is used with high-performance liquid chromatography (HPLC) is electrospray ionization (ESI). In an HPLC/ESI inlet system and ionization source, the effluent coming from the HPLC column is transferred into the ESI capillary that has a high voltage applied to it inducing the ESI process. In this configuration, the inlet system and ionization source are located at atmospheric pressure outside of the mass spectrometric instrumentation that is under vacuum. The spray that is produced passes through a tiny orifice

Analytical Chemistry: A Chemist and Laboratory Technician's Toolkit, First Edition. Bryan M. Ham and Aihui MaHam.

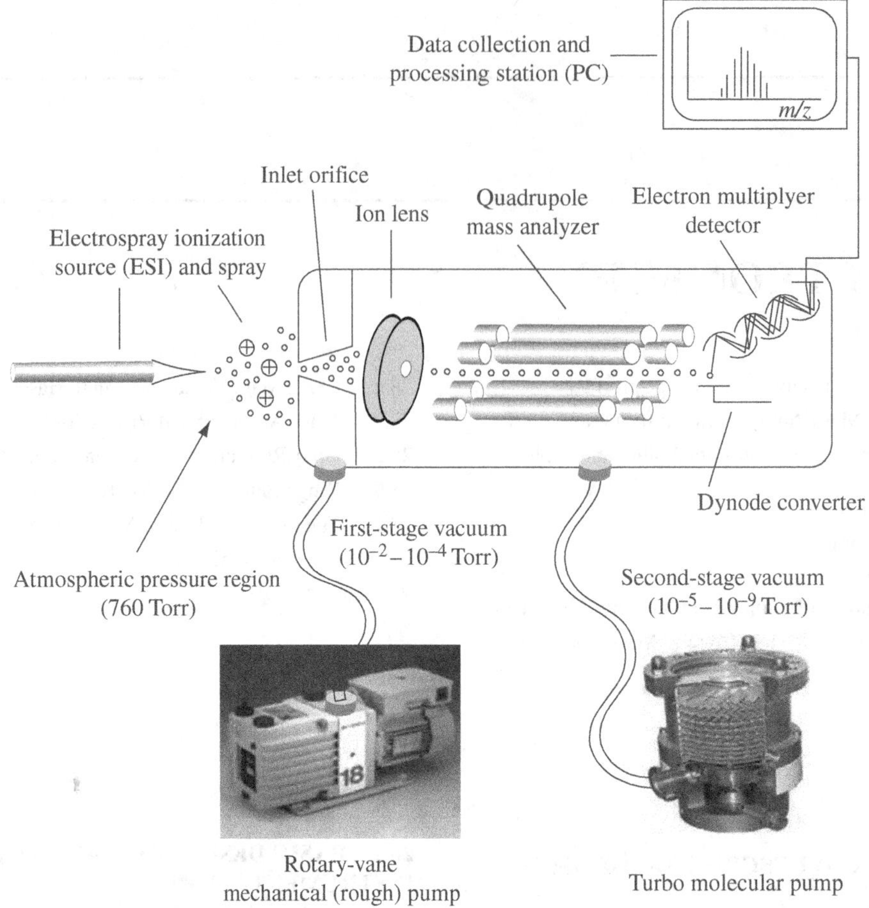

FIGURE 28.1 The 6 components that make up the fundamental configuration of mass spectrometric instrumentation consisting of (1) inlet and ionization system, (2) inlet orifice (source), (3) mass analyzer, (4) detector, (5) vacuum system, and (6) data collection and processing station (PC). (Turbomolecular pump reprinted with permission from the Wiki Creative Commons Attribution-Sharealike 3.0 Unported License.)

that separates the internal portion of the mass spectrometer that is under vacuum from its ambient surroundings that are at atmospheric pressure. This orifice is also often called the *inlet* and/or the *source*. In the case of the coupling of a gas chromatograph to the mass spectrometer, the capillary column of the gas chromatograph is inserted through a heated transfer capillary directly into the internal portion of the mass spectrometer that is under vacuum. This is possible due to the fact that the species eluting from the capillary column are already in the gas phase making their introduction into the mass spectrometer more straightforward as compared to the liquid eluant from an HPLC where analytes must be transferred from the solution phase to the gas phase. An example of an ionization process that takes place under vacuum in the front end of the mass spectrometer is a process called matrix-assisted laser desorption ionization (MALDI). In this ionization technique, a laser pulse is directed toward a MALDI target that contains a mixture of the neutral analytes and a strongly UV absorbing molecule, often times a low molecular weight organic acid such as dihydroxybenzoic acid (DHB). The analytes are lifted off of the MALDI target plate directly into the gas phase in an ionized state. This is due to transference of the laser energy to the matrix and then to the analyte. The MALDI technique takes place within a compartment that is at the beginning of the mass spectrometer instrument and is under vacuum. The compartment that this takes place is often called the ionization source thus combining the *inlet system* and the *ionization source* together into one compartment. As illustrated in Figure 28.1, the analyte molecules (small circles), in an ionized state, pass from atmospheric conditions to the first stage of vacuum in the mass spectrometer through an inlet orifice that separates the mass spectrometer that is under vacuum from ambient conditions. The analytes are guided through a series of ion lenses into the mass analyzer. The *mass analyzer* is the heart of the system, which is a separation device that separates positively or negatively charged ionic species in the gas phase according to their respective mass-to-charge ratios. The mass analyzer gas-phase ionic species separation can be performed by an external field such as an electric field or a magnetic field or by a field-free region such as within a drift tube. For the detection of the gas-phase separated ionic species, electron multipliers are often used as the *detector*. Electron multipliers are mass impact detectors that convert the impact of the gas-phase separated ionic species into a cascade of electrons thereby multiplying the signal of the impacted ion many times fold.

The *vacuum system* ties into the inlet, the source, the mass analyzer, and the detector of the mass spectrometer at different stages of increasing vacuum as movement goes from the inlet to the detector (left to right in Figure 28.1). It is very important for the mass analyzer and detector to be under high vacuum as this removes ambient gas, thereby reducing the amount of unwanted collisions between the mass-separated ionic species and gas molecules present. As illustrated in Figure 28.1, ambient, atmospheric conditions are generally at a pressure of 760 Torr. The first-stage vacuum is typically at or near 10^{-3} Torr immediately following the inlet orifice and around the first ion transfer lenses. This stage of vacuum is obtained using two-stage rotary vane mechanical pumps that are able to handle high pressures such as atmospheric and large variation in pressures but are not able to obtain the lower pressures that are required further into the mass spectrometer instrument. The two-stage rotary vane mechanical pump has an internal configuration that utilizes a rotating cylinder that is off-axis within the pump's hollow body. The off-axis-positioned rotor contains two vanes that are opposed and directed radially and are spring controlled to make pump body contact. As the cylinder rotates the volume between the pump's body and the vanes changes, the volume increases behind each vane that passes a specially placed gas inlet port. This will cause the gas to expand behind the passing vane, while the trapped volume between the exhaust port and the forward portion of the vane will decrease. The exhaust gas is forced into a second stage and then is released by passing through the oil that is contained within the pump's rear oil reservoir. This configuration is conducive for starting up at atmospheric pressure and working toward pressures usually in the range of 10^{-3}–10^{-4} Torr.

The lower stages of vacuum are obtained most often using turbo molecular pumps as illustrated in Figure 28.1. Turbo molecular pumps are not as rugged as the mechanical pumps described previously and need to be started in a reduced pressure environment. Typically, a mechanical pump will perform the initial evacuation of an area. When a certain level of vacuum is obtained, the turbo molecular pumps will then turn on and bring the pressure to higher vacuum. Using a mechanical vane pump to provide a suitable forepump pressure for the turbo molecular pump is known as roughing or "rough out" the chamber. Therefore, two-stage rotary vane mechanical pumps are often referred to as rough pumps. As illustrated in Figure 28.1, the turbo molecular pump contains a series of rotor/stator pairs that are mounted in multiple stages. The principle of turbomolecular pumps is to transfer energy from the fast rotating rotor (turbo molecular pumps operate at very high speeds) to the molecules that make up the gas. After colliding with the blades of the rotor, the gas molecules gain momentum and move to the next lower stage of the pump and repeat the process with the next rotor. Eventually, the gas molecules enter the bottom of the pump and exit through an exhaust port. As gas molecules are removed from the head or beginning of the pump, the pressure before the pump is continually reduced as the gas is removed through the pump, thus achieving higher and higher levels of vacuum. Turbo molecular pumps can obtain much higher levels of vacuum (up to 10^{-9} Torr) as compared to the rotary vane mechanical pumps (up to 10^{-4} Torr).

The final component of the mass spectrometer is a data processing system. This is typically a personal computer (PC) allowing the mass spectrometric instrumentation to be software controlled enabling precise measurements of carefully designed experiments and the collection of large amounts of data. Commercially bought mass spectrometers will come with its own software that is used to set the operating parameters of the mass spectrometer and to collect and interpret the data, which is in the form of mass spectra.

28.3 MASS SPECTROMETRY OF PROTEIN, METABOLITE, AND LIPID BIOMOLECULES

Proteins and lipids are two important classes of biological compounds that are found in all living species. Proteomics (the identification and characterization of proteins in cells, organisms, etc.) has now become a very important field of research in mass spectrometry, which is often used to describe a cell's or organism's proteome or to investigate a response to a stress upon a system through a change in protein expression. Global metabolic profiles (metabolomics) have also become a quickly advancing area of study utilizing mass spectrometric techniques. Lipidomics is a subclass of metabolomics that is also experiencing an increase in mass spectral applications. Lipids are often associated with proteins and act as physiological activators. Lipids constitute the bilayer components of biological membranes and directly participate in membrane protein regulation and function. In the past decade, mass spectrometry has experienced increasing applications to the characterization and identification of these two important biological compound classes. Using proteomics, metabolomics, and lipidomics as examples of the application of mass spectrometry to biomolecular analysis, we will briefly look at these areas in the following three sections.

28.3.1 Proteomics

The area of proteomics has been applied to a wide spectrum of physiological samples often based upon comparative studies where a specific biological system's protein expression is compared to either another system or the same system under stress. Often in the past, the comparison is made using 2-dimensional electrophoresis where the gel maps for the two systems are compared looking for changes such as the presence or absence of proteins and the up- or downregulation of proteins. Proteins of interest are cut from the gel and identified by mass spectrometry. Figure 28.2 illustrates a 1-dimensional (1D) sodium dodecylsulfate polyacrylamide gel electrophoresis (SDS-PAGE) gel (12% acrylamide) of rabbit tear from a normal versus a dry eye diseased state model and a protein molecular weight marker [1]. The technique of SDS-PAGE accomplishes the linear separation of proteins according to their molecular weights. In this process, the proteins are first denatured and sulfide bonds are cleaved, effectively unraveling the tertiary and secondary structure of the protein. Sodium dodecylsulfate, which is negatively charged, is then used to coat the protein in a fashion that is proportional to the proteins' molecular weight. The proteins are then separated within a polyacrylamide gel by placing a potential difference across the gel. Due to the potential difference across the gel, the proteins will experience an electrophoretic movement through the gel, thus separating them according to their molecular weight with

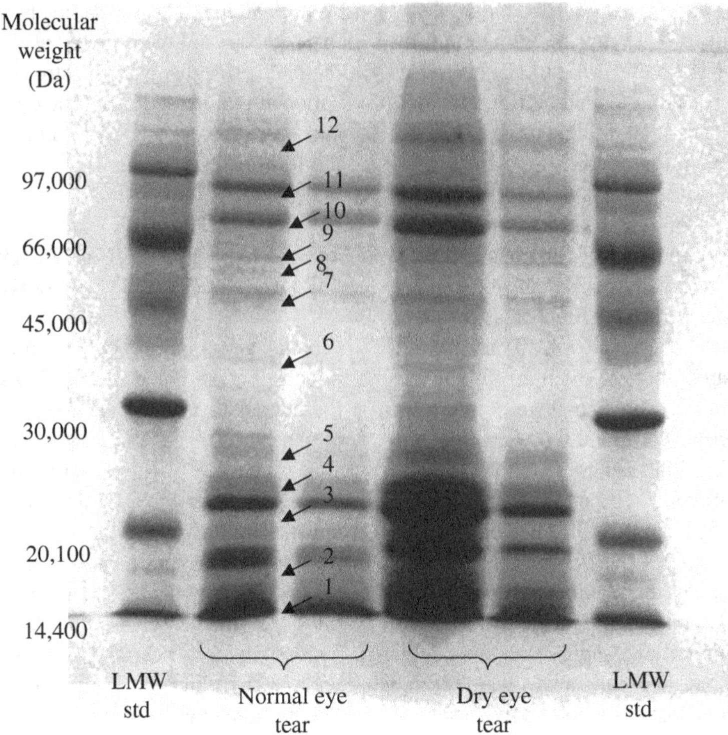

FIGURE 28.2 1D SDS-PAGE of rabbit tear proteins from a study of normal eye versus a dry eye model. Arrows and numbers represent bands of separated proteins that are used for protein identifications. (Reprinted with kind permission from Springer Science and Business Media from Ref. [1], Figure 1, p. 891.)

the lower molecular weight proteins having a greater mobility through the gel and the higher molecular weight proteins having a lower mobility through the gel. There are 12 clear bands observed and designated in Figure 28.2 as indicated by the numbering and associated arrows. These bands represent proteins of different molecular weights that have been separated electrophoretically. Often, bands such as these are cut from the gel, and the proteins are digested with an enzyme such as trypsin, which is an endopeptidase that cleaves within the polypeptide chain of the protein at the carboxyl side of the basic amino acids arginine and lysine (the trypsin enzyme has optimal activity at a pH range of 7–10 and requires the presence of Ca^{+2}). The enzyme-cleaved peptides are then extracted from the gel and analyzed by mass spectrometry for their identification.

Figure 28.3 shows the identification of one of the proteins as lipophilin CL2 [1] recovered from band #1 in Figure 28.2 using the mass spectrometric technique of MALDI time-of-flight (TOF) post-source decay (PSD) mass spectrometry. A recent example of a 2D SDS-PAGE approach was reported by Nabetani et al. [2] in a study of patients with primary hepatolithiasis, an intractable liver disease, where the authors reported the upregulation of 12 proteins and the downregulation of 21 proteins. Another recent approach for proteomic studies by mass spectrometry uses a technique known as multidimensional protein identification technology (MudPIT) [3]. This is a gel-free approach that utilizes multiple HPLC-mass spectrometry analysis of in-solution digestions of protein fractions. Kislinger et al. [4] demonstrated the effectiveness of this approach in a study of biochemical (mal)adaptations involving heart tissue that was both healthy and diseased. Other examples include the spatial profiling of proteins and peptides on brain tissue sections [5], the analysis of viruses' capsid proteins for viral identification and posttranslational modifications [6], and structural immunology studies including antibody and antigen structures, immune complexes, and epitope sequencing and identification [7]. The discipline of proteomics and the applications of mass spectrometry to protein analysis will be extensively covered in Chapter 31.

28.3.2 Metabolomics

The study of global metabolite profiles that are contained within a system at any one time representing a set of conditions is metabolomics [8]. The system can be comprised of a cell, tissue, or an organism. Metabolomics has also been described as the measurement of all metabolite concentrations in cells, tissues, and organisms, while metabonomics involves the quantitative measurement that pathophysiological stimuli or genetic modification has upon the metabolic responses of multicellular systems [9]. The system's genomic interaction with the environmental system condition results in the production of metabolites. Metabolites represent an extremely diverse and broad set of biomolecule classes that include lipids, amino acids, nucleotides, and organic acids and cover widely different concentration ranges [10]. It has been estimated that there are from 2000 major metabolites [11] to 5000, with over 20,000 genes and 1,000,000 proteins in humans [12]. Figure 28.4 illustrates the investigative strategies for two major approaches of mass spectrometry-based metabolomics. The first strategy of the figure (left) is quantitative

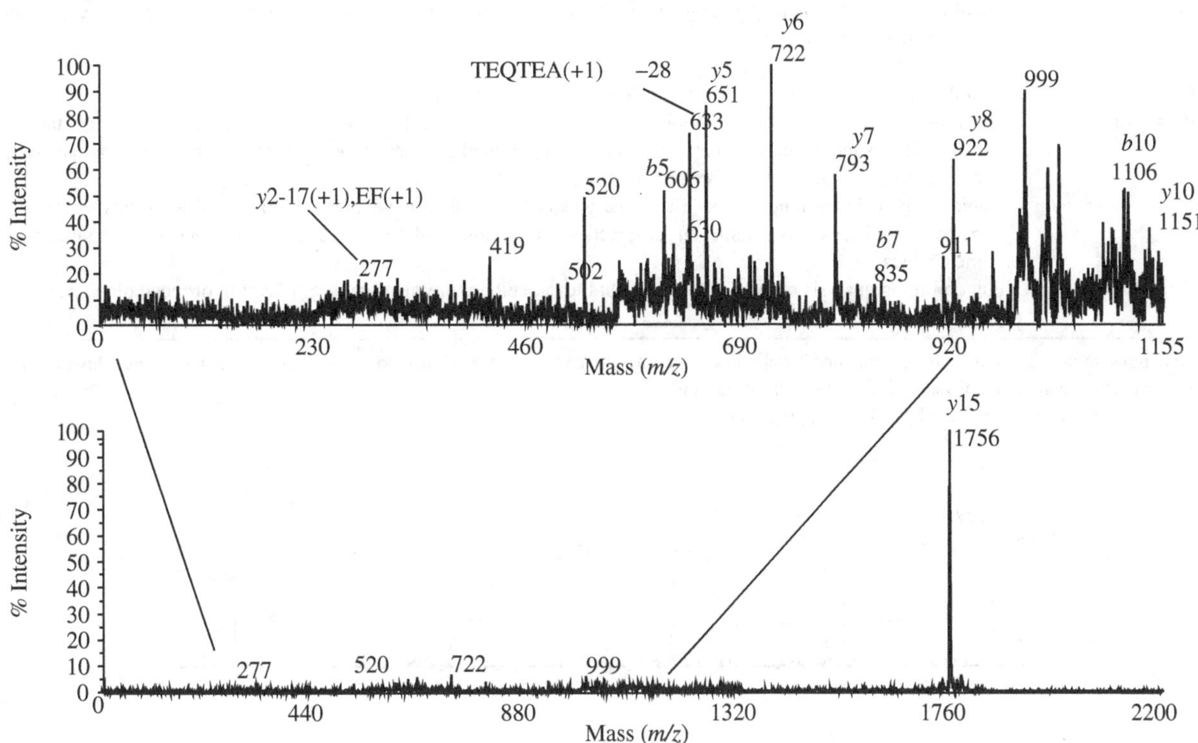

FIGURE 28.3 Matrix-assisted laser desorption ionization (MALDI) time-of-flight (TOF) post-source decay (PSD) mass spectrum illustrating the identification of the protein lipophilin CL2 recovered from the #1 band in Figure 28.2. (Reprinted with kind permission from Springer Science and Business Media from Ref. [1], Figure 5, p. 895.)

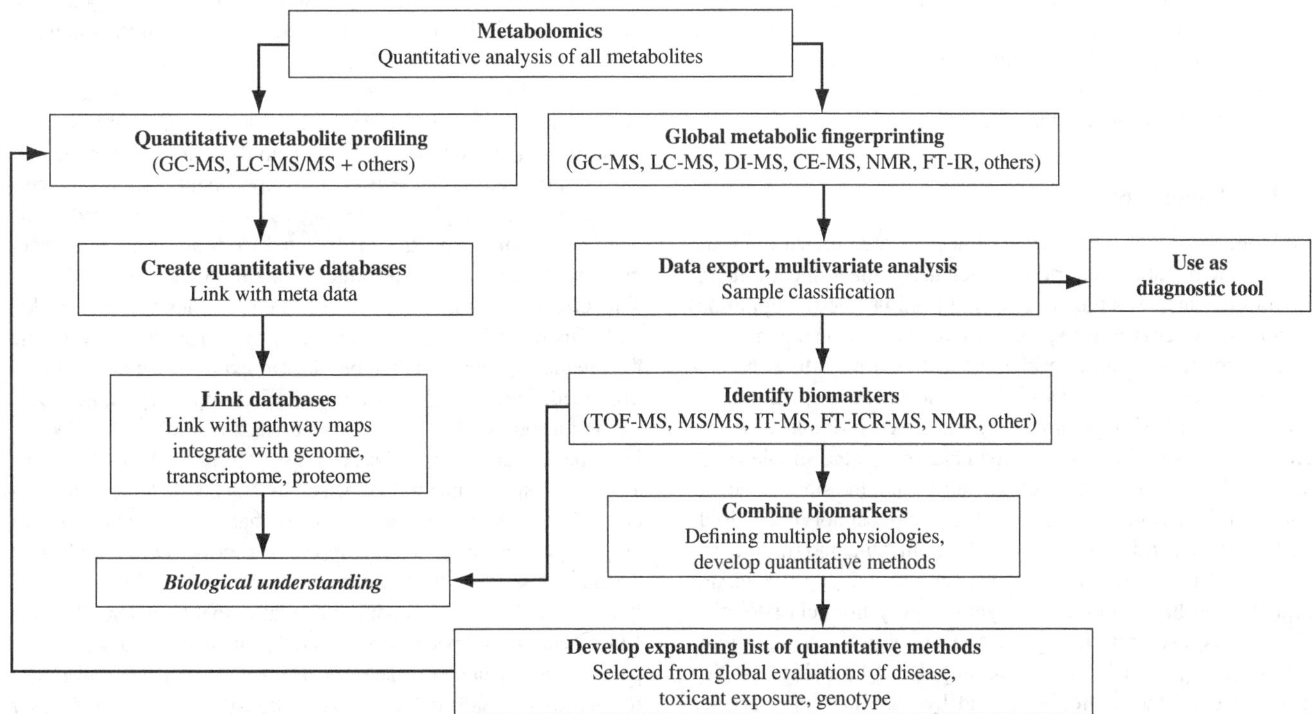

FIGURE 28.4 Strategies for metabolomic investigations. (Reprinted with permission from Ref. [10]. Copyright 2007.)

TABLE 28.1 Metabolomics-Related Definitions.

Metabolite	Small molecules that participate in general metabolic reactions and that are required for the maintenance, growth, and normal function of a cell*
Metabolome	The complete set of metabolites in an organism
Metabolomics	Identification and quantification of all metabolites in a biological system
Metabolic profiling	Quantitative analysis of set of metabolites in a selected biochemical pathway or a specific class of compounds. This includes target analysis, the analysis of a very limited number of metabolites, for example, single analytes as precursors or products of biochemical reactions
Metabolic Fingerprinting	Unbiased, global screening approach to classify samples based on metabolite patterns or "fingerprints" that change in response to disease, environmental, or genetic perturbations with the ultimate goal to identify discriminating metabolites
Metabolic footprinting	Fingerprinting analysis of extracellular metabolites in cell culture medium as a reflection of metabolite excretion or uptake by cells

*This is in contrast to xenobiotic or foreign compound metabolites, which may overshadow natural metabolites in an analytical procedure. However, they can be very valuable in evaluating the physiological status of the organism.

Source: Reprinted with permission from Ref. [10]. Copyright 2007.

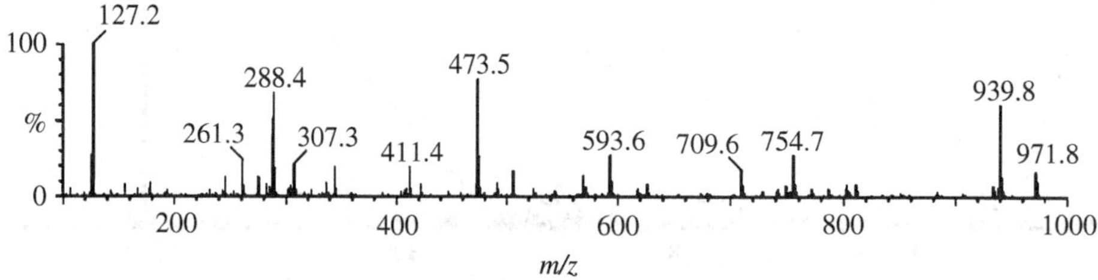

FIGURE 28.5 Electrospray ionization quadrupole–hexapole–quadrupole (ESI/QhQ) mass spectrum of the lithium adducts of a complex biological extract illustrating a number of biomolecules that are effectively separated in the gas phase.

metabolic profiling where generally predefined metabolites are targeted and measured to answer a biological question(s). The second strategy of the figure (right) is global metabolic fingerprinting where the patterns or "fingerprints" of metabolites of a system are compared to measure changes due to a response of the system to disease, environmental changes, etc. Table 28.1 lists some metabolomics-related definitions.

28.3.3 Lipidomics

Lipidomics has also experienced a dramatic increase in the amount of literature-reported studies using mass spectrometry. Lipidomics involves the study of the lipid profile (or lipidome) of cellular systems and the processes involved in the organization of the protein and lipid species present [13] including the signaling processes and metabolism of the lipids [14]. Lipids are a diverse class of physiologically important biomolecules that are often classified into three broad classes: (i) the simple lipids such as the fatty acids, the acylglycerols, and the sterols such as cholesterol; (ii) the more complex polar phosphorylated lipids such as phosphatidylcholine and sphingomyelin; and (iii) the isoprenoids, terpenoids, and vitamins. Recently, the LIPID MAPS organization has proposed an eight-category lipid classification (fatty acyls, glycerolipids, glycerophospholipids, sphingolipids, sterol lipids, prenol lipids, saccharolipids, and polyketides) that is based upon the hydrophobic and hydrophilic characteristics of the lipids [15]. Past methodology for the analysis of lipids often relied heavily on thin-layer chromatography that generally requires a substantial amount of sample and where the detection is nonspecific done either visually or by spray phosphorus. For specific identifications, the spot is usually scrapped from the plate, and the lipids extracted and then either analyzed directly or reacted with other reagents to increase volatility or detection ability with chromophore or fluorophore labeling. Often, mass spectrometry can be used to separate biomolecules as illustrated in a single-stage spectrum that shows the molecular species present. Mass spectrometric experiments can then be performed where a single analyte species of interest observed in the single-stage mass spectrum is isolated from all of the other species present and collision-induced decomposition is performed. This effectively fragments the molecular species that is recorded and illustrated in a product ion spectrum. The structure of the fragmented species can then be ascertained from the product ions produced. Figure 28.5 illustrates an ESI single-stage mass spectrum of a complex biological extract where a number of molecular ion species are being observed as lithium adducts $[M+Li]^+$. The mass spectrometer has effectively separated these species according to their respective mass-to-charge (*m/z*) ratios, producing a spectrum of peaks that represents individual biomolecular species. Using a type of mass spectrometer known as a quadrupole–hexapole–quadrupole mass spectrometer, the *m/z* 631 precursor ion species was isolated from all of the other species present, and a tandem mass spectrometry experiment was performed as collision-induced dissociation (CID) of the precursor *m/z* 631 ion to produce the product ion spectrum illustrated in Figure 28.6. The quadrupole–hexapole–quadrupole mass

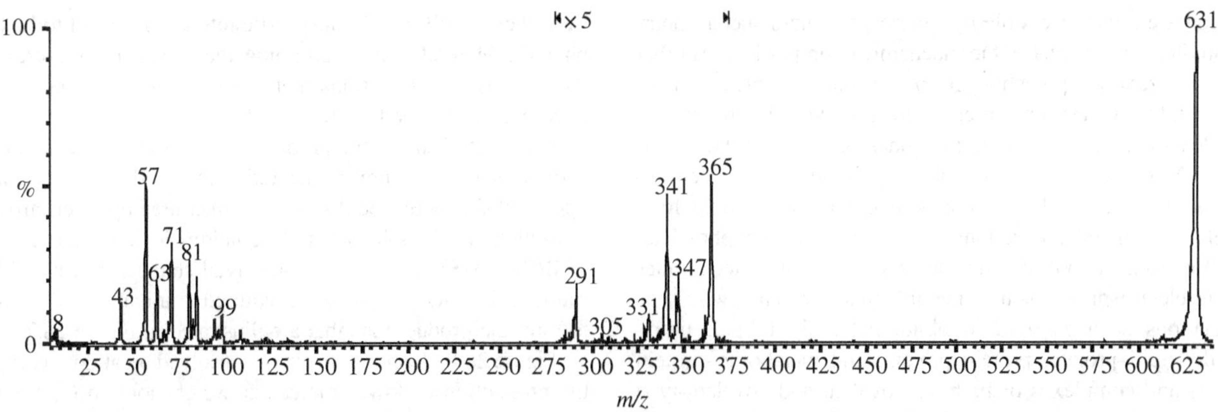

FIGURE 28.6 Electrospray ionization quadrupole–hexapole–quadrupole (ESI/QhQ) product ion mass spectrum of the lithium adduct precursor ion *m/z* 631 illustrated in Figure 1.5. Two product ion ranges, a high and a low, are observed that are used for structural identification of the unknown biomolecule. The *m/z* 631 species was identified as 1,3-distearin.

spectrometer uses the inherent ability of a quadrupole mass analyzer to filter out all precursor ions except the one of interest, thus allowing it to pass through the system for the ensuing CID fragmentation study. Due to their filtering effect, quadrupoles are often referred to as mass filters. As illustrated in the product ion spectrum in Figure 28.6, there are two ranges of product ions, for the low range approximately *m/z* 43 to *m/z* 99 and for the high range approximately *m/z* 291 to *m/z* 365, that are used to deduce the structure and identity of the unknown biomolecule *m/z* 631. Using the mass spectrometric technique of CID and subsequent fragmentation, the unknown *m/z* 631 species was identified as 1,3-distearin.

Electrospray ionization quadrupole–hexapole–quadrupole (ESI/QhQ) product ion mass spectrum of the lithium adduct precursor ion *m/z* 631 is illustrated in Figure 28.5. Two product ion ranges, a high and a low, are observed and are used for structural identification of the unknown biomolecule. The *m/z* 631 species was identified as 1,3-distearin.

A recent review by Pulfer and Murphy [16] offers an informative discussion of the analysis of the phosphorylated lipids by mass spectrometry with a focus upon electrospray as the ionization technique, while Griffiths [17] covers the mass spectrometric analysis of the simple class of lipids (fatty acids, triacylglycerols, bile acids, and steroids). Recent examples of the application of mass spectrometry to biological sample lipid analysis include a study by Lee et al. [18] where the relatively new technique of electron-capture atmospheric pressure chemical ionization mass spectrometry was used in the quantitative work of rat epithelial cell lipidomes, the analysis of human cerebellum gangliosides by nanoelectrospray tandem mass spectrometry [19], the analysis of lyso-phosphorylated lipids in ascites from ovarian cancer patients [20], the analysis of Amadori-glycated phosphatidylethanolamine in plasma samples from individuals with and without diabetes using a quadrupole ion trap mass spectrometer [21], and the identification of nonpolar lipids and phosphorylated lipids in tear [22,23].

A primary direction illustrated by these studies is for the understanding of biological systems in both normal and diseased states using the investigative technique of mass spectrometry. Proposed biological systems investigated can include studies of the expression of biomolecules from diseased states and stress models searching for relevant markers, the study of biological fluids for characterization, the study of biological processes intra- and extracellular, and the study of biomolecule interactions. Areas of mass spectrometric study of biomolecular systems include the characterization of the specific classes of proteins, lipids, carbohydrates, sugars, etc.; the development of extraction, purification, and enrichment of biomolecules from complex biological samples; the development of quantitative analysis of biomolecules; the development of identification methods for biomolecules; and the identification and quantification of biomolecules in normal and diseased state expression studies and their function and importance.

28.4 FUNDAMENTAL STUDIES OF BIOLOGICAL COMPOUND INTERACTIONS

The study of thermodynamic properties of complexes such as bond dissociation energies (BDE) by CID tandem mass spectrometry for the measurement of the dissociation of gas-phase complexes has been a useful approach. For the study of the BDE of a cobalt carbene ion, Armentrout and Beauchamp [24] introduced an ion beam apparatus and have since refined the measurement of noncovalent, collision-induced threshold bond energies of metal–ligand complexes by mass spectrometry. There are extensive examples of the use of CID tandem mass spectrometry [25] for the measurement of thermodynamic properties of gas-phase complexes including investigations of noncovalent interactions [26–29]; BDE [16]; BDE using guided ion beam tandem mass spectrometry [30]; gas-phase dissociative electron transfer reactions [31]; critical energies for ion dissociation using a quadrupole ion trap [26], an electrospray triple quadrupole [32,33], or a flowing afterglow-triple quadrupole [27,28]; and gas-phase equilibria [34–36].

During the ionization process, electrospray [37–40] is an ionization method that is well known to produce intact gas-phase ions with minimal fragmentation. Species originating from solution are quite accessible by electrospray mass spectrometry for studies of gas-phase thermochemical properties. Another "soft" ionization approach to gas-phase ionic complex analysis is MALDI; however, during crystallization with the matrix, the complex's character in the solution phase can be lost. However, using a MALDI ion source coupled to a hybrid mass spectrometer, Gidden et al. [41] derived experimental cross sections for a set of alkali metal ion cationized poly(ethylene terephthalate) (PET) oligomers. With the advent of soft ionization techniques such as electrospray, it is now possible to study intact, wild-type associations such as lipid–metal/nonmetal, lipid–lipid, lipid–peptide, lipid–protein, protein–protein, and metabolite–protein adducts and complexes of both noncovalent and covalent-type bonding. Electrospray effectively allows the transfer of intact associations within solution phase (physiological/aqueous or organic solvent modified) to the gas phase without interruption of the weak associations, thus preserving the original interaction (however, specific and nonspecific interactions need to be deciphered). Methods for determining the dissociation energies of these types of associations that can be applied to these studies have been developed [42].

Figure 28.7 shows the product ion spectrum of the diacylglycerol 1-stearin, 2-palmitin illustrating the product ions produced upon collision-activated dissociation measured by an electrospray ionization quadrupole–hexapole–quadrupole mass spectrometer (ESI/QhQ-MS). Using energy-resolved studies, it is possible to estimate the energies that were required to initiate bond breakage to form the product ions that are illustrated in Figure 28.7.

Figure 28.8 illustrates an energy-resolved breakdown graph of the production of lower molecular weight ions in CID experiments of monopentadecanoin showing increased production with increasing internal energy. The experiments used to construct the plot in Figure 28.8 are performed by measuring the products produced by CID such as those illustrated in Figure 28.7 at

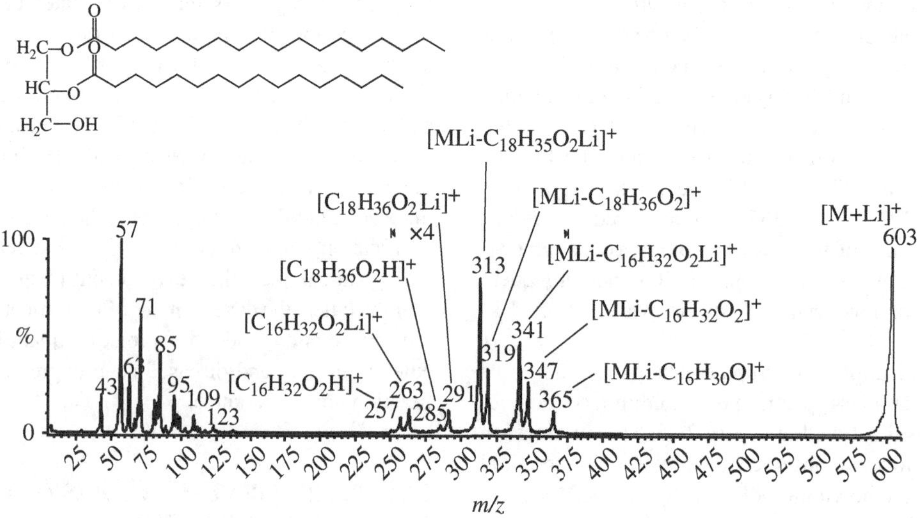

FIGURE 28.7 Product ion spectrum of the lithium adduct of the diacylglycerol 1-stearin,2-palmitin at *m/z* 603 illustrating the product ions generated by collision-induced dissociation.

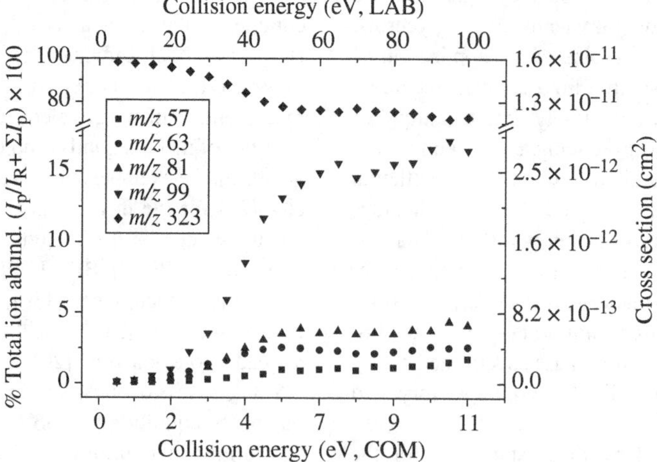

FIGURE 28.8 Energy-resolved breakdown graph illustrating product ion formation with increasing internal energy for the collision-induced dissociation of monopentadecanoin.

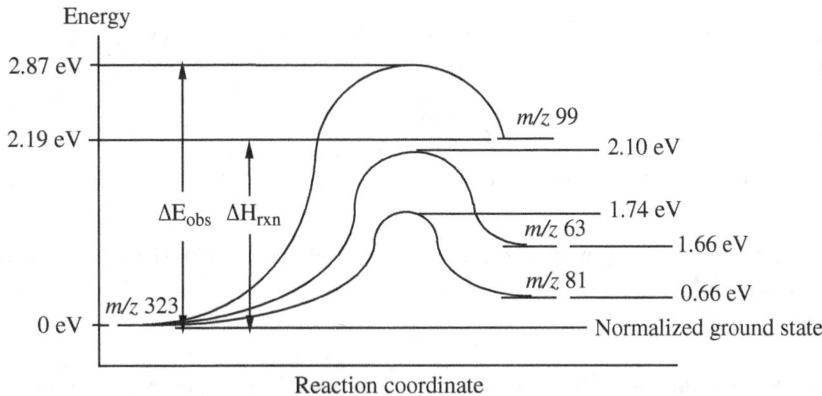

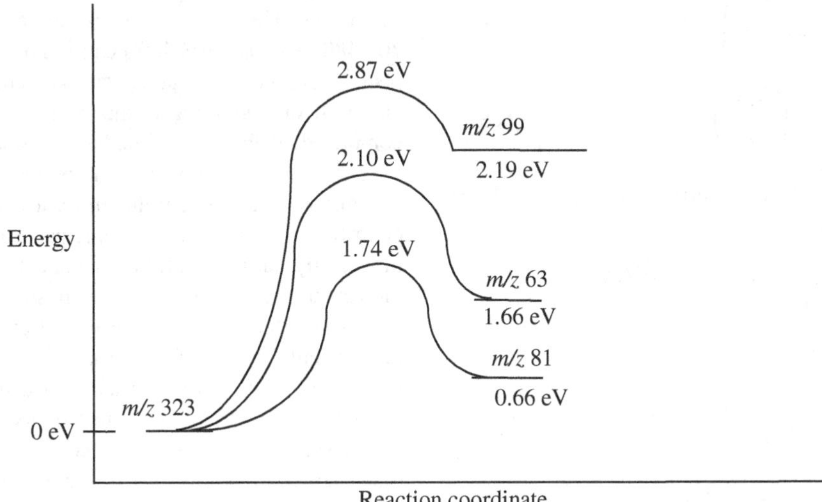

FIGURE 28.9 Energy diagram describing the formation of selected product ions from the collision-induced dissociation of monopentadecanoin.

increasing internal energies. These results can then be used to relate the measured products of the CID experiments to the dissociation energies involved in the bond cleavages. Figure 28.9 displays an energy diagram describing the production of a selected number of low molecular weight product ions from the CID of monopentadecanoin.

28.5 MASS-TO-CHARGE (*m/z*) RATIO: HOW THE MASS SPECTROMETER SEPARATES IONS

For an analyte to be separated by a mass analyzer and subsequently measured by a detector, it must first be in an ionized state. Often, the analyte is in a neutral state whether in a solid or liquid matrix (e.g., in a soil sample, solid tissue sample, an aqueous solution, or as an organic solvent extract). Using various ionization techniques, the analyte is ionized and transferred into the gas phase in preparation for introduction into the mass analyzer. The ionization of the analytes and transfer into the gas phase are typically done in an apparatus called a source that acts as a front-end preparation stage for the mass analysis of the analyte as a positively or negatively charged ion. An analyte "M" can be in the form of a protonated molecule giving the analyte species a single positive charge $[M+H]^+$, a metal adduct such as sodium giving the molecular species a single positive charge $[M+Na]^+$, a chloride adduct giving the molecular species a single negative charge $[M+Cl]^-$, a deprotonated state giving the molecular species a single negative charge $[M-H]^-$, through electron loss giving the molecular species a single positive charge as a radical $M^{+\cdot}$ (also known as a molecular ion), and so on. All mass analyzers that are used in mass spectrometric analysis possess a characteristic means of relating an analyte species' mass to its respective charge (*m/e*), where m is the species mass and *e* is the fundamental charge constant of 1.602×10^{-19} C. The TOF mass analyzer affords a simple and direct example of the mass analyzer's physical relationship of analyte mass to analyte charge. The TOF mass analyzer separates analytes by time that have drifted through a preset described length drift tube. In the ionization source, the analytes have all been given the same initial kinetic energy (KE); therefore, their physical description of $1/2mv^2$ is virtually constant for all of the ions for their drift through the drift tube. Using the relationship of $KE = zeV = 1/2mv^2$, where V is the acceleration potential and v is the velocity of the ion, an analyte with a larger mass m will possess a lower velocity v. The time (t)

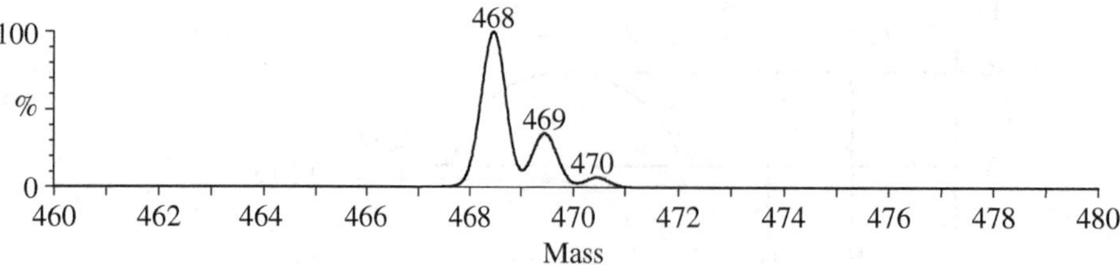

FIGURE 28.10 Typical mass spectrum illustrating three peaks at *m/z* 468, *m/z* 469, and *m/z* 470. The y-axis is in percentage response and the *x*-axis represents the mass-to-charge (*m/z*) ratio.

for the ions to arrive at the detector at the end of the drift tube is measured and is converted to a mass scale:

$$\mathrm{KE} = zeV = \frac{1}{2}mv^2 \qquad (28.1)$$

$$v = \left(\frac{2zeV}{m}\right)^{\frac{1}{2}} \qquad (28.2)$$

$$t = \frac{L}{v}, \quad L = \text{length of drift tube} \qquad (28.3)$$

$$t = L\left(\frac{m}{2zeV}\right)^{\frac{1}{2}}, \quad V \text{ and } L \text{ are fixed} \qquad (28.4)$$

Solving for *m/z*,

$$t^2 = L^2\left(\frac{m}{2zeV}\right) \qquad (28.5)$$

$$\frac{m}{z} = \frac{2eVt^2}{L^2} \qquad (28.6)$$

Thus, we have related the mass-to-charge (*m/z*) ratio of the analyte species as a positively or negatively charged ionic species to the drift time (t) of the ion through the drift tube while keeping the acceleration potential (V) and the length (L) of the drift tube fixed. This effectively allows the TOF mass analyzer to separate ionized species according to their mass-to-charge ratio. In general, an ion with a smaller mass and a larger charge will have the shorter drift time t. All mass analyzers must possess the physical characteristic of separating ionized species according to their mass-to-charge ratio.

An example of a typical mass spectrum illustrating one usual design used is illustrated in Figure 28.10. The *x*-axis of the mass spectrum represents the mass-to-charge (*m/z*) ratio measured by the mass analyzer, and the *y*-axis is a percentage scale that has been normalized to the most abundant species in the spectrum. Often, the *y*-axis may also be presented as an intensity scale of arbitrary units. In this particular mass spectrum, there are three peaks illustrated at *m/z* 468, *m/z* 469, and *m/z* 470.

28.6 EXACT MASS VERSUS NOMINAL MASS

Mass analyzers with high enough resolution have the ability to separate the naturally occurring isotopes of the elements. Therefore, in mass spectra, there is often observed a series of peaks for the same species. This is illustrated in Figure 28.11 for an *m/z* 703.469 species (the farthest most left peak with lowest mass). The isotopic peaks are at *m/z* 704.465 and *m/z* 705.480. The *m/z* 703.469 peak is known as the monoisotopic peak where the particular species is made up of ^{12}C, and if containing hydrogen, oxygen, nitrogen, phosphorus, etc., it will be comprised of the most abundant isotope for that element: ^{1}H, ^{16}O, ^{14}N, ^{31}P, etc. Organic compounds are primarily comprised of carbon and hydrogen; therefore, it is these two elements that contribute the most to the isotopic pattern observed in mass spectra (mostly carbon contribution) and to the mass defect of the molecular weight of the species (mostly hydrogen contribution). The mass defect of a species is the difference between the molecular formula's nominal value and its exact mass. A molecular formula nominal mass is calculated using the integer value of the most abundant isotope, and a molecular formula exact mass (monoisotopic mass) is calculated using the most abundant isotopes' exact masses. For a formula of $C_{10}H_{20}O_3$, the nominal mass is 10×12 amu $+ 20\times 1$ amu $+ 3\times 16$ amu $= 188$ amu, and the exact monoisotopic mass is 10×12 amu $+ 20\times 1.007825035$(12) amu $+ 3\times 15.99491463(5)$ amu $= 188.141245$ amu. Therefore, the mass defect is equal to 188.141245 amu $- 188$ amu $= 0.141245$ amu. For the mass spectrum shown in Figure 28.11, the second peak at *m/z* 704.465 is due to the inclusion of one ^{13}C into the structure of the measured species. The peak at *m/z* 705.480 is present due to the inclusion of two ^{13}C into the structure of the measured species.

The periodic table that the student is most familiar with lists the average weighted mass for the various elements listed in the table. For example, carbon in the periodic table is listed as 12.0107 amu, while in exact mass notation used in mass spectrometry, carbon has an atomic weight of 12 amu. The abundances of the natural isotopes of carbon are ^{12}C (12.000000 amu by definition) [43] at 98.93(8)% and ^{13}C (13.003354826(17) amu) at 1.07(8)%. Therefore, the weighted average of the two carbon isotopes gives a mass of 12.0107 amu as is found in periodic tables. This is true of all the elements where the average atomic weight values in the periodic table represent the sum of the mass of the isotopes for that particular element according to their natural abundances (see Appendix I for a listing of the naturally occurring isotopes).

Some of the elements such as chlorine or bromine if present will also have a significant influence on the isotopic pattern observed in mass spectra. For example, chlorine has two isotopes—^{35}Cl (34.968852721(69) amu at 75.78(4)% natural

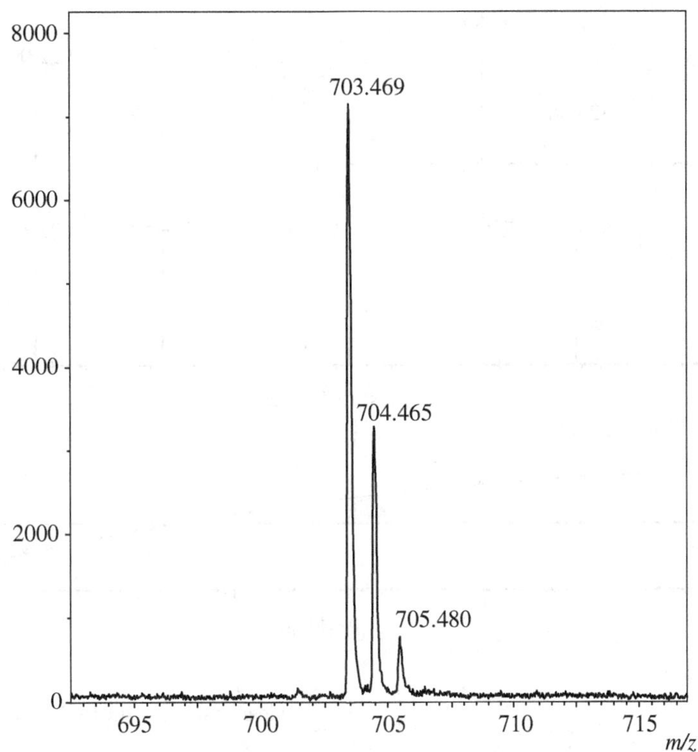

FIGURE 28.11 Mass spectrum of isotopically resolved peaks where *m/z* 703.469 species (the farthest most left peak of lowest mass) contains only ^{12}C and is known as the monoisotopic peak, the *m/z* 704.465 peak contains one ^{13}C, and the *m/z* 705.480 peak contains two ^{13}C.

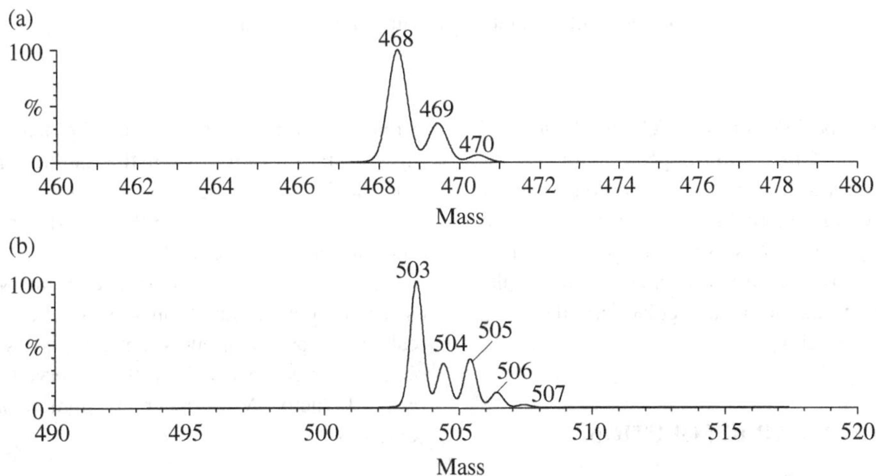

FIGURE 28.12 Isotope patterns for molecular formulas $C_{30}H_{60}O_3$ (top with *m/z* 468 nominal mass) and $C_{30}H_{60}O_3Cl$ (bottom with *m/z* 503 nominal mass).

abundance) and ^{37}Cl (36.96590262(11) amu at 24.22(4)% natural abundance)—that produce a more complex isotope pattern than that illustrated in Figure 28.9. The contribution of a chlorine atom to the isotopic pattern seen in mass spectra is illustrated in Figure 28.12 where the spectrum in Figure 28.12a is the isotope pattern for a molecular formula of $C_{30}H_{60}O_3$ (nominal mass of 468). In this isotope pattern, there are three peaks observed for the ^{12}C monoisotopic peak at *m/z* 468, the inclusion of one ^{13}C at *m/z* 469 and the inclusion of two ^{13}C at *m/z* 470. If we take the same formula and add one chlorine atom giving a molecular formula of $C_{30}H_{60}O_3Cl$ (nominal mass of 503), we will see a more complex isotope pattern. This is illustrated in the mass spectrum of Figure 28.12b where there are now five clear isotopic peaks present. The *m/z* 503 peak represents the ^{12}C- and ^{35}Cl-containing species, the *m/z* 504 peak represents the ^{12}C- and ^{35}Cl-containing species with one ^{13}C, the *m/z* 505 peak represents a mixture of a ^{12}C- and ^{35}Cl-containing species with two ^{13}C- and a ^{12}C- and ^{37}Cl-containing species, the *m/z* 506 peak represents a

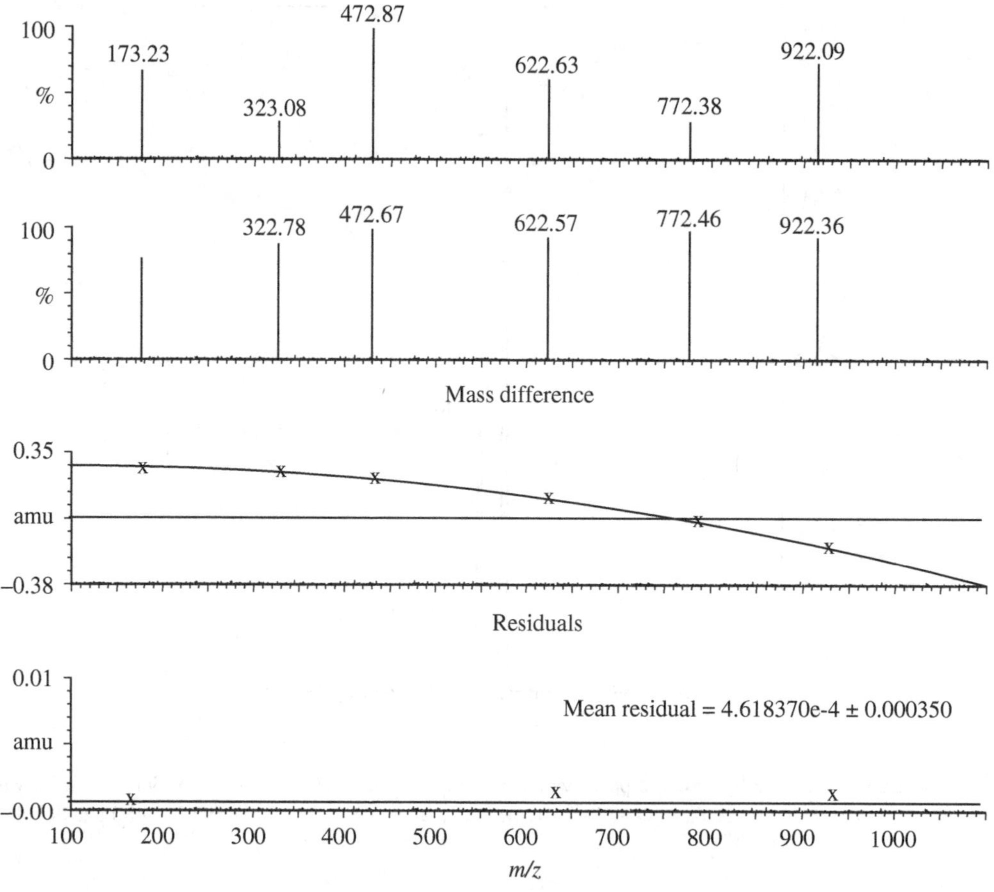

FIGURE 28.13 Calibration output for NaI clusters.

^{12}C- and ^{37}Cl-containing species with one ^{13}C, and the *m/z* 507 peak represents a ^{12}C- and ^{37}Cl-containing species with two ^{13}C. This can in fact be a useful tool when a more complex isotopic series is observed that contains more than three peaks in the identification of elements present. Most software packages used with mass spectrometers have the ability to calculate and display isotopic series for a given atomic or molecular formula and to perform isotopic pattern matching.

28.7 MASS ACCURACY AND RESOLUTION

Mass accuracy in mass spectrometry is related to the calibration of the mass analyzer to properly assign the true mass-to-charge ratio to a detected ion and to the resolution of the detector response, which is in the form of an intensity spike or peak within a mass spectrum. In calibrating a mass spectrometer, a series of standard compounds, usually in the form of a multicomponent standard, are measured and related to the fundamental properties of the mass analyzer such as the drift time within a TOF mass spectrometer drift tube. An example of an electrospray TOF mass analyzer calibration is illustrated in Figure 28.13 for sodium iodide clusters (NaI). The top spectrum is the experimental spectrum obtained on the instrument in the laboratory. The second to top spectrum is a reference spectrum of a sodium iodide standard. A mass difference plot and residual plot are also displayed showing the statistical results of the calibration fit. Calibrations of mass analyzers can be linear or polynomial; however, if more than two points are used for the calibration of the mass analyzer, a polynomial fit is generally used.

The mass accuracy in mass spectrometry is usually calculated as a parts-per-million (ppm) error where the theoretical mass (calculated as the monoisotopic mass) is subtracted from the observed mass, divided by the observed mass and multiplied by a 10^6 factor. Mass accuracy (ppm error) can thus be represented as

$$\text{Mass accuracy (ppm)} = \frac{m_{\text{observed}} - m_{\text{theoretical}}}{m_{\text{observed}}} \times 10^6 \quad (28.7)$$

In mass spectrometry, mass resolution is generally calculated as $m/\Delta m$ where m is the *m/z* value obtained from the spectrum and Δm is the full peak width at half maximum (FWHM). This is illustrated in Figure 28.14 for the *m/z* 703.469 ion, which has a full peak width at half maximum of 0.1797 *m/z*. The resolution of the *m/z* 703.469 would be calculated as $m/\Delta m = 703.469/0.1797 = 3914$. The narrower and sharper a peak is will result in a higher-resolution value and a better estimate of the apex of the Gaussian-shaped peak, which thus results in a better estimate of the true value of the mass-to-charge ratio and therefore

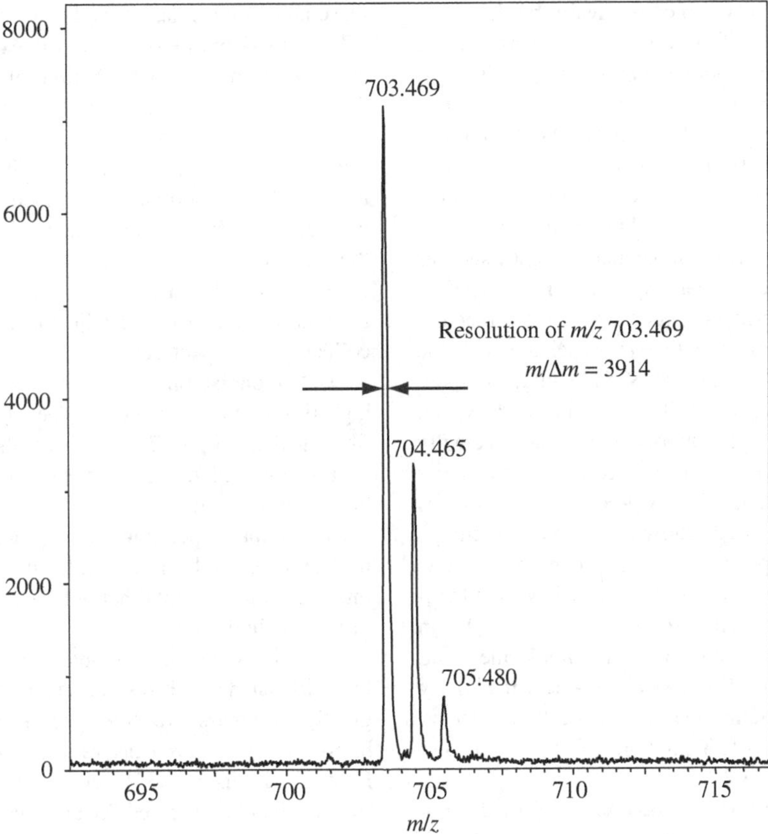

FIGURE 28.14 Full width at half maximum (FWHM) calculation of the resolution of m/z 703.469 as $m/\Delta m$ = 3914.

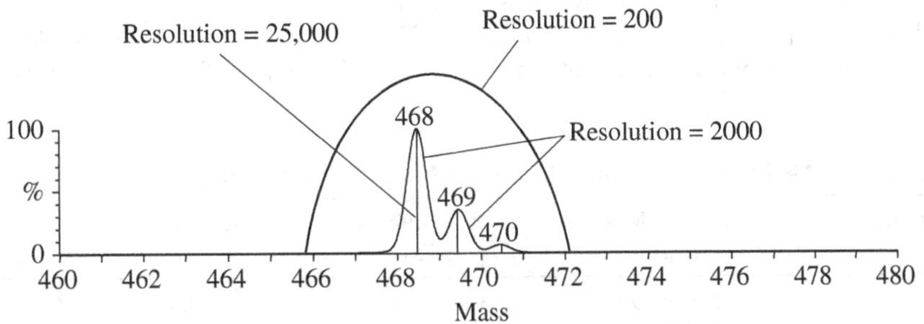

FIGURE 28.15 Effect of mass resolving power on the mass spectrum.

higher accuracy. In contrast to this, as a peak becomes broader and wider, the apex of the peak becomes more distorted, resulting in a lower-resolution value and thus reducing the accuracy of the estimate of the mass-to-charge (m/z) ratio.

In order to get a better idea of the impact that the mass resolving power of a mass analyzer has upon mass spectra, the illustration in Figure 28.15 shows a comparison of a mass resolution of approximately 200 for the overall curve where none of the isotopic peaks have been resolved. A resolution of 200 will give an average mass value for the species being measured with a high degree of error for the exact mass (>500 ppm). The three isotope peaks are resolved at a resolution of 2000, and an intermediate accurate exact mass is obtainable. At a resolution of 25,000, the three isotopic peaks are completely baseline resolved, and a mass accuracy typically less than 5 ppm can be obtained.

28.8 HIGH-RESOLUTION MASS MEASUREMENTS

There is a useful property that can be used when measuring the monoisotopic peak that was previously discussed in Section 28.5. For low molecular weight species (usually less than 2000 amu) measured as mass-to-charge (m/z) ratio ions in the gas phase using mass spectrometry, the monoisotopic peak will be the leftmost peak (lowest m/z value) of the isotopic peaks for that particular species, will be the most abundant peak in the isotopic series, and will contain only ^{12}C isotopes. Again, by definition and used as a standard, the ^{12}C isotope has been assigned a mass of 12.00000000 amu. In bioorganic molecules, the isotopic species most often encountered are carbon 12 as ^{12}C = 12.00000000 amu, hydrogen 1 as ^{1}H = 1.007825035 amu, nitrogen 14 as ^{14}N = 14.00307400 amu, phosphorus 31 as ^{31}P = 30.973763 amu, and oxygen 16 as ^{16}O = 15.99491463 amu. When adding up the isotopes' mass contributions for a particular molecule (known as the theoretical molecular weight), such as a lithium adduct of *cis*-9-octadeceamide as $[C_{18}H_{35}NOLi]^+$, a cation in positive ion mode mass spectrometry, the molecular weight will include the mass error that is associated with the masses just listed for hydrogen (+0.007825035 amu), nitrogen (+0.00307400 amu), phosphorus (−0.026237 amu), and oxygen (−0.0.00508537 amu) where phosphorus and oxygen actually have a negative mass defect. The theoretical molecular weight for the lithium adduct of *cis*-9-octadeceneamide as $[C_{18}H_{35}NOLi]^+$ is *m/z* 288.2879 (shown as a mass-to-charge *m/z* ion measured by mass spectrometry). To a certain accuracy, we can derive the only molecular formula that will fit the theoretical mass of 288.2879 Da. At a mass accuracy of 50 ppm and using the most common atoms found in biomolecules (and also lithium 7 at ^{7}Li = 7.016005 Da as we already have the luxury of knowing what type of cationized species it is), there exist six molecular formulas for a mass of 288.2879 Da. The six molecular formulas are listed in Table 28.2.

As can be seen in Table 28.1, a mass accuracy of 25 ppm will include the top four entries; at a mass accuracy of 10 ppm, the top three will be included; and finally, at a mass accuracy of 5 ppm, only the top formula will be included, which is in fact the correct one. In this example, we knew the true formula beforehand. However, often times the mass spectrometrist will be working with unknown samples. In this case, the most accurate measurement of the monoisotopic peak is essential for reasonable molecular formula determination or confirmation. A benchmark or rule-of-thumb value of ≤5 ppm for the mass error is often used for consideration of an unknown's molecular formula derived from a high-resolution mass measurement. Figure 28.16 shows the results of a high-resolution mass measurement of the lithium adduct of *cis*-9-octadecenamide. The spectrum has been smoothed, centroided, and a lock mass technique has been used for fine adjustment of the calibration curve within the vicinity of the monoisotopic peak of interest. In this particular case, the NaI cluster peak at *m/z* 322.7782 has been used as an internal standard for fine adjustment of the calibration used in collecting this spectrum.

The monoisotopic peak of *cis*-9-octadecenamide has a high-resolution mass measurement value of *m/z* 288.2876 as illustrated in Figure 28.16. Using the mass accuracy equation of Equation 28.7, we calculate an error of 1.0 ppm for the high-resolution mass measurement of *cis*-9-octadecenamide. Using the mass spectrometric approach of high-resolution mass measurement and an error tolerance of ≤5 ppm, there is only one molecular formula that will match demonstrating the utility of this methodology.

At higher molecular weights, the possible molecular formulas that will match a high-resolution mass measurement will increase greatly for a mass error tolerance of ≤10 ppm. For example, for a high-resolution mass measurement of *m/z* 489.3110, illustrated in Figure 28.17, using the NaI cluster at *m/z* 472.6725 as the internal standard calibrant gives 23 possible molecular formulas. The 23 molecular weight possibilities for the unknown species at *m/z* 489.3110 are listed in Table 28.3 for a mass error tolerance of ≤10 ppm. The most abundant atoms that make up biomolecules were included in the molecular formula search and are C, H, N, O, P, and Na.

If the mass error tolerance is decreased to ≤5 ppm, it can be seen in Table 28.2 that there are still 14 possibilities for the

TABLE 28.2 Possible Formulas for *m/z* 288.2879 at a Mass Accuracy of 50 ppm.

Theoretical Mass (Da)	Calculated Mass (Da)	Mass Difference (Da)	Mass Error (ppm)	Formula
288.2879	288.2879	0.0	0.0	$C_{18}H_{35}LiNO$
	288.2903	−2.4	−8.2	$C_{17}H_{38}NO_2$
	288.2855	2.4	8.4	$C_{19}H_{32}Li_2N$
	288.2815	6.4	22.3	$C_{14}H_{32}Li_2N_3O_2$
	288.2991	−11.2	−38.9	$C_{17}H_{35}LiN_3$
	288.3015	−13.6	−47.1	$C_{16}H_{38}N_3O$

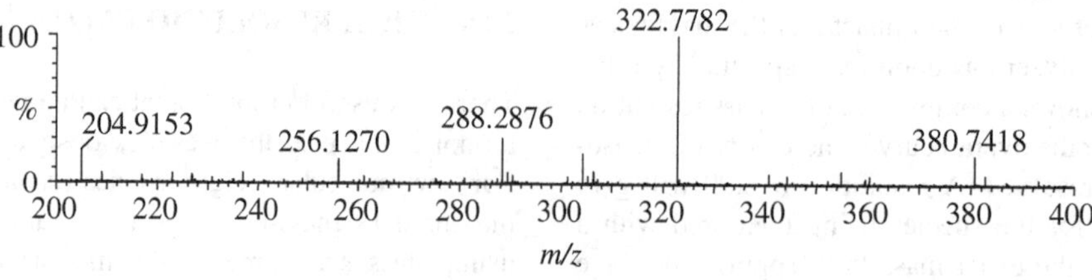

FIGURE 28.16 High-resolution mass measurement of *cis*-9-octadecenamide at *m/z* 288.2876 as the lithium adduct $[C_{18}H_{35}NOLi]^+$. The NaI cluster peak at *m/z* 322.7782 was used for the internal calibration standard.

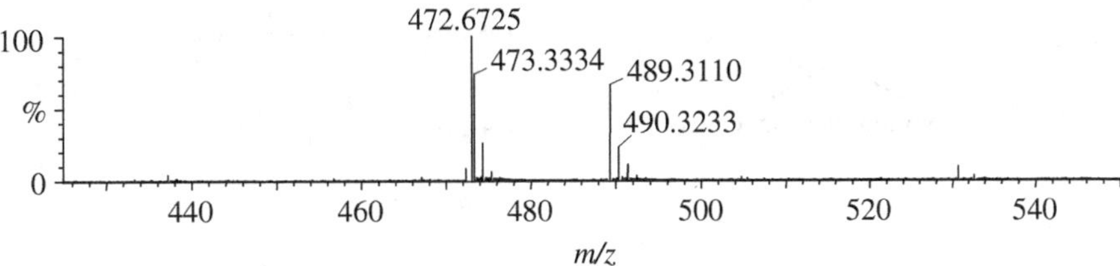

FIGURE 28.17 High-resolution mass measurement of unknown species at m/z 489.3110 as the sodium adduct $[M+Na]^+$. The NaI cluster peak at m/z 472.6725 was used for the internal calibration standard.

TABLE 28.3 Possible Formulas for *m/z* 489.3110 at a Mass Accuracy of 10 ppm.

Theoretical Mass (Da)	Calculated Mass (Da)	Mass Difference (Da)	Mass Error (ppm)	Formula
489.3110	489.3110	0.0	0.0	$C_{22}H_{51}O_7P_2$
	489.3110	0.0	0.1	$C_{27}H_{47}O_4PNa$
	489.3109	0.1	0.1	$C_{32}H_{43}ONa_2$
	489.3116	−0.6	−1.1	$C_{24}H_{49}N_2ONaP_2$
	489.3117	−0.7	−1.5	$C_{31}H_{41}N_2O_3$
	489.3099	1.1	2.2	$C_{21}H_{48}N_4O_3NaP_2$
	489.3099	1.1	2.2	$C_{26}H_{44}N_4Na_2P$
	489.3123	−1.3	−2.7	$C_{28}H_{43}N_4NaP$
	489.3123	−1.3	−2.7	$C_{23}H_{47}N_4O_3P_2$
	489.3094	1.6	3.4	$C_{24}H_{46}N_2O_6P_2$
	489.3093	1.7	3.4	$C_{29}H_{42}N_2O_3Na$
	489.3133	−2.3	−4.8	$C_{34}H_{42}ONa$
	489.3134	−2.4	−4.9	$C_{29}H_{46}O_4P$
	489.3086	2.4	5.0	$C_{25}H_{48}O_4Na_2P$
	489.3136	−2.6	−5.3	$C_{19}H_{45}N_4O_{10}$
	489.3140	−3.0	−6.1	$C_{26}H_{48}N_2ONaP_2$
	489.3077	3.3	6.8	$C_{26}H_{41}N_4O_5$
	489.3147	−3.7	−7.6	$C_{30}H_{42}N_4P$
	489.3069	4.1	8.3	$C_{22}H_{47}N_2O_6NaP$
	489.3069	4.1	8.4	$C_{27}H_{43}N_2O_3Na_2$
	489.3152	−4.2	−8.6	$C_{22}H_{46}N_2O_8Na$
	489.3064	4.6	9.5	$C_{25}H_{45}O_9$
	489.3157	−4.7	−9.7	$C_{36}H_{41}O$

identification of the unknown species molecular weight. If the mass error tolerance is decreased to ≤2 ppm, there are now still five possibilities illustrating the utility of a high mass accurate measurement for decreasing the number of possible molecular formulas of an unknown species. Finally, the inclusion of the theoretical isotopic distribution can help to reduce the number of possible formulas associated with an unknown's molecular weight even further.

28.9 RINGS PLUS DOUBLE BONDS ($r + db$)

Due to the valences of the elements that make up most of the biomolecules that are analyzed by mass spectrometry, there is a very useful tool known as the "total rings plus double bonds" ($r + db$) that can be applied to molecular formulas such as $C_{12}H_{24}O_2$. The most general representation of the rings plus double bonds relationship is

$$r+db=\sum \text{Group IVA}-\frac{1}{2}\sum(H+\text{Group VIIA}) +\frac{1}{2}\sum \text{Group VA}+1 \tag{28.8}$$

In mass spectrometry, this expression is often simplified to only include the most common elements that typically make up the molecular ion species measured. This is generally presented as

$$r+db=x-\frac{1}{2}y+\frac{1}{2}z+1 \tag{28.9}$$

for the molecular formula $C_xH_yN_zO_n$

For example, normal hexane has no rings plus double bonds, cyclohexane contains 1 ring plus double bonds, and benzene contains 4 rings plus double bonds as illustrated in Figure 28.18.

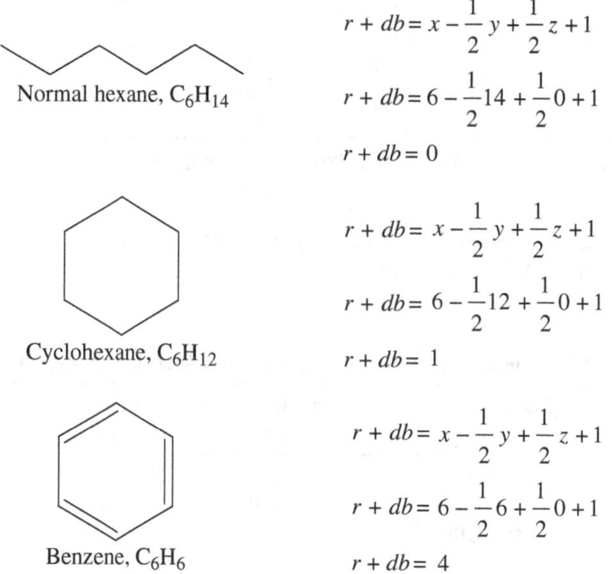

FIGURE 28.18 Application of the rings plus double bonds association for normal hexane, cyclohexane, and benzene using the simplified form of the equation for molecular formulas represented by $C_xH_yN_zO_n$.

The rings plus double bonds relationship presented in Equations 28.8 and 28.9 are applicable to organic compounds and to odd electron (OE) molecular ions ($M^{+\cdot}$) that are produced by electron ionization mass spectrometry EI-MS. However, for even electron (EE) ions that are produced by electrospray ionization mass spectrometry (ESI-MS) or matrix-assisted laser desorption ionization mass spectrometry (MALDI-MS) where the species has been ionized in the form of an acid adduct $[M+H]^+$, a metal adduct such as sodium $[M+Na]^+$, or a chloride adduct $[M+Cl]^-$, Equations 28.8 and 28.9 must be followed by the subtraction of ½. This is also a useful tool for interpreting whether a molecular ion is EE or OE, as will be applied in the next section "The Nitrogen Rule."

28.10 THE NITROGEN RULE IN MASS SPECTROMETRY

Another useful tool used in mass spectrometry of biomolecule analysis is the so called "nitrogen rule." The nitrogen rule for organic compounds states that if a molecular formula has a molecular weight that is an even number, then the compound contains an even number of nitrogen (e.g., 0N, 2N, 4N, etc.). The same applies to an odd molecular weight species, which will thus contain an odd number of nitrogen (e.g., 1N, 3N, 5N, etc.). The nitrogen rule is due to the number of valences that the different elements possess. The number of valences is the number of hydrogen that can be bonded to the element. Most of the elements encountered in mass spectrometry either have an even mass and subsequent even valence or an odd mass and an odd valence. For example, the even mass/even valence elements are:

Carbon ^{12}C has a valence of 4, CH_4 = 16 Da (even mass and even number of $N = 0$)

Oxygen ^{16}O has a valence of 2, H_2O = 18 Da (even mass and even number of $N = 0$)

Silicon ^{28}Si has a valence of 4, H_4Si = 32 Da (even mass and even number of $N = 0$)

Sulfur ^{32}S has a valence of 2, H_2S = 34 Da (even mass and even number of $N = 0$)

and the odd mass/odd valence elements are:

Hydrogen ^{1}H has a valence of 1, H_2 = 2 Da (even mass and even number of $N = 0$)

Fluorine ^{19}F has a valence of 1, HF = 20 Da (even mass and even number of $N = 0$)

Phosphorus ^{31}P has a valence of 3 or 5, H_3P = 34 Da (even mass and even number of $N = 0$)

Chlorine ^{35}Cl has a valence of 1, HCl = 36 Da (even mass and even number of $N = 0$).

However, nitrogen has an even mass but odd valence:

Nitrogen ^{14}N has a valence of 3, NH_3 = 17 Da (odd mass and odd number of $N = 1$).

An odd number of nitrogen results in an odd mass such as ammonia NH_3, which has a formula weight of 17 Da. An example of a two-nitrogen-containing compound would be urea, CH_4N_2O with a nominal mass of 60 Da. The nitrogen rule can be directly applied to OE molecular ion species ($M^{+\bullet}$) that are generated with electron ionization (EI) because the ionization process has not changed the species molecular formula or molecular weight. However, this is not the case with processes such as ESI-MS or MALDI-MS where the species has been ionized in the form of an acid adduct $[M+H]^+$, a metal adduct such as sodium $[M+Na]^+$, or a chloride adduct $[M+Cl]^-$. In these cases, the precursor ion is an EE ion, and the weight of the original molecule has changed, while the valence has not. Here, the exact opposite to the above stated nitrogen rule that applied to organic compounds and OE molecular ions is the case. In other words, the nitrogen rule for EE precursor ions states that an ion species with an odd mass will have an even number of nitrogen (e.g., 0N, 2N, 4N, etc.) and an ion species with an even mass will have an odd number of nitrogen (e.g., 1N, 3N, 5N, etc.).

28.11 CHAPTER PROBLEMS

28.1 What are the six components that comprise a mass spectrometer?

28.2 The mass analyzer separates charged ionic species in the gas phase according to what?

28.3 Why is it important for the mass analyzer and detector to be under high vacuum?

28.4 What two types of pumps are used with mass spectrometric instrumentation, and what pressures do they obtain approximately?

28.5 What are three areas that mass spectrometry has currently found an increased application to?

28.6 Give a definition of proteomics, metabolomics, and lipidomics.

28.7 How is it that the soft ionization technique of electrospray allows gas-phase analysis of solution-phase associations? What are some examples of these associations?

28.8 What are a couple of examples of ionized analytes that are measured using mass spectrometry?

28.9 Given an acceleration voltage (KE) of 2000 eV, calculate the mass-to-charge (*m/z*) ratio of a biomolecule measured by a TOF mass spectrometer with a drift tube length of 2 m and a flight time of 68 μs.

28.10 A biomolecule has a mass of 129 amu and a charge of 2+, if it is given an acceleration voltage of 1800 eV, what would be its drift time through a drift tube of length 2.3 m?

28.11 Calculate the exact mass (using the "Relative Atomic Mass" of the highest abundance isotope in Appendix I) and the weighted average mass (using the "Standard Atomic Weight" in Appendix I) for each of the following molecular formulas. What is the associated mass defect for each of the molecular formulas?

a. $C_{10}H_{32}O_2$
b. $C_{15}H_{28}N_2OPNa$
c. $C_{29}H_{42}N_2O_3K$
d. $C_{40}H_{75}N_2OP_3$
e. $C_5H_{10}O_3Cl_3$

28.12 Calculate the mass accuracy of the following observed masses for the following biomolecules according to their proposed formula:

a. $C_{24}H_{45}O_3N$ observed at *m/z* 395.4511
b. $C_{64}H_{120}O_{10}N_4P_2$ observed at *m/z* 1166.8398
c. $C_{48}H_{72}O_6$ observed at *m/z* 744.5310
d. $C_{78}H_{130}O_{12}N_8$ observed at *m/z* 1370.9932.

28.13 Using the following data, calculate the resolution of the following observed mass spectral peaks:

a. *M/z* 342.4766 with peak width of *m/z* 0.6624
b. *M/z* 1245.6624 with peak width of *m/z* 0.0573
c. *M/z* 245.1191 with peak width of *m/z* 1.2444.

28.14 Using the rings plus double bonds formula, decide whether the following ions are even electron or odd electron ions and draw their formula as such (as an example, the first one is done). Using the nitrogen rule, list whether you would expect each ion to have an even or odd mass-to-charge (*m/z*) ratio value. What is that value?

a. CH_4
$r + db = 0$, odd electron (OE) as $CH_4^{+\cdot}$, even mass ion with *m/z* = 16 Th
b. NH_3
c. NH_4
d. H_3O
e. C_2H_7O
f. $C_{14}H_{28}O_2$
g. $C_6H_{14}NO$
h. $C_4H_{10}SiO$
i. $C_{29}H_{60}N_2O_6P$.

REFERENCES

1. Ham, B.M.; Jacob, J.T.; Cole, R.B. *Anal. Bioanal. Chem.* 2007, **3**, 889–900.
2. Nabetani, T.; Tabuse, Y.; Tsugita, A.; Shoda, J. *Proteomics* 2005, **5**, 1043–1061.
3. Wolters, D.A.; Washburn, M.P.; Yates, J.R. *Anal. Chem.* 2001, **73**, 5683–5690.
4. Kislinger, T.; Gramolini, A.O.; MacLennan, D.H.; Emili, A. *J. Am. Soc. Mass Spectrom.* 2005, **16**, 1207–1220.
5. Pierson, J.; Norris, J.L.; Aerni, H.R.; Svenningsson, P.; Caprioli, R.M.; Andren, P.E. *J. Proteome Res.* 2004, **3**, 289–295.
6. Siuzdak, G. *J. Mass Spectrom.* 1998, **33**, 203–211.
7. Downard, K.M. *J. Mass Spectrom.* 2000, **35**, 493–503.
8. Rochfort, S. *J. Nat. Prod.* 2005, **68**, 1813–1820.
9. Nicholson, J.K.; Wilson, I.D. *Nat. Rev. Drug Discov.* 2003, **2**, 668–676.
10. Dettmer, K.; Aronov, P.A.; Hammock, B.D. *Mass Spectrom. Rev.* 2007, **26**, 51–78.

11. Beecher, C.W.W. In: *Metabolic Profiling: Its Role in Biomarker Discovery and Gene Function Analysis, Springer, New York, NY, 2003.*
12. Ginsburg, G.S.; Haga, S.B. *Expert Rev. Mol. Diagn.* 2006, **6**, 179–191.
13. Han, X.; Gross, R.W. *Mass Spectrom. Rev.* 2005, **24**, 367–412.
14. Han, X.; Gross, R.W. *J. Lipid Res.* 2003, **44**, 1071–1079.
15. Fahy, E. et al. *J. Lipid Res.* 2005, **46**, 839–862.
16. Pulfer, M.; Murphy, R.C. *Mass Spectrom. Rev.* 2003, **22**, 332–364.
17. Griffiths, W.J. *Mass Spectrom. Rev.* 2003, **22**, 81–152.
18. Lee, S.H.; Williams, M.V.; DuBois, R.N.; Blair, I.A. *Rapid Commun. Mass Spectrom.* 2003, **17**, 2168–2176.
19. Zamfir, A.D. et al. *J. Am. Soc. Mass Spectrom.* 2004, **15**, 1649–1657.
20. Xiao, Y.; Schwartz, B.; Washington, M.; Kennedy, A.; Webster, K.; Belinson, J.; Xu, Y. *Anal. Biochem.* 2001, **290**, 302–313.
21. Nakagawa, K.; Oak, J.H.; Higuchi, O.; Tsuzuki, T.; Oikawa, S.; Otani, H.; Mune, M.; Cai, H.; Miyazawa, T. *J. Lipid Res.* 2005, **46**, 2514–2524.
22. Ham, B.M.; Jacob, J.T.; Keese, M.M.; Cole, R.B. *J. Mass Spectrom.* 2004, **39**, 1321–1336.
23. Ham, B.M.; Jacob, J.T.; Cole, R.B. *Anal. Chem.* 2005, **77**, 4439–4447.
24. Armentrout, P. B.; Beauchamp, J. L. *J. Chem. Phys.* 1981, **74**, 2819–2826.
25. Shukla, A. K.; Futrell, J. H. *J. Mass Spectrom.* 2000, **35**, 1069–1090.
26. Daniel, J.M.; Friess, S.D.; Rajagopala, S.; Wendt, S.; Zenobi, R. *Int. J. Mass Spectrom.* 2002, **216**, 1–27.
27. Graul, S.T.; Squires, R.R. *J. Am. Chem. Soc.* 1990, **112**, 2517–2529.
28. Sunderlin, L.S.; Wang, D.; Squires, R.R. *J. Am. Chem. Soc.* 1993, **115**, 12060–12070.
29. Pramanik, B.N.; Bartner, P.L.; Mirza, U.A.; Liu, Y.H.; Ganguly, A.K. *J. Mass Spectrom.* 1998, **33**, 911–920.
30. Ervin, K.M.; Armentrout, P.B. *J. Chem. Phys.* 1985, **83**, 166–189.
31. Dougherty, R.C. *J. Am. Soc. Mass Spectrom.* 1997, **8**, 510–518.
32. Anderson, S.G.; Blades, A.T.; Klassen, J.; Kebarle, P. *Int. J. Mass Spectrom. Ion Process.* 1995, **141**, 217–228.
33. Klassen, J.S.; Anderson, S.G.; Blades, A.T.; Kebarle, P. *J. Phys. Chem.* 1996, **100**, 14218–14227.
34. Nielsen, S.B.; Masella, M.; Kebarle, P. *J. Phys. Chem. A* 1999, **103**, 9891–9898.
35. Kebarle, P. *Int. J. Mass Spectrom.* 2000, **200**, 313–330.
36. Peschke, M.; Blades, A.T.; Kebarle, P. *J. Am. Chem. Soc.* 2000, **122**, 10440–10449.
37. Whitehouse, C.M.; Dreyer, R.N.; Yamashita, M.; Fenn, J.B. *Anal. Chem.* 1985, **57**, 675–679.
38. Fenn, J.B. *J. Am. Soc. Mass Spectrom.* 1993, **4**, 524–535.
39. Cole, R.B. *J. Mass. Spectrom.* 2000, **35**, 763–772.
40. Cech, N.B.; Enke, C.G. *Mass Spec. Rev.* 2001, **20**, 362–387.
41. Gidden, J.; Wyttenbach, T.; Batka, J.J.; Weis, P.; Jackson, A.T.; Scrivens, J.H.; Bowers, M.T. *J. Am. Soc. Mass Spectrom.* 1999, **10**, 883–895.
42. Ham, B.M.; Cole, R.B. *Anal. Chem.* 2005, **77**, 4148–4159.
43. Rosman, K.J.R.; Taylor, P.D.P. Commission on Atomic Weights and Isotopic Abundances report for the International Union of Pure and Applied Chemistry in *Isotopic Compositions of the Elements 1997, Pure Appl. Chem.*, 1998, **70**, 217. [Copyright 1998 IUPAC].

29

IONIZATION IN MASS SPECTROMETRY

29.1 Ionization Techniques and Sources
29.2 Chemical Ionization (CI)
29.2.1 Positive CI
29.2.2 Negative CI
29.3 Atmospheric Pressure Chemical Ionization (APCI)
29.4 Electrospray Ionization (ESI)
29.5 Nanoelectrospray Ionization (Nano-ESI)
29.6 Atmospheric Pressure Photo Ionization (APPI)
29.6.1 APPI Mechanism
29.6.2 APPI VUV Lamps
29.6.3 APPI Sources
29.6.4 Comparison of ESI and APPI
29.7 Matrix Assisted Laser Desorption Ionization (MALDI)
29.8 FAB
29.8.1 Application of FAB versus EI
29.9 Chapter Problems
References

29.1 IONIZATION TECHNIQUES AND SOURCES

There are numerous ionization techniques for the production of molecular cations (e.g., $M^{+\bullet}$, a radical cation) or adducted ionized species (e.g., $[M + H]^+$ or $[M + Cl]^-$) from neutral biomolecules in use today. Closely related are the instrumental sources used for the introduction of the ionized species into mass spectrometers that are available to the mass spectrometrist and both will be covered in this chapter. Before an analyte can be measured by a mass analyzer, it must first be in an ionized state. Biomolecules for mass analysis are often in a neutral state and must be converted to an ionized state that may exist as a cation or as an anion depending upon the molecule to be mass analyzed. This is a very important fundamental aspect of mass spectrometry, and much developmental work has been devoted to processes that can be used to convert a neutral molecule into an ionized species. The source, which is used to transfer the newly formed ionized analyte into the mass spectrometer, is the second important step in mass analysis of ionized analyte species. The source is often considered synonymous with the ionization technique. This is often due to the ionization technique taking place within what would be considered the source. Source/ionization systems include electron ionization (EI), electrospray ionization (ESI), chemical ionization (CI), atmospheric pressure chemical ionization (APCI), atmospheric pressure photo ionization (APPI), and matrix-assisted laser desorption ionization (MALDI). These ionization techniques produce ions of analyte molecules (often designated as "M" for molecule), which includes molecular ions $M^{+\cdot}$(from EI), protonated molecules ($[M + H]^+$), deprotonated molecules ($[M - H]-$), and metal ($[M + metal]^+$, e.g., $[M + Na]^+$) or halide ($[M + halide]-$, e.g., $[M + Cl]-$) adduct (all possible from ESI, CI, APCI, APPI, and MALDI). All of these ionization techniques and sources are used to effectively ionize a neutral molecule and transfer the formed ionized molecule into the gas phase in preparation to introduction into the low-vacuum environment of the mass analyzer.

29.2 CHEMICAL IONIZATION (CI)

A complimentary ionization technique to EI is CI. CI often can be chosen to be used as a softer ionization technique rather than electron impact (EI), which is known as a hard ionization technique often depleting the molecular ion produced during the ionization. Therefore, a greater abundance of the analyte in an ionized state is observed in CI as compared to EI. CI in turn can add confidence in the determination of the molecular weight of the ionized analyte, while EI gives structural information through extensive fragmentation of the molecular ion. This is illustrated in Figure 29.1 for the CI mass spectrum of methyl oleate. Here, we see the precursor ion at *m/z* 297.2 as a protonated species $[M + H]^+$, giving MW information for the analyte. In CI, the source acts to create a very reactive, high-pressure region containing the ionized reagent gas known as the plasma region. It is in this plasma region that the analyte passes through allowing ionization and subsequent mass analysis and detection. The source design for CI is similar to that used for EI however with a few operational differences required for

Analytical Chemistry: A Chemist and Laboratory Technician's Toolkit, First Edition. Bryan M. Ham and Aihui MaHam.

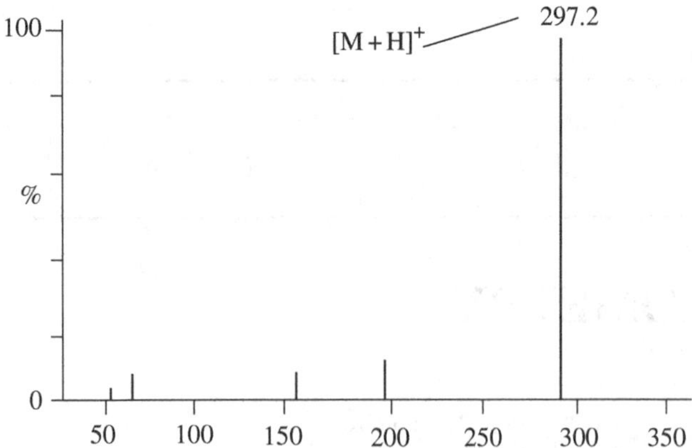

FIGURE 29.1 Chemical Ionization (CI) mass spectrum of methyl oleate at *m/z* 297.2 as a protonated species $[M+H]^+$, giving MW information for the analyte without extensive fragmentation.

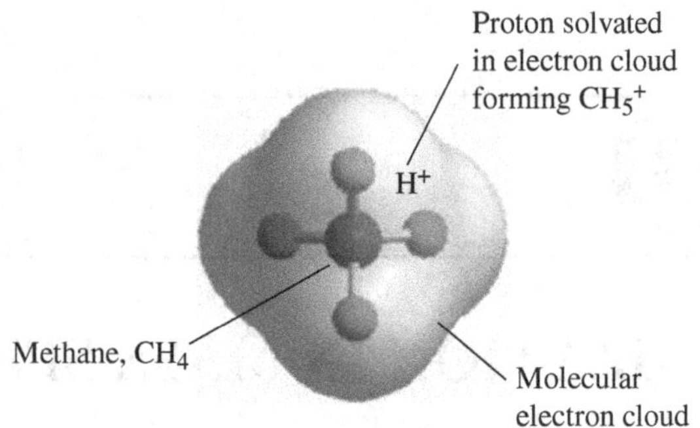

FIGURE 29.2 Protonated form of methane. The proton is solvated by the sigma bonds of the covalently bound hydrogen producing a highly reactive and unstable species. The protonated methane will transfer a proton to a neutral analyte (exothermic reaction) producing the ionized form of the analyte as a protonated species $[M+H]^+$.

the technique. The slits and apertures are smaller in the CI source to help maintain the high pressure that is created in the source due to the addition of a reactive gas into the source used for the CI. In CI, the emission current leaving the filament is measured instead of the target current as in EI. The repeller voltage is also kept lower in CI as compared to EI in order to maximize the number of collisions between the analyte and the CI gas. In the CI source design, a reagent gas is introduced into the "tight" source and maintained at a pressure that is higher than the mass analyzer but lower than atmospheric pressure. A maintained higher pressure in the source of the reagent gas serves to maximize the number of collisions between the reagent gas and the analyte. The higher pressure also serves in promoting dampening collisions between the analyte and the reagent gas. Dampening collisions are slow, low-energy collisions that help to relax the ionized analyte molecules by transferring excess energy from the ionized analyte to the gas molecule. These dampening collisions help to decrease the amount of fragmentation that can take place due to excess energy from the ionization process. This helps to maintain a softer ionization producing a greater abundance of intact ionized analyte molecules. However, the amount of energy transferred in the ionization step can also be controlled by the analyst through the use of different chemical reagent gasses as will be discussed next.

In the CI process, the first step is the ionization of the reagent gas by the electron beam from the filament/target setup, which is similar to that found in EI sources. The ionization of the reagent gas by the electron beam often produces a protonated reagent gas that then transfers a proton to the analyte producing the ionized form of the analyte, $[M+H]^+$. This gives even electron (EE) analyte ions that are generally more stable then the radical, odd electron (OE) molecular ions ($M^{+\bullet}$) that are produced with EI. In positive CI that results in protonation, the analyte ion is 1 mass unit greater than the neutral analyte (M + H). In negative CI that results in proton abstraction, the analyte ion is 1 mass unit less than the neutral analyte (M − H):

$$\begin{aligned}&\text{Positive ion: } GH^+ + M \rightarrow MH^+ + G\\&\text{Negative ion: } (G-H)^- + M \rightarrow (M-H)^- + G\end{aligned} \quad (29.1)$$

where G is the reagent gas and M is the analyte molecule.

29.2.1 Positive CI

CI processes are ion–molecule reactions between the analyte molecules (M) and the ionized reagent gas ions (G) that produce the analyte ions. These are gas-phase acid–base reactions according to the Bronsted–Lowry theory. In general, these are exothermic reactions taking place in the gas phase. In positive CI, the three most common reagent gasses used are methane (CH_4), isobutane (i-C_4H_{10}), and ammonia (NH_3). The first step in the process is EI of the reagent gas by the filament-produced electron beam in the source. This is illustrated for the reactive gas formation of methane as follows (the species designated with "∗" are the most reactive forms of the ionized gas molecules):

$$\begin{aligned}&CH_4 + e^- \rightarrow [CH_4]^{+\bullet} + 2e^-\\&\qquad\qquad\quad \hookrightarrow CH_3^+ + H^\bullet\\&CH_4^{+\bullet} + CH_4 \rightarrow (CH_5^+)^* + CH_3^\bullet\\&CH_4^{+\bullet} + CH_4 \rightarrow (C_2H_5^+)^* + H_2 + H^\bullet\end{aligned} \quad (29.2)$$

In the ionization of the reagent gas, a very reactive, high-pressure region in the source containing the reagent ionized gas has been created that is called the plasma region. The protonated form of methane, $(CH_5^+)^*$, is a very reactive reagent gas that transfers the proton to the analyte with a high degree of excess energy. This is a very exothermic gas-phase reaction that is a "hard" ionization of the analyte producing an ionized analyte that often to some degree fragments. Methane is able to accommodate the extra proton in the gas phase through a solvation of the proton by the sigma electrons of the four covalent hydrogen bonds (illustrated in Figure 29.2). The protonated form of methane however is very unstable and highly reactive.

A second example of a reactive CI gas used is isobutane. Again, the first step in the process of creating the reactive gas is the EI of the reagent gas by an electron-emitting filament. The EI produces the radical cation form of the isobutane reagent gas:

$$\begin{aligned}&i\text{-}C_4H_{10} + e^- \rightarrow i\text{-}C_4H_{10}^{+\bullet} + 2e^-\\&i\text{-}C_4H_{10}^{+\bullet} + i\text{-}C_4H_{10} \rightarrow \left(i\text{-}C_4H_9^+\right)^* + C_4H_9^\bullet + H_2\end{aligned} \quad (29.3)$$

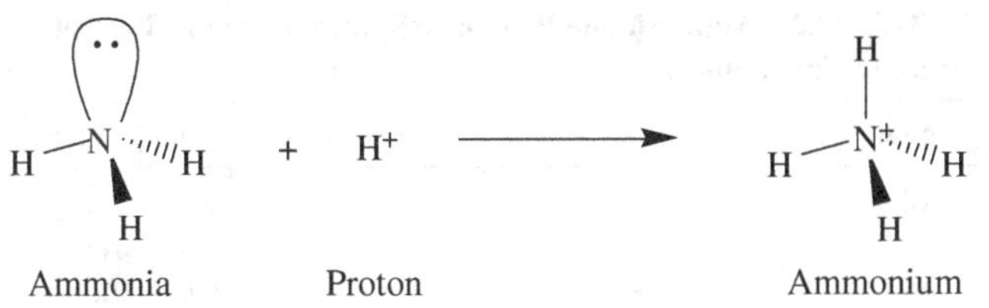

FIGURE 29.3 Reaction for the protonation of ammonia. Ammonia has available a lone pair of electrons in which to donate to the proton, forming the protonated reactive reagent gas species ammonium (NH_4^+). The proton shares the lone pair of electrons on the nitrogen producing a reactive species that is more stable than protonated methane. The protonated ammonia will transfer a proton to a neutral analyte (exothermic reaction) producing the ionized form of the analyte as a protonated species.

The second step is the same as observed for methane in Equation 29.2, the reactive radical cation form of the reagent gas reacts with other reagent gas molecules present to form the reactive CI species. Unlike methane though, isobutane produces primarily only one reactive species as compared to two that are produced when using methane as the reagent gas. In the case of isobutane, however, the reactive species is not a protonated form of isobutane, but the cation gaseous species i-$C_4H_9^+$ that is highly reactive and able to transfer a proton (H^+) for analyte ionization.

Finally, we have ammonia as a third example of a gaseous reagent used in CI. Like methane, ammonia also produces two reactive species that may be used in the CI of neutral analyte molecules. In the second step of the ionization of ammonia to produce the reactive reagent gas species, a protonated form of ammonia is produced. This species is more stable than the protonated form of methane:

$$\begin{aligned} &NH_3 + e^- \rightarrow NH_3^{+\bullet} + 2e^- \\ &NH_3^{+\bullet} + NH_3 \rightarrow \left(NH_4^+\right)^* + NH_2^\bullet \\ &NH_4^+ + NH_3 \rightarrow \left(N_2H_7^+\right)^* \end{aligned} \tag{29.4}$$

due to the lone pair of electrons available on the nitrogen atom in the ammonia molecule, which helps to delocalize the positive charge (see Figure 29.3). The two reactive species produced from ammonia are both protonated species that have an extra proton that is available to transfer to a neutral analyte during the CI process.

The reactivity in CI is dependent upon the proton affinities (PA) of the ionized reagent gas (RG) and the analyte (M). The transfer of the proton from the reagent gas to the analyte must be an exothermic reaction for this to occur spontaneously in the CI source. The change in enthalpy (ΔH^0) for the following reaction must be less than 0 for the transfer of the proton to take place spontaneously:

$$RGH^+ + M \rightarrow MH^+ + RG \quad \Delta H^0 < 0 \tag{29.5}$$

The PA of the analyte molecule must be greater than the PA of the reactive reagent gas; $PA_{molecule} > PA_{reagent\ gas}$. This is almost always the general case as the reagent gas is a highly reactive species with a very low relative PA (in other words,

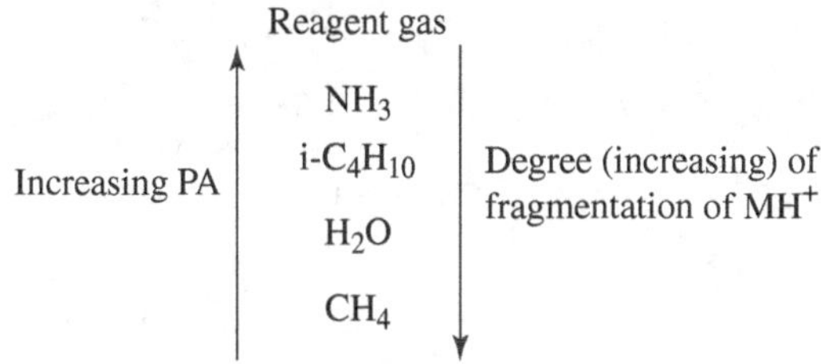

FIGURE 29.4 Illustration of the inverse relationship between the proton affinity of reagent gasses versus the degree of fragmentation observed upon protonation of the analyte molecule within the chemical ionization source. Ammonia (NH_3) has the greatest proton affinity producing the lowest amount of fragmentation of the protonated analyte species.

the reagent gas is looking to lose the extra proton it possesses). The PA is equal to the change in enthalpy of the deprotonation reactions of the reagent gas and the analyte:

$$\begin{array}{ll} RGH^+ \rightarrow RG + H^+ & \Delta H = PA_{reagent\ gas} \\ \underline{M + H^+ \rightarrow MH^+} & \underline{\Delta H = PA_{molecule}} \\ RGH^+ + M \rightarrow MH^+ + RG & \Delta H = PA_{RG} - PA_M \end{array} \tag{29.6}$$

If $PA_M > PA_{RG}$ than the change in enthalpy is negative, the reaction is exothermic and can proceed spontaneously to the right, thus protonating the analyte molecule.

The reagent gasses used in CI possess different PA; therefore, the proper choice of reagent gas must be made according to the amount of fragmentation desired. The greater the exothermicity of the proton transfer reaction will dictate the amount of fragmentation that will take place in the CI process. The following diagram illustrates the relationship between the PA of the reagent gas and the degree of fragmentation that is observed upon ionization (proton transfer) of the analyte molecule.

As illustrated in Figure 29.4, there is an inverse relationship between PA of the reagent gas and the degree of fragmentation of the analyte molecule. Ammonia (NH_3) has the greatest PA owing to its lone pair of electrons on the nitrogen. This in turn equates to a lesser degree of fragmentation of the analyte molecule during the proton transfer reaction in the CI source. At the bottom of the figure, we see that methane has the lowest PA and therefore produces the greatest amount of fragmentation of the analyte molecule during the proton transfer reaction in the CI source. The effect of the choice of reagent gas is also illustrated in Figure 29.5 for the mass analysis of 4-hydroxy-pentanoic acid methyl ester ($C_6H_{12}O_3$). In Figure 29.5a, ammonia (NH_3) was chosen as the reagent gas where very little fragmentation is observed to be taking place. Figure 29.5b illustrates substantial fragmentation with the use of methane as the reagent gas (CH_4).

The PA for typical CI reagent gasses are listed in Table 29.1. The PA for a number of organic compounds are listed in Table 29.2. The protonated form of diatomic hydrogen (H_3^+) has a PA value of 418.4 kJ/mol, while the protonated form of ammonia has a PA value of 857.7 kJ/mol. Again, the PA of a substance is the amount of energy required to pull off the positively charged proton (H^+). It takes much less energy to remove the proton from protonated diatomic hydrogen than it does for protonated ammonia. As discussed previously, the lone pair of

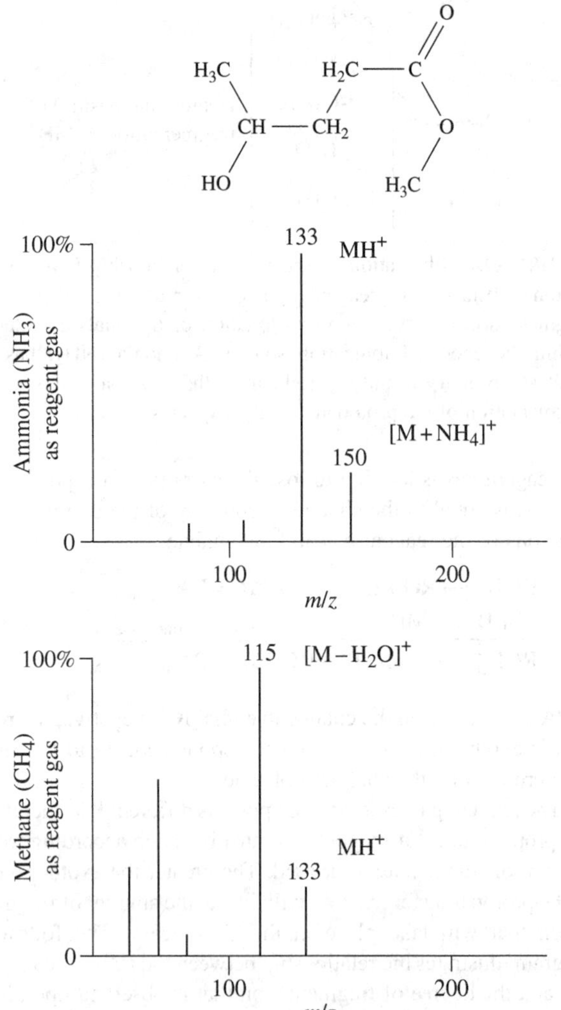

FIGURE 29.5 Effect of the choice of reagent gas for the mass analysis of 4-hydroxy-pentanoic acid methyl ester ($C_6H_{12}O_3$). (a) Ammonia (NH_3) chosen as the reagent gas where very little fragmentation is observed to be taking place. (b) Substantial fragmentation with the use of methane as the reagent gas (CH_4).

TABLE 29.1 Proton Affinities of Chemical Ionization Reagent Gasses.

Reagent Gas	Ionized Reagent Gas	Proton Affinity (kJ/mol)
H_2	H_3^+	418.4
CH_4	CH_5^+	531.4
H_2O	H_3O^+	690.4
i-C_4H_{10}	i-$C_4H_9^+$	815.9
NH_3	NH_4^+	857.7

electrons on the nitrogen of ammonia is available to delocalize the charge on the proton, thus stabilizing the ion. The charge on the proton adducted to diatomic hydrogen only has available to it the sigma (σ) electrons of the hydrogen–hydrogen covalent bonds making this species unstable and highly reactive. From Table 29.2, we see that the PA of alcohols in general is greater than the PA of the protonated form of methane (CH_5^+); therefore, CH_5^+ will always protonate alcohols. For example, if we select the middle of the PA range for alcohols as 774.0 kJ/mol, the overall change in enthalpy for the CI reaction with protonated methane would be

TABLE 29.2 Approximate Proton Affinities for Some Typical Organic Compounds.

Neutral Species	Proton Affinity (kJ/mol)
Alcohols	732.2–815.9
Acids	736.4–811.7
Aldehydes	740.6–807.5
Esters	795.0–845.2
Ethers	807.5–841.0
Ketones	836.8

$$\begin{array}{ll} CH_5^+ \rightarrow CH_4 + H^+ & \Delta H = 531.4\ \text{kJ/mol} \\ \text{Alcohols} + H^+ \rightarrow \text{AlcoholsH}^+ & \Delta H = -774.0\ \text{kJ/mol} \\ \hline CH_5^+ + \text{Alcohols} \rightarrow \text{AlcoholsH}^+ + CH_4 & \Delta H = -242.6\ \text{kJ/mol} \end{array} \quad (29.7)$$

The change in enthalpy for the reaction is −242.6 kJ/mol indicating that the reaction is exothermic and will proceed to the right spontaneously.

29.2.2 Negative CI

Negative CI, while less common than positive CI presented in Section 29.2.1, can also be used as a method for analyte ionization in preparation for mass analysis. As we saw earlier, the most common form of positive CI is the protonation of a neutral analyte molecule by a highly reactive reagent gas. In contrast to this, one form of negative CI is the production of a negatively charged analyte species through proton exchange or abstraction. In this process, a reactive ion that has a large affinity for a proton is used to remove a proton from the analyte. Some examples of commonly used reactive ions are fluoride (F^-), chloride (Cl^-), oxide radical ($O^{-\bullet}$), hydroxide (OH^-), and methoxide (CH_3O^-). This form of CI is an acid–base reaction that can generally be represented as

$$X^- + M \rightarrow XH + [M-H] \quad (29.8)$$

where X^- represents the reactive reagent gas ion and M the neutral analyte.

Due to the high degree of negative charge on the reactive reagent gas ion, the proton exchange from the analyte to the reactive ion is exothermic and will proceed to the right spontaneously. An actual example is as follows where a proton is abstracted from the commonly used ketone solvent acetone by the reactive chloride anion:

$$Cl^- + H_3C{-}\overset{\overset{\Large O}{\|}}{C}{-}CH_3 \longrightarrow HCl + H_2C^-{-}\overset{\overset{\Large O}{\|}}{C}{-}CH_3 \quad (29.9)$$

Electron capture has also been used in negative CI. In this technique, electrons emitted from a filament are captured by neutral analyte molecules present in the source producing a radical anion species. These bare electrons emitted from the filament are unstable by themselves and will readily attach to the analyte. The introduction of a nonreactive gas into the source will help to slow down the electrons through collisions and enhance the ionization of the neutral analyte molecules:

$$N_2 + e^-(70\ \text{eV}) \rightarrow N_2 + e^-(\gg 70\ \text{eV}) \qquad (29.10)$$

Some examples of nonreactive gasses used are argon (Ar), nitrogen (N_2), methane (CH_4), and isobutane (i-C_4H_{10}). Fast-moving electrons that have not undergone any collisional cooling have cross sections (probabilities) that are small for capture to take place. This often meant a very low yield in negatively ionized analytes resulting in poor sensitivity:

$$\text{Neutral compound (M)} + e^-(\text{slow}) \xrightarrow{\text{electron capture}} \text{Ionized compound } (M^{-\bullet}) \qquad (29.11)$$

The slower (relaxed or cooled) electrons thus have an increased probability of being captured resulting in higher yields and greater sensitivity for the negative CI technique. It is often observed though that certain groups have a higher affinity for capturing electrons as compared to others. Compounds such as pesticides that contain chlorine atoms often will exhibit a very high affinity for electron capture.

Another process used for negative CI is known as dissociative electron capture. In this process, enough energy has been absorbed by the analyte during the electron-capturing process to initiate a small amount of fragmentation to take place. This process generally does not give a large amount of structural information and can be a problem if the initial analyte is an unknown compound.

In general, CI gives a much higher abundance of the intact analyte ion versus the process of EI. The intact analyte ions produced by CI exist as, for example, in positive ion mode, as protonated molecules $[M+H]^+$ (sometimes referred to as acid adducts), and in the negative ion mode as either deprotonated molecules $[M-H]^-$ or radical anions $M^{-\bullet}$. The ionized analyte species formed using CI in the positive ion mode are almost exclusively EE ions. In contrast, EI almost exclusively produces OE ions ($M^{+\bullet}$) during the in-source analyte ionization process. Oftentimes, the sensitivity of the detection of the ionized analyte species can be functional group dependent when using CI. Electron capture can be difficult without the presence of a moiety within the molecular structure of the analyte that possesses a high electron affinity. Often, the CI spectrum will contain a single peak representing the *m/z* value of the intact precursor ion allowing the determination of the molecular weight of the analyte species. The EI spectrum usually does not contain the precursor *m/z* value to any appreciable amount but contains a significant amount of product ions giving structural information. The combination of the two techniques can be complimentary in determining a species molecular weight and structure.

29.3 ATMOSPHERIC PRESSURE CHEMICAL IONIZATION (APCI)

The previous sections introducing EI and CI as two techniques for transforming an analyte molecule from a neutral state to an ionized state both shared a common design where the ionization takes place within a source that is either at a much lower pressure (EI, 10^{-5} Torr) than ambient (760 Torr) or at a somewhat slightly lower pressure than ambient (CI, 0.1–2 Torr) due to input of reactive reagent gas. Figure 29.6 illustrates the basic design of an atmospheric pressure ionization (API) source.

As can be seen in Figure 29.6, the API source is directly suitable for a liquid stream such as that obtained from a separation science technique like high-performance liquid chromatography (HPLC). In the APCI technique, the ions are formed at atmospheric pressure using either a beta particle (β^-) emitting source (e.g., ^{63}Ni foil, old technique that is not in use much anymore) or a corona discharge needle. In corona discharge, a needle is held near the source with a very high negative voltage applied to it. Right on the sharp tip of the needle, there will be a very high electron (e^-) density buildup. The electrons will jump off of the needle tip and be accelerated forward by the negative field associated with the discharge needle. The discharged electrons then ionize the surrounding ambient species present in the surrounding air such as oxygen or water. The ionized ambient gaseous species (H_2O^- and $O^{-\bullet}$) will then undergo true ion–molecule reactions with the analyte species present. This process is illustrated in the expanded view of the corona needle in Figure 29.6. In this way, an introduced reactive reagent gas is not necessary for the ionization of the analyte species to take place; thus, a low-pressure source is not needed. During the ion–molecule reactions taking place, an adduct is formed between the neutral analyte species and the surrounding ionized ambient gasses. The reactive intermediate ions produced will then interact further with other neutral analyte molecules present producing ionized species. There are a number of ion–molecule reactions that are possible during the APCI process as follows:

$$\begin{array}{ll}
M + e^- \rightarrow M^{-\bullet} & \text{(associative electron capture)} \\
MX + e^- \rightarrow M^\bullet + X^- & \text{(dissociative electron capture)} \\
R^{+/-} + M \rightarrow R + M^{+/-} & \text{(charge transfer)} \\
MH + R^+ \rightarrow M^+ + RH & \text{(hydride abstraction)} \\
MH + R^- \rightarrow M^- + RH & \text{(proton transfer)} \\
RH^+ + M \rightarrow R + MH^+ & \text{(protonation)} \\
M + R^{+/-} \rightarrow MR^{+/-} & \text{(adduct formation)} \\
MX + R^+ \rightarrow M^+ + X + R & \text{(charge transfer with dissociation)}
\end{array} \qquad (29.12)$$

where M represents the neutral analyte, X represents any species, and R represents the reactive species formed during the APCI process.

The ionization takes place at atmospheric pressure making this technique well suited for a liquid stream such as that eluting from a liquid chromatography column. There is an inherent incompatibility between liquids and low pressure though if the liquid is directly introduced into the low-pressure inlet region,

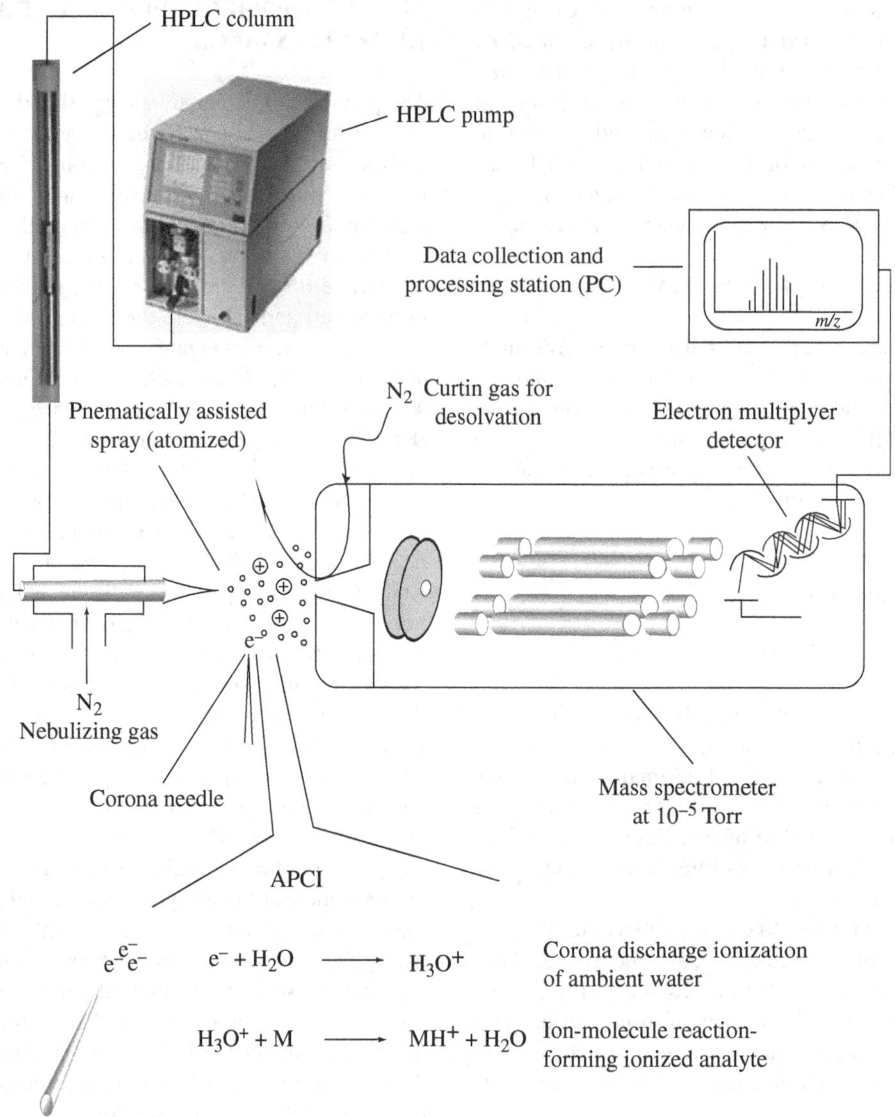

FIGURE 29.6 Design of a general atmospheric pressure chemical ionization (APCI) inlet system for mass spectrometric analysis.

the liquid will expand and vaporize instantaneously, thus producing high pressure within the source. This is too difficult to maintain and control when the desired pressure in the source inlet region is to be kept at or near 10^{-5} Torr. It has been observed that near 100% ionization efficiency can be obtained with APCI for compounds that are present at very low concentrations, thus increasing the sensitivity of the method. However, oftentimes, adduct and cluster ions of the ionized ambient air and water species are observed to be formed during the APCI process causing a high chemical background noise reducing the analyte signal-to-noise (S/N) ratios (decreased sensitivity). To overcome this problem, a dry nitrogen (N_2) curtain gas (see Figure 29.6) is used to break up the clusters through collisions between the N_2 gas and the clusters. The clusters will then tend to fall apart producing less interference at higher masses. The nitrogen curtain gas also aids in the evaporation of the solvent prior to introduction into the instrument source. This helps to reduce the problem of a large liquid stream evaporating and expanding within the source producing higher pressures. Finally, as can be seen in Figure 29.6, nitrogen is also used as a nebulizing gas that helps to atomize the eluant spray.

29.4 ELECTROSPRAY IONIZATION (ESI)

ESI is a process that enables the transfer of compounds in solution phase to the gas phase in an ionized state, thus allowing their measurement by mass spectrometry. The use of ESI coupled to mass spectrometry was pioneered by Fenn et al. [1,2] in 1985, by extending the work of Dole et al. [3] in 1968, who demonstrated the production of gas-phase ions by spraying macromolecules through a steel capillary that was electrically charged and subsequently monitoring the ions with an ion-drift spectrometer. The process by which ESI works has received much theorization,

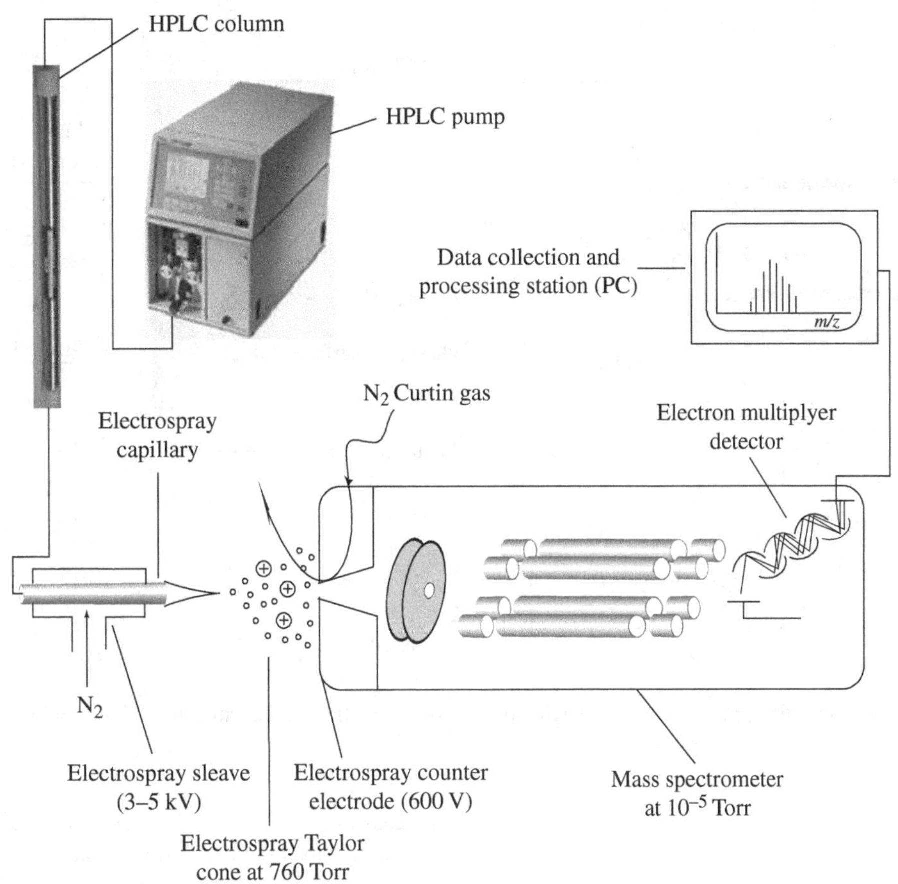

FIGURE 29.7 General setup for ESI when measuring biomolecules by electrospray mass spectrometry.

study, and debate [4–10], in the scientific community, especially the formation of the ions from the Taylor [11] cone droplets and offspring droplets. Figure 29.7 shows the general setup for ESI when measuring biomolecules by electrospray mass spectrometry. The ESI process is done at atmospheric pressure, and the system is quite similar to that shown for APCI in Figure 29.6. Both APCI and ESI are shown in-line with HPLC and a mass analyzer, however, the two differ extensively in the area of the spray right before introduction into the mass analyzer. The electrospray process is achieved by placing a potential difference between the capillary and a flat counter electrode. This is illustrated in Figure 29.8 where the "Spray Needle" is the capillary and the "Metal Plate" is the flat counter electrode. The generated electric field will penetrate into the liquid meniscus and create an excess abundance of charge at the surface. The meniscus becomes unstable and protrudes out forming a Taylor cone. At the end of the Taylor cone, a jet of emitting droplets (number of drops estimated at 51,250 with radius of 1.5 μm) will form that contain an excess of charge. Pictures of jets of offspring droplets are illustrated in Figure 29.9. As the droplets move toward the counter electrode, a few processes take place. The drop shrinks due to evaporation, thus increasing the surface charge until columbic repulsion is great enough that offspring droplets are produced. This is known as the Rayleigh limit producing a columbic explosion. The produced offspring droplets have 2% of the parent droplets mass and 15% of the parent droplets charge. This process will continue until the drop contains one molecule of analyte and charges that are associated with basic sites (positive ion mode). This is referred to as the "charged residue model" that is most important for large molecules such as proteins. This process is illustrated in Figure 29.10. As the droplets move toward the counter electrode, a second process also takes place known as the "ion evaporation model." In this process, the offspring droplet will allow evaporation of an analyte molecule from its surface along with charge when the charge repulsion of the analyte with the solution is great enough to allow it to leave the surface of the drop. This usually takes place for droplets with a radius that is less than 10 nm. This type of ion formation is most important for small molecules.

In the ensuring years since its introduction, electrospray mass spectrometry has been used for structural elucidation and fragment information [13–15], and noncovalent complex studies [16,17], just to name a few recent examples of its overwhelmingly wide range of applications.

Electrospray [1,2,5,18] is an ionization method that is now well known to produce intact gas-phase ions with very minimal, if any, fragmentation being produced during the ionization process. In the transfer process of the ions from the condensed phase to the gas phase, several types of "cooling" processes of the ions are taking place in the source: (i) cooling during the desolvation

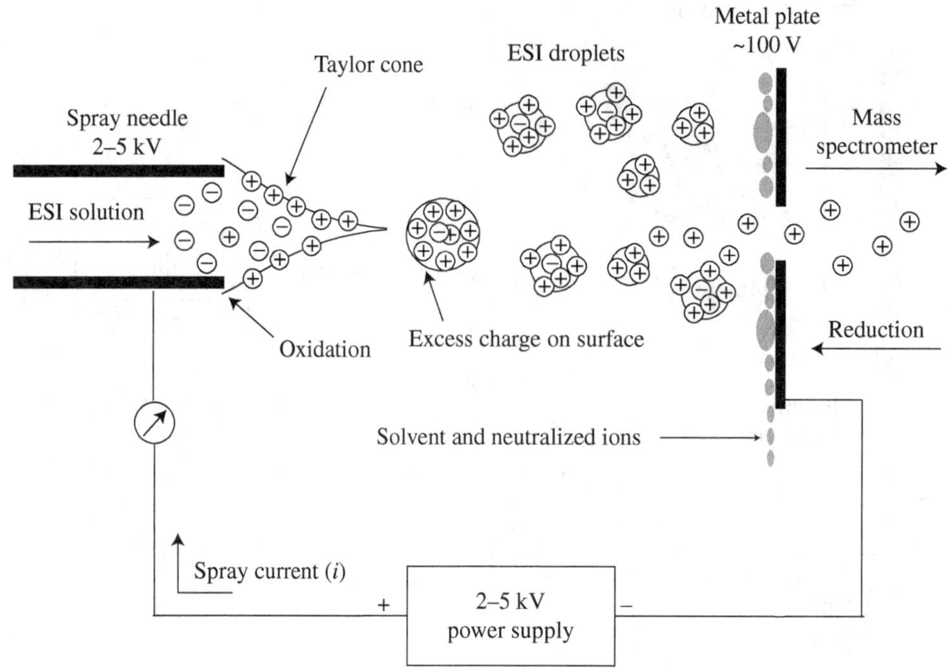

FIGURE 29.8 Electrospray ionization process [6] illustrated in positive ion mode. (Reprinted with permission of John Wiley & Sons, Inc. [6]. Copyright 2001.)

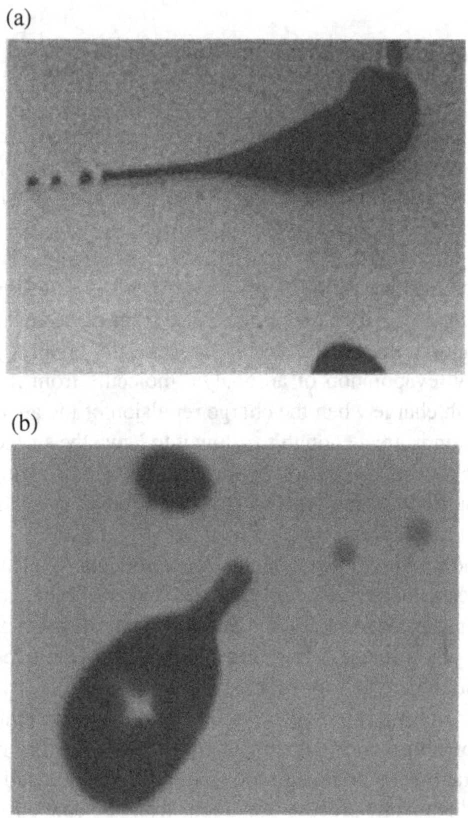

FIGURE 29.9 Pictures illustrating the jet production of offspring droplets [12]. (Reused with permission from Ref. [12]. Copyright 1994, American Institute of Physics.)

process through vibrational energy transfer from the ion to the departing solvent molecules, (ii) adiabatic expansion of the electrospray as it enters the first vacuum stage, (iii) evaporative cooling, and (iv) cooling due to low-energy dampening collisions with ambient gas molecules. The combination of these effects, and the fact that electrospray can effectively transfer a solution-phase complex to the gas phase with minimal interruption of the complex, makes the study of noncovalent complexes from solution by ESI mass spectrometry attractive.

29.5 NANOELECTROSPRAY IONIZATION (NANO-ESI)

A major application of biomolecule analysis using mass spectrometry has been the ability to allow liquid flows to be introduced into the source of the mass spectrometer. This has enabled the coupling of HPLC to mass spectrometry where HPLC is used for a wide variety of biomolecule analysis. Normal ESI, introduced in the preceding section, typically has flow rates on the order of microliters per min (~1–500 μl/min). Traditional analytical HPLC systems designed with UV/Vis detectors generally employ flow rates in the range of milliliters per minute (~0.1–1 ml/min). A recent advancement in the ESI technique has been the development of nanoelectrospray where the flows employed are typically in the range of nanoliters per minute (~1–500 nl/min). Following the progression of the development of electrospray from Dole's original reporting in 1968 through Fenn's work reported in 1984 and 1988, a more efficient electrospray process was reported by Wilm et al. [19] employing flows in the range of 25 nl/min. This early reporting of low flow rate electrospray was initially termed as microelectrospray by Wilm

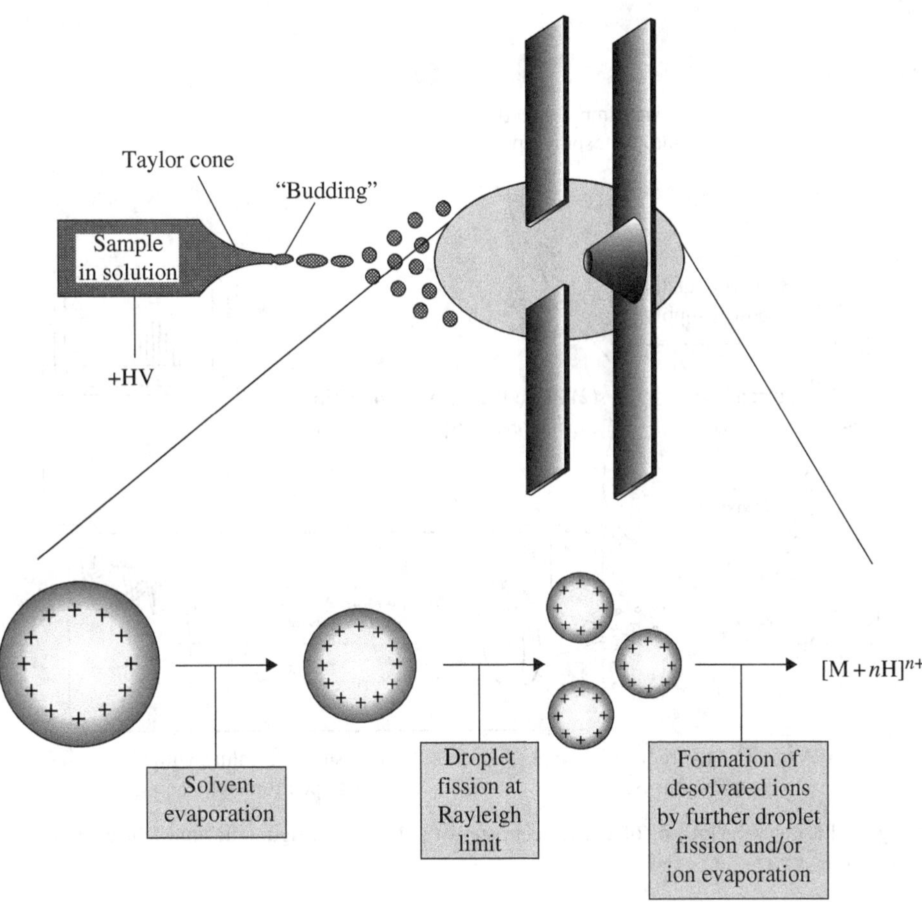

FIGURE 29.10 Gas-phase ion formation process from electrospray droplets [7]. (Reproduced with permission from Ref. [7].)

et al. but was later changed to nanoelectrospray [20]. At the same time that Wilm et al. [19] had reported the microelectrospray, Caprioli et al. [21] had also reported a miniaturized ion source that they had named microelectrospray. The name "nanoelectrospray" for Wilm's source is actually more descriptive due to flow rates used in the nanoliter per minute range and the droplet sizes that are produced in the nanometer range. Conventional electrospray sources before the introduction of nanoelectrospray produced droplets on the order of 1–2 μm. The nanoelectrospray source produces droplets in the size range of 100–200 nm, which is 100–1000 smaller in volume. When spraying standard solutions at concentrations of 1 pmol/μl, it is estimated that droplets of the nanometer size contain only one analyte molecule per droplet.

The original nanoelectrospray sources that were used were comprised of pulled fused silica capillary tips 3–5 cm long with orifices of 1–2 μm in diameter. The tips also have a thin gold plating that allows current flow. The tips are loaded with 1–5 μl of sample directly using a pipette [22] and coupled to the electrospray source completing the closed circuit required for the production of the applied voltage electrospray Taylor cone generation. This is illustrated in Figure 29.11 where in the top portion of the figure a sample is being loaded into the nanospray tip using a pipette. The tip is then placed into the closed circuit system for the electrospray to take place. The sample flow rate is very low using the nanospray tips allowing the measurement of a very small sample size over an extended period of time. It has also been observed that nanospray requires a lower applied voltage for the production of the electrospray that helps to reduce problems with corona electrical discharges that will interrupt the electrospray. In nanoelectrospray, the flow rate is lower than in conventional electrospray and is felt to have a direct impact on the production of the droplets within the spray and the efficiency of ion production. The lower flow rate produces charged droplets that are reduced in size as compared to conventional electrospray. This has been described in detail by Wilm et al. [19], Fernandez de la Mora et al. [23], and Pfeifer and Hendricks [24]. There are fewer droplet fission events required with smaller initial droplets in conjunction with less solvent evaporation taking place before ion release into the gas phase [25,26]. A result of this is that a larger amount of the analyte molecule is transferred into the mass spectrometer for analysis. Though the efficiency of ionization is increased with nanoelectrospray, the process is also influenced by the size and shape of the orifice tip [27,28]. Pictures of nanoelectrospray orifice tips are illustrated in Figure 29.12.

Figure 29.13 gives a good picture of an array of Taylor cones formed from a microelectrospray emitter. In the picture, multiple cones can be seen along with their associated spray produced from the electrospray process.

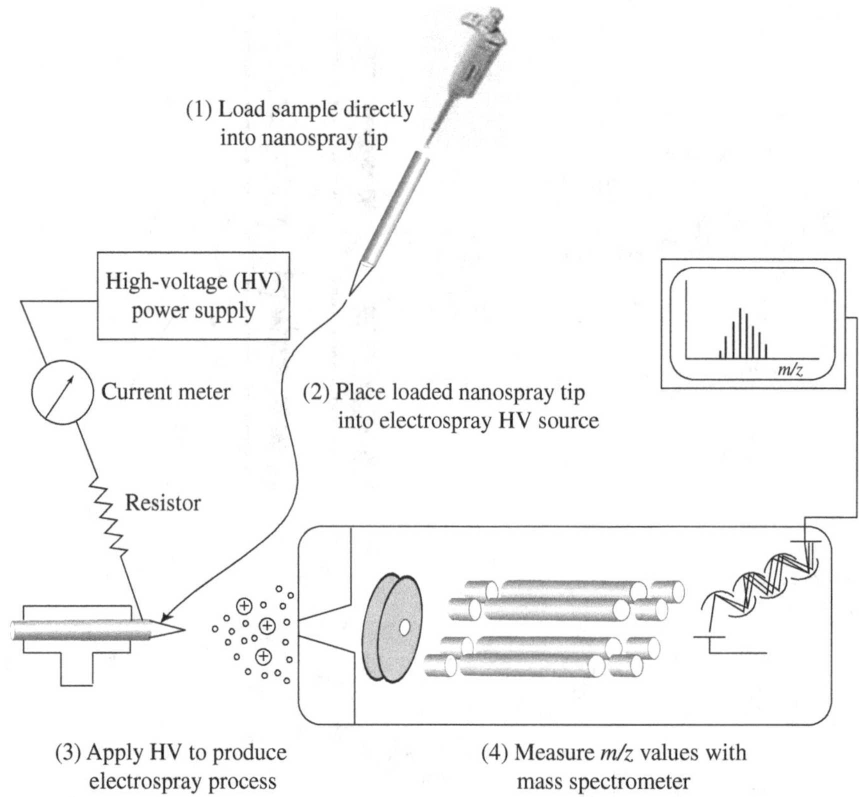

FIGURE 29.11 Top of figure illustrates the loading of a nanoelectrospray tip. Bottom of figure illustrates the coupling of the nanoelectrospray tip to the closed circuit system.

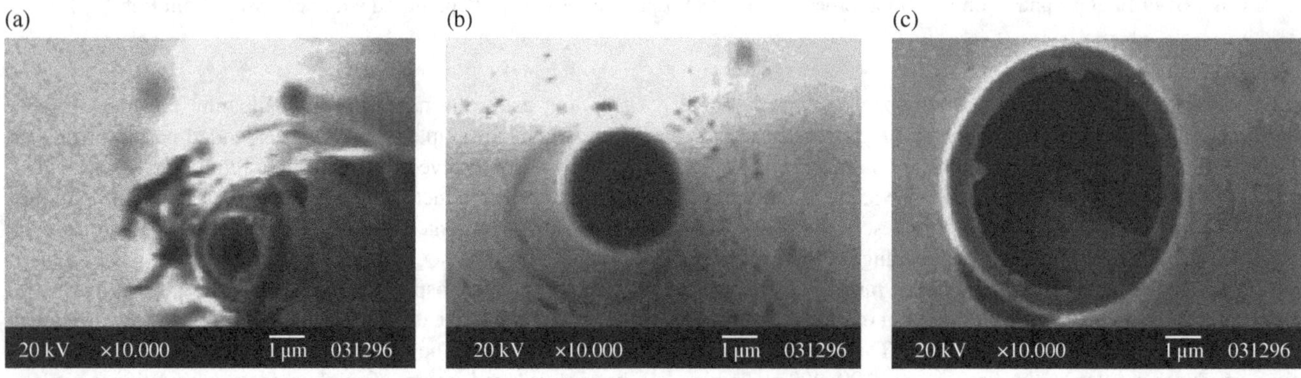

FIGURE 29.12 Illustration of different nanoelectrospray tip orifice diameters. Scanning electron microscopy images of employed nanospray emitters: (a) 1-, (b) 2-, and (c) 5-μm tip. Images were obtained after 2 h of use. (Reprinted with permission from Ref. [27]. Copyright 2003 American Chemical Society.)

As mentioned previously, nano-HPLC is increasingly being coupled to nanoelectrospray for biomolecule analysis. A nano-HPLC–ESI system is illustrated in Figure 29.15. The flow involved in nano-HPLC–ESI often ranges between 10 and 100 nl/min. The fused silica capillary columns that are used in nano-HPLC have very small diameters often around 50 μm. These small diameter columns can often create high back pressures in the HPLC system. One way to achieve the very low flow rate through the fused silica nano-HPLC column is to use a flow splitter that is located in-stream between the column and the HPLC pump as illustrated in Figure 29.14. The tubing from the splitter to waste is called a restrictor and is used to regulate the flow through the nanocolumn. A smaller diameter restrictor used will increase the back pressure forcing more mobile phase through the nanocolumn. If a larger diameter restrictor is used, the back pressure will be lower resulting in less flow being directed through the column. The nanocolumns have a nano-ESI tip coupled to them (diameters can range from 1 to 100 μm) to produce the electrospray. Another difference observed here as compared to the atmospheric pressure source is the absence of a nebulizing gas or a drying gas. These are not needed or used in nano-ESI.

29.6 ATMOSPHERIC PRESSURE PHOTO IONIZATION (APPI)

A recent addition to ionization sources that are useful in mass spectrometric measurement of biomolecules is the APPI source. As the name indicates, the photoinduced ionization of analytes is taking place under atmospheric conditions. As we will see, the technology has been applied to analytes that are within a liquid effluent such as that coming out of an HPLC column making the ionization methodology directly applicable to biomolecule analysis by mass spectrometry. Photoionization of analytes is actually a technique that has been in use for a number of years. The photoionization technique was primarily used in conjunction with gas chromatography (GC) using a detection system

FIGURE 29.13 Photograph of nine stable electrosprays generated from the nine-spray emitter array. (Reprinted with permission from Ref. [29]. Copyright 2001 American Chemical Society.)

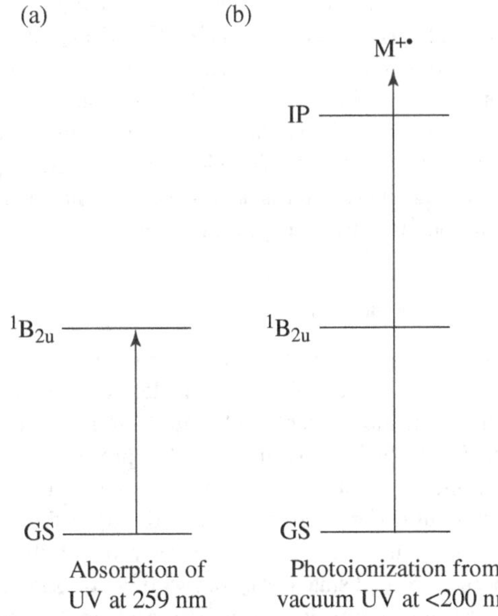

FIGURE 29.15 Single-photon processes for a molecule of benzene: (a) a photon is absorbed with an energy $E < \text{IP}$ resulting in the elevation of an electron from the ground state (GS) to an excited state ($^1B_{2u}$) but with no ionization in the form of electron ejection. (b) A photon is absorbed with an energy $E > \text{IP}$ resulting in electron ejection from the benzene molecule forming a radical cation ($M^{+\bullet}$) of the benzene molecule.

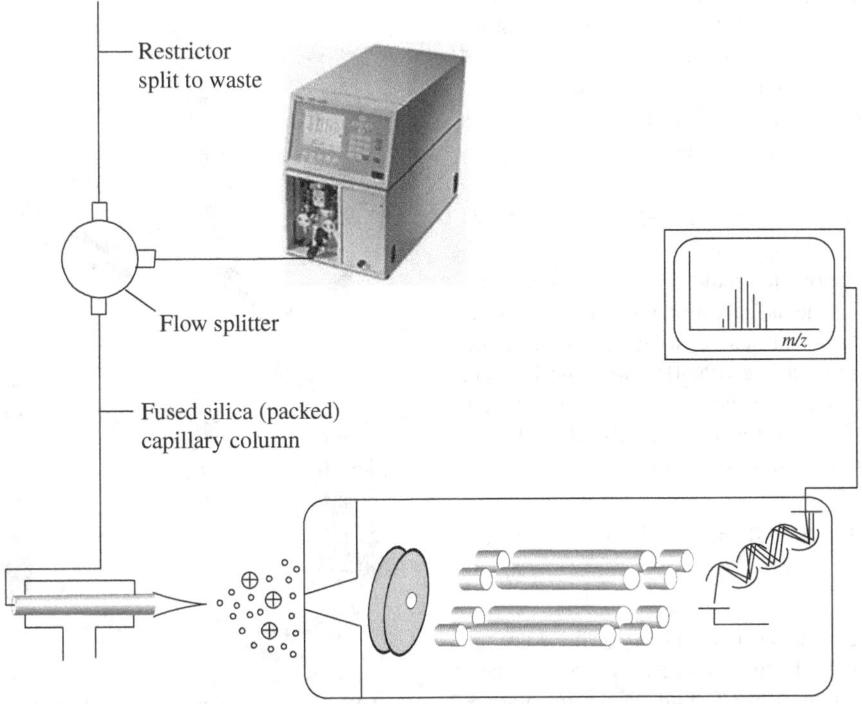

FIGURE 29.14 Design of a nano-HPLC nano-ESI system for mass spectrometric analysis of biomolecules.

measuring a change in current through a collection electrode induced by the photoionized analyte species eluting from the GC column [30,31]. The photoionization was initiated using discharge lamps that produce vacuum ultraviolet (VUV) photons. The technique of photoionization for the detection of analyte species has also been explored with the use of liquid chromatography [32–35]. However, with both GC and LC analyses of postcolumn photoionized biomolecular species, there is no information obtained regarding the analyte's mass and no structural information either. This is where the mass spectrometer contributes a clear advantage with the addition of measured information regarding analyte mass and structural information as compared to the other two analytical techniques.

29.6.1 APPI Mechanism

The process of photoionization involves the absorption of radiant energy from a UV source where the incident energy is greater than the first ionization potential (IP) of electron loss from the analyte. In Figure 29.15, two single-photon processes are illustrated for a molecule of benzene. In process (a), a photon is absorbed with an energy that is less than the IP of the benzene molecule resulting in the elevation of an electron from the ground state (GS) to an excited state ($^1B_{2u}$) but with no ionization in the form of electron ejection. In process (b), a photon is absorbed with an energy that is greater than the first IP of the benzene molecule resulting in electron ejection from the benzene molecule. Photoionization in this example is a single-photon process that results in the liberation of an outer valence shell electron forming a radical cation ($M^{+\cdot}$) of the benzene molecule. The process of photoionization at atmospheric pressure usually results in the production of molecular ions ($M^{+\bullet}$) often with minimal fragmentation taking place during the ionization. Most organic compounds have IP that range between 7 and 10 eV; therefore, photon sources are needed that will be able to supply photons with sufficient energy to induce photoionization. Two ionization processes are common in photoionization [36]. The first involving the absorption of radiant energy with subsequent liberation of an electron producing the radical cation:

$$M + h\nu \rightarrow M^{+\bullet} + e^- \quad \Delta H_{PI} = IP(M) - h\nu \tag{29.13}$$

where M is the neutral analyte molecule, $h\nu$ is the incident photoionization energy, $M^{+\cdot}$ is the radical cation formed, and e^- is the liberated electron. The change in enthalpy for the photoionization is equal to the IP of the molecule (IP(M)) minus the incident photoionization energy ($h\nu$). The second process involves the abstraction of a proton from the surrounding solvent producing a positively charged, protonated analyte species:

$$M^{+\bullet} + S \rightarrow MH^+ + S(-H), \quad \Delta H = IP(H) - IP(M) - PA(M) + D_H(S) \tag{29.14}$$

where PA(M) is the PA of the analyte and D_H is the hydrogen bond energy. Finally, a third process is also observed to take place with photoionization in the form of analyte fragmentation, which when controlled can give structural information.

29.6.2 APPI VUV Lamps

Most organic molecules have first IP between 7 and 10 eV; thus, a VUV lamp source is needed that can supply this energy. There are three such VUV lamps that can supply energies in this range: the xenon (Xe) lamp that generates light at 8.4 eV, the krypton (Kr) lamp that generates light at 10.0 and 10.6 eV, and the argon (Ar) lamp that generates light at 11.7 eV [37]. The Kr lamp is the most common lamp used between the three VUV lamp choices. Examples of typical VUV lamps are illustrated in Figure 29.16.

29.6.3 APPI Sources

There are two types of APPI sources that are generally in use at this time and are available commercially. This includes the design originally reported by Robb et al. [38] (available from Sciex) and is illustrated in Figure 29.17. A second APPI source configuration is based off of the original design by Syage et al. [39]. In both sources, the analytes undergo photoionization using a 10 eV krypton discharge lamp, which can be followed by gas-phase reactions. The gas-phase reaction approach is used in the APPI source shown in Figure 29.17 where we see the addition of a dopant (used to promote ionization of the neutral analyte). Both sources are also similar in that nebulization and high-temperature desolvation are used to evaporate the liquid solution containing the analyte. The second Syage APPI source contains an open area where the gaseous analyte is ionized, while in the APPI source illustrated in Figure 29.17, the ionization takes place within the closed quartz tube.

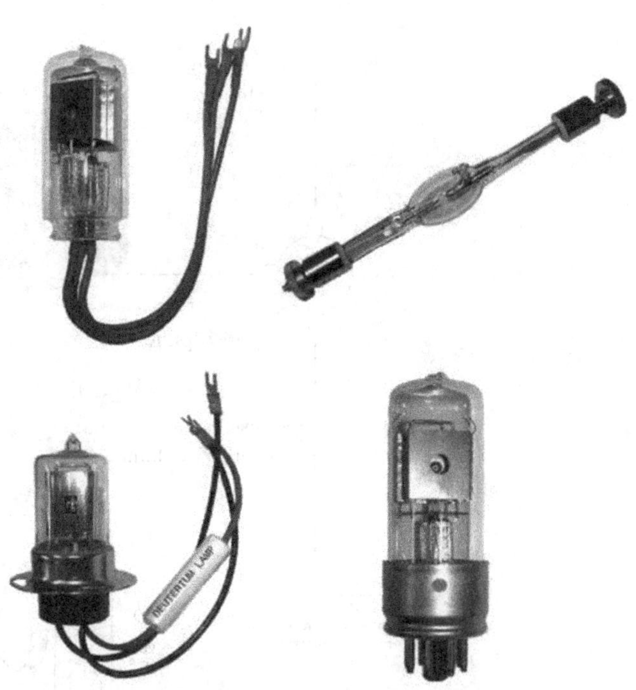

FIGURE 29.16 Examples of common designs of vacuum ultraviolet (VUV) lamp sources.

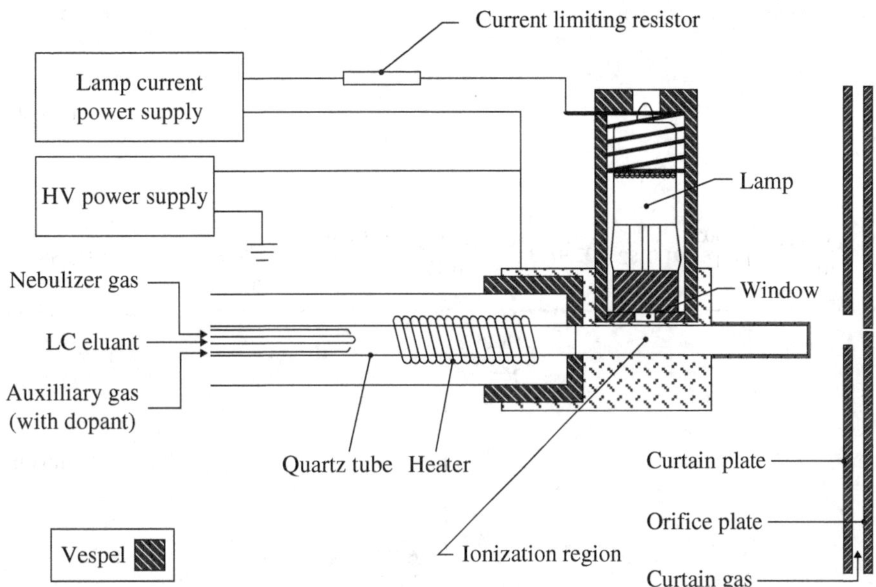

FIGURE 29.17 Schematic of the APPI ion source, including the heated nebulizer probe, photoionization lamp, and lamp mounting bracket. (Reprinted with permission from Ref. [38]. Copyright 2000 American Chemical Society.)

29.6.4 Comparison of ESI and APPI

In the preceding two sections of Chapter 2, we looked at the ionization process involving electrospray, both normal electrospray and nanoelectrospray. At this point, we are at an advantageous place to compare APPI to electrospray affording us an opportunity to get a closer look at the two ionization techniques and also some examples of the application of APPI to biomolecule analysis. When analyzing neutral biomolecules by electrospray, the mechanism of ionization is primarily dependent upon the PA (also directly related to the metal cation affinity) of the analyte. For example, the reactions involved during ESI for the production of a proton adduct (acidic solution) or a metal (sodium, Na^+) adduct are

$$\begin{aligned} &M + H^+ \rightarrow MH^+ \quad \text{(proton adduct)} \\ &M + Na^+ \rightarrow MNa^+ \quad \text{(sodium adduct)} \end{aligned} \tag{29.15}$$

Because the ionization process is dependent upon the PA of the analyte in ESI, often nonpolar species have very low detection. With APPI, the ionization process is dependent upon the incident photon possessing greater energy than the IP of the analyte. This in effect can allow the detection of nonpolar species by APPI that are difficult to detect using ESI. Figure 29.18 illustrates the versatility of APPI for the detection of nonpolar compounds by comparing spectra obtained with ESI and APPI [36]. The spectra in Figure 29.18 were collected using a dual source that has the ability to apply APPI in conjunction with ESI. In the top spectrum, the APPI lamp source is turned off during the spraying of a 100 ng/μl solution of progesterone. With the APPI lamp off, very little of the analyte is observed. In the bottom spectrum, the APPI lamp has been turned on, and an abundant $[M+H]^+$ peak for the protonated form of progesterone is observed at *m/z* 315.1678. Figure 29.19 illustrates the structure of progesterone, the biomolecule that is being analyzed by ESI and APPI in Figure 29.18. The proton adduct that is being observed is formed by the help of the dopant toluene that is being included in the spray liquid.

The ionization process in ESI is a solution-based process, while the ionization process in APPI is a gas-phase process; this is where the two ionization processes fundamentally differ. This difference is the case for direct photoionization; however, often the addition of a dopant can also be used in APPI, which brings about a more complicated mechanism for the production of an ionized analyte. The process of APPI ionization with the inclusion of a dopant has similarities to the solution-phase mechanism of ESI where the PA of species now has an influence on the ionization process. The mechanism for dopant-assisted ionization in APPI involves first the ionization of the dopant to a radical cation ($M^{+\bullet}$). Often, the dopant is a chemical species that possesses a relatively low IP that allows its ionization to take place with high efficiency. An example of a common dopant used in APPI is toluene, which has an IP of 8.82 eV. Table 29.3 lists some common solvents and dopants used in APPI along with their respective PA and IP.

The next step in the APPI process involving a dopant is the protonation of the solvent molecules by the dopant radical cation. The solvent molecules that are protonated can next transfer a proton to the neutral analyte molecule producing an ionized species detectable by mass spectrometry. It is also possible for the dopant radical cation to directly ionize the neutral analyte through the process of charge transfer [40,41].

Other examples of the use of APPI and its comparison to ESI include a study of the analysis of dinitropyrene and aminonitropyrene by LC–MS/MS [42], the analysis of aflatoxins in cow milk [43], and a study of the acylglycerols in edible oils, which

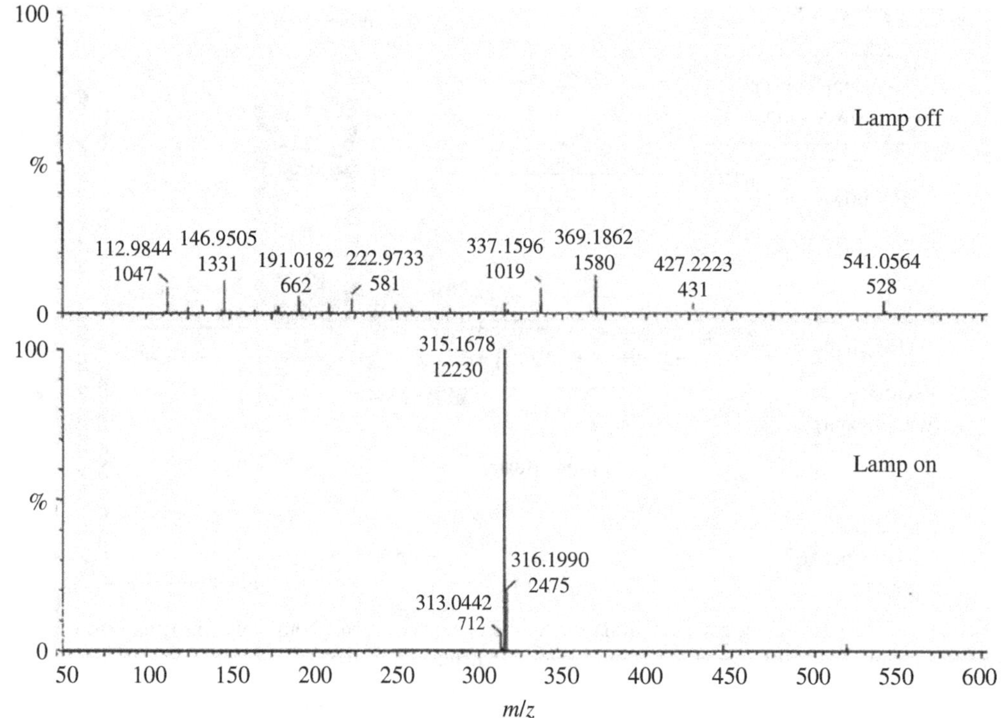

FIGURE 29.18 Positive ion mass spectra of progesterone (100 ng/μL) recorded for the dual ESI/APPI source on the Waters LCT. The top spectrum is in ESI-only mode and the bottom spectrum is in dual mode. Spectra were recorded by flow injection analysis (5 μL sample injection) for mixture of methanol–toluene (95:5) solvent. Both mass spectra are on the same absolute intensity scale. (Reprinted with permission from Ref. [36]. Copyright Elsevier 2004.)

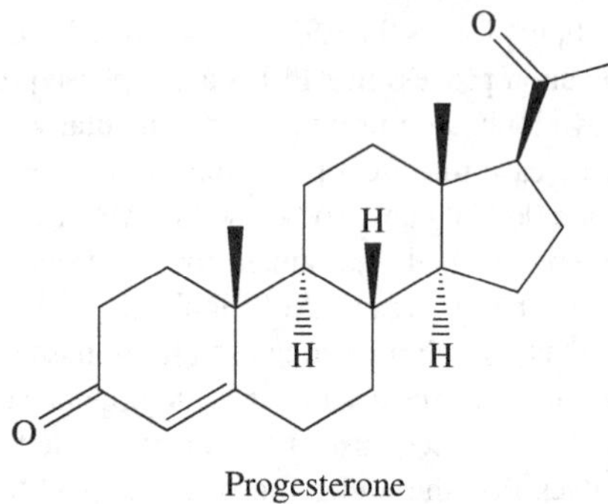

FIGURE 29.19 Structure of neutral progesterone, the biomolecule that is being analyzed by ESI and APPI in Figure 29.18.

TABLE 29.3 Compilation of Thermochemical Data for Some Common Solvents and Dopants Used in APPI.

Compound	IP	PA
Methanol	10.85	7.89
H_2O	12.61	7.36
Acetonitrile	12.19	8.17
DMSO	9.01	9.16
Naphthalene	8.14	8.44
Benzene	9.25	7.86
Phenol	8.47	8.51
Aniline	7.72	9.09
m-Chloroaniline	8.09	8.98
1-Aminonaphthalene	7.1	9.41
Toluene	8.82	8.23
TNT	10.59	

Source: Redrawn with permission from Ref. [36].

we shall take a closer look at [44]. In their comparison of edible oils, it was observed that the ESI spectra of the oils primarily gave the spectra of the triacylglycerols. This is illustrated in Figure 29.20 of the ESI spectra of (a) sunflower oil and (b) of corn oil. In comparison to this, the spectra in Figure 29.21 were obtained for the same oils using APPI where there is observed a much higher abundance of the diacylglycerols and the monoacylglycerols. This was attributed to fragmentation taking place of the triacylglycerols during the APPI process. In Figure 29.20, the majority of biomolecules that are being observed are in the range of *m/z* 800–900. This is illustrated in the *m/z* regions that are expanded in the spectra. The *m/z* 800–900 region is in the range of the molecular weights of the triacylglycerols. In Figure 29.21, the *m/z* 800–900 region is now much lower in intensity as compared to the *m/z* 300–700 region. The *m/z* 250–400 region contains biomolecules such as free fatty acids and monoacylglycerols in these edible oil samples. In the *m/z* 500–700 region, the biomolecules are primarily diacylglycerols. Both the monoacyls and the diacyls are thought to be produced by APPI process from the original triacylglycerols present in the oil samples. Edible oil samples are typically comprised of 80–98% triacylglycerols.

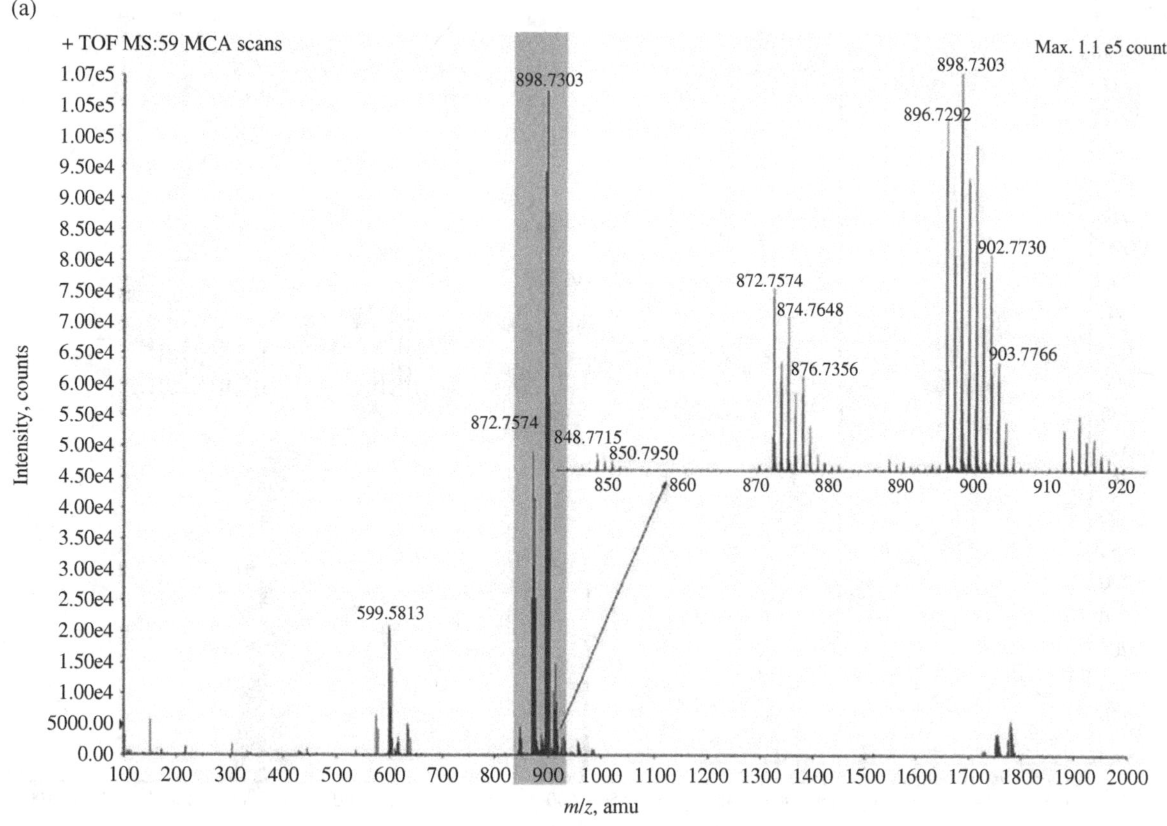

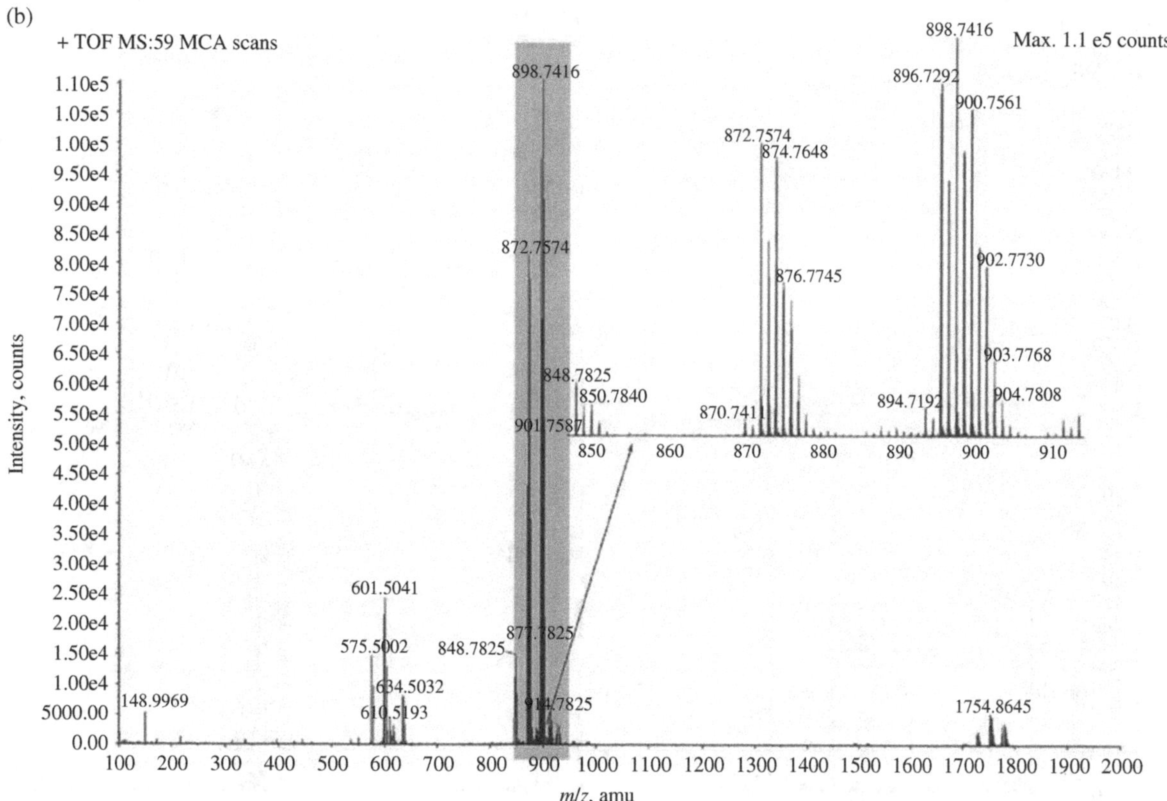

FIGURE 29.20 Full-scan ESI-MS: (a) sunflower oil and (b) corn oil. The region between *m/z* 800 and 900 is magnified. (Reprinted with permission from Ref. [44]. Copyright Elsevier 2006.)

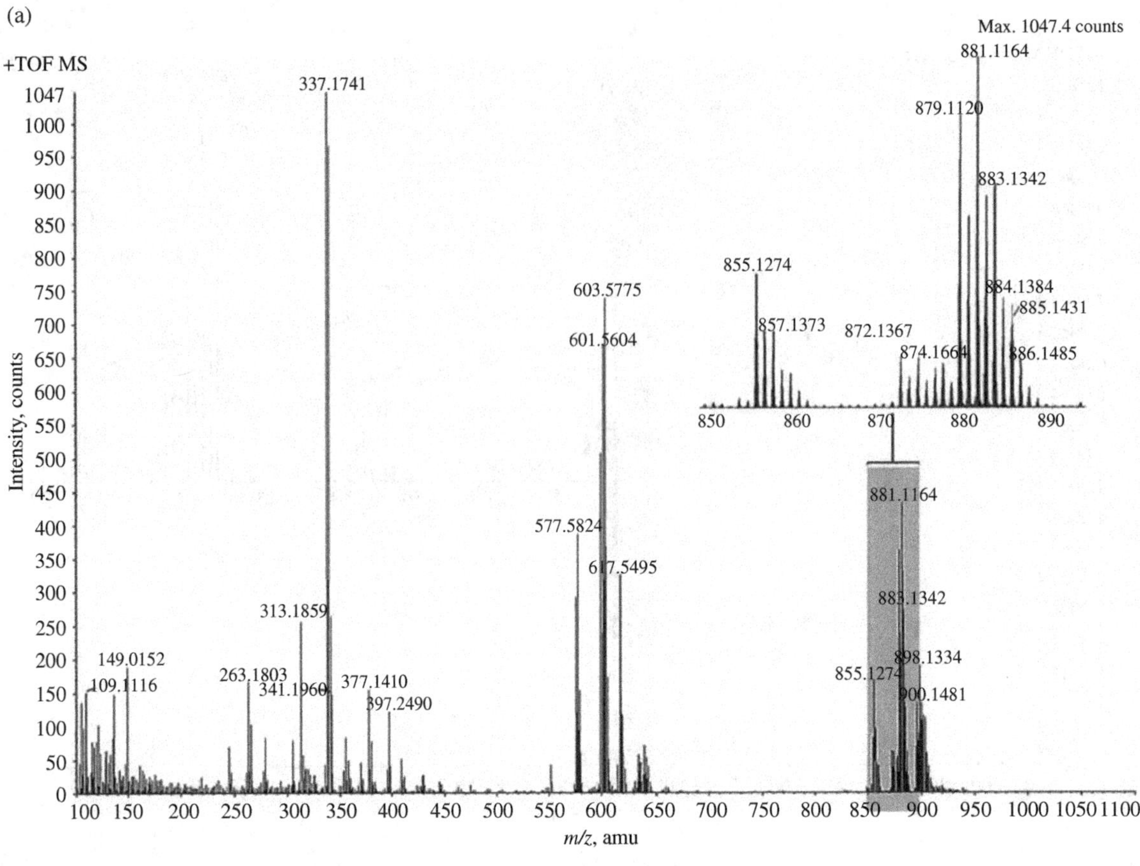

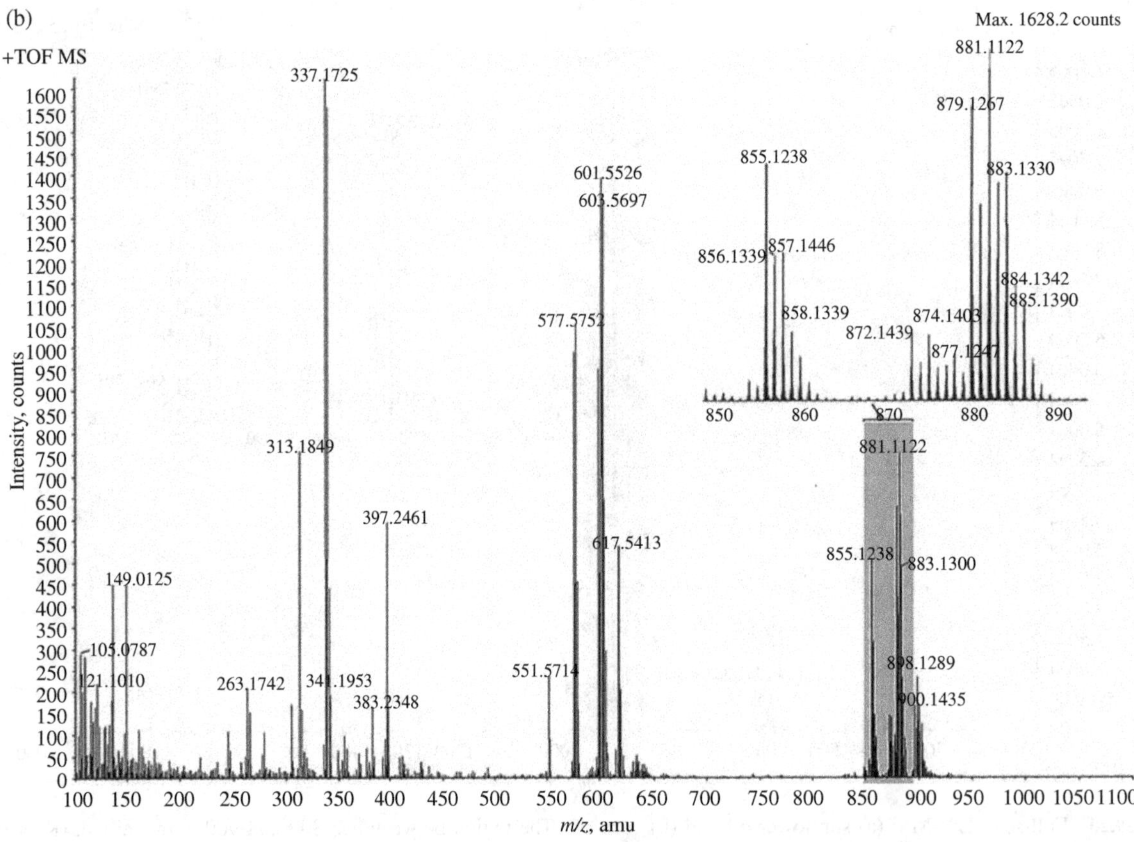

FIGURE 29.21 Full-scan APPI-MS: (a) sunflower oil and (b) corn oil. The region between *m/z* 800 and 900 is magnified. (Reprinted with permission from Ref. [44]. Copyright Elsevier 2006.)

29.7 MATRIX ASSISTED LASER DESORPTION IONIZATION (MALDI)

Matrix assisted laser desorption ionization (MALDI) is a process that enables the transfer of compounds in a solid, crystalline phase to the gas phase in an ionized state, thus allowing their measurement by mass spectrometry [45–47]. The process involves mixing the analyte of interest with a strongly ultraviolet (UV)-absorbing organic compound, applying the mixture to a target surface (MALDI plate) and then allowing it to dry (illustrated in Figure 29.22). Examples of typical organic matrix compounds used are 2,5-dihydroxybenzoic acid (DHB), 3,5-dimethoxy-4-hydroxy-trans-cinnamic acid (sinapic or sinapinic acid), and α-cyano-4-hydroxy-trans-cinnamic acid (α-CHCA) (structures illustrated in Figure 29.23). There are many techniques described for applying the matrix to the target such as the "dried droplet" method [47,48] where the analyte and matrix are mixed usually in an approximate 1:1000 ratio and then 0.5–2 μl of the analyte/matrix mixture are spotted onto the target and allowed to dry forming a crystalline solid. The "thin-film" method [49] is done by depositing a thin polycrystalline film of a homogenous 1:1 mixture of the matrix and analyte onto the target plate and allowed to dry. There are layer or sandwich methods where a layer of matrix may be applied and allowed to dry followed by a layer of the analyte then followed by another layer of matrix. Often, it is best to try a few approaches to find which method works best for the analytes being measured. The target plate after spot drying is placed into the source of the mass spectrometer, and the source is evacuated. The dried crystalline mixture "film" or "spot" is then irradiated with a nitrogen laser (337 nm) or an Nd:YAG laser (266 nm). The strongly UV-absorbing matrix molecules accept energy from the laser and desorb from the surface carrying along any analyte that is mixed with it. The desorbed matrix and analyte molecules, ions, and neutrals form a gaseous plume above the target and within the source. The analyte is cationized in the plume above the crystalline surface with, in the positive mode, either a hydrogen proton $[M+H]^+$ transferred from the acidic matrix or a metal cation present such as sodium $[M+Na]^+$. Figure 29.24 illustrates the MALDI process where a pulsed laser is irradiated upon the crystalline matrix and analyte mixture creating the plume above the target. There is often a delayed extraction period allowing a decrease in the spatial distribution of the analytes within the plume. There are two known problems with MALDI ionization in the gaseous plume of analytes above the target that cause a decrease in the resolution of the detected analytes. The first resolution decreasing effect is the initial spatial distribution where not all of the desorbed analytes are at the exact same distance from the detector at the start of their flight (placement in crystal structure may attribute to this). The second resolution decreasing effect involves the initial velocity distribution of the analytes within the desorbed gaseous plume above the target. Not all of the analytes may have exactly the same velocity at the start of their flight toward the detector. Together, these have attributed to the low-resolution ability of the MALDI–TOF/MS technique where spectra are typically generated at a resolution of approximately $m/\Delta m = 500$. We will see later in Chapter 30 that improvements in time-of-flight instrumentation and source design have helped to effectively address these problems [delayed extraction in the source and an electrostatic mirror (reflectron) after the drift tube] and greatly increase spectral resolution up to an $m/\Delta m$ value of 10,000 or higher. The desorbed, ionized compounds from the MALDI process are then introduced into a mass spectrometer for analysis by an extraction grid known as an accelerating voltage or drawout pulse.

Recently, there has been the addition to the MALDI ionization technique two new types of matrixes: liquid ionic matrixes [50–54] and solid ionic matrixes [55,56]. These new types of MALDI matrixes have been introduced to address inherent problems that have been associated with the traditional solid MALDI matrixes. For example, when analyzing biological lipid extracts by MALDI prompt fragmentation involving cleavage at the head group of phosphorylated lipids is a common problem. There are also known difficulties when using solid MALDI matrixes pertaining to the ability to obtain quantitative [53] and spatial [57] information. The matrix layer also causes variations in responses due to so-called "hot" or "sweet" spots versus

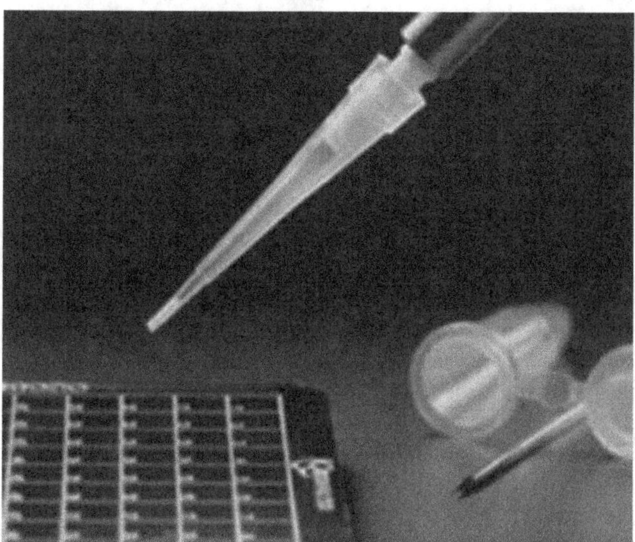

FIGURE 29.22 Spotting of a MALDI target plate. (Courtesy of EMD Millipore Corporation, Billerica, MA.)

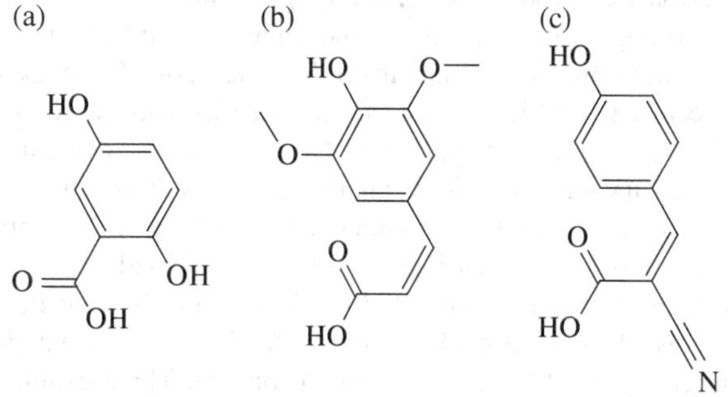

FIGURE 29.23 Examples of typical organic matrix compounds used are (a) 2,5-dihydroxybenzoic acid (DHB), (b) 3,5-dimethoxy-4-hydroxy-trans-cinnamic acid (sinapic or sinapinic acid), and (c) α-cyano-4-hydroxy-trans-cinnamic acid (α-CHCA).

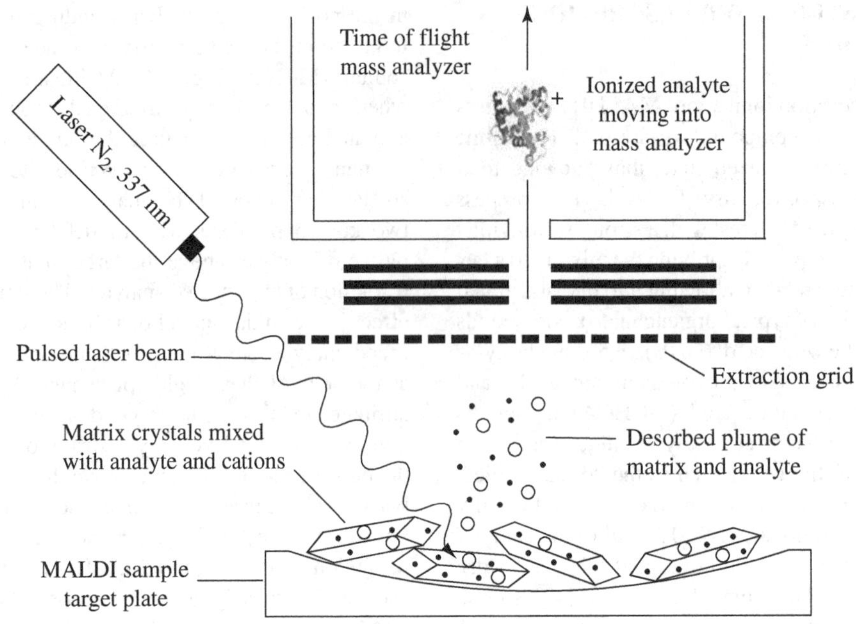

FIGURE 29.24 Matrix-assisted laser desorption ionization "MALDI" process of desorption and ionization.

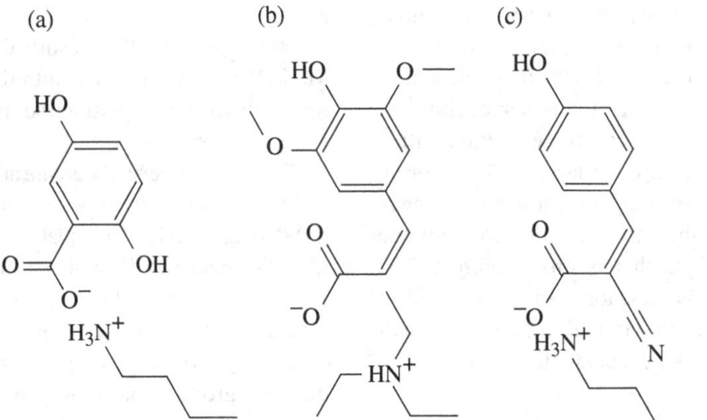

FIGURE 29.25 Ionic liquid matrixes used for improved shot-to-shot reproducibility, and a reduction in fragmentation induced by MALDI. (a) 2,5-dihydroxybenzoic acid butylamine, (b) 3,5-dimethoxy-4-hydroxycinnamic acid triethylamine, and (c) alpha-cyano-4-hydroxycinnamic acid butylamine.

the areas that produce lower-level signals that are the results of inhomogeneous deposits of the analyte within the matrix and nonuniformities in the matrix/analyte spot. There exists between the MALDI spectral signal intensity and the amount of the measured analyte present in the spot [57] a rather complex relationship. The use of ionic liquid matrixes is demonstrated to have improved shot-to-shot reproducibility of signal intensities over traditional solid matrixes. This has also enabled more accurate quantitative analysis by MALDI. When using the ionic liquid matrixes, there is also observed a reduction in fragmentation induced by the MALDI ionization technique. Three examples of ionic liquid matrixes are illustrated in Figure 29.25. Through combining the appropriate viscous liquid amines with the crystal MALDI matrix, after having dissolved both in methanol, the ionic liquid matrixes are formed. Finally, the methanol and free amine are removed producing the ionic pair, which is then mixed with a small amount of ethanol to reduce the viscosity of the liquid matrix. The ionic liquid matrix enhances the MALDI ionization technique by more uniformly dissolving the analyte within the liquid matrix versus the crystalline state that the traditional matrixes produce upon drying. Furthermore, the enhancement may also result from an action that is similar to that observed with fast atom bombardment (FAB) (covered in the next section) where the analyte is mixed in glycerol. As the analyte is depleted from the surface of the liquid during the desorption/ionization step, the action of the liquid may allow the immediate replenishment of the analyte to the liquid surface, thus greatly enhancing the response. The ionic liquids have been observed to be stable at both atmospheric pressure and reduced pressure within the source. Solid ionic crystal matrixes

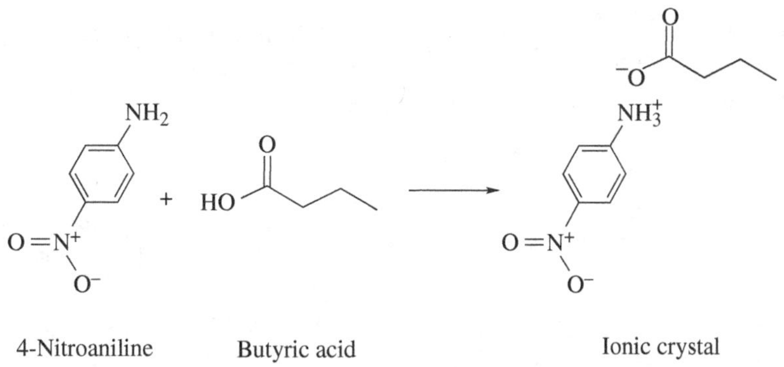

FIGURE 29.26 Synthesis of the solid ionic crystal matrix for MALDI upon the addition of the matrix modifier butyric acid to 4-nitroaniline.

have also recently demonstrated an enhancement in MALDI ionization for certain applications as compared to the traditional crystalline matrixes illustrated in Figure 29.23. One example is the ionic pair produced between the addition of the matrix modifier butyric acid to 4-nitroaniline as illustrated in Figure 29.26.

The resultant matrix mixture contrasts with those in Figure 29.25 in that an ionic liquid was not produced but upon analyte mixing and drying, a solid ionic crystal is formed that acts as a powerful gas-phase proton donor and enhances phosphorylated lipid response. This new matrix preparation is crystalline and shows deposition (spotting) behavior similar to other widely used solid MALDI matrixes such as sinapinic acid, DHB, or α-CHCA. The new solid crystal ionic matrix was observed to give an enhanced response for phosphorylated lipid analysis by MALDI–TOF mass spectrometry. Primarily protonated analyte ions $[\text{Phospholipid+H}]^+$ were observed in the mass spectra with a reduced degree of prompt ion fragmentation, which was earlier discussed as a major drawback of the more traditional MALDI matrixes illustrated in Figure 29.23. The combination of mostly protonated molecular ions and reduction in the amount of prompt ion fragmentation greatly facilitates the interpretation of unknown spectra.

Difficult applications will be more likely to succeed when a solid ionic crystal matrix is used in preference to an ionic liquid matrix, such as depositing the matrix upon tissue or other surfaces for MALDI analyses revealing two-dimensional spatial distributions [58]. In contrast to the acidic MALDI matrixes, DHB or α-CHCA acidification of the analyte environment should not be severe with the solid ionic crystal matrix (where butyric acid is largely ion paired with a *p*-nitroaniline). This property can facilitate the analysis of certain compounds that are prone to hydrolysis in an acidic environment, such as plasmalogens.

Another recent application of solid ionic MALDI matrixes has been in the analysis of peptides [59]. The direct measurement of peptides was obtained by application of solid ionic crystal matrixes to tissue samples. It was demonstrated that three solid ionic crystal matrixes involving α-cyano-4-hydroxycinnamic acid (CHCA) were observed to give an enhanced peptide response for rat brain tissue analysis as compared to CHCA only. Three of the CHCA matrix modifiers used in the study were aniline, *N*,*N*-dimethylaniline, and 2-amino-4-methyl-5-nitropyridine. In this study, the enhanced properties of using a solid ionic MALDI matrix were also observed including an enhancement in the quality of the spectra obtained in the form of sensitivity, resolution, and high tolerance to contamination. It was also observed that there was a better ability for a homogeneous crystallization of the solid ionic matrix upon the tissue to be analyzed. The solid ionic MALDI matrixes were also observed to have satisfactory stability in the high-vacuum source, good analyte response in both positive and negative modes, and sufficient molecular ion intensities to perform postsource decay (PSD) prompt ion fragmentation studies, thus allowing structural information to be collected and analyzed.

29.8 FAB

A similar ionization technique to MALDI is FAB where both techniques involve the use of a matrix. In MALDI, the matrix is used to transfer energy from the laser to the analyte inducing desorption of the analyte. The matrix also serves to transfer a charge to the analyte converting it from the neutral state to the ionized state, thus allowing its measurement by mass spectrometry. In FAB, a matrix is also used in the liquid state such as glycerol or *m*-nitrobenzyl alcohol (NBA). This is in contrast to MALDI that often uses a matrix in the crystalline state (excluding however the more recent room temperature ion pair liquid matrixes currently being used with the MALDI technique, e.g., see Section 29.6). In FAB, the analyte is dissolved and dispersed within the glycerol liquid and is bombarded with a high-energy beam of atoms typically comprised of neutral argon (Ar) or xenon (Xe) atoms or charged atoms of cesium (Cs^+). As the high-energy atom beam (6 keV) strikes the FAB matrix/analyte mixture, the kinetic energy from the colliding atom is transferred to the matrix and analyte effectively desorbing them into the gas phase. The analyte can already be in a charged state or may become charged during the desorption process by surrounding ionized matrix. Figure 29.27 illustrates the ionization mechanism of FAB where Figure 29.27a shows the overall process and Figure 29.27b is a closeup view of the desorption process. FAB is another soft ionization technique where there is the observance of a high yield of cationized analyte species with minimal fragmentation taking place. During the desorption and ionization process, the matrix

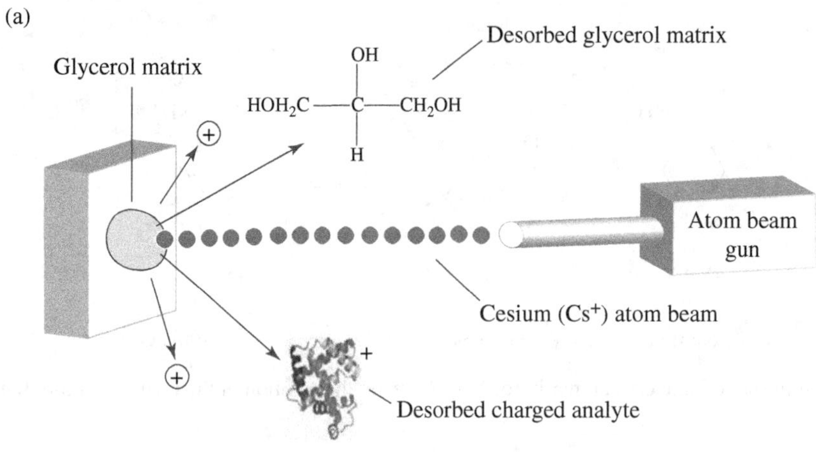

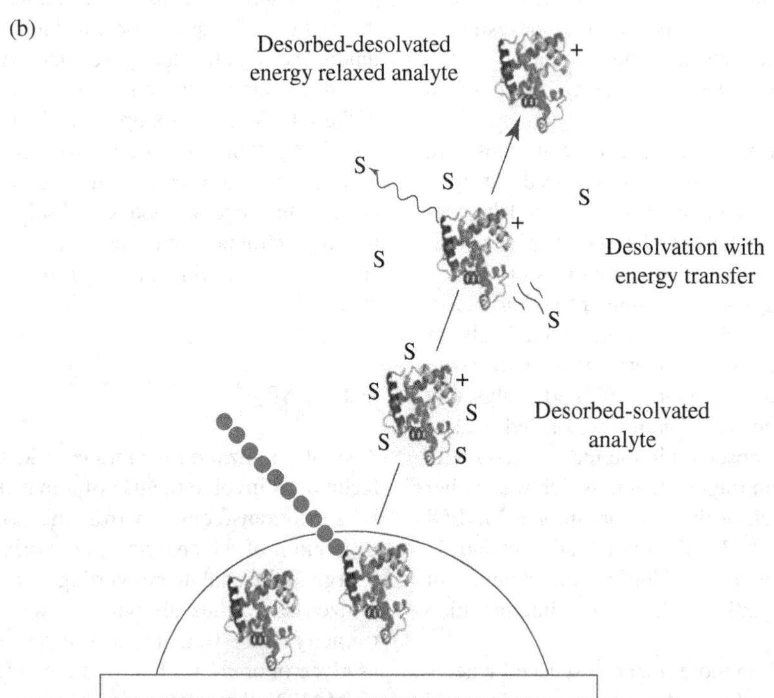

FIGURE 29.27 Fast atom bombardment (FAB) desorption and ionization process. (a) A cesium (Cs^+) atom beam gun directs a beam at the FAB analyte/matrix mixture desorbing both analyte and matrix. (b) Desorbed analyte is solvated with matrix ions. Transference of excess energy takes place between analyte and matrix relaxing the desorbed analyte.

absorbs the largest amount of the available kinetic energy from the incoming high-energy atom beam, thus sparing the analyte from unwanted decomposition. When decomposition does take place, the product ions derived from the precursor analyte are EE ions owing to the softer ionization process taking place. The matrix serves to constantly replenish the analyte to the surface of the liquid matrix for desorption and to limit analyte fragmentation. As shown in Figure 29.27b, the desorbed analyte goes through a desolvation mechanism where excess energy obtained during the desorption process is transferred to the matrix solvent molecules, thus relaxing the desorbed analyte ion. This effectively prevents excess fragmentation of the desorbed analyte and subsequently increased observation of intact cationized analyte species. This is of course important when the analyst is attempting to determine the molecular weight of an unknown analyte species. If the mass spectrum contains primarily only one major m/z peak versus a complicated spectrum of many m/z peaks of analyte and fragments, the determination of the molecular weight of the unknown is simple and straightforward. The FAB technique can be useful for the measurement of nonvolatile compounds that can also be thermally labile.

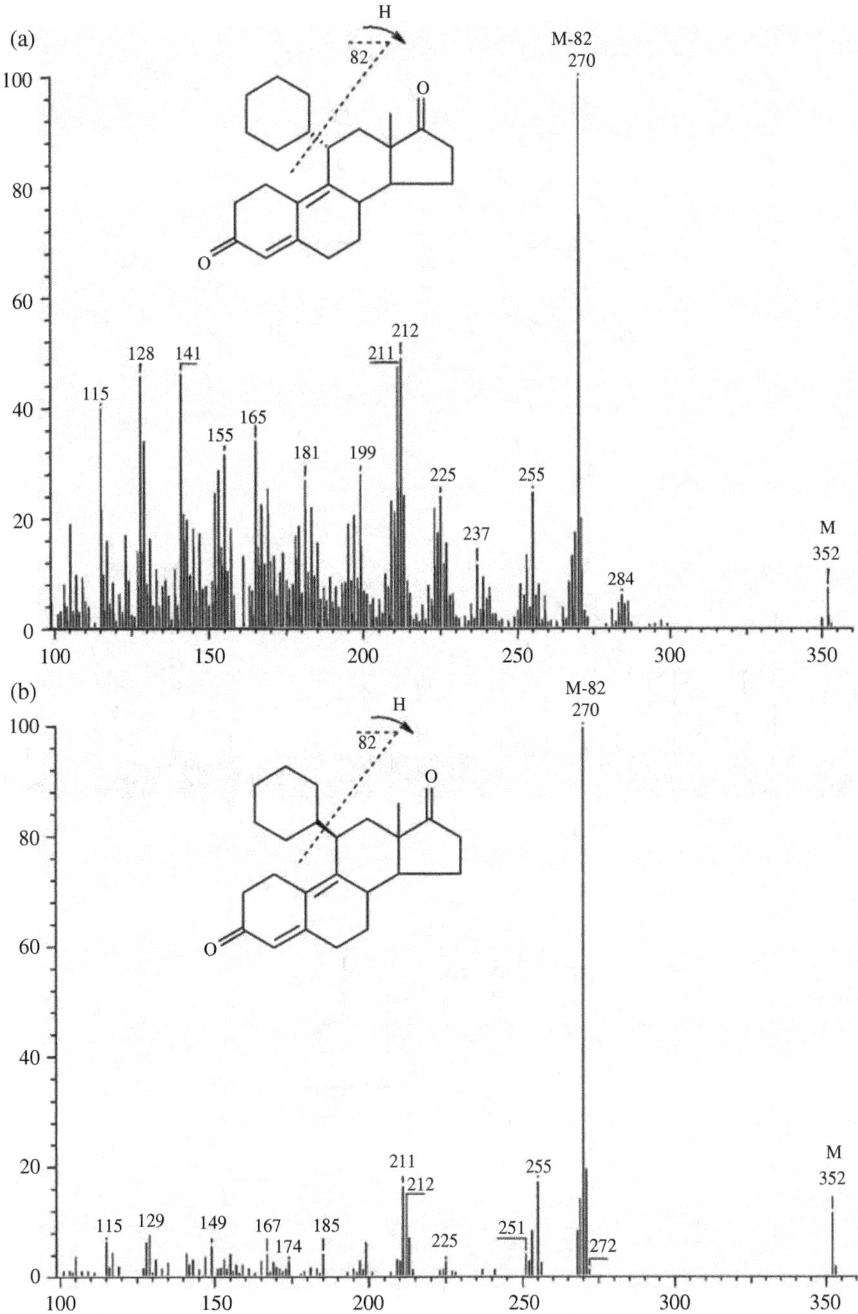

FIGURE 29.28 The electron ionization mass spectra of 1a and 1b epimers. (Reprinted with permission from Ref. [60]. Copyright Elsevier 2004.)

29.8.1 Application of FAB versus EI

As an example of the ionization technique of FAB, we will look at the comparative analysis of steroids by both FAB and by EI. Steroids, especially the unhydroxylated, tend to be very nonpolar and thus do not have a very predominant spectrum when analyzed using ESI. An alternative approach to their measurement has been the application of FAB, which is conducive to nonpolar analytes. Figure 29.28 illustrates the EI mass spectra of two epimers of estrans where Figure 29.28a is the 11α-cyclohexyl estran and Figure 29.28b is the 11β-cyclohexyl estran [60]. Notice that the product ions, such as the *m/z* 270 product ion for the loss of the cyclohexyl substituent, are the predominant ions in the spectra versus the precursor molecular ion ($M^{+\bullet}$) at *m/z* 352. The spectra for the two same steroid species are illustrated in Figure 29.29 collected using FAB. Notice in these spectra that the predominant ion is the *m/z* 353 proton adduct of the precursor (MH^+), while the peaks at *m/z* 270 and 271 are very minor peaks. This illustrates that the FAB ionization is a soft ionization technique as compared to the EI approach.

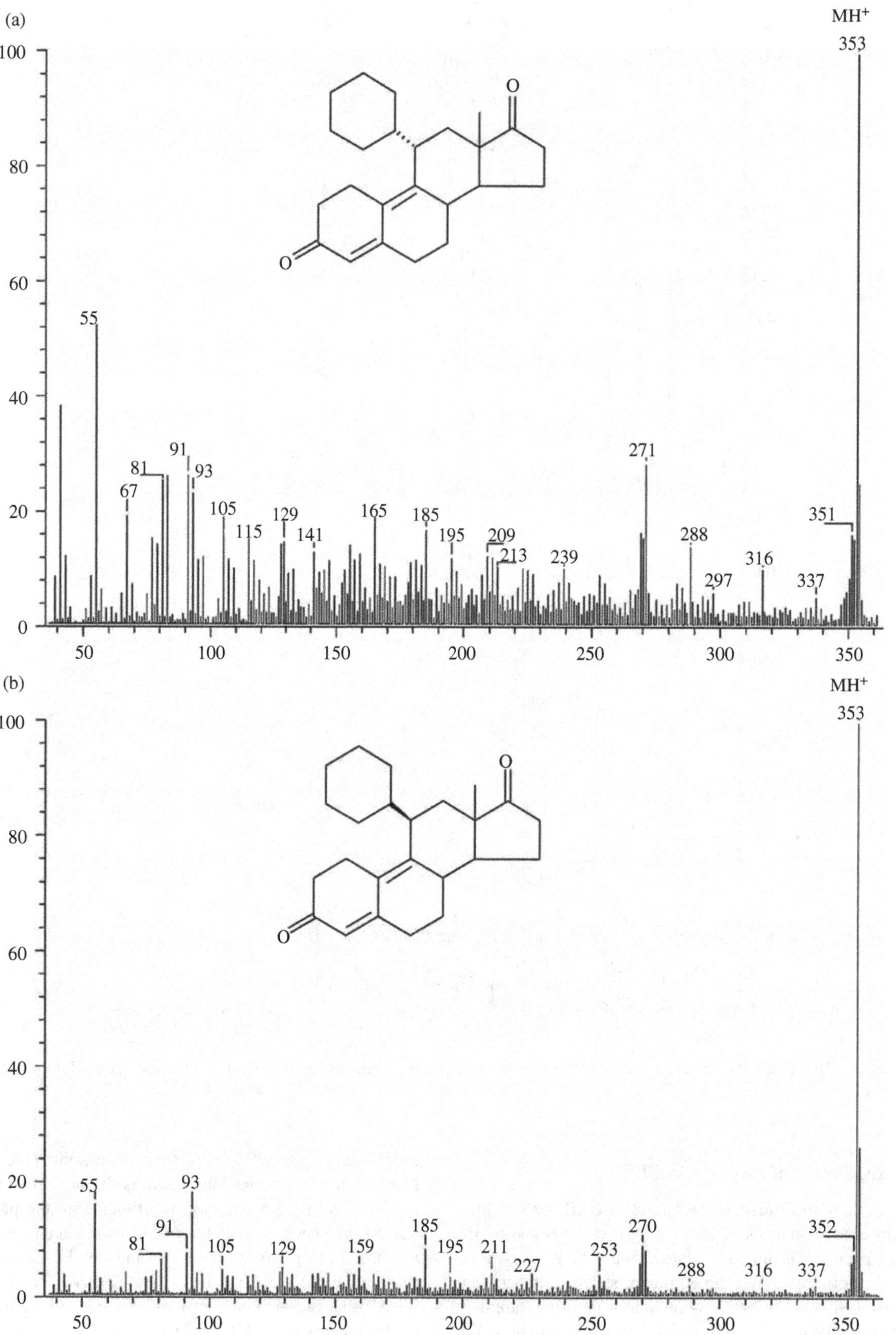

FIGURE 29.29 The FAB mass spectra of 1a and 1b epimers. (Reprinted with permission from Ref. [60]. Copyright Elsevier 2004.)

29.9 CHAPTER PROBLEMS

29.1 List some source/ionization systems that are typically used in mass spectrometry. Why is a source/ionization system important and a necessary component of mass spectrometric instrumentation?

29.2 What parameter is kept constant in the EI source that allows standardization of EI-generated mass spectra?

29.3 Small magnets are included as part of the EI source. Are these used to remove neutral analytes from the source? If not, what is their purpose in the EI source?

29.4 Chemical ionization and electron ionization source designs are similar but posses some key differences. What are they?

29.5 What are the most common reagent gasses used in positive chemical ionization? Explain the differences in reactivity between the reagent gasses used.

29.6 If an alcohol had a value of 895.0 kJ/mol for its proton affinity (PA), will the use of ammonia as the chemical ionization reagent gas result in the protonated form? What is the change in enthalpy for the reaction?

29.7 Demonstrate by calculating the change in enthalpy what other CI reagent gas may and may not form the protonated alcohols.

29.8 What reagent gasses will protonate ketones? What reagent gasses will not?

29.9 How can the two ionization techniques electron ionization and chemical ionization be used in a complementary fashion?

29.10 In the APCI source, a dry nitrogen curtain gas is used. Explain what its inclusion in the source is used to achieve.

29.11 Describe and explain what takes place to produce a Taylor cone and initiate the electrospray process.

29.12 What is the difference between the "charged residue" model and the "ion evaporation" model?

29.13 What are some of the cooling processes taking place during the electrospray process? What do these tend to promote?

29.14 Describe how flow rate enhances nanoelectrospray as compared to normal electrospray.

29.15 Describe the APPI mechanism including the ionization process and the photon source needed.

29.16 Why is it easier at times to detect nonpolar analytes with APPI than with electrospray?

29.17 Though ESI is a solvent-based process and APPI is a gas-phase process, how does the inclusion of a dopant to APPI affect the ionization process? Is APPI still a gas-phase ionization process?

29.18 For the compounds listed in Table 29.3, which could be ionized by the use of a xenon (Xe) lamp? Are there any that could not be analyzed by an argon (Ar) lamp?

29.19 What are some of the physical and chemical characteristics needed in order to make a good MALDI matrix? What are some examples of MALDI matrixes used?

29.20 What are the three general MALDI plate spotting techniques that are used?

29.21 Briefly describe the desorption/ionization process that takes place with the MALDI technique.

29.22 What are the two known problems with the MALDI ionization technique that affects resolution of the analytes? What instrumental designs have been made to reduce their effect?

29.23 List some differences that are found when using liquid ionic matrixes as compared to solid crystal matrixes.

29.24 What are some of the advantages of using a solid ionic crystal MALDI matrix?

29.25 Briefly describe the ionization mechanism of FAB.

29.26 What processes take place that tend to relax the analyte and produce a soft ionization technique in FAB.

REFERENCES

1. Whitehouse, C.M.; Dryer, R.N.; Yamashita, M.; Fenn, J.B. *Anal. Chem.* 1985, **57**, 675.
2. Fenn, J.B. *J. Am. Soc. Mass Spectrom.* 1993, **4**, 524.
3. Dole, M.; Hines, R.L.; Mack, R.C.; Mobley, R.C.; Ferguson, L.D.; Alice, M.B. *J. Chem. Phys.* 1968, **49**, 2240.
4. Kebarle, P.; Ho, Y. In *Electrospray Ionization Mass Spectrometry*; Cole, R.B., Ed.; Wiley: New York, 1997; p. 17.
5. Cole, R.B. *J. Mass Spectrom.* 2000, **35**, 763–772.
6. Cech, N.B.; Enke, C.G. *Mass Spectrom. Rev.*, 2001, **20**, 362–387.
7. Gaskell, S.J. *J. Mass Spectrom.*, 1997, **32**, 677–688.
8. Cech, N.B.; Enke, C.G. *Anal. Chem.*, 2000, **72**, 2717–2723.
9. Sterner, J.L.; Johnston, M.V.; Nicol, G.R.; Ridge, D.P. *J. Mass Spectrom.* 2000, **35**, 385–391.
10. Cech, N.B.; Enke, C.G. *Anal. Chem.*, 2001, **73**, 4632–4639.
11. Taylor, G.I. *Proc. R. Soc. London, Ser. A*, 1964, **280**, 383.
12. Gomez, A.; Tang, K. *Phys. Fluids* 1994, **6**, 404.
13. Cao, P.; Stults, J.T. *Rapid Commun. Mass Spectrom.* 2000, **14**, 1600–1606.
14. Ho, Y.P.; Huang, P.C.; Deng, K.H. *Rapid Commun. Mass Spectrom.* 2003, **17**, 114–121.
15. Kocher, T.; Allmaier, G.; Wilm, M. *J. Mass Spectrom.* 2003, **38**, 131–137.
16. Lorenz, S.A.; Maziarz, E.P.; Wood, T.D. *J. Am. Soc. Mass Spectrom.* 2001, **12**, 795–804.
17. Daniel, J.M.; Friess, S.D.; Rajagopalan, S.; Wendt, S.; Zenobi, R. *Int. J. Mass Spectrom.* 2002, **216**, 1–27.
18. Cech, N.B.; Enke, C.G. *Mass Spec. Rev.* 2001, **20**, 362.

19. Wilm, M. S.; Mann, M. *Int. J. Mass Spectrom. Ion Processes* 1994, **136**, 167–180.
20. Wilm, M.; Mann, M. *Anal. Chem.* 1996, **68**, 1–8.
21. Caprioli, R.M.; Emmett, M.E.; Andren, P. Proceedings of the 42nd ASMS Conference on Mass Spectrometry and Allied Topics, Chicago, IL, May 29–June 3, 1994, p. 754.
22. Qi, L.; Danielson, N.D. *J. Pharm. Biomed. Anal.* 2005, **37**, 225–230.
23. Fernandez de la Mora J.; Loscertales, I.G. *J. Fluid Mech.* 1994, **260**, 155–184.
24. Pfiefer, R.J.; Hendricks, C.D., Jr. *AIAA J.* 1968, **6**, 496–502.
25. Juraschek, R.; Dulcks, T.; Karas, M. *J. Am. Soc. Mass Spectrom.* 1999, **10**, 300–308.
26. Schmidt, A.; Karas, M. *J. Am. Soc. Mass Spectrom.* 2003, **14**, 492–500.
27. Li, Y.; Cole, R.B. *Anal. Chem.* 2003, **75**, 5739–5746.
28. El-Faramawy, A.; Siu, K.W.M.; Thomson, B.A. *J. Am. Soc. Mass Spectrom.* 2005, **16**, 1702–1707.
29. Tang, K.; Lin, Y.; Matson, D.W.; Kim, T.; Smith, R.D. *Anal. Chem.* 2001, **73**, 1658–1663.
30. Driscoll, J.N.; Spaziani, F.F. *Res./Dev.* 1976, **27**(5), 50–54.
31. Langhorst, M.L. *J. Chromatogr. Sci.* 1981, **19**, 98–103.
32. Schermund, J.T.; Locke and D.C. *Anal. Lett.* 1975, **8**, 611–625.
33. Locke, D.C.; Dhingra, B.S.; Baker, A.D. *Anal. Chem.* 1982, **54**, 447–450.
34. Driscoll, J.N.; Conron, D.W.; Ferioli, P.; Krull, I.S.; Xie, K.H. *J. Chromatogr.* 1984, **302**, 43–50.
35. De Wit, J.S.M.; Jorgenson, J.W. *J. Chromatogr.* 1987, **411**, 201–212.
36. Syage, J.A.; Hanold, K.A.; Lynn, T.C.; Horner, J.A.; Thakur, R.A. *J. Chromatogr. A* 2004, **1050**, 137–149.
37. Short, L.C.; Cai, S.-S.; Syage, J.A. *J. Am. Soc. Mass Spectrom.* 2007, **18**(4), 589–599.
38. Robb, D.B.; Covey, T.R.; Bruins, A.P. *Anal. Chem.* 2000, **72**, 3653–3659.
39. Syage, J.A.; Evans, M.D.; Hanold, K.A. *Am. Lab.* 2000, **32**, 24–29.
40. Kauppila, T.J.; Bruins, A.P.; Kostianen, R. *J. Am. Soc. Mass Spectrom.* 2005, **16**, 1399–1407.
41. Kauppila, T.J.; Kotiaho, T.; Kostianen, R.; Bruins, A.P. *J. Am. Soc. Mass Spectrom.* 2004, **15**, 203–211.
42. Straube, E.A.; Dekant, W.; Volkel, W. LC-MS/MS, *J. Am. Soc. Mass Spectrom.* 2004, **15**, 1853–1862.
43. Cavaliere, C.; Foglia, P.; Pastorini, E.; Samperi, R.; Lagana, A. *J. Chromatogr. A* 2006, **1101**, 69–78.
44. Gomez-Ariza, J. L.; Arias-Borrego, A.; Garcia-Barrera, T.; Beltran, R. *Talanta*, 2006, **70**, 859–869.
45. Karas, M.; Bachmann, D.; Bahr, U.; Hillenkamp, F. *Int. J. Mass Spectrom. Ion Processes* 1987, **78**, 53.
46. Tanaka, K.; Waki, H.; Ido, Y.; Akita, S.; Yoshida, Y.; Yoshida, T. *Rapid Commun. Mass Spectrom.* 1988, **2**, 151–153.
47. Karas, M.; Hillenkamp, F. *Anal. Chem.* 1988, **60**, 2299.
48. Xiang, F.; Beavis, R.C. *Org. Mass Spectrom.* 1993, **28**, 1424.
49. Xiang, F; Beavis, R.C. *Rapid Commun. Mass Spectrom.* 1994, **8**, 199–204.
50. Armstrong, D.W.; Zhang, L.-K.; He, L.; Gross, M.L. *Anal. Chem.* 2001, **73**, 3679–3686.
51. Carda-Broch, S.; Berthod, A.; Armstrong, D.W. *Rapid Commun. Mass Spectrom.* 2003, **17**, 553–560.
52. Li, Y.L.; Gross, M.L. *J. Am. Soc. Mass Spectrom.* 2004, **15**, 1833–1837.
53. Mank, M.; Stahl, B.; Boehm, G. *Anal. Chem.* 2004, **76**, 2938–2950.
54. Li, Y.L.; Gross, M.L.; Hsu, F.-F. *J. Am. Soc. Mass Spectrom.* 2005, **16**, 679–682.
55. Ham, B.M.; Jacob, J.T.; Cole, R.B. *Anal. Chem.* 2005, **77**, 4439–4447.
56. Lemaire, R.; Tabet, J.C.; Ducoroy, P.; Hendra, J.B.; Salzet, M.; Fournier, I. *Anal. Chem.* 2006, **78**, 809–819.
57. Aebersold, R.; Mann, M. *Nature* 2003, **422**, 198–207.
58. Luxembourg, S.L.; McDonnell, L.A.; Duursma, M.C.; Guo, X.; and Heeren, R.M.A. *Anal. Chem.* 2003, **75**, 2333–2341.
59. Lemaire, R.; Wisztorski, M.; Desmons, A.; Tabet, J.C.; Day, R.; Salzet, M.; Fournier, I. *Anal. Chem.* 2006, **78**, 7145–7153.
60. Mak, M.; Francsics-Czinege, E.; Tuba, Z. *Steroids* 2004, **69**, 831–840.

30

MASS ANALYZERS IN MASS SPECTROMETRY

30.1 Mass Analyzers
30.2 Magnetic and Electric Sector Mass Analyzer
30.3 Time-of-Flight Mass Analyzer (TOF/MS)
30.4 Time-of-Flight/Time-of-Flight Mass Analyzer (TOF–TOF/MS)
30.5 Quadrupole Mass Filter
30.6 Triple Quadrupole Mass Analyzer (QQQ/MS)
30.7 Three-Dimensional Quadrupole Ion Trap Mass Analyzer (QIT/MS)
30.8 Linear Quadrupole Ion Trap Mass Analyzer (LTQ/MS)
30.9 Quadrupole Time-of-Flight Mass Analyzer (Q-TOF/MS)
30.10 Fourier Transform Ion Cyclotron Resonance Mass Analyzer (FTICR/MS)
30.10.1 Introduction
30.10.2 FTICR Mass Analyzer
30.10.3 FTICR Trapped Ion Behavior
30.10.4 Cyclotron and Magnetron Ion Motion
30.10.5 Basic Experimental Sequence
30.11 Linear Quadrupole Ion Trap Fourier Transform Mass Analyzer (LTQ–FT/MS)
30.12 Linear Quadrupole Ion Trap Orbitrap Mass Analyzer (LTQ–Orbitrap/MS)
30.13 Chapter Problems
References

30.1 MASS ANALYZERS

The mass analyzer is the heart of the mass spectrometric instrumentation used in the separation of molecular ions ($M^{+\cdot}$) and analyte ions (e.g., $[M+H]^+$) in the gas phase. The most fundamental aspect of the mass analyzer is the ability to separate ions according to their mass-to-charge (*m/z*) ratio (see Chapter 28, Section 28.4). In this chapter, we will take a close look at the most common mass analyzers used today including their basic design and construction and the theory behind their ability to separate ionized analyte species according to their mass-to-charge ratio. Figure 30.1 illustrates the basic design and layout of most of the mass analyzers that are in use today. Included in Figure 30.1 are (a) an electric and magnetic sector mass analyzer, (b) a time-of-flight mass analyzer (TOF/MS), (c) a time-of-flight/time-of-flight mass analyzer (TOF–TOF/MS), (d) the hybrid (hybrids are mass analyzers that couple together two separate types of mass analyzers) quadrupole time-of-flight mass analyzer (Q-TOF/MS), (e) a triple quadrupole or linear ion trap mass analyzer (QQQ/MS or LIT/MS), (f) a three-dimensional quadrupole ion trap mass analyzer (QIT/MS), (g) a Fourier transform ion cyclotron resonance mass analyzer (FTICR/MS), and finally (h) the linear ion trap–Orbitrap mass analyzer (IT–Orbitrap/MS). Also included later in the chapter are discussions of the two more recently introduced hybrid mass analyzers in use in laboratories today: the linear quadrupole ion trap Fourier transform mass spectrometer (LTQ–FT/MS) and the linear quadrupole ion trap–Orbitrap mass spectrometer (LTQ–Orbitrap/MS). As discussed in Chapter 28 (Section 28.2), the most general configuration for mass spectrometric instrumentation is an ion source, a mass analyzer, and finally a detector (see Figure 28.1). As we also learned in Chapter 29, two very common techniques of analyte ionization are electrospray ionization (ESI) and matrix-assisted laser desorption ionization (MALDI). We will see in this chapter how these ionization sources are used in conjunction with the common mass analyzers illustrated in Figure 30.1. We will also look at a few newer, state-of-the-art mass analyzers that have come out recently such as the linear ion trap Fourier transform mass analyzer and the Orbitrap mass analyzer. Finally, we will take a look at a few examples of applications of the LTQ–FT/MS and the LTQ–Orbitrap/MS for the measurement of biomolecules at the end of the chapter.

30.2 MAGNETIC AND ELECTRIC SECTOR MASS ANALYZER

When charged particles enter a magnetic field, they will possess a circular orbit that is perpendicular to the poles of the magnet.

Analytical Chemistry: A Chemist and Laboratory Technician's Toolkit, First Edition. Bryan M. Ham and Aihui MaHam.

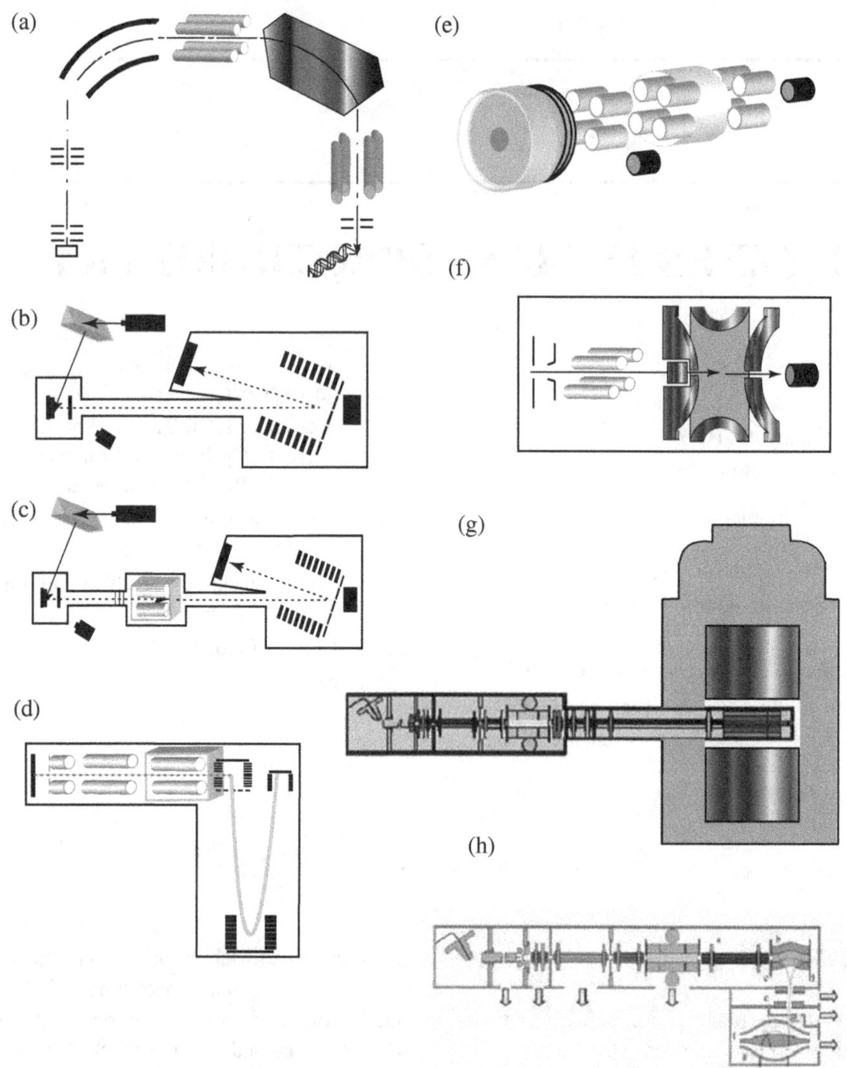

FIGURE 30.1 Some of the most common mass analyzers in use today: (a) electric and magnetic sector mass analyzer, (b) time-of-flight mass analyzer (TOF/MS), (c) time-of-flight/time-of-flight mass analyzer (TOF–TOF/MS), (d) quadrupole time-of-flight mass analyzer (Q-TOF/MS), (e) triple quadrupole or linear ion trap mass analyzer (QQQ/MS or LIT/MS), (f) three-dimensional quadrupole ion trap mass analyzer (QIT/MS), (g) Fourier transform ion cyclotron resonance mass analyzer (FTICR/MS), and (h) linear ion trap–Orbitrap mass analyzer (IT–Orbitrap/MS).

This phenomenon has been applied to a magnetic sector mass analyzer, which is a momentum separator. Figure 30.2 illustrates the basic design of the magnetic sector mass analyzer. Notice that the slits are normally placed collinear with the apex of the magnet. The ions enter a flight tube (first field-free region) from a source through a source exit slit and travel into the magnetic field. The accelerating voltage in the source will determine the kinetic energy (KE) that is imparted to the ions:

$$\mathrm{KE} = zeV = \frac{1}{2}mv^2 \qquad (30.1)$$

where V is the accelerating voltage in the source, e is the fundamental charge of an electron (1.60×10^{-19} C), m is the mass of the ion, v is the velocity of the ion, and z is the number of charges. The magnetic field will deflect the charged particles according to the radius of curvature of the flight path (r) that is directly proportional to the mass-to-charge (m/z) ratio of the ion. The centripetal force that is exerted upon the ion by the magnetic field is given by the relationship

$$F_{\mathrm{M}} = Bzev \qquad (30.2)$$

where B is the magnetic field strength. The centrifugal force acting upon the ion from the initial momentum is given by

$$F_{\mathrm{C}} = \frac{mv^2}{r} \qquad (30.3)$$

For the ion to reach the detector, the centripetal force acting upon the ion from the magnet (down pushing force) must equal the centrifugal force acting upon the ion from the initial momentum of the charged ion (upwardly pushing force). This equality is represented by

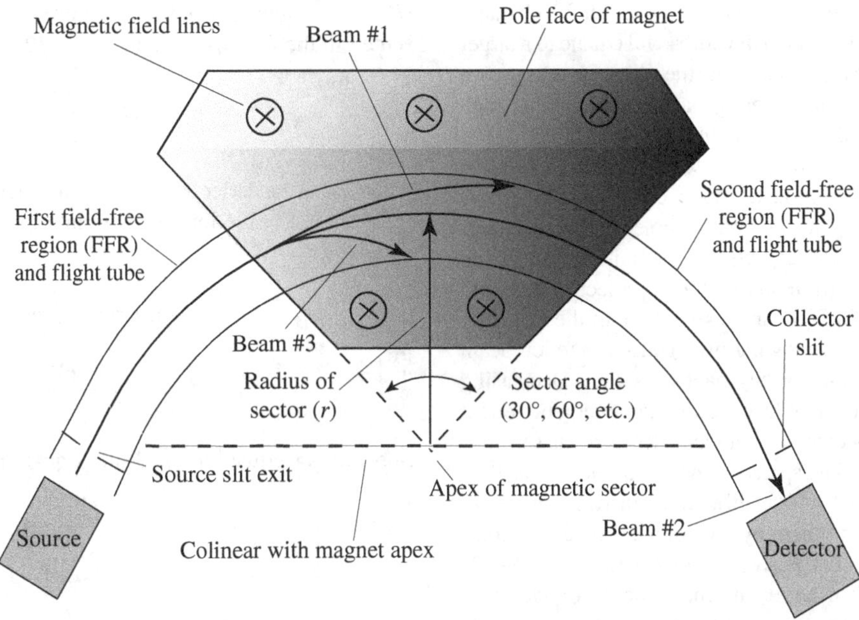

FIGURE 30.2 Basic components of a magnetic sector mass analyzer. The magnetic field lines are directed into the plane of the paper for negative ion mode that follows the right-hand rule. For positive ion mode, the polarity is switched and the magnetic field lines are directed out of the plane of the paper.

$$F_M = F_C \tag{30.4}$$

$$Bzev = \frac{mv^2}{r} \tag{30.5}$$

where B is the magnetic field strength, e is the electron fundamental charge, m is the mass of the ion, v is the velocity of the ion, z is the number of charges, and r is the radius of the curvature of the sector. Solving Equation 30.5 for v and substituting into Equation 30.1, we can derive the relationship between the mass-to-charge (m/z) ratio of the ion to the force of the magnetic field strength:

$$Bzev = \frac{mv^2}{r} \tag{30.5}$$

$$v = \frac{Bzer}{m} \tag{30.6}$$

$$KE = zeV = \frac{1}{2}mv^2 \tag{30.1}$$

$$zeV = \frac{1}{2}m\left(\frac{Bzer}{m}\right)^2 \tag{30.7}$$

$$zeV = \frac{1}{2}m\frac{B^2z^2e^2r^2}{m^2} \tag{30.8}$$

$$V = \frac{1}{2}\frac{B^2zer^2}{m} \tag{30.9}$$

$$\frac{m}{z} = \frac{B^2er^2}{2V} \tag{30.10}$$

This relationship is often referred to as the scan law. In normal experimental operation, the radius of the sector is fixed, and the accelerating voltage V is held constant, while the magnetic field strength B is scanned. From Equation 30.10, it can be seen that a higher magnetic field strength equates to a higher mass-to-charge ratio stability path. The magnetic field strength B is typically scanned from lower strength to higher strength for m/z scanning in approximately 5 s. In exact mass measurements and when calibrating the magnetic sector mass analyzer, the magnetic field strength B of the magnet is kept fixed, while the accelerating voltage is scanned. This is done to avoid the slight error that is inherent in the magnet known as hysteresis where after the magnet has scanned, its original strength is slightly changed from its initial condition before the scanning was commenced. Scanning the accelerating voltage does not have this problem. The calibration of the magnetic sector mass analyzer is a nonlinear function; therefore, several points are often required for calibration. In calibrating, a standard mass is scanned by scanning the acceleration voltage and holding the magnetic field strength constant. When the mass is detected, the values are set for that set of acceleration voltage versus magnetic field strength. The resolution of the magnetic sector mass analyzer is directly proportional to the magnet radius and the slit width of the source exit slit and the slit width of the collector slit:

$$\text{Resolution} \propto \frac{r}{S_1 + S_2} \tag{30.11}$$

where

r	=	Radius of the magnet
S_1	=	Slit width of the source exit slit
S_2	=	Slit width of the collector slit

It can be seen from the relationship in Equation 30.11 that a larger magnet radius and smaller slit widths will equate to a larger resolution value. However, decreasing the slit widths causes a loss in sensitivity, as the amount of ions allowed to pass through will decrease. The typical resolution obtained under normal scan mode for a magnetic sector mass analyzer is ≤5000. This means that the magnetic sector mass analyzer can separate an *m/z* 5000 molecular ion from an *m/z* 5001 molecular ion, an *m/z* 500.0 molecular ion from an *m/z* 500.1 molecular ion, an *m/z* 50.00 molecular ion from an *m/z* 50.01 molecular ion, and so on. Two factors that limit the mass resolution of the magnetic sector mass analyzer are the angular divergence of the ion beam and the KE spread of the ions as they leave the source. Decreasing the source exit slit width will reduce the angular divergence effect upon the resolution; however, this will also decrease the sensitivity of the mass analyzer. The spread in the KE of the ions as they leave the source is caused by slight differences in the initial velocity of the ions imparted to them by the source. In electron ionization (EI), the initial KE spread is approximately 1–3 eV. To address the KE spread distribution in a magnetic sector mass analyzer, an electric sector that acts as a KE separator is used in combination with the magnetic sector. The electric sector shown in Figure 30.3 is used as an energy filter where only one energy of ions will pass through the electric field. The electric sector will only pass ions that have the exact energy match as that with the acceleration voltage in the source. In combination with slits, the electric sector can be called an energy focuser. The electric sector is constructed of two cylindrical plates with potentials of $+1/2E$ and $-1/2E$, where the total field $= eE = (+1/2E) + (-1/2E)$. For ions to pass through the electric sector, the deflecting force (K_E) must equal the centrifugal force (K_C). This equality is represented in the following Equation 30.12 as

$$eE = \frac{mv^2}{r_E} \tag{30.12}$$

where eE is the deflecting force (K_E) and mv^2/r_E is the centrifugal force (K_C). For the ions arriving into the electric sector, the KE equals

$$\text{KE} = zeV = \frac{1}{2}mv^2 \tag{30.1}$$

$$v^2 = \frac{2zeV}{m} \tag{30.13}$$

Substituting Equation 30.13 into Equation 30.12, we get

$$eE = \frac{m\left(\dfrac{2zeV}{m}\right)}{r_E} \tag{30.14}$$

$$E = \frac{2zV}{r_E} \tag{30.15}$$

$$r_E = \frac{2zV}{E} \tag{30.16}$$

From this relationship, we see that the radius of the curvature will equal two times the accelerating voltage divided by the total electric field strength. What ions pass through the electric sector does not depend upon the mass or charge of the ions but upon how well the ion's KE matches the field according to the acceleration

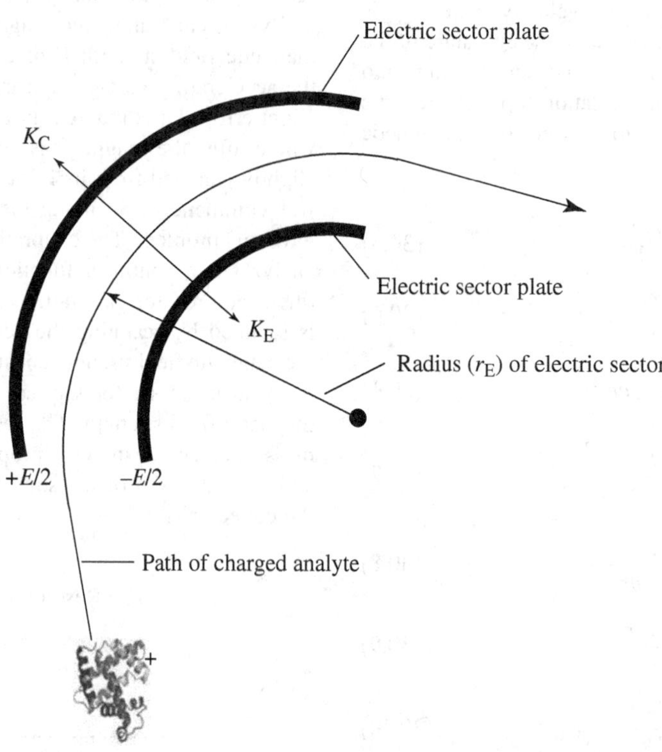

FIGURE 30.3 Path of charged particle through the electric sector energy filter. Forces acting upon the particle include the centrifugal force (K_C) and the deflecting force (K_E).

voltage within the source. This relationship demonstrates that the electric sector is not a mass analyzer like the magnetic sector but rather is an energy filter.

By combining the magnetic sector with the electric sector, a double-focusing effect can be achieved where the magnetic sector does directional focusing, while the electric sector does energy focusing. Two types of double-focusing geometries are illustrated in Figures 30.4 and 30.5. The double-focusing geometry illustrated in Figure 30.4 is known as the Mattauch–Herzog double-focusing geometry. The electric sector (electrostatic analyzer) located before the magnetic analyzer in the instrumental design has effectively been added to gain in the resolution of the mass-analyzed ions. The Mattauch–Herzog double-focusing geometry results in a focal plane within the mass analyzer. This is a static analyzer plane where a photoplate or microchannel array detectors can be placed parallel to the focal plane. In this design, the different mass-to-charge ratios will be focused at different points on the microchannel plate. The detection of the separated mass-to-charge ratio ions will require no scanning increasing the sensitivity in the instrumental response. The second design in Figure 30.5 is the Nier–Johnson double-focusing mass spectrometer. Here, also the electrostatic analyzer is placed before the magnetic analyzer. In this design, ions are focused onto a single point and the magnetic analyzer must be scanned.

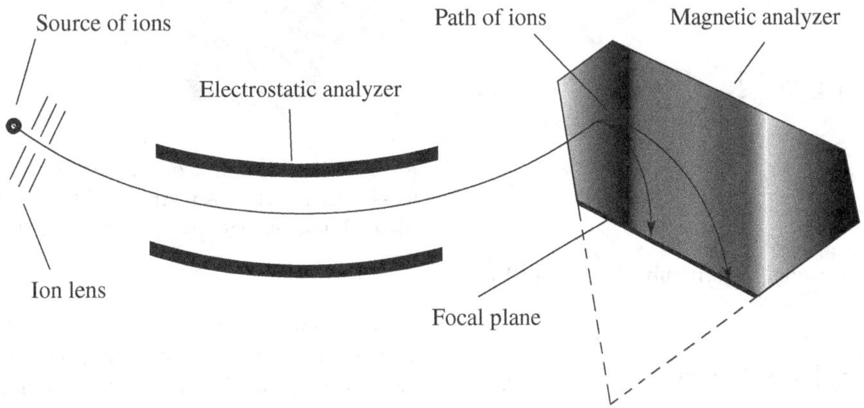

FIGURE 30.4 Mattauch–Herzog double-focusing geometry mass spectrometer.

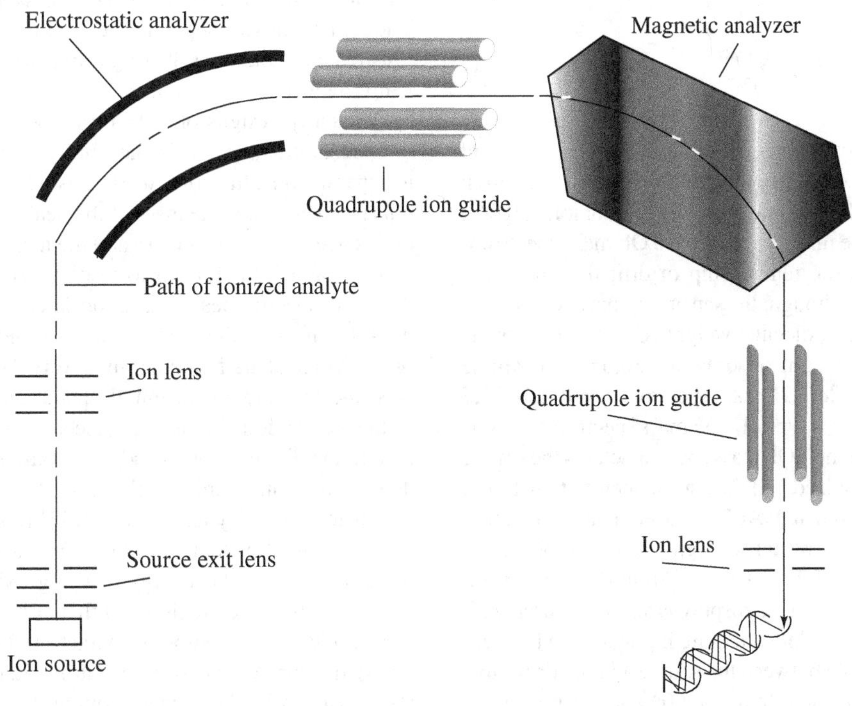

FIGURE 30.5 Nier–Johnson double-focusing geometry mass spectrometer.

30.3 TIME-OF-FLIGHT MASS ANALYZER (TOF/MS)

The most common mass spectrometer that is coupled to the MALDI ionization technique is the time-of-flight (TOF) mass spectrometer (TOF/MS) [1,2]. The TOF mass spectrometer separates compounds according to their mass-to-charge (*m/z*) ratios through a direct relationship between a compound's drift time through a predetermined drift path length and the analyte ion's mass-to-charge (*m/z*) ratio. Initially, all the ions have similar KE imparted to them from the draw-out pulse (representing time zero), which accelerates them into the flight tube. Because the compounds have different masses, their velocities will be different according to the relationship between KE and mass represented by $KE = zeV = 1/2mv^2$. From this expression, the mass-to-charge ratio is related to the ion's flight time by the following expression: $m/z = 2eVt^2/L^2$:

$$KE = zeV = \frac{1}{2}mv^2 \quad (30.1)$$

$$v = \left(\frac{2zeV}{m}\right)^{\frac{1}{2}} \quad (30.17)$$

$$t = \frac{L}{v}, \quad L = \text{length of drift tube} \quad (30.18)$$

$$t = L\left(\frac{m}{2zeV}\right)^{\frac{1}{2}}, \quad V \text{ and } L \text{ are fixed} \quad (30.19)$$

Solving for *m/z*,

$$t^2 = L^2\left(\frac{m}{2zeV}\right) \quad (30.20)$$

$$\frac{m}{z} = \frac{2eVt^2}{L^2} \quad (30.21)$$

There is a high transmission efficiency of the ions into and through the drift tube that equates to very low levels of detection limits that are in the femtomole (10^{-15}) to attomole (10^{-18}) ranges. Theoretically, the mass range of the TOF mass spectrometer is unlimited due to the relationship of drift time for mass measurement. In practice though, the sensitivity needed to detect a very slow-moving large molecular weight compound limits the TOF/MS to 1–2 million daltons or so. Figure 30.6 is an example of a commercially available TOF mass spectrometer. Figure 30.7 shows the different parts of a typical TOF mass spectrometer. The first major component of the TOF mass spectrometer is the source where the analytes are ionized and then subsequently transferred into the mass analyzer. When MALDI is used as the ionization technique, the source will be under high vacuum typically at an approximate pressure of 1×10^{-6} torr. An atmospheric pressure ionization source such as electrospray is also often used with TOF mass analyzers, which do not require high vacuum but generally require a quadrupole between the source and the drift tube. These quadrupoles can be radio frequency (rf) only acting as ion guides, or they can be fully functional mass filters used to isolate analyte ions before transfer into the drift tube. This is often done for tandem mass spectrometric analysis of collision-induced dissociation product ion spectra generation (this will be covered in Section 30.8). At this point, we will consider a MALDI ionization source. Upon generation of gas-phase ions by the laser and matrix, the ions are drawn into the second component of the TOF mass analyzer by an extraction grid that imparts equal KE to all of the analyte ions present. The analyte ions then transfer into the second part of the TOF mass analyzer, the drift tube. The analyte ions are separated according to the relationship that is illustrated in Equations 30.17 through 30.21 and are detected typically by electron multipliers.

FIGURE 30.6 Example of a commercially available time-of-flight mass spectrometer. An Applied Biosystems Voyager DE STR MALDI time-of-flight mass spectrometer (MALDI-TOF/MS).

The early designs of TOF mass spectrometers suffered from poor resolution, which is the ability of the mass spectrometer to separate ions. In earlier work, this was expressed as resolution $= m/\Delta m$ where m is the mass of the peak of interest and Δm is the difference between this mass and the next closest peak (more of a chromatography approach to resolution). Typically used today, the mass spectral peak resolution is calculated as $m/\Delta m$ where m is the mass of the peak of interest and Δm is the full width of the peak at half maximum (FWHM). The poor resolution was due to nonuniform initial spatial and energy distributions of the formed ions in the mass spectrometer's ionization source. Due to the distribution spreads, the mass resolution was directly dependent upon the initial velocity of the formed ions. Early work conducted by Wiley and McLaren [3] reported upon the correction for initial velocity distributions using a technique they described as "time-lag energy focusing" where the ions were produced in a field-free region and then, after a preset time, a pulse was applied to the region to extract the formed ions. They also reported upon a way to correct for initial special distributions through a two-field pulsed ion source. They demonstrated though that correction can only be made for one of the distributions at a time. In MALDI, the spatial distribution spread is not a

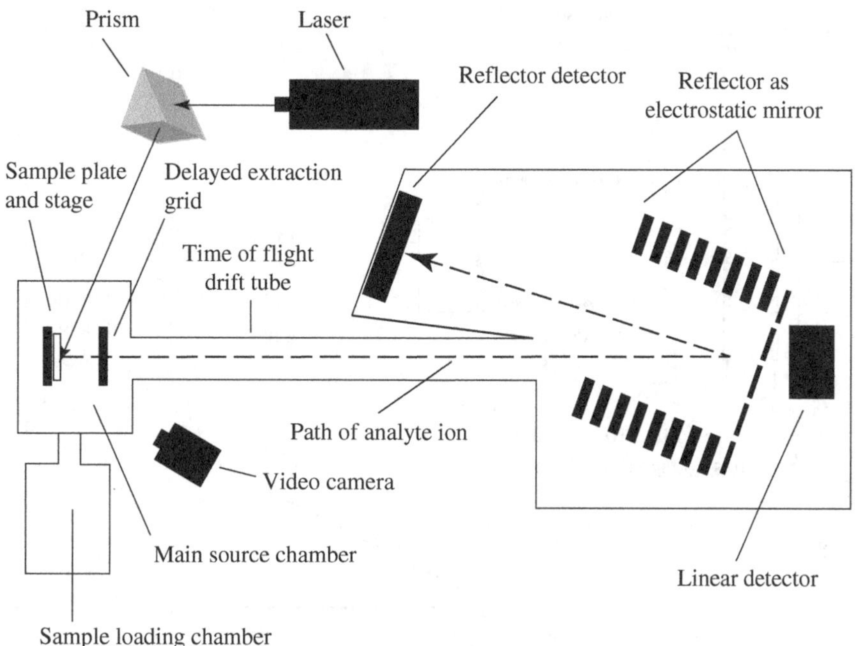

FIGURE 30.7 Components of a time-of-flight mass spectrometer illustrating the major sections including the source, drift tube, reflectron electrostatic mirror, and detector.

significant problem; therefore, "delayed extraction" of the ions from the field-free source region should correct for the initial energy distribution spread of the formed ions [4]. With this correction, the resolution should be directly dependent upon the total flight time ratio to the error in the time measurement; thus, an increase in the length of the flight path should equate to a higher resolution. The resolution enhancement due to the delayed extraction initial energy focusing and flight path was demonstrated in a study reported by Vestal et al. [4] using a MALDI-TOF mass spectrometer. Early work reported by Lennon et al. in 1995 (see Fig. 30.8 caption) gave a good description of a pulsed ion extraction system applied to MALDI-TOF mass spectrometry. Figure 30.8 illustrates a schematic representation of the pulsed ion extraction apparatus they used and the location of it within the source section of the TOF mass spectrometer. It is described as a three-grid system (G1, G2, and a third grounded grid) where the grids are constructed as a mesh of wire for uniform electrostatic field formation. In the delayed extraction mechanism, a nitrogen laser is pulsed onto the MALDI matrix/sample mixture effectively desorbing the matrix and the sample into a gas-phase plume above the G1 and between G2. After a brief delay period (340 ns to 4 μs) for optional plume formation, a short pulse (0–3 kV for ~10 μs) of voltage is applied to G1 while maintaining G2 at the prepulse voltage (bias voltage). This creates a potential field gradient between G1 and G2 highest at G1 and decreasing toward G2. Slower moving ions within the plume will be closer to G1 with respect to faster moving ions and will experience a stronger repelling force from G1. The result will be that the ions that were initially moving slower will effectively catch up to ions with the same *m/z* value that possessed slightly higher initial velocities. By combining an optimized delay time and extraction pulse, the flight time spread contribution to resolution loss due to different initial velocities can be corrected for.

A second modification of TOF mass spectrometers involves time focusing of the ions while they are in flight through the use of an electrostatic mirror at the end of the flight tube [5,6]. The electrostatic mirror focuses the ions with the same mass-to-charge ratio but slightly different KE by allowing a slightly longer path for the higher KE-containing ion as compared to a slightly lower KE ion, thus allowing the ions to catch up to one another. The electrostatic mirror also increases the flight path length of the mass spectrometer, thus providing a double-focusing effect. All recent TOF mass spectrometers employ both delayed extraction and the reflectron electrostatic mirror for analysis of compounds below 10 kDa. For compounds greater than 10 kDa, a linear mode is typically used where the electrostatic mirror is turned off. Figure 30.9 illustrates the focusing of the two ions that have the same *m/z* value but slightly different KE. Ion path 1 represents the path of the ion with the slightly higher KE in relation to ion path 2. Ion path 1 travels slightly farther into the electrostatic mirror field gradient, thus focusing its flight time to match that of ion path 2. After traveling through the second field-free region, the two ions are focused and arrive at the detector at the same time, producing a narrower peak representing its detection and thus increasing the resolution.

30.4 TIME-OF-FLIGHT/TIME-OF-FLIGHT MASS ANALYZER (TOF–TOF/MS)

A modification of the reflectron TOF mass analyzer is the coupling of essentially two TOF mass analyzers together with the chief addition of a central floating collision cell between them.

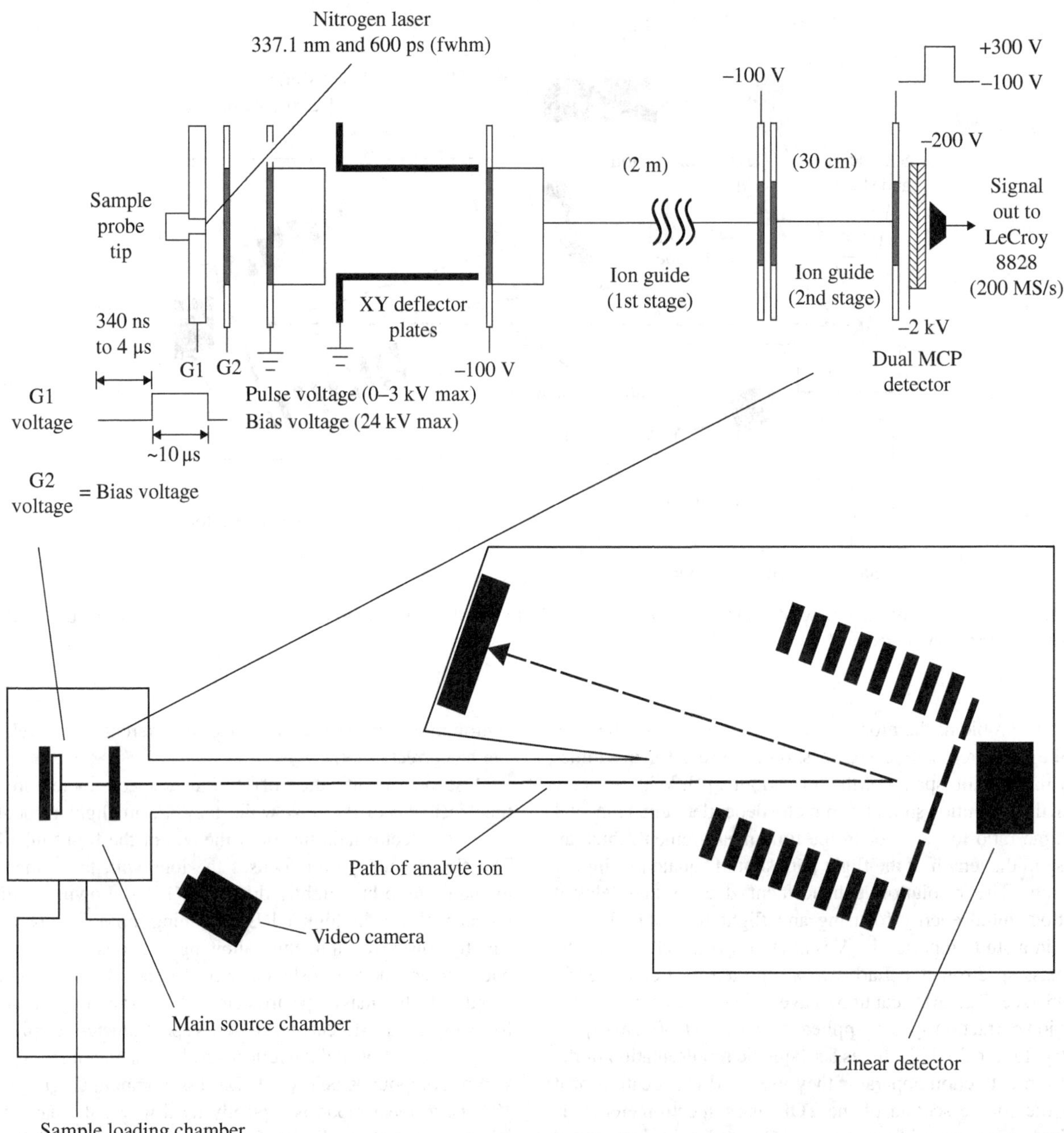

FIGURE 30.8 Expanded view of the delayed extraction setup within the source of the time-of-flight mass spectrometer. (Reprinted with permission from Brown, R.S.; Lennon, J.J. Mass Resolution Improvement by Incorporation of Pulsed Ion Extraction in a Matrix-Assisted Laser Desorption/Ionization Linear Time-of-Flight Mass Spectrometer. *Anal. Chem.* **1995**, *67*, 1998–2003. Copyright 1995 American Chemical Society.)

This effectively allows tandem mass spectrometric analyses to be performed that consist of collision-induced dissociation experiments of precursor analyte ions for structural information and identification. The design of this system primarily dictates though that a MALDI source must be used. The MALDI ionization technique almost exclusively produces singly charged analyte ions that are then mass analyzed. One drawback of singly charged ions, particularly in the mass spectrometric analysis of peptides, is the sometimes observed low efficiency in the number of fragmentation pathways that give useful product ions. ESI is a technique that often produces multiply charged analyte ions (peptides and proteins) that tend to have a higher degree in the number of fragmentation pathways that produce product ions. This is due to the higher degree of charging by electrospray that tends to activate a higher degree of fragmentation in CID experiments. Figure 30.10 illustrates the basic components that made up a time-of-flight/time-of-flight (TOF–TOF) mass spectrometer. A tandem TOF–TOF mass analyzer, such as the one illustrated in Figure 30.10, produces a product ion spectrum by first isolating the gas-phase analyte ion species of interest separated from the other species present by the first drift tube, which is then allowed to pass out of the drift tube and into a central rf-only

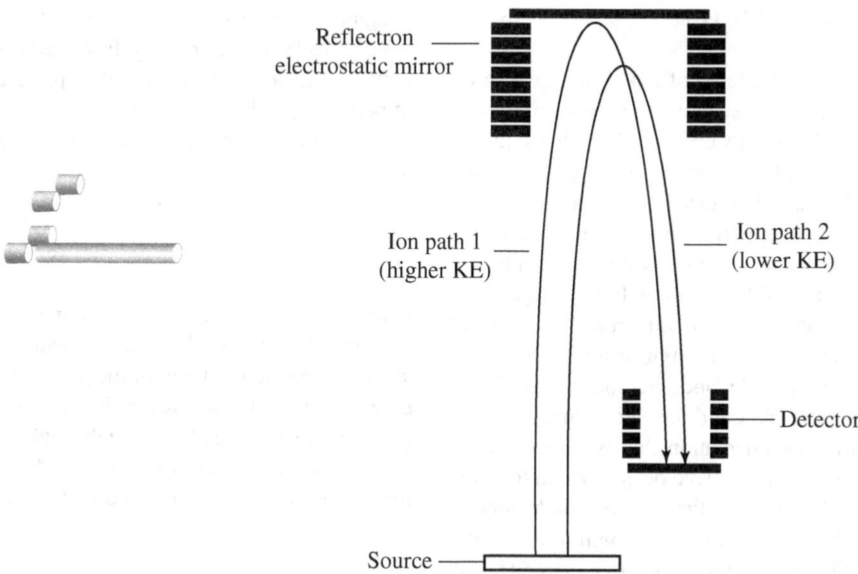

FIGURE 30.9 Electrostatic mirror focusing of two ions that have the same *m/z* value but slightly different kinetic energies. Ion path 1 possesses slightly higher kinetic energy in relation to ion path 2. Ion path 1 travels slightly farther to match that of ion path 2. The two ions are focused and arrive at the detector at the same time.

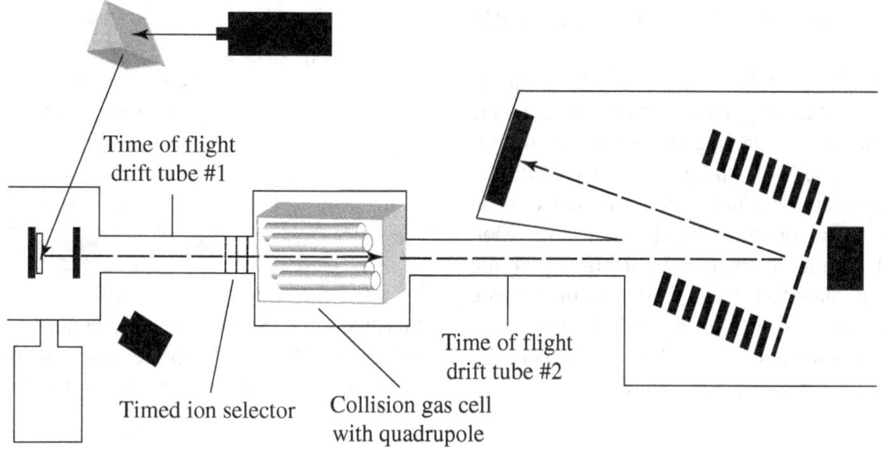

FIGURE 30.10 Basic components that make up a time-of-flight/time-of-flight mass spectrometer (TOF–TOF MS).

quadrupole or hexapole collision cell with the use of an ion gate acting as a timed ion selector. The ion gate is comprised of a series of wires that alternating voltages can be applied to usually as ±1000 V. When the ion gate is switched on, no ions are allowed to pass through the gate, thus preventing the transmission of any ions through the remainder of the instrument. When the gate is switched off, all ions will be able to pass through the gate. By switching off the gate at a predetermined flight time, a specific *m/z* value will be allowed to pass through the gate. In this way, the timed ion selector is turned on and off to allow the passage of the desired *m/z* species according to its predetermined drift tube flight time. The central rf-only quadrupole or hexapole collision cell is filled with a stationary target gas such as argon. The ion is induced to collide with the gas, thus activating the ion for dissociation. The central rf-only quadrupole or hexapole collision cell is not a mass analyzer itself but functions to refocus the product ions before exiting the cell. The products produced by the collision-induced dissociation are then transferred to the second TOF mass analyzer by a series of ion guides. To this point, the series of events include the production of gas-phase ionized analyte species in the source phase, the delayed extraction pulsing of the ions into the first TOF mass analyzer, the separation of the different *m/z* values according to their drift times in the first flight tube, and finally the selection and allowed passage of a single *m/z* species by the timed ion selector into the central collision cell. The *m/z* species that was allowed passage into the collision cell undergoes product ion activating collision with the stationary argon target gas. The refocused product ions are then transferred into the second TOF drift tube where they are separated into different *m/z* values according to their respective drift times. The TOF–TOF/MS instrumentation also include the reflector (electrostatic mirror) used to focus the product ions prior to detection. Finally, the mass-analyzed gas-phase ions are detected, and their spectra stored for processing.

30.5 QUADRUPOLE MASS FILTER

A quadrupole mass analyzer is made up of four cylindrical rods that are placed precisely parallel to each other. One set of opposite poles has a DC voltage (U) supply connected, while the other set of opposite poles of the quadrupole has an rf voltage (V) connected. The quadrupole DC and rf voltage configuration is illustrated in Figure 30.11. Ions are accelerated into the quadrupole by a small voltage of 5 eV, and under the influence of the combination of electric fields, the ions follow a complicated trajectory path. If the oscillation of the ions in the quadrupole has finite amplitude, it will be stable and pass through. If the oscillations are infinite, they will be unstable and the ion will collide with the rods. These path descriptions are illustrated in Figure 30.12. In the particular orientation that is illustrated in Figure 30.12, the DC and rf voltages have been selected to give an *m/z* value of 100 a stable trajectory through the quadrupoles. An *m/z* value of 10, which is a less massive ion, will have a very unstable trajectory and will collide with the quadrupole rods at an early stage. An *m/z* 1000 species will be a more massive ion of greater momentum, which tends to travel further through the quadrupole field but still possesses an unstable trajectory and will also suffer collision with the rods and thus also be effectively filtered out. The construction of the quadrupoles is based off of the equipotential curves of a quadrupole field, which are illustrated in Figure 30.13.

As Figure 30.13 shows, the equipotential curves of the quadrupole field are comprised of rectangular hyperbolae. The field is created in the mass analyzer by selecting one set of the equipotential curve rectangular hyperbolae and placing the quadrupole electrodes in a configuration that follows the dimensions of the rectangular hyperbolae. The optimum field and transmission through the quadrupole would be obtained by using hyperbolic rods; however, for ease of manufacturing, the rods are often made cylindrical (notice that the inside of the hyperbolae curves follows a semicircular path). The construction of an ideal hyperbolic field is achieved by orienting the quadrupole electrodes in an imaginary square so that the opposite poles lay 1/1.148 times the electrode diameter away from each other [7]. The inherent faults in the electric field caused by using cylindrical rods instead of perfect hyperbolae can be corrected by placing the electrodes 1/1.148 times the electrode diameter according to the following relationship:

$$r_0 = \frac{r}{1.148} \tag{30.22}$$

where r_0 is the radius of the field and r is the electrode radius.

The ratio of 1/1.148 is theoretically derived to produce an ideal hyperbolic field within the geometric center of the quadrupole. The mathematical derivation of the description of the stable trajectory of a charged particle through the quadrupole has been well characterized, and the reader is directed to other references for a more thorough coverage of this subject if interested [8].

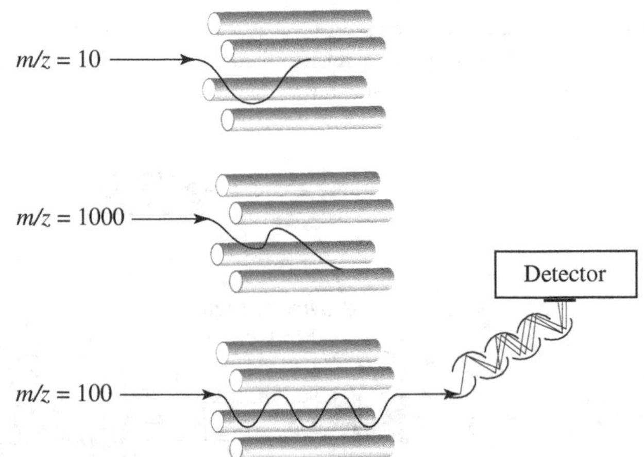

FIGURE 30.12 Stable and unstable trajectories of ions through the quadrupole. The *m/z* 100 species has been selected for stable path and transmission through the quadrupole for detection.

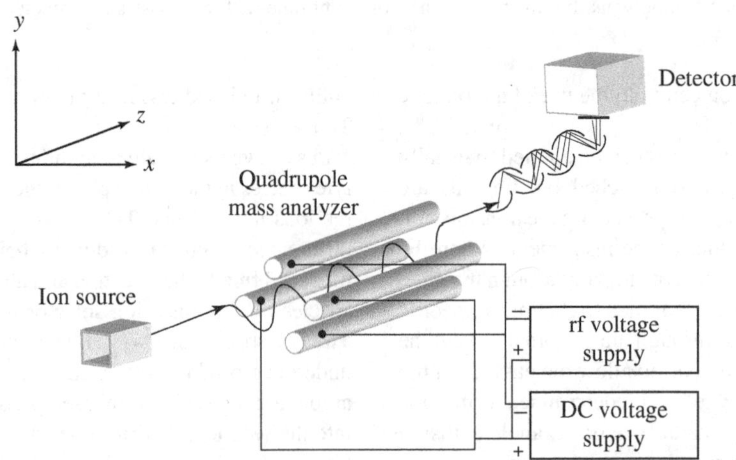

FIGURE 30.11 Quadrupole orientation and the configuration for the connections of the DC voltage (*U*) and radio frequency (rf) voltage (*V*). Ions are accelerated into the quadrupole by a small voltage of 5 eV, and under the influence of the combination of electric fields, the ions follow a complicated trajectory path.

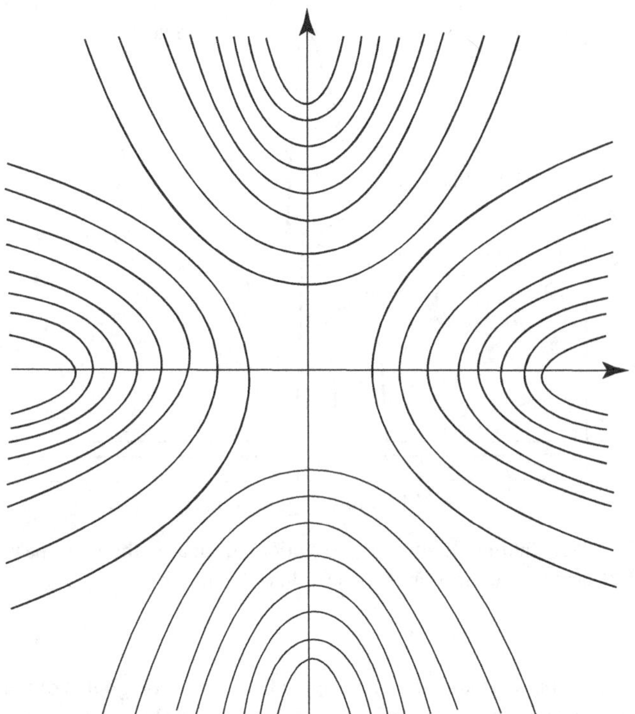

FIGURE 30.13 Equipotential curves of the quadrupole field comprised of rectangular hyperbolae where the inside of the hyperbolae curves follows a semicircular path.

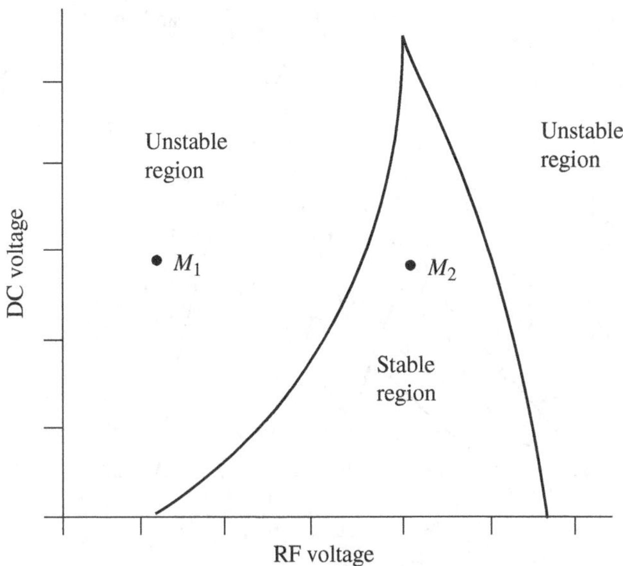

FIGURE 30.14 Stability diagram relating the DC voltage amplitude versus the rf voltage amplitude for masses m_1 and m_2 where m_2 lies within a designated stable area for transmission through the quadrupole.

However, a brief description will be offered; the derivation of stability starts with the simple $F = ma$ force equation and results in a second order differential. The canonical form of the Mathieu equation [9] that describes the stability of a charged particle's trajectory through the quadrupole, in a Cartesian coordinate system, is as follows:

$$\frac{d^2u}{d\xi^2} + (a_u - 2q\cos 2\xi)u = 0 \tag{30.23}$$

with u representing either x or y:

$$a_u = a_x = -a_y = \frac{4eU}{m\omega^2 r_0^2}, \quad q = q_x = -q_y = \frac{2eV}{m\omega^2 r_0^2} \tag{30.24}$$

where e is the charge on an electron, U is the applied DC voltage, V is the applied zero-to-peak rf voltage, m is the mass of the ion, ω is the angular frequency, and r is the effective radius between the quadrupole electrodes. Intuitively, the student can instantly recognize that the quadrupole possesses the fundamental dependence of mass-to-charge (m/z) ratio to the effective voltages applied to the electrodes, which is required of a mass analyzer.

Stability diagrams can be constructed from the interdependencies of the stable trajectories of the ions through the quadrupole. The quadrupole field produced by the electrodes focuses the mass-analyzed ion down to the center of the quadrupole by the alternating biases applied to the oppositely aligned electrodes. Usually, the rf amplitude is kept constant, while the polarity of the DC amplitude is switched. For a positively charged ion, the focusing force will exist in the plane of the positively biased electrodes in the form of $U + V_0 \cos \omega t$. The two rods opposite each other will possess the same exact combination of fields. The two opposed rods to the positively biased electrodes that destabilize the trajectory of the positive ion will possess a negative bias in the form of $-U - V_0 \cos \omega t$. The rapidly changing biases producing rapidly changing fields will cause the ion to oscillate back and forth within the quadrupole. A potential difference applied across the length of the quadrupole will draw the ion through the oscillating fields. Ions with unstable trajectories will collide with the rods. This mass filtering takes place primarily near the beginning of the quadrupole, and often, quadrupole mass analyzer designs will include small rods making up a prequadrupole filter. The plot in Figure 30.14 illustrates an ion stability diagram relating the dependence between U and V. At a certain range of DC voltage U and rf potential V, the ion of interest will have a stable region (ion m_1 vs. ion m_2). Due to this functionality, quadrupoles are called mass filters; however, they are still dependent upon a mass-to-charge ratio. The plot in Figure 30.15 illustrates the ability to scan mass-to-charge (m/z) ratios with the quadrupole allowing the measurement of a range of masses, typically up to 4000 Da. The slope of the line is a fixed ratio of U to V. As U and V are varied, the scan follows the scan line and subsequent m/z values are recorded. The k value of the x-axis represents an instrument calibration constant. In the plot of Figure 30.15, there are two scan lines labeled I and II. The scan lines demonstrate that the resolution of the quadrupole is inversely proportional to the sensitivity. Scan line I has a higher resolution than scan line II, but the sensitivity is lower in scan line I versus scan line II. Scan line II intersects a greater area under the curves for masses M_1 and M_2 thus allowing a greater proportion of the ions to transmit through the quadrupole and be detected increasing the sensitivity of the instrument. The trade-off is that a greater

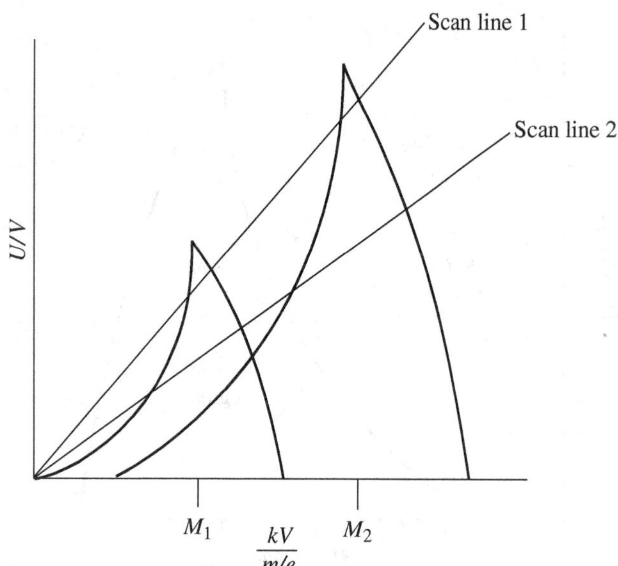

FIGURE 30.15 Scanning mass-to-charge (*m/z*) ratios with the quadrupole allowing the measurement of a range of masses, typically up to 4000 Da. As the fixed ratio of *U* and *V* are varied, the scan follows the scan line and subsequent *m/z* values are recorded. The *k* value of the *x*-axis represents an instrument calibration constant. Scan line I represents higher resolution and lower sensitivity than scan line II.

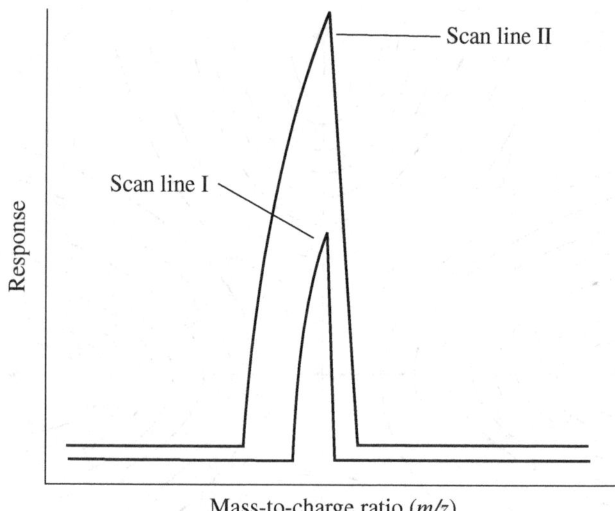

FIGURE 30.16 Result of resolution decrease through peak broadening and mass shift for scan line II versus scan line I.

amount of ion transmission coincides with a greater amount of energy spread within the ions that manifests itself as peak broadening, thus lowering the resolution of the quadrupole mass analyzer. It can be seen that the curves in the stability diagrams rise three times slower than they fall. Therefore, as the scan line is lowered, the frontal portion of the peak in the mass spectrum moves toward the lower mass value three times faster than the trailing part of the curve moves toward higher mass values. This results in a shift of the apex of the mass peak to a slightly lower mass value. Figure 30.16 shows the effect of peak shifting and broadening due to scan line I and scan line II.

The quadrupole mass analyzer possesses some unique advantages such as a high acceptance angle capability for the incoming ion particle beam (angle of incidence). As the ion particles enter the quadrupole, they are very effectively focused into a coherent path if they possess the chosen stable trajectory. In this way, the quadrupole is highly tolerant to a wider range of incidence angle where the incoming ion particles can be off focus up to a certain critical angle. In relation to this, the quadrupole also effectively focuses the position of the incoming ion particle incidence through the field oscillations (position of incidence). The quadrupole is also relatively insensitive to the spread in the KE of the incoming ion particle beam (KE spread). As stated earlier, 5 eV is an optimal energy for ions to enter the quadrupole. The ability of the quadrupole to analyze nonideal beams successively increases with the diameter and the rod length of the electrodes. The quadrupole mass analyzer is also relatively easy to operate and calibrate; they are mechanically rugged and are relatively inexpensive to produce. Some limitations of the quadrupole are related to the imperfections in the quadrupole field. If the KE of the ion is too low, it may be lost and not pass through the quadrupole filter even though it is of the correct mass-to-charge ratio. If the KE of an ion is too high, there is the probability that it may pass through the quadrupole mass filter and be incorrectly detected even though it is not of the proper mass-to-charge ratio. This tends to increase the width of the peak and therefore reduce the resolution of the proper *m/z* peak. Quadrupoles can also be easily contaminated due to buildup near the front end of the quadrupole. Instruments often possess short rod prefilter quadrupoles directly before the main quadrupoles to decrease this effect and allow ease of cleaning. Quadrupoles are low-resolution mass analyzers typically possessing unit mass resolution at each mass ($1/\Delta m = 1$ for $R = m/\Delta m$). Finally, transmission of the ions that are formed at the source through the quadrupole mass filter is mass dependent, and there exists a slight biased discrimination against high mass ions.

30.6 TRIPLE QUADRUPOLE MASS ANALYZER (QQQ/MS)

In the preceding section, a mass analyzer was described comprised of four cylindrical rods placed precisely parallel to one another making up a quadrupole. One set of opposite poles has a DC voltage (*U*) supply connected, while the other set of opposite poles of the quadrupole has an rf voltage (*V*) connected. Ions are accelerated into the quadrupole by a small voltage of 5 eV, and under the influence of the combination of electric fields, the ions follow a complicated trajectory path. The single quadrupole setup can be used as a mass spectrometer scanning a range of *m/z* values (typically from *m/z* 1 to 4000) or can be used as a mass filter where all other ions besides the ion of interest will be given unstable trajectories and removed, thus filtering out all except for a single *m/z* species. Besides the single quadrupole mass filter, a combination of quadrupoles can be constructed in tandem to form a more complex mass spectrometer known as the triple quadrupole mass spectrometer. The triple quadrupole mass spectrometer

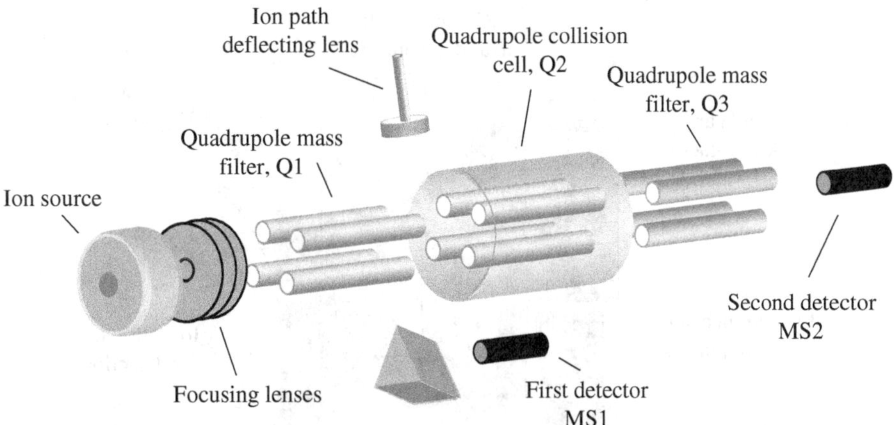

FIGURE 30.17 Illustration of arrangement of tandem quadrupoles in space making up a triple quadrupole mass spectrometer. The central quadrupole is contained within a floating gas cell used for fragmentation studies.

is quite a versatile instrument in its ability for structural analysis of biomolecules. The alignment of the three quadrupoles in tandem allows a combination of unique processes where the quadrupoles are scanned or held static. Two of the quadrupoles, the first called Q1 and the third called Q3, have individual detectors (MS1 and MS2), while the middle quadrupole, Q2, is surrounded by a gas cell used for collision-induced product ion generation and is rf only with no detector. Figure 30.17 shows the arrangement in space of the quadrupoles. Having the quadrupoles aligned in tandem allows approaches to compound isolation and subsequent identification by using the quadrupoles as mass filters. The mass spectrometry methods used utilizing the triple quadrupole in biomolecule analyses include single-stage analysis (ESI-MS) where the first quadrupole (Q1) is scanned and the first detector (MS1) is used. This type of scanning is for the measurement of biomolecule molecular weights as mass-to-charge (*m/z*) ratios. A second method used in biomolecule triple quadrupole mass spectrometry is neutral loss scanning. This is used to determine the initial biomolecule that losses a certain neutral mass upon fragmentation induced within the central quadrupole (Q2) collision cell. In this procedure, the first and third quadrupoles are scanned at a differential equal to the neutral loss value. For example, if the NL value is 50, then when Q1 is at *m/z* 200, Q3 is scanning at *m/z* 150. When *m/z* 150 is registered by the second detector MS2, the Q1 *m/z* value at that particular scan time is recorded along with the intensity recorded with detector MS2. A third method used is precursor ion scanning. This is used to determine the initial biomolecule that produces a specific product ion generated in Q2 and detected at MS2. Here, Q3 is held static at the ion fragment *m/z* value of interest, and Q1 is scanned over an *m/z* range. When the ion is detected by Q3, the *m/z* value in Q1 is recorded in conjunction with the intensity recorded by detector MS2. An example of precursor ion scanning is in the analysis of phosphatidylcholines where the protonated form of the phosphatidylcholine head group at *m/z* 184 is used. The biomolecules are scanned in Q1 and allowed one by one to pass through Q2 where collisions take place and the generation of product ions. During the collision activation in Q2, the phosphatidylcholine biomolecule allowed to pass through Q1 will lose the head group and generate the *m/z* 184 product ion. The specific *m/z* value that represents the phosphatidylcholine biomolecule that was allowed to pass through Q1 will be associated with the event of the detection of *m/z* 184 by MS2, thus identifying that this particular species contains the phosphatidylcholine head group. Finally, a fourth method used is product ion scanning where a species is filtered and isolated in Q1, allowed to pass through Q2 with collisions generating product ions. The product ions that are generated in Q2 are scanned in Q3 and recorded by MS2. This type of scan is often used for structural information of biomolecules.

30.7 THREE-DIMENSIONAL QUADRUPOLE ION TRAP MASS ANALYZER (QIT/MS)

A second design utilizing the equipotential curves of the quadrupole field is the three-dimensional (3D) Paul quadrupole ion trap. This was a second mass analyzer invented and described by Paul [10] in addition to the quadrupole mass filter discussed in the preceding section (see Fig. 30.11 for design of quadrupole). Initially, the ion trap mass analyzer was not utilized to the extent the quadrupole mass filter was due to its complexity in design and operation. The early quadrupole ion traps were operated in stability scan mode the same as the quadrupole mass filters are. This is where the amplitude of the dc and rf components of the ring electrode is ramped at a certain ratio, which would subsequently stabilize higher and higher mass-to-charge ratio ions for detection. In the design of the quadrupole ion trap, a potential well is formed by wrapping a hyperboloidal electrode into a ring that has two electrodes on each side, creating a trapping space that are also of hyperboloidal shape. The basic design and construction of the quadrupole ion trap mass analyzer are illustrated in Figure 30.18. The two opposite electrodes, known as end caps, are characterized by passages or slits used for ion introduction into the trap (Figure 30.18a) and for detection of ions exiting the trap (Figure 30.18b). Probably the most striking ability that distinguishes the quadrupole ion trap from the quadrupole mass analyzer illustrated in Figure 30.11 is the capability of the quadrupole ion trap to perform multiple stages of product ion

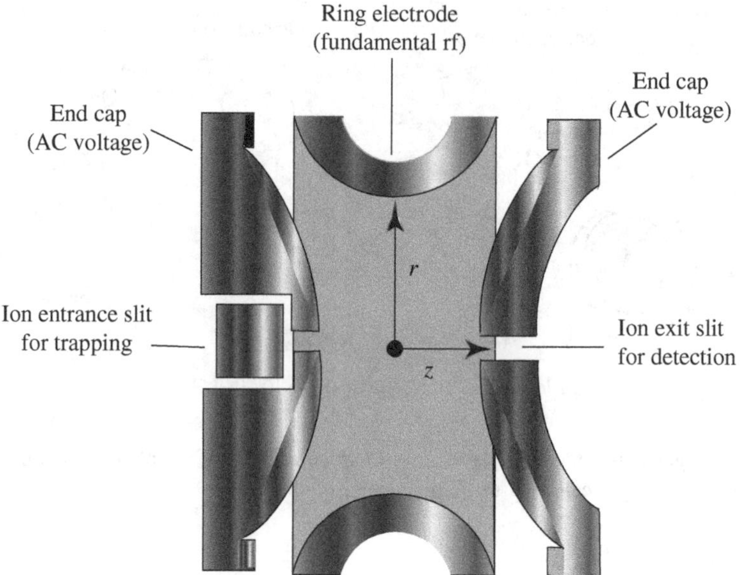

FIGURE 30.18 Basic construction of a quadrupole ion trap mass analyzer including the end cap electrodes and the ring electrode.

fragmentation as MS^n. The quadrupole ion trap has been able to achieve up to 12 stages of product ion fragmentation, MS^{12}. This can allow the investigation of structural aspects of compounds to a very exact and specific level.

An ideal quadrupole field is characterized according to the following relationship between the radius of the ring electrode and the distance to the end cap from the center of the trap:

$$r_0^2 = 2z_0^2 \tag{30.25}$$

where r_0 is the ring electrode radius and z_0 is the end cap electrodes' axial distance from the trap's center. Using this relationship when r_0 has been set, then the distance of the placing of the end caps is predetermined. However, recent work has demonstrated that moving the end caps to a distance slightly greater than the relationship of Equation 30.25 dictates allows access to greater multiple fields, thus increasing the controlling of the efficiency of ion ejection and excitation while also increasing resolution. Typically, the size of a quadrupole ion trap used in most mass analyzers employs an r_0 value of 1.00 or 0.707 cm, equating to an z_0 value of 0.707 or 0.595 cm, respectively.

The derivation of the useful form of the Mathieu equation for quadrupole ion trap mass analyzers is similar to the quadrupole discussed in the preceding section except that for the 3D quadrupole ion trap, the derivations for a_z and q_z are multiplied by a factor of 2, giving the following expressions:

$$a_z = \frac{-8eU}{m\omega^2 r_0^2}, \quad q_z = \frac{4eV}{m\omega^2 r_0^2} \tag{30.26}$$

The dimensionless parameter q_z describes the dependency of the motion of the trapped ion upon the parameters of the quadrupole ion trap that includes the mass (m) and charge (e) of the ion, the ion trap radial size (r_0), the fundamental rf oscillating frequency (ω), and the voltage amplitude on the ring electrode (V). Related to q_z, the parameter a_z in Equation 30.26 describes the DC potential effect upon the motion of the ions trapped within the trap when applied to the ring electrode. Ions are introduced into the trap through slits in one of the end caps. The operation of the trap usually includes the presence of a bath or dampening gas such as helium at a pressure of approximately 1 mTorr. The purpose of the bath gas is to induce low-energy collisions with the ions to thermally cool the ions. These dampening collisions will decrease the KE of the trapped ions. This has the effect of focusing the ion packets into tighter ion trajectories near the center of the trap. Figure 30.19 illustrates a simplified ion trajectory within the ion trap in the presence of helium bath gas. A second focusing of the ions toward the center of the trap is achieved through the fundamental rf voltage that is applied to the ring electrode. This is an oscillating voltage whose amplitude

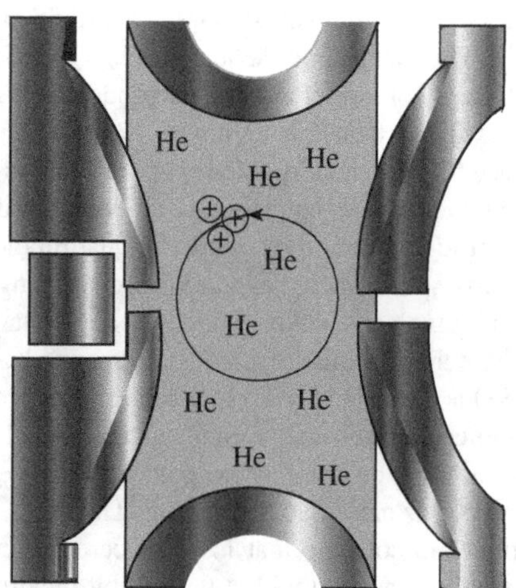

FIGURE 30.19 Simplified trajectory of ions trapped within the ion trap in the presence of helium bath gas.

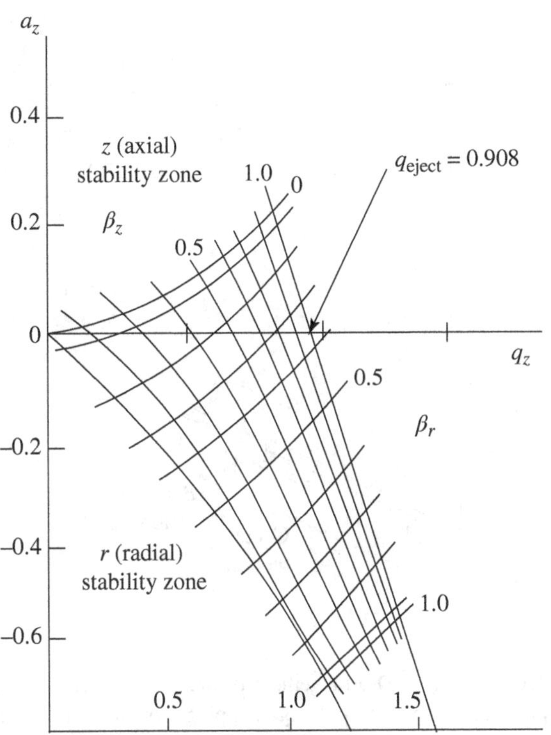

FIGURE 30.20 Stability region diagram for the radial and axial trajectories as a function of a_z and q_z.

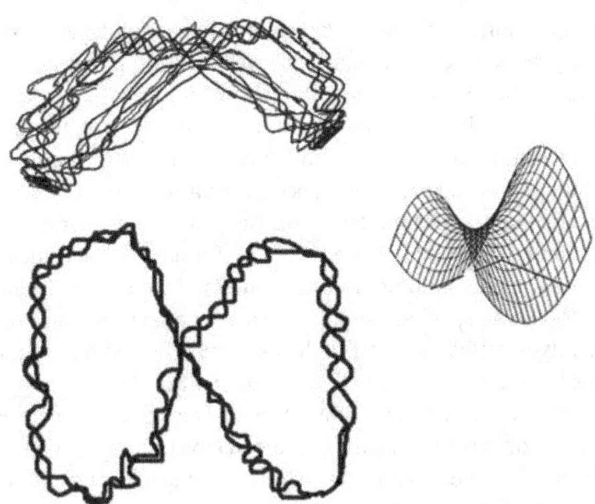

FIGURE 30.21 Trajectory path of the secular motion of an ion trapped within the quadrupole ion trap. The ion follows a Figure 8 saddle path.

will determine the mass-to-charge range of ions that are trapped. The terminology of "fundamental" implies that there are other higher-order frequencies, but they are not of practical use. In trapping operation and scanning for masses, the end caps are held at ground potential; thus, the field that confines the ions is oscillatory only from the frequency of the fundamental rf ring electrode voltage. The method for scanning the ion trap for the mass-to-charge *m/z* species present is known as "mass-selective axial instability mode." The amplitude of the rf applied to the ring electrode is ramped in a linear fashion called an analytical scan. This will cause increasingly unstable trajectories of the ions from lower *m/z* values to higher *m/z* values; in other words, lower *m/z* values that have trajectories closer to the center of the trap will be ejected first before larger *m/z* values that have larger orbital trajectories. To store ions in the ion trap, their trajectories must be stable in both the *r*- and the *z*-directions. The theoretical region of radial (*r*-direction) and axial (*z*-direction) stability overlap is illustrated in the stability diagram of Figure 30.19. Quadrupole ion trap mass analyzers are typically operated where the a_z parameter is equal to zero in the stability diagram of Figure 30.20. The region in the stability diagram where the *z*-axial stability zone and the *r*-radial stability zone intersect is known as the A region and is closest to the origin. There is a second *z*-stable/*r*-stable intersection zone known as region B, but we will not discuss this second one further. In the stability diagram of Figure 30.20, an ion will have a q_z value according to the amplitude of the rf voltage (*V*) applied to the ring electrode. If the ion's q_z value is less than the q_z value of 0.908, then the ion will have a stable secular motion. The motion of the ions in the trap is described by two secular frequencies that are axial (ω_z) and radial (ω_r). The ion trajectory within the quadrupole ion trap follows a Figure 8 "saddle" path as illustrated in Figure 30.21. The sloping path is due to the quadrupole potential field surface that resembles the form of a saddle. The small curving oscillations along the path are due to the oscillating quadrupole potential surface that has the effect of spinning the surface of the saddle. In Figure 30.20, the value of 0.908 for q_z at $a_z = 0$ is known as the low-mass cut-off (LMCO) and is the mass-to-charge lowest value that can be stored within the trap for a given rf voltage (*V*) applied to the ring electrode. In most quadrupole ion trap mass analyzers with r = 1 cm, the working applied rf voltage to the ring electrode ranges from 0 to 7500 V due to limitations of high voltages and circuitry. This has a specific consequence: namely, at 7500 V, a mass-to-charge value of *m/z* 1500 will only have a q_z value of 0.404 (unitless) and thus cannot be ejected from the ion trap. This severely limits the mass range that can be trapped and subsequently measured by the quadrupole ion trap mass analyzer. A technique known as resonance ejection has been developed that allows the extension of the mass-to-charge range that can be measured by the quadrupole ion trap. Resonance ejection is obtained through the application of an ac voltage applied to the end caps of the quadrupole ion trap. Resonance of the ion is induced by applying an ac voltage (usually a few hundred millivolts) across the end caps and then adjusting the q_z value to match the secular frequency of the ion to the frequency of the applied ac voltage. This effectively uses the axial secular frequencies of the ions to induce resonant excitation. To scan the quadrupole ion trap for the mass-to-charge species trapped within the ion trap, the resonant excitation is set at a low applied resonance voltage and then ramped up to high resonance voltage, subsequently ejecting the ions from the trap. If a sufficiently large enough amplitude resonance signal is applied, the ions will be lost or ejected from the ion trap. Smaller amplitude resonance signals can be used to excite the ions to greater KE values that will produce more energetic collisions with the bath gas. This can be used to induce unimolecular dissociation reactions within the quadrupole ion trap for structural elucidation. Another advantage to using resonance excitation is the inversely proportional relationship between q_z and mass that allows the ejection of larger molecular weight ions trapped within the quadrupole ion trap, leaving lower molecular weight ions still trapped within the ion trap. This is known as resonance ejection and is used to isolate a specific mass-to-charge ion within the quadrupole ion trap. Forward and reverse voltage sweeps are performed on both sides of the ion of interest in order to isolate the ion by ejecting all other mass-to-charge species present. When ions are ejected and subsequently detected, one half of the ions are collected through the ion exit slit in the end cap allowing one half of the ions trapped within the cell to be lost and not detected. This has the deleterious effect of reducing the sensitivity of the 3D ion trap, a problem that has been eliminated with the introduction of the linear quadrupole ion trap (LTQ) covered in the next section.

30.8 LINEAR QUADRUPOLE ION TRAP MASS ANALYZER (LTQ/MS)

The linear ion trap is comprised of rod-shaped quadrupole electrodes similar to those that are used in the quadrupole mass analyzer and often placed with the same geometry in space as four oppositely opposed rods. Again, as discussed previously, the design of cylindrical rods is easier to manufacture as compared to rods with a truer hyperbolic shape. Ions in a linear quadrupole are confined by a two-dimensional (2D) rf field derived from the four electrodes and stopping potentials induced from two end caps. The rf field defines the radial motion of the trapped ions, while the end caps define the axial motion. In contrast to the 3D quadrupole ion traps, linear ion traps can store a greater amount of ions, and the introduction of ions into the trap is more efficient. The ion density in the linear ion trap can be increased by increasing the length of the quadrupoles making a longer ion trap, thus allowing the storage of a greater number of ions. In 2D quadrupole fields, the motion of ions has the same form of the Mathieu equation as the 3D quadrupole ion trap (Eq. 30.26) except the directions are in the x–y plane:

$$a_x = -a_y = \frac{8eU}{m\omega^2 r_0^2}, \quad q_x = -q_y = \frac{4eV_{rf}}{m\omega^2 r_0^2}, \quad \xi = \frac{\omega t}{2} \qquad (30.27)$$

where m is the mass of the ion, e is the charge of the ion, r_0 is the ion trap radial size, ω is the rf oscillating frequency, and V_{rf} is the rf voltage amplitude on the electrodes. Figure 30.23 shows the construction of the four quadrupole electrode rods aligned in space. The axial direction through the quadrupole is along the z-axis, which is perpendicular to the plane of the page. The ions trapped within the quadrupole will have oscillations in the x–y-direction as shown in the direction of ion movement in the graph. At the back of the poles, there is a stopping plate aperture used to trap the ions axially. Usually, a second aperture is also placed at the front end of the quadrupole system acting as a stopping plate. In quadrupole fields, the motion in the x-direction is independent of the motion in the y-direction. In higher multipoles such as hexapoles and octopoles, the two

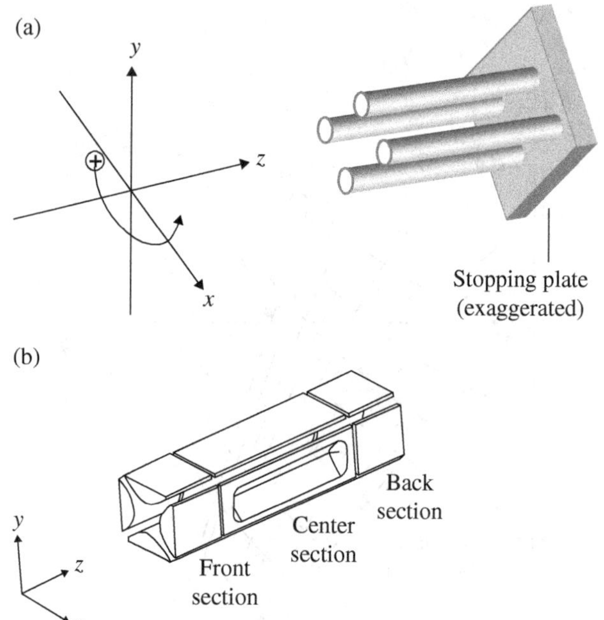

FIGURE 30.22 (a) 3D arrangement of quadrupole electrodes with end plate aperture for applying stopping potential. Plot shows basic motion of ions in the x–y plane. (b) Illustration of the hyperbolic rod-shaped quadrupole used in the two-dimensional ion trap.

directional movements are not independent of one another, and therefore, it is not possible to construct stability diagrams. In a description of the construction and application of a 2D quadrupole ion trap mass spectrometer, Schwartz et al. [11] show the use of quadrupole rods that have a hyperbolic rod design of the 2D linear ion trap as illustrated in Figure 30.22b. The radius of the field (r_0) has been set to 4 mm, and each rod has been cut into 12, 37, and 12 mm length axial sections. This helps to avoid problems with fringe field distortions to the resonance excitation and trapping fields. This quadrupole ion trap design also contains end plates on the front and back sections used for axial trapping (not showing in Figure 30.22). For ion ejection from the trap, a 0.25 mm high slot was cut into the middle electrode.

In the configuration of the 2D ion trap, there is a successive decrease in pressures moving from the electrospray source at ambient pressure (760 Torr) to the ion trap that is at 2.0×10^{-5} Torr. The ion beam produced from the electrospray process passes through a series of ion lenses and rf-only quadrupole and octopole systems to be finally introduced into the ion trap axially. The detector, located radically to the ion trap, is comprised of a conversion dynode and a channeltron electron multiplier.

The Mathieu equations of motion for the ions in the 2D quadrupole field are

$$\frac{d^2x}{d\xi^2} + (a_x - 2q_x \cos 2\xi)x = 0 \qquad (30.28)$$

$$\frac{d^2y}{d\xi^2} + (a_y - 2q_y \cos 2\xi)y = 0 \qquad (30.29)$$

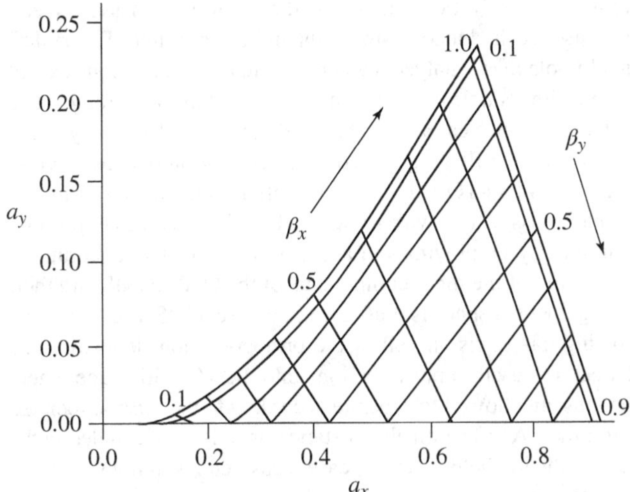

FIGURE 30.23 Stability diagram for the motion of ions within a 2D quadrupole ion trap.

The solutions to these equations give zones of stability and zones of instability allowing the construction of a stability diagram as illustrated in Figure 30.23.

The frequencies with which the ions will oscillate within the 2D LTQ are given by

$$\omega_n = (2n+\beta)\frac{\omega}{2} \quad 0 \le \beta \le 1 \quad n = 0, \pm 1, \pm 2, \ldots \tag{30.30}$$

where β is described as a function of a and q.

As was observed in the 3D quadrupole ion trap, the presence of a bath gas within the LTQ improved performance by way of enhancing the transmission of the ions axially through the end trapping aperture. This is due to the cooling of the ions through low-energy collisions with the neutral bath gas where the ion's KE is slightly dissipated through the bath gas collisions. The ions with lowered KE will focus closer to the middle of the quadrupole trap where the quadrupole field effective energy is lower. Since the ions are more focused at the center of the quadrupole, they can be transmitted through the end stopping plate aperture more efficiently for detection.

There are three methods that can be used to excite ions within the linear ion trap including resonant excitation, dipole excitation, and quadrupole excitation. When using dipole excitation, an auxiliary voltage is applied to two opposite rods that is at a frequency that matches the motion of the ion to be excited. The amplitude of the ion's oscillation will increase with the application of the dipole excitation that can be used for ion ejection or ion fragmentation studies using multiple collisions with the bath gas. The frequency used in dipole excitation is obtained using Equation 30.30 where n is usually set at zero ($n = 0$). With quadrupole excitation, there is an excitation waveform that is applied to the electrodes. To scan a range of mass-to-charge values for ejection, the dipole or quadrupole excitation is set, and the trapping rf voltage is changed to coincide with different ion frequencies for ejection. There are other approaches for ion excitation and ejection such as applying multiple frequencies as a frequency scan or chirp or using broadband waveforms. Finally, there are two designs for ion ejection, axial and radial. Axial ejection is obtained through an exit aperture leading to an electron multiplier for signal detection. Radial ejection is obtained through modified rod electrodes. Designs have included two detectors enabling the collection of all of the ejected ions greatly increasing sensitivity as compared to 3D quadrupole ion traps.

30.9 QUADRUPOLE TIME-OF-FLIGHT MASS ANALYZER (Q-TOF/MS)

A second variation of the TOF mass analyzer is the quadrupole TOF mass analyzer that is a hybrid instrument. Mass spectrometers that are a coupling of two types of mass analyzers are known as hybrid mass spectrometers. This is different from the TOF instrument that contains an initial rf-only quadrupole that acts as an ion guide and not a mass filter. This type of instrumentation is often used with an ESI source for exact mass measurements of analyte ions. An internal standard is used to obtain a high level of accuracy in the determination of the mass-to-charge ratio of other ions within the mass spectrum, often with errors well below 10 ppm. This type of analysis is used for compound identification and molecular formula confirmation. Another source that has been successfully used for the rf-only quadrupole TOF mass analyzer is the atmospheric pressure MALDI ionization source. A chief advancement with the design in this instrumentation lies in the orthogonal arrangement of the TOF mass analyzer to the source. Following the quadrupole or hexapole ion guide is a pulsed ion storage chamber where the ions momentarily reside before being introduced into the TOF mass analyzer. A pulsed extraction grid will draw the ions out of the ion storage chamber imparting equal KE to all of the ions present. The ions will then be separated according to their mass-to-charge ratio drift times within the TOF mass analyzer. This configuration has made it possible to couple a quadrupole to a TOF mass analyzer. The KE of the ions can be controlled by the pulsed ion storage chamber and extraction grid design allowing introduction into the TOF mass analyzer. A second benefit of this design is the ability of the orthogonal acceleration to effectively eliminate the nonuniform spatial distribution of the ions from the source. When the ions are pulsed out of the ion storage chamber into the TOF mass analyzer, the drift time is set to the zero starting point. The initial KE distribution imparted to the ions from the source is greatly reduced by the pre-TOF ion optics, skimmers, and the quadrupole. The spatial distribution produced by the ion storage chamber is the result of collimation. The TOF mass analyzer's reflectron will compensate for the KE distribution that was formed from the spatial distribution in the storage chamber when the ions were extracted. These instrumental reductions in spatial and KE distributions coupled with internal standard calibration allow the orthogonal acceleration design to achieve high resolutions that are used in high mass accuracy exact mass measurements (usually well below 10 ppm error).

In the quadrupole time-of-flight hybrid mass analyzer (Q-TOF/MS) instrumental design, a fully functional quadrupole mass analyzer is placed in tandem to a TOF mass analyzer. This type of configuration allows sources at atmospheric pressure to be used such as an ESI source in place of a source that requires a vacuum such as a MALDI source. Secondly, a floating collision cell is also incorporated into the path of the ion particle beam that can be used for collision-induced formation product ion. The fully functional quadrupole TOF mass analyzer also incorporates the orthogonal acceleration design described previously, allowing the reduction in spatial and KE distribution spread before introduction into the TOF mass analyzer increasing resolution and mass accuracy.

A commercially available Q-TOF mass spectrometer is illustrated in Figure 30.24. The design of this mass spectrometer includes an electrospray source inlet that can be used to couple the instrument to liquid eluant-containing samples from high-performance liquid chromatography (HPLC), capillary electrophoresis, or direct infusion from syringe pumps. The instrument includes a source inlet, which is most often electrospray (but can also be MALDI). The ions that are introduced into the Q-TOF will pass through an rf–DC quadrupole that is a fully functional mass analyzer. Notice that prior to the quadrupole mass analyzer, there is also included a small quadrupole prefilter. The rf–DC quadrupole mass analyzer can be used to filter out all ions except an ion with a single *m/z* value that is allowed to pass through. The isolated *m/z* species can then be accelerated into the rf-only quadrupole with the floating gas collision cell for product ion formation. Recent Q-TOF designs also offer functionality that was primarily associated with triple quadrupoles such as neutral loss scanning by employing the abilities of the rf–DC quadrupole. A drawing of the inner components of the Q-TOF /MS and their arrangement in space is illustrated in Figure 30.25. Following the floating gas collision cell is the orthogonal ion deflector. The deflector is used to pulse the ions into the TOF drift tube where the ions are allowed to separate according to their *m/z* associated drift times. A reflection electrostatic mirror is also included in the path of the ion helping to increase measured resolutions. Typical detectors used are conversion dynodes and channel electron multipliers.

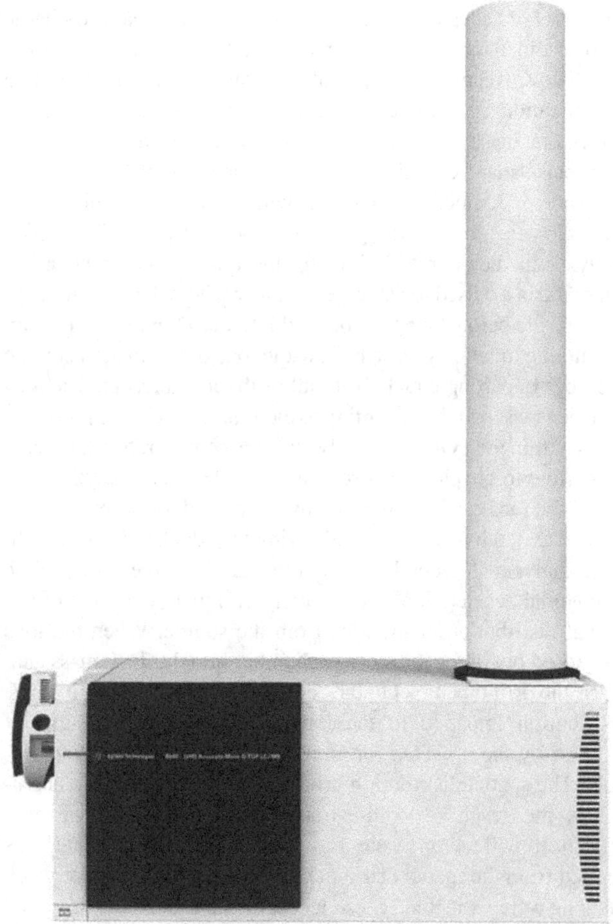

FIGURE 30.24 An Agilent 6500 Series Accurate-Mass Quadrupole Time-of-Flight (Q-TOF) LC/MS. The instrumental setup illustrated in the picture includes an electrospray source inlet used for coupling to liquid eluants such as from high-performance liquid chromatography, capillary electrophoresis, or direct infusion using a syringe pump. (© Agilent Technologies, Inc. 2014. Reproduced with permission. Courtesy of Agilent Technologies, Inc.)

30.10 FOURIER TRANSFORM ION CYCLOTRON RESONANCE MASS ANALYZER (FTICR/MS)

30.10.1 Introduction

Out of all the mass analyzers, the Fourier transform ion cyclotron resonance mass analyzer (FTICR/MS) has the ability to achieve mass resolutions and mass accuracies that are unsurpassed by all other mass analyzers. Routinely, mass accuracies of <5 ppm are obtained with resolutions as high as 1,000,000 using Fourier transform ion cyclotron resonance mass spectrometry. FTICR mass spectrometers are trapping mass analyzers that use the phenomenon of ion cyclotron resonance in the presence of a homogenous, static magnetic field. When an ionized particle enters a strong magnetic field, it will undergo a circular motion that is perpendicular to the magnetic field lines known as cyclotron motion. The cyclotron motion that the ions exhibit has a resonance frequency that is specific to the ions' mass-to-charge (*m/z*) ratio. Therefore, mass analysis is achieved in FTICR mass analyzers by detecting the cyclotron frequencies of the trapped ions that are specifically unique to each *m/z* value. Unlike the quadrupole, TOF, and magnetic sector mass analyzers that perform a separate spatial ion formation, separation, and detection according to their mass-to-charge ratio, the FTICR mass analyzer can perform a temporal ion formation, separation, and subsequent ion detection all within the ICR cell. All of the ions trapped within the cell undergo a simultaneous time-domain cyclotron motion that allows the measurement of the signals produced by each mass-to-charge ratio present without the need for scanning. The need for scanning a mass analyzer reduces the amount of signal that can be measured. In the FTICR mass analyzer, the entire signal produced by the resonating ions is measured at the same time, thus increasing the signal detection and therefore increasing the sensitivity. By keeping the amount of charge in the ion cyclotron cell at an optimum, usually less than 10^7 to insure that spatial charge interactions are minimized, the ability to measure the signal without scanning allows the FTICR mass analyzer to achieve a high degree of sensitivity. Permanent magnets and

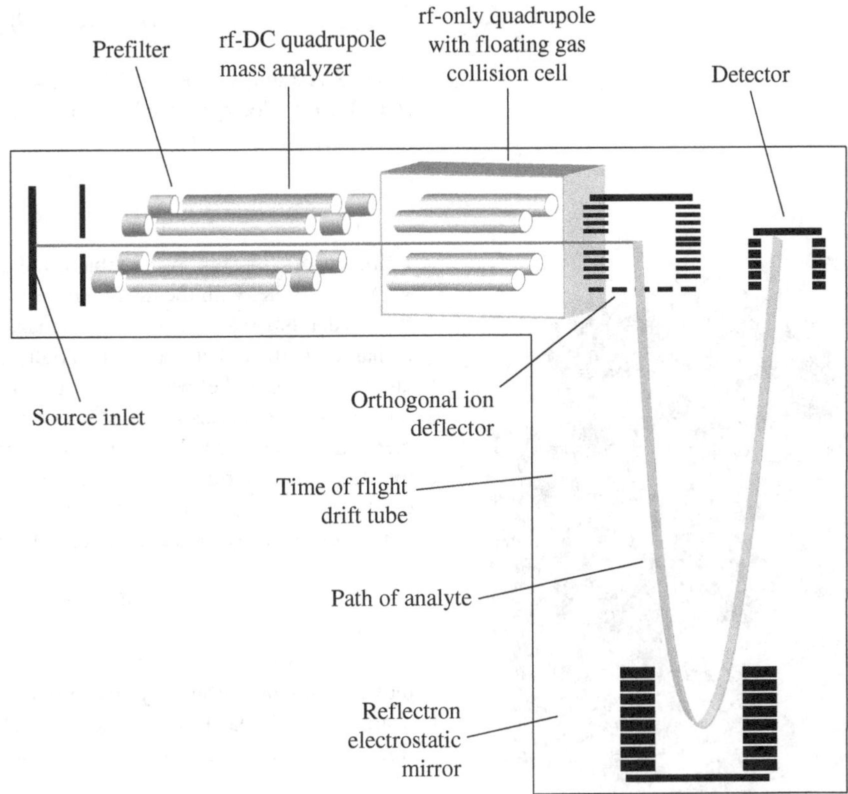

FIGURE 30.25 A schematic diagram of the internal components of a Q-TOF mass spectrometer. Important components include the quadrupole mass filter, the floating collision cell, the orthogonal ion deflector, the time-of-flight drift tube, the reflectron electrostatic mirror, and the detector.

electromagnets are typically not used due to the low field strengths obtainable, usually less than 2 Tesla (T). In general, superconducting magnets are used that range in magnetic field strengths from 3.5, 7, 9.4, and 11.5 T where a 20 T has also been constructed, but such high strengths are not often used. It will be shown later that the performance of the FTICR mass analyzer is directly proportional to the magnetic field strength.

30.10.2 FTICR Mass Analyzer

The basic components of an FTICR mass analyzer include (i) a magnet capable of producing a stable and uniform magnetic field that is constant in space distribution and in time, (ii) a trapping cell located within the middle of the magnet where the ions are measured (can be cubicle, cylindrical, or other shapes), (iii) an extremely high vacuum system capable of levels down to 10^{-9} or 10^{-10} Torr (required for high-resolution measurements as will be discussed later), and a data system that is capable of collecting and storing large amounts of signal responses required for high-resolution mass measurements. Figure 30.26 illustrates the basic components of an FTICR mass spectrometer. Though ions can be formed within the FTICR cell itself by processes such as EI or photoionization (PI), the standard technique is to produce the ions using an external source such as MALDI or ESI prior to injection into the cell. Often, there is an ion trap in tandem with the FTICR cell located between the cell and the source where the source-generated ions can be collected and accumulated. Once a sufficient amount of ions have been accumulated, they are guided into the FTICR cell by ion optics for mass analysis. Figure 30.27 illustrates the basic design of a cubic Penning trap FTICR cell, which is useful in understanding the components of the cell used for mass analysis. Cubic cells fit well within the shape of the lower strength permanent and electromagnetic magnets; however, for the higher field strength superconducting magnets, a cylindrical-shaped cell fits more optimally within the central bore of the magnet. To trap the ions within the cell, an electrostatic field is produced by two opposing plates where the direction of the electrostatic field is parallel to the magnetic field. The cylindrical cell functions similarly to the cubic cell where two outer cylinders are used for trapping the ions, while the inner cylinder is comprised of four electrodes used for excitation and signal measurement.

30.10.3 FTICR Trapped Ion Behavior

The trapped ions within the cell have a force imposed upon them by the strong magnetic field. This is described by Newton's force equation:

$$\text{Force} = \text{mass} \times \text{acceleration} \tag{30.31}$$

$$F = m\frac{dv}{dt} \tag{30.32}$$

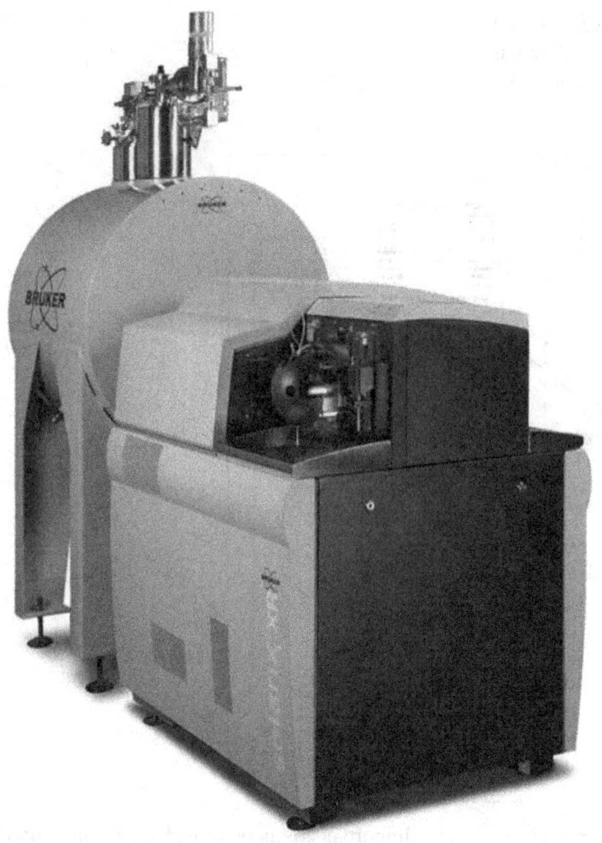

FIGURE 30.26 A Bruker Daltonics solariX XR Fourier transform ion cyclotron resonance mass spectrometer (FTICR/MS). The large circular portion of the mass spectrometer is the shielded magnet, which comes in 7.0, 9.4, 12, or even 15 T magnets that are actively shielded. The ion cyclotron cell is located within the central bore of the magnet. The front, right portion of the mass spectrometer houses an inlet source and a trapping quadrupole for ion accumulation prior to entry into the ICR cell. (Reprinted with permission from Bruker Daltonics.)

$$F = q\bar{v} \otimes \bar{B} \tag{30.33}$$

where q is the charge of the ion, v is the velocity of the ion, and B is the magnetic field strength. The magnetic component of the Lorentz force (using the right-hand rule, the force on the ion is both perpendicular to the magnetic field and the direction of its velocity) will cause the circular path of the ion shown in Figure 30.28 (positive ion). The circular path is in the x, y plane and perpendicular to the z plane. The magnetic field lines are toward the negative direction (opposed to) of the z-axis (into the plane of the paper). Note that the path of a negative ion will be in the opposite direction, but will experience the exact same force magnitude.

For the magnetic force equation,

$$\text{Force} = q\bar{v} \otimes \bar{B} \tag{30.34}$$

Due to the definition of the cross product, the magnetic force is in the direction given by

$$q\bar{v} \otimes \bar{B} = (qvB\sin\theta)\hat{n} \tag{30.35}$$

where $\hat{n}$ is a unit vector that is perpendicular to both B and v. If $\theta = 90°$ (the velocity vector is perpendicular to the magnetic field lines), then magnetic force is

$$F = qvB \tag{30.36}$$

From this relationship, we see that if the direction of the velocity vector is parallel with the magnetic field, then $\theta = 0°$ and $\sin\theta = 0$. This demonstrates that there is no magnetic force exerted upon a charged particle that is moving axially with the magnetic field and will therefore be lost from the ICR cell. This is why a trapping electric field is also applied in conjunction with the imposed magnetic field to the ICR cells whose combination produces a 3D ion trap. The electric and magnetic forces from the applied field can be added as vectors; therefore, the net force exerted upon the ion is described by the Lorentz force law:

$$F = q(E + v \otimes B) \tag{30.37}$$

The imposed strong magnetic field induces the ions to precess or move in a circular orbit at cyclotron frequencies that are uniquely characteristic to each mass-to-charge (*m/z*) ratio. The cyclotron frequency may be derived from Newton's second law:

$$\sum F = ma \tag{30.38}$$

$$qvB = m\left(\frac{v^2}{R}\right) \tag{30.39}$$

where circular motion velocity can be described as

$$v = 2\pi\omega_c R \tag{30.40}$$

and substituting into Equation 30.39,

$$q(2\pi\omega_c R)B = m\left(\frac{(2\pi\omega_c R)^2}{R}\right) \tag{30.41}$$

$$qB = 2\pi m\omega_c \tag{30.42}$$

$$\omega_c = \frac{qB}{2\pi m} \tag{30.43}$$

Equation 30.43 demonstrates that smaller ions are moving with a greater cyclotron velocity as compared to larger ions, which precess with slower cyclotron velocity. Also demonstrated by Equation 30.43 is that the cyclotron frequency for ions of the same mass-to-charge ratio is independent from the initial position or velocity of the ions. Because the cyclotron frequency is independent from the initial velocity, it is independent from the initial KE of the ion. This is one direct reason why FTICR mass spectrometry is able to achieve such high mass resolutions where other mass analyzers are required to focus the translational energy of the ions to achieve resolution of the *m/z* values. The FTICR independence from initial position is also in contrast to the MALDI-TOF mass spectrometer where initial spatial

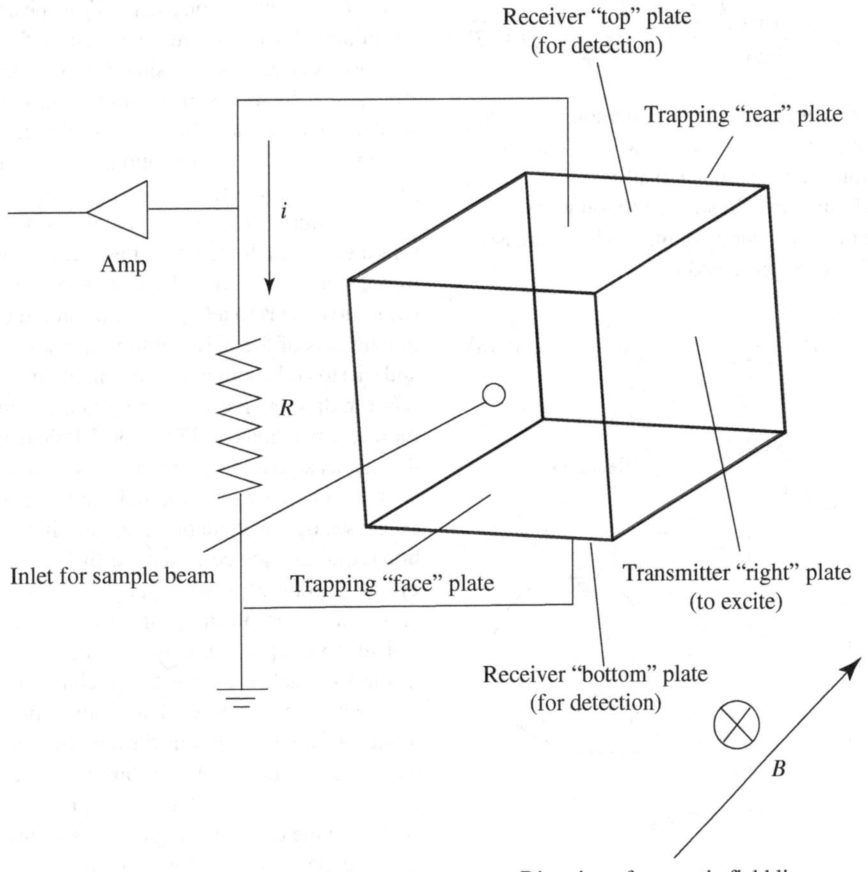

FIGURE 30.27 Cubic Penning trap design of an ICR cell.

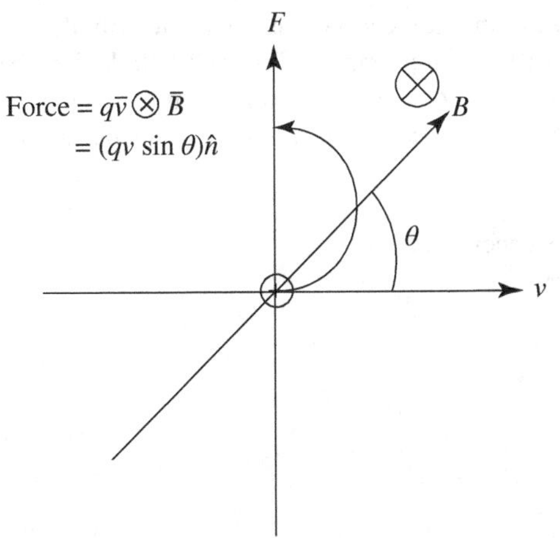

FIGURE 30.28 Circular path of an ion within a strong magnetic field.

distribution affects resolution and the quadrupole mass analyzer where a critical angle exists for introduction and transmission of the ion through the quadrupole.

The ions are generally trapped with low KE so their cyclotron radius are low resulting in orbits near the middle or center of the ICR cell, thus making the detection of their cyclotron frequencies unavailable. As illustrated in Figure 30.27, the ICR cell design includes two opposing plates that are used as rf transmitter plates that excite the *m/z* ion packets as they move in their circular orbits. The excitation of the ion packets is performed by applying an oscillating electric field that has the same frequency as the motion of the *m/z* ion packet. By keeping the applied rf oscillating electric field in resonance with the ion packet, the *m/z* ions in resonance with the imposed field will absorb energy in the form of KE. The frequency of motion will not change as illustrated in Equation 30.43; however, the velocity of the ions will increase according to the resonant rf electric field. Taking Equation 30.39 and rearranging, we obtain

$$qvB = m\left(\frac{v^2}{R}\right) \tag{30.39}$$

$$R = \frac{mv}{qB} \tag{30.44}$$

and by substituting in the relationship between KE (*eV*) and *mv*:

$$\text{KE} = eV = \frac{1}{2}mv^2 \tag{30.45}$$

$$mv = (2eVm)^{\frac{1}{2}} \tag{30.46}$$

we obtain

$$R = \frac{(2eVm)^{\frac{1}{2}}}{qB} \qquad (30.47)$$

The relationship illustrated in Equation 30.47 demonstrates that the radius (R) of the ion's cyclotron rotation will increase with increasing KE (eV) as imposed by the resonant rf electric field. However, the cyclotron frequency of resonant motion will remain the same also indicating that the velocity of motion has increased. Finally, Equation 30.47 can be presented as

$$R = \frac{V_P t}{2dB} \qquad (30.48)$$

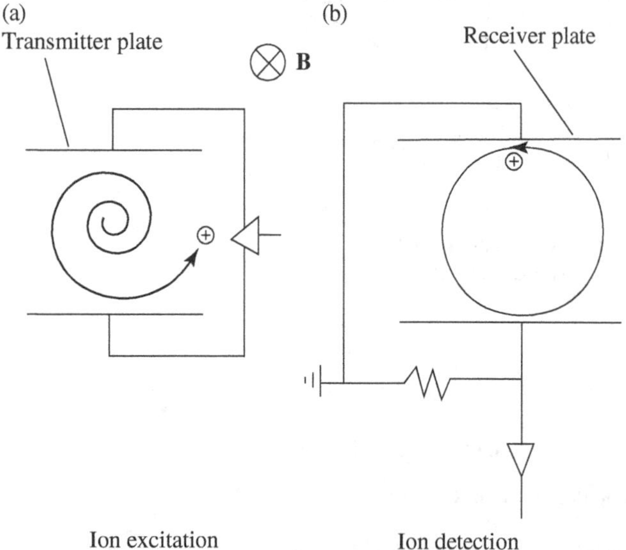

FIGURE 30.29 (a) Ions are excited to a larger cyclotron rotation radius through a resonant rf electric field induced by the transmitter plates. (b) Ions are detected by the receiver plates acting as electrodes.

where R is the cyclotron radius, V_P is the amplitude of the rf excitation signal, t is the time duration of the rf pulse, and d is the distance between the transmitter plates. Equation 30.48 demonstrates that the radius of the ions' cyclotron motion after excitation is independent of the mass-to-charge ratios. This means that there is no mass discrimination in the excitation of the ions into greater orbital radius.

Excitation of the *m/z* ion packet is performed to (i) drive the ions away from the center of the ICR cell and subsequently closer to the plate boundary of the ICR cell for signal measurement, (ii) to drive ions to a larger cyclotron radius than the constrained dimensions of the ICR cell to remove (eject) them from the cell, and (iii) to add enough internal energy in the form of translational KE for dissociative collision product ion generation and ion-molecule reactions. In Figure 30.28, there are two opposed plates that act as signal receiver plates used for the detection of the ion packets trapped within the ICR cell. Figure 30.29 illustrates the processes of ion excitation resulting in increased cyclotron rotation radius and ion detection. As the ion packet moves through its cyclotron rotation, it will approach one of the receiver plates and induce a current in the form of a flow of electrons through the resistor to the plate (this is in positive ion mode where the ions in the ion packet are positively charged). Subsequently, as the ion packet moves toward the opposing receiver plate, it will induce a flow of electrons through the resistor, producing a current in the form of electron buildup. The excited ions coherent motion will produce a transient signal, which is called an image current in the two opposing receiver plates. This in turn produces a time dependent waveform by the excited ions, which are resonating at a specific cyclotron frequency where the ion abundances are directly proportional to the magnitude of the frequency components.

A graphical plot of an image (transient) current for a group of ions present in the ICR cell is illustrated in the bottom of Figure 30.30. The top graph of Figure 30.30 depicts the rf burst used to excite the ions trapped within the ICR cell to larger radius cyclotron orbits (unchanged cyclotron frequencies). The top of

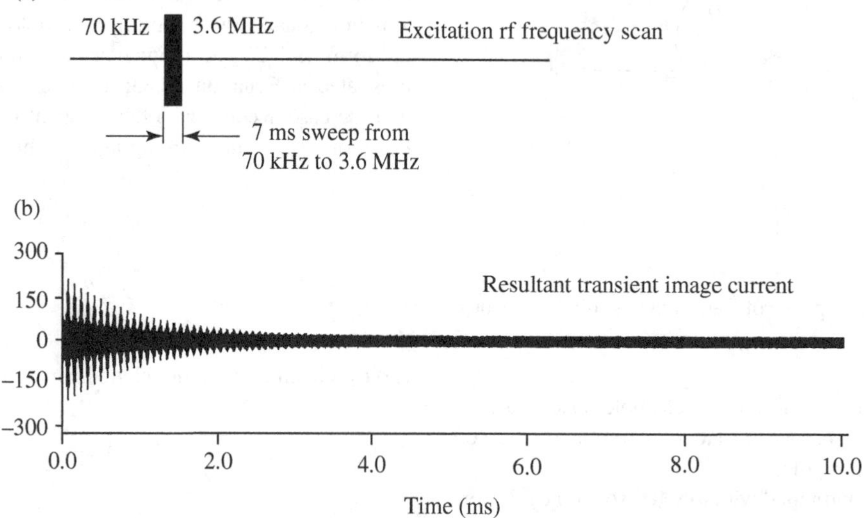

FIGURE 30.30 (a) Excitation rf burst to excite trapped ions within the ICR cell to larger radius cyclotron orbits. (b) Resultant transient image current produced by resonant cyclotron frequencies of ion packets trapped and excited within the ICR cell.

Figure 30.30 shows a burst of 7 ms where the excitation frequency is scanned from 70 kHz to 3.6 MHz. The bottom of Figure 30.30 shows the transient image current produced by the excited ion packets within the ICR cell. In the excitation process, a single frequency can also be applied in the rf excitation process where only a single *m/q* ion packet that is in resonant with the rf will be excited to a larger radius cyclotron orbit. A range of frequencies can also be swept to enable broadband detection of many different *m/q* ions present in the ICR cell by using a rapid frequency sweep known as an rf chirp. This is what is illustrated in the bottom plot of Figure 30.30, a composite of different sinusoids of varying frequencies and amplitudes. With time, the magnitude of the cyclotron frequency amplitudes is observed to steadily decrease as is illustrated in the composite transient signal in the bottom of Figure 30.30. This is due to collisions taking place between the trapped ions and between the trapped ions with neutrals. This is a process called collision-mediated radial diffusion where the coherence of the ion packets is reduced. Collisions will drive the ions toward the outer dimensions of the cell where they will be lost to neutralizing collisions with the cell walls. Optimal conditions require high vacuum in the order of 10^{-9} torr where collisions are greatly reduced. To better understand what the transient image represents and therefore what the transient image contains, we will look at a single frequency rf excitation of the ICR cell followed by the addition of many rf through an rf chirp. Figure 30.31 shows the resultant transient plot for a single *m/z* ion trapped within the ICR cell after coherent rf excitation by the transmitter plate. The first plot shows the sinusoidal behavior of the cyclotron frequency (ω_c) of the trapped *m/z* ion. The magnitude of the ω_c amplitude equates to the charge density of the orbiting ion packet that is proportional to the number of ions present and therefore proportional to the ion concentration within the cell. Using the technique of Fourier transform (FT), the time-domain transient signal can be converted to a frequency-domain plot as illustrated in the right half of Figure 30.31. With calibration of the mass analyzer, the frequency domain can be converted to a mass spectrum as illustrated in the bottom of Figure 30.31.

In continuation of the demonstration in the construction of the composite transient signal illustrated in Figure 30.31, Figure 30.32 represents the measurement of two orbiting ion packets within the ICR cell that have undergone a coherent rf excitation from the cell's transmitter plates. The transient signal for M_1 indicates an ion packet with a greater (higher) cyclotron frequency equating to a lower mass as compared to M_2. The ion packet represented by M_2 has a lower (less) cyclotron frequency value indicating a higher mass ion packet as compared to M_1. The right transient plot in Figure 30.32 illustrates the effect of the combination of the simultaneous detection of the cyclotron frequencies of M_1 and M_2. We now begin to get a feel for the complexity that is illustrated in the composite transient signal of Figure 30.30. Figure 30.33 ties together all of the aspects discussed in Figures 30.30 through 30.32. In the top of Figure 30.33, there is the illustration of the construction of the complex composite transient signal plot of the cyclotron frequencies for a large number of orbiting ion packets $M_1 + M_2 + \cdots + M_n$. Next, FT is

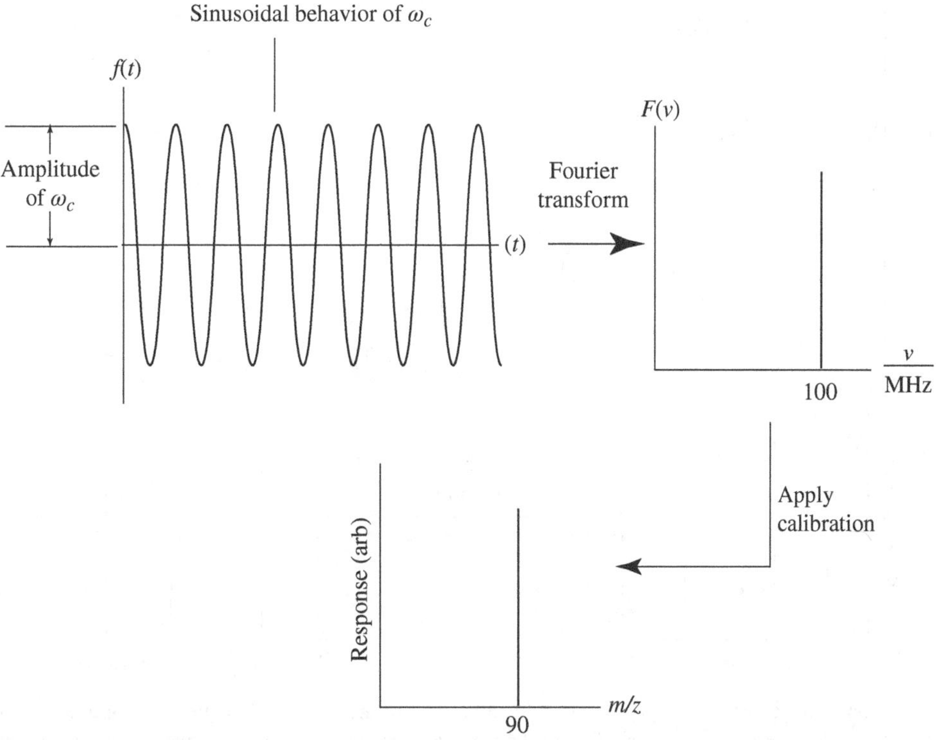

FIGURE 30.31 Progression of the production of a mass spectrum from transient signal measurement. Top left is the transient signal recorded for a single ion trapped within the cell. Fourier transform is applied to change from a time domain to a frequency domain as plotted to the right. A calibration is applied that allows the conversion of the frequency-domain plot to be converted to a mass spectrum as illustrated by the bottom mass spectrum.

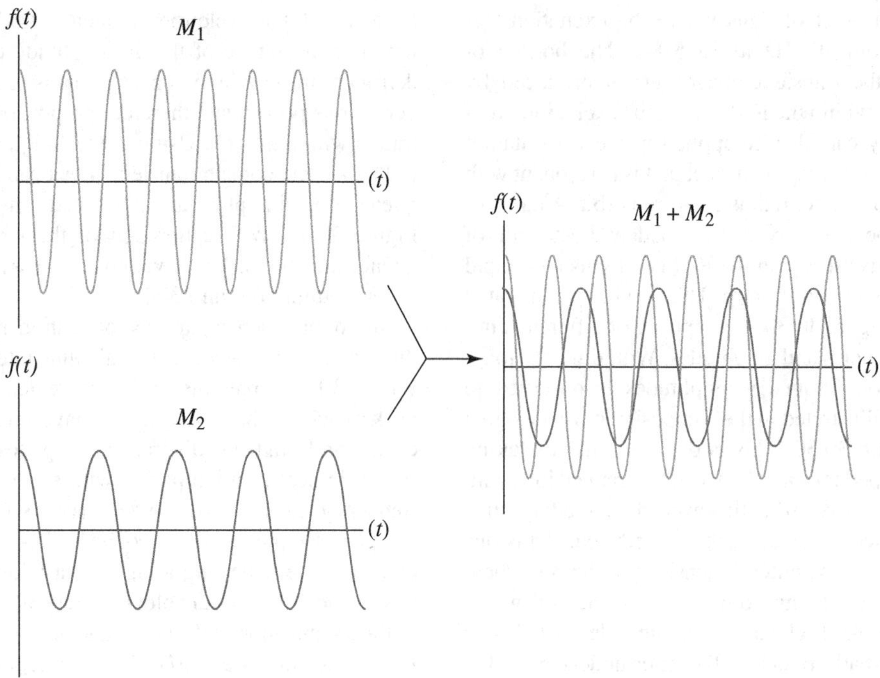

FIGURE 30.32 Illustration of the transient signal composite of the two orbiting ion packets M_1 and M_2. As illustrated by the cyclotron frequencies, the transient signal for M_1 indicates an ion packet with a greater (higher) cyclotron frequency equating to a lower mass as compared to M_2. The ion packet represented by M_2 has a lower (less) cyclotron frequency value indicating a higher mass ion packet as compared to M_1.

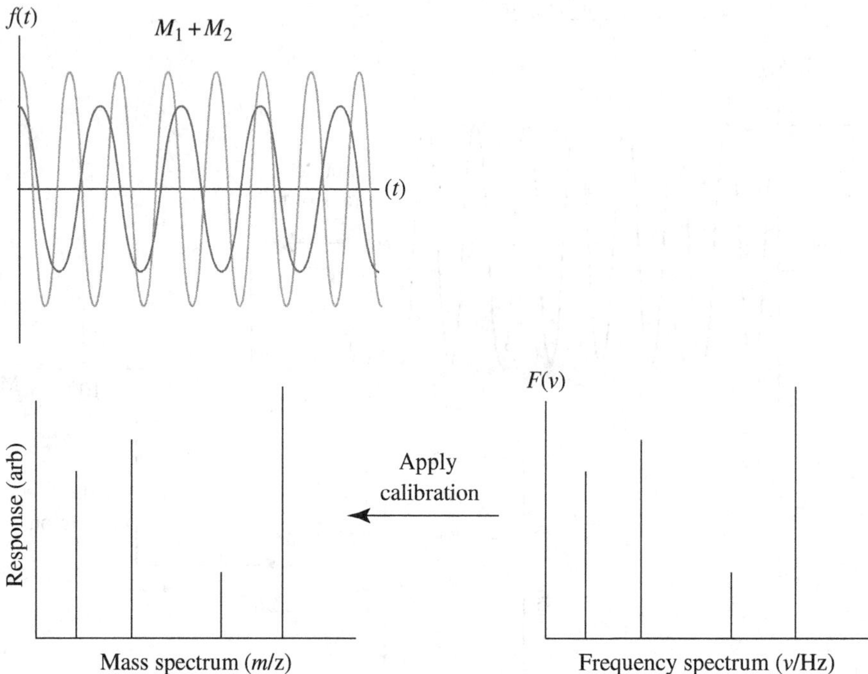

FIGURE 30.33 Construction of the complex composite transient signal plot of the cyclotron frequencies for a large number of orbiting ion packets $M_1 + M_2 + \cdots + M_n$. Fourier transform is applied to the time-domain transient signal converting it to a frequency-domain signal containing multiple signals at specific frequencies. A calibration is applied to the frequency-domain signals converting them to their respective mass-to-charge ratio values, which are used to construct the multiple peak mass spectrum.

applied to the time-domain transient signal converting it to a frequency-domain signal containing multiple signals at specific frequencies. A calibration is applied to the frequency-domain signals converting them to their respective mass-to-charge ratio values, which are used to construct the multiple peak mass spectrum illustrated in the bottom of Figure 30.33. The mass spectral representation of the orbiting ion packets within the ICR cell is the typical type of plot used when performing experiments using the FTICR mass analyzer.

30.10.4 Cyclotron and Magnetron Ion Motion

Within the ICR cell, the ions undergo three types of motion that consist of cyclotron motion due to the magnetic field, trapping motion due to the electric field produced by the trapping plates, and magnetron motion that is attributed to a combination of the magnetic and electric fields acting upon the trapped ions. This is illustrated in the Lorentz force law of Equation 30.37 where the total force acting upon the ions is the electric field (E) times the charge (q) upon the ion plus the cross product relationship between the magnetic field (B) and the ion's velocity. As presented earlier, the magnetic field induces a circular motion of the ions in an orbit that is perpendicular to the direction of the magnetic field. In the z-axis direction (direction of the magnetic field), there is no force acting upon the ion by the magnetic field. Therefore, an electric field is produced by two opposite trapping plates that are located perpendicular to the magnetic field to confine the ions within the ICR cell. A positive trapping potential is applied to the plates to trap positive ions, and a negative potential is applied to trap negative ions. This produces an oscillation motion for the ions back and forth between the trapping plates that follows simple harmonic motion. The combination of the two fields however produces a third magnetron motion, which is a circular motion that effectively follows the contours of one of the isopotential curves of the electric field. The combination of cyclotron and magnetron motion is illustrated in Figure 30.34 where the larger overall motion is due to the cyclotron orbit with the smaller tube producing motion due to the magnetron orbit. A closer look at the ion motion within the trap reveals that the ion's circular oscillation is actually a complex composition of three difficult motions whose frequencies we can describe and compare. These three motions are comprised of an axial motion (z) and two radial motions of magnetron (−) and cyclotron (+). The frequencies of these oscillatory motions are

$$\omega_z^2 = \frac{qV_0}{md^2} \text{ (axial)} \tag{30.49}$$

$$\omega_{\pm} = \frac{\omega_c}{2} \pm \sqrt{\frac{\omega_c^2}{4} - \frac{\omega_z^2}{2}} \text{ (radial, (+)cyclotron, (−)magnetron)} \tag{30.50}$$

where ω_c is the cyclotron frequency in Equation 30.43, V_0 is the depth of the ICR trap, m is the mass of the ion, q is the charge, and d is a trap characteristic parameter. Figure 30.34 illustrates the three motions that make up the total cyclotron motion of the ion within the ICR cell. Space charge effects can also influence the motion of the ions within the ICR cell. When ions are accumulated above the trapping optimization of the ICR cell, space charge effects can broaden the ion packet, producing a wider peak shape of the signal. At higher ion charge densities, peak coalescence can occur where the coherent oscillation of closely spaced ion packets becomes joined into a single cyclotron orbit. To obtain high resolution, a transient signal often must be collected for relatively long periods of time, thus allowing the peak coalescence to occur within this time frame. Peak coalescence can often be controlled by reducing the charge density within the ICR cell along with adjustment of other parameters such as the trapping voltage. A breakdown of the three components of the cyclotron and magnetron motion of the ions within the ICR cell is illustrated in Figure 30.35.

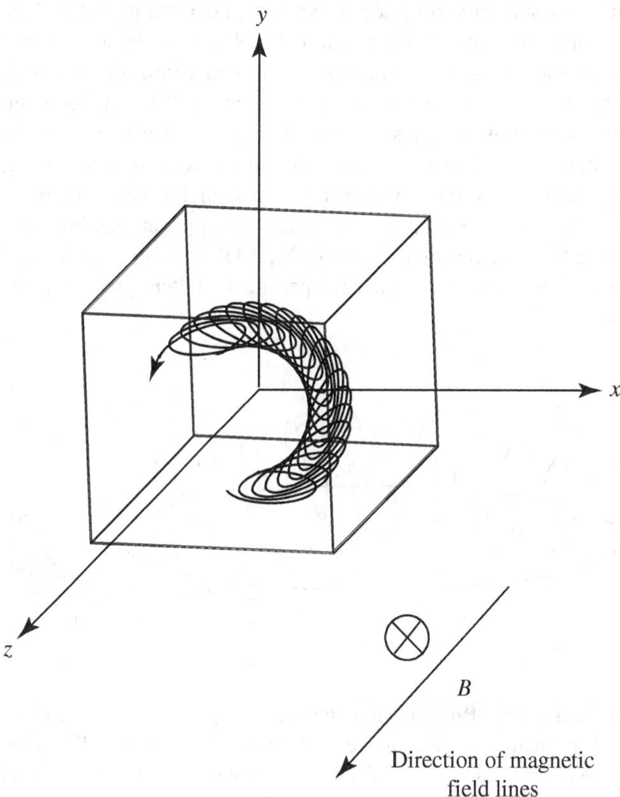

FIGURE 30.34 Cyclotron and magnetron motion of the ions within the ICR cell. The larger overall circular motion is due to the cyclotron orbit, and the smaller circular tube producing motion is due to the magnetron orbit.

30.10.5 Basic Experimental Sequence

The basic experimental sequence using FTICR mass spectrometry is different from other mass spectrometers where the sequence of events happens in time instead of in space. The sequence is basically comprised of cleaning out the ICR cell by removing all of the ions present, an ionization process that can be either internal or external, excitation of the ions, and then subsequent detection of the ions. The purging of the ICR cell in preparation for introduction of new ions into the cell is usually accomplished by applying an asymmetric potential to the opposed trapping plates. One trapping plate will be given a positive potential, while the opposite trapping plate possesses a negative potential. The asymmetric potential within the ICR cell causes the ions to exit along the z-axis, thus ejecting all the ions within the ICR cell within a very short period of time. Once the ICR cell has been purged of ions, a new set of ions is introduced into the cell. In earlier work, EI was used to form ions directly within the cell by passing a stream of electrons through the cell producing odd electron molecular ions. Recently, the application of external ionization sources such as ESI or MALDI has been widely used in FTICR mass spectrometry experiments. Often, preceding the ICR cell will be an ion trap that allows accumulation of ions prior to introduction into the ICR cell when using ionization techniques such as electrospray. When MALDI is used, the source is usually also under high vacuum, and the generated ions are guided into the ICR cell with lenses and ion guides. Figure 30.37 illustrates the steps involved in the experimental sequence. On the left, the ions are purged from the cell, essentially leaving the cell empty of ions and neutrals that have been removed by high vacuum. The cell is then filled with ions that are either generated within the cell or externally. The ions are then excited to higher orbit radius and detected as shown in the right of Figure 30.36.

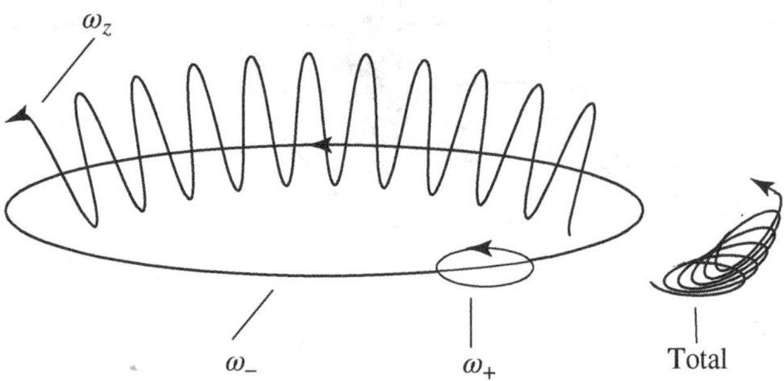

FIGURE 30.35 Breakdown of the three components of the cyclotron and magnetron motion of the ions within the ICR cell. The three motions are comprised of an axial motion (z) and two radial motions of magnetron (−) and cyclotron (+).

The next process to take place is the input of KE into the orbiting ion packets to increase the cyclotron orbit radius of the ions. Initially, the ions within the ICR cell usually have low KE and low cyclotron orbital radius. This is the case for both externally ionized species and internally ionized species. The ions are orbiting near the center of the ICR cell, which is too far away from the receiver plates to generate the transient signal used for mass detection and analysis. To excite the ions, a sinusoidal voltage is applied to the transmitter plates whose frequency is in resonance with the cyclotron frequency of an orbiting ion packet. By scanning a frequency range, all ions present within a certain mass-to-charge ratio range that is constrained by the complete FTICR mass analyzer system will be excited to a greater orbital radius, thus allowing detection of the ion packets present within the ICR cell.

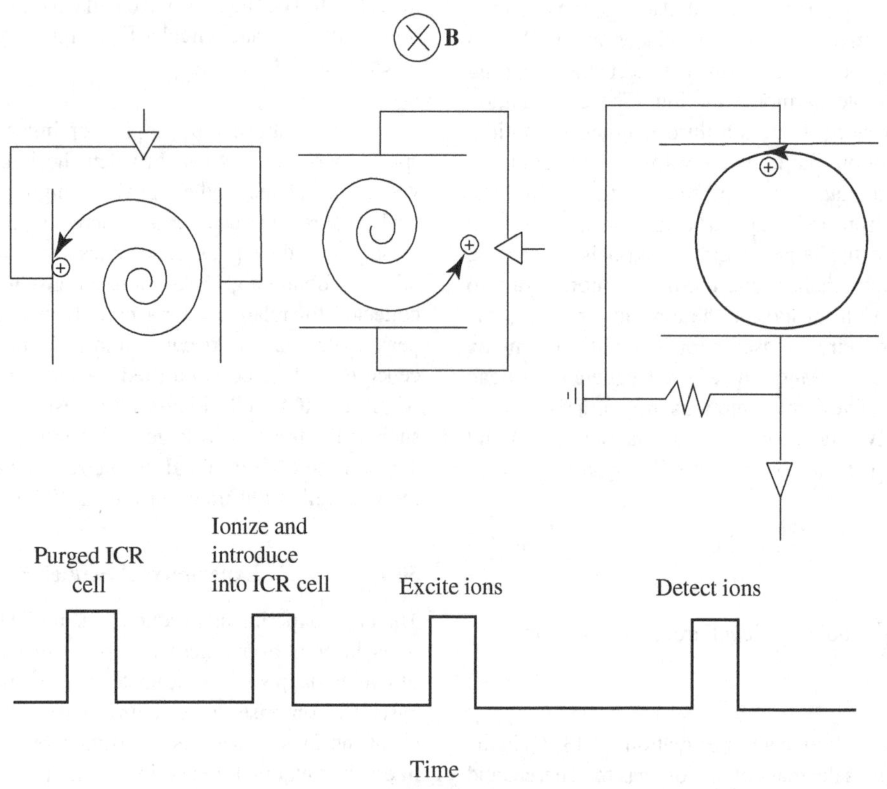

FIGURE 30.36 Progression of the event sequence in time for the FTICR mass analyzer. Purging of the ICR cell in preparation for introduction of new ions into the cell, ionization of species, input of kinetic energy into the orbiting ion packets to increase the cyclotron orbit radius of the ions, and lastly detection of the orbiting ions.

Finally, as shown in the right cell of Figure 30.36, the orbiting ions are detected by the transient electric current induced in the receiver plates. The mass range that can be detected using FTICR mass spectrometry, typically up to 10,000 m/z, is directly dependent upon the magnetic field strength and constrained by the dimensions of the ICR cell. The size of the ICR cell will physically allow only so great of a cyclotron radius diameter; however, in practical use, the largest radius diameter is smaller than the upper limit of the cell dimension. Typically, ICR cells are usually a few centimeters in length upon each side.

As a consequence of the relationships derived from Equation 30.43, we see that many aspects of the functionality of the FTICR mass spectrometer are directly influenced by the magnetic field strength B. For example, the maximum amount of ions that can be trapped within the ICR cell is proportional to the square of the magnetic field (B^2). Increasing the concentration of ions within the ICR cell without deleterious space charge effects will directly increase the sensitivity of the mass analyzer. Another important relationship is the increase in maximum ion KE with increasing magnetic field strength (B^2). In the relationship of increasing energy with increasing magnetic field strength, the experimenter can put more energy into the orbiting ions before they are removed by cell wall neutralization. This gives access to higher collision energies that can be used to utilize fragmentation pathways that require larger activation energies. Finally, when using a higher magnetic field strength, the radius of the orbiting ion packet will be smaller for the same KE as compared to a lower field strength. This allows the trapping of a larger mass-to-charge range and increases the resolution and dynamic range of the mass analyzer.

30.11 LINEAR QUADRUPOLE ION TRAP FOURIER TRANSFORM MASS ANALYZER (LTQ–FT/MS)

Increasingly in today's research, industrial method development, and pharmaceutical drug discovery, there is a need to couple separation science as a front-end analytical methodology to mass spectrometric instrumentation that is subsequently used as an extremely versatile detector. The front-end separation science used today is often HPLC but can also be gas chromatography (GC) or capillary zone electrophoresis (CZE). The HPLC is used to separate complex mixtures of analytes prior to their introduction into the mass spectrometer. A recent and important example of this is in the field of proteomics where a set of proteins isolated from a biological system (the proteome) such as a culture of cells, tissue, or bodily fluids is digested into peptides creating an extremely complex mixture of molecules. For example, a cell's proteome may contain as a hypothetical estimate 3000 proteins that when digested to the peptide level turns into 10,000 individual peptides. It is utterly impossible to infuse this complex mixture directly into the mass spectrometric instrumentation for analysis to any reasonable extent of identification of many of the individual species; often, only the most abundant species present will be observed obscuring most of the other peptides present. Therefore, HPLC is used in proteomic studies to chromatographically separate the peptides into less complex mixtures that can be detected and identified using mass spectrometry. There has also grown an ever-increasing need for the mass spectrometer to possess the ability to scan the incoming analytes in a rate that is comparable to the complexity and speed of the analytes eluting from the HPLC column. A mass spectrometer that samples an ion beam such as a triple quadrupole mass analyzer must use the level of the analyte within the ion beam for its mass measurement and product ion experiments. The ion trap on the other hand does have the ability to trap and accumulate the incoming analytes from the inlet ion stream, thus allowing the enrichment of the sample (increased sensitivity and detection). The relatively recent introduction of the linear ion trap has also allowed a much faster scan rate of the trapped ions as compared to the 3D quadrupole. While the dynamic range and fragmentation efficiency of these instrumentation are quite good, these instruments possess a very low-resolution power, often confined to unit mass resolution with no better than a 20 ppm mass accuracy. The advent and introduction of the orthogonal quadrupole TOF hybrid mass analyzer did allow much higher resolution and mass accuracy, but the instrumentation is still sampling an ion beam where analytes are mass filtered in the first-stage quadrupole and measured in the second-stage TOF mass analyzer. This arrangement suffers from low transmission of the ions and a limited dynamic range for mass accuracy measurements.

The next hybrid mass spectrometer that has been constructed is the combination of the high scan rate and trapping ability of the linear quadrupole ion trap mass spectrometer (LTQ-IT/MS) with the ultrahigh-resolution Fourier transform ion cyclotron resonance mass analyzer (FTICR/MS). With this hybrid, the linear ion trap accumulates ions externally to the ICR cell and can isolate and introduce a single analyte to the ICR cell or fragment an analyte and transmit the product ions into the ICR cell. The linear ion trap also has the ability for MS^n fragmentation studies. The FTICR mass spectrometer possesses the highest obtainable mass resolution (routinely up to 1,000,000) thus allowing routine mass accuracies in the 1–2 ppm range with a relatively broad dynamic range.

The first hybrid LTQ–FTICR mass spectrometer was designed and reported by Syka et al. [12] at the University of Virginia and is illustrated in Figures 30.37 and 30.38. Figure 30.37 is a picture of the LTQ–FTICR mass spectrometer where the linear ion trap is located in the front of the system and the superconducting magnet is the large cylindrically enclosed apparatus in the back of the picture. In this system, the superconducting magnet has a 3 tesla (3 T) magnetic field strength. An illustration of the ion optics and trapping cells used in the construction of the mass spectrometer is illustrated in Figure 30.38. The sample is introduced into the mass spectrometric instrumentation through the inlet located to the far left of the drawing. The analytes will pass through an inlet orifice into a lower-pressure region under vacuum and guided into the LTQ by ion guide lenses and an rf-only quadrupole. The analytes enter the LTQ where they are trapped and accumulated. Here, the ions can be isolated and fragmented for structural information and elucidation. Beyond the LTQ are a series of ion guide lenses and rf-only quadrupoles used to guide the ions from the LTQ into the Penning trap of the FTICR mass spectrometer.

The sensitivity of the system was demonstrated at 550 zmol (10^{-21}) of angiotensin I and is illustrated in Figure 30.39. The sample was analyzed by nano-HPLC, and both a single stage and a product ion spectrum were recorded. The dynamic range of the system was demonstrated at 4000:1 with routine mass accuracies of 1–2 ppm.

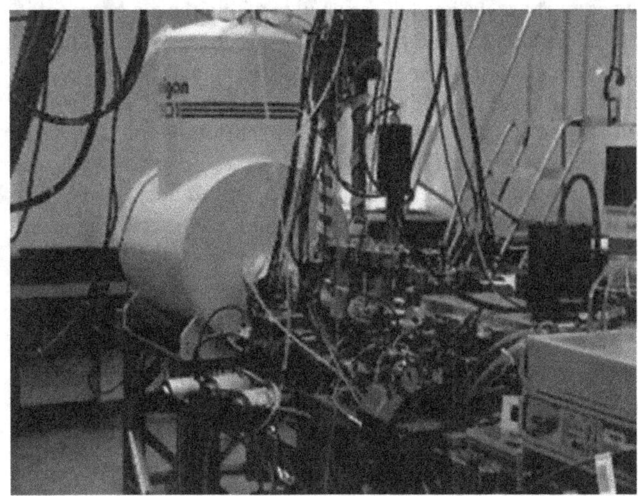

FIGURE 30.37 Linear quadrupole ion trap Fourier transform ion cyclotron resonance mass spectrometer (LTQ–FTICR/MS) developed and designed by Syka et al. at the University of Virginia. (Reprinted with permission from Ref. [12]. Copyright 2004 American Chemical Society.)

Figure 30.40 is a picture of a commercially available hybrid mass spectrometer that combines a linear ion trap mass spectrometer with a Fourier transform ion cyclotron resonance mass spectrometer. Also included in the figure is an HPLC system for sample separation prior to the introduction into the ion trap via ESI. A schematic diagram of the instrument is illustrated in Figure 30.41 showing the linear ion trap, ion optics, rf-only multipoles, and the ICR cell located in the middle of a 7 T superconducting magnet. The electron-capture dissociation (ECD) assembly and the infrared multiphoton dissociation (IRMPD) laser assembly are features added to the instrument designed to enhance fragmentation of ions within the ICR cell.

30.12 LINEAR QUADRUPOLE ION TRAP ORBITRAP MASS ANALYZER (LTQ–ORBITRAP/MS)

The most recent addition to the mass analyzer family is the Orbitrap mass analyzer developed and reported by Makarov [13]. As we will see, this mass analyzer achieves very high mass resolution (up to 150,000) without the need of a supercooled superconducting magnet with magnetic field strengths starting at 7 T and going higher as we saw in Section 30.10 for FTICR mass spectrometry. This eliminates the need for the manufacturing of a complicated magnetic system that requires both the refilling of liquid nitrogen and liquid helium reservoirs to maintain the

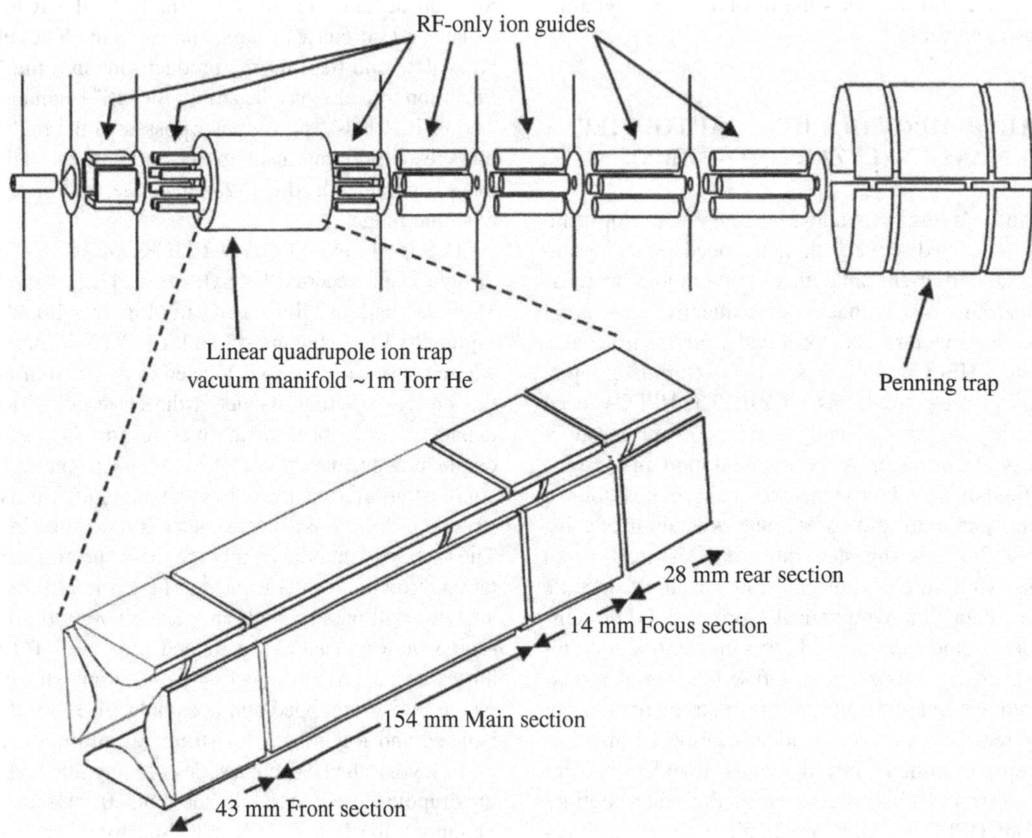

FIGURE 30.38 Ion optics and trapping cell used in the construction of the mass spectrometer. The superconducting magnet has a 3 Tesla (3 T) magnetic field strength. (Reprinted with permission from Ref. [12]. Copyright 2004 American Chemical Society.)

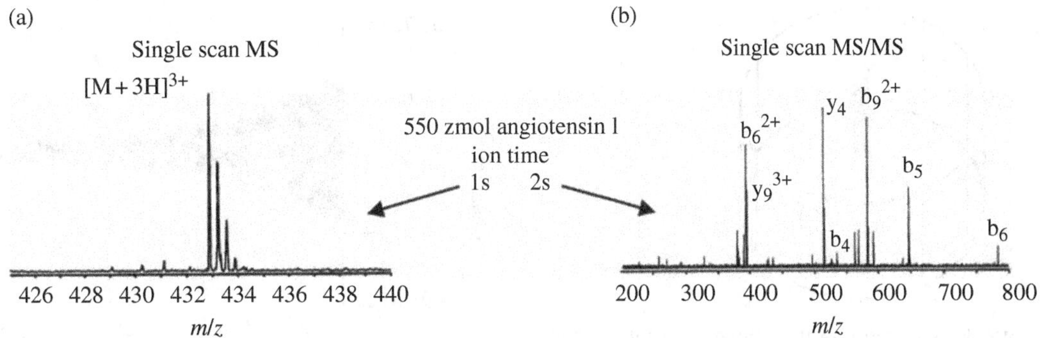

FIGURE 30.39 Sample levels for the detection and sequence analysis of peptides on the prototype QLT/FTMS instrument; (a) MS spectrum recorded on 550 zmol (550×10^{-21} mol) of angiotensin 1, (b) MS/MS spectrum recorded on $(M + 3H)^{+3}$ ions generated from angiotensin 1 at the 550 zmol sample level. (Reprinted with permission from Ref. [12]. Copyright 2004 American Chemical Society.)

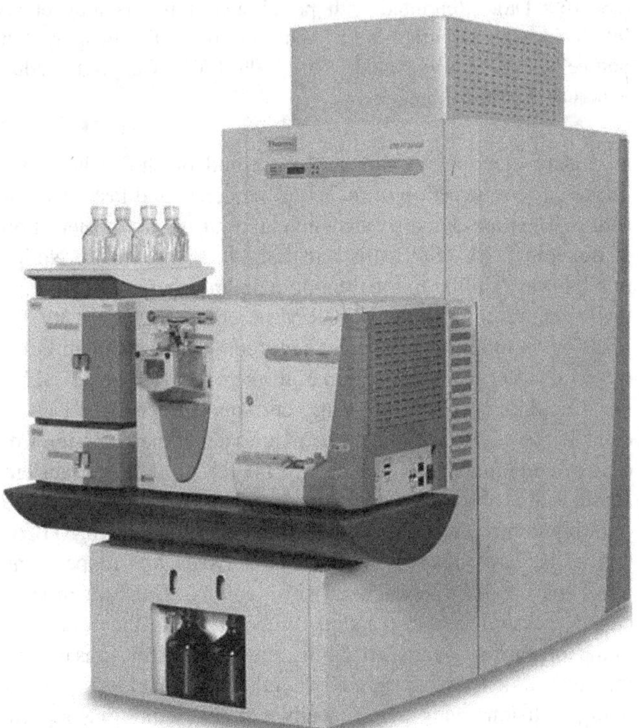

FIGURE 30.40 Picture of the LTQ–FT Ultra™ mass spectrometer. (Reprinted with permission from Ref. [12]. Copyright 2004 American Chemical Society.)

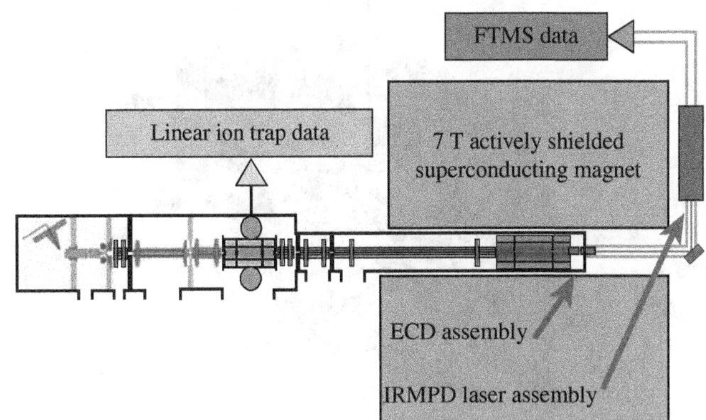

FIGURE 30.41 Schematic diagram of 7 T LTQ–FTICR mass spectrometer showing the linear ion trap, ion optics, rf-only multipoles, and the ICR cell located in the middle of a 7 T superconducting magnet. The electron-capture dissociation (ECD) assembly and the infrared multiphoton dissociation (IRMPD) laser assembly are features added to the instrument designed to enhance fragmentation of ions within the ICR cell. (Reprinted with permission from Thermo Scientific.)

low temperatures needed for superconductivity and the active shielding required in order to reduce the strong magnetic field surrounding the superconducting magnet. The Orbitrap mass analyzer has the ability to achieve high mass resolution because its detection system is based off of an orbiting packet of ions very similar to those detected in FTICR where an oscillating transient current is produced and can be fast Fourier transformed (fast FT) from a transient time-domain signal into an *m/z* versus intensity mass spectrum. However, there are significant differences between the construction, motion, and mass detection in the Orbitrap mass analyzer as compared to the FTICR mass analyzer. The physicist Kingdon first introduced orbital trapping of ions in the gas phase in 1923 [14]. His design, which was subsequently named the Kingdon trap, consisted of an outer open cylinder that contained a central wire running axial to the cylinder and end flanges that were used to enclose the trapping volume. The orbiting of the ions was achieved by applying a voltage between the outer cylinder and the central axial wire. The ions were attracted by the central wire and would move toward it. If the ions possessed enough KE (tangential velocity), they would not be stopped by the central wire but would begin to orbit around the wire. The ion's axial movement in relation to the wire was constrained by the fields induced from the flanges that restrained the ions from escaping the Kingdon trap axially. Figure 30.42 shows the basic design and principle of the Kingdon trap where a centrally located wire runs axial to the outer cylinder.

The Orbitrap does not use the simple design that is illustrated in Figure 30.42 for the Kingdon trap but rather uses outer and inner electrodes that are curvature in nature. The basic design of the Orbitrap is illustrated in Figure 30.43. Both the outer and inner electrodes are curved, and the electrostatic field

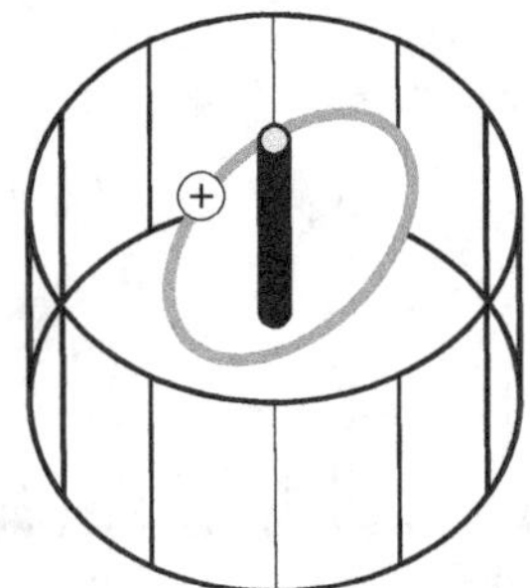

FIGURE 30.42 Simple Kingdon trap illustrating the central axial electrode (wire) surrounded by the outer cylindrical electrode (Reprinted with permission from Thermo Scientific.).

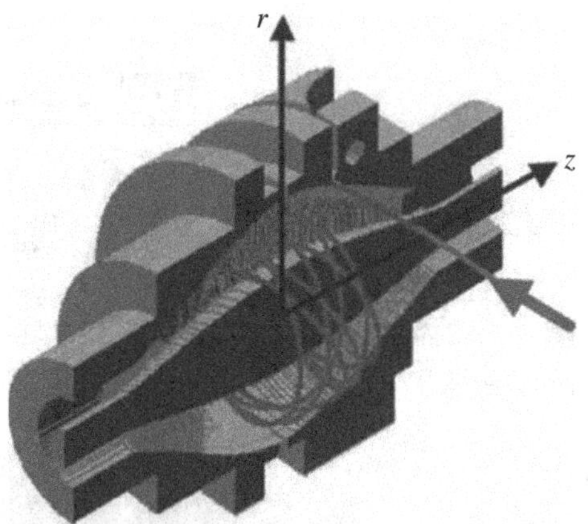

FIGURE 30.43 Cutaway view of the Orbitrap mass analyzer. Ions are injected into the Orbitrap at the point indicated by the arrow. The ions are injected with a velocity perpendicular to the long axis of the Orbitrap (the z-axis). Injection at a point displaced from $z = 0$ gives the ions potential energy in the z-direction. Ion injection at this point on the z-potential is analogous to pulling back a pendulum bob and then releasing it to oscillate. (Reproduced with permission from Ref. [16]. Copyright John Wiley & Sons, Ltd, 2005.)

produces ion movement both in a circular path around the inner electrode and an axial oscillation. The axial oscillation back and forth along the central electrode was what was finally used for the ion detection instead of the frequency of the ion rotation around the central electrode. Remember in the FTICR cell that the frequency of the ion rotation is measured, in the form of a transiently induced oscillating current, as the ions orbit in their cyclotron radius within the cell induced by the strong surrounding magnetic field, by opposing detector plates. It was observed for the Orbitrap that the frequency of ion rotation was greatly dependent upon the initial radius of the ion rotation and upon the ion velocity. This type of dependency will result in poor mass resolution (e.g., in FTICR mass spectrometry, the frequency of the orbiting ion packet is independent of the ions' initial KE or orbital radius affording extremely high levels of mass resolution). In the application of the Orbitrap, the harmonic ion oscillation frequencies along the field of axis are used to derive mass-to-charge (m/z) ratios. While mass-selective instability can be used in the Orbitrap to eject and subsequently detect the trapped ions using a

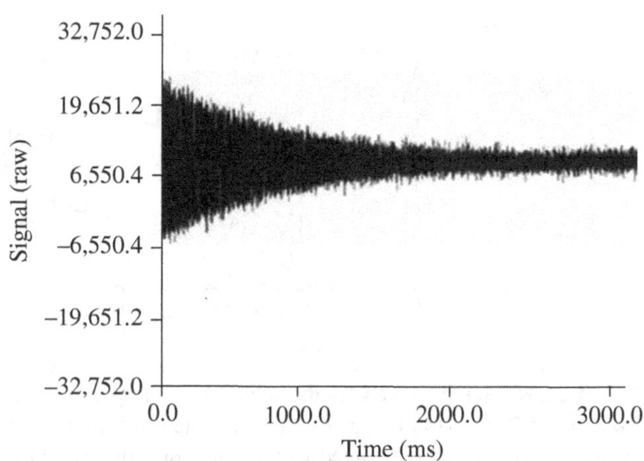

FIGURE 30.44 Image current transient from ions of doxepin (280.1696 Da). (Reprinted with permission from Makarov et al. Interfacing the Orbitrap Mass Analyzer to an Electrospray Ion Source. *Anal. Chem.* **2003**, *75*, 1699–1705. Copyright 2003 American Chemical Society.)

secondary electron multiplier, the method of choice is to use image current detection of the axial frequency and fast FT algorithms. The time-domain transient image current for the detection of doxepin (280.1696 Da) is illustrated in Figure 30.44. Notice that it is very similar to the time-domain transient image currents that were presented in the FTICR/MS section (see Figure 30.29). The time-domain transient image currents measured in an Orbitrap also suffer from dissipation of the signal as was observed for the time-domain transient image currents measured in FTICR cells. If you recall, with time, the magnitude of the cyclotron frequency amplitudes measured in an FTICR time-domain transient signal is also observed to steadily decrease as is illustrated in the Orbitrap composite transient signal in Figure 30.44. In FTICR, this is due to collisions taking place between the trapped ions and between the trapped ions with neutrals. This is a process called collision-mediated radial diffusion where the coherence of the ion packets is reduced. Collisions will drive the ions toward the outer dimensions of the cell where they will be lost to neutralizing collisions with the cell walls. Optimal conditions require high vacuum in the order of 10^{-9} torr where collisions are greatly reduced. This is also the same case within the Orbitrap where ultrahigh vacuum pressures are incorporated in the Orbitrap mass spectrometer in excess of 2×10^{-10} mbar. The high resolving power achievable by the Orbitrap mass analyzer is demonstrated in Figure 30.45, which illustrates an expanded view of the frequency spectrum of the doxepin time-domain transient signal shown in Figure 30.44. The frequency resolving power is in excess of 300,000 that equates to a mass resolving power of 150,000. This high mass resolving power clearly rivals the FTICR in the fact of the relative simplicity of the construction of the Orbitrap mass analyzer, which does not require the incorporation of a superconducting magnet.

The description of the Orbitrap's potential field has similarities to a 3D quadrupole; however, the design of the Orbitrap requires an additional potential description. The Orbitrap induces trapping within static electrostatic fields, while the quadrupole ion trap uses an electrostatic field that is dynamic with an

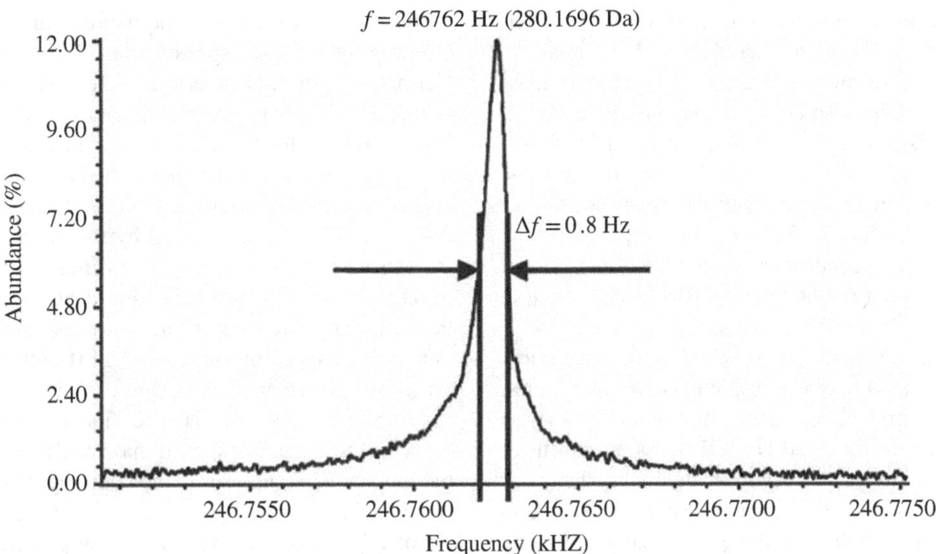

FIGURE 30.45 Expansion of the frequency spectrum demonstrating frequency resolving power of ~300,000 and mass resolving power of 150,000. The spectrum was obtained from the transient in Figure 30.45 by fast Fourier transform with double zero filling. Due to high magnification, only the main isotopic peak of doxepin fits into the window. (Reprinted with permission from Ref. [15].)

oscillation that is approximately 1 MHz. The Orbitrap obtains an electrostatic field orbital trapping with a potential distribution that is described by the following function [15]:

$$U(r,z) = \frac{k}{2}\left(z^2 - \frac{r^2}{2}\right) + \frac{k}{2}(R_m)^2 \ln\left[\frac{r}{R_m}\right] + C \qquad (30.51)$$

where r and z are the cylindrical coordinates ($z = 0$ as the plane of the field symmetry), R_m is the characteristic radius, k is the field curvature, and C is a constant. The electrostatic field of the Orbitrap comprises two independent fields that include the ion trap's quadrupole field and the cylindrical capacitor's logarithmic field. The field for the Orbitrap has been described by combining the two into what is known as a quadrologarithmic field.

The physical design and dimensions of the trap as illustrated in Figure 30.43 include a central electrode with a spindle-like shape that runs through the axis of the outer barrel-like-shaped electrode. In the illustration, R_1 is the radius of the inner spindle-like electrode, R_2 is the radius of the outer barrel-like electrode, and r is the cylindrical coordinate with $z = 0$ denoting the plane of symmetry. The shape of the two electrodes has been derived by Makarov from Equation 30.51 as

$$z_{1,2}(r) = \sqrt{\frac{r^2}{2} - \frac{(R_{1,2})^2}{2} + (R_m)^2 \ln\left[\frac{R_{1,2}}{r}\right]} \qquad (30.52)$$

This equation allows the determination of the maximum radius of the inner (R_1) electrode, the outer (R_2) electrode, the characteristic radius (R_m), and the corresponding r polar coordinate with respect to the $z = 0$ plane of symmetry. Makarov has also described that the stable trajectories of trapped ions within the Orbitrap possess both oscillations around the central axis combined with rotation around the central electrode. This results in movement that is a repeating intricate spiral back and forth within the trap. The polar coordinates (r, φ, z) equation of motion has shown to (including mass-to-charge (m/z) ratio) be

$$\ddot{r} - r\dot{\varphi}^2 = -\frac{q}{m}\frac{k}{2}\left[\frac{(R_m)^2}{r} - r\right] \qquad (30.53)$$

$$\frac{d}{dt}(r^2\dot{\varphi}) = 0 \qquad (30.54)$$

$$\ddot{z} = -\frac{q}{m}kz \qquad (30.55)$$

It is deduced from these relationships (Eq. 30.53) that if $r < R_m$, there is an attraction of ions by the electric field to the central axis. If $r > R_m$, then there is a repulsion of the ions. Therefore, in the physical design of the Orbitrap, only radii below the R_m are useful. Also deduced from these equations (Eqs. 30.53, 30.54, 30.55) is that the equation of motion along z is a simple harmonic oscillator with an exact solution of

$$z(t) = z_0 \cos(\omega t) + \sqrt{\left(\frac{2E_z}{k}\right)} \sin(\omega t) \qquad (30.56)$$

From this relationship, the frequency of axial oscillations (in rad/s) can be derived as

$$\omega = \sqrt{\left(\frac{q}{m}\right)k} \qquad (30.57)$$

Within the Orbitrap's quadrologarithmic field, each respective packet of m/q ions will possess a distinctive frequency of axial oscillation (ω). This relationship allows the transient time-domain signal measurement of the oscillating ions according to their mass-to-charge (*m/z*) ratios that is in effect a mass analyzer. Again, for the same reasons that an FTICR mass analyzer can achieve such high-resolution values, the Orbitrap mass analyzer can also obtain resolution values in the order of 100,000–200,000. Like the cyclotron frequency in FTICR, the Orbitrap's frequency of axial oscillations for ions of the same mass-to-charge ratio is independent from the initial position or velocity of the ions. Because the frequency is independent from the initial velocity, it is independent from the initial KE of the ion. This is one very important reason why the FTICR and Orbitrap mass spectrometers are able to achieve such high mass resolutions where other mass analyzers are required to focus the translational energy of the ions to achieve resolution of the *m/z* values. The FTICR and Orbitrap's independence from initial position is also again in contrast to the MALDI-TOF mass spectrometer where initial spatial distribution affects resolution and the quadrupole mass analyzer where a critical angle exists for introduction and transmission of the ion through the quadrupole that directly affects resolution. As a consequence of the design of the Orbitrap based on the optimal radii dimensions, the Orbitrap has a larger trapping volume that the FTICR traps and the quadrupole Paul traps. Because the Orbitrap's field can be defined to a very high accuracy, high mass resolution is achievable routinely to 150,000. Also, due to the independence of the mass-to-charge ratio to the trapping potential, an increased space charge capacity at higher masses is possible.

Figure 30.46 is a picture of a commercially available hybrid linear ion trap–Orbitrap mass spectrometer. The linear ion trap is located in the front, while the Orbitrap is located in the back

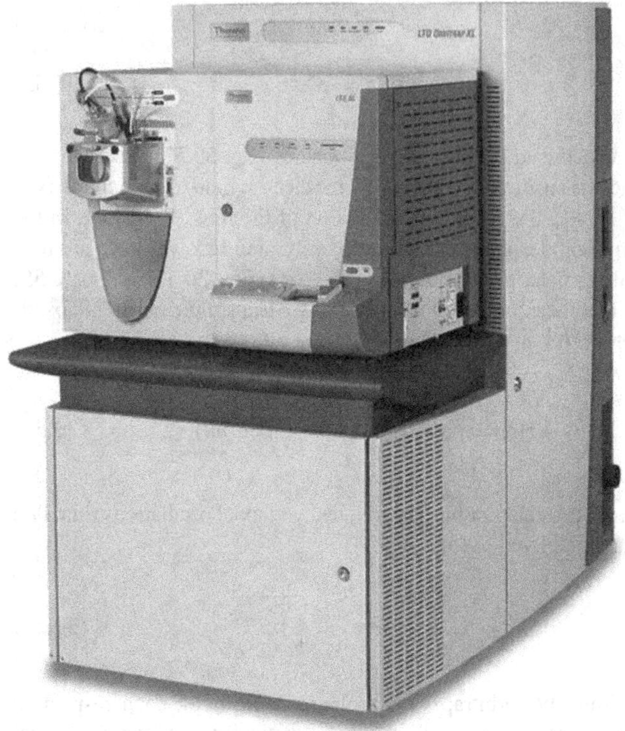

FIGURE 30.46 Picture of the LTQ Orbitrap^{TM} mass spectrometer. (Reprinted with permission from Ref. [15].)

portion of the instrument. The major components of the LTQ–Orbitrap hybrid mass spectrometer (Thermo Electron, Bremen, Germany) are illustrated in the schematic diagram of Figure 30.47. Also illustrated is an experimental sequence that is performed with the Orbitrap for ion mass analysis. For experimental work that has been performed thus far with the Orbitrap, an ESI source has been used and is depicted at the far left of Figure 30.47. Ions are formed by the electrospray process and transferred into the first stages of differential vacuum pumping through an inlet orifice. The desolvated ions introduced into the instrument are now in the gas phase and are subsequently guided by a series of ion lenses and rf-only multipoles into the LTQ mass analyzer. The LTQ in itself is also a fully functional ion trap mass spectrometer, and this is why the LTQ–Orbitrap mass spectrometer is a hybrid mass analyzer (as you remember, the hybrid mass analyzer is the coupling of two different, fully functional mass analyzers into one instrument). The recent models of LTQ mass analyzers also now possess an electronically controlled procedure known as automatic gain control (AGC) where the ion current within the predefined mass range trapped within the LTQ is scanned using a prescan prior to the full analytical scan. This allows the storage of a targeted number of ions within the trap producing an enhanced signal due to an optimal signal without the degradation of signal from space charge effects of trap overpopulation. The design of the Orbitrap has many similarities to the LTQ–FT mass spectrometer where the ability to analyze ions in the second mass analyzer is possible (at very high mass resolutions). Following the LTQ mass analyzer is a transfer octopole (dimensions include 300 mm long and an inscribed diameter of 5.7 mm, see "a" in Figure 30.47) that guides the ions ejected from the LTQ into a curved rf-only C-trap. The C-trap is a recently added feature of the instrumentation as compared to the first Orbitrap mass analyzer that was reported in the literature [16]. The first Orbitrap mass analyzer was a developmental mass spectrometer that was not commercially available and differed from the present commercially available one. The first system coupled a storage quadrupole with transfer lenses to introduce ions into the Orbitrap. The storage quadrupole is necessary because the Orbitrap works on a pulsed set of ions introduced into the trap and not upon a constant beam of ions like a triple quadrupole mass spectrometer. The recent Orbitrap instrumental design includes a C-trap after the LTQ mass analyzer and before the Orbitrap mass analyzer. The axis of the C-trap follows an arc that is C shaped composed of rods with hyperbolic surfaces. On the two ends of the C-trap are plates used for ion introduction (plate located between the octopole and the C-trap) and trapping (plate located at other end of C-trap). The C-trap is also filled with nitrogen bath gas at a pressure of approximately 1 mTorr. The bath gas is used for collisional dampening of the ions trapped within the C-trap. The collisions between the nitrogen bath gas and the trapped ions are at energies low enough not to activate fragmentation of the trapped ions. The ions experience enough collisional cooling in the C-trap to where they form a stable, thin thread along the curved axis of the C-trap. The trapping plates at the two ends of the C-trap are given a positive potential of 200 V that compresses the trapped ions axially. Using a combination of pullout potentials, the ions are removed from the C-trap and introduced into the Orbitrap. This is achieved in a pulsed fashion

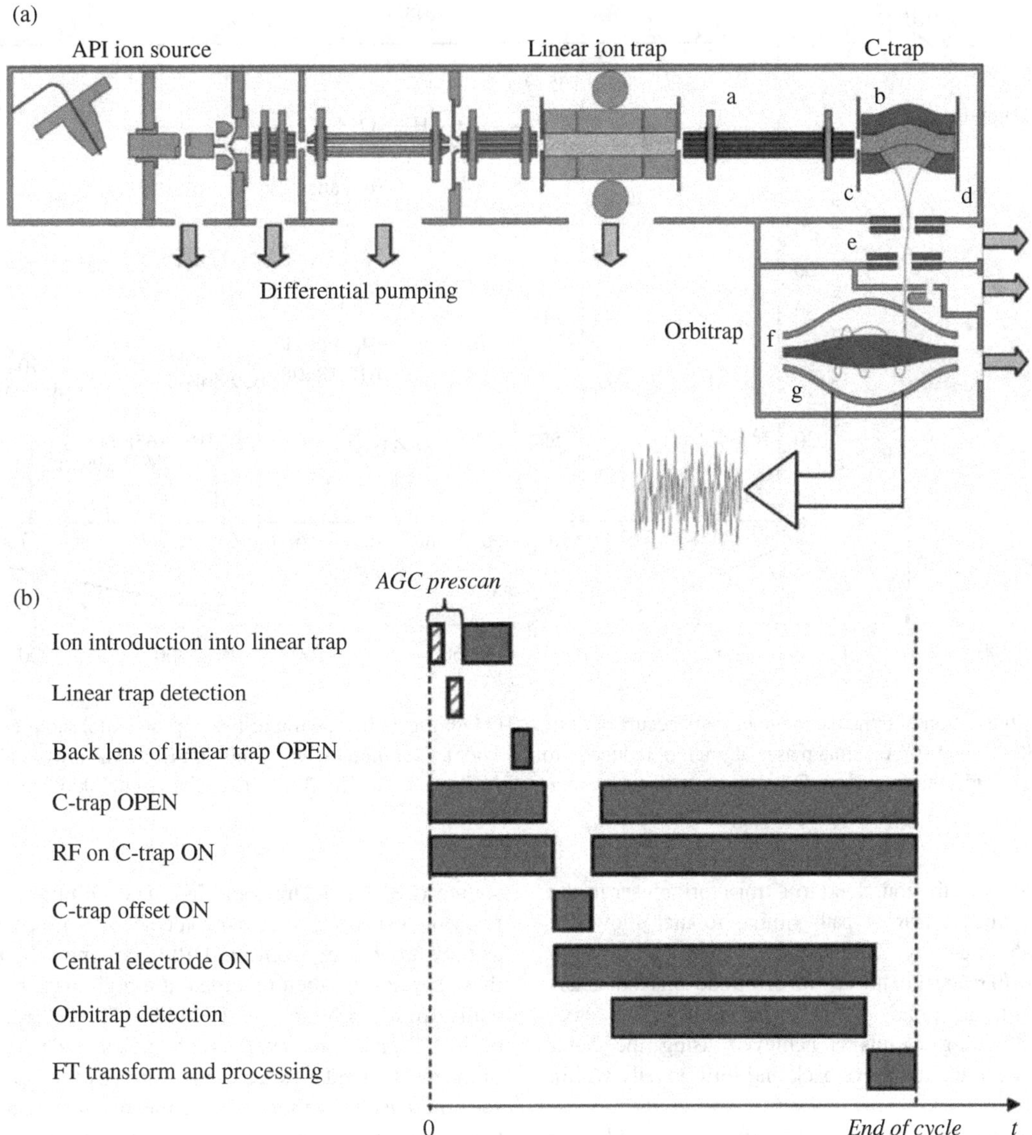

FIGURE 30.47 (A) Schematic layout of the LTQ–Orbitrap mass spectrometer: (a) transfer octopole; (b) curved rf-only quadrupole (C-trap); (c) gate electrode; (d) trap electrode; (e) ion optics; (f) inner Orbitrap electrode; (g) outer Orbitrap electrode. (B) Simplest operation sequence of the LTQ–Orbitrap mass spectrometer (not shown are the following: optional additional injection of internal calibrant; additional MS or MS^n scans of linear trap during the Orbitrap detection). (Reprinted with permission from Thermo Scientific.)

where large ion populations are transferred from the C-trap into the Orbitrap in a fast and uniform fashion. As illustrated in Figure 30.48, the ions pass through slightly curved ion optics from the C-trap to the Orbitrap. The packet of ions will also pass through three different stages of vacuum upon arrival into the Orbitrap. This is done because in order to achieve the high mass resolution within the Orbitrap, there must be a very low incidence of collisions between the trapped ions and any form of ambient gas. As you remember, it is the collisions within the Orbitrap that decay the inherent oscillations, thus decaying the transient time-domain signal over time. The ions that are trapped within the C-trap are at a relatively high background pressure (~1 mTorr) as compared to the optimal operating pressure of the Orbitrap (~10^{-9} torr). The distance between the C-trap and the Orbitrap is very small; thus, any TOF separation is kept to a negligibly small unwanted negative effect.

Upon exiting the C-trap, the ions are moved through a series of curved ion optics (as depicted in Figure 30.47) and are also simultaneously accelerated to high KE. This has the effect of compressing the ion packet into a tight cloud that is able to transverse an entrance aperture that is relatively small and is offset tangentially to the center of the Orbitrap. The ions are injected into the trap at a position that is offset from the center of the Orbitrap at a distance of 7.5 mm from the equatorial center of the trap. In this manner, the ions will start coherent axial oscillations due to the present fields in the trap, thus not requiring any type of initial excitation. The capturing of the ions within the trap is obtained by applying a rapidly increasing electric field that squeezes or

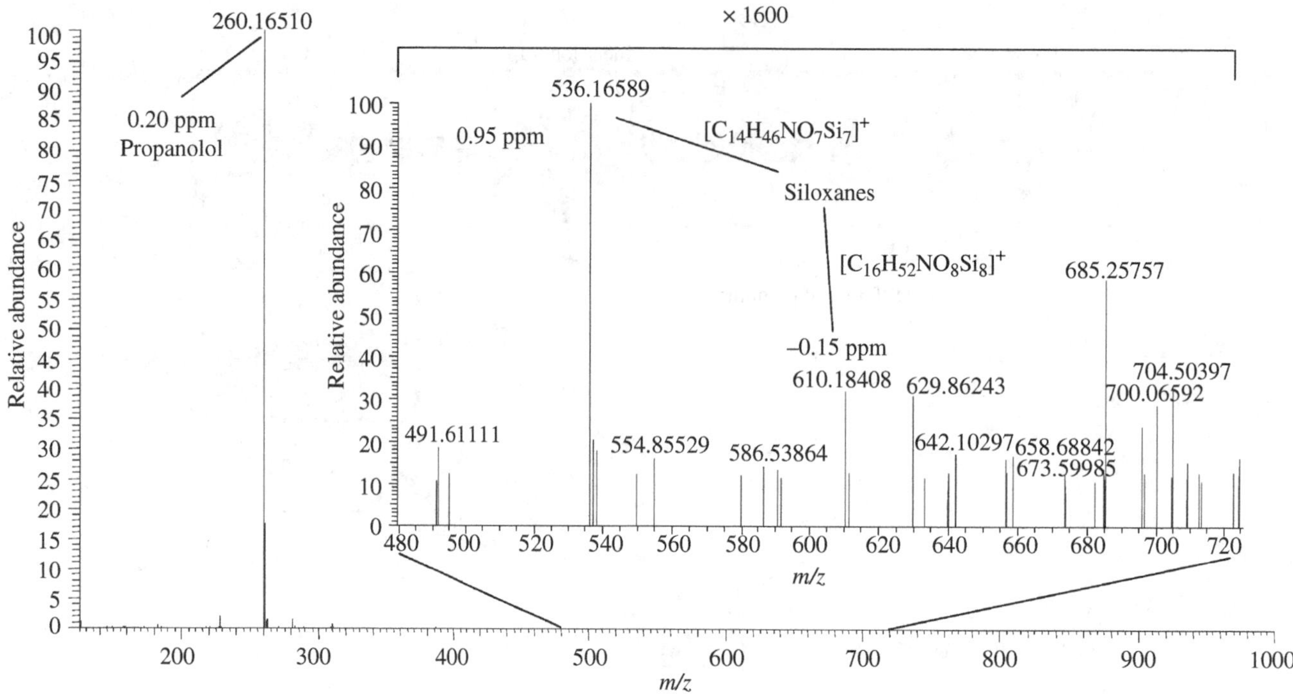

FIGURE 30.48 Illustration of dynamic range of mass accuracy of the LTQ Orbitrap (siloxane impurities in propanolol sample) in a single 1-s scan ($R = 60\ 000$, $N = 2 \times 10^6$, external mass calibration, reduced profile mode). (Reprinted with permission from Makarov et al. Performance Evaluation of a Hybrid Linear Ion Trap/Orbitrap Mass Spectrometer. *Anal. Chem.* **2006**, *78*, 2113–2120. Copyright 2006 American Chemical Society.)

contracts the radius of the ion cloud to a trajectory closer to the axis. The ion clouds follow a path similar to that shown in Figure 30.43 where clouds with higher *m/z* values will have a radius of orbit with respect to the central electrode axis that is larger as compared to ion clouds of smaller *m/z* values. The detection of the oscillating clouds is achieved using the outer electrodes as the ion clouds move back and forth axially within the Orbitrap.

The LTQ–Orbitrap hybrid mass spectrometer also, like the LTQ–FT hybrid mass spectrometer, possesses multifunctional capabilities including high mass resolution and accuracy of precursor ions and precursor ion dissociation through collision-induced dissociation (CID) for structural elucidation. The Orbitrap mass analyzer has been demonstrated to possess a wide dynamic range for mass accuracy determinations such as that illustrated in Figure 30.49 where the measurement of impurities and propanolol in a propanolol sample demonstrates mass accuracy. The impurity species of siloxanes being present in the sample enabled the comparison of mass accuracies for both the most abundant species present being propanolol at *m/z* 260.16510 with a mass accuracy of 0.20 ppm and siloxane impurities at *m/z* 536.16589 with a mass accuracy of 0.95 ppm and the *m/z* 610.18408 siloxane species with a mass accuracy of −0.15 ppm.

The LTQ–Orbitrap mass analyzer also has the capability of performing data-dependent scanning for precursor fragmentation studies of species as they elute from an HPLC column. In this type of experiment, species are eluting from an HPLC column and are converted to gas-phase ions using the ESI source (ESI-LTQ–Orbitrap/MS). The Orbitrap mass analyzer is used to scan the precursor species at a given point in time as they are eluting from the HPLC column. The *m/z* value for these species are then recorded at a high mass resolution (typically from $R = 50{,}000$ to 100,000). The controlling software of the LTQ mass analyzer is set to pick a predetermined number of the most abundant species measured by the Orbitrap and subsequently isolates them within the linear trap and fragments them. The product ions can then either be detected using the linear trap or the Orbitrap. The linear trap will collect the spectra at a much lower resolution than the Orbitrap, but the use of the linear trap for product ion spectral measurements is often used when a large number of data-dependent spectra are to be collected (such as a setting of the collection of the top 10 most intense precursor ions, a setting that is often used when analyzing a complex peptide mixture by nano-HPLC LTQ–Orbitrap/MS). Figure 30.49 illustrates the use of data-dependent scans for a limited amount of precursors (top 3) where the Orbitrap was used for product ion spectral collection at a resolution of $R = 7500$, which is still much higher than that achievable by the LTQ mass analyzer (typical $R = 4000$). Notice however that in Figure 30.49a the resolutions actually range from $R = 72{,}923$ (with a mass accuracy of 2.01 ppm) for buspirone at *m/z* 386.25583 to $R = 90{,}556$ (with a mass accuracy of 1.48 ppm) for propanolol at *m/z* 260.164,89. The spectra in Figure 30.49b–d are the product ion spectra of the precursors in Figure 30.50a, all demonstrating quite high mass accuracies themselves, which can be used in confirmation of structural elucidation studies.

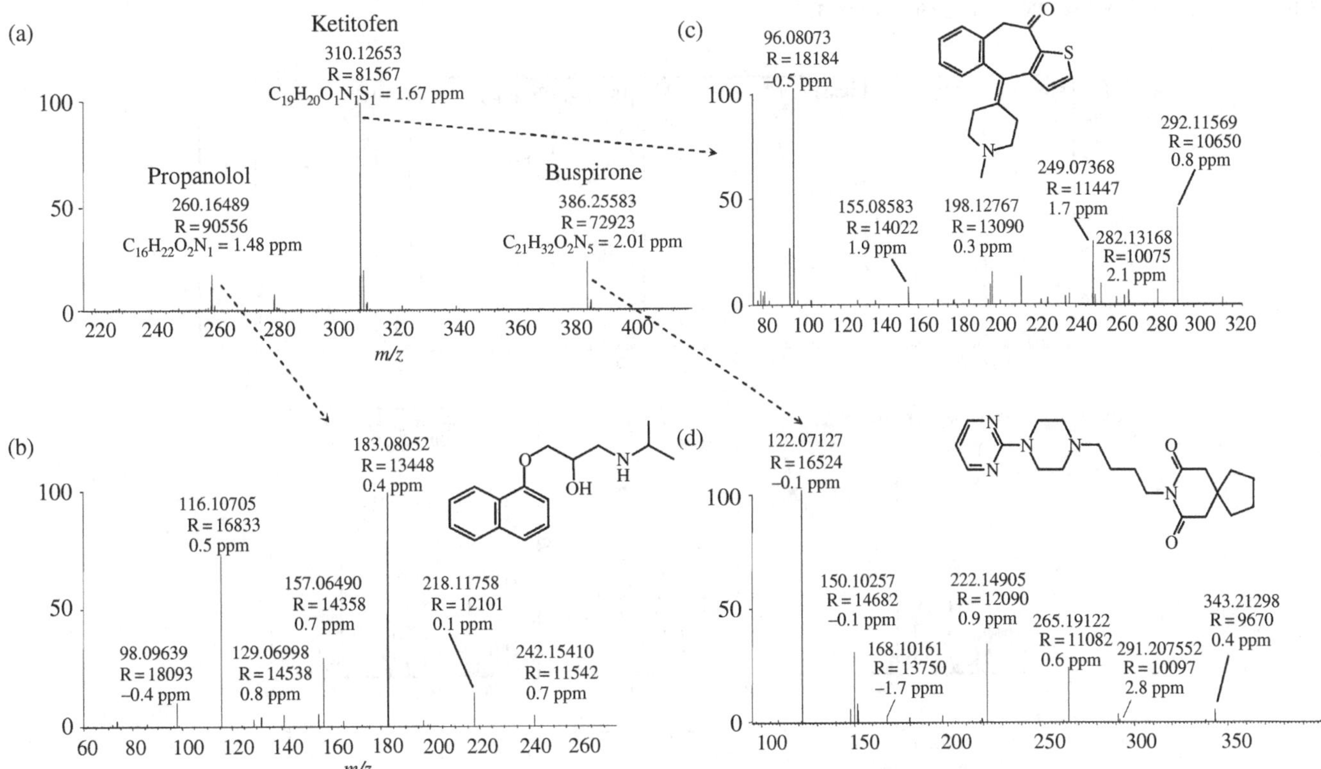

FIGURE 30.49 Example of data-dependent acquisition with external mass calibration for a sample containing small molecules, with one high-resolution mass spectrum recorded of the precursors at R = 60,000 and n = 500,000 (a) followed by three data-dependent MS/MS spectra at R = 7500 and N = 30,000 (b) for precursor at m/z = 260, (c) for precursor at m/z 310, and (d) for precursor at m/z = 386. (Reprinted with permission from Makarov et al. Performance Evaluation of a Hybrid Linear Ion Trap/Orbitrap Mass Spectrometer. *Anal. Chem.* **2006**, 78, 2113–2120. Copyright 2006 American Chemical Society.)

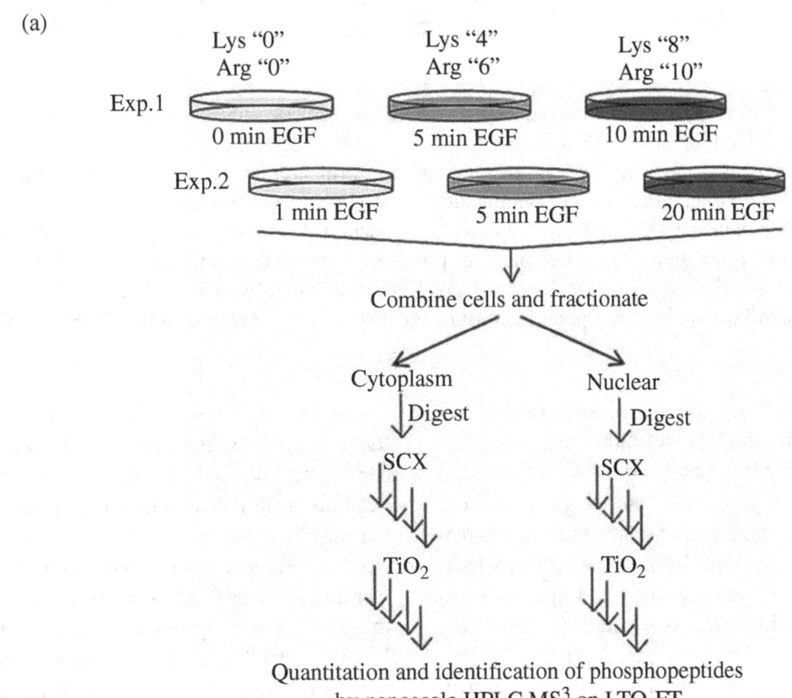

FIGURE 30.50 Quantitative and time-resolved phosphoproteomics using SILAC. (a) Three cell populations are SILAC encoded with normal and stable isotope-substituted arginine and lysine amino acids, creating three stages distinguished by mass. Each population is stimulated for a different length of time with EGF, and the experiment is repeated to yield five time points. Cells are combined, lysed, and enzymatically digested, and phosphopeptides are enriched and analyzed by mass spectrometry.

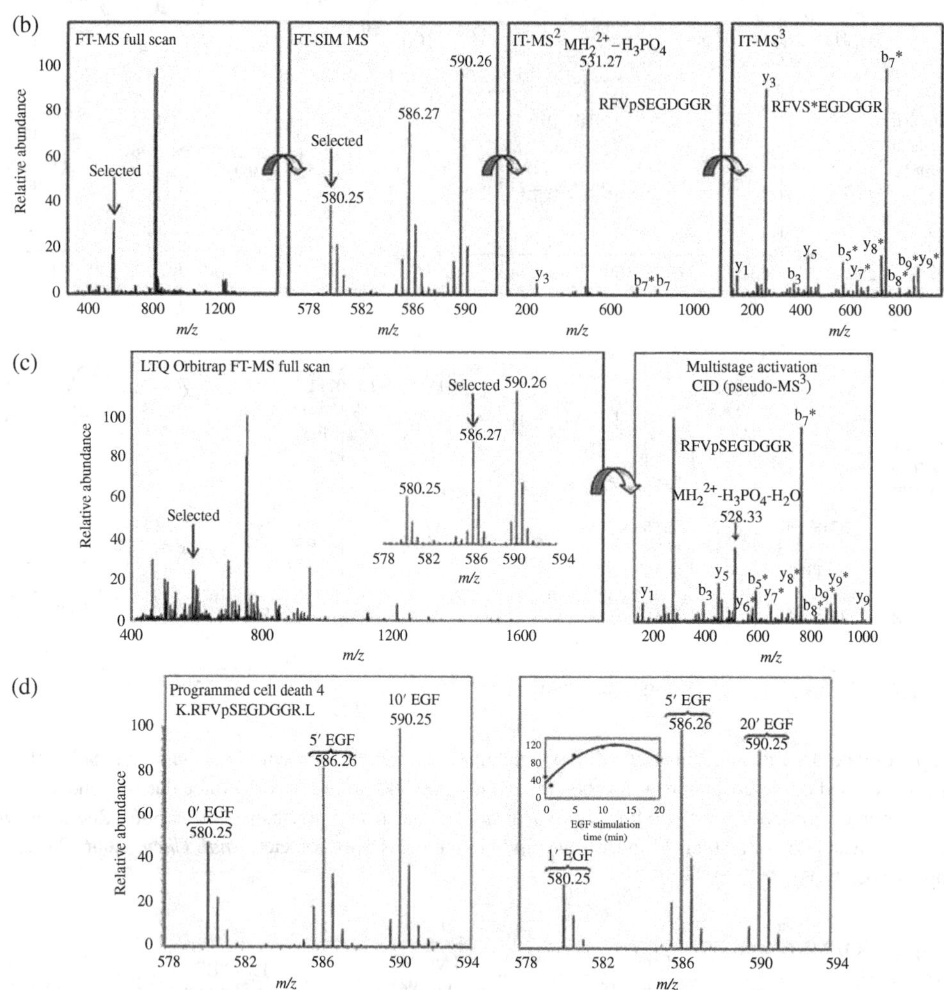

FIGURE 30.50 (Continued) (b) Mass spectra of eluting peptides reveal SILAC triplets (same peptide from the three cell populations), and these triplets are remeasured in selected ion monitoring (SIM) scans for accurate mass determination. Phosphopeptides are identified by loss of the phospho group in a first fragmentation step followed by sequence-related information from a second fragmentation step. (c) Same peptides as in (b) but measured on the LTQ–Orbitrap. Inset shows a magnification of the SILAC peptide selected for fragmentation. Right-hand panel shows the result of multistage activation of the peptide. (d) Raw data of a phosphopeptide from the protein programmed cell death 4. The three peptide intensities in the two experiments are combined using the 5 min time point, resulting in the quantitative profile shown in the inset. (Reprinted with permission. This article was published in *Cell*, Jesper V. Olsen, Blagoy Blagoev, Florian Gnad, Boris Macek, Chanchal Kumar, Peter Mortensen, Matthias Mann Global, in vivo, and site-specific phosphorylation dynamics in signaling networks. **2006**, *127*, 635–648. Copyright Elsevier 2006.)

Lastly, there is one more very interesting feature that has been recently introduced with the LTQ–Orbitrap mass analyzer that allows the collection of MS^3 spectra into a single spectral result. This capability is known as multistage activation (MSA) and is now actually available on the most recently introduced LTQ mass analyzers. Before the introduction of MSA, MS^3 spectra were actually separately collected spectra from the second-stage MS^2 product ion spectra. The traditional MS^n spectra collected by an ion trap were performed by first isolating the precursor ion within the trap and then activating it, thus inducing fragmentation and product ion production. Isolating the product ion of interest by excluding all other ions from the ion trap and then activating it inducing fragmentation and product ion production would then perform the next stage of MS^3. This would be a separately collected spectrum from the MS and the MS^2 spectra. With MSA, the spectral result contains the product ions from the second-stage MS^2 and the third-stage MS^3 experiments. In MSA, instead of removing all of the product ions from the second-stage activation of the precursor ion, the product ion of interest is activated while retaining the entire previously produced product ions and subsequently fragmented, creating a product ion spectrum that contains the second-and third-stage product ions. The capability of MSA is very useful when studying the product ion spectra of posttranslational modifications of proteins such as phosphorylation. Often, when a product ion spectrum is collected of phosphorylated peptides, the predominant product ion observed in the spectrum is a peak formed through a neutral loss of the phosphate group, with very little other information in the spectrum in the form of product ions. A typical approach for studying the product ion spectra of phosphorylated peptides was to look for this predominant phosphate neutral loss product ion and then isolate it for MS^3 spectral collection. The MS^3 spectra obtained could either be interpreted manually by inspecting it in conjunction with the MS^2 product ion spectrum of the precursor or by offline combining the spectra into a single spectrum. With MSA, the composited spectrum is obtained in real time during the spectral acquisition. Presented in Figure 30.50 is an excellent example of the use of MSA in the study of posttranslational modification of proteins by phosphorylation. The experiments were collected on an LTQ–FT mass spectrometer using the approach described previously where separate spectra are obtained for the MS^2 and the MS^3 product ion fragmentation studies (see Fig. 30.50b). The same capability is available on the LTQ–Orbitrap mass spectrometer along with the capability of collecting MS^3 spectra from MSA studies (see Fig. 30.50c). Figure 30.50 also affords the opportunity to see what type of biological experiments can be studied using mass spectrometry.

30.13 CHAPTER PROBLEMS

30.1 List some common mass spectrometers in use today.

30.2 With a source exit slit width of 4 μm and a collector slit width of 6 μm, what magnet radius is needed to obtain a resolution of 10,000?

30.3 Is the electric sector a mass analyzer? If not, explain what it is.

30.4 In the TOF–TOF mass analyzer, what is the ion gate and what is it used for?

30.5 By what energy are ions accelerated into the quadrupole?

30.6 How do the quadrupoles induce stable and unstable trajectories?

30.7 What is it in the Mathieu equations of 30.24 that demonstrates the quadrupole to be a mass analyzer?

30.8 Explain the relationship between sensitivity and resolution for the quadrupole mass analyzer.

30.9 What are some advantages of quadrupole mass analyzers?

30.10 What are some limitations of quadrupole mass analyzers?

30.11 What are the four methods used in scanning with the triple quadrupole mass analyzer? Briefly describe their usefulness.

30.12 What is the purpose of the bath gas used in ion trap mass analyzers?

30.13 What is mass-selective axial instability mode?

30.14 How is resonance ejection obtained in ion traps?

30.15 What are the three methods to excite ions within the ion trap?

30.16 What is the chief difference between a time-of-flight mass analyzer and a quadrupole time-of-flight mass analyzer?

30.17 What kind of spatial configuration is used in quadrupole time-of-flight mass analyzers?

30.18 What two analyses routinely done sets FTICR mass analyzers apart from all others?

30.19 What are the strengths of permanent and electromagnets and superconducting magnets?

30.20 What vacuum pressures are needed in the FTICR cell and why?

30.21 What are three reasons why excitation of the *m/z* ion packet is performed in FTICR mass spectrometry?

30.22 (a) What is collision-mediated radial diffusion? (b) How is it observed?

30.23 What functionalities of the FTICR mass spectrometer are influenced by the magnetic field strength?

30.24 What advantages are there for the Orbitrap mass analyzer not requiring a strong magnet?

REFERENCES

1. Hillenkamp, F.; Unsold, E.; Kaufmann, R.; Nitsche, R. *Appl. Phys.* 1975, **8**, 341–348.
2. Van Breemen, R.B.; Snow, M.; Cotter, R.J. *Int. J. Mass Spectrom. Ion Phys.* 1983, **49**, 35–50.
3. Wiley, W.C.; McLaren, I.H. *Rev. Sci. Instrum.* 1955, **26**, 1150–1156.
4. Vestal, M.L.; Juhasz, P.; Martic, S.A. *Rapid Commun. Mass Spectrom.* 1995, **9**, 1044–1050.
5. Mamyrin, B.A.; Karateev, V.I.; Shmikk, D.V.; Zagulin, V.A. *Sov. Phys. JETP* 1973, **37**, 45.
6. Della Negra, S.; Le Beyec, Y. *Int. J. Mass Spectrom. Ion Process.* 1984, **61**, 21.
7. Johnstone, R.A.W.; Rose, M.E. *Mass Spectrometry for Chemists and Biochemists,* 2nd ed.; Cambridge University Press: New York, 1996; Chapter 2.
8. March, R.E.; Hughes, R.J. *Quadrupole Storage Mass Spectrometry,* Wiley Interscience: New York, 1989; Chapter 2: "Theory of Quadrupole Mass Spectrometry", Pages 31–110.
9. White, F.A. *Mass Spectrometry in Science and Technology.* John Wiley and Sons, 1968.
10. Paul, W. *Angew. Chem. Int. Ed. Engl.* 1990, **29**, 739.
11. Schwartz J.C.; Senko, M.W.; Syka, J.E.P. *J. Am. Soc. Mass Spectrom.* 2002, **13**, 659–669.
12. Syka, J.E.P.; Marto, J.A.; Bai, D.L.; Horning, S.; Senko, M.W.; Schwartz, J.C.; Ueberheide, B.; Garcia, B.; Busby, S.; Muratore, T.; Shabanowitz, J.; Hunt, D.F. *J. Proteome Res.*2004, **3**, 621–626.
13. Makarov, A. *Anal. Chem.* 2000, **72**, 1156–1162.
14. Kingdon K.H. *Phys. Rev.* 1923, **21**, 408–418.
15. Hardman, M., Makarov, A.A. *Anal. Chem.* 2003, **75**, 1699–1705.
16. Hu, Q.; Noll, R.J.; Li, H.; Makarov, A.; Hardman, M.; Cooks, G.R. *J. Mass Spectrom.* 2005, **40**, 430–443.

31

BIOMOLECULE SPECTRAL INTERPRETATION: SMALL MOLECULES

31.1 Introduction
31.2 Ionization Efficiency of Lipids
31.3 Fatty Acids
  31.3.1 Negative Ion Mode Electrospray Behavior of Fatty Acids
31.4 Wax Esters
  31.4.1 Oxidized Wax Esters
  31.4.2 Oxidation of Monounsaturated Wax Esters by Fenton Reaction
31.5 Sterols
  31.5.1 Synthesis of Cholesteryl Phosphate
  31.5.2 Single-Stage and High-Resolution Mass Spectrometry
  31.5.3 Proton Nuclear Magnetic Resonance (^{1}H-NMR)
  31.5.4 Theoretical NMR Spectroscopy
  31.5.5 Structure Elucidation
31.6 Acylglycerols
  31.6.1 Analysis of Monopentadecanoin
  31.6.2 Analysis of 1,3-Dipentadecanoin
  31.6.3 Analysis of Triheptadecanoin
31.7 ESI-Mass Spectrometry of Phosphorylated Lipids
  31.7.1 Electrospray Ionization Behavior of Phosphorylated Lipids
  31.7.2 Positive Ion Mode ESI of Phosphorylated Lipids
  31.7.3 Negative Ion Mode ESI of Phosphorylated Lipids
31.8 Chapter Problems
References

31.1 INTRODUCTION

In this chapter, we will look at the use of mass spectrometry for the analysis of biomolecules as small molecules, which has become an important and increasingly used tool in numerous fields of research and study. Today, when one considers small molecule analysis by mass spectrometry, it is generally associated with the fields of lipidomics and metabolomics, disciplines that study the compliment of small biomolecules that can be associated with biological processes, cells, tissue, and physiological fluid samples such as plasma or urine. Examples of small biomolecules are steroids, fatty acid amides, fatty acids, wax esters, fatty alcohols, phosphorylated lipids, and acylglycerols. Biomolecules referred to as small molecule are contrasted to the larger molecular weight biopolymers such as peptides, proteins, polysaccharides, and nucleic acids, all covered in Chapters 32 and 33.

Lipidomics involves the study of the lipid profile (or lipidome) of living systems and the processes involved in the organization of the protein and lipid species present [1] including signaling processes and metabolism [2]. Mass spectrometry has lead to a dramatic increase in our understanding of lipidomics over the past several years in a variety of cellular systems. Lipids are a diverse class of physiologically important biomolecules that are often classified into three broad classes: (i) the simple lipids such as the fatty acids, the acylglycerols, and the sterols such as cholesterol; (ii) the more complex polar phosphorylated lipids such as phosphatidylcholine and sphingomyelin; and (iii) the isoprenoids, terpenoids, and vitamins. Recently, the Lipid Maps organization has proposed an eight category lipid classification (fatty acyls, glycerolipids, glycerophospholipids, sphingolipids, sterol lipids, prenol lipids, saccharolipids, and polyketides) that is based upon the hydrophobic and hydrophilic characteristics of the lipids [3].

31.2 IONIZATION EFFICIENCY OF LIPIDS

We will begin with a look at the ionization behavior of lipids when using electrospray as the ionization source. Ionization efficiency studies were performed on an ion trap mass spectrometer and a Q-TOF mass spectrometer of a five-component lipid standard consisting of *cis*-9-octadecenamide (oleamide, a primary fatty acid amide, molecular weight (MW) of 281.3 Da), palmityl oleate (an unsaturated wax ester, MW of 506.5 Da), palmityl behenate (a saturated wax ester, MW of 564.6 Da), cholesteryl

Analytical Chemistry: A Chemist and Laboratory Technician's Toolkit, First Edition. Bryan M. Ham and Aihui MaHam.

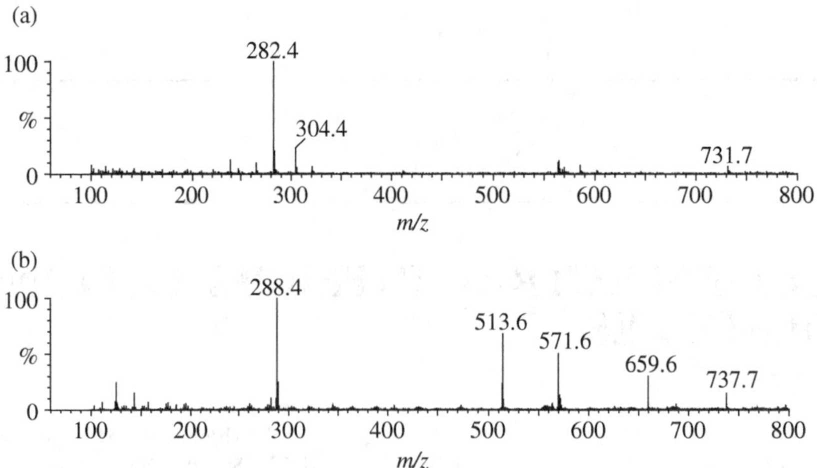

FIGURE 31.1 Single-stage positive ion mode mass spectra of a five-component lipid standard collected on a Q-TOF mass spectrometer for (a) an acidic solution showing the ionization of oleamide at *m/z* 282.4 as $[M+H]^+$, the sodium adduct of oleamide at *m/z* 304.4 as $[M+Na]^+$, and sphingomyelin as an acid adduct at *m/z* 731.7 and (b) a lithium solution showing the ionization of the lithium adduct of oleamide at *m/z* 288.4 as $[M+Li]^+$, at *m/z* 513.6 for the lithium adduct of palmityl oleate as $[M+Li]^+$, at *m/z* 571.6 for the lithium adduct of palmityl behenate as $[M+Li]^+$, at *m/z* 659.6 for the lithium adduct of cholesteryl stearate as $[M+Li]^+$, and at *m/z* 737.7 for the lithium adduct of sphingomyelin as $[M+Li]^+$.

stearate (a saturated cholesterol ester, MW of 652.6 Da), and sphingomyelin (a phosphorylated lipid, MW of 730.6 Da). The ionization solutions consisted of either 10 mM LiCl in 1:1 chloroform/methanol or 1% acetic acid in 1:1 chloroform/methanol. Figure 31.1a illustrates a positive ion mode single-stage mass spectrum of the five-component lipid standard in the acidic solution. The standard was run on a Q-TOF mass spectrometer with direct infusion nano-flow electrospray as the ionization source. The acidic solution did not promote ionization of all of the lipids present in the standard. The species that are being ionized and observed in the mass spectrum include the acid adduct of oleamide at *m/z* 282.4 as $[M+H]^+$, the sodium adduct of oleamide at *m/z* 304.4 as $[M+Na]^+$, and sphingomyelin as an acid adduct at *m/z* 731.7 as $[M+H]^+$ which was observed to ionize to a limited extent. Figure 31.1b illustrates a positive ion mode single-stage mass spectrum of the five-component lipid standard in the lithium solution. The LiCl ionization solution is observed to promote ionization of all five of the lipids in the standard at *m/z* 288.4 for the lithium adduct of oleamide as $[M+Li]^+$, at *m/z* 513.6 for the lithium adduct of palmityl oleate as $[M+Li]^+$, at *m/z* 571.6 for the lithium adduct of palmityl behenate as $[M+Li]^+$, at *m/z* 659.6 for the lithium adduct of cholesteryl stearate as $[M+Li]^+$, and at *m/z* 737.7 for the lithium adduct of sphingomyelin as $[M+Li]^+$.

The addition of Li^+ is the most common method to cationize lipids using electrospray in the positive ion mode [4–6]. Many lipids are not readily protonated, while sodiated ions are not preferred as sodium cations give very little fragmentation information. Lithium cations give rich fragmentation patterns, and thus, it is fairly standard in the analytical field to use this approach. Further, adding a specific cation will often alleviate variation in the metal ion adducts. For example, a mass spectrum run without the addition of a cationizing reagent will show a distribution of $M+H^+$, $M+Na^+$, and $M+K^+$. When a cation is added to the sample, it will primarily shift all the species to that cation. The addition of Na^+ will shift the equilibrium to form almost exclusively $M+Na^+$ and makes interpretation of unknowns easier. Finally, many compounds ionize according to a cation affinity. Some compounds have a higher affinity to sodium than a proton, for example. In order to fully observe all lipid species present, researchers will use multiple cations in order to confirm the assignments of the compounds and to discover new species that might not have a high proton affinity or sodium affinity (e.g., oleamide is seen at 282 Th as $M+H^+$, 304 Th as $M+Na^+$, and 288 Th as $M+Li^+$).

Figure 31.2 illustrates the same effect for the five-component lipid standard when run on a three-dimensional quadrupole ion trap mass spectrometer. In Figure 31.2a, the acidic solution shows the ionization of oleamide at *m/z* 282.4 as $[M+H]^+$, the potassium adduct of oleamide at *m/z* 323.3 as $[M+K]^+$, and sphingomyelin as an acid adduct at *m/z* 731.7. The other species that are present are primarily chemical noise in the form of contaminants. The spectrum in Figure 31.2b is in a lithium solution showing the ionization of the lithium adduct of oleamide at *m/z* 288.4 as $[M+Li]^+$, the lithium adduct of palmityl oleate at *m/z* 513.6 as $[M+Li]^+$, the lithium adduct of palmityl behenate at *m/z* 571.6 as $[M+Li]^+$, the lithium adduct of cholesteryl stearate at *m/z* 659.7 as $[M+Li]^+$, and the lithium adduct of sphingomyelin at *m/z* 737.8 as $[M+Li]^+$.

31.3 FATTY ACIDS

Fatty acids are often measured in the negative mode when analyzing by electrospray mass spectrometry [7]. Studies of fatty acids by electrospray mass spectrometry include very long-chain fatty acid analysis [8], fingerprinting of vegetable oils [9], and the analysis of fatty acid oxidation products [10]. An excellent coverage of the mass spectrometric analysis of the simple class of lipids (fatty acids, triacylglycerols, bile acids, and steroids) is given by Griffiths in a recent review [11].

Mass spectrometric analysis performed on free fatty acid samples in acid solution do not show a good response for the free fatty acids as acid adducts, $[M+H]^+$. Figure 31.3 shows a mass spectrum of a free fatty acid containing biological sample in 1 mM ammonium acetate 50:50 methanol/chloroform solution, collected in the

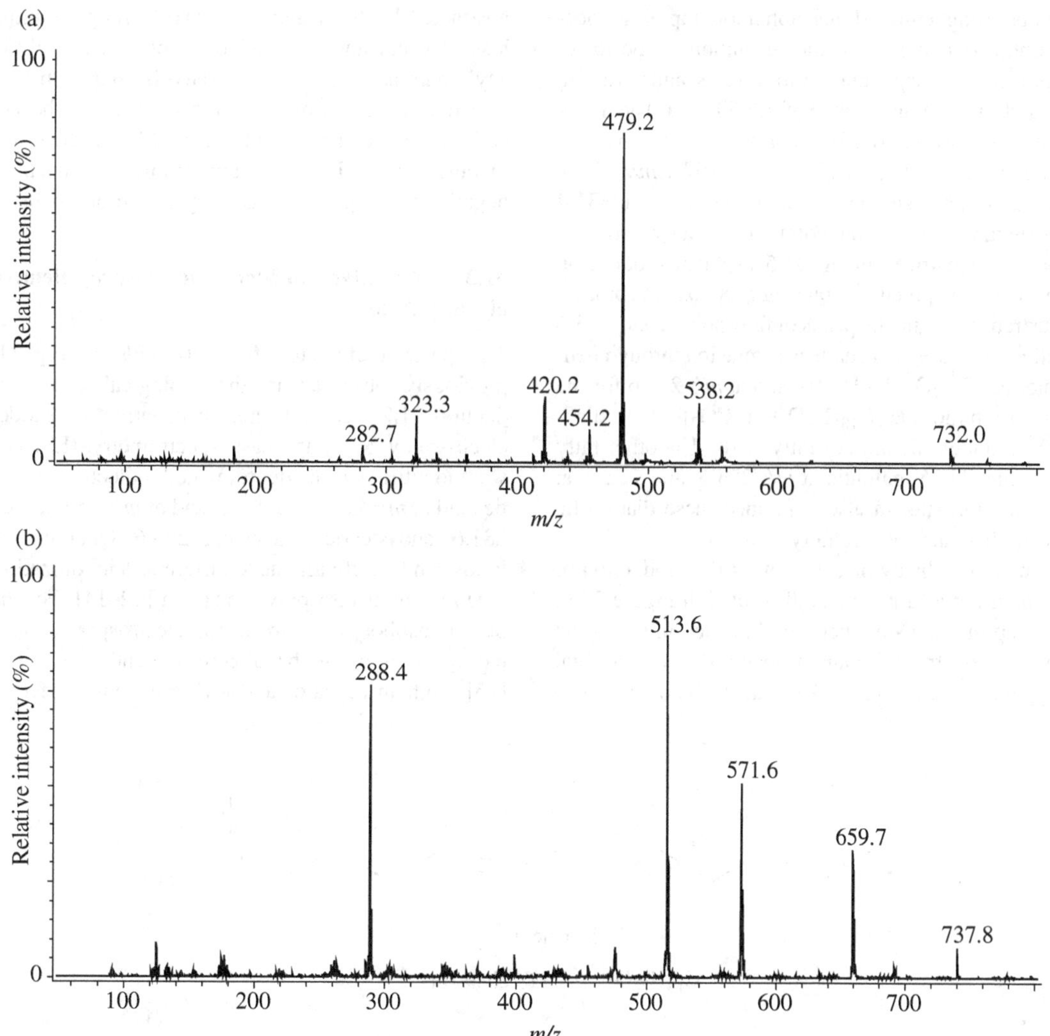

FIGURE 31.2 Single-stage positive ion mode mass spectra of a five-component lipid standard collected on a three-dimensional quadrupole ion trap mass spectrometer for (a) an acidic solution showing the ionization of oleamide at *m/z* 282.4 as $[M+H]^+$, the potassium adduct of oleamide at *m/z* 323.3 as $[M+K]^+$, and sphingomyelin as an acid adduct at *m/z* 731.7 and (b) a lithium solution showing the ionization of the lithium adduct of oleamide at *m/z* 288.4 as $[M+Li]^+$, at *m/z* 513.6 for the lithium adduct of palmityl oleate as $[M+Li]^+$, at *m/z* 571.6 for the lithium adduct of palmityl behenate as $[M+Li]^+$, at *m/z* 659.7 for the lithium adduct of cholesteryl stearate as $[M+Li]^+$, and at *m/z* 737.8 for the lithium adduct of sphingomyelin as $[M+Li]^+$.

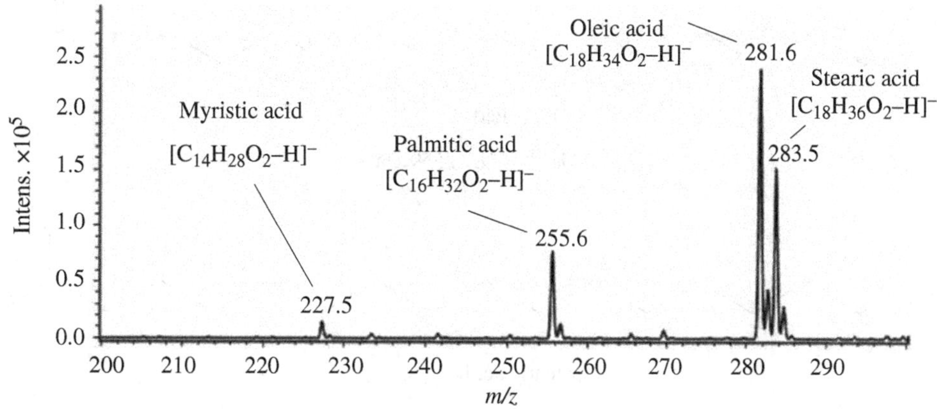

FIGURE 31.3 Negative ion mode mass spectrum of a biological extract in 1 mM ammonium acetate 50:50 methanol/chloroform solution. Peaks in the spectrum are deprotonated ions $[M-H]^-$ of myristic acid at *m/z* 227.5, palmitic acid at *m/z* 255.6, oleic acid at *m/z* 281.6, and stearic acid at *m/z* 283.6.

negative ion mode using a three-dimensional ion trap mass spectrometer. For optimized sensitivity and resolution of the lower molecular weight lipid species, the ion trap was scanned during spectral accumulation for a mass range of *m/z* 50–400. The major peaks in the spectrum are comprised of deprotonated ions $[M-H]^-$ identified as myristic acid at *m/z* 227.5, palmitic acid at *m/z* 255.6, oleic acid at *m/z* 281.6, and stearic acid at *m/z* 283.6. Figure 31.4 illustrates the structures of these four common free fatty acids that are found in biological matrices. Figure 31.5 displays a product ion spectrum of the oleic acid precursor ion at *m/z* 281.6 in the biological sample illustrating diagnostic peaks of fatty acids at *m/z* 263.6 for the production of a fatty acyl chain as ketene ion through neutral loss of water $[C_{18}H_{34}O_2\text{–}H\text{–}H_2O]-$ and at *m/z* 249.6 for the neutral loss of carbon dioxide $[C_{18}H_{34}O_2\text{–}H\text{–}CO_2]-$, both diagnostic losses for the identification of fatty acids. The other fatty acid (myristic at *m/z* 227.5, palmitic at *m/z* 255.6, and stearic at *m/z* 283.6) product ion spectra also contained these diagnostic losses identifying them also as free fatty acids.

The fragmentation pathway mechanism for the production of the fatty acyl chain ion as a ketene is illustrated in Figure 31.6. This is a three-step mechanism where the first step involves the transfer of an α-proton from the alpha position to the terminal fatty acid oxygen. In step two, a fatty acyl chain proton is abstracted by the leaving hydroxyl group, resulting in neutral loss of water and the production of a negatively charged fatty acyl chain as ketene, as illustrated by step 3 in the mechanism. The fragmentation pathway mechanism for the neutral loss of carbon dioxide producing the *m/z* 249.6 product ion is illustrated in Figure 31.7. This is a simpler one-step mechanism where a negatively charged fatty acyl hydrocarbon chain is produced.

31.3.1 Negative Ion Mode Electrospray Behavior of Fatty Acids

The spectrum of the four free fatty acids (see Fig. 31.3) that were previously observed in the biological extract presented in Section 31.2 were collected in negative ion mode analysis by electrospray ion trap mass spectrometry (ESI-IT/MS). These were identified as the deprotonated molecules $[M-H]^-$ of myristic acid at *m/z* 227.5, palmitic acid at *m/z* 255.6, oleic acid at *m/z* 281.6, and stearic acid at *m/z* 283.6. Quantitative analysis of lipids can be difficult due to differences in ionization efficiencies when using electrospray ionization [12–14]. For the more complex phosphorylated lipids, the electrospray ionization process has been shown to be directly dependent upon concentration [15]. Both unsaturation and acyl chain length affect the efficiency

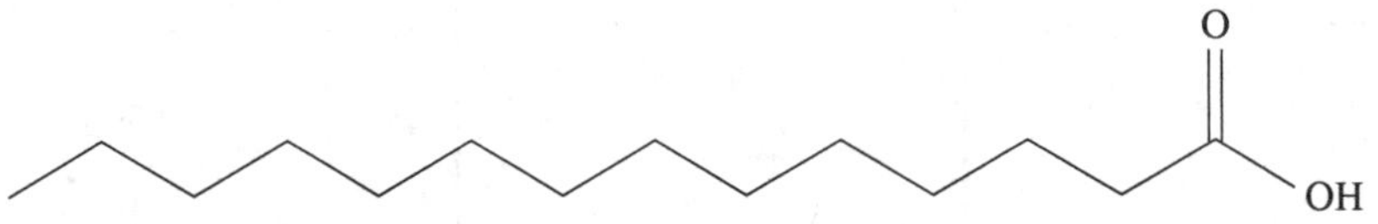

O
OH
Myristic acid
$C_{14}H_{28}O_2$ 228.2089 Da

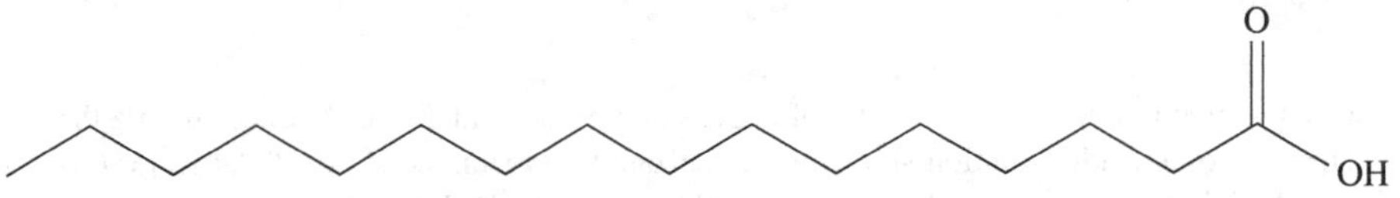

O
OH
Palmitic acid
$C_{16}H_{32}O_2$ 256.2402 Da

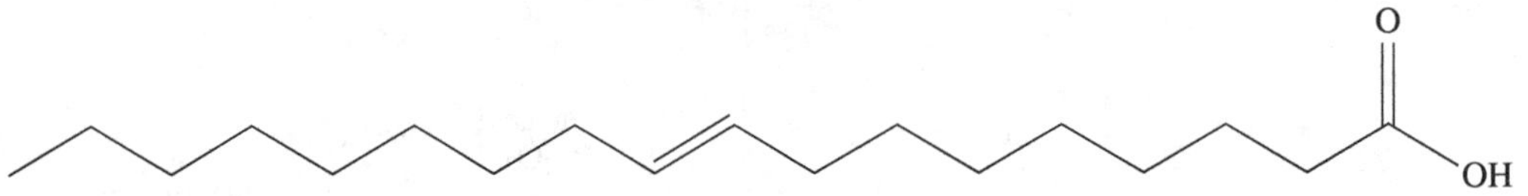

O
OH
Oleic acid
$C_{18}H_{34}O_2$ 282.2559 Da

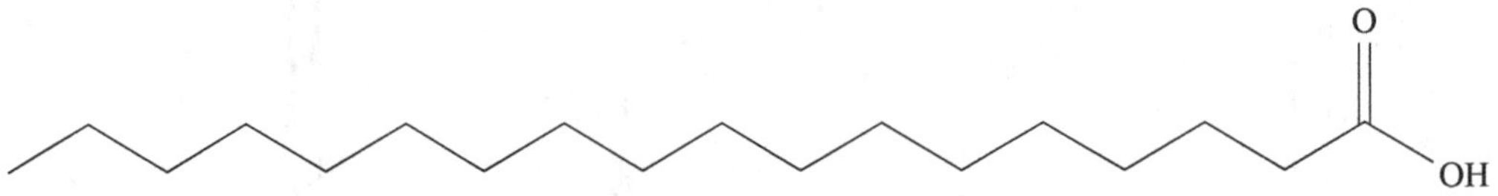

O
OH
Stearic acid
$C_{18}H_{36}O_2$ 284.2715 Da

FIGURE 31.4 Structures of four common free fatty acids found in biological matrices.

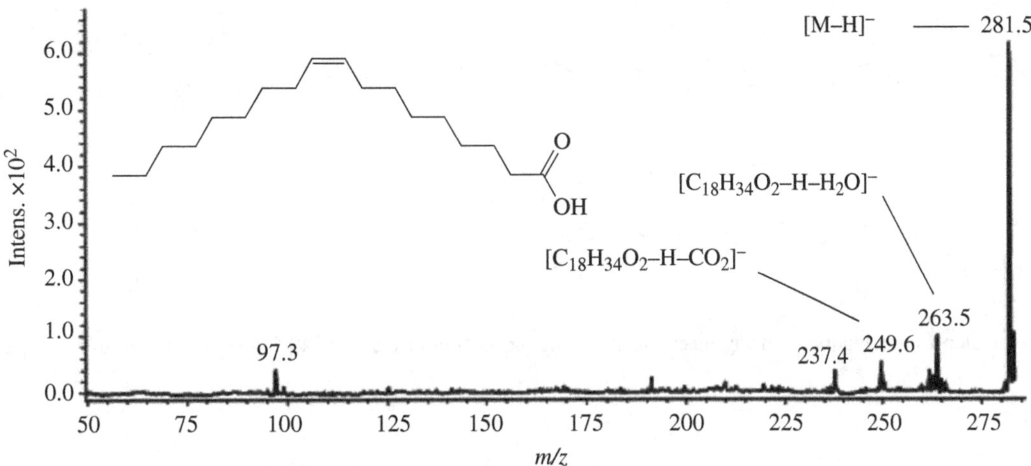

FIGURE 31.5 Product ion spectrum of the oleic acid precursor ion at *m/z* 281.6 in the biological extract illustrating diagnostic peaks of fatty acids at *m/z* 263.6 for the production of a fatty acyl chain as ketene ion through neutral loss of water $[C_{18}H_{34}O_2–H–H_2O]^-$ and at *m/z* 249.6 for the neutral loss of carbon dioxide $[C_{18}H_{34}O_2–H–CO_2]^-$.

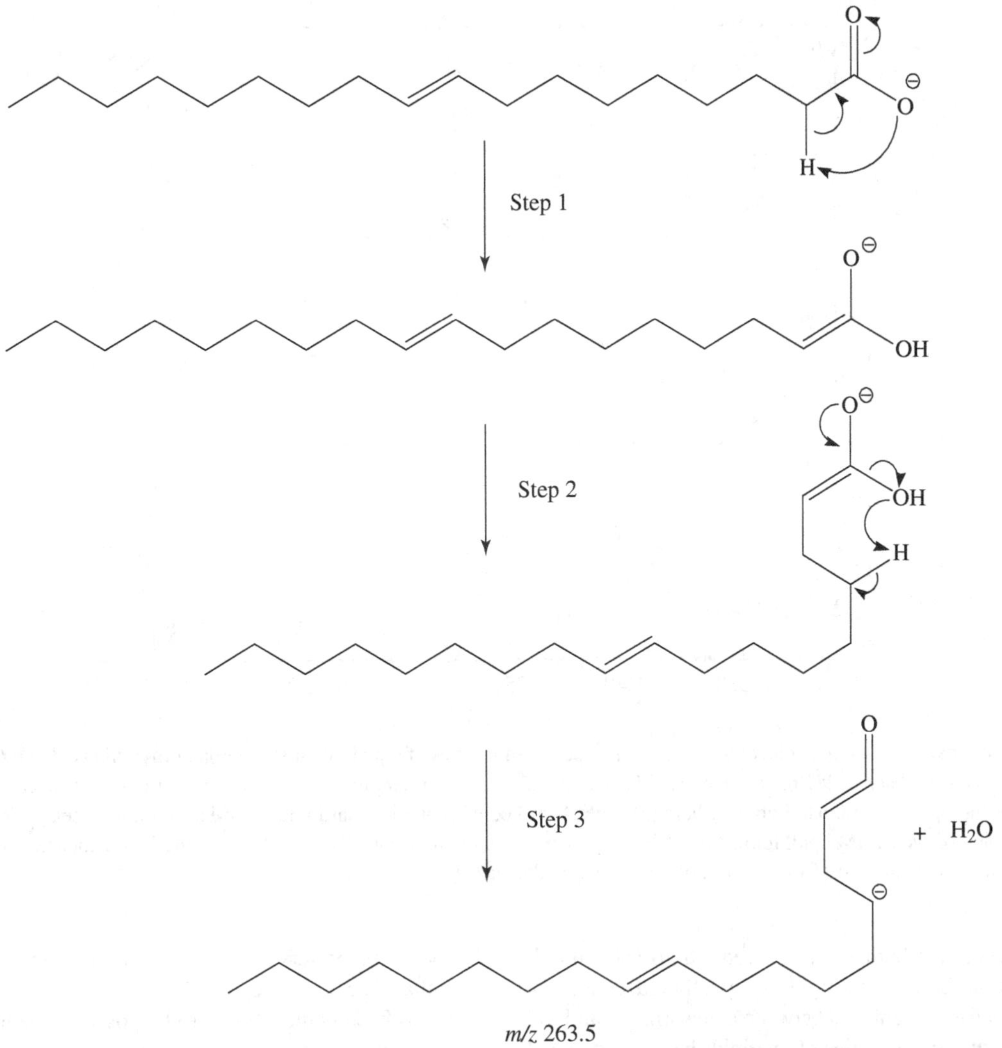

O
O⊖
H
Step 1
O⊖
OH
Step 2
O⊖
OH
H
Step 3
O
⊖
+ H_2O

m/z 263.5

FIGURE 31.6 Fragmentation pathway mechanism for the neutral loss of water producing a negatively charged fatty acyl chain as ketene at *m/z* 263.5.

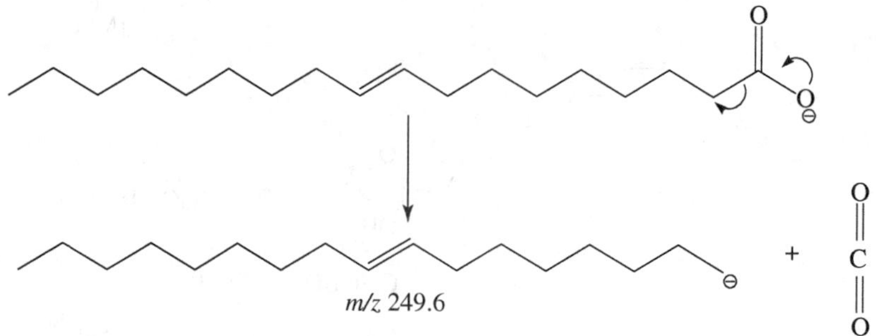

FIGURE 31.7 One-step fragmentation pathway mechanism for the production of the *m/z* 249.6 product ion through neutral loss of carbon dioxide.

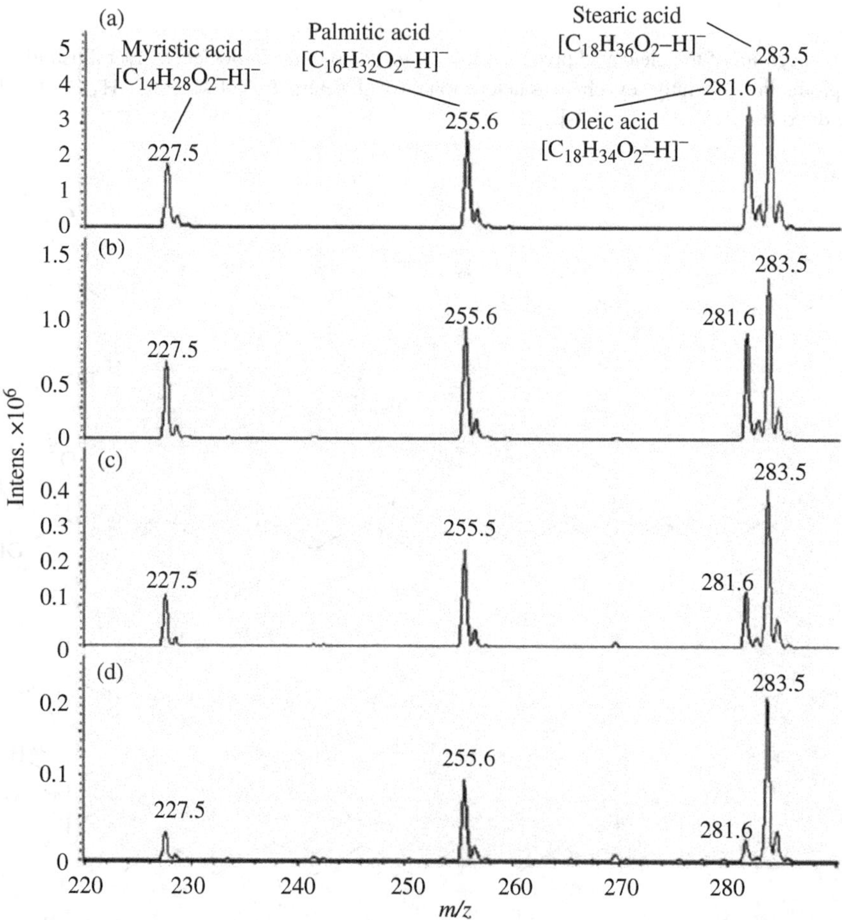

FIGURE 31.8 Electrospray ion trap negative ion mode mass spectra of a series of equal molar standards of myristic acid at *m/z* 227.5, palmitic acid at *m/z* 255.6, oleic acid at *m/z* 281.6, and stearic acid at *m/z* 283.5. The spectra represent scans of serial dilutions at successively decreasing concentrations of the lipids from top to bottom where (a) is a 0.1 mM equal molar standard with 1 mM ammonium acetate, (b) is 10 μM lipid, (c) is 1 μM lipid, and (d) is 0.1 μM lipid. Ionization efficiency increases with increasing fatty acid chain length and apparent suppression effect for oleic acid as the concentration of the equal molar standard is decreased.

of the lipid ionization at higher lipid concentrations (>0.1 nmol/μl). The interactions between the lipids are greatly reduced, and a linear correlation can be achieved between concentration and ion intensities when the concentration of the lipids has been reduced [15]. Due to differences in ionization efficiencies of lipids (and subsequent suppression), internal standard spiking of lipids known to be of very low concentration is performed to directly quantitate the lipid species of interest in the biological extract. This has been demonstrated to be possible due to a direct linear correlation between lipid concentration and lipid response [12–15]. Figure 31.8 contains negative ion mode spectra of a series of equal molar standards of myristic acid at *m/z* 227.5, palmitic acid at *m/z* 255.6, oleic acid at *m/z* 281.6, and stearic acid at *m/z* 283.5 at successively decreasing concentrations of the lipids from top to bottom. Figure 31.8a illustrates a 0.1 mM equal molar standard with 1 mM ammonium acetate, while Figure 31.8b shows 10 μM lipid, Figure 31.8c 1 μM lipid, and Figure 31.8d 0.1 μM lipid. The spectrum illustrated in Figure 31.8a might indicate that the ionization efficiency increases with increasing fatty acid chain length as demonstrated by the lower-intensity peak for myristic acid at *m/z* 227.5, which has a 14 carbon chain, as compared to stearic acid at *m/z* 283.5, which has an 18 carbon chain. However, there is also observed in the Figure 31.8 spectra an apparent suppression effect for oleic acid as the concentration of the equal molar standard is being decreased. This decrease in ionization efficiency for oleic acid at lower concentrations may be attributed to the single unsaturation in its hydrocarbon chain. It does not appear that the fatty acids measured in negative ion mode using the ion trap possess the same linear correlation that the more complex phosphorylated lipids have exhibited in the past.[15] To ascertain whether the single unsaturation in the oleic fatty acyl chain has an effect upon the ionization efficiency, a series of oleic acid standards at successively decreasing concentration were analyzed using the electrospray quadrupole time-of-flight (Q-TOF/MS) mass spectrometer with the results presented in Figure 31.9a–g. Single-stage negative ion mode mass spectra were collected of oleic acid at concentrations of 90 μM (Fig. 31.9a), 50 μM (Fig. 31.9b), 10 μM (Fig. 31.9c), 1 μM (Fig. 31.9d), 100 nM (Fig. 31.9e), 10 nM (Fig. 31.9f), and 1 nM (Fig. 31.9g), all in 1 mM ammonium acetate 1:1 $CHCl_3$/MeOH solutions. In the spectra, oleic acid was observed to decrease in its response, while the emergence and subsequent spectral dominance of myristic acid, palmitic acid, and stearic acid was noted to take place. In the negative ion mode spectrum of Figure 31.9g for 1 nM oleic acid, the spectrum is being dominated primarily by chemical noise from contaminants. Adjustment of the source parameters appeared to have little effect; the emergence in the spectra of the *m/z* 227.5 (myristic acid) and *m/z* 255.6 (palmitic acid) species was also observed. The adjustment of source parameters entailed varying the sample cone voltage between approximately 20 and 80 V and varying the capillary voltage between approximately 2000 and 3500 V. This behavior of the suppression of oleic acid or the reduction of oleic acid to stearic acid and the emergence of the other two fatty acid species (myristic acid and palmitic acid) would explain the observation made in Figure 31.8. With decreasing concentration of the lipids, the oleic acid is either suffering from suppression or is being converted to stearic acid with lesser amounts of myristic and palmitic being produced. This is illustrated in Figure 31.8d where stearic acid at *m/z* 283.5 is the predominant peak in the spectrum, followed by palmitic acid at *m/z* 255.6, myristic acid at *m/z* 227.5, and lastly oleic acid at *m/z* 281.6.

To further investigate the suppression of oleic acid or the reduction of oleic acid during the electrospray process, a number of experiments were performed. The double-bond position was first studied by electrospray experiments which included 6-octadecenoic and 11-octadecenoic standards. Both of the different positional C18:1 unsaturated fatty acid lipids (6- and 11-) showed similar behavior to the C18:1 9-octadecenoic fatty acid illustrated in Figure 31.8 (i.e., the C18:1 fatty acid decreased significantly with the significant increase in the C18:0 fatty acid, followed by the palmitic and myristic fatty acids). This indicates that the position of the double bond within the acyl chain does not influence this ESI behavior. A trans C18:1 fatty acid, 9-*trans*-octadecenoic, was analyzed to determine whether a conformational effect would be observed. It was found that 9-*trans*-octadecenoic also had similar behavior as is illustrated in Figure 31.8 for C18:1 9-octadecenoic. This indicates that location of the double bond is not a factor nor is the conformation of the fatty acid. A C20:1 11-eicosenoic lipid standard was analyzed to determine the electrospray effect upon a longer-chain fatty acid lipid. At decreasing concentrations, the production of a C20:0 fatty acid by reduction of the C20:1 fatty acid was clearly not observed to any extent that was observed for the C18:1 electrospray experiments. However, the emergence of the other three fatty acids was observed which included myristic acid at *m/z* 227.1, palmitic acid at *m/z* 255.2, and stearic acid at *m/z* 283.2. Finally, 9-hexadecenoic (C16:1 fatty acid) was analyzed by electrospray in the negative ion mode to determine whether a lower molecular weight singly unsaturated fatty acid would exhibit the same behavior as the C18:1 fatty acid. This in fact was observed where at decreasing concentrations palmitic acid (C16:0) was observed to emerge along with a small amount of myristic acid. Stearic acid was also observed at lower concentrations to become a major peak in the spectra; however, it was also observed as a contaminant in the standard. The fact that the singly unsaturated C20:1 fatty acid did not significantly produce the C20:0 saturated fatty acid during the electrospray process shows that the reduction of the double bond is not taking place. Also, that a C18:0 fatty acid was observed as a major species in the C16:1 electrospray experiments further shows that this is an ionization suppression effect that is taking place with the singly unsaturated fatty acids at decreasing concentrations. Small amounts of the fatty acids are observed in all of the standards as contaminants, except a C20:0 fatty acid. This suggests that at decreasing concentration the singly unsaturated fatty acids are suffering from significant suppression during the electrospray process.

To rule out the contribution of contaminant myristic acid, palmitic acid, and stearic acid presence (e.g., from the oleic acid standard or internal instrument carryover), influencing the spectra, a carbon 13 (^{13}C) algal fatty acid standard mix consisting of 51.0% 16:0, 8.7% 16:1, 2.3% 18:0, 18.3% 18:1, 17.7% 18:2, and 2.0% 18:3 was measured at decreasing concentrations. Single-stage negative ion mode mass spectra of dilutions of the ^{13}C fatty acid standard mixture were collected by ESI-Q-TOF/MS in solutions of 1 mM ammonium acetate 1:1 $CHCl_3$/MeOH. Figure 31.10 illustrates the results of the negative ion mode studies where Figure 31.10a contains 940 μM ^{13}C16:0, Figure 31.10b contains 470 μM ^{13}C16:0, Figure 31.10c contains 240 μM ^{13}C16:0, Figure 31.10d contains 9.4 μM ^{13}C16:0, and Figure 31.10e contains 0.94 μM ^{13}C16:0. In the figure, the ^{13}C16:0 lipid is at *m/z* 271.3, the ^{13}C16:1 lipid is at *m/z* 269.3, the ^{13}C18:1 lipid is at *m/z* 299.3, and the ^{13}C18:2 lipid is at *m/z* 297.3. Close inspection of the ^{13}C16:0 lipid at *m/z* 271.3 and the ^{13}C16:1 lipid at *m/z* 269.3 reveals the decreasing response for the unsaturated ^{13}C16:1 lipid at *m/z* 269.3 as compared to the saturated ^{13}C16:0 lipid at *m/z* 271.3. To better illustrate the observed

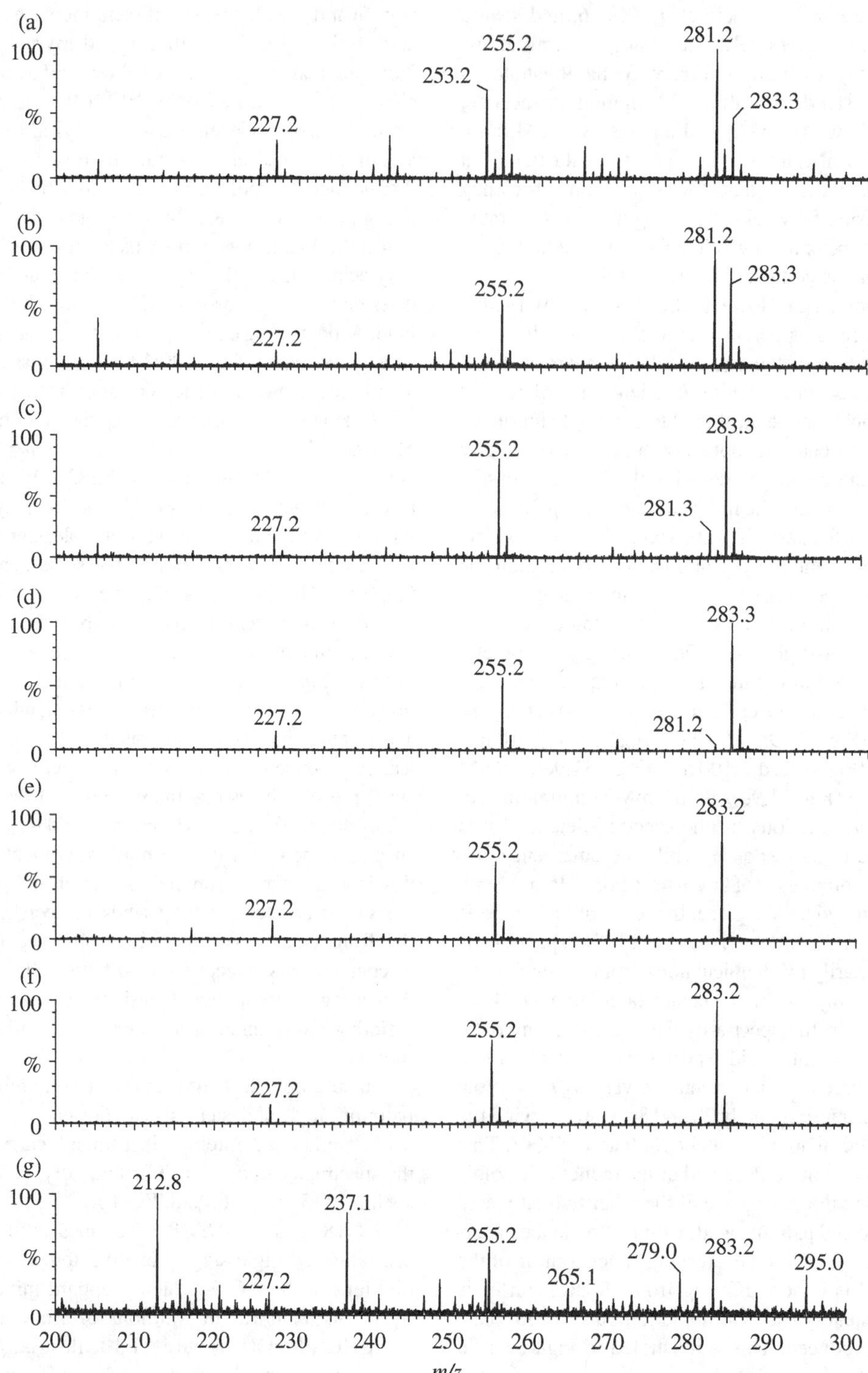

FIGURE 31.9 Single-stage negative ion mode electrospray quadrupole time-of-flight (Q-TOF/MS) mass spectra of a series of oleic acid standards at successively decreasing concentrations of (a) 90 μM, (b) 50 μM, (c) 10 μM, (d) 1 μM, (e) 100 nM, (f) 10 nM, and (g) 1 nM, all in 1 mM ammonium acetate 1:1 $CHCl_3$/MeOH solutions.

effect visually, ratio graphs were constructed. Figure 31.11 contains graphs illustrating the behavior of the unsaturated fatty acids C16:1, C18:1, and C18:2 contained within the ^{13}C fatty acid standard mixture analyzed in Figure 31.10. Figure 31.11a shows the ratio of the mass spectral response of the singly unsaturated C16:1 fatty acid to the mass spectral response to its unsaturated counterpart C16:0 fatty acid plotted versus concentration. The graphs are from seven standards that cover a concentration range from 95 μM C16:0 to 950 μM C16:0 and 16 μM C16:1 to 160 μM C16:1. The graphs are presented with the *x*-axis represented by the concentration of the C16:0 fatty acid. The graph demonstrates the change in

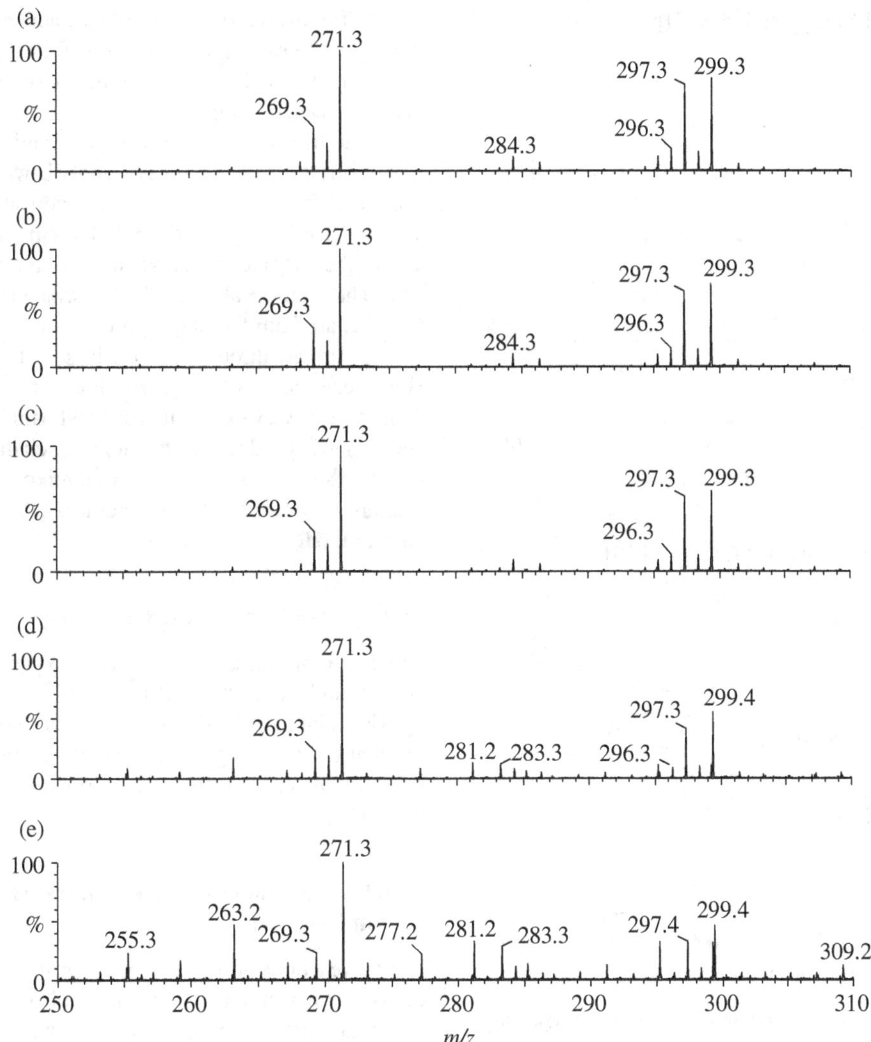

FIGURE 31.10 Negative ion mode electrospray single-stage mass spectra collected on a Q-TOF mass spectrometer illustrating the results of decreasing the concentration of the carbon 13 lipids. (a) Contains 940 μM ^{13}C16:0, (b) contains 470 μM ^{13}C16:0, (c) contains 240 μM ^{13}C16:0, (d) contains 9.4 μM ^{13}C16:0, and (e) contains 0.94 μM ^{13}C16:0. The ^{13}C16:0 lipid is at *m/z* 271.3, the ^{13}C16:1 lipid is at *m/z* 269.3, the ^{13}C18:1 lipid is at *m/z* 299.3, and the ^{13}C18:2 lipid is at *m/z* 297.3. The ^{13}C16:0 lipid at *m/z* 271.3 and the ^{13}C16:1 lipid at *m/z* 269.3 are decreasing in response for the unsaturated ^{13}C16:1 lipid at *m/z* 269.3 as compared to the saturated lipid ^{13}C16:0 lipid at *m/z* 271.3.

response (suppression) of the C16:1 fatty acid as compared to the C16:0 fatty acid at successively decreasing concentrations. This behavior is also observed in the suppression of C18:1 and C18:2 at successively lower concentrations as illustrated in Figure 31.11b. Due to the relationship that the response of a lipid using ESI is linearly dependent upon concentration, it is also apparent that the suppression effect can also be linearly dependent upon concentration as is illustrated in Figure 31.11a and b. Also of note is that Figure 31.11b includes C18:1 and C18:2 as a comparison of the influence of the number of unsaturation to suppression. As stated previously, the carbon 13 algal fatty acid standard mix contained 18.3% of the C18:1 fatty acid and 17.7% of the C18:2 fatty acid (giving a 3% difference in concentration). The curves in Figure 31.11b illustrate that there appears to be a reduced response for the C18:2 unsaturated lipid as compared to the C18:1 unsaturated lipid as the unsaturation increases from 1 to 2.

The significance of the observance of the suppression of the singly unsaturated fatty acids (oleic acid and palmitoleic acid) illustrates the difficulty in quantitating the amounts of the free fatty acids in biological samples using electrospray as the ionization technique. Care must be taken in assigning relative amounts to the free fatty acids, as well as to all the other lipids present in biological extracts when using an electrospray source.

31.4 WAX ESTERS

Wax esters are lipid species that are quite common in both plants and animals and have been studied using mass spectrometric techniques. Wax ester and oxidized wax ester analysis has primarily been performed using the chromatographic coupled mass spectrometric technique of GC-MS [16]. Recently, a study was

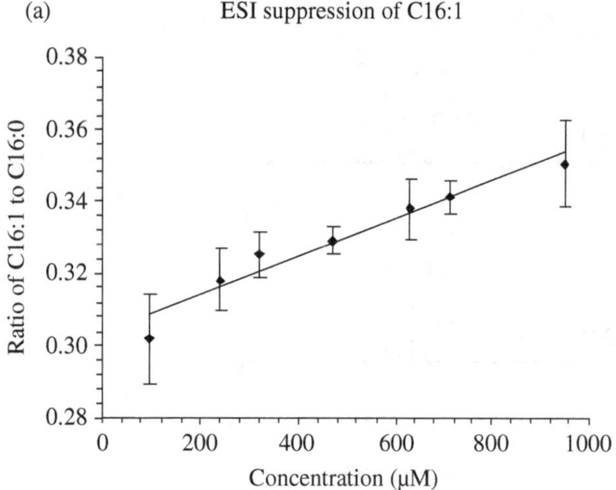

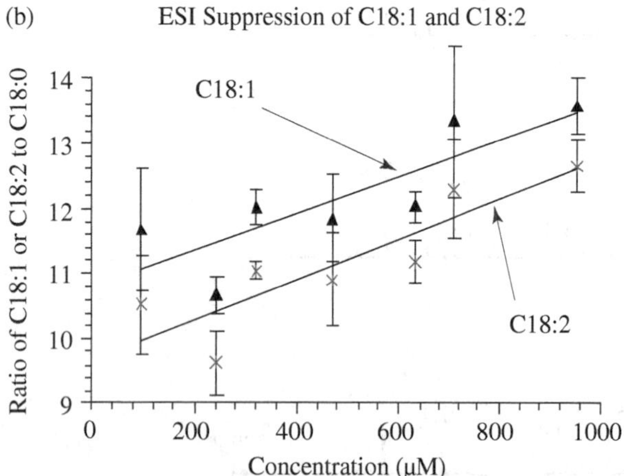

FIGURE 31.11 ESI-Q-TOF/MS plotted results of single-stage negative ion mode mass spectral carbon 13 fatty acid responses for decreasing concentrations. The graphs are from seven standards that cover a concentration range from 95 to 950 μM. Each standard spectrum was collected for 1 min, in triplicate, for each standard concentration (the average plotted as data points) and standard deviation (*y*-error bars). (a) Palmitoleic acid (C16:1) graphical results of the response of C16:1 divided by the response of C16:0 at each standard concentration. (b) Oleic acid (C18:1) and linoleic acid (C18:2) graphical results of the response of C18:1 and C18:2 divided by the response of stearic acid (C18:0) at each standard concentration.

reported in the literature where the wax esters were analyzed by electrospray mass spectrometry [17].

Figure 31.12 illustrates the product ion spectrum of the lithium adduct of palmityl behenate, a representative fatty acid/fatty alcohol wax ester. The product ion spectrum was collected on a triple quadrupole mass spectrometer by collision-induced dissociation (CID). In the low mass region, only the hydrocarbon fragments are observed at *m/z* 43, *m/z* 57, *m/z* 71, *m/z* 83 (unsaturated), and *m/z* 85. In the *m/z* 350 region, there is one predominant peak appearing at *m/z* 347. This represents the neutral loss of the carbon chain of the fatty alcohol $[MLi–C_{16}H_{32}]^+$, leaving the lithium adduct of the fatty acid portion of the wax ester at *m/z* 347 $[C_{22}H_{44}O_2Li]^+$.

The fragmentation pathway mechanism for the production of the *m/z* 347 product ion is illustrated in Figure 31.13.

Figure 31.14 shows the fragmentation behavior of a standard unhydroxylated, single unsaturation containing wax ester oleyl palmitate (the saturated fatty acid palmitic acid esterified to the singly unsaturated fatty oleyl alcohol, wax ester structure also shown in figure) as the lithium cation at *m/z* 513.8 $[M+Li]^+$. The major product ion at *m/z* 263.5 is produced through neutral loss of the fatty alcohol alkyl chain forming the lithiated palmitic acid. The minor peak at *m/z* 245.5, which is 18 atomic mass units (amu) apart from the major product ion at *m/z* 263.5, is the C16 ketene formed through neutral loss of the oleyl fatty alcohol. However, this is a very minor fragmentation pathway. A mixture of wax esters such as oleyl palmitate and palmityl oleate primarily produce product ions which differ by 26 amu (spectrum not shown). In other words, a mixture of wax esters would primarily differ by C_nH_{2n} if saturated such as a difference of 56 amu or C_4H_8.

31.4.1 Oxidized Wax Esters

Oxidation of lipids occurs when oxygen attacks the double bond found in unsaturated fatty acids to form peroxide and or epoxide linkages and can result in the destruction of the original lipid leading to the loss of function and is believed to be a significant contributor to coronary artery disease and diabetes [18].

31.4.2 Oxidation of Monounsaturated Wax Esters by Fenton Reaction

Fenton reactions were performed on the monounsaturated wax esters oleyl palmitate and palmityl oleate for product ion fragmentation pathway determinations. The Fenton reaction is used for *in vitro* studies of the oxidation of lipids where the addition of hydrogen peroxide in the presence of Fe^{2+} induces oxidative conditions [19]. The oxidation of oleyl palmitate and palmityl oleate was performed using the following procedure [19]. Approximately 5 nmols of lipid in 50 μl of chloroform was pipetted into a 1.5 ml Eppendorf tube and brought just to dryness with a stream of dry nitrogen. The lipid was then reconstituted into 332 μl of a 0.1 M ammonium bicarbonate buffer solution (pH 7.4). For the Fenton reaction, 15 μmols of $FeCl_2$ and 158 μ mols of hydrogen peroxide (H_2O_2) were added to the reconstituted lipid in ammonium bicarbonate. The mixture was then left to react for periods of time (20, 22, and 26 h) coupled with sonication by immersing the Eppendorf tube in a sonicator bath. The oxidized and unoxidized lipid species in the reaction mixture were then extracted using a modified Folch method consisting of 2:1 chloroform/methanol (vol/vol) [20,21]. The extracted lipid species were brought just to dryness with a stream of dry nitrogen and reconstituted in a 10 mM LiCl chloroform/methanol solution (1:1, vol/vol) for electrospray ionization mass spectrometric analysis.

The Fenton reaction for monounsaturated wax esters was found to produce the desired oxidation effect with extended reaction periods. The molecular oxygen addition to the monounsaturated wax esters was observed and is illustrated in the single-stage mass spectrum of Figure 31.15 where the *m/z* 529.7 peak

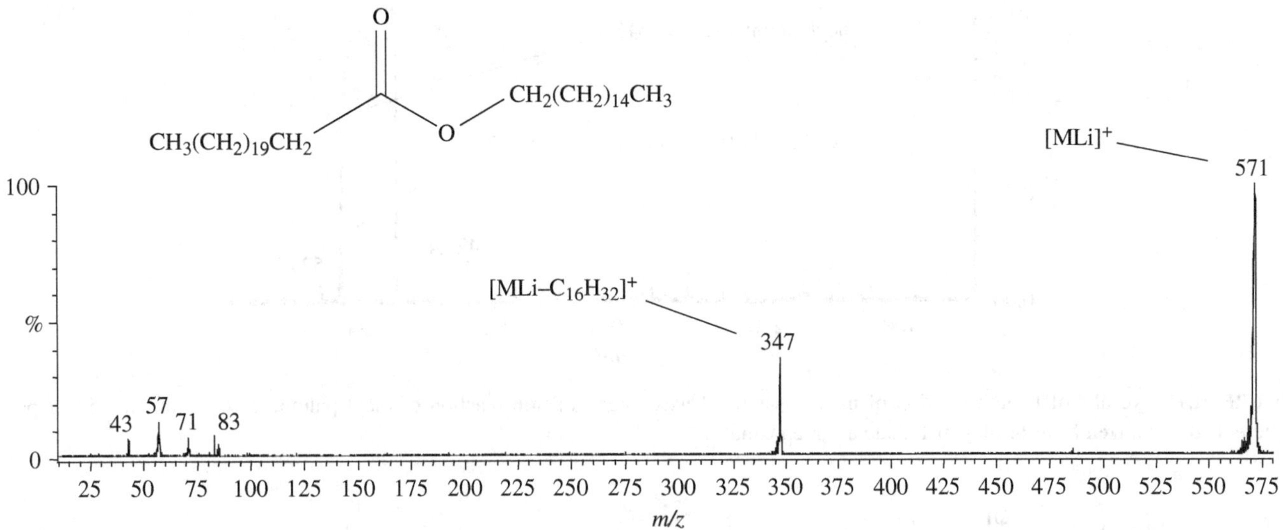

FIGURE 31.12 Collision-induced dissociation product ion spectrum of the lithium adduct of palmityl behenate at *m/z* 571 collected on a triple quadrupole mass spectrometer. Product ions include *m/z* 43, *m/z* 57, *m/z* 71, *m/z* 83, and *m/z* 347.

FIGURE 31.13 *M/z* 347 product ion fragmentation pathway mechanism.

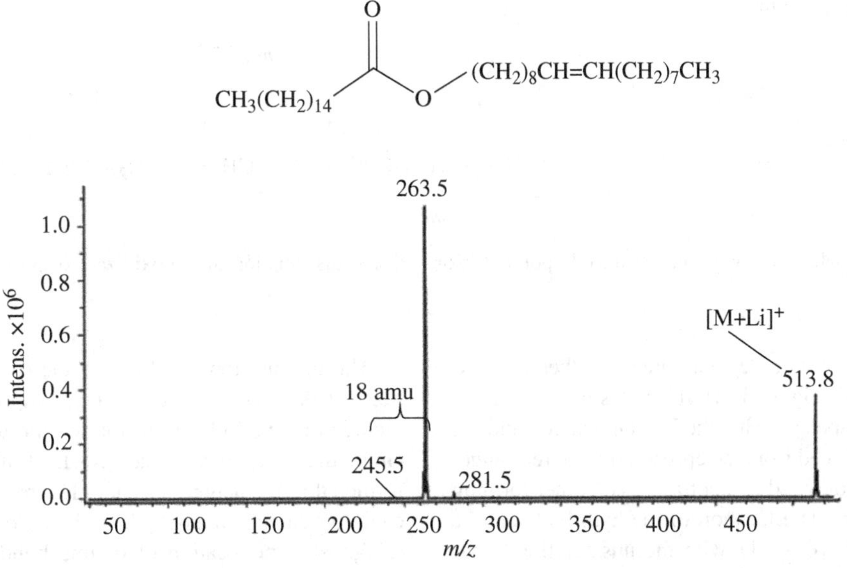

FIGURE 31.14 Fragmentation behavior of an unhydroxylated, single unsaturation containing wax ester oleyl palmitate (the saturated fatty acid palmitic acid esterified to the singly unsaturated fatty oleyl alcohol) as the lithium adduct at *m/z* 513.8 $[M+Li]^+$. Product ions include *m/z* 263.5 through neutral loss of the fatty alcohol alkyl chain forming the lithiated palmitic acid and *m/z* 245.5 (18 amu apart from *m/z* 263.5), a C16 ketene formed through neutral loss of the oleyl fatty alcohol.

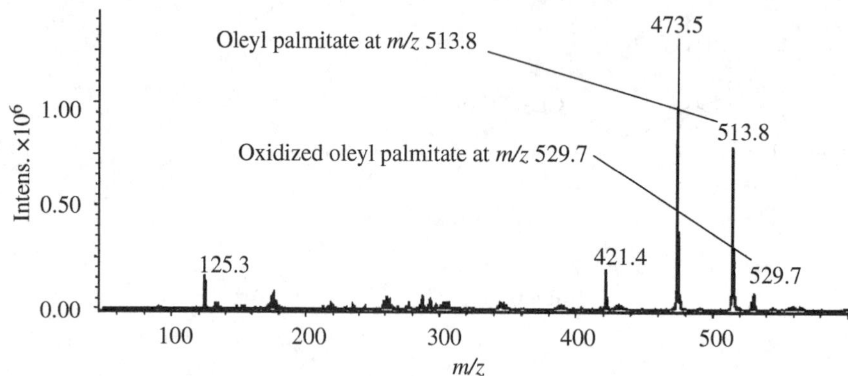

FIGURE 31.15 Results of Fenton reaction of monounsaturated wax esters. Fenton reaction of oleyl palmitate where the *m/z* 529.7 peak represents the oxidized form of oleyl palmitate as an epoxide.

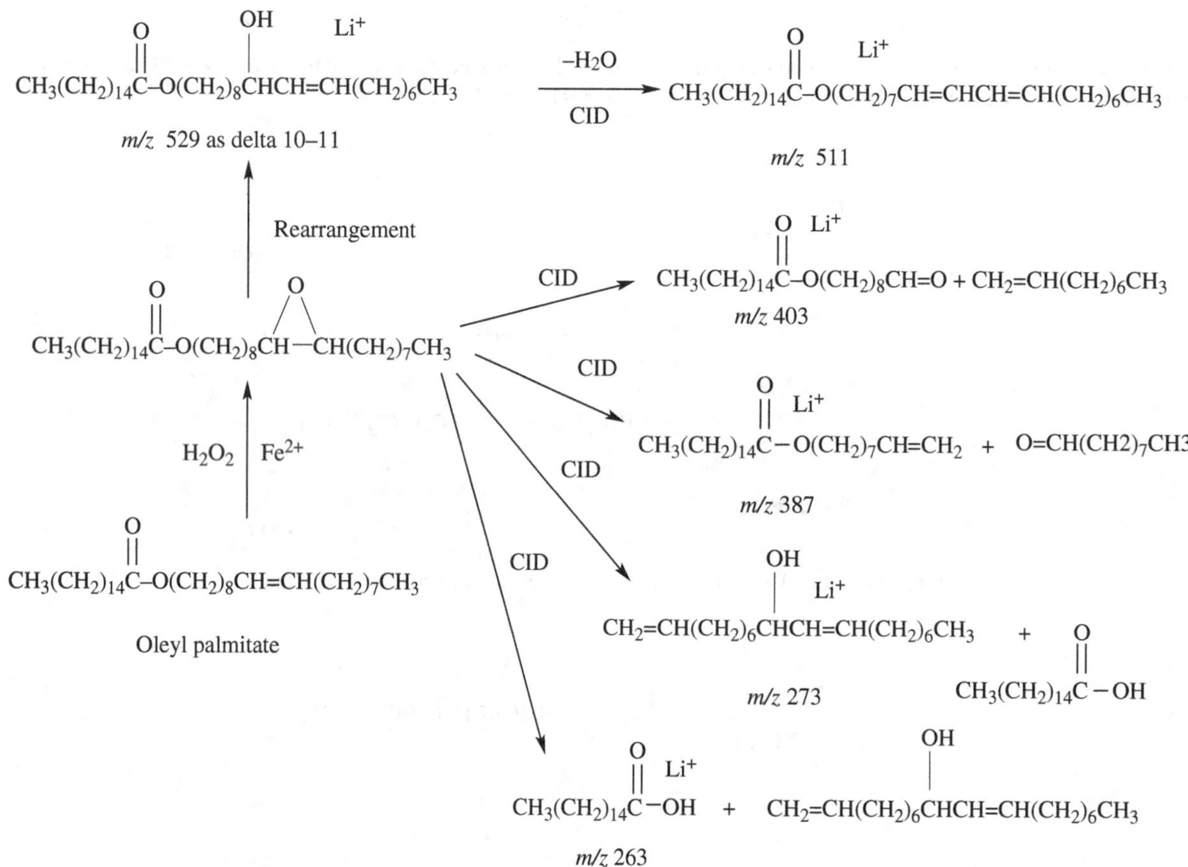

FIGURE 31.16 Major product ions produced upon collision-induced dissociation of the oxidized form of oleyl palmitate.

represents the oxidized form of oleyl palmitate as either an epoxide [22] or hydroxylated. Figure 31.16 illustrates the production of the epoxidated lipid species using the Fenton reaction and the major product ions produced from the epoxide state or rearrangement to the hydroxide followed by neutral loss of water forming the *m/z* 511 product ion. The location of the hydroxide can be either at positions 8, 9, 10, or 11 with the unsaturation either remaining at the Δ9-10 original position or either at the two adjacent positions (Δ8–9 or Δ10–11) [23].

The product ion spectrum of the oxidized wax ester at *m/z* 529.7 is illustrated in Figure 31.17. Important product ion peaks include the *m/z* 511.7 peak formed through loss of H_2O producing a diene of the wax ester $[M+Li-H_2O]^+$, the *m/z* 403.6 peak formed through neutral loss of C_nH_{2n} from the fatty alcohol alkyl chain producing an aldehyde at the C_9 position $[M+Li-C_9H_{18}]^+$ (diagnostic for location of double bond) [24,25], and the *m/z* 387.6 product ion formed through neutral loss of a nine carbon aldehyde from the fatty alcohol substituent $[M+Li-C_9H_{18}O]^+$

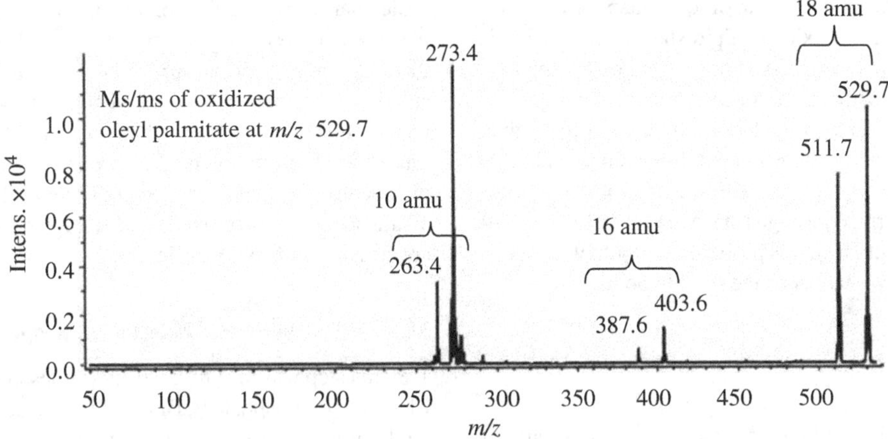

FIGURE 31.17 ESI-ion/trap product ion spectrum of the oxidized wax ester at *m/z* 529.7. Important product ion peaks include the *m/z* 511.7 peak formed through loss of H_2O, *m/z* 403.6 peak formed through neutral loss of C_nH_{2n} $[M+Li–C_9H_{18}]^+$ (diagnostic for location of double bond), *m/z* 387.6 product ion formed through neutral loss of a nine carbon aldehyde from the fatty alcohol substituent $[M+Li–C_9H_{18}O]^+$ (diagnostic for location of double bond), *m/z* 273.4 peak formed through neutral loss of a C16:0 fatty acid forming a lithiated C18:1 fatty alcohol $[M+Li–C_{16}H_{32}O_2]^+$, and *m/z* 263.4 product ion (10 amu difference from *m/z* 273.4) formed through neutral loss of the fatty alcohol alkyl chain $[M+Li–C_{18}H_{36}]^+$.

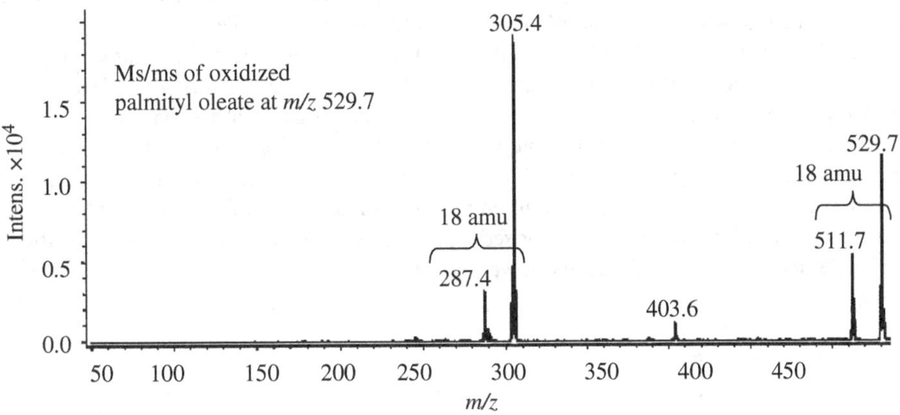

FIGURE 31.18 Product ion spectrum of the oxidized epoxide of palmityl oleate.

(diagnostic for location of double bond) [26,27]. A predominant pair of product ions in the middle of the spectrum are the *m/z* 273.4 peak formed through neutral loss of a C16:0 fatty acid forming a lithiated C18:1 fatty alcohol $[M+Li–C_{16}H_{32}O_2]^+$ and the *m/z* 263.4 product ion (10 amu difference from *m/z* 273.4) formed through neutral loss of the fatty alcohol alkyl chain $[M+Li-C_{18}H_{36}]^+$. The *m/z* 263.4 product ion is the predominant pathway observed for the unoxidized oleyl palmitate wax ester where a loss analogous to *m/z* 273.4 is not observed. The presence of the epoxide is thus apparently allowing these two pathways to take place with a much higher efficiency for the *m/z* 273.4 pathway as compared to the *m/z* 263.4 pathway. In comparison, Figure 31.18 displays the product ion spectrum of the oxidized epoxide of palmityl oleate where the unsaturation is now located on the fatty acid substituent of the wax ester as opposed to the fatty alcohol substituent as is the case with the oleyl palmitate wax ester of Figure 31.17. The same loss of H_2O at *m/z* 511.7 and the loss of C_9H_{18} at *m/z* 403.6 indicative of the location of the double bond are also observed as discussed previously for the oxidized epoxide of oleyl palmitate. The major product ions in the middle of the spectrum however differ from that of oleyl palmitate due to the location of the epoxide. The *m/z* 305.4 product ion is formed through neutral loss of the fatty alcohol chain which is also the major product ion formed with unoxidized wax esters. The product ion at *m/z* 287.4 is formed through neutral loss of the C14:0 fatty alcohol producing a C18 ketene with a difference of 18 amu from the *m/z* 305.4 product ion. These types of losses can be used in diagnostic identification of wax ester unknowns whether saturated or unsaturated and oxidized or unoxidized. Low levels of the peroxides and diols of the wax esters were observed, and the product ion spectra were similar to the epoxides but slightly more complex. Loss of H_2O is observed along with double-bond location, and the major substituent losses differing by either 10 amu for oleyl palmitate or 18 amu for palmityl oleate were also contained in the product ion spectra. A number of biological sample extracted lipids observed in the electrospray ionization mass spectra produce product ion spectra that indicated that they are oxidized lipid species by a characteristic neutral loss of either 16 or 18 amu from the precursor ion. Water loss from the precursor often indicates an oxidized lipid species [22]. Often in lipid work, researchers have observed the 16 and/or 18 amu neutral loss associated with oxidation. As an example, in the study of Reis et al. [19] of linoleic acid oxidation, the inspection of their product ion spectra all shows spectra containing a neutral loss of 18 amu associated with water loss. Also included in the product ion spectra is a peak derived through a neutral loss of 16 amu. The loss of atomic oxygen at 16 amu is probably attributable as a loss from an epoxide that reestablishes the double bond.

31.5 STEROLS

In metabolomics studies, often an unknown biomolecule is observed in a system under study by mass spectrometry. The structural elucidation and subsequent identification of an unknown small biomolecule is a process that often involves a number of steps. Figure 31.19 shows a single-stage ESI-Q-TOF mass spectrum of a biological extract containing unknown biomolecules in a solution of 1:1 chloroform/methanol with 10 mM LiCl. The spectrum contains primarily lithium adducts of the lipid species in the biological extract as $[M+Li]^+$. The predominant biomolecule observed in the biological extract at *m/z* 473.5 (also observed at *m/z* 489 as the sodiated species) will be the subject of study concerning its identification in the following example of identifying an unknown biomolecule.

Figure 31.20 indicates an ESI-Q-TOF single-stage negative ion mode mass spectrum of the same biological extract as illustrated in Figure 31.19 in a 1:1 chloroform/methanol solution containing 1 mM ammonium acetate. The spectrum is comprised of deprotonated biomolecule species as $[M-H]^-$. In the lower molecular weight region of the spectrum, there are four peaks observed representing the free fatty acids myristic acid at *m/z* 227.3, palmitic acid at *m/z* 255.4, oleic acid at *m/z* 281.4, and stearic acid at *m/z* 283.4. The peak at *m/z* 465.5 is suspected to be the same biomolecule as that of *m/z* 473.5 observed in positive ion mode. The unknown species was suspected to be cholesteryl phosphate. To verify this, the cholesteryl phosphate species was synthesized, characterized, and compared to the unknown for confirmation of its identification.

31.5.1 Synthesis of Cholesteryl Phosphate

The first step was to synthesize the cholesteryl phosphate species. Cholesteryl phosphate (CP) was synthesized following the methodology outlined by Sedaghat et al. [28] and Gotoh et al. [29] where the major modification consisted of changing the methodology from a gram scale to a milligram scale. Briefly, 0.67 g of cholesterol was dissolved in 3.5 ml of dry pyridine. The solution was then cooled in ice water, and phosphorus oxychloride (0.175 ml dissolved in 3.5 ml acetone) was added slowly with stirring. A precipitate was formed immediately, and the solution was allowed to cool for 10 min. in the ice water. The precipitate (cholesteryl phosphorochloridate) was filtered and washed with 15 ml cold dry acetone. The solid was dissolved in tetrahydrofuran (THF) and refluxed for 2 h after adding aqueous NaOH (2.1 equivalent). The resulting cholesteryl phosphate precipitate was filtered and washed with dry cold acetone and dried in a Vacufuge™ (Eppendorf). High-resolution mass measurements (including both the sodium adduct and the lithium adduct) of the synthesized cholesteryl phosphate and H-NMR results are presented for verification of its identification in the next two sections.

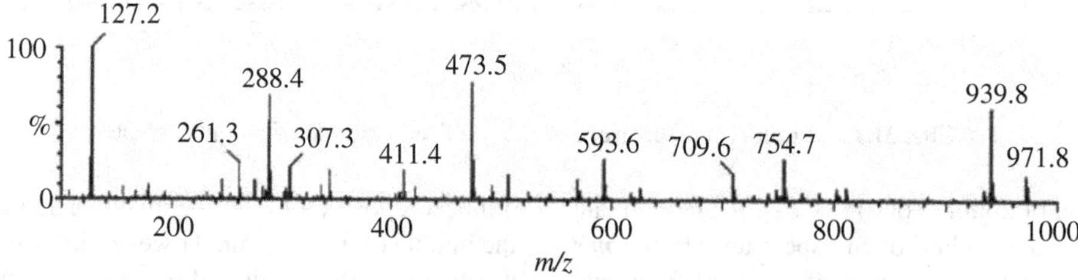

FIGURE 31.19 Single-stage ESI-Q-TOF mass spectrum of a biological extract in a solution of 1:1 chloroform/methanol with 10 mM LiCl containing primarily lithium adducts as $[M+Li]^+$. *M/z* 473.5 (also observed at *m/z* 489 as the sodiated species) is the subject of study concerning its identification. (Reprinted with permission from Ref. [19]. Copyright Springer 2003.)

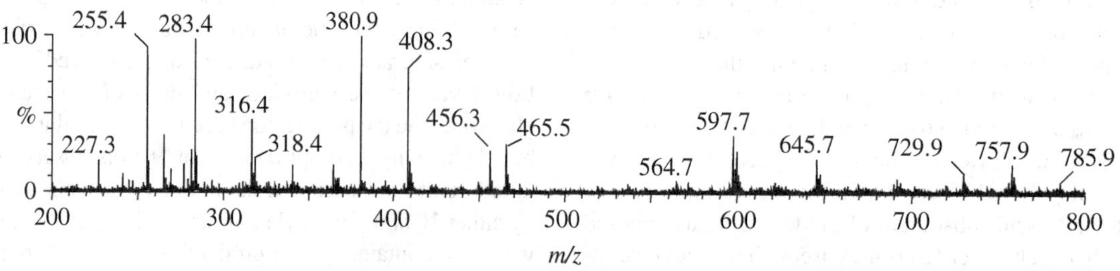

FIGURE 31.20 ESI-Q-TOF single-stage mass spectrum of the biological extract collected in negative ion mode in a 1:1 chloroform/methanol solution containing 1 mM ammonium acetate. Spectrum is comprised of deprotonated lipid species as $[M-H]^-$ illustrating the free fatty acids myristic acid at *m/z* 227.3, palmitic acid at *m/z* 255.4, oleic acid at *m/z* 281.4, stearic acid at *m/z* 283.4, and an unknown species at *m/z* 465.5. (Reprinted with permission from Ref. [19]. Copyright Springer 2003.)

31.5.2 Single-Stage and High-Resolution Mass Spectrometry

The synthesized cholesteryl phosphate (CP) was next characterized using high-resolution mass spectrometry. Figure 31.21a depicts a single-stage mass spectrum in positive ion mode acquired on the Q-TOF mass spectrometer of the synthesized cholesteryl phosphate in an acidic solution containing NaI (1% acetic acid in 1:1 $CHCl_3$/MeOH). The two major peaks in the spectrum are at *m/z* 489.4 for cholesteryl phosphate as the sodium adduct and *m/z* 955.8 for a cholesteryl phosphate dimer. Figure 31.21b shows a single-stage mass spectrum in negative ion mode acquired on the Q-TOF mass spectrometer of the synthesized cholesteryl phosphate in a 10 mM ammonium hydroxide 1:1 $CHCl_3$/MeOH solution. The major peak in the spectrum is at *m/z* 465.3 for the deprotonated form of cholesteryl phosphate as $[C_{27}H_{47}O_4P{-}H]^-$. Figure 31.21c displays the high-resolution mass measurement of the synthesized CP as the lithium adduct $[C_{27}H_{47}O_4PLi]^+$ at *m/z* 473.3362 (*m/z* 472.6725 is the internal standard NaI peak used for the exact mass measurement). The theoretical mass of CP as the lithium adduct is *m/z* 473.3372 equating to a calculated −2.1 ppm error. Figure 31.21d illustrates the high-resolution mass measurement of the synthesized CP as the sodium adduct $[C_{27}H_{47}O_4PNa]^+$ at *m/z* 489.3103 (*m/z* 472.6725 is the internal standard NaI peak used for the exact mass measurement). The theoretical mass of CP as the sodium adduct is *m/z* 489.3110 equating to a calculated −1.4 ppm error.

31.5.3 Proton Nuclear Magnetic Resonance (^{1}H-NMR)

The next step was to characterize the synthesized cholesteryl phosphate using nuclear magnetic resonance (NMR) to confirm its synthesis. Single-stage proton (^{1}H-NMR) spectra were obtained on a Bruker DRX 800 (800 MHz) nuclear magnetic resonance spectrometer (Bruker, Bremen, Germany). The spectra of the synthesized cholesteryl phosphate were obtained at 800 MHz (shifts in ppm) with deuterated chloroform ($CDCl_3$) used as solvent. Figure 31.22 contains the structure of cholesteryl phosphate with the carbon atoms numbered and the experimental proton NMR results of the synthesized cholesteryl phosphate.

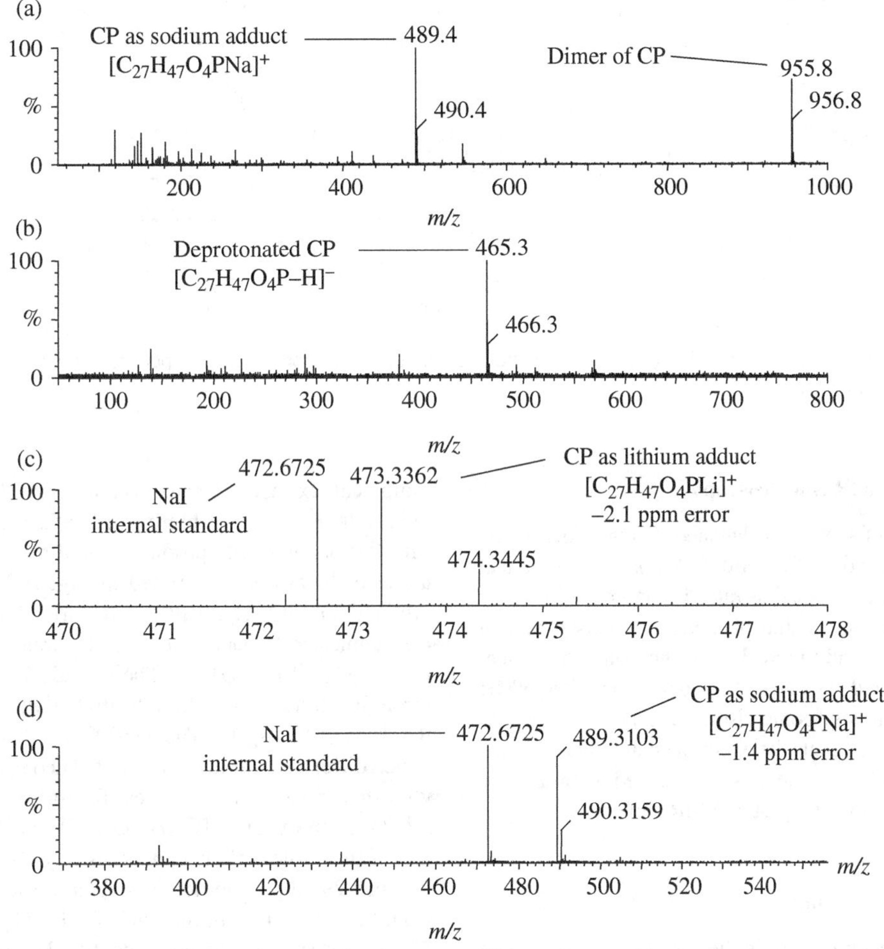

FIGURE 31.21 (a) Single-stage mass spectrum in positive ion mode of the synthesized cholesteryl phosphate in an acidic solution containing NaI (1% acetic acid in 1:1 $CHCl_3$/MeOH). (b) Single-stage mass spectrum in negative ion mode of the synthesized cholesteryl phosphate in a 10 mM ammonium hydroxide 1:1 $CHCl_3$/MeOH solution. (c) High-resolution mass measurement of the synthesized CP as the lithium adduct $[C_{27}H_{47}O_4PLi]^+$ at *m/z* 473.3362 (calculated −2.1 ppm error). (d) High-resolution mass measurement of the synthesized CP as the sodium adduct $[C_{27}H_{47}O_4PNa]^+$ at *m/z* 489.3103 (calculated −1.4 ppm error).

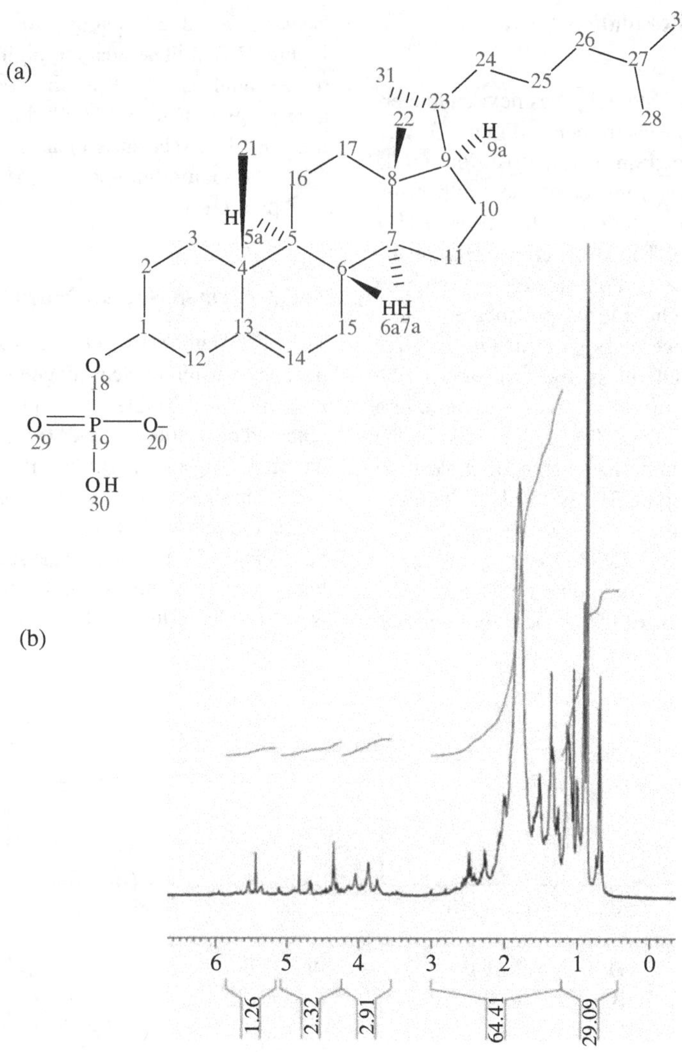

FIGURE 31.22 (a) Structure of cholesteryl phosphate with the carbon atoms numbered. (b) Experimental proton NMR results of the synthesized cholesteryl phosphate.

31.5.4 Theoretical NMR Spectroscopy

The NMR proton shifts were calculated using ChemDraw (Cambridge Soft Corporation, Cambridge, MA). The theoretical proton shifts for the structure are illustrated in Figure 31.23 where Figure 31.23a represents the structure of cholesteryl phosphate with the proton shifts labeled, and Figure 31.23b shows the theoretical NMR spectrum generated by ChemDraw using the ChemNMR feature. Table 31.1 illustrates a comparison of the theoretically generated proton NMR shifts versus the experimentally obtained proton NMR shifts. Good agreement was observed between the theoretical and the experimental proton NMR shifts.

31.5.5 Structure Elucidation

Product ion mass spectra were collected of the unknown biomolecule to compare to the product ion spectra of the synthesized cholesteryl phosphate. Figure 31.24 illustrates a product ion spectrum of the *m/z* 473.5 biomolecule, as $[M+Li]^+$, in the biological extract collected on a Q-TOF mass spectrometer in a solution of 1:1 chloroform/methanol with 10 mM LiCl. The major products produced and the associated fragmentation pathways are illustrated in Figure 31.25. The *m/z* 255.3 product ion, $[C_{11}H_{21}O_4PLi]^+$, is formed through cleavage of the cholesteryl backbone as illustrated in Figure 31.25, $[C_{27}H_{47}O_4P{+}Li{-}C_{16}H_{26}]^+$. The *m/z* 237.2 product ion is derived through neutral water loss from the *m/z* 255.3 product ion, $[C_{11}H_{21}O_4PLi{-}H_2O]^+$. The *m/z* 293.3 and *m/z* 311.3 product ions which differ by 18 amu (H_2O) are formed in an analogous fashion with their respective structures illustrated in Figure 31.25. The *m/z* 311.3 product ion, $[C_{15}H_{29}O_4PLi]^+$, is formed through cleavage of the cholesteryl backbone as $[C_{27}H_{47}O_4P + Li{-}C_{12}H_{18}]^+$. The *m/z* 293.3 product ion is derived through neutral water loss from the *m/z* 311.3 product ion, $[C_{15}H_{29}O_4PLi{-}H_2O]^+$. The *m/z* 473.4 lipid species has been identified as the lithium adduct of cholesteryl phosphate with the ionic formula of $[C_{27}H_{47}O_4P{+}Li]^+$ and a theoretical mass of 473.3372 Da. High-resolution mass measurements of the *m/z* 473.4 lipid species in meibum

(a)

(b)

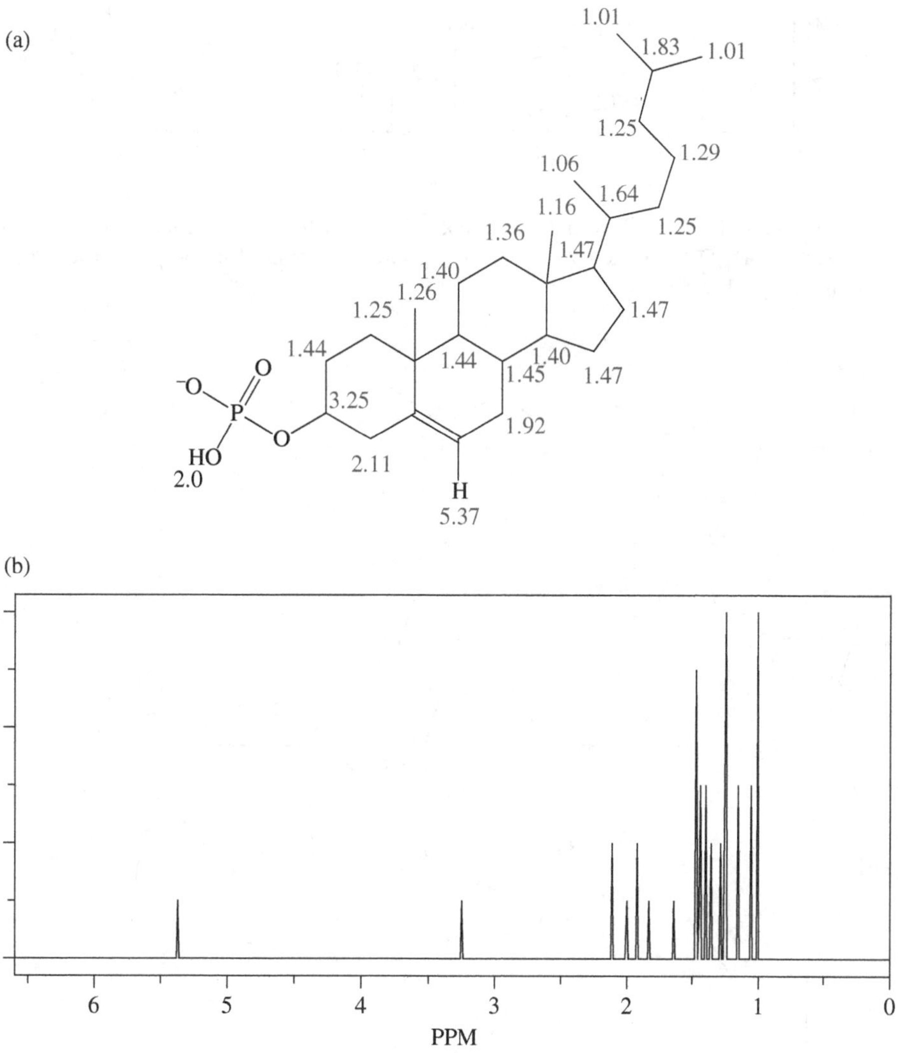

FIGURE 31.23 (a) Structure of cholesteryl phosphate with the proton shifts labeled. (b) Theoretical NMR spectrum of cholesteryl phosphate generated by ChemDraw using the ChemNMR feature.

TABLE 31.1 Proton NMR Spectral Results.

Node	Theoretical Shift (ppm)	Experimental Shift (ppm)
CH3(18)		0.68–0.73 (0.69) [28]
CH3(26), CH3(27)		0.88–0.89 (0.86–0.87) [28]
CH3(21)		0.92 (0.91) [28]
CH3	1.01	1.01 (1.0) [28]
CH3	1.01	1.02
CH3	1.06	1.08
CH3	1.16	1.15–1.18
CH2	1.25	
CH2	1.25	
CH2	1.25	
CH3	1.26	1.275
CH2	1.29	1.29
CH2	1.36	1.36
CH	1.40	1.40

TABLE 31.1 (*Continued*)

Node	Theoretical Shift (ppm)	Experimental Shift (ppm)
CH2	1.40	1.40
CH2	1.44	
CH	1.44	
CH	1.45	
CH2	1.47	1.46
CH2	1.47	1.47
CH	1.47	1.48
CH	1.64	1.64
CH	1.83	1.83
CH2	1.92	1.90
OH	2.0	2.01
CH2	2.11	2.11
CH	3.25	
C=C–H	5.37	5.46 (5.3) [28]

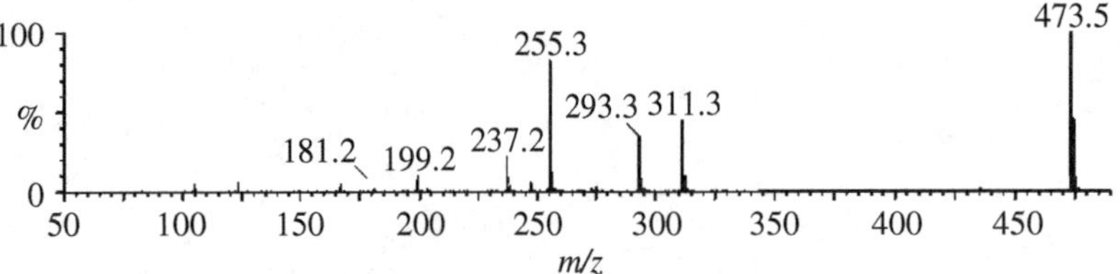

FIGURE 31.24 Product ion spectrum of *m/z* 473.7 identified as cholesteryl phosphate. Product ion spectrum of synthesized cholesteryl phosphate as the lithium adduct $[C_{27}H_{47}O_4P{+}Li]^+$ at *m/z* 473.4. The spectrum is identical to the product ion spectrum of the unknown *m/z* 473.5 species observed in biological extract.

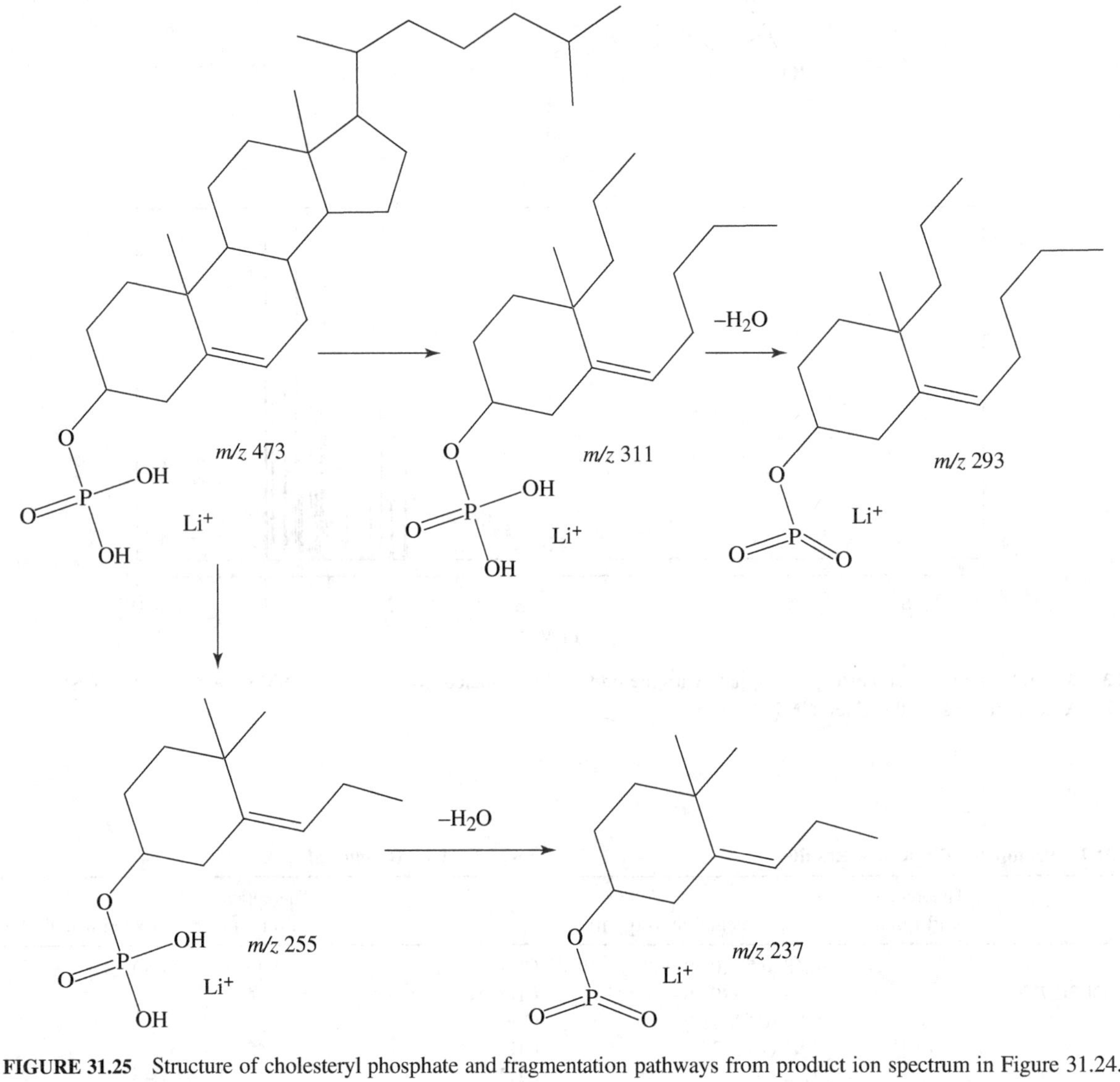

FIGURE 31.25 Structure of cholesteryl phosphate and fragmentation pathways from product ion spectrum in Figure 31.24.

was determined at *m/z* 473.3361 equating to a mass error of −2.3 ppm. The *m/z* 489.3 lipid species has been identified as the sodium adduct of cholesteryl phosphate with the ionic formula of $[C_{27}H_{47}O_4P{+}Na]^+$ and a theoretical mass of 489.3110 Da. High-resolution mass measurements of the *m/z* 489.3 lipid species in meibum was determined at *m/z* 489.3099 equating to a mass error of −2.2 ppm.

Figure 31.26 illustrates the product ion spectrum of the *m/z* 465.5 species observed in the biological extract obtained using the Q-TOF mass spectrometer with an electrospray ionization source collected in negative ion mode. In Figure 31.26, there are two major product ion peaks observed at *m/z* 127.0 and 97.0. The *m/z* 465.5 lipid species has been identified as the deprotonated form of cholesteryl phosphate with the ionic formula $[M{-}H]^-$ of

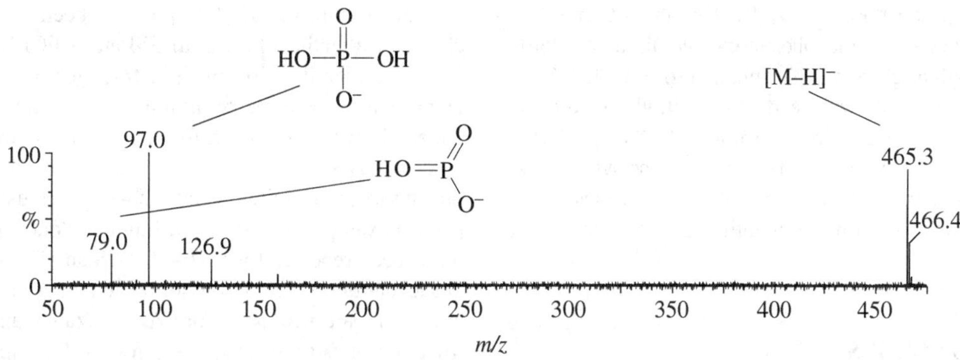

FIGURE 31.26 Q-TOF MS negative ion mode product ion spectrum of the *m/z* 465.5 species in the biological extract identified as the deprotonated form of cholesteryl phosphate with the ionic, deprotonated formula [M–H]⁻ of $[C_{27}H_{47}O_4P–H]^-$.

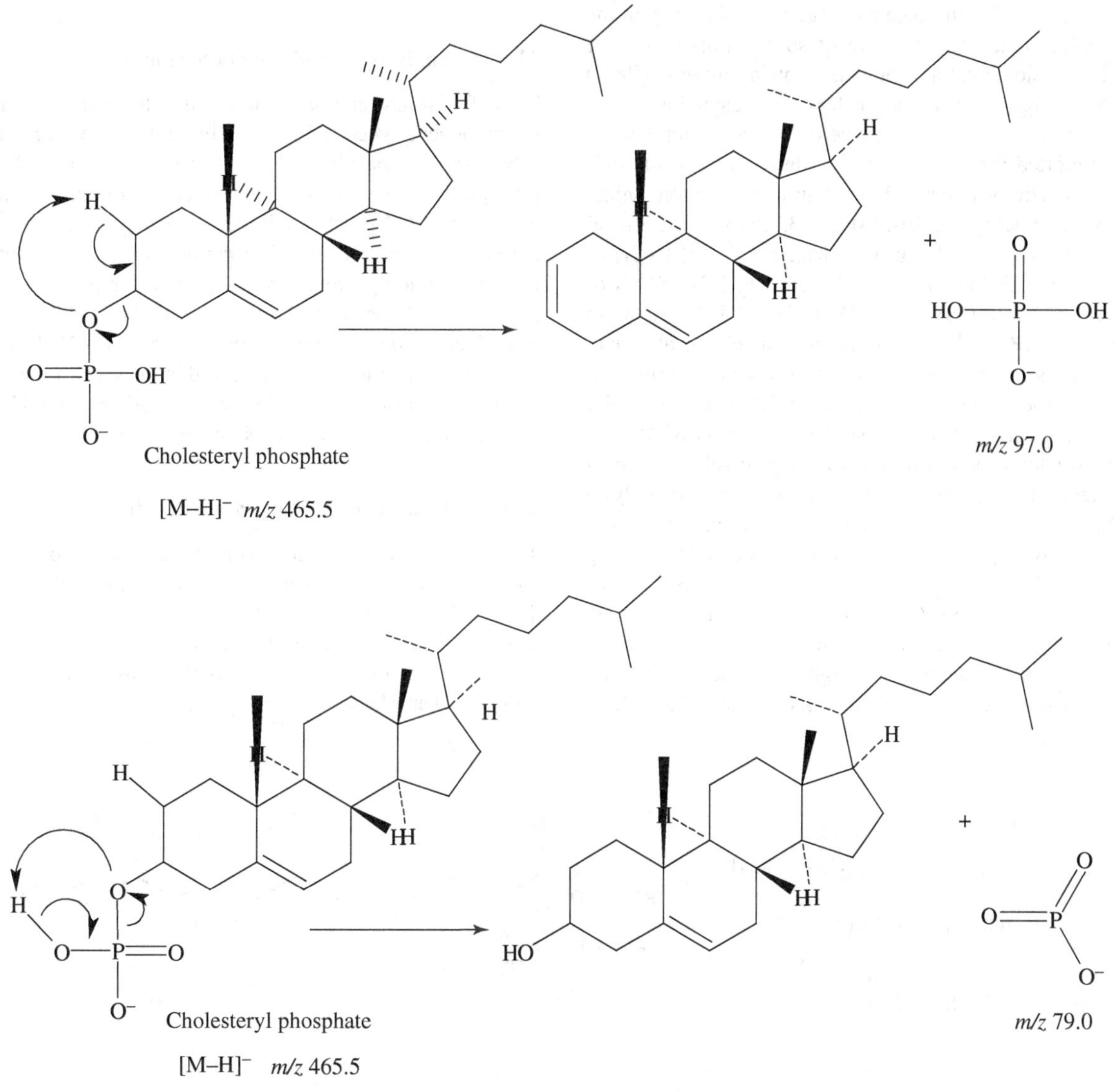

Cholesteryl phosphate

[M–H]⁻ *m/z* 465.5

m/z 97.0

Cholesteryl phosphate

[M–H]⁻ *m/z* 465.5

m/z 79.0

FIGURE 31.27 Structure of the *m/z* 465.5 lipid species and fragmentation pathways.

$[C_{27}H_{47}O_4P–H]^-$ and a theoretical mass of 465.3134 Da. High-resolution mass measurements of the *m/z* 465.5 lipid species in the biological extract was determined at *m/z* 465.3140 equating to a mass error of 1.3 ppm. The structure of the *m/z* 465.5 lipid species is illustrated in Figure 31.27 along with proposed fragmentation pathways that describe the production of the *m/z* 97.0 and *m/z* 79.0 product ions. Both the *m/z* 97.0 and 79.0 product ions represent the H_2PO_4 phosphate portion of the head group. This same

H_2PO_4 phosphate product ion at *m/z* 97.0 is also observed in the product ion spectra of the phosphorylated lipid standards 1-palmitoyl-2-oleoyl-*sn*-glycero-3-phosphate (16:0–18:1 PA or POPA, spectrum not shown) and 1-palmitoyl-2-oleoyl-*sn*-glycero-3-[phospho-*rac*-(1-glycerol)] (16:0–18:1 PG or POPG, spectrum not shown). The product ion spectrum of the synthesized cholesteryl phosphate is identical to the *m/z* 465.5 species in the biological extract and confirms its identification as cholesteryl phosphate.

31.6 ACYLGLYCEROLS

Lipids are very important biomolecules that are found in all living species. These include nonpolar lipids such as the acylglycerols and the more polar phosphorylated lipids. Of the nonpolar lipids, triacylglycerols are thought to be a storage form of energy in the cells [30], while diacylglycerols are of special importance for their role as physiological activators of protein kinase C (PKC) [31,32]. The acylglycerols are a subclass of lipids and are comprised of mono-, di-, and triacylglycerols. Various approaches have been reported for the mass spectrometric analysis of acylglycerols such as derivative gas chromatography–electron ionization mass spectrometry (GC/EI-MS) [33–35], positive chemical ionization [33], negative chemical ionization [36–38] (NCI) of chloride adducts, ES-MS/MS [39,12] including ES-MS/MS of ammonium adducts [40,41], MALDI-TOF [42,43], fast atom bombardment (FAB) [44], atmospheric pressure chemical ionization (APCI) mass spectrometry [40,45], and APCI liquid chromatography/mass spectrometry [46,26,27] (LC/MS). The nonpolar lipids in general have also been studied by ESI-MS/MS and include the structural characterization of acylglycerols [41], direct qualitative analysis [47], regiospecific determination of triacylglycerols [48], analysis of triacylglycerol oxidation products [49], quantitative analysis [12], and acylglycerol mixtures [50]. Using ratios of diacylglycerol fragment ions versus precursor ions for quantifying triacylglycerols and fragmentation schemes has been reported for single-stage APCI/MS experiments [51–54]. Even in acidified solutions, saturated acylglycerols do not readily form protonated molecules; therefore, alkali metal salt (Na^+) and ammonium acetate (NH_4^+) have been used as additives in chloroform/methanol for both ESI and APCI LC/MS.

In the identification of lipid fractions in biological samples using mass spectrometry, lithium (Li^+) is sometimes chosen as the alkali metal for adduct formation for wax esters, for the nonpolar lipids such as the acylglycerols, and for the more polar phosphorylated lipids, which have previously been reported [5]. As examples of the use of lithium adducts for lipid analysis, it has been reported for FAB-MS/MS and ES-MS studies of oligosaccharides [55,56]; for ionization of aryl 1,2-diols by fast atom bombardment [44]; for FAB ionization and structural identification of fatty acids [4]; and for the ES ionization and structural determination of triacylglycerols [12,6]. In this section, we will take a detailed look at the major fragments produced from ES-MS/MS studies employing CID of lithium adducts of monoacylglycerols, diacylglycerols, and triacylglycerols.

31.6.1 Analysis of Monopentadecanoin

We will first have a look at the simplest form of an acylglycerol, the monoacylglycerol. This lipid biomolecule has one fatty acyl substituent on the glycerol backbone. Figure 31.28 depicts the CID product ion spectrum of lithiated monopentadecanoin at *m/z* 323 as $[M{+}Li]^+$. At the lower molecular weight region, there are observed three peaks characteristic of the fragmentation of lithiated monoglycerols. The *m/z* 99 major product ion is the glycerol backbone derived from the neutral loss of the C15:0 fatty acyl chain as ketene from the parent ion $[MLi{-}C_{15}H_{28}O]^+$. The next two product ion peaks observed in the spectrum, *m/z* 81 and *m/z* 63, are the subsequent loss of one and then two H_2O molecules from the glycerol backbone (Figure 31.29).

31.6.2 Analysis of 1,3-Dipentadecanoin

The diacylglycerol 1,3-dipentadecanoin is a disubstituted glycerol where the glycerol backbone contains two fatty acyl chains. The product ion spectrum of 1,3-dipentadecanoin is illustrated in Figure 31.30. Similar to the monoacylglycerol presented earlier, the 1,3-diacylglycerol also contains hydrocarbon fragment ions, two water losses from the glycerol backbone at *m/z* 63, and

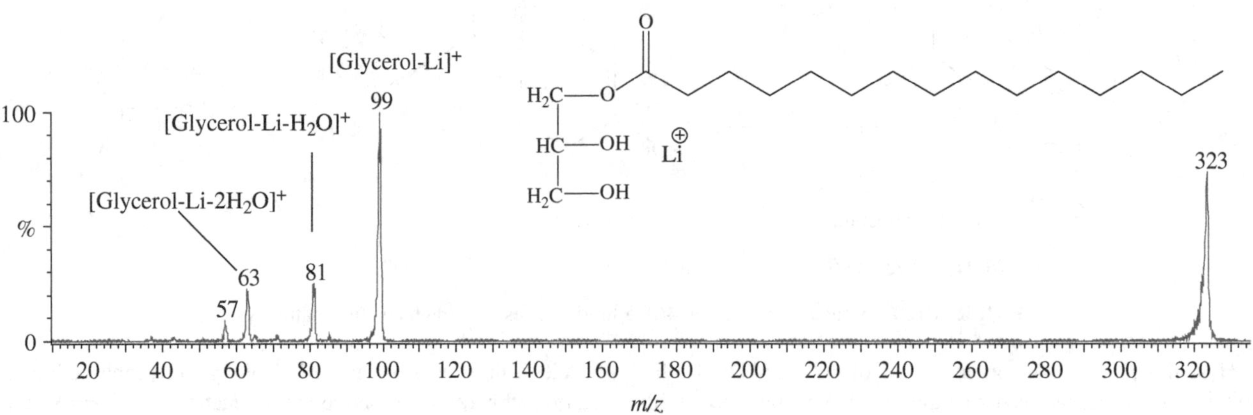

FIGURE 31.28 Positive ion mode ESI-MS/MS product ion spectra of the lithium adducts of monopentadecanoin at *m/z* 323 as $[MLi]^+$.

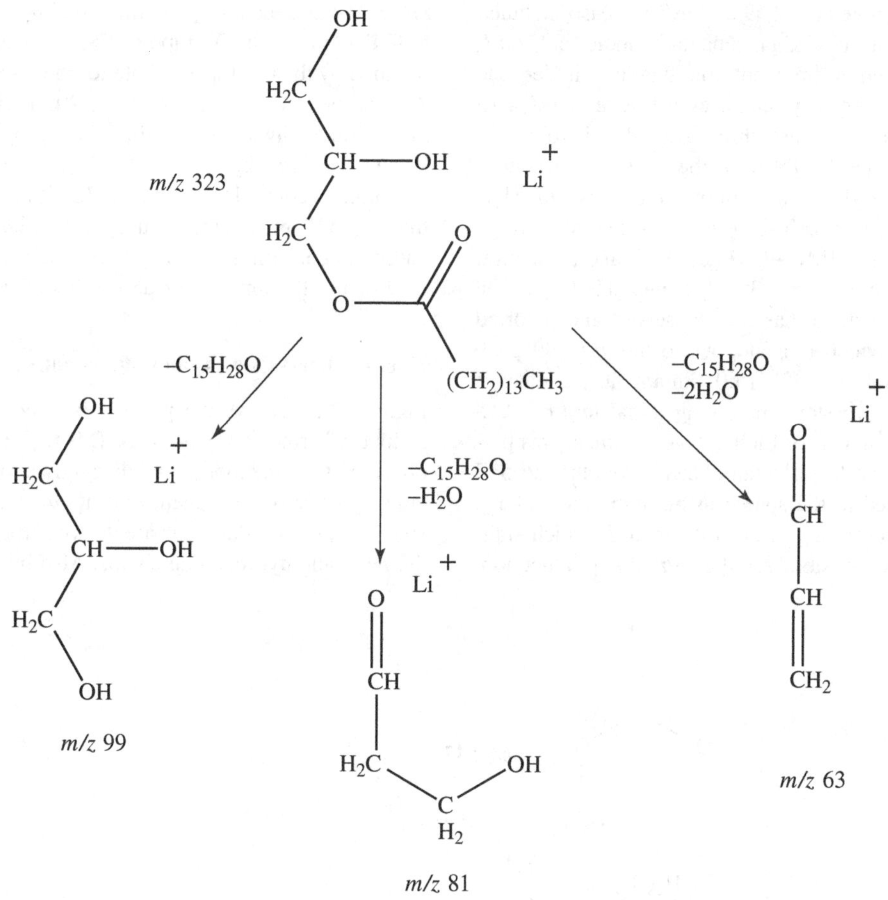

m/z 323

$-C_{15}H_{28}O$

$-C_{15}H_{28}O$
$-H_2O$

$-C_{15}H_{28}O$
$-2H_2O$

m/z 99

m/z 81

m/z 63

FIGURE 31.29 Major fragment ions produced from the collision-induced dissociation of monopentadecanoin *m/z* 323 $[MLi]^+$. Ion at *m/z* 99 is the glycerol backbone derived from the neutral loss of the C15:0 fatty acyl chain as a ketene from the precursor ion $[MLi–C_{15}H_{28}O]^+$. *M/z* 81 and *m/z* 63 are the subsequent loss of one and then two H_2O molecules from the lithiated glycerol backbone.

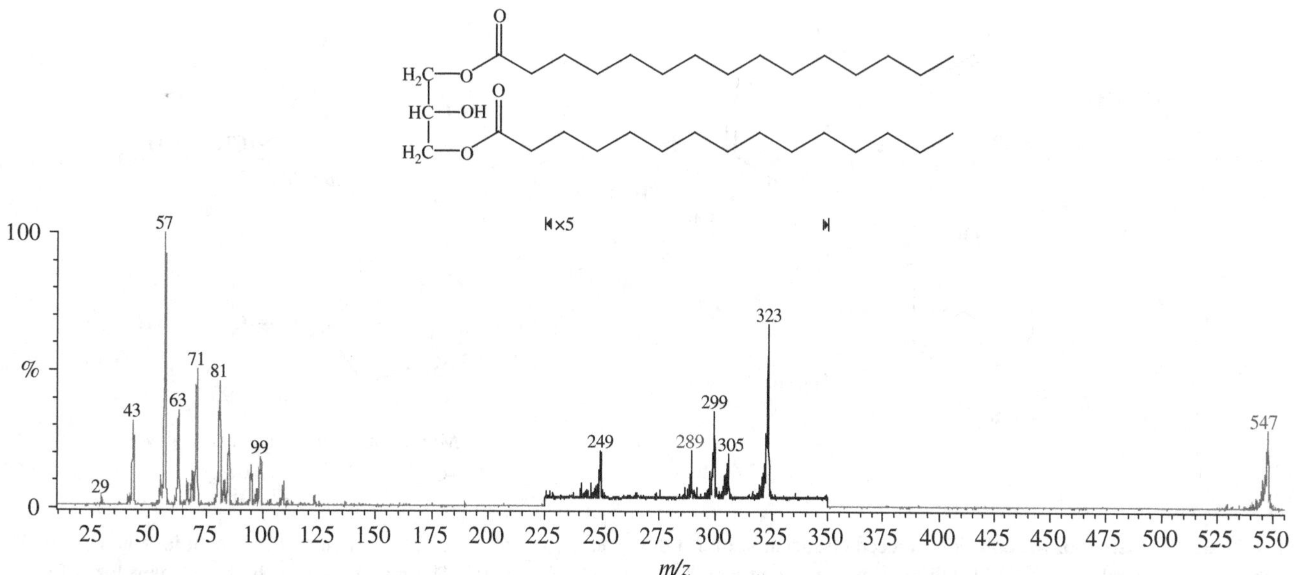

FIGURE 31.30 Positive ion mode ESI-MS/MS product ion spectra of the lithium adducts of 1,3-dipentadecanoin *m/z* 547.

also the glycerol backbone at *m/z* 99 and *m/z* 81. Also included in the product ion spectrum is lithiated pentadecanoic acid at *m/z* 249 $[C_{15}H_{30}O_2Li]^+$ and a fragment ion that has undergone neutral loss of C15:0 fatty acyl chain as a ketene at *m/z* 323 $[MLi–C_{15}H_{28}O]^+$. There are also three loses that involve the fatty acyl substituent at *m/z* 289 from the consecutive neutral loss of C15:0 fatty acyl ketene, followed by loss of H_2O $[MLi–C_{15}H_{28}O–H_2O]^+$; at *m/z* 299 for the neutral loss of C15:0 lithium fatty acetate $[MLi–C_{15}H_{29}O_2Li]^+$; and the neutral loss of C15:0 fatty acid at *m/z* 305 $[MLi–C_{15}H_{30}O_2]^+$. The loss of NH_4OH, then ketene, has been observed and reported by LC-APCI/MS[33], which is analogous to the *m/z* 299 peak formed through neutral loss of C15:0 lithium fatty acetate if the process occurs in two steps, involving initial loss of C:15 fatty ketene, followed by loss of LiOH. An experiment was performed where the loss of the C:15 fatty ketene from the *m/z* 547 precursor was produced in the source using high cone voltage (i.e., "in-source" or "upfront" CID) to distinguish which fragment pathway was responsible for the *m/z* 299 product ion. *m/z* 323 was isolated by the first quadrupole and then subjected to CID in the central hexapole. The product ions generated were scanned by the third quadrupole for the loss of LiOH at *m/z* 299. A small peak was observed at *m/z* 299 indicating that this fragmentation pathway is possible. The proposed loss of a fatty ketene followed by loss of LiOH is much less favored than the direct neutral loss of the C:15 lithium fatty acetate from the *m/z* 547 precursor. Figure 31.31 illustrates fragmentation pathways concerning the major and minor losses associated with CID of the lithium adduct of 1,3-dipentadecanoin.

31.6.3 Analysis of Triheptadecanoin

Figure 31.32 shows the product ion spectrum of the lithiated adduct of triheptadecanoin as $[M+Li]^+$ at *m/z* 855. As was observed with the mono- and diacylglycerols, the spectrum contains hydrocarbon fragment ions at *m/z* 43, *m/z* 57, *m/z* 71, and *m/z* 85. These product ions are derived from σ-bond cleavage of the fatty acid hydrocarbon chains. Also included are the product

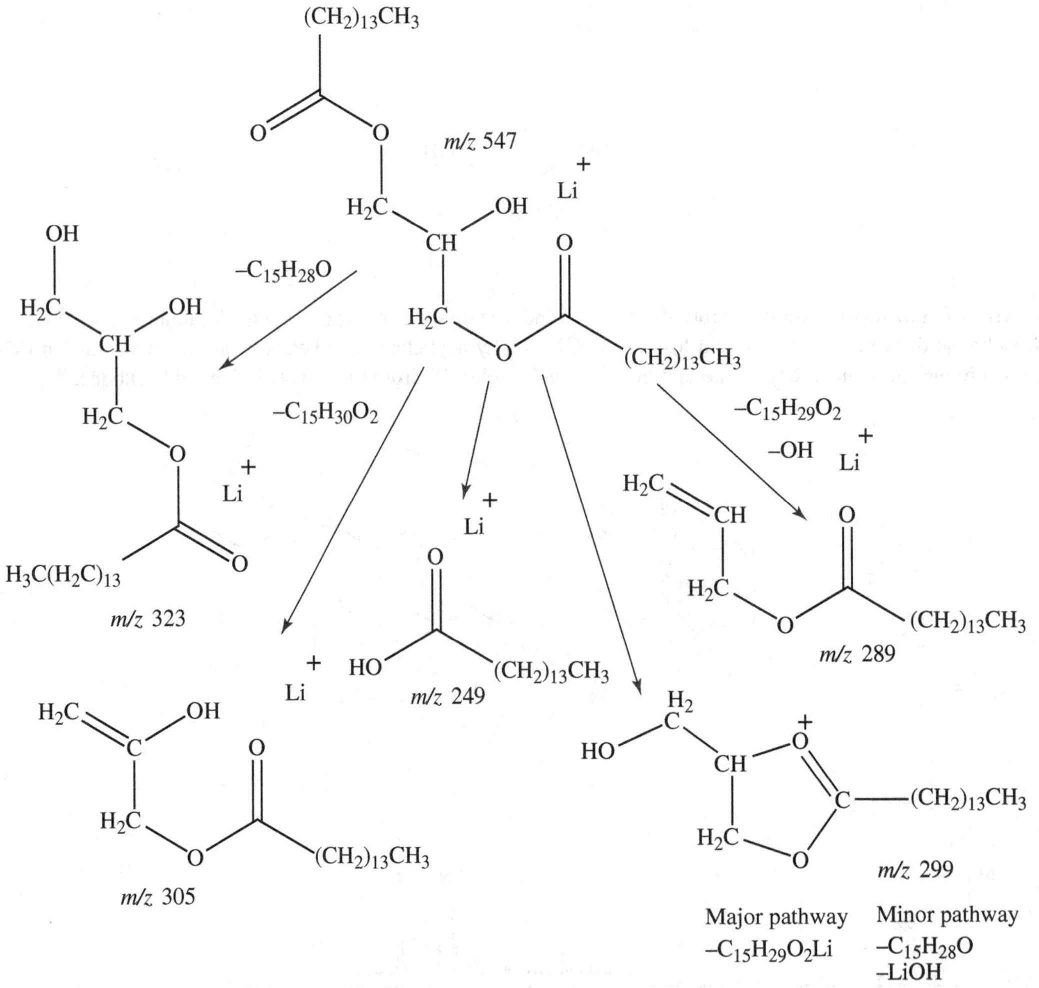

FIGURE 31.31 Listing of the collision-induced dissociation of 1,3-dipentadecanoin at *m/z* 547 $[MLi]^+$ major fragment ions generated. The major path of neutral loss of C15:0 lithium fatty acetate at *m/z* 299 $[MLi–C_{15}H_{29}O_2Li]^+$. The minor path of the subsequent loss of C15:0 fatty acyl chain as ketene from the precursor ion followed by LiOH at *m/z* 299 $[MLi–C_{15}H_{28}O–LiOH]^+$. The precursor ion peak minus the neutral loss of C15:0 fatty acyl chain as ketene at *m/z* 323 $[MLi–C_{15}H_{28}O]^+$. The neutral loss of C15:0 fatty acid at *m/z* 305 $[MLi –C_{15}H_{30}O_2]^+$, the consecutive loss of C15 fatty acyl ketene, followed by loss of H_2O at *m/z* 289 $[MLi–C_{15}H_{28}O–H_2O]^+$ and pentadecanoic acid at *m/z* 249 $[C_{15}H_{30}O_2Li]^+$.

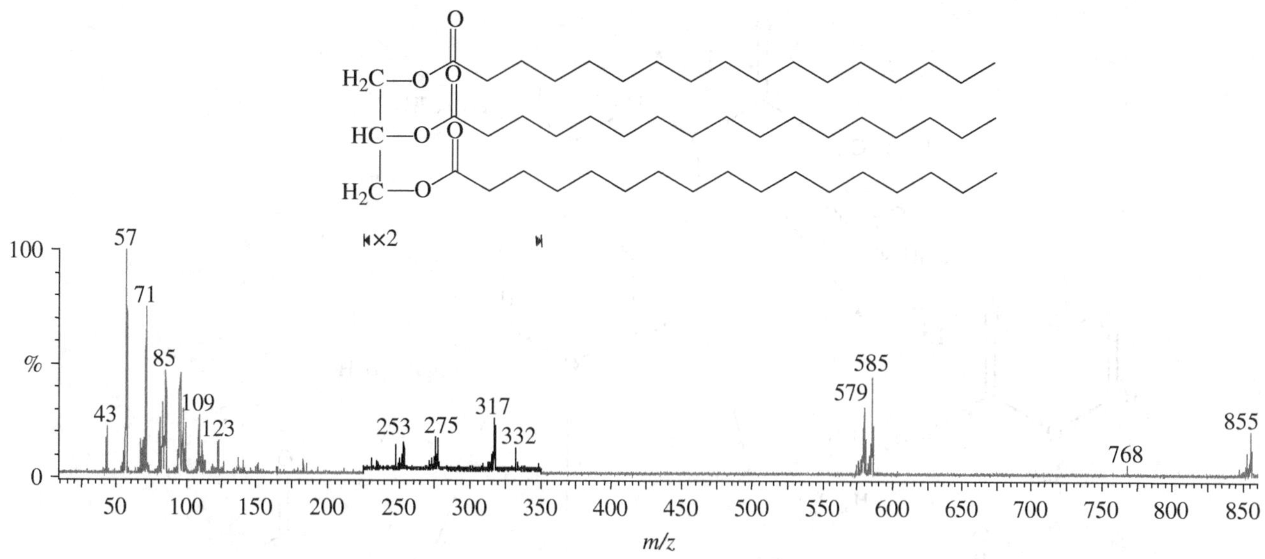

FIGURE 31.32 Positive ion mode ESI–MS/MS product ion spectra of the lithium adducts as $[M+Li]^+$ of triheptadecanoin at *m/z* 855.

ions at *m/z* 99 for the lithium adduct of the glycerol backbone and at *m/z* 81 derived from water loss from this glycerol backbone. The product ion spectrum includes peaks that are indicative of the triacylglycerol at *m/z* 253 for the C17:0 acylium ion $[C_{17}H_{33}O]^+$; at *m/z* 277 for the lithium adduct of heptadecanoic acid $[C_{17}H_{34}O_2Li]^+$, the *m/z* 317 product ion from the neutral loss of heptadecanoic acid, and then the neutral loss of a C17:1 alpha-beta unsaturated fatty acid, to give $[MLi–C_{17}H_{34}O_2–C_{17}H_{32}O_2]^+$; at *m/z* 579 for the neutral loss of lithiated heptadecanoate $[MLi–C_{17}H_{33}O_2Li]^+$; and the precursor ion minus the neutral loss of heptadecanoic acid at *m/z* 585 $[MLi–C_{17}H_{34}O_2]^+$. The product ions of *m/z* 317 and *m/z* 585, which both involve the neutral loss of a fatty acid, have not been observed in ESI tandem mass spectrometry studies with ammonium adducts [43], but the *m/z* 579 product ion from the neutral loss of a catonized fatty acetate. Figure 31.33 illustrates fragmentation pathways concerning the major and minor losses associated with CID of the lithium adduct of triheptadecanoin.

31.7 ESI-MASS SPECTROMETRY OF PHOSPHORYLATED LIPIDS

Phosphorylated lipids are also very important biomolecules that are found in all living species. Of the phosphorylated lipids that are polar lipids, those containing phosphoryl head groups, such as the phosphatidylcholines (PC), the phosphatidylethanolamines (PE), and the phosphatidylserines (PS) (see Fig. 31.34 for structures), constitute the bilayer components of biological membranes, determine the physical properties of these membranes, and directly participate in membrane protein regulation and function. The phosphatidylcholines and phosphatidylethanolamines are zwitterionic lipid species that contain a polar head group, although the lipid is neutral overall. Other types of phosphorylated lipids are the sphingomyelins and the glycosphingolipids, which have been reported to be involved in different biological processes such as growth and morphogenesis of cells [57]. It is postulated that sphingomyelins, similar to the other zwitterionic phosphorylated lipids, are directly involved in the anchoring of the tear lipid layer to the aqueous portion of the tear film. The polar head group of sphingomyelins may be associated with the aqueous layer of the tear film, while the nonpolar portion of the lipid (the fatty acyl chains of the substituents) is probably associated with the other nonpolar lipids, creating an interface between the two. Phosphatidylserine is a membrane phospholipid in prokaryotic and eukaryotic cells and is found most abundantly in brain tissue. Phosphatidylserine is also a precursor biological molecule in the biosynthesis of phosphatidylethanolamine through a reaction that is catalyzed by PS decarboxylase. At physiological pH (7.38), phosphatidylserine carries a full negative charge and is often referred to as an anionic phosphorylated lipid. Thus, analysis of this special class of biomolecules has been of great interest to medical researchers, biologists, and chemists.

Recent analyses of phospholipids have employed HPLC [58], MEKC [58,59], CE-ES-MS [60], ES-MS/MS [61,62,5,63–67], and to a lesser extent MALDI-FTICR [68] and MALDI-TOF [69–73]. The mass spectrometric analyses of nonvolatile lipids have been reviewed by Murphy et al. [74], and it is pointed out that the advent of soft ionization techniques such as ES-MS and MALDI-MS has obviated the need for derivatization of lipids. Biological samples are complex mixtures of a wide variety of compounds that can have a negative influence on the ability to identify and quantify lipids by mass spectrometry; effects include signal suppression and interferences from isomeric and isobaric compounds.

31.7.1 Electrospray Ionization Behavior of Phosphorylated Lipids

The phosphorylated lipids will ionize differently depending on the ion mode that the analysis is performed in and upon the ionic species (Na^+, H^+, NH_4^+, and Li^+ in positive ion mode and –H or Cl^- in

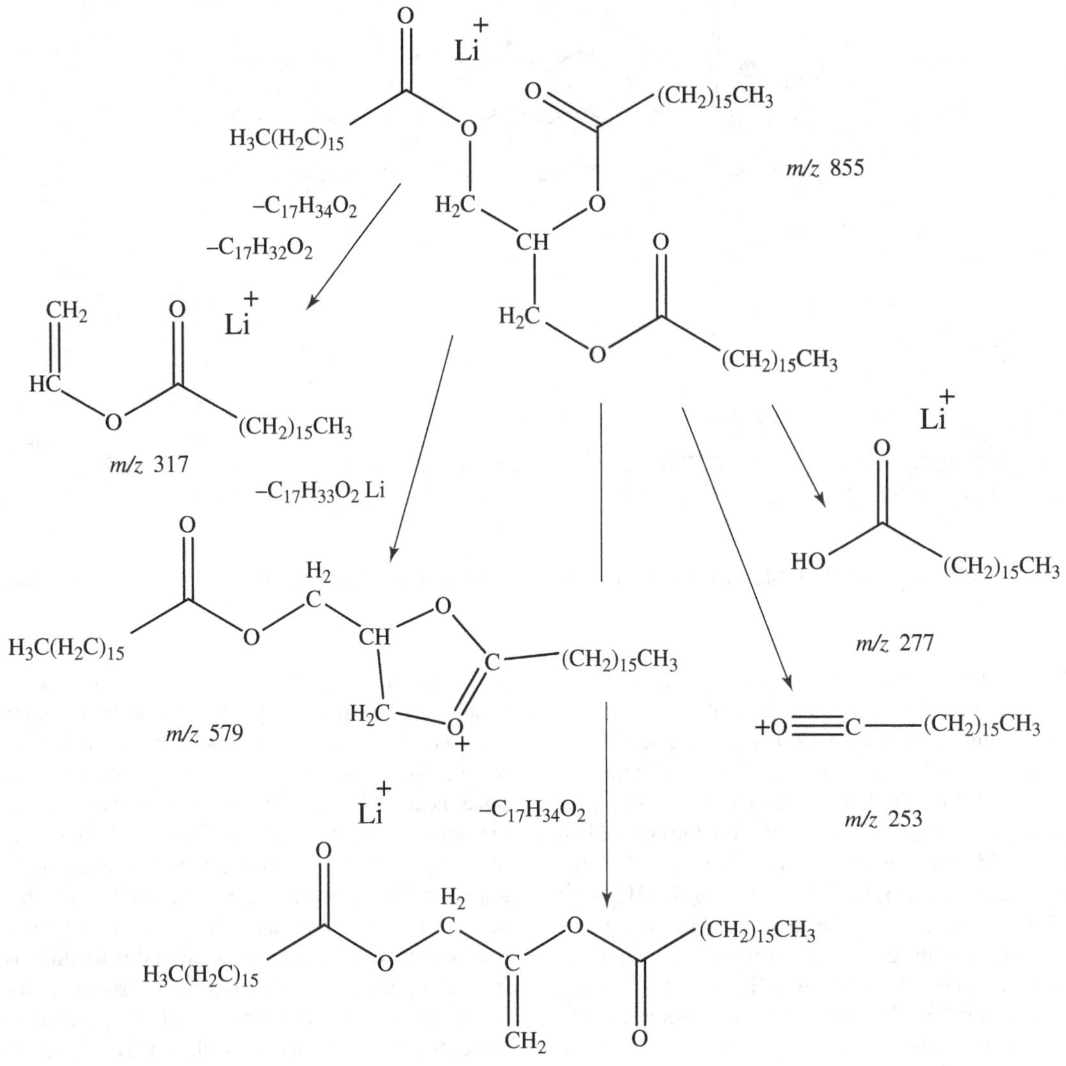

FIGURE 31.33 Major fragment ions produced from the collision-induced dissociation of triheptadecanoin at *m/z* 855 $[MLi]^+$. The C17:0 acylium ion at *m/z* 253 $[C_{17}H_{33}O]^+$; the lithium adduct of heptadecanoic acid at *m/z* 277 $[C_{17}H_{34}O_2Li]^+$; the neutral loss of heptadecanoic acid and then the neutral loss of a C17:1 alpha-beta unsaturated fatty acid, to give the *m/z* 317 product ion $[MLi{-}C_{17}H_{34}O_2{-}C_{17}H_{32}O_2]^+$; the precursor ion minus lithiated heptadecanoate at *m/z* 579 $[MLi{-}C_{17}H_{33}O_2Li]^+$; and the precursor ion minus the neutral loss of heptadecanoic acid at *m/z* 585 $[MLi{-}C_{17}H_{34}O_2]^+$.

negative ion mode) that are present during the ionization process. Figure 31.35 illustrates the structure of the phosphatidic acid 1-palmitoyl-2-oleyl-sn-glycero-3-phosphate (POPA, 16:0–18:1 PA) as a neutral sodium salt adduct. There are two acidic hydrogens on the phosphate head group of POPA. The pK_a for the removal of the first acidic hydrogen (pK_a1) is 3.0; therefore, at a pH of 3.0, the phosphate head group is 50% ionized. At a pH of approximately 5, the head group is 100% ionized. The second acidic hydrogen has a pK_a of 8.0 (pK_a2); thus, at physiological pH (7.36), the POPA species is ionized. In general, the common phosphorylated lipids all have a pka1 that is ≤3 and are subsequently ionized in solutions that are not highly acidic. The common phosphorylated lipids also tend to associate with sodium during the ionization process. The neutral sodium salt adduct molecular formula for POPA is $C_{37}H_{70}O_8PNa$ and has a mass of 696.4706 Da. In negative ion mode analysis, the POPA lipid is in the desodiated, deprotonated form of $[M{-}H]^-$, as $C_{37}H_{70}O_8P$ with a mass of 673.4808 Da. In positive ion mode analysis, the phosphorylated lipids can exhibit a more complex ionization behavior. For example, let's look at the positive ion mode electrospray mass spectrum of the POPA phosphorylated lipid in an ionization solution containing 0.1% acetic acid (H^+), 1 mM lithium (Li^+), and 1 mM sodium (Na^+) in a 1:1 methanol/chloroform solution. The single-stage mass spectrum of the POPA lipid in the multi-ionic solution was collected using nanoelectrospray with a Q-TOF mass spectrometer and is illustrated in Figure 31.36. In the mass spectrum is observed four major peaks for the deprotonated di-lithium adduct at *m/z* 687.7 as $[C_{37}H_{70}O_8P{+}2Li]^+$, the protonated (this term is used loosely here as the protonated form is actually the neutral molecule) lithium adduct at *m/z* 697.7 as $[C_{37}H_{70}O_8P{+}H{+}Li]^+$, the deprotonated lithium sodium adduct at *m/z* 703.7 as $[C_{37}H_{70}O_8P{+}Li{+}Na]^+$,

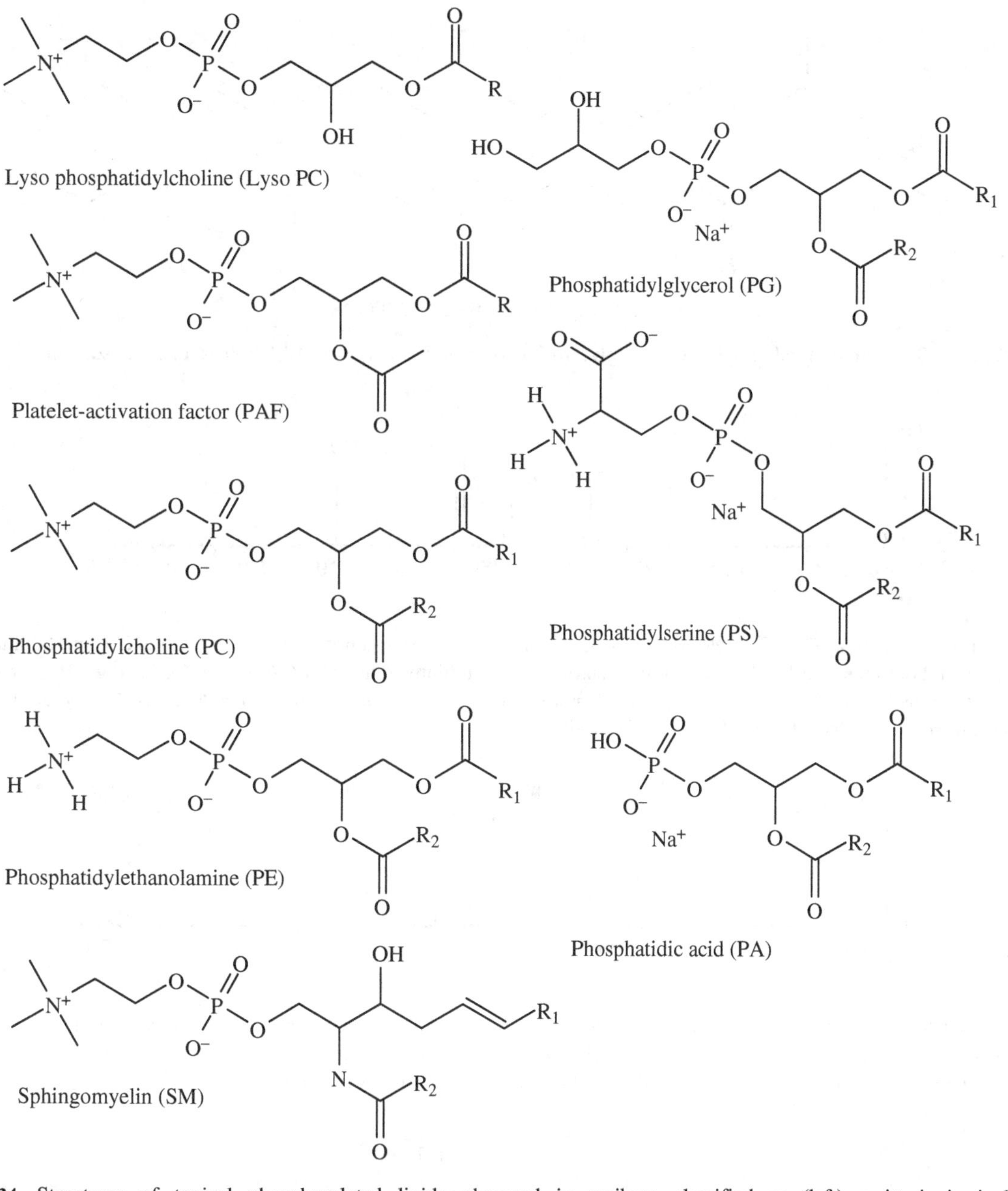

Lyso phosphatidylcholine (Lyso PC)

Phosphatidylglycerol (PG)

Platelet-activation factor (PAF)

Phosphatidylcholine (PC)

Phosphatidylserine (PS)

Phosphatidylethanolamine (PE)

Phosphatidic acid (PA)

Sphingomyelin (SM)

FIGURE 31.34 Structures of typical phosphorylated lipids observed in meibum classified as (left) zwitterionic including lyso phosphatidylcholine (lyso PC), phosphatidylcholine (PC), phosphatidylethanolamine (PE), and sphingomyelin (SM) and (right) anionic including phosphatidylglycerol (PG), phosphatidylserine (PS), and phosphatidic acid (PA).

and the deprotonated di-sodium adduct at *m/z* 719.7 as $[C_{37}H_{70}O_8P{+}2Na]^+$. This ionization behavior of the phosphorylated lipids can be used to help determine whether an unknown species is in fact a phosphorylated lipid.

31.7.2 Positive Ion Mode ESI of Phosphorylated Lipids

The product ion mass spectra can be used to determine the fatty acid substituents of the phosphorylated lipids. Figure 31.37 illustrates the product ion mass spectrum for the deprotonated di-lithium adduct of 1-palmitoyl-2-oleyl-sn-glycero-3-phosphate at *m/z* 687.7 as $[C_{37}H_{70}O_8P{+}2Li]^+$ collected using nanoelectrospray on a Q-TOF mass spectrometer. The two major peaks in the mid mass range of the spectrum are associated with the neutral loss of the fatty acid substituents. The *m/z* 431.5 product ion is generated from the neutral loss of the C16:0 fatty acid substituent, and the *m/z* 405.4 product ion is generated from the neutral loss of the C18:1 fatty acid substituent. These two product ions can be used to identify the fatty acid substituents of the phosphorylated lipid. Another observation of interest is that the fatty acid substituent neutral losses that produce the *m/z* 431.5 and *m/z* 405.4 product ions do not involve either of the lithium cations. This directly indicates that both of the lithium cations are associated with the phosphatidic head group rather than with the fatty acid substituent

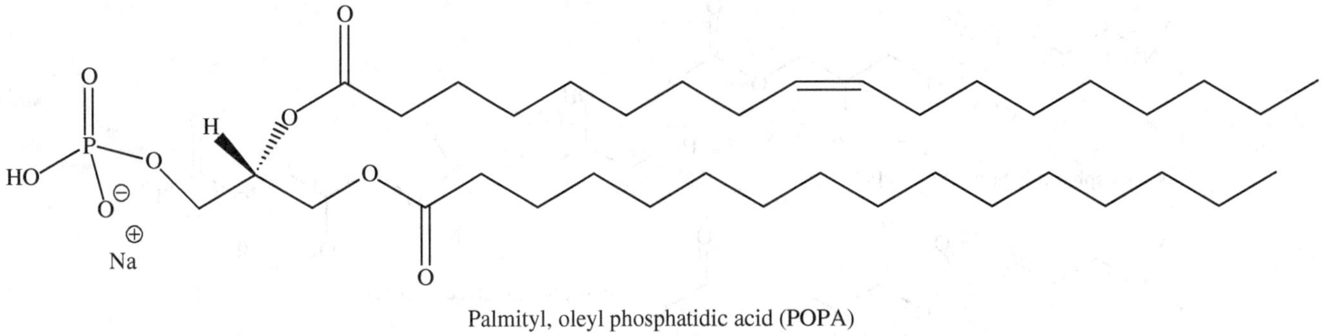

Palmityl, oleyl phosphatidic acid (POPA)

FIGURE 31.35 Structure of 1-palmitoyl-2-oleyl-sn-glycero-3-phosphate (POPA, 16:0–18:1 PA) as a neutral sodium salt adduct.

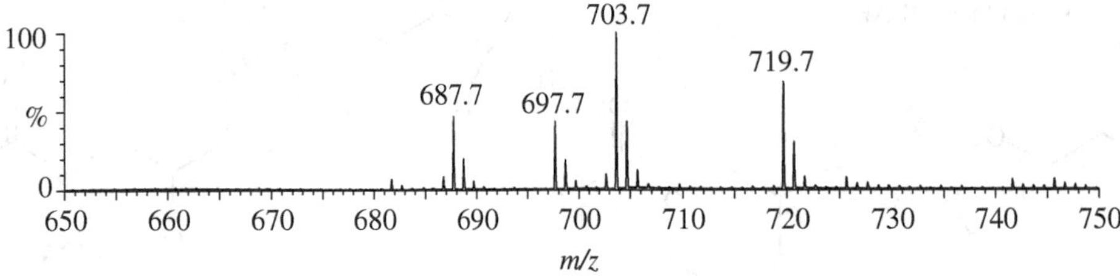

FIGURE 31.36 Nanoelectrospray Q-TOF single-stage mass spectrum of POPA in a solution of 0.1% acetic acid (H^+), 1 mM lithium (Li^+), and 1 mM sodium (Na^+). Four major peaks observed for the deprotonated di-lithium adduct at *m/z* 687.7 as $[C_{37}H_{70}O_8P{+}2Li]^+$, the protonated lithium adduct at *m/z* 697.7 as $[C_{37}H_{70}O_8P{+}H{+}Li]^+$, the deprotonated lithium sodium adduct at *m/z* 703.7 as $[C_{37}H_{70}O_8P{+}Li{+}Na]^+$, and the deprotonated di-sodium adduct at *m/z* 719.7 as $[C_{37}H_{70}O_8P{+}2Na]^+$.

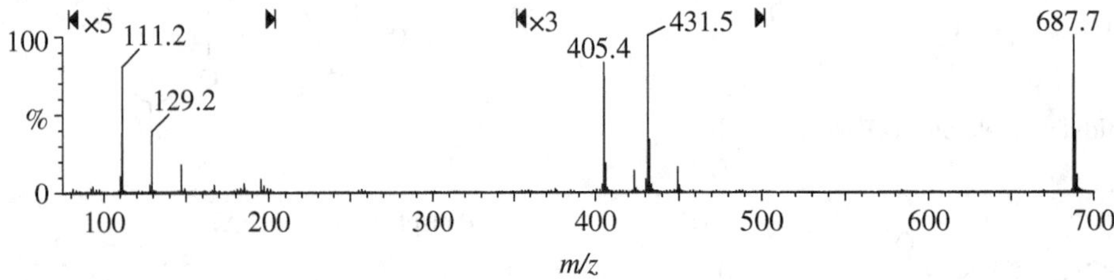

FIGURE 31.37 Nanoelectrospray Q-TOF/MS/MS product ion mass spectrum for the deprotonated di-lithium adduct of 1-palmitoyl-2-oleyl-sn-glycero-3-phosphate at *m/z* 687.7 as $[C_{37}H_{70}O_8P{+}2Li]^+$. The *m/z* 431.5 product ion is generated from the neutral loss of the C16:0 fatty acid substituent, and the *m/z* 405.4 product ion is generated from the neutral loss of the C18:1 fatty acid substituent.

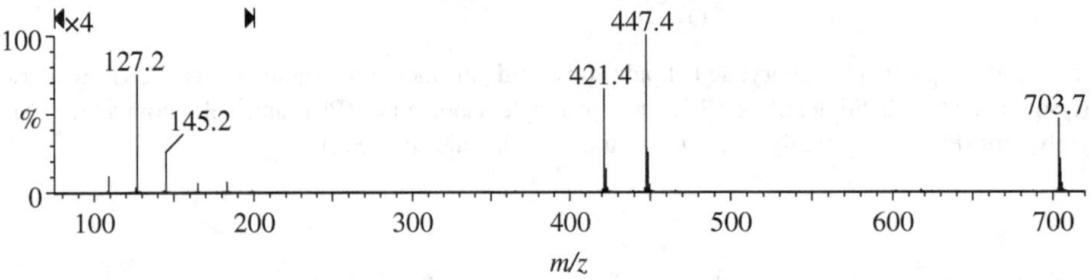

FIGURE 31.38 Nanoelectrospray Q-TOF/MS/MS product ion mass spectrum for the deprotonated lithium sodium adduct of 1-palmitoyl-2-oleyl-sn-glycero-3-phosphate at *m/z* 703.7 as $[C_{37}H_{70}O_8P{+}Li{+}Na]^+$. The *m/z* 447.4 product ion is generated from the neutral loss of the C16:0 fatty acid substituent, and the *m/z* 421.4 product ion is generated from the neutral loss of the C18:1 fatty acid substituent.

portion of the phosphorylated lipid. This is also seen with the *m/z* 111.2 product ion that represents the di-lithium adduct of the phosphate head group as $[H_2PO_4Li_2]^+$. The *m/z* 129.2 product ion also represents the phosphatidic head group as $[C_2H_3PO_4Li]^+$; however, one of the lithium ions has been transferred to one of the fatty acid substituents.

This same behavior is also observed for the product ion mass spectrum for the deprotonated lithium sodium adduct of 1-palmitoyl-2-oleyl-sn-glycero-3-phosphate at *m/z* 703.7 as $[C_{37}H_{70}O_8P{+}Li{+}Na]^+$ collected using nanoelectrospray on a Q-TOF mass spectrometer illustrated in Figure 31.38. The two major peaks in the mid mass range of the spectrum are associated

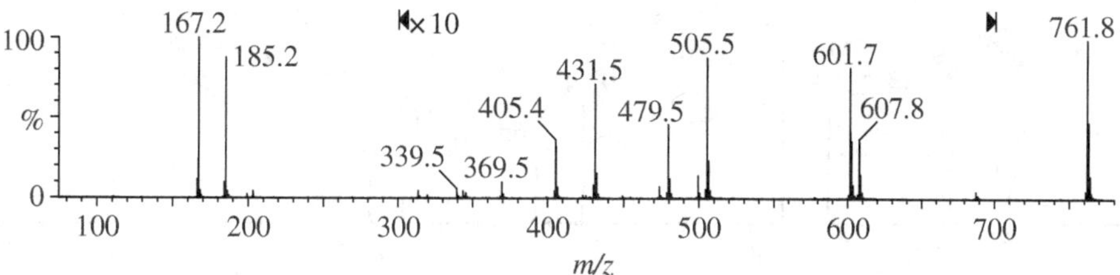

FIGURE 31.39 Nanoelectrospray Q-TOF/MS/MS product ion mass spectrum of the deprotonated di-lithium adduct of 1-palmitoyl-2-oleoyl-sn-glycero-3-[phospho-rac-(1-glycerol)] at *m/z* 761.8 as $[C_{40}H_{76}O_{10}P{+}2Li]^{+}$.

m/z 761.8

m/z 167.2

m/z 185.2

m/z 601.7

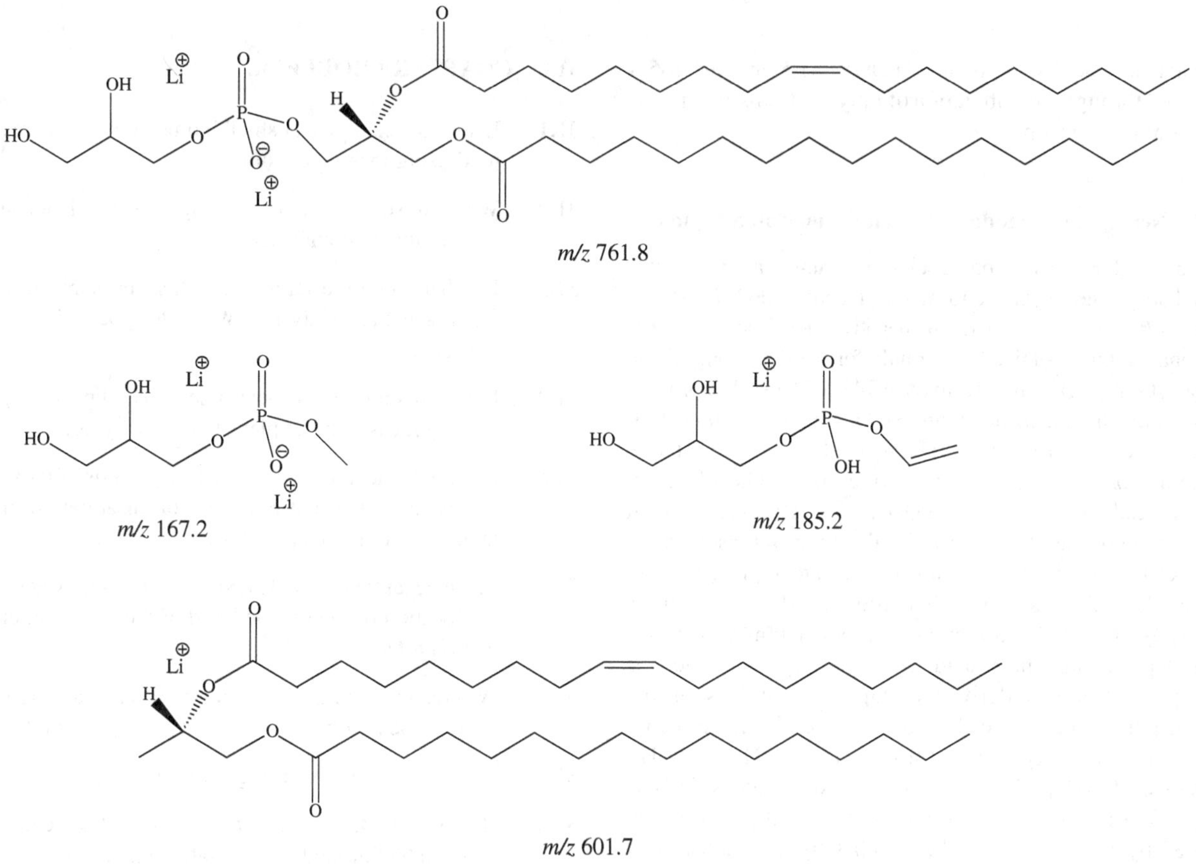

FIGURE 31.40 Some of the major product ions formed from the dissociation of POPG.

with the neutral loss of the fatty acid substituents. The *m/z* 447.4 product ion is generated from the neutral loss of the C16:0 fatty acid substituent, and the *m/z* 421.4 product ion is generated from the neutral loss of the C18:1 fatty acid substituent. Here too, the lithium and sodium cations are not observed to be lost with the fatty acid substituents but stay associated with the phosphatidic head group. The *m/z* 127.2 product ion represents the lithium sodium adduct of the phosphate head group as $[H_2PO_4LiNa]^{+}$, similar to what was observed previously. Interestingly though, the product ion at *m/z* 145.2 represents the phosphatidic head group as $[C_2H_3PO_4Na]^{+}$; however, the lithium ion has been transferred to one of the fatty acid substituents instead of the sodium ion.

The product ion mass spectrum of 1-palmitoyl-2-oleoyl-sn-glycero-3-[phospho-rac-(1-glycerol)] (POPG, 16:0–18:1 PG) however is more complex than that of POPA. Figure 31.39 illustrates the nanoelectrospray Q-TOF/MS/MS product ion mass spectrum for the deprotonated di-lithium adduct of 1-palmitoyl-2-oleoyl-sn-glycero-3-[phospho-rac-(1-glycerol)] at *m/z* 761.8 as $[C_{40}H_{76}O_{10}P{+}2Li]^{+}$. There are a number of product ions in this spectrum that are associated with the head group as compared to that of POPA earlier.

The *m/z* 167 product ion is the di-lithiated head group and its structure is illustrated in Figure 31.40 along with the structure for *m/z* 185.2. The *m/z* 601.7 product ion is formed through the neutral loss of the head group and one lithium as $[C_{40}H_{76}O_{10}P{+}Li{-}C_4H_{10}PO_4Li]^{+}$ and is illustrated in Figure 31.40. The *m/z* 505.5 product ion is generated from the neutral loss of the C16:0 fatty acid substituent, and the *m/z* 479.5 product ion is generated from the neutral loss of

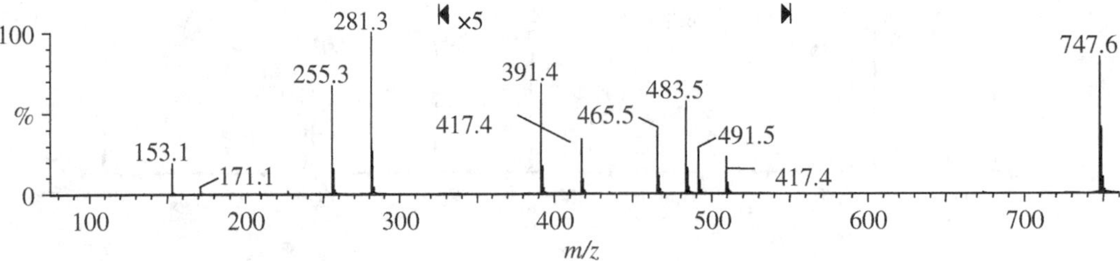

FIGURE 31.41 Product ion mass spectrum for POPG standard reference at *m/z* 747.6. The product ion mass spectrum contain product ions at *m/z* 253.4 (deprotonated form of a C16:1 palmitic acid [$C_{16}H_{30}O_2$–H]–) and *m/z* 281.4 (deprotonated form of a C18:1 oleic acid [$C_{18}H_{34}O_2$–H]–) that are diagnostic for the identification of the fatty acid substituents of the lipids.

the C18:1 fatty acid substituent. The *m/z* 431.5 and *m/z* 405.4 are formed through a combination of fatty acid substituent loss and partial head group loss.

31.7.3 Negative Ion Mode ESI of Phosphorylated Lipids

Figure 31.41 illustrates the product ion spectrum of a 1-palmitoyl-2-oleoyl-*sn*-glycero-3-[phospho-*rac*-(1-glycerol)] (POPG) standard at *m/z* 747.6 collected in negative ion mode as the deprotonated ion $[M–H]^-$. The formula for the deprotonated ion is $[C_{40}H_{76}O_{10}P]^-$ with an exact mass of 747.5176 Da. The product ion spectrum illustrated in Figure 31.41 contains a number of diagnostic peaks used for unknown identification. These include the *m/z* 153.1 product ion, which is the phosphatidylglycerol head group as $[C_3H_6PO_5]^-$, and *m/z* 171.1, which is also a phosphatidylglycerol head group represented as $[C_3H_8PO_6]^-$. The fatty acid substituents are represented by *m/z* 255.3, which is a C16:0 deprotonated fatty acid as $[C_{16}H_{32}O_2–H]^-$, and the *m/z* 281.3 product ion, which is a C18:1 deprotonated fatty acid as $[C_{18}H_{34}O_2–H]^-$. The *m/z* 391.4 product ion is derived through neutral loss of the glycerol portion of the head group followed by neutral loss of the C18:1 fatty acid substituent $[C_{40}H_{77}O_{10}P–H–C_3H_6O_2–C_{18}H_{34}O_2]^-$. The *m/z* 417.4 product ion is derived through neutral loss of the glycerol portion of the head group followed by neutral loss of the C16:0 fatty acid substituent $[C_{40}H_{77}O_{10}P–H–C_3H_6O_2–C_{16}H_{32}O_2]^-$. The *m/z* 465.5 product ion is derived through neutral loss of the C18:1 fatty acid substituent, $[C_{40}H_{77}O_{10}P–H–C_{18}H_{34}O_2]^-$, while *m/z* 483.5 is derived through neutral loss of the C18:1 fatty acyl chain as a ketene $[C_{40}H_{77}O_{10}P–H–C_{18}H_{32}O]^-$. The C16:0 fatty acid substituent also produces product ions that are analogous to those produced by the C18:1 fatty acid substituent, such as the *m/z* 491.5 product ion, which is derived through neutral loss of the C16:0 fatty acid substituent, $[C_{40}H_{77}O_{10}P–H–C_{16}H_{32}O_2]^-$, and the *m/z* 509.5 product ion, which is derived through neutral loss of the C16:0 fatty acyl chain as a ketene $[C_{40}H_{77}O_{10}P–H–C_{16}H_{30}O]^-$. The difference between the product ions for neutral loss of fatty acid substituent versus fatty acyl chain as ketene is 18 amu. When this pattern is observed in product ion spectra such as Figure 31.41, it can be used as an indication of ester linkages as fatty acid substituents in the structures and can be used in unknown lipid species identification.

31.8 CHAPTER PROBLEMS

31.1 List some examples of small biomolecules that are measured using mass spectrometry.

31.2 What are some benefits of using a metal adduct such as lithium for lipid analysis?

31.3 In what mode are fatty acids often measured in during mass spectral analysis? Why do you think this is the case?

31.4 In electrospray mass spectrometry, can lipids be quantitatively measured? Explain why or why not.

31.5 During the negative ion mode ESI analysis of fatty acids, it was suspected that reduction of unsaturation may be taking place. Is this possible?

31.6 In the negative ion mode ESI behavior of fatty acids, was the location of the double bond or the structural conformation a factor?

31.7 Was reduction of the fatty acids in ESI negative ion mode analysis being observed, or was it suppression?

31.8 What do the graphs in Figure 8.11 demonstrate?

31.9 How was contamination ruled out in the ESI negative ion mode behavior study of the fatty acids?

31.10 What is the significance of the ESI negative ion mode behavior study?

31.11 What is found to allow quantitative analysis by GC/EI mass spectrometry?

31.12 Briefly describe the internal standard method.

31.13 Why are different sample dilutions measured when quantitating species by mass spectrometry?

31.14 In Figure 8.16, what fragment is the lithium cation associated with and why?

31.15 What are a couple of neutral losses in the CID of wax esters that are used as diagnostic for location of a double bond?

31.16 What form of loss from a lipid often indicates an oxidized species?

31.17 In the sterols section 31.5, what steps were performed in order to identify the unknown biomolecule?

31.18 List some approaches that have been used for mass spectrometric analysis of acylglycerols.

31.19 What are zwitterionic lipid species?

31.20 Describe the types of ions that can be produced in an acidic solution that also contains metal salts such as Li^+, Na^+, and K^+.

31.21 What can product ion spectra be used for in analysis of lipids?

31.22 With the phosphorylated lipids, what part is the metal cation generally found associated with?

31.23 What is the difference in neutral loss for loss of fatty acid substituents versus fatty acyl chain as ketene? What can this be useful in determining?

REFERENCES

1. Han, X.; Gross, R.W. *Expert Rev. Proteomics* 2005, **2**, 253–264.
2. Han, X.; Gross, R.W. *J. Lipid Res.* 2003, **44**, 1071–9.
3. Fahy, E.; Subramaniam, S.; Brown, H.A. *J. Lipid Res.* 2005, **46**, 839–62.
4. Adams, J.; Gross, M.L. *Anal. Chem.* 1987, **59**, 1576–1582.
5. Hsu, F.F.; Bohrer, A.; Turk, J. *J. Am. Soc. Mass Spectrom.* 1998, **9**, 516–526.
6. Hsu, F.F.; Turk, J. *J. Am. Soc. Mass Spectrom.* 1999, **10**, 587–599.
7. Kerwin, J.L.; Wiens, A.M.; Ericsson, L.H. *J. Mass Spectrom.* 1996, **31**, 184.
8. Valianpour, F.; Selhorst, J.J.M.; van Lint, L.E.M.; van Gennip, A.H.; Wanders, R.J.A.; Kemp, S. *Mol. Genet. Metab.* 2003, **79**, 189.
9. Catharino, R.R.; Haddad, R.; Cabrini, L.G.; Cunha, I.B.S.; Sawaya, A.C. H.F.; Eberlin, M.N. *Anal. Chem.* 2005, **77**, 7429.
10. Moe, M.K.; Strom, M.B.; Jensen, E.; Claeys, M. *Rapid Commun. Mass Spectrom.* 2004, **18**, 1731.
11. Griffiths, W.J. *Mass Spectrom. Rev.* 2003, **22**, 81.
12. Han, X.; Gross, R.W. *Anal. Biochem.* 2001, **295**, 88.
13. Gross, M.L. *Int. J. Mass Spectrom.* 2000, **200**, 611.
14. Brugger, B.; Erben, G.; Sandhoff, R.; Wieland, F.T.; Lehmann, W.D. *Proc. Natl. Acad. Sci. U. S. A..* 1997, **94**, 2339.
15. Han, X.; Gross, R.W. *Mass Spectrom. Rev.* 2005, **24**, 367.
16. Wakeham, S.G.; Frew, N.M. *Lipids* 1982, **17**, 831.
17. Alveraz, H.M.; Luftmann, H.; Silva, R.A.; Cesari, A.C.; Viale, A.; Waltermann, M.; Steinbuchel, A. *Microbiology* 2002, **148**, 1407.
18. Gurr, M.I.; Harwood, J.L.; Frayn, K.N. In *Lipid Biochemistry: An Introduction. 5th Edition* Culinary and Hospitality Industry Publications Services: Weimer, Texas 2002.
19. Reis, A.; Domingues, R.M.; Amado, F.M.L.; Ferrer-Carreia, A.J.V.; Domingues, P. *J. Am. Soc. Mass Spectrom.* 2003, **14**, 1250.
20. Folch, J.; Lees, M.; Stanly, G.H.S. *J. Biol. Chem.* 1957, **226**, 497.
21. Bligh, E.G.; Dyer, W.J. *Can. J. Biochem. Physiol.* 1959, **37**, 911.
22. Wilcox, A.L.; Marnett, L.J. *Chem. Res. Toxicol.* 1993, **6**, 413.
23. Porter, N.A.; Caldwell, S.E.; Mills, K.A. *Lipids* 1995, **30**, 277.
24. Bierl-Leonhardt, B.A.; DeVilbuss, E.D.; Plimmer, J.R. *J. Chromatogr. Sci.* 1980, **18**, 364.
25. Tomer, K.B.; Crow, F.W.; Gross, M.L. *J. Am. Chem. Soc.* 1983, **105**, 5487.
26. Holcapek, M.; Jandera, P.; Fischer, J. *Crit. Rev. Anal. Chem.* 2001, **31**, 53.
27. Mottram, H.R.; Woodbury, S.E.; Evershed, R.P. *Rapid Commun. Mass Spectrom.* 1997, **11**, 1240.
28. Sedaghat, S.; Désaubry, L.; Streiff, S.; Ribeiro, N.; Michels, B.; Nakatani, Y.; Ourisson, G. *Chem. Biodivers.* 2004, **1**, 124–128.
29. Gotoh, M.; Ribeiro, N.; Michels, B.; Elhabiri, M.; Albrecht-Gary, A.M.; Yamashita, J.; Hato, M.; Ouisson, G.; Nakatani, Y. *Chem. Biodivers.* 2006, **3**, 198–209.
30. Ohlrogge, J.; Browse, J. *Plant Cell* 1995, **7**, 957–970.
31. Hodgkin, M.N.; Pettitt, T.R.; Martin, A.; Wakelam, M.J.O. *Biochem. Soc. Trans.* 1996, **24**, 991–994.
32. Pettitt, T. R.; Martin, A.; Horton, T.; Liassis, C.; Lord, J.M.; Wakelam, M.J.O. *J. Biol. Chem.* 1997, **272**, 17354–17359.
33. Murphy, R.C. In *Mass Spectrometry of Lipids.* Plenum Press: New York 1993; 189.
34. Liu, Q.T.; Kinderlerer, J.L. *J. Chromatogr. A* 1999, **855**, 617.
35. Harvey, D.J.; Tiffany, J.M. *J. Chromatogr.* 1984, **301**, 173.
36. Kuksis, A.; Marai, L.; Myher, J.J. *J. Chromatogr.* 1991, **588**, 73.
37. Marai, L.; Kukis, A.; Myher, J.J.; Itabashi, Y. *Biol. Mass Spectrom.* 1992, **21**, 541.
38. Cole, R.B.; Zhu, J. *Rapid Commun. Mass Spectrom.* 1999, **13**, 607.
39. Duffin, K.L.; Henion, J.D. *Anal. Chem.* 1991, **63**, 1781.
40. Byrdwell, W.C.; Neff, W.E. *Rapid Commun. Mass Spectrom.* 2002, **16**, 300.
41. Marzilli, L.A.; Fay, L.B.; Dionisi, F.; Vouros, P. *J. Am. Oil Chem. Soc.* 2003, **80**, 195.
42. Waltermann, M.; Luftmann, H.; Baumeister, D.; Kalscheuer, R.; Steinbuchel, A. *Microbiology* 2000, **146**, 1143.
43. Schiller, J.; Arnhold, J.; Benard, S.; Muller, M.; Reichl, S.; Arnold, K. *Anal. Biochem.* 1999, **267**, 46.
44. Leary, J.A.; Pederson, S.F. *J. Org. Chem.* 1989, **54**, 5650.
45. Byrdwell, W.C. *Lipids* 2001, **36**, 327.
46. Mu, H.; Sillen, H.; Hoy, C.E. *J. Am. Oil Chem. Soc.* 2000, **77**, 1049.
47. McAnoy, A.M.; Wu, C.C.; Murphy, R.C. *J. Am. Soc. Mass Spectrom.* 2005, **16**, 1498.
48. Kalo, P.; Kemppinen, A.; Ollilainen, V.; Kuksis, A. *Lipids* 2004, **39**, 915.
49. Giuffrida, F.; Destaillats, F.; Skibsted, L.H.; Dionisi, F. *Chem. Phys. Lipids* 2004, **131**, 41.
50. Ham, B.M.; Jacob, J.T.; Keese, M.M.; Cole, R.B. *J. Mass Spectrom.* 2004, **39**, 1321.
51. Holcapek, M.; Jandera, P.; Fischer, J.; Prokes, B. *J. Chromatogr. A* 1999, **858**, 13.
52. Holcapek, M.; Jandera, P.; Zderadicka, P.; Hruba, L. *J. Chromatogr. A* 2003, **1010**, 195.
53. Jakab, A.; Jablonkai, I.; Forgacs, E. *Rapid Commun. Mass Spectrom.* 2003, **17**, 2295.
54. Fauconnot, L.; Hau, J.; Aeschlimann, J.M.; Fay, L.B.; Dionisi, F. *Rapid Commun. Mass Spectrom.* 2004, **18**, 218.
55. Zhou, Z.; Ogden, S.; Leary, J.A. *J. Org. Chem.* 1990, **55**, 5444.
56. Striegel, A.M.; Timpa, J.D.; Piotrowiak, P.; Cole, R.B. *Int. J. Mass Spectrom. Ion Process.* 1997, **162**, 45.
57. Gu, M.; Kerwin, J. L.; Watts, J. D.; Aebersold, R. *Anal. Biochem.* 1997, **244**, 347–356.
58. Szucs, R.; Verleysen, K.; Duchateau, G.S.M.J.E.; Sandra, P.; Vandeginste, B.G.M. *J. Chromatogr. A* 1996, **738**, 25–29.
59. Verleysen, K.; Sandra, P. *J. High Resol. Chromatogr.* 1997, **20**, 337–339.
60. Raith, K.; Wolf, R.; Wagner, J.; Neubert, R.H.H. *J. Chromatogr. A* 1998, **802**, 185–188.

61. Han, X.; Gross, R.W. *J. Am. Chem. Soc.* 1996, **118**, 451–457.
62. Hoischen, C.; Ihn, W.; Gura, K.; Gumpert, J. *J. Bacteriol.* 1997, **179**, 3437–3442.
63. Khaselev, N.; Murphy, R.C. *J. Am. Soc. Mass Spectrom.* 2000, **11**, 283–291.
64. Hsu, F.F.; Turk, J. *J. Mass Spectrom.* 2000, **35**, 596–606.
65. Liebisch, G.; Drobnik, W.; Lieser, B.; Schmitz, G. *Clin. Chem.* 2002, **48**, 2217–2224.
66. Ho, Y.P.; Huang, P.C.; Deng, K.H. *Rapid Commun. Mass Spectrom.* 2003, **17**, 114–121.
67. Hsu, F.F.; Turk, J. *J. Am. Soc. Mass Spectrom.* 2004, **15**, 1–11.
68. Marto, J.A.; White, F.M.; Seidomridge, S.; Marshall, A.G. *Anal. Chem.* 1995, **67**, 3979–3984.
69. Schiller, J.; Arnhold, J.; Benard, S.; Muller, M.; Reichl, S.; Arnold, K. *Anal. Biochem.* 1999, **267**, 46–56.
70. Ishida, Y.; Nakanishi, O.; Hirao, S.; Tsuge, S.; Urabe, J.; Sekino, T.; Nakanishi, M.; Kimoto, T.; Ohtani, H. *Anal. Chem.* 2003, **75**, 4514–4518.
71. Al-Saad, K.A.; Zabrouskov, V.; Siems, W.F.; Knowles, N.R.; Hannan, R.M.; Hill Jr., H.H. *Rapid Commun. Mass Spectrom.* 2003, **17**, 87–96.
72. Rujoi, M.; Estrada, R.; Yappert, M.C. *Anal. Chem.* 2004, **76**, 1657–1663.
73. Woods, A.S.; Ugarov, M.; Egan, T.; Koomen, J.; Gillig, K.J.; Fuhrer, K.; Gonin, M.; Schultz, J.A. *Anal. Chem.* 2004, **76**, 2187–2195.
74. Murphy, R.C.; Fiedler, J.; Hevko, J. *Chem. Rev.* 2001, **101**, 479–526.

32

MACROMOLECULE ANALYSIS

32.1 Introduction
32.2 Carbohydrates
32.2.1 Ionization of Oligosaccharides
32.2.2 Carbohydrate Fragmentation
32.2.3 Complex Oligosaccharide Structural Elucidation
32.3 Nucleic Acids
32.3.1 Negative Ion Mode ESI of a Yeast 76-mer $tRNA^{Phe}$
32.3.2 Positive Ion Mode MALDI Analysis
32.4 Chapter Problems
References

32.1 INTRODUCTION

The three main types of biological macromolecules (or biopolymers) are proteins, polysaccharides, and nucleic acids. The mass spectrometric analysis of proteins, peptides, and their associated posttranslational modifications will be covered extensively in Chapter 33 and will not be covered here. In this chapter, we will take a close look at the mass spectrometric analysis of polysaccharides and nucleic acids, both having their own set of fragmentation pathway schemes and naming systems quite similar to that of the peptides to be presented in Chapter 33. The biological macromolecules all consist of repeating units of monomers with a variety that can include a single monomeric repeating unit up to perhaps 20 different monomeric units involved as is the case with the amino acids. The biopolymers are synthesized through condensation reactions between the monomeric units that join the monomers with the elimination of water. The chemical splitting apart of the monomeric units of a biopolymer is a hydrolysis reaction where the addition of water is used to separate the monomers, which is the reverse of the condensation reaction. These two processes are constantly taking place within a living organism such as the synthesis of starch for energy storage (condensation) and the process of digestion for breaking down large molecules (hydrolysis).

Mass spectrometry has been applied to the gas-phase measurement, characterization, and identification through structural elucidation of the other macromolecules polysaccharides and nucleic acids. This is an important area of study, and though much work has already been accomplished in the area of macromolecule mass spectrometric analysis, there is room for added information and insight.

32.2 CARBOHYDRATES

Polysaccharides are biopolymers that are comprised of long chains of repeating sugar units, usually either the same repeating monomer or a pattern of two alternating units. The sugar monomers that make up the polysaccharide chain are called monosaccharides from the Greek word "*mono*" meaning "single" (the Greek "*poly*" means "many") and "*saccharide*" meaning "sugar." Glycogen and starch are two forms of storage polysaccharides, while cellulose is a structural polysaccharide. It is the monosaccharide glucose that makes up these three polysaccharides as a single repeating unit. It is the bonding between the repeating glucose monosaccharide and the structural design through branching that gives each one of these its special properties designed for specific functions. Starch is the storage form of polysaccharide that is found in plants, while it is glycogen that is the storage form found in animals. Glycogen is primarily found in the liver where it is used to maintain blood sugar levels and in muscle tissue where it is used for energy in muscle contraction. These polysaccharides are comprised of α-d-glucose units that are linked together by α-glycosidic bonds at the 1 and 4 positions of the glucose unit. Cellulose, found in the cell walls of plants, is a polysaccharide comprised of the repeating glucose monomer β-d-glucose with a β-glycosidic bond between the 1 and 4 carbons of the sugar. Figure 32.1 illustrates the structures of α-d-glucose and β-d-glucose. Starting with carbon number one in the numbering of the ring to the right of the structure, the difference between the two is that in α-d-glucose the hydroxyl group points downward and in β-d-glucose it points upward. Glucose, also known as aldohexose d-glucose, is a six-carbon structure with a molecular formula of $C_6H_{12}O_6$. The sugars are of the

Analytical Chemistry: A Chemist and Laboratory Technician's Toolkit, First Edition. Bryan M. Ham and Aihui MaHam.

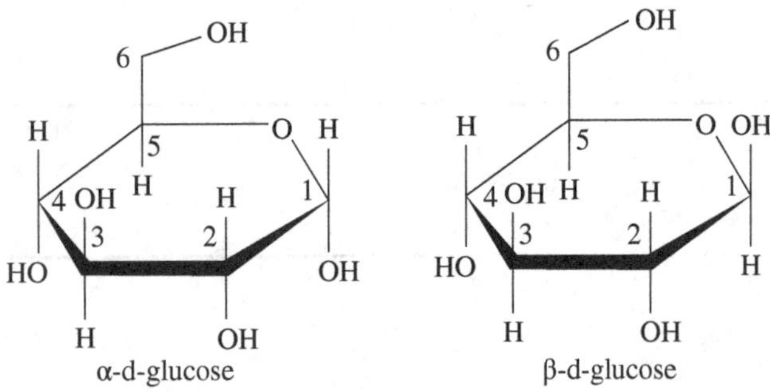

FIGURE 32.1 Structures of α-d-glucose (the unit that repeats in starch and glycogen) where the hydroxyl group points downward and of β-d-glucose (the unit that repeats in cellulose) where the hydroxyl group points upward.

carbohydrate group with a general formula of $C_nH_{2n}O_n$ where the carbon to hydrogen to oxygen ratio is 1:2:1. Carbohydrate is actually an archaic nomenclature from early chemist studies that in general termed compounds such as sugars as a hydrate of carbon represented as $C_n(H_2O)_n$. There are two types of linkages that connect the monosaccharides forming the polysaccharides, the α-glycosidic bond and the β-glycosidic bond. These two types of bonds are illustrated in Figure 32.2 for the disaccharides maltose and lactose. In the disaccharide maltose structure illustrated in Figure 32.2a, there are two α-d-glucose units connected forming a α-glycosidic bond, while in lactose (Fig. 32.2b), there is a β-d-galactose unit connected to a β-d-glucose forming a β-glycosidic bond. It is also possible to have a α-sugar connected to a β-sugar as illustrated in Figure 32.2c between α-d-glucose and β-d-fructose forming the common

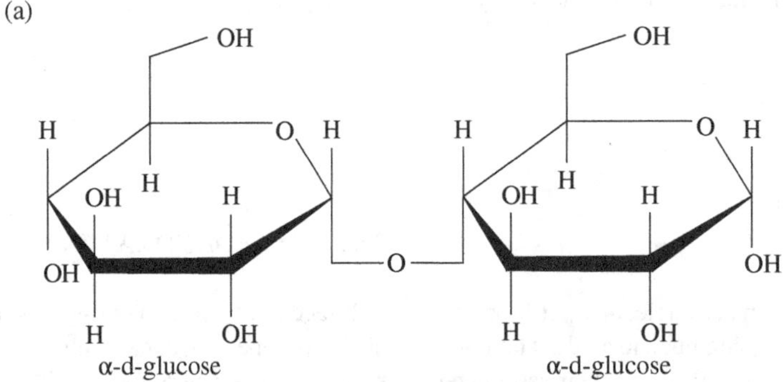

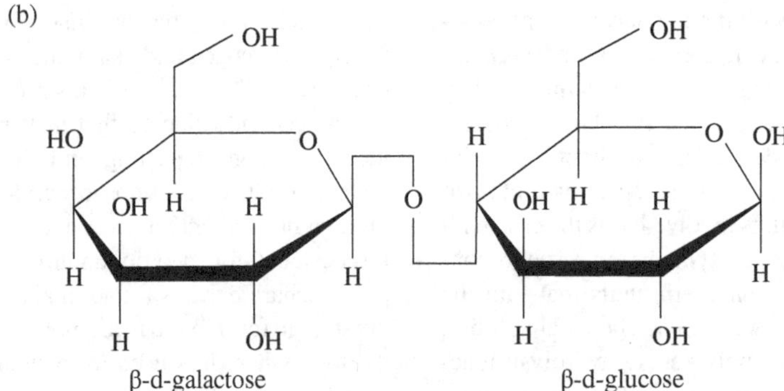

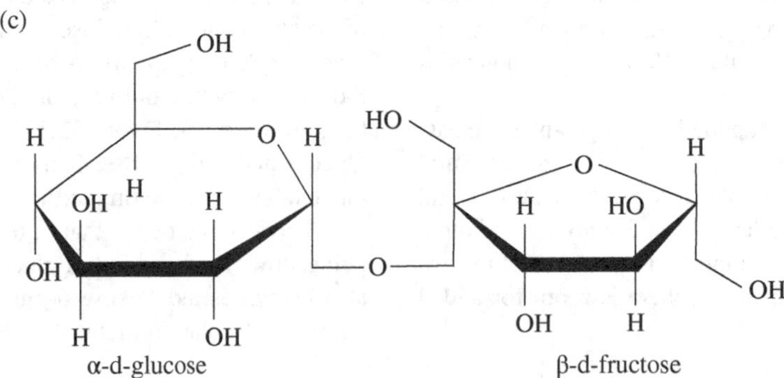

FIGURE 32.2 (a) Maltose structure comprised of two α-d-glucose units connected forming a α-glycosidic bond. (b) Lactose structure comprised of a β-d-galactose unit connected to a β-d-glucose forming a β-glycosidic bond. (c) A α-sugar connected to a β-sugar as α-d-glucose and β-d-fructose, the common table sugar sucrose forming a α-glycosidic bond.

table sugar sucrose with a α-glycosidic bond. While the three polysaccharides glycogen, starch, and cellulose are the most abundant forms of polysaccharides in biological systems, there are other examples of polysaccharides that contain different sugar units than glucose that the mass spectrometrist may encounter. Examples include derivatives of β-glucosamine, a sugar where an amino group has replaced a hydroxyl group on carbon atom 2, found in the cell wall of bacteria and chitin, which is found in the exoskeletons of insects. There are also the oligosaccharides, shorter chains of sugar units than the polysaccharides, found as modifications on proteins, which will be covered in Chapter 33.

32.2.1 Ionization of Oligosaccharides

Normal electrospray that is pneumatically assisted has been observed to produce a low response for oligosaccharides that are not modified when analyzed by mass spectrometry. The use of nanoelectrospray however has shown to induce an enhanced response for the analysis of oligosaccharides by mass spectrometry that is comparable to that observed of the ionization of peptides [1]. This is due to the decrease in the size of the nanoelectrospray drops, which helps to increase the surface activity of the hydrophilic oligosaccharides. With electrospray, the response of the oligosaccharides tends to decrease with increasing size of the oligosaccharides [2]. This is not the case when using MALDI as the ionization source as the response of the oligosaccharides basically stays constant with increasing oligosaccharide size. The electrospray process though is a softer ionization process as compared to MALDI, which has been observed to promote metastable fragmentation of the oligosaccharides. The inclusion of metastable fragmentation can increase the complexity of the mass spectra obtained.

Similar to the mass spectrometric analysis of most biomolecules, there are three choices for the ionized species as either protonated $[M + nH]^{n+}$ or metalized $[M + Na]^{+}$ or $[M + Li]^{+}$ both measured in positive ion mode or deprotonated $[M - nH]^{n-}$ measured in negative ion mode. Product ion mass spectra of protonated oligosaccharides tend to dominate in B_m- and Y_n-type ions. Also, when CID is performed on larger oligosaccharides, the product ions are primarily derived through glycosidic bond cleavage giving sequence information but little information concerning branching. When CID is performed on smaller oligosaccharides, the product ions are produced through both glycosidic bond cleavage and cross-ring cleavage. This is due to the greater number of vibrational degrees of freedom in the larger oligosaccharides that are able to dissipate the internal energy imparted to them from the collisions through vibrational relaxation. In high-energy CID studies such as those obtained with a sector mass analyzer or the MALDI TOF–TOF mass analyzer, the use of protonated species was observed to have more efficient fragmentation of the glycosidic bonds as compared to metal-adducted oligosaccharides. Negative ion mode mass spectrometric analysis of deprotonated oligosaccharides does not have a very efficient or high response for neutral oligosaccharides. The action of hydroxylic hydrogen migration has been suggested as a limiting factor for the observance of abundant cross-ring cleavage ions in negative ion mode analysis of deprotonated ions while supporting glycosidic bond cleavages [3]. Sialylated and sulfated oligosaccharides can have an enhanced response and informative product ion spectra as deprotonated species analyzed in negative ion mode.

32.2.2 Carbohydrate Fragmentation

Oligosaccharides are often observed in nature to be very complex mixtures of species that are very closely related isomeric compounds. This and the fact that they contain many labile bonds and can be highly branched make the analysis of oligosaccharides by mass spectrometry difficult. In mass spectrometric analysis of the polysaccharides using collision-induced dissociation for the production of product ion spectra, the polysaccharides fragment according to specific pathways that have a standardized naming scheme. Figure 32.3 illustrates the types of fragmentation that can

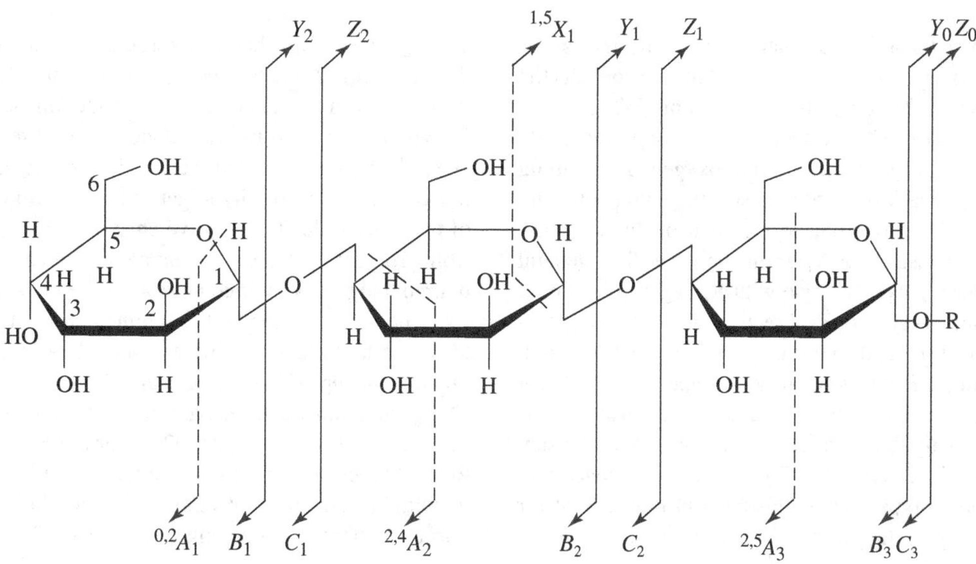

FIGURE 32.3 Fragmentation pathways and naming scheme for polysaccharides.

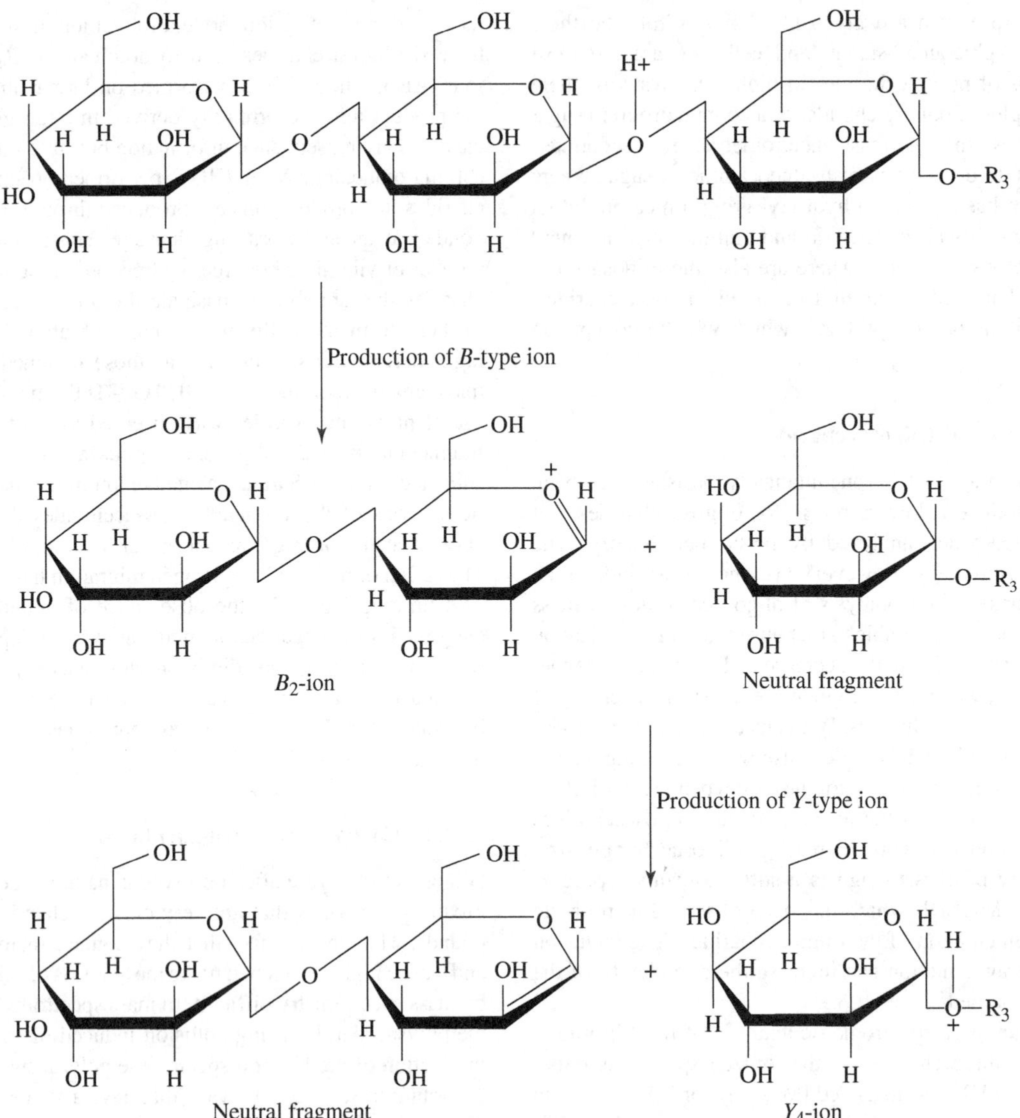

FIGURE 32.4 Production of B_2-type and Y_4-type ions for a six-sugar unit polysaccharide in positive ion mode.

take place with the polysaccharide chain (adapted from Costello et al. [4]). The sugar ring structure on the far left also contains the numbering scheme for the sugar ring carbon atoms. Whether it is a five-membered ring or a six-membered ring, the counting starts with the first carbon to the left of the ring oxygen. The naming scheme in Figure 32.3 as illustrated is for a three sugar containing oligosaccharide, with an R-group to the right of the structure; therefore, the naming starts at Y_0 from the right. The general naming scheme for a polysaccharide with R_n sugar units would result in a naming of Y_n and Z_n directly left of the R_n, Y_{n+1} and Z_{n+1} for the next set, and so forth. We will take a closer look at the generation of the *B*-type ions and the *Y*-type ions in positive ion mode to better understand the generation of the product ions and their naming using the more general scheme for a polysaccharide based off of the scheme in Figure 32.3. The generation of the B_2-type and Y_4-type ions is illustrated in Figure 32.4 for a six-sugar unit polysaccharide, obtained in positive ion mode. The B_2-type ion is a charged oxonium ion that is contained within the sugar ring structure. The Y_4-type ion carries a positive charge in the form of protonation. The production of the C_2-type and Z_4-type ions in positive ion mode is illustrated in Figure 32.5. Here, the C_2-type ion carries a positive charge in the form of protonation derived from hydrogen transfer from the reducing side of the saccharide. The positive charge of the Z_4-type ion is also from protonation, but the ion has also suffered loss of water to the nonreducing carbohydrate portion of the polysaccharide.

In negative ion mode, the fragmentation of the polysaccharides also takes place with the glycosidic bonds; however, ring opening and epoxide formation are observed in the product ions. The production of the *B*-type and *Y*-type ions in negative ion mode is illustrated in Figure 32.6. For the *B*-type ion, the deprotonated precursor will dissociate where the hydroxyl group will go with the reducing portion of the sugar, while the carbohydrate portion suffers ring opening and epoxide formation. A similar

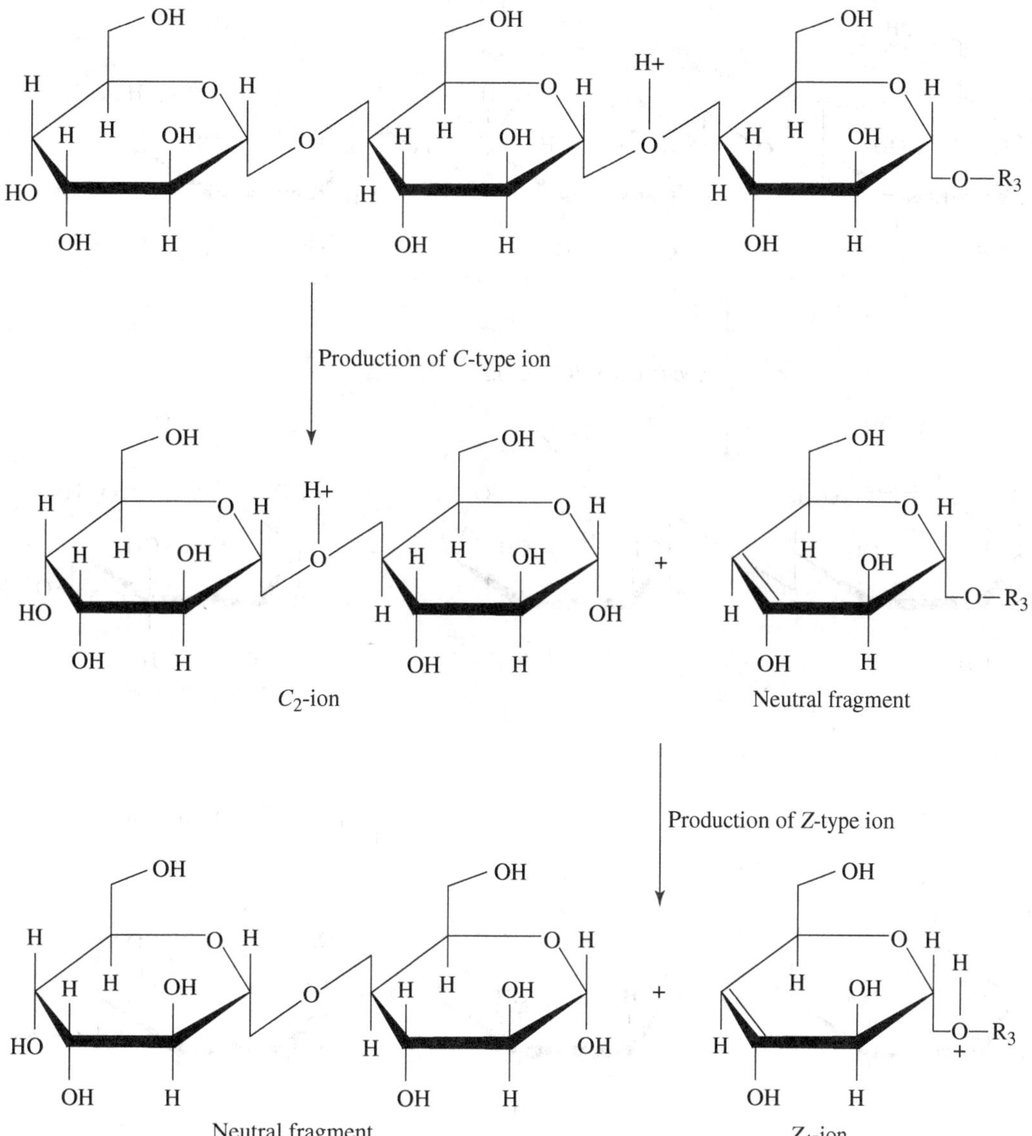

FIGURE 32.5 Production of C_2-type and Z_4-type ions for a six-sugar unit polysaccharide in positive ion mode.

mechanism is observed for the *Y*-type ion except there is a proton transfer from the reducing portion of the structure to the carbohydrate portion, leaving the reducing end deprotonated and negatively charged. The production of the *C*-type and *Z*-type ions in negative ion mode is illustrated in Figure 32.7. In these pathways, the epoxide is forming on the reducing end of the polysaccharide, and there is no ring opening taking place.

In positive ion mode, cleavage of the ring structure producing the *A*-type ion is not often observed. In negative ion mode, cleavage of the ring structure producing the *A*-type ion is observed. In both positive and negative ion modes, the cleavage of the sugar ring producing the *X*-type ions is observed and produces a number of different types of product ion structures.

When generating productions of the oligosaccharides using high-energy CID, the products produced are different for protonated oligosaccharides than that of metal ion adducts. It has been observed that the production of cross-ring cleavages is more enhanced in the metal ion adduct oligosaccharides. In low-energy CID, the amount of glycosidic bond cleavage is low for the metal ion adduct species. This can be explained due to the types of bonding that is taking place between the oligosaccharide and the proton or metal ion. In protonation, the proton is associated with the glycosidic oxygen, which is the most basic site in the structure. This destabilizes the glycosidic bond, resulting in charge-localized and charge-driven fragmentation of the glycosidic bond. However, with the metal ion, the bonding is more complex where the charge of the metal ion is not directly associated with the glycosidic bond oxygen but is distributed to a number of local oxygen atoms. This results in a higher activation barrier for the induced fragmentation of the oligosaccharide. This effect is illustrated in Figure 32.8 where Figure 32.8a is for the low activation barrier protonated oligosaccharide and Figure 32.8b shows the high activation barrier metal ion adduct.

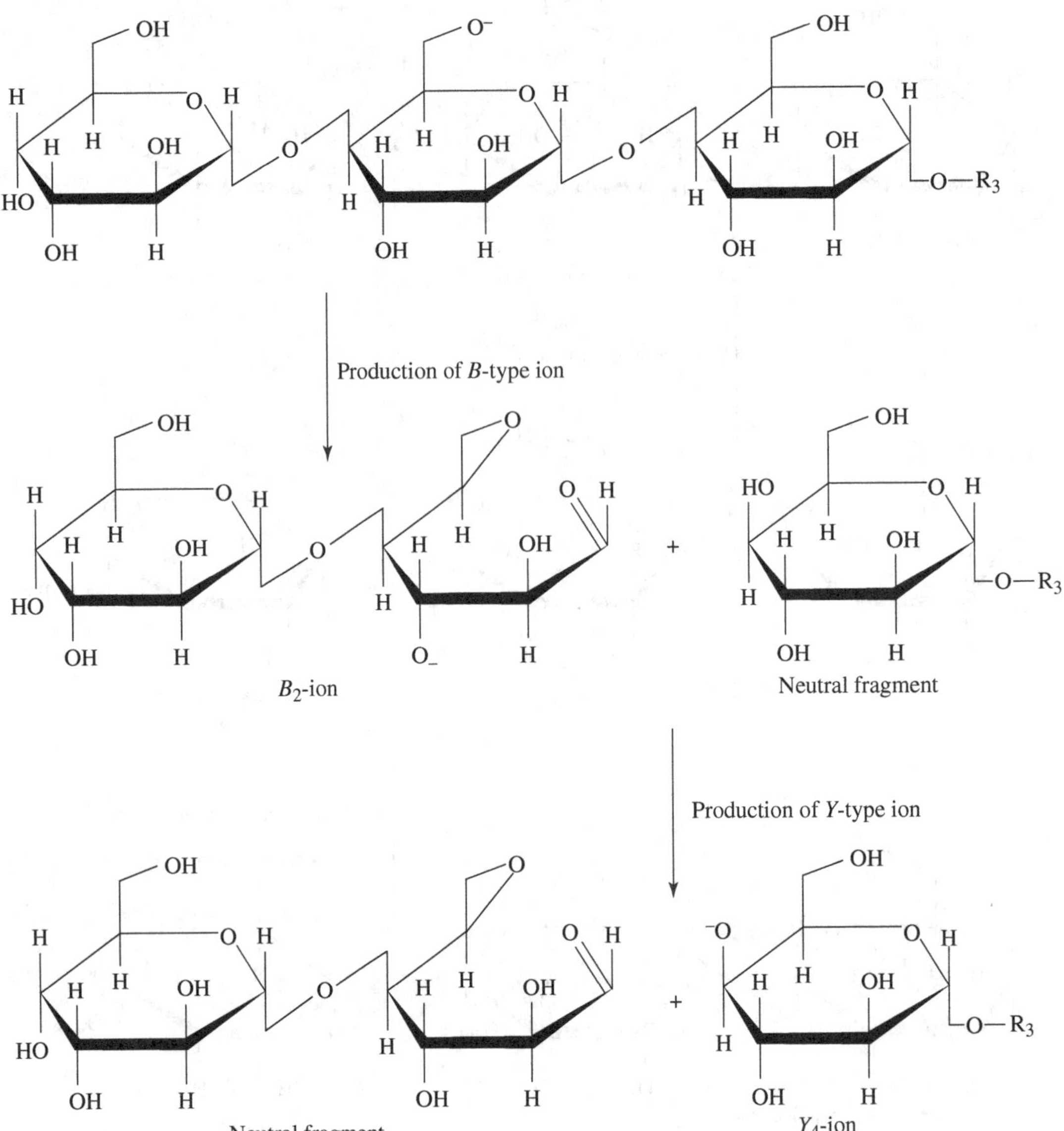

FIGURE 32.6 Production of B_2-type and Y_4-type ions for a six-sugar unit polysaccharide in negative ion mode.

32.2.3 Complex Oligosaccharide Structural Elucidation

It should be noted that it is actually quite difficult to completely sequence a highly branched and/or substituted polysaccharide that are often many units in length. The possible combinations are quite large, and a systematic approach to sequencing complicated polysaccharides does not exist. For neutral carbohydrates, the best approach is the product ion mass spectral generation using metal adducts in positive ion mode with either ESI or MALDI as the ionization source. Acidic oligosaccharides are measured in negative ion mode as deprotonated species using either ESI or MALDI as the ionization source. Typically, one cannot very readily interpret and assign a structure of a complicated oligosaccharide from its tandem mass spectrum. The best approach is the use of experience in combination with product ion spectral libraries to solve the structure of complicated, branched, and/or substituted oligosaccharides. There are some examples where with careful consideration a complex oligosaccharide structure can be either partially or sometimes fully elucidated with product ion spectral interpretation. High-energy product ion spectra of the hybrid glycan $(Man)_5(GlcNAc)_4$ was obtained using a magnetic sector mass spectrometer with a MALDI ionization source, a collision cell, and an orthogonal TOF mass analyzer by Bateman and coworkers [6] that allowed sequencing of the complex oligosaccharide. Figure 32.9 is the product ion spectrum of the high-energy collision-induced dissociation of the hybrid glycan $(Man)_5(GlcNAc)_4$. The spectrum has been split into two sections to better illustrate the rich abundance of product ions generated. The product ion spectrum contains

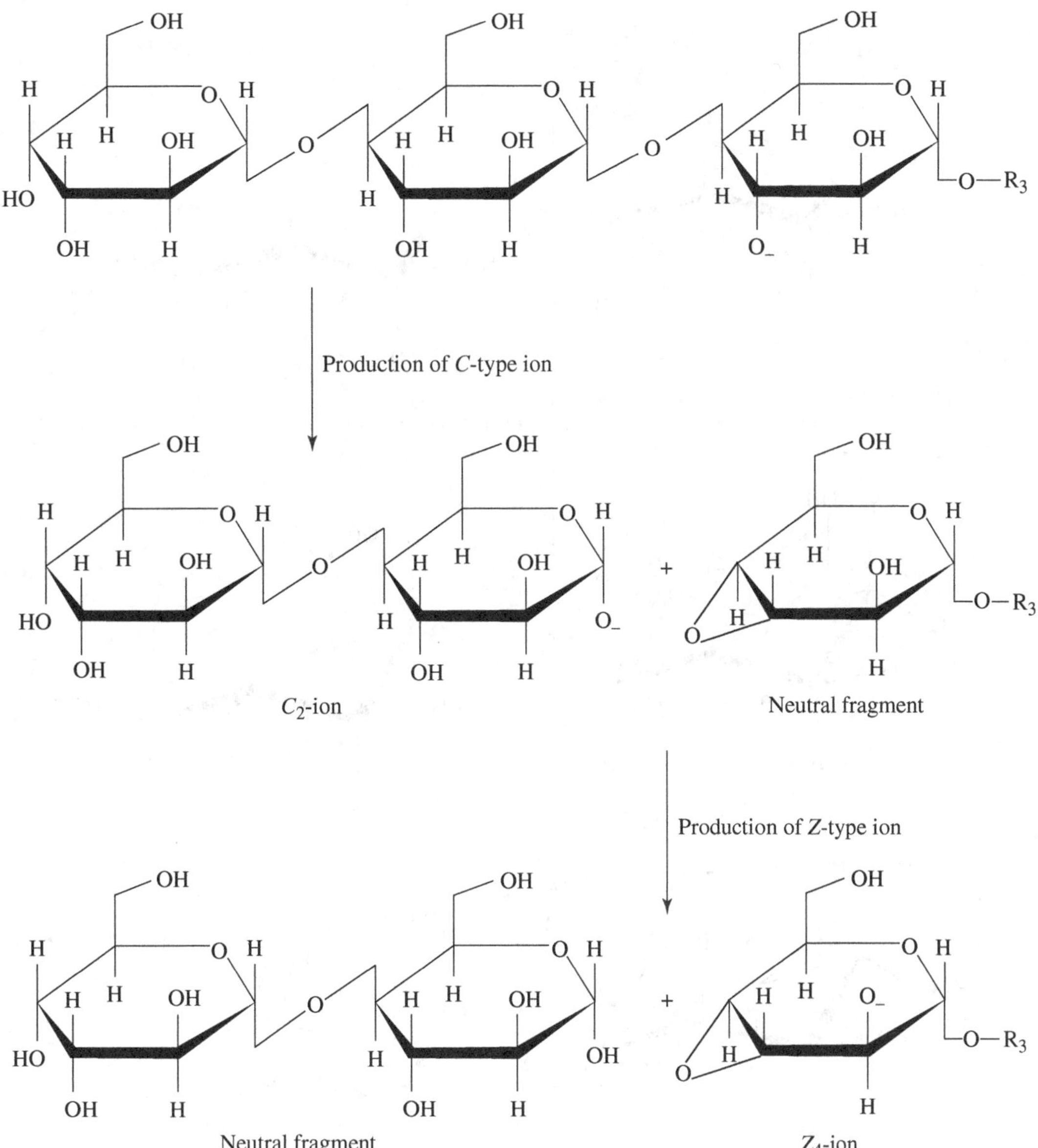

FIGURE 32.7 Production of C_2-type and Z_4-type ions for a six-sugar unit polysaccharide in negative ion mode.

B-, *Y*-, and *X*-type ions that allowed the structural elucidation of the oligosaccharide. Figure 32.10a and b illustrate the assignment of the product ions contained within the Figure 32.9 spectrum to the structure of the oligosaccharide.

32.3 NUCLEIC ACIDS

Nucleic acids are also analyzed using mass spectrometric techniques, and like the polysaccharide section just covered, we will start with a background look at the makeup of nucleic acids before looking at the mass spectrometry. Unlike polysaccharides and like the proteins, nucleic acids are specifically directional in their makeup and contain nonidentical monomers that have a distinct sequence that produce informational macromolecules. The nucleic acids reside in the nucleus of the cell and are the storage, expression, and transmission of genetic information of living species. The two types of nucleic acids are deoxyribonucleic acid (DNA) and ribonucleic acid (RNA). There are two distinct parts of their chemical structure and makeup that differentiate the two: (i) DNA contains the 5-carbon sugar deoxyribose, while RNA contains ribose, and (ii) DNA contains the base thymine (T), while RNA contains uracil (U). The molecules that make up the DNA and RNA structures are illustrated in Figure 32.11. This comprises the purine bases adenine (A) and guanine (G) and the pyrimidine bases cytosine (C), uracil (U), and thymine. Also illustrated in Figure 32.11 are the two sugars d-deoxyribose and d-ribose and finally the phosphate group

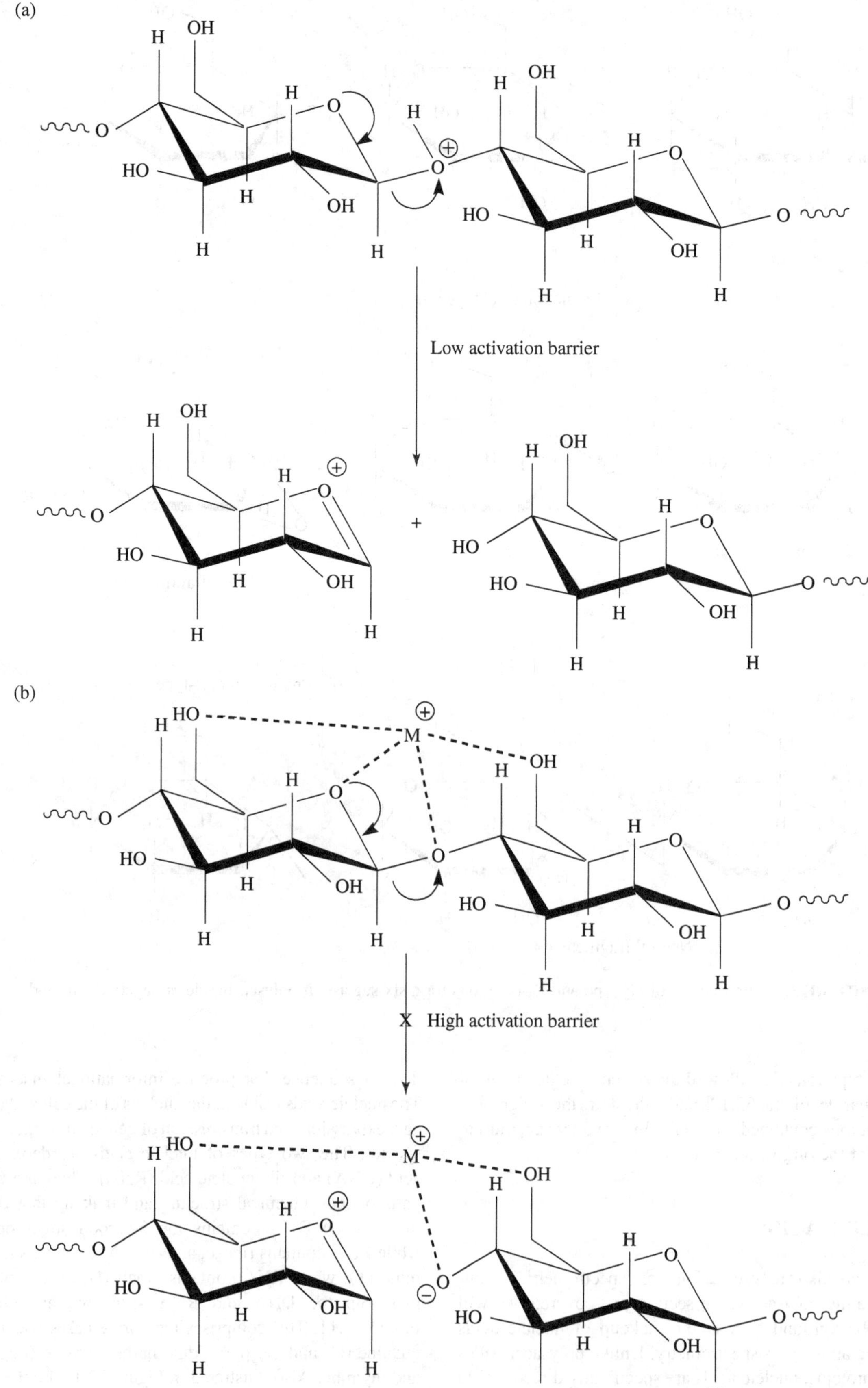

FIGURE 32.8 Fragmentation pathway mechanisms for the cleavage of the glycosidic bond through (a) protonation and (b) metal ion adduct. (Modified and reprinted with permission from Cancilla et al. [5].)

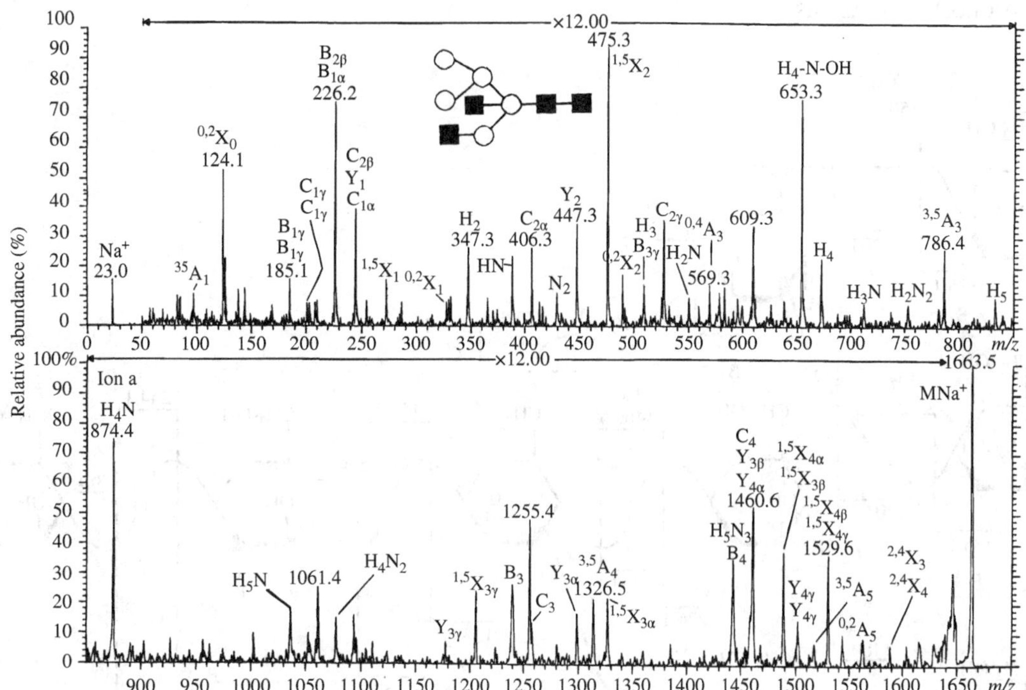

FIGURE 32.9 (a) High-energy (800 eV) MALDI±CID spectrum of the hybrid glycan $(Man)_5(GlcNAc)_4$ recorded from 2,5-DHB. The type D-ion is at *m/z* 874.4, and the presence of the bisecting GlcNAc residue is indicated by the ion 221 mass units lower at *m/z* 653.3. (Reprinted with permission. Harvey, D.J.; Bateman, R.H.; Green, M.R. High-energy collision-induced fragmentation of complex oligosaccharides ionized by matrix-assisted laser desorption/ionization mass spectrometry. *J. Mass Spectrom.* **1997**, 32, 167–187. Copyright 1997. Copyright John Wiley & Sons, Inc.)

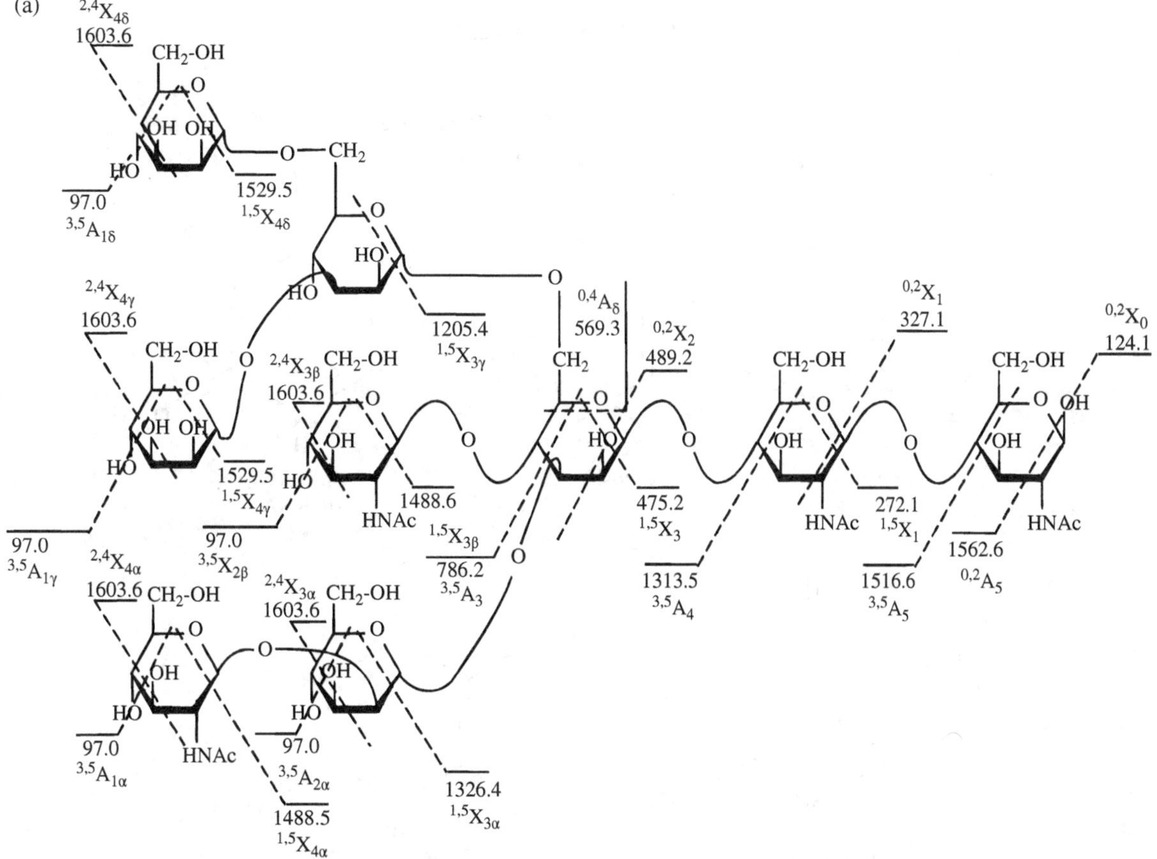

FIGURE 32.10 Scheme to show the formation of the cross-ring (a) and glycosidic (b) fragment ions for the spectrum shown in Figure 32.9a. (Reprinted with permission. Harvey, D.J.; Bateman, R.H.; Green, M.R. High-energy collision-induced fragmentation of complex oligosaccharides ionized by matrix-assisted laser desorption/ionization mass spectrometry. *J. Mass Spectrom.* **1997**, 32, 167–187. Copyright 1997. Copyright John Wiley & Sons, Inc.)

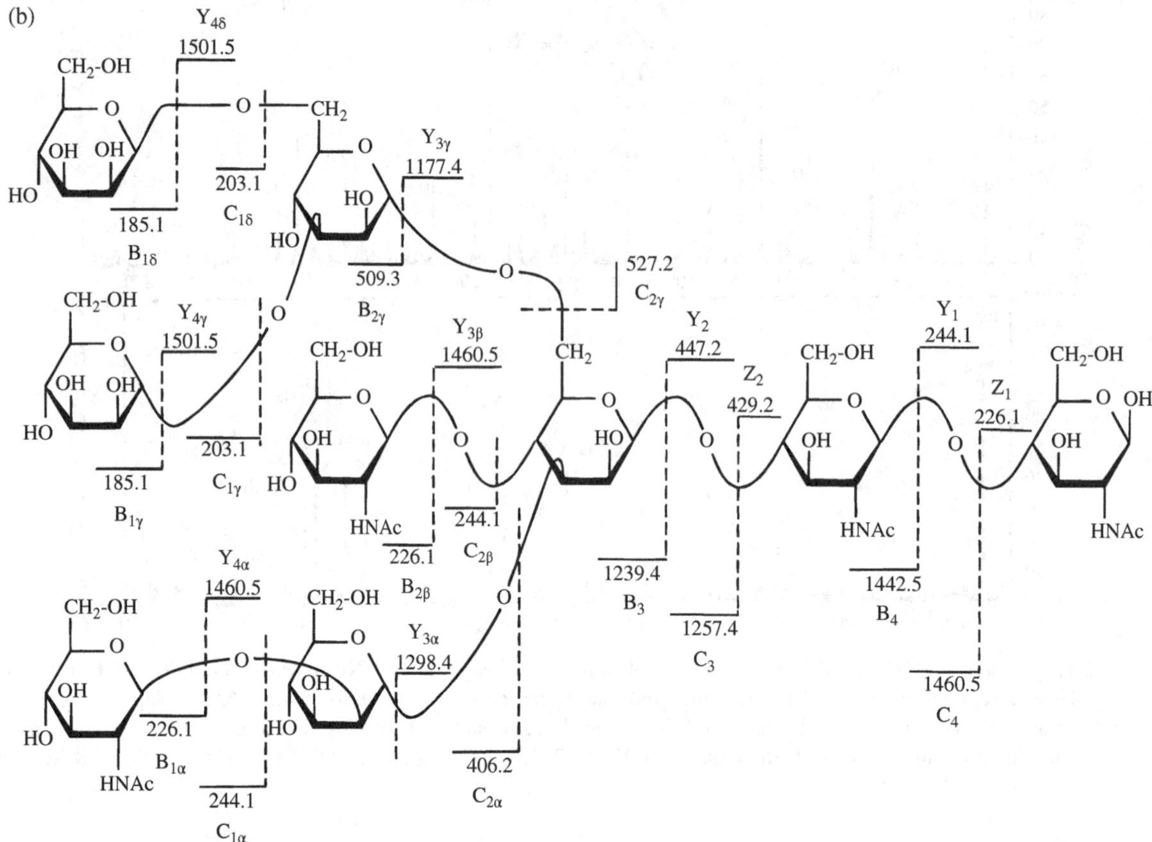

FIGURE 32.10 (Continued)

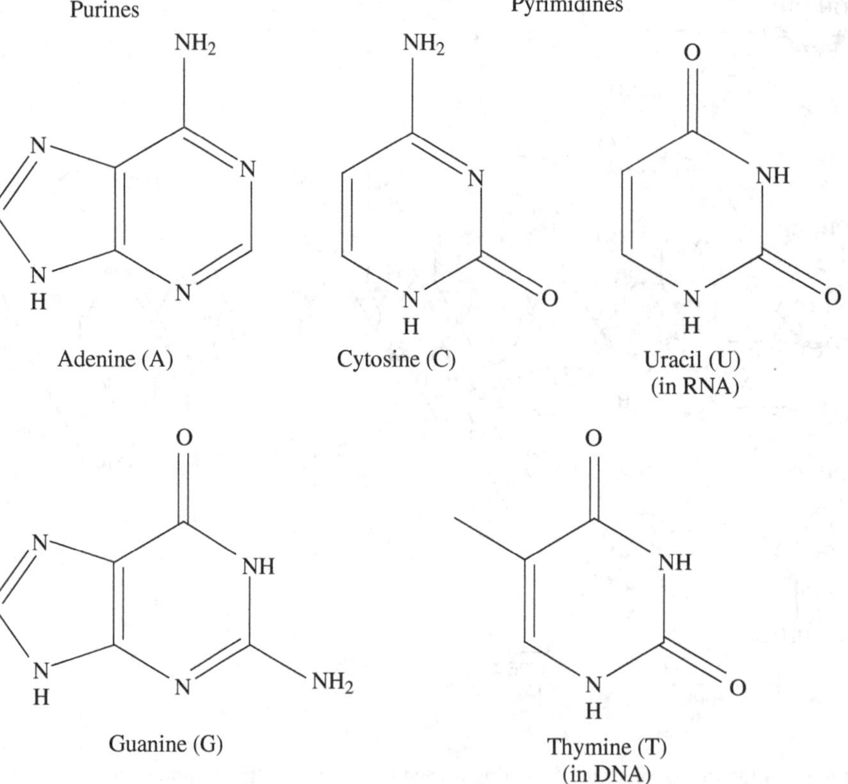

FIGURE 32.11 Structures of the molecules that make up the nucleic acids DNA and RNA. The purine bases adenine (A) and guanine (G); the pyrimidine bases cytosine (C), uracil (U), and thymine; the sugars d-deoxyribose and d-ribose; and the phosphate group.

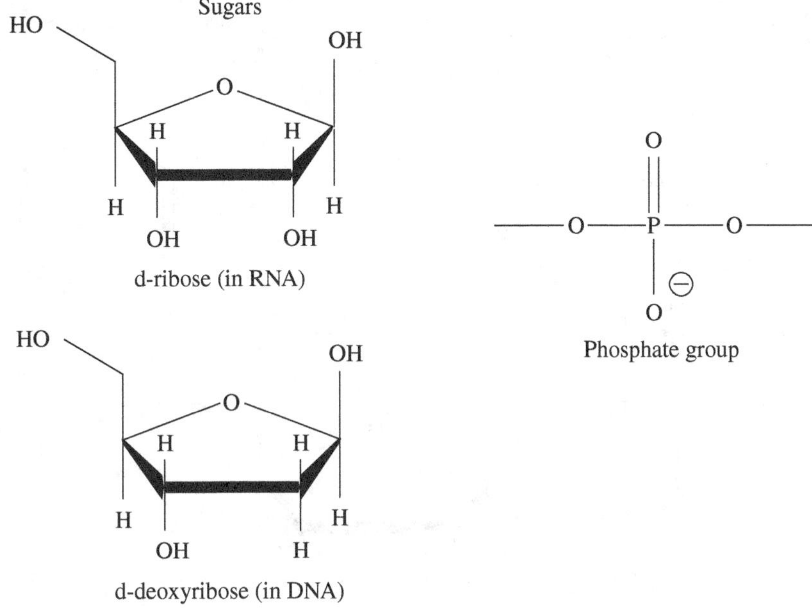

FIGURE 32.11 (Continued)

that acts as the backbone of the nucleic acids linking the nucleotides together. Nucleotides are the monomeric units that make up the nucleic acids. There are actually only four nucleotides that make up DNA and RNA, a much smaller number than the twenty amino acids found in proteins. Examples of nucleotides found in DNA and RNA are illustrated in Figure 32.12. Figure 32.12a is a DNA nucleotide where the number 2′ carbon in the sugar ring contains a hydrogen atom for d-deoxyribose. One of the bases will be attached to the 1′ carbon of the sugar through an aromatic nitrogen, and the phosphate will be attached to the number 5′ sugar carbon with a phosphoester bond. The RNA nucleotide illustrated in Figure 32.12b has the same types of bonding as illustrated for the DNA nucleotide but to a d-ribose sugar. In the case that the phosphate group is removed from the nucleotide, the remaining base sugar structure is called a nucleoside.

The nucleotides are linked to each other through the phosphate group forming a linear polymer. The nucleotides undergo a condensation reaction through the linking of the phosphate group on the 5′ carbon to the 3′ carbon of the next nucleotide known as a 3′,5′-phosphodiester bond. The resulting polynucleotide therefore has a 5′-hydroxyl group at the start (by convention) and a 3′-hydroxyl group at the end (by convention). Representative linear nucleotide structures are illustrated for (a) RNA and (b) DNA in Figure 32.13.

A similar naming scheme that is used for the fragmentation ions generated by CID of peptides was proposed by Glish and coworkers [7] for nucleic acids and is illustrated in Figure 32.14. There are four cleavage sites producing fragmentation along the phosphate backbone from CID. When the product ion contains the 3′-OH portion of the nucleic acid, the naming includes the letters *w*, *x*, *y*, and *z* where the numeral subscript is the number of bases from the associated terminal group. When the product ion contains the 5′-OH portion of the nucleic acid, the naming includes the letters *a*, *b*, *c*, and *d*. Losses are also more complicated than that shown in Figure 32.14 due to the neutral loss of base moieties. Figure 32.15 illustrates an actual structural cleavage at the w_2/a_2 site of a 4-mer nucleic acid's phosphate backbone according to the naming scheme of Figure 32.14. Numerous mechanisms have also been reported for the fragmentation pathways leading to charged base loss and also neutral base loss. These are losses that are observed in product ion spectra other than the cleavage along the phosphate backbone that is illustrated in Figures 32.14 and 32.15. Neutral and charged base losses add to the complexity of the product ion spectra but also add information concerning the makeup of the oligonucleotide. Figure 32.16 illustrates a couple examples of proposed fragmentation pathway mechanisms for neutral and charged base losses. In Figure 32.16a, a simple nucleophilic attack on the C-1′ carbon atom by the phosphodiester group results in the elimination of a charged base [8]. Figure 32.16b illustrates a two-step reaction where in the first step there is a neutral base loss followed by breakage of the 3′-phosphoester bond [9]. There are other proposed fragmentation pathways for a number of other possible mechanisms for the production of the product ions observed in tandem mass spectra of the nucleic acids.

32.3.1 Negative Ion Mode ESI of a Yeast 76-mer tRNA^{Phe}

In electrospray ionization, it has been observed that the nucleic acids posses an enhanced response in the negative ion mode due to the presence of the phosphate groups in the nucleic acid

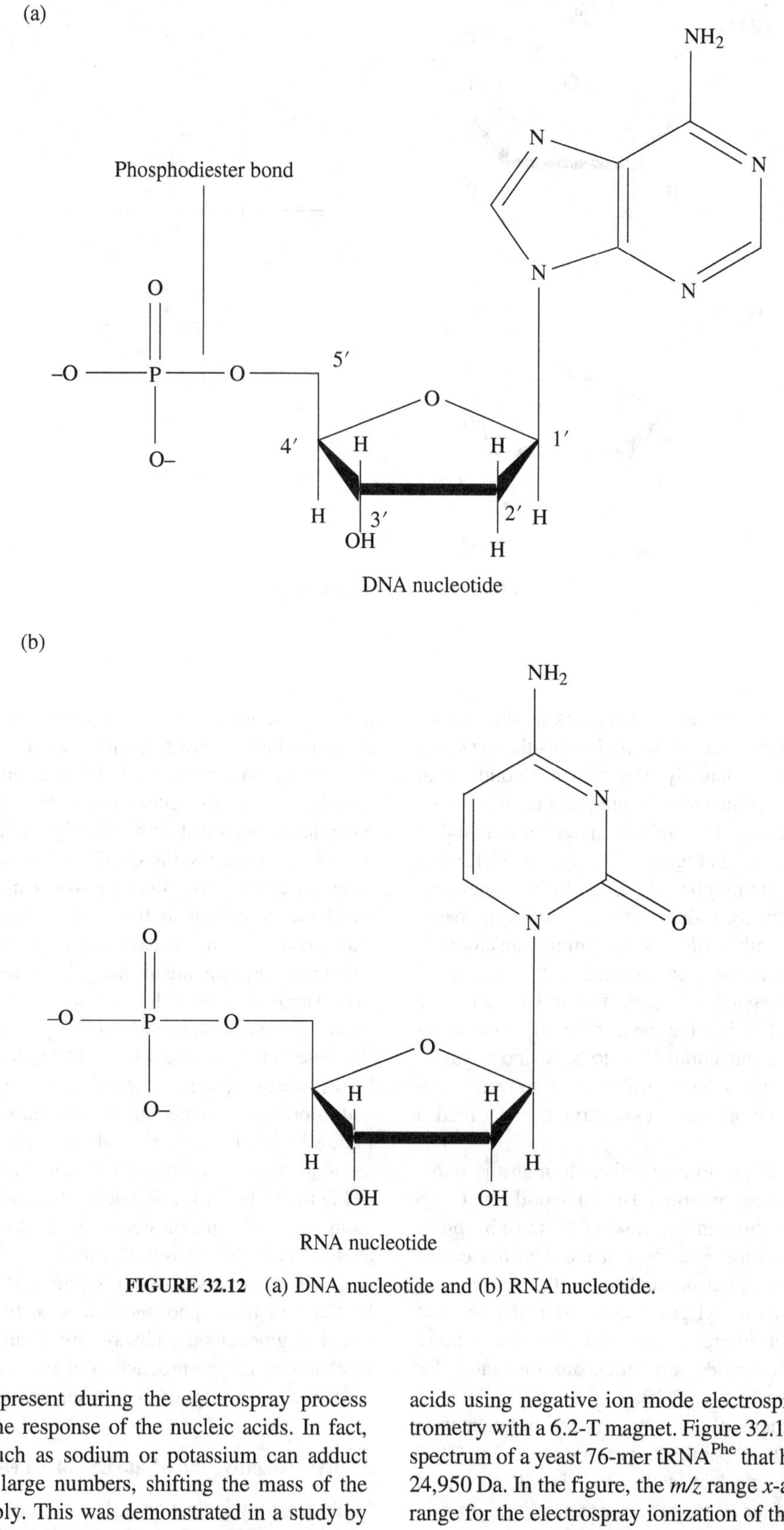

(a)

(b)

FIGURE 32.12 (a) DNA nucleotide and (b) RNA nucleotide.

backbone. Salt being present during the electrospray process though can suppress the response of the nucleic acids. In fact, many metal species such as sodium or potassium can adduct to the nucleic acid in large numbers, shifting the mass of the nucleic acid considerably. This was demonstrated in a study by McLafferty and coworkers [10] that looked at different nucleic acids using negative ion mode electrospray FTICR mass spectrometry with a 6.2-T magnet. Figure 32.17 is a single-stage mass spectrum of a yeast 76-mer $tRNA^{Phe}$ that has a molecular mass of 24,950 Da. In the figure, the *m/z* range *x*-axis is in the 26,000 Da range for the electrospray ionization of the $tRNA^{Phe}$. This is due to the species having between 34 and 55 Na adducts changing its

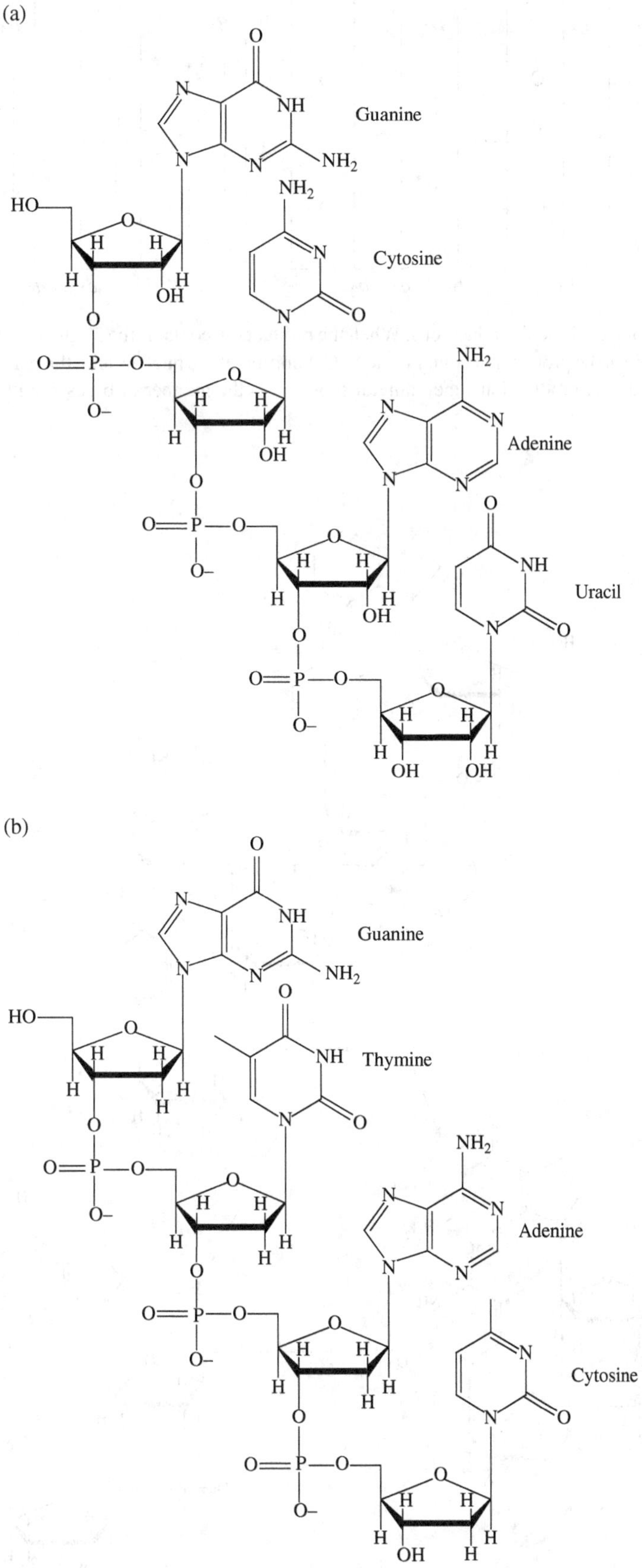

(a)

(b)

FIGURE 32.13 Linear nucleic acid structures for (a) RNA and (b) DNA.

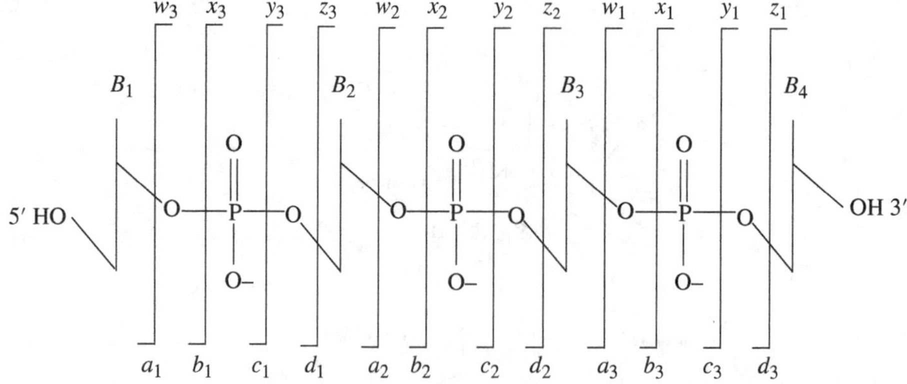

FIGURE 32.14 Naming scheme for nucleic acid product ions. When the product ion contains the 3′-OH portion of the nucleic acid, the naming includes the letters *w*, *x*, *y*, and *z*. When the product ion contains the 5′-OH portion of the nucleic acid, the naming includes the letters *a*, *b*, *c*, and *d*. In both 3′-OH- and 5′-OH-containing product ions, the numeral subscript is the number of bases from the associated terminal group.

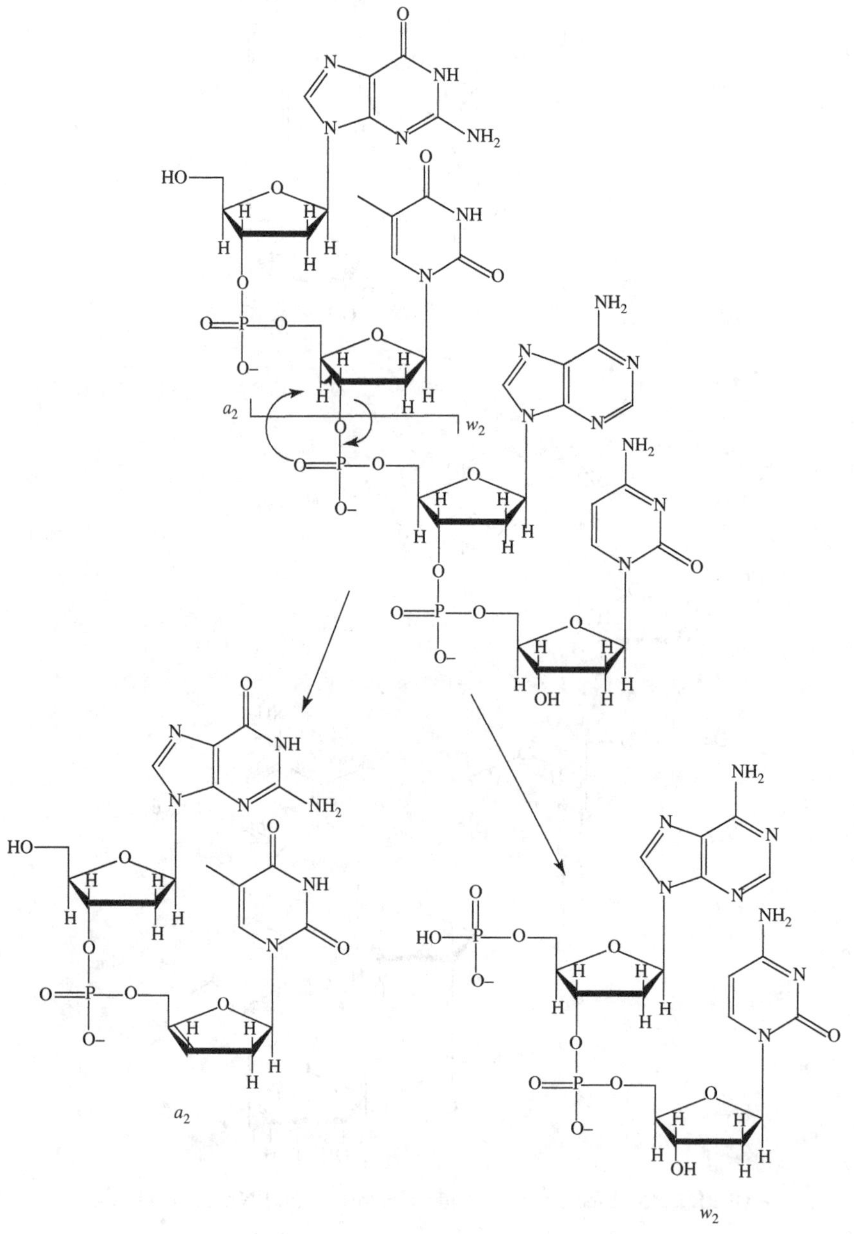

FIGURE 32.15 Example of structural cleavage at the w_2/a_2 site of a 4-mer nucleic acid's phosphate backbone according to the naming scheme of Figure 9.14.

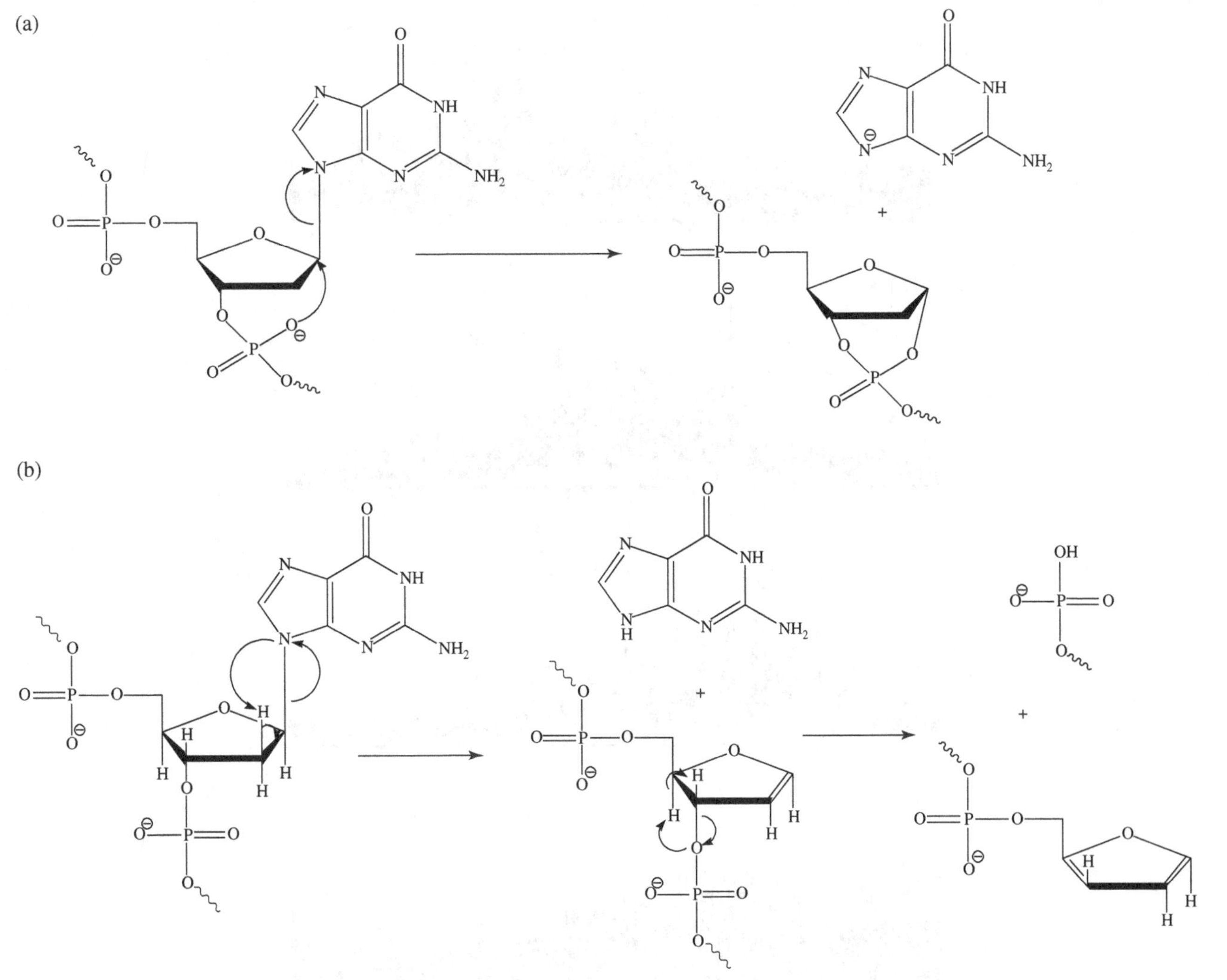

(a)

(b)

FIGURE 32.16 Proposed fragmentation pathways associated with the base substituent groups. (a) Nucleophilic attack on the C-1′ carbon atom by the phosphodiester group results in the elimination of a charged base. (b) Two-step reaction mechanism where in the first step there is neutral base loss followed by breakage of the 3′-phosphoester bond.

mass by 750–1200 Da. The sodium metal atoms will associate with the phosphate groups in place of the hydrogen protons. The higher the charge state of the nucleic acid, the higher the amount of the adduct sodium will be. The sample was desalted using HPLC, and the single-stage electrospray negative ion mode mass spectra were recollected. Figure 32.18 illustrates the same yeast $tRNA^{Phe}$ after the HPLC desalting. In the lower spectrum, a number of different charge states are now observed with a much improved response in the signal-to-noise (S/N) ratio. The phosphate groups are now deprotonated and are giving the $tRNA^{Phe}$ a series of high negative charge states from 27^- to 18^- in the figure. The upper portion of Figure 32.18 is an expanded view of the 24^- charge state peak revealing that the peak in the lower spectrum is actually comprised of a series of peaks. This is the isotopic distribution of the $tRNA^{Phe}$ species. The FTICR mass spectrometer used possesses the resolution high enough to resolve these isotopic peaks to give the upper spectrum.

32.3.2 Positive Ion Mode MALDI Analysis

Oligonucleotides are also analyzed using the MALDI ionization technique that allows the measurement of a singly charged species. The MALDI ionization technique results in the production of ionized species that carry a single charge from an adducted proton or a metal cation such as sodium. This can allow studies of fragmentation pathways of species that have undergone hydrogen/deuterium exchange (H/DX), a method where exchangeable hydrogen such as hydroxyl hydrogen are replaced with deuterium (H = 1.007825 Da, D = 2.01565 Da). Hydrogen/deuterium studies of oligonucleotides have been extensively analyzed using mass spectrometry to characterize the fragmentation pathways of oligonucleotides [11–15]. Using H/DX in conjunction with mass spectrometric analysis of oligonucleotide product ion generation, it has been proposed that for DNA every fragmentation pathway is initiated by loss of a nucleobase

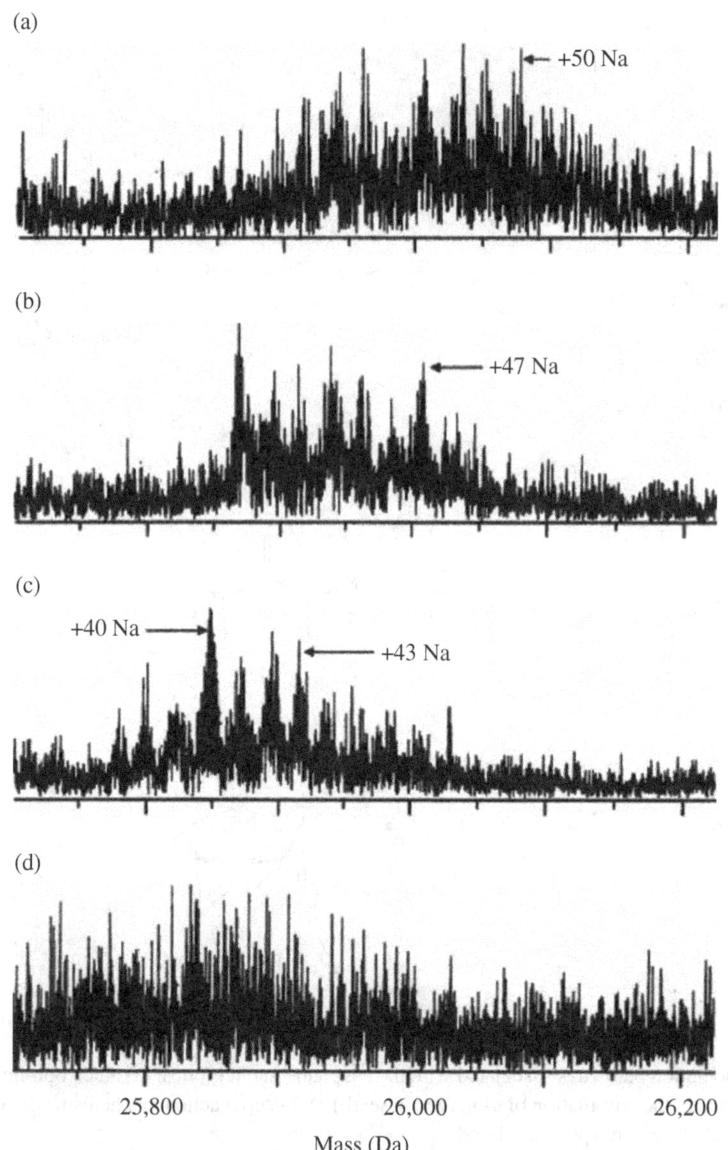

FIGURE 32.17 ESI FTMS spectra for the 16^- (a), 18^- (b), 20^- (c), and 22^- (d) anions of yeast $tRNA^{Phe}$ without desalting. (Reprinted with permission from Ref. [10]. Copyright 1995 National Academy of Sciences, U.S.A.)

that does not involve simple water loss or 3′- and 5′-terminal nucleoside/nucleotides.

In a study conducted by Anderson et al. [16], tandem mass spectra of H/D-exchanged (i) UGUU and (ii) UCUA, small RNA oligonucleotides were collected. The mass spectra were collected on a 4700 Proteomics Analyzer (Applied Biosystems, Framingham, MA) that is a tandem in space time-of-flight (TOF)–time-of-flight (TOF) mass spectrometer with MALDI as the ionization source (MALDI TOF–TOF/MS). As described in Chapter 30, Section 30.4, the MALDI TOF–TOF mass spectrometer contains a gas-filled collision cell located in between the two TOF mass analyzers that allows collision-induced dissociation generation of product ions. The MALDI TOF–TOF mass analyzer is one of the few commercially available mass analyzers that are capable of high-energy collision-induced dissociation product ion generation. Collision-induced dissociation and PSD analysis using MALDI TOF mass spectrometry has been observed to produce similar product ions, indicating that the fragmentation mechanisms are the same. The structures of the UGUU and UCUA deuterated RNA oligonucleotides used in the experiments to generate the product ion spectra are illustrated in Figure 32.19. In both spectra, a number of *a*-type, *z*-type, *y*-type, and *w*-type product ions are observed. In Figure 32.19, the exchangeable hydrogens, the sugar hydroxyl hydrogen and the base nitrogen bound hydrogen, are replaced by deuterium. The exchange of deuterium for hydrogen allows the interpretation of the fragmentation pathways that generate the observed product ions. An example of mechanism determination is illustrated in Figure 32.20 for the charged cytosine base loss from the UCUA oligonucleotide.

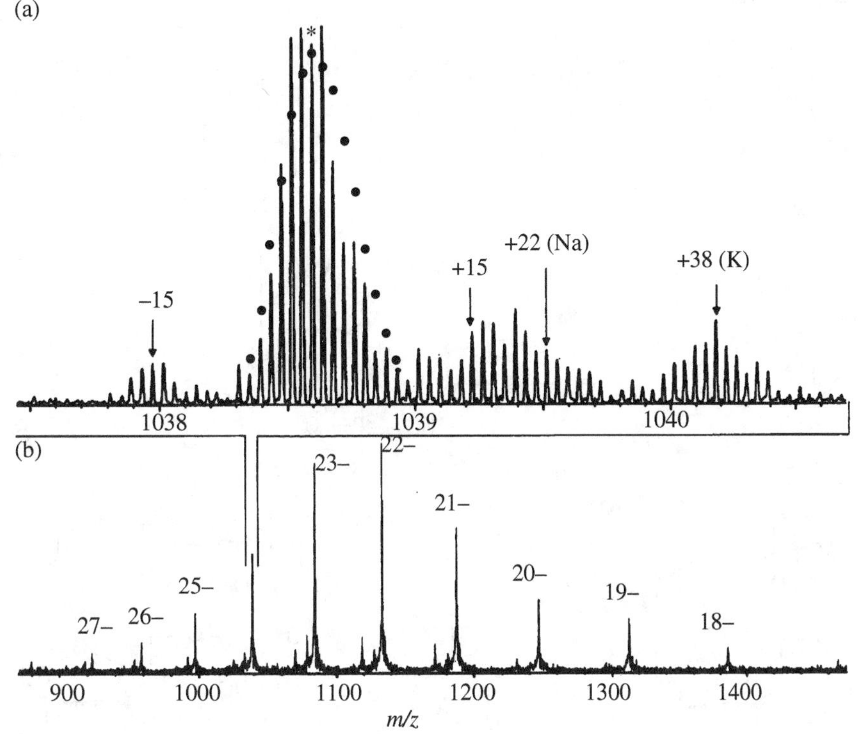

FIGURE 32.18 (b) ESI FTMS spectrum of yeast tRNAPhe after desalting. (a) The 24- anion region [heterodyne scan, SWIFT (34) isolation of 23- to 25-region]; after fitting the $(M{-}24H)^{24-}$ isotopic peak abundances to those expected theoretically (indicated with dots), the asterisk indicates the peak that corresponds to the most abundant of the theoretical distribution. Peaks at approximately −15 and +15 Da are from minor variants. (Reprinted with permission from Ref. [10]. Copyright 1995 National Academy of Sciences, U.S.A.)

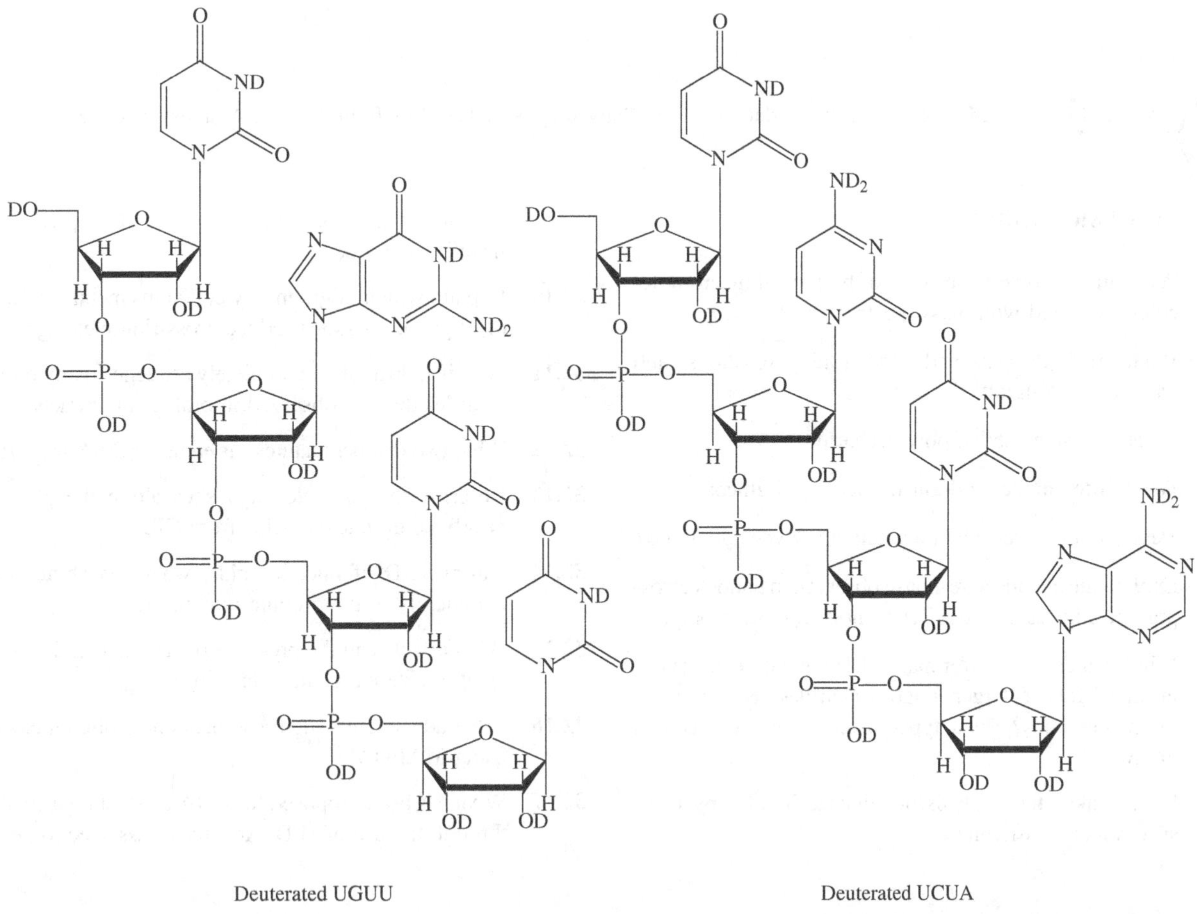

Deuterated UGUU

Deuterated UCUA

FIGURE 32.19 Structures of the RNA oligonucleotides UGUU and UCUA used in the study for the product ions.

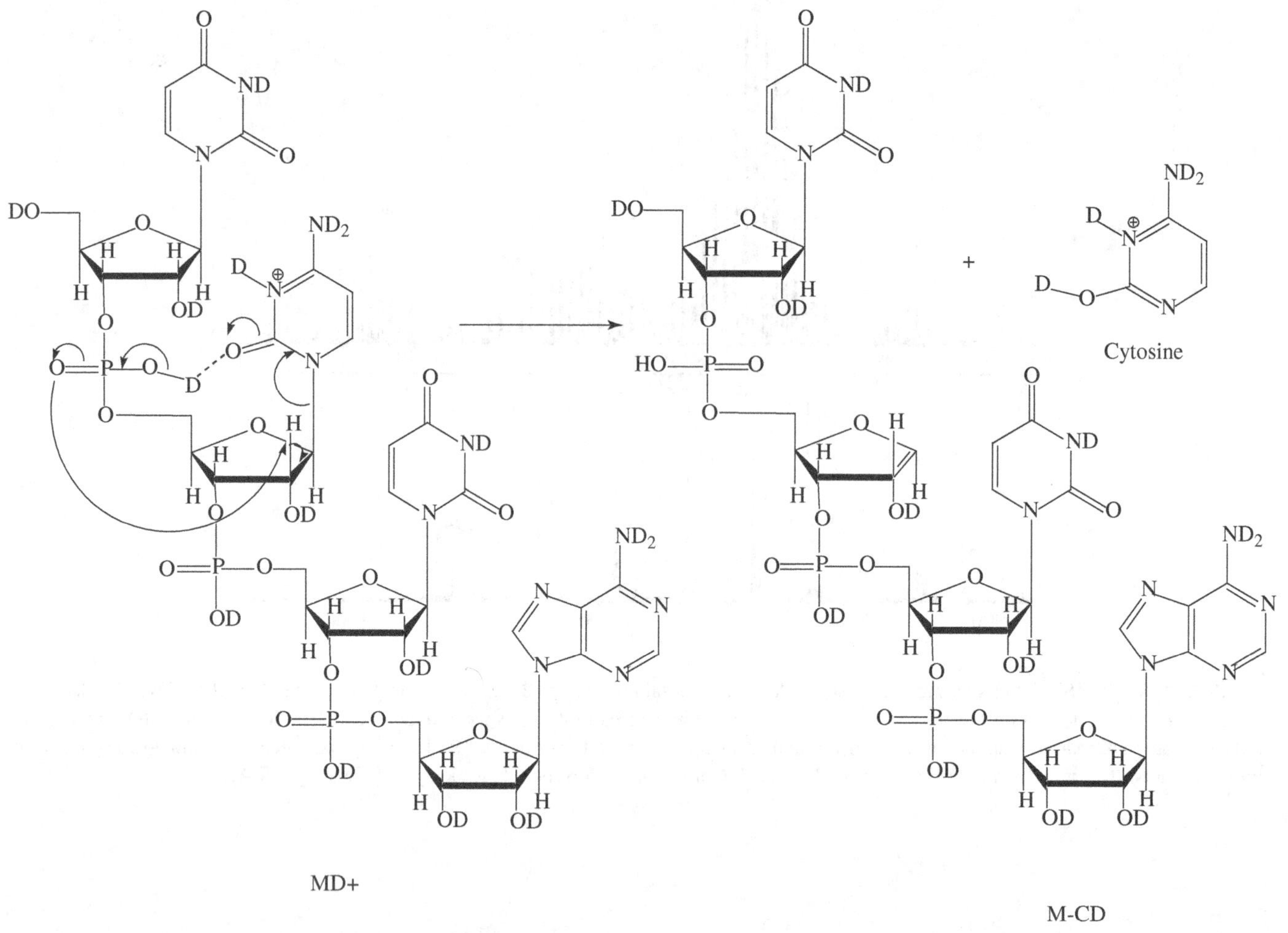

MD+

M-CD

FIGURE 32.20 Mechanism for the production of the charged cytosine base loss from the UCUA oligonucleotide.

32.4 CHAPTER PROBLEMS

32.1 What are the three main types of biological macromolecules measured with mass spectrometry?

32.2 What are hydrolysis and condensation reactions, and what are they involved in?

32.3 What are three uses of polysaccharides?

32.4 What differentiates α-d-glucose from β-d-glucose?

32.5 Describe a α-glycosidic bond and a β-glycosidic bond?

32.6 Explain the enhanced response observed in nanoelectrospray of oligosaccharides versus normal electrospray.

32.7 Why is there less informative fragmentation observed in the CID of larger oligosaccharides as compared to smaller ones? What types of cleavages are seen in the two?

32.8 What makes the analysis of oligosaccharides by mass spectrometry difficult?

32.9 Explain what happens for production of the β-type ion in negative ion mode.

32.10 Explain why in high-energy collisions metal ion adducts of oligosaccharides produce cross-ring cleavages.

32.11 Why is it difficult to completely sequence many polysaccharides that are observed in biological extracts?

32.12 What two distinct features differentiate DNA from RNA?

32.13 Describe the four cleavage sites along the phosphate backbone on nucleic acids from CID.

32.14 During CID of nucleic acids, what contributes to the complexity of the product ion spectra?

32.15 What effects can the presence of salt have in the electrospray ionization of nucleic acids?

32.16 What advantage exits for ionization of oligonucleotides using MALDI?

32.17 What has been proposed for DNA product ion generation through the use of H/D exchange mass spectrometry?

REFERENCES

1. Bahr, U.; Pfenninger, A.; Karas, M.; Stahl, B. *Anal. Chem.* 1997, **69**, 4530–4535.
2. Harvey, D.J. *Rapid Commun. Mass Spectrom.* 1993, **7**, 614–619.
3. Hofmeister, G.E.; Zhou, Z.; Leary, J.A. *J. Am. Chem. Soc.* 1991, **113**, 5964–5970.
4. Domon, B.; Costello, C.E. *Glycoconjugate J.* 1988, **5**, 397–409.
5. Cancilla, M.T.; Penn, S.G.; Carroll, J.A.; Lebrilla, C.B. *J. Am. Chem. Soc.* 1996, **118**, 6736–6745.
6. Clayton, E.; Bateman, R.H. *Rapid Commun. Mass Spectrom.* 1992, **6**, 719–720.
7. McLuckey, S.A.; Van Berkel, G.J.; Glish, G.L. *J. Am. Soc. Mass Spectrom.* 1992, **3**, 60–70.
8. Cerny, R.L.; Gross, M.L.; Grotjahn, L. *Anal. Biochem.* 1986, **156**, 424.
9. McLuckey, S.A.; Habibi-Goudarzi, S. *J. Am. Chem. Soc.* 1993, **115**, 12085.
10. Little, D.P.; Thannhauser, T.W.; McLafferty, F.W. *Proc. Natl. Acad. Sci. U. S. A.* 1995, **92**, 2318–2322.
11. Wu, J.; McLuckey, S.A. *Int. J. Mass Spectrom.* 2004, **237**, 197–241.
12. Phillips, D.R.; McCloskey, J.A. *Int. J. Mass Spectrom. Ion Process.* 1993, **128**, 61–82.
13. Gross, J.; Leisner, A.; Hillenkamp, F.; Hahner, S.; Karas, M.; Schafer, J.; Lutzenkirchen, F.; Nordhoff, E. *J. Am. Soc. Mass Spectrom.* 1998, **9**, 866–878.
14. Wan, K.X.; Gross, J.; Hillenkamp, F.; Gross, M.L. *J. Am. Soc. Mass Spectrom.* 2001, **12**, 193–205.
15. Gross, J.; Hillenkamp, F.; Wan K.X.; Gross, M.L. *J. Am. Soc. Mass Spectrom.* 2001, **12**, 180–192.
16. Anderson, T.E.; Kirpekar, F.; Haselmann, K.F. *J. Am. Soc. Mass Spectrom.* 2006, **17**, 1353–1368.

33

BIOMOLECULE SPECTRAL INTERPRETATION: PROTEINS

33.1 Introduction to Proteomics
33.2 Protein Structure and Chemistry
33.3 Bottom-up Proteomics: Mass Spectrometry of Peptides
  33.3.1 History and Strategy
  33.3.2 Protein Identification through Product Ion Spectra
  33.3.3 High-Energy Product Ions
  33.3.4 De Novo Sequencing
  33.3.5 Electron Capture Dissociation
33.4 Top-Down Proteomics: Mass Spectrometry of Intact Proteins
  33.4.1 Background
  33.4.2 GP Basicity and Protein Charging
  33.4.3 Calculation of Charge State and Molecular Weight
  33.4.4 Top-Down Protein Sequencing
33.5 PTM of Proteins
  33.5.1 Three Main Types of PTM
  33.5.2 Glycosylation of Proteins
  33.5.3 Phosphorylation of Proteins
  33.5.4 Sulfation of Proteins
33.6 Systems Biology and Bioinformatics
  33.6.1 Biomarkers in Cancer
33.7 Chapter Problems
References

33.1 INTRODUCTION TO PROTEOMICS

Proteomics is the study of a biological system's complement of proteins (e.g., from cell, tissue, or a whole organism) at any given state in time and has become a major area of focus for research and study in many different fields and applications. In proteomic studies, mass spectrometry can be employed to analyze both the intact, whole protein and the resultant peptides obtained from enzyme-digested proteins. The mass spectrometric analysis of whole intact proteins is often called top-down proteomics where the measurement study starts with the analysis of the intact protein in the gas phase (GP) and subsequently investigates its identification and any possible modifications through collision-induced dissociation (CID) measurements. The mass spectrometric analysis of enzyme-digested proteins that have been converted to peptides is known as bottom-up proteomics. Finally, mass spectrometry is also used to study posttranslational modifications (PTM) that have taken place with the proteins such as glycosylation, sulfation, and phosphorylation. We shall begin with a look at bottom-up proteomics, the most common approach, followed by top-down proteomics, which is seeing more applications and study lately, and finally the PTM of glycosylation, sulfation, and phosphorylation. Bioinformatics has become an important tool used in the interpretation of results obtained from mass spectrometry studies. In the last part of this chapter, we will briefly look at what bioinformatics is and what it can be used for in relation to mass spectrometry and proteomic studies. Due to the enormous impact of proteomics on research into biological processes, organisms, diseased states, tissues, and so on, we will begin this section by starting with a brief overview of proteins including their structure and makeup.

33.2 PROTEIN STRUCTURE AND CHEMISTRY

Of all biological molecules, proteins are one of the most important, next only to the nucleic acids. All living cells contain proteins, and their name is derived from the Greek word proteios, which has the meaning of "first" [1]. There are two broad classifications for proteins related to their structure and functionality: water-insoluble fibrous proteins and water-soluble globular proteins. The three-dimensional configuration of a protein is described by its primary, secondary, tertiary, and quaternary structures. Figure 33.1 shows a three-dimensional ribbon representation of the protein RNase. The primary structures of proteins are made up of a sequence of amino acids forming a polypeptide chain. Typically, if the chain is less than 10,000 Da, the compound is called a polypeptide, and if greater than 10,000 Da, the compound is called a protein. There are 20 amino acids that make up the protein chains through carbon to nitrogen peptide bonds. Figure 33.2 illustrates the 20 amino acid structures that make up the polypeptide backbone chain of proteins. Amino acids possess an amino group (NH_2) and a carboxyl group (COOH) that are bonded to the same carbon atom that is alpha to

Analytical Chemistry: A Chemist and Laboratory Technician's Toolkit, First Edition. Bryan M. Ham and Aihui MaHam.

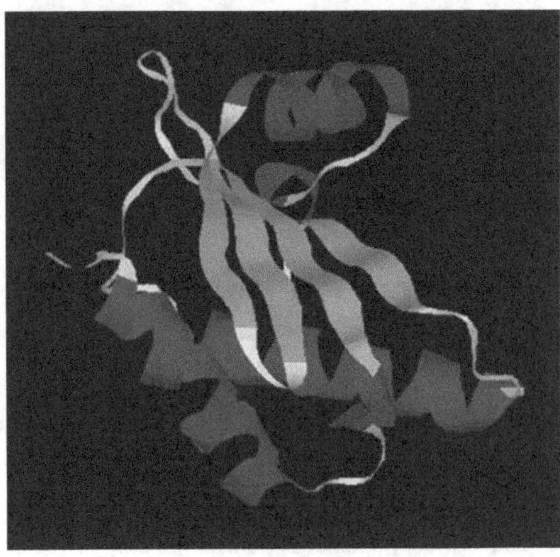

FIGURE 33.1 Ribbon structure representation of the RNase protein illustrating substructures of alpha-helixes and beta-sheets.

both groups; therefore, amino acids are called alpha amino acids (α-amino acid). At physiological pH (~7.36), the amino acids can be subdivided into four classes according to their structure, polarity, and charge state: (i) negatively charged consisting of aspartic acid (Asp) and glutamic acid (Glu); (ii) positively charged consisting of lysine (Lys), arginine (Arg), and histidine (His); (iii) polar consisting of serine (Ser), threonine (Thr), tyrosine (Tyr), cysteine (Cys), glutamine (Gln), and asparagine (Asn); and (iv) nonpolar consisting of glycine (Gly), leucine (Leu), isoleucine (Ile), alanine (Ala), valine (Val), proline (Pro), methionine (Met), tryptophan (Trp), and phenylalanine (Phe). The carbon to nitrogen peptide bonds are formed through condensation reactions between the carboxyl and amino groups. An example condensation reaction between the amino acids leucine and tyrosine is illustrated in Figure 33.3. The peptide C—N bonds are found to be shorter than most amine C—N bonds due to a double bond nature that contributes to 40% of the peptide bond [2]. This double bond character lessens the free rotation of the bond, thus affecting the overall structure of the protein [3]. The secondary structure of the protein is described by two different configurations and turns. The two configurations are alpha-helixes (first proposed by Pauling and Corey in 1951) and beta-sheets (parallel and antiparallel) and are illustrated in Figure 33.4. The alpha-helix is described as a right-hand-turned spiral that has hydrogen bonding between oxygen and the hydrogen of the nitrogen atoms of the chain backbone. This hydrogen bonding stabilizes the helical structure. The R-group side chains that make up the amino acid residues extrude out from the helix. The beta-sheet is a flat structure that also has hydrogen bonding between oxygen and the hydrogen of the nitrogen atoms but from different beta-sheets (parallel and antiparallel) that run alongside each other. These hydrogen bonds also work to stabilize the structure. The R-group side chains alternatively extrude out flat with the sheet from the sides of the sheet. The third secondary structure, the turn, basically changes the direction of the polypeptide strand. The tertiary structure, which includes the disulfide bonds, is comprised of the ordering of the secondary structure, which is stabilized through side chain interactions. The quaternary structure is the arrangement of the polypeptide chains into the final working protein. All of these structures describe what is actually a folded protein, where the apolar regions of the protein are tucked away inside the structure, away from the aqueous medium they are found in naturally, and more polar regions are on the surface.

33.3 BOTTOM-UP PROTEOMICS: MASS SPECTROMETRY OF PEPTIDES

33.3.1 History and Strategy

The proteomic approach comprised of measuring the enzymatic products of the protein digestion (after protein extraction from the biological sample), namely, the peptides, using mass spectrometry is known as bottom-up proteomics. In the bottom-up approach using nano-electrospray ionization high-performance liquid chromatography–mass spectrometry (nano-ESI-HPLC/MS), the peptides are chromatographically separated and subjected to CID in the GP. The product ion spectra thus obtained of the separated peptides are then used to identify the proteins present in the biological system being studied. Prior to the use of nano-ESI-HPLC/MS for peptide measurement, Edman degradation was used to sequence unknown proteins. The method of Edman sequencing involves the removal of each amino acid residue one by one from the polypeptide chain starting from the N-terminus of the peptide or protein [4]. The method worked well for highly purified protein samples that contained a free amino N-terminus, but the analysis was slow usually taking a day to analyze the sequence of one protein. Mass spectrometry was first coupled with Edman sequencing in 1980 by Shimonishi et al. [5] where the products of the Edman degradation were measured using field desorption (FD) mass spectrometry. FD, introduced in 1969 by Beckey, is an ionization technique not commonly in use today. FD consists of depositing the sample, either solid or dissolved in solvent, onto a needle and applying a high voltage. The processes of desorption and ionization are obtained simultaneously. The analyte ions produced from the FD are then introduced into the mass spectrometer for mass analysis. Fast atom bombardment was also used as an ionization technique to measure peptides obtained from the Edman sequencing approach [6].

Another early approach to proteomics using mass spectrometry was the application of matrix-assisted laser desorption ionization (MALDI) time-of-flight (TOF) mass spectrometry (MALDI-TOF/MS) to the measurement of peptides obtained from in-gel digestions of proteins separated by gel electrophoresis. This technique was reported by several groups and called peptide mass fingerprinting (PMF) [7–9]. In the PMF approach, proteins are first separated using two-dimensional gel electrophoresis (2-DE), a protein separation technique first introduced in the 1970s [10]. The gel used in electrophoresis is a rectangular gel comprised of polyacrylamide. The protein sample is loaded onto the gel, and the proteins are separated according to their isoelectric point (pH where the protein has a zero charge). This is the first dimension of the separation. The second dimension is a linear separation of the proteins according to their molecular weights. In preparation for sodium dodecyl sulfate polyacrylamide gel electrophoresis (SDS-PAGE), the proteins are first denatured (usually with 8 M urea and boiling), and sulfide bonds are cleaved effectively, unraveling the tertiary and secondary structure of the protein. Sodium dodecyl sulfate, which is negatively charged, is then used to coat the protein in a fashion that is proportional to the proteins' molecular weight. The proteins are then separated within a polyacrylamide gel by placing a potential difference across the gel. Due to the potential difference across the gel, the

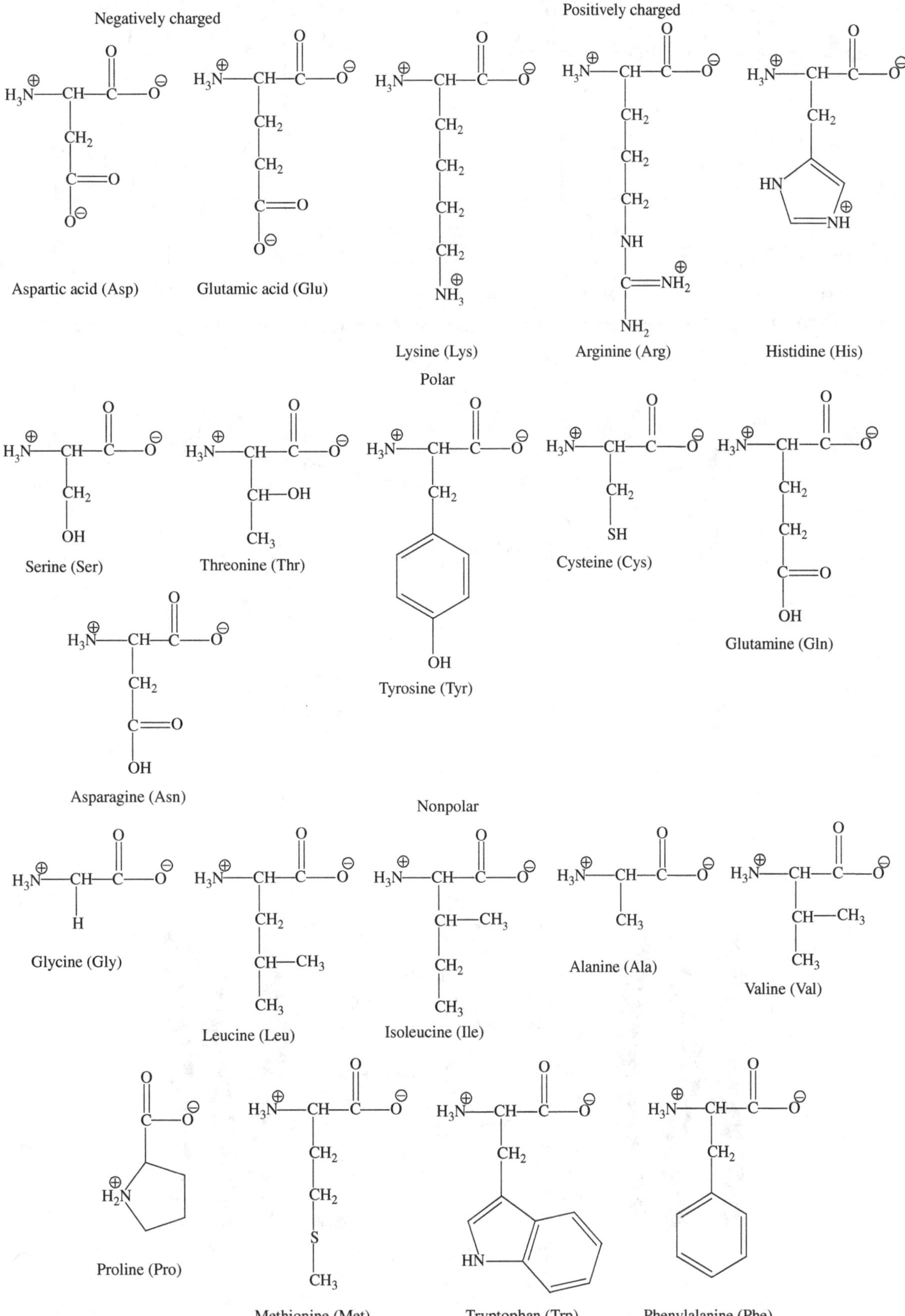

FIGURE 33.2 Structures of the 20 amino acids that make up the polypeptide backbone of proteins. Divisions include negatively charged, positively charged, polar, and nonpolar.

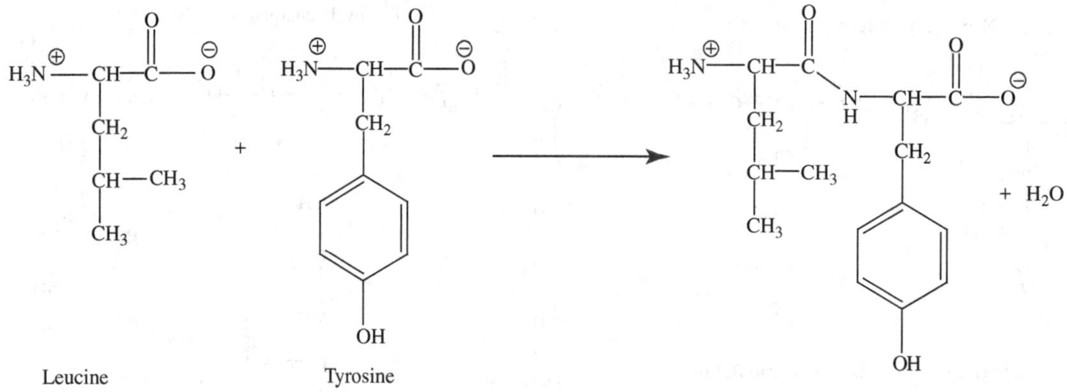

Leucine Tyrosine

FIGURE 33.3 Condensation reaction between the amino acids leucine and tyrosine forming a peptide bond.

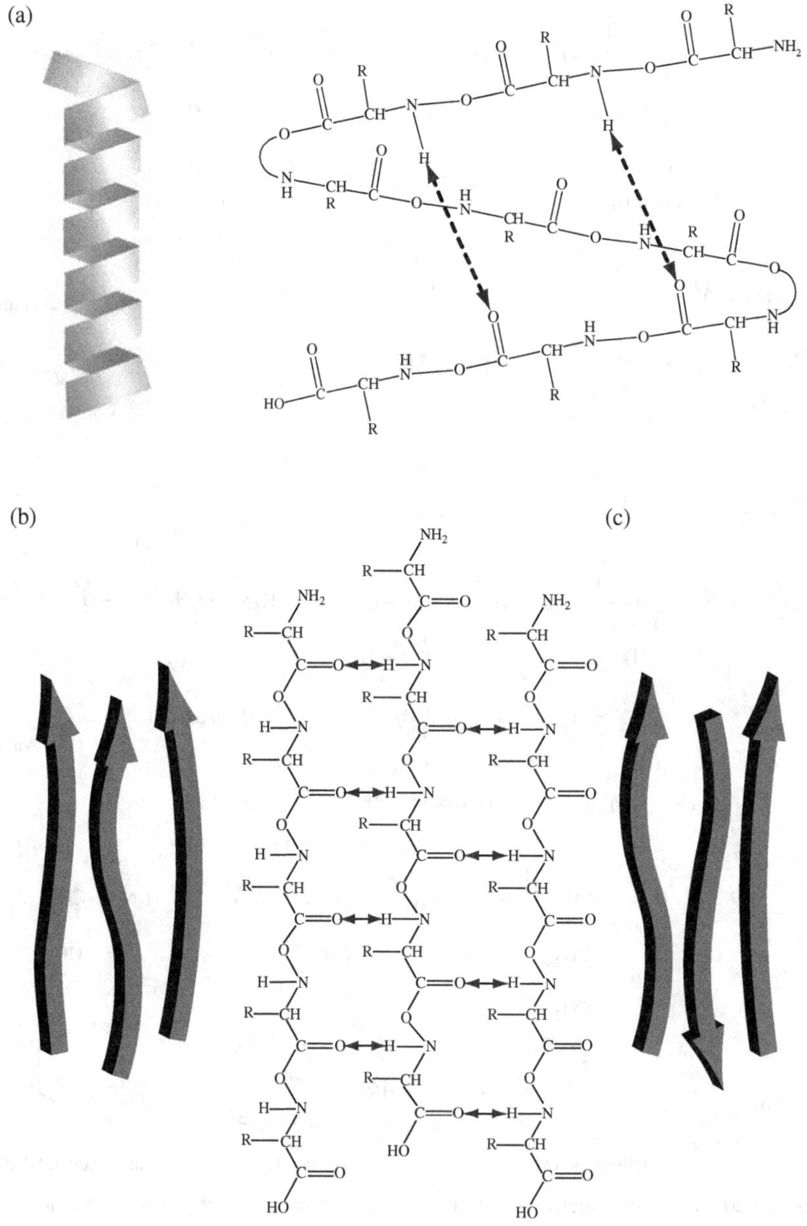

FIGURE 33.4 (a) Alpha-helixes. Beta-sheets, (b) parallel and (c) antiparallel.

proteins will experience an electrophoretic movement through the gel, thus separating them according to their molecular weight with the lower molecular weight proteins having a greater mobility through the gel and the higher molecular weight proteins having a lower mobility through the gel. The resultant 2-DE separation is a collection of spots on the gel that can be up to a few thousand in number. In two-dimensional sodium dodecyl sulfate polyacrylamide gel electrophoresis (2D SDS-PAGE), the proteins have been essentially separated into single protein spots. This allows the digestion of the protein within the spot (excised from the gel) using a protease with known cleavage specificity into subsequent peptides that are unique to that particular protein. The peptides extracted from the in-gel digested proteins separated by 2D SDS-PAGE are then measured by MALDI-TOF/MS, creating a spectrum of peaks that represent the molecular weight of the protein's enzymatic generated peptides. This list of measured peptides can be compared to a theoretical list according to the specificity of the enzyme used for digestion. There is an extensive list of references and searching software that has been introduced for the PMF approach to proteomics that has been reviewed [11]. The 2D SDS-PAGE and PMF approach to proteomics is illustrated in Figure 33.5.

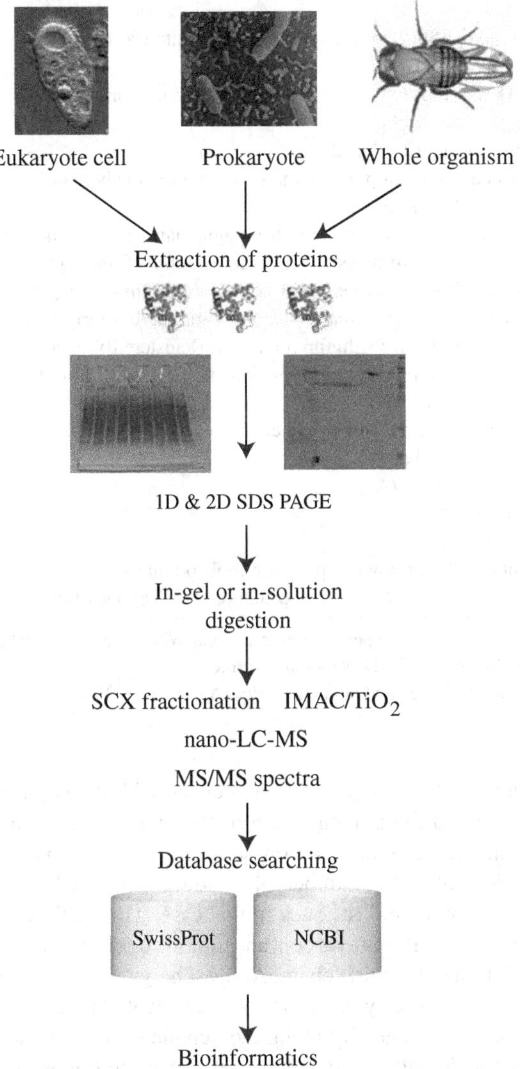

FIGURE 33.5 General strategy and sample flow involved in proteomics.

In bottom-up proteomics, the proteins are generally extracted from the sample of interest, which can include a sample of cultured cells, bacterium, tissue, or a whole organism. A general scheme for the extraction and peptide mass fingerprint mass spectrometric analysis typically followed in early proteomic studies is illustrated in Figure 33.5. The initial sample is lysed and the proteins are extracted and solubilized. The proteins can then be separated using one-dimensional sodium dodecyl sulfate polyacrylamide gel electrophoresis (1D SDS-PAGE) or 2D SDS-PAGE. Proteins can be digested in the gels, or the proteins in solution are digested using a protease such as trypsin. Trypsin is an endopeptidase that cleaves within the polypeptide chain of the protein at the carboxyl side of the basic amino acids arginine and lysine (the trypsin enzyme has optimal activity at a pH range of 7–10 and requires the presence of Ca^{+2}). It has been observed though that trypsin does not efficiently cleave between the residues Lys–Pro or Arg–Pro. Tryptic peptides are predominantly observed as doubly or triply charged when using electrospray as the ionization source. This is due to the amino-terminal residue being basic in each peptide, except for the C-terminal peptide. There exists a number of protease that are available to the mass spectrometrist when designing a digestion of proteins into peptides. These can be used to target cleavage at specific amino acid residues within the polypeptide chain. Examples of available protease and their cleavage specificity are listed in Table 33.1. The enzymes will cleave the proteins into smaller chains of amino acids (typically from 5 amino acid residues up to 100 or so). These short-chain amino acids are mostly water soluble and can be directly analyzed by mass spectrometry. However, often, a lysis and extract from a biological system will comprise a very complex mixture of proteins that requires some form of separation to decrease the complexity prior to mass spectral measurement.

TABLE 33.1 Examples of Protease Available for Polypeptide Chain Cleavage.

Protease	Polypeptide Cleavage Specificity
Trypsin	At carboxyl side of arginine and lysine residues
Chymotrypsin	At carboxyl side of tryptophan, tyrosine, phenylalanine, leucine, and methionine residues
Proteinase K	At carboxyl side of aromatic, aliphatic, and hydrophobic residues
Factor Xa	At carboxyl side of Glu–Gly–Arg sequence
Carboxypeptidase Y	Sequentially cleaves residues from the carboxy (C)-terminus
Submaxillary Arg-C protease	At carboxy side of arginine residues
S. aureus V-8 protease	At carboxy side of glutamate and aspartate residues
Aminopeptidase M	Sequentially cleaves residues from the amino (N)-terminus
Pepsin	Nonspecifically cleaves at exposed residues favoring the aromatic residues
Ficin	Nonspecifically cleaves at exposed residues favoring the aromatic residues
Papain	Nonspecifically cleaves at exposed residues

33.3.2 Protein Identification through Product Ion Spectra

More recently, nano-electrospray ionization high-performance liquid chromatography–tandem mass spectrometry (nano-ESI-HPLC-MS/MS) has been employed using reversed-phase C18 columns to initially separate the peptides prior to introduction into the mass spectrometer. If a highly complex complement of digested proteins are being analyzed such as those obtained from eukaryotic cells or tissue, a greater degree of complexity reduction is employed such as strong cation exchange (SCX) fractionation, which can separate the complex peptide mixture into up to 25 fractions or more. The coupling of online SCX with nano-ESI C18 reversed-phase HPLC-MS/MS has also been employed and is called 2D HPLC and multidimensional protein identification technology (MudPIT) [12]. This is a gel-free approach that utilizes multiple HPLC-mass spectrometry analysis of in-solution digestions of protein fractions. The separated peptides are introduced into the mass spectrometer, and product ion spectra are obtained. The product ions within the spectra are assigned to amino acid sequences. A complete coverage of the amino acid sequence within a peptide from the product ion spectrum is known as de novo sequencing. This can unambiguously identify a protein (except for a few anomalies that will be covered shortly) according to standard spectra stored in protein databases. Two examples of protein databases are NCBInr, a protein database consisting of a combination of most public databases compiled by the National Center for Biotechnology Information (NCBI), and Swiss-Prot, a database that includes an extensive description of proteins including their functions, PTM, domain structures, and so on. The correlation of peptide product ion spectra with theoretical peptides was introduced by Yates and coworkers [13]. At the same time, Mann and Wilm [14] proposed a partial sequence error-tolerant database searching for protein identifications from peptide product ion spectra. There exists now a rather large choice of searching algorithms that are available for protein identifications from peptide product ion spectra. A list of identification algorithms and their associated URLs is illustrated in Table 33.2. The final step in the proteomic analysis of a biological system is the interpretation of the identified proteins, which has been called bioinformatics. Bioinformatics attempts to map and decipher interrelationships between observed proteins and the genetic description. Valuable information can be obtained in this way concerning biomarkers for diseased states, the descriptive workings of a biological system, biological interactions, and so on.

TABLE 33.2 List of Identification Algorithms.

MS Identification Algorithms and URLs PMF
Aldente—http://www.expasy.org/tools/aldente/
Mascot—http://www.matrixscience.com/search_form_select.html
MOWSE—http://srs.hgmp.mrc.ac.uk/cgi-bin/mowse
MS-Fit—http://prospector.ucsf.edu/ucsfhtml4.0/msfit.htm
PeptIdent—http://www.expasy.org/tools/peptident.html
ProFound—http://65.219.84.5/service/prowl/profound.html
MS/MS Identification Algorithms and URLs PFF
Phenyx—http://www.phenyx-ms.com/
Sequest—http://fields.scripps.edu/sequest/index.html
Mascot—http://www.matrixscience.com/search_form_select.html
PepFrag—http://prowl.rockefeller.edu/prowl/pepfragch.html
MS-Tag—http://prospector.ucsf.edu/ucsfhtml4.0/mstagfd.htm
ProbID—http://projects.systemsbiology.net/probid/
Sonar—http://65.219.84.5/service/prowl/sonar.html
TANDEM—http://www.proteome.ca/opensource.html
SCOPE—N/A
PEP_PROBE—N/A
VEMS—http://www.bio.aau.dk/en/biotechnology/vems.htm
PEDANTA—N/A
De Novo Sequencing
SeqMS—http://www.protein.osaka-u.ac.jp/rcsfp/profiling/SeqMS.html
Lutefisk—http://www.hairyfatguy.com/Lutefisk
Sherenga—N/A
PEAKS—http://www.bioinformaticssolutions.com/products/peaksoverview.php
Sequence Similarity Search
PeptideSearch—http://www.narrador.embl-heidelberg.de/GroupPages/Homepage.html
PepSea—http://www.unb.br/cbsp/paginiciais/pepseaseqtag.htm
MS-Seq—http://prospector.ucsf.edu/ucsfhtml4.0/msseq.htm
MS-Pattern—http://prospector.ucsf.edu/ucsfhtml4.0/mspattern.htm
Mascot—http://www.matrixscience.com/search_form_select.html
FASTS—http://www.hgmp.mrc.ac.uk/Registered/Webapp/fasts/
MS-Blast—http://dove.embl-heidelberg.de/Blast2/msblast.html
OpenSea—N/A
CIDentify—http://ftp.virginia.edu/pub/fasta/CIDentify/
Congruence Analysis
MS-Shotgun—N/A
MultiTag—N/A
Tag Approach
Popitam—http://www.expasy.org/tools/popitam/
GutenTag—http://fields.scripps.edu/GutenTag/index.html

Source: Reprinted with permission from John Wiley & Sons, Hernandez, P., Muller, M., Appel, R. D. Automated protein identification by tandem mass spectrometry: Issues and strategies, *Mass Spectr. Rev.* **2006**, *25*, 235–254.

In the identification of proteins from peptides, CID using mass spectrometry is performed to fragment the peptide and identify its amino acid residue sequence. In most mass spectrometers used in proteomic studies such as the ion trap, the quadrupole time of flight, the triple quadrupole, and the Fourier transform ion cyclotron resonance (FTICR), the collision energy is considered low (5–50 eV), and the product ions are generally formed through cleavages of the peptide bonds. According to the widely accepted nomenclature of Roepstorff and Fohlman [15], when the charge is retained on the N-terminal portion of the fragmented peptide, the ions are depicted as *a*, *b*, and *c*. When the charge is retained on the C-terminal portion, the ions are denoted as *x*, *y*, and *z*. The description of the dissociation associated with the peptide chain backbone and the nomenclature of the produced ions are illustrated in Figure 33.6. The ion subscript, for example, the "2" in y_2, indicates the number of residues contained within the ion, two amino acid residues in this case. The weakest bond is between the carboxyl carbon and the nitrogen located directly to the left in the peptide chain. At low-energy CID of the peptide in mass spectrometry, the primary breakage will take place at the weakest bond, generally along the peptide backbone chain, and produce *a*, *b*, and *y* fragments. Notice that the *c* ions and the *y* ions contain an extra proton that they have abstracted from the

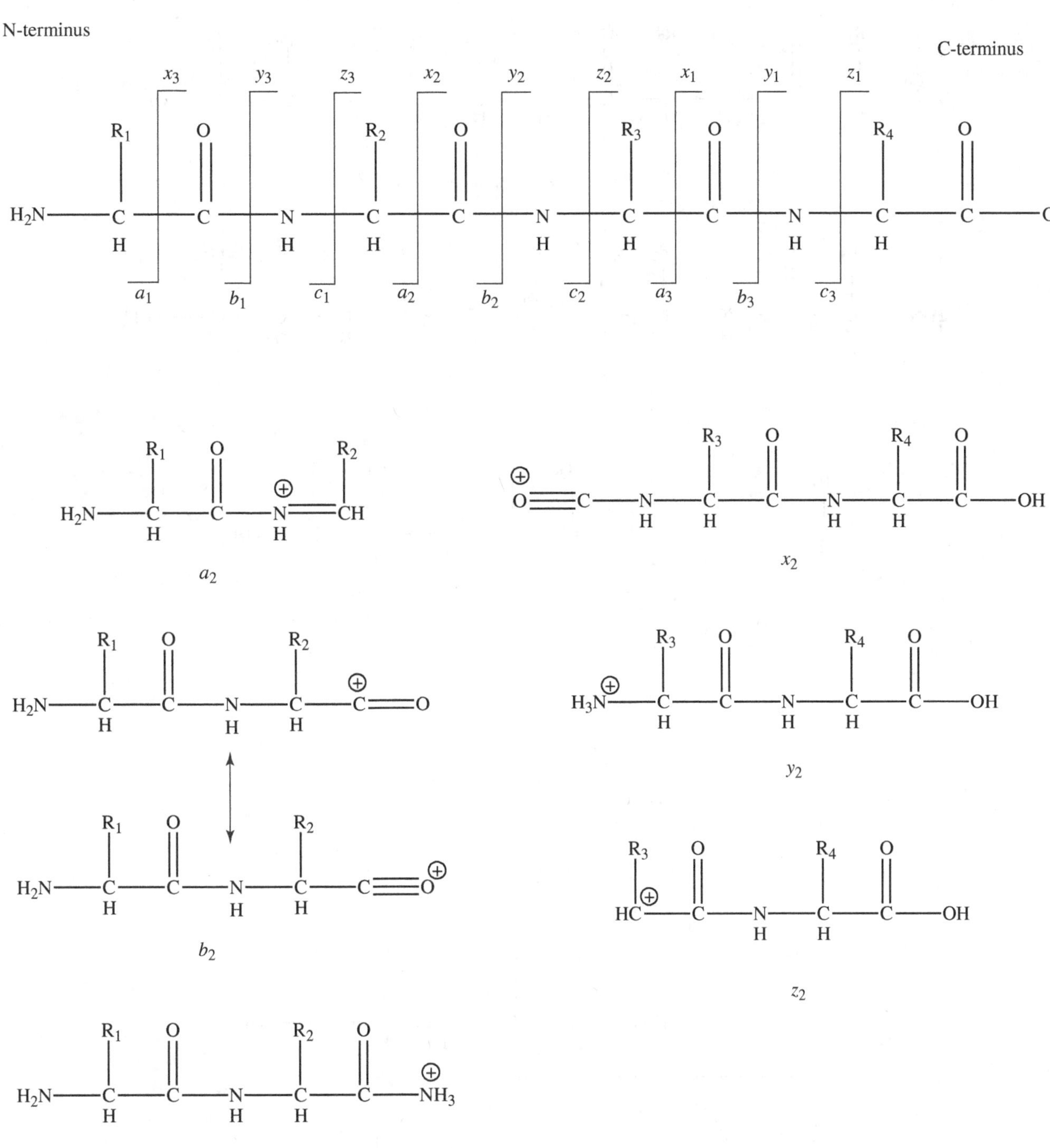

FIGURE 33.6 Dissociation associated with the peptide chain backbone and the nomenclature of the produced ions. Charges retained on the N-terminal portion of the fragmented peptide ions are depicted as *a*, *b*, and *c*. Charges retained on the C-terminal portion of the ions are denoted as *x*, *y*, and *z*. Ion subscript, for example, "2" in y_2, indicates the number of residues (two) contained within the ion.

precursor peptide ion. There has also been proposed a third structure for the *b* ion that is formed as a protonated oxazolone, which is suggested to be more stable through cyclization [16] (see b_2 ion in Figure 33.7). The stability of the *y* ion can be attributed to the transfer of the proton that is producing the charge state to the terminal nitrogen, thus inducing new bond formation and a lower energy state. The model that describes the dissociation of protonated peptides during low-energy CID is called the "mobile proton" model [17]. Peptides fragment primarily from charge-directed reactions where protonation of the peptide can take place at side chain groups, amide oxygen and nitrogen, and at the terminal amino acid group. On the peptide chain

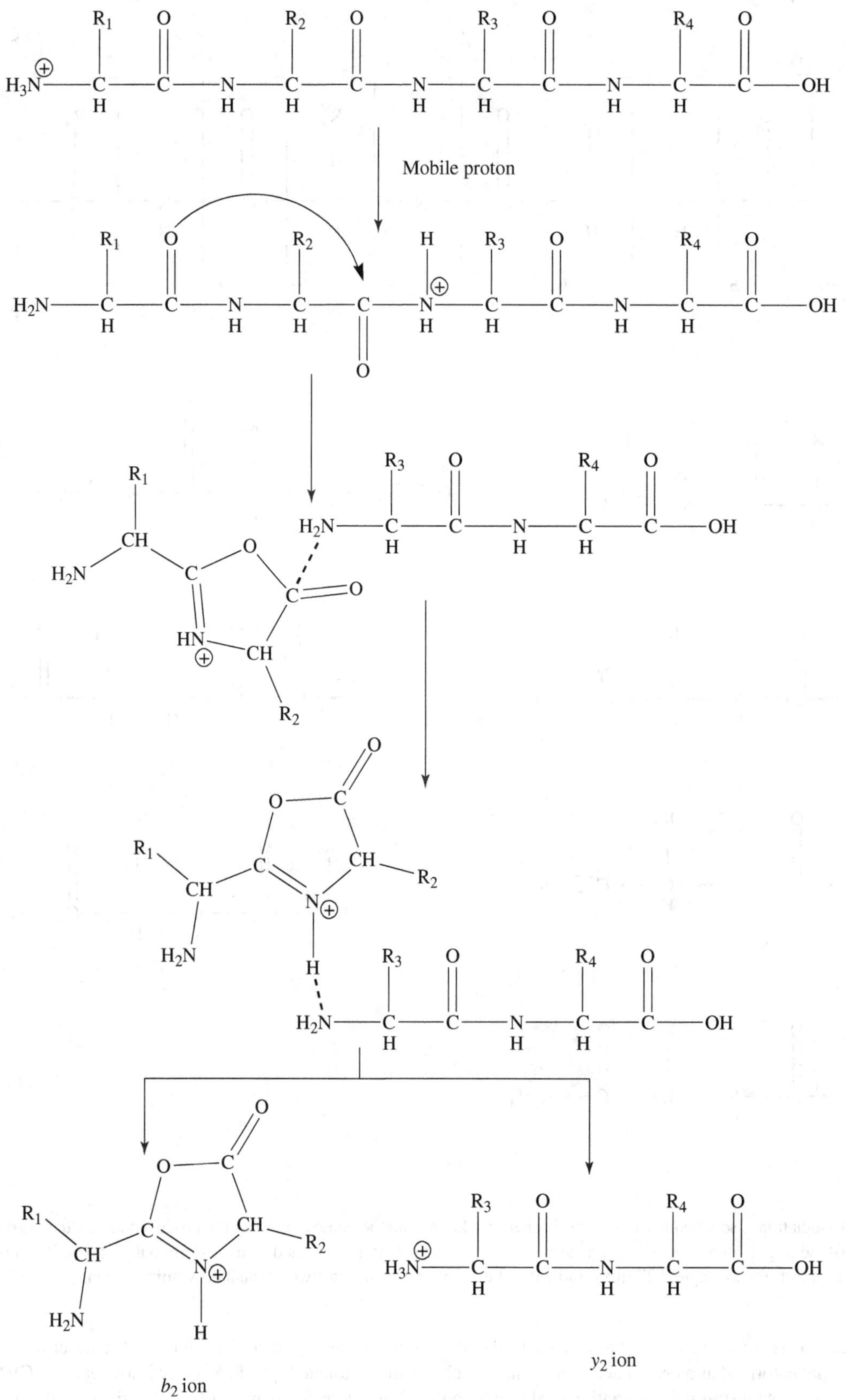

Mobile proton

b_2 ion

y_2 ion

FIGURE 33.7 Fragmentation pathway leading to the production of the *b* and *y* ions from collision-induced dissociation from the polypeptide backbone chain.

backbone, protonation of the amide nitrogen will lead to a weakening of the amide bond inducing fragmentation at that point. However, it is more thermodynamically favored, as determined by molecular orbital calculations [17,18], for protonation to take place on the amide oxygen, which also has the effect of strengthening the amide bond. Inspection of peptide product ion fragmentation spectra has demonstrated though that the protonating of the amide nitrogen is taking place over the protonating of the amide oxygen. This is in contrast to the expected site of protonation from a thermodynamic point of view that indicates the amide oxygen protonation and not the amide nitrogen. This discrepancy has been explained by the "mobile proton model," introduced by Wysocki and coworkers [17,19], which describes that the proton(s) added to a peptide, upon excitation from CID, will migrate to various protonation sites, provided they are not sequestered by a basic amino acid side chain prior to fragmentation. The fragmentation pathway leading to the production of the *b* and *y* ions is illustrated in Figure 33.7. The protonation takes place first on the N-terminus of the peptide. The next step is the mobilization of the proton to the amide nitrogen of the peptide chain backbone where cleavage is to take place. The protonated oxazolone derivative is formed from nucleophilic attack by the oxygen of the adjacent amide bond on the carbon center of the protonated amide bond. Depending upon the location of the retention of the charge, either a *b* ion or a *y* ion will be produced.

Besides the amide bond cleavage producing the *b* and *y* ions that are observed in low-energy collision product ion spectra, there are also a number of other product ions that are quite useful in peptide sequence determination. Ions that have lost ammonia (−17 Da) in low-energy collision product ion spectra are denoted as $a*$, $b*$, and $y*$. Ions that have lost water (−18 Da) are denoted as a^{o}, b^{o}, and y^{o}. The *a* ion illustrated in Figure 33.6 is produced through loss of CO from a *b* ion (−28 Da). Upon careful inspection of the structures in Figure 33.6 for the product ions, it can be seen that the *a* ion is missing CO as compared to the structure of the *b* ion. When a difference of 28 is observed in product ion spectra between two *m/z* values, an *a–b* ion pair is suggested and can be useful in ion series identification. Internal cleavage ions are produced by double backbone cleavage, usually by a combination of *b*- and *y*-type cleavage. When a combination of *b*- and *y*-type cleavage takes place, an amino-acylium ion is produced. When a combination of *a*- and *y*-type internal cleavage takes place, an amino-immonium ion is produced. The structures of an amino-acylium ion and an amino-immonium ion are illustrated in Figure 33.8. These types of product ions that are produced from internal fragmentation are denoted with their one-letter amino acid code. Though not often observed, *x*-type ions can be produced using photodissociation.

33.3.3 High-Energy Product Ions

Thus far, the product ions that have been discussed, the *a*-, *b*-, and *y*-type ions, are produced through low-energy collisions such as those observed in ion traps. The collision-induced activation in ion traps is a slow heating mechanism, produced through multiple collisions with the trap bath gas that favors lower-energy fragmentation pathways. High-energy collisions that are in the keV range such as those produced in MALDI TOF-TOF mass spectrometry produce other product ions in addition to the types that have been discussed so far. Side chain cleavage ions that are produced by a combination of backbone cleavage and a side chain bond are observed in high-energy collisions and are denoted as *d*, *v*, and *w* ions. Figure 33.9 contains some illustrative structures of *d*-, *v*-, and *w*-type ions.

Immonium ions are produced through a combination of *a*-type and *y*-type cleavage that results in an internal fragment that contains a single side chain. These ions are designated by the one-letter code that corresponds to the amino acid. Immonium ions are not generally observed in ion trap product ion mass spectra but are in MALDI TOF-TOF product ion mass spectra. The structure of a general immonium ion is illustrated in Figure 33.10. Immonium ions are useful in acting as confirmation of residues suspected to be contained within the peptide backbone. Table 33.3 presents a compilation of the amino acid residue information that is used in mass spectrometry analysis of peptides. The table includes the amino acid residue's name, associated codes, residue mass, and immonium ion mass.

33.3.4 De Novo Sequencing

An example of de novo sequencing is illustrated in Figure 33.11. The product ion spectrum in Figure 33.11a is for a peptide comprised of seven amino acid residues. The peptide product ion spectrum in Figure 33.11b is also comprised of seven amino acid residues; however, the serine residue (Ser, $C_3H_5NO_2$, 87.0320 amu) in (a) has been replaced by a threonine residue (Thr, $C_4H_7NO_2$, 101.0477 amu) in (b). The product ion spectra are very similar, but a difference can be discerned with the b_5 ion and the y_3 ions where a shift of 14 Da is observed due to the difference in amino acid residue composition associated with the serine and threonine.

Though the sequencing of the amino acids contained within a peptide chain can be discerned by de novo mass spectrometry as just illustrated, there is a problem associated with isomers and isobars. Isomers are species that have the same molecular formula but differ in their structural arrangement, while isobars are species with different molecular formulas that possess similar (or

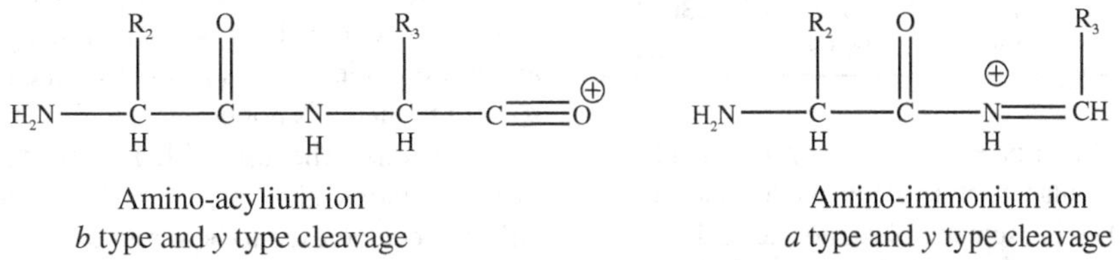

Amino-acylium ion
b type and *y* type cleavage

Amino-immonium ion
a type and *y* type cleavage

FIGURE 33.8 Structure of (left) an amino-acylium ion produced through a combination of *b*- and *y*-type cleavage and (right) an amino-immonium ion through a combination of *a*- and *y*-type internal cleavage.

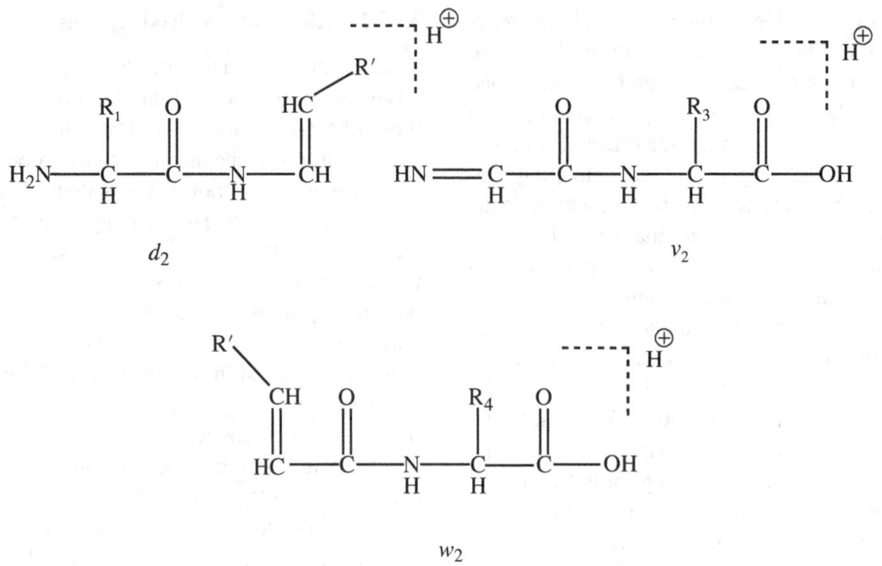

FIGURE 33.9 Structures of d-, v-, and w-type ions produced by a combination of backbone cleavage and a side chain bond observed in high-energy collision product ion spectra.

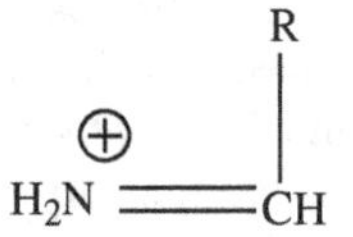

FIGURE 33.10 Structure of a general immonium ion.

TABLE 33.3 Amino Acid Residue Names, Codes, Masses, and Immonium Ion m/z Values.

Residue	1-Letter Code	3-Letter Code	Residue Mass	Immonium Ion (m/z)
Alanine	A	Ala	71.04	
Arginine	R	Arg	156.10	129
Asparagine	N	Asn	114.04	87.09
Aspartic acid	D	Asp	115.03	88.04
Cysteine	C	Cys	103.01	76
Glutamic acid	E	Glu	129.04	102.06
Glutamine	Q	Gln	128.06	101.11
Glycine	G	Gly	57.02	30
Histidine	H	His	137.06	110.07
Isoleucine	I	Ile	113.08	86.1
Leucine	L	Leu	113.08	86.1
Lysine	K	Lys	128.09	101.11
Methionine	M	Met	131.04	104.05
Phenylalanine	F	Phe	147.07	120.08
Proline	P	Pro	97.05	70.07
Serine	S	Ser	87.03	60.04
Threonine	T	Thr	101.05	74.06
Tryptophan	W	Trp	186.08	159.09
Tyrosine	Y	Tyr	163.06	136.08
Valine	V	Val	99.07	72.08

the same) molecular weights. For example, it is not possible to determine whether a particular peptide contains leucine (Leu, $C_6H_{11}NO$, 113.0841 amu) or its isomer isoleucine (Ile, $C_6H_{11}NO$, 113.0841 amu) both at a residue mass of 113.0841 amu. Furthermore, even though the remaining 18 amino acid residues each contain distinctive elemental compositions and thus distinct molecular masses, some combinations of residues will actually equate to identical elemental compositions. This produces an isobaric situation where different peptides will possess either very similar or identical sequence masses. If every single peptide amide bond cleavage is not represented within the product ion spectrum, then it is not possible to discern some of these possible combinations. The use of high-resolution/high mass accuracy instrumentation such as the FTICR mass spectrometer or the Orbitrap can be used to help reduce this problem when complete de novo sequencing is not possible. Table 33.4 shows a listing of some of the amino acid combinations that may arise that can contribute to unknown sequence determination when complete de

TABLE 33.4 Examples of Combinations of Amino Acid Residues Where Isobaric Peptides Can Be Observed.

Amino Acid Residue	Residue Mass (Da)	Δ Mass (Da)
Leucine	113.08406	
Isoleucine	113.08406	0
Glutamine	128.05858	
Glycine + alanine	128.05858	
	(57.02146 + 71.03711)	0
Asparagine	114.04293	
2 × glycine	114.04293 (2 × 57.02146)	0
Oxidized methionine	147.03540	
Phenylalanine	147.06841	0.03301
Glutamine	128.05858	
Lysine	128.09496	0.03638
Arginine	156.10111	
Glycine + valine	156.08987	
	(57.02146 + 99.06841)	0.01124
Asparagine	114.04293	
Ornithine	114.07931	0.03638
Leucine/Isoleucine	113.08406	
Hydroxyproline	113.04768	0.03638
2 × valine	198.13682 (2 × 99.06841)	
Proline + threonine	198.10044	
	(97.05276 + 101.04768)	0.03638

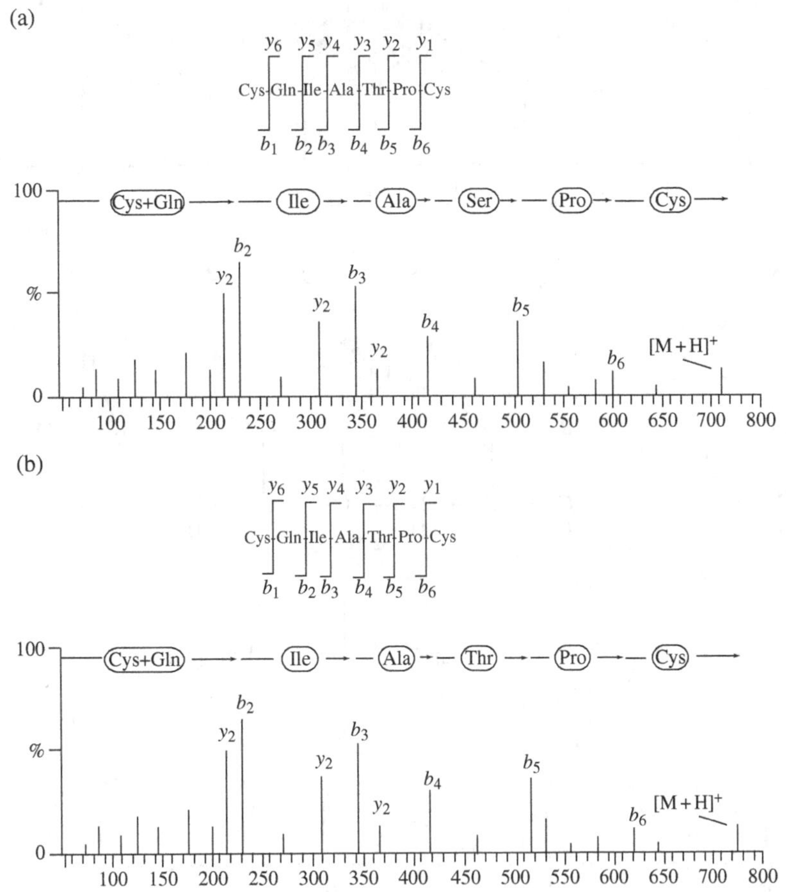

FIGURE 33.11 Example of de novo sequencing using product ion spectra collected by collision-induced dissociation mass spectrometry. (a) Peptide comprised of seven amino acid residues. (b) Peptide comprised of seven amino acid residues with the serine residue (Ser, $C_3H_5NO_2$, 87.0320 amu) replaced by a threonine residue (Thr, $C_4H_7NO_2$, 101.0477 amu). The product ion spectra are very similar, but a difference can be discerned with the b_5 ion and the y_3 ions where a shift of 14 Da is observed due to the difference in amino acid residue composition associated with the serine and threonine.

novo sequencing is not being obtained. For a peptide with a mass of 800 Da, the differences in the table for, for example, the glutamine versus lysine difference at 0.03638 Da, it would take a mass accuracy of better than 44 ppm to distinguish the two. For the arginine versus glycine + valine at 0.01124 Da, it would require a mass accuracy of 14 ppm to distinguish the two. For the FTICR mass spectrometers and the hybrid mass spectrometers such as the LTQ-FT or the LTQ-Orbitrap, this is readily achievable, but often for ion traps, this is not always achievable.

33.3.5 Electron Capture Dissociation

Other techniques such as electron capture dissociation (ECD) [20] and electron transfer dissociation (ETD, discussed in Section 33.5.3 Phosphorylation of Proteins) have also been used to alleviate the problem of isobaric amino acid combinations by giving complementary product ions (such as *c* and *z*• ions) that help to obtain complete sequence coverage. The technique of ECD tends to promote extensive fragmentation along the polypeptide backbone, producing *c*- and *z*•-type ions, while also preserving modifications such as glycosylation and phosphorylation. The general *z*-type ion that is shown in Figure 33.6 is different though from the *z*•-type ion that is produced in ECD, which is a radical cation. Peptide cation radicals are produced by passing or exposing the peptides, which are already multiply protonated by electrospray ionization (ESI), through low-energy electrons. The mixing of the protonated peptides with the low-energy electrons will result in exothermic ion–electron recombinations. There are a number of dissociations that can take place after the initial peptide cation radical is formed. These include loss of ammonia, loss of H atoms, loss of side chain fragments, cleavage of disulfide bonds, and most importantly peptide backbone cleavages. The *c*-type ion is produced through homolytic cleavage at the N—C peptide bond, and charges are present in the amino-terminal fragment. The *z*•-type ion is produced when charges are present in the carboxy-terminal fragment. The mechanism that has been given for the promotion of fragmentation of the peptides is due to electron attachment to the protonated sites of the peptide. The now cation radical intermediate that has formed will release a hydrogen atom. A nearby carbonyl group will capture the released hydrogen atom, and the peptide will dissociate by cleavage of the adjacent N—C peptide bond. The mechanism for the production of an α-amide radical of the peptide C-terminus, a *z*•-type ion, and the enolamine of the N-terminus portion of the peptide, a *c*-type ion, is illustrated in Figure 33.12.

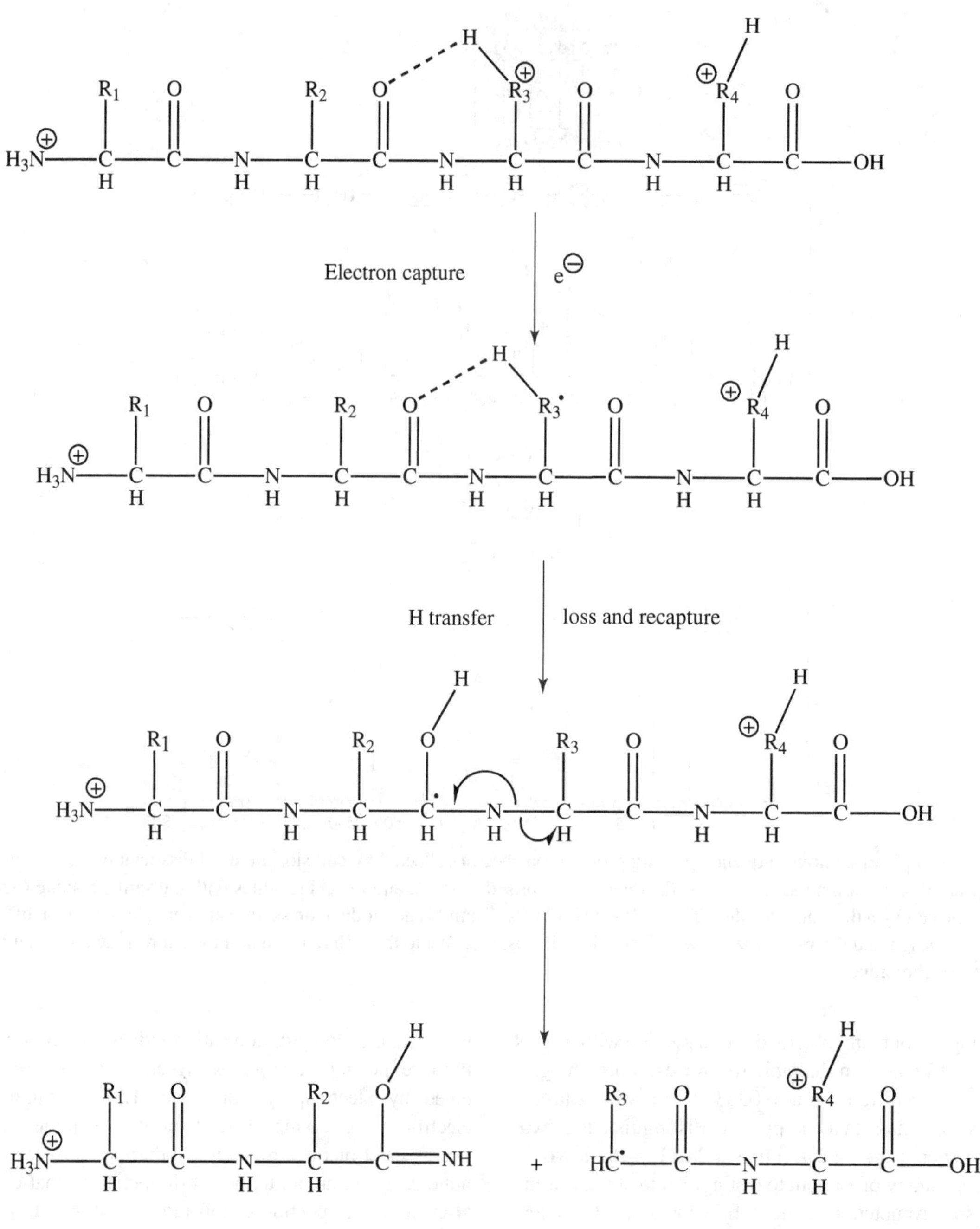

FIGURE 33.12 Mechanism for the production of *c*- and *z*-type ions observed in ECD.

33.4 TOP-DOWN PROTEOMICS: MASS SPECTROMETRY OF INTACT PROTEINS

33.4.1 Background

Measuring the whole, intact protein in the GP using mass spectrometric methodologies is known as "top-down" proteomics. Top-down proteomics measures the intact protein's mass followed by CID of the whole protein breaking it into smaller parts. A vital component of top-down proteomics is the accuracy in which the masses are measured. Often, high-resolution mass spectrometers such as the FTICR mass spectrometer are used to accurately measure the intact protein's mass and the product ions produced during CID experiments. In early top-down experiments though, this was not the case. Mass spectrometers such as the triple quadrupole coupled with electrospray were first used to measure intact proteins in the GP [21,22]. However, the triple quadrupole mass spectrometer does not allow the resolving of the isotopic distribution of the product ions being generated in the top-down approach. The use of FT-MS/MS was later reported with high enough resolution to resolve isotopic peaks [23,24]. An

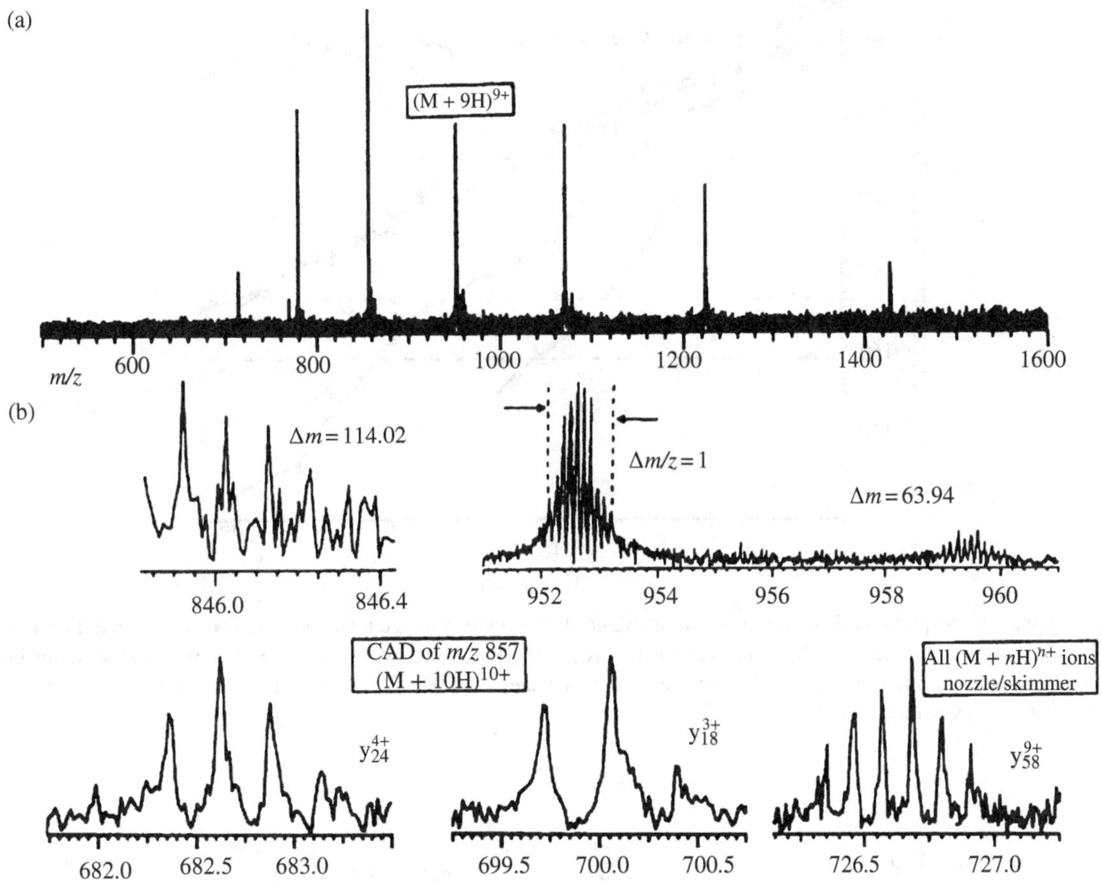

FIGURE 33.13 (a) ESI mass spectrum of ubiquitin (sum of 10 scans, 64,000 data). (Center) Regions expanded to show the presence of impurities. (b) MS/MS fragment ions from collisionally activated dissociation of $(M+10H)^{10+}$ and from placing 200 V between the nozzle and skimmer of the ESI source. (Reprinted with permission from Ref. [23].)

example of these early top-down experiments utilizing FT mass spectrometry is illustrated in Figure 33.13. Extensive initial, pre-mass analysis sample preparation such as cleanup, digestion, desalting, enriching, etc., all often incorporated in "bottom-up" proteomics, is not necessarily required in top-down approaches. The dynamic range in top-down proteomics can be limited by the number of analytes that can be present during analysis, but this is usually overcome by using some type of separation prior to introduction into the mass spectrometer. The separation of complex protein mixtures can be obtained using techniques such as reversed-phase high-performance liquid chromatography (RP-HPLC), gel electrophoresis, anion exchange chromatography, and capillary electrophoresis. Typically, in bottom-up analysis, the digested protein peptides are <3 kDa, and the complete description of the original intact protein is not possible. With top-down analysis, there is often 100% coverage of the protein being analyzed. This allows the determination of the N- and C-termini, the exact location of modifications to the protein such as phosphorylation, and the confirmation of DNA-predicted sequences.

33.4.2 GP Basicity and Protein Charging

It is the process of ESI that allows the measurement of high molecular weight intact proteins. As a rule of thumb, for each 1000 Da of the protein, there is associated one charge state. For example, a 30 kDa protein, as an approximation, will have a charge state of 30+. This brings the measured mass of the protein down into the range of many mass spectrometers that typically scan between *m/z* 100 and *m/z* 4000 (*m/z* = 30 kDa/30+ = 1000 Th). Multiple charging of peptides and proteins is achieved during the ESI process due to the presence of amino acid residue basic sites. There is a limit to the number of charges that can be placed onto a peptide or protein during the ESI process as was demonstrated by Schnier et al. [25]. In their study, the apparent GP basicity as a function of charge state was measured using cytochrome c. A graphical plot of apparent GP basicity versus charge state is illustrated in Figure 33.14. The curves in Figure 33.14 have a negative trend (go down) as each charge state is increased. As a new charge state is added (increasing *x*-axis), the apparent GP basicity decreases (*y*-axis). In the graph, the dashed line represents the GP basicity of methanol, which is included as a reference of a species present during the ESI process that can also accept a charge. The charging of large molecules during the ionization process is thought to follow the charged residue mechanism. In this ionization process, the solvent molecules evaporate from around the protein leaving charges that associate to basic sites within the protein. What the intersection in the graph between the charging of

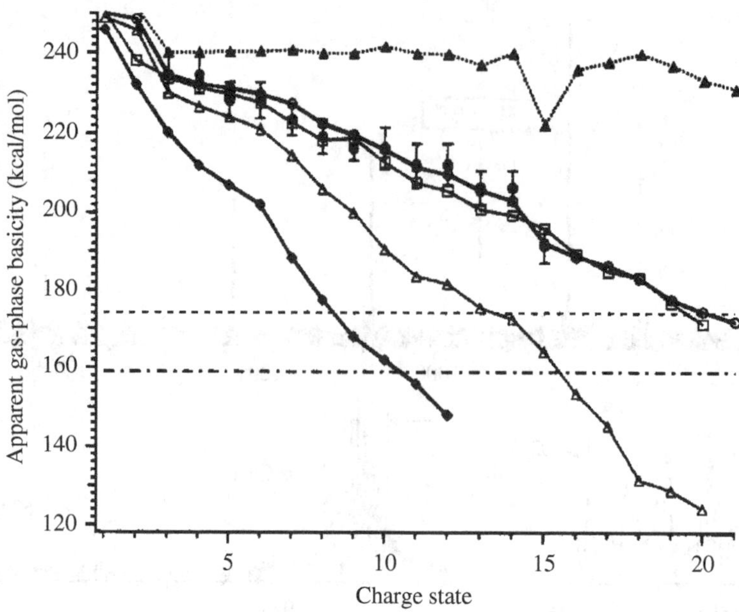

FIGURE 33.14 Apparent gas-phase basicity as a function of charge state of cytochrome c ions, measured (•); calculated, linear (ϵ_r = 1.0, Δ; best fit ϵ_r = 2.0, O); intrinsic, ▲; calculated, X-ray crystal structure (ϵ_r = 2.0, ♦); calculated, a-helix (ϵ_r = 4.1, □). The dashed line indicates GB of methanol (174.1 kcal/mol) and the dash-dot line indicates GB of water (159.0 kcal/mol). (Reprinted with permission from Ref. [25]. Copyright 1995 American Chemical Society.)

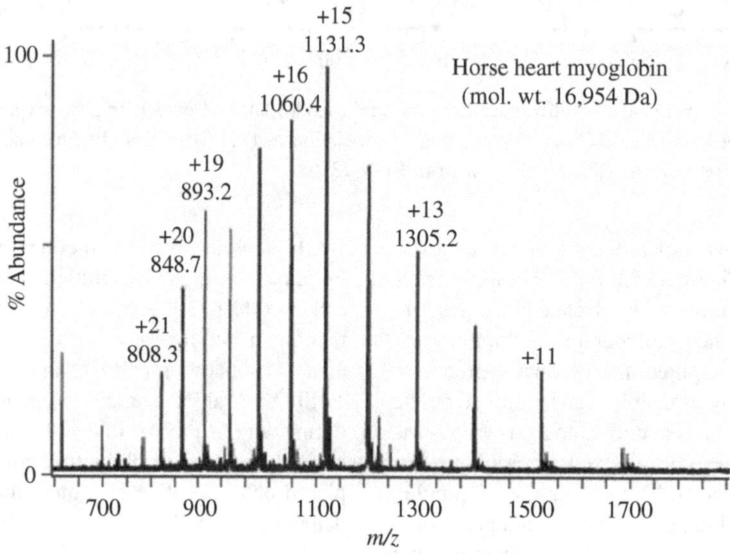

FIGURE 33.15 Electrospray mass spectrum of horse heart myoglobin at a molecular weight of 16,954 Da. The spectrum illustrates an envelope of peaks of different *m/z* values and charge states for the protein.

cytochrome c and the GP basicity of methanol is demonstrating is that there is a limit to the amount of charges that can be placed upon a species during ionization. At some point (intersection), it becomes thermodynamically favorable to put the next charge (proton) onto a methanol molecule than onto the protein. Here, a state has been reached where coulombic repulsion between the charges and a loss of basic sites does not allow further charging. At this point, the maximum charge state has been reached for the protein molecule.

33.4.3 Calculation of Charge State and Molecular Weight

Another interesting feature of the electrospray charging of proteins is the ability to calculate the charge state and molecular weight of an unknown protein from its single-stage mass spectrum. Figure 33.15 illustrates the electrospray mass spectrum of horse heart myoglobin at a molecular weight of 16,954 Da. If this were an unknown protein species, we would only have the respective *m/z* values from the mass spectrum. Using the following two simultaneous equations with two unknowns, the charge states and the molecular weight of the unknown protein can be calculated using only the information obtained within the mass spectrum:

$$m/z = \frac{m + z(1.0079)}{z} \text{ higher } m/z \text{ peak from spectrum} \quad (33.1)$$

$$m/z = \frac{m + (z+1)(1.0079)}{z+1} \text{ lower } m/z \text{ peak from spectrum} \quad (33.2)$$

If we solve the higher mass/lower charge state peak (*m/z* = 1131.3 Th) for *m* by taking Equation 33.1 and solving for m, we obtain

$$m = 1131.3z - 2.0079z \quad (33.3)$$

$$m = 1130.2921z \quad (33.4)$$

If we next substitute this mass into the lower mass/higher charge state (*m/z* = 1060.4 Th) in Equation 33.2, we can calculate the charge state of the *m/z* 1131.3 peak:

$$m/z = \frac{m + (z+1)(1.0079)}{z+1} \quad (33.2)$$

$$1060.4 = \frac{m + 1.0079z + 1.0079}{z+1} \quad (33.5)$$

$$1060.4 = \frac{1130.2921m + 1.0079z + 1.0079}{z+1} \quad (33.6)$$

$$z = 15 \quad (33.7)$$

Therefore, we have determined that the charge state of the *m/z* 1131.3 peak is +15. To calculate the molecular weight of the unknown protein, we can substitute the charge state value into Equation 33.1 and solve for *m*:

$$m/z = \frac{m + z(1.0079)}{z} \quad (33.1)$$

$$1131.3 = \frac{m + 15(1.0079)}{15} \quad (33.8)$$

$$m = 16954 \quad (33.9)$$

This process is known as deconvolution where a distribution (mass spectral envelope) of a protein's *m/z* values and associated charge states is collapsed down to a single peak representing the molecular weight of the protein. The deconvolution of the horse heart myoglobin protein is illustrated in Figure 33.16. The width of the deconvoluted peak indicates the variability in the calculation of the molecular weight of the protein from the mass spectral peaks. While it is possible to deconvolute protein peaks by hand by solving two simultaneous equations with two unknowns, mass spectral computer software is typically used to perform this task.

33.4.4 Top-Down Protein Sequencing

In bottom-up proteomics, the mass spectral identification of proteins through sequence determination of separated peptides requires the isolation of a single peptide for fragmentation experiments. This is typically done by removing from an ion trap mass spectrometer all *m/z* species present except for one that is of interest for fragmentation and subsequent sequencing. It has been demonstrated though that multiple proteins can be simultaneously fragmented and identified in top-down proteomics. In a study reported by Patrie et al. [26], the authors used a hybrid mass spectrometer that coupled a quadrupole mass analyzer to a FTICR mass spectrometer that uses a 9.4 T magnet. The instrumental design is illustrated in Figure 33.17. Prior to the FTICR/MS, there is a resolving quadrupole that can act as either an rf-only ion guide or as a fully functional mass analyzer. Following this in the instrumental design is an accumulation octopole that was used to accumulate and store ions prior to introduction into the FTICR mass spectrometer. Nitrogen or helium gas at a pressure of approximately 1 mTorr was introduced into the accumulation octopole to help improve the accumulation. The FTICR cell located within the 9.4 T magnet is an open-ended capacity

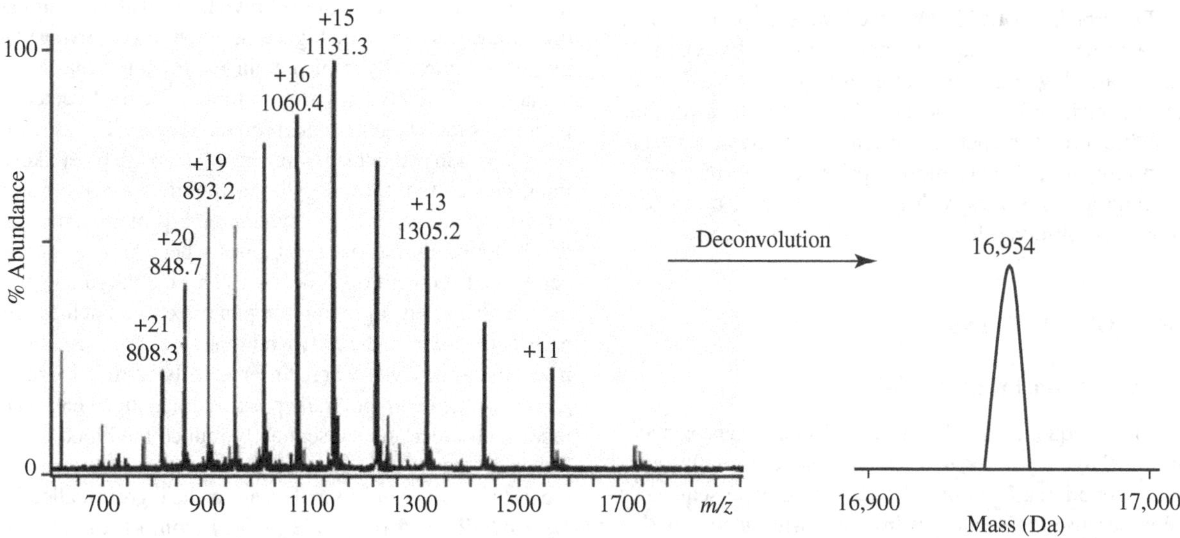

FIGURE 33.16 Deconvoluted, computer-generated spectrum of horse heart myoglobin at a molecular weight of 16,954 Da.

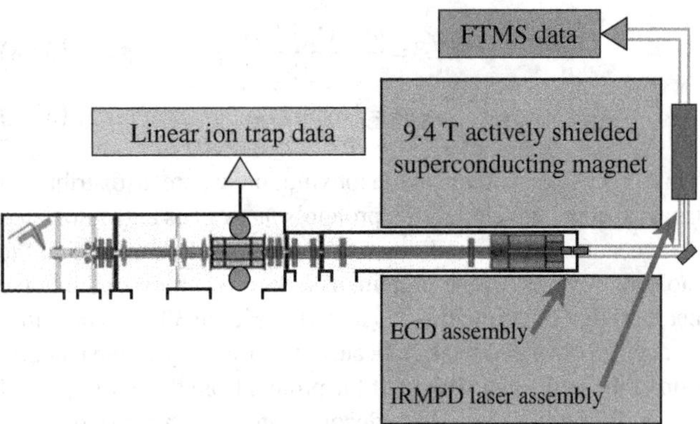

FIGURE 33.17 Schematic representation of the quadrupole/Fourier transform ion cyclotron resonance hybrid mass spectrometer for versatile MS/MS and improved dynamic range by means of *m/z*-selective ion accumulation external to the superconducting magnet bore.

coupled cell that is cylindrical and divided axially into five segments. At the end of the instrument (far right side) is located a laser that is used for infrared multiphoton dissociation (IRMPD) experiments. For fragmentation experiments, CID could be performed in the accumulation octopole, IRMPD could be performed within the ICR cell by irradiating the trapped species with the laser, or finally the instrumental design also included ECD capabilities. A top-down experiment of a mixture of proteins collected as a fraction eluting from a reversed-phase LC (RPLC) separation was studied by the authors. First, a broadband spectrum of the RPLC fraction was collected where a very low response is observed for the proteins present. The same broadband spectrum was collected after using the accumulation octopole to increase the amount of sample that is being introduced into the FTICR cell and thus increase the sensitivity of the mass measurement of the seven proteins present. Fragmentation results of an IRMPD experiment of the seven proteins that were present were then reported. Of the seven proteins present, three were identified by top-down proteomics, listed as X, a 19,431.8 Da protein MJ0543; O, a 20,511.3 Da protein MJ0471; and Z, a 17,263.0 Da protein MJ0472. Observed were a large number of amino acid residues contained within the *b*- and *y*-type ions and an associated high number of charges (e.g., the Oy_{63}^{8+} product ion that contains 63 amino acid residues and 8 charges). This is quite different from the peptides that are normally observed in bottom-up proteomics where most peptides contain between 7 and 25 amino acid residues with mostly 2 charges, but 3 or 4 charges are also observed for the longer-chain peptides.

33.5 PTM OF PROTEINS

33.5.1 Three Main Types of PTM

In the genomic sequencing field, the use of robotic gene sequencers allowed large-scale sequencing that was essentially automated. The robotic automation of determining gene sequences is possible because the sequences involved with genes involve only four bases (see Section 32.3), and there are no variations induced in the form of postmodification. This has resulted in the well-publicized entire sequencing of the human genome (Human Genome Project, *Nature*, February 2001). This is not the case with proteins where there is not only the observance of spliced variants from alternative splicing from the messenger RNA (mRNA) but where there are also PTM that can take place with the amino acids contained within the protein. There are over 200 PTM that can take place with proteins as has been described by Wold [27]. As examples, here are 22 different types of PTM that can take place with proteins: acetylation, amidation, biotinylation, C-mannosylation, deamidation, farnesylation, formylation, flavinylation, gamma-carboxyglutamic acids, geranylgeranylation, hydroxylation, lipoxylation, myristoylation, methylation, N-acyl diglyceride (tripalmitate), O-GlcNAc, palmitoylation, phosphorylation, phosphopantetheine, pyrrolidone carboxylic acid, pyridoxal phosphate, and sulfation [28]. There are also artifactual modifications such as oxidation of methionine. Of these, the three types of PTM that are primarily observed and studied using mass spectrometric techniques are glycosylation, sulfation, and phosphorylation. The observance of PTM is increasingly being used in expression studies where a normal state proteome is being compared to a diseased state proteome. However, the PTM of a protein during a biological of physiological change within an organism may take place without any change in the abundance of the protein involved and often is one piece of a complex puzzle. Methods that measure PTM using mass spectrometric methodologies often focus on the degree (increase or decrease or, alternatively, upregulation or downregulation) of PTM for any given protein or proteins. We shall briefly look at glycosylation and sulfation, which are less involved in cellular processing than phosphorylation, a major signaling cascade pathway for the response to a change in cellular condition(s).

33.5.2 Glycosylation of Proteins

Glycosylation is the covalent addition of a carbohydrate chain to amino acid side chains of a protein producing a glycoprotein. The carbohydrate side chains can be anywhere from 1 to 70 sugar units in length, branched, or straight chained and are most commonly comprised of mannose, galactose, *N*-acetylglucosamine, and sialic acid. The structures and abbreviations of these four carbohydrates are illustrated in Figure 33.18. Protein glycosylation is the most common type of protein modification that is found in eukaryotes. The glycosylation of proteins helps to determine the proteins' structure, is crucial in cell–cell recognition, and may also be involved in cellular signaling events, though most likely not as important as the phosphorylation modification is in signaling. Glycoproteins are primarily membrane proteins and are abundantly found in plasma membranes where they are involved in cell-to-cell recognition processes. The carbohydrate groups of the membrane glycoproteins are positioned so that they are externally extruded from the cell membrane surface. An example of this involves erythrocytes where the externally extruded carbohydrate groups of the cell membrane possess a negative charge and thus repel each other and subsequently reduce the blood's viscosity. The two types of linkage of the carbohydrates to the amino acids are (i) to the nitrogen atom of an amino acid group called *N-linked glycosylation* and (ii) to the oxygen atom of a hydroxyl group called *O-linked glycosylation*. In the N-linked glycosylation, the carbohydrates are attached to the asparagine (Asn) side chain

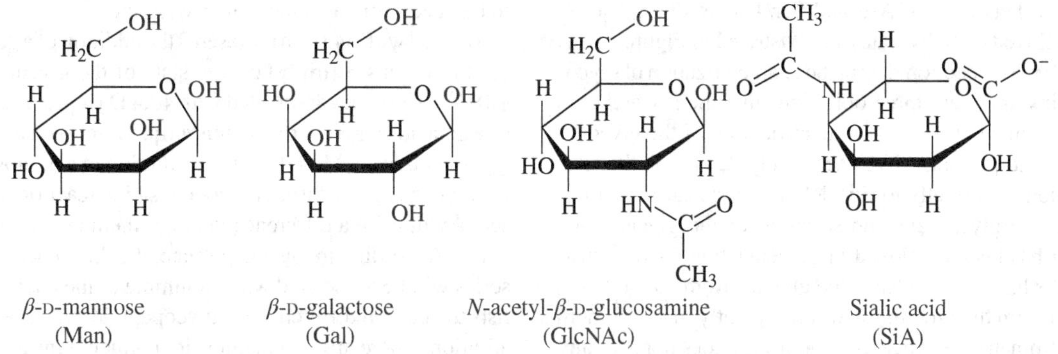

FIGURE 33.18 The most common four carbohydrates found in glycoproteins: mannose (Man), galactose (Gal), *N*-acetylglucosamine (GlcNAc), and sialic acid (SiA).

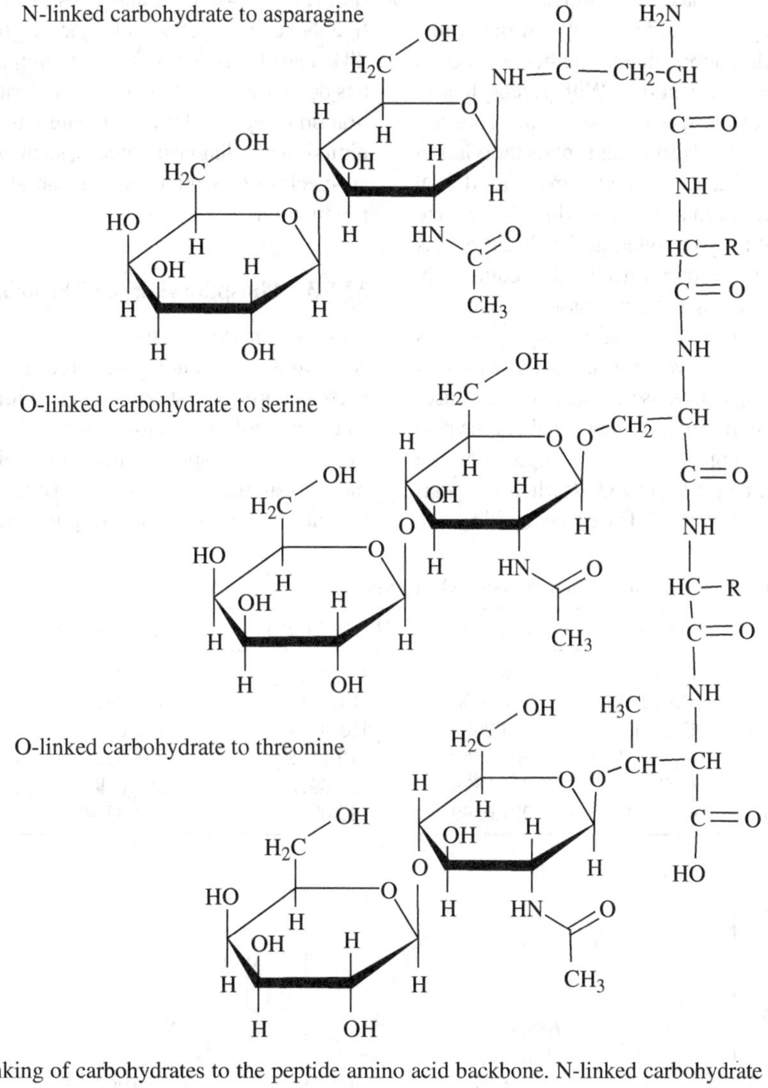

FIGURE 33.19 Covalent linking of carbohydrates to the peptide amino acid backbone. N-linked carbohydrate to asparagine. Notice the one amino acid residue between the next amino acid residue serine (could also be threonine). O-linked carbohydrate to serine. O-linked carbohydrate to threonine.

amino group. In the O-linked glycosylation, the carbohydrates are attached to the serine (Ser) or threonine (Thr) side chain hydroxyl group. The current specific amino acid sequences involved with glycosylation state that the N-linked carbohydrates are linked through asparagine (Asn) and *N*-acetylglucosamine. The associated amino acid sequence with the N-linked carbohydrate consists of a serine (Ser) or threonine (Thr) separated by one amino acid residue (any one of 19 residues excluding proline), both located toward the C-terminus of the peptide chain. The proline (Pro) amino acid residue however does not participate as the one amino acid between the Asn and Ser/Thr residues. The N-linked and O-linked carbohydrates are illustrated in Figure 33.19.

In applying mass spectrometry in the characterization of glycosylated proteins, there are three objectives that the researcher is attempting to achieve: (i) to get an identification of the glycosylated peptides and proteins, (ii) to accurately determine the sites of glycosylation, and finally to (iii) determine the carbohydrates that make up the glycan and the structure of the glycan. One approach that has been employed in glycosylation characterization of proteins has been to cleave the glycans from the proteins, separate them from the proteins, and subsequently measure them by mass spectrometry. However, this approach does not give any information as to what proteins, and associated sites, belonged to what glycans. Increasingly, glycosylated proteins are being digested with an endoprotease producing glycopeptides. The glycopeptides are then analyzed by mass spectrometry. The mass spectrometric analysis of the glycopeptides is not as well defined as is the case with other modifications due to the heterogeneous nature of the oligosaccharide modifications. With glycosylation, the mass shift associated with the modification is not constant as compared to acetylation, oxidation, and phosphorylation (e.g., phosphorylation adds 80 Da to the peptide mass as HPO_3). Sometimes, mass pattern recognition can be used in single-state precursor ion scans to identify glycosylation. Table 33.5 lists the monoisotopic masses of the monosaccharides commonly found in glycosylated peptides along with their associated residue masses formed through water loss. Glycosylation heterogeneity is observed in precursor mass spectra when a repeating pattern is observed. This type of repeating pattern is indicative of the subsequent addition of a monosaccharide to the glycan chain of the glycosylated peptide. The repeating pattern for a high-galactose glycosylation pattern is illustrated in Figure 33.20 where a repeating value of 54.0 and 40.4 Da is observed. The glycopeptide in the mass spectrum has a molecular weight of 2709.3 Da. The series associated with 54.0 Da represents the addition of a galactose moiety to the plus 3 (triply) charge state of the glycopeptide as $[M+3H]^{3+}$. The series associated with 40.4 Da represents the addition of a galactose moiety to the plus 4 (quadruply) charge state of the glycopeptide as $[M+4H]^{4+}$. Each addition of a monosaccharide to the glycan chain is through a condensation reaction; therefore, the series will have a difference value of the monosaccharide minus water. According to the charge state, the difference between each series will be associated with a multiple of the charge. Table 33.6 lists the pattern difference for glycopeptide heterogeneity residue addition. There are also diagnostic fragment ions that can be produced during CID product ion spectral collection of glycated peptides known as oxonium ions. Hexose (Hex, generic name for galactose and mannose) has an oxonium ion at *m/z* 163, fucose (Fuc) at *m/z* 147, sialic acid (SiA) at *m/z* 292, *N*-acetylhexosamine (HexNAc) at *m/z* 204, and *N*-acetylglucosamine (GlcNAc) at *m/z* 204. The observance of these oxonium ions in product ion spectra has been used to help in the identification of the glycan modification on a peptide. However, care must be observed in using oxonium ions as diagnostic peaks, such as in precursor ion scanning, as species other than glycans can also generate similar isobaric product ions.

33.5.3 Phosphorylation of Proteins

As stated previously, the genomic DNA sequencing that has been accomplished cannot give direct information concerning PTM such as glycosylation (see previous section) and sulfation (section to follow). Another major form of PTM that we are looking at is the phosphorylation of proteins, a significant regulatory mechanism that controls a variety of biological functions in most organisms. Examples of phosphorylation regulating mechanisms

TABLE 33.5 Common Monosaccharides and Their Associated Masses.

Monosaccharide	Formula	Mass (Da)	$[M+H]^+$ (*m/z*)	Residue Mass (Da)	$[M+H]^+$ (*m/z*) Oxonium Ion
Mannose (Man)	$C_6O_6H_{12}$	180.0634	181.0712	162.0528	163.0607
Galactose (Gal)	$C_6O_6H_{12}$	180.0634	181.0712	162.0528	163.0607
Fucose (Fuc)	$C_6O_5H_{12}$	164.0685	165.0763	146.0579	147.0657
Sialic acid (SiA)	$C_{11}O_9NH_{19}$	309.1060	310.1138	291.0954	292.1032
N-acetylglucosamine (GlcNAc)	$C_8O_6NH_{15}$	221.0899	222.0978	203.0794	204.0872
N-acetylgalactosamine (GalNAc)	$C_8O_6NH_{15}$	221.0899	222.0978	203.0794	204.0872

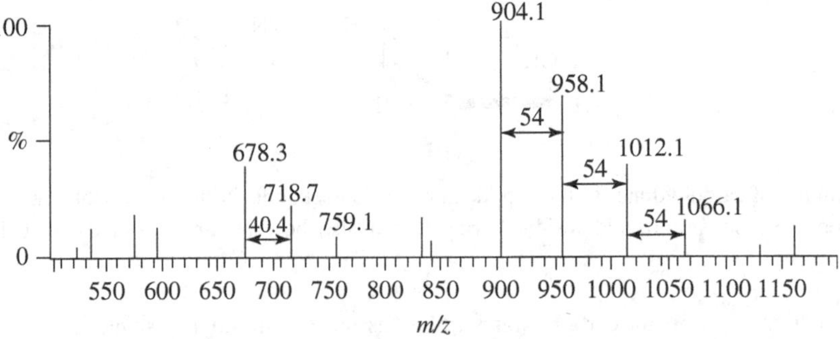

FIGURE 33.20 The repeating mass pattern for a high-galactose glycosylation for a glycopeptide with molecular weight of 2709.3 Da. Series associated with a 54.0 Da difference represents the addition of a galactose moiety to the plus 3 (triply) charge state of the glycopeptide as $[M+3H]^{3+}$. The series associated with a 40.4 Da difference represents the addition of a galactose moiety to the plus 4 (quadruply) charge state of the glycopeptide as $[M+4H]^{4+}$.

TABLE 33.6 Pattern Difference in Mass for Glycopeptide Heterogeneity Residue Addition.

Sugar	Formula	+1	+2	+3	+4	+5
Hexose	$C_6O_5H_{10}$	162.1	81.0	54.0	40.4	32.4
dHexose	$C_6O_4H_{10}$	146.1	73	48.7	36.5	29.2
HexNac	$C_8O_5NH_{13}$	203.2	101.6	67.7	50.8	40.6
SiA	$C_{11}O_8NH_{17}$	291.3	145.6	97.1	72.8	58.2

include gene expression, cell cycle processes, apoptosis, cytoskeletal regulation, and signal transduction. It has been estimated that up to 30% of all of the proteins in humans exist in the phosphorylated form where 2% of the human genome encode for protein kinases (>2000 genes) [29]. In eukaryotic cells, the protein phosphorylation takes place with the serine, threonine, and tyrosine residues. The reversible protein phosphorylation of the serine, threonine, and tyrosine residues is an integral part of cellular processes involving signal transduction [30,31]. The identification of the phosphorylation sites is important in understanding cellular signal transduction.

Protein phosphatases are enzymes responsible for the removal of phosphate groups from a target (i.e., reversible protein phosphorylation), and protein kinases are enzymes responsible for the addition of phosphate groups to a target. These two enzymes work together to control cellular processes and signaling pathways. Greater attention has been given in the literature to the study of signaling pathways primarily involved with protein kinase as compared to specific types of phosphatase [32–37]. However, the importance of studying protein phosphatase enzymes and their targets has been demonstrated in recently reported disease state studies where the abnormal condition has been attributed at least in part to malfunctioning protein phosphatase enzymes [38–40]. In covalent modification of proteins such as phosphorylation, the activity of the modified enzyme has been altered in the form of activated, inactivated, or to otherwise regulate its activity upwardly or downwardly. The most common mechanism for phosphorylation is the transfer of a phosphate group from adenosine triphosphate (ATP) to the hydroxyl group of either serine, threonine, or tyrosine within the protein. Figure 33.21 illustrates the general cellular mechanism involving protein kinase phosphorylation and protein phosphatase dephosphorylation. In the top portion of Figure 33.21, a cellular signal is received in a kinase/phosphatase cycle, often in the form of a messenger biomolecule such as lipid (diacylglycerol shown), that initiates the phosphorylation of the target protein by the protein kinase–ATP action. The enzyme has been phosphorylated resulting in its participation in the signaling cycle. Usually, the phosphorylated enzyme does not permanently stay in its modified form but will undergo dephosphorylation through interaction with protein phosphatase. Figure 33.22 illustrates the nonphosphorylated structure of the serine, threonine, and tyrosine residues, each having a hydroxyl moiety, and the phosphorylated form of the amino acid residue.

While the serine, threonine, and tyrosine residues all have a side group hydroxyl moiety available for PTM, over 99% of the phosphorylated modification takes place with the serine and threonine amino acid residues in eukaryotic cells [41]. Recently, Olsen et al. [31] have reported a slight variation to the widely referred study of Hunter and Sefton [42] where the relative abundances of amino acid residue phosphorylation were assigned as 0.05% for phosphotyrosine (pY), 10% for phosphothreonine (pT), and 90% for phosphoserine (pS). In Olsen's recent study, this has been adjusted to 1.8% pY, 11.8% pT, and 86.4% pS where the larger percentage value allocated to pY was attributed to more sensitive methodology being employed, thus allowing the characterization of lower abundant phosphorylated proteins. However, the stoichiometry of the phosphorylated proteome (in the form of tryptic peptides) is small (≤ 1) in relation to the nonphosphorylated proteome (in the form of tryptic peptides), thus requiring that the sensitivity of the phosphorylated proteome analysis to be as optimal as possible.

Mass spectrometry is often used to identify the sites of phosphorylation in the protein backbone when studying cellular signaling pathways [43–46]. Prior to the introduction of mass spectrometric methodology for phosphorylated protein analysis, researchers studied protein phosphorylation using ^{32}P labeling, followed by 2D polyacrylamide gel electrophoresis and finally Edman sequencing. This methodology was time consuming and involved the handling of radioactive isotopes. This has prompted a considerable amount of methodology development using mass spectrometric techniques that are able to measure whole proteomes from complex biological systems and not just single proteins that the Edman sequencing approach is most suited for. There are numerous studies reported in the literature of signaling pathways primarily involved with protein kinase [40,47–51]. The study of protein phosphatase enzymes and their targets has also gained importance where there are recently reported numerous disease states that have been attributed at least in part to malfunctioning protein phosphatase enzymes [33,52–54].

The mass spectrometric instrumentation in use today possesses the sensitivity needed for PTM analysis; therefore, often the limiting factor in phosphorylated proteome analyses lies in the sample treatment prior to mass spectral measurement. Immobilized metal affinity chromatography (IMAC) is the methodology of choice for phosphorylated proteome cleanup and enrichment [55–57]. Labeling is also the most widely used approach for relative quantitation of the phosphorylated proteome when comparing a normal state to a perturbed state. It typically involves using stable isotopes with either ^{2}D-methanol or ^{13}C-SILAC or iTRAQ reagent [58–60]. However, often labeling procedures can cause an increase in sample complexity, can be cumbersome or incomplete, and can ultimately result in sample losses.

The ideal methodology for PTM studies would entail both high sample recovery and sample specificity while avoiding any additional modifications to the proteins being studied. The first critical step in sample preparation is the lysis and solubilization of the sample's complement of proteins. This is followed by additional steps for sample cleanup and treatment that often are accompanied by protein/peptide loss at each step. Another limiting factor is the presence of nucleic acids during the IMAC enrichment step, which are known to poison or compete with the phosphorylated peptides for binding sites on the IMAC bed. Alternative approaches exist for the removal of RNA and DNA such as the addition of RNase and DNase to the whole cell lysate buffer [61], passing the lysate repeatedly through a tuberculin syringe fitted with a 21 gauge needle to shear the RNA and DNA mechanically [62], or using the QIAshredder (Qiagen) [63]. The use of RNase and DNase is often avoided in an effort to reduce the amount of additives and the production of low

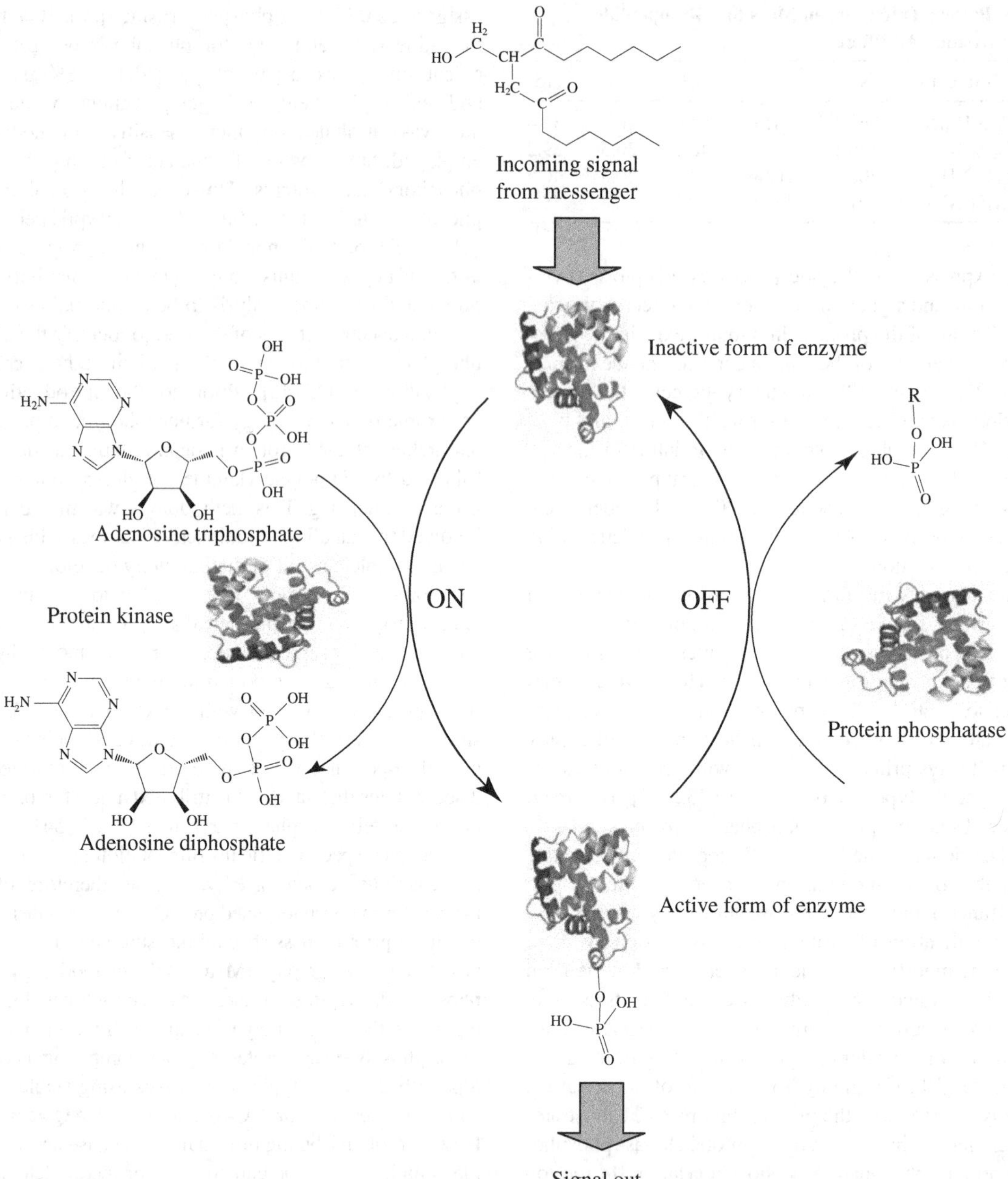

FIGURE 33.21 Example of the general cellular mechanism involving protein kinase phosphorylation and protein phosphatase dephosphorylation. (Top) A cellular signal is received in the form of the messenger biomolecule diacylglycerol lipid, which initiates the phosphorylation of the target protein by the protein kinase–ATP action. The phosphorylated enzyme propagates the signaling cycle and then is recycled through dephosphorylation by a phosphatase.

molecular weight nucleic acid-lysed products, which are difficult to remove and also poison the IMAC bedding. The choice of ultracentrifugation can be made over mechanical shearing for similar reasons that are associated with the production of low molecular weight nucleic acid species.

The use of 1D SDS-PAGE is another alternative approach for whole cell lysate intact protein cleanup that includes removal of the nucleic acids. Current examples include a study by Gygi and coworkers [33] who used preparatory gels for HeLa cell nuclear protein separation followed by whole gel digestion (cut into 10 regions) where 967 proteins revealed 2002 phosphorylation sites. A recent study by Mann and coworkers [64] reported that five times more proteins from tear fluid were identified after gel electrophoresis as compared to in-solution digestion. Other potential advantages for the use of 1D SDS-PAGE are the ability to target specific molecular weight ranges that can be identified and excised from various band regions within the gels and a more efficient tryptic digest due to the enhanced accessibility of the protein backbone denatured into a linear orientation locked within the gel. Recoveries from gels can be an added concern though as all steps in protein/peptide preparation can contribute to an overall loss of protein.

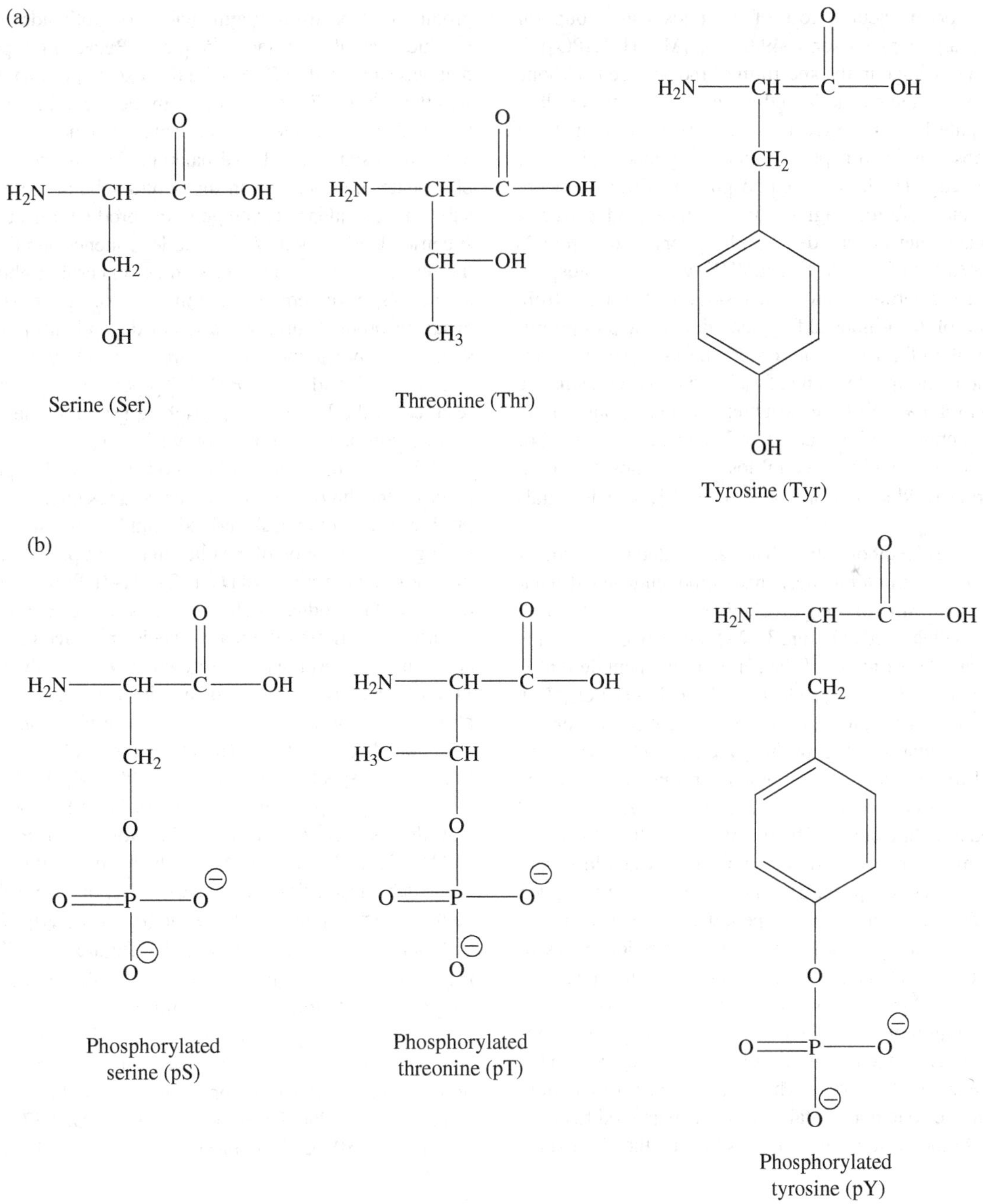

(a)

(b)

FIGURE 33.22 Structures of the nonphosphorylated (a) and phosphorylated (b) states of the amino acids serine, threonine, and tyrosine.

Mass spectrometric studies of phosphorylation PTM typically employ tandem mass spectrometry to generate product ion spectra of phosphorylated peptides. This usually entails a premass spectrometric separation of a complex mixture of phosphorylated peptides by reversed-phase carbon-18 stationary-phase high-performance liquid chromatography with an electrospray ionization source (RP C18 HPLC ESI-MS/MS). Fragmentation pathway studies using ion trap mass spectrometry of the phosphorylation of the three possible sites in peptides—serine, threonine, and tyrosine—have been studied and characterized revealing different mechanisms. Peptides that contain phosphoserine generally will lose the phosphate group as the predominant product ion in the spectrum. This often results in limited information about the sequence of the peptide where other peptide backbone fragmentation is not well observed. Neutral loss of the phosphate group for the plus one charge state (+1 CS) is observed at 98.0 Da, $[M+H-H_3PO_4]^+$; for the plus two charge state (+2 CS) is observed at 49.0 Da, $[M+2H-H_3PO_4]^{2+}$; and for the plus 3 charge state (+3 CS) is observed at 32.7 Da, $[M+3H-H_3PO_4]^{3+}$. Let's consider hypothetical product ion spectra for a phosphoserine-containing peptide at *m/z* 987.5 (monoisotopic mass of 1972.98 Da). In the product ion spectrum, the predominant product ion is observed at *m/z* 938.7 for the neutral loss of the phosphate group for the doubly charged precursor (–49.0 Da), $[M+2H-H_3PO_4]^{2+}$. There is often coverage in the spectrum of the peptide backbone sequence as illustrated by the *b*- and *y*-type ions; however, their response is quite low and often is not observed in product ion spectra of phosphorylated peptides when using tandem ion trap mass spectrometry. The loss of the phosphate modification from the peptide is the preferred fragmentation pathway and is usually the predominant one observed. The phosphorylated peptide's sequence is RApSVVGTTYWMAPEVVK, where the phosphorylation is on the serine amino acid residue that is third from the left. In the collision-induced fragmentation of the phosphorylated peptide, all of the *y*-type ions observed (y_5–y_{14}) do not contain the phospho group. Only one of the *b*-type ions contains the phospho group at b_{12}. This is due to a preferential cleavage taking place on the peptide backbone at the proline residue, while all of the other *b*-type ions include neutral loss of the phospho group from the serine residue and are denoted as b_n^{Δ}, which equals b_n—H_3PO_4.

The loss of the phosphorylation that is associated with serine is through a β-(beta) elimination mechanism producing dehydroalanine. The mechanism for dehydroalanine production through β-elimination is illustrated in Figure 33.23 for phosphoserine. This would represent the structure of the serine in the peptide that is represented by the *m/z* 938.7 product ion, $[M+2H-H_3PO_4]^{2+}$.

Application of MS^3 product ion spectral collection for the enhanced fragmentation of phosphorylated peptides can often be employed to help solve this problem. In this approach, the main product ion collected from MS^2, *m/z* 938.7, is isolated and subjected to a third stage of fragmentation. In the MS^3 product ion spectrum, there is a predominant peak for water loss from the *m/z* 938.7 precursor peak. This also does not afford much information for the sequence of the peptide; however, there are also observed in the spectrum numerous product ion peaks of the *b* and *y* types that have a much greater response than that observed in the MS^2 product ion spectrum. It is often noticed that in the fragmentation of the phosphorylated peptide, product ions derived through cleavage from both sides directly adjacent to the serine residue did not take place. This means that a positive identification of the residue that contains the phosphate modification, namely, the serine residue, is not possible. In the third-stage product ion spectrum, fragmentation on both sides of the serine residue does often in fact take place. Because the precursor ion that was subjected to CID in the third-stage product ion spectrum was the *m/z* 938.7 product ion from the second-stage fragmentation, the serine residue has been replaced in the sequence by a "B" ion, which stands for dehydroalanine. Due to the absence of the phosphate modification on the peptide chain, there is observed substantial sequence coverage in the product ions, thus allowing specific identification of the peptide sequence and the location of the phosphorylation. Further studies concerning phosphorylated serine (pS) have demonstrated that loss of the phosphate H_3PO_4 group in product ion spectra is not dependent upon the charge state of the precursor ion. All three commonly observed charge states, +1, +2, and +3, for ESI CID product ion spectral collection resulted in the loss of the phosphate group as the predominant product ion peak with other minor losses.

When the phosphorylation modification of a peptide takes place on the threonine (Thr) amino acid residue, the predominant product ion spectral peak derived from CID is also observed to be through β-elimination of the phosphate group. The neutral loss of the phosphate group, −98 Da as $[M+H-H_3PO_4]^+$, from the threonine residue produces dehydroaminobutyric acid. A product ion spectrum illustrating the major species produced from a peptide containing a phosphorylated threonine is RASVVGTpTYWMAPEVVK, which has a monoisotopic mass of 1972.98 Da. The phosphorylation of the peptide has taken place on the threonine amino acid residue (pT). The charge state of the peptide in the product ion spectrum is +2 at *m/z* 987 as $[M+2H]^{2+}$. Loss of the H_3PO_4 phosphate group is the primary fragmentation pathway that is observed in the product ion spectrum at *m/z* 939.0 as $[M+2H-H_3PO_4]^{2+}$. In the product ions produced from CID of the *m/z* 987 species, there are a number of *b*-type ions that both contain the phosphate modification ($b_8, b_9, b_{10}, b_{11}, b_{12}, b_{14}, b_{15}$) and that do not $(b_8^{\Delta}, b_9^{\Delta}, b_{10}^{\Delta}, b_{12}^{\Delta}, b_{14}^{\Delta})$. Figure 33.24 illustrates the fragmentation pathway mechanism for the production of the dehydroaminobutyric acid from the neutral loss—β-elimination of the phosphate group from phosphorylated threonine amino acid. One difference in the product ion spectrum for fragmentation of phosphorylated threonine as compared to phosphorylated serine is the product ion peak observed at *m/z* 947.5 for the neutral loss of 80 Da. The product ion at *m/z* 947.5 represents

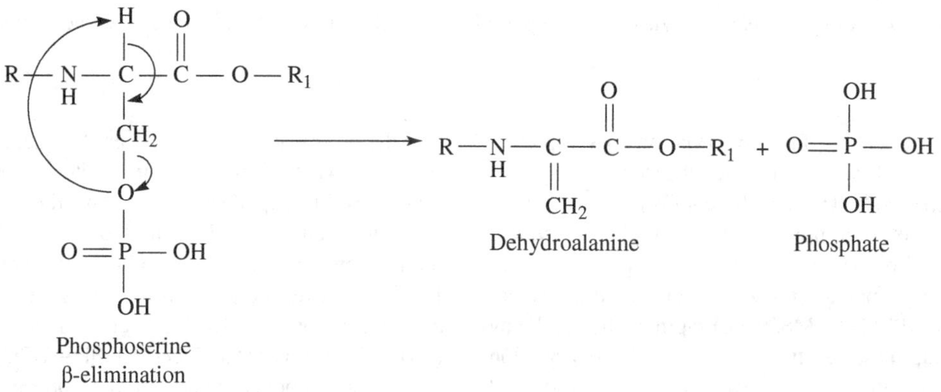

FIGURE 33.23 β-Elimination mechanism for the fragmentation pathway producing dehydroalanine for the phosphorylated serine amino acid residue.

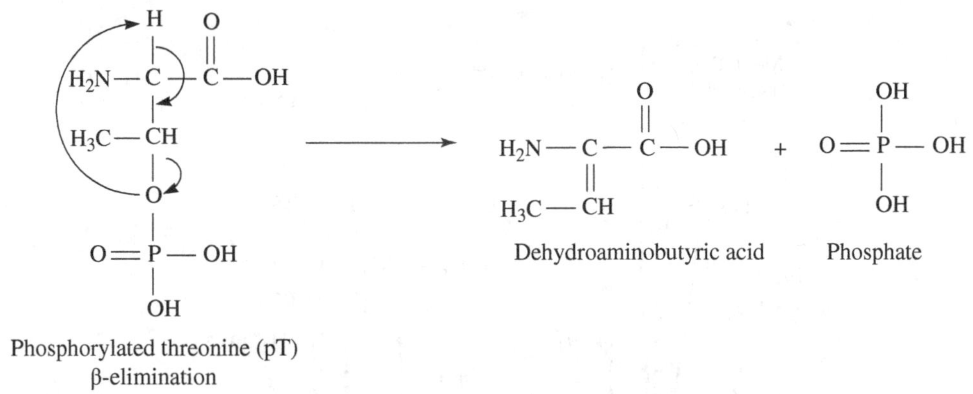

Phosphorylated threonine (pT)
β-elimination

Dehydroaminobutyric acid

Phosphate

FIGURE 33.24 Mechanism for the production of the dehydroaminobutyric acid from phosphorylated threonine through neutral loss—β-elimination of the phosphate group.

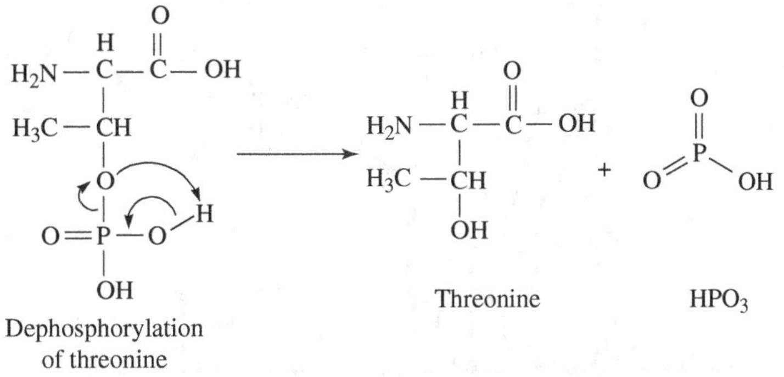

Dephosphorylation of threonine

Threonine

HPO_3

FIGURE 33.25 Mechanism for dephosphorylation of the threonine amino acid resulting in the structure of the original amino acid residue of threonine.

Phosphorylated tyrosine (pY)

Dephosphorylated tyrosine

HPO_3

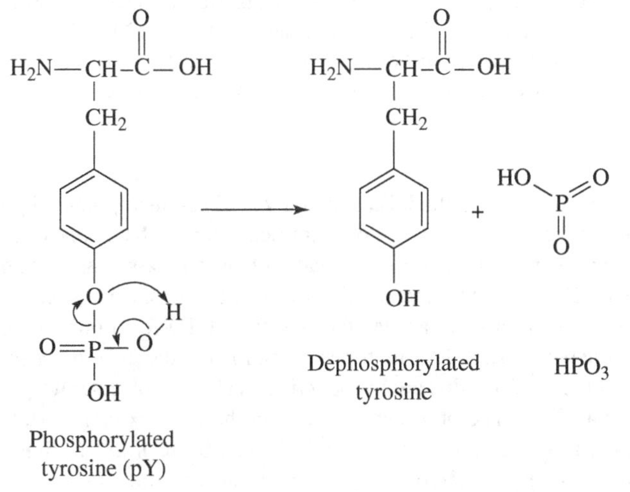

FIGURE 33.26 Mechanism for the dephosphorylation of tyrosine.

dephosphorylation through neutral loss of HPO_3 from the precursor peak as $[M+2H-HPO_3]^{2+}$. This particular loss is not observed in the product ion spectrum of phosphorylated serine. The mechanism for dephosphorylation of the threonine amino acid is illustrated in Figure 33.25. The dephosphorylation results in the structure of the original amino acid residue of threonine.

Product ion spectra of phosphorylated tyrosine (pY)-containing peptides also illustrate losses associated with 80 and 98 Da, which would appear to be similar to the losses observed with phosphorylated threonine peptides. The 80 Da loss is due to dephosphorylation of the tyrosine residue resulting in the original structure of the tyrosine residue. The mechanism is illustrated in Figure 33.26. However, due to the structure of tyrosine, a similar mechanism of β-elimination for loss of the phosphate (−98 Da, H_3PO_4) group is most probably not likely. The neutral loss of phosphate at 98 Da has also been proposed to not happen through a two-step mechanism involving both water (H_2O) loss of 18 Da and HPO_3 loss of 80 Da, in any order. The neutral loss of 98 Da is more than likely associated with some form of rearrangement in the fragmentation pathway mechanism.

Other mass spectrometric methods are used in phosphorylated peptide analysis that has been investigated to help increase the efficiency of the fragmentation that takes place. This can allow more direct approaches for modification location without the need of third-stage fragmentation experiments that require the interpretation of two individual spectra. One example of an alternative mass spectrometric approach has been the use of the FTICR mass spectrometer using IRMPD and ECD. Figure 33.27 illustrates a comparison of these two dissociation approaches. In the top figure, IRMPD was used to excite and dissociate the phosphorylated peptide. The phosphorylation is

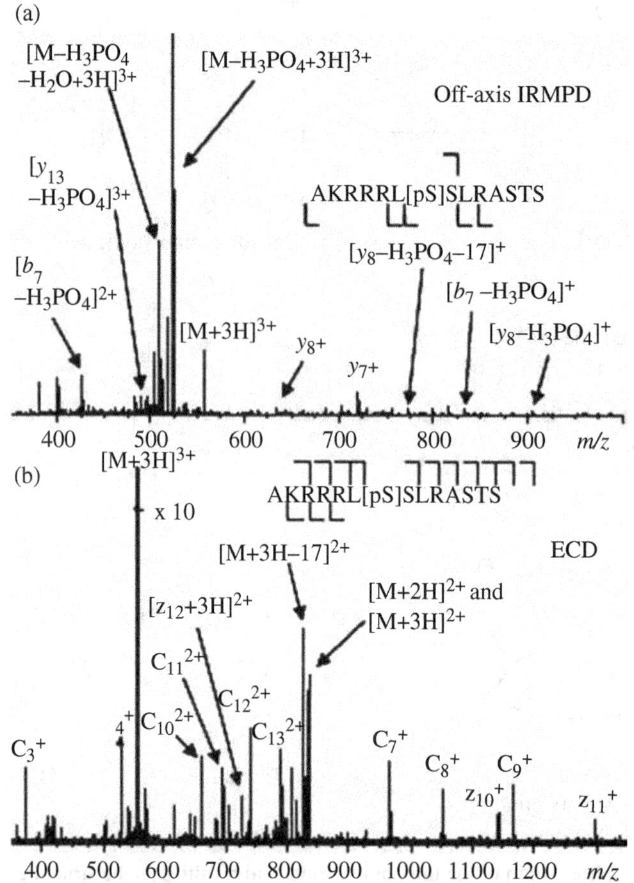

FIGURE 33.27 (a) Product ion spectrum obtained from off-axis IRMPD FTICR MS/MS of a population of quadrupole- and SWIFT-isolated $[AKRRRL(pS)SLRASTS+3H]^{3+}$ phosphopeptide ions. The spectrum is dominated by ions resulting from neutral losses of H_3PO_4, NH_3, and H_2O. Five (out of 13) peptide backbone bonds are broken, and the location of the phosphorylation site is identified only by observation of the ($y8$–H_3PO_4–NH_3) ion (present at very low abundance) and the singly and doubly charged ($b7$–H_3PO_4) ions. Irradiation was for 500 ms at ~36 W laser power, and the data represent a sum of 10 scans. (b)_Product ion spectrum obtained from ECD (20 ms irradiation) FTICR MS/MS of the same quadrupole- and SWIFT-isolated phosphopeptide. Twelve out of 13 peptide backbone bonds are cleaved, and the location of the phosphate is readily assigned by observation of the abundant $c7$ ions. (Reprinted with permission from Ref. [30]. Copyright Elsevier 2004.)

located on a serine (pS) residue within the middle of the peptide chain having the sequence: AKRRRL(pS)SLRASTS. In the product ion spectrum collected using IRMPD, the major product ions that were observed were all associated with the neutral loss of the phosphate group as $[M+3H-H_3PO_4]^{3+}$ and also phosphate and water loss as $[M+3H-H_3PO_4-H_2O]^{3+}$. As also can be observed in the top product ion spectrum of Figure 33.27, very little information is given concerning the peptide chain's sequence. The bottom product ion spectrum of Figure 33.27 illustrates the effectiveness of using ECD for phosphorylated peptide sequence determinations. In this spectrum, there is an appreciable amount of peptide backbone fragmentation ions in the form of *c*-type and *z*-type ions. The major peak in the spectrum is the triply protonated precursor ion as $[M+3H]^{3+}$. Also, notice that the phosphorylation was maintained within the structure of the product ions. The two product ion spectra are complementary where the top spectrum is diagnostic for the determination of a phosphorylation of the peptide, while the bottom spectrum gives very good sequence coverage of the peptide.

More recently, the ETD capability has been incorporated into linear ion traps that also allow significant peptide backbone cleavages similar to the ECD capability usually associated with FTICR mass spectrometers. Figure 33.28 illustrates the extensive fragmentation that can be achieved with ETD for a quite long sequence of peptide. The top spectrum is the dissociation spectrum for a 35-residue phosphopeptide that has a molecular weight of 4093 Da. The phosphopeptide is in the plus six charge state (+6) at *m/z* 683.3 for $[M+6H]^{6+}$. As can be seen in the figure for this very long sequence phosphorylated peptide, essentially complete coverage of the peptide chain has been achieved along with the determination of the phosphorylation site. The ETD product ion spectrum illustrated in Figure 33.28b is for a phosphorylated histidine (pH) peptide. Phosphorylation on the histidine residue is an important type of PTM that is observed in prokaryotic proteomes and will be discussed next.

33.5.3.1 Phosphohistidine as PTM The phosphoryl modification of serine, threonine, and tyrosine is a PTM primarily

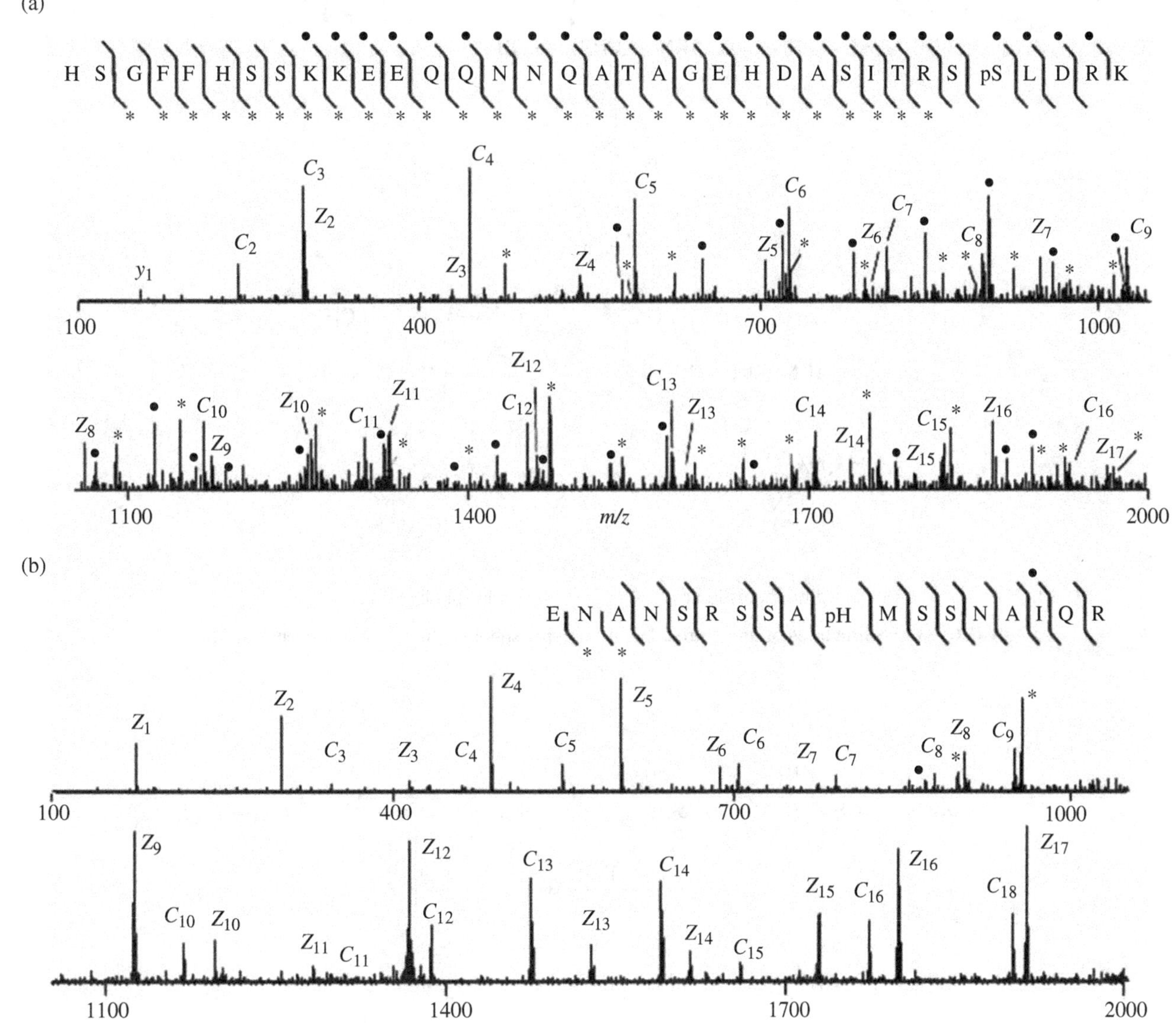

FIGURE 33.28 Phosphopeptide mass spectra. ETD mass spectra recorded on $(M + 6H)^{+6}$ ions at m/z 683.3 for a 35-residue phosphopeptide of MW 4093 (*A*) and $(M + 3H)^{+3}$ ions from a pHis-containing peptide at the C-terminus of the septin protein, Cdc10 (*B*). Observed c and z• ions are indicated on the peptide sequence by] and [, respectively. Observed doubly charged c and z• ions are indicated by an additional label, circle and asterisk, respectively. (Reprinted with permission from Chi, A., Huttenhower, C., Geer, L.Y., Coon, J.J., Syka, J.E.P., Bai, D.L., Shabanowitz, J., Burke, D.J., Troyanskaya, O.G., Hunt, D.F. Analysis of phosphorylation sites on proteins from *Saccharomyces cerevisiae* by electron transfer dissociation (ETD) mass spectrometry *PNAS,* **2007**, *104*, 2193–2198. Copyright 2007 National Academy of Sciences, U.S.A.)

associated with regulating enzyme activity in eukaryotic systems. The phosphoester linkage resulting with the PTM of serine, threonine, and tyrosine is relatively stable at physiological pH (~7.36) and thus generally requires a phosphatase for the removal of the modification. The phosphorylation and subsequent dephosphorylation of proteins are activities associated in biological systems with respect to cellular signaling pathways. In prokaryotic biological systems, the histidine group is phosphorylated primarily for the purpose of transferring the phosphate group from one biomolecule to another. These biomolecules are known as phosphodonor and phosphoacceptor molecules where the phosphohistidine is acting as a high-energy intermediate in some type of biological process on the molecular level [65]. The phosphohistidine transfer potential (ΔG^{o} of transfer) of the phosphate is estimated at −12 to −14 kcal/mol, reflecting a relatively high-energy system [66]. The phosphorylation of the histidine residue produces a phosphoramidate that contains a large standard free energy of hydrolysis, making them the most unstable form of phosphoamino acids. The stability of the phosphohistidine is largely influenced by its local amino acid residues and the nature and makeup of the associated protein. It is speculated due to this instability that often a histidine phosphatase is not required. It has been estimated that up to 6% of the phosphorylation that takes place in eukaryotes [67] and prokaryotes [68] is with the histidine amino acid residue. It is also estimated that the abundance of phosphohistidine is 10- to 100-fold greater than

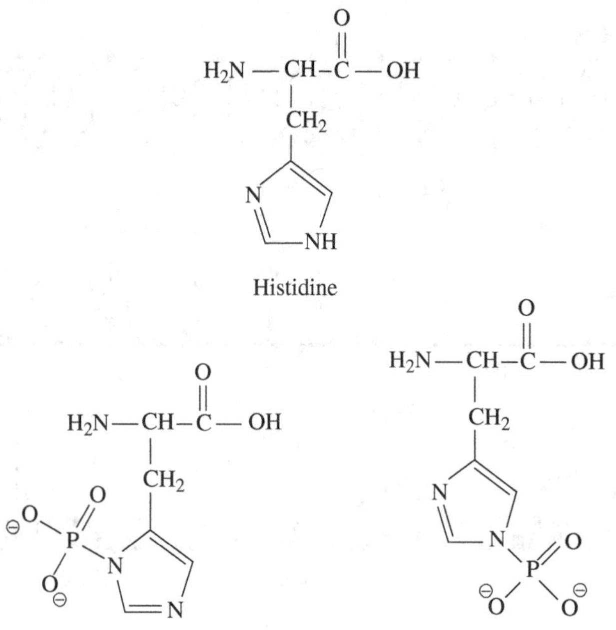

Histidine

1-phosphohistidine

3-phosphohistidine

FIGURE 33.29 Structures of unmodified histidine, 1-phosphohistidine, and 3-phosphohistidine.

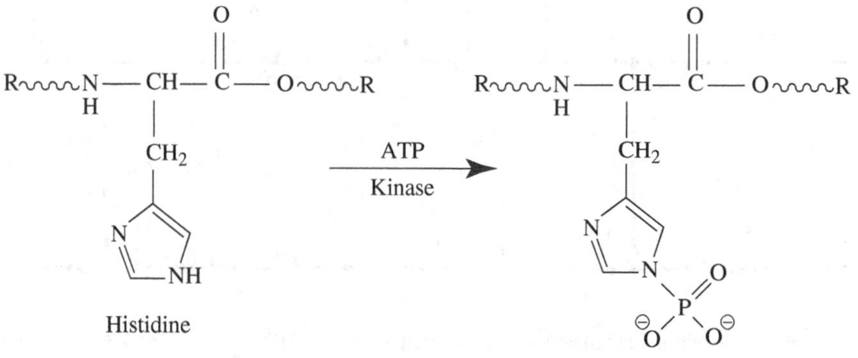

ATP

Kinase

Histidine

3-phosphohistidine

FIGURE 33.30 Kinase phosphorylation of the histidine amino acid residue producing 3-phosphohistidine.

that of phosphotyrosine but much less than phosphoserine [67]. The phosphorylation of the histidine amino acid residue can take place on either the nitrogen in the 1-position or the nitrogen in the 3-position. This is illustrated in Figure 33.29 for the phosphorylation of the histidine amino acid residue. Studies have demonstrated that the 3-phosphohistidine is more stable than the 1-phosphohistidine [69] and is therefore more likely to be the positional isomer observed in biological systems. The histidine amino acid residue is phosphorylated by kinase using the reactive intermediate ATP. This process is illustrated in Figure 33.30 where the modification produces the 3-phosphohistidine residue.

The bacterial histidine kinases of the two-component system are an example of phosphohistidines found in prokaryotic systems. Figure 33.31 shows a drawing of a two-component signaling system that involves histidine phosphorylation and dephosphorylation. The figure illustrates a membrane-bound protein that contains a carboxy-terminal histidine kinase domain. The subsequent steps involved include the binding of a ligand, which is followed by dimerization of two kinases. Phosphorylation of the histidine residue is done through the reactive intermediate ATP. The next step is the binding of a cytosolic response regulator protein that contains an aspartate residue in the amino-terminal domain. The aspartate residue in the response regulator is then phosphorylated by the first membrane-bound protein and released to perform its downstream function. There are various bacterial two-component histidine kinase signaling systems that have been observed and characterized. Table 33.7 shows a listing of some of the bacteria that have been observed to have the two-component system.

The major difficulty in measuring the phosphorylation of a histidine residue in a peptide is due to the instability of the

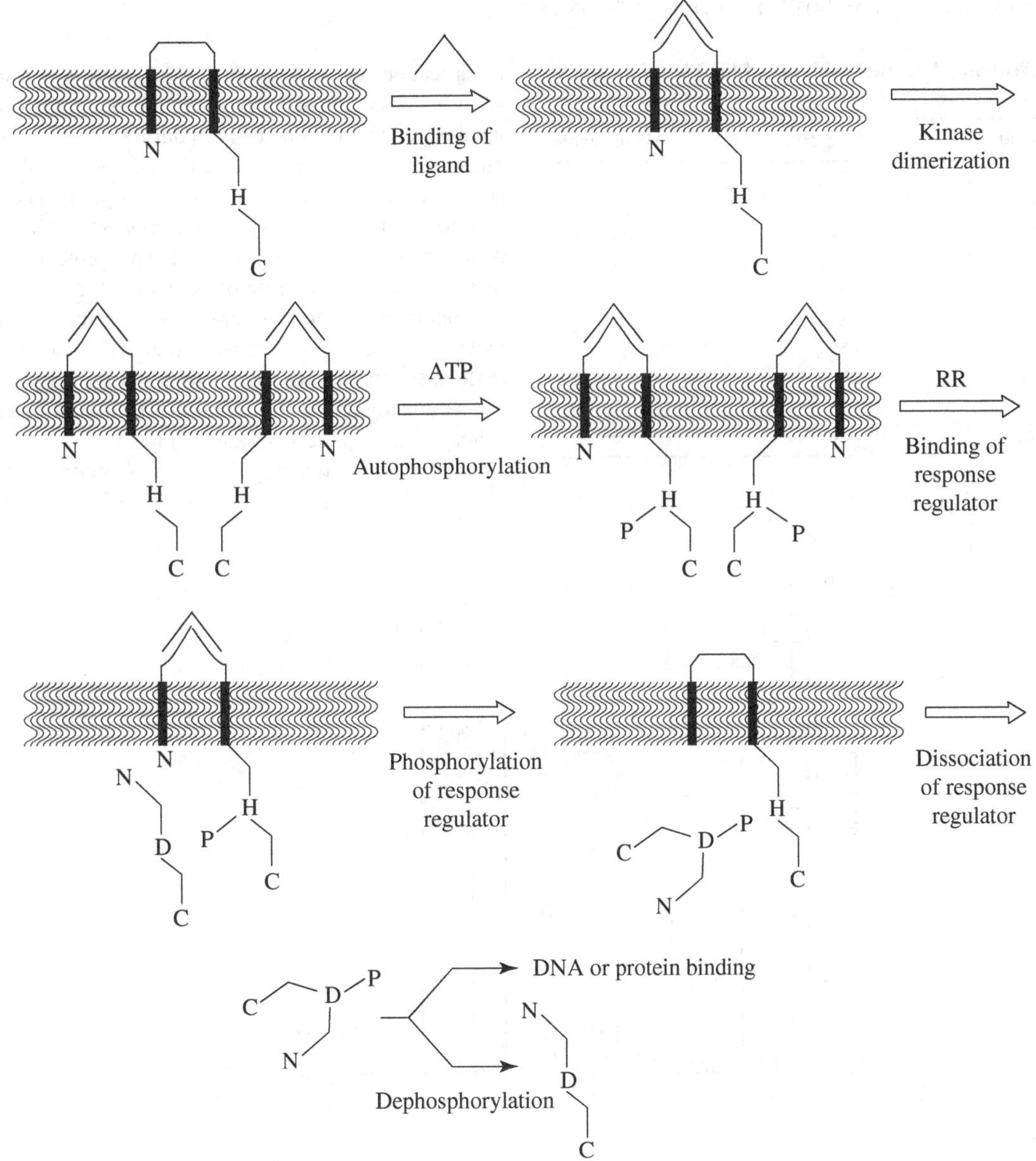

FIGURE 33.31 A model for two-component signaling systems involving the phosphorylation of the histidine amino acid residue. (Reprinted with permission. This article was published in *Chemistry & Biology,* Pirrung, M.C. Histidine kinases and two-component signal transduction systems. **1999**, *6*, R167–R175. Copyright Elsevier 1999.)

TABLE 33.7 Some Bacterial Two-Component Histidine Kinase Signaling System.

Histidine Kinase	Control Function	Source	Receiver
CheA	Chemotaxis, flagellar motor	*E. coli*	CheY
EnvZ	Osmosensing, outer membrane proteins	*E. coli*	OmpR
VanS	Cell wall biosynthesis	*Enterococcus faecium*	VanR
KinA	Osmosensing	*E. coli*	SpoOF
PhoR	Phosphate metabolism	*E. coli*	PhoB
FrzE	Fruiting body formation	*Myxococcus xanthus*	FrzE (internal receiver)
RscC	Cell capsule synthesis	*E. coli*	RscC (internal receiver)
VirA	Host recognition, transformation	*Agrobacterium tumefaciens*	VirA (internal receiver)
ArcB	Anaerobiosis	*E. coli*	ArcB (internal receiver) and ArcA
BvgS	Virulence	*Bordetella pertussis*	BvgS (internal receiver)

Source: Reprinted with permission. This article was published in *Chemistry & Biology*, Pirrung, M.C. Histidine kinases and two-component signal transduction systems. **1999**, *6*, R167–R175. Copyright Elsevier 1999.

TABLE 33.8 Acid and Alkaline Stabilities of the Phosphorylated Amino Acids.

Phosphoamino Acid	Acid Stable	Alkali Stable
N-Phosphates		
Phosphoarginine	No	No
Phosphohistidine	No	Yes
Phospholysine	No	Yes
O-Phosphates		
Phosphoserine	Yes	No
Phosphothreonine	Yes	Partial
Phosphotyrosine	Yes	Yes
Acyl phosphate		
Phosphoaspartate	No	No

modification. The phosphorylated histidine is very unstable in an acidic environment that is typically the matrix that the phosphorylated peptide is contained within during normal proteomic preparation steps prior to mass spectral analyses. Table 33.8 lists the acidic or alkaline stabilities of the phosphorylated amino acids. Due to the acid lability of the phosphorylated histidine residue, it is not observed during most phosphoproteome studies using mass spectrometric techniques. Current studies of histidine-phosphorylated peptides are incorporating neutral-level pH methodologies to preserve the modification in conjunction with enrichment approaches such as immobilized copper (II) ion affinity chromatography [70]. Figure 33.32a illustrates a nonphosphorylated peptide's product ion mass spectrum collected using ESI and a quadrupole ion trap. Very good coverage of

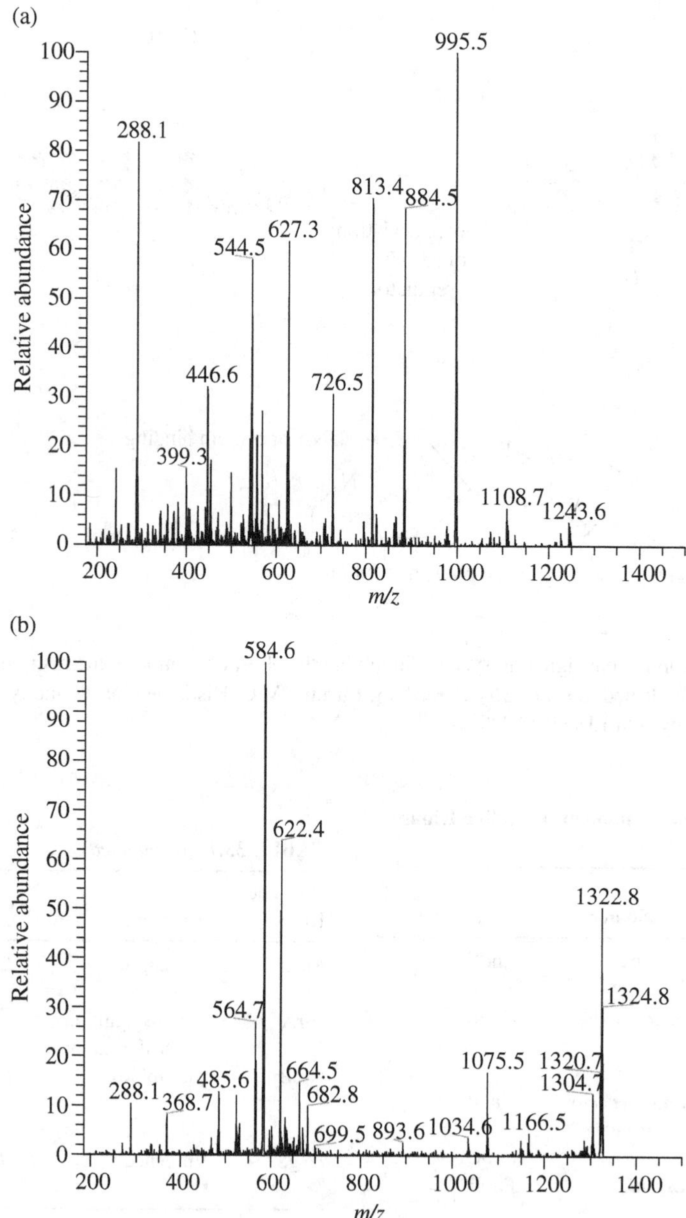

FIGURE 33.32 (a) Product ion mass spectrum of a nonphosphorylated peptide at m/z 642.4 as $[M+2H]^{2+}$. Good coverage of the peptide backbone is observed with the *y*-series ions allowing the sequencing of the peptide. (b) Product ion spectrum of the same peptide after phosphorylation of a histidine residue at m/z 682.0 as $[M+2H]^{2+}$. Major product ion observed is for the neutral loss of the phosphate moiety and water at m/z 633.1 as $[M+2H-HPO_3-H_2O]^{2+}$.

the peptide backbone is observed with the *y*-series ions allowing the sequencing of the peptide. The precursor was a doubly protonated species at *m/z* 642.4 as $[M+2H]^{2+}$. Figure 33.32b shows the product ion spectrum of the same peptide after phosphorylation of a histidine residue contained in the peptide chain. The mass spectrum was collected on the doubly protonated species at *m/z* 682.0 as $[M+2H]^{2+}$. The major product ion observed in the spectrum is for the neutral loss of the phosphate moiety and water at *m/z* 633.1 as $[M+2H–HPO_3–H_2O]^{2+}$. There are a couple of important considerations when considering product ion spectra of phosphorylated amino acid residues. When measuring these species in positive ion mode, the phosphorylation of the histidine residue, as illustrated in Figures 33.29 and 33.30, actually results in the addition of 79.9663 Da (HPO_3) to the mass of the peptide. The ATP is contributing, in a sense, HPO_3 to the peptide as illustrated in Figure 33.33a. In physiological conditions, the addition is more likely to be PO_3^-, but measured by mass spectrometry in positive ion mode, it is HPO_3. During CID product ion spectral accumulation, the loss due to the phosphoryl modification is also at 79.9663 Da for HPO_3. This would suggest that in positive ion mode, the GP fragmentation mechanism would be something like that illustrated in Figure 33.33b. The phosphoryl group on the histidine amino acid residue is presented as being fully protonated. The channel pathway during the fragmentation process results in the neutral loss of HPO_3 and the reprotonated, original form of the histidine residue without a change in the charge state of the peptide.

Similar to the product ion spectra of the phosphorylated serine, threonine, and tyrosine (also sulfated) amino acid residues, the phosphorylated histidine product ion spectra are primarily dominated by the neutral loss of the modification. Future work utilizing mass spectrometric methodologies such as ECD and ETD will also be beneficial in the GP analysis of phosphorylated histidine peptides.

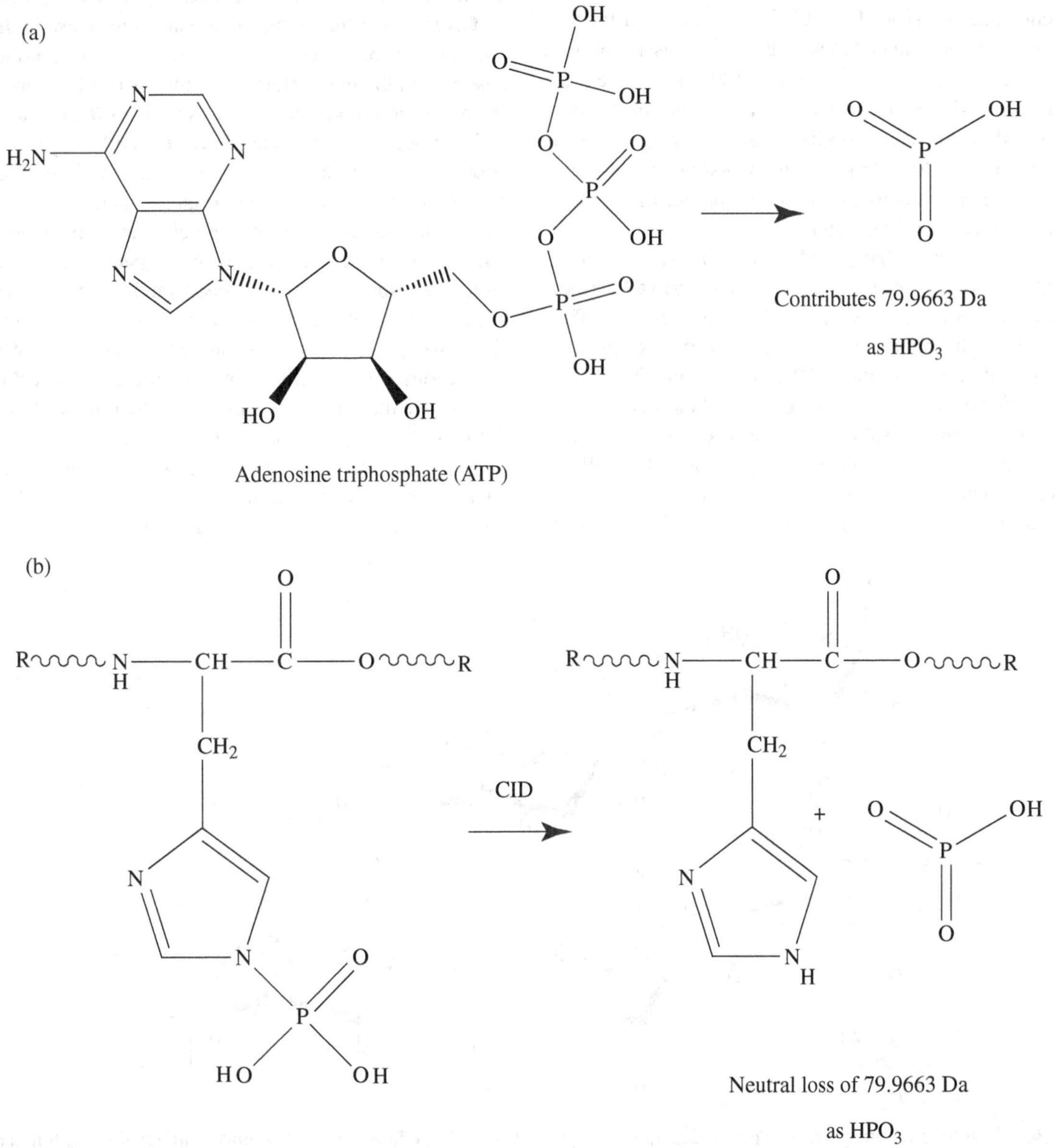

(a)

Adenosine triphosphate (ATP)

Contributes 79.9663 Da as HPO_3

(b)

CID

Neutral loss of 79.9663 Da as HPO_3

FIGURE 33.33 (a) ATP contributing HPO_3 to the peptide when considered mass spectrometrically in positive ion mode. (b) Collision-induced dissociation product ion spectra illustrate the loss due to the phosphoryl modification at 79.9663 Da for HPO_3.

33.5.4 Sulfation of Proteins

33.5.4.1 Glycosaminoglycan Sulfation There are two primary types of sulfation that occur involving protein PTM: carbohydrate sulfation of cell surface glycans and sulfonation of protein tyrosine amino acid residues. We will begin with a look at carbohydrate sulfation, which is an important process that occurs in conjunction with extracellular communication. The glycosaminoglycan modification presented previously in Section 33.5.2 occurs with the portion of the membrane-bound protein that is protruding out of the surface of the cell. The predominant enzyme carbohydrate sulfotransferase is the protein responsible for adding the sulfation modification to the cell surface proteins and is found in the extracellular matrix. The process of glycosaminoglycan sulfation is a precise process and has been found to be associated for the activation of cytokines and growth factors and for inflammation site endothelium adhesion [72–75]. The biological modification of protein-bound carbohydrate sulfation is a process that has been associated with normal cellular processes, arthritis, cystic fibrosis, and pathogenic diseased states [75–79]. Sulfation is used for the removal of proteins and other biomolecules, such as metabolic end products, steroids, and neurotransmitters [80], from the extracellular matrix or body due to the increased solubility of the sulfated species and the reduced bioactivity. Specific glycoproteins that have been observed to undergo sulfation modification include the gonadotropin hormones follitropin, lutropin, and thyrotropin [81]; mucins [71]; and erythropoietin [82].

Previously, we observed that phosphorylation PTM of peptides predominantly takes place on the free hydroxyl moiety of the serine, threonine, and tyrosine amino acid residues by kinase enzymes. The amino acid residue phosphorylation mechanism is through the use of the activated donor molecule ATP (see Section 33.5.3 and Figure 33.24). The eukaryotic cell process of sulfation is through a class of enzymes known as sulfotransferase. The sulfotransferase class of enzymes comprises two general types that either sulfate small molecules such as steroids and metabolic products: the cytosolic sulfotransferase or the membrane-bound, Golgi-localized sulfotransferase that will sulfate glycoproteins or the tyrosine residue of proteins (to be covered in the next section). The transfer of the sulfonate (SO_3^-) group to the glycoprotein (or amino acid hydroxyl group) is through the activated donor phosphoadenosine phosphosulfate (PAPS) that is produced through the combination of SO_4^{2-} and ATP synthesized by the protein PAPS synthase [83]. An example of a sulfated glycoprotein is illustrated in Figure 33.34. The glycan has been O-linked to a general Ser/Thr amino acid residue within the protein's peptide chain. The mechanism for the sulfation of a saccharide is illustrated in Figure 33.35.

When glycated peptides are fragmented by tandem mass spectrometry, the product ions produced are classified according to a naming nomenclature that is similar to the peptide nomenclature presented in Section 33.3.2 (see Figure 33.6). The fragmentation nomenclature for carbohydrates is covered extensively in Section 32.1 [84]. However, a brief coverage is presented here to illustrate the products that are produced upon CID of sulfated and glycated mucins. Figure 33.36 illustrates the general types of fragmentation observed in tandem mass spectrometry of carbohydrates.

Glycated peptides have an enhanced response in the negative ion mode. It has been observed that different product ions are observed in the two different ion modes (positive and negative). In positive ion mode, product ions of types *B* and *Y* are observed, while in negative ion mode, *Y*-, *B*-, *C*-, and *Z*-type ions are often observed. The product ion spectra illustrated in Figure 33.37 are examples of sulfated mucin oligosaccharides that have been chemically released from the mucins and converted to alditols prior to HPLC separation and mass spectral analysis. The two product ion spectra were obtained from isomers of an *m/z* 975 oligosaccharide collected in negative ion mode as deprotonated species $[M-H]^-$. The spectra illustrate the production of *B*-, *Y*-, and *Z*-type ions from the fragmentation of the oligosaccharide. Also observed in the spectra is a peak at *m/z* 97 produced from the neutral loss of HSO_4^- indicative of sulfation.

Examples of fragmentation pathway mechanisms for the production of a select number of product ions illustrated in Figure 33.37 are presented in Figure 33.38.

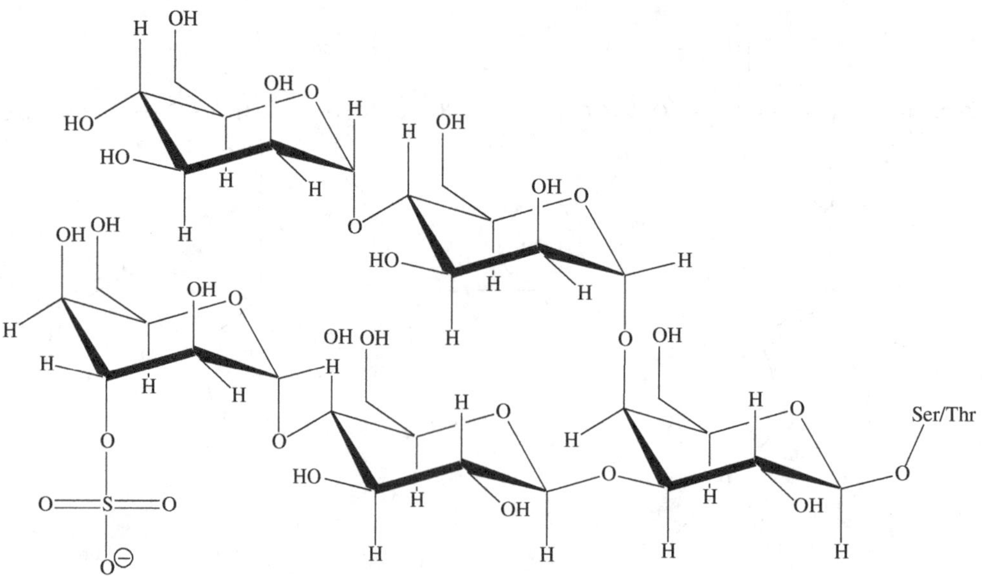

FIGURE 33.34 A sulfated glycoprotein. The glycan has been O-linked to a general Ser/Thr amino acid residue within the protein's peptide chain.

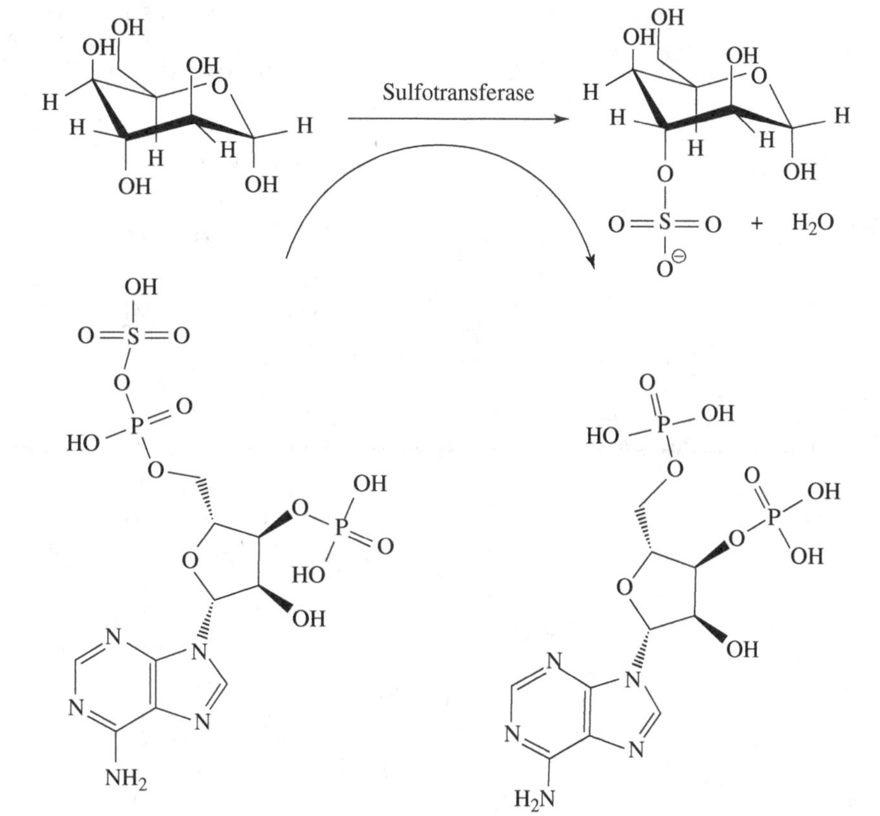

Sulfotransferase

3-phosphoadenosine-5-phosphosulfate (PAPS) 3,5-ADP

FIGURE 33.35 Mechanism for biosynthesis of sulfated saccharide.

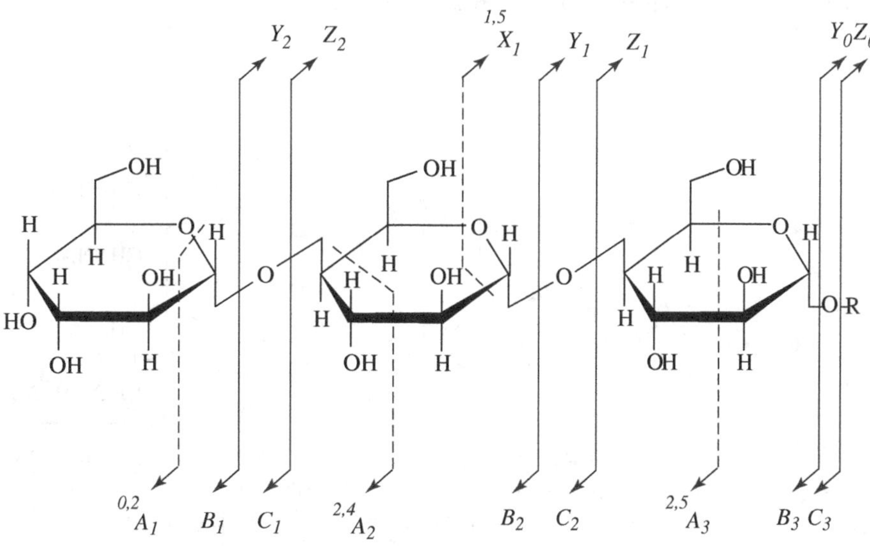

FIGURE 33.36 Carbohydrate fragmentation ion types and associated nomenclature [91].

33.5.4.2 Tyrosine Sulfation We saw previously that the tyrosine amino acid residue can be phosphorylated though it is the least modified as compared to serine and threonine. The PTM in the form of sulfation also takes place with tyrosine, and it is the most common form of PTM that involves tyrosine. Within the total protein content of an organism, it is estimated that up to 1% of all tyrosine residues may be sulfated [85]. The first reporting of the observance of the sulfate to tyrosine covalent

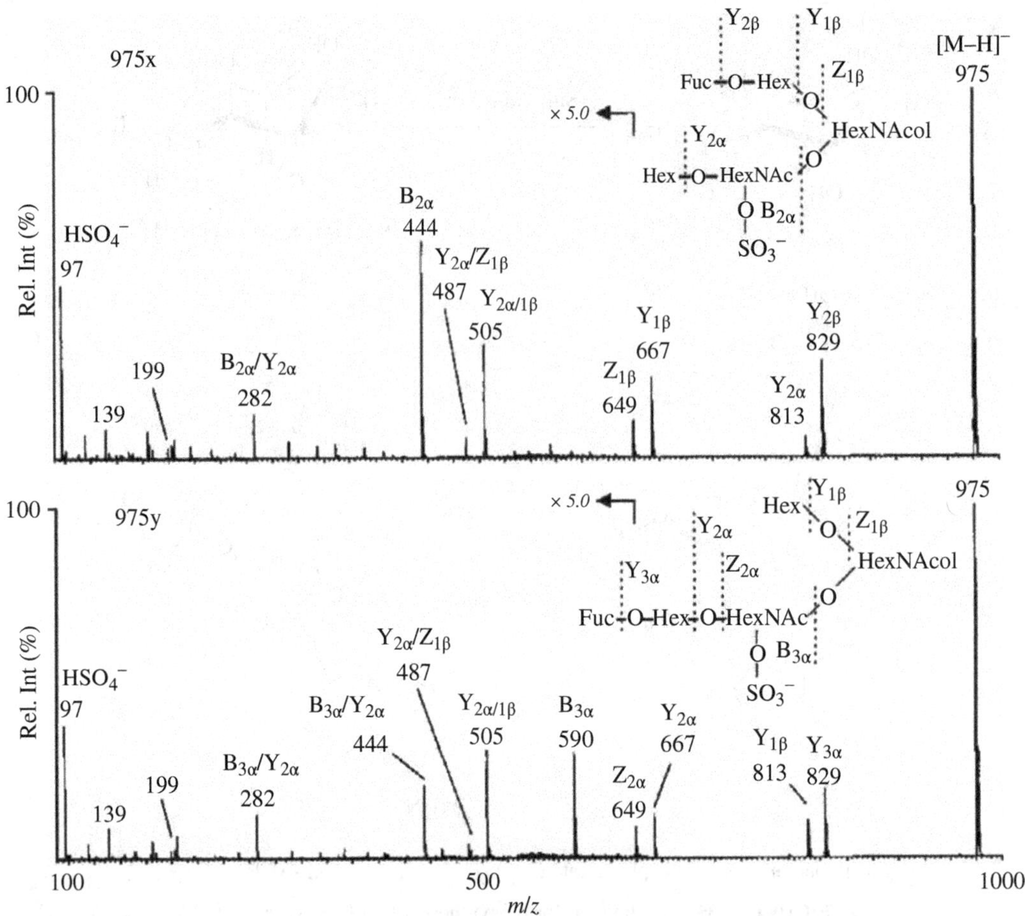

FIGURE 33.37 ESI tandem mass spectra of two isomers with [M–H]– *m/z* 975 that eluted at 46.5 (975x) and 49.5 min (975y) obtained from porcine stomach mucins. The fragmentation nomenclature is according to Figure 33.41. The 5-marked region has been magnified 5 times. (Reprinted with permission from Ref. [71]. Copyright 2000 American Chemical Society.)

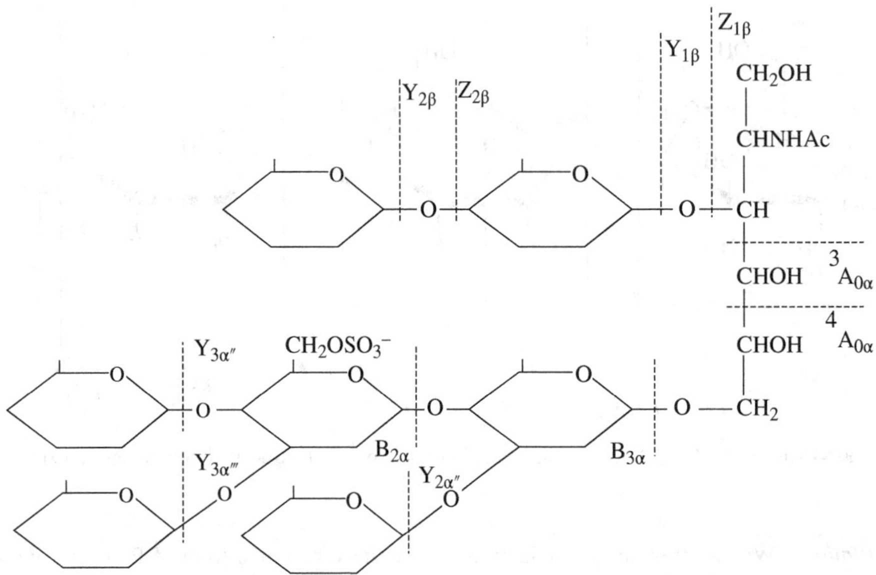

FIGURE 33.38 Fragment annotations applied in this study based upon the suggested nomenclature by Domon and Costello [85] and our previous report. (Reprinted with permission from Ref. [71]. Copyright 2000 American Chemical Society.)

modification was in 1954 by Bettelheim [86]. Sulfation takes place on transmembrane-spanning and secreted proteins. The sulfation mechanism within eukaryotic cells takes place in the trans-Golgi where the membrane-bound enzyme tyrosylprotein sulfotransferase catalyzes the modification (sulfation) of proteins synthesized in the rough endoplasmic reticulum (RER). The mechanism for sulfation of the tyrosine residue forming tyrosine *O*-sulfate esters (aryl sulfate) is the same as for glycosaminoglycan sulfation through the activated donor $3'$-phosphoadenosine-$5'$-phosphosulfate (PAPS). There are two different tyrosylprotein sulfotransferases that have been isolated and identified by researchers, TPST1 and TPST2 [87,88]. Others have observed that N-sulfation and S-sulfation can occur [89] and that on rare occasions serine and threonine can be sulfated [90]. The biosynthesis pathway for the sulfation of the tyrosine amino acid residue by tyrosylprotein sulfotransferases through the activated donor PAPS is illustrated in Figure 33.39.

Sulfation of the tyrosine amino acid residue is directly involved in the recognition processes of protein-to-protein interactions of membrane-bound and secreted proteins. A recent article from Woods et al. [91] at the National Institutes of Health (NIH) gives a listing of proteins that have been observed to be sulfated that includes plasma membrane proteins, adhesion proteins, immune components, secretory proteins, and coagulation factors, to name a few. A listing of O-sulfated human proteins currently included in the Swiss-Prot database is shown in Table 33.9. The UniProt database currently lists 275 proteins that are tyrosine sulfated (http://www.uniprot.org). There has not at

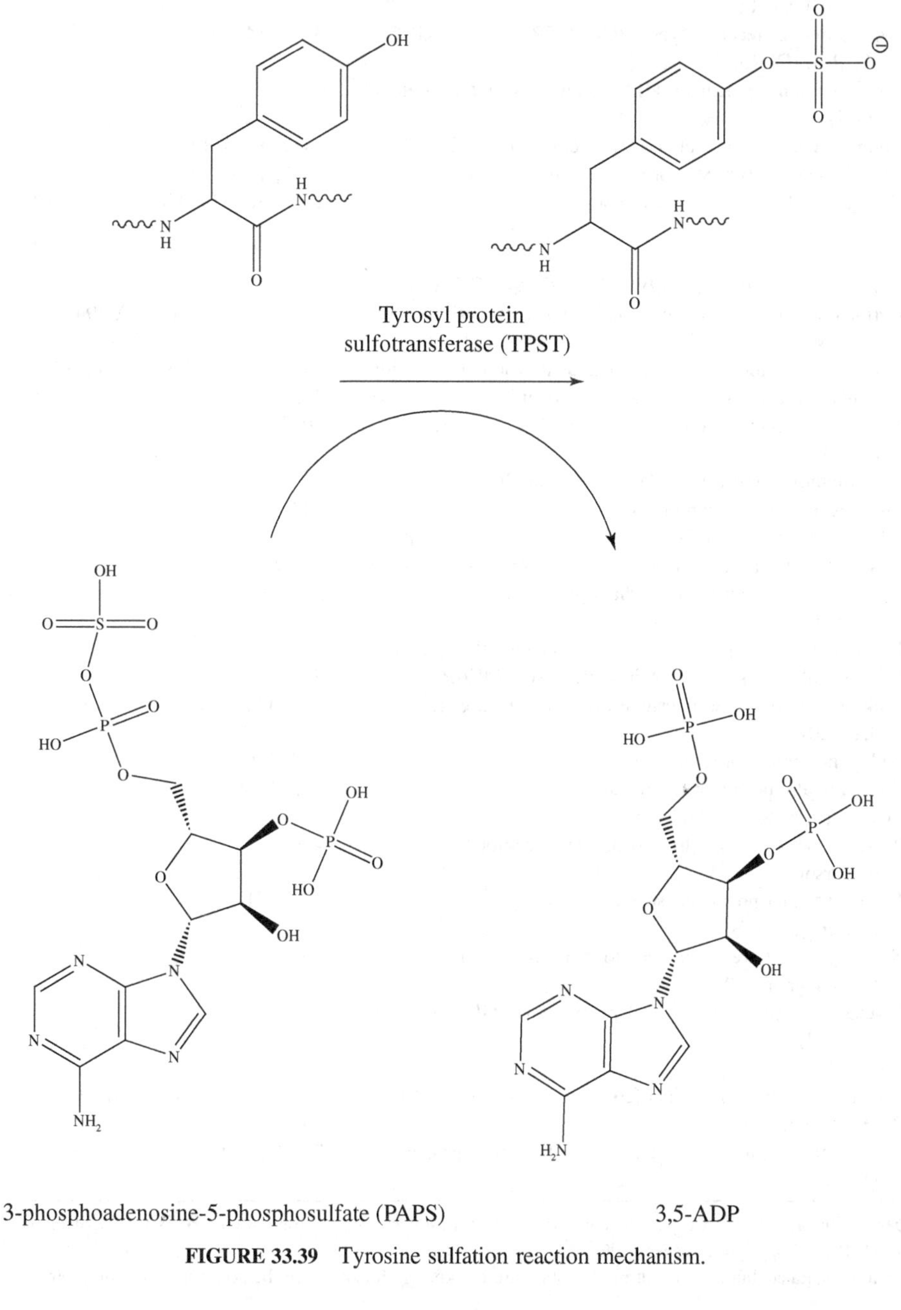

Tyrosyl protein sulfotransferase (TPST)

3-phosphoadenosine-5-phosphosulfate (PAPS)

3,5-ADP

FIGURE 33.39 Tyrosine sulfation reaction mechanism.

TABLE 33.9 Proteins Listed in Swiss-Prot That Are Annotated as Having at Least One *O*-Sulfated Amino Acid Residue.

Swiss-Prot Name	Description	Sulfation Site(s)	Sequence Length
1A01_HUMAN	HLA class I histocompatibility antigen, A-1 alpha chain precursor		
(MHC class I antigen A∗1)	83	365	
7B2_HUMAN	Neuroendocrine protein 7B2 precursor	–	212
A2AP_HUMAN	Alpha-2-antiplasmin precursor (alpha-2-plasmin inhibitor) (alpha-2-PI)	484	491
AMD_HUMAN	Peptidyl-glycine alpha-amidating monooxygenase precursor	961	973
AMPN_HUMAN	Aminopeptidase N (EC 3.4.11.2) (hAPN) (alanyl aminopeptidase)	175, 418, 423, 912	966
C3AR_HUMAN	C3a anaphylatoxin chemotactic receptor (C3a-R) (C3AR)	174, 184, 318	482
C5AR_HUMAN	C5a anaphylatoxin chemotactic receptor (C5a-R) (C5aR) (CD88 antigen)	11, 14	350
CCKN_HUMAN	Cholecystokinin precursor (CCK)	111, 113	115
CCR2_HUMAN	C\C chemokine receptor type 2 (C\C CKR-2) (CC-CKR-2) (CCR-2) (CCR2)	26	374
CCR5_HUMAN	C\C chemokine receptor type 5 (C\C CKR-5) (CC-CKR-5) (CCR-5) (CCR5)	3, 10, 14, 15	352
CMGA_HUMAN	Chromogranin A precursor (CgA) (pituitary secretory protein I) (SP-I)	–	457
CO4A_HUMAN	Complement C4-A precursor (acidic complement C4)	1417, 1420, 1422	1744
CO4B_HUMAN	Complement C4-B precursor (basic complement C4)	1417, 1420, 1422	1744
CO5A1_HUMAN	Collagen alpha-1(V) chain precursor	234, 236, 338, 340, 346, 347, 416, 417,	
420, 421, 1601,1604	1838		
CXCR4_HUMAN	C–X–C chemokine receptor type 4 (CXC-R4) (CXCR-4)	21	352
DERM_HUMAN	Dermatopontin precursor (tyrosine-rich acidic matrix protein) (TRAMP)	23, 162, 164, 166, 167, 194	201
FA5_HUMAN	Coagulation factor V precursor (activated protein C cofactor)	693, 724, 726, 1522, 1538, 1543, 1593	2224
FA8_HUMAN	Coagulation factor VIII precursor (procoagulant component)	365, 414, 426, 737, 738, 742, 1683, 1699	2351
FA9_HUMAN	Coagulation factor IX precursor (EC 3.4.21.22) (Christmas factor)	201	461
FETA_HUMAN	Alpha-fetoprotein precursor (alpha-fetoglobulin)	–	609
FIBG_HUMAN	Fibrinogen gamma chain precursor	444, 448	453
FINC_HUMAN	Fibronectin precursor (FN) (cold-insoluble globulin) (CIG)	876, 881	2386
GAST_HUMAN	Gastrin precursor [contains gastrin 71 (Component I); gastrin	52 87	101
GP1BA_HUMAN	Platelet glycoprotein Ib alpha chain precursor (glycoprotein Ibalpha)	292, 294, 295	626
HEP2_HUMAN	Heparin cofactor 2 precursor (heparin cofactor II) (HC-II)	79, 92	499
MFAP2_HUMAN	Microfibrillar-associated protein 2 precursor (MFAP-2)	47, 48, 50	183
MGA_HUMAN	Maltase-glucoamylase, intestinal [includes maltase (EC 3.2.1.20)	415, 424, 1281	1856
NID1_HUMAN	Nidogen-1 precursor (entactin)	289, 296	1247
OMD_HUMAN	Osteomodulin precursor (osteoadherin)	25, 31, 39	421
OPT_HUMAN	Opticin precursor (oculoglycan)	71	332
ROR2_HUMAN	Tyrosine-protein kinase transmembrane receptor ROR2 precursor	469, 471	943
SCG1_HUMAN	Secretogranin-1 precursor (secretogranin I) (SgI) (chromogranin B)	173, 341	677
SCG2_HUMAN	Secretogranin-2 precursor (secretogranin II) (SgII) (chromogranin C)	151	617
SELPL_HUMAN	P-selectin glycoprotein ligand 1 precursor (PSGL-1) (selectin P ligand)		
(CD162 antigen).	46, 48, 51	412	
SUIS_HUMAN	Sucrase-isomaltase, intestinal [contains sucrase (EC 3.2.1.48)	236, 238, 390, 399, 666, 762, 764	1826
THYG_HUMAN	Thyroglobulin precursor	24	2768
VTNC_HUMAN	Vitronectin precursor (serum spreading factor) (S-protein) (V75)	75, 78, 282	478

Source: Reprinted with permission. This article was published in *Biochimica et Biophysica Acta,* Monigatti, F., Hekking, B., Steen, H. Protein sulfation analysis – A primer. **2006**, *1764*, 1904–1913. Copyright Elsevier 2006.

Some proteins are known to be sulfated, but the sites of modification are not known. It can clearly be seen that sulfation preferentially occurs in clusters.

this point been observed any type of enzymatic mechanism that would desulfate the modified proteins, and because sulfation is pH stable, proteins containing a sulfation modification can be observed in urine excretion.

Product ion mass spectra of sulfated peptides by CID are similar to that observed for phosphorylated peptides in that the major product ion observed is for the neutral loss of sulfur trioxide SO_3 (−80 Da) from the precursor ion, often with little other information in the way of product ions. This is illustrated in the product ion mass spectrum of Figure 33.40. In the figure, there are two predominant peaks at *m/z* 647 for the doubly protonated, sulfated precursor ion $[M+2H]^{2+}$ and at *m/z* 607 for the neutral loss of the sulfate modification $[M+2H-SO_3]^{2+}$. Loss of 80 Da from the precursor ion is also observed with the neutral loss of HPO_3 from the phosphorylated tyrosine and histidine amino acid residues. The exact mass of HPO_3 is 79.9663 Da, while the exact mass of SO_3 is 79.9568 Da, a difference of 119 parts per million (ppm). The fragmentation pathway mechanisms for the loss of the phosphate or sulfate modifications are the same, resulting with the original, unmodified tyrosine residue, as contrasted with the production of dehydroalanine produced by loss of H_3PO_4 (98 Da). The GP fragmentation pathway mechanism for the neutral loss of the sulfate modification of tyrosine during CID product ion production is illustrated in Figure 33.41.

Increasing the collision energy will generate further product ions of the *m/z* 647 sulfated species that is illustrated in Figure 33.40, giving sequence information of the peptide in the form of *b*- and *y*-type ions. This can allow the sequencing of the peptide and subsequent identification of the protein where it came from. The higher-energy collision-induced product ions though are generated from the precursor after the loss of the labile sulfo moeity; thus, the ability to identify the modification site is often not possible. The loss of the labile sulfo group is proton or charge driven. Conditions that allow or favor charge remote fragmentation can induce the production of product ions that still retain the sulfo moeity. This condition will give the opportunity to determine the site of the sulfo modification of the peptide. One approach to product ion spectral generation that has demonstrated this is the use of ECD in FTICR mass spectrometers. As we saw previously, the ECD fragmentation mechanism is not based on the activation of the precursor through repeated collisions but through the capture of thermal electrons that tend to promote backbone dissociations while still retaining the labile modifications (sulfation, phosphorylation, etc.). Figure 33.42 illustrates

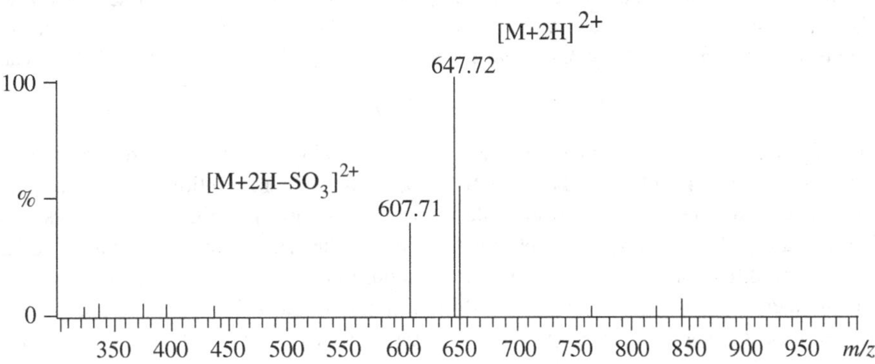

FIGURE 33.40 Product ion mass spectrum of a sulfated peptide at *m/z* 647 for $[M+2H]^{2+}$. The mass spectrum is dominated by the precursor ion and the associated SO_3 neutral loss product ion at *m/z* 607.

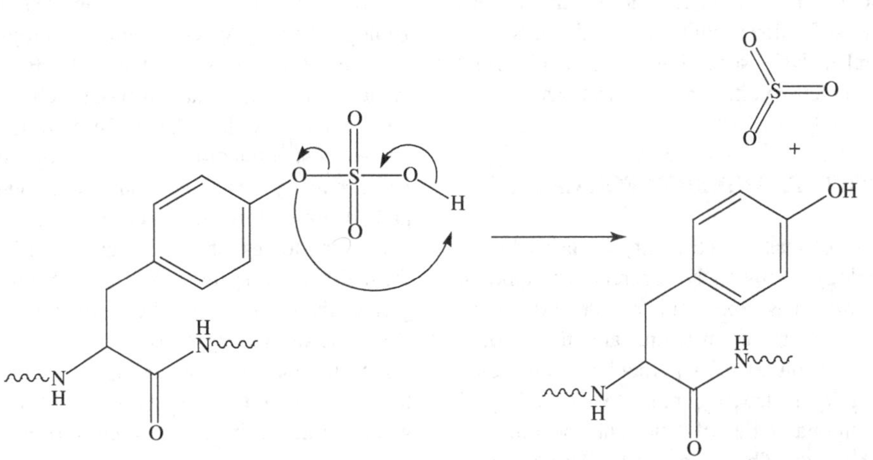

FIGURE 33.41 Gas-phase fragmentation pathway mechanism for the neutral loss of the sulfate modification of tyrosine during CID product ion production.

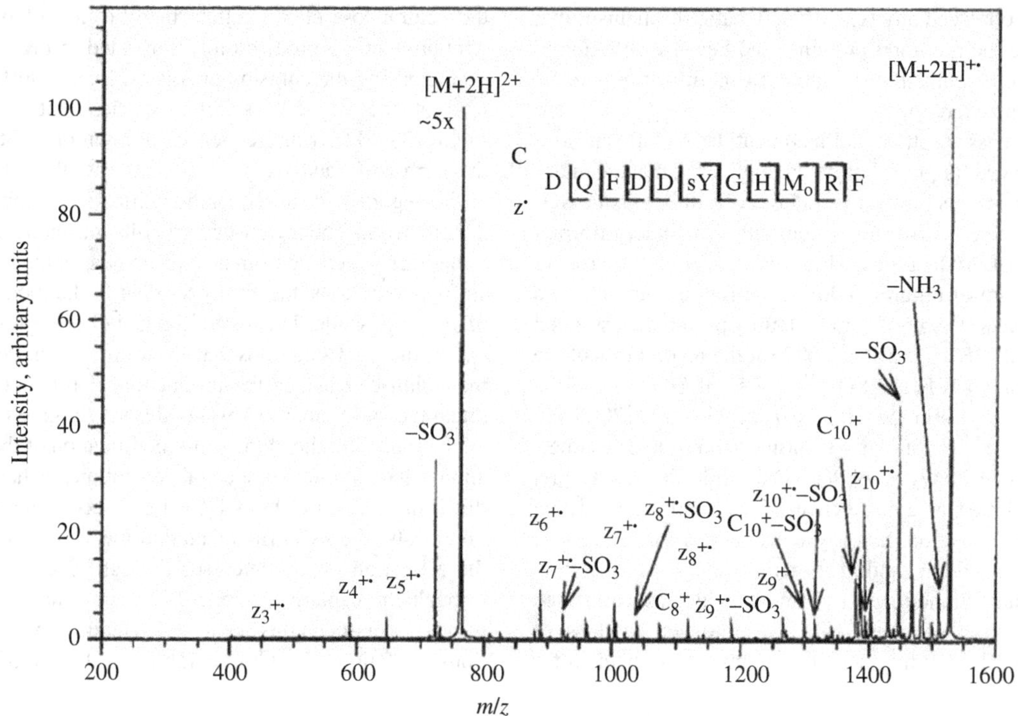

FIGURE 33.42 ECD product ion mass spectrum of drosulfakinin. (Reprinted with permission from Haselmann, K.F., Budnik, B.A., Olsen, J.V., Nielsen, M.L., Reis, C.A., Clausen, H., Johnsen, A.H., Zubarev, R.A. Advantages of External Accumulation for Electron Capture Dissociation in Fourier Transform Mass Spectrometry *Anal. Chem.* **2001**, *73*, 2998–3005. Copyright 2001 American Chemical Society.)

a product ion spectrum of drosulfakinin, a polypeptide that is known to have a sulfated tyrosine residue, in the +2 CS $[M+2H]^{2+}$. The product ion spectrum contains a considerable amount of backbone fragmentation allowing sequence determination in conjunction with modification site determination. Again, one drawback to the technique of ECD is the required use of an FTICR mass spectrometer. As was presented previously, the technique of ETD used in ion traps will most likely contribute to studies of sulfation PTM of peptides. A comparison of the biology, biochemistry, and mass spectrometry of sulfation versus phosphorylation is given in Table 33.10. The table gives a brief description of the subcellular protein locations associated with PTM, modification stabilities, and behaviors associated with mass spectral analysis such as ionization stabilities and masses corresponding to the two different modifications.

33.6 SYSTEMS BIOLOGY AND BIOINFORMATICS

The application and use of mass spectrometry as an analytical tool to the field of biology is obviously apparent. The amount of information given from mass spectrometric methodologies, that is, the molecular weight, the structure, and the amount (can be relative and/or absolute) of a particular biomolecule extracted from a biological matrix, is similar to the biological analytical approach traditionally used of "one gene" or "one protein" at a time study. The study of biological systems is now moving toward more encompassing analysis such as the sequencing of an organism's genome or proteome. The trend now is to study both the single components of a system in conjunction with the particular system's entire complement of components. Of special interest is how the various components of the system interact with one another under normal conditions and some type of perturbed condition such as a diseased state or a change in the systems environment (e.g., lack of oxygen, food, water, etc.). Thus, systems biology is the study of the processes and complex biological organizational behavior using information from its molecular constituents [92]. This is quite broad in the sense that the biological organization may go to the level of tissue up to a population or even an ecosystem. The hierarchical levels of biological information were summed up by L. Hood at the Institute for Systems Biology (Seattle, WA) showing the progression from DNA to a complex organization as illustrated in Figure 33.43. The idea is to gather as much information about each component of a system to more accurately describe the system as a whole. This encompasses most of the disciplines of biology and incorporates analytical chemistry (mass spectrometry) through the need to measure and identify individual species on the molecular level. When attempting to describe the behavior of a biological system, often, the reality of the system is that the behavior of the whole system is greater than what would be predicted from the sum of its parts [93]. The study of systems biology incorporates multiple analytical techniques and methodologies, which includes mass spectrometry, to study the components of a biological system (e.g., genes, proteins, metabolites, etc.) to better model their interactions [94].

The experimental workflow involved in systems biology that centers around mass spectrometric analysis is illustrated in

TABLE 33.10 Summary of the Relevant Characteristics of Phosphorylation and Sulfation as Protein Posttranslational Modifications.

		Phosphorylation	Sulfation
Biology	Location	Intracellular	Extracellular (membrane bound)
	Reversibility	Reversible	Irreversible
	Location of modifying enzyme	Cytosolic and nuclear	Membrane bound in trans-Golgi network
Function	Activation, inactivation, modulation of protein interaction	Modulation of protein interaction	
Biochemistry	Radioactive isotopes	^{32}P and ^{33}P	^{35}S
	Chemical stability	pY is stable pS/pT are alkaline labile pE/pD/pH are acid labile	sY is acid labile
	Edman compatibility	pY compatibility limited solubility	sY is hydrolyzed during TFA cleavage step
		pS undergoes β-elimination to dehydroalanine pT undergoes β-elimination, giving rise to many different side products	
	Removal	Phosphatases pS/pT: harsh alkaline treatment, many side products	Arylsulfatases of limited use Quantitative hydrolysis: 1 M HCl, 95 °C, 5 min
	Enrichment	α-pY and anti-phospho domain antibodies available	No antibody
		IMAC, TiO2	No affinity reagents
Mass spectrometry	Property	Acidic (2−)	Acidic (1−)
	Mass difference	+80 Da (79.9663 Da)	+80 Da (79.9568 Da)
	Stability during Ionization	ESI	ESI
		pS/pT: good pY: stable MALDI pS/pT partial loss possible pY stable	sY easily lost under standard conditions MALDI +ve: complete loss −ve: partial loss
	Stability during CID	pY: stable pS/pT: sequence dependent loss of H3PO4	Complete loss Multiple sY's are more stable
	Signature after CID	pS: dehydroalanine (69 Da) pT: dehydroaminobutyric acid (83 Da)	None
	Characteristic neutral loss	−H3PO4 (−98 Da) −(H3PO4+H2O) (−116 Da)	−SO3 (− 80 Da)
	Characteristic fragment ion	PO3 − (−79 Da)	SO3 − (−80 Da)

Source: Reprinted with permission. This article was published in *Biochimica et Biophysica Acta,* Monigatti, F., Hekking, B., Steen, H. Protein sulfation analysis – A primer. **2006**, *1764*, 1904–1913. Copyright Elsevier 2006.

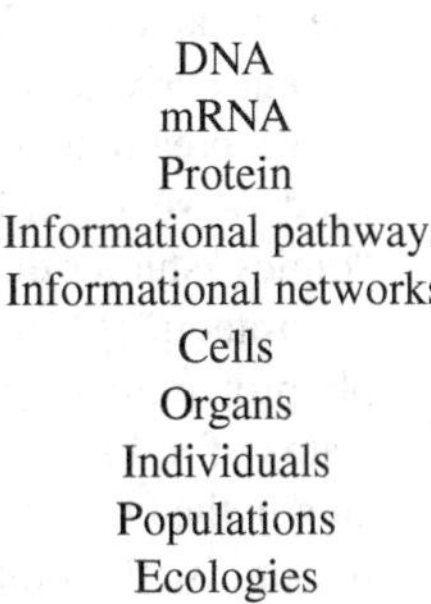

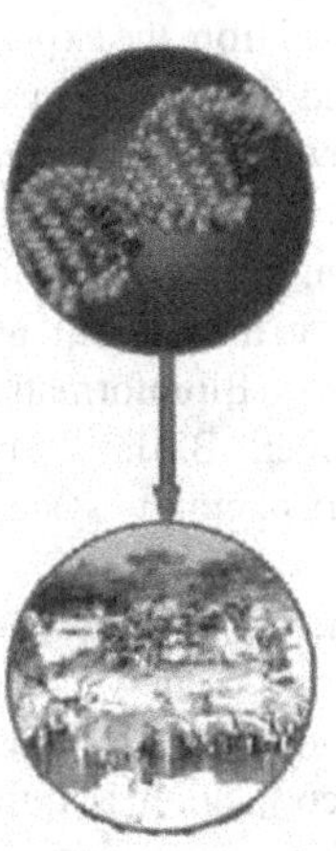

FIGURE 33.43 Hierarchical levels of biological information. (Reprinted with permission from Hood, L. A Personal View of Molecular Technology and How It Has Changed Biology *J. Proteome Res.* **2002**, *1*, 399–409. Copyright 2002 American Chemical Society.)

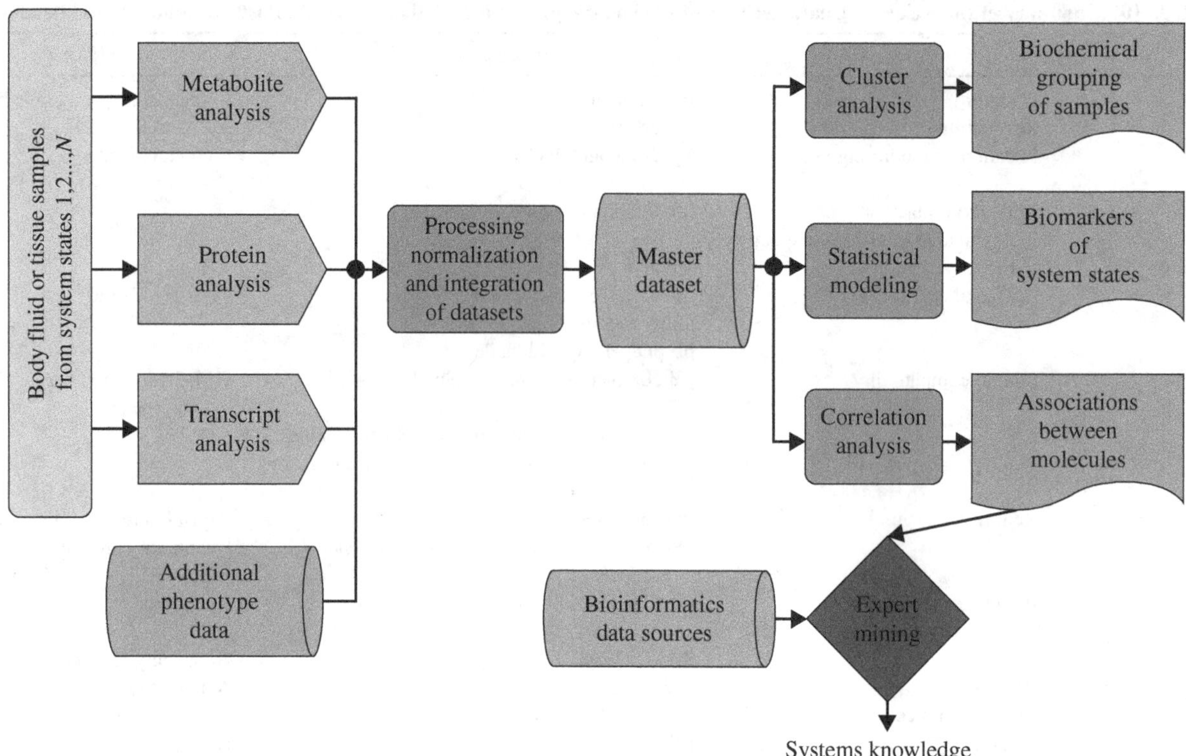

FIGURE 33.44 Systems biology workflow. Data are produced by different platforms (transcriptomics, proteomics, and metabolomics), followed by integration into a master data set. Different biostatistical strategies are pursued: clustering, modeling, and correlation analysis. Integration with extensive bioinformatics tools and expert biological knowledge is key to the creation of meaningful knowledge. (Reprinted with permission from van der Greef, J., Martin, S., Juhasz, P., Adourian, A., Plasterer, T., Verheij, E.R., McBurney, R.N. The Art and Practice of Systems Biology in Medicine: Mapping Patterns of Relationships *J. Proteome Res.* **2007**, *6*, 1540–1559. Copyright 2007 American Chemical Society.)

Figure 33.44. Specifically, the metabolite analysis and the protein analysis, illustrated in the second step of the flow, are increasingly performed using highly efficient separation methodology such as nano-HPLC (1D and 2D) coupled with high-resolution mass spectrometry such as FTICR-MS and LTQ-Orbitrap-MS. Much development has been done, and is still ongoing, in the processing and data extraction of the information obtained from high-peak capacity nano-HPLC-ESI mass spectrometry. A tremendous amount of information is obtained from the experiments that need to be processed, validated, and statistically evaluated. Bioinformatics for proteomics has developed with great speed and complexity with many open-source software available with statistical methods and filtering algorithms for proteomic data validation. Some examples at the present are Bioinformatics.org; SourceForge.net; Open Bioinformatics Foundation that features toolkits such as BioPerl, BioJava, and BioPython; and BioLinux, an optimized Linux operating system for bioinformaticians.

Finally, the visualization of the data obtained from mass spectrometric analysis has also been developed significantly in the past few years. The ability to take different complements of analyses and integrate them with the goal of correlation is quite daunting when hundreds to thousands of biomolecules are involved. Programs like Cytoscape allow the visualization of molecular interaction networks. Figure 33.45 shows a correlation network between data obtained from proteomics, metabolomics, and transcriptomics from the liver tissue and plasma. The correlation network found biomarker analytes in the plasma that may be useful in monitoring processes occurring in the organ.

33.6.1 Biomarkers in Cancer

Mass spectrometry is having a substantial impact on systems biology-type studies such as those of biomarker discovery in cancer diagnostics. Figure 33.46 illustrates a scheme for the search for biomarkers in patient-derived samples through mass spectral analyses. From the patient, the figure lists nine types of patient-derived samples that include blood, urine, sputum, saliva, breath, tear fluid, nipple aspirate fluid, and cerebrospinal fluid. Also described in the figure are the various types of "omics" studies done using mass spectrometric techniques such as proteomics, metabonomics, peptidomics, glycomics, phosphoproteomics, and lipidomics. Indicative of the progress or status of a disease a biomarker is a biologically derived molecule in the body that is measured along with many other species present by the omics methods. Bioinformatics analysis is performed on the mass spectral data to determine the presence of potential biomarkers through expression studies and response differentials. Once identified, the biomarkers can be used for clinical diagnostics such as early detection before the onset of a serious disease like cancer. This allows medical intervention that may have a substantial influence upon the success of an

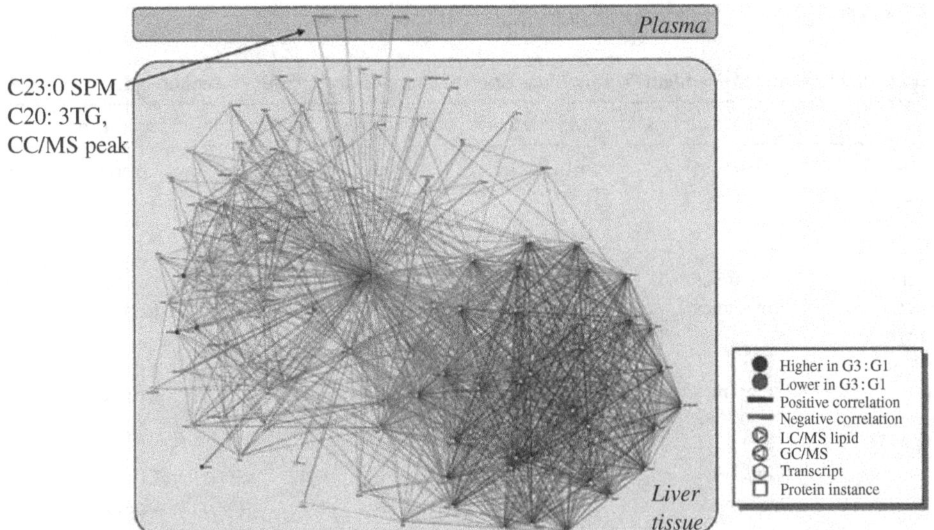

FIGURE 33.45 Correlation network of analytes across blood plasma (top of figure) and liver tissue (bottom of figure). Analytes include proteins, endogenous metabolites, and gene transcripts. Not only is structure evident among analytes profiled from liver tissue, but there are also a number of correlations to analytes profiled in plasma in this case. Such analytes can serve as useful circulating biomarkers for the tissue-based biochemical processes occurring in the organ. (Reprinted with permission from van der Greef, J., Martin, S., Juhasz, P., Adourian, A., Plasterer, T., Verheij, E.R., McBurney, R.N. The Art and Practice of Systems Biology in Medicine: Mapping Patterns of Relationships *J. Proteome Res.* **2007**, *6*, 1540–1559. Copyright 2007 American Chemical Society.)

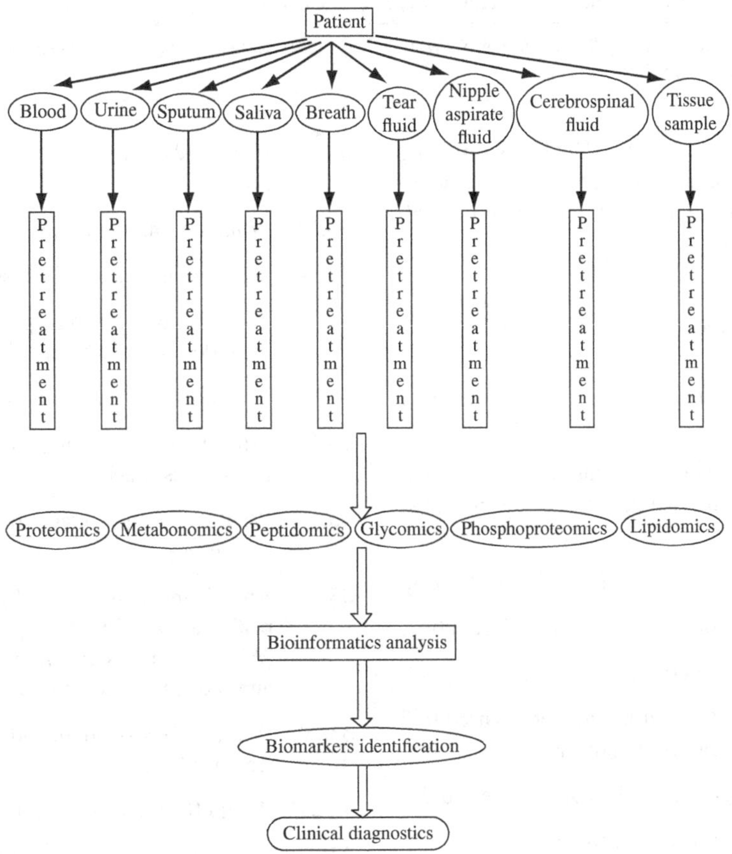

FIGURE 33.46 Scheme for mass spectrometry-based "omics" technologies in cancer diagnostics. Proteomics is the large-scale identification and functional characterization of all expressed proteins in a given cell or tissue, including all protein isoforms and modifications. Metabonomics is the quantitative measurement of metabolic responses of multicellular systems to pathophysiological stimuli or genetic modification. Peptidomics is the simultaneous visualization and identification of the whole peptidome of a cell or tissue, that is, all expressed peptides with their posttranslational modifications. Glycomics is the identification and study of all the glycan molecules produced by an organism, encompassing all glycoconjugates (glycolipids, glycoproteins, lipopolysaccharides, peptidoglycans, proteoglycans). Phosphoproteomics is the characterization of phosphorylation of proteins. Lipidomics is systems-level analysis and characterization of lipids and their interacting partners. (Reprinted with permission from John Wiley & Sons, Inc. Mass spectrometry based "omics" technologies in cancer diagnostics, Zhang, X., Wei, D., Yap, Y., Li, L., Guo, S., Chen, F. *Mass Spectrometry Reviews* **2007**, *26*, 403–431.)

TABLE 33.11 Potential Cancer Biomarkers Identified by Mass Spectrometry-Based "Omics" Technologies.

Biomarkers	"omics" Platforms	MS Methods	Sample Source	Cancer Type	References
Apolipoprotein A1, inter-α-trypsin inhibitor Haptoglobin-a-subunit Transthyretin	Proteomics	SELDI-TOF	Serum	Ovarian	Ye et al., 2003; Zhang et al., 2004
Vitamin D-binding protein	Proteomics	SELDI-TOF	Serum	Prostate	Hlavaty et al., 2003
Stathmin (Op18), GRP 78 14-3-3 isoforms. Transthyretin Protein disulfide Isomerase	Proteomics	ESI-MS	Tissue	Lung	Chen et al., 2003
Peroxiredoxin, Enolase Protein disulfide Isomerase HSP 70, α-1-antitrypsin	Proteomics	MALDI-TOF LC-MS	Tissue	Breast	Somiari et al., 2003
HSP 27	Proteomics	MALDI-TOF	Serum	Liver	Feng et al., 2005
Annexin 1, Cofilin, GST Superoxide dismutase Peroxiredoxin, Enolase Protein disulfide Isomerase	Proteomics	MALDI-TOF, ESI-MS, Q-TOF	Tissue	Colon	Seike et al., 2003; Stierum et al., 2003
Neutrophil peptides 1-3	Proteomics	SELDI-TOF	Nipple aspirate fluid	Breast	Li et al., 2005b
PCa-24	Proteomics	MALDI-TOF	Tissue	Prostate	Zheng et al., 2003
Alkanes, Benzenes	Metabonomics	GC-MS	Breath	Lung	Phillips et al., 1999
Decanes, Heptanes	Metabonomics	GC-MS	Breath	Breast	Phillips et al., 2003
Hexanal, Heptanal	Metabonomics	LC-MS	Serum	Lung	Deng et al., 2004
Pseu, m1A, m1l	Metabonomics	HPLC, LC-MS	Urine	Liver	Yang et al., 2005b

Source: Reprinted with permission from John Wiley & Sons, Zhang, X., Wei, D., Yap, Y., Li, L., Guo, S., Chen, F. Mass spectrometry based "omics" technologies in cancer diagnostics, *Mass Spectr. Rev.* **2007**, *26*, 403–431.

early treatment and subsequent cure. Table 33.11 lists a number of identified potential cancer biomarkers that have been discovered through mass spectrometric analyses, primarily through proteomics and metabonomics.

33.7 CHAPTER PROBLEMS

33.1 What is the difference between top-down and bottom-up proteomics?

33.2 What are the three forms of posttranslational modifications (PTM) of proteins that are studied using mass spectrometry?

33.3 List as many descriptive features of proteins as you can.

33.4 List the four structures of proteins and their description.

33.5 What is peptide mass fingerprinting?

33.6 In Table 33.1, what are these different proteins used for? What makes one different from another?

33.7 What is de novo sequencing, and what is it used for?

33.8 What does bioinformatics attempt to do?

33.9 Describe the nomenclature of Roepstorff for fragmented peptides according to charge retainment.

33.10 What is the mobile proton model, and what discrepancy does it address?

33.11 What other types of losses can be observed besides *b*- and *y*-type ions?

33.12 In high-energy collisions, how are d, v, and w ions produced?

33.13 What are immonium ions, and what is their usefulness?

33.14 What are isomers and isobars?

33.15 What can be used to help reduce the problem of incomplete de novo sequencing?

33.16 How does ETD differ from CID?

33.17 What is the vital component of top-down proteomics, and how is it achieved?

33.18 What are two advantages of top-down proteomics versus bottom-up?

33.19 An ESI mass spectrum of a protein has an envelope of peaks at *m/z* 2250.7, *m/z* 2344.5, *m/z* 2446.4, and *m/z* 2557.5. What is the molecular weight of the protein and the associated charge states of the peaks?

33.20 What is glycosylation, and what is it used for in biological systems?

33.21 What effect does glycosylation have on erythrocytes?

33.22 What three aspects of glycosylated proteins can be obtained using mass spectrometry?

33.23 If two series in a single-stage mass spectrum consisted of an increase of 145.6 Da for one series and 97.1 Da for the second, what could be ascertained from this?

33.24 What amino acid residues in eukaryotic cells can be phosphorylated?

33.25 Are the three amino acid residues phosphorylated to an even extent? If not, what are the relative percentages?

33.26 Explain why the stoichiometry of the phosphorylated proteome is so small (≤ 1%).

33.27 What are known to interfere with IMAC columns, and what are some ways to remove them?

33.28 What are the neutral loss masses of the phosphate group at charge states +1, +2, and +3?

33.29 Neutral loss of the phosphate group from serine produces what form of the amino acid residue? What about from phosphorylated threonine?

33.30 Describe the neutral losses associated with phosphorylated tyrosine.

33.31 What sets apart ECD and ETD from CID? Is this an advantage?

33.32 Briefly describe the difference between eukaryotic and prokaryotic phosphorylation.

33.33 If up to 6% of phosphorylation in eukaryotes and prokaryotes is with the histidine residue, why is it not typically observed?

33.34 What are some stabilities of the other phosphorylated amino acid residues?

33.35 What is the structure of HPO_3?

33.36 What are the two types of protein sulfation that occur?

33.37 Are the neutral loss masses of a phosphate modification similar to that of a sulfate modification?

33.38 Give a brief definition of systems biology.

33.39 List some of the patient samples that are used in cancer biomarker studies.

REFERENCES

1. Morrison, R.T.; Boyd, R.N. *Organic Chemistry,* Fifth Edition; Allyn and Bacon: Massachusetts, 1987; p. 1345.
2. Schulz, G.E.; Schirmer, R.H., In *Principles of Protein Structure, Springer Advanced Texts in Chemistry;* Cantor, C.R., Ed.; Springer-Verlag: New York, 1990.
3. Franks, F. *Biophys. Chem.* 2002, **96**, 117–127.
4. Edman, P. *Acta Chem. Scand.* 1950, **4**, 283–293.
5. Shimonishi, Y.; Hong, Y.M.; Kitagishi, T.; Matsuo, H.; Katakuse, I. *Eur. J. Biochem.* 1980, **112**, 251–264.
6. Morris, H.R.; Panico, M.; Barber, M.; Bordoli, R.S.; Sedgwick, R.D.; Tyler, A. *Biochem. Biophys. Res. Commun.* 1981, **101**, 623–631.
7. Mann, M.; Hojrup, P.; Roepstorff, P. *Biol. Mass Spectrom.* 1993, **22***(6)*, 338–345.
8. Pappin, D.D.J.; Hojrup, P.; Bleasby, A.J. *Curr. Biol.* 1993, **3**, 327–332.
9. Henzel, W.J.; Billeci, T.M.; Stults, J.T.; Wong, S.C.; Grimley, C.; Watanabe, C. *Proc. Natl. Acad. Sci. U. S. A.* 1993, **90***(11)*, 5011–5015.
10. Kenrick, K.G.; Margolis, J. *Anal. Biochem.* 1970, **33**, 204–207.
11. Gras, R.; Muller, M. *Curr. Opin. Mol. Ther.* 2001, **3***(6)*, 526–532.
12. Wolters, D.A.; Washburn, M.P.; Yates, J.R. *Anal. Chem.* 2001, **73**, 5683–5690.
13. Eng, J.K.; McCormack, A.L.; Yates, J. III *J. Am. Soc. Mass Spectrom.* 1994, **5**, 976–989.
14. Mann, M.; Wilm, M. *Anal. Chem.* 1994, **66**, 4390–4399.
15. Roepstorff P.; Fohlman, J. *Biomed. Mass Spectrom.* 1984, **11***(11)*, 601.
16. Yalcin, T.; Csizmadia, I.G.; Peterson, M.R.; Harrison, A.G. *J. Am. Soc. Mass Spectrom.* 1996, **7**, 233–242.
17. Dongre, A.R.; Jones, J.L.; Somogyi, A.; Wysocki, V.H. *J. Am. Chem. Soc.* 1996, **118**, 8365–8374.
18. McCormack, A.L.; Somogyi, A.; Dongre, A.R.; Wysocki, V.H. *Anal. Chem.* 1993, **65**, 2859–2872.
19. Somogyri, A.; Wysocki, V.H.; Mayer, I. *J. Am. Soc. Mass Spectrom.* 1994, **5**, 704–717.
20. Zubarev, R.A.; Kelleher, N.L.; McLafferty, F.W. *J. Am. Chem. Soc.* 1998, **120**, 3265–3266.
21. Loo, J.A.; Edmonds, C.G.; Smith, R.D. *Science* 1990, **248**, 201–204.
22. Feng, R.; Konishi, Y. *Anal. Chem.* 1993, **65**, 645–649.
23. Loo, J.A., et al. *Proc. Natl. Acad. Sci. U. S. A.* 1992, **89**, 286–289.
24. Senko, M.W.; Beu, S.C.; McLafferty, F. W. *Anal. Chem.* 1994, **66**, 415–417.
25. Schnier, D.S.; Gross, E.R.; Williams, E.R. *J. Am. Chem. Soc.* 1995, **117**, 6747–6757.
26. Patrie, S.M.; Charlebois, J.P.; Whipple, D.; Kelleher, N.L.; Hendrickson, C.L.; Quinn, J.P.; Marshall, A.G.; Mukhopadhyay, B. *J. Am. Soc. Mass Spectrom.* 2004, **15**, 1099–1108.
27. Wold, F. *Ann. Rev. Biochem.* 1981, **50**, 783–814.
28. Wilkins, M.R.; Gasteiger, E.; Gooley, A.A.; Herbert, B.R.; Molloy, M. P.; Binz, P.A.; Ou, K.; Sanchez, J.C.; Bairoch, A.; Williams, K.L.; Hochstrasser, D.F. *J. Mol. Biol.* 1999, **280**, 645–657.
29. Manning, G.; Whyte, D.B.; Martinez, R.; Hunter, T.; Sudarsanam, S. *Science* 2002, **298**, 1912–1934.
30. Chalmers, M.J.; Kolch, W.; Emmett, M.R.; Marshall, A.G.; Mischak, H. *J. Chromatogr. B* 2004, **803**, 111–120.
31. Olsen, J.V.; Blagoev, B.; Gnad, F.; Macek, B.; Kumar, C.; Mortensen, P.; Mann, M. *Cell* 2006, **127**, 635–648.
32. Zhou, H.; Watts, J.D.; Aebersold, R. *Nat. Biotechnol.* 2001, **19**, 375–378.
33. Beausoleil, S.A.; Jedrychowski, M.; Schwartz, D.; Elias, J.E.; Villen, J.; Li, J.; Cohn, M.A.; Cantley, L.C.; Gygi, S.P. *Proc. Natl. Acad. Sci. U. S. A.* 2004, **101**, 12130–12135.
34. Amanchy, R.; Kalume, D.E.; Iwahori, A.; Zhong, J.; Pandey, A. *J. Proteome Res.* 2005, **4**, 1661–1671.
35. Ballif, B.A.; Roux, P.P.; Gerber, S.A.; MacKeigan, J.P.; Blenis, J.; Gygi, S.P. *Proc. Natl. Acad. Sci. U. S. A.*, 2005, **102**, 667–672.
36. Hoffert, J.D.; Pisitkun, T.; Wang, G.; Shen, R.F.; Knepper, M.A. *Proc. Natl. Acad. Sci. U. S. A.*, 2006, **103**, 7159–7164.
37. Yang, F.; Stenoien, D.L.; Strittmatter, E.F.; Wang, J.H.; Ding, L.H.; Lipton, M.S.; Monroe, M.E.; Nicora, C.D.; Gristenko, M.A.; Tang, K.Q.; Fang, R.H.; Adkins, J.N.; Camp, D.G.; Chen, D.J.; Smith, R.D. *J. Proteome Res.*, 2006, **5**, 1252–1260.
38. Arroyo, J.D.; Hahn, W.C. *Oncogene* 2005, **24**, 7746–7755.
39. Brady, M.J.; Saltiel, A.R. *Recent Prog. Horm. Res.* 2001, **56**, 157–173.
40. Oliver, C.J.; Shenolikar, S. *Front. Biosci.* 1998, **3**, D961–972.
41. Brown, L.; et al. *Biochim. Biophys. Acta*, 2000, **1492**, 470–476.
42. Hunter, T.; Sefton, B.M. *Proc. Natl. Acad. Sci. U. S. A.* 1980, **77**, 1311–1315.
43. McLachlin, D.T.; Chait, B.T. *Curr. Opin. Chem. Biol.* 2001, **5**, 591–602.
44. Qian, W.J.; Goshe, M.B.; Camp II, D.G.; Yu, L.R.; Tang, K.; Smith, R.D. *Anal. Chem.* 2003, **75**, 5441–5450.
45. Garcia, B.A.; Shabanowitz, J.; Hunt, D.F. *Methods* 2005, **35**, 256–264.

46. Salih, E. *Mass Spectrom. Rev.* 2005, **24**, 828–846.
47. Smith, R.D.; Anderson, G.A.; Lipton, M.S.; Pasa-Tolic, L.; Shen, Y.; Conrads, T.P.; Veenstra, T.D.; Udseth, H.R. *Proteomics* 2002, **2**, 513–523.
48. Janssens, V.; Goris, J. *Biochem. J.* 2001, **353**, 417–439.
49. Ceulemans, H.; Bollen, M. *Physiol. Rev.* 2004, **84**, 1–39.
50. Ficarro, S.B.; McCleland, M.L.; Stukenberg, P.T.; Burke, D.J.; Ross, M.M.; Shabanowitz, J.; Hunt, D.F.; White, F.M. *Nat Biotechnol* 2002, **20**, 301–305.
51. Kim, J.E.; Tannenbaum, S.R.; White, F.M. *J. Proteome Res.* 2005, **4**, 1339–1346.
52. Moser, K.; White, F.M. *J. Proteome Res.* 2006, **5**, 98–104.
53. Ballif, B.A.; Villen, J.; Beausoleil, S.A.; Schwartz, D.; Gygi, S.P. *Mol. Cell. Proteomics* 2004, **3**, 1093–1101.
54. Gentile, S.; Darden, T.; Erxleben, C.; Romeo, C.; Russo, A.; Martin, N.; Rossie, S.; Armstrong, D.L. *Proc. Natl. Acad. Sci. U. S. A.* 2006, **103**, 5202–5206.
55. Posewitz, M.C.; Tempst, P. *Anal. Chem.* 1999, **71**, 2883–2892.
56. Cao, P.; Stults, J.T. *Rapid Commun. Mass Spectrom.* 2000, **14**, 1600–1606.
57. Kocher, T.; Allmaier, G.; Wilm, M. *J. Mass Spectrom.* 2003, **38**, 131–137.
58. Stover, D.R.; Caldwell, J.; Marto, J.; Root, R.; Mestan, J.; Stumm, M.; Ornatsky, O.; Orsi, C.; Radosevic, N.; Liao, L.; Fabbro, D.; Moran, M.F. *Clin. Proteomics* 2004, **1**, 069–080.
59. Gruhler, A.; Olsen, J.V.; Mohammed, S.; Mortensen, P.; Faergeman, N. J.; Mann, M.; Jensen, O.N. *Mol. Cell. Proteomics* 2005, **4**, 310–327.
60. Zhang, Y.; Wolf-Yadlin, A.; Ross, P.L.; Pappin, D.J.; Rush, J.; Lauffenburger, D.A.; White, F.M. *Mol. Cell. Proteomics* 2005, **4**, 1240–1250.
61. Garrels, J. *J. Biol. Chem.* 1979, **254**, 7961–7977.
62. Leimgruber, R.M. In: *The Proteomics Protocols Handbook,* 2005, Walker, J.M. ed., Humana Press, New Jersey, p. 4.
63. Leimgruber, R.M. *Proteomics* 2002, **2**, 135–144.
64. de Souza, G.A.; Godoy, L.M.F.; Mann, M. *Genome Biol.* 2006, **7**:R72.
65. Stock, J.; Ninfa, A.; Stock, A.M. *Microbiol. Rev.* 1989, **53**, 450–490.
66. Stock, J.B.; Stock, A.M.; Mottonen, J.M. *Nature* 1990, **344**, 395–400.
67. Mathews, H.R. *Pharmacol. Ther.* 1995, **67**, 323–350.
68. Waygood, E.B.; Mattoo, R.L.; Peri, K.G. *J. Cell Biochem.* 1984, **25**, 139–159.
69. Pirrung, M.C.; James, K.D.; Rana, V.S. *J. Org. Chem.* 2000, **65**, 8448–8453.
70. Napper, S.; Kindrachuk, J.; Olson, D.J.H.; Ambrose, S.J.; Dereniwsky, C.; Ross, A.R.S. *Anal. Chem.* 2003, **75**, 1741–1747.
71. Thomsson, K.A.; Karlsson, H.; Hansson, G.C. *Anal. Chem.* 2000, **72**, 4543–4539.
72. Hooper, L.V.; Manzella, S.M.; Baenziger, J.U. *FASEB J.* 1996, **10**, 1137–1146.
73. Bowman, K.G.; Bertozzi, C. R. *Chem. Biol.* 1996, **6**, R9–R22.
74. Habuchi, O. *Biochim. Biophys. Acta* 2000, **1474**, 115–127.
75. Bowman, K.G.; Cook, B.N.; de Graffenried, C.L.; Bertozzi, C.R. *Biochemistry* 2001, **40**, 5382–5391.
76. Plaas, A.H.K.; West, L.A.; Wong Palms, S.; Nelson, F.R.T. *J. Biol. Chem.* 1998, **273**(20), 12642–12649.
77. Bayliss, M.T.; Osborne, D.; Woodhouse, S.; Davidson, C. *J. Biol. Chem.* 1999, **274**(22), 15892–15900.
78. Desaire, H.; Sirich, T.L.; Leary, J.A. *Anal. Chem.* 2001, **73**(15), 3513–3520.
79. Jiang, H.; Irungu, J.; Desaire, H. *J. Am. Soc. Mass Spectrom.* 2005, **16**, 340–348.
80. Falany, C.N. *FASEB J.* 1997, **22**, 206–216.
81. Green, E.D.; Baenziger, J.U. *J. Biol. Chem.* 1988, **26**, 36–44.
82. Kawasaki, N.; Haishima, Y.; Ohta, M.; Satsuki, I.; Hyuga, M.; Hyuga, S.; Hayakawa, T. *Glycobiology* 2001, **11**, 1043–1049.
83. Hemmerich, S.; Rosen, S.D. *Glycobiology* 2000, **10**, 849–856.
84. Domon, B.; Costello, C.E. *Glycoconjugate J* 1988, **5**, 397–409.
85. Baeuerle, P.A.; Huttner, W.B. J. *Biol. Chem.* 1985, **260**, 6434–6439.
86. Bettelheim, F.R. *J. Am. Chem. Soc.* 1954, **76**, 2838–2839.
87. Beisswanger, R.; Corbeil, D.; Vannier, C.; Thiele, C.; Dohrmann, U.; Kellner, R.; Ashman, K.; Niehrs, C.; Huttner, W.B. *Proc. Natl. Acad. Sci. U. S. A.* 1998, **95**, 11134–11139.
88. Ouyang, Y.B.; Moore, K.L. *J. Biol. Chem.* 1998, **273**, 24770–24774.
89. Huxtable, R.J. *Biochemistry of Sulfur,* Plenum, New York, 1986.
90. Medzihradszky, K.F.; Darula, Z.; Perlson, E.; Fainzilber, M.; Chalkley, R.J.; Ball, H.; Greenbaum, D.; Bogyo, M.; Tyson, D.R.; Bradshaw, R.A.; Burlingame A.L. *Mol. Cell. Proteomics* 2004, **3**, 429–440.
91. Woods, A.S.; Wang, H.Y.J.; Jackson, S.N. *J. Proteome Res.* 2007, **6**, 1176–1182.
92. Kirschner, M.W. *Cell* 2005, **121**, 503–504.
93. Srivastava, R.; Varner, J. *Biotechnol. Prog.* 2007, **23**, 24–27.
94. Smith, J.C.; Lambert, J.P.; Elisma, F.; Figeys, D. *Anal. Chem.* Published on Web 05/04/2007.

APPENDIX I

CHAPTER PROBLEM ANSWERS

4.1 a. Four, b. three, c. five, d. six, e. three.
4.2 a. 5.17, b. 196.65, c. 3.2, d. 10111, e. 712, f. −41.5, g. 1.7161, h. 5.58.
4.3 134 in.
4.4 2230 cm^3.
4.5 4.83 lb.
4.6 0.562 qt.
4.7 12.4 oz.
5.1 0.34.
5.2 1.88.
5.3 1.5.
5.4 1.47.
5.5 0.035 M.
5.6 0.351% ± 0.003%.
5.7 Systematic error may exist.
5.8 Not a significant difference.
5.9 Cannot be removed.
5.10 Can be removed.
6.1 No answer.
6.2 No answer.
6.3 $m = 1.25$, $y \approx 2.4$.
6.4 $y = 1.25x + 2.4$.
6.5 $m = 1.269$, $y = 2.414$.
6.6 $y = 0.01438x + 4.967$.
6.7 $y = 0.0560x + 7.073$, $R^2 = 0.9963$.
6.8 $y = -0.9538x + 307.6445$, $R^2 = 0.9980$.
8.1 a. 221.81093 $g{\cdot}mol^{-1}$, b. 110.11064 $g{\cdot}mol^{-1}$, c. 197.169341 g mol^{-1}, 342.44664 g mol^{-1}.
8.2 a. 220.30954, b. 384.144148, c. 56.10632, d. 220.0567, e. 354.7978, f. 292.24272.
8.3 $m = 0.02516$.
8.4 $m = 0.0530$.
8.5 0.0340.
8.6 0.02513.
8.7 0.6001.
8.8 ppm lactose = 16.4, ppb lactose = 16,400, % lactose = 0.001,64.
8.9 130 mg Cu_2SO_4 needed.
8.10 Concentration KNO_3 = 0.04787 M.
8.11 10.0747 g $KMnO_4$ needed.
8.12 Concentration = 1.4168 N.
8.13 10.420 g H_2SO_4.
9.1 a. Cl^-. b. HSO_4^-, SO_4^-. c. $C_2H_2(COOH)COO^-$, $C_2H_2(COO^-)_2$, d. CH_3COO^-, e. $HCOO^-$.
9.2 a. $HCl + NH_3 \leftrightarrow NH_4^+ + Cl^-$,
b. $H_2SO_4 + NH_3 \leftrightarrow NH_4^+ + HSO_4^-$ and $HSO_4^- + NH_3 \leftrightarrow NH_4^+ + SO_4^{2-}$,
c. $C_2H_2(COOH)_2 + NH_3 \leftrightarrow NH_4^+ + C_2H_2(COOH)COO^-$ and $C_2H_2(COOH)COO^- + NH_3 \leftrightarrow NH_4^+ + C_2H_2(COO^-)_2$,
d. $CH_3COOH + NH_3 \leftrightarrow NH_4^+ + CH_3COO^-$,
e. $HCOOH + NH_3 \leftrightarrow NH_4^+ + HCOO^-$.
9.3 a. $K_a = \dfrac{[NH_4^+][Cl^-]}{[HCl][NH_3]}$

b. $K_{a1} = \dfrac{[NH_4^+][HSO_4^-]}{[H_2SO_4][NH_3]}$, $K_{a2} = \dfrac{[NH_4^+][SO_4^{2-}]}{[HSO_3^-][NH_3]}$

c. $K_{a1} = \dfrac{[NH_4^+][C_2H_2(COOH)COO^-]}{[C_2H_2(COOH)_2][NH_3]}$, $K_{a2} = \dfrac{[NH_4^+][C_2H_2(COO^-)_2]}{[C_2H_2(COOH)COO^-][NH_3]}$

d. $K_a = \dfrac{[NH_4^+][CH_3COO^-]}{[CH_3COOH][NH_3]}$

e. $K_a = \dfrac{[NH_4^+][HCOO^-]}{[HCOOH][NH_3]}$

Analytical Chemistry: A Chemist and Laboratory Technician's Toolkit, First Edition. Bryan M. Ham and Aihui MaHam.

9.4 a. $K_c = \frac{[H^+][HCO_3^-]}{[H_2CO_3]}$

b. $K_c = \frac{[Ni(CN)_4^{2-}]}{[Ni^{2+}][CN^-]^4}$

c. $K_c = \frac{[H^+][CO_3^{2-}]}{[HCO_3]}$

d. $K_c = \frac{Ag(NH_3)_2^+}{[Ag][NH_3]^2}$

e. $K_c = \frac{[H^+]^2[CO_3^{2-}]}{[H_2CO_3]}$

9.6 $x = 9.3 \times 10^{-4} = [H^+]$.
9.7 $x = 8.1 \times 10^{-6} = [H^+]$.
9.8 $[H^+] = 4.13 \times 10^{-5}$.
9.9 4.8% and 1.6%.
9.10

$$H_2NCH_2CH_2NH_2 + H_2O \Leftrightarrow H_2NCH_2CH_2NH_3^+ + OH^-$$
$$H_2NCH_2CH_2NH_3^+ + H_2O \Leftrightarrow {}^+H_3NCH_2CH_2NH_3^+ + OH^-$$

9.11

$$K_{b1} = \frac{[H_2NCH_2CH_2NH_3^+][OH^-]}{[H_2NCH_2CH_2NH_2]}$$
$$K_{b2} = \frac{[{}^+H_3NCH_2CH_2NH_3^+][OH^-]}{[H_2NCH_2CH_2NH_3^+]}$$

9.12 2.7% and 0.1%.
9.15 5.44×10^{-4} M.
9.16 2.73×10^{-2} M.
9.17 1.185×10^{-7} M.
9.18 a. 1.60.
9.19 a. 12.2.
9.20 2.0.
9.21 12.
9.22 4.4.
9.23 a. 3.85, b. the pH has changed from 3.85 to 3.55, c. the pH has changed from 3.85 to 4.15.
9.24 6.35:1.
9.25 The pH has changed from 3.14 to 3.11, only 0.03 units.
9.26 412 and 588 ml.
10.1 a. 2735 mg.
10.2 a. 6.9 ml, 13.9 ml.
10.3 Sulfuric acid $M = 17.8$, $N = 35.6$.
10.4 12.5 M, 27.3 ml.
10.5 15.27 g.
10.6 0.1126 M.
10.7 215.70 ml.
10.8 0.1167 M.
10.9 0.2587 g, 3.45%.
10.10 19.87%.
10.11 500 ppm, 5.1×10^{-3} M, 1.02×10^{-2} N.
10.12 17.83%.
10.13 13.12%.
10.14 45.01, 90.03.
10.15 2021 ml.
10.16 287 ml.
10.17 4520 ppm, 0.0390 M.
10.18 10.1%.
10.19 2.54%.
10.20

$$Co^{2+} + H_2Y^{2-} \rightleftharpoons CoY^{2-} + 2H^+$$
$$Cr^{3+} + H_2Y^{2-} \rightleftharpoons CrY^- + 2H^+$$
$$La^{3+} + H_2Y^{2-} \rightleftharpoons LaY^- + 2H^+$$
$$Th^{4+} + H_2Y^{2-} \rightleftharpoons ThY + 2H^+$$

11.1 a. $K_{E_l}|K^+(1\,M)||Br_2(aq)|_{E_r}2Br^-$

b. $Co_{E_l}|Co^{2+}(1\,M)||Cu^{2+}(1\,M)|_{E_r}Cu$

c. $Na_{E_l}|Na^+(1\,M)||N_2O_4(1\,M)|_{E_r}NO_2^-$

d. $Mg_{E_l}|Mg^{2+}(1\,M)||Au^{3+}(1\,M)|_{E_r}Au$

e. $2I^-(aq)_{E_l}|I_2(s)||Zn^{2+}(1\,M)|_{E_r}Zn$

11.2 a.
$$2K^+ + 2e^- \rightarrow 2K(s)\ (-2.931\,V)$$
$$Br_2(aq) + 2e^- \rightarrow 2Br^-(+1.066\,V)$$
$$Br_2(aq) + 2K(s) \rightarrow 2Br^- + 2K^+$$

b.
$$Co^{2+} + 2e^- \rightarrow Co(s)(-0.28\,V)$$
$$2Cu^{2+} + 2e^- \rightarrow 2Cu^+(+0.159\,V)$$
$$2Cu^{2+} + Co(s) \rightarrow 2Cu^+ + Co^{2+}$$

c.
$$2Na^+ + e^- \rightarrow 2Na(s)(-2.71\ V)$$
$$N_2O_4 + 2e^- \rightarrow 2NO_2^-(+0.867\,V)$$
$$2Na(s) + N_2O_4 \rightarrow 2Na^+ + 2NO_2^-$$

d.
$$Mg^{2+} + 2e^- \rightarrow Mg(s)(-2.372\ V)$$
$$Au^{3+} + 3e^- \rightarrow Au(s)(+1.52\,V)$$
$$3Mg(s) + 2Au^{3+} \rightarrow 3Mg^{2+} + 2Au(s)$$

e.
$$I_2(s) + 2e^- \rightarrow 2I^-(+0.54\,V)$$
$$Zn^{2+} + 2e^- \rightarrow Zn(s)(-0.7618\,V)$$
$$2I^- + Zn^{2+} \rightarrow I_2 + Zn$$

11.3 a. −1.865 V, not spontaneous, b. +0.439 V, spontaneous, c. +3.577 V, spontaneous, d. +3.892 V, spontaneous, e. −1.3018 V, not spontaneous.

11.4 a.
$$Ni + 2Ag^+ \rightarrow Ni^{2+} + 2Ag$$
$$Ni(s)_{E_l}|Ni^{2+}(1\,M)|\ |Ag^+(1\,M)|_{E_r}Ag(s)$$

b.
$$Ca + Cl_2 \rightarrow Ca^{2+} + 2Cl^-$$
$$Ca(s)_{E_l}|Ca^{2+}(1\,M)|\ |Cl_2(1\,M)|_{E_r}Cl^-(aq)$$

c.
$$Au^{3+} + 3Ag \rightarrow Au + 3Ag^+$$
$$Ag(s)_{E_l}|Ag^+(1\,M)|\ |Au^{3+}(1\,M)|_{E_r}Au(s)$$

d.
$$Pb^{2+} + Fe^{2+} \rightarrow Pb + Fe^{3+}$$
$$Fe^{2+}{}_{E_l}|Fe^{3+}(1M)|\ |Pb^{2+}(1\,M)|_{E_r}Pb(s)$$

e.
$$Sn^{4+} + 2Br^- \rightarrow Sn^{2+} + Br_2$$
$$Br^-{}_{E_l}|Br_2(1\,M)|\ |Sn^{4+}(1\,M)|_{E_r}Sn^{2+}$$

11.5 a.
$$Ni^{2+}(aq) + 2e^- \rightarrow Ni(s)(E^o = -0.26\,V)$$
$$Ag^+(aq) + e^- \rightarrow Ag(s)(E^o = 0.80\,V)$$

b.
$$Ca^{2+} + 2e^- \rightarrow Ca(s)(-2.868\,V)$$
$$Cl_2(g) + 2e^- \rightarrow 2Cl^-(+1.36)$$

c.
$$Ag^+ + e^- \rightarrow Ag(s)(+0.7996\,V)$$
$$Au^{3+} + 3e^- \rightarrow Au(s)(+1.52\,V)$$

d. $Fe^{3+} + e^- \rightarrow Fe^{2+}(+0.77\,V)$
$Pb^{2+} + 2e^- \rightarrow Pb(s)(-0.13\,V)$

e. $Sn^{4+} + 2e^- \rightarrow Sn^{2+}(+0.15\,V)$
$Br_2(aq) + 2e^- \rightarrow 2Br^-(+1.0873\,V)$

11.6 a. 1.06 V, spontaneous, b. 4.228 V, spontaneous, c. 0.7204 V, spontaneous, d. 0.90 V, spontaneous, e. 0.9373 V, spontaneous.
11.7 a. −0.2743 V, b. −2.909 V, c. 1.436 V, d. −0.1478 V, e. −0.2000 V.
11.8 a. 0.255 M, b. 2.87×10^{-3} M, c. 8.90×10^{-2} M, d. 1.54×10^{-5} M, e. 2.13×10^{-4} M.
11.9 $E_{cell} = 0.652$, not, or slightly significant.
11.10 $E_{cell} = 1.08$, significant.
11.11 $E_{cell} = 2.81$, significant.
11.12 a. 2.27×10^{10}, b. 2.27×10^{41}, c. 5.79×10^{105}.
11.13 1.7×10^{37}.
11.14 a. 4.96×10^{179}, c. 5.3×10^{59}.
11.15 a. 1.22 V, b. 2.3×10^{41}.
11.16 0.2601 N.
11.17 a. 0.04945 M, b. +0.54, +0.56, c. 3+ to 5+, d. −0.02 V.
11.18 b. 0.06815 M, c. 0.1758 M, d. 25.26%, e. 52.40%.
13.1 0.14.
13.2 a. 0.20, b. 0.013, c. 0.49, d. 0.15.
13.3 a. 2.26, b. 0.162, c. 0.0829, d. 0.022.
13.4 70–80%.
13.5 $6.23\,l^{-1}\,mol^{-1}\,cm$.
13.6 0.013 M.
13.7 0.265.
13.8 0.0796 M.
13.9 390 nm.
13.10 $0.337\,l^{-1}\,mol^{-1}\,cm$.
13.11 1.175 M.
13.12 Yes.
13.14 $0.0541\,mg^{-1}\,cm^{-1}$, $2164\,l^{-1}\,mol^{-1}\,cm^{-1}$.
13.15 5.45 ppm, 1.36×10^{-4} M.
13.17 0.65%, 0.022,85 M.
13.18 Not on curve.
14.1 5.03×10^{-19} J.
14.2 209 nm (209×10^{-9} m).
14.3 0.738.
14.4 187.9.
14.6 Intensity of fluorescence emission directly proportional to incident radiation.
14.10 $k_{fluor} = 7.46 \times 10^{-12}$.
14.11 $\Phi_{fluor} = 0.55$, quantum yield has decreased 27%.
14.12 7.40×10^{10}.
14.13 74 s, or 1 min and 14 s. Most likely phosphorescence due to long lifetime.
14.14 $\tau_M = 1.01 \times 10^{11}$. Lifetime increased to 101 s, or 1 min and 41 s. Lifetime has increased 36%.
14.15 They are directly proportional.
14.16 0.1075 M.
14.17 15.94 1/M s.
14.18 5000.
14.19 1.645 M.
14.20 480 nm.
14.21 620 nm.
14.23 $537.5\,l\,mg^{-1}\,cm^{-1}$, $2.58 \times 10^{8}\,l\,mole^{-1}\,cm^{-1}$.
14.24 1.45×10^{-5} M.
14.26 1.40×10^{-3} M.
14.27 3.00%.
14.28 0.051%, 2.86×10^{-3} M.
15.1 Ambient carbon dioxide and water.
15.2 Changes in dipole moments.
15.3 a. Inactive, b. active, c. inactive.
15.4 a. Inactive, b. inactive, c. active, d. active.
15.5 a. 3468, 2923, 2854, 1735, 1463, 1169, and 723 cm^{-1}, c. vegetable fat.
15.6 Includes ester moiety.
15.7 Calcium carbonate.
15.8 555 and 1019 cm^{-1}—phosphate—$(PO_4)^3$.
15.9 Aromatic.
15.10 Aromatic and alcohol.
15.11 Carbonyl carbon.
15.12 Shift in carbonyl band from 1690 to 1721 cm^{-1}.
15.14 Fumaric acid.
15.15 Ionic, not covalent.
15.16 IR spectrum due to O–H bonds of water.
26.1 *m/z* 296, 264.1, 222.2, 180.2, and 74.0.
28.1 Inlet, ionization source, mass analyzer, detector, data processing system, and vacuum system.
28.2 Their mass-to-charge ratio *m/z*.
28.3 To reduce the amount of unwanted collisions.
28.4 Two-stage rotary vane mechanical pump (rough pump) at ~10^{-4} Torr and turbo molecular pump at ~ 10^{-7}–10^{-9} Torr.
28.5 Proteomics, metabolomics, and lipidomics.
28.7 Allows the transfer of intact associations within solution phase to the gas phase without interruption of the weak associations thus preserving the original interaction. Lipid:metal/nonmetal, lipid:lipid, lipid:peptide, lipid:protein, protein:protein, and metabolite:protein adducts.
28.8 Protonated molecule $[M+H]^+$, sodium metal adduct as $[M+Na]^+$, a chloride adduct as $[M+Cl]^-$, a deprotonated species as $[M-H]^-$, through electron loss as a radical $M^{+\cdot}$.
28.9 446 amu.
28.10 31 μs.
28.11 a. 184.2402, 184.3653, 0.2402. b. 306.1837, 306.3668, 0.1837. c. 505.2833, 505.7717, 0.2833. d. 692.5092, 692.9780, 0.5092. e. 222.9696, 224.4927, −0.0304.
28.12 a. 281 ppm. b. −7 ppm. c. −3 ppm. d. 9 ppm.
28.13 a. 517. b. 21,739. c. 197.
28.14 a. Odd electron (OE) as $CH_4^{+\cdot}$, even mass ion with *m/z* = 16 Th. b. OE as $NH_3^{+\cdot}$, odd mass ion with *m/z* = 17 Th. c. even electron (EE) as NH_4^+, even mass ion with *m/z* = 18 Th. d. EE as H_3O^+, odd mass ion with *m/z* = 19 Th. e. EE as $C_2H_7O^+$, odd mass ion with *m/z* = 47 Th. f. OE as $C_{14}H_{28}O_2^{+\cdot}$, even mass ion with *m/z* = 228 Th. g. EE as $C_6H_{14}NO^+$, even mass ion with *m/z* = 116 Th. h. OE as $C_4H_{10}SiO^{+\cdot}$, even mass ion with *m/z* = 102 Th. i. EE as $C_{29}H_{60}N_2O_6P^+$, odd mass ion with *m/z* = 563 Th.
29.1 Electron ionization (EI), electrospray ionization (ESI), chemical ionization (CI), atmospheric pressure chemical

ionization (APCI), atmospheric pressure photo ionization (APPI), and matrix-assisted laser desorption ionization (MALDI).

29.2 To ionize a neutral molecule and transfer into the gas phase in preparation to introduction into the low vacuum environment of the mass analyzer.

29.3 Potential difference between the filament and the target producing electron beam energy of 70 eV.

29.4 To induce the electrons in the beam to follow a spiral trajectory from filament to target causing the electrons to follow a longer path from filament to target producing a greater chance of ionizing the neutral analyte molecules passing through the electron beam.

29.5 Slits and apertures are smaller to maintain the higher pressure in the source, emission current leaving the filament is measured, and repeller voltage is kept lower.

29.6 Methane (CH_4), isobutane (i-C_4H_{10}), and ammonia (NH_3).

29.7 Yes. -37.3 kJ mol^{-1}.

29.8 Though not exothermically apparent, most will be protonated.

29.9 All will except ammonia may not result in a high abundance.

29.10 Chemical ionization for molecular weight and electron ionization for structural information.

29.11 Breakup clusters, aid in evaporation of the solvent, and a nebulizing gas that helps to atomize the eluant spray.

29.12 Generated electric field penetrates into the liquid meniscus and creates an excess abundance of charge at the surface. The meniscus becomes unstable and protrudes out forming a Taylor cone.

29.13 Charged residue for larger molecules, ion evaporation for smaller.

29.14 Desolvation, adiabatic expansion, evaporative cooling, and low-energy dampening collisions. Soft ionization.

29.15 Reduced size droplets resulting in more efficient ion production.

29.16 Absorption of radiant energy from a UV source where the incident energy is greater than the first ionization potential of electron loss from the analyte.

29.17 Not a solution-phase process based upon proton affinity.

29.18 Dopant is a chemical species that possesses a low ionization potential (IP) that allows its ionization to take place with high efficiency. Yes.

29.19 Naphthalene, benzene, phenol, aniline, m-chloroaniline, 1-aminonaphthalene. Water and acetonitrile.

29.20 UV absorber and proton donor. 2,5-dihydroxybenzoic acid (DHB), 3,5-dimethoxy-4-hydroxy-trans-cinnamic acid (sinapic or sinapinic acid), and α-cyano-4-hydroxy-trans-cinnamic acid (α-CHCA).

29.21 Dried droplet, thin film, and layer or sandwich.

29.22 The UV absorbing matrix molecules accepts energy from the laser and desorbs from the surface carrying along any analyte that is mixed with it forming a gaseous plume where the analyte is ionized.

29.23 Initial spatial and energy distributions. Delayed extraction and the reflectron.

29.24 Drying behavior and analyte distribution.

29.25 Drying behavior, spatial distributions, and gas-phase proton donor.

29.26 Analyte is dissolved in glycerol and is bombarded with a high-energy beam of atoms where the kinetic energy from the colliding atom is transferred to the matrix and analyte effectively desorbing them into the gas phase with ionization.

29.27 Matrix absorbs the largest amount of kinetic energy, and the desorbed analyte goes through a desolvation mechanism where excess energy is transferred to the matrix solvent molecules.

30.1 Electric and magnetic sector mass analyzer, time-of-flight mass analyzer (TOF/MS), time-of-flight/time-of-flight mass analyzer (TOF–TOF/MS), quadrupole time-of-flight mass analyzer (Q-TOF/MS), triple quadrupole or linear ion trap mass analyzer (QQQ/MS or LIT/MS), three-dimensional quadrupole ion trap mass analyzer (QIT/MS), Fourier transform ion cyclotron mass analyzer (FTICR/MS), and linear ion trap-Orbitrap mass analyzer (IT-Orbitrap/MS).

30.2 100 mm.

30.3 The electric sector is not a mass analyzer but an energy filter.

30.4 The ion gate acts as a timed ion selector by switching on and off to allow ions to pass.

30.5 5 eV.

30.6 Under the influence of the combination of electric fields comprised of DC voltage (U) and radio frequency (rf) voltage (V).

30.7 The mass-to-charge relationship (m/e).

30.8 The plot of Figure 3.16 demonstrates that the resolution of the quadrupole is inversely proportional to the sensitivity.

30.9 Relatively insensitive to kinetic energy spread, easy to operate and calibrate, mechanically rugged, and inexpensive to produce.

30.10 Imperfections in the quadrupole field's influence upon ion detection.

30.11 i. Single stage analysis, ii. neutral loss scanning, iii. precursor ion scanning, iv. product ion scanning.

30.12 To induce low-energy collisions with the ions to thermally cool them, decrease their kinetic energy, and focus the ion packets into tighter ion trajectories near the center of the trap.

30.13 Method for scanning mass-to-charge species in the trap.

30.14 Through the application of an ac voltage applied to the end caps of the quadrupole ion trap.

30.15 Resonant excitation, dipole excitation, and quadrupole excitation.

30.16 The inclusion of a quadrupole mass analyzer in tandem prior to the time-of-flight mass analyzer.

30.17 Orthogonal, which allows a uniform spatial distribution of ions from the source.

30.18 Mass accuracies of <5 ppm with resolutions as high as 1,000,000.

30.19 Permanent magnets and electromagnets are less than 2 Tesla (T), while superconducting magnets are from 3.5, 7, 9.4, and 11.5 T.

30.20 10^{-9} or 10^{-10} Torr to reduce collision-mediated radial diffusion.

30.21 i. Drive the ions away from the center of the ICR cell for signal measurement, ii. drive ions to a larger cyclotron radius to remove (eject) them from the cell, and iii. to add internal energy for dissociative collision product ion generation and ion–molecule reactions.

30.22 a. Collisions taking place between the trapped ions and between the trapped ions with neutrals will drive the ions toward the outer dimensions of the cell where they will be lost to neutralizing collisions with the cell walls. b. Cyclotron frequency amplitudes are observed to steadily decrease.

30.23 The maximum amount of ions that can be trapped within the ICR cell is proportional to the square of the magnetic field (B^2), increase in maximum ion kinetic energy with increasing magnetic field strength (B^2), and the radius of the orbiting ion packet will be smaller for the same kinetic energy.

30.24 No refilling of liquid nitrogen and liquid helium reservoirs to maintain the low temperatures needed for superconductivity, and no active shielding of the magnet.

31.1 Steroids, fatty acid amides, fatty acids, wax esters, fatty alcohols, phosphorylated lipids, and acylglycerols.

31.2 Promote ionization, rich fragmentation patterns, alleviate variations in adducts.

31.3 Negative ion mode. The organic acid is easily deprotonated.

31.4 Yes they can. Due to a direct linear correlation between lipid concentration and lipid response, however, there are differences in ionization efficiencies of lipids and subsequent suppression that can interfere with quantitation.

31.5 It is possible in negative ion mode due to reduction taking place at the surface of the ESI capillary.

31.6 No they were not.

31.7 Suppression.

31.8 The suppression effect being observed.

31.9 A carbon 13 (^{13}C) algal fatty acid standard mix was analyzed.

31.10 The difficulty in quantitating free fatty acids in biological samples using electrospray as the ionization technique.

31.11 The response of ion detectors coupled with single quadrupole mass spectrometers is linear in regions.

31.12 Dissolving the biological sample in a solvent solution containing an internal standard that is then quantitated against a series of fatty acid amide standards ratioed to the internal standard (IS).

31.13 To measure the biomolecules at responses that are within the linear response of the associated calibration curves.

31.14 The fatty acid portion of the biomolecule due to an electrostatic noncovalent interaction.

31.15 The *m/z* 403.6 aldehyde peak and the *m/z* 387.6 product ion formed from the fatty alcohol substituent.

31.16 An 18 Da neutral loss of water.

31.17 1. Exact mass measurements.

2. Synthesis and analysis of suspected unknown.
3. Proton NMR.
4. Structural elucidation of unknown by CID in positive and negative ion mode.

31.18 GC-EI-MS, positive chemical ionization, negative chemical ionization of chloride adducts, ES-MS/MS, MALDI-TOF, fast atom bombardment, atmospheric pressure chemical ionization (APCI) mass spectrometry, and APCI liquid chromatography/mass spectrometry (LC/MS).

31.19 Species that contain polarized regions such as positive and negative but are neutral overall.

31.20 Deprotonated dimetal(1) adduct, protonated metal adduct, deprotonated metal(1)-metal(2) adduct, and the deprotonated dimetal(2) adduct.

31.21 Identification of substituents and head groups.

31.22 Typically the head group.

31.23 18 atomic mass units (amu). Can be used as an indication of ester linkages as fatty acid substituents in the structures and in unknown identification.

32.1 Proteins, polysaccharides, and nucleic acids.

32.2 Hydrolysis reactions cleave apart monomers with the addition of water. Condensation reactions join monomers with the elimination of water. Synthesis of biopolymers.

32.3 Glycogen and starch for storage and cellulose for structural use.

32.4 In α-d-glucose, the hydroxyl points downward, while for β-d-glucose, it points upward.

32.5 Two α-d-sugar units connected form a α-glycosidic bond and two β-d-sugar units connected form a β-glycosidic bond.

32.6 The decrease in the size of the nanoelectrospray drops increases the surface activity of the hydrophilic oligosaccharides.

32.7 The greater number of vibrational degrees of freedom in the larger oligosaccharides enables the dissipation of the internal energy. B_m- and Y_n-type ions. Larger oligosaccharides through glycosidic bond cleavage, smaller oligosaccharides through both glycosidic bond cleavage and cross-ring cleavage.

32.8 Complex mixtures of very closely related isomeric compounds that contain many labile bonds and can be highly branched.

32.9 The deprotonated precursor will dissociate with the hydroxyl group going to the reducing portion of the sugar, while the carbohydrate portion suffers ring opening and epoxide formation.

32.10 The charge of the metal ion is not directly associated with the glycosidic bond oxygen but is distributed to a number of local oxygen atoms.

32.11 Polysaccharides are often highly branched and/or substituted.

32.12 DNA contains the 5-carbon sugar deoxyribose, while RNA contains ribose. DNA contains the base thymine (T), while RNA contains uracil (U).

32.13 When the product ion contains the 3′-OH portion of the nucleic acid, the naming includes the letters *w*, *x*, *y*, and *z*. When the product ion contains the 5′-OH portion of the nucleic acid, the naming includes the letters *a*, *b*, *c*, and *d*.

32.14 Neutral and charged base losses.

32.15 Suppression and shifting of the mass of the nucleic acid.

32.16 Measurement of a singly charged species.

32.17 Every fragmentation pathway is initiated by loss of a nucleobase that does not involve simple water loss or 3′- and 5′-terminal nucleoside/nucleotides.

33.1 Top-down is the analysis of whole intact proteins, while bottom-up is the analysis of enzyme-digested proteins that have been converted to peptides.

33.2 Glycosylation, sulfation, and phosphorylation.

33.3 Soluble and insoluble; primary, secondary, tertiary, and quaternary structure; 20 amino acids; four classes of amino acids; beta-sheets and alpha-helix.

33.4 Primary consists of sequence of amino acids, secondary of alpha-helixes and beta-sheets, tertiary of the ordering of the secondary, and quaternary is the arrangement of the polypeptide chains into the working protein.

33.5 A comparison of a protein's enzymatic generated peptides measured by MALDI-TOF/MS against a theoretical list.

33.6 Protease used to digest proteins. The cleavage location within the polypeptide chain.

33.7 A complete coverage of the amino acid sequence within a peptide from the product ion spectrum. Can unambiguously identify a protein according to standard spectra stored in protein databases.

33.8 To map and decipher interrelationships between observed proteins and the genetic description.

33.9 When the charge is retained on the N-terminal portion of the fragmented peptide, the ions are depicted as *a*, *b*, and *c*. When the charge is retained on the C-terminal portion, the ions are denoted as *x*, *y*, and *z*.

33.10 Upon excitation of the peptide from CID, the proton will migrate to various protonation sites prior to fragmentation. The expected site of protonation from a thermodynamic point of view indicates amide oxygen protonation and not the amide nitrogen.

33.11 Loss of ammonia and water.

33.12 Side chain cleavage ions that produced by a combination of backbone cleavage and a side chain bond.

33.13 Immonium ions are produced through a combination of *a*-type and *y*-type cleavage that results in an internal fragment that contains a single side chain. Immonium ions are useful in acting as confirmation of residues suspected to be contained within the peptides backbone.

33.14 Isomers are species that have the same molecular formula but differ in their structural arrangement, while isobars are species with different molecular formulas that posses similar (or the same) molecular weights.

33.15 The use of high-resolution/high mass accuracy instrumentation.

33.16 Promotes extensive fragmentation along the polypeptide backbone while preserving modifications such as glycosylation and phosphorylation.

33.17 The accuracy in which the masses are measured. High-resolution mass spectrometers to accurately measure the intact protein's mass and the product ions.

33.18 A decreased need for extensive sample precleanup and up to 100% coverage of protein including modifications.

33.19 56,243 Da. *M/z* 2250.7 at +25, *m/z* 2344.5 at +24, *m/z* 2446.4 at + 23, and *m/z* 2557.5 at +22.

33.20 The covalent addition of a carbohydrate chain to amino acid side chains. To determine the proteins' structure, is crucial in cell–cell recognition, and may also be involved in cellular signaling events.

33.21 Reduces the viscosity of the blood.

33.22 (1) Get an identification of the glycosylated peptides and proteins, (2) accurately determine the sites of glycosylation, and (3) determine the carbohydrates that make up the glycan and the structure of the glycan.

33.23 Repeated addition of SiA at a charge state of +2 (146.5 Da) and of +3 (97.1 Da).

33.24 Serine, threonine, histidine, and tyrosine.

33.25 No. Approximately 2% pY, 12% pT, and 86% pS.

33.26 For a given protein, there may be only a couple of phosphorylated tryptic peptides as compared to many nonphosphorylated.

33.27 Nucleic acids. Ultracentrifugation, RNase and DNase, QIAShredder.

33.28 +1 at 98.0 Da, +2 at 49.0 Da, and +3 32.7 Da?

33.29 Dehydroalanine and dehydroaminobutyric acid.

33.30 Loss of 80 Da as HPO_3 and loss of 98 Da as H_3PO_4.

33.31 Promotes extensive fragmentation along the polypeptide backbone while preserving modifications such as glycosylation and phosphorylation. Yes.

33.32 In eukaryotic systems, the phosphoryl modification is associated with regulating enzyme activity. In prokaryotic systems, the histidine group is phosphorylated to transfer the phosphate group from one biomolecule to another.

33.33 Histidine phosphorylation is labile in acidic conditions, the conditions encountered during sample preparation steps.

33.34 See Table 33.8.

33.35 Neutral with a valence of 5 for the phosphorus.

33.36 Carbohydrate sulfation of cell surface glycans and sulfonation of protein tyrosine amino acid residues.

33.37 Yes, HPO_3 has a mass of 79.9663 Da and SO_3 has a mass of 79.9568 Da.

33.38 The study of the processes and complex biological organizational behavior using information from its molecular constituents.

33.39 Blood, urine, sputum, saliva, breath, tear fluid, nipple aspirate fluid, and cerebrospinal fluid.

APPENDIX II

ATOMIC WEIGHTS AND ISOTOPIC COMPOSITIONS

Symbol	Relative Atomic Mass	Abundance	Standard Atomic Weight
^{1}H	1.0078250321(4)	99.9885(70)	1.00794(7)
^{2}D	2.0141017780(4)	0.0115(70)	
^{3}T	3.0160492675(11)		
^{3}He	3.0160293097(9)	0.000137(3)	4.002602(2)
^{4}He	4.0026032497(10)	99.999863(3)	
^{6}Li	6.0151223(5)	7.59(4)	6.941(2)
^{7}Li	7.0160040(5)	92.41(4)	
^{9}Be	9.0121821(4)	100	9.012182(3)
^{10}B	10.0129370(4)	19.9(7)	10.811(7)
^{11}B	11.0093055(5)	80.1(7)	
^{12}C	12.0000000(0)	98.93(8)	12.0107(8)
^{13}C	13.0033548378(10)	1.07(8)	
^{14}C	14.003241988(4)		
^{14}N	14.0030740052(9)	99.632(7)	14.0067(2)
^{15}N	15.0001088984(9)	0.368(7)	
^{16}O	15.9949146221(15)	99.757(16)	15.9994(3)
^{17}O	16.99913150(22)	0.038(1)	
^{18}O	17.9991604(9)	0.205(14)	
^{19}F	18.99840320(7)	100	18.9984032(5)
^{20}Ne	19.9924401759(20)	90.48(3)	20.1797(6)
^{21}Ne	20.99384674(4)	0.27(1)	
^{22}Ne	21.99138551(23)	9.25(3)	
^{23}Na	22.98976967(23)	100	22.989770(2)
^{24}Mg	23.98504190(20)	78.99(4)	24.3050(6)
^{25}Mg	24.98583702(20)	10.00(1)	
^{26}Mg	25.98259304(21)	11.01(3)	
^{27}Al	26.98153844(14)	100	26.981538(2)
^{28}Si	27.9769265327(20)	92.2297(7)	28.0855(3)
^{29}Si	28.97649472(3)	4.6832(5)	
^{30}Si	29.97377022(5)	3.0872(5)	
^{31}P	30.97376151(20)	100	30.973761(2)
^{32}S	31.97207069(12)	94.93(31)	32.065(5)
^{33}S	32.97145850(12)	0.76(2)	
^{34}S	33.96786683(11)	4.29(28)	
^{36}S	35.96708088(25)	0.02(1)	
^{35}Cl	34.96885271(4)	75.78(4)	35.453(2)
^{37}Cl	36.96590260(5)	24.22(4)	
^{36}Ar	35.96754628(27)	0.3365(30)	39.948(1)
^{38}Ar	37.9627322(5)	0.0632(5)	
^{40}Ar	39.962383123(3)	99.6003(30)	
^{39}K	38.9637069(3)	93.2581(44)	39.0983(1)
^{40}K	39.96399867(29)	0.0117(1)	
^{41}K	40.96182597(28)	6.7302(44)	
^{40}Ca	39.9625912(3)	96.941(156)	40.078(4)
^{42}Ca	41.9586183(4)	0.647(23)	
^{43}Ca	42.9587668(5)	0.135(10)	
^{44}Ca	43.9554811(9)	2.086(110)	
^{46}Ca	45.9536928(25)	0.004(3)	
^{48}Ca	47.952534(4)	0.187(21)	
^{45}Sc	44.9559102(12)	100	44.955910(8)
^{46}Ti	45.9526295(12)	8.25(3)	47.867(1)
^{47}Ti	46.9517638(10)	7.44(2)	
^{48}Ti	47.9479471(10)	73.72(3)	
^{49}Ti	48.9478708(10)	5.41(2)	
^{50}Ti	49.9447921(11)	5.18(2)	
^{50}V	49.9471628(14)	0.250(4)	50.9415(1)
^{51}V	50.9439637(14)	99.750(4)	
^{50}Cr	49.9460496(14)	4.345(13)	51.9961(6)
^{52}Cr	51.9405119(15)	83.789(18)	
^{53}Cr	52.9406538(15)	9.501(17)	
^{54}Cr	53.9388849(15)	2.365(7)	
^{55}Mn	54.9380496(14)	100	54.938049(9)
^{54}Fe	53.9396148(14)	5.845(35)	55.845(2)
^{56}Fe	55.9349421(15)	91.754(36)	
^{57}Fe	56.9353987(15)	2.119(10)	
^{58}Fe	57.9332805(15)	0.282(4)	

(continued)

Analytical Chemistry: A Chemist and Laboratory Technician's Toolkit, First Edition. Bryan M. Ham and Aihui MaHam.

Symbol	Relative Atomic Mass	Abundance	Standard Atomic Weight
^{59}Co	58.9332002(15)	100	58.933200(9)
^{58}Ni	57.9353479(15)	68.0769(89)	58.6934(2)
^{60}Ni	59.9307906(15)	26.2231(77)	
^{61}Ni	60.9310604(15)	1.1399(6)	
^{62}Ni	61.9283488(15)	3.6345(17)	
^{64}Ni	63.9279696(16)	0.9256(9)	
^{63}Cu	62.9296011(15)	69.17(3)	63.546(3)
^{65}Cu	64.9277937(19)	30.83(3)	
^{64}Zn	63.9291466(18)	48.63(60)	65.409(4)
^{66}Zn	65.9260368(16)	27.90(27)	
^{67}Zn	66.9271309(17)	4.10(13)	
^{68}Zn	67.9248476(17)	18.75(51)	
^{70}Zn	69.925325(4)		
^{69}Ga	68.925581(3)		69.723(1)
^{71}Ga	70.9247050(19)	39.892(9)	
^{70}Ge	69.9242504(19)	20.84(87)	72.64(1)
^{72}Ge	71.9220762(16)	27.54(34)	
^{73}Ge	72.9234594(16)	7.73(5)	
^{74}Ge	73.9211782(16)	36.28(73)	
^{76}Ge	75.9214027(16)	7.61(38)	
^{75}As	74.9215964(18)	100	74.92160(2)
^{74}Se	73.9224766(16)	0.89(4)	78.96(3)
^{76}Se	75.9192141(16)	9.37(29)	
^{77}Se	76.9199146(16)	7.63(16)	
^{78}Se	77.9173095(16)	23.77(28)	
^{80}Se	79.9165218(20)	49.61(41)	
^{82}Se	81.9167000(22)	8.73(22)	
^{79}Br	78.9183376(20)	50.69(7)	79.904(1)
^{81}Br	80.916291(3)	49.31(7)	
^{78}Kr	77.920386(7)	0.35(1)	83.798(2)
^{80}Kr	79.916378(4)	2.28(6)	
^{82}Kr	81.9134846(28)	11.58(14)	
^{83}Kr	82.914136(3)	11.49(6)	
^{84}Kr	83.911507(3)	57.00(4)	
^{86}Kr	85.9106103(12)	17.30(22)	
^{85}Rb	84.9117893(25)	72.17(2)	85.4678(3)
^{87}Rb	86.9091835(27)	27.83(2)	
^{84}Sr	83.913425(4)	0.56(1)	87.62(1)
^{84}Sr	85.9092624(24)	9.86(1)	
^{84}Sr	86.9088793(24)	7.00(1)	
^{84}Sr	87.9056143(24)	82.58(1)	
^{89}Y	88.9058479(25)	100	88.90585(2)
^{90}Zr	89.9047037(23)	51.45(40)	91.224(2)
^{91}Zr	90.9056450(23)	11.22(5)	
^{92}Zr	91.9050401(23)	17.15(8)	
^{94}Zr	93.9063158(25)	17.38(28)	
^{96}Zr	95.908276(3)	2.80(9)	
^{93}Nb	92.9063775(24)	100	92.90638(2)
^{92}Mo	91.906810(4)	14.84(35)	95.94(2)
^{94}Mo	93.9050876(20)	9.25(12)	
^{95}Mo	94.9058415(20)	15.92(13)	
^{96}Mo	95.9046789(20)	16.68(2)	
^{97}Mo	96.9060210(20)	9.55(8)	
^{98}Mo	97.9054078(20)	24.13(31)	
^{100}Mo	99.907477(6)	9.63(23)	
^{97}Tc	96.906365(5)	[98]	
^{98}Tc	97.907216(4)		
^{99}Tc	98.9062546(21)		
^{96}Ru	95.907598(8)	5.54(14)	101.07(2)

Symbol	Relative Atomic Mass	Abundance	Standard Atomic Weight
^{98}Ru	97.905287(7)	1.87(3)	
^{99}Ru	98.9059393(21)	12.76(14)	
^{100}Ru	99.9042197(22)	12.60(7)	
^{101}Ru	100.9055822(22)	17.06(2)	
^{102}Ru	101.9043495(22)	31.55(14)	
^{104}Ru	103.905430(4)	18.62(27)	
^{103}Rh	102.905504(3)	100	102.90550(2)
^{102}Pd	101.905608(3)	1.02(1)	106.42(1)
^{104}Pd	103.904035(5)	11.14(8)	
^{105}Pd	104.905084(5)	22.33(8)	
^{106}Pd	105.903483(5)	27.33(3)	
^{108}Pd	107.903894(4)	26.46(9)	
^{110}Pd	109.905152(12)	11.72(9)	
^{107}Ag	106.905093(6)	51.839(8)	107.8682(2)
^{109}Ag	108.904756(3)	48.161(8)	
^{106}Cd	105.906458(6)	1.25(6)	112.411(8)
^{108}Cd	107.904183(6)	0.89(3)	
^{110}Cd	109.903006(3)	12.49(18)	
^{111}Cd	110.904182(3)	12.80(12)	
^{112}Cd	111.9027572(30)	24.13(21)	
^{113}Cd	112.9044009(30)	12.22(12)	
^{114}Cd	113.9033581(30)	28.73(42)	
^{116}Cd	115.904755(3)	7.49(18)	
^{113}In	112.904061(4)	4.29(5)	114.818(3)
^{115}In	114.903878(5)	95.71(5)	
^{112}Sn	111.904821(5)	0.97(1)	118.710(7)
^{114}Sn	113.902782(3)	0.66(1)	
^{115}Sn	114.903346(3)	0.34(1)	
^{116}Sn	115.901744(3)	14.54(9)	
^{117}Sn	116.902954(3)	7.68(7)	
^{118}Sn	117.901606(3)	24.22(9)	
^{119}Sn	118.903309(3)	8.59(4)	
^{120}Sn	119.9021966(27)	32.58(9)	
^{122}Sn	121.9034401(29)	4.63(3)	
^{124}Sn	123.9052746(15)	5.79(5)	
^{121}Sb	120.9038180(24)	57.21(5)	121.760(1)
^{123}Sb	122.9042157(22)	42.79(5)	
^{120}Te	119.904020(11)	0.09(1)	127.60(3)
^{122}Te	121.9030471(20)	2.55(12)	
^{123}Te	122.9042730(19)	0.89(3)	
^{124}Te	123.9028195(16)	4.74(14)	
^{125}Te	124.9044247(20)	7.07(15)	
^{126}Te	125.9033055(20)	18.84(25)	
^{127}Te	127.9044614(19)	31.74(8)	
^{130}Te	129.9062228(21)	34.08(62)	
^{127}I	126.904468(4)	100	126.90447(3)
^{124}Xe	123.9058958(21)	0.09(1)	131.293(6)
^{126}Xe	125.904269(7)	0.09(1)	
^{128}Xe	127.9035304(15)	1.92(3)	
^{129}Xe	128.9047795(9)	26.44(24)	
^{130}Xe	129.9035079(10)	4.08(2)	
^{131}Xe	130.9050819(10)	21.18(3)	
^{132}Xe	131.9041545(12)	26.89(6)	
^{134}Xe	133.9053945(9)	10.44(10)	
^{136}Xe	135.907220(8)	8.87(16)	
^{133}Cs	132.905447(3)	100	132.90545(2)
^{130}Ba	129.906310(7)	0.106(1)	137.327(7)
^{132}Ba	131.905056(3)	0.101(1)	
^{134}Ba	133.904503(3)	2.417(18)	

Symbol	Relative Atomic Mass	Abundance	Standard Atomic Weight
^{135}Ba	134.905683(3)	6.592(12)	
^{136}Ba	135.904570(3)	7.854(24)	
^{137}Ba	136.905821(3)	11.232(24)	
^{138}Ba	137.905241(3)	71.698(42)	
^{138}La	137.907107(4)	0.090(1)	138.9055(2)
^{139}La	138.906348(3)	99.910(1)	
^{136}Ce	135.907140(50)	0.185(2)	140.116(1)
^{136}Ce	137.905986(11)	0.251(2)	
^{136}Ce	139.905434(3)	88.450(51)	
^{136}Ce	141.909240(4)	11.114(51)	
^{141}Pr	140.907648(3)	100	140.90765(2)
^{142}Nd	141.907719(3)	27.2(5)	144.24(3)
^{143}Nd	142.909810(3)	12.2(2)	
^{144}Nd	143.910083(3)	23.8(3)	
^{145}Nd	144.912569(3)	8.3(1)	
^{146}Nd	145.913112(3)	17.2(3)	
^{148}Nd	147.916889(3)	5.7(1)	
^{150}Nd	149.920887(4)	5.6(2)	
^{145}Pm	144.912744(4)	[145]	
^{147}Pm	146.915134(3)		
^{144}Sm	143.911995(4)	3.07(7)	150.36(3)
^{147}Sm	146.914893(3)	14.99(18)	
^{148}Sm	147.914818(3)	11.24(10)	
^{149}Sm	148.917180(3)	13.82(7)	
^{150}Sm	149.917271(3)	7.38(1)	
^{152}Sm	151.919728(3)	26.75(16)	
^{154}Sm	153.922205(3)	22.75(29)	
^{151}Eu	150.919846(3)	47.81(3)	151.964(1)
^{153}Eu	152.921226(3)	52.19(3)	
^{152}Gd	151.919788(3)	0.20(1)	157.25(3)
^{154}Gd	153.920862(3)	2.18(3)	
^{155}Gd	154.922619(3)	14.80(12)	
^{156}Gd	155.922120(3)	20.47(9)	
^{157}Gd	156.923957(3)	15.65(2)	
^{158}Gd	157.924101(3)	24.84(7)	
^{160}Gd	159.927051(3)	21.86(19)	
^{159}Tb	158.925343(3)	100	158.92534(2)
^{156}Dy	155.924278(7)	0.06(1)	162.500(1)
^{158}Dy	157.924405(4)	0.10(1)	
^{160}Dy	159.925194(3)	2.34(8)	
^{161}Dy	160.926930(3)	18.91(24)	
^{162}Dy	161.926795(3)	25.51(26)	
^{163}Dy	163 162.928728(3)	24.90(16)	
^{164}Dy	163.929171(3)	28.18(37)	
^{165}Ho	164.930319(3)	100	164.93032(2)
^{162}Er	161.928775(4)	0.14(1)	167.259(3)
^{164}Er	163.929197(4)	1.61(3)	
^{166}Er	165.930290(3)	33.61(35)	
^{167}Er	166.932045(3)	22.93(17)	
^{168}Er	167.932368(3)	26.78(26)	
^{170}Er	169.935460(3)	14.93(27)	
^{169}Tm	168.934211(3)	100	168.93421(2)
^{168}Yb	167.933894(5)	0.13(1)	173.04(3)
^{170}Yb	169.934759(3)	3.04(15)	
^{171}Yb	170.936322(3)	14.28(57)	
^{172}Yb	171.9363777(30)	21.83(67)	
^{173}Yb	172.9382068(30)	16.13(27)	
^{174}Yb	173.9388581(30)	31.83(92)	

Symbol	Relative Atomic Mass	Abundance	Standard Atomic Weight
^{176}Yb	175.942568(3)	12.76(41)	
^{175}Lu	174.9407679(28)	97.41(2)	174.967(1)
^{176}Lu	175.9426824(28)	2.59(2)	
^{174}Hf	173.940040(3)	0.16(1)	178.49(2)
^{176}Hf	175.9414018(29)	5.26(7)	
^{177}Hf	176.9432200(27)	18.60(9)	
^{178}Hf	177.9436977(27)	27.28(7)	
^{179}Hf	178.9458151(27)	13.62(2)	
^{180}Hf	179.9465488(27)	35.08(16)	
^{180}Ta	179.947466(3)	0.012(2)	180.9479(1)
^{181}Ta	180.947996(3)	99.988(2)	
^{180}W	179.946706(5)	0.12(1)	183.84(1)
^{182}W	181.948206(3)	26.50(16)	
^{183}W	182.9502245(29)	14.31(4)	
^{184}W	183.9509326(29)	30.64(2)	
^{186}W	185.954362(3)	28.43(19)	
^{185}Re	184.9529557(30)	37.40(2)	186.207(1)
^{187}Re	186.9557508(30)	62.60(2)	
^{184}Os	183.952491(3)	0.02(1)	190.23(3)
^{186}Os	185.953838(3)	1.59(3)	
^{187}Os	186.9557479(30)	1.96(2)	
^{188}Os	187.9558360(30)	13.24(8)	
^{189}Os	188.9581449(30)	16.15(5)	
^{190}Os	189.958445(3)	26.26(2)	
^{192}Os	191.961479(4)	40.78(19)	
^{191}Ir	190.960591(3)	37.3(2)	192.217(3)
^{193}Ir	192.962924(3)	62.7(2)	
^{190}Pt	189.959930(7)	0.014(1)	195.078(2)
^{192}Pt	191.961035(4)	0.782(7)	
^{194}Pt	193.962664(3)	32.967(99)	
^{195}Pt	194.964774(3)	33.832(10)	
^{196}Pt	195.964935(3)	25.242(41)	
^{198}Pt	197.967876(4)	7.163(55)	
^{197}Au	196.966552(3)	100	196.96655(2)
^{196}Hg	195.965815(4)	0.15(1)	200.59(2)
^{198}Hg	197.966752(3)	9.97(20)	
^{199}Hg	198.968262(3)	16.87(22)	
^{200}Hg	199.968309(3)	23.10(19)	
^{201}Hg	200.970285(3)	13.18(9)	
^{202}Hg	201.970626(3)	29.86(26)	
^{204}Hg	203.973476(3)	6.87(15)	
^{203}Tl	202.972329(3)	29.524(14)	204.3833(2)
^{205}Tl	204.974412(3)	70.476(14)	
^{204}Pb	203.973029(3)	1.4(1)	207.2(1)
^{206}Pb	205.974449(3)	24.1(1)	
^{207}Pb	206.975881(3)	22.1(1)	
^{208}Pb	207.976636(3)	52.4(1)	
^{209}Bi	208.980383(3)	100	208.98038(2)
^{209}Po	208.982416(3)		[209]
^{210}Po	209.982857(3)		
^{210}At	209.987131(9)		[210]
^{211}At	210.987481(4)		
^{211}Rn	210.990585(8)		[222]
^{220}Rn	220.0113841(29)		
^{222}Rn	222.0175705(27)		
^{223}Fr	223.0197307(29)		[223]
^{223}Ra	223.018497(3)		[226]
^{224}Ra	224.0202020(29)		

(continued)

Symbol	Relative Atomic Mass	Abundance	Standard Atomic Weight
^{226}Ra	226.0254026(27)		
^{228}Ra	228.0310641(27)		
^{227}Ac	227.0277470(29)		[227]
^{230}Th	230.0331266(22)		232.0381(1)
^{232}Th	232.0380504(22)	100	
^{231}Pa	231.0358789(28)	100	231.03588(2)
^{233}U	233.039628(3)		238.02891(3)
^{234}U	234.0409456(21)	0.0055(2)	
^{235}U	235.0439231(21)	0.7200(51)	
^{236}U	236.0455619(21)		
^{238}U	238.0507826(21)	99.2745(106)	

Symbol	Relative Atomic Mass	Abundance	Standard Atomic Weight
^{237}Np	237.0481673(21)		[237]
^{239}Np	239.0529314(23)		
^{238}Pu	238.0495534(21)		[244]
^{239}Pu	239.0521565(21)		
^{240}Pu	240.0538075(21)		
^{241}Pu	241.0568453(21)		
^{242}Pu	242.0587368(21)		
^{244}Pu	244 244.064198(5)		
^{241}Am	241.0568229(21)		[243]
^{243}Am	243.0613727(23)		

APPENDIX III

FUNDAMENTAL PHYSICAL CONSTANTS

Quantity	Symbol	Value	Unit
Electron mass	m_e	$9.10938215(45) \times 10^{-31}$	kg
Proton mass	m_p	$1.672621637(83) \times 10^{-27}$	kg
Proton–electron mass ratio	m_p/m_e	1836.15267247(80)	
Speed of light in vacuum	c	2.99792458×10^{8}	$m\ s^{-1}$
Magnetic constant	μ_0	$4\pi \times 10^{-7}$	$N\ A^{-2}$
Rydberg constant $\alpha^2 m_e c/2h$	R_∞	10,973 731.568527(73)	m^{-1}
Avogadro constant	N_A, L	$6.02214179(30) \times 10^{23}$	mol^{-1}
Faraday constant $N_A e$	F	96,485.3399(24)	$C\ mol^{-1}$
Molar gas constant	R	8.314472(15)	$J\ mol^{-1}\ K^{-1}$
Boltzmann constant R/N_A	k	$1.3806504(24) \times 10^{-23}$	$J\ K^{-1}$
Stefan–Boltzmann constant $(\pi^2/60)k^4/\hbar^3c^2$	σ	$5.670400(40) \times 10^{-8}$	$W\ m^{-2}\ K^{-4}$
Electric constant $1/\mu_0 c^2$	ϵ_0	$8.854187817 \times 10^{-12}$	$F\ m^{-1}$
Newtonian gravitation constant	G	$6.67428(67) \times 10^{-11}$	$m^3\ kg^{-1}\ s^{-2}$
Electron volt (e/C) J	eV	$1.602176487(40) \times 10^{-19}$	J
Unified atomic mass unit	u	$1.660538782(83) \times 10^{-27}$	kg
Plank constant	h	$6.62606896(33) \times 10^{-34}$	J s
Plank constant $h/2\pi$	$\hbar$	$1.054571628(53) \times 10^{-34}$	J s
Elementary charge	e	$1.602176487(40) \times 10^{-19}$	C
Magnetic flux quantum $h/2e$	Φ_0	$2.067833667(52) \times 10^{-15}$	Wb
Conductance quantum $2e^2/h$	G_0	$7.7480917004(53) \times 10^{-5}$	S

Analytical Chemistry: A Chemist and Laboratory Technician's Toolkit, First Edition. Bryan M. Ham and Aihui MaHam.

APPENDIX IV

REDOX HALF REACTIONS

Half Reaction			
Oxidant	⇌	Reductant	$E°$ (V)
$Sr^+ + e^-$	⇌	Sr	−3.80
$Ca^+ + e^-$	⇌	Ca	−3.80
$Pr^{3+} + e^-$	⇌	Pr^{2+}	−3.1
$^3/_2 N_2(g) + H^+ + e^-$	⇌	$HN_3(aq)$	−3.09
$Li^+ + e^-$	⇌	Li(s)	−3.0401
$N_2(g) + 4H_2O + 2e^-$	⇌	$2NH_2OH(aq) + 2OH^-$	−3.04
$Cs^+ + e^-$	⇌	Cs(s)	−3.026
$Ca(OH)_2 + 2e^-$	⇌	$Ca + 2\ OH^-$	−3.02
$Er^{3+} + e^-$	⇌	Er^{2+}	−3.0
$Ba(OH)_2 + 2e^-$	⇌	$Ba + 2\ OH^-$	−2.99
$Rb^+ + e^-$	⇌	Rb(s)	−2.98
$K^+ + e^-$	⇌	K(s)	−2.931
$Ba^{2+} + 2e^-$	⇌	Ba(s)	−2.912
$La(OH)_3(s) + 3e^-$	⇌	$La(s) + 3OH^-$	−2.90
$Fr^+ + e^-$	⇌	Fr	−2.9
$Sr^{2+} + 2e^-$	⇌	Sr(s)	−2.899
$Sr(OH)_2 + 2e^-$	⇌	$Sr + 2\ OH^-$	−2.88
$Ca^{2+} + 2e^-$	⇌	Ca(s)	−2.868
$Eu^{2+} + 2e^-$	⇌	Eu(s)	−2.812
$Ra^{2+} + 2e^-$	⇌	Ra(s)	−2.8
$Ho^{3+} + e^-$	⇌	Ho^{2+}	−2.8
$Bk^{3+} + e^-$	⇌	Bk^{2+}	−2.8
$Yb^{2+} + 2e^-$	⇌	Yb	−2.76
$Na^+ + e^-$	⇌	Na(s)	−2.71
$Mg^+ + e^-$	⇌	Mg	−2.70
$Nd^{3+} + e^-$	⇌	Nd^{2+}	−2.7
$Mg(OH)_2 + 2e^-$	⇌	$Mg + 2\ OH^-$	−2.690
$Sm^{2+} + 2e^-$	⇌	Sm	−2.68
$Be_2O_3^{2-} + 3\ H_2O + 4e^-$	⇌	$2\ Be + 6\ OH^-$	−2.63
$Pm^{3+} + e^-$	⇌	Pm^{2+}	−2.6
$Dy^{3+} + e^-$	⇌	Dy^{2+}	−2.6
$No^{2+} + 2e^-$	⇌	No	−2.50

Half Reaction			
Oxidant	⇌	Reductant	$E°$ (V)
$HfO(OH)_2 + H_2O + 4e^-$	⇌	$Hf + 4\ OH^-$	−2.50
$Th(OH)_4 + 4e^-$	⇌	$Th + 4\ OH^-$	−2.48
$Md^{2+} + 2e^-$	⇌	Md	−2.40
$Tm^{2+} + 2e^-$	⇌	Tm	−2.4
$La^{3+} + 3e^-$	⇌	La(s)	−2.379
$Y^{3+} + 3e^-$	⇌	Y(s)	−2.372
$Mg^{2+} + 2e^-$	⇌	Mg(s)	−2.372
$ZrO(OH)_2(s) + H_2O + 4e^-$	⇌	$Zr(s) + 4OH^-$	−2.36
$Pr^{3+} + 3e^-$	⇌	Pr	−2.353
$Ce^{3+} + 3e^-$	⇌	Ce	−2.336
$Er^{3+} + 3e^-$	⇌	Er	−2.331
$Ho^{3+} + 3e^-$	⇌	Ho	−2.33
$H_2AlO_3^- + H_2O + 3e^-$	⇌	$Al + 4\ OH^-$	−2.33
$Nd^{3+} + 3e^-$	⇌	Nd	−2.323
$Tm^{3+} + 3e^-$	⇌	Tm	−2.319
$Al(OH)_3(s) + 3e^-$	⇌	$Al(s) + 3OH^-$	−2.31
$Sm^{3+} + 3e^-$	⇌	Sm	−2.304
$Fm^{2+} + 2e^-$	⇌	Fm	−2.30
$Am^{3+} + e^-$	⇌	Am^{2+}	−2.3
$Dy^{3+} + 3e^-$	⇌	Dy	−2.295
$Lu^{3+} + 3e^-$	⇌	Lu	−2.28
$Tb^{3+} + 3e^-$	⇌	Tb	−2.28
$Gd^{3+} + 3e^-$	⇌	Gd	−2.279
$H_2 + 2e^-$	⇌	$2H^-$	−2.23
$Es^{2+} + 2e^-$	⇌	Es	−2.23
$Pm^{2+} + 2e^-$	⇌	Pm	−2.2
$Tm^{3+} + e^-$	⇌	Tm^{2+}	−2.2
$Dy^{2+} + 2e^-$	⇌	Dy	−2.2
$Ac^{3+} + 3e^-$	⇌	Ac	−2.20
$Yb^{3+} + 3e^-$	⇌	Yb	−2.19
$Cf^{2+} + 2e^-$	⇌	Cf	−2.12
$Nd^{2+} + 2e^-$	⇌	Nd	−2.1

(continued)

Analytical Chemistry: A Chemist and Laboratory Technician's Toolkit, First Edition. Bryan M. Ham and Aihui MaHam.

Half Reaction			
Oxidant	⇌	Reductant	$E°$ (V)
$Ho^{2+} + 2e^-$	⇌	Ho	−2.1
$Sc^{3+} + 3e^-$	⇌	Sc(s)	−2.077
$AlF_6^{3-} + 3e^-$	⇌	$Al + 6F^-$	−2.069
$Am^{3+} + 3e^-$	⇌	Am	−2.048
$Cm^{3+} + 3e^-$	⇌	Cm	−2.04
$Pu^{3+} + 3e^-$	⇌	Pu	−2.031
$Pr^{2+} + 2e^-$	⇌	Pr	−2.0
$Er^{2+} + 2e^-$	⇌	Er	−2.0
$Eu^{3+} + 3e^-$	⇌	Eu	−1.991
$Lr^{3+} + 3e^-$	⇌	Lr	−1.96
$Cf^{3+} + 3e^-$	⇌	Cf	−1.94
$Es^{3+} + 3e^-$	⇌	Es	−1.91
$Pa^{4+} + e^-$	⇌	Pa^{3+}	−1.9
$Am^{2+} + 2e^-$	⇌	Am	−1.9
$Th^{4+} + 4e^-$	⇌	Th	−1.899
$Fm^{3+} + 3e^-$	⇌	Fm	−1.89
$Np^{3+} + 3e^-$	⇌	Np	−1.856
$Be^{2+} + 2e^-$	⇌	Be	−1.847
$H_2PO_2^- + e^-$	⇌	$P + 2\,OH^-$	−1.82
$U^{3+} + 3e^-$	⇌	U	−1.798
$Sr^{2+} + 2e^-$	⇌	Sr/Hg	−1.793
$H_2BO_3^- + H_2O + 3e^-$	⇌	$B + 4\,OH^-$	−1.79
$ThO_2 + 4H^+ + 4e^-$	⇌	$Th + 2\,H_2O$	−1.789
$HfO^{2+} + 2\,H^+ + 4e^-$	⇌	$Hf + H_2O$	−1.724
$HPO_3^{2-} + 2\,H_2O + 3e^-$	⇌	$P + 5\,OH^-$	−1.71
$SiO_3^{2-} + H_2O + 4e^-$	⇌	$Si + 6\,OH^-$	−1.697
$Al^{3+} + 3e^-$	⇌	Al(s)	−1.662
$Ti^{2+} + 2e^-$	⇌	Ti(s)	−1.63
$ZrO_2(s) + 4H^+ + 4e^-$	⇌	$Zr(s) + 2H_2O$	−1.553
$Zr^{4+} + 4e^-$	⇌	Zr(s)	−1.45
$Ti^{3+} + 3e^-$	⇌	Ti(s)	−1.37
$TiO(s) + 2H^+ + 2e^-$	⇌	$Ti(s) + H_2O$	−1.31
$Ti_2O_3(s) + 2H^+ + 2e^-$	⇌	$2\,TiO(s) + H_2O$	−1.23
$Zn(OH)_4^{2-} + 2e^-$	⇌	$Zn(s) + 4OH^-$	−1.199
$Mn^{2+} + 2e^-$	⇌	Mn(s)	−1.185
$Fe(CN)_6^{4-} + 6H^+ + 2e^-$	⇌	$Fe(s) + 6HCN(aq)$	−1.16
$Te(s) + 2e^-$	⇌	Te^{2-}	−1.143
$V^{2+} + 2e^-$	⇌	V(s)	−1.13
$Nb^{3+} + 3e^-$	⇌	Nb(s)	−1.099
$Sn(s) + 4H^+ + 4e^-$	⇌	$SnH_4(g)$	−1.07
$SiO_2(s) + 4H^+ + 4e^-$	⇌	$Si(s) + 2H_2O$	−0.91
$B(OH)_3(aq) + 3H^+ + 3e^-$	⇌	$B(s) + 3H_2O$	−0.89
$Fe(OH)_2(s) + 2e^-$	⇌	$Fe(s) + 2OH^-$	−0.89
$Fe_2O_3(s) + 3H_2O + 2e^-$	⇌	$2Fe(OH)_2(s) + 2OH^-$	−0.86
$TiO^{2+} + 2H^+ + 4e^-$	⇌	$Ti(s) + H_2O$	−0.86
$2H_2O + 2e^-$	⇌	$H_2(g) + 2OH^-$	−0.8277
$Bi(s) + 3H^+ + 3e^-$	⇌	BiH_3	−0.8
$Zn^{2+} + 2e^-$	⇌	Zn(Hg)	−0.7628
$Zn^{2+} + 2e^-$	⇌	Zn(s)	−0.7618
$Ta_2O_5(s) + 10H^+ + 10e^-$	⇌	$2\,Ta(s) + 5H_2O$	−0.75
$Cr^{3+} + 3e^-$	⇌	Cr(s)	−0.74
$[Au(CN)_2]^- + e^-$	⇌	$Au(s) + 2CN^-$	−0.60
$Ta^{3+} + 3e^-$	⇌	Ta(s)	−0.6
$PbO(s) + H_2O + 2e^-$	⇌	$Pb(s) + 2OH^-$	−0.58
$2\,TiO_2(s) + 2H^+ + 2e^-$	⇌	$Ti_2O_3(s) + H_2O$	−0.56
$Ga^{3+} + 3e^-$	⇌	Ga(s)	−0.53
$U^{4+} + e^-$	⇌	U^{3+}	−0.52
$H_3PO_2(aq) + H^+ + e^-$	⇌	$P(white) + 2H_2O$	−0.508
$H_3PO_3(aq) + 2H^+ + 2e^-$	⇌	$H_3PO_2(aq) + H_2O$	−0.499

Half Reaction			
Oxidant	⇌	Reductant	$E°$ (V)
$H_3PO_3(aq) + 3H^+ + 3e^-$	⇌	$P(red) + 3H_2O$	−0.454
$Fe^{2+} + 2e^-$	⇌	Fe(s)	−0.44
$2\,CO_2(g) + 2H^+ + 2e^-$	⇌	HOOCCOOH(aq)	−0.43
$Cr^{3+} + e^-$	⇌	Cr^{2+}	−0.42
$Cd^{2+} + 2e^-$	⇌	Cd(s)	−0.40
$GeO_2(s) + 2H^+ + 2e^-$	⇌	$GeO(s) + H_2O$	−0.37
$Cu_2O(s) + H_2O + 2e^-$	⇌	$2\,Cu(s) + 2OH^-$	−0.360
$PbSO_4(s) + 2e^-$	⇌	$Pb(s) + SO_4^{2-}$	−0.3588
$PbSO_4(s) + 2e^-$	⇌	$Pb(Hg) + SO_4^{2-}$	−0.3505
$Eu^{3+} + e^-$	⇌	Eu^{2+}	−0.35
$In^{3+} + 3e^-$	⇌	In(s)	−0.34
$Tl^+ + e^-$	⇌	Tl(s)	−0.34
$Ge(s) + 4H^+ + 4e^-$	⇌	$GeH_4(g)$	−0.29
$Co^{2+} + 2e^-$	⇌	Co(s)	−0.28
$H_3PO_4(aq) + 2H^+ + 2e^-$	⇌	$H_3PO_3(aq) + H_2O$	−0.276
$V^{3+} + e^-$	⇌	V^{2+}	−0.26
$Ni^{2+} + 2e^-$	⇌	Ni(s)	−0.25
$As(s) + 3H^+ + 3e^-$	⇌	$AsH_3(g)$	−0.23
$AgI(s) + e^-$	⇌	$Ag(s) + I^-$	−0.15224
$MoO_2(s) + 4H^+ + 4e^-$	⇌	$Mo(s) + 2H_2O$	−0.15
$Si(s) + 4H^+ + 4e^-$	⇌	$SiH_4(g)$	−0.14
$Sn^{2+} + 2e^-$	⇌	Sn(s)	−0.13
$O_2(g) + H^+ + e^-$	⇌	$HO_2\cdot(aq)$	−0.13
$Pb^{2+} + 2e^-$	⇌	Pb(s)	−0.13
$WO_2(s) + 4H^+ + 4e^-$	⇌	$W(s) + 2H_2O$	−0.12
$P(\mathit{red}) + 3H^+ + 3e^-$	⇌	$PH_3(g)$	−0.111
$CO_2(g) + 2H^+ + 2e^-$	⇌	HCOOH(aq)	−0.11
$Se(s) + 2H^+ + 2e^-$	⇌	$H_2Se(g)$	−0.11
$CO_2(g) + 2H^+ + 2e^-$	⇌	$CO(g) + H_2O$	−0.11
$SnO(s) + 2H^+ + 2e^-$	⇌	$Sn(s) + H_2O$	−0.10
$SnO_2(s) + 2H^+ + 2e^-$	⇌	$SnO(s) + H_2O$	−0.09
$WO_3(aq) + 6H^+ + 6e^-$	⇌	$W(s) + 3H_2O$	−0.09
$P(\mathit{white}) + 3H^+ + 3e^-$	⇌	$PH_3(g)$	−0.063
$Fe^{3+} + 3e^-$	⇌	Fe(s)	−0.04
$HCOOH(aq) + 2H^+ + 2e^-$	⇌	$HCHO(aq) + H_2O$	−0.03
$2H^+ + 2e^-$	⇌	$H_2(g)$	0.0000
$AgBr(s) + e^-$	⇌	$Ag(s) + Br^-$	+0.07133
$S_4O_6^{2-} + 2e^-$	⇌	$2\,S_2O_3^{2-}$	+0.08
$Fe_3O_4(s) + 8H^+ + 8e^-$	⇌	$3\,Fe(s) + 4H_2O$	+0.085
$N_2(g) + 2H_2O + 6H^+ + 6e^-$	⇌	$2\,NH_4OH(aq)$	+0.092
$HgO(s) + H_2O + 2e^-$	⇌	$Hg(l) + 2OH^-$	+0.0977
$Cu(NH_3)_4^{2+} + e^-$	⇌	$Cu(NH_3)_2^+ + 2NH_3$	+0.10
$Ru(NH_3)_6^{3+} + e^-$	⇌	$Ru(NH_3)_6^{2+}$	+0.10
$N_2H_4(aq) + 4H_2O + 2e^-$	⇌	$2\,NH_4^+ + 4OH^-$	+0.11
$H_2MoO_4(aq) + 6H^+ + 6e^-$	⇌	$Mo(s) + 4H_2O$	+0.11
$Ge^{4+} + 4e^-$	⇌	Ge(s)	+0.12
$C(s) + 4H^+ + 4e^-$	⇌	$CH_4(g)$	+0.13
$HCHO(aq) + 2H^+ + 2e^-$	⇌	$CH_3OH(aq)$	+0.13
$S(s) + 2H^+ + 2e^-$	⇌	$H_2S(g)$	+0.14
$Sn^{4+} + 2e^-$	⇌	Sn^{2+}	+0.15
$Cu^{2+} + e^-$	⇌	Cu^+	+0.159
$HSO_4^- + 3H^+ + 2e^-$	⇌	$SO_2(aq) + 2H_2O$	+0.16
$UO_2^{2+} + e^-$	⇌	UO_2^+	+0.163
$SO_4^{2-} + 4H^+ + 2e^-$	⇌	$SO_2(aq) + 2H_2O$	+0.17
$TiO^{2+} + 2H^+ + e^-$	⇌	$Ti^{3+} + H_2O$	+0.19
$SbO^+ + 2H^+ + 3e^-$	⇌	$Sb(s) + H_2O$	+0.20
$AgCl(s) + e^-$	⇌	$Ag(s) + Cl^-$	+0.22233
$H_3AsO_3(aq) + 3H^+ + 3e^-$	⇌	$As(s) + 3H_2O$	+0.24
$GeO(s) + 2H^+ + 2e^-$	⇌	$Ge(s) + H_2O$	+0.26

Half Reaction			
Oxidant	⇌	Reductant	$E°$ (V)
$UO_2^+ + 4H^+ + e^-$	⇌	$U^{4+} + 2H_2O$	+0.273
$Re^{3+} + 3e^-$	⇌	Re(s)	+0.300
$Bi^{3+} + 3e^-$	⇌	Bi(s)	+0.308
$VO^{2+} + 2H^+ + e^-$	⇌	$V^{3+} + H_2O$	+0.34
$Cu^{2+} + 2e^-$	⇌	Cu(s)	+0.340
$[Fe(CN)_6]^{3-} + e^-$	⇌	$[Fe(CN)_6]^{4-}$	+0.36
$O_2(g) + 2H_2O + 4e^-$	⇌	$4OH^-(aq)$	+0.40
$H_2MoO_4 + 6H^+ + 3e^-$	⇌	$Mo^{3+} + 2H_2O$	+0.43
$CH_3OH(aq) + 2H^+ + 2e^-$	⇌	$CH_4(g) + H_2O$	+0.50
$SO_2(aq) + 4H^+ + 4e^-$	⇌	$S(s) + 2H_2O$	+0.50
$Cu^+ + e^-$	⇌	Cu(s)	+0.520
$CO(g) + 2H^+ + 2e^-$	⇌	$C(s) + H_2O$	+0.52
$I_3^- + 2e^-$	⇌	$3I^-$	+0.53
$I_2(s) + 2e^-$	⇌	$2I^-$	+0.54
$[AuI_4]^- + 3e^-$	⇌	$Au(s) + 4I^-$	+0.56
$H_3AsO_4(aq) + 2H^+ + 2e^-$	⇌	$H_3AsO_3(aq) + H_2O$	+0.56
$[AuI_2]^- + e^-$	⇌	$Au(s) + 2I^-$	+0.58
$MnO_4^- + 2H_2O + 3e^-$	⇌	$MnO_2(s) + 4OH^-$	+0.59
$S_2O_3^{2-} + 6H^+ + 4e^-$	⇌	$2S(s) + 3H_2O$	+0.60
$Fc^+ + e^-$	⇌	Fc(s)	+0.641
$H_2MoO_4(aq) + 2H^+ + 2e^-$	⇌	$MoO_2(s) + 2H_2O$	+0.65
O, O $+ 2H^+ + 2e^-$	⇌	HO, OH	+0.6992
$O_2(g) + 2H^+ + 2e^-$	⇌	$H_2O_2(aq)$	+0.70
$Tl^{3+} + 3e^-$	⇌	Tl(s)	+0.72
$PtCl_6^{2-} + 2e^-$	⇌	$PtCl_4^{2-} + 2Cl^-$	+0.726
$H_2SeO_3(aq) + 4H^+ + 4e^-$	⇌	$Se(s) + 3H_2O$	+0.74
$PtCl_4^{2-} + 2e^-$	⇌	$Pt(s) + 4Cl^-$	+0.758
$Fe^{3+} + e^-$	⇌	Fe^{2+}	+0.77
$Ag^+ + e^-$	⇌	Ag(s)	+0.7996
$Hg_2^{2+} + 2e^-$	⇌	2Hg(l)	+0.80
$NO_3^-(aq) + 2H^+ + e^-$	⇌	$NO_2(g) + H_2O$	+0.80
$2FeO_4^{2-} + 5H_2O + 6e^-$	⇌	$Fe_2O_3(s) + 10OH^-$	+0.81
$[AuBr_4]^- + 3e^-$	⇌	$Au(s) + 4Br^-$	+0.85
$Hg^{2+} + 2e^-$	⇌	Hg(l)	+0.85
$[IrCl_6]^{2-} + e^-$	⇌	$[IrCl_6]^{3-}$	+0.87[4]
$MnO_4^- + H^+ + e^-$	⇌	$HMnO_4^-$	+0.90
$2Hg^{2+} + 2e^-$	⇌	Hg_2^{2+}	+0.91
$Pd^{2+} + 2e^-$	⇌	Pd(s)	+0.915
$[AuCl_4]^- + 3e^-$	⇌	$Au(s) + 4Cl^-$	+0.93
$MnO_2(s) + 4H^+ + e^-$	⇌	$Mn^{3+} + 2H_2O$	+0.95
$[AuBr_2]^- + e^-$	⇌	$Au(s) + 2Br^-$	+0.96
$[HXeO_6]^{3-} + 2H_2O + 2e^- +$	⇌	$[HXeO_4]^- + 4OH^-$	+0.99
$H_6TeO_6(aq) + 2H^+ + 2e^-$	⇌	$TeO_2(s) + 4H_2O$	+1.02
$Br_2(l) + 2e^-$	⇌	$2Br^-$	+1.066
$Br_2(aq) + 2e^-$	⇌	$2Br^-$	+1.0873
$IO_3^- + 5H^+ + 4e^-$	⇌	$HIO(aq) + 2H_2O$	+1.13
$[AuCl_2]^- + e^-$	⇌	$Au(s) + 2Cl^-$	+1.15

Half Reaction			
Oxidant	⇌	Reductant	$E°$ (V)
$HSeO_4^- + 3H^+ + 2e^-$	⇌	$H_2SeO_3(aq) + H_2O$	+1.15
$Ag_2O(s) + 2H^+ + 2e^-$	⇌	$2Ag(s) + H_2O$	+1.17
$ClO_3^- + 2H^+ + e^-$	⇌	$ClO_2(g) + H_2O$	+1.18
$[HXeO_6]^{3-} + 5H_2O + 8e^-$	⇌	$Xe(g) + 11OH^-$	+1.18
$Pt^{2+} + 2e^-$	⇌	Pt(s)	+1.188
$ClO_2(g) + H^+ + e^-$	⇌	$HClO_2(aq)$	+1.19
$2IO_3^- + 12H^+ + 10e^-$	⇌	$I_2(s) + 6H_2O$	+1.20
$ClO_4^- + 2H^+ + 2e^-$	⇌	$ClO_3^- + H_2O$	+1.20
$O_2(g) + 4H^+ + 4e^-$	⇌	$2H_2O$	+1.229
$MnO_2(s) + 4H^+ + 2e^-$	⇌	$Mn^{2+} + 2H_2O$	+1.23
$[HXeO_4]^- + 3H_2O + 6e^-$	⇌	$Xe(g) + 7OH^-$	+1.24
$Tl^{3+} + 2e^-$	⇌	Tl^+	+1.25
$Cr_2O_7^{2-} + 14H^+ + 6e^-$	⇌	$2Cr^{3+} + 7H_2O$	+1.33
$Cl_2(g) + 2e^-$	⇌	$2Cl^-$	+1.36
$CoO_2(s) + 4H^+ + e^-$	⇌	$Co^{3+} + 2H_2O$	+1.42
$2NH_3OH^+ + H^+ + 2e^-$	⇌	$N_2H_5^+ + 2H_2O$	+1.42
$2HIO(aq) + 2H^+ + 2e^-$	⇌	$I_2(s) + 2H_2O$	+1.44
$Ce^{4+} + e^-$	⇌	Ce^{3+}	+1.44
$BrO_3^- + 5H^+ + 4e^-$	⇌	$HBrO(aq) + 2H_2O$	+1.45
ß-$PbO_2(s) + 4H^+ + 2e^-$	⇌	$Pb^{2+} + 2H_2O$	+1.460
a-$PbO_2(s) + 4H^+ + 2e^-$	⇌	$Pb^{2+} + 2H_2O$	+1.468
$2BrO_3^- + 12H^+ + 10e^-$	⇌	$Br_2(l) + 6H_2O$	+1.48
$2ClO_3^- + 12H^+ + 10e^-$	⇌	$Cl_2(g) + 6H_2O$	+1.49
$MnO_4^- + 8H^+ + 5e^-$	⇌	$Mn^{2+} + 4H_2O$	+1.51
$HO_2^{\bullet} + H^+ + e^-$	⇌	$H_2O_2(aq)$	+1.51
$Au^{3+} + 3e^-$	⇌	Au(s)	+1.52
$NiO_2(s) + 4H^+ + 2e^-$	⇌	$Ni^{2+} + 2OH^-$	+1.59
$2HClO(aq) + 2H^+ + 2e^-$	⇌	$Cl_2(g) + 2H_2O$	+1.63
$Ag_2O_3(s) + 6H^+ + 4e^-$	⇌	$2Ag^+ + 3H_2O$	+1.67
$HClO_2(aq) + 2H^+ + 2e^-$	⇌	$HClO(aq) + H_2O$	+1.67
$Pb^{4+} + 2e^-$	⇌	Pb^{2+}	+1.69
$MnO_4^- + 4H^+ + 3e^-$	⇌	$MnO_2(s) + 2H_2O$	+1.70
$AgO(s) + 2H^+ + e^-$	⇌	$Ag^+ + H_2O$	+1.77
$H_2O_2(aq) + 2H^+ + 2e^-$	⇌	$2H_2O$	+1.78
$Co^{3+} + e^-$	⇌	Co^{2+}	+1.82
$Au^+ + e^-$	⇌	Au(s)	+1.83
$BrO_4^- + 2H^+ + 2e^-$	⇌	$BrO_3^- + H_2O$	+1.85
$Ag^{2+} + e^-$	⇌	Ag^+	+1.98
$S_2O_8^{2-} + 2e^-$	⇌	$2SO_4^{2-}$	+2.010
$O_3(g) + 2H^+ + 2e^-$	⇌	$O_2(g) + H_2O$	+2.075
$HMnO_4^- + 3H^+ + 2e^-$	⇌	$MnO_2(s) + 2H_2O$	+2.09
$XeO_3(aq) + 6H^+ + 6e^-$	⇌	$Xe(g) + 3H_2O$	+2.12
$H_4XeO_6(aq) + 8H^+ + 8e^-$	⇌	$Xe(g) + 6H_2O$	+2.18
$FeO_4^{2-} + 3e^- + 8H^+$	⇌	$Fe^{3+} + 4H_2O$	+2.20
$XeF_2(aq) + 2H^+ + 2e^-$	⇌	$Xe(g) + 2HF(aq)$	+2.32
$H_4XeO_6(aq) + 2H^+ + 2e^-$	⇌	$XeO_3(aq) + H_2O$	+2.42
$F_2(g) + 2e^-$	⇌	$2F^-$	+2.87
$F_2(g) + 2H^+ + 2e^-$	⇌	2HF(aq)	+3.05

APPENDIX V

PERIODIC TABLE OF ELEMENTS

Analytical Chemistry: A Chemist and Laboratory Technician's Toolkit, First Edition. Bryan M. Ham and Aihui MaHam.

PERIODIC TABLE OF THE ELEMENTS

http://www.ktf-split.hr/periodni/en/

Key: GROUP IUPAC (13) · GROUP CAS (IIIA) · RELATIVE ATOMIC MASS (1) (10.811) · ATOMIC NUMBER (5) · SYMBOL (B) · ELEMENT NAME (BORON)

Metal · Semimetal · Nonmetal

1 Alkali metal · 2 Alkaline earth metal · Transition metals · Lanthanide · Actinide · 16 Chalcogens element · 17 Halogens element · 18 Noble gas

STANDARD STATE (25 °C; 101 kPa): Ne - gas · Fe - solid · Ga - liquid · Tc - synthetic

PERIOD	1 IA	2 IIA	3 IIIB	4 IVB	5 VB	6 VIB	7 VIIB	8 (VIIIB)	9 VIIIB	10 (VIIIB)	11 IB	12 IIB	13 IIIA	14 IVA	15 VA	16 VIA	17 VIIA	18 VIIIA
1	1 1.0079 H HYDROGEN																	2 4.0026 He HELIUM
2	3 6.941 Li LITHIUM	4 9.0122 Be BERYLLIUM											5 10.811 B BORON	6 12.011 C CARBON	7 14.007 N NITROGEN	8 15.999 O OXYGEN	9 18.998 F FLUORINE	10 20.180 Ne NEON
3	11 22.990 Na SODIUM	12 24.305 Mg MAGNESIUM											13 26.982 Al ALUMINIUM	14 28.086 Si SILICON	15 30.974 P PHOSPHORUS	16 32.065 S SULPHUR	17 35.453 Cl CHLORINE	18 39.948 Ar ARGON
4	19 39.098 K POTASSIUM	20 40.078 Ca CALCIUM	21 44.956 Sc SCANDIUM	22 47.867 Ti TITANIUM	23 50.942 V VANADIUM	24 51.996 Cr CHROMIUM	25 54.938 Mn MANGANESE	26 55.845 Fe IRON	27 58.933 Co COBALT	28 58.693 Ni NICKEL	29 63.546 Cu COPPER	30 65.39 Zn ZINC	31 69.723 Ga GALLIUM	32 72.64 Ge GERMANIUM	33 74.922 As ARSENIC	34 78.96 Se SELENIUM	35 79.904 Br BROMINE	36 83.80 Kr KRYPTON
5	37 85.468 Rb RUBIDIUM	38 87.62 Sr STRONTIUM	39 88.906 Y YTTRIUM	40 91.224 Zr ZIRCONIUM	41 92.906 Nb NIOBIUM	42 95.94 Mo MOLYBDENUM	43 (98) Tc TECHNETIUM	44 101.07 Ru RUTHENIUM	45 102.91 Rh RHODIUM	46 106.42 Pd PALLADIUM	47 107.87 Ag SILVER	48 112.41 Cd CADMIUM	49 114.82 In INDIUM	50 118.71 Sn TIN	51 121.76 Sb ANTIMONY	52 127.60 Te TELLURIUM	53 126.90 I IODINE	54 131.29 Xe XENON
6	55 132.91 Cs CAESIUM	56 137.33 Ba BARIUM	57-71 La-Lu Lanthanide	72 178.49 Hf HAFNIUM	73 180.95 Ta TANTALUM	74 183.84 W TUNGSTEN	75 186.21 Re RHENIUM	76 190.23 Os OSMIUM	77 192.22 Ir IRIDIUM	78 195.08 Pt PLATINUM	79 196.97 Au GOLD	80 200.59 Hg MERCURY	81 204.38 Tl THALLIUM	82 207.2 Pb LEAD	83 208.98 Bi BISMUTH	84 (209) Po POLONIUM	85 (210) At ASTATINE	86 (222) Rn RADON
7	87 (223) Fr FRANCIUM	88 (226) Ra RADIUM	89-103 Ac-Lr Actinide	104 (261) Rf RUTHERFORDIUM	105 (262) Db DUBNIUM	106 (266) Sg SEABORGIUM	107 (264) Bh BOHRIUM	108 (277) Hs HASSIUM	109 (268) Mt MEITNERIUM	110 (281) Uun UNUNNILIUM	111 (272) Uuu UNUNUNIUM	112 (285) Uub UNUNBIUM		114 (289) Uuq UNUNQUADIUM				

LANTHANIDE

57 138.91 La LANTHANUM	58 140.12 Ce CERIUM	59 140.91 Pr PRASEODYMIUM	60 144.24 Nd NEODYMIUM	61 (145) Pm PROMETHIUM	62 150.36 Sm SAMARIUM	63 151.96 Eu EUROPIUM	64 157.25 Gd GADOLINIUM	65 158.93 Tb TERBIUM	66 162.50 Dy DYSPROSIUM	67 164.93 Ho HOLMIUM	68 167.26 Er ERBIUM	69 168.93 Tm THULIUM	70 173.04 Yb YTTERBIUM	71 174.97 Lu LUTETIUM

ACTINIDE

89 (227) Ac ACTINIUM	90 232.04 Th THORIUM	91 231.04 Pa PROTACTINIUM	92 238.03 U URANIUM	93 (237) Np NEPTUNIUM	94 (244) Pu PLUTONIUM	95 (243) Am AMERICIUM	96 (247) Cm CURIUM	97 (247) Bk BERKELIUM	98 (251) Cf CALIFORNIUM	99 (252) Es EINSTEINIUM	100 (257) Fm FERMIUM	101 (258) Md MENDELEVIUM	102 (259) No NOBELIUM	103 (262) Lr LAWRENCIUM

(1) Pure Appl. Chem., 73, No. 4, 667-683 (2001)
Relative atomic mass is shown with five significant figures. For elements have no stable nuclides, the value enclosed in brackets indicates the mass number of the longest-lived isotope of the element.

However three such elements (Th, Pa, and U) do have a characteristic terrestrial isotopic composition, and for these an atomic weight is tabulated.

Editor: Aditya Vardhan (adivar@nettlinx.com)

APPENDIX VI

INSTALLING AND RUNNING PROGRAMS

Installing the ChemTech Program from DVD

1. Open up the DVD in Windows Explorer to illustrate the files on the disk.
2. There is one folder "ChemTech."
3. Copy the ChemTech folder to your C: drive creating the directory C:\ChemTech\.
4. Double click to open the C:\ChemTech folder.
5. There will be five folders (ChemTech, LIMS, MoleWeightCalc, Programs, and QuickTime) and two other files "NewDB.mdb and weights.xls."
6. Double click the ChemTech folder to open the folder.
7. Double click the setup.exe program to install the ChemTech program.
8. Follow the instructions using the suggested default setup.
9. When setup is complete, the ChemTech program can be opened by going to the start menu and open all programs. There should be a ChemTech program to run.

Installing the Molecular Weight Calculator Program

1. Double click to open the C:\ChemTech\MoleWeightCalc folder.
2. Double click the MolecularWeightCalculator.msi program to install the Molecular Weight Calculator program.
3. Follow the instructions using the suggested default setup.
4. When setup is complete, the MolecularWeightCalculator program can be run from within the ChemTech program.

Installing the QuickTimeInstaller Program

1. Double click to open the C:\ChemTech\QuickTime folder.
2. Double click the QuickTimeInstaller.exe program to install QuickTime 7 player.
3. Follow the instructions using the suggested default setup.

Running the LIMS Program from the Stand-Alone File

1. The LIMS program can be opened from the file under the C:\ChemTech folder.
2. Double click the LIMS folder.
3. There is currently only one version of the LIMS program in the directory for Access 2010, which is distributed with Windows 7.
4. Double click the file ChemTechLIMS2010.accdb to open up the LIMS program.
   i. NOTE: If the LIMS database ever becomes corrupted, it can be recopied from the DVD to the C:\ChemTech\LIMS folder.

Analytical Chemistry: A Chemist and Laboratory Technician's Toolkit, First Edition. Bryan M. Ham and Aihui MaHam.

INDEX

abscissa, 110
absolute error, 94, 107
absorbance, 111–116, 118–119, 124, 135–138, 141, 228–231, 235–242, 252, 257, 270–271, 278, 373–374, 396, 437
absorption frequency, 278t
absorption spectra, 221, 228
absorptivity, 229, 238–239, 295
accelerating voltage, 328–329, 331, 492–494
accuracy, 93–94, 101–102, 107, 183, 191, 194, 249, 449, 460–463, 465, 507–508, 517, 522, 524, 588, 626
accurate mass, 508, 526f
acetone, 22, 157, 288, 470, 542
acid, 4, 7, 21–22, 25–26, 106, 108, 149, 150t, 151–152, 155–161, 162t, 167–184, 185t, 186–197, 204–206, 208, 211, 213, 224, 225t, 230, 239, 248, 267t, 270, 275–276, 279, 283–289, 302, 370, 371t, 375, 378t, 395–397, 415
acid adduct, 464, 471, 530, 531f
acid-base theory, 4, 159, 162t
acidic pesticides, 375, 377, 378f, 378t
acidic solution, 152, 161, 186, 189–190, 205, 206t, 248, 288f, 289, 435, 479, 530, 530f, 531f, 543, 543f, 557
acid ionization constant (K_a), 159, 161, 162t, 164, 171, 179, 181
activation barrier, 563, 566f
acylglycerols, 417, 454–455, 479–480, 529–530, 548, 550, 557, 625
acylium ion, 551, 552f, 587, 587f
adenine, 239, 565, 568f, 571f
adenosine triphosphate (ATP), 597, 598f, 607f
adiabatic expansion, 474, 624
aerosol spray, 296, 305
affinities, 351, 366, 368–369, 395, 435–436, 443, 469, 470t
aliquot, 147, 177, 181, 185–186, 195, 208, 260
alkali metal, 168, 190, 299, 456, 548
alpha-beta unsaturated, 552f
alpha-bond, 424
alpha cleavage, 423–424
alpha-cyano-4-hydroxycinnamic acid butylamine, 484f
alpha-cyano-4-hydroxy-trans-cinnamic acid (α-CHCA), 483
alpha-helixes, 580, 582f
alumina, 395
amalgam, 205
ambient, 185, 263–264, 273, 325, 412, 421, 436, 450–451, 471, 472f, 474, 506, 523, 623
amide bond, 587, 588
amino acids, 239, 452, 525f, 559, 569, 579–580, 581f, 583, 587, 594, 599f, 606t, 626
amino-acylium ion, 587
amino-immonium ion, 587
1 2-amino-3-methyl-pentanoic acid, 283
aminopropyl-bonded silica (LC-NH_2), 369
ammonia, 150t, 159, 161, 163, 163t, 167f, 169, 172, 187, 188f, 192, 194, 328, 425–427, 464, 468–470, 489, 587, 589, 624, 626
ammonium adduct $[M + NH_4]^+$, 422, 548, 551
ammonium ion, 162t, 169, 187, 188f, 267t
ammonium phosphate, 165t, 271, 272f, 273t
analysis ToolPak, 143, 145
analytes, 153, 177, 296, 351, 352f, 354–355, 357–359, 361, 364, 368, 397, 414
analytical balance, 15f, 16
analytical chemist, xxiii, 1, 89
angiotensin I, 517
angular divergence, 330, 494
angular momentum, 241, 277, 292
aniline, 163t, 248, 480t, 485, 624
anion, 161f, 164, 175, 188t, 189–190, 369, 381, 467, 470, 551, 553f, 574f
anion exchange chromatography (AEC), 369, 371, 372f, 372t, 591
anisole, 248
anode, 199–201, 207, 251, 297, 298f, 335, 373–374, 418f, 419, 441, 443f, 446f
anthracene, 247, 247f, 259, 260
applications of mass spectrometry, 449, 452

Analytical Chemistry: A Chemist and Laboratory Technician's Toolkit, First Edition. Bryan M. Ham and Aihui MaHam.

arbitrary units, 111f, 112f, 113, 458
archived samples, 214, 220
argon plasma torch, 303–305, 306f
arithmetic mean, 94–96
artifactual modification, 594
ash, 11
ashing, 11, 14
atmospheric conditions, 450, 451, 477
atmospheric pressure, 15, 325, 421, 449, 450f, 451, 468, 471, 473, 476, 478, 484, 507
atmospheric pressure chemical ionization (APCI), 5, 455, 467, 471, 472f, 548, 625
atmospheric pressure ionization (API), 471, 496
atmospheric pressure photo ionization (APPI), 5, 467, 624
atomic absorption spectroscopy, 141, 296, 297f, 303
atomic emission spectroscopy, 303
atomic mass spectrometry, 325
atomizer, 296, 298f, 304
A-type ion, 563
automatic gain control (AGC), 522
autosampler, 375, 376f, 407
average deviation, 94, 96–98, 100–102, 107
average mass, 461, 465
Avogadro's number, 147, 157
axial motion, 506, 515, 516f
axial oscillation, 520–523
axial torch, 304
azo dyes, 221, 225, 237

back titration, 175, 189, 195, 197
balances, xxiv, 6, 7, 15, 15f, 16, 16f
base, 21–22, 83, 149, 150t, 151, 157, 159–161, 163, 168–174, 176–182, 185t, 192, 197, 204
base ionization constant (K_b), 163, 163t, 164, 171
baseline absorption, 354
baseline resolved, 357, 357f, 461
basic solution, 159, 161, 171, 196, 205, 435
bath gas, 504, 505, 507, 522, 527, 587
beakers, 7, 9f, 11, 12f
beam splitter, 244, 253f, 261, 263
bed packing, 381
Beer's law, 119–121, 228–230, 237, 239, 257, 373
bending, 261, 271
benzene, 247, 266t, 275, 463, 464f, 477–478, 480t, 618t, 624
beryllium window, 335, 336f
best fit line, 111–115, 118f, 119, 124, 138
beta-(β)-carotene, 246
bile acids, 455, 530
bioinformatics, 6, 579, 583f, 584, 614, 616, 617f, 618
bioluminescence, 239
biomarkers, 453f, 584, 616, 617f, 618t
biomolecules, 287, 451, 454–455, 462–463, 465, 467, 473, 477, 479–480, 491, 503, 529, 542, 548, 551
biopolymers, 529, 559, 625
biphenyl, 247, 247f
borosilicate, 344
bottom-up proteomics, 579–580, 583, 591, 593–594, 618
Bragg equation, 338, 340f, 345
Bragg parameters, 341, 342t
Bragg's law, 340, 345
Bremsstrahlung effect, 335
broken glass boxes, 29
Brønsted–Lowry acid-base theory, 160
Brønsted–Lowry definition, 160
b-type ion, 600
B-type ion, 562
buffer, 4, 153, 159, 168–173, 185, 192, 194, 298f, 365f, 369, 371–372, 436, 441, 443, 538, 597
Bunsen burner, 7, 11, 13f
butyl, 22, 379t
butyric acid, 162t, 485, 600

calculators hand held, computer (Window's), xxiii, 2, 3, 86, 89, 97, 110, 171
calibrated test electrode, 201
calibrating, 329, 460, 493
calibration, 8, 10, 93, 98, 102, 104, 106, 109, 111, 113–114, 124, 135, 147, 157, 168, 206, 249, 302f, 329, 429–430, 431f, 460, 462, 493, 502
cancer, 234–235, 255, 256f, 436, 438, 440–441, 447f, 448, 455, 616, 617f, 618t, 619
capacity, 29, 85t, 169, 381
capacity factor, 361, 399
capillary, 297f, 361–362, 368
  column, 402f, 406, 413, 429f
  effects, 395
  inlet, 411, 414
capillary zone electrophoresis (CZE), 435, 517
carbohydrate, 276, 381, 449, 455, 559–560
  fragmentation, 561, 609f
C-NH_2 aminopropyl bonded silica, 369, 381
carbon C, 2, 4f, 147–148, 158
carbon isotope, 147
carbon-13 NMR, 283–286
carotenoids, 351
carrier gas, 361, 401, 402f, 403, 410, 412–413, 416f, 420
Cartesian coordinate system, 109–110, 123, 501
cathode, 199–201, 207, 251, 297, 335, 441, 443f, 446f
  lamp, 298f, 300–301
cation, 190
cation exchange chromatography (CEC), 369, 370f, 371t, 436, 448t
cationize, 456
cationizing agent, 426
C-18 packing, 381
cell potential, 200–201, 204t, 207–208
cells in Excel®, 126f, 127, 129
cellulose, 559, 560f, 561, 625
central tendency of data, 94–95
centrifugal force, 328–330, 492, 494
centripetal force, 328, 492
cerebrospinal fluid, 425, 616, 626
certificate of analysis (C of A), xxiii, 1, 5, 209, 220
change in enthalpy (ΔH^o), 469–470, 478, 489
channeltron, 506
charge-coupled device (CCD), 252
charge injection device (CID), 310
charge remote fragmentation, 613
charge state, 573, 580, 585, 591, 592f, 593f
charged residue model, 473, 489
Chart Wizard, 130–132, 136–137
ChemDraw, 289, 290f, 544, 545f
chemical ionization (CI), 5, 422, 455, 467–472, 489
chemicals, 2, 19–23, 27, 346
chemical shift, 281–282, 284–287, 293
chemiluminescence, 239, 257
chemist, xxiii, 1–2, 4–7, 18–19, 83, 89, 115, 119, 127, 147, 160
ChemNMR, 289, 290f, 544, 545f
ChemTech—The Chemist and Technician Toolkit Companion, xxiii, 1–4, 6, 83–84, 86, 88, 98, 101, 115
chloride adduct $[M + Cl]^-$, 422, 457, 464, 548, 623, 625

chlorophylls, 351, 362
cholesteryl phosphate, 288, 289f, 290f, 291f, 292f, 542–548
cholesteryl stearate, 529–530, 531f
chromatogram, 352f, 353–356, 357f, 399, 412, 415, 417, 429, 431f, 440f
chromatography, xxiii, 5–6, 104, 142, 235, 351, 354f
chromatography column, 351, 355, 363, 421, 471
chromophore, 221, 224, 236, 238, 454
cis-9-octadeceamide, 462
cleavage, 289, 423–428, 457, 483, 544, 550, 561, 563, 566f, 569, 572f, 576, 583, 583f
cobalt carbene ion BDE, 455
coefficient for *longitudinal diffusion (B),* 358, 359f
coefficient for *mass transfer between phases* (C_S for stationary and C_M for mobile), 358, 359f
coefficient for *multiple flow paths (A, eddy diffusion)*, 358, 359f
coefficient of variation (CV), 97
coelution, 356
collision cell, 328, 497, 499, 503, 508, 509f, 564, 574
collision gas, 328, 499f
collision induced dissociation (CID), 423, 427f, 454, 456f, 496, 498–499, 538, 539f
collision mediated radial diffusion, 513, 520, 527, 624
collision/reaction cell, 325, 328
collisional cooling, 471, 522
collisional excitation, 303
collisionally activated dissociation (CAD), 591f
colorimetric analysis, 221
color wheel, 224, 228, 237
column flow, 414, 420f
column headings, 127, 129f
column length (*L*), 357, 358f, 406
column liquid chromatography (LC), 5, 351
complex formation titration, 190, 191f, 193
complimentary color, 224, 237
compressed gas cylinders, 403
computer modeling, 438, 439f
computers, xxiii, 1, 110, 334
concentration, 4, 22, 86, 95, 98, 103–105, 107, 109, 114–124, 132, 141, 148, 149t, 150–158
condensation reaction, 559, 569, 576, 580, 582f, 596, 625
conductometric titration, 186, 187f, 188, 188f, 194
confidence limits, 101, 102t, 103–104, 107, 142
conjugated double bond, 230f, 246, 397
conjugate salt, 159
constructive interference, 345
consumables, 368, 375, 411
continuous spectrum, 295, 296f, 373–374
continuous wave NMR, 278, 279f
conversion dynode, 506, 508
conversion tool, 89, 91
converting units *unit-factor method*, the *factor-label method*, or *dimensional analysis,* 84
cooled cone interface (CCI), 308
copper sulfate, 111, 114, 271, 272f, 273t
corona discharge, 471, 472f
coulombic, 370
coulombic repulsion/explosion, 592 *see also* Rayleigh limit
coupling constant (J), 282, 283f, 293
crucible, 11, 13f, 14, 16
crystal lattice spacing, 345
crystalline lattice, 345
C-trap, 522–523
c-type ion, 563f, 565f
C-type ion, 589, 590f
cubic cells, 509
curve fitting, 110, 114–115, 119–120, 121f
cuvettes, 231, 237, 245–246, 257
CW-NMR, 278
cyanopropyl bonded endcapped silica (LC-CN), 369
cyclohexane, 463, 464f
cyclohexyl estran, 487
cyclonic spray chamber, 304
cyclotron motion, 508, 512, 515
cyclotron velocity, 510
cytosine, 239, 565, 568f, 571f, 574, 576f
dalton (Da), 496
dampening collisions, 468, 474, 504, 624
dampening gas, 504 *see also* bath gas
data outliers, 105
data processing system, 449, 451, 623
de novo sequencing, 584, 584t, 587–589, 618
d-deoxyribose, 565, 568f, 569
dead time (t_m), 354–355, 398, 401–402
decay lifetimes, 244
deconvolution, 593
degrees of freedom (*n*–1), 97, 102–105, 141, 143, 145, 561, 625
dehydroalanine, 600, 613, 615t, 626
dehydroaminobutyric acid, 600, 601f, 615t, 626
delayed extraction (DE), 483, 497f, 498–499, 624
delta (δ), 281, 282f, 283f, 540f
density functional theory (DFT), 255, 256f
deoxyribonucleic acid (DNA), 565
Department of Labor Occupational Safety and Health Administration (OSHA), 23, 29, 30
deprotonated molecule $[M–nH]^{n-}$, 5, 467, 471, 532
derivatization, 551
derivatized, 361, 369, 370f, 371f
desalting, 381, 573, 574f, 575f, 591
deshielding, 281
desolvation, 296, 303, 472f, 473, 478, 486, 624
destructive interference, 345
detection limit, 303, 304f, 308, 310, 496
detector, 93, 120, 202, 228, 233f, 235, 244–245, 248–254, 261–263, 265f, 273, 298f, 310, 325
developing solvent, 395, 396f
developing tank, 395, 397f
deviation factors (k), 96
Dewar, 280
diacylglycerol, 456, 480, 548, 550
dichromic mirror, 253–254
difference in means, 104
diffraction, 222, 254, 261, 273, 338, 340f, 342, 345–346
diffraction grating, 254, 261
diffraction patterns, 345–346
diffractive optics, 337
diffusion, 198, 358–359, 361–362, 403, 513
digital fluorescence microscopy, 252
2,5-dihydroxybenzoic acid (DHB), 483–484, 624
2,5-dihydroxybenzoic acid butylamine, 484
3,5-dimethoxy-4-hydroxycinnamic acid trimethylamine, 484
3,5-dimethoxy-4-hydroxy-trans-cinnamic acid, 624
dimethyl sulfoxide (DMSO), 480t
diode array detector (DAD), 372, 436
diol-bonded silica (LC-diol), 369
1,3-dipentadecanoin, 529, 548–550
dipole excitation, 507, 624
disposable pipettes, 10, 11f
1,3-distearin, 455

dopant, 478–479, 480t
double-focusing mass spectrometer, 332, 495
doublet, 241, 283t, 286
doxepin, 520, 521f
d-ribose, 565, 568f, 569
dried droplet method, 483
drift tube, 450, 457–458, 460, 465, 483, 496–499, 508–509
drying
  dishes, 11
  gas, 476
  sample, 14
  spot, 483–485
  stage, 299
drying oven, 15
duties
  analyst, 19
dynamic range, 308, 310, 342, 418, 517–518, 524, 591, 594f
dynode, 251, 450, 506, 508

ease of ionization, 423
effect
  data point, 95
  health, 22
  meniscus, 8
efficiency of a column, 357
electric and magnetic sector mass analyzer, 5, 491–492, 624
electricity, 198, 277, 421
electric sector, 330–332, 491, 494–495, 527, 624
electrochemical cell, 5, 198, 202, 207
electrochemistry, 5
electrode, 167–168, 186–187, 190, 198
electrode potential, 198–201, 207
electroluminescence, 239
electromagnetic radiation, 5, 221–223, 236–237, 261, 278
electromagnetic spectrum, 5, 221–222, 236–237
electromagnets, 509, 527, 624
electromotive force (emf), 198
electron
  EI mass spectrometry, 425
  EI processes, 422
  electron capture detector (ECD), 418
  electron capture dissociation (ECD), 589
  electron ionization (EI), 421
  spin states, 240–241
electron affinity, 418, 471
electron beam, 421–422, 468, 624
electron capture detector (ECD), 418
electron capture dissociation (ECD), 589
electron donation tendency, 424
electron ionization (EI), 421
electronic configuration, 228
electronic excited state (first, second), 223, 237, 262
electronic transition, 228, 262
electronic laboratory notebook, xxiii, 1, 5
electron multiplier, 327, 450, 496, 506–507, 520
electron spin *see* spin states
electron transfer dissociation (ETD), 589, 603
electron volt (eV), 631
electroosmotic flow (EOF), 436, 446
electrophoretic movement, 451, 582–583
electropositive, 281, 293
electrospray ionization (ESI), 5, 234, 290, 422, 426, 436, 449, 454, 464, 467, 472, 491, 532, 538, 541, 546, 551
  nano-, 474
electrostatic analyzer, 331–332, 495
electrostatic attraction, 8
electrostatic field, 497, 509, 519–521
electrostatic mirror, 483, 497, 499, 508–509 *see also* reflectron
electrothermal atomization, 299
elements, 2–3, 147–148, 157, 224, 225t, 278t, 295–296, 299–300, 303–305, 325, 328, 333
elimination
  beta, 600–601, 615t
  of charged base, 569, 573
  of water, 559
elution, 352–356, 359–360, 369, 371–372, 398, 402, 435, 441
emergency eye wash station, 23–24
emergency face wash station, 23–24
emergency safety shower, 24
emission spectra, 240, 242–243
empirical formula, 283
end cap electrode, 504
endopeptidase, 452, 583
endpoint, 176, 178, 181–186, 188–191, 194–197, 202, 205, 207–208, 252
energy
  energy diagram, 261, 457
  energy dispersive X-ray fluorescence (EDXRF), 334, 338
  energy resolved breakdown graph, 456
energy-dispersive X-ray fluorescence (EDXRF)
  applications, 341
  detectors, 335–336
  hand held, 339
  instrumentation, 334, 337
  process, 334
enthalpy, 469–470, 478, 489
Environmental Health and Safety (EH&S) Departments, 21
enzyme, 5, 169, 452, 579, 583, 597–598, 603, 611, 615t, 626
epoxide, 538, 540–542, 562–563, 625
equation
  average deviation, 96
  confidence limits, 103
  linear regression, 114
  nernst, 5
  null hypothesis, 105
  point-slope, 112
  Q test, 105
  redox, 5
  standard deviation, 97
  T_n test, 106
equilibrium constant (K_c), 161, 164, 171, 203, 207–208
equipotential curves, 500–501, 503
equivalence point, 176, 178–182, 186–191, 193–195, 203–205
equivalent weight, 151–152, 156–157, 181–183
ergonomic sample introduction system, 309
Erlenmeyer flasks, 7, 9
error curve, 94, 107
errors
  absolute, 94
  detector, 233
  Q-TOF/MS, 507
  random, 93
  relative, 94
  systematic, 93, 106
  XRD, 346
erucamide, 429–432
ethanol, 158, 281–282, 286, 351
ethylenediaminetetraacetic acid (EDTA)

for cations, 190
chelating agent, 299
complex titration, 191
mg complex, 192
structure, 190
table of titration ions, 192
evaporating dishes, 11–12
evaporative cooling, 474, 624
even electron ion (EE), 422, 426, 464, 468, 623
exact mass, 289, 329, 458, 461, 465, 493, 507, 543, 556, 613
excitation wavelength, 244, 246, 248, 250, 252, 254
excited state, 223–224, 228, 237, 240–243, 248, 257, 262, 295, 303, 374, 422, 477
exothermic, 152, 468–470, 589, 624
experimental shift, 287t, 545t
eye protection, 19, 152

Fahrenheit, 213
fast atom bombardment (FAB), 422, 484, 486, 548, 580, 625
fatty acids, 211, 287, 454–455, 529–530, 537, 556, 625
femtomole, 496
Fenton reaction, 538, 540
FID decay signal, 280
field desorption (FD), 580
field free region (FFR), 328–329, 450, 492–493, 496–497
filament, 335, 373–374, 420–422, 468, 471, 624
filter cube, 252–254
filters, 245, 253, 326, 328, 419, 455, 496, 501, 503
fingerprinting, 453–454, 530, 580
fire extinguishers
class A, 24
class B, 25
class C, 25
class D, 25
class K, 25
first aid kits, 27
Fisher Scientific, 20, 231, 263–264, 309, 332, 347, 363, 375, 377
fishhook arrow, 424
flame cabinets, 19, 27
flame ionization detector (FID), 418
flame photometric detector (FPD), 419
flame temperatures, 297t
flammable liquids, 19, 25, 27
flash chromatography, 351, 363
floating collision cell, 328, 497, 509
fluorene, 247
fluorescein, 246
fluorescence
continuous variation method, 256
excel example, 128, 131
factors affecting, 244–248
instrumentation, 249–254
introduction to, 239
Jablonski diagram, 240
linear data example, 118–124
quantum yield rate constants, 243
quantitative analysis, 248
special topic, 235
Stokes shift, 242
theory, 240
fluorimetry, 221
fluorometers, 5, 221, 236
fluorophore, 221, 236, 240, 244
focal plane, 331, 495
focusing lenses, 298, 422, 503
footprinting, 454t
formal (*F*) solutions, 149
formula weight, 147–152, 154
Fourier transform infrared spectrometers (FTIR), 5, 221, 236, 261, 263–271
Fourier transform ion cyclotron mass analyzer (FTICR/MS), 5, 491–492, 508, 510, 517, 584, 624
fraction collecting, 363
fraction collector, 364–366, 374–375
fragment ion, 425, 548–550, 552, 567, 591, 615t
frame-transfer CCD, 254
free fatty acid (FFA), 213, 287, 480, 530, 532, 537, 542, 625
frequency, 95, 98, 103, 222–223, 237, 240–242, 247–248, 257, 270–271, 277, 278t, 279–282, 285, 303, 328, 496, 500
frequency domain, 280–281, 513–515
frequency of collision, 248
fructose, 560–561
F-test, 144–145
FTIR correlation table, 266–269
FT-NMR, 280, 292–293
Full-frame CCD, 254–255
full peak width at half maximum (FWHM), 460
fume hoods, 7, 14
functional group, 264, 270–276, 471
fused silica capillary columns, 362, 406, 476
fuzziness, 346

galactose, 560, 594–595, 596t
galvanic cell, 198–207
gamma-hydrogen (γ-H) rearrangement, 424
gamma rays, 222, 240, 257, 333
gas chromatography (GC), 5, 104, 351, 359, 361, 401
columns and stationary phases, 404, 405t
oven, 415
packed columns, 406
partition factors, 403
solvating GC, 361
Stationary phases, 405t
theory, 401
gas chromatography electron ionization mass spectrometry (GC/EI-MS), 425
quantitative analysis, 429
acylglycerols, 548
gas chromatography mass spectrometry (GC-MS), 421
gas cylinders, 28–29, 402–404
gaseous metal atoms, 303
gas liquid chromatography (GLC), 5, 351, 361, 401
gas phase, 235, 256, 325, 402, 421, 426, 436, 438
gas phase basicity, 592f
gas phase complexes, 455
gas solid chromatography (GSC), 404
Gaussian peak shape, 353–354, 358
gel-free approach, 452, 584
general glove rules, 22
glass electrode, 167–168, 206
glass (SiO_2), 344
glassware, 2, 6–11, 16, 19
volumetric, 181
titration, 183
pyrex, 344
glass wool, 404–405, 410–412
globar, 261, 273

gloves, 20–22
glucose, 104, 276, 559
glycoprotein, 594–595, 608, 612t
glycosaminoglycan, 608, 611
glycosidic bond, 276, 559–563, 566
glycosylation, 579, 589, 594–596, 618
Goggles, 19–20
gold seal, 410–411, 414
goniometer, 341–342, 346–348
gradient run, 364
graduated cylinders, 8–9
grams per volume, 149t
graph, 4, 109
graph of equations, 111
graphite ferrule, 412, 417
graphite furnace, 299–300
graphite tube, 299–300
grating, 252–254, 261, 273, 441
gravimetric factor, 150, 158
gravity convection, 15
grid, 483–484, 496
gridline, 110, 131, 137
ground energy state, 223, 237
ground state (GS), 223, 228, 240–243
guanine, 239, 565, 568, 571
guard column, 368, 372

half reaction, 198, 199t, 200–204, 206, 633t
halogenated aromatic, 24
hand held EDXRF, 339
hard ionization, 422, 467–468
hazardous chemical waste, 20
headgroup, 291
heavy metal contamination, 295
height equivalent of a theoretical plate (HETP/H), 355, 357, 398–399
hertz (Hz), 222, 237, 278, 282
heterocyclic, 247
heterolysis (heterolytic cleavage), 424, 427
hexose, 559, 596t
high energy collision induced dissociation, 564, 574
highest occupied molecular orbital (HOMO), 228
high mass resolution, 510, 518–519, 522–524
high performance liquid chromatography (HPLC), 5, 235, 351, 361, 363, 435, 449, 471, 508, 580
high-purity silicon wafers, 336
high resolution ICP-MS, 328, 332
high resolution mass spectrometry, 288, 543, 616
hollow cathode lamp, 297–301
homolysis (homolytic cleavage), 424
hot plate, 11–12
human genome project, 594
hybrid mass spectrometer, 456, 507, 517, 522, 524, 589
hydride vapor generator, 304
hydrocarbon series, 428–429
hydrogen bonding, 369, 580
hydrogen/deuterium exchange, 573
hydrogen nuclei, 292
hydronium ion, 160–161
hydrophilic attractions, 369
hydroxide ion concentration, 167
4-hydroxy-pentanoic acid methyl ester, 469–470
hygroscopic, 262t
hyperbolae, 500–501 *see also* equipotential curves

i-cleavage (inductive cleavage), 423–425, 427
ICP-OES, 303
identification algorithms, 584t
immobilized metal affinity chromatography (IMAC), 597
immonium ion, 587, 588t, 618, 626
incandescent lamps, 244
indicators
  acid/base, 180
  dyes, 229
  redox titrations, 204–205
  visual, 180, 193
indole, 247
inductively coupled plasma, 303
inductively coupled plasma mass spectrometer (ICP-MS), 18, 325
inductively coupled plasma optical emission spectroscopy (ICP-OES), 303
infrared, 5, 221
infrared multi photon dissociation (IRMPD), 518
infrared region, 223
injection
  loop, 366–368, 436
  peak, 354–355, 398, 402
  port, 406–408, 412, 414
  port liner, 408
  port sleeve, 408
  valve, 354, 364
inlet, 366, 381–382, 407
inorganic elemental analysis, 224, 225t
inorganic salts used for IR optics, 262t
Institute for Systems Biology, 614
interactive Periodic Table, 2–3, 148
interface, 115, 308–310, 325, 361
interferences
  chemical, 205, 299, 302, 325, 326t, 551
  general, 306, 308
  spectral, 299
interferogram, 263–264, 273
interferometer, 261, 263, 273
interline transfer CCD, 254
intersystem crossing, 240, 242–243, 258
inverse centimeters, 261
inverted fluorescence microscopy, 252
ion-drift spectrometer, 472
ion evaporation model, 473, 489
ion exchange (IEC), 168, 368
ion exchange HPLC (IEX-HPLC), 369
ion gate, 499, 527, 624
ionization
  constant, 161, 163
  efficiency, 472, 529, 534–535
  potential (IP), 422, 478, 624
  process, 422, 450, 455, 464, 468, 471, 473–474, 479
  source, 5, 290, 421, 449
ion-molecule reaction(s), 468, 471–472, 625
ion product for water (K_w), 164
ion-selective electrodes, 190, 197, 201, 206
irradiation, 240, 257, 333
isobar, 551, 587–588, 596, 618
isobutane, 468–469, 471, 624
isocratic run, 364
isoelectric point, 580
isomer, 551, 561, 587–588, 604, 608
isotope, 147, 158
isotopic abundance, 278t

isotopic peak, 458–459, 461

Jablonski diagram, 240, 242, 257
Jones redactor, 205t

ketene, 425, 532
kinase, 548, 597
kinetic energy, 328, 330–331, 457, 485, 492, 499, 516
Kingdon trap, 519–520

in the laboratory, xxiii, 1
laboratory coats, 20
laboratory glassware, 2, 7
Laboratory Information Management Systems (LIMS), 1–2, 5, 209
lag time, 361
Lambert–Beer Law *see* Beer's Law
laser desorption ionization, 5, 422, 450, 464, 467, 483
lasers, 20, 244
least-squares method, 114
Le Chatelier's principle, 169, 172
light absorption transitions, 223
light emission, 239–240, 257
light emitting diodes (LEDs), 244
light source, 244, 249–250, 252, 297, 352, 373, 443
limit of detection (LOD), 374t
LIMS *see* Laboratory Information Management Systems
linear ion trap mass analyzer (QQQ/MS or LIT/MS), 5, 491, 506
linear ion trap–Orbitrap mass analyzer (IT-Orbitrap/MS), 5, 491, 518
linear regression, 114, 116
linear velocity, 358–362, 403
linoleic acid, 211, 538, 542
lipidome, 454–455, 529
lipidomics, 451, 454, 465, 529, 616
lipids, 425, 428, 449, 451, 454
lipophilin, 452–453
liquid chromatography (LC), 5, 351
liquid helium, 280, 293, 519, 625
liquid ionic matrix, 483, 489
liquid nitrogen, 280, 293, 336–337, 519, 625
lithium adducts $[M + Li]^+$, 287–288, 454, 542
lithium-drifted silicon crystals, 336
litmus paper, 159–160, 171, 283
litmus test, 159
Lorentz force, 510, 515
loss of ammonia, 425–427, 589, 626
lowest unoccupied molecular orbital (LUMO), 228
low resolution ICP-MS, 325
lyso phosphatidylcholine (Lyso PC), 553f *see also* phosphorylated lipids

macromolecules, 472, 559, 565
magnetic field, 241, 277–281, 282f, 284, 287
magnetic moment, 241, 277, 278f, 281
magnetic quantum numbers, 277
magnetic sector, 328, 329–332, 491–495, 508, 564
magnetogyric ratio, 277, 278
magnetron motion, 515, 516f
main menu, 2, 3f, 84, 86, 88f, 89, 90f, 91f, 98, 99f, 101, 115, 118, 125, 132, 135, 141, 147, 148, 153, 154f, 155, 170, 177, 182, 185, 188, 190, 209, 210, 211, 212, 214, 218, 220
mannose, 594, 595f, 596
mass accuracy, 460, 461, 462, 463t
mass analyzers, 2, 325, 457, 458, 460, 467, 468, 473, 484f, 491–511, 513, 515, 516, 517, 518, 519, 520f, 522, 524, 526, 574
mass defect, 458, 462
mass filter, 328, 421, 455, 496, 501, 502, 503, 507, 509f, 517
mass pattern recognition, 596
mass resolution, 330, 460, 461, 494, 496, 498f, 502, 508, 510, 517, 518, 519, 520, 522, 523, 524
mass spectrometry, 5, 6, 18, 234, 287, 288, 289, 290, 325, 328, 331f, 332, 368, 374t, 375, 397, 404, 421, 422f, 423f, 425, 426, 429, 436, 441, 449–452, 454–458, 460–464, 467, 472–474, 472f, 473f, 474, 475, 476f, 477–479, 483, 485, 491, 495, 496, 497, 498, 499f, 502, 503, 503f, 506, 507, 508, 509, 510, 515, 517, 518, 519, 520, 522, 523f, 524, 525f, 527, 529, 530, 531f, 532, 535, 537f, 538, 539f, 543, 544, 546, 552, 553, 554, 564, 573, 574, 580, 583, 584, 589, 590, 591, 593, 594, 601, 602, 613, 614
mass spectrum, 287–289, 422, 425f, 426, 427, 431, 453f, 454, 455, 458–461, 467, 468, 486, 502, 507, 513f, 514f, 515, 519, 525f, 530, 531f, 538, 542, 543, 552–556, 564, 570, 591, 592, 596, 606, 607, 613, 614
mass-to-charge ratio (*m/z*), 287–292, 325, 328–330, 421–433, 502f *see also* mass-to-charge ratio
material safety data sheet (MSDS), 22
math, 2, 4, 83, 89, 109, 114, 263, 346, 500, 501, 504, 506
Mathieu, 501, 504, 506
matrix assisted laser desorption ionization (MALDI), 5, 422, 450, 453f, 467, 483, 491, 580
McLafferty rearrangement, 424, 425
mean, 22, 94–98, 102–105, 142, 143, 145, 159, 183, 224, 240, 243, 330
measure of dispersion, 97
mechanical convection, 15
mechanical pump, 451
mechanism, 24, 420, 424–427, 429, 433, 478, 479, 485, 486, 497, 532, 533f, 534f, 538, 539f, 563, 566f, 569, 573f, 574, 576f, 587, 589, 590f, 591, 597–601, 608, 609f, 611, 613
median, 95
megahertz (MHz), 278–280, 289, 512f, 513, 521, 543
meniscus, 8, 10f, 473
mercury vapor emission line spectrum, 244
mercury vapor lamp, 244
mercury vapor unit, 303, 304f
Merlin Microseal, 407–409
metabolic, 451, 452, 453f, 454, 608, 617f
metabolite, 421, 452, 453f, 454, 456, 614, 616, 617f
metabolome, 454
metal adduct, 457, 564
metals analysis, 153, 300, 301f
methane, 7, 185t, 266t, 286, 362, 468, 469–471
method of continuous variation, 255, 438
methoxide, 470
methylene blue, 234, 235, 255, 256f, 436, 437f, 444f, 448t
methyl oleate, 423, 429, 430f, 431, 432, 467, 468
metric system, 4, 83, 84, 85t
Michelson interferometer, 263
microchannel plate, 331, 495
microelectrospray, 474, 475
Microsoft Excel®, 4, 120, 125, 126f, 127, 132, 135, 138, 141, 142f
microwaves, 221
midpoint, 169, 179, 181, 182, 185
migrated, 398, 399
mobile phase, 5, 351, 352, 353, 354, 355, 357, 358, 359, 361, 362f, 363, 364, 365f, 366, 368, 369, 370, 371, 372, 373f, 377, 378, 395, 396, 398, 401, 402f, 403, 435, 436, 441, 442f, 448, 476

mobile phase linear velocity in centimeters per second (F, cm/sec), 358
mobile proton model, 585
mode, 148, 168, 242, 249, 252, 254, 265, 287, 288, 289, 291, 328, 329f, 330, 422, 426, 436, 438, 439f, 440, 451, 452f, 455, 462, 471, 473, 474f, 480f, 483, 485, 493, 494, 497, 503, 505, 512, 522, 524f, 530, 531f, 532, 534, 535, 536f, 537f, 538f, 542, 543, 546, 547f, 548f, 549f, 551, 552, 553, 556, 561, 562, 563, 564, 569, 570, 573, 607, 608
molal, 149
molar absorptivity, 229
molar extinction coefficient, 244, 248, 373
molar hydrogen ion concentration, 161
molarity, 4, 95, 98, 103, 150–155, 185
molar titrations, 178
mole, 83, 84t, 147–152, 154, 175, 176, 182, 190, 255
molecular electron cloud, 468
molecular formula, 193, 280, 431, 458–460, 462–464, 507, 552, 559, 588, 591
molecular ion, 330, 421–425, 454, 457, 463–465, 467, 468, 478, 485, 487, 491, 494
molecular luminescence, 239
molecular orbital diagram, 246
molecular vibrations, 240, 261
molecular weight, 148, 154, 177, 178, 193, 436, 450–452, 454, 457, 458, 461–464, 467, 471, 480, 486, 496, 503, 506, 529, 532, 535, 542, 548, 580, 583, 588, 591–593, 596, 598, 602, 614, 639
molecular weight calculator, 148, 154, 177, 178, 639
molecular weight marker, 451
mole fraction, 149, 438
monochromators, 245, 249, 250
monoisotopic mass, 458, 460, 596, 599, 600
monoisotopic peak, 458, 459, 461, 462
monomer, 366, 559, 565, 569
monopentadecanoin, 432–433, 456, 457, 548, 549f
monosaccharide, 559, 560, 596
MS^n, 427, 504, 517, 523, 526
MS^3, 427, 525f, 526, 527, 600
muffle furnace, 14, 15
multidetection microplate reader, 252
multidimensional protein identification technology (MUDPIT), 452, 584
multi-element spectrum, 340f
multiplets, 282, 283t
multiplicity, 241, 282, 283t
multistage activation (MSA), 526
myristamide, 429, 430f, 431, 432t
myristic acid, 287, 531f, 532, 534, 535, 542f

N-acetylglucosamine, 594, 595f, 596
naphthalene, 193, 247, 480t
National Center for Biotechnology Information (NCBI), 584
natural isotope, 147, 458
natural rubber (latex), 21
n-bonding, 228
near infrared (NIR) spectroscopy, 245
nebulizer, 296, 297f, 304, 305, 307f, 308f, 309, 325, 479f
nebulizer gas, 296, 479f
negative chemical ionization, 548
negative ion mode, 287, 288f, 289, 291, 329f, 422, 471, 493f, 531f, 532, 534, 535, 536f, 537, 538f, 542, 543, 546, 547, 552, 556, 561–565, 569, 570, 573, 608,
negative slope, 119f, 186, 339
neoprene, 21, 22
Nernst equation, 5, 200, 201, 203
Nernst glower, 261
neutral, 26, 152, 159–161, 166, 176, 177, 187, 188, 193, 198, 248, 289, 299, 325, 328f, 369, 371, 421–427, 429, 436, 441, 449, 450, 457, 467–471, 478–480, 483, 485, 503, 507, 508, 513, 516, 520, 526, 532, 533f, 538, 540–542, 544, 548, 549f, 550–556, 561–565, 569, 573f, 599–602, 606–608, 613, 615t
neutral loss scan, 503
Newton's force equation, 509
nitrile, 22, 267t
4-nitroaniline, 485
m-nitrobenzyl alcohol (NBA), 485
nitrogen laser, 483, 497, 498f
nitrogen phosphorus detector (NPD), 419
nitrogen rule, 464–465
N-linked glycosylation, 594
NMR sample probe, 280
nominal mass, 458, 459, 464 *see also* exact mass
non-linear regression, 121
normal distribution curve, 94, 95f, 97, 98
normal hexane, 369, 463, 464f
normality, 4, 151–153, 155–157, 181, 185, 213
normal phase (NP), 368, 369f
normal phase extraction, 381
normal phase HPLC (NP-HPLC), 368
normal titrations, 182, 183
nuclear interaction, 277, 281
nuclear magnetic resonance (NMR), 277, 397, 543
nucleic acid, 529, 559, 565, 568f, 569, 570, 572f, 573, 579, 597, 598
nucleoside, 569, 574
nucleotide, 239, 452, 569, 570, 574
Nujol oil, 271
null hypothesis, 105, 143, 145

objectives, 252–254, 265, 596
Occupational Safety and Health Administration (OSHA), 23, 404
offspring droplet, 473, 474f
ohmic heating, 303
oleamide, 425–432, 529–531
oleic acid, 213, 287, 531f, 532, 533f, 534f, 535, 538, 542, 556f
oleyl palmitate, 538, 540, 541
oligosaccharide, 548, 561, 562–565, 567f, 596, 608
O-linked glycosylation, 594
optics, 231, 233f, 245, 249, 250f, 261, 262t, 263, 265, 308, 325, 337, 346, 507, 509, 517, 518f, 523
optimal flow rate, 359
orbitrap, 5, 491, 492, 518–527, 588, 616
organic dyes, 224
orthogonal ion deflector, 508, 509f
oxidant, 199, 296, 297f, 633t–635t
oxidation, 5, 197–200, 202, 205, 206, 474f, 530, 538, 542, 548, 594, 596
oxidation-reduction (redox), 197
oxide radical, 470
oxidized wax esters, 538, 541
oxidizing agent, 197, 205, 206
oxonium ions, 596

packed columns, 381, 404, 406f, 413
palmitamide, 429–431, 432t
palmitic acid, 287f, 531f, 532, 534, 535, 538, 539f, 542, 556
palmityl behenate, 529–531, 538, 539f
palmityl oleate, 529–531, 538, 541
pants, 20

paper chromatography, 351, 395, 396f
PAPS, 608, 609f, 611
paramagnetic, 241
  particle emitting source, 471
particle theory, 221–222
partitioning coefficients, 354
partitioning dynamics, 358
parts per billion (ppb), 153
parts per hundred (pph), 94, 97
parts per million (ppm), 4, 104, 132, 149, 281, 460, 613
  ppm error, 288f, 289, 460, 507, 543 *see also* mass accuracy
path length, 119, 228–230, 232, 234f, 295, 372, 373, 496, 497
Pauli exclusion principle, 241
peak broadening, 357, 358, 360, 362, 399, 502
peak coalescence, 515
Peltier cooled silicon drift detectors (SDD), 335
Peltier cooled Si(Li) detectors, 335
Penning trap, 509, 511f, 517, 518
peptide bond, 579–581, 584, 589
peptide mass fingerprinting (PMF), 580
peptides, 5, 369, 375, 449, 452, 485, 498, 517, 519f, 525f, 526, 527, 529, 559, 561, 569, 579, 580, 582–584, 588–591, 593–597, 599–601, 606–608, 613, 614, 617, 618t
percentile values, 102
percent ionization, 164
percent transmittance, 230, 231, 264, 266f, 272f
periodic table, 303, 637
peristaltic pump, 296, 304, 305, 307f, 309, 325
peroxide, 211, 538, 541
personal protective equipment (PPE), 19
pH, 4, 7, 86
phase interface, 361
phenol, 162t, 180f, 181, 182, 184, 185, 248, 267, 369, 381, 480
phosphatase, 597–598, 603, 615t
phosphate group, 526, 565, 568f, 569, 573, 597, 599, 600, 601f, 602, 603
phosphate neutral loss, 527
phosphoester bond, 569, 573f
phosphohistidine, 602–604, 606t
phosphoramidate, 603
phosphorescence, 239–244, 248, 250f
phosphorylated lipids, 455, 483, 529, 532, 535, 548, 551, 552, 555, 556
  phosphatidic acid (PA), 553
  phosphatidylcholine (PC), 553
  phosphatidylethanolamine (PE), 553
  phosphatidylglycerol (PG), 553
  phosphatidylserine (PS), 553
  phosphatidylsphingomyelin (SM), 454
phosphorylation, 526, 527, 579, 589, 591, 594, 596–608, 613, 614, 615t, 617
photocathode, 251
photoelectrons, 254
photographic plate, 345, 346f
photoionization, 374t, 477–479, 509
photoionization energy, 478
photomultiplier tube (PMT), 249, 251
photon, 223, 228, 235, 239, 245, 251, 252, 254, 328f, 333–337, 436, 477f, 478, 479, 518, 519, 594
photoplate, 331, 495
Pi antibond orbital, 245
Pi (π) bonds, 246
Pi bond orbital, 245
picket fence series, 428
pigments, 224, 227t, 351
pipettes, 8, 11f
  auto, 12f
  class A, 8
  class B, 8
pixels, 254
Planck's equation, 278
plane chromatography, 395
plasma region, 305f, 467, 468
plasma torch, 303–306, 308f, 325
plate theory, 357, 398, 401
  platelet activation factor (PAF), 553
plotting, 4, 109–112, 114, 116, 119, 122f, 129, 130f, 131, 132, 135, 138, 141, 179f, 186, 204
point-slope equation, 112, 114, 115, 118, 120
polychromator, 310
polyether ether ketone (PEEK), 366
polynucleotide, 569
polyprotic acid, 181
polysaccharide, 529, 559–565, 617f
polyvinyl alcohol (PVA), 22
polyvinyl chloride (PVC), 22
population mean, 94, 97, 142
positive ion mode, 493, 512, 530, 531f, 543, 548f, 549f, 551–553, 561–564, 573, 607, 608
post source decay (PSD), 452, 453f
potential gradient, 422
potentiometer, 202, 203
potentiometric titration, 5, 202, 204f, 206
precessional, 278
precipitation titration, 188
precision, 16, 94, 96, 97, 101, 102, 145
precursor ion, 426–429, 433, 454, 455, 465, 467, 471, 503, 524, 526, 532, 533f, 542, 548, 549f, 550f, 551, 552f, 596, 600, 602, 613
precursor ion scan, 503, 596
primary acid, 185, 186f
primary base, 185
primary standards, 184
priming the pump, 366
probability curve, 94
probability level, 103
product ion scan, 503
product ion, 289–291, 422–429, 431–433, 454–457, 471, 486, 487, 496, 498, 499, 503, 504, 508, 512, 517, 518, 524, 526, 527, 532–534, 538–542, 544, 546–556, 561–565, 569, 572–574, 580, 584–587, 587–590, 594, 596, 599–602, 606–608, 613–614
  profiling, 452, 453f, 454
  scan, 503
  spectrum, 289–291, 423, 426–428, 431–433, 454–456, 498, 518, 526, 527, 533f, 538–541, 544, 546–548, 550, 551, 556, 564, 583, 588, 599–602, 606, 614
progesterone, 479, 480f
prompt ion fragmentation, 485
propanolol, 524, 525f
protease (table), 583
protein charging, 591
protein(s), 5, 21, 239, 369, 449, 451–456, 473, 498, 517, 526, 529, 548, 551, 559, 561, 565, 569, 579–584, 589–605, 608, 611, 612t, 613, 614, 615t, 616, 617f, 618t
proteomics, 5, 6, 375, 451, 452, 517, 574, 579–580, 583–584, 590, 591, 593, 594, 616, 617f, 618t
proton decoupled NMR spectrum, 286
proton NMR, 279–280, 282f, 284, 285f, 287t, 289, 543–545
pulsed field Fourier transform NMR, 280

pump head, 364–366
purge syringe, 366
purge valve, 366
purine(s), 565, 568f
Pyrex, 344, 402, 404, 406f
pyrimidine(s), 565, 568f

Q test, 95, 105, 106
quadrant, 109, 110
quadrupole excitation, 507
quadrupole ion deflector (QID), 325, 328f
quadrupole mass spectrometer, 5, 421, 429, 454, 456, 502, 503f, 522, 538, 539f, 590
quadrupole time-of-flight mass analyzer (Q-TOF/MS), 5, 491, 492f, 507
quantitative analysis, 111, 248, 264, 341, 421, 429, 453f, 454t, 455, 484, 532, 548
quantum efficiency, 240, 243, 245, 248
quantum number, 240, 243, 245, 248
quantum of energy, 240, 298
quantum yield, 243, 244, 248
quenching, 248, 249
Quick Access Toolbars, 127
quinolone, 247

rabbit tear, 451, 452
radial torch, 304
radical anion, 471
radioisotopes, 334
radioluminescence, 239
radio waves, 271
Raman scattering, 248
random (indeterminate) error, 93, 106
range, 5, 8, 16, 27, 94, 96–98, 100, 101, 103–105, 130–132, 134–136, 138–139, 142, 144, 145, 160, 170, 171, 180–182, 185–186, 189, 191, 205, 216, 217f, 222, 224, 228, 230, 231, 233f, 244–246, 249, 252, 253, 261, 262t, 265f, 270, 277–280, 284, 286, 287, 308, 310, 336–338, 342, 362, 367, 368, 373–375, 399, 403, 414, 418, 420, 429, 448, 451, 452, 455, 470, 473–476, 478, 480, 496, 501–503, 505, 507, 509, 513, 516–518, 522, 524, 532, 536, 553, 554, 570, 583, 587, 591, 594, 598
rate constants, 243, 244
rate theory, 357–358, 398
Rayleigh limit, 473, 475f
Rayleigh scattering, 248
Rayleigh–Tyndall Scattering, 248
reagent gas, 467–471
receiver plate, 512, 516, 517
reconstituting, 354
redox equilibria, 198
redox reactions, 197, 198, 202, 205
redox titrations, 5, 202, 204, 205
reducing agent, 198, 205
reductant, 199t, 633t–635t
reduction, 5, 197–201, 205, 206, 474, 484, 485, 507, 508, 535, 584
reference electrode, 167, 168, 199–202, 206
reflectrons, 346, 483, 497, 499f, 507, 509f
refraction, 222
refractive index, 354, 374
refrigerators and freezers, 16
relative error, 94
repeller, 422, 468
resistor, 186, 187f, 420, 476f, 479f, 512
resolution (R_s), 233f, 255, 265f, 278, 310, 329–331, 336, 338, 346, 356, 357, 357f, 362, 399, 403, 413, 458, 460, 461f, 483, 485, 493, 494–498, 501, 502, 504, 508, 510, 511, 517, 518, 522, 523, 525, 532, 573
resolving power, 461, 520, 521f
resonance ejection, 505, 506
resonance frequency, 281, 508
resonant excitation, 505, 507
restrictor, 413f, 476, 477f
retardation factor (R_F), 398
retention factor (k'), 361, 362, 398, 399
retention time (t_R), 235, 354, 355f–358, 361, 362, 398, 399, 401, 402, 438
retro-Diels–Alder, 423–424
reverse phase (RP), 368, 375
reverse phase HPLC (RP-HPLC), 369, 435, 584
rheodyne injection valve, 366–368
ribonucleic acid (RNA), 565
right hand rule, 329, 510
rigidity, 247
ring electrode, 503–505
rings plus double bonds (R+DB), 428, 463, 464
rocking, 261
root mean square (RMS), 97
rotary vane mechanical pump, 451
rotovap, 18
rough pump, 325, 450–451
ROY G BIV, 222
rubber septum, 407

safety, 2, 6–7, 19, 21–28
safety glasses, 19, 20, 25
salt bridge, 167, 168f, 198f, 200–202, 206
sample loop, 354, 366f
scan law, 329, 493
scan line, 501–502
scatter, 94, 96, 102, 109, 114, 130, 132, 133f, 135, 137, 338, 345, 346f
scattering, 248, 374
scientific method, 7
scissoring, 261
Search Wizard, 209, 214
secondary standards, 184, 185
septa, 407–409, 411
septum, 407–409, 412–415
septum purge flow, 413f–415f
sequential WDXRF, 340–342
sharps containers, 29
SHE half reaction, 199
  sheet, 199
shoes, 20
sialic acid, 594, 595f, 596t
sigma antibond orbital, 244
Sigma-Aldrich, 20, 150, 229, 230, 375
sigma (σ) bond, 244, 423, 426–428, 468f
sigma bond orbital, 244
sigma (σ) cleavage, 423–425
signaling, 425, 454, 526f, 529, 594, 597, 598, 603–605
signal-to-noise ratio (S/N), 280
significant figures, 4, 84–86, 88, 89, 100, 118
silica, 271, 344, 362, 369, 370, 377, 381, 395, 402f, 406, 408, 429, 435, 436, 441, 443, 445f, 475, 476f
silica gel, 395
silicon oxide, 346, 349f
silicon semiconductors, 336

silicone diode photosensors
siloxane, 402, 405t, 524
Silver Shield/4H, 22
simultaneous WDXRF, 338–341
sinapinic acid, 483, 485
single element lamps, 299
single photon, 477, 478
single stage analysis, 503
single stage mass spectrum, 287f, 288, 454, 530, 538, 542, 543, 552, 554, 570, 592
singlet, 240–243, 283t, 286
slits, 233, 244, 249, 253, 302, 328–330, 348, 414, 422, 468, 492–494, 504, 506
slope (*m*), 112
slope-intercept equation, 113, 138
sodium adduct $[M + Na]^+$, 288–290, 422, 463
sodium dodecylsulfate (SDS), 451
sodium iodide (NaI) cluster(s), 460
soft ionization, 456, 465, 485, 551
solid ionic matrix, 483
solid phase extraction (SPE), 5, 351, 377, 381
solubility product constant (K_{sp}), 164
solute, 4
solutions, 147
solvents, 4
source data, 130
sources
  atmospheric pressure photo ionization (APPI), 477–478
  atomic absorption, 296
  atomic emission, 296
  dye references, 225
  electron ionization (EI), 423, 468
  electrospray, 475
  ionization, 5
  infrared radiation, 261, 273
  light, 244
  mass spectrometry, 467
  plasma, 303
  ultraviolet (UV), 373
  X-ray, 334, 338
Soxhlet extractions, 16
space charge effects, 515, 517, 522
spargers, 363
SPE *see* solid phase extraction
spectral lines, 280, 292, 341, 343
spectrofluorometer, 249–252, 281
spectroscopy
  atomic absorption (AA), 295
  atomic emission (AE), 303
  fluorescence, 239
  Fourier transform infrared (FTIR), 261
  instrumental analysis, 5
  nuclear magnetic resonance (NMR), 277
  ultraviolet and visible (UV/VIS), 221
sphingomyelin (SM), 454, 529–531, 551–553 *see also* phosphorylated lipids
spill cleanup kits, 25
spin pairing, 241
spin quantum number, 277, 292
spin-spin coupling, 283t
spin-spin splitting, 281, 293
split flow, 412, 414–415
split injection, 414
splitless injection port, 414–415
split ratio, 414, 429
splitter, 244, 253, 261–263, 410, 476
spontaneous
  chemical reaction, 198, 206
  exothermic reaction, 469–470
  redox reaction, 198, 622
spray needle, 473–474
spreadsheet, 4, 83, 125
sputter, 298
sputtering, 152
stability constant (K_{MY}), 191
stability diagram, 501, 502, 505
standard deviation, 86, 94
standard electrode potentials, 199t
standard hydrogen electrode (SHE), 199
standardization, 185t
starch, 276, 559–561, 625
state of equilibrium, 160
stationary phase, 5, 351–353
stearamide, 429
stearic acid, 106, 287, 531
1-stearin, 2-palmitin, 456
steroid(s), 369
sterol(s), 454, 529, 542
Stokes shift, 241
stopping plate, 506–507
stretching, 261, 270
strong cation exchange (SCX), 375, 435
structural determination, 449
structure elucidation, 289, 544
student's distribution *t*, 101, 102t
sulfation, 579, 594, 596, 608
sulfotransferase, 608–609, 611
sulfuric acid charring, 397
superconducting magnets, 280, 293, 509
superconducting solenoid, 280, 293
supercritical fluid chromatography (SFC), 351
suppression, 233, 534–535, 537–538, 551, 625
Swiss Prot, 584, 611, 612t
systematic (determinate) error, 93, 95, 98, 102
systems biology, 614, 616

tandem mass spectrometry, 234, 436, 454–455, 584, 599, 608
target plate, 450, 483–484
Taylor cone, 473–475, 489, 624
technician, xxiii, 1–2, 5–7, 18–20, 83
temperature, 14–16, 20, 22, 24, 84t, 150, 164, 168, 200, 212, 239, 241, 248, 261, 296, 297t, 299, 374, 402, 413, 415–417, 429
terpenoid, 246, 454, 529
tesla, 509, 517–518, 624
test tubes, 16–17, 381–382
tetrahedra, 346
tetramethylsilane, 281, 293
theoretical plate number (N), 355, 357, 399, 401
theoretical shift, 545t
theory of chromatography, 351, 362, 401
thermal conductivity detector (TCD), 420
thermochemistry, 449
thermoplastic, 366
theta (θ), 338, 340, 345
thin film method, 483
thin Layer Chromatography (TLC), 5, 351, 395
Thompson, J. J., 421
Thompson (Th), 421

three dimensional quadrupole ion trap mass analyzer (QIT/MS), 5, 491–492, 503, 624
thymine, 239, 371–372, 565, 568, 571, 625
tight source, 468
timed ion selector, 499, 624
time-of-flight mass analyzer (TOF/MS), 5, 491–492, 497
time-zero (t_0), 353–355, 402, 417, 496
titrand, 176, 184–185, 188–192, 194–196, 202, 205
titrant, 169, 175–184, 185t
titration, 4, 151, 153, 167, 175
Titrimetric analysis, 4, 175, 194
TLC *see* thin layer chromatography
TMS, 281–283, 293–294, 519
T_n Test, 106
Tongs, 11, 14
Toolbar ribbon, 132, 139
top-down proteomics, 579, 590
top loading balance, 16
total flow, 414–415
total ion chromatogram (TIC), 429, 431
transient image current, 512–513, 520
transmission, 238, 245, 257, 261, 310, 496, 500
transmitter plate, 511–513, 516
trapping plate, 515, 522
trendline, 138, 141
triheptadecanoin, 550–552
triple cone interface, 325
triple quadrupole mass analyzer, 502, 517, 527
triplet, 240–243, 247, 258, 283
trypsin, 452, 583, 583t, 618t
t-test, 143–145
tungsten anode, 297
tungsten-halogen lamp, 244
turbomolecular pump, 325, 327, 450–451
twisting, 261, 271
two-component system, 604
two-dimensional electrophoresis (2DE), 5
two-stage regulator, 403–404
Tyndall scattering, 248
types of fires, 24–25
tyrosine sulfation, 609, 611

ubiquitin, 591
ultraviolet, 5, 20, 221
ultraviolet and visible (UV/Vis) spectroscopy, 221
  spectrophotometers, 5, 221
unimolecular dissociation, 424, 505
units, 83
universal resource locators (URLs), 584t
uracil, 239, 565, 568, 571, 625
urea, 118t, 119, 121t, 123, 127, 162t, 464, 580
UV/Vis detector, 235, 354, 372, 374, 474

vacuum degasser, 363–364, 366, 375, 436
vacuum drying oven, 15
vacuum manifold
  linear quadrupole ion trap, 518
  SPE, 381
vacuum pumps, 18, 325
vacuum system, 449–450, 509, 623
valence shell electron, 223, 237, 241, 422, 478
Van Deemter equation, 358, 361
vanillin, 270, 271
velocity of light (c), 223
vendors, xxiii, 1, 149, 305, 363
vertical θ–θ Bragg–Brentano setup, 346, 348
vertical/horizontal θ–2θ geometry, 346
vibrational degrees of freedom, 561, 625
vibrational level, 224, 241
visible spectrum, 221, 244
visual indicators, 180, 193, 204
viton, 22
volatile, 14
volatile organic compounds, 401
Volhard method for anions, 189
volt, 198, 200
voltaic cell, 201, 207
volume percent, 149t
volumetric flasks, 7–8, 183
volumetric method of analysis, 4, 175
VUV lamps (APPI), 478
VWR International, 20

wagging, 261, 271
watch glasses, 11–12
water insoluble, 262t, 579
wavelength, 221–224, 228
wavelength dispersive X-ray fluorescence (WDXRF), 337
wavenumbers, 264, 266, 271
wave theory, 222, 237
wax esters, 529, 537–538, 540–541, 548
weight percent, 149t
wetting effects, 395
Whatman filter paper, 395–396
Wheatstone bridge circuit, 420
white light, 221–222, 228, 237, 244, 250
W-type ion, 587

x-axis, 94, 107, 110
xenon (Xe)
  arc lamp, 244–245
  lamp, 250, 478, 489
X-ray crystallography, 344
X-ray diffraction (XRD), 340, 342
X-ray fluorescence (XRF), 18, 333
X-ray tube, 334, 338, 341–342, 348
X-rays, 221–222, 237, 240, 333
X-type ion, 563, 565

y-axis, 94, 110
y intercept, 112
y-type ion, 587, 594, 600, 613
Y-type ion, 562–564

Z-flow cells, 373
zone broadening, 358–361
z-type ion, 590, 602
Z-type ion, 563, 565, 590, 608
zwitterionic, 551, 553, 557